VALENCE INSTABILITIES

VALENCE INSTABILITIES

Proceedings of the International Conference held in
Zürich, Switzerland, April 13–16, 1982

Editors:

P. Wachter and H. Boppart

Eidgenössische Technische Hochschule Zürich
Laboratorium für Festkörperphysik
CH-8093 Zürich, Switzerland

1982

NORTH-HOLLAND PUBLISHING COMPANY – AMSTERDAM · NEW YORK · OXFORD

Publishers:
NORTH-HOLLAND PUBLISHING COMPANY – AMSTERDAM · NEW YORK · OXFORD

Sole distributors for the U.S.A. and Canada:
ELSEVIER SCIENCE PUBLISHING COMPANY, INC.
52 VANDERBILT AVENUE, NEW YORK, N.Y. 10017

ISBN: 0 444 86475 x

Library of Congress Cataloging in Publication Data

Main entry under title:

Valence instabilities.

Proceedings of the International Conference on Valence Instabilities held at the Physics Dept. of the Swiss Federal Institute of Technology under the auspices of the European Physical Society and sponsored by the Institute and the Swiss National Science Foundation.
Includes index.
1. Valence (Theoretical chemistry)--Congresses. 2. Chemical systems--Congresses. 3. Solid-state chemistry--Congresses. I. Wachter, P. (Peter), 1932- . II. Boppart, H. (Heinz), 1951- . III. International Conference on Valence Instabilities (1982 : Physics Dept., Swiss Federal Institute of Technology) IV. European Physical Society. V. Eidgenössische Technische Hochschule Zürich. VI. Schweizerscher Nationalfonds zur Förderung der Wissenschaft-lichen Forschung.
QD469.V34 1982 541.2'24 82-14438
ISBN 0-444-86475-X

PRINTED IN THE NETHERLANDS

PREFACE

The International Conference on Valence Instabilities was held in Zürich, Switzerland, from April 13th till 16th, 1982 at the Physics Department of the Swiss Federal Institute of Technology (ETHZ), Hönggerberg. The conference was under the auspices of the European Physical Society and was sponsored by the Swiss Federal Institute of Technology (ETHZ) and the Swiss National Science Foundation, which is gratefully acknowledged. The conference was organized by members of the Laboratorium für Festkörperphysik, ETH Zürich.

The place and the timing of the conference has been chosen because a wealth of information has been collected in the years after the first conference on this topic, the Rochester meeting in 1976. The St. Barbara conference 1981 was planned with emphasis on theory and the Zürich conference with emphasis on experiment, although in all conferences there was a good mixture of theory and experiment. This first conference on valence instabilities outside the USA reflects the present great activity in this field in Europe. About 200 scientist from 17 countries attended this meeting.

The topics of the conference were: Ground State Properties, Lattice and Valence, Valence Changes, Photoemission, Spectroscopies, Kondo, Transport Properties, Magnetic Properties and Mixed Valence Compounds. A panel session on photoemission on cerium and on gaps in intermediate valence materials was included. The proceedings contain all the 119 papers presented at the conference.

I gratefully acknowledge the manifold and valuable help of the International Advisory Committee who selected the invited speakers and the Local Organizing Committee who established the program and chose the scientific contributions. Special thanks are due to the session chairmen for running the sessions and to the referees of the papers.

I would like to thank in particular Heinz Boppart for his service as secretary to the conference who was responsible for the flawless organization of the social and scientific activities and to Miss Anne Bütikofer whose friendliness, efficiency and ability to converse in several languages is well remembered by the conferees. Special thanks are also due to H.R. Ott for help with the technical part of the discussion recordings.

Last but not least, I gratefully acknowledge the excellent cooperation with all authors, referees and the publishers of North Holland who enabled a fast publication and complete documentation of the topics presented at the International Conference on Valence Instabilities.

P. Wachter
Chairman

INTERNATIONAL ADVISORY COMMITTEE

P.F. de Châtel	University of Amsterdam, Amsterdam, The Netherlands
F. Holtzberg	IBM T.J. Watson Research Center, Yorktown Heights, USA
A. Jayaraman	Bell Laboratories, Murray Hill, New Jersey, USA
E. Kaldis	Laboratorium für Festkörperphysik, ETH Zürich, Switzerland
T. Kasuya	Tohoku University, Sendai, Japan
R. Martin	Xerox Corporation, Palo Alto, California, USA
Sir N. Mott	University of Cambridge, Cambridge, England
I.A. Smirnov	Academy of Sciences of the USSR, Leningrad, USSR
K.W.H. Stevens	University of Nottingham, Nottingham, England
R. Tournier	Centre de Recherches sur les Très Basses Températures, Grenoble, France
D. Wohlleben	Universität zu Köln, Köln, Germany
P. Wachter	Laboratorium für Festkörperphysik, ETH Zürich, Switzerland

LOCAL ORGANISATION

P. Wachter chairman

H. Boppart secretary

H.R. Ott

J. Schoenes

J. Sierro

SPONSORS

ETH Zürich

European Physical Society

Swiss National Science Foundation

PANEL DISCUSSION

Discussion Leader:

P.F. de Châtel	University of Amsterdam, Amsterdam, The Netherlands

Panelists:

S. Hüfner	Universität des Saarlandes, Saarbrücken, Germany
N. Martensson	Linköping University, Linköping, Sweden
R.D. Parks	Polytechnic Institute, New York, USA
M. Schlüter	Bell Laboratories, Murray Hill, New Jersey, USA
K.W.H. Stevens	University of Nottingham, Nottingham, England
P. Wachter	Laboratorium für Festkörperphysik, ETH Zürich, Switzerland

LIST OF CHAIRMEN

GROUND STATE PROPERTIES	P.F. de Châtel, Amsterdam, The Netherlands
LATTICE AND VALENCE	R. Monnier, Zürich, Switzerland
VALENCE CHANGES I	D. Wohlleben, Köln, Germany
PHOTOEMISSION	M. Campagna, Jülich, Germany
SPECTROSCOPIES	K.W.H. Stevens, Nottingham, England
KONDO	T.M. Rice, Zürich, Switzerland
TRANSPORT PROPERTIES	T. Kasuya, Sendai, Japan
VALENCE CHANGES II	A. Jayaraman, Murray Hill, USA
MAGNETIC PROPERTIES	R.D. Parks, New York, USA
MIXED VALENCE COMPOUNDS	E. Kaldis, Zürich, Switzerland

CONTENTS

SPECTROSCOPIES

KONDO

TRANSPORT PROPERTIES

MAGNETIC PROPERTIES

MIXED-VALENCE COMPOUNDS

Valence Instabilities
P. Wachter and H. Boppart (eds.)

DENSITIES OF STATES IN VALENCE FLUCTUATORS

John W. Wilkins

Laboratory of Atomic and Solid State Physics
Cornell University, Ithaca, NY 14853

INTRODUCTION

This paper is not a record of the talk given at the conference. That talk concentrated on current confusions in the experimental literature and was intended to stimulate discussion. It seems inappropriate to dignify with a printed record the rantings of a theorist about experiments. Instead this paper attempts to summarize what is known theoretically about the density of states of local moment systems and valence fluctuators. Prime attention is paid to the arguments for "resonant levels" near the Fermi level. As it turns out, surprisingly little is known. Accordingly a set of theoretical questions close the paper.

The narrow focus of this paper prevents discussion of important topics some of which were the subject of an earlier paper more concerned with experiments (see ref. 1). There have been several long reviews[2-6] which give a broader perspective of valence fluctuators than this paper.

2. DILUTE LIMIT

There are two reasons for studying a dilute concentration of rare earth ions in a metal. First there do appear to be dilute valence fluctuators,[7,8] although the experiments are relatively difficult. Second, and more to the point, it is only for dilute magnetic impurities that we have any solidly (and not so solidly) based theoretical results.

2.1 Kondo Model

In the Kondo model of a dilute magnetic impurity in a metal, the local antiferromagnetic interaction between the impurity spin and the spin density of the conduction electrons is characterized by the interaction strength ρJ. For temperatures at or below the Kondo temperatures $T_K \propto D/\rho J \exp(-1/\rho J)$, the electronic properties are strongly perturbed. One electron is bound to the impurity at low temperature. The structure of this bound state is such that it induces interactions between the remaining electrons.[9] Accordingly some[10,11,12] have devised Fermi liquid descriptions for the very low temperature properties, such as the zero temperature limit of $\chi T/C$ and the low temperature resistivity.

The renormalization group approach,[9] which leads to a Fermi liquid description, is capable of calculating the full temperature dependence at least for the susceptibility χ and specific heat C. In particular the low temperature specific heat and susceptibility are well described by aresonant level model whose Lorentzian density of states has a height of $1/\Gamma_{eff}$ and width Γ_{eff}, and hence a unit area. Figure 1 shows how well the susceptibility and specific heat for this resonant level model, with $\Gamma_{eff} = 1.6\ T_K$, agree with those calculated with the numerical renormalized group approach.[9,13] These results,[13] recently confirmed[14,15] by the Bethe Ansatz approach, further support a simple appealing picture[15] that one would like to carry over to valence fluctuations; namely, the entire effect of the many body interactions is to place a "resonance" at the Fermi level with width ~ T_K.

2.2 Symmetric Anderson Model

The Anderson model provides a microscopic under pinning of the Kondo model and also a point of view for discussing valence fluctuations. The phenomenological Hamiltonian H_A is illustrated in Fig. 2:

$$H_A = \sum_s \varepsilon_k n_{ks} + \sum (V c_k^+ f + \text{h.c.})$$

$$+ \varepsilon_f \sum_s n_{fs} + U n_{f\uparrow} n_{f\downarrow} . \qquad (1)$$

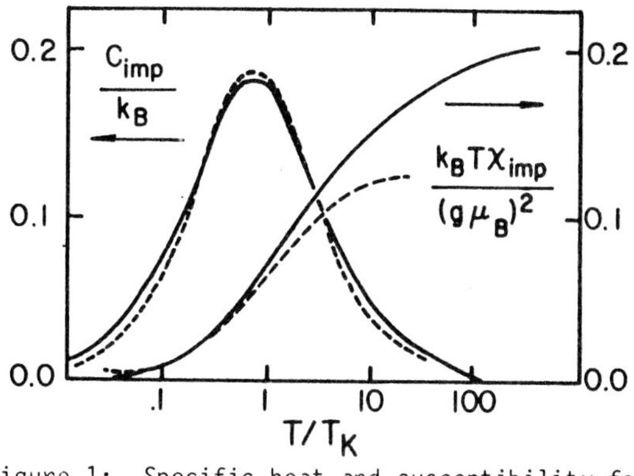

Figure 1: Specific heat and susceptibility for the Kondo model. The numerical renormalization group results (solid line) are well mimicked by those of a Lorentzian resonant level centered at the Fermi level with unit weight and width ~ 1.6 T_K (dashed line).

Here ε_f appears to be the energy of a single "f" orbital of the impurity, U is the Coulomb energy for putting two electrons into the impurity orbital, and $\Gamma = \pi \rho V^2$ is the level width of the impurity orbit arising from its mixing with conduction states whose energies ε_k lies in band of width of 2D. In the symmetric case $\varepsilon_f = -U/2$.

2.2a Many Body Nature of the Parameters

We must think of ε_f and U as many body energies. Consider Figure 2 showing the manifold of states of the asymmetric Anderson model when $\Gamma = 0$. The thick lines represent the ground states corresponding to the conduction electrons being in the lowest lying states for the impurity occupancy $n_f = 0,1$, or 2. For each impurity configuration, the conduction band can be in any of its continuum of many body states. Inclusion of Γ causes transitions between the states.

Now we can envisage a XPS absorption from $f^1 \to f^0$ as one in which the conduction electrons adjust around the f hole to keep the local environment charge neutral. Then the total energy difference

$$\Delta_-(f^1 \to f^0) = E\,(F^0 v^{m+1}) - E\,(F^1 v^m) = -\varepsilon_f \quad (2)$$

for the Anderson model with $\Gamma = 0$. (Here v denotes the valence electrons and $m \simeq 3$ in most cases.) From this perspective ε_f is not a single-particle energy but the difference between the (fully-relaxed) ground state energies of the neutral f^1 and f^0 configurations. In the same way the BIS excitation energy Δ_+ defined by

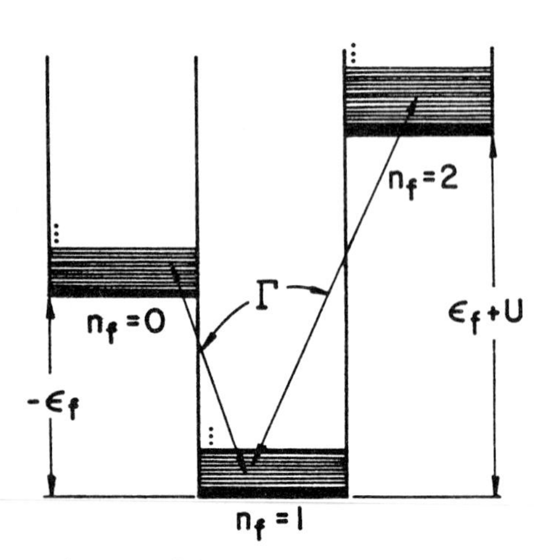

Figure 2: Manifold of the states of the asymmetric Anderson model for f occupancies n_f = 0, 1 and 2.

$$\Delta_+(f^1 \to f^2) = E(f^2\, v^{m-1}) - E(f^1\, v^m) = \varepsilon_f + U\ .$$

The obvious generalization of these are

$$\Delta_-(f^{\,n} \to f^{n-1}) = -\varepsilon_f \quad \text{and} \quad (3a)$$

$$\Delta_+(f^{\,n} \to f^{n+1}) = \varepsilon_f + U\ , \quad (3b)$$

and hence the Anderson model parameters ε_f and U are related to ground state energy differences of various 4f configurations. In this paper ε_f will always have the meaning (3a). This is convenient because then $\varepsilon_f \simeq 0$ will correspond to two valence configurations being degenerate. On the other hand if ε_f were an orbital energy (in the Hartree-Fock sense) then the valence degeneracy condition of Δ_- or Δ_+ being zero would occur when ε_f lay far below the Fermi level.[16]

This rather satisfactory[18] state of affairs for ε_f and U does not extend to the mixing parameter V. To date, no one has devised a satisfactory way of imbedding discrete atomic levels in a continuum that permits anything remotely close to a first-principles calculation of V or equivalently Γ. Instead Γ must be viewed as a parameter to be determined by experiment. How difficult that is can be seen by comparing the results of neutron scattering[19] which suggests $\Gamma \simeq 0.02$ eV and XPS measurements[20] giving Γ values at least an order of magnitude larger.

2.2b Properties of the Resonant Level

There are three sorts of theoretical evidences that point to a narrow "resonance" at the Fermi level for the Anderson model. The first is that the low temperature susceptibility of the symmetric Anderson model is known[21a,b] to be identical to that of the Kondo model whose properties we have already noted are consistent with a resonance of unit intensity at the Fermi level with width ~ T_K. The connection between the parameter of the two models is given[22] by

$$\rho J = \frac{2\Gamma}{\pi}\left(\frac{-1}{\varepsilon_f} + \frac{1}{\varepsilon_f + U}\right), \text{ and} \quad (4a)$$

$$T_K \sim .1\, U \sqrt{\rho J}\, \exp\,(-1/\rho J)\,. \quad (4b)$$

A more explicit demonstration of the resonance is provided by the calculation of the electron self energy[23] and other properties[24] as a power series in U. In particular for the symmetric Anderson model, the f-density of states (Fig. 3-5, ref. 23) has, for $U \sim 15\,\Gamma$, broad peaks at ± U/2 with width Γ, and a narrow peak of $1/\pi\Gamma$ at the Fermi level. While the width of the Fermi level resonance is much less that Γ, it cannot be of order T_K simply because low order perturbation theory in U cannot reproduce (4b). The bulk of the f density of states is associated with the f^1 level at - U/2 and the f^2 level at +U/2. Only a small part of the f level density is associated with the Fermi level resonance.

These perturbation - in - U results can be understood more generally by considering the implications of a Friedel-type sum rule[25] for the Anderson model. In particular it follows that the density of f states at the Fermi level at T=0 K is

$$\rho_{fs}(\mu,0) = \sin^2[\pi\langle n_{fs}\rangle]/\pi\Gamma. \quad (5)$$

This result is valid for any value of ε_f, U and Γ. For the symmetric Anderson model (ε_f = - U/2) with U >> Γ, several conclusions follows. At T = 0 K, there is no moment, hence $n_{f\uparrow} = n_{f\downarrow} = 1/2$. For the symmetric Anderson model the resonance must be centered at the Fermi level. Further the only low energy scale[21a,b] is T_K. Hence the f-level resonance has a peak density of states $1/\pi\Gamma$ at the Fermi level and a width ~ T_K. This f-resonance has weight $T_K/\Gamma << 1$, and the bulk of the density lies within a few Γ of ± U/2. Unfortunately, there are no detailed calculations directly verifying this picture although the numerical renormalization group calculations of the susceptibility are consistent with it. The Bethe Ansatz approach[14b,c] should be capable of calculating the density of f-states and the properties of the Anderson model. One unfortunate feature of a system with interactions, such as the Anderson model, is that one cannot go directly from the density of f states to properties such as the susceptibility or specific heat. A substantial theoretical effort is required to show that a non-interacting model of a resonance with unit weight and width ~ T_K mimics the properties of the symmetric Anderson model in which the f-state density has a peak of weight T_K/Γ and width ~ T_K.

2.3 Asymmetric Anderson Model

2.3a The "Resonance"

Two valence configurations become nearly degenerate as ε_f approaches zero. For this fascinating case there are few exact results. Only the susceptibility[21c,d] has been calculated. However the perturbation in U calculations[26] suggest the following features. For $-\varepsilon_f >> \Gamma$, there are broad bumps (width ~ Γ) at $-\varepsilon_f$ and ε_f + U, corresponding to the f^1 and f^2 configurations, and a narrow "resonance" lying within Γ of the Fermi level. As $-\varepsilon_f$ decreases to of order Γ, one might suspect that the f^1 peak and the "resonance" merge giving rise to one bump of width ~ Γ lying within Γ of Fermi level. This possibility is consistent with the result (5), arising from the Friedel sum rule. As $-\varepsilon_f$ moves up, $\langle n_{fs}\rangle$ should decrease from 1/2 which would occur if the peak, still with magnitude $\sim(\pi\Gamma)^{-1}$, had moved away from the Fermi level.

While the precise details of the scenario as $-\varepsilon_f \to 0$ may differ[27] from one approximate theory to another, most would probably agree that there is a peak of nearly unit weight near the Fermi level with width Γ. In the valence fluctuation regime ($|\varepsilon_f|/\Gamma \sim 1$), the f density of states near the Fermi level is comparable to that of the conduction electrons. One tends to think of the f and conduction electron as hybridizing and in concentrated system a natural model is that of d-f bands. Finally, it is possible, using scaling theory[28] or the renormalization group approach[21c], to describe how the width of the "resonance" change from $T_K \sim \sqrt{\Gamma U}\exp\{-\pi|\varepsilon_f|/2\Gamma\}$ for the case $U >> |\varepsilon_f| >> \Gamma$ to Γ for $|\varepsilon_f|/\Gamma \sim 1$.

2.3b Putting "Realism" into the Anderson Model

It has been recognized, even from the early 1960s, that orbital degeneracies, Hund's rules and crystal electric fields (CEF) might be important for calculating the actual properties. Consider the relatively simple case of the f^1 configuration in cerium. The spin orbit interaction split terms $^2F_{5/2}$ and $^2F_{7/2}$ some 0.2 eV apart, so that only the six-fold degenerate $^2F_{5/2}$ need be considered for most properties. Finally a crystal field (cubic in most systems) will split $^2F_{5/2}$ into a quartet and a doublet by about 20 meV (the number varies greatly depending on the environment). There is a tendency to overlook this further lowering of symmetry because the CEF is seldom seen in most valence fluctuations[29] due most likely to decay rates of at least that magnitude.

Features of atomic physics are important[30] not only for the ground state but also for excited states that might be probed thermally or with XPS and BIS. For example, XPS will see all the terms of f^{n-1} and BIS those of f^{n+1}. Further, the relative intensities of XPS and BIS from configuration f^n are in the ratio n/(14-n) which is 1/13 for Ce and 13 for Yb.

2.3c The Effective Position of the f Level

The interaction V of the local f level with the conduction band gives rise to a shift in the f level, and a broadening, if the new level overlaps the band. The shifted level E_f in lowest order Brillouin perturbation theory[28,31,32,21c] is

$$E_f = \varepsilon_f + \Delta g \frac{\Gamma}{\pi} \ln\left(\frac{U}{|E_f|}\right). \quad (6)$$

where Δg is the difference in the degeneracy of the two configurations being considered. For the single orbital Anderson model (1) $\Delta g = 1$ and a solution of (6) is given in Fig. 12 of ref. 21c. For the case of Ce, the two configurations are f^1 and f^0 and $\Delta g = 13$; the spin-orbit interaction reduces Δg to 5 and crystal electric field to 3 (if the quartet lies lowest).

The effect of the second term in (6) is to stabilize the non-magnetic state (i.e., the one of lower degeneracy); an effect which is particularly important in the valence fluctuation regime where $|\varepsilon_f| \sim \Gamma$. Then one can calculate the properties in perturbation theory[31,21c]. At very low temperature $\chi \sim C/T \sim \Gamma/E_f^2$. At finite temperature there are additional corrections due to the fact that ε_f depends on temperature. A large enough Δg can stabilize the nonmagnetic state ($E_f > 0$) over a large range (compared to Γ) of initial ε_f values, which situation would reduce the complexity of the theory of valence fluctuations compared to that of the Kondo or Anderson model (with $|E_f| >> \Gamma$). Unfortunately, it seems that for real rare earth alloys Δg is fairly small.

2.4 Two-Impurity Calculations

Most valence fluctuators contain a high concentration of rare earth ions. As a result any single-impurity calculation must be modified to include (i) the interactions between rare earth ions and (ii) the shift in the chemical potential (relative to ε_f) due to a finite concentration of rare earth impurities. Several calculations[32,33,34,35a] have been performed, and a consistent picture is starting to emerge.

The strength of the interaction between the impurities depends on the position of f level. For $-\varepsilon_f << \Gamma$ as in the Kondo model there is a RKKY interaction[35b] $[\sim \rho J^2 \cos(2k_F R)/(k_F R)^3]$ that for some inter-impurity spacing R binds the two impurities into a spin-one object for $T > T_K$. As the temperature is lowered the Kondo effect freezes out the spins. For the asymmetric Anderson model when $\Gamma << -\varepsilon_f << U$, the RKKY interaction can acquire an antiferromagnetic contribution[32,35a,36] [falling off as $(k_F R)^{-2}$ for a restricted range of the inter-impurity spacing]. Finally for $|\varepsilon_f| \sim \Gamma$, the interactions between the impurities is quite weak, and the low temperture properties should be dominated by single-impurity results scaled by the concentration of impurities.

Nonetheless there is one additional result which clouds the simple view that single-impurity properties dominate the physics of concentrated valence fluctuation materials. The susceptibility for a single impurity has a contribution proportional to $\Gamma \ln(D/T)$. This "Kondo singularity" is identically cancelled[33,34] by a contribution to the chemical potential. A similar cancellation occurs at all orders in Γ[32]. This result suggests that it may be possible to devise a simpler theory for many rare earth ions than is possible in the dilute limit. To date, however, there is no similar theoretical effort for the density of states or any other property for the two-impurity problem.

3. CONCENTRATED CASE

Although most valence fluctuators are rare earth compounds, as opposed to the relatively few dilute alloy examples, there are no rigorous theoretical results for concentrated systems. The only guidance offered[37] is that the Luttinger sum rule[38] must apply. That sum rule says the Fermi surface volume is equal to the number of conduction electrons per unit cell. In the case of a valence fluctuator for which presumably ε_f lies in the conduction band, the number of electrons must include the f electrons. (For example, if one believed that some form of Ce metal were a valence fluctuator, the number of electrons per unit cell would be four.) The Luttinger sum rule can be thought of as just the Friedel sum rule applied to the crystal as whole. But as in the case of the Friedel sum rule, which led to (5), one needs to know more in order to reach any useful conclusions. To be specific, what is the density of states and is there some "resonance" near the Fermi level? Are the f electrons extended or localized, and are they strongly hybridized to the other conduction electrons? All we have in the way of answers are a variety of model calculations. Two popular models, the band and fermi liquid models, illustrate some of the problems that confront the field.

3.1 Band Model of a Valence Fluctuation

De Haas-van Alphen measurements[39] in $CeSn_3$ of Fermi orbits and masses strongly support a band picture. The anomalously high (4.2-9.2) effective masses and the presence of orbits not found in $LaSn_3$ imply that f electrons are strongly hybridized with the other conduction electrons. This view is further supported by the increase in the magnetic form factor[40] at small q as the temperature is lowered below 40 K indicating a possible delocalization of the f electrons or admixture of 5d electrons. A self-consistent band calculation[39] yields a ~ eV wide f band at the Fermi level containing 1.1 electrons. These band features are similar to those found in band calculations for γ- and α-cerium.[41,42] The principal unease one might have with a band picture is its considerable differences from the rather atomic picture associated with the Anderson model discussed at the start of this paper. For example, all self-consistent calculations to date find slightly more than one f electron per cerium atom. That it is even close to one is impressive given the imperfections of the potential construction scheme. But in the absence of any detailed calculations of properties such as the susceptibility and the x-ray photoemission spectra, it is impossible to say an itinerant view won't work. In fact its partial success may lend support to a more phenomenological model.

3.2 Fermi Liquid Model

The simplest form of this model, which usually omits all interactions, is to throw out everything but non-interacting f electrons with a density of states modeled by a Lorentzian resonant level of width Δ centered at ε_r with respect to the Fermi level. The proponents[43] of this scheme propose to treat a valence fluctuator as group of these independent resonances, but one could just view the whole thing as a very simplified band model. The important assumption, which drives the model, is to treat ε_r as a function of temperature, $\varepsilon_r(T)$, chosen to keep the number of f electrons n_f fixed (as required by the Luttinger sum rule if there are only f electrons).

The density of states of the model is given by

$$\rho(\varepsilon) = \frac{\Delta/\pi}{(\varepsilon - \varepsilon_r(T))^2 + \Delta^2} \qquad (7)$$

in which $\varepsilon_r(T)$ (always measured from the Fermi level) is determined by

$$n_f = (2J+1) \int d\varepsilon\, \rho(\varepsilon)\, f(\varepsilon) \qquad (8)$$

where (2J+1) is the multiplicity of the f level and $f(\varepsilon)$ is the Fermi function. Then the specific heat and susceptibilities are given by

$$\chi(T)/\chi(0) = \int d\varepsilon\, \rho(\varepsilon)\, (-\partial f(\varepsilon)/\partial\varepsilon)/\rho(0) \qquad (9)$$

where $\chi(0) = (g_J \mu_B)^2 (2J+1) J(J+1) \rho(0)/3$ and

$$CT = (2J+1) \int d\varepsilon\, \varepsilon[\varepsilon - T\{\partial\varepsilon_r(T)/\partial T\}] \qquad (10)$$
$$\rho(\varepsilon)(-\partial f(\varepsilon)/\partial\varepsilon).$$

In Fig. 3 we show the results of calculations for the case of J=7/2 and $n_f = 1$ which fixes $\varepsilon_r(0)/\Delta \approx 2.4$. The solid curves are for the case that ε_r is frozen at its zero temperture value so that n_f increases with temperature as the Fermi function expands into the higher density of states about ε_r. Accordingly the susceptibility and specific heat are similar to those in Fig. 1, especially at high temperature, with the exception that χ has a peak at $T \sim \varepsilon_r$ as does C. Note in particular the high-temperature Curie law form for χ occurs naturally for any narrow band model. The peak in $\chi(T)$ is much larger than is found in most valence fluctuators, reflecting the fact that, for ε_r constant, n_f is increasing.

The effect of keeping n_f constant is dramatic: for $T > \Delta$, $\varepsilon_r(T)$ increase linearly with temperature moving the resonance far from the Fermi level. (This case is shown by dashed lines in Fig. 3.) As a result the susceptibility is greatly reduced so that χ more closely resembles the measurements for valence fluctuators (namely, dull χs). Similarly there is a substantial reduction in the high temperature part ($T>\Delta$) of the specific heat due to keeping n_f constant. These effects can, of course, be modulated by postulating some additional band states which can soak up the increasing n_f and permit ε_f to increase less rapidly. Such refinements increase the number of parameters and diminish the appealing simplicity of the model.

Even more to the point, what is this resonance? To fit experimental data requires Δ and $\varepsilon_r(0)$ on the order of a few meV. If this resonance were an f level, it is much too narrow to be consistent with a band model or

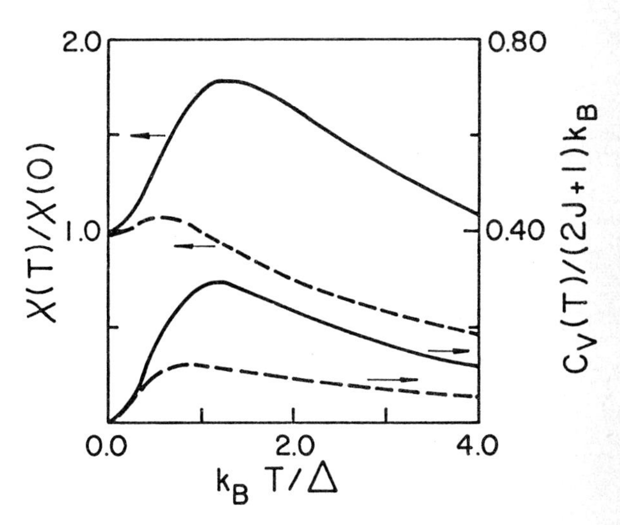

Figure 3: Specific heat and susceptibility of a Lorentzian resonant level model[43] centered at $\varepsilon_r(0) = 2.4\,\Delta$ and degeneracy 2J + 1 = 8. The strong features seen in constant $\varepsilon_r(T)$ case (solid line) are ameliorated when $\varepsilon_r(T)$ is allowed to vary in order to keep $n_f = 1$ (dashed lines).

even with XPS estimates of Γ for dilute systems. In fact a Δ that small should permit the observation of crystal field levels. On the other hand, one might interpret this as a many-body resonance such as is seen in the dilute Kondo and Anderson models. Then of course it is much less clear how the position of the resonance ε_r should be fixed since presumably the bulk of the f electron lies at lower energy as is suggested by resonant photoemission experiments. But in either case, we know too little to properly calculate a wide range of properties.

4. FUTURE DIRECTIONS FOR THE THEORY

There are several theoretical tasks, the successful completion of which would be very helpful. (1) Within the context of the Anderson model, what is the first-principle prescription for calculating the mixing parameter V or equivalently the resonant width Γ? How does that prescription merge with band theory for concentrated rare earth alloys? (2) How important are the interactions between Anderson model impurities? Is it possible that a concentrated system may be easier computationally than a dilute system? (3) How important is it to consider a more realistic model for a magnetic impurity and, for example, to include orbital degeneracy and crystal electric field effect. Are there qualitative differences associated with these complications on the single-orbit Anderson model? (4) What is the density of states for an asymmetric Anderson model; how does it change with ε_f? How also does the position and width of "resonance" depend on the presence of nearby impurities?

This list of questions is limited to those that have naturally arisen from the topics discussed in this review. There are many others that come to mind. First among these is need to develop a theory of XPS that allows one to predict the relative intensities and line widths of main and satellite peaks.

ACKNOWLEDGEMENTS

My long term collaborators, Jan Herbst, C. Jayaprakash, H.R. Krishna-murthy, Luiz Oliveira and Ken Wilson, have largely determined my present attitudes. In addition the following have specifically assisted me with one or more points in the paper or talk: J.W. Allen, H.U. Baranger, A.E. Carlsson, D.L. Cox, C. Jayaprakash, J.F. Herbst and J. Weaver. I am extremely grateful to them all. My portion of any of the cited papers has been supported by the NSF, both through grants (presently DMR-80-20429) and the Material Science Center at Cornell University, and by Nordita in Copenhagen.

REFERENCES

[1] Int. Conf. on Valence Fluctuations in Solids, eds. L.M. Falicov, W. Hanke, and M.B. Maple (North Holland, Amsterdam, 1981).

[2a] D.K. Wohlleben and B.R. Coles, Magnetism V, ed. H. Suhl (Academic, New York, 1973).

[2b] M. Campagna, G.K. Wertheim and Y. Baer, Topics in Applied Physics v 27, Photemission in Solids II, eds. L. Ley and M. Cardona (Springer-Verlag, Berlin, 1978) p. 217.

[2c] M. Campagna, G.K. Wertheim, and E. Bucher, Structure and Bonding 30, 99 (1976).

[3] L.L. Hirst, Adv. In Phys. 27, 231 (1978).

[4] M.B. Maple. L.E. DeLong, and B.C. Sales, Handbook on the Physics and Chemistry of Rare Earths 1, eds. K.A. Gschneider, Jr. and L. Eyring (North Holland, Amsterdam, 1978) p. 797.

[5] N. Grewe, H.J. Leder, and P. Entel, Festkörperprobleme XX, ed. J. Treusch (Vieweg, Braunschweig, 1980), [Theory Review].

[6] J.M. Lawrence, P.S. Riseborough, and R.D. Parks, Rep. Prog. Phys. 44, 1 (1981).

[7] For table of such systems with references see the paper by J.W. Wilkins, ref. 1, p. 459.

[8] Almost all the data is for slow measurements; see, however, D.K. Wohlleben (ref. 1, p. 1) who reports $L_{2,3}$ -edge results for dilute Eu in $ScAl_2$. Furthermore XPS measurements by N. Martensson, B. Reihl, R.A. Pollak, F. Holtzberg, G. Kaindl, and D.E. Eastman, Bull Amer. Phys. Soc. 27, 277 (1982) indicates the presence of 2^+ and 3^+ Tm in $Tm_x Y_{1-x}$ Se for x = 0.05 and 0.20.

[9] K.G. Wilson, Collective Properties of Physical Systems, Nobel Symposium 24 (Academic Press, 1974); Rev. Mod. Phys. 47, 773 (1975).

[10] P. Nozieres, J. Low Temp. Phys. 17, 31 (1974).

[11] A more microscopic effort for the Anderson model has evolved, using Ward identities, from H. Shiba, Prog. Theor. Phys. 54, 967 (1975) and A. Yoshimori, Prog. Theor. Phys. 55, 67 (1976) to L. Mihaly and A. Zawadowski, J. Phys. Lett. 39, L483 (1978).

[12] Scaling arguments have been used on the degenerate Anderson model including CEF and spin-orbit interactions to deduce the effective low temperature Hamiltonians and

the zero temperature ratios $\chi T/C$: P. Nozieres and A. Blandin, J. de Phys. 41, 193 (1980).

[13] L.N. Oliveira and J.W. Wilkins, Phys. Rev. Lett. 47, 1553 (1981).

[14] (a) The temperature dependent properties were reported by V.T. Rajan, J.H. Lowenstein, and N. Andrei, Bull. Amer. Phy. Soc. 27, 305 (1984), (b) N. Andrei and J.H. Lowenstein, Phys. Rev. Lett. 46, 356 (1981), and (c) V.M. Filyov, A.M. Tzvelik, and P.B. Wiegmann, Phys. Lett. 81A, 175 (1981).

[15] The resonant level model has the difficulty that without some tinkering, see ref. 13, it gives a low temperature ratio $R = (3\pi^2/2\ g^2\ \mu_B^2)\ T\chi/C$ of unity instead of two. The present state of the experiments on these properties for valence fluctuators does not warrant further discussion of this point.

[16] J.F. Herbst, R.E. Watson, and J.W. Wilkins, (a) Phys. Rev. B13, 1439 (1976); (b) Phys. Rev. B17, 3089 (1978).

[17] This point of view is further expanded in the Appendix to ref. 16b.

[18] Even the actual values of ε_f and U calculated in ref. 16 for the rare earth elements agree well with the XPS and BIS measured of J.K. Lang, Y. Baer, and P.A. Cox, J. Phys. F11, 121 (1981); esp. see p. 133 and Fig. 5.

[19] M. Loewenhaupt and E. Holland-Moritz, J. Appl. Phys. 50, 1 (1979); J. Magn. Magn. Mat. 14, 227 (1979); E. Holland-Moritz, et al. Phys. Rev. Lett. 17, 983 (1977).

[20] J.W. Allen et al., Phys. Rev. Lett. 46, 1100 (1981) and M. Croft et al., ibid, p. 1104.

[21] H.R. Krishna-murthy, J.W. Wilkins and K.C. Wilson, (a) Phys. Rev. Lett. 35, 1101 (1975); (b) Phys. Rev. B21, 1003 (1980); (c) ibid, p. 1044; (d) Valence Instabilities and Related Narrow-Band Phenomena, ed. R.D. Parks (Plenum, New York, 1977), p. 177.

[22] J.R. Schrieffer and P.A. Wolff, Phys. Rev. 149, 491 (1966).

[23] K. Yamada, Prog. Theor. Phys. 53, 970 (1975).

[24] K. Yosida and K. Yamada, Prog. Theor. Phys. Suppl. 46, 241 (1970).

[25] D.C. Langreth, Phys. Rev. 150, 516 (1966).

[26] B. Horvatic and V. Zlatic, Phys. Stat. Sol. (b) 99, 251 (1980).

[27] See for example, F.D.M. Haldane in ref. 1, p. 153. One of the confusions is whether, when $-\varepsilon_f$ is still several Γ away from the Fermi level, the resonance might be centered at the Fermi level.

[28] F.D.M. Haldane, Phys. Rev. Lett. 40, 416, 911(E) (1978).

[29] (a) S.M. Shapiro, J. Appl. Phys. 52, 2129 (1981), (b) A. Furrer, W. Bührer, and P. Wachter, Solid State Comm. 40, 1011 (1981).

[30] P.A. Cox, J.K. Lang and Y. Baer, J. Phys. F11, 113 (1981) work out the theory. Experiments showing the relative intensities of XPS and BIS for the rare earth meals are shown in ref. 19.

[31] T.V. Ramakrisnan, ref. 1, p. 13; T.V. Ramakrisnan and K. Sur, to be published.

[32] H. Keiter and N. Grewe, ref. 1, p. 129; N. Grewe and H. Keiter, Phys. Rev. B24, 4420 (1981).

[33] A.C. Hewson, J. Phys. C10, 4973 (1972).

[34] T.K. Lee and S. Chakravarty, Phys. Rev. B22, 3609 (1980).

[35] (a) C. Jayaprakash, H.R. Krishna-murthy, and J.W. Wilkins, J. Appl. Phys. 53, xxx (1982); (b) same authors, Phys. Rev. Lett. 47, 737 (1981).

[36] A.M. Tsvelik, Sov. Phys. JETP 49, 1142 (1979).

[37] R.M. Martin, Phys. Rev. Lett. 48, 362 (1982).

[38] J.M. Luttinger, Phys. Rev. 119, 1153 (1960).

[39] W.R. Johanson, G.W. Crabtree, D.D. Koelling, A.S. Edelstein, and O.D. Masters, J. Appl. Phys. 52, 2134 (1981); W.R. Johanson, et al., Phys. Rev. Lett. 46, 504 (1981); G.W. Crabtree et al., ref. 1, p. 93.

[40] C. Stassis, et al., J. Appl. Phys. 50, 7567 (1979). W.E. Pickett, A.J. Freeman, and D.D. Koelling, Phys. Rev. B23, 1266 (1981).

[41] W.E. Pickett, A.J. Freeman, and D.D. Koelling, Phys. Rev. B23, 1266 (1981).

[42] D. Glötzel, Physics of Solids Under High Pressure, eds. J.S. Schilling and R.N. Shelton, (North Holland, Amsertdam, 1981),

p. 263. See refs. 31-33 for other calculations.

[43] D.M. Newns and A.C. Hewson, J. Phys. F10, 2429 (1980).

D. RIEGEL: From what experiment do you have taken the information that the line width from an XPS experiment is larger than 0.2 eV?

J.W. WILKINS: I said it was on the scale of two tenth of an eV. Some of the data analysis was done for a group of Ce compounds and from this information I said it was definitely an order of magnitude larger than the 0.02 eV from neutron scattering experiments.

D. RIEGEL: They can't measure that. They only have the information that it is less than 1 eV or 0.5 eV.

J.W. Wilkins: You say this because there is no reliable way of extracting the line widths from the data. I couldn't agree more.

A. JAYARAMAN: You left off the volume changes in this phase transition ($\gamma - \alpha$ transition in Ce). I think this is very important because in some way there is a drastic change in the screening, as indicated by volume.

J.W. WILKINS: I wouldn't disagree with that. There have been calculations of the bands for both of those cases. A start to take the volume into account is to scale the spectrum of X-ray absorption with the lattice constant. Another shocking feature which I didn't even elude to is that the spectra hardly change at all when you go through the $\gamma - \alpha$ transition.

COMMENT: B. LENGELER: Can I make a comment on that. One change that can be seen is the difference in peak heights for γ-and α-Cerium.

GROUND STATE PROPERTIES

Valence Instabilities
P. Wachter and H. Boppart (eds.)

PERIODIC MODELS FOR ANOMALOUS RARE-EARTH COMPOUNDS

R. Jullien

Laboratoire de Physique des Solides, Bât. 510
Université Paris-Sud, 91405 Orsay, France

Theoretical investigations by means of real-space renormalization-group and finite cell calculations are presented for the one dimensional Kondo lattice and Anderson lattice hamiltonians. It is shown that for two electrons per site, the ground state is both non-magnetic and insulating in the whole range of parameters including the large Coulomb repulsion limits corresponding to mixed valence and Kondo-effect. Quantitative results are given for the insulating charge gap and for the magnetic gap. The extension to the more realistic three dimensional case is discussed.

1. INTRODUCTION

In regular rare-earth compounds, the magnetic properties are governed by the well defined ionic configuration of the f-shell which is well localized at energies far below the Fermi energy. At low temperatures the f-moments order magnetically due to their indirect interactions via the conduction electron band (R.K.K.Y. interaction).(1) In many other compounds, mainly with Ce, Yb, Sm, Tm... this regular behavior is not observed. These "anomalous" rare-earth compounds have recently received a great deal of experimental and theoretical attention.(2) The anomalous low temperature properties of these compounds are generally attributed to the presence of the f level states in the vicinity of the Fermi level. When the f level is close to the Fermi level, the d-f mixing cannot be neglected. The ionic configuration of the f shell is not well defined and can be considered as intermediate between two configurations : this is the well known mixed-valence effect. When the f level is quasi-localized not too far below the Fermi energy (typically of the order of few of its resonance width) an effective antiferromagnetic interaction between the f-moments and the conduction electron spins can win the direct ferromagnetic interaction favorizing the formation of a non-magnetic singlet ground state for the magnetic excitations of the system : this is the well known Kondo effect. All these phenomena have been extensively studied in a case of one impurity with the Anderson hamiltonian as a starting model.(3)

In many cases the one impurity theories apply fairly well to the anomalous rare-earth compounds which can be considered as a collection of independent rare-earth impurities imbeded in a broad s-d conduction band. In most of these one-impurity theories, the highly correlated electron gas is considered as a Fermi liquid at low temperatures. In the mixed valence case, the large effective mass of the electrons comes directly from the fact that there are effectively nearly magnetic f-electrons in the vicinity of the Fermi level and the spin-fluctuation theory applies fairly well. In the Kondo case the many body resonance at the Fermi level is formed by the conduction electrons themselves and has nothing to do with the f-resonance which is located below the Fermi level. If the origin of the resonance is intrinsically different in both cases, some common features are shared. The magnetic susceptibility has a Curie-like behavior at high temperature and saturates to a high constant value at low temperatures without magnetic order. The electrical resistivity has a very large T^2 term at low temperature and then saturates and eventually decreases at high temperatures. The slope γ of the specific heat versus T at low temperatures is very large. Unless there is no precise evidence for an integer or intermediate valence it is sometimes difficult to distinguish between mixed valence (or spin fluctuations) and Kondo effect. Photoemission experiments are very usefull because they locate the f-levels, however their theoretical interpretation is not always obvious.

However some peculiar properties of anomalous rare-earth compounds are not explained within one impurity theories. To describe theoretically coherence effects such as the reduction (or disappearance) of the ordering temperature by Kondo effect (in $CeAl_2$, $CeAl_3$...), the occurence of an insulating-like character at low temperatures (in SmS under pressure, TmSe...), the existence of a well defined Fermi surface for the f-electrons (in $CeSn_3$), it is essential to start from periodic models where the rare-earth atoms are regularly spaced on a lattice. The Anderson lattice model and its quasi-localized limit, the Kondo lattice model, which are the natural extensions of the Anderson and Kondo hamiltonian to the concentrated case of one impurity per site, are basic ingredients to understand such peculiar properties. Those models are very hard to treat theoretically especially to account for coherence effects and correlation effects together. It is generally necessary to go beyond mean-field like approximations or usual decouplings.

In this paper we review some theoretical investigations of the Kondo lattice and Anderson lattice hamiltonians. In part II we present the models. In part III we present real-space renormalization group and finite cell calculations on the one dimensional Kondo lattice. In part IV we present a recent finite cell calculation on the one-dimensional Anderson lattice. In part V we discuss extensions to the more realistic three dimensional case.

2. THE MODELS

Let us write generally :

$$\mathcal{H} = H - \mu N_p \tag{1}$$

where $\mathcal{H}$ and H represent the hamiltonian respectively written in the grand-canonical and canonical representation ; μ and N_p are respectively the Fermi energy (or chemical potential) and the total number of particles.

In the case of the Anderson lattice we will consider H and N_p given by :

$$H=\varepsilon_f \sum_{i,\sigma} f^+_{i\sigma} f_{i\sigma} + \frac{t}{z} \sum_{<i,j>\sigma} (d^+_{i\sigma} d_{j\sigma} + d^+_{j\sigma} d_{i\sigma}) \tag{2}$$

$$+V \sum_{i,\sigma} (d^+_{i\sigma} f_{i\sigma} + f^+_{i\sigma} d_{i\sigma}) + U \sum_i f^+_{i\uparrow} f_{i\uparrow} f^+_{i\downarrow} f_{i\downarrow}$$

$$N_p = \sum_{i,\sigma} (d^+_{i\sigma} d_{i\sigma} + f^+_{i\sigma} f_{i\sigma}) \tag{3}$$

The narrow f-band with zero width is centered at energy ε_f. U is a large Coulomb repulsion between f-electrons. The broad s-d band with half-width t is centered at energy $\varepsilon_d = 0$. In first approximation, we consider only d hopping between next-neighbour <i,j> sites. z is the number of next neighbors for a given site. V is a constant hybridization parameter between the two bands.

Both bands are considered to be only spin-degenerated and the real orbital degeneracies have been completely neglected. This is obviously the main approximation which does not allow to describe each rare-earth separately. However, it is well known that, apart from this restriction, the hamiltonian describes all the main physical situations of the anomalous rare-earth compounds including mixed valence and Kondo regime. When applying the model to the physical situation we must consider U and t of the same order of magnitude, typically 1 eV, while V must be considered as one or two order of magnitude smaller. The ε_f and μ must be chosen to fit the best the physical situation and to sketch the d and f fillings.

The mixed valence regime occurs when one of the atomic f levels, ε_f or ε_f + U lies near the Fermi energy, i. e. when $\varepsilon_f \sim \mu$ or $\varepsilon_f + U \sim \mu$, while the d-band is partially occupied. The case $\varepsilon_f \sim \mu$ (resp. $\varepsilon_f + U \sim \mu$) correspond to a mixed valence state in which the occupancy of f level is between 1, a magnetic state, and 0 (resp. 2), a non magnetic state. So, only interconfiguration between magnetic and non-magnetic valence states are described here. By neglecting the real orbital degeneracies, as well as spin orbit splitting and crystalline-field effects we forget all other possibilities.

The Kondo regime occurs when the atomic f levels ε_f and ε_f + U are both away from the Fermi level: $\varepsilon_f < \mu < \varepsilon_f + U$. In this limit we shall assume that the Schrieffer-Wolff transformation can apply in the case of a lattice and that the Anderson lattice hamiltonian can be replaced by the Kondo lattice hamiltonian for which H and N_p are given by :

$$H= \frac{t}{z} \sum_{<i,j>,\sigma} (d^+_{i\sigma} d_{j\sigma} + d^+_{j\sigma} d_{i\sigma}) + \tag{4}$$

$$J\sum_i \{2(d^+_{i\uparrow} d_{i\downarrow} S^-_i + d^+_{i\downarrow} d_{i\uparrow} S^+_i) + (d^+_{i\uparrow} d_{i\downarrow} - d^+_{i\downarrow} d_{i\downarrow}) S^z_i\}$$

$$N_p = \sum_{i,\sigma} d^+_{i\sigma} d_{i\sigma} \tag{5}$$

In the Schrieffer-Wolf transformation, the positive constant J, which corresponds to an antiferromagnetic coupling, is given by :

$$J = \frac{V^2}{\mu - \varepsilon_f} + \frac{V^2}{\varepsilon_f + U - \mu} \tag{6}$$

In this hamiltonian the f electrons are well localized and are no longer considered as particles ; conveniently they are not counted in N_p. They are sketched as spin 1/2 regularly disposed on the lattice and interacting antiferromagnetically with the spins of the conduction electrons. The operators S^+_i, S^-_i and S^z_i are spin 1/2 Pauli matrices :

$$S^+_i = \begin{pmatrix} 0 & 1 \\ 0 & 0 \end{pmatrix}, \; S^-_i = \begin{pmatrix} 0 & 0 \\ 1 & 0 \end{pmatrix}, S^z_i = \begin{pmatrix} 1 & 0 \\ 0 & -1 \end{pmatrix} \tag{7}$$

We do not discuss here the validity of the Schrieffer-Wolf transformation in the case of a lattice. This is the subject of another paper of this conference.(4) It has been shown that other terms could appear which could modify the physical results on the Kondo lattice. We will neglect these terms here.

Before discussing those models in the one dimensional case let us describe simply the coherence effects when there is no electron-electron interaction, i. e. U = 0. In that case the Anderson lattice hamiltonian can easily be diagonalized in the k-space after a simple Bogoliubov rotation. The main feature in one dimension is the existence of a non-zero indirect energy gap near ε_f and a very narrow resonance near the gap edges. The $G_{o\pi}$ indirect gap as well as the width of the resonance are of the order $\Delta=2V^2/t$ when ε_f lies near the center of the d band. These well-known results must be compared with the one impurity case. In that case one observes also a resonance of width Δ at ε_f : the socalled virtual bound state. The great difference is the existence of a gap at the center of the resonance in the periodic case resulting from the coherent hybridization between the d and f bands. In higher dimensions this gap may disappear ; it remains however that the f electron have a well defined Fermi surface in contrast with the one impurity case.

A physical situation of peculiar interest, which will be considered all along this paper is the case where there are two electrons per site in Anderson lattice $N_p = 2N$, (which corresponds to one electron per site in the Kondo lattice). In that case for $U = 0$, the ground state consists in filling up completely the lower subband so that the system is insulating at $T = 0$. One important problem, which is one of the motivation of the following study is to know how this insulating phase is modified in presence of electron-electron interactions in both the highly correlated Kondo and mixed valence regimes.

3. THE KONDO LATTICE

3. 1. Historical

The Kondo lattice model has been introduced by Doniach (5) in order to explain the anomalous low temperature properties of $CeAl_2$ and $CeAl_3$. Two competing effects are present in the same model : the R.K.K.Y. mechanism which tends to favorize long range magnetic order at low temperatures and the Kondo effect which tends to favorize a non magnetic ground state. Doniach simplified the problem by considering a one dimensional analog consisting in two coupled spin chains : the "Kondo necklace". This hamiltonian is supposed to be a modelisation of the magnetic degrees of freedom of the original Kondo lattice hamiltonian while the charge degrees of freedom are not considered. So, from the beginning, this model excludes the possibility of coherence effects. The Kondo necklace has been studied by mean field (5) and real-space renormalization group techniques.(6) We have shown that this simple model exhibits a transition at $T = 0$ between a small coupling phase with long range magnetic order and a large coupling phase with a non magnetic singlet ground state. However from simple arguments it is obvious that charge degrees of freedom cannot be neglected in the Kondo lattice. In the special case of one conduction electron per site, the formation of the Kondo singlet must be accompanied by a localization of the conduction electrons on each site leading to an insulating ground state for the system. This makes the Kondo lattice very different from the usual Kondo impurity effect. To describe such coherence effect, it is necessary to start directly from the full Kondo lattice hamiltonian.(1)(4)(5) This theoretical study is considerably more difficult. Lacroix and Cyrot (7) have used a decoupled scheme with functional integral technique. They have recovered a magnetic-non magnetic transition by increasing the Kondo coupling for a general filling of the conduction band N_p/N different from 1. For $N_p = N$ they found that the ground state is an insulator for all values fo the Kondo coupling J and that the insulating gap varies as $\exp(-t/J)$. In the mean time we have performed real-space renormalization group calculations on the Kondo lattice in one dimension in the half filled case $N_p=N$ ($\mu = 0$). While in a first calculation (8) we found a transition at a very small value of J/t in a more clever second calculation (9) we found an insulating non magnetic singlet ground state for every values of J/t. In fact, by using a slightly different technique near $J = 0$ (10), we have given some arguments to show that the gap opens exponentially when J is small. This last method has the great advantage to be analytical near $J = 0$. The previous numerical studies had difficulties in reproducing the exponential behavior and instead was leading to an erroneous transition or to a power law behavior for the gap. This exponential opening of the gap in one dimension must be compared with the exact solution of the half-filled Hubbard model in one dimension. The magnetic degrees of freedom in the Kondo lattice behave very similarly to the charge degrees of freedom in the Hubbard model. (10) However we must notice that the spins and charges are not coupled in the same manner in both models ; the half-filled Hubbard model is magnetic and insulating while here the half-filled Kondo model is both non-magnetic and insulating. We have chosen to describe here the argument which tends to conclude for an exponential behavior for the gap as well as finite cell calculations which support also this result.

3. 2. Real space renormalization-group near J=0

We want to focuze here on the behavior of the Kondo lattice in one dimension for small values of the Kondo coupling constant. We shall use a perturbation expansion in J/t combined with a blocking renormalization-group method in one dimension. The real-space blocking renormalization group method is now a well known procedure (11) which can be applied directly on fermion systems. The method proceeds as follows :

1) The lattice is split into adjacent blocks of n_s sites. The hamiltonian for each block is diagonalized exactly.

2) We select only n_L low-lying states among the states obtained in the block diagonalization. For the choice we are guided by the symetries in order that the block hamiltonian can be written as an original one site hamiltonian with

renormalized parameters.

3) The original interblock interaction is rewritten in term of the block states in order that the interblock part takes the same form as the original intersite interaction but with renormalized parameters.

The whole procedure allows to rewrite the old hamiltonian defined on a lattice of N sites as a new hamiltonian of the same form defined on a contracted lattice of N/n_s. The new parameters can be obtained as a function of the old ones after a length scale by n_s.

Before considering the spin degree of freedom of the Kondo lattice chain let us consider the half-filled non interacting Fermi see given by hamiltonian :

$$H = t \sum_i c_i^+ c_{i+1} \tag{8}$$

In that case the method is particularly simple and can be done entirely analytically for all values of n_s by taking $n_L = 2$. For n_s odd the ground state of the block is a doublet at energy $- t \cos(\frac{n_s-1}{n_s+1} \frac{\pi}{2})$. Considering the two components of the doublet as new vacuum and one particle state for the block, we get the renormalization group equation for the hopping constant :

$$t' = \frac{2}{n_s + 1} t \tag{9}$$

This formula tells us that we are at a fixed point in which the energy excitations scale with size ($E \sim N^{-z}$) with a "dynamical" exponent

$$z = \ln \frac{n_s + 1}{2} / \ln n_s \tag{10}$$

The exact value $z = 1$ is only recovered when $n_s \to \infty$. The approximative result for n_s finite comes from the approximation (neglecting the excited levels). However the essential result, the fact that we are at a critical point, is well described by the method. We want now to show how we go away from this fixed point when applying $J \neq 0$.

Before considering the full Kondo-lattice hamiltonian for a finite block, let us consider what gives the Kondo hamiltonian on one site for $\mu = 0$. We get eight levels forming a singlet-triplet quadruplet system. The singlet at energy $- 3J$ and the triplet at energy $+ J$ correspond to the one particle states. The quadruplet at zero energy correspond to the zero particle and two particle states which are degenerated due to the electron-hole symmetry (insured by $\mu = 0$). The quadruplet is at the barycenter of the singlet-triplet system. If we would have considered an electron-hole-symmetric Coulomb-like term of the form :

$$U_d \sum_i (d_{i\uparrow}^+ d_{i\uparrow} - \tfrac{1}{2})(d_{i\downarrow}^+ d_{i\downarrow} - \tfrac{1}{2}) \tag{11}$$

in addition to the other terms of (4), the energies of the singlet and triplet would have been lowered by $U_d/4$ while the energy of the quadruplet would have been increased by $U_d/4$.

Let us consider now a finite block of $n_s = 3$ sites with free ends. For $J = 0$, the degeneracy of the ground state obtained above for the spin less Fermi see must be multiplied by the spin degeneracies of the conduction electrons (2) and of the localized spins (2^{n_s}, i. e. 8 for $n_s = 3$). We get a 32 fold degenerate ground state. This high degeneracy is partially lifted to first order in J. We get three different energy levels. The lowest level with energy $-t\sqrt{2} - 2J$ is four-fold degenerate, the second with energy $- t\sqrt{2}$ is twenty-fold degenerate and the third with energy $-t\sqrt{2} + J$ is eight-fold degenerate. The four-fold degeneracy of the ground state is lifted only up to second order in J and gives rise to a singlet-triplet system. In the numerical calculation of reference (9) we were taking this singlet-triplet system and a quadruplet coming from the twenty fold degenerate level to map onto a single site Kondo hamiltonian and to construct the renormalization group transformation. This choice is not very good in the limit $J \to 0$ because the new intersite coupling is tending to zero when $J \to 0$ instead of tending to a constant value, and we do not recover the simple relation (9) of the $J = 0$ fixed point. In fact it is more convenient to consider instead an excited triplet coming from the twenty-fold degenerate level because this triplet is more strongly coupled with the quadruplet on the other sites and contributes better in forming the ground state of the system. Making this coice of levels we map on to a single site hamiltonian containing a J' term and a Hubbard-like U' term because the quadruplet is not at the barycenter of the singlet-triplet system. J' and U' are given by :

$$\begin{aligned} J' &= J/2 \\ U' &= J + U/2 \end{aligned} \tag{12}$$

The second term in the expression of U' is obtained by considering such kind of Hubbard term in the original hamitlonian and corresponds to the renormalization of U obtained in the direct study of the Hubbard hamiltonian by the same method.(12)

To lowest order, the hopping constant is renormalized as for $J = 0$, i. e. for $n_s = 3$:

$$t' = t/2 \tag{13}$$

For the parameters U/t and J/t we get :

$$\begin{aligned} (J/t)' &= (J/t) + \ldots \\ (U/t)' &= (U/t) + 2(J/t) + \ldots \end{aligned} \tag{14}$$

We find a "marginal" behavior in both J/t and U/t.

The consequence is that both gaps, the "charge" gap between the singlet and the quadruplet, and the "magnetic" gap between the singlet and the triplet open exponentionnally when J is applied :

$$G \sim \exp\left(- A\left(\frac{t}{J}\right)^{\sigma}\right) \tag{15}$$

To find the marginal exponent σ, it is necessary to have the higher order terms in the expansion of $(J/t)'$ and $(U/t)'$ and we have not done this expansion.

The preceeding calculation can be extended for larger n_s, it has been done for $n_s = 5$ leading to the same result. It might be possible to show that marginalism will appear in this method for all n_s as it is the case for the Hubbard model. (12)

3.3 Finite size calculations.

Finite size calculations can give some help in understanding the phase transitions in one dimensional quantum hamiltonians.(13) We have diagonalized exactly the Kondo lattice hamiltonian (for $\mu = 0$) for open chains (with free ends) up to size N = 5. We have determined the S = 0 lowest singlet, the S = 1 lowest triplet and we have verified that they belong to the subspace $N_p = N$ while the lowest S = 1/2 quadruplet belongs to the subspaces $N_p = N \pm 1$. As already found above by perturbation expansion in J, the lowest magnetic gap G^m between the singulet and the triplet is quadratic in J while the lowest electronic gap G^{el} between the singulet and the quadruplet is linear in J. Determining the coefficient A by :

$$G^m/t = A(J/t)^2 \tag{16}$$

we have found A = 1.98, 0.96, 1.33, 0.59 respectively for N = 2, 3, 4, 5. Regardless to the odd-even oscillations, it seems that A tends to zero when N tends to infinity.

In order to be more quantitative, let us assume that J = 0 is a "regular" fixed point where the coherence length diverges as $(J/t)^{-\nu}$. Then we would have the following scaling form for G^m :

$$\frac{G^m}{t} \sim \left(\frac{J}{t}\right)^{\nu z} F\left(N\left(\frac{J}{t}\right)^{\nu}\right) \tag{17}$$

where z is the dynamical exponent and where F is a regular function. We deduce that A must follow the following asymptotic behavior when $N \to \infty$

$$A \sim \left(\frac{\partial^2 G^m}{\partial J^2}\right)_{J=0} \sim N^{-z+\frac{2}{\nu}} \tag{18}$$

Assuming that this asymptotic law is verified for N small and assuming z = 1 (which is a reasonable assumption) we find $\nu = 4.7$ when comparing N = 2 and N = 4 and $\nu \simeq 40.$ when comparing N = 3 and N = 5. This method in determining ν shows that ν converges rapidly to an infinite value for $N \to \infty$. This is again an argument for an essential singularity at J = 0 with an exponential opening for the gap.

4. THE ANDERSON LATTICE : FINITE SIZE CALCULATIONS

4.1 Principles of the calculation

If we forget very recent treatments, the Anderson lattice has been only studied within mean-field like approximations (14) which are generally unable to recover the Kondo effect correctly. Here we present exact calculations done on finite on dimensional cells so that correlation effects are exactly taken into account. Let us give the expressions of H and N_p entering (1) written in k space, for a one-dimension chain of N sites :

$$H = \varepsilon_f \sum_{k,\sigma} f^+_{k\sigma} f_{k\sigma} + t \sum_{k,\sigma} \cos k \; d^+_{k\sigma} d_{k\sigma}$$

$$+ V \sum_{k,\sigma} (d^+_{k\sigma} f_{k\sigma} + f^+_{k\sigma} d_{k\sigma}) \tag{19}$$

$$+ \frac{U}{N} \sum_{k_1 k_2 k_3} f_{k_1\uparrow} f_{k_2\uparrow} f_{k_3\downarrow} f_{(k_3+k_1-k_2)\downarrow}$$

$$N_p = \sum_{k,\sigma} (d^+_{k\sigma} d_{k\sigma} + f^+_{k\sigma} f_{k\sigma}) \tag{20}$$

We are more interested in the case $N_p = 2N$ for which coherent effects are supposed to be important at low temperature. We have numerically solved this hamiltonian for finite cells of N sites up to N = 4.

In the case of a finite cell of N sites, the regular periodic boundary conditions allow only discrete values for the wave vector k : k = 2 n/N, with n = 0, 1, ... N - 1. In our case the most interesting k points,where the $\varepsilon_f = 0$ level crosses the d-band are located near $k \approx \pm \frac{\pi}{2}$. They are only considered for cells with N = 4, 8, ... All the properties of the system will thus vary with N with oscillations of period 4. This makes very difficult the extrapolation to $N \to \infty$ when comparing only few small cells. To avoid this difficulty, we have introduced "modified" periodic boundary conditions by considering an arbitrary phase ϕ so that $kN = 2n\pi + \phi$. This phase is irrelevant in the large N limit but for small N it produces a simple shift of the k points. By varying ϕ any k point can be in principle reached, even for a small cell. However, there are now artificial oscillations in the excitation energies with period $2\pi/N$ in k ; the amplitude of these oscillations vanishes when $N \to \infty$. The most exact results are obtained for the special value of ϕ, $\phi = \phi_o(N) = N\,\pi/2$, when the Fermi wave vector of the infinite system $k_F = \pi/2$ is included in the set of k points. Then, for this special choice of ϕ, the results for different N values can be best compared.

N and ϕ being given, we have diagonalized exactly H in each subspace corresponding to the possible values of the good quantum numbers : the total number of electrons N_p, the projection of the total spin over the z axis Σ^z and the total wave vector K. We have use the Lanczös procedure (15) to solve the large matrices obtained. We have first determined the absolute ground state of H for $N_p = 2N$. We have found that it is always a non magnetic singlet with $\Sigma^z = 0$. We have determined its energy E^o_{2N} and its total wave vector K^o. Then we have considered the k-dependent electronic and magnetic excitation energies by calculating the ground state energies for $N_p = 2N \pm 1$, $\Sigma^z = \pm 1/2$ and $K = K° + k$, $E_{2N\pm1}(k)$, and the ground state energy for $N_p = 2N$, $\Sigma^z = \pm 1$ and $K=K°+k$, $E^1_{2N}(k)$. The electronic and hole excitation energies are then defined as :

$$\begin{aligned} \varepsilon_{2N+1} - \varepsilon_{2N} &= \varepsilon^+(k) - \mu \\ \varepsilon_{2N} - \varepsilon_{2N-1} &= \varepsilon^-(k) - \mu \end{aligned} \qquad (21)$$

with

$$\begin{aligned} \varepsilon^+(k) &= E_{2N+1}(k) - E^o_{2N} \\ \varepsilon^-(k) &= E^o_{2N} - E_{2N-1}(k) \end{aligned} \qquad (22)$$

We define the "electronic" gaps

$$G^{el}_{kk'} = \varepsilon^+(k') - \varepsilon^-(k) = E_{2N+1}(k') + E_{2N-1}(k) - 2E^o_{2N} \qquad (23)$$

and the "magnetic" gaps

$$G^{mag}_k = E^1_{2N}(k) - E^o_{2N} \qquad (24)$$

Let us point out that the k-dependence of these excitation energies keep their physical meaning in the infinite system limit only if the smallest electronic gap stays positive and non zero when $N \to \infty$. In that case the ground state for $N_p = 2N$ corresponds to an insulating completely filled up subband, so that the extramomentum k corresponds really to the momentum of the added electron or hole.

4. 2 Results

Let us consider the non-interacting limit $U \to 0$. For $U = 0$ and $\varepsilon_f = 0$ the electronic dispersion curbes are analytically given by :

$$\varepsilon^{\pm}(k) = \frac{t\cos k}{2} \pm \frac{1}{2}(t^2\cos^2 k + 4V^2)^{1/2} \qquad (25)$$

and are given by the dashed curves of fig. 1. For U infinitesimal, but non-zero, the spectrum is strongly modified. States far from the Fermi energy can decay by excitation of electron-hole pairs (as in the Auger effect). This is illustrated in figure 1 where we have shown, by the continuous curves, the smallest values of $E_{2N\pm1}(k) - E^o_{2N}$ in the k-region where this mechanism occurs. The modification is more important when $N \to \infty$ due to the greater number of k points available. However, the dispersion curves near

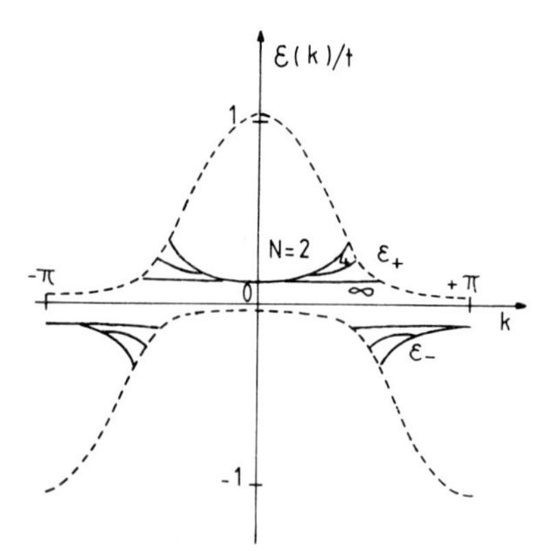

Figure 1 : lower limit of the electron and hole excitation spectrums for U infinitesimal (continuous curves) for N = 2, 4, ∞, compared with the one electron dispersion curves for U = 0 (dashed curves) for $\varepsilon_f = 0$, $V = 0.2t$.

the lowest gap edges are not affected and the lowest gap remains the indirect one $G_{o\pi} \simeq 2V^2/t = \Delta$ while the lowest direct gap changes from $G_{\pi/2,\ \pi/2} = 2V$ for $U = 0$ to $G_{oo} \approx 2\Delta$ for $U \neq 0$.

Let us consider now the case where U is large. First, we present the results for the electron-hole symmetric case $\varepsilon_f = -U/2$ which correspond

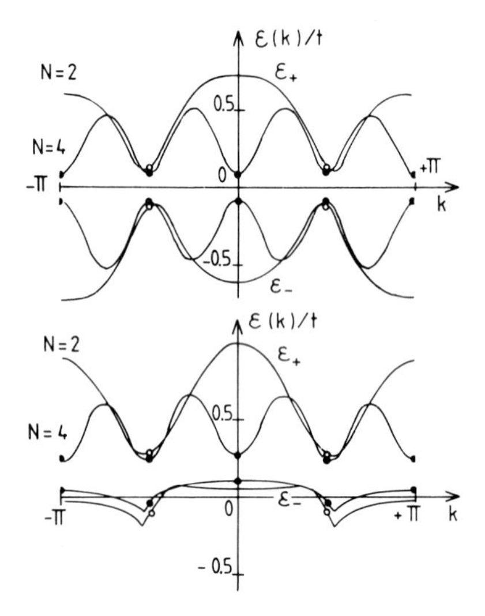

Figure 2 : results for the electron and hole excitation spectrums for N = 2, 4 when U=1.5t, V = 0.2t in the symmetric Kondo lattice case $\varepsilon_f = -U/2$ (top) and in a mixed valence case $\varepsilon_f = 0$ (bottom). The dots show the best set of points for $\phi = \phi_o$

to the Kondo lattice limit. On top of fig. 2, we have reported the lowest electronic excitations $\varepsilon^{\pm}(k)$ for U = 1.5t and V = 0.2t. The points for $\phi = \phi_o(N)$ show that there is an energy gap with a very small dispersion and little dependence upon N. Evidently we observe large artificial oscillations in k when varying ϕ. These oscillations are consequences of the Kondo effect and result from the difficulty in constructing the narrow many body resonance near the Fermi energy from the highly dispersive d-states. These oscillations must disappear in the limit $N \to \infty$ leading to a very flat dispersion near the band edges.

The gaps $G^{el}_{o\pi}$, $G^{el}_{\pi/2,\ \pi/2}$, G^{el}_{oo}, G^{mag}_{o}, G^{mag}_{π} are given, on top of fig. 3, as a function of U^2 (keeping always $\varepsilon_f = -U/2$). For small U, the

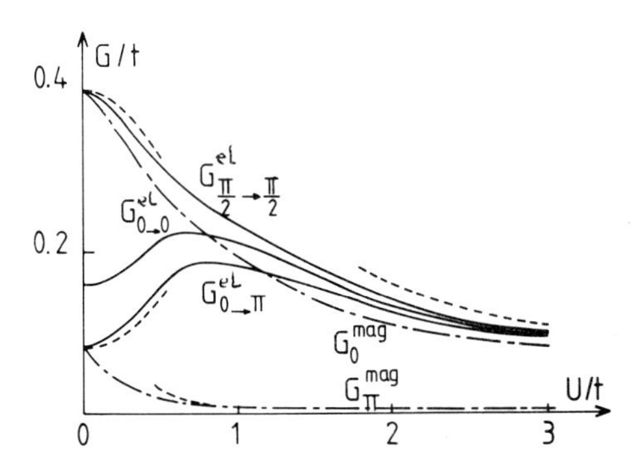

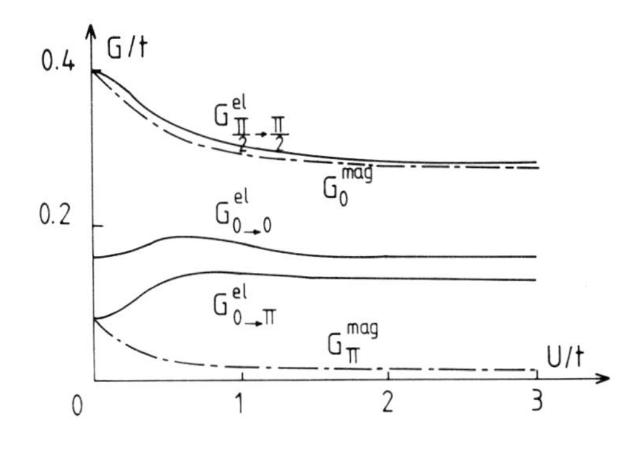

Figure 3 : Variation of the gaps with U for N=4, V=0.2t in the symmetric Kondo lattice case ε_f=-U/2 (top and in a mixed valence case ε_f=0 (bottom). Solid and dash-dot lines refer to electronic and magnetic gaps respectively. In the symmetric case the N = ∞ low U expansions of ref. 16 and the R. G. results on the Kondo lattice (for n_s=3) of ref. 9 are represented by dotted lines.

first two gaps vary as U^2 and agree well with recent perturbation expansions on the infinite one dimensional lattice done by Yamada and Yoshida.(16) For large U we recover that the electronic and magnetic gaps scale respectively as J and J^2/t (see § 3.3) if $J = 4V^2/U$ is the Kondo coupling as deduced from the Schrieffer-Wolff transformation. Note that from the analysis done for the Kondo lattice, one must expect that this large U behaviors must be changed into exponential (exp - aU) behaviors when $N \to \infty$. All the electronic gaps converge into the same curve showing that there is a very small dispersion of the electronic excitations in the large U Kondo limit. For intermediate values of U the calculations give a smooth interpolation. In particular there is no crossing of $G^{el}_{o\pi}$ and $G^{el}_{\pi/2,\ \pi/2}$, as has been suggested.(16) The dispersion of the magnetic gaps also tends to zero as $U \to \infty$, however the gap G^{mag}_{π} is smaller indicating an insipient antiferromagnetic instability due to the $0\to\pi$ nesting of the one dimensional Fermi surface.

The corresponding results in a mixed valence case are shown in the lower parts of fig. 2 and 3 where we have considered the same U and V values as before but now $\varepsilon_f = 0$. We observe that there is always a gap and that we must choose $\mu > 0$ in the gap to have the absolute ground state for $N_p = 2N$. This means that the f-level has been effectively shifted as found in Hartree-Fock and CPA calculations.(14) The hole dispersion curve depends little on N and looks like the simple $U \approx 0$ regime shown in fig. 1. while the electron dispersion curve shows the same kind of oscillations as in the Kondo case. As shown in the bottom of fig. 3, the electronic gaps does not vary too much with U and tend to constant values of order Δ when $U \to \infty$. This means that the correlations do not modify strongly the coherence effect. Note that the magnetic gap G^{mag}_{π} is smaller, of order Δ^2/t. This is due, as in the Kondo case, to the $0\to\pi$ nesting of the Fermi surface.

4.3 Discussion

The most important question concerning our results is how to extrapolate to the infinite one dimensional lattice. It is essential to check if the lowest gap stay finite in the limit $N\to\infty$. The comparison of our N = 4 results with the perturbation expansions done on the infinite lattice in the low U region is already a strong support. Also, the modified periodic boundary conditions help to extrapolate our results. For example, we can compare $G^{el}_{\pi/2,\ \pi/2}$ for N = 2 with antiperiodic boundary conditions with the same gap for N = 4 but with regular periodic boundary conditions. Assuming a form $G = G^{\infty} + A/N$ leads to $G^{\infty} \neq 0$. However this is not a demonstration and it would be better either to reach larger cells (N = 6 is not impossible) or to perform a renormalization group transformation. The present work which analyzes the levels for a small cell, is an essential starting point for a future carefull real-space renormalization-group study.

5. EXTENSION OF THREE DIMENSIONS AND CONCLUSION

So, we have found a complete continuity between the Kondo lattice regime and the mixed valence regime. In all case we have found an insulating gap in the half-filled case as resulting from the coherence of states on different sites in a periodic system.

These one dimensional results cannot be extended straightforwardly in higher dimensions. The insulating gap which appears always in the half-filled case in one dimension can disappear in some range of parameters. The exponential opening of the gap in the Kondo lattice case is certainly typical of one dimension, as for the Hubbard model. It could be that in higher dimensions the gap opens at a critical value $(J/t)_c \neq 0$. This is suggested by an extension of the arguments of § 3.3 for an hypercubic lattice in dimension $d > 1$. Considering hypercubic blocks of 3^d sites, by renormalizing in each direction successively the relations(12) and (13) are changed into

$$J' = J/2^d \qquad t' = t/2 \tag{26}$$

so that

$$J'/t' = (J/t)/2^{d-1} \tag{27}$$

The fixed point $J/t = 0$ becomes stable for $d > 1$ suggesting that the metallic and magnetic phases $J = 0$ extends up to a critical value of J/t. However, this renormalization procedure which renormalizes in each direction successively breaks the cubic symmetry of the lattice. The hypercubic lattice in d dimensions has an important nesting in the half-filled case which is not taken into account. Extension to other three dimensional structures is absolutely not obvious. Note that the more simple Hubbard model itself has not been studied seriously even in the half-filled case for $d > 1$. To give a quite general conclusion we could say that the gap could vanish especially in the range of parameters where it has been found to be very small in one dimension. But it is not obvious that both magnetic and electronic gap should close together ; it could appear a magnetic an insulating phase for some range of parameters. The nature of the three dimensional lattice plays certainly a role in determining the boundaries of the different phases.

Even, if there may be no gap spanning the entire Brillouin zone in three dimensions, the main conclusion of the present work remains : coherence effects are certainly very important at sufficiently low temperature and at T = 0 the f-electrons form a well-defined Fermi surface in contrast with the one-impurity case. All one impurity theories which consider the rare-earth compound as a collection of independent impurities become wrong at sufficiently low temperature.

Coherence effects can be invoked to explain the low temperature behavior of TmSe. This is strongly supported by the variation of the low temperature resistivity with stoechiometry which shows an insulating-like character at concentration one strictly.(17) However in the TmSe case the low temperature phase is both insulating and antiferromagnetically ordered. This could be accounted for by taking into account mixed valence between two magnetic configurations. It would be interesting to extend our calculation in that case.

Another example of evidence for low temperature coherence effects is given by the recent de Haas van Alphen experiments on $CeSn_3$.(18) Some parts of the Fermi surface are similar to its normal analog $LaSn_3$ and other parts are qualitatively different showing large effective masses. This shows that f-electrons have a well-defined Fermi surface in this compound.

We are presently pursuing our theoretical investigations, one of our main aims being to perform a carefull renormalization group analysis of the three dimensional Anderson lattice hamiltonian which takes care of coherence effects.

REFERENCES

(1) For a review on normal rare-earth, see the book of B. Coqblin : "The electronic structure of rare-earth metals and alloys : the magnetic heavy rare-earths", 1977, Academic press, New York.
(2) For a review on anomalous rare-earths and mixed valence, see the review papers : Jefferson J. H. and Stevens K. W. H., J. Phys. C. Solid State Physics, 11, 3919 (1978) ; Lawrence J. M., Riseborough P. S. and Parks R. D., Reports on progress in Physics 44, 1 (1981) ; Coqblin B., Les Houches course (1980) and the conference proceedings of Rochester : "Valence instabilities and related narrow band phenomena" ed. by R. D. Parks, 1977, Plenum Press, New York ; Santa Barbara : "Valence fluctuations in solids" ed. by I. M. Falikov, W. Hanke and M. P. Maple, 1981, North Holland, Amsterdam
(3) Gruner G., and Zawadowski A., Rep. Prog. Phys. 37, 1497 (1974)
(4) Lopez L. C., Jullien R., Bhattacharjee A.K. and Coqblin B., Phys. Rev. B, to be published and this conference
(5) Doniach S., Physica 91B, 231 (1977) and in "Valence instabilities and related narrow band phenomena" ed. by R. D. Parks 2977, Plenum press, New York
(6) Jullien R., Fields J. N. and Doniach S., Phys. Rev. Lett. 38, 1500 (1977)
(7) Lacroix C. and Cyrot M., Phys. Rev. B 20, 1969 (1979)
(8) Jullien R., Pfeuty P., Fields J. N. and Doniach S., Journal de Physique (France) 40, C5-293 §1979)

(9) Jullien R., Pfeuty P., Bhattacharjee A. K. and Coqblin B., J. Appl. Phys. 50, 7555 (1979)
(10) Jullien R. and Pfeuty P., J. Phys. F 11, 353 (1981)
(11) Jullien R., Can. J. of Physics 59, 605 (1980)
Pfeuty P., Jullien R. and Penson K. A., in "Real space renormalization" ed. by T. W. Burkhardt and J. M. J. Van Leuuwen, Springer Verlag (1982)
(12) Dasgupta C. and Pfeuty P., J. Phys. C 14, 717 (1981)
(13) Hamer C. J. and Barber M. N., 14, 241 and 259 (1981)
(14) Leder H. J. and Mühlschleger B., Z. Phys. B 29, 341 (1978) ; Martin R. M. and Allen J. W., J. Applied Phys. 50, 7561 (1979) ; Grewe N., Leder H. J. and Entel P., in 'Feskörperprobleme' (Advances in solid state physics), vol. XX, p. 413, J. Trensh ed. Vieweg. Braunschweig, 1980.
(15) Whitehead R., in "Theory and applications of moment method in many fermion systems" ed. by J. B. Dalton, S. M. Grimes, J. P. Vary and S. A. Williams, Plenum (1980)
(16) Yamada K. and Yoshida K., I. S. S. P. report A 1100, Dec. 80
(17) Haen P., Lapierre F., Mignot J. M., Tournier R. and Holtzberg F., Phys. Rev. Lett. 43, 304 (1979)
(18) Johansson W. R., Grabtree G. W., Edelstein A. S. and Mc Masters O. D., Phys. Rev. Lett. 46, 504 (1981).

C.M. VARMA; You fixed the ground state charge in both the Anderson lattice in the Kondo limit and the Mixed-valence limit to be 2 per site. Did you vary the chemical potential μ in going from one to the other ? I ask this because in calculations by M. Schlüter and me, if we fix $\mu = E_d$ (middle of d-band) $= E_f$ (position of f-level) $= 0$, we find the ground state to be 1 ½ electrons/ site in the large U limit. Incidentally, this is not in violation of Luttinger's theorem, if correctly interpreted.

R. JULLIEN: In one case we have taken E_f and E_f+U symmetric compared to $E_d = 0$. In that case there is no shift of the Fermi level ($\mu=0$). In a second case we have taken $E_f = 0$ and E_f+U increasing. In this case we were obliged to take into account a finite μ in order to keep $N_p=2N$. This corresponds to an effective shift of the f-level compared to the atomic position which have been already found in CPA and other approximation. We have not considered the case $N_p = (1+1/2)N$ for which certainly (as you found) the ground state is magnetic.

N.F. MOTT: Do you have any criterion about when the Kondo energy kT_K should be greater than the RKKY ?

R. JULLIEN: In the 3D Kondo lattice the transition should appear when the strength of the Kondo coupling J becomes the order of the bandwidth of the conduction electrons t, i.e. $(J/t)_c \sim 1$. However, it must be noticed that the critical J value can be too large so that we have already entered the mixed valence regime. It is then to use a criterion in terms of the Anderson lattice parameters. This criterion is $|E_f - E_F| \sim \Delta$.

K.W.H. STEVENS: In an exact solution of the 3-site problem we had approximately 500 eigenstates. How are your gaps defined in terms of the eigenvalue spectrum ?

R. JULLIEN: We have really determined all the levels of the 4-sites problem. The subspace corresponding to 8 particles ($N_p=2N$) was more than one thousand by one thousand (even reduced by spin symmetries). The gaps are really between the absolute ($N_p=2N$, $S=0$) singlet ground state and the first excited state ($N_p=2N\pm1$, $S= \pm 1/2$) and ($N_p=2N$, $S=1$).

Valence Instabilities
P. Wachter and H. Boppart (eds.)

RENORMALIZED PERTURBATION EXPANSION FOR GREEN'S FUNCTIONS OF INTERMEDIATE VALENCE SYSTEMS

Norbert Grewe

Institut für Theoretische Physik der Universität zu Köln,
Zülpicher Str. 77, D-5000 Köln 41, FRG

The Anderson Hamiltonian in the limit of infinite Coulomb repulsion is studied in a perturbation expansion with respect to the hybridization. A short review is given on results for the impurity problem, which have been obtained with Brilloin-Wigner-type summations of divergent perturbational contributions. In particular, linked single f-electron parts are discussed, which play the central role in the subsequent treatment of an IV-compound. It is shown that in the compound problem there actually exists some substitute for Wick's theorem at low temperatures which allows for a practicable cumulant and linked cluster expansion of free energy and Green's functions respectively. Explicit formulas are derived which express these compound quantities in terms of the linked single f-electron parts of the impurity problem at T=0, and a low temperature expansion is indicated. Some consequences of this reduction and possible results are pointed out.

I. INTRODUCTION

In recent years it has turned out that there exists a well defined renormalized perturbation expansion for physical quantities of Intermediate Valence (IV) systems in terms of the hybridization matrix elements between 4f-and band electron states /1/. The difficulties which had to be overcome were twofold: (1) The dynamics of 4f-shells is largely determined by strong local Coulomb interactions, which can not well be treated perturbatively and should be included into the zeroth order description. This poses a technical problem, since then Wick's theorem and the standard many body procedures are not available. (2) In the unperturbed situation a high degree of degeneracy exists due to a nonzero angular momentum of the participating Hund's rule groundstates of ionic 4f-shells and to their very small excitation energy against charge transfer to the conduction band, both combined with the finite density of band electron states at the Fermi-level. This leads to vanishing energy denominators which in thermodynamic perturbation expansions show up as divergencies in the limit of vanishing temperature.

A way to cope with these difficulties consists in using time-ordered local processes with a corresponding description of perturbational contributions to physical quantities in terms of Goldstone-diagrams, and in summing infinite classes of these processes in a systematic way. It leads to a renormalized perturbation expansion which is completely well behaved and even may converge very fast under favourable circumstances such as a large orbital degeneracy /2/. These ideas have first been developed by Keiter and Kimball /3/, who studied the partition function of the impurity Anderson model, and have subsequently been applied to a calculation of the static susceptibility of a realistic Rare Earth impurity in the IV-regime by Bringer and Lustfeld /4/. An extensive treatment of the impurity partition function and derived quantities is due to Ramakrishnan /5/, who particularly stresses the importance of a high degeneracy in one of the ionic multiplets involved for the convergence of the renormalized perturbation series.

A generalization of this theory to the Green's functions of the Rare Earth impurity and to the case of concentrated systems has been worked out by Grewe and Keiter /6/. The low order results for the single f-electron Green's function will be reviewed in Sec.II together with some necessary features of the formalism. Essentially, all (many particle) f-electron Green's functions of the impurity problem constitute the building blocks for processes in concentrated systems involving several sites. Interactions between these sites are mediated by single band electron Green's functions which merge with the local parts of a multi-site process in form of external lines.

Any attempt of calculating properties of concentrated IV-systems using this approach has met up to now specific difficulties, which have their origin in the lack of Wick's theorem for the local parts. In general, an (impurity-) many particle f-Green's function can not be seperated into one particle contributions. This implies that any temporary admixture of a band electron into a particular 4f-shell is heavily correlated with similar events at the same site simply by the specific (renormalized) initial state of the 4f-shell which must be the same for all of these events. The higher Green's functions however are difficult to calculate and pose serious topological problems regarding a practicable classification of multi-site processes. Moreover, since a site may not appear more than once in any process - i.e., if all the external lines at this site are used to connect it with certain other sites, no band electron ejected from a further site can be admixed - there appear exclusions in site summations, which prevent any natural exponentiation of sums of processes with unconnected contributions and a cancellation of unlinked factors in thermodynamic expectation values.

The main purpose of this paper is to show that in the IV-situation there exists, at least in an expansion around T=0, a way to cope with these technical difficulties. In Sec.III a factorization of local parts with external lines into irreducible contributions, which as a substitute for Wick's theorem is exactly valid at T=0, is used to derive rigorously a practicable cumulant sum for the free energy and a linked cluster expansion of Green's functions in the compound problem. The cumulants can graphically be identified with generalized topological rings and the linked clusters for the single f-electron Green's function as generalized chains. If only irreducible contributions with up to two external lines are taken into account, the sums of all rings and of all chains can explicitely be calculated, as shown in Sec.IV. The corresponding expressions for the free energy and all the one particle Green's functions reduce these quantities to certain pieces of the impurity f-Green's function. Sec.V contains a short discussion of the physical results to be expected and a comparison to the trivially solvable case without Coulomb-interaction.

II. GENERAL DESCRIPTION AND THE IMPURITY PROBLEM

For simplicity one can restrict the following discussion to the Anderson model in the extremely asymmetric situation $-E_\sigma \approx U \gg W \gg \Delta$ $(\Delta = \pi \mathcal{N}_F V^2$, W=bandwidth),

$$H-\mu N = \sum_{\underline{k}\sigma} \epsilon_{\underline{k}\sigma} d^+_{\underline{k}\sigma} d_{\underline{k}\sigma} + \sum_\nu [E_2 X^\nu_{22} + \sum_\sigma E_{1\sigma} X^\nu_{1\sigma 1\sigma}] + V \frac{1}{\sqrt{N}} \sum_{\nu \underline{k}\sigma} [e^{i\underline{k}\underline{R}_\nu} d^+_{\underline{k}\sigma} X^\nu_{1-\sigma 2} + h.c.] \quad , \qquad (1)$$

which generally is assumed to bear essential features of the IV-phenomenon. The site index ν for a compound runs over all lattice points whereas in the impurity case ν denotes one single site placed at the origin $\underline{R}_\nu=0$. The Hamiltonian has been written in the spirit of the ionic picture which is particularly convenient in the limit $U \to \infty$ considered here. It uses transfer operators $X^\nu_{MM'}$ acting between the possible ionic states $|M\rangle \equiv |n^f J_z\rangle = |1\tfrac{1}{2}\rangle, |1-\tfrac{1}{2}\rangle, |20\rangle$, which can be expressed in terms of f-electron operators $f_{\nu\sigma}$, and ionic energies E_M which are connected in an obvious way with the original one particle levels E_σ, chemical potential μ included /6/.

A perturbation expansion of the model (1) in terms of the hybridization V meets difficulties resulting from the degeneracies in the unperturbed part of the Hamiltonian, i.e. the first and second terms. Keiter and Kimball have shown how one can sum classes of logarithmically divergent terms in order to obtain a completely well behaved renormalized perturbation series for the free energy of the impurity problem /3/. Their result for the partition function $Z^{(1)}_{\nu=0}$ reads:

$$Z^{(1)} = \sum_M e^{-\beta(E_M + \tilde{E}_M)} \quad , \qquad \tilde{E}_M = \Gamma^0_M(\tilde{E}_M) \quad . \qquad (2)$$

It expresses the influence of interactions on the unperturbed ionic levels E_M, as a simple energy shift $\tilde{E}_M$, which has to be determined as the fixed point of a properly regularized irreducible self energy function Γ^0_M of real argument. Γ^0_M is the sum of all possible different parts without external lines of a local process, which start and end with state $|M\rangle$ and do not possess intermediate energies identically equal to E_M. The simple form of Eq.(2) suggests the notion of renormalized ionic energy levels and of statistical quasiparticles with regard to corresponding ficticious single particle excitations.

Unfortunately, the free particle structure of the partition function Eq.(2) apparently does not find an equally simple analogue in dyramical quantities like the single f-electron Green's function. It turned out recently /6/, that the perturbative method devised for the pertition function can be extended to all the impurity Green's functions. These are not only interesting quantities in themselves, but also serve as a kind of local input for calculations in concentrated systems. It is convenient to express the general result in the following form /6/:

$$C_M(s_1|\ldots|s_\ell) = -\beta \tilde{P}_M [e^{\beta\tilde{E}_M} (\varsigma_M \frac{d}{d\tilde{E}_M})^{\ell-1} (\varsigma_M e^{-\beta\tilde{E}_M} \Gamma^{s_1}_M(\tilde{E}_M) \cdot \ldots \cdot \Gamma^{s_\ell}_M(\tilde{E}_M))] \, . \qquad (3)$$

Eq.(3) gives an explicit expression for the total contribution C_M of all local processes based on the initial state M, which contain (a) exactly l irreducible parts in any temporal sequence, each with a definite nonempty sequence of distinct external lines, which is specified in the symbol s_j (j= ,...,l) including the information about spin σ, Fermi-frequency $i\omega_n$ and direction carried by each line, and additionally (b) all possible combinations of irreducible parts without external lines. In analogy to the quantity Γ^0_M defined above one introduces the sums $\Gamma^{s_j}_M$ for irreducible pieces with the external lines specified in s_j. These sums appear explicetely on the r.h.s. of Eq.(3), whereas Γ^0_M implicitely serves to determine the argument $\tilde{E}_M$ and thereby takes the renormalization of the initial state $|M\rangle$ into account, wherever it appears between the irreducible parts with external lines. This is also the reason for the appearence of the Z-factors ς_M and the changed initial probability $\tilde{P}_M$ given by:

$$\varsigma_M = [1 - \frac{d\Gamma^0_M}{d\tilde{E}_M}(\tilde{E}_M)]^{-1} \, , \quad \tilde{P}_M = Z^{(1)^{-1}} e^{-\beta(E_M+\tilde{E}_M)} = Z^{(1)^{-1}} P_M Z^{(1)}_0 e^{-\beta\tilde{E}_M} \, . \qquad (4)$$

Using the properly normalized $\tilde{P}_M$ instead of $P_M e^{-\beta\tilde{E}_M}$ and leaving out the matrix element V at vertices with external lines gives the most direct connection between the C_M's and local Green's functions. For example in the one particle case one obtains:

$$F^{(1)}_\sigma(i\omega_n) \equiv -\frac{1}{2}\int_{-\beta}^{\beta} d\tau \, e^{i\omega_n\tau} \langle T(X^\nu_{1\sigma 2}(\tau) X^\nu_{21\sigma}) \rangle = \sum_{M=1\sigma,1-\sigma,2} \tilde{P}_M F^{i\sigma}_M(i\omega_n,\sigma) \, , \qquad (5)$$

$$F^{i\sigma}_M(i\omega_n,\sigma) = \varsigma_M [\Gamma^{s_1}_M(\tilde{E}_M) - \Gamma^{s_2}_M(\tilde{E}_M)] \quad , \quad \begin{cases} s_1 = (in, i\omega_n, -\sigma; out, i\omega_n, -\sigma) \\ s_2 = (out, i\omega_n, -\sigma; in, i\omega_n, -\sigma) \end{cases}$$

Approximations can best be formulated using a graphical language, in which the various processes are represented by Goldstone diagrams drawn on an imaginary time axis. An example contributing to the three particle f-Green's function with two different irreducible parts L_1 and L_2 with external lines characterized by $s_2 = (out, i\omega_n, -\sigma;\ in, i\omega_n, -\sigma)$ and $s_1 = (in, i\omega_m, -\sigma;\ in, i\omega_{m'}, -\sigma;\ out, i\omega_{m''}, -\sigma;\ out, i\omega_{m+m'+m''}, -\sigma)$ is:

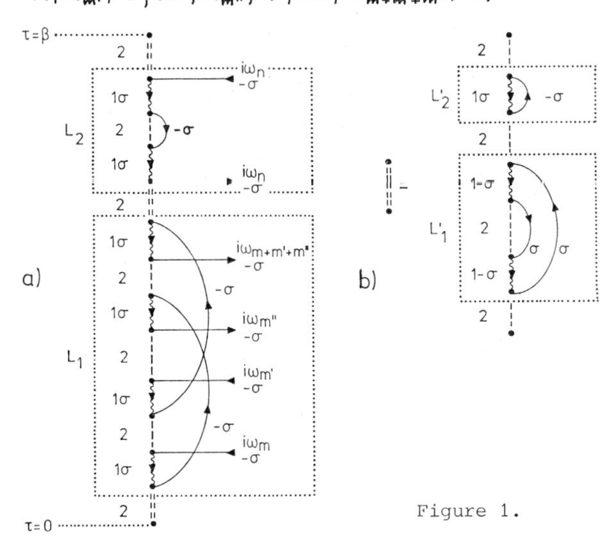

Figure 1.

In Fig.1b a possible insertion for the double hatched lines of Fig.1a is shown. It contains two different irreducible parts L_1' and L_2' without external lines, which contribute to the self-energy $\Gamma_2^o(z)$. The simple buckle L_2' inserted into Γ^o generates the most divergent approximation for $\tilde{E}$, which has extensively been studied by several authors /3-6/ and gives quite a good qualitative explanation for some measured static properties of IV-systems like the magnetic susceptibility. L_1' defines the next divergent approximation if the inner buckle itself is treated as a self energy insertion, too, and proper regularization provided /7/. In the same sense L_2 in Fig. 1a represents the most divergent approximation for the second term, abbreviated as $\Gamma_2^{\sigma i}(i\omega_n, -\sigma)$ on the r.h.s. of Eq. (5) with M=2.

In the most divergent Approximation the following results have been derived in Ref./6/

$$\Gamma_{1\sigma}^{i\sigma}(i\omega_n, -\sigma) = [i\omega_n + \tilde{E}_{1\sigma} - \Delta E_\sigma - \Gamma_2^o(i\omega_n + \tilde{E}_{1\sigma} - \Delta E_\sigma)]^{-1} ,$$

$$\Gamma_2^{\sigma i}(i\omega_n, -\sigma) = -[i\omega_n - \tilde{E}_2 - \Delta E_\sigma + \Gamma_{1\sigma}^o(-i\omega_n + \tilde{E}_2 + \Delta E_\sigma)]^{-1} . \quad (6)$$

where $\Delta E_\sigma = E_2 - E_{1\sigma}$, and the self-energies are given by:

$$\Gamma_{1\sigma}^o(z) = \frac{V^2}{N} \sum_{\underline{k}} \frac{f_{\underline{k}-\sigma}}{z - \Delta E_\sigma + \epsilon_{\underline{k}-\sigma}} , \quad \Gamma_2^o(z) = \frac{V^2}{N} \sum_{\underline{k}\sigma'} \frac{1 - f_{\underline{k}-\sigma'}}{z + \Delta E_{\sigma'} - \epsilon_{\underline{k}-\sigma'}} . \quad (7)$$

In the continuous limit these functions Γ_M^o have a branch cut on the real axis, where $-z + \Delta E_\sigma$ and $z + \Delta E_{\sigma'}$ respectively lie inside the band $\epsilon_{\underline{k}-\sigma}$. For T=0 and $\mathrm{Im}\, z \equiv \delta \to 0$ they furnish imaginary parts of magnitude Δ and 2Δ respectively and sign equal to $-\mathrm{sgn}\,\delta$, if only Re $z > 0$ (and not too large). In the extreme IV-situation $\Delta E_\sigma \approx 0$ one finds $\tilde{E}_2 \approx 2\tilde{E}_{1\sigma} < 0$. Therefore $\Gamma_{1\sigma}^{i\sigma}$ and $\Gamma_2^{\sigma i}$ at T=0, if analytically continued to the real ω-axis, both have a simple pole at $\omega_\sigma = \Delta E_\sigma + \tilde{E}_2 - \tilde{E}_{1\sigma} < 0$ and smooth nonzero imaginary parts for $\omega > -\tilde{E}_{1\sigma} > 0$ and $\omega < \tilde{E}_2 < \omega_\sigma$ respectively.

This picture however and its implications for the Green's function Eq. (5) are certainly incomplete. The irreducible part L_2 considered so far may be obtained from L_1' by cutting the outer band electron line. If one cuts the inner line instead (and adds the most divergent self energy insertions in the intermediate state 2) one arrives at the following first approximation for $\Gamma_2^{i\sigma}$:

$$\Gamma_2^{i\sigma}(i\omega_n, \sigma) = \frac{V^2}{N} \sum_{\underline{k}} (\tilde{E}_2 + \Delta E_{-\sigma} - \epsilon_{\underline{k}\sigma})^{-2} (1 - f_{\underline{k}\sigma}) \times \Gamma_{1-\sigma}^{i\sigma}(i\omega_n + \omega_{-\sigma} - \epsilon_{\underline{k}\sigma}, \sigma) . \quad (8)$$

When analytically continued, this apparently furnishes for $\omega > 0$ a smooth nonzero imaginary part to $F_M^{i\sigma}(i\omega_n, \sigma)$, Eq. (5), beside the one for $\omega < \tilde{E}_2$ and the pole at ω_σ coming from $\Gamma_2^{\sigma i}$. A more extensive discussion with further results and their consequences for the properties of an IV-compound will be published elsewhere /8/. It is suggested that the final result has the form

$$F_2^{i\sigma}(\omega + i\delta, \sigma) = \xi [\omega - \omega_\sigma + i\,\mathrm{sgn}\,\delta \cdot \Delta - \Sigma_+^f(\omega + i\delta)]^{-1} , \quad (9)$$

where the residual self-energy Σ_+^f has a smooth imaginary part, which for T=0 vanishes or is very small in a neighborhood of ω_σ.

III. LOW TEMPERATURE TREATMENT OF CONCENTRATED SYSTEMS

The following discussion of an IV-compound is based on the periodic Anderson model (1) and focuses on the partition function and the one particle Green's functions, which now are briefly introduced. It is convenient to consider the reduced partition function $\tilde{Z} = Z/Z_{band} \cdot \prod_\nu Z_\nu^{(1)}$ normalized in order to give directly the increment $\tilde{\Omega} = -\frac{1}{\beta} \ln \tilde{Z}$ in thermodynamic potential due to intersite interactions. The single particle excitation spectrum may be obtained from the band electron Green's function

$$G_{\underline{k}\sigma}(i\omega_n) = -\frac{1}{2} \int_{-\beta}^{\beta} d\tau\, e^{i\omega_n \tau} \langle T(d_{\underline{k}\sigma}(\tau) d_{\underline{k}\sigma}^+) \rangle . \quad (10)$$

It is connected with the 4f-Green's function F_σ and the T-matrix $T_{\underline{k}\sigma}$ defined by

$$F_\sigma(\underline{R}_{\nu\mu}, i\omega_n) = -\frac{1}{2} \int_{-\beta}^{\beta} d\tau\, e^{i\omega_n \tau} \langle T(X_{1\sigma 2}^{\mu}(\tau) X_{21\sigma}^{\nu}) \rangle$$

$$= V^{-2} \frac{1}{N} \sum_{\underline{k}} e^{-i\underline{k}\underline{R}_{\nu\mu}} T_{\underline{k}-\sigma}(i\omega_n) \quad (11)$$

via a sequence of two exact equations of motion leading to

$$G_{\underline{k}\sigma}(i\omega_n) = G_{\underline{k}\sigma}^o(i\omega_n) + G_{\underline{k}\sigma}^o(i\omega_n) T_{\underline{k}\sigma}(i\omega_n) G_{\underline{k}\sigma}^o(i\omega_n) . \quad (12)$$

In Eq. (11) $\underline{R}_{\nu\mu} = \underline{R}_\mu - \underline{R}_\nu$ has been introduced and unlike in Eq. (5) the two site indices ν and μ need not be equal. $G_{\underline{k}\sigma}^o(i\omega_n) = [i\omega_n - \epsilon_{\underline{k}\sigma}]^{-1}$ is the unperturbed band electron Green's function.

The lack of Wick's theorem discussed in Sec.I has as a consequence $\lim_{R_{\nu\mu}\to\infty} F_\sigma(R_{\nu\mu}, i\omega_n) \neq F_\sigma(0, i\omega_n)$ where both sides are defined via Eq. (5). In the diagrammatic example shown in Fig.2 this deficiency in general prevents a factorization into the contributions of the two closed rings. The two rings with circulating independent frequencies $i\omega_n$ and $i\omega_m$ respectively join the same site ν_2 -although in different irreducible parts L_1, L_2- and therefore are correlated to have the same initial and final states M and the same local shift $\tilde{E}_M$ at this point.

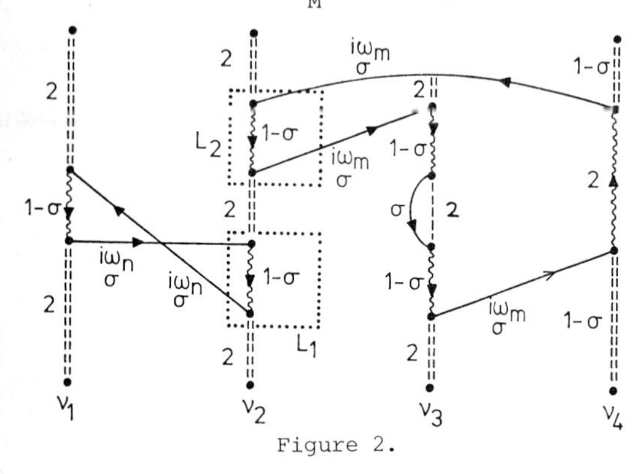

Figure 2.

Additionally, it is essential for the same reason to maintain in all summations over sites that $\nu_1, \nu_2, \nu_3, \nu_4$ are actually different, which leads to complicated site exclusions and prevents cumulant and linked cluster expansions of being useful.

There exist however particular circumstances in which these difficulties can be overcome. A first important condition is just realized in the IV-situation, where only one of the possible ionic states contributes at low temperatures, namely the singlet M=2. More precisely, it is well known from renormalized perturbation theory /5,6/ and from renormalization group analysis /9,10/ that there exists a regime of parameters of the original Anderson model, typically the most asymmetric case with $\Delta E_\sigma \approx 0$, where the singlet by hybridization effects is always renormalized to a lower energy than the multiplet (here the doublet $M=1\sigma$). The reason simply lies in the larger number of intermediate states available for the singlet. In the present context this implies $E_2 + \tilde{E}_2 < E_{1\sigma} + \tilde{E}_{1\sigma}$ so that the local quasiparticle occupation probabilities introduced in Eq.(4) have the following low temperature behaviour:

$$\text{IV}: \quad \tilde{P}_2 \to 1 \ , \quad \tilde{P}_{1\sigma} \to 0 \quad (T\to 0) \quad . \qquad (13)$$

This constitutes an essential simplification insofar, as only diagrams with the (renormalized) singlet as initial state have to be considered at T=0. In the frame of the present treatment condition (13) has a self-consistent character. In principle, it has to be checked after a low temperature calculation of the grand canonical potential Ω has been carried through, the chemical potential been determined via electron conservation, $N_{el} = -\partial\Omega/\partial\mu|_{T,B}$, and inserted into $E_2, E_{1\sigma}$.

A second essential point can directly be seen from Eq.(3) for the contribution of a local process with initial state M(=2) and irreducible pieces with external lines. Each of the derivatives can be performed on the factors $\mathcal{G}_M$, $e^{-\beta\tilde{E}_M}$ and the various Γ_M's. Since it is to be expected that $\mathcal{G}_M$ and the Γ_M's are well behaved functions for all temperatures in a real neighborhood of $\tilde{E}_M$, the most important terms at low temperatures arise from the differentiation of $e^{-\beta\tilde{E}_M}$, giving factors $-\beta$. Then Eq.(3) factorizes according to

$$C_M(s_1|\ldots|s_\ell) \to \tilde{P}_M \prod_{j=1}^{\ell} [-\beta\mathcal{G}_M \Gamma_M^{s_j}(\tilde{E}_M)] \quad (T\to 0) \qquad (14)$$

into the contributions of its irreducible parts $\Gamma_M^{s_j}$. Altogether, in a rigorous way at T=0, for a multisite process all sites now become equivalent in that they all have the same initial state M=2, and they all can be revisited from everywhere without a correlation of the corresponding irreducible local part with others on the same site. This fact constitutes, as will be explained below, a powerful substitute for Wick's theorem.

It is now a straightforward task to develop a cumulant expansion of the thermodynamic potential $\tilde{\Omega}$ and linked cluster expansions for the Green's functions. As will be discussed in detail elsewhere /8/ the cumulant can be identified with the contributions R_λ of generalized (closed) rings and the linked clusters with those of generalized (open) chains S_λ. A diagrammatic example for such a generalized ring is shown in Fig.3.

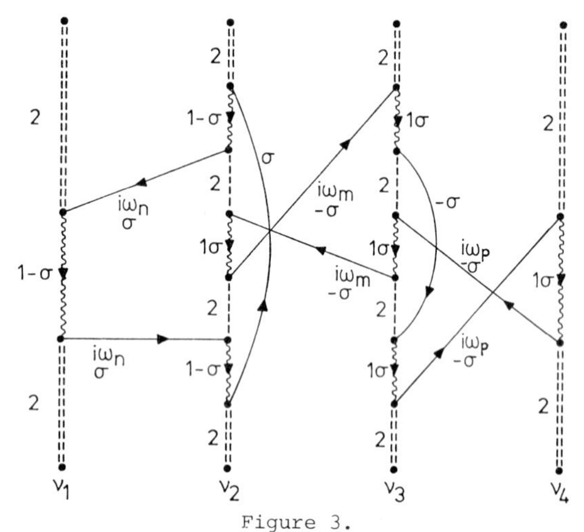

Figure 3.

One recognizes a quite complicated topological structure with three subrings carrying different Fermi-frequencies $i\omega_n, i\omega_m, i\omega_p$ the contribution of which do not factorize because they are glued together in the irreducible parts at sites ν_2 and ν_3 with four external lines each. If only irreducible parts with two external lines would be taken into account the resulting rings would be simple with only one circulating frequency $i\omega_n$. If one of the intersite connections in Fig. 3 would be cut open and the corresponding ends of this line considered as external lines of the whole diagram, one would obtain a generalized chain contributing to the f-Green's function (11).

As a rule a generalized ring R is identified by its specific different $\varkappa$ sites $\{\nu_1,\ldots,\nu_\varkappa\}$, by a tupel $(s_1^{(j)},\ldots,s_{\ell_j}^{(j)})$ of symbols for each site j, which characterize the external lines of possible irreducible parts Γ_2^s, and the specific network of bonds (band electron lines) between the parts on different sites. All different Fermi-frequencies $i\omega_n$ and spin indices σ in a ring and one wave vector for each bond are summed over. As far as the detailed diagrammatic rules are concerned, the reader is referred to Ref./6/. A generalized chain is obtained from a ring by cutting a bond and dropping the corresponding summations. It is then possible to derive the following results /8/

$$\tilde{\Omega} = -\frac{1}{\beta}\sum_\lambda R_\lambda \quad , \quad T_{\underline{k}\sigma}(i\omega_n) = -\frac{1}{\beta}\sum_\lambda \frac{1}{N} S_\lambda \quad , \quad (15)$$

which are exactly valid at T=O. The summations are over all different generalized rings and chains respectively.

At this stage, a further simplification is achieved by the introduction of topologically distinct rings $\tilde{R}_\alpha$ (chains $\tilde{S}_\alpha$), which sum classes of different, but similar, rings (chains). $\tilde{R}_\alpha$ is characterized by the number ℓ of site-indices ν_j involved, by a set of symbols $\{s_1,\ldots,s_\ell\}$ -one for each site-index- and the specific network of bonds between the irreducible local parts belonging to these symbols. Sums are performed as in R_λ, but an additional summation over the site-indices ν_j is also implied with the restriction, that each single bond runs between different sites. It is essential that there is only one symbol s_j for each site-index ν_j, since all contributions with more than one irreducible part at a site are automatically generated by this summation prescription. Each sequence of bonds may return to the same site after a minimum of two steps. Since topological rings have certain invariance properties against the interchange of site-symbols, an additional symmetry factor t_α has to be introduced, leading to:

$$\tilde{\Omega} = -\frac{1}{\beta}\sum_\alpha t_\alpha \tilde{R}_\alpha \quad , \quad T_{\underline{k}\sigma}(i\omega_n) = -\frac{1}{\beta}\sum_\alpha \frac{1}{N}\tilde{S}_\alpha \quad . \quad (16)$$

Analogous results exist for all higher Green's functions. The meaning of Eq.(16) will become more explicit in the next section, where the r.h.s. are evaluated using irreducible parts with two external lines only.

Concluding this section, it is pointed out that the foregoing formulas for thermodynamic potential and T-matrix keep their rigorous character also for nonzero T in the sense of a low temperature expansion. Besides the temperature dependence already contained in the expressions Γ_2^s for local irreducible parts one has to include systematically (renormalized) doublets $M=I\sigma$ as initial states and those terms generated by differentiating the $\mathcal{G}$'s and Γ's instead of $\exp(-\beta\tilde{E}_M)$ in Eq. (3) /8/.

IV. EXPLICIT RESULTS FOR IV-COMPOUNDS

The remaining task is to calculate sums of topologically different rings and chains. It turns out that all simple rings and chains, which are obtained by using only local irreducible parts with up to two external lines, can be summed in closed form. The general structure of a simple ring is shown in Fig.4, where it is understood that subsequent site indices are different, $\nu_j \neq \nu_{j+1}$ $(j=1,\ldots,\ell;\ \nu_{\ell+1}\equiv\nu_1)$, but otherwise may coincide. The boxes stand for all possible irreducible parts with two external lines, which are collected in $\mathcal{G}_2^{-1} F_2^{i\sigma}(i\omega_n,\sigma)$.

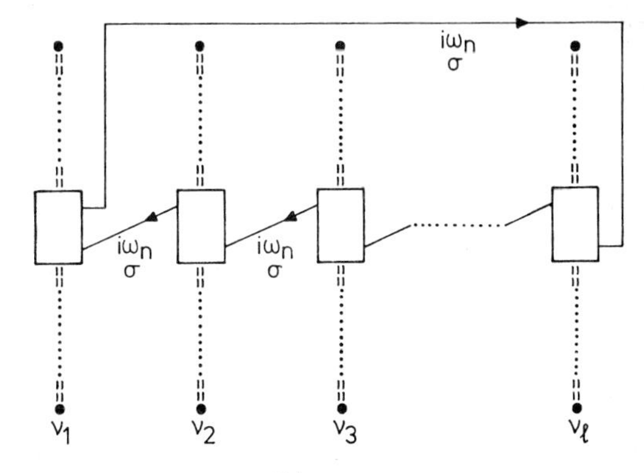

Figure 4.

A simple chain again is obtained by cutting one of the bonds between sites. It is obvious that a topologically distinct simple ring is completely characterized by the number l of its site indices, and that its symmetry number t_ℓ is $\frac{1}{\ell}$, since one may start with any of the irreducible parts shown in Fig.4. The corresponding contribution according to the diagrammatic rules stated in Ref./6/ is:

$$t_\ell \tilde{R}_\ell = -\frac{1}{\ell}\sum_{\omega_n,\sigma}\ \sum_{\substack{\nu_1,\ldots,\nu_\ell \\ \nu_1\neq\nu_2\neq\ldots\neq\nu_\ell\neq\nu_1}}\ \prod_{j=1}^{\ell}[V^2F_2^{i\sigma}(i\omega_n,\sigma) \times G^0_{-\sigma}(\underline{R}_{\nu_{j+1}\nu_j}, i\omega_n)] \ . \ (17)$$

Here the convention $\ell+1\equiv 1$ is used and coherence factors and $\underline{k}$-sums have been included into the band propagator:

$$G^0_{-\sigma}(\underline{R}_{\nu\mu}, i\omega_n) = \frac{1}{N}\sum_{\underline{k}} e^{i\underline{k}(\underline{R}_\nu - \underline{R}_\mu)} G^0_{\underline{k}-\sigma}(i\omega_n) \quad . \quad (18)$$

With a definite enumeration of lattice sites it is convenient to introduce a matrix-notation:

$$(\hat{F}(i\omega_n,\sigma))_{\nu\mu} = \delta_{\nu\mu}V^2F_2^{i\sigma}(i\omega_n,\sigma) \ , \ (\hat{G}^0(i\omega_n,\sigma))_{\nu\mu} = G^0_\sigma(\underline{R}_{\mu\nu}, i\omega_n) \ ,$$

$$(\hat{D}^0(i\omega_n,\sigma))_{\nu\mu} = \delta_{\nu\mu}G^0_\sigma(0,i\omega_n) = \delta_{\nu\mu}\frac{1}{N}\sum_{\underline{k}}\frac{1}{i\omega_n - \epsilon_{\underline{k}\sigma}} \quad . \quad (19)$$

The trivial ring l=1 with contribution 1 to Z has been accounted for already by the exponential form of Z, see Eq.(16), so that the sum of all topologically different simple rings gives the following result for the thermodynamic potential:

$$\tilde{\Omega} = \frac{1}{\beta}\sum_{\ell=2}^{\infty}\frac{1}{\ell}\sum_{\omega_n,\sigma} \mathrm{Tr}\{[\hat{F}(i\omega_n,\sigma)(\hat{G}^{\circ}-\hat{D}^{\circ})(i\omega_n,-\sigma)]^{\ell}\}$$

$$= -\frac{1}{\beta}\sum_{\omega_n,\sigma} \mathrm{Tr}\{\ln[\hat{I}-\hat{F}(i\omega_n,\sigma)(\hat{G}^{\circ}-\hat{D}^{\circ})(i\omega_n,-\sigma)]\}. \quad (20)$$

$\hat{I}$ is the unit matrix and $\mathrm{Tr}\{\hat{F}(\hat{G}^{\circ}-\hat{D}^{\circ})\}=0$ has been used. The quantities $\tilde{R}_\ell$ can be viewed as the cumulants for the partition function with site exclusions, which already in the present approximation deviate from their standard form /8/. Eq.(20) reduces the ground state energy of an IV-compound to the trace of a matrix in lattice space, the only input of which is the function $F_2^{i\sigma}$ to be derived from the impurity problem. Physically, it expresses the gain in binding energy brought about by the interactions between renormalized singlets at all lattice points occupied by a Rare Earth ion, which are mediated by band electrons travelling from site to site.

In a simple chain the last site index ν_ℓ is -in contrast to a ring- free to sum over. Since the first site in a chain can uniquely be fixed by the outgoing external line, no symmetry factor is necessary. One obtains:

$$\tilde{S}_\ell = -\beta V^2 F_2^{i\sigma}(i\omega_n,-\sigma)\sum_{\substack{\nu_1,\dots,\nu_\ell\\ \nu_1\neq\nu_2\neq\dots\neq\nu_\ell}} e^{i\underline{k}(\underline{R}_{\nu_1}-\underline{R}_{\nu_\ell})}\prod_{j=1}^{\ell-1}[V^2F_2^{i\sigma}(i\omega_n,-\sigma) \times G^{\circ}_{\sigma}(\underline{R}_{\nu_{j+1}\nu_j}, i\omega_n)]. \quad (21)$$

Now the restricted summation over sites of the product of coherence factors on the r.h.s. of Eq.(21) can explicitely be carried out with the result

$$\sum_{\substack{\nu_1,\dots,\nu_\ell\\ \nu_1\neq\nu_2\neq\dots\neq\nu_\ell}} e^{i\underline{k}(\underline{R}_{\nu_1}-\underline{R}_{\nu_\ell})}e^{i\underline{k}_1(\underline{R}_{\nu_2}-\underline{R}_{\nu_1})}\cdot\dots\cdot e^{i\underline{k}_{\ell-1}(\underline{R}_{\nu_\ell}-\underline{R}_{\nu_{\ell-1}})} = N\prod_{j=1}^{\ell-1}(N\delta_{\underline{k}_j\underline{k}}-1), \quad (22)$$

so that the sum of all topologically different simple chains is geometric and leads to the following result for the T-matrix:

$$T_{\underline{k}\sigma}(i\omega_n) = V^2F_2^{i\sigma}(i\omega_n,-\sigma)[1-V^2F_2^{i\sigma}(i\omega_n,-\sigma) \times (G^{\circ}_{\underline{k}\sigma}(i\omega_n)-D^{\circ}_{\sigma}(i\omega_n))]^{-1}. \quad (23)$$

In accord with Eq.(19), $D^{\circ}_{\sigma}(i\omega_n) = \frac{1}{N}\sum_{\underline{k}}[i\omega_n-\epsilon_{\underline{k}\sigma}]^{-1}$ has been introduced. Eq.(23) describes multiple scattering of a band electron from the system of renormalized ionic singlets, but without irreducible merging of two scattering events. It is left to write down the final results for the two single particle Green's functions, Eq.'s (11) and (12), which can be done in the following form:

$$F_\sigma(\underline{R}_{\nu\mu}, i\omega_n) = \frac{1}{N}\sum_{\underline{k}} e^{-i\underline{k}\underline{R}_{\nu\mu}}[F_2^{i\sigma}(i\omega_n,\sigma)^{-1}-\Sigma^{f}_{\underline{k}\sigma}(i\omega_n)]^{-1},$$

$$G_{\underline{k}\sigma}(i\omega_n) = [G^{\circ}_{\underline{k}\sigma}(i\omega_n)^{-1}-\Sigma^{d}_{\underline{k}\sigma}(i\omega_n)]^{-1}. \quad (24)$$

The two self-energies are:

$$\Sigma^{f}_{\underline{k}\sigma}(i\omega_n) = V^2[G^{\circ}_{\underline{k}-\sigma}(i\omega_n)-D^{\circ}_{-\sigma}(i\omega_n)],$$

$$\Sigma^{d}_{\underline{k}\sigma}(i\omega_n) = V^2[F_2^{i\sigma}(i\omega_n,-\sigma)^{-1}+V^2D^{\circ}_{\sigma}(i\omega_n)]^{-1}. \quad (25)$$

As Eq.(20) for the thermodynamic potential, these results for the Green's functions reduce properties of the compound to those of the impurity problem. Some direct conclusions may be drawn by inserting the suggested form (9) for $F_2^{i\sigma}$ and using $V^2D^{\circ}_{\sigma}(\omega+i\delta) = -i\,\mathrm{sgn}\,\delta\cdot\Delta$ in Eq.(24). Then one obtains:

$$T_{\underline{k}\sigma}(\omega+i\delta) = \xi V^2[\omega-\omega_{-\sigma}+i\,\mathrm{sgn}\,\delta\cdot(1-\xi)\Delta - \Sigma^{f}_{+}(\omega+i\delta)-\xi V^2 G^{\circ}_{\underline{k}\sigma}(\omega+i\delta)^{-1}]^{-1},$$

$$G_{\underline{k}\sigma}(\omega+i\delta) = [G^{\circ}_{\underline{k}\sigma}(\omega+i\delta)^{-1}-\xi V^2(\omega-\omega_{-\sigma} + i\,\mathrm{sgn}\,\delta\cdot(1-\xi)\Delta-\Sigma^{f}_{+}(\omega+i\delta))^{-1}]^{-1}. \quad (26)$$

Were it not for the still unknown, but possibly small residual self-energy Σ^{f}_{+}, Eq.(26) would describe a band $\epsilon_{\underline{k}\sigma}$ hybridized with a broadened local level $\omega_{-\sigma}$ for each spin direction. The effective hybridization is $\sqrt{\xi}\cdot V$ and the level-width $(1-\xi)\Delta$. (Notice: $\xi<1$.) This would lead to a small density of states inside the original gap. It must however be pointed out that no definite conclusions are possible at the moment. Moreover, as one can convince oneself very easily by inserting the known low order Brillouin-Wigner results for $F_2^{i\sigma}$, see for instance Eq.'s (6) and (8) or the results of Ref./6/, the spectral functions derived from Eq.'s (26) are even not positive semi-definite everywhere. Apparently one needs a much larger effort to derive meaningful results for dynamical quantities of the impurity problem in contrast to the partition function, which is given in terms of three single numbers $\tilde{E}_{1\sigma}$, $\tilde{E}_2$ only. It may be necessary to resort to integral equations, which locally sum infinite classes of terms in renormalized perturbation theory, as has been suggested by Keiter /3,7/.

V. CONCLUSION

In order to review the main line of thought and to shed an additional light on the difficulties encountered in the foregoing treatment of an IV-compound at low temperatures it is useful to study the case without Coulomb-repulsion, U=0. The corresponding Hamiltonian

$$\bar{H} = \sum_{\underline{k}\sigma}\epsilon_{\underline{k}\sigma}d^{+}_{\underline{k}\sigma}d_{\underline{k}\sigma} + \sum_{\nu\sigma}E_\sigma f^{+}_{\nu\sigma}f_{\nu\sigma} + \frac{V}{\sqrt{N}}\sum_{\nu\underline{k}\sigma}[e^{i\underline{k}\underline{R}_\nu}d^{+}_{\underline{k}\sigma}f_{\nu\sigma}+\mathrm{h.c.}] \quad (27)$$

is bilinear and can trivially be solved leading to the well known picture of a broad band hybridized with a flat one for each spin direction /11/. It is instructive however, to see how this result comes out when a route analogous to the treatment of the U=∞ -case of this paper is chosen. From the equations of motion one easily convinces oneself that the quantity $\bar{F}_\sigma(\underline{R}_{\nu\mu}, i\omega_n) = \langle\langle f_{\mu\sigma}; f^{+}_{\nu\sigma}\rangle\rangle_{i\omega_n}$ (Zubarev's notation) now plays the role of the 4f-Green's function $F(\underline{R}_{\nu\mu}, i\omega_n)$

defined in Eq.(11).

Sorting out only local processes at a particular fixed site $\nu=0$ is -apart from the role played by the shift in chemical potential, which is not explicitely considered here- equivalent to calculating the local Green's function $\bar{F}^{(1)}_{\sigma}(i\omega_n)$ of the impurity problem at this site. $\bar{F}^{(1)}_{\sigma}$ also is exactly known /11/,

$$\bar{F}^{(1)}_{\sigma}(i\omega_n) = [i\omega_n - E_{\sigma} - V^2 \frac{1}{N}\sum_{\underline{k}} G^{\circ}_{\underline{k}\sigma}(i\omega_n)]^{-1} \quad . \quad (28)$$

It corresponds to the result (5), which at low temperatures and in the IV-situation reduces to the quantity $F_2^{i\sigma}(i\omega_n,\sigma)$. The resonant levelform of Eq.(28) can be recognized in the final version (9) for $F_2^{i\sigma}$ (using $V^2\frac{1}{N}\sum_{\underline{k}} G^{\circ}_{\underline{k}\sigma}(\omega + i\delta) - i\,\mathrm{sgn}\,\delta\cdot\Delta$), in which because of $\xi<1$ spectral weight from the pole at ω_{σ} is shifted to regions where the residual self-energy Σ^f_{σ} is nonzero. Physically, Eq.(28) describes a broad resonance of width Δ centered around $\omega = E_{\sigma}$. In the compound problem, the corresponding pole at z=E-$i\,\mathrm{sgn}\,\mathrm{Im}\,z\cdot\Delta$ on the unphysical sheet has to be cancelled by the interactions between these resonant levels at each lattice point mediated by travelling band electrons. This part of the calculation resembles a random-walk problem.

The inclusion of all diagrams with bonds between different sites into the calculation of the full 4f-Green's function $\bar{F}_{\sigma}(\underline{R}_{\nu\mu},i\omega_n)$ is straightforward, since Wick's theorem and the linked cluster theorem are available in the noninteracting case. Local parts always factor into one-particle mixing events described by the impurity Green's function $\bar{F}^{(1)}_{\sigma}(i\omega_n)$ and disconnected closed (Feynman-) diagrams are cancelled by the partition function in the denominator. To prevent overcounting one only has to make sure that in a nonlocal contribution bonds really are present only between different sites. This leads to

$$\begin{aligned}\bar{T}_{\underline{k}\sigma}(i\omega_n) &= V^2\bar{F}^{(1)}_{\sigma}(i\omega_n)\{1+V^2\bar{F}^{(1)}_{\sigma}(i\omega_n)[G^{\circ}_{\underline{k}\sigma}(i\omega_n)-\frac{1}{N}\sum_{\underline{k}'}G^{\circ}_{\underline{k}'\sigma}(i\omega_n)] \\ &\qquad + \ldots\} \\ &= V^2\{\bar{F}^{(1)}_{\sigma}(i\omega_n)^{-1} - V^2[G^{\circ}_{\underline{k}\sigma}(i\omega_n)-\frac{1}{N}\sum_{\underline{k}'}G^{\circ}_{\underline{k}'\sigma}(i\omega_n)]\}^{-1} \\ &= V^2[i\omega_n - E_{\sigma} - V^2 G^{\circ}_{\underline{k}\sigma}(i\omega_n)]^{-1} \qquad (29)\end{aligned}$$

where $\bar{T}_{\underline{k}\sigma}$ as before is V^2 times the Fourier-transform of $\bar{F}_{\sigma}(\underline{R}_{\nu\mu})$ and Eq.(28) has been inserted in the last line. Eq.(29) exhibits the structure of sharp, hybridized bands mentioned before, and this comes about because the exclusion term $\frac{1}{N}\Sigma_{\underline{k}'}G^{\circ}_{\underline{k}'\sigma}(i\omega_n)$ in the random walk form of Eq.(29) (first line) cancels the resonant level pole in the impurity Green's function (28).

This exact cancellation signals possible difficulties in the problem with Coulomb-interactions. Presumably, most of the approximations one is forced to do in the impurity-problem, i.e. the "local part" of the compound problem, are not fully consistent with those made in treating all possible kinds of bonds in the concentrated system. An example for a possible effect of such inconsistencies are the aforementioned one particle spectral densities which may become negative in some region of the ω -axis, if a low order Brillouin-Wigner summation of local parts is combined with the approximation of linked clusters in nonlocal diagrams by simple chains. One can hope that these deficiencies may be remedied by summing selected classes of renormalized local diagrams, for example via integral equations. Actual results about the one-particle spectrum of the compound can only be obtained, after more is known about the residual local self-energy defined in Eq.(9). (First approximations are obtained by using Eq.'s (6) to (8).) If this quantity is set to zero, Eq.'s (26) describe the effect of mixing the bands $\epsilon_{\underline{k}\sigma}$ with a lattice of broadened local levels, which only leads to a pseudogap. The reason for this outcome would be that only a part $\xi<1$ of each local resonance is coherently bound via propagation through the band, whereas the other part $1-\xi$ is smeared out over the region of the original gap.

A final word concerns the role of the new results derived in Sec.'s III and IV of this paper for an actual calculation of compound properties with the diagrammatic perturbation technique developed in Ref./6/. It seems now as if these properties in the IV-regime -at least at low temperatures- are directly related to those of the IV-impurity via Eq.'s (20) and (26) and a self-consistency-loop for the chemical potential at fixed particle number.

ACKNOWLEDGEMENTS

The author would like to thank Professors H.Keiter and E.Müller-Hartmann for valuable discussions.

This work was performed in the program of the Sonderforschungsbereich 125 "Magnetische Momente und Unordnungsphänomene in Metallen".

REFERENCES

/1/ For recent review articles see:
J.M.Jefferson and K.W.Stevens, J.Phys.C11, 3919 (1978)
N.Grewe, H.J.Leder and P.Entel, in "Festkörperprobleme XX", edited by J.Treusch (Vieweg, Braunschweig, 1980)
J.M.Lawrence, P.S.Riseborough and R.D.Parks, Rep.Progr.Phys.44,1 (1981)

/2/ P.W.Anderson, in "Valence Fluctuatons in Solids", edited by L.M.Falicov, W.Hanke and M.B.Maple (North-Holland, Amsterdam, 1981)

/3/ H.Keiter and J.C.Kimball: Int.J.Magnetism 1, 233 (1971)

/4/ A.Bringer and H.J.Lustfeld: Z.Physik B22, 213 (1977)

/5/ T.V.Ramakrishnan, in "Valence Fluctuations in Solids", edited by L.M.Falicov, W.Hanke and M.B.Maple (North-Holland, Amsterdam, 1981)
T.V.Ramakrishnan, preprint

/6/ H.Keiter and N. Grewe, in "Valence Fluctuations in Solids", edited by L.M. Falicov, W. Hanke and M.B. Maple (North Holland, Amsterdam, (1981)
N. Grewe and H.Keiter, Phys.Rev.B24, 4420 (1981)

/7/ See the article of H.Keiter et al. in this same volume.

/8/ N. Grewe, to be published in Z.Physik

/9/ H.R. Krishna-murthy, J.W.Wilkins and K.G. Wilson, Phys.Rev.B21, 1003; 21, 1044 (1980)

/10/ J.H. Jefferson, J.Phys,C10, 3589 (1977)
F.D.M. Haldane, Phys.Rev.Lett.40, 416 (1978)

/11/ See for example: W.A. Harrison, "Solid State Theory", McGraw-Hill Book Company, New York, 1970

COMMENT: H. KEITER: If your f states were real singlets with occupancy = 1, how is a non-integer valence obtained ?

N. GREWE: At first it must be stressed that the description presented does primarily use the concept of "statistical quasiparticles". These are fictitious entities, whose energies completely determine the partition function of the impurity problem. At vanishing temperature the important diagrammatic contributions are based on the renormalized (quasiparticle-) singlet. Renormalization means admixture of the degenerate configuration and thus produces a non-integer valence. This valence can, in principle, vary continuously with a variation of the chemical potential which influences the degree of admixture via the quasiparticle energy shifts E_M.

A.C. HEWSON: I should like to draw your attention to the fact that there is a generalized Wick's theorem for the transfer operators X_{pg} in the sense that a many time Green's function can be factorized into sums of products of two time Green's functions. This can be used as a basis for a perturbation theory. The resulting diagrammatic formalism is complicated but may be simpler than the approach used here. It does have the advantage that certain types of excitations are associated with corresponding propagators.

N. GREWE: I have seen this non-standard reduction scheme but was not able to make direct use of it. The advantage of the locally time ordered perturbation expansion is clearly that one uses Goldstone-diagrams representing physical processes and that one can actually perform the infinite order resummations in a systematic way which are necessary to remove the divergencies inherent in perturbation theory. I hope it also became clear from my talk that with the present scheme one may calculate all the necessary Green's functions and obtain the excitation spectrum.

Valence Instabilities
P. Wachter and H. Boppart (eds.)

SOLUTION OF THE MIXED-VALENCE LATTICE PROBLEM BY A NUMERICAL RENORMALIZATION-GROUP METHOD, AND VARIATIONAL RESULTS FOR A MIXED-VALENT IMPURITY WITH BOTH STATES MAGNETIC

C. M. Varma, M. Schlüter and Y. Yafet

Bell Laboratories
Murray Hill, New Jersey 07974

The major part of this paper deals with a new numerical renormalization group method for the one-dimensional mixed-valent lattice model, developed by M. Schlüter and C. M. Varma. This model is equivalent in a certain range of parameters to a model for a one dimensional rare-earth metal lattice (sometimes called the Kondo Lattice[1]). We will present excerpts of the results of the calculations using the new method for various choices of parameters. The results include ground state spin and charge correlations, low lying excitation energies and low temperature specific heat and susceptibility.

We also present a summary of the results (obtained by C. M. Varma and Y. Yafet) for the energies of the ground and a few important excited states for a mixed-valence impurity with both states magnetic. These calculations are variational, of the type used earlier[2] to obtain similar quantities for a mixed-valence impurity with one of the states non-magnetic.

The One-Dimensional Anderson Lattice

We consider the Hamiltonian,

$$H = \epsilon_\alpha \sum_{i,\sigma} C^+_{i\alpha\sigma} C_{i\alpha\sigma} + U \sum_i n_{i\alpha\sigma} n_{i\alpha-\sigma}$$

$$+ \epsilon_\beta \sum_{i,\sigma} C^+_{i\beta\sigma} C_{i\beta\sigma} + \frac{1}{2} V_\beta \sum_i (C^+_{i+1,\beta\sigma} C_{i\beta\sigma} + \text{c.c.})$$

$$+ V_{\alpha\beta} \sum_{i,\sigma} (C^+_{i\alpha\sigma} C_{i\beta\sigma} + \text{c.c.})$$

$$- \mu \sum_{i,\sigma} (n_{i\alpha\sigma} + n_{i\beta\sigma}) \quad , \tag{1}$$

describing localized orbitals α on a one dimensional lattice of sites i hybridized on the same site to orbitals β by the matrix element $V_{\alpha\beta}$. The orbitals β have the nearest neighbor hopping matrix element V_β. We will work in the limit that Coulomb repulsion U for two orbitals α on the same site is infinity. Then a given site either has no f electron or an f-electron with up or down spin. (One of the mixed-valence states is therefore "non-magnetic"). Of interest is the case $V_{\alpha\beta} \ll V_\beta$. We obviously have the f-orbitals in mind as α and the s–d orbitals as β although in the present work orbital degeneracy is not considered. μ is the chemical potential. Without loss of generality we take $\epsilon_\beta = 0$.

For $(\mu - \epsilon_\alpha)/V_{\alpha\beta} \gg 1$, but μ still in the range $-2V_\beta < \mu < 2V_\beta$, the rare-earth metals are described. This is equivalent to the recently discussed Kondo Lattice:

$$H + \sum_i J\vec{S}_{\alpha i} \cdot \vec{S}_{\beta i} + \frac{1}{2} V_\beta \sum_i (C^+_{i+1\beta\sigma} C_{i\beta\sigma} + \text{c.c.}) \tag{2}$$

with

$$J = -|V_{\alpha\beta}|^2 / |\mu - \epsilon_\alpha| \tag{3}$$

We combine the real space renormization group (RB) technique with statistical methods of treating spectra in complex quantum-mechanical problems and take advantage of the one small parameter, $|V_{\alpha\beta}/V_\beta|$, in the present problem. The RG method consists in itertatively expressing the Hamiltonian for the 2N site problem in a basis set which is formed of the products of the eigenstates of the N site problems. The method works straight-forwardly in principle for one-dimensional problems with short-range interactions. The Hamiltonian for 2N sites $H^{(2N)}$ can be written

$$H^{(2N)} = H^{(N),L} + H^{(N),R} + H^{(N),LR} \quad , \tag{4}$$

where L and R stand for left and right and $H^{(N),LR}$ is the interaction term mixing the solutions of the N site problems. Let the solutions of the N site problem be $\Psi^{(N)}_{\{\lambda_1\}}$ where $\{\lambda_1\}$ are the set of quantum numbers. The quantum numbers are the conserved set of quantities, charge, total spin, z-component of spin, and parity about the mid-point, at any stage of iteration. The basis set $\Phi^{(2N)}_{\{\lambda_1\},\{\lambda_2\}}$ of the 2N site problem is

$$\Phi^{(2N)}_{\{\lambda_1\},\{\lambda_2\}} = \Psi^{(N),L}_{\{\lambda_1\}} \Psi^{(N),R}_{\{\lambda_2\}} \quad . \tag{5}$$

In this basis set the diagonal elements of $H^{(2N)}$ are $E^{(N)}_{\{\lambda_1\}} + E^{(N)}_{\{\lambda_2\}}$. Using the orthonormal properties of $\Psi^{(N)}_{\{\lambda\}}$, the other elements can be shown to be given by

$$\langle \Phi^{(2N)}_{\lambda_1,\lambda_2} | H | \Phi^{(2N)}_{\lambda_1'\lambda_2'} \rangle$$

$$= \sum a^{(N)}_{\lambda_1;\lambda_{1\ell},\lambda_{1r}} \sum a^{(N)}_{\lambda_2;\lambda_{2\ell},\lambda_{2r}} \sum a^{(N)}_{\lambda_1';\lambda_{1\ell}',\lambda_{1r}'} \sum a^{(N)}_{\lambda_2';\lambda_{2\ell}',\lambda_{2r}'}$$

$$\times \langle \Phi^{(N)}_{\lambda_{1r},\lambda_{2\ell}} | H | \Phi_{\lambda_{1r}',\lambda_{2\ell}'} \rangle \delta_{\lambda_{1\ell},\lambda_{1\ell}'} \delta_{\lambda_{2r},\lambda_{2r}'} \tag{6}$$

where the coefficients $a_{\{\ldots\}}$ are given by the eigenvectors found in the solution of the N-site problem:

$$\Psi^{(N)}_\lambda = \sum_{\lambda_\ell,\lambda_r} a^{(N)}_{\lambda;\lambda_\ell,\lambda_r} \Phi^{(N)}_{\lambda_\ell,\lambda_r} \quad . \tag{7}$$

Starting from the solution of the 2 site problem, the solution of the N site problem requires $\ell n_2 N$ iteration. We are in general interested in the ground state and the low lying excited states only. For these states an energy scaling in principle will yield good results. In an energy renormalization, the low lying eigenvalues and eigen-vectors of $H^{(2N)}$ are found by eliminating the coupling between the matrix $H^{(2N)}_K$ formed from the low-lying eigenstates states of $H^{(N)}$ and the rest by one scheme or another.

At this point the renormalization scheme is in principle similar to the one employed by Drell *et al.*[3] in some field theory problems, by Jullien *et al.*[4] in the Kondo lattice problem and by Chui and Bray[5] for the one dimensional Hubbard model. This scheme, however, is unworkable for the mixed-valence problem. There are two problems: 1) the sheer size of the matrices to be diagonalized, even approximately and 2) the enormous number of terms to be added to obtain any given matrix element even after the great reduction introduced by the iteration procedure of Eq. (6). At the single site level the number of eigenstates per site is 12. For an N site problem the general matrix is therefore of size $12^N \times 12^N$. Using the conservation laws the size of matrices to be diagonalized can be reduced at the 8 site level to be of order $10^6 \times 10^6$. Good results require going to the 16 site level! Secondly, already at the 8 site level the number of terms to be summed in evaluating *each* matrix element through Eq. (6) is 10^5; and this must be done $0(10^{10})$ times just to obtain the 8 site matrix, which can then be perturbatively treated to obtain the ground and low-lying excited states.

These large numbers and the presence of the small parameter $|V_{\alpha\beta}/V_\beta| \ll 1$ can be turned to advantage by adopting a statistical approach. Imagine we have the large matrices $H_{\{\lambda\}}^{(N)}$ for any given set of quantum numbers λ and have divided it into $H_{\lambda,K}^{(N)}$, $H_{\lambda,G}^{(N)}$ and $H_{\lambda,KG}^{(N)}$ as shown in Fig. (1). $H_{\lambda,K}^{(N)}$ is to be treated exactly; we need it to be typically of $0(10^3)$ which can be handled. Imagine that we apply a unitary transformation which, as shown, leaves $H_{\lambda,K}^{(N)}$ invariant, but diagonalizes $H_{\lambda,G}^{(N)}$ to $\hat{H}_{\lambda,G}^{(N)}$ and transforms the $H_{\lambda,KG}^{(N)}$ to $\hat{H}_{\lambda,KG}^{(N)}$. Our interest in $\hat{H}_{\lambda,G}^{(N)}$ and $\hat{H}_{\lambda,KG}^{(N)}$ is simply to renormalize the terms in $H_{\lambda,K}^{(N)}$. For this purpose we need only the integrals over the distribution of eigenvalues (weighted by elements of $\hat{H}_{\lambda,KG}^{(N)}$) as far as $\hat{H}_{\lambda,G}^{(N)}$ is concerned. It is our contention that, for $|V_{\alpha\beta}/V_\beta| \ll 1$, no change is introduced in these integrals (for a dense enough distribution) if we put $V_{\alpha\beta} = 0$. With $V_{\alpha\beta} = 0$, the problem is exactly soluble; the eigenstates correspond to propagating d-electrons with localized f-electrons. This spectrum has considerable degeneracies because of the different configurations of putting a given number of f-electrons with given spin etc. The spectrum is spread on the scale of V_β. For finite $V_{\alpha\beta} \ll V_\beta$, the degeneracies of each level are lifted due to different phases for different configurations but only on the scale of $V_{\alpha\beta}$. As long as the number of states for each λ in $H_{\lambda,G}^{(N)}$ are properly counted, (as is done for $V_{\alpha\beta}=0$) the degeneracies on the scale of $V_{\alpha\beta}$ do not make any difference in integral quantities, and, therefore, to the renormalization of $H_{\lambda,K}^{(N)}$.

That leaves us with the problem of $\hat{H}_{\lambda,GK}^{(N)}$. Clearly the matrix elements are too numerous to be calculated. Having "diagonalized" $H_{\lambda,KG}^{(N)}$, the renormalization of $H_{\lambda,K}^{(N)}$ only requires sums of squares of rows and columns of $\hat{H}_{\lambda,KG}^{(N)}$ and therefore, only information on the statistical distribution of the elements of such matrices. The distribution arises at any N-site stage because of the statistical variation in the phases of the wave-functions at the mid-point; it is energy independent and a function solely of V_β, λ, and of N. We have in the calculations to be presented characterized it by the root mean square of the distributions estimated from Eq. (6) and checked it by explicit calculation of the distribution of elements in $H_{\lambda,K}^{(N)}$. The idea of treating matrix elements statistically arose from Wigner's statistical treatment of spectra of complex nuclei.[6] The basic physical point here is that because of the enormous number of levels involved, a statistical description of the higher lying states with a proper care for the quantum numbers λ is sufficient for the renormalization of the low lying states.

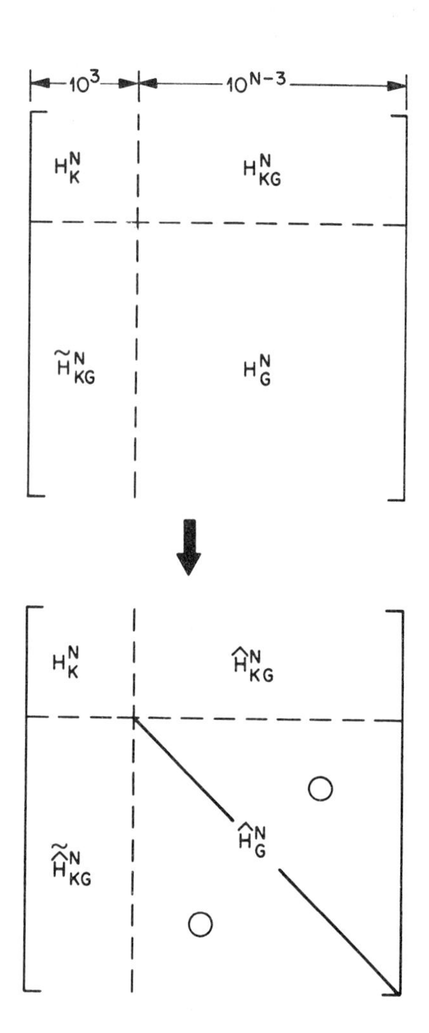

Fig. 1: Schematics of the matrix reduction procedure outlined in the text.

With the two steps described above, the matrices $H_{\lambda,K}^{(N)}$ can be renormalized to any level of accuracy desired.

Results Of Calculations

We present excerpts of the numerical results using the above method. Throughout $V_{\alpha\beta}/V_\beta = 0.1$. All the results presented are for 8 sites. The two simplifications described above are made in going from the 4 site to the 8 site problems, but not in going from the 2 site to the 4 sites, for which the whole matrix is explicitly calculated.

Results for the solution of the 16 sites problem will be presented elsewhere.

A. Rare-Earth Metal (Kondo-Lattice): Results are presented for $\mu = \epsilon_\beta = 0$ and $\epsilon_\alpha = -V_\beta/\sqrt{2}$. For these parameters, the model is equivalent to the Kondo-lattice. Using the Schrieffer-Wolff transformation for $U \to \infty$

$$J/V_{\alpha\beta} = V_{\alpha\beta}/\epsilon_\alpha = -0.1/\sqrt{2} \quad .$$

As expected for these parameters, we find in the ground state $\langle n_f \rangle = \langle n_d \rangle = 1$ per site, and a total spin zero. The ground state nearest-neighbor correlations

$$C_{0\alpha\sigma,1\beta\sigma'} = \frac{1}{N} \sum_i \langle C^+_{i\alpha\sigma} C_{i+1\beta\sigma'} \rangle$$

are shown in Fig. (2a) on a relative scale. What is most striking are the very strong antiferromagnetic correlations both in the α(f) orbitals and in the β(d) orbitals and the lack of significant correlations among the α and the β orbitals. The antiferromagnetic correlations can be traced to the fact that in the ground state wavefunction at any stage of iteration, about 90% of the contribution is from a specific wave-function of the previous stage of iteration. Ultimately at the 2 site level, this wavefunctions is the linear combination of (↑↓,0) (0,↑↓) (↑,↓) and (↓,↑) forming the S = 0 state. This may simply be understood as the hydrogen molecule effect. This S = 0 state, of course, hybridizes only with the S = 0 state of the f-orbitals. hence the antiferromagnetic correlation among the f-orbitals also. We believe the very strong antiferromagnetic tendency in the model is primarily a one-dimensional effect. The lack of significant α-β correlations can be traced to the large value of $V_\beta/V_{\alpha\beta}$.

The low-lying excited states are all spin-waves, which exhibit dispersion. In this connection note that due to the finite size good results are to be expected only for excitation energies above the characteristic bandwidth of the excitation divided by N, (and limited above by the region where the eigenvalues of $H_K^{(N)}$ extend). This effectively means that eigenvalues with primarily β(d) character are not given well; this is alright since that is in any case the trivial part of the problem. We have calculated the low temperature specific heat and the susceptibility. They are shown in Figs. (3) and (4). The characteristic temperature at which the antiferromagnetic correlations set in is of $0(\Gamma = \frac{V^2_{\alpha\beta}}{V_\beta})$. More studies are required to study the variation of this temperature with J. However, we can say that it is not proportional to J^2 as expected on a naive RKKY basis.

B. Mixed-Valence Semi-Conductor: We fix $\mu = \epsilon_\alpha = \epsilon_\beta = 0$. We find in this case that the ground state has $\langle n_\alpha \rangle = ½$, $\langle n_\beta \rangle = 1$. In this connection note that Luttinger's theorem has not much to say about the Mixed-Valence problem. In normal rare-earth metals, the very large U and the very small hopping of the f-orbitals, already leads in effect to a Mott transition among the f-orbitals, restricting f-charge fluctuations while the conduction band electrons are quite delocalized. The Mixed-Valence problem is in effect the study of how the f-orbitals get around the Mott transition when they are near the chemical potential by using the "one-sided" degree of charge fluctuation this affords and develop long-range spatial coherence.

The ground state still exhibits strong antiferromagnetic correlations and for the same reason as (A); see Fig. (2b). The f-holes are mobile, however, as evidenced by the dispersion in the particle hole spectrum. The low-lying excited states are quasi-particle and quasi-hole states which have predominantly an f-character. They exhibit a gap. We have calculated only the states that are of even and odd parity about the middle. The even gap (k=0, direct) is lower, approximately equal to 5Γ while the odd, Brillouin zone gap $(k = \frac{\pi}{2a})$ is 7Γ.

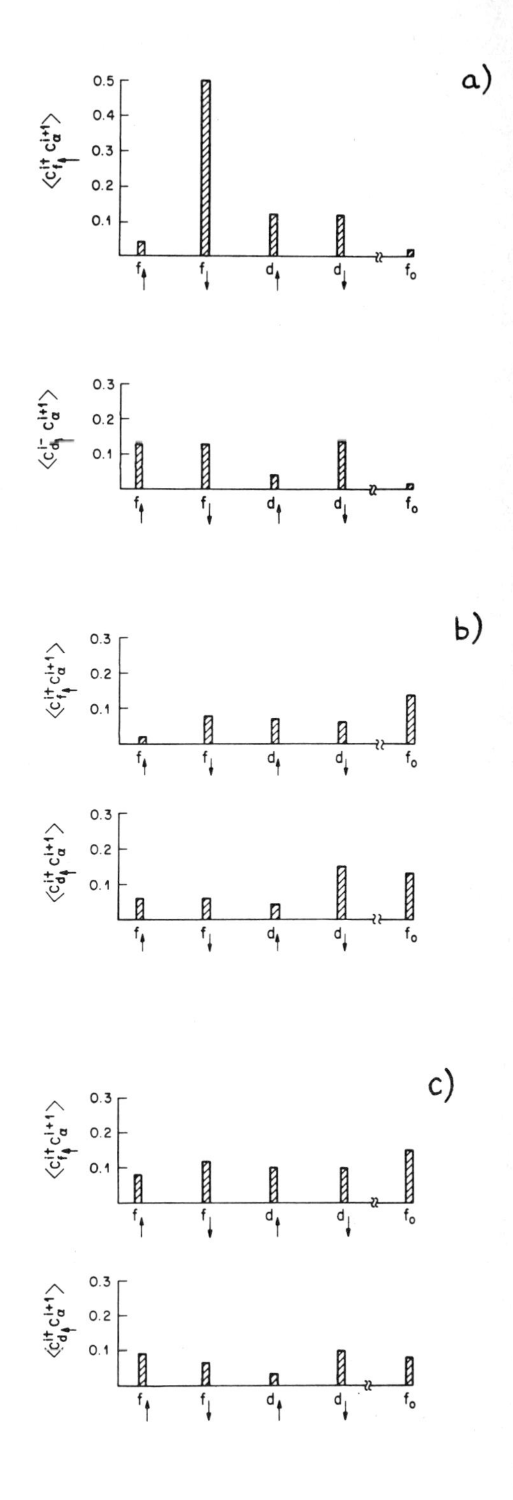

Fig. 2: Ground state correlations. Also indicated is the probability of an f-hole given an f↑ or d↑ at a neighboring site.

C. Mixed-Valence Metal: We fix $\epsilon_\alpha = \epsilon_\beta = 0$, and let $\Gamma \lesssim \mu \lesssim 2\Gamma$, (or $-2\Gamma \lesssim \mu \lesssim \Gamma$). While the variation of μ in $-\Gamma \lesssim \mu \lesssim \Gamma$ produces no change in the ground state occupation, it produces rapid variation in $\langle n_f \rangle$, between $\frac{1}{2}$ and 1, leaving $\langle n_d \rangle \approx 1$ in the range indicated above. Different charge states become nearly degenerate with the energy separation limited by the finite lattice size effects. We consider this characteristic of having passed from the "semi-conducting" to the "metallic" regime as a function of μ. Corresponding to the near degeneracy, the wave-function at any stage of iteration is an admixture with comparable weight of the basis states which are products of eigenstates of the previous iteration with different charges. These charge fluctuations severely suppress the spin-correlations, compared to cases A and B, as shown in Fig. (2c).

From the low-lying excitations, we have computed the low temperature specific heat and the magnetic susceptibility. These are shown in Figs. (3) and (4). The very low temperature results are artifacts of finite lattice size effect. The expected linear specific heat and the Pauli susceptibility are not easy to discern; to establish these, calculations with a larger number of lattice sites are required.

Conclusions: Using the presence of the small parameter $|V_{\alpha\beta}/V_\beta|$ and taking advantage of the statistical nature of the problem, due to the fact that a very large number of eigenstates with random phases renormalize the low-lying states, we have presented a method by which the one dimensional mixed-valent lattice problem can be solved to a desired level of accuracy. Excerpts of *preliminary* numerical calculations for various ranges of parameters have also be presented.

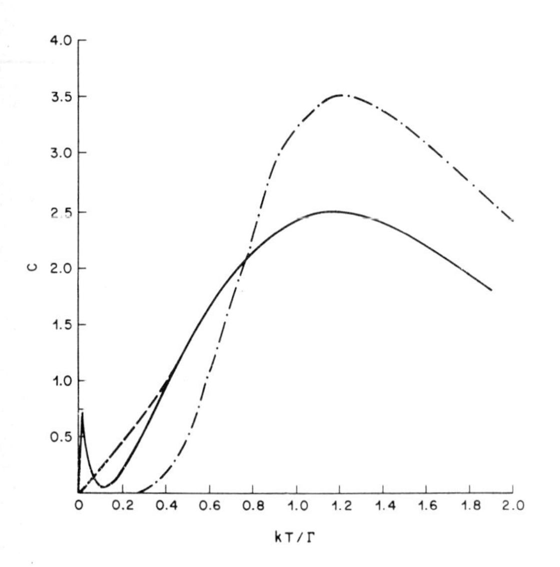

Fig. 3: Calculated specific heat for the Kondo-lattice (dashed-line) and the mixed-valence metal (full-line). Results are for 8 sites only.

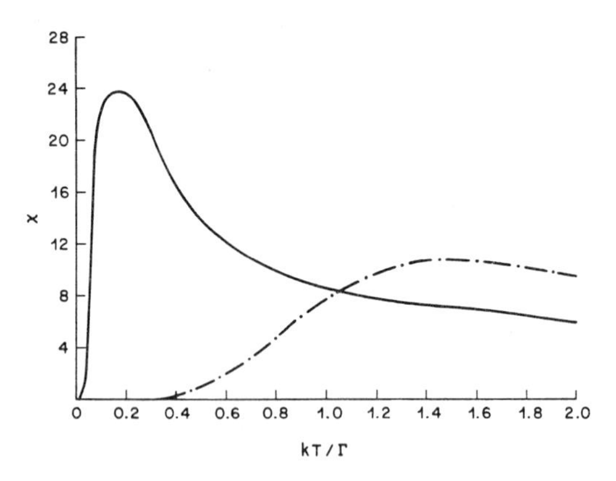

Fig. 4: Calculated magnetic susceptibility for the Kondo-lattice (dahsed-line) and the mixed-valence metal (full-line). Results are for 8 sites only.

Results Of Variational Calculations Of M.V. Impurities

Homogeneous Mixed-Valence materials appear to fall in two distinct classes,[1] one in which one of the valence states has no magnetic moment as in Ce and Yb compounds, and the other in which both valence states have magnetic moments, as in TmSe.[8] For significant valence ratios the former remain a normal Fermi-liquid i.e., show no long range magnetic order while the latter order magnetically at low temperatures. A significant question in this connection is whether this difference arises from the nature of an isolated mixed-valent ions or whether it arises due to a difference in the interaction between them in the two different cases.

Some years ago Varma and Yafet[2] constructed a variational wavefunction for Mixed-Valence impurity of class I and calculated the energy and the magnetic susceptibility of the ground state, which is a singlet. This energy is given in Table I. This result has subsequently been obtained in a number of different ways.[9] It is worthwhile to extend the variational approach to Mixed-Valence impurities of class II. Such a calculation was recently attempted by Mazzaferro *et al*;[10] unfortunately they did not examine the different possible spin states of the impurity-Fermi-sea system and concluded erroneously that the ground state is magnetic (doublet). We have performed the calculations for singlet, doublet and triplet cases and the results are shown in Table I. For class II, analytic expressions could not be obtained and numerical calculations were resorted to.

The significant point about the results in Table II are: (i) In class II also the ground state is a singlet, (ii) the energy separation between the ground state singlet and the excited magnetic states in class I is of order Γ whereas it is of order $10^{-1}\Gamma$ for class II. This is, of course, of great relevance to the fact that class II magnetically orders. The fact that in both cases the ground state is a singlet means that interaction effects for concentrated systems are essential to obtain the correct ground state.

Table I: |Binding energy| for the *lowest* singlet, doublet and triplet states for the case when one of the valence states is non-magnetic, class I and the case when both states are magnetic, class II. For the latter case results are presented for $W/\Gamma = 20$, although there is very little sensitivity of the correction indicated to this parameter.

	|Binding Energy: |ω|	
	Class I Valence States 0; ↑, ↓	Class II Valence States ↑, ↓; ↑↑, ↓↓
Doblet	$\frac{2\Gamma}{\pi} \ln \left\|\frac{W}{\omega}\right\|$	$\frac{3\Gamma}{2\pi} \ln \left\|\frac{W}{\omega}\right\|$
Triplet	$\frac{\Gamma}{\pi} \ln \left\|\frac{W}{\omega}\right\|$	$\frac{3\Gamma}{2\pi} \ln \left\|\frac{W}{\omega}\right\| + 1.6 \times 10^{-3}\Gamma$, (for W/Γ=20)
Singlet	$\frac{\Gamma}{\pi} \ln \left\|\frac{W}{\omega}\right\|$	$\frac{3\Gamma}{2\pi} \ln \left\|\frac{W}{\omega}\right\| + 9 \times 10^{-2}\Gamma$, (for W/Γ=20)

Acknowledgements

Ms. Barbara Jones (Harvard University), while a summer visiting student at Bell Laboratories, wrote some of the computer codes for the renormalization calculations. We are also very grateful to E. I. Blount for frequent discussions.

References

1. S. Doniach, Physica B *91*, 231 (1977).
2. C. M. Varma and Y. Yafet, Phys. Rev. B *13*, 2950 (1976).
3. S. D. Drell, M. Weinstein and S. Yankielowicz, Phys. Rev. D *14*, 487 (1976).
4. R. Jullien, J. N. Fields, S. Doniach, Phys. Rev. Letters *38*, 1500 (1977).
5. S. T. Chui and J. W. Bray, Phys. Rev.
6. E. Wigner, see various preprints in Statistical Theories of Spectra, Edited by C. E. Porter, Academic Press, New York (1965).
7. C. M. Varma, Rev. Mod. Phys. *48*, 219 (1976).
8. B. Batlogg, H. R. Ott, E. Kaldis, W. Thöni and P. Wachter, Phys. Rev. *19*, 247 (1979).
9. A. Bringer and H. Lustfeldt, Z. Physik B *28*, 213 (1977). Also see papers by F. D. Haldane, T. V. Ramakrishnan, and H. Keiter and N. Grewe in *Valence Fluctuations in Solids*, Edited by L. Falicov, W. Hanke and M. Maple, North Holland, New York (1981).
10. A. Mazzaferro, C. A. Balseiro and B. Alascio, Phys. Rev. Letters. *47*, 274 (1981).

S. VON MOLNAR: You said that you had calculated the specific heat and other thermodynamic properties of those that are semiconducting mixed valence materials, but you didn't show a sketch. Could you give us an idea of how they look like.

C.M. VARMA: I was just running out of time. I can show you the graphs after the session.

R. JULLIEN: Are you sure that the "Probability trick" that you use to treat the off diagonal elements do not distroy the coherence effects ?

C.M. VARMA: I am sure they do not because we do find coherence effects. Also, we can vary the size of the matrix, we treat exactly and confirm that the "probability trick" works.

COMMENT: R. JULLIEN: When we found a gap in the symmetric Kondo lattice in one dimension, it was so tiny (varying as exp(-t/J)) that we must be careful with the approximations not to distroy it.

C.M. VARMA: I am very conscious of this. We are very careful. We go to the extent of keeping 10^3 states precisely at each iteration. I believe you treat only states of order 10 precisely in your Kondo lattice renormalization work and for that problem find rather consistent results.

T.A. KAPLAN: Did you calculate the long-range correlations ? You should be able to and the results would be interesting - it is really needed to discern antiferromagnetism from nonmagnetism.

C.M. VARMA: We have calculated only the nearest neighbour correlations so far. I agree with you that further neighbour correlations are required to decide definitively on long-range order.

Valence Instabilities
P. Wachter and H. Boppart (eds.)

DIAGRAMMATIC EXPANSIONS IN I.V.: EXACT OR APPROXIMATE?

K. W. H. Stevens

Department of Physics, University of Nottingham,
University Park, Nottingham NG7 2RD, England

The Keiter and Kimball separation of the partition function of the Anderson Hamiltonian into a product of the conduction electron partition function and the partition function of a localised "effective Hamiltonian" is examined using a simple example and a different version of their treatment. It is shown that the final formulae of the two treatments, though quite similar in appearance, differ on a term-by-term comparison. The question of whether or not the two series can be rearranged to agree is still unanswered.

The Anderson Hamiltonian provides a way of describing the interaction between conduction electrons and a localised impurity, which can have a complicated structure. Keiter and Kimball [1] investigated the partition function by means of perturbation theory and showed that it could be factorised into the partition function of the conduction electrons alone multiplied into the partition function of an "effective" localised impurity. The "effective" impurity would have temperature dependent energy levels in one-to-one correspondence with those of the actual impurity. In other words the coupled system of conduction electrons and a localised moment could be decoupled into the conduction electrons alone together with an "effective" local impurity. They gave formulae for the "effective" energy levels. The same formalism has recently been used by Ramakrishnan [2] for an intermediate valence impurity, and Keiter and Grewe [3] have extended the formalism to intermediate valence systems with many impurities. The discussion in the original Keiter and Kimball paper (to be referred to as KK) is quite complicated, as is the discussion in Grewe and Keiter and it is not easy to see whether it contains approximations, though the impression is given that it is exact. To examine this point several courses are open. One is to repeat the analysis in detail, another is to find a simpler method of arriving at the important formulae and the third is to test the KK formalism on a simple model which can be handled in some quite different way. In this paper the second and third approaches will be discussed, with the second approach being little more than a reversal of the KK treatment.

The simple model is that of a single impurity, with a Hubbard U, interacting with a conduction band which consists of a single Bloch function.

The Hamiltonian is: $\mathscr{H} = \mathscr{H}_C + \mathscr{H}_L + V$

$$\text{where } \mathscr{H}_L = E(a_+^* a_+ + a_-^* a_-) + U\, a_+^* a_+ a_-^* a_-$$

$$\mathscr{H}_C = \varepsilon(C_+^* C_+ + C_-^* C_-)$$

$$\text{and } V = V_0(a_+^* C_+ + C_+^* a_+ + a_-^* C_- + C_-^* a_-)$$

Its eigenvalues are readily found, for the five possible numbers of electrons, 0, 1, 2, 3 and 4. For N = 0 the eigenvalue is zero. For N = 1 there are two doublets at $(E \pm V)$. For N = 2 there is a triplet at 2E, and three singlets

$$\text{at } U + 2E,\ \tfrac{1}{2}[4E + U) \pm (U^2 + 16V^2)^{\frac{1}{2}}]$$

N = 3 has two doublets at

$$\tfrac{1}{2}[(3E + 3\varepsilon + U) \pm \{(E + U - \varepsilon)^2 + 4V^2\}^{\frac{1}{2}}$$

and N = 4 has a singlet at $(2E + U + 2\varepsilon)$. The complete partition function is readily constructed and division by the partition function for $\mathscr{H}_C$ alone will give the partition function for the effective local impurity. The KK theory suggests that this will take the form

$$\exp(-\beta\tilde{E}_0) + 2\exp(-\beta\tilde{E}_1) + \exp(-\beta\tilde{E}_2)$$

where $\tilde{E}_0$, $\tilde{E}_1$ and $\tilde{E}_2$ will be temperature dependent and correspond to local occupancies of zero, one and two electrons. There is no obvious way of making a unique such decomposition within the context of the direct evaluation of the eigenvalues, and however it is done it appears that the temperature dependences of the effective energies will be quite complicated.

We turn therefore to the KK method, but using an approach that while differing from theirs is immediately suggested by some of their observations. We use the B-W perturbation method [4]. The only slightly awkward point is that the method assumes that the unperturbed Hamiltonian, $\mathscr{H}_L + \mathscr{H}_C$, has non-degenerate energy levels. Ours has degeneracies. The point is readily circumvented by adding suitable infinitesimals to remove the degeneracies and then letting them tend to zero at the end. The theory shows that if $|i\rangle$ is an eigenstate of $\mathscr{H}_L + \mathscr{H}_C$, with eigenvalue E_i, the change in its energy, ΔE_i on adding the perturbation V is given by that solution of

$$z = \Gamma_i(z) = \langle i|V + VQV + VQVQV + \ldots |i\rangle$$

which tends to zero as $V \to 0$, where $Q = (1 - P_i)/(z - Ho)$ with $P_i = |i\rangle\langle i|$. It should be noted that Q depends on i because of the projection operator in its definition, and on z.

The Lagrange formula (see KK) can then be used to obtain the contribution

$\exp[-\beta(E_i + \Delta E_i)]$ to the partition function,

for $\exp[-\beta(E_i + \Delta E_i)] = \exp(-\beta E_i) \times$

$$\left\{1 - \beta \sum_{m=0} \frac{1}{(m+1)!} \left[\frac{d^m}{dz^m} e^{-\beta z}\left\{\Gamma_i(z)\right\}^{m+1}\right]_{z=0}\right\}$$

The R.H.S. is entirely in terms of the eigenstates of the unperturbed Hamiltonian, which are products of conduction and localised states, so a contribution to the partition function of the effective local Hamiltonian is obtained by taking a partial trace, over the conduction states, and dividing out the conduction electron partition function. The result is then written in the form

$\exp[-\beta\tilde{E}_i]$ where $\tilde{E}_i$ is an eigenvalue of the effective local Hamiltonian.

KK would arrive at a corresponding point by a quite different route, but more particularly the formula they would obtain, though looking very similar, is different. If $\langle\ \rangle$ denotes a thermal average over the conduction states a term in (1), such as:

$$\left[\frac{d^m}{dz^m} e^{-\beta z}\left\{\Gamma_i(z)\right\}^{m+1}\right]_{z=0}$$

becomes

$$\left\langle \left[\frac{d^m}{dz^m} e^{-\beta z}\left\{\Gamma_i(z)\right\}^{m+1}\right]_{z=0} \right\rangle$$

whereas they would have

$$\left[\frac{d^m}{dz^m} e^{-\beta z}\left\{\langle\Gamma_i(z)\rangle\right\}^{m+1}\right]$$

That they differ can readily be seen, because $\Gamma_i(z)$ will contain $C_+^* C_+$ and

$\langle(C_+^* C_+)^{m+1}\rangle = \langle C_+^* C_+\rangle$, and is not equal

to $\{\langle C_+^* C_+\rangle\}^{m+1}$. So on a term by term comparison the two formulae differ. However this does not automatically imply that at least one of them is incorrect for it may be possible to rearrange one of them to give the other, but so far this has not proved possible. Until it is done a question must hang over the formalisms, and some care should be exercised in their use.

Conclusions

With the simple Hamiltonian we have been using it is straightforward to obtain the eigenvalues and partition function. It is also possible, though rather tedious, to verify that the B-W form of perturbation theory obtains the eigenvalues correctly, and that the Lagrange formula gives the exponentials correctly. The problems arise in manipulating the traces over the conduction states, and further study seems desirable, for the technique of using the B-W expansion by-passes a good deal of the complication of the original KK work, and offers considerable promise for extension to systems with many impurities. Finally we note that Ramakrishnan [2] used an approximate form of the KK expression, retaining only the first term (m = 0) in the expansion, and to this approximation the two formalisms are identical.

References

1 Keiter, H. and Kimball, J.C. Int J Magn 1, 233 (1971)

2 Ramakrishnan, T.V. Valence Fluctuations in Solids, (North-Holland, Amsterdam, 1981) p. 13

3 Grewe, N. and Keiter, H. Phys Rev B24, 4420 (1981)

4 Brandon, B. Rev Mod Phys 39, 771 (1967)

Valence Instabilities
P. Wachter and H. Boppart (eds.)

GREEN'S FUNCTIONS OF THE ANDERSON LATTICE HAMILTONIAN*

B. H. Brandow

Theoretical Division
Los Alamos National Laboratory
University of California
Los Alamos, NM 87545, USA

Diagrammatic analysis provides an exact general form for the conduction-electron and f-electron Green's functions; both are determined by a single function $\mathcal{G}$. Zubarev-Hubbard equations of motion are then used to obtain a "Hubbard III" type approximation for this $\mathcal{G}$. The latter analysis incorporates some self-consistency requirements, to enforce consistency with the general diagrammatic result. This new approximation goes beyond CPA by inclusion of "resonance broadening" type processes, and may possibly provide a well-defined Fermi surface.

1. INTRODUCTION

For valence fluctuation (VF) systems, there is now a strong presumption that an adequate theory of the elementary excitations (quasiparticles) should exhibit two important features: (1) For systems which would be insulating if their intra-atomic repulsions U were absent, there should be an insulating gap. (2) For "essentially metallic" systems (intermetallic compounds) there should be a well-defined Fermi surface, i.e., the quasiparticle lifetime should become infinite as $(k, E) \to (k_F, \varepsilon_F)$. It has recently been emphasized[1] that both of these features are consequences of Luttinger's formal analysis[2] of Fermi liquid systems. However, it is equally important to recognize that only an exact solution of the assumed Hamiltonian can determine theoretically whether the system actually satisfies the necessary condition [adiabatic equivalence to a suitable "noninteracting" (U=0) system] for these Luttinger properties to be manifested. The present belief that periodic (concentrated, stoichiometric) valence fluctuators have these Luttinger properties is based on experimental evidence (particularly from SmB_6--insulating gap, and $CeSn_3$--Fermi surface), and a variety of approximate theoretical analyses.[1]

The most systematic theoretical treatments of quasiparticle spectra to date, for periodic VF systems, involve approximations of either the "Hubbard I" or the CPA type. The Hubbard I-type approximation[3-5] trivially provides a Fermi surface, since its quasiparticles have infinite lifetimes for all (k, E), and its spectrum always exhibits a gap. Unfortunately, however, in versions appropriate for the insulating compound SmB_6 its Fermi level does not fall in this gap.[4] A modified form[6] of this Hubbard I approximation also fails to place ε_F in the gap.[7] In contrast, the CPA or alloy-analogy approximation[8-10] does have the important virtue of exhibiting a gap and placing ε_F exactly in this gap. But this fails to produce a Fermi surface for the essentially metallic compounds, and the gap also vanishes throughout a portion of the valence-fluctuation regime. In particular, for parameters that seem appropriate to SmB_6, the CPA gap is either nonexistent or at least an order of magnitude too small.[7] Since no available approximation exhibits all of the desired features, a more fundamental study is clearly desirable.

Our study is based on a combination of diagrammatic and algebraic methods. A simple but exact diagrammatic analysis provides a general form which the Green's functions of the (unknown) exact theory must satisfy. This is quite significant. In practice, this can be used as a consistency check on proposed approximations, and also as a guide in the development of new approximations. The latter aspect is a key feature of this work. The algebraic analysis is based on the Green's function methods of Zubarev[11] and Hubbard.[12] The Zubarev formalism has the important virtue that formal analyses for zero temperature and finite temperatures are essentially identical; the thermal aspects enter in a very simple and convenient way. Moreover, the equation-of-motion approach readily incorporates effects of d-f hybridization to infinite order; the "noninteracting" (U = 0) case of hybridized bands is handled very simply and exactly. The projection-operator technique of Hubbard is convenient for large-U systems, especially for the present case where $U \to \infty$ is assumed. We have tried to analyze the Hamiltonian strictly on its own merits, without reference to the specific approximations of Hubbard. In particular, we pay close attention to the requirement of lattice translational invariance. Nevertheless, this study is sufficiently similar to Hubbard's that use of his terminology may be instructive.

The main outcome of this study is a new approximation which is analogous to that in the third paper[12] of Hubbard's series on narrow-band systems, i.e., a "Hubbard III" type of approximation. We recall the schematic relation[12]

$$\text{Hubbard III} = \text{Hubbard I} + \text{scattering correction} + \text{resonance broadening.} \quad (1)$$

The CPA approximation presumes that the opposite-spin ("down spin") localized electrons are frozen in some disordered static configuration, and it thereby neglects the analog of Hubbard's "resonance broadening." The latter is included here. Although we have not yet achieved a fully self-consistent numerical solution, our efforts so far suggest that this new approximation should exhibit a well-defined (Luttinger) Fermi surface.

A noteworthy feature of this study is that we recover the CPA in a purely deductive manner, as a direct and unforced consequence of the formal analysis. This contrasts with the previous studies[8-10] of CPA for VF systems, which have all been based on the intuitive analogy to a disordered binary alloy. This result ends any doubts that the randomness assumption of the alloy analogy may have generated some otherwise-avoidable errors with respect to the actual lattice translational invariance, and it also tends to increase confidence in the present new approximation.

We use the simplest model Hamiltonian that appears to contain the essence of VF phenomena, the Anderson lattice Hamiltonian[13]

$$H = \sum_{k\sigma} \varepsilon_k \hat{n}_{k\sigma} + \varepsilon_f \sum_{j\sigma} \hat{n}_{j\sigma} + U \sum_j (1-\hat{n}_{j\uparrow})(1-\hat{n}_{j\downarrow}) + \sum_{kj\sigma} (\upsilon_{kj} \eta^\dagger_{k\sigma} \eta_{k\sigma} + \text{h.c.}) \quad , \tag{2}$$

where $\hat{n} = \eta^\dagger \eta$ is the number operator. The 5d conduction bands are represented here by a single "5s" band, with Bloch eigenvalues ε_k. The localized 4f's are represented by "4s" Wannier functions with energy ε_f and site index j, where these sites constitute a Bravais lattice. Thus, orbital degeneracy and any intrinsic 4f bandwidth are neglected. We set $\varepsilon_f = 0$ to define the origin of the energy scale, and assume $U \to \infty$ so that only f^1 and f^2 (actually s^2) configurations are permitted at each site.

2. GENERAL DIAGRAMMATIC ANALYSIS

We use a diagrammatic convention in which vertical lines represent occupied f orbitals, and sloping lines represent occupied conduction orbitals. Since U is the largest physical parameter, it is inappropriate to regard this as a perturbation; only the last term of (2) is to be treated perturbatively. It follows that the familiar Feynman-Dyson concepts of "propagator" and "self energy" are not strictly applicable here, at least not in their usual forms. With this warning, we now define an "irreducible f-electron propagator"[14] $\mathcal{G}^\sigma_{jj'}$ as the sum of all linked diagrams which connect the creation operator $\eta^\dagger_{j'\sigma}$ to the annihilation operator $\eta_{j\sigma}$, with the proviso that none of these diagrams can be separated into two disconnected pieces by cutting only a single conduction-orbital line. [Note that the second and succeeding diagrams of

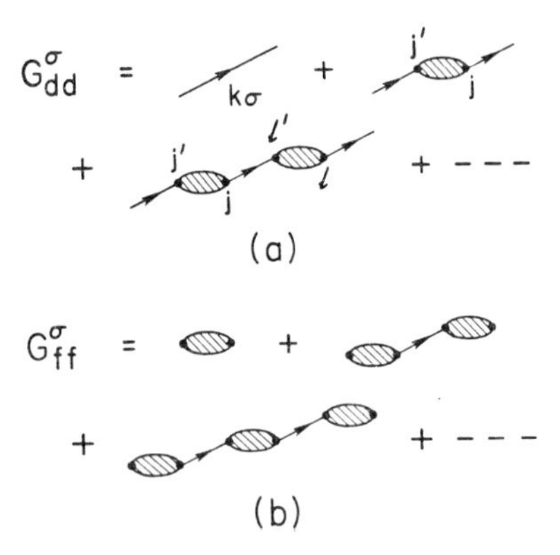

Figure 1: Schematic form of perturbation diagrams included in (a) G^σ_{cc} and (b) G^σ_{ff}.

Fig. 1(b) provide counterexamples; these are "reducible" f-propagator diagrams.] This irreducible $\mathcal{G}$ is represented schematically by a shaded blob with dots j' and j indicating the entrance and exit f orbitals. With these conventions, we see that the conduction electron Green's function G^σ_{cc} is represented by the sum of diagrams shown in Fig. 1(a). The $\mathcal{G}^\sigma_{jj'}$ blobs are drawn "horizontally," as a reminder that $\eta^\dagger_{j'\sigma}$ may occur either before or after $\eta_{j\sigma}$ [as shown explicitly in Fig. 2(a)]. Each internal black dot is accompanied by a factor υ_{kj} or υ_{jk}. We therefore find that the summation of these diagrams leads to the general Dyson-like result

$$G^\sigma_{cc}(k,E) = [E-\varepsilon_k - V_k^2 \mathcal{G}^\sigma_k(E)]^{-1} \quad , \tag{3}$$

where V_k and $\mathcal{G}_k$ are the lattice Fourier transforms of υ_{kj} and $\mathcal{G}_{jj'}$. Similarly, G^σ_{ff} is given by the diagrams of Fig. 1(b), whose sum is

$$G^\sigma_{ff}(k,E) = \mathcal{G}^\sigma_k(E) + \mathcal{G}^\sigma_k(E) V_k G^\sigma_{cc}(k,E) V_k \mathcal{G}^\sigma_k(E)$$

$$= (E-\varepsilon_k)\mathcal{G}^\sigma_k(E) G^\sigma_{cc}(k,E) \quad . \tag{4}$$

Actually, the transition from diagrams to the results (3)-(4) requires a further and quite important step. If one draws individual time-ordered (Goldstone) perturbation diagrams which contribute to the various general schematic diagrams of Fig. 1, it becomes apparent that there are restrictions on the summations over the f-orbital site indices j, j', ℓ, ℓ', etc. Many of these labels must be distinct. Grewe and Keiter[15] have emphasized the importance of such restrictions, and have shown that their precise treatment can lead to very complicated expressions. What we have done in (3) and (4) is to add and subtract terms wherever necessary in order to remove these restrictions, thereby modifying and redefining the operator $\mathcal{G}^\sigma_{jj'}$, and its transform $\mathcal{G}^\sigma_k$. This step is obviously necessary in order that the consequence of translational invariance (k-conservation) be manifest in a simple form. An important point of this report is to demonstrate that this is easy to do in practice, at least to a limited but very reasonable extent [see (9) below].

We note that the "Hubbard I" approximation[3-5] exhibits the general form (3), (4), with $\mathcal{G}^{\sigma}_{jj'} \to \delta_{jj'} n_{f,-\sigma}/E$, $n_{f\sigma} = \langle \hat{n}_{j\sigma} \rangle$. This $\mathcal{G}$ includes just the two time-ordered diagrams of Fig. 2(a). The very similar approximation of Ref. 6 also exhibits this form, but with E shifted by a constant ($E \to E - \Delta$) within $\mathcal{G}$. To compare (3) and (4) with the CPA studies one must average these expressions over k, $\langle j|G|j\rangle = N^{-1}\Sigma_k G(k)$, to obtain corresponding results for the single-site G_{cc} and G_{ff}. The G's of Refs. 8 and 10 are easily seen to manifest the resulting general form, and this can also be demonstrated for Ref. 9. (Note, however, that Ref. 8 employs a different model Hamiltonian.)

3. ALGEBRAIC DEVELOPMENT

We now seek a solution of the Zubarev equations of motion, subject to two simplifying restrictions: (a) $\mathcal{G}$ must be site-diagonal, and (b) only terms with no more than one nontrivial k-summation will be retained. Self-consistency will also be imposed where appropriate. We find it convenient to represent $\mathcal{G}$ as a generalization of the Hubbard I result,

$$\mathcal{G}_{jj'} \to \delta_{jj'} \mathcal{G}_{jj} \equiv \delta_{jj'} \tilde{n}(E)/E \quad , \tag{5}$$

thus our goal is to determine the function $\tilde{n}(E)$.

We focus on the $|E| \ll U$ region of $\langle\langle \eta_{j\uparrow}; \eta^{\dagger}_{j'\uparrow} \rangle\rangle$, and introduce the notation $\langle\langle \alpha\beta\gamma\delta \rangle\rangle \equiv \langle\langle \eta^{\dagger}_{\alpha\downarrow}\eta_{\beta\downarrow}\eta_{\gamma\uparrow}; \eta^{\dagger}_{\delta\uparrow} \rangle\rangle$, where α etc. can signify any of the f or conduction orbital labels. For $|E| \ll U$ we need consider only the $\hat{n}_{j\downarrow}$ Hubbard projection of $\langle\langle \eta_{j\uparrow}; \eta^{\dagger}_{j'\uparrow} \rangle\rangle$, namely $\langle\langle jjjj' \rangle\rangle$, whose equation of motion is

$$E\langle\langle jjjj'\rangle\rangle = \frac{\delta_{jj'} n_{f\downarrow}}{2\pi} + \sum_k \upsilon_{jk}(\langle\langle jkjj'\rangle\rangle + \langle\langle jjkj'\rangle\rangle) \quad . \tag{6}$$

An additional term, $-\Sigma_k \upsilon_{kj} \langle\langle kjjj'\rangle\rangle$, does not contribute here in the $U \to \infty$ limit. Similarly,

$$(E-\varepsilon_k) \langle\langle jjkj'\rangle\rangle = \Sigma_m \upsilon_{km} \langle\langle jjmj'\rangle\rangle \quad , \tag{7}$$

where additional terms are ignored for reason (b) above. For $m \neq j$ we replace $\eta_{m\uparrow}$ by $\hat{n}_{m\downarrow}\eta_{m\uparrow}$, thus

$$E\langle\langle jjmj'\rangle\rangle_{m\neq j} \to \frac{\delta_{mj'}}{2\pi} n^2_{f\downarrow} + \tilde{n} \sum_k \upsilon_{mk} \langle\langle jjkj'\rangle\rangle \, . \tag{8}$$

Here we have introduced a self-consistency ansatz: any processes that contribute to $\mathcal{G}^{\uparrow}_{jj}$ must likewise contribute to $\mathcal{G}^{\uparrow}_{mm}$, thereby generating the present $\tilde{n}$ [see (10) below]. This is quite evident from the corresponding diagrams. (Note, however, that we are now neglecting summation restrictions between processes internal to $\mathcal{G}^{\uparrow}_{mm}$ and the remaining processes within $\langle\langle \eta_{j\uparrow}; \eta^{\dagger}_{j'\uparrow} \rangle\rangle$.) The $m \neq j$ restriction is now removed by adding and subtracting terms, thus

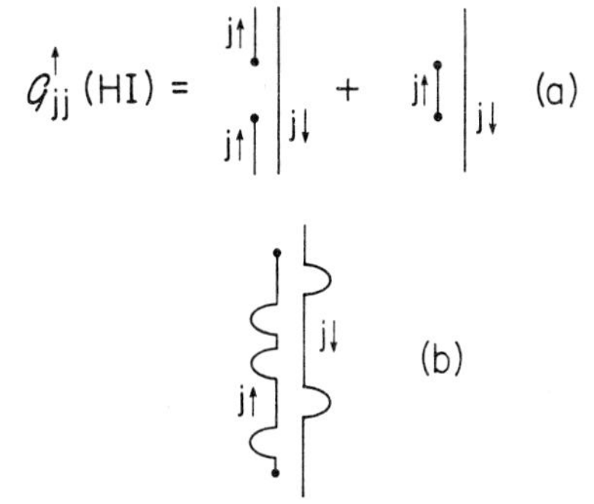

$$E\langle\langle jjmj'\rangle\rangle = \frac{\delta_{mj'}}{2\pi} n^2_{f\downarrow} + \tilde{n} A_{mjj'} + \delta_{mj}[E\langle\langle jjjj'\rangle\rangle - \frac{\delta_{jj'}}{2\pi} n^2_{f\downarrow} - \tilde{n} A_{jjj'}] \quad , \tag{9}$$

where we have introduced $A_{mjj'} = \Sigma_k \upsilon_{mk} \langle\langle jjkj'\rangle\rangle$. We combine (7) and (9), and Fourier transform to find

$$A_{kjj'} = \Sigma_m A_{mjj'} e^{-ik\cdot(R_m - R_j)} = \frac{V_k^2[\;] + (n^2_{f\uparrow} V_k^2/2\pi E) e^{ik\cdot(R_j - R_{j'})}}{E - \varepsilon_k - V_k^2(\tilde{n}/E)} \quad , \tag{10}$$

where [ ] represents the bracket in (9). The denominator here is just the present approximation (5) for the $G^{\uparrow}_{cc}$ of (3); this is what initially motivated our introduction of $\tilde{n}$ in (8). The inverse Fourier transform now gives

$$A_{jjj'} = \frac{[\langle\langle jjjj'\rangle\rangle - \frac{\delta_{jj'} n^2_{f\downarrow}}{2\pi E}]F_o + \frac{n^2_{f\downarrow}}{2\pi E} F_{jj'}}{1 + \tilde{n}x} \quad , \tag{11}$$

$$F_{jj'} = \frac{1}{N} \sum_k V_k^2 G^{\uparrow}_{cc}(k,E) e^{ik\cdot(R_j - R_{j'})} \quad , \tag{12}$$

$F_o = F_{jj}$, and $x = F_o/E$. This $A_{jjj'}$ is our

Figure 2: (a) Lowest order contributions to $G^{\uparrow}_{jj}$; these constitute the Hubbard I approximation. (b) Example of type of perturbation diagrams included in the present approximation for $G^{\uparrow}_{jj}$.

result for the last term in (6).

The remaining term in (6) is obtained similarly, using $B_{\ell jj'} = \Sigma_k \upsilon_{\ell k} \langle\langle jkjj'\rangle\rangle$. The result is

$$B_{jjj'} = [\langle\langle jjjj'\rangle\rangle F_o + \frac{\delta_{jj'}}{2\pi}\{F\}]/(1 + \tilde{n}x) \quad , (13)$$

$$\{F\} = F_{f\uparrow c} + \frac{1}{E}[F_{f\uparrow nf} - \langle \eta^{\dagger}_{j\downarrow}\hat{n}_{j\uparrow}\eta_{j\downarrow}\rangle F_o] \quad , (14)$$

$$F_{f\uparrow c} = \frac{1}{N}\sum_k V_k \langle f^{\dagger}c\rangle_k \, G^{\downarrow}_{cc}(k,E) \quad , (15)$$

$$F_{f\uparrow nf} = \frac{1}{N^2}\sum_k V_k^2 \langle f^{\dagger}\hat{n}f\rangle_k G^{\downarrow}_{cc}(k,E) \quad , (16)$$

where $\langle f^{\dagger}c\rangle_k$, $\langle f^{\dagger}\hat{n}f\rangle_k$ are the respective Fourier transforms of $\langle \eta^{\dagger}_{j\downarrow}\eta_{k\downarrow}\rangle$ and $\langle \eta^{\dagger}_{j\downarrow}\hat{n}_{\ell\uparrow}\eta_{\ell\downarrow}\rangle$. Inserting these A and B results in (6), a final Fourier transformation gives

$$G^{\uparrow}_{ff}(k,E) = \frac{y}{Ez} + \frac{n^2_{f\downarrow}V_k^2}{E^2 z}[E-\varepsilon_k - V_k^2 \frac{\tilde{n}}{E}]^{-1} \quad , (17)$$

$$y = n_{f\downarrow}[1 + (\tilde{n} - n_{f\downarrow})x] + \{F\} \quad , (18)$$

$$z = 1 + (\tilde{n} - 2)x \quad . (19)$$

Comparing (17) with (3), we are tempted to identify y/z with $\tilde{n}$, except for the problem that the second term of (17) would then not fully agree with its counterpart in (3). However, diagrammatic analysis and the equations of motion both convince us that there must be additional terms, whose effect is to replace $n^2_{f\downarrow}/z$ by $\tilde{n}^2$. [All processes which contribute to the first (single $\mathcal{G}$) term in Fig. 1(b) must likewise contribute to the other $\mathcal{G}$'s within G_{ff}.] In view of the addition and subtraction within (9), the same replacement must also be made within the first term of (17), whereby we finally arrive at our defining equation for $\tilde{n}$,

$$\tilde{n} = z^{-1}[n_{f\downarrow}(1 + \tilde{n}x) + \{F\}] - \tilde{n}^2 x \quad (20)$$

An example of the type of processes included in the present $\mathcal{G}^{\dagger}_{jj}$ is illustrated in Fig. 2(b). Each semicircular line now denotes G_{cc}, and leads to a factor of $x = F_{jj}/E$.

Repeating the entire foregoing analysis for $|E+U| \ll U$, we find that the deep poles (at $E \approx -U$) contribute $(1-n_{f,-\sigma})$ and 0, respectively, to the occupation numbers $n_{f\sigma}$, $n_{k\sigma}$, just as in the Hubbard I approximation.[4]

Our A and B correspond respectively to the "scattering correction" and "resonance broadening" of Hubbard.[12] It is thus of interest to see what happens when $B_{jjj'}$ is omitted. The above arguments then lead to $\tilde{n} = n_{f\downarrow}[1+(\tilde{n}-1)x]^{-1}$, which is exactly the CPA result[10] for (2), in present notation.

The present set of equations has been programmed, using a semi-elliptic state density for the conduction band, but a fully self-consistent solution has not yet been achieved. We observe, however, that {F} has a strong dependence on the Fermi level ε_F, via the standard Green's function prescription[11] for the <> terms in (14)-(16) and that this can lead to interesting behavior for $E \approx \varepsilon_F$. Our efforts so far suggest that $\tilde{n}$ may become purely real at $E = \varepsilon_F$, thus giving infinite quasiparticle lifetimes and a well-defined Fermi surface.

I thank R. M. Martin and J. W. Allen for helpful discussions about their CPA studies.

*Work support by the US Department of Energy.

REFERENCES

[1] R. M. Martin, Phys. Rev. Lett. 48, 362 (1982), and references therein.
[2] J. M. Luttinger, Phys. Rev. 119, 1153 (1960).
[3] C. M. Varma and Y. Yafet, Phys. Rev. B 13, 2950 (1976).
[4] B. H. Brandow, Int. J. Quantum Chem. Symp. 13, 423 (1979).
[5] M. Roberts and K. W. H. Stevens, J. Phys. C 13, 5941 (1980).
[6] S. K. Sinha and A. J. Fedro, J. de Physique 40, C4-214 (1979); A. J. Fedro and S. K. Sinha, in "Valence Fluctuations in Solids," L. M. Falicov, W. Hanke, and M. P. Maple, eds. (North-Holland, 1981).
[7] B. H. Brandow, unpublished calculations.
[8] O. Sakai, S. Seki, and M. Tachiki, J. Phys. Soc., Japan 45, 1465 (1978).
[9] R. M. Martin and J. W. Allen, J. Appl. Phys. 50, 7561 (1979).
[10] H. J. Leder and G. Czycholl, Z. Physik B 35, 7 (1979).
[11] D. N. Zubarev, Sov. Phys. Usp. 3, 320 (1960) (English trans.).
[12] J. Hubbard, Proc. R. Soc. (London) A 281, 401 (1964).
[13] C. M. Varma, Rev. Mod. Phys. 48, 219 (1976); J. H. Jefferson and K. W. H. Stevens, J. Phys. C 11, 3919 (1978).
[14] This is actually an off-diagonal generalization of the "locator" concept; see for example J. A. Blackman, D. M. Esterling, and N. F. Berk, Phys. Rev. B4, 2412 (1971).
[15] N. Grewe and H. Keiter, Phys. Rev. B24, 4420 (1981); N. Grewe, this volume.

Valence Instabilities
P. Wachter and H. Boppart (eds.)

A STUDY OF THE STEVENS-BRANDOW VARIATIONAL GROUND STATE WAVE FUNCTION

Patrik Fazekas+

Institut für Theoretische Physik der Universität zu Köln
Zülpicher Str.77, D-5000 Köln 41, West Germany

The variational wave function suggested by Stevens and Brandow for the ground state of the periodic nondegenerate Anderson model is reconsidered in the case of f^1-f^2 valence mixing. Non-orthogonality problems are avoided by expanding the trial state in an orthonormal basis. A single variational parameter τ is used which turns out to play the role of a "hybridization temperature". The ground state energy is a well-behaved function of the hybridization strength V. The valence and the shift of the chemical potential are calculated. The band occupation numbers follow a Fermi distribution at "temperature" $\tau \sim V$.

1. INTRODUCTION

It is a long-standing problem to construct a trial ground state wave function which incorporates the strong local correlations characteristic of the mixed valence state /1/, shows long-range phase coherence and is still simple enough to make the variational procedure practicable. While neglecting the orbital degeneracy of the f-state leads to losing some essential physics /2/, one still hopes to gain useful experience from considering first the periodic Anderson model with orbitally non-degenerate d- and f-states

$$H=\sum_{\underline{k}\sigma} \epsilon(\underline{k})\, d^+_{\underline{k}\sigma} d_{\underline{k}\sigma} + E_f \sum_{\underline{g}\sigma} f^+_{\underline{g}\sigma} f_{\underline{g}\sigma} + U\sum_{\underline{g}} n_{\underline{g}\uparrow} n_{\underline{g}\downarrow} - \frac{V}{\sqrt{N}}\sum_{\underline{g}\underline{k}\sigma} (e^{-i\underline{k}\underline{g}}\, d^+_{\underline{k}\sigma} f_{\underline{g}\sigma} + \text{c.c.}) \qquad (1)$$

with $V > 0$. The on-site form of the hybridization has been chosen for convenience. $d^+_{\underline{k}\sigma}$ creates a d-electron with spin σ in the Bloch-state $\underline{k}$; $f^+_{\underline{g}\sigma}$ an f-electron at the site $\underline{g}$; $n_{\underline{g}\sigma} = f^+_{\underline{g}\sigma} f_{\underline{g}\sigma}$ and N is the number of lattice sites. U is thought to be very large.

In what follows, we assume that there are two electrons per atom and that the V=0 position of the Fermi level

$$E_F^0 = E_f + U \qquad (2)$$

lies in the lower half of the d-band, so it is the valencies f^1 and f^2 which mix. To describe this situation, Stevens /3/ and Brandow /4/ suggested the following Ansatz

$$|G\rangle_B = \prod_{\underline{g}} \left[1 + \frac{1}{\sqrt{N}} \sum_{\underline{k}} a(\underline{k})\, e^{-i\underline{k}\underline{g}} (d^+_{\underline{k}\uparrow} f_{\underline{g}\uparrow} + d^+_{\underline{k}\downarrow} f_{\underline{g}\downarrow})\right] |v\rangle \qquad (3)$$

where the coefficients $a(\underline{k})$ are variational parameters, and in the vacuum state $|v\rangle$ all atoms have the f^2 configuration. $|G\rangle_B$ is obviously homogeneous and nonmagnetic which makes it a promising candidate for modelling the ground state of compounds with an even number of electrons per rare earth atom, such as SmB_6.

The disadvantage of $|G\rangle_B$ is that it is rather awkward to handle. To evaluate any expecta tion value, $|G\rangle_B$ should be expanded into a multiple sum over $\underline{g}$ and $\underline{k}$ indices. This is made difficult by the requirements of the exclusion principle: if $d^+_{\underline{g}\sigma}$ acts at a site $\underline{j}$, it should be omitted from the $\underline{k}$-sums at all other sites. This difficulty which Brandow /4/ described as the collision of spin compensation clouds in $\underline{k}$-space, has been circumvented either by intuitive reasoning as to how the exclusion principle can be enforced, or by invoking diagram techniques /3, 4/.

The aim of the present work is to take the variational problem defined by (3) and (1) literally, and carry out the minimization procedure with as few extra assumptions as possible. The essential steps are: first $|G\rangle_B$ is expanded in the orthonormal set formed by the eigenstates of the unperturbed (V=0) Hamiltonian. Then a specific form for the $a(\underline{k})$ will be chosen by expressing all the N variational parameters through a single one denoted by τ. We may call τ the "hybridization temperature" since, in agreement with the intuitive arguments by Stevens /3, 5/ and Brandow /4, 6/, the effect of switching on hybridization will appear to be "heating" the electrons to a temperature $\tau \sim V$.

2. THE EXPANSION OF THE TRIAL STATE

A clue as to how $|G\rangle_B$ can be expanded in an orthonormal basis, can be obtained by considering first the configuration in which the d-band contains two up-spin electrons $\underline{k}_1$ and $\underline{k}_2$, with the corresponding f-holes located at $\underline{g}_1$ and $\underline{g}_2$. This arises in the expansion (3) in two ways: first, when $d^+_{\underline{k}_1\uparrow}$ acts at $\underline{g}_1$ and second, when it acts at $\underline{g}_2$:

$$N^{-1} a(\underline{k}_1) a(\underline{k}_2) \Big(e^{-i\underline{k}_1\underline{g}_1 - i\underline{k}_2\underline{g}_2} d^+_{\underline{k}_1\uparrow} f_{\underline{g}_1\uparrow} d^+_{\underline{k}_2\uparrow} \cdot f_{\underline{g}_2\uparrow} + e^{-i\underline{k}_2\underline{g}_1 - i\underline{k}_1\underline{g}_2} d^+_{\underline{k}_2\uparrow} f_{\underline{g}_1\uparrow} d^+_{\underline{k}_1\uparrow} f_{\underline{g}_2\uparrow} \Big) |v\rangle =$$

$$= N^{-1} a(\underline{k}_1) a(\underline{k}_2) \left\| e^{-i\underline{k}\underline{g}} \right\|_1^2 \left| \begin{matrix} \underline{g}_1 & \underline{g}_2 \\ \underline{k}_1 & \underline{k}_2 \end{matrix} \right\rangle \qquad (4)$$

where we used a short-hand notation for the determinant

$$\left\| e^{-i\underline{k}\underline{g}} \right\|_1^2 = \begin{vmatrix} e^{-i\underline{k}_1\underline{g}_1} & e^{-i\underline{k}_2\underline{g}_1} \\ e^{-i\underline{k}_1\underline{g}_2} & e^{-i\underline{k}_2\underline{g}_2} \end{vmatrix} \qquad (5)$$

and introduced a standard configuration corresponding to a fixed standard ordering of the operators

$$\left| \begin{matrix} \underline{g}_1 & \underline{g}_2 \\ \underline{k}_1 & \underline{k}_2 \end{matrix} \right\rangle = d^+_{\underline{k}_1\uparrow} f_{\underline{g}_1\uparrow} d^+_{\underline{k}_2\uparrow} f_{\underline{g}_2\uparrow} |v\rangle \qquad (6)$$

The general basis state

$$\left| \begin{matrix} \underline{g}_1 \ldots \underline{g}_p ; & \underline{h}_1 \ldots \underline{h}_{r-p} \\ \underline{k}_1 \ldots \underline{k}_p ; & \underline{q}_1 \ldots \underline{q}_{r-p} \end{matrix} \right\rangle \qquad (7)$$

contains r electrons in the d-band: p with spin up, with wavevectors $\underline{k}_1 \ldots \underline{k}_p$, and r-p with spin down, with wavevectors $\underline{q}_1 \ldots \underline{q}_{r-p}$. The corresponding p up-spin f-holes are located at sites $\underline{g}_1 \ldots \underline{g}_p$, while the r-p down-spin f-holes are at $\underline{h}_1 \ldots \underline{h}_{r-p}$. The basis vector (7) is created with the standard ordering of the operators: the f are arranged according to a fixed order of site indices; each f is preceded by a d^+ with the same spin and, within these constraints, the wave vectors follow a standard ordering for each of the two spin directions.

Obviously, (7) appears in the expansion of (3) p!(r-p)! times, with different phase factors. When the operators are commuted to the standard order, the Bloch phase factors combine into a determinant, as we have seen in (4 - 6). Thus, (3) can be rewritten as

$$|G\rangle_B = \sum_{r=0}^{N} N^{-\frac{r}{2}} \sum_{p=0}^{r} \sum_{\{\underline{k}_1^p\}} \sum_{\{\underline{q}_1^{r-p}\}} \prod_{i=1}^{p} a(\underline{k}_i) \prod_{j=1}^{r-p} a(\underline{q}_j) \sum_{\{\underline{g}_1^p\}} \sum_{\{\underline{h}_1^{r-p}\}} \left\| e^{-i\underline{k}\underline{g}} \right\|_1^p \left\| e^{-i\underline{q}\underline{h}} \right\|_1^{r-p} \left| \begin{matrix} \underline{g}_1 \ldots \underline{g}_p ; & \underline{h}_1 \ldots \underline{h}_{r-p} \\ \underline{k}_1 \ldots \underline{k}_p ; & \underline{q}_1 \ldots \underline{q}_{r-p} \end{matrix} \right\rangle \qquad (8)$$

The notations of the determinants are obvious generalizations of (5). The summation over

$$\{\underline{k}_1^p\} = \{\underline{k}_1, \underline{k}_2 \ldots \underline{k}_p\}$$

is over all the possible ways of choosing p different wave vectors from N; the other short notations have analogous meanings. While the $\underline{k}$- and $\underline{q}$-sums are independent, the exclusion of the f^0 configuration demands that the $\underline{g}$- and $\underline{h}$-sets have no common element.

In (8), the overall phase coherence of the ground state is described by the determinant coefficients. The phase relationship between different configurations is not influenced by the choice of the $a(\underline{k})$, and is thus the same as in the completely localised Kaplan-Mahanti ground state /7/, which can be obtained from (3) by choosing $a(\underline{k})$ $\underline{k}$-independent:

$$|G\rangle_{KM} = \prod_{\underline{g}} \left[1 + \frac{a}{\sqrt{N}} \sum_{\underline{k}} e^{-i\underline{k}\underline{g}} (d^+_{\underline{k}\uparrow} f_{\underline{g}\uparrow} + d^+_{\underline{k}\downarrow} f_{\underline{g}\downarrow}) \right] |v\rangle$$

$$= \prod_{\underline{g}} \left[1 + a (d^+_{\underline{g}\uparrow} f_{\underline{g}\uparrow} + d^+_{\underline{g}\downarrow} f_{\underline{g}\downarrow}) \right] |v\rangle \qquad (9)$$

Here $d^+_{\underline{g}\sigma}$ creates an electron in a d-band Wannier state at site $\underline{g}$.

While the simple limiting case (9) may be a good choice for strong enough f-d interaction /8/, it gives far too high an energy with (1) /5/. Therefore, we have to consider a strongly $\underline{k}$-dependent $a(\underline{k})$. However, then the sensitive dependence of the determinant coefficients on the choice of the $\{\underline{k}\}$ -, etc. sets threatens to make further progress impossible. Fortunately, for the evaluation of the expectation value of (1), it is not really necessary to know the determinants one by one; it is sufficient to know a few sum rules for them. In fact, we need only two. To evaluate diagonal matrix elements, such as those of f^+f, or d^+d, we have to use

$$\sum_{\{\underline{k}_1^p\}} \left| \left\| e^{-i\underline{kq}} \right\|_1^p \right|^2 = N^p \tag{10}$$

(10) can be satisfied by replacing the determinants by their real averages

$$\left\| e^{-i\underline{kg}} \right\|_1^p \longrightarrow (C_p^N)^{-1/2}\, N^{p/2} \tag{11}$$

where C is a binomial coefficient.

If we substitute (11) in (3), the vital phase coherence seems to be lost. We can restore as much of it as is needed for the minimization procedure, by introducing an extra rule for the non-diagonal matrix element:

$$\left\langle \begin{matrix} \underline{g}_1 \ldots \underline{g} \ldots \underline{g}_p ; \underline{h}_1 \ldots \underline{h}_{r-p} \\ \underline{k}_1 \ldots \underline{k} \ldots \underline{k}_p ; \underline{q}_1 \ldots \underline{q}_{r-p} \end{matrix} \right| d^+_{\underline{k}\sigma} f_{\underline{g}\sigma} \tag{12}$$

$$\left| \begin{matrix} \underline{g}_1 \ldots \underline{g}_p ; \underline{h}_1 \ldots \underline{h}_{r-p} \\ \underline{k}_1 \ldots \underline{k}_p ; \underline{q}_1 \ldots \underline{q}_{r-p} \end{matrix} \right\rangle = e^{i\underline{kg}} \left[\frac{N}{(p+1)(N-p)} \right]^{1/2}$$

As it is shown in detail elsewhere /9/, the use of the simplifying assumptions (11,12) is justified by recovering exact results in the limiting case (9).

3. THE MODIFIED TRIAL STATE

In (3), the trial state is expanded in terms of the eigenstates of the unperturbed (V=O) part of the Hamiltonian (1). The energy corresponding to (7) is

$$\sum_{i=1}^{p} \epsilon(\underline{k}_i) + \sum_{j=1}^{r-p} \epsilon(\underline{q}_j) - r(E_f+U) + N(2E_f+U) \tag{13}$$

For the minimization of the ground state energy, it is obviously advantageous to choose the $a(\underline{k})$ in such a way that the weight of a configuration (7) will be the smaller, the higher its zeroth-order energy (13). This is achieved by choosing

$$a(\underline{k}) = \exp\left\{ -\frac{1}{2\tau}\left[\sum_1^p \epsilon(\underline{k}_i) + \sum_1^{r-p} \epsilon(\underline{q}_j) - rE_F^O \right] \right\} \tag{14}$$

where $\tau \geq 0$, and E_F^O is given by (2). (14) is like a Boltzmann factor, with the variational parameter τ playing the role of temperature. The weight factor (14), as well as the appearance of the determinant phase factors, are reminiscent of the Gutzwiller treatment /10/ of the Hubbard model.

Using (11) and (14), we can now write down the final form of the Ansatz

$$|G\rangle = \sum_{r=0}^{N} \exp\left(\frac{rE_F^O}{2\tau}\right) \sum_{p=0}^{r} (C_p^N C_{r-p}^N)^{-1/2} \sum_{\{\underline{k}_1^p\}} \cdot$$

$$\exp\left(-\frac{1}{2\tau}\sum_1^p \epsilon(\underline{k}_i)\right) \sum_{\{\underline{q}_1^{r-p}\}} \exp\left(-\frac{1}{2\tau}\sum_1^{r-p} \epsilon(\underline{q}_j)\right)$$

$$\sum_{\{\underline{g}_1^p\}} \sum_{\{\underline{h}_1^{r-p}\}} \left| \begin{matrix} \underline{g}_1 \ldots \underline{g}_p ; \underline{h}_1 \ldots \underline{h}_{r-p} \\ \underline{k}_1 \ldots \underline{k}_p ; \underline{q}_1 \ldots \underline{q}_{r-p} \end{matrix} \right\rangle \tag{15}$$

While (15) suffices for the evaluation of the diagonal matrix elements, the extra rule (12) has to be invoked to calculate the contribution of the hybridization term.

A detailed evaluation of the expectation value of (1) with (15) has to be reserved for a forthcoming publication /9/; here we have to go straight to the results.

4. DISCUSSION

The minimization procedure yields

$$\tau = \pi V \left[\frac{2(1-\xi_o)}{2-\xi_o} \right]^{1/2} \left[\ln^2 \frac{2(1-\xi_o)}{2-\xi_o} + \frac{\pi^2}{3} \right]^{-1} \tag{16}$$

justifying the term "hybridization temperature". ξ_o is the zeroth-order value of the fractional valence (concentration of f^1 sites). The ground state energy

$$E_o = E_o(V{=}O) - \pi^2 V^2 g(E_F^O) \frac{\dfrac{2(1-\xi_o)}{2-\xi_o}}{\ln^2 \dfrac{2(1-\xi_o)}{2-\xi_o} + \dfrac{\pi^2}{3}} \tag{17}$$

is a well-behaved function of V, as suggested by Brandow/4/. The fractional valence

$$\xi = \xi_o + 2\tau g(E_F^O) \ln \frac{2(1-\xi_o)}{2-\xi_o} \tag{18}$$

is less than ξ_o. This can be understood by an argument due to Stevens /3/: "heating up" the band electrons to temperature τ makes it energetically favourable to transfer electrons back to the f-states.

The d-band occupation numbers are found to follow a Fermi distribution at temperature τ, with the chemical potential

$$\mu = E_F^o + \tau \ln \frac{2(1-\xi_o)}{2-\xi_o} \qquad (19)$$

shifted downwards from E_F^o in a manner resembling Brandow's suggestion /6/

$$\mu - E_F^o \sim \ln(1-\xi_o).$$

Looking just at the ground state one can not say whether there is a gap in the spectrum of current-carrying excitations, but circumstantial evidence suggests that we are dealing with an insulating ground state. (15) can be looked upon as the Fourier expanded form of the wave function suggested by Stevens /3/:

$$|G\rangle = \prod_{\underline{g}} \left[1 + (\chi^+_{\underline{g}\uparrow} f_{\underline{g}\uparrow} + \chi^+_{\underline{g}\downarrow} f_{\underline{g}\downarrow}) \right] |v\rangle \qquad (20)$$

where χ^+ creates an electron in a generalized d-band Wannier state whose envelope $b(\underline{r})$ is just the Fourier transform of our $a(\underline{k})$

$$b(\underline{r}) \sim \exp\left(-\frac{r^2 m^* \tau}{\hbar^2}\right) \qquad (21)$$

where m^* is the effective mass at the bottom of the d-band. (With V=0.01 eV and m^*=m(free), $b(\underline{r})$ has a radius of about 30 Å). - The form (20), taken together with (21), makes it apparent that in $|G\rangle$ the d-electron is still localized around its parent f-hole. This seems to indicate a narrow gap semiconductor, in accordance with arguments /11/ based on the Luttinger theorem.

Since in the present treatment minimizing the ground state energy somewhat resembles minimizing a free energy at "temperature" τ, one may ask if this could provide microscopic justification for heuristic thermodynamic arguments /12/ which use a compound quantity composed of the actual temperature and a "quantum mechanical" temperature T_f. Could our τ be that T_f? For a model in which the valence is primarily determined by band filling, the answer could be a tentative yes. One should note, however, that Wohlleben's /12/ work aims at explaining why the valence seems to be actually rather independent of the concentration of rare earth atoms, so the models are not really comparable.

ACKNOWLEDGEMENT

The author is indebted to N.Grewe, B.Mühlschlegel and D.Wohlleben for useful discussions and encouragement, and to SFB 125 for financial support.

REFERENCES

+Permanent address: Central Research Institute for Physics, H-1525 Budapest 114, POB 49, Hungary.

/1/ For recent reviews see:
N.Grewe, H.J.Leder and P.Entel, in: Festkörperprobleme XX, Ed.J.Treusch Vieweg, Braunschweig, 1980

J.M.Lawrence, P.S.Riseborough and R.D.Parks, Rep.Progr.Phys.44,1 (1981)

/2/ P.W.Anderson, in "Valence Fluctuations in Solids", ed. by L.M.Falicov, W.Hanke and M.B.Maple (North Holland, Amsterdam, 1981)

/3/ K.W.H.Stevens, J.Phys.C 11,985 (1978)

/4/ B.H.Brandow, Int.J.Quantum Chem., Symp.Vol.13, 423 (1979)

/5/ K.W.H.Stevens, in: "Valence Fluctuations in Solids", ed. L.M.Falicov. W.Hanke amd M.B.Maple (North Holland, Amsterdam, 1981)

/6/ B.H.Brandow, Physica 102 B,368 (1980)

/7/ T.A.Kaplan and S.D.Mahanti, Phys.Lett. 51A, 265 (1975)

/8/ M.Iwamatsu, Physica 106B, 415 (1981)

/9/ P.Fazekas, to be published in Z.Physik

/10/ M.C.Gutzwiller, Phys.Rev. 137, A 1726 (1965)

/11/ R.M.Martin, Phys.Rev.Lett. 48,362 (1982)

/12/ D.K.Wohlleben, in: "Valence Fluctuations in Solids", ed. L.M.Falicov, W.Hanke and M.B.Maple (North Holland, Amsterdam, 1981)

Valence Instabilities
P. Wachter and H. Boppart (eds.)

INFINITE ORDER BRILLOUIN-WIGNER PERTURBATION THEORY OF THE MIXED VALENT IMPURITY

G. Czycholl, H. Keiter, E. Niebur

Institut für Physik, Universität Dortmund
D-4600 Dortmund 50, F.R.G.

A perturbative approach to the mixed valent lattice in terms of Feynman-Goldstone diagrams needs as input the statistical quasiparticle energies and Green's functions for the f-electrons of a mixed valent impurity. Infinite order partial summations of such diagrams lead to a coupled set of non-linear equations, which become integral equations when performing naively the continuum limit. Their study, however, may be misleading unless vanishing energy denominators are properly regularized.

1. INTRODUCTION

Recently, systematic perturbation techniques with respect to the mixing were applied to model Hamiltonians for the electronic properties of intermediate valence compounds [1-4]. The terms of the expansion can be represented by diagrams, which reflect the elementary excitation processes, i.e. the emission and absorption of delocalized electrons at specific sites. On-site processes (the emitted electrons being reabsorbed at the same site) are calculated from finite temperature Goldstone-diagrams. Inter-site processes are calculated from mixed Feynman-Goldstone diagrams, in which the different sites are connected via band-electron Green functions. Resummations of the diagrams can be performed but they encounter an excluded site problem [4] for the inter-site processes, which is discussed in a paper by N. Grewe at this conference.

The topic of the present paper is the summation of higher order on-site processes. Using the results for the single impurity Anderson-model [5] which follow from the application of Lagrange's formula [6], the energy shifts of the statistical f-quasi-particles are obtained from Brillouin-Wigner equations. For stronger hybridization and (or) for f-level positions below the Fermi-level the perturbative treatment of these equations becomes insufficient. As long as one does not perform the continuum limit in sums over d-electron states, the large class of Brillouin-Wigner diagrams with universal (i.e. band structure independent) logarithmic divergences can be summed [5]. A coupled system of "integral"-equations is obtained, which, at zero temperature, has also been recently derived by Inagaki [7] within the Yosida-Yoshimori-theory. From a comparison with the Brillouin-Wigner result, one finds an interesting inconsistency, which is discussed in section 2 together with Inagaki's approximate treatment and a numerical solution of the system of integral equations. In section 3 the difficulties are traced back to the problem of regularizing energy denominators when going to the continuum limit of d-electron energies. To sum regularized diagrams turns out to be quite cumbersome, as is seen from examples. For some special values of the temperature and the position of the f-level, the results of the unregularized and the regularized approach may coincide.

2. PROBLEMS CAUSED BY NAIVE RESUMMATION OF ON-SITE PROCESSES.

In Fig. 1a (two-site-)process has been drawn, which is an example for the excitation at the two sites caused by two band-electrons with spin σ running between them. Among the excitations at site ν_1, the ones leading to $\Gamma_0^{(2)}$ and $\Gamma_0^{(4)}$ respectively, restore the initial f-state (here labled "0") after an even number of interactions. These "unlinked" structures can be eliminated from all diagrams in favour of an "energy shift

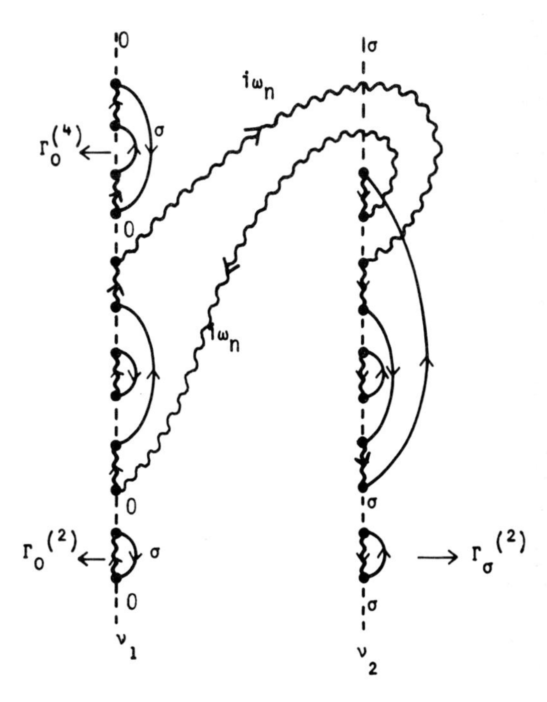

Fig.1: Excitation process, involving two sites

for statistical quasiparticles", $\tilde{E}_0$, which has to be added to all energy-denominators of the remaining "linked" parts, and which also modifies the thermal occupation probability of the initial f-configuration. It furthermore produces quasi-particle renormalization factors, corresponding to the number of linked parts with external lines. (For all technical details, including the rules for calculation of the excitation processes, Ref. [4] should be consulted). The energy-shift $\tilde{E}_0$ is determined from a type of Brillouin-Wigner condition

$$\tilde{E}_0=\Gamma_0(\tilde{E}_0)=\Gamma_0^{(2)}(\tilde{E}_0)+\Gamma_0^{(4)}(\tilde{E}_0)+\ldots, \quad (1)$$

in which Γ_0 is obtained diagrammatically, the first contributions shown in Fig. 1. For the processes starting from an f-configuration with one additional electron of spin σ present in the initial state, the analog of Eq. (1) reads

$$\tilde{E}_\sigma=\Gamma_\sigma(\tilde{E}_\sigma) \quad (2)$$

If the r.h.s. of Eqs. (1) and (2) is approximated by the lowest order terms the starting point of Ramakrishnan's theory [3] is obtained. To go beyond this, one has to perform partial summations within the diagrams for Γ_0 or Γ_σ and within the linked parts with external connections. As long as one does not go over to the continuum limit for the sums on d-electron states, the class of diagrams for Γ_0 and Γ_σ without crossing of lines can be naively summed by the observation that Γ_σ acts as an energy shift for the processes starting from the f-state "0" and vice versa. The result is

$$\Gamma_0(\tilde{E}_0)=\sum_{\vec{k}\sigma}|V_{\vec{k}}|^2 f_{\vec{k}\sigma}\frac{1}{\tilde{E}_0-\varepsilon_{f\sigma}+\varepsilon_{\vec{k}\sigma}-\Gamma_\sigma(\tilde{E}_0-\varepsilon_{f\sigma}+\varepsilon_{\vec{k}\sigma})} \quad (3)$$

$$\Gamma_\sigma(\tilde{E}_\sigma)=\sum_{\vec{k}}|V_{\vec{k}}|^2(1-f_{\vec{k}\sigma})\frac{1}{\tilde{E}_\sigma+\varepsilon_{f\sigma}-\varepsilon_{\vec{k}\sigma}-\Gamma_0(\tilde{E}_\sigma+\varepsilon_{f\sigma}-\varepsilon_{\vec{k}\sigma})} \quad (4)$$

Here, $V_{\vec{k}}$ represents the Fourier-component of the hybridization potential, $f_{\vec{k}\sigma}$ stands for the Fermi distribution function and $\varepsilon_{\vec{k}\sigma}$, $\varepsilon_{f\sigma}$ are the single particle energies of the d- and f-electrons, reckoned from the chemical potential. This system of equations, which was first written down as a single equation for $\Gamma_\sigma(\tilde{E}_\sigma)$[5], has been discussed in various approximations. If one replaces the Γ's on the r.h.s. of Eqs. (3) and (4) by their lowest order approximations, the decoupled system can be studied numerically [2]. An even simpler approximation is useful, when the degeneracy in one channel (here due to the spin) is large. Then one may neglect the Γ's on the r.h.s. of Eqs. (3) and (4) altogether [3]. A much more ambitious treatment of the system was published by Inagaki [7]. He derived the zero temperature version of the system within the frame of the Yosida-Yoshimori-theory. There is a remarkable difference, however, in the interpretation of $\tilde{E}_0$ in this theory and the Brillouin-Wigner-approach. In the latter the energy shifts enter the partition function of the single impurity problem in the following way

$$Z = Z_{Band}(\exp[-\beta\tilde{E}_0]+\sum_\sigma\exp[-\beta(\varepsilon_{f\sigma}+\tilde{E}_\sigma)]) \quad (5)$$

whereas in Inagaki's approach $\tilde{E}_0$ furnishes the ground state energy! At T=0, and for a constant density of states N_0, cut off at $\pm D$, the $\tilde{E}_0$ and $\tilde{E}_\sigma$ in Eqs. (3) and (4) can be shifted into the limits of integration, if one naively proceeds to the continuum limit, with the result

$$\Gamma_0(\tilde{E}_0) = V^2N_0\sum_\sigma\int_{-D+\tilde{E}_0}^{\tilde{E}_0} d\varepsilon\,\frac{1}{\varepsilon-\varepsilon_{f\sigma}-\Gamma_\sigma(\varepsilon-\varepsilon_{f\sigma})} \quad (6)$$

$$\Gamma_\sigma(\tilde{E}_\sigma) = V^2N_0\int_{-\tilde{E}_\sigma}^{D-\tilde{E}_\sigma} d\varepsilon\,\frac{1}{\varepsilon_{f\sigma}-\varepsilon-\Gamma_0(\varepsilon_{f\sigma}-\varepsilon)} \quad (7)$$

A system of difference-differential equations is obtained by taking the derivatives with respect to $\tilde{E}_0$ and $\tilde{E}_\sigma$. From this system, or directly from Eqs. (6) and (7), the effect of zeroes in the denominators can be studied. These zeroes correspond to solutions of the Brillouin-Wigner conditions (1) and (2). A singularity in the integrand of Eq. (6), e.g. leads to an infinite derivative of Γ_0. (More precisely: If, near x_0, $\Gamma_0(\tilde{E}_0)\sim|\tilde{E}_0-x_0|^{2/3}$, then $\Gamma_\sigma(\tilde{E}_\sigma)\sim|\tilde{E}_\sigma-x_0|^{1/3}$, and for a model with λ-fold degeneracy in one f-level, the exponents are changed into $\lambda/\lambda+1$ and $1/\lambda+1$ respectively). This nonanalytic behavior, which leads to branching points of the solution, is also a characteristic feature of Inagaki's approximate solution of the system of Eqs. (6) and (7). He - in a somewhat changed notation - neglects $\tilde{E}_0$ besides D, and solves the resulting approximate system exactly from the equivalent differential equations [7]. The result for $\tilde{E}_0$, obtained from his beautiful mathematics, is highly nonanalytic, however: For $V\to0$, $\tilde{E}_0\to\varepsilon_f$ for zero magnetic field, while, from the perturbation expansion, one would expect $\tilde{E}_0\to0$. Furthermore, for zero magnetic field and for $\varepsilon_f=0$, we found that the $\tilde{E}_\sigma$ obtained within the Inagaki approximation, fulfills $\tilde{E}_0=\tilde{E}_\sigma$. Similar solutions can be obtained already from the lowest order Brillouin-Wigner approximations. But in that case one also finds others, which, as required, continuously go to zero, if $V\to0$. We therefore had to look for other solutions of the integral equations (6) and (7) beyond Inagaki's approximation. For a numerical solution of the integral equations (6) and (7) we added a small imaginary part to the denominators in order to get no numerical problems because of the poles of the integrands. Then the real part of the resulting integral must be taken. In other words we regularized the singular integrals in the most obvious way by numerically calculating their principal values. Only by this means a single valued solution of the coupled equations (6, 7) can be obtained. The Γ's for zero and for finite

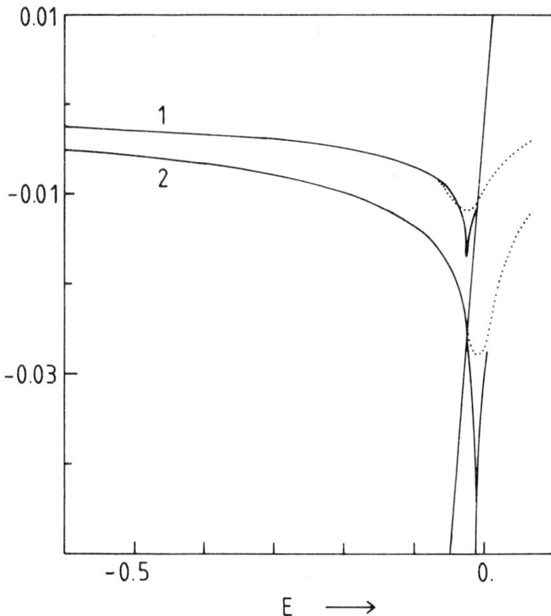

Fig. 2: Numerical result for the Γ's, 1: $\Gamma_\sigma(E)$, 2: $\Gamma_0(E)$, full line: zero temperature, dotted: finite T, for V=0.07, D=1, $\varepsilon_f=0$

temperature are displayed in Fig. 2, together with the intersecting line. These results show the expected behaviour, in particular concerning the asymptotics; furthermore for T=0 the Γ's have the power law singularities with powers of about 1/3 and 2/3, as discussed above. These singularities are smeared over thermally for finite temperatures and we observe only a corresponding minimum in the Γ's. The resulting $\tilde{E}_0$ and $\tilde{E}_\sigma$ can be pursued numerically and go to zero with vanishing hybridization. Because of the tip or minimum structure of the Γ's, however, there may be more than one intersecting point between the Γ-curves and the identity line, i.e. it may happen that there is not a single valued solution for the $\tilde{E}_{0,\sigma}$, at least for low temperatures.

In view of all these problems we decided to reconsider the analytic properties of the diagrams leading to Eqs. (3) and (4).

3. SUMMATIONS OF REGULARIZED DIAGRAMS.

It is a standard problem of any perturbation expansion to avoid vanishing energy-denominators. Usually these denominators are produced automatically, when one performs the continuum limit of discrete quantum states, e.g. of the $\vec{k}$'s in Eqs. (3) and (4). In a given order of the expansion parameter, however, the sum of all contributions behaves regularly, though individual contributions may have vanishing energy denominators. In fact, the ambiguities disappear already within the contribution of a "family of diagrams, rotated on a cylinder" [4-6]. So, the diagrams of these families must be dealt with on the same footing. At a first glance back at the naive summation of all diagrams without crossings of lines, which was performed in section 2, we indeed have summed up all the members of the corresponding families. At a closer look, however, there is a problem: By eliminating first the "unlinked" structures in favour of an energy shift etc. in the remaining diagrams, we did not treat all the individual members of a family on the same footing. This is why a regularization procedure is required. It amounts to assigning small imaginary parts to the independant energy-variables and average over their sign and their relative sizes. For details we refer to [5] and [6]. To understand the meaning of the averaging process, one should note, e.g. that the regularized contributions of $\Gamma_0^{(2)}$ and $\Gamma_0^{(4)}$ in Fig. 1 have to factor from each other and from the rest of the diagram, before they can be summed. The simplest diagrammatic contributions ($\Gamma_\sigma^{(2)}$ and $\Gamma_0^{(2)}$) contain only one denominator, to which the small imaginary part $i\eta$ has to be added and the average over its sign has to be taken. This results in a principal value. It is precisely the same regularization procedure, which was used in the numerical treatment of the integral equation. Unfortunately, taking just principal values, is wrong in higher order diagrams. What happens, may be seen from the regularized contribution $\Gamma_0^{(4)}$

$$\Gamma_0^{(4)}(\tilde{E}_0) \to (2^2.2!)^{-1} \sum{}' \sum\!\!\!\!\!\!\int_{\vec{k},\sigma} |V_{\vec{k}}|^2 f_{\vec{k}\sigma} \frac{1}{(\tilde{E}_0-\varepsilon_{f\sigma}+\varepsilon_{\vec{k}\sigma}+i\eta)^2}$$
$$\times \sum\!\!\!\!\!\!\int_{\vec{k}'} |V_{\vec{k}'}|^2 (1-f_{\vec{k}'\sigma}) \frac{1}{\tilde{E}_0-\varepsilon_{\vec{k}'\sigma}+\varepsilon_{\vec{k}\sigma}+i\eta-i\eta'} \quad (8)$$

in which the prime on the first sum stands for the $2^2.2!$ possibilities of averaging over the signs of η and η', as well as over the two possibilities $|\eta| \gg |\eta_1|$ and $|\eta_1| \gg |\eta|$. The second denominator in Eq. (8) is determined by $i\eta$, if $|\eta| \gg |\eta_1|$, indicating a hierarchy of electron lines. So the $\vec{k}$-integral has to be evaluated as a principal value integral added to a retarded or advanced integral, depending on the sign of η. As a drastic consequence of this, the approximation to the system of Eqs. (3) and (4) used by Bringer and Lustfeld [2] reads in the regularized version

$$\Gamma_0(\tilde{E}_0) = \sum\!\!\!\!\!\!\int_{\vec{k}\sigma} |V_{\vec{k}}|^2 f_{\vec{k}\sigma} \frac{1}{\mathrm{Im}\Gamma_\sigma^{(2)}(\tilde{E}_0+i\eta-\varepsilon_{f\sigma}+\varepsilon_{\vec{k}\sigma})}$$
$$\times \arctan \frac{\mathrm{Im}\Gamma_\sigma^{(2)}(\ldots)}{\tilde{E}_0-\varepsilon_{f\sigma}+\varepsilon_{\vec{k}\sigma}-\mathrm{Re}\,\Gamma_\sigma^{(2)}(\ldots)} \quad (9)$$

(The equation for $\Gamma_\sigma(\tilde{E}_\sigma)$ follows from symmetry, see Eqs. (3) and (4)). As long as the denominator of the arctan is large compared with the numerator, Eq. (9) goes over into the unregularized result. In the opposite case (e.g. near zeros of the denominator) the analytic structure of Eq. (9) differs markedly from the unregularized version. Our numerical results for $\tilde{E}_0$ calculated from Eqs. (9) and (1) are too premature to allow for conclusions about the differences. Moreover, for the intermediate valence compound, if it is treated in the single-site approximation, the shift of the chemical potential probably prevents a zero in the denominator of the arctan in Eq. (9) within the range of integration. The difficulties come up, however, if one proceeds towards the local moment region [8].

The regularization also influences the structure of the single site Green functions, as may be seen from the following example. For the one-particle f-electron Green function we consider those diagrams, in which in the irreducible parts there is only one particle-hole pair in the intermediate states not containing the external "energy" (or Matsubara frequency) $i\omega_n$. (The particle-hole pair may consist either of an f-particle + a d-hole, and vice versa, or a d-particle-hole pair). The regularization problem drops out in energy denominators containing $i\omega_n$.

The classes of diagrams, in which the external line enters at the bottom and leaves at the top (or vice versa) are then readily summed as a geometric series - see e.g. the first contribution on the r.h.s. of Eq. (10). In all the other diagrams there are accidentally vanishing energy denominators which have to be regularized in the way described before. It is a little bit painstaking to sum the regularized diagrams then, but the result is simple

$$F_\sigma(i\omega_n) = \tilde{P}_0 z_0 \{ \frac{1}{i\omega_n + E_0 - \varepsilon_{f\sigma} - \Gamma_\sigma(i\omega_n + \tilde{E}_0 - \varepsilon_{f\sigma})}$$

$$- \oint_{\vec{k}} |V_{\vec{k}}|^2 f_{\vec{k}\sigma} \frac{1}{-i\omega_n + \tilde{E}_0 + \varepsilon_{k\sigma} - \Gamma_0^{(2)}(\tilde{E}_0 + \varepsilon_{\vec{k}\sigma} - i\omega_n)} \times$$

$$\times \frac{1}{\mathrm{Im}\,\Gamma_\sigma^{(2)}(\tilde{E}_0 + \varepsilon_{\vec{k}\sigma} - \varepsilon_{f\sigma} + i\eta)} \times$$

$$\times \mathrm{Im} \frac{1}{\tilde{E}_0 - \varepsilon_{f\sigma} + \varepsilon_{\vec{k}\sigma} - \Gamma_\sigma^{(2)}(\tilde{E}_0 + \varepsilon_{\vec{k}\sigma} - \varepsilon_{f\sigma} + i\eta)} \}$$

$$+ \tilde{P}_\sigma z_\sigma \ldots \qquad (10)$$

Here the three dots stand for the contributions of those diagrams with an initial f state containing an electron with spin σ, z_0 and z_σ are the renormalization factors for the statistical quasi-particles, e.g.

$$z_0 = [1 - \frac{d}{d\tilde{E}_0} \Gamma_0(\tilde{E}_0)]^{-1} \qquad (11)$$

and $\tilde{E}_0$ has to be computed from the Brillouin-Wigner conditions Eq. (1) with $\Gamma_0(\tilde{E}_0)$ obtained from Eq. (9). Γ_σ of the complex argument follows from Eq. (4), the unregularized result. For vanishing imaginary part of $\Gamma_\sigma^{(2)}(\ldots)$ Eq. (10) goes over into the unregularized result, which in a slightly more general form was given in [4]. Results like Eq. (10), taking the correct regularization of diagrams into account, are hoped to overcome the difficulties which arose in using unregularized versions of Green's function approximations.

[1] Proceedings of the International Conference on Valence Fluctuations in Solids, Santa Barbara, 1981 (L. Falicov, W. Hanke, and M. B. Maple, EDS.) North Holland Publ. Co. 1981
[2] Bringer, A., and Lustfeld, H., Z. Physik B 28, 213 (1977)
[3] Ramakrishnan, T.V. in Ref. [1], and Sur, K. preprint
[4] Keiter, H., and Grewe, N., in Ref. [1], and Phys. Rev. B 24, 4420 (1981)
[5] Keiter, H., and Kimball, J. C., Int. J.Magn. 1, 233 (1971)
[6] Balian, R., and De Dominicis, C., Ann. Phys. (N.Y.) 62, 229 (1971)
[7] Inagaki, S., Progr. Theor. Phys. 62, 1441 (1979) and Progr. Theor. Phys. Supplement 69, 232 (1980)
[8] Lustfeld, H. private communication

Valence Instabilities
P. Wachter and H. Boppart (eds.)

BETHE-ANSATZ SOLUTION OF THE COQBLIN-SCHRIEFFER MODEL

J.W. Rasul

Department of Mathematics
Imperial College
London SW7 2B7

The Coqblin-Schrieffer model is solved using the Bethe-Ansatz as formulated by Andrei [1]. Finite temperature equations are obtained and discussed.

The Bethe-Ansatz has recently been used to obtain an exact solution to the spin - ½ Kondo problem [1-4] in agreement with previous renormalisation group results [5]. An important generalisation of the Kondo problem to the case of orbital degeneracy is that of the Coqblin-Schrieffer (CS) Hamiltonian [6]. This Hamiltonian is appropriate to impurities with strong spin-orbit interaction such as Ce, Yb and Sm in the absence of significant crystal-field splitting. The CS Hamiltonian describes the integral valence limit of an impurity described by the orbitally degenerate asymmetric Anderson model. It is therefore appropriate to discuss nearly integral valent intermediate valent dilute alloys, such as $La_{1-x}Sm_xSn_3$ or $Yb_xY_{1-x}CuAl$ in terms of the CS model. Indeed the applicability of a single impurity model to concentrated IV systems has been stressed by some authors [7,8], raising the possibility that the CS model may be useful for understanding some aspects of concentrated, nearly integral valent, IV systems.

In this paper we derive an exact solution of the CS model using the Bethe-Ansatz. Such an approach has been investigated by Tsvelick and Wiegmann [9] who considered ground state properties and outlined a finite temperature treatment leading to the known value of the Wilson ratio. We generalise the problem in the spirit of Andrei [1] and consider a two-component one dimensional (2j + 1) - channel Fermi gas. The conduction electrons correspond to one component (with channel index m) and the other component represents impurities with infinite mass and channel label m. For the case of a single impurity we can make a connection with the CS model by interpreting the channel label m as the magnetic quantum number. A linear dispersion relation is used for conduction electrons since only states near the Fermi level play an essential role. Denoting electron (impurity) operators by c(d) and defining $N_c = 2j+1$ we have

$$H = \int dx \Big[-i \sum_m c_m^+ \frac{d}{dx} c_m + J \sum_{mm'} \big[c_{m'}^+ d_m^+ d_{m'} c_m - \frac{1}{N_c} c_m^+ d_{m'}^+ d_{m'} c_m \big] \Big] \qquad (1)$$

where the coupling constant J has been normalised by the Fermi velocity. Electrons (impurities) can be distinguished by a "purity" label β (=1,0). From the antisymmetry of the N-particle wave function, interchanging channel labels between 2 particles is equivalent to interchanging their purity labels (with a factor -1). (1) can therefore be written in terms of purity exchange operators P_{ij}. The N-particle wave function can be factorised into a product of a purity-dependent part $F(x \ldots x_n : \beta_1 \ldots \beta_n)$ and a channel dependent part described by a complimentary Young diagram to that of F. F satisfies hF = EF where

$$h = -i \sum_{i=1}^{N} \beta_i \frac{d}{dx_i} - J \sum_{i>j}^{N} \delta(x_i - x_j) \big[P_{ij} + 1/N_c \big] (\beta_i - \beta_j)^2 \qquad (2)$$

We now use the Bethe Ansatz

$$F(x_1 \ldots x_N : \beta_1 \ldots \beta_N) = \sum_P \xi_P(Q) \exp\Big(i \sum_{j=1}^{N} K_{P_j} x_{Q_j}\Big) \prod_{l=1}^{N} \delta_{\alpha_{P_l}, \beta_{Q_l}} \qquad (3)$$

F is a linear combination of plane waves labelled by momenta $K_1 \ldots K_n$ and purities $\alpha_1 \ldots \alpha_n$. $\xi_P(Q)$ is the coefficient of the Pth permutation of plane waves in the Qth region defined by $x_{\alpha_1} < x_{\alpha_2} \ldots < x_{\alpha_n}$. Substitution into the Schrodinger equation leads to a set of relations between the ξ in neighbouring regions Q and Q′ corresponding to an interchange of coordinate labels a and b. If i and j label the momenta multiplying these coordinates and P, P′ denote the permutations distinguished by interchanging i and j then

$$\xi_{P'} = [1 - i(\alpha_i - \alpha_j)J/N_c + J^2/4(1 - 1/N_c^2)]^{-1}[i(\alpha_i - \alpha_j)J + (1 - J^2/4(1 - 1/N_c^2))P_{ij}^{ab}]\xi_P, \quad \alpha_i \neq \alpha_j$$

$$\xi_{P'} = \xi_P, \quad \alpha_i = \alpha_j \qquad (4)$$

where P^{ab} interchanges components $\xi p(Q)$ and $\xi p(Q')$ of the vector ξp. We impose periodic boundary conditions which can be written in the form [10,11]

$$X'_{j+1,j} \ldots X'_{N,j} X'_{1,j} \ldots X'_{j-1,j} \Phi = e^{ik_j L} \Phi$$

$$X'_{ij} = M_{ij} \frac{(i(\alpha_i - \alpha_j) + \bar{c} P^{ij})}{(i(\alpha_i - \alpha_j) + \bar{c})} \qquad (5)$$

Where Mij is a constant and the effective coupling constant $\bar{c}$ is given by

$$\bar{c} = J[1 - \tfrac{J^2}{4}(1 - N_c^{-2})]^{-1} \qquad (6)$$

A generalised Bethe-Yang hypothesis[11] leads to a relation between the momentum eigenvalues and a set of numbers $\Lambda_{1,\delta}$ ($\delta = 1 \ldots M_1$)

$$k_j L = 2\pi I_j - i \sum_{i \neq j}^{N} \ln M_{ij} + \sum_{s=1}^{M_1} [\Theta(2\alpha_j - 2\Lambda_{1s}) - \pi] \qquad (7)$$

where $\Theta(x) = 2 \tan^{-1}(x/c)$, and is the first of $N_c - 1$ sets of numbers related by

$$\sum_{\mu=1}^{M_{r-1}} \Theta(2\Lambda_{r,p} - 2\Lambda_{r-1,\mu}) = \sum_{v=1}^{M_r} \Theta(\Lambda_{r,p} - \Lambda_{r,v}) - \sum_{s=1}^{M_{r+1}} \Theta(2\Lambda_{r,p} - 2\Lambda_{r+1,s}) + 2\pi J_{r,p} \qquad (8)$$

where the Is and Js are integers (or half integers) and are quantum numbers of the system. The number of particles in channel r is given by $M_{r-1} - M_r$ and we have defined $M_0 = 0$, $M_{nc} = 0$, $\Lambda_{0,\mu} = \alpha_\mu$, $\Lambda_{N_c} = 0$. The total energy of the system $E = \sum_j^N \alpha_j K_j$. In the ground state the $N \to \infty$ limit is taken by replacing sums by integrals over densities $\sigma_r(\Lambda)$ from some limit $-B_r$ (say) to infinity. Inverting the above equations yields

$$\sigma_r(\Lambda) = \sigma_r^0(\Lambda) - \sum_{s=1}^{N_c - 1} \int_{-\infty}^{-B_s} (R_{rs}(\Lambda - \Lambda') - \delta_{rs}\,\delta(\Lambda - \Lambda'))\,\sigma_s(\Lambda')\,d\Lambda' \qquad (9)$$

where R is defined by its Fourier transform

$$R_{rs}(\omega) = e^{c|\omega|/2} \frac{\sinh \frac{\bar{c}\omega}{2}(N_c - r)\,\sinh \frac{\bar{c}\omega s}{2}}{\sinh \frac{\bar{c}\omega N_c}{2}\,\sinh \frac{\bar{c}\omega}{2}}, \quad r \geq s \qquad (10)$$

and σ_r^0 is the solution in the case of all $B_r = \infty$

$$\sigma_r^0(\Lambda) = \frac{1}{\bar{c}N_c} \sin\left(\frac{\pi r}{N_c}\right)\left[N_e\left[\cosh \frac{2\pi(\Lambda - 1)}{\bar{c}N_c} - \cos\frac{\pi r}{N_c}\right]^{-1} + N_i\left[\cosh \frac{2\pi\Lambda}{\bar{c}N_c} - \cos\left(\frac{\pi r}{N_c}\right)\right]^{-1}\right] \qquad (11)$$

In order to derive the magnetisation and susceptibility for finite magnetic fields in cases of interest equations (9) have to be solved numerically . We now discuss finite temperature properties and make the usual conjecture [3,12]

$$\Lambda_{r,\alpha} \to \Lambda_{r,\alpha}^{(n)} + i(n + 1 - 2j)\bar{c}/2 + O(e^{-N})$$

$$j = 1 \ldots n \qquad (12)$$

Thus the set $\Lambda_{r,\alpha}$ which took only real values in the ground state have been generalised to include discrete imaginary parts. For a particular "n-string" these take on the values

$$i(n-1)\bar{c}/2 \;\ldots\ldots\; -i(n-1)\bar{c}/2 \qquad (13)$$

Equations (7) and (8) are thus generalised to include sums over j and string label n. At finite temperatures the sets of quantum numbers $J^n_{r,p}$ and their corresponding $\Lambda^n_{r,p}$ are not all occupied. In the thermodynamic limit $N \to \infty$ we introduce particle (hole) densities ρ ($\tilde{\rho}$) such that the total number of $\Lambda^n_{r,p}$ is given by

$$N \int d\lambda \left(\rho_n^{(r)}(\lambda) + \tilde{\rho}_n^{(r)}(\lambda)\right) \qquad (14)$$

The sums in (7) and (8) become integrals

from $-\infty$ to $+\infty$. Taking the Fourier transform of (8) yields

$$2\pi \tilde{\rho}_n^{(r)}(\omega) = \sum_{m=1}^{\infty} A_{nm}(\omega)\Big[\big(\rho_m^{(r+1)}(\omega) + \rho_m^{(r-1)}(\omega)\big)\frac{\mathrm{sech}(c\omega/2)}{2} - \rho_m^{(r)}(\omega)\Big] \quad (15)$$

where

$$A_{n,m}(\omega) = 2\pi \coth\left|\frac{c\omega}{2}\right|\Big[\exp\Big(-|(n-m)|\frac{c\omega}{2}\Big) - \exp\Big(-(n+m)\left|\frac{c\omega}{2}\right|\Big)\Big]$$

Inverting these equations we obtain particle densities in terms of hole densitites

$$\rho_n^{(r)}(\omega) = \sum_{s=1}^{N_c-1} R_{rs}(\omega)\Big[-(1+e^{-|c\omega|})\tilde{\rho}_n^{(s)}(\omega) + e^{-|c\omega/2|}\big(\tilde{\rho}_{n-1}^{(s)}(\omega) + \tilde{\rho}_{n+1}^{(s)}(\omega)\big)\Big], \quad (16)$$

$$\tilde{\rho}_0^{(r)}(\omega) = \delta_{1,r}(N_e e^{i\omega} + N_i)/N$$

The energy can be written in terms of particle densities using (7). It is more convenient to express this in terms of hole densities

$$E = E_0 + E_c(T) + \frac{2N_e N}{L}\sum_{s=1}^{N_c-1}\int_{-\infty}^{+\infty} d\lambda\, \tilde{\rho}_1^{(s)}(\lambda) \times \tan^{-1}\Big[\frac{\sin(\pi s/N_c)}{e^{-2\pi(\lambda-1)/cN_c} - \cos(\frac{\pi s}{N_c})}\Big] \quad (17)$$

where E_0 is the ground state energy and E_c arises from the charge degrees of freedom. In the CS model these are uncoupled from the spin densities. Consequently minimisation of the free energy with respect to variations in the densities only involves spin degrees of freedom. The entropy can be found by generalising to the case of N_c channels and an infinite number of n-strings the arguments of Yang and Yang[13].

$$S = N\sum_{n=1}^{\infty}\sum_{r=1}^{N_c-1}\int_{-\infty}^{+\infty} d\lambda\,\Big[\big(\rho_n^{(r)}(\lambda) + \tilde{\rho}_n^{(r)}(\lambda)\big) \times \ln\big(\rho_n^{(r)}(\lambda) + \tilde{\rho}_n^{(r)}(\lambda)\big) - \rho_n^{r}(\lambda)\ln\rho_n^{r}(\lambda) - \tilde{\rho}_n^{r}(\lambda)\ln\tilde{\rho}_n^{r}(\lambda)\Big] \quad (18)$$

Taking variations of the free energy $E - TS$ with respect to densities and substituting for ρ we obtain

$$\frac{2N_e}{LT}\tan^{-1}\Big[\frac{\sin(\pi r/N_c)}{e^{-2\pi(\lambda-1)/cN_c} - \cos(\pi r/N_c)}\Big] = \ln(1 + 1/\eta_n^r(\lambda)) + \sum_{s=1}^{N_c-1}\Big[g_{rs} * \ln(1+\eta_n^s(\lambda)) + f_{rs} * [\ln(1+\eta_{n-1}^s(\lambda)) + \ln(1+\eta_{n+1}^s(\lambda))]\Big],$$

$$f * g(\lambda) = \int_{-\infty}^{+\infty} d\lambda'\, f(\lambda-\lambda')\, g(\lambda') \quad (19)$$

where f and g are defined by their Fourier transforms

$$g_{rs}(\omega) = -(1 + e^{-|c\omega|})R_{rs}(\omega)$$
$$f_{rs}(\omega) = e^{-c|\omega|/2} R_{rs}(\omega) \quad (20)$$

and $\eta_n^r = \tilde{\rho}_n^{(r)}/\rho_n^{(r)}$ is related to the field H by the boundary condition

$$\lim_{n\to\infty} \eta_n^r/n = H/T \quad (21)$$

The free energy is related to $\eta_1^r(\lambda)$ by

$$F = E_0 + F_c(T) - \frac{TN}{cN_c}\sum_{r=1}^{N_c-1}\int d\lambda\,\ln(1+\eta_1^r(\lambda)) \Big[\frac{N_e \sin(\pi r/N_c)}{\cosh\frac{2\pi(\lambda-1)}{cN_c} - \cos(\frac{\pi r}{N_c})} + \frac{N_i \sin \pi r/N_c}{\cosh\frac{2\pi\lambda}{cN_c} - \cos(\frac{\pi r}{N_c})}\Big] \quad (22)$$

where $F_c(T)$ is the free energy of a gas of spinless fermions

$$F_c(T) = \frac{-\pi T^2 L}{12} \quad (23)$$

In the high temperature limit the driving term in (19) vanishes. The η_n^r are then independent of λ and satisfy a set of non-linear difference equations. Solving these leads to the expected free energy for a free N_c-fold degenerate impurity. This provides a check on the consistency of the n-string conjecture.

At low temperatures $T \ll H$[3,14,15] we introduce "pseudoenergies" $\varepsilon_n^{\Gamma} = T \ln h_n^{\Gamma}$ and assume that the I - string pseudoenergies ε_1^{Γ} are decreasing functions of λ with zeroes at $\lambda = -B_r$ (say). Expanding the ε_1^{Γ} in powers of T introduces equations similar to (9). Taking the limit $H \to 0$ ($B_r \to \infty$) we find the T^2 term in the free energy

$$-\frac{\pi T^2 L}{12}\left[N_c + (N_i/N_e)(N_c - 1) e^{2\pi/\varepsilon N_c} \right]$$

Since the ratio of impurity susceptibility to electron susceptibility is [9]

$$\chi_{imp}/\chi_{el} = e^{2\pi/\varepsilon N_c}$$

it follows that the Wilson ratio has the known value

$$R = N_c/(N_c - 1) \qquad (24)$$

The author is grateful to A.C. Hewson and D.M. Newns for stimulating discussions and to the S.R.C. for financial support.

References:

(1) Andrei, N., Phys. Rev. Lett. 45 (1980) 379.

(2) Wiegmann, P.B., JETP Lett. 31 (1980) 364 and J. Phys. C 14 (1981) 1463.

(3) Filyov, V.M., Tsvelick, A.M., and Wiegmann, P.B., Phys. Lett. 81A (1981) 175.

(4) Andrei, N., and Lowenstein, J.H., Phys. Rev. Lett. 46 (1981) 356.

(5) Wilson, K.G., Rev. Mod. Phys. 67 (1975) 773.

(6) Coqblin, B., and Schrieffer, J.R., Phys. Rev. 185 (1969) 847.

(7) Newns, D.M., and Hewson, A.C., J. Phys. F 10 (1980) 2429 and A local Fermi liquid theory of intermediate valence systems, in Falicov, L.M. Hanke, W. and Maple, M.B. (eds.), Valence Fluctuations in Solids (North - Holland, Amsterdam, 1981).

(8) Ramakrishnan, T.V., Perturbative theory of mixed valence systems, in Falicov, L.M., Hanke, W., and Maple, M.B. (eds.), Valence Fluctuations in Solids (North - Holland, Amsterdam, 1981).

(9) Tsvelick, A.M. and Wiegmann, P.B., J. Phys. C 15 (1982) 1707.

(10) Yang, C.N., Phys. Rev. Lett. 19 (1967) 1312.

(11) Sutherland, B., Phys. Rev. Lett. 20 (1967) 98.

(12) Takahashi, M., Prog. Theor. Phys. 46 (1971) 401.

(13) Yang, C.N., and Yang, C.P., J. Math. Phys. 10 (1969) 1115.

(14) Johnson, J.D. and McCoy, B.M., Phys. Rev. A6 (1972) 1613.

(15) Takahashi, M. Prog. Theor. Phys. 50 (1973) 1519.

Valence Instabilities
P. Wachter and H. Boppart (eds.)

A NEW DESCRIPTION OF THE INTERMEDIATE VALENCE GROUND STATE

O.L.T. de Menezes and A.Troper

Centro Brasileiro de Pesquisas Físicas
Rua Xavier Sigaud 150, 22.290 Rio de Janeiro, Brasil

It is suggested that an attractive interaction between f- and d- conduction electrons mediated by phonon exchange can provide a ground state for intermediate valence systems. This theoretical assumption gives a qualitatively good description of some experimental features observed in these systems.

The intermediate valence (IV) phenomena exhibited by some systems containing rare earths (Ce, Sm, Eu, Tm and Yb) has been under active study due to its anomalous behaviour in respect to normal rare earth systems. In spite of this interest, a complete understanding of IV remains still open, due to the absence of a model describing adequately the main features of IV.

Several theoretical models have been proposed, the majority of them based on the Anderson hamiltonian, within its purely electronic nature. On the other hand, some approaches, supported by increasing experimental evidences of phonon anomalies in IV [1], have incorporated the electron-phonon (EP) interaction [2], argued to play an important role in the understanding of these systems. Such approaches put the question whether EP mechanisms are relevant due the thermal phonon excitations, which are damped out at very low temperatures, or if their role could be deeper, acting as virtual mechanisms changing the nature of the ground state.

In this work one analyses the existence of an attractive interaction between f- and d- electrons at the Fermi level due to the exchange of virtual phonons. This approache gives a completely different insight into the understanding of IV phenomena and, within a simple picture, is in qualitative agreement with some of the most characteristic anomalous features of IV.

Starting from a generalized version of the Fröhlich hamiltonian, the renormalizations on the electronic contributions can be written in a quite general form as [3]:

$$H^{r} = - \sum_{\alpha\beta\alpha'\beta'} \sum_{\underset{\sim}{k}\underset{\sim}{k}'\underset{\sim}{q}'} \sum_{\sigma\sigma'} \frac{g_{\alpha\beta}^{\underset{\sim}{k}\underset{\sim}{q}}\, g_{\alpha'\beta'}^{\underset{\sim}{k}'\underset{\sim}{q}'}}{h\omega_{\underset{\sim}{q}}} \times \alpha^{\dagger}_{\underset{\sim}{k}+\underset{\sim}{q}\sigma} \beta_{\underset{\sim}{k}\sigma} \alpha'^{\dagger}_{\underset{\sim}{k}'+\underset{\sim}{q}'\sigma'} \beta_{\underset{\sim}{k}'\sigma'} \delta_{\underset{\sim}{q}',-\underset{\sim}{q}'} \qquad (1)$$

where $\alpha,\beta(\alpha',\beta')$ can be either f- or d- electronic states wavevectors $\underset{\sim}{k}(\underset{\sim}{k}')$ and $\sigma(\sigma')$.

The energy of phonons with wavevector $\underset{\sim}{q}$ is $h\omega_{\underset{\sim}{q}}$ and $g_{\alpha\beta}$ are EP couplings between α- and β- electrons.

Several scattering processes are included in eq.(1). If one supposes that IV systems are characterized by the simultaneous presence of f- and d- electrons at the Fermi level, within a n=2 model, and that the most relevant EP scatterings are those involving opposite spins, eq. (1) turns out to be:

$$H^{int} = - \Lambda \sum_{\underset{\sim}{k}\underset{\sim}{q}\sigma} f^{\dagger}_{\underset{\sim}{k}+\underset{\sim}{q}\sigma} f_{\underset{\sim}{k}\sigma} d^{\dagger}_{-\underset{\sim}{k}-\underset{\sim}{q}-\sigma} d_{-\underset{\sim}{k}-\sigma} \qquad (2)$$

where Λ denotes the effective EP couplings involving f- and d- electrons, considered for simplicity as being constant and proportional to $g_{ff}g_{dd}$, if spin-flip scattering-like terms are neglected.

So, the model hamiltonian reads:

$$H = H^{el} + H^{int} = \sum_{\underset{\sim}{k}\sigma} \varepsilon^{d}_{\underset{\sim}{k}} d^{\dagger}_{\underset{\sim}{k}\sigma} d_{\underset{\sim}{k}} + \sum_{\underset{\sim}{k}\sigma} \varepsilon^{f}_{\underset{\sim}{k}} f^{\dagger}_{\underset{\sim}{k}\sigma} f_{\underset{\sim}{k}\sigma} + V \sum_{\underset{\sim}{k}\sigma} (f^{\dagger}_{\underset{\sim}{k}\sigma} d_{\underset{\sim}{k}\sigma} + d^{\dagger}_{\underset{\sim}{k}\sigma} f_{\underset{\sim}{k}\sigma}) + H^{int}. \qquad (3)$$

In eq.(3) d-f Coulomb interaction between f- and d-parallel spins is neglected whereas d-f Coulomb interaction among opposite spins is overcomed by EP interaction and henceforth Λ may be regarded as an effective d-f coupling. Moreover, it is assumed that 4f- electrons show, at least in the case of Ce, a certain degree of itineracy [4].

The trial wave function (TWF) for the ground state of H, within the n=2 model, involves the simultaneous creation of f- and d- electrons, and according with the suggested role played by EP interaction, is written as:

$$|\psi\rangle = \sum_{\underset{\sim}{k}\sigma} A(\underset{\sim}{k})\, f^{\dagger}_{\underset{\sim}{k}\sigma} d^{\dagger}_{-\underset{\sim}{k}-\sigma} |\phi\rangle \qquad (4)$$

$|\phi\rangle$ denoting the "core states".

Note that this TWF has some similarities with those proposed by Kaplan and Mahanti [5] (KM), Stevens [6] and Brandow [7]. In KM model, the mechanism responsible for the IV ground state, assumes that f- and d- electrons are localized. An attractive term between f- and d- electrons arises from a van der Waals-type interaction, which however has been argued to be very small [8]. Moreover the presence of a dipole moment in IV phase was not until now experimentaly observed. In Refs. 6 and 7 the variational ground state wave function, as also in KM model, is not affected by EP interactions.

From eqs. (3) and (4) one can calculate the energy $E = \langle\psi|H|\psi\rangle$ and minimizing E respect to $A(\underset{\sim}{k})$'s, under the constraint $\sum_{\underset{\sim}{k}} |A(\underset{\sim}{k})|^2 = 1$, one gets:

$$E = E_F - \Delta \,, \tag{5}$$

where E_F is the Fermi energy without EP interaction:

$$E_F = \sum_{\underset{\sim}{k}} \left(\varepsilon^f_{\underset{\sim}{k}} + \varepsilon^d_{\underset{\sim}{k}} \right) \tag{6}$$

and Δ is the energy which stabilizes the ground state:

$$\Delta = 2\hbar\omega \, \exp\left[-\Lambda \rho(E_F) \right] \tag{7}$$

$\rho(E_F)$ being the density of states at E_F, expected to be high due to the presence of f- electrons around the Fermi level.

In spite of similarities with BCS results, the present model does not imply in a superconducting behaviour, since d- conduction states are not available around the Fermi level to be paired among them. Moreover the symmetry of $|\psi_A\rangle$ does not allow Bose-Einstein condensation.

Eq.(4) could be written under a symmetric ($|\psi_s\rangle$) and antisymmetric ($|\psi_A\rangle$) form:

$$|\psi_s\rangle = \frac{1}{\sqrt{2}} \sum_{\underset{\sim}{k}} \left(f^\dagger_{\underset{\sim}{k}\uparrow} d^\dagger_{-\underset{\sim}{k}\downarrow} + f^\dagger_{-\underset{\sim}{k}\downarrow} d^\dagger_{\underset{\sim}{k}\uparrow} \right) |\phi\rangle$$
$$|\psi_A\rangle = \frac{1}{\sqrt{2}} \sum_{\underset{\sim}{k}} \left(f^\dagger_{\underset{\sim}{k}\uparrow} d^\dagger_{-\underset{\sim}{k}\downarrow} - f^\dagger_{-\underset{\sim}{k}\downarrow} d^\dagger_{\underset{\sim}{k}\uparrow} \right) |\phi\rangle \tag{8}$$

both states being equally effected by H^{int}.

For a $n = 2$ model, the possible states involving f- and d- electrons, as a tentative basis may be written as:

$$|\psi_A\rangle = \sum_{\underset{\sim}{k}\underset{\sim}{k}'} \sum_{\sigma\sigma'} A(\underset{\sim}{k},\underset{\sim}{k}') \, \alpha^\dagger_{\underset{\sim}{k}\sigma} \beta^\dagger_{\underset{\sim}{k}'\sigma'} |\phi\rangle \,, \tag{9}$$

Clearly, $|\psi_A\rangle$ and $|\psi_s\rangle$ are included in eq. (8), when one makes $\underset{\sim}{k}' = -\underset{\sim}{k}$ and $\sigma' = -\sigma$ with $\alpha = f$ and $\beta = d$. One can see that hybridization admixes $|\psi_s\rangle$ and other states involving opposite spins whereas $|\psi_A\rangle$ is not affected by hybridization. Therefore $|\psi_A\rangle$ and $|\psi_s\rangle$ are split in energy, and $E_A < E_s$, implying that $|\psi_A\rangle$ may be indeed a good TWF for the ground state. Moreover the other possible states apart from $|\psi_s\rangle$ are degenerate ($E = E_F$), if one neglects Coulomb interactions. Once one takes into account these interactions some of these excited states will rise up in energy.

One of the most interesting consequences of the existence of the energy gap Δ is that the magnetic susceptibility at zero temperature should be finite. This follows from a van Vleck type behaviour at low temperatures assigned to the presence of degenerate excited magnetic states with energy E_F. As the temperature increases, $\Delta(T)$ is expected to decrease smoothly and a maximum in the magnetic susceptibility should be expected around a temperature T*, corresponding to $\Delta(T^*) = 0$. For $T > T^*$, the system exhibits a Curie-Weiss type behaviour and the role of f-d hybridization becomes important. Moreover, above T*, EP interaction still remains acting as thermal excitations, renormalizing the purely electronic interactions [3].

Electron-Phonon interaction at low temperatures and the structure of the ground state are expected to modify the electronic scattering channels as compared as to normal systems. Such modification could be, for instance, detected in ESR relaxation measurements. In spite of the controversial effects of IV on the relaxation rate [9-10] one should expect that for $T < T^*$ the relaxation rate should be almost constant, whereas for $T > T^*$ it becomes faster as seems to be the case of $CePd_3$.

One could expect also that several anomalous behaviours are to be found T*, e.g., resitivity, specific heat etc. Detailed calculation should provide a more realistic and even quantitative information about the consequence of the model.

One of the authors (O.L.T.M.) would like to thank Alexander von Humboldt-Stiftung for financial Support.

REFERENCES

[1] Valence Fluctuation in Solids, Falicov, L.M., Hanke, W. and Maple, M.P., editors (North-Holland, Amsterdam, 1981).

[2] Lawrence, J.M., Riseborough, P.S. and Parks, R.D., Rep. Progr. Phys. 44 (1981)1.

[3] Menezes, O.L.T. de and Troper, A., Phys. Rev. B22 (1980) 2127.

[4] Pickett, W.E., Freeman, A.J. and Koellings, D.D., Phys. Rev. B2 (1981) 1266.

[5] Kaplan, T.A. and Mahanti, S.D., Phys.Lett. 51A (1975) 265.

[6] Stevens, K.W., J.Phys. C (Solid St.Phys.) 11 (1978) 985.

[7] Brandow, B.H., Int. J. Quantum Chem. Symposium 13 (1979) 423.

[8] Khomskii, D.I., Sov. Phys. Usp. 22 (1979) 879.

[9] Gambke, T., Elschner, B. and Hirst, L.L., Phys. Rev. Lett. 40 (1978) 1290.

[10] Barberis, G.E., Davidov, D., Rettori, C., Donoso, J.P., Torriani, I. and Gandra, F.C., Phys. Rev. Lett. 45 (1980) 1966.

Valence Instabilities
P. Wachter and H. Boppart (eds.)

ONE PARTICLE PROPAGATORS OF MIXED VALENCE COMPOUNDS

H.G.Baumgärtel and E.Müller-Hartmann

Institut für Theoretische Physik, Universität zu Köln
Zülpicher Str.77, D-5000 Köln 41

For various reasons it is difficult to formulate a systematic theory of one-particle propagators of mixed valence compounds which takes the intraionic Coulomb repulsion serious and is at the same time correct to even the lowest order in the hybrisization. We have obtained such a theory, correct to second order, by carefully decoupling equations of motion. In this theory, the one-particle propagator for the 4f-electrons in momentum representation has the structure

$$F_{\underline{k}}(z) = \mathcal{N}(z)/(z-E-\mathcal{M}(z)-\mathcal{N}(z)\frac{V^2}{z-\varepsilon_{\underline{k}}}) \quad .$$

The self-energy $\mathcal{M}(z)$ and the renormalisation factor $\mathcal{N}(z)$ satisfy self-consistent non-linear integral equations.

The 4f-shell of rare earth ions has unique properties: This shell is only partially occupied but lies in the interior of the Xe-core of the ions. It's radius is much smaller than that of the 6s-orbital.

The pecularity of mixed valence compounds is that the position of the Fermi energy allows transitions between the 4f-shell and the conduction band without cost of energy. This leads to the coexistence of two configurations, $4f^n$ and $4f^{n+1}$. The 4f-orbitals overlap with the 5d and 6s-states of the neighbouring atoms. Therefore the 4f-electrons and the conduction band hybridizes.

The 4f-electrons of a RE-ion are in highly correlated n or n+1 electron states. The correlation energy is much larger than the hybridization energy. That is the reason why the starting point for a reasonable description of mixed valence compounds has to be a model for the RE-ions which includes at least the Hund's rule correlations of the two degenerate configurations exactly. The hybridization V can then treated as a perturbation.

The resultant perturbation theory shows several difficulties. The degeneracy of the $4f^n$ and the $4f^{n+1}$ configurations causes an infrared divergence. The lowest order self-energy:

$$\Delta = |V|^2 \cdot N_F \ln(D/k_B T)$$

(D=band-width, N_F=density of states at the Fermi surface) shows a logarithmic divergence. Also the occupation number of the 4f-electrons $n_{4f} = n^o_{4f} - 2\cdot N_F \cdot \Delta$ and the thermal expansion $\alpha \propto |V|^2 \cdot N_F/T$ diverge at low temperatures /1/.

The pertubation theory is therefore to be considered a high-temperature approximation. To obtain results for the interesting low-temperature region one has to set up some sort of infinite order summation or a self-consistent treatment of the perturbation expansion /1/. Since the major correlations in the RE-ions are caused by two-body forces a perturbation theory can not take advantage of Wick's theorem. This leads to complications of the diagrammatic perturbation theory /2/,/3/.

The aim of the present work is the formulation of a self-consistent theory which describes the low-temperature cut-off of the divergent self-energy mentioned above. We are considering the hierarchy of the equations of motion for the 4f-electron propagator. To obtain a closed chain of equations we decouple all contributions with an explicit factor V^2.

The two configurations $4f^n$ and $4f^{n+1}$ have ground states with angular momentum J and $\bar{J}$. For the model Hamiltonian of the RE-ions we use the transfer-operators $X_{PQ} = |P\rangle\langle Q|$ (P and Q may be one of the n or n+1 electron states $|JM\rangle$ or $|\bar{J}\bar{M}\rangle$) introduced by Hubbard /4/:

$$H_L = E_n \sum_{iM} X_{iMM} + E_{n+1} \sum_{i\bar{M}} X_{i\bar{M}\bar{M}}$$

Here E_n and E_{n+1} are the energies of a RE-ion with n and n+1 4f-electrons respectively. The i-summation runs over all sites with a RE-ion. Since we consider her a stochiometric compound, we suppose H_L to be lattice invariant. For the conduction band we have:

$$H_c = \sum_{\underline{k}\sigma} \varepsilon_{\underline{k}} c^+_{\underline{k}\sigma} c_{\underline{k}\sigma}$$

The hybridization $\hat{V}$ describes the transition of an electron with orbital angular momentum m and spin σ from the the 4f-shell into the conduction band as an electron with momentum $\underline{k}$ and spin σ and the reverse process:

$$\hat{V} = \frac{1}{\sqrt{N}} \sum_{\underline{k}\sigma i m} (V^*_{\underline{k}m} e^{-i\underline{k}\cdot\underline{R}_i} c^+_{\underline{k}\sigma} f_{im\sigma} + \text{h.c.})$$

Because of their high energies the exited states of the RE-ions are not involved in the processes considered here. Therefore it is reasonable to restrict the state space to the ground-states of the two configurations $4f^n$ and $4f^{n+1}$:

$$\hat{V} = \sum_{PQ} |P\rangle\langle P|\hat{V}|Q\rangle\langle Q|$$

Here the P and Q-summations run over the restricted state space:

$$\hat{V} = \frac{1}{\sqrt{N}} \sum_{\underline{k}\sigma i M\bar{M}} (V^{M\bar{M}*}_{\underline{k}m} e^{-i\underline{k}\cdot\underline{R}_i} c^+_{\underline{k}\sigma} X_{iM\bar{M}} + \text{h.c.})$$

with the matrix elements:

$$V^{M\bar{M}}_{\underline{k}\sigma} = \sum_m V_{\underline{k}m} \langle M | f^+_{im\sigma} | \bar{M} \rangle .$$

To take optimum advantage of the translational invariance we use the Fourier transformed transfer-operators:

$$X_{\underline{k}M_1M_2} = \frac{1}{\sqrt{N}} \sum_i e^{-i\underline{k}\cdot\underline{R}_i} X_{iM_1M_2} .$$

with the commutator rule:

$$\{X_{\underline{k}M_1M_2}, X_{\underline{k}'M_3M_4}\}_{\mp} = \frac{1}{\sqrt{N}} (\delta_{M_2M_3} X_{\underline{k}+\underline{k}'M_1M_4} \pm \delta_{M_4M_1} X_{\underline{k}+\underline{k}'M_3M_2}) .$$

Here "+" applies if M_1+M_2 and M_3+M_4 are half-integral and "-" otherwise. With the Hamiltonian:

$$H = H_L + H_c + \hat{V}$$

we set up the equation of motion for the propagator ($E = E_n + E_{n+1}$)

$$(z-E)\langle\langle X_{\underline{k}M\bar{M}} | X_{-\underline{k}\bar{M}'M'} \rangle\rangle = \frac{1}{\sqrt{N}} \left(\delta_{MM'} \langle X_{\underline{0}MM} \rangle + \delta_{\bar{M}\bar{M}'} \langle X_{\underline{0}\bar{M}\bar{M}} \rangle \right)$$
$$+ \frac{1}{\sqrt{N}} \sum_{\substack{\underline{k}_1\sigma_1 \\ M_1\bar{M}_1}} V^{M_1\bar{M}_1}_{\underline{k}_1\sigma_1} \Big[\delta_{\bar{M}\bar{M}_1} \langle\langle X_{\underline{k}-\underline{k}_1 MM_1} c_{\underline{k}_1\sigma_1} | X_{-\underline{k}\bar{M}'M'} \rangle\rangle$$
$$+ \delta_{MM_1} \langle\langle X_{\underline{k}-\underline{k}_1 \bar{M}_1\bar{M}} c_{\underline{k}_1\sigma_1} | X_{-\underline{k}\bar{M}'M'} \rangle\rangle \Big]$$

The new propagators on the right hand site satisfay the equations:

$$(z-\varepsilon_{\underline{k}_1}) \langle\langle X_{\underline{k}-\underline{k}_1 MM_1} c_{\underline{k}_1\sigma_1} | X_{-\underline{k}\bar{M}'M'} \rangle\rangle$$
$$= \frac{\delta_{MM'}}{\sqrt{N}} \langle X_{-\underline{k}_1\bar{M}M_1} c_{\underline{k}_1\sigma_1} \rangle + \sum_{M_2\bar{M}_2} V^{M_2\bar{M}_2*}_{\underline{k}_1\sigma_1} \langle\langle X_{\underline{k}-\underline{k}_1 MM_1} X_{\underline{k}_1 M_2\bar{M}_2} | X_{-\underline{k}\bar{M}'M'} \rangle\rangle$$
$$+ \frac{1}{\sqrt{N}} \sum_{\underline{k}_2\sigma_2\bar{M}_2} \Big[-V^{M_1\bar{M}_2*}_{\underline{k}_2\sigma_2} \langle\langle c^+_{\underline{k}_2\sigma_2} c_{\underline{k}_1\sigma_1} X_{\underline{k}-\underline{k}_1+\underline{k}_2 M\bar{M}_2} | \rangle\rangle + V^{M\bar{M}_2}_{\underline{k}_2\sigma_2} \langle\langle c_{\underline{k}_1\sigma_1} c_{\underline{k}_2\sigma_2} X_{\underline{k}-\underline{k}_1-\underline{k}_2 \bar{M}_2M_1} | \rangle\rangle$$

and:

$$(z-\varepsilon_{\underline{k}_1}) \langle\langle X_{\underline{k}-\underline{k}_1 \bar{M}_1\bar{M}} c_{\underline{k}_1\sigma_1} | X_{-\underline{k}\bar{M}'M'} \rangle\rangle$$
$$= \frac{\delta_{\bar{M}\bar{M}'}}{\sqrt{N}} \langle X_{-\underline{k}_1\bar{M}_1M} c_{\underline{k}_1\sigma_1} \rangle + \sum_{M_2\bar{M}_2} V^{M_2\bar{M}_2*}_{\underline{k}_1\sigma_1} \langle\langle X_{\underline{k}-\underline{k}_1\bar{M}_1\bar{M}} X_{\underline{k}_1M_2\bar{M}_2} | \rangle\rangle$$
$$+ \frac{1}{\sqrt{N}} \sum_{\underline{k}_2\sigma_2M_2} \Big[\langle\langle c^+_{\underline{k}_2\sigma_2} c_{\underline{k}_1\sigma_1} X_{\underline{k}-\underline{k}_1+\underline{k}_2 M_2\bar{M}} | \rangle\rangle V^{M_2\bar{M}_1*}_{\underline{k}_2\sigma_2} + V^{M_2\bar{M}}_{\underline{k}_2\sigma_2} \langle\langle c_{\underline{k}_2\sigma_2} c_{\underline{k}_1\sigma_1} X_{\underline{k}-\underline{k}_1-\underline{k}_2 \bar{M}_1M_2} | \rangle\rangle$$

These equations are still exact, but on the right-hand site there are new propagators again. Our decoupling approximation sets in at this point: We decouple all propagators which contribute to $\langle\langle X_{\underline{k}MM} | X_{-\underline{k}M'\bar{M}'} \rangle\rangle$ in order V^2 into thermal averages and one-particle propagators. In the decoupling special care has to be taken not to violate the strong intraionic correlations: only transfer-operators referring to different RE-ions may be decoupled. In the Fourier transformed notation this leads to a decoupling procedure which is examplified by the following equation:

$$\langle\langle X_{\underline{k}-\underline{k}_1\bar{M}_1\bar{M}} X_{\underline{k}_1M_2\bar{M}_2} | X_{-\underline{k}\bar{M}'M'} \rangle\rangle \approx$$
$$(\delta_{\underline{k}\underline{k}_1} N - 1) \langle X_{\underline{0}\bar{M}_1\bar{M}} \rangle \langle\langle X_{\underline{k}M_2\bar{M}_2} | X_{-\underline{k}\bar{M}'M'} \rangle\rangle$$
$$- \frac{1}{\sqrt{N}} \langle\langle X_{\underline{k}M_0\bar{M}} | X_{-\underline{k}\bar{M}'M'} \rangle\rangle \Big[\langle X_{-\underline{k}_1\bar{M}_1M_0} X_{\underline{k}_1M_2\bar{M}_2} \rangle - \frac{1}{N} \sum_{\underline{k}_3} \langle X_{-\underline{k}_3\bar{M}_1M_0} X_{\underline{k}_3M_2\bar{M}_2} \rangle \Big]$$

For the mixed propagators we use the identity:

$$\langle\langle c_{\underline{k}\sigma} | X_{-\underline{k}\bar{M}'M'} \rangle\rangle = \frac{1}{z-\varepsilon_{\underline{k}}} \sum_{M\bar{M}} V^{M\bar{M}*}_{\underline{k}\sigma} \langle\langle X_{\underline{k}M\bar{M}} | X_{-\underline{k}\bar{M}'M'} \rangle\rangle .$$

With this we have obtained a closed system of equations, the decoupling equations for the one-particle propagator $\langle\langle X_{\underline{k}M\bar{M}} | X_{-\underline{k}\bar{M}'M'} \rangle\rangle$. It is possible to take into account cristal fields and a magnetic field. Futhermore, the equations are able to describe mixed valence of compounds with RE-ion, in particular Tm-compounds with two magnetic configurations. For this general case the decoupling equations are, however, rather complex and their solution requires a carefull grouptheoretical analysis. To avoid such problems we restrict here to the isotrop case. We assume that anisotropies of the conduction band and the hybridization may be neglected. Futhermore we assume one configuration to be nomagnetic. If J=0 we obtain:

$$\langle\langle X_{\underline{k}0\bar{M}} | X_{\underline{k}\bar{M}0} \rangle\rangle = \frac{\mathcal{N}_{\bar{M}}}{z - E - \mathcal{M}_{\bar{M}}(z) - \mathcal{N}_{\bar{M}} \cdot \frac{\sum_{\sigma_1} |V^{0\bar{M}}_{\underline{k}\sigma_1}|^2}{z - \varepsilon_{\underline{k}}}}$$

where:

$$\mathcal{N}(z) = \langle X_{00} \rangle + \langle X_{\bar{M}\bar{M}} \rangle + \frac{1}{N} \sum_{\substack{\underline{k}_1\sigma_1 \\ \bar{M}_1 \neq \bar{M}}} \frac{V^{0\bar{M}}_{\underline{k}_1\sigma_1} \langle X_{-\underline{k}_1\bar{M}0} c_{\underline{k}_1\sigma_1} \rangle}{z - \varepsilon_{\underline{k}_1}}$$

and:

$$\mathcal{M}(z) = \frac{1}{N} \sum_{\underline{k}_1\sigma_1} \left[\frac{V^{0\bar{M}_1 *}_{\underline{k}_1\sigma_1}}{z - \varepsilon_{\underline{k}_1}} \left(V^{0\bar{M}}_{\underline{k}_1\sigma_1} \langle 1 - n_{\underline{k}_1\sigma_1} \rangle + V^{0\bar{M}_1}_{\underline{k}_1\sigma_1} \langle n_{\underline{k}_1\sigma_1} \rangle \right) \right.$$
$$\left. + \frac{|V^{0\bar{M}}_{\underline{k}_1\sigma_1}|^2 \left(\langle X_{00} \rangle + \langle X_{\bar{M}\bar{M}} \rangle \right)}{z - \varepsilon_{\underline{k}_1}} - \sum_{\bar{M}_1 \neq \bar{M}} \frac{|V^{0\bar{M}_1}_{\underline{k}_1\sigma_1}|^2 \left(\langle X_{-\underline{k}_1\bar{M}0} X_{\underline{k}_1 0\bar{M}} \rangle - \langle X_{\bar{M}\bar{M}} \rangle \right)}{z - \varepsilon_{\underline{k}_1}} \right]$$

In the metal approximation (broad conduction band) the self-consistency equations are non-linear integral equations in terms of the variable z.

The numerical solution of these equations is in progress. The solution will provide information about the one particle excitations of mixed valence compounds at low temperatures. It also will allow to calculate thermodynamic properties like anomalies of the thermal expansion and of the specific heat.

/1/ E.Müller-Hartmann,
Solid State Comm. 31(1979)113

/2/ H.G.Baumgärtel, Störungstheorie des periodischen Anderson-Modells, Thesis 1981

/3/ N.Grewe and H.Keiter,
Phys.Rev.B24(1981)4420

/4/ J.Hubbard
Proc.R.Soc. London,Ser. A 285 (1965) 542

Valence Instabilities
P. Wachter and H. Boppart (eds.)

VALENCE TRANSITION IN THE ANDERSON MODEL

C. LACROIX

Laboratoire Louis Néel,
C.N.R.S. 166 X,
38042 Grenoble-Cédex, France

The single impurity Anderson model is studied using a procedure of Green's function decoupling, as a function of the parameters U, E_o-E_F, Δ and the temperature T. In this method it is possible to describe both the Kondo regime ($U\to\infty$, $|E_o-E_F|\gg\Delta$) and the mixed valence regime ($|E_o-E_F|\lesssim\Delta$). In the mixed valence regime the number of d electrons in the bound state varies discontinuously due to the renormalisation of the position of the bound state. These results are compared with recent experiments on the valence change in the compounds $Ce(Ni_{1-x}Cu_x)_5$.

1. INTRODUCTION

The Anderson model (1) was first proposed as a model for a magnetic impurity in a non magnetic metal host. Schrieffer and Wolff (2) have shown that in the limit of strong Coulomb interaction this model is related to the s-d model or Kondo model. More recently the Anderson model has been widely used in the mixed valence problem (3). In this paper we present a decoupling procedure which can describe both the Kondo and the mixed valence impurity.

The method has been described in a preceding paper (4) and we give here only the final expression for the Green's function : in the case of infinite U, the impurity Green's function $G_d(\omega)$ is obtained as :

$$G_d(\omega+i\alpha) = \frac{1-\langle n_d\rangle/2 - A(\omega+i\alpha)}{\omega-E_o+i\Delta+\frac{\Delta}{\pi}\int_{-D}^{D}\frac{f(\omega')d\omega'}{\omega'-\omega-i\alpha} - 2i\Delta A(\omega+i\alpha)} \quad (1)$$

where $\langle n_d\rangle$ is the total number of d electrons on the impurity :

$$\langle n_d\rangle = -\frac{2}{\pi}\int f(\omega')\,\mathrm{Im}G_d(\omega'+i\alpha)\,d\omega' \quad (2)$$

and $A(\omega)$ must be calculated selfconsistently :

$$A(\omega+i\alpha) = -\frac{\Delta}{\pi}\int\frac{G_d^{\star}(\omega'+i\alpha)\,f(\omega')\,d\omega'}{\omega'-\omega-i\alpha} \quad (3)$$

In these expressions, E_o is the position of the d level, Δ the width of the bound state ($\Delta=\pi V^2/2D$) and the conduction electrons density of states is taken as $\rho(\varepsilon)=1/2D$ if $-D<\varepsilon<D$.

A similar, but a little more complicated expression can be obtained for finite U. In particular in the symetric case (i.e. $2E_o+U=0$, $E_F=0$) $G_d(E_F)$ is independant of U at T=0 K : $G_d(E_F)=-i/\Delta$. Thus the density of states at the Fermi level does not depend on U for this case ($\rho_d(E_F)=1/\pi\Delta$) in agreement with Haldane's (5) and Yamada's (6) results.

In the following we give the solution of equation 1 in the two cases : Kondo and mixed valence impurity.

2. KONDO IMPURITY

We summarize briefly the results for the case of a Kondo impurity, which are described in more details in ref 7.

At low temperature it is possible to solve equations 1 and 3 in an approximate way : supposing that $G_d(\omega')$ varies more smoothly near E_F than $f(\omega')/\omega'-\omega-i\alpha$, is possible to calculate $A(\omega+i\alpha)$ for $\omega\simeq E_F$; in fact it is in this region that $A(\omega+i\alpha)$ has an important contribution to the Green's function, because it diverges logarithmically at the Fermi level. At zero temperature we find :

$$A(\omega+i\alpha) \simeq -\frac{\Delta}{\pi}G_d^{\star}(\omega+i\alpha)\,\mathrm{Ln}\left|\frac{\omega-E_F}{D}\right| \quad (4)$$

Using this approximate solution we have shown (4,7) that the density of states at zero temperature consists of three peaks : the two resonances at E_o and E_o+U and a third one at the Fermi level E_F. The width of this third peak is very small (of the order of the Kondo temperature $kT_K=D\exp\pi(E_o-E_F)/\Delta$) and the height is in agreement with Haldane's exact results (5) $\rho(E_F)=1/\pi\Delta$.

At high temperatures we can make a perturbation expansion in powers of Δ in equation 3 : the Kondo peak disappears above the Kondo temperature. A numerical calculation (7) has shown that this disappearance is gradual : there is no discontinuity at the Kondo temperature. The numerical results show that there is still a small peak at the Fermi level well above T_K. This is related to the fact that the Kondo effect can be still detected well above T_K.

When the impurity level E_o comes near the Fermi level the Kondo resonance disappears. The critical value is :

$$E_F-E_o - \frac{\Delta}{\pi} - \frac{\Delta}{\pi}\,\mathrm{Ln}\,\frac{\Delta}{\pi D} \tag{5}$$

If E_F-E_o is smaller than this critical value the impurity is in the mixed valence regime.

3. MIXED VALENCE IMPURITY

For a mixed valence impurity, the expansion in powers of Δ of equation 3 is a good approximation (4). To the first order in Δ we obtain

$$A(\omega+i\alpha) = -\frac{\Delta}{\pi}\,\frac{1-n_d/2}{\omega-E_o}\left[\psi\left(\frac{1}{2} - i\beta\,\frac{\omega-E_F}{2\pi}\right) - \psi\left(\frac{1}{2} - i\beta\frac{E_o-E_F}{2\pi}\right)\right] \tag{6}$$

Then to the first order in Δ we can write $G_d(\omega)$ as :

$$G_d(\omega) = \frac{1-n_d/2}{\omega-E_o + i\Delta + \frac{\Delta}{\pi}\int_{-D}^{D}\frac{f(\omega')\,d\omega'}{\omega'-\omega-i\alpha} + A(\omega+i\alpha)\frac{\omega-E_o}{1-n_d/2}} \tag{7}$$

which can be written as :

$$G_d(\omega) = \frac{1-n_d/2}{\omega-E_o'+i\Delta'} \tag{8}$$

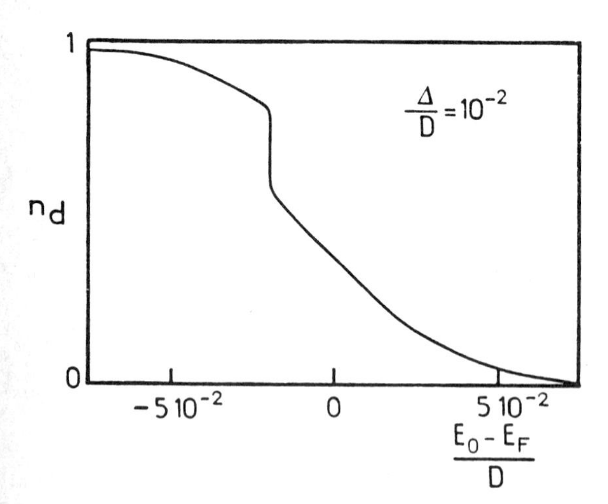

Figure 1 : Variation of the valence with the position of the d level.

where the renormalised position of the bound state E_o' is given by ;

$$E_o' = E_o - \frac{\Delta}{2\pi}\,\mathrm{Ln}\,\frac{(\pi kT)^2 + (E_o'-E_F)^2}{D^2} \tag{9}$$

and $\Delta' = 2\Delta$ if $E_o<E_F$ and $\Delta' = \Delta$ if $E_o>E_F$.

Equation 9 shows that the position of the bound state is modified and depends on temperature. The same expression was obtained by other authors (8,9,10) using different techniques.

Figure 1 shows the variation of the valence with E_o-E_F at $T=0$ K : there is a discontinuity at a critical value of E_o-E_F ; this is due to the fact that the renormalized value of E_o, E_o' varies discontinuously with E_o-E_F when $E_o-E_F\to 0$.

Figures 2 and 3 show the variation of the valence with temperature. As E_o' depends on temperature, we find a strong variation of n_d with temperature : this temperature variation is the result of two effects

- the variation of E_o' with temperature : E_o' decreases when the temperature increases and this gives an increase of n_d with temperature.

- when $E_o'-E_F$ is of the order of the width of the level Δ, there is a strong variation of n_d due to the effect of the Fermi function $f(\omega)$ in equation 2. This gives as decrease of n_d if $E_o'-E_F<0$ and an increase if $E_o'-E_F>0$.

The numerical results (figures 2 and 3) show that the first effect dominates at low temperature, and the second one at higher temperatures.

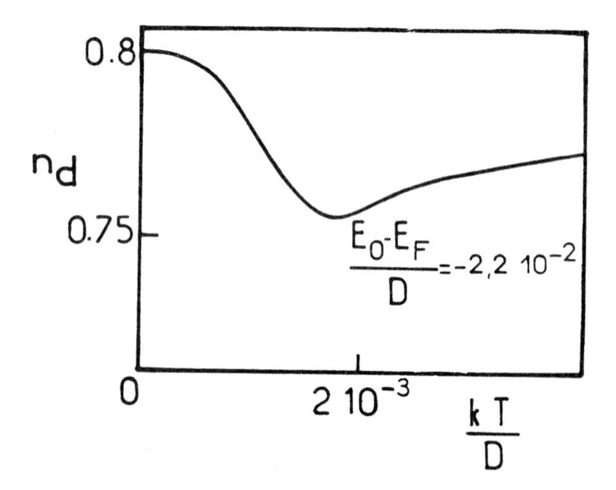

Figure 2 : Variation of the valence with temperature for $n_d=0.8$.

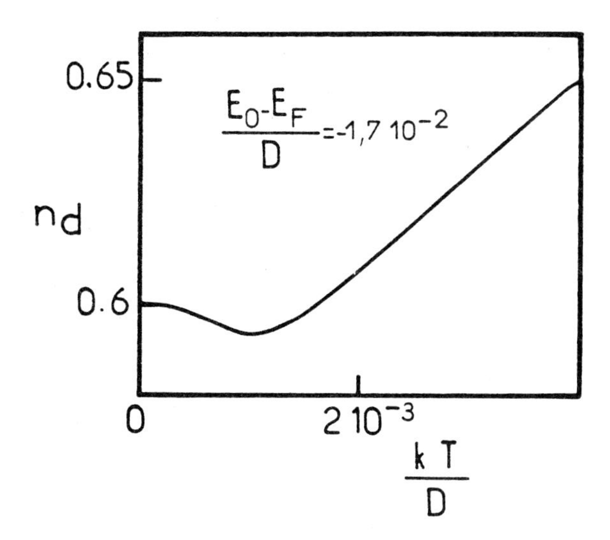

Figure 3 : Variation of the valence with temperature for n_d=0.6.

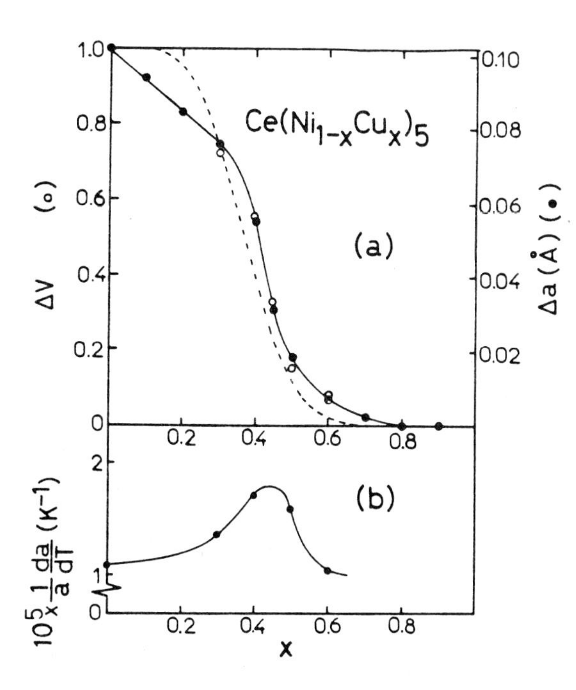

Figure 4 :
a) Deviation of the lattice parameter Δa and change of valence ΔV in the Ce(Ni$_{1-x}$Cu$_x$)$_5$ compounds

b) Dilatation coefficient in the Ce(Ni$_{1-x}$Cu$_x$)$_5$ compounds

(from reference 11)

4. CONCLUSION

We have shown that at zero temperature there is a discontinuous variation of n_d with E_F-E_o in the Anderson model. Near the discontinuity the valence has a strong temperature dependence.

These results can be compared with the recent experiments on the compounds Ce(Ni$_{1-x}$Cu$_x$)$_5$ by Gignoux et al (11) : in these compounds the valence state of Ce deduced from the variation of the lattice parameter decreases with x : in CeNi$_5$ the valence is almost 4^+ and in CeCu$_5$ the valence of the Ce ions is 3^+ (figure 4-a), and there is a strong decrease of the valence for x≃0.4-0.5. In fact an increase of the concentration corresponds to a decrease of the total electron number and to a lowering of the Fermi level. Thus the behaviour of the valence change in these compounds can be qualitatively explained by our model (for quantitative comparison on should use an Anderson lattice model). The rapid variation of the valence at the critical concentration can be due to the renormalisation of the f level.

The strong variation of the dilatation coefficient (figure 4-b) around the critical concentration corresponds to a strong variation of the valence with temperature in this region. In fact a variation of 2% of the lattice parameter Δa/a, corresponds to a valence change of about 1. Thus if we consider that the dilatation coefficient contains a lattice term (almost independant of x) and a term due to the valence change, we can estimate, from figure 4-b : dV/dT=0.5 10^{-3} K^{-1} for x≃0.4, which is in agreement with our results : if the parameter D is taken as 2 eV we find that dV/dT is of the order of 10^{-3} K^{-1} ; thus we find the correct order of magnitude. But in order to compare more quantitatively with experimental results, the calculation has to be generalized to the Anderson lattice.

REFERENCES

1. ANDERSON P.W.
Phys. Rev. 124, 41 (1961)

2. SCHRIEFFER J.R., WOLF P.A.
Phys. Rev. 149, 491 (1966)

3. Proceedings of the conference on Valence fluctuations in solids, ed. L. Falicov, W Hanke and MB Maple (Amsterdam, North Holland 1981)

4. LACROIX C.
J. Phys. F 11, 2389 (1981)

5. HALDANE F.D.M.
Thesis, University of Cambridge (1978)

6. YAMADA K.
Prog. Theor. Phys. 53, 970 (1975)

7. LACROIX C.
Proceedings of the MMM Conference (Atlanta 1981)

8. KRISHNA-MURTHY H.R., WILKINS J.W., WILSON K.G.
Phys. Rev. B 21, 1044 (1980)

9. UEDA K.
J. Phys. Soc. Jap. : 47, 811 (1979)

10. HALDANE F.D.M.
Phys. C 11, 5015 (1978)

11. GIGNOUX D., GIVORD F., LEMAIRE R., LAUNOIS H., SAYETAT F.
J. Physique 43, 173 (1982)

LATTICE AND VALENCE

Valence Instabilities
P. Wachter and H. Boppart (eds.)

PHONON ANOMALIES AND PHASE TRANSITIONS IN MIXED-VALENT COMPOUNDS

H. Bilz, H. Büttner, G. Güntherodt, W. Kress and M. Miura

Max-Planck-Institut für Festkörperforschung,
Heisenbergstrasse 1
7000 Stuttgart - 80, Federal Republic of Germany

The interaction of electronic f(d,p) transitions with the ionic lattice in mixed-valent compounds is treated in a 'breathing' shell model. The isotropic breathing deformability of the ions with f electrons leads, in the self-energy approximation, to a description of phonon anomalies in mixed-valent compounds[1]. (e.g. $Sm_xY_{1-x}S$) while the light-induced breathing fluctuations are responsible for the phonon Raman spectra[2]. In a non-linear extension, third-and fourth- order terms of the breathing potential are considered. A pseudo-one dimensionnal model corresponds to the complete factorization of the dynamical matrix in the main symmetry direction. Several phase transitions are possible dependent upon the intra- and intersite coupling parameters. In particular one obtains

i) a homogeneous transition (wave-vector q=o) with an isotropic charge-density contraction at every lattice site. This is a purely electronic transitions;

ii) an 'anti'-type phase transition with two different valencies alternating along the chain. This couples strongly to the longitudinal optic phonon at the zone bounrady and to the longitudinal acoustic ones in the elastic regime (long wavelength).

The different types of mixed-valent systems are discussed in view of the model.

Valence Instabilities
P. Wachter and H. Boppart (eds.)

EVIDENCE FOR A LOCALIZED POLARONIC CHARGE FLUCTUATION MODE: "BOUND FLUCTUON"

N. Stüßer and G. Güntherodt

II. Physikalisches Institut,[+] Universität zu Köln, 5000 Köln 41, FRG

A. Jayaraman

Bell Laboratories, Murray Hill, N.J. 07974, USA

K. Fischer

IFF der KFA Jülich, 5170 Jülich 1, FRG

F. Holtzberg

IBM T.J. Watson Research Center, Yorktown Heights, N.Y. 10598, USA

An inelastic excitation is identified within the gap of acoustic and optic phonons of the intermediate valence materials $Sm_{1-x}R_xS$ ($0.15<x<0.5$) (R=Y,La,Pr,Gd,Tb,Dy,Tm), $SmS_{1-x}As_x$ (x=0.1, 0.2, 0.6), TmSe, $TmSe_{.50}S_{.50}$, and $TmSe_{.85}Te_{.15}$ using Raman scattering. This "gap mode" is a first order process, has full symmetric (A_{1g}) character, shifts parallel to the frequency of the LO(L) phonons and disappears for dilute Sm concentrations ($x\gtrsim 0.50$) in $Sm_{1-x}R_xS$. The "gap mode is attributed to a coupling of the incoherently fluctuating 4f charge density of intermediate valence ions to local lattice distortions, stabilizing a "bound fluctuon" mode. This mode should be contrasted with the phonon induced f-d charge fluctuation, giving rise to charge density waves and phonon frequency renormalizations.

1. INTRODUCTION

Light scattering spectroscopy is a versatile tool for the study of low frequency elementary excitations of solids. The scattering cross section is determined by the spatial and temporal fluctuations in the electronic contributions to the electric susceptibility.

Raman scattering experiments in intermediate valence (IV) materials have determined the strength and symmetry of the electron-phonon interaction[1-3] and provide an experimental check for the lattice dynamical models[4,5]. These experiments have given evidence for the anomalous phonon softening of the longitudinal optic (LO) phonons near the L point of the Brillouin zone (BZ) and of the longitudinal acoustic (LA) phonons in the (111) direction of the IV systems $Sm_{.75}Y_{.25}S$[1,2] and TmSe[3,6], consistent with the phonon dispersion curves measured by neutron scattering[7,8].

[+]Work supported by Deutsche Forschungsgemeinschaft, SFB 125

Strong Raman scattering intensities are observed from these phonon anomalies due to the anomalously large electron-phonon matrix element entering the scattering cross section. This matrix element can be described by phonon induced intra-ionic charge deformabilities of the valence fluctuating ions[1,2,5].

So far Raman scattering experiments have only been concerned with the investigations of charge deformabilities of the valence fluctuating ions which are induced by the phonons and no particular search has been carried out for the charge density fluctuations inherent to the valence-fluctuating phenomenon. In general, the light scattering cross section for charge density fluctuations is proportional to the density-density correlation function $\langle\rho(-q,t)\rho(q,0)\rangle$ which should give rise to quasielastic light scattering and should give a measure of the charge relaxation rate of IV materials[9]. While such measurements will become feasible in the near future by means of advanced Brillouin scattering techniques, a possible coupling of these quasielastic charge density fluctuations with phonons may shift them onto the phonon frequency scale, thus resulting in an inelastic excitation.

In neutron scattering experiments on IV $Sm_{.75}Y_{.25}S$ at room temperature a dispersionless mode at about ω=175 cm^{-1} (21.9 meV) is found which has been assigned to a localized vibration mode of the lighter Y ions compared to Sm[7]. The frequency position of this "localized" mode agrees well with a simple mass scaling argument:

$$\omega = (m_{Sm}/m_Y)^{1/2}\ \omega_{LA(L)}$$

where $\omega_{LA(L)}$ denotes the frequency of the longitudinal acoustic phonons at the zone boundary L. Consistent with the neutron scattering data there appears in the Raman spectra of $Sm_{.75}Y_{.25}S$ a pronounced shoulder at nearly the same frequency of 185 cm^{-1} (23.1 meV)[1]. To test this local mode assignment by its mass and concentration dependence we have performed Raman experiments on a series of single crystals of $Sm_{1-x}R_xS$ (R=Y,La,Pr,Gd,Tb,Dy,Tm) and of $SmS_{1-x}As_x$.

2. EXPERIMENTAL RESULTS

The unpolarized Raman spectrum of $Sm_{.65}Y_{.35}S$ at room temperature in Fig.1 (bottom) shows similar to $Sm_{.75}Y_{.25}S$ a pronounced shoulder near 200 cm^{-1} (25 meV). In substituting the Y ions by the much heavier rare earth elements we observe in the unpolarized Raman spectra of $Sm_{.75}Pr_{.25}S$, $Sm_{.78}Gd_{.22}S$ and $Sm_{.80}Gd_{.10}Pr_{.10}S$ at 300 K in Fig.1 also an excitation near 190 cm^{-1} (23.8 meV) - 200 cm^{-1} (25 meV) in the gap between the acoustic and optic phonons. This is surprising since the absence of a mass defect between Sm and Pr or Gd should not give rise to a localized vibration mode in these materials. Hence we conclude that the "gap mode" in the Pr- and Gd-substituted $Sm_{1-x}R_xS$ samples cannot be due to a local vibration mode. Consequently, the previous assignment of a local vibration mode in $Sm_{.75}Y_{.25}S$ near 175 cm^{-1} (21.9 meV) has to be revised.

In order to further characterize the "gap mode" we have carried out the symmetry analysis of the scattering intensity which is easily obtained by using at least three different scattering geometries[1-3]. The analysis for $Sm_{.78}Gd_{.22}S$ at 300 K in terms of the observable three even symmetry components (A_{1g}, E_g, T_{2g}) of the scattering

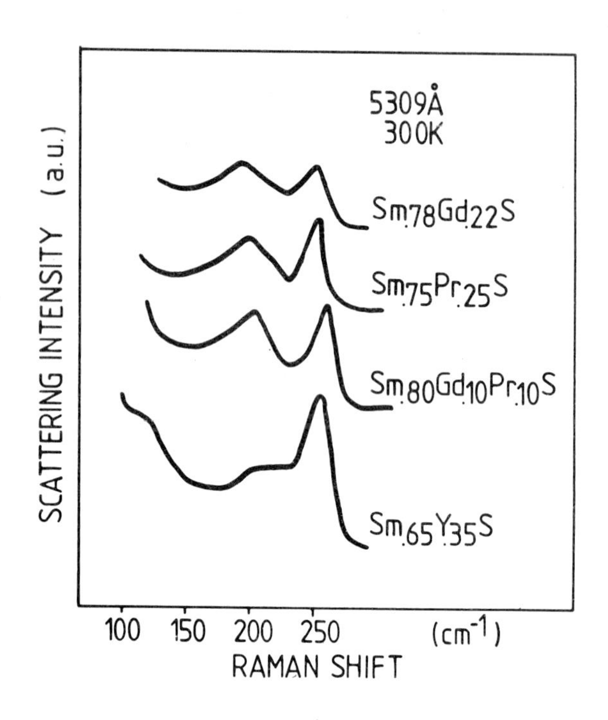

Fig. 1 : Unpolarized Raman spectra of cleaved (100) faces of $Sm_{.78}Gd_{.22}S$, $Sm_{.75}Pr_{.25}S$, $Sm_{.80}Gd_{.10}Pr_{.10}S$ and $Sm_{.65}Y_{.35}S$ single crystals at 300 K using 5309 Å laser excitation.

tensor in the NaCl structure is shown in Fig.2 (bottom). Obviously, the scattering intensity is dominated by the full symmetric A_{1g} component. For the maximum near 260 cm^{-1} the A_{1g} component is related to the breathing charge deformability induced by the LO phonons near the L point of the BZ[1,2,5].

The "gap mode" does not originate from two-phonon processes of acoustic phonons. This has been proved by the temperature dependence of the scattering intensity (see Fig.2, top), which is nearly constant upon cooling from 300 K to 80 K, but should decrease by a factor of 5 for a two-phonon process.

In $Sm_{.65}Y_{.35}S$ the frequency positions of the "gap mode" as well as of the LO phonons near the L point are both shifted towards higher frequencies by 15cm^{-1} as compared to $Sm_{.75}Y_{.25}S$. The same parallel shift of the "gap mode" with the LO(L)-phonon modes is observed in the Raman spectra of the Gd-, Pr-substituted $Sm_{1-x}R_xS$ samples in Fig.1.

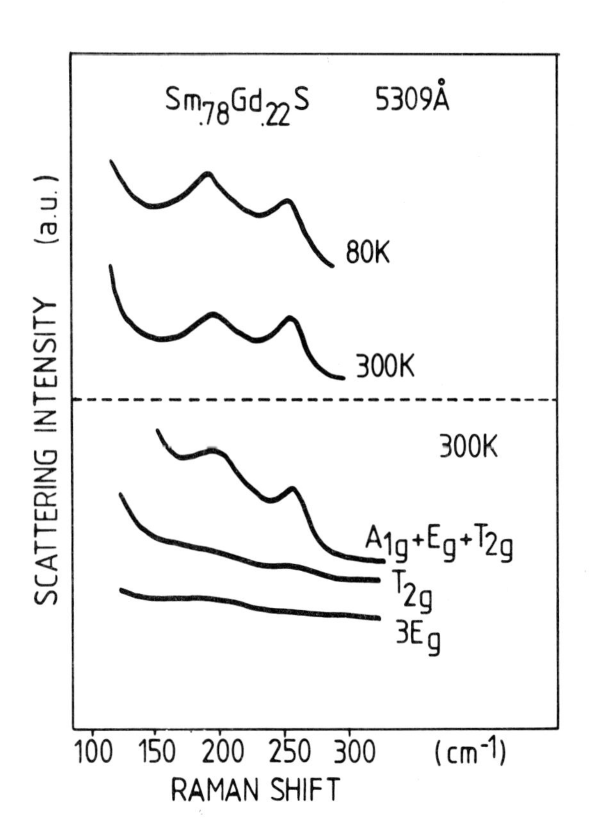

Fig. 2 : Symmetry-analyzed Raman spectra at 300 K (bottom) and unpolarized Raman spectra at 300 K and 80 K (top) of a cleaved $Sm_{.78}Gd_{.22}S$ single crystal.

One of the most interesting features of the $Sm_{1-x}R_xS$ solid solution system is that the "gap mode" is observable only in the concentrated intermediate valence crystals with $0.15 < x < 0.50$ whereas in the dilute cases ($x \gtrsim 0.50$), i.e. for instance in $Sm_{.50}Tm_{.50}S$, $Sm_{.50}La_{.50}S$ and $Sm_{.25}Dy_{.75}S$ (see Fig.3), no first order excitation in the frequency region around 170 cm^{-1} - 210 cm^{-1} has been observed. Instead, in these dilute systems only second order scattering from acoustic phonons has been detected within the optic-acoustic phonon gap which is analogous to the behaviour of integralvalent materials, like for instance GdS[10] and TmS (see Fig.6). For $Sm_{.25}Dy_{.75}S$ the second order scattering intensity is indicated by the dashed line in Fig.3 and identified as such by its disappearence in the spectrum at 80 K. On the contrary, the spectrum of $Sm_{.76}Dy_{.24}S$ at 300 K in Fig.3 remains practically unchanged upon cooling to 80 K similar to the case of $Sm_{.78}Gd_{.22}S$ in Fig.2.

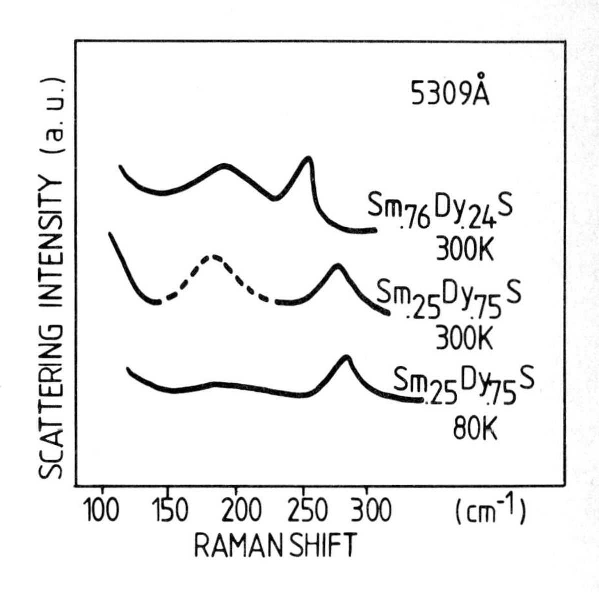

Fig. 3 : Unpolarized Raman spectra of cleaved single crystals of $Sm_{.76}Dy_{.24}S$ at 300 K and $Sm_{.25}Dy_{.75}S$ at 300 K and 80 K.

In order to further corroborate the above arguments in favor of a new type of "gap mode" we have also investigated the $SmS_{1-x}As_x$ solid solution system. The unpolarized room temperature Raman spectra are shown for x=0.10, 0.20, 0.60 and 1.0 in Fig.4. Again, a "gap mode" is found between the optic phonons near 250 cm^{-1} and the acoustic phonons below 150 cm^{-1} for all values of x, except for the integral valent $Sm^{3+}As$ (x=1.0). The persistence of the "gap mode" for all values of x in the IV phase of $SmS_{1-x}As_x$ ($0.10 < x < 1.0$) with a fixed Sm concentration gives evidence that the disappearance of the "gap mode" in the $Sm_{1-x}R_xS$ system with increasing x is not connected with alloy effects due to Gd- or Pr-substitution, but rather with reduced Sm-Sm interactions. There is also a clear indication of a frequency shift of the "gap mode" which parallels that of the maximum of the scattering intensity near 250 cm^{-1} due to the LO(L) phonons. In Fig.5 we give proof of the dominant A_{1g} symmetry character of the "gap mode" (bottom) as well as of its first order scattering intensity behaviour (top) for $SmS_{.90}As_{.10}$. Simple mass scaling between S and As for the LO(L) phonon frequency gives a local As mode near 165 cm^{-1} contrary to the "gap mode" position near 200 cm^{-1}. Furthermore, a local As mode is ruled out by the absence of any broadening

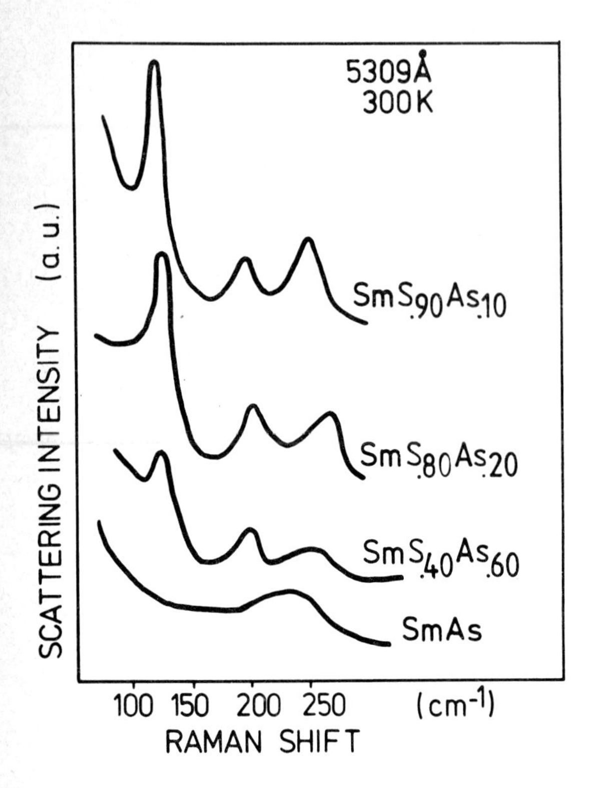

Fig. 4 : Unpolarized Raman spectra of cleaved single crystals of $SmS_{1-x}As_x$ (x=0.10, 0.20, 0.60, 1.0) at 300 K.

of the "gap mode" with increasing As concentration in Fig.4.

The above results shed new light on the inelastic excitation near 145 cm^{-1} in TmSe[3,6] which we had attributed to a two-phonon process of the anomalous longitudinal acoustic phonons. Due to the expectedly strong anharmonicity in IV materials a two-phonon bound state may occur, presumably accounting also for the anomalous temperature dependence of the 145 cm^{-1} mode which is typical for a first order process. However, since no two-phonon scattering intensity from the anomalous LO(L) phonons has been detected, such an interpretation is questionable.

On the other hand, for the 145 cm^{-1} excitation of TmSe the first order nature[3,6], the dominant A_{1g} symmetry[3] as well as its frequency shift in parallel to that of the LO(L) phonons in $TmSe_{.85}Te_{.15}$ (Fig.6) provide an identification analogously to that of the "gap mode" in the IV phases of $Sm_{1-x}R_xS$ and of $SmS_{1-x}As_x$.

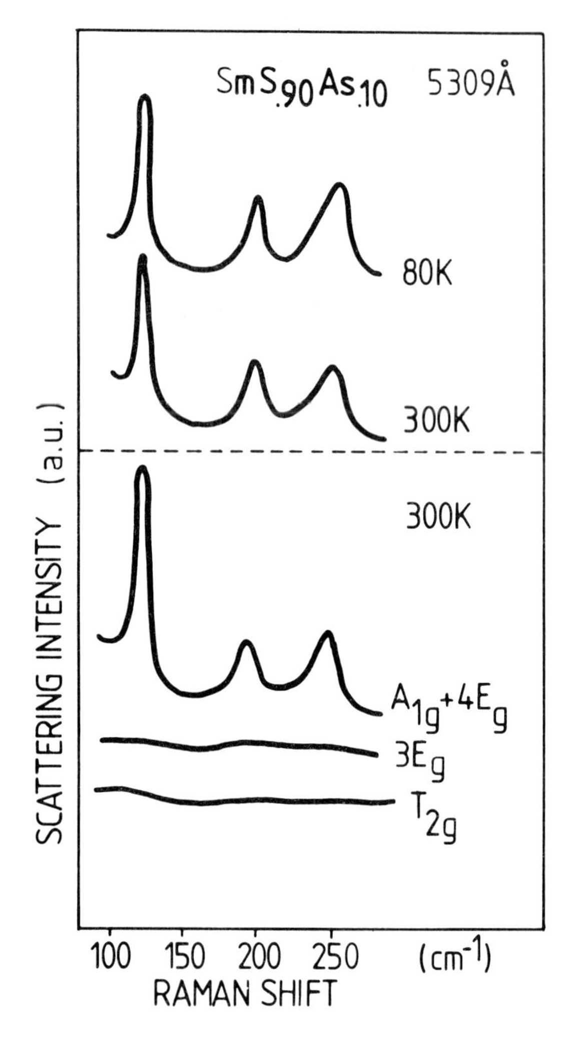

Fig. 5 : Symmetry-analyzed Raman spectra at 300 K (bottom) and unpolarized Raman spectra at 300 K and 80 K (top) of a cleaved $SmS_{.90}As_{.10}$ single crystal.

In Fig.6 we show the unpolarized Raman spectra of $Tm_{1.01}Se$, $TmSe_{.85}Te_{.15}$ and of $TmSe_{1-x}S_x$ for x=0.5 and 1.0. The "gap mode" near 145 cm^{-1} (18.1 meV) in $Tm_{1.01}Se$ is shifted to 123 cm^{-1} (15.4 meV) in $TmSe_{.85}Te_{.15}$ which parallels the softening of the LO(L) phonons due to a Tm valence change from 2.75 towards 2.6. The persistence of the "gap mode" in $TmSe_{.50}S_{.50}$ is similar to the case of $SmS_{1-x}As_x$. In nominally integral valent TmS we observe second order scattering (dashed line) within the gap of acoustic phonons below 120 cm^{-1} and optic phonons near 290 cm^{-1}. Thus we believe that the "gap mode" is a quite general, intrinsic feature of valence fluctuations in at least the NaCl-structure type IV rare earth chalcogenides.

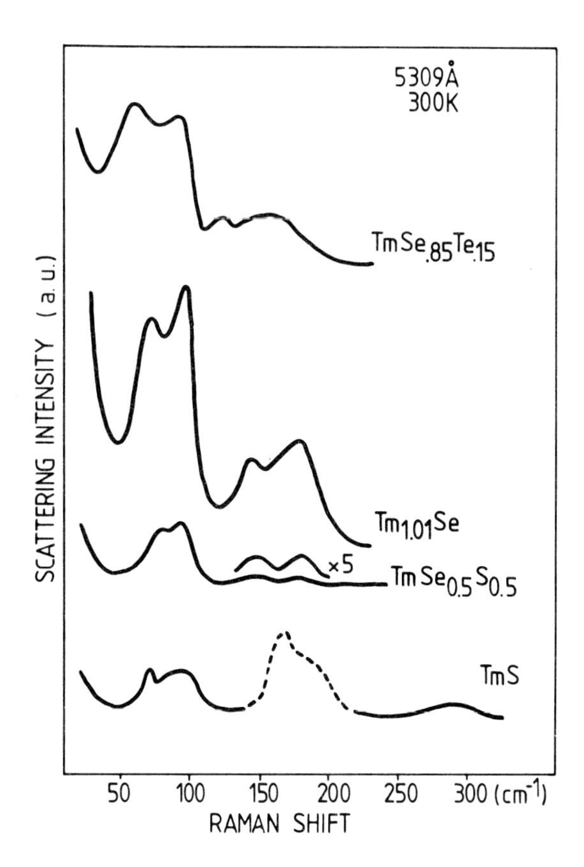

Fig. 6 : Unpolarized Raman spectra of cleaved single crystals of $TmSe_{1-x}S_x$ (x=1.0, 0.5), $Tm_{1.01}Se$ and $TmSe_{.85}Te_{.15}$ at 300 K.

The A_{1g} symmetry of the "gap mode" in IV $Sm_{1-x}R_xS$ rules out any contribution from the J=0→J=1 excitation of the Sm $4f^6$ (^7F_J) configuration. This excitation can only appear in the antisymmetric T_{1g} (Γ_{15}^+) component[11] and could not be detected in Raman scattering of single crystals of $Sm_{.75}Y_{.25}S$[11]. The J=0→J=1 excitation has been identified in neutron scattering on powder samples of $Sm_{.75}Y_{.25}S$ near 31 meV (250 cm^{-1})[12]. Neutron scattering experiments on $Sm_{.75}Y_{.25}S$ single crystals at 20 K and q=(002) show a broad (12 meV at FWHM) distribution of magnetic scattering with two maxima near 28 meV and 25 meV[13]. We do not assume that the "gap mode" in IV $Sm_{1-x}R_xS$ is associated with the J=0→J=1 excitation, since a similar magnetic excitation is absent near the "gap mode" frequency in TmSe.

3. DISCUSSION

On the basis of our above experimental characterization of the "gap mode" we can rule out any contribution from excitations of the 4f J-multiplet levels or from phonons. On the other hand, the A_{1g} symmetry of the "gap mode" excitation is consistent with scattering by electron density fluctuations[14,15]. The shift of the "gap mode" frequency in parallel with that of the LO phonons may suggest a coupled plasmon-LO phonon mode. Such a plasmon in these metallic IV materials, with a frequency comparable to that of the LO phonons, could only be due to the predicted acoustic plasmon[16], i.e. to the out-of-phase motion of the two component f-d plasma. Theoretically the dispersion of such an acoustic plasmon is described by[16]

$$(\omega_q^{ac})^2 \sim N_f\, q^2/(q^2+k_s^2)$$

where N_f is the density of the 4f electrons, q is the wavevector and k_s denotes the screening wavevector, which is determined by the Fermi wavevector $k_f \sim n^{1/3}$, with n the conduction electron concentration. For small q values the frequency is proportional to $\omega_q^{ac} \sim n^{-1/3}$. Thus the frequency position of the "gap mode" in $Sm_{1-x}R_xS$ relative to the LO phonons should be strongly dependent on the concentration of the substituted R^{3+} ions, which contribute one conduction electron each. No such evidence is found in the spectra for various concentrations x of the $Sm_{1-x}R_xS$ and the $SmS_{1-x}As_x$ systems. In the latter n decreases at 300 K by about an order of magnitude in going from x=0.10 to x=0.60[17]. No change in the "gap mode" of $SmS_{.90}As_{.10}$ is observed upon cooling from 300 K to below 2 K, with an order of magnitude decrease of n[17]. Analogously the "gap mode" of stoichiometric TmSe is not affected by cooling to 1.5 K, despite the strong increase in resistivity below the Néel temperature T_N=3.5 K[19]. The existence of an acoustic plasmon at large q values is questionable because of Landau damping.

In the light of our experimental results we interpret the "gap mode" in all of the above investigated IV materials as a coupling of the incoherently fluctuating localized 4f charge density of IV rare earth ions to local lattice distortions. The latter are dominantly determined by LO(L) phonons which sense most effectivly the ionic size effect exhibited in occupying the $4f^n$ configuration of larger ionic volume compared to $4f^{n-1}5d^1$. This local, full symmetric (A_{1g}) coupling of bound electron-hole pair excitations to LO(L) phonons in real space gives rise to a dispersionless excitation in k-space, i.e. a "polaronic bound exciton" or "bound fluctuon". This is consistent with the dispersionless mode observed in neutron scattering of $Sm_{.75}Y_{.25}S$. The magnetic scattering observed in neutron experiments around 25 meV[13] is consistent with a 4f form factor of the localized charge density of the "bound fluctuon".

The localized, incoherent charge fluctuation mode should be contrasted with the periodic modulation of the charge density of the valence fluctuating ions induced by propagating lattice vibrations (coherent lattice distortions), giving rise to pinned charge density waves. These more delocalized, periodic charge fluctuation modes have been described in terms of phonon induced ionic charge deformabilities, which successfully account for the anomalies of the phonon dispersion curves[4,2,5]. Based on experimental Raman data the renormalization of the LO phonons near the L point of the BZ amounts to 6 meV in $Sm_{.75}Y_{.25}S$[1] and to 2 meV in TmSe, respectively, compared to the reference YS[1] and GdSe[19].

On the other hand, the "bound fluctuon" mode leads to a LO(L) renormalization of 12 meV in $Sm_{.75}Y_{.25}S$ and 6 meV in TmSe. The difference in renormalization of the LO(L) phonons due to the pinned charge density wave and the "bound fluctuon" is 6 meV in $Sm_{.75}Y_{.25}S$ and 4 meV in TmSe and gives a measure of the polaronic electron-hole binding energy.

Interestingly enough, an electron-hole binding energy has been deduced from electron tunneling spectra of IV SmS (2-3 meV) and TmSe (1-2 meV)[20]. Assuming that these electron-hole binding energies are dominantly due to the Coulomb interaction explains why the above polaronic electron-hole binding energies are by roughly a factor of two larger. Hence we deduce a polaronic lattice deformation energy of about 3 meV in $Sm_{.75}Y_{.25}S$ and 2 meV in TmSe. We would like to point out that the disappearance of the "gap mode" in $Sm_{1-x}R_xS$ for $x \gtrsim 0.5$ suggests that a nearest neighbor Sm-Sm interaction and presumably "bipolaronic" effects are of importance for the existence of the "bound fluctuon". Thus the localized charge fluctuation at one lattice site may be mediated within the exciton radius of the bound electron (e^-)-hole ($4f^{n-1}$) pair onto the neighboring site. The exciton radius would be of the order of roughly $a\sqrt{2}$, where a is the lattice constant.

In the displaced harmonic oscillator model of Sherrington and von Molnar[21], shifts in the electronic excitation spectrum of IV materials result by allowing the lattice to locally adjust appropriately to the individual ionic 4f configuration. Such polaronic effects are shown to lower the energy of purely electronic interconfiguration fluctuations for which the lattice is not allowed to relax. The "relaxation energy" is $E=4\lambda^2/\hbar\omega$, where λ is the electron-phonon coupling constant and ω the frequency of a local, optic phonon.

With respect to our discussion we note that the modulation of the charge density of the IV ions imposed by the propagating lattice modes does not allow for local lattice distortions. Instead it leads to a dynamic hybridization between f and d states (phonon induced f-d transitions) and to a renormalization of the phonon frequencies. On the other hand, starting from the purely electronic, incoherent fluctuations of an IV ion and allowing the lattice to relax locally, leads to a stabilization into a localized fluctuation mode, i.e. the "bound fluctuon" and to a reduction of the f-d hybridization matrix element V_{fd} by $V_{fd}\cdot\exp(-(\lambda/\hbar\omega)^2)$ [21]. Thus the width of the "gap mode" observed in our experiments gives a measure of the renormalized charge relaxation rate Γ_c' in contrast to that (Γ_c) observable in quasielastic light scattering around the exciting laser frequency (zero Raman shift).

The difference in frequency between the renormalized LO phonon branch near the L point and the "bound fluctuon" mode is due to the lattice deformation energy of order $4\lambda^2/\hbar\omega$ and the electron-hole Coulomb binding energy. Assuming, as discussed above, that the lattice deformation energy and Coulomb energy each contribute half of the renormalization energy of 6 meV in $Sm_{.75}Y_{.25}S$ and 4 meV in TmSe, we can estimate the electron-phonon coupling constant λ=5 meV for $Sm_{.75}Y_{.25}S$ and λ=3 meV for TmSe.

ACKNOWLEDGEMENT

We would like to thank H.Bilz, W.Kress, P.Entel and T.V.Ramakrishnan for fruitful discussions.

REFERENCES

1. Güntherodt,G., Jayaraman,A., Kress, W., Bilz,H., Phys.Lett.82A,26(1981)
2. Güntherodt,G., Jayaraman,A., Bilz, H., Kress,W., in Valence Fluctuations in Solids, ed.by L.M.Falicov, W.Hanke and M.B.Maple, North-Holland, Amsterdam, 1981,p.121
3. Stüßer,N., Barth,M., Güntherodt,G., Jayaraman,A., Solid State Commun.39, 965(1981)
4. Bilz,H., Güntherodt,G., Kleppmann, W., Kress,W., Phys.Rev.Lett.43,1998 (1979)
5. Kress,W., Bilz,H., Güntherodt,G., Jayaraman,A., J.de Physique 42, Coll.C6,3(1981)
6. Treindl,A., Wachter,P., Solid State Commun. 32,573(1979)
7. Mook,H.A., Nicklow,R.M., Penney,T., Holtzberg,F., Shafer,M.W., Phys.Rev. B18,2925(1978)
8. Mook,H.A., Holtzberg,F., in Valence Fluctuations in Solids, ed. by L.M. Falicov,W.Hanke,M.B.Maple, North-Holland, Amsterdam, 1981, p.113
9. Lopez,A., Balseiro,C., Phys.Rev. B17,99(1978)
10. Güntherodt,G., Grünberg,P., Anastassikis,E., Cardona,M., Hackfort,H., Zinn,W., Phys.Rev.B16,3504(1977)
11. Güntherodt,G., Jayaraman,A., Anastassakis,E., Bucher,E., Bach,H., Phys.Rev.Lett.46,885(1981)
12. Mook,H.A., Penney,T., Holtzberg,F., Shafer,M.W., J. de Physique 8, Coll.C6,837(1978)
13. Mook,H.A., private commun.
14. Platzmann,P.M., Phys.Rev.A139,379 (1965)
15. McWorther,A.L.,in Physics of Quantum Electronics, ed. by P.L. Kelly, B.Lax, P.E.Tannenwald, McGraw-Hill, p.111(1966)
16. Varma,C.M., Rev.Mod.Phys.48,219 (1976)
17. von Molnar,S.,Penney,T., Holtzberg, F., J.de Physique37,Coll.C4,241(1976)
18. Holtzberg,F., Penney,T., Tournier,R. , J.de Physique40,Coll.C5,314(1979)
19. Treindl,A., Wachter,P.,Phys.Lett.64A 147(1977)
20. Güntherodt,G., Thompson,W.A., Holtzberg,F., Fisk,Z., to be published
21. Sherrington,D., von Molnar,S., Solid State Commun.16,1341(1975)

COMMENT: T.M. HOLDEN: There is a much more prosaic explanation of the hatched region on the dispersion relation of uranium sulphide. This originates in the polycristalline vibrational scattering from the thin aluminium sample holder.

G. GUENTHERODT: That is too bad. Unfortunately, we were not successful so far to measure reproducibly the Raman spectrum of US because of surface oxidation problems in order to check on this point independently.

K.W.H. STEVENS: Have you considered the possibility that your interband mode is an excitation between R.E. spin-orbit levels ?

G. GUENTHERODT: Yes, but from our Raman scattering experiments we can clearly exclude any involvement of the $J=0 \rightarrow J=1$ excitation of the Sm^{2+} $4f^6$ (^7F_J) spin-orbit multiplet. In our polarized Raman experiments we can distinguish between magnetic scattering from J multiplet levels and from phonons. In the solid solution system $Sm_{1-x}Y_xS$ we see that the $J=0 \rightarrow 1$ excitation in semiconducting SmS (x=0) is strongly broadened with increasing x upon approaching the phase transition near x=0.15. We can no more identify this excitation in the mixed valence phase for x>0.15. In the case of TmSe no spin-orbit multiplet excitation exists near the "gap-mode" at 18 meV.

M. CROFT: Have you monitored the localized excitation you speak of as a function of temperature in the immediate vicinity of the gold to black transition (near 200 K) in the $Sm_{1-x}R_xS$ alloys ?

G. GUENTHERODT: Yes, we have and it shows no effect.

J.W. WILKINS: Since I understand you to say that this new mode is dispersionless and should be universal among the rare earth valence fluctuators, shouldn't it be prominent in inelastic processes and hence show up in the far infrared and tunneling spectroscopies ?

G. GUENTHERODT: Yes, it should, since it is a localized charge fluctuation coupled to phonons. The question is just what weight this mode has as a function of q, i.e. magnetic for small q's and phononic for large q's towards the zone boundary. The 18 meV gap mode of TmSe has been seen in tunneling spectroscopy, but not in a reproducible way.

Valence Instabilities
P. Wachter and H. Boppart (eds.)

CONDUCTION BAND-PHONON COUPLING IN I.V. MATERIALS

D. G. Cantrell and K. W. H. Stevens

Department of Physics,
University of Nottingham,
Nottingham, England

Intermediate Valence materials are very sensitive to lattice strains. In this paper a study is made of the effect on the conduction bands of a strain having the symmetry of a zone boundary longitudinal mode directed along a four-fold axis in SmB_6. The conduction band is distorted, a rearrangement of conduction electrons occurs and also a transfer from the localised levels. There is an energy lowering proportional to the square of the strain, which manifests itself as a decrease of the phonon frequency. Estimates suggest the effect is far from negligible.

INTRODUCTION

In many materials the transverse phonon branches are below the corresponding longitudinal branches, but this is not so in a number of I.V. materials [1]. It has become clear that the I.V. phenomenon can have a profound effect on lattice vibrations and distortions. There are a number of ways in which the special electronic properties of I.V. can interact with lattice vibrations. In this paper we shall report on a study of one of them. The work has been stimulated by results on SmB_6 [2] and it is this material we have in mind throughout. The boron octahedra will be considered as rigid and carrying two negative charges, in which case the crystal can be regarded as a simple cubic lattice of Sm ions with the boron clusters at the centre of the cells. The Sm electrons will be distributed between localised 4f states and conduction orbitals. We shall slightly modify this and assume that the lower lying band states are built solely from the $5e_g$ orbitals (these would be split off from the $5t_{2g}$ in a crystal field model) and that the conduction electrons only occupy e_g - type Bloch functions. In the absence of lattice distortions all Sm sites are regarded as identical and the intermediate valence is taken to be due to a coherent superposition, at each site, of states of type $(4f)^n$ and $(4f)^{n-1}c$. Being inside the Sm ion we take the 4f states to be relatively insensitive to lattice distortions.

Suppose now that the crystal is given a static distortion of the same symmetry as that of a longitudinal phonon mode at the zone boundary in the (0,0,k) direction. The unit cell is thereby doubled in the Oz direction and the samarium ions now become subjected to a crystal field of tetragonal symmetry from the boron groups. This can be expected to split the local degeneracy of the e_g orbitals and so alter the band structure. If the number of conduction electrons could be held constant one might expect that there would be a distortion of the Fermi surface as electrons move from regions of $\underline{k}$-space where the energies have been increased by the distortion to regions where there has been a decrease. The Fermi energy is likely to be altered. In an I.V. material if it is lowered electrons previously in 4f orbitals will migrate to the conduction band to restore the Fermi energy to its original value. If it is raised then conduction electrons will migrate to the 4f states. Either way there is a coupling to this particular lattice distortion which is not present in most materials. If the magnitude of the lattice distortion is denoted by δ it is clear that reversing δ leads to the same behaviour so any energy change will be even in δ. For small displacements we may take it as $-g\delta^2$ per Sm ion. If one now considers an actual phonon oscillation of this symmetry δ will be time dependent. Its period can be assumed to be long compared with the times involved in electronic readjustments. The effect we have outlined will appear as a change in elastic energy. That is the angular frequency (Ω) of such a zone boundary phonon should relate to that of a similar mode, of angular frequency ω, in a non I.V. material by

$$\Omega = \omega\left(1 - \frac{g}{m\omega^2}\right)$$

where m is the mass of a B_6 octahedron.

DETAILED THEORY

The original cell is doubled by the distortion. It is convenient to use the new cell as basis with wavevectors denoted by $\underline{K}$. In the doubled cell there are two Sm ions each with two e_g -type orbitals. Four Bloch functions of a given $\underline{K}$ can be constructed. The strain we consider is an alternate expansion and contraction of the cubic unit cell caused by displacement of planes of the boron octahedra in the Oz and -Oz directions. This leaves the Sm-Sm nearest neighbour distances unchanged. The e_g states at one site in the unit cell have a decrease in mean energy and split into $3z^2-r^2$ and x^2-y^2 types. At the other site the mean energy is raised and the splitting is inverted. It is assumed that the periodic Hamiltonian only has transfer matrix elements between adjacent sites. Some of these vanish on symmetry grounds e.g. between $3z^2-r^2$ on one site and x^2-y^2 on an adjacent site in the Oz direction. The rest can be expressed in terms of two quantities (assumed independent of charge transfers): the matrix element from $(3z^2-r^2)$ to $(3z^2-r^2)$ along Oz (-A) and from (x^2-y^2) to (x^2-y^2) along Oz (-B). The 4x4 matrix can be readily diagonalised if A is assumed to equal B. With this approximation the band energies corresponding to (K_x, K_y, K_z) are

$$6A - 2A(\cos K_x a + \cos K_y a) \pm (4A^2\cos^2 K_z a + \varepsilon_1^2)^{\frac{1}{2}}$$

and $$6A - 2A(\cos K_x a + \cos K_y a) \pm (4A^2\cos^2 K_z a + \varepsilon_2^2)^{\frac{1}{2}}$$

where A is positive, ε_1 and ε_2 are the energy changes of $(3z^2-r^2)$ and (x^2-y^2) at a typical site. The zero of energy is taken as the bottom of the band when $\varepsilon_1=\varepsilon_2=0$. The presence of the distortion splits the previously doubly degenerate bands.

We take a simple model for the 4f shell, replacing it by a simple localised orbital with energy E and a Hubbard U which restricts the occupancy to be either one or two. With periodic boundary conditions corresponding to N cells in each direction (in the distorted configuration) and with the conduction band so high in energy that it is not occupied each local orbital will have two electrons. So the total number of electrons is $4N^3$. When the conduction band is lowered to give an I.V. system there will always be at least one electron per localised orbital which accounts for $2N^3$ electrons, leaving $2N^3$ to be distributed between conduction and localised states. In a first approximation it is convenient to assume that the local energy is close to the bottom of the conduction band so that relatively few electrons are in the conduction band. Then K_x, K_y, and K_z will be small and the two lowest bands, which are the only ones which will be populated, have energies

$$a^2A(K_x^2 + K_y^2 + K_z^2) - \frac{\varepsilon_1^2}{4A} \quad \text{and}$$

$$a^2A(K_x^2 + K_y^2 + K_z^2) - \frac{\varepsilon_2^2}{4A}$$

Both of these bands are filled to energy E.

Writing these as

$$\varepsilon_o + \Delta\varepsilon_1 \quad \text{and} \quad \varepsilon_o + \Delta\varepsilon_2$$

We see that ΔE the total change in electronic energy compared to the $\varepsilon_1 = \varepsilon_2 = 0$ case is

$$\Delta E = \int_{-k_o-\Delta k_1}^{k_o+\Delta k_1} (\varepsilon_o+\Delta\varepsilon_1)8\pi k^2\rho dk + \int_{-k_o-\Delta k_2}^{k_o+\Delta k_2} (\varepsilon_o+\Delta\varepsilon_2)8\pi k^2\rho dk - \int_{-k_o}^{k_o} 2\varepsilon_o 8\pi k^2\rho dk + E(N_L'-N_L)$$

where $$(N_L'-N_L) = \int_{-k_o}^{k_o} 2\rho 8\pi k^2 dk - \int_{-k_o-\Delta k_1}^{k_o+\Delta k_1} 8\pi k^2\rho dk - \int_{-k_o-\Delta k_2}^{k_o+\Delta k_2} 8\pi k^2\rho dk$$

where k_o = Fermi wavevector of $\varepsilon_1 = \varepsilon_2 = 0$ bands

$k + \Delta k_1$ = Fermi wavevector of ε_1 band

$k + \Delta k_2$ = Fermi wavevector of ε_2 band

ρ = density of states

Neglecting terms $O(\delta^4)$ and $O(\delta^2k^2)$ we find a decrease in energy

$$\Delta E = \int_{-k_o}^{k_o} (\Delta\varepsilon_1 + \Delta\varepsilon_2)8\pi k^2\rho dk$$

or

$$\Delta E = \frac{4\pi k_o^3}{3}\rho(\varepsilon_1^2 + \varepsilon_2^2)$$

since $$\rho = \frac{a^3N^3}{4\pi^3}$$

and $E = a^2Ak_o^2$ where k_o is the Fermi wavevector of the $\varepsilon_1 = \varepsilon_2 = 0$ band it follows that

$$\Delta E = \frac{1}{3}\frac{N^3}{\pi^2}\left(\frac{E}{A}\right)^{\frac{3}{2}}\frac{1}{A}(\varepsilon_1^2 + \varepsilon_2^2)$$

The next step is to relate ε_1 and ε_2 to the displacement δ. We have used standard crystal field theory following the method outlined in [3]. The cubic field splitting is assumed to be much greater than the effects due to the displacement. We have thus neglected the t_{2g} states. Knowing the cubic field splitting to be of the order of electron volts [4] we use

$$\langle r^4 \rangle_{5d} = 4.1 \times 10^{-39}\ m^4$$

$$\langle r^2 \rangle_{5d} = 5.4 \times 10^{-20}\ m^2$$

which gives $\varepsilon_1^2 + \varepsilon_2^2 = 1.9 \times 10^{22}\ e^2\ \delta^2$ (e is the electronic charge), and the corresponding lattice mode will have angular frequency

$$\omega\left(1 - \frac{1}{m\omega^2}\ \frac{1}{6\pi^2}\ \left(\frac{E}{A}\right)^{\frac{3}{2}}\ \frac{1}{A}\ 1.9 \times 10^{22}\ e^2\right)$$

Taking the e_g conduction bandwidth to be 0.5 eV([4] calculate ∿0.1 eV which seems small) and E/A ∿ 10^{-2} we obtain

$$\frac{g}{m\omega^2} \sim 0.1$$

for phonons of energy 200 cm^{-1} (6 x 10^{12} Hz)

Thus we see that even a small estimate for E/A (which puts few electrons in the conduction band) gives an appreciable softening of the phonon frequency.

REFERENCES

[1] Mook, H.A. Nicklow, R.M., Penney, T., Holtzberg, F. and Schafer, M.W., Phonon dispersion in Intermediate Valence $Sm_{0.75}\ Y_{0.25}S$, Phys. Rev. B 18 (1978) 2925-2928.

[2] Mörke, I., Dvořák, V., and Wachter, P., Raman scattering in Intermediate Valent SmB_6, Solid State Commun. 40 (1981) 331-334.

[3] Hutchings, M.T., Point-charge calculations of energy levels of magnetic ions in crystalline electric fields, in Seitz, F. and Turnbull, D. (eds.), Solid State Physics vol. 16 (New York Academic Press, 1964).

[4] Nickerson, J.C., White, R.M., Lee, K.N., Bachmann, R., Geballe, T.H. and Hull Jr, G.W., Physical properties of SmB_6, Phys. Rev. B 3 (1971) 2030-2042.

A similar mechanism has been proposed by Penney et al (see below) to account for the softening of the bulk modulus of SmS in the gold phase. As here a softening could only be obtained if the Fermi energy was in a region of overlap of f and d states.

Penney, T., Melcher, R.L., Holtzberg, F., and Güntherodt, G. Soft bulk modulus at the configurational phase transition in $Sm_{1-x}\ Y_x\ S$, AIP Conf. Proc. (USA) no. 29 (1976) 392-393.

Valence Instabilities
P. Wachter and H. Boppart (eds.)

EVOLUTION OF INTERMEDIATE VALENCY IN $TmSe_{0.32}Te_{0.68}$ UNDER PRESSURE

H. Boppart,[1] W. Rehwald,[2] E. Kaldis,[1] and P. Wachter[1]

[1]Laboratorium für Festkörperphysik, ETH Zürich, 8093 Zürich

[2]Laboratories RCA Ltd, 8048 Zürich, Switzerland

In semiconducting $TmSe_{0.32}Te_{0.68}$ a pressure induced continuous transition to the metallic state takes place around 1.5 GPa. In order to study the variation of the elastic behavior through the transition we measured the ultrasonic sound velocities in the [100] and [110] directions as well as the specific volume up to 1.7 GPa. The softening of the longitudinal acoustic phonons and the bulk modulus is mainly caused by the negative elastic constant c_{12}. It appears that an appreciable mixture of 4f and 5d states must be present already in the semiconducting range leading to an "intermediate valent semiconductor".

1. INTRODUCTION

The elastic properties of intermediate valent materials are still of great interest because of their anomalous behavior manifesting itself in phonon anomalies, a negative elastic constant c_{12} and a soft bulk modulus. These features are thought to be typical for intermediate valent compounds like TmSe (1) and $Sm_{0.75}Y_{0.25}S$ (2).

In the pressure induced semiconductor to metal transition (SMT) of various rare earth chalcogenides a softening of the bulk modulus is observed already in the semiconducting state (3,4). So the aim of this work is to show how the elastic constants change through a continuous transition from the semiconducting to the metallic state and in which region the elastic constant c_{12} becomes negative.

Macroscopically a negative c_{12} means in an uniaxial pressure experiment a shortening not only in the direction of compression but also in the orthogonal directions (1). Microscopically an isotropic change of the ionic diameter takes place due to the "breathing" deformability of the ion (5,6).

In the $TmSe_{1-x}Te_x$ system the compounds with x>0.4 are semiconducting (7-10). Starting from semiconducting $Tm^{2+}Te$ with an energy gap of ~0.3 eV one observes a reduction of the gap with increasing incorporation of Se. This gap is determined by the separation in energy of the $4f^{13}$ and the $4f^{12}(5d6s)^1$ configurations. In these semiconducting compositions a SMT can be induced by external pressure. This SMT is mainly driven by an increase of the ligand field splitting of the 5d-conduction band as the anion-cation distance decreases. Consequently, the $4f^{13}$-$4f^{12}(5d6s)^1$ energy gap is reduced and finally 4f electrons flow into the conduction band and the valence of the Tm ions changes.

Because of the availability of large single crystals we chose the composition $TmSe_{0.32}Te_{0.68}$ for our investigations. In this compound the energy gap is about 0.2 eV. The measurement of the electrical resistivity under pressure reveals an exponential behavior as it is well known for these compounds and a transition into the metallic state at 1.5 GPa (=15 kbar) (4).

In this work we measured under pressure up to 1.7 GPa both longitudinal and transverse sound velocities in [100] and [110] directions as well as the length change by means of the strain gauge technique on the same single crystal (4x5x5 mm^3). This permits the determination of the changes in all three second-order elastic constants c_{11},c_{12} and c_{44} with two internal checks.

2. EXPERIMENTAL RESULTS

Hydrostatic pressure has been generated in a conventional piston cylinder device with petroleum ether as pressure transmitting medium. The longitudinal and the transverse sound velocities v_L and v_T were obtained at 20 MHz by the pulse echo overlap method. The quartz transducers were bonded to the sample using Nonaq.

In a cubic material the sound velocities are related to the elastic constants c_{11},c_{12} and c_{44} by $v_L^2= c_{11}/\rho$, $v_T^2= c_{44}/\rho$ in the [100] direction and by $v_L^2= (c_{11}+c_{12}+2c_{44})/2\rho$, $v_{T_1}^2= c_{44}/\rho$ and $v_{T_2}^2= (c_{11}-c_{12})/2\rho$ in the [110] direction. We measured both the longitudinal and the transverse sound velocity in the [100] direction as well as the longitudinal and one transverse (v_{T2}) sound velocity in [110] direction.

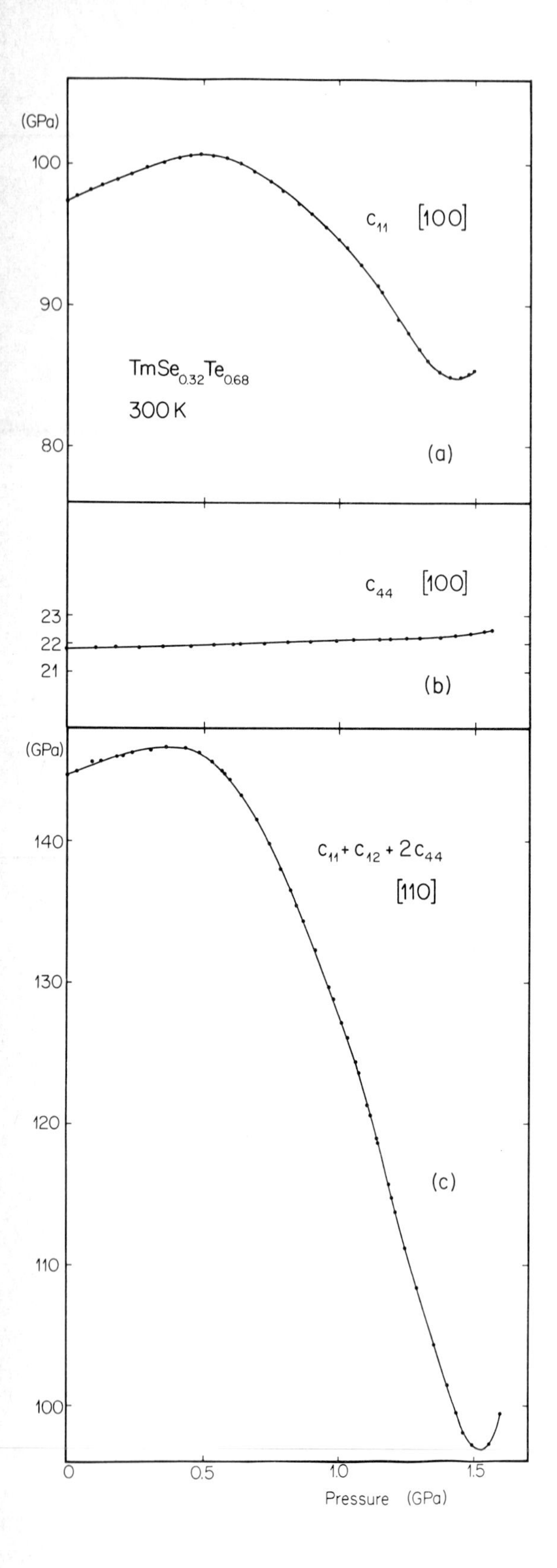

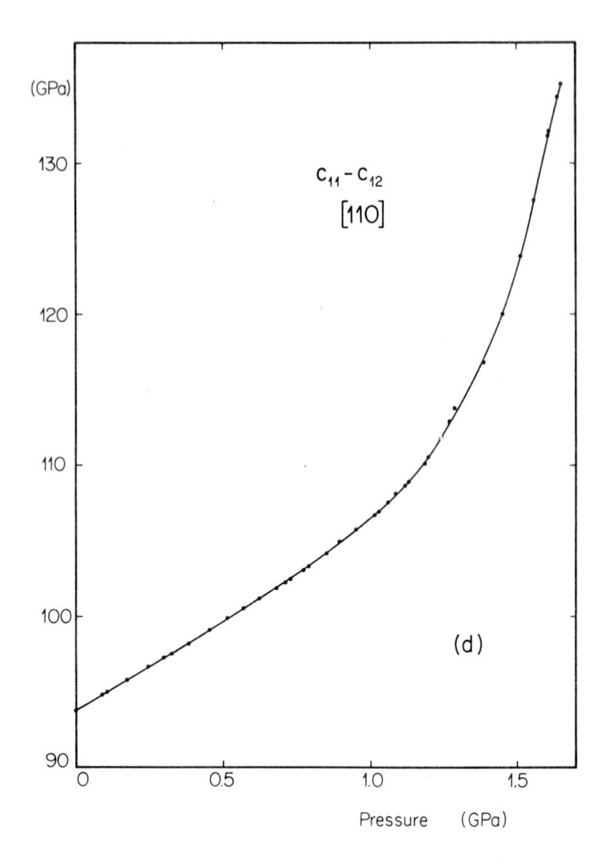

Fig. 1: Pressure dependence of the elastic constants of $TmSe_{0.32}Te_{0.68}$, a) c_{11}, b) c_{44}, c) $c_{11}+c_{12}+2c_{44}$ and d) $c_{11}-c_{12}$

Due to the large volume change under pressure in going through a SMT we measured the length change under pressure by means of the strain gauge technique. Taking this into account for the density and the velocity change we calculated the elastic moduli c_{ij} from

$c_{ij}(p)= \rho_o\ l_o^3[l(p)\cdot t^2]^{-1}$, where ρ_o is the density at ambient pressure (9), l_o is the initial path length and t is the transit time. From the volume-pressure measurement the pressure dependence of the bulk modulus $B= (c_{11}+2c_{12})/3$ can be derived. In this way we have five measured quantities for the three unknown elastic constants.

In Fig. 1 a-d) the pressure dependence of the elastic constants c_{11},c_{44}, $c_{11}+c_{12}+2c_{44}$ and $c_{11}-c_{12}$ is shown. The elastic constant c_{11} increases almost linearly up to about 0.4 GPa and decreases by ∿15% up to the minimum at 1.42 GPa (Fig. 1a), c_{44} increases smoothly (∿3%) with increasing pressure (Fig. 1b). For the longitudinal mode $c_{11}+c_{12}+2c_{44}$ a similar behavior as for c_{11} is observed but with a stronger decrease (∿30-40%) and a minimum at 1.54 GPa (Fig. 1c). The shear mode $c_{11}-c_{12}$ shows initially a linear behavior and then a strong but smooth increase up to 1.7 GPa (Fig. 1d).

The linear increase in the pressure range up to 0.4 GPa reflects the same behavior as it has been observed for several alkali halides (11). The deviation from the linear behavior indicates the onset of a phase transition due to the valence change of the rare earth ion. Comparing the curves in Fig. 1 only the longitudinal modes c_{11} and $c_{11}+c_{12}+2c_{44}$ show a minimum at about 1.5 GPa because they are relevant with regard to volume deformations (at constant pressure). Both shear modes c_{44} and $c_{11}-c_{12}$ conserve the volume and therefore change smoothly over the phase transition.

The results of the volume-pressure measurement by means of the strain gauge technique are shown in Fig. 2 a). On the right hand scale the change of the lattice constant with pressure is given too. From these data the length and the density change of the crystal were calculated. The dashed line in Fig. 2 a) shows the expected behavior of $TmSe_{0.32}Te_{0.68}$ as variation of pressure with purely divalent Tm. The deviation from this curve around 0.4 GPa indicates the beginning of the valence change of the rare earth ion.

The compressibility or its inverse, the bulk modulus vs pressure as a derivative of the volume-pressure curve is shown in Fig. 2 b) (full line). The initial value of the isothermal bulk modulus B_T is 37±1 GPa. With increasing pressure the bulk modulus decreases and at the transition pressure of 1.54 GPa B_T is only 6 GPa. Above this pressure the crystal stiffens and B_T increases again. The dashed line in Fig. 2 b) represents the adiabatic bulk modulus $B= (c_{11}+2c_{12})/3$ as calculated from the elastic constants c_{11} and c_{12}. The initial value is 35±2 GPa. The isothermal bulk modulus B_T is related to the adiabatic bulk modulus via the transformation $B_T= 1/3\ (c_{11}+2c_{12})\cdot(1+\alpha\ \gamma_G\ T)$, where α is the thermal coefficient of volume expansion, γ_G is the Grüneisen constant and T is the temperature in Kelvin. For the alkali halides the initial isothermal bulk modulus is about 4-6% higher than the adiabatic bulk modulus (12). In principal our results reflect the same behavior but it is worth mentioning that this difference in the bulk moduli lies within the measuring uncertainty.

From all our ultrasonic data a value for c_{12} can be calculated by using different equations. All values are equal within 10% and so the overall behavior can be seen in Fig. 3. The most striking point is the negative sign of c_{12} above 0.6 GPa far below the transition pressure to the metallic state at about 1.5 GPa. The fact that the sum of the square of the sound velocities in each direction is equal to $v^2= (c_{11}+2c_{44})/\rho$ provides a further check of the reliability of our ultrasonic data. For the [100] and [110] directions both sums are within 0.4% at lower pressures and 2% at higher pressures.

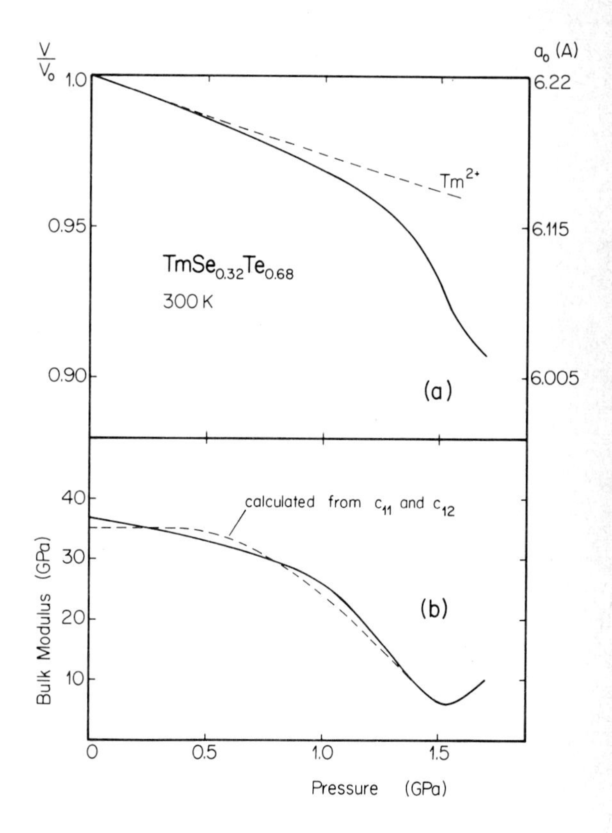

Fig. 2: Pressure dependence of a) the specific volume and b) the bulk modulus as a derivative of the specific volume (full line) and as calculated from the elastic constants (dashed line)

To a first approximation the second-order elastic moduli of most solid materials vary in a linear fashion with pressure. This behavior is mainly described by the third-order elastic constants which relate stress to strain in nonlinear finite elasticity theory. Their knowledge allows an evaluation of first-order anharmonic terms of the interatomic potential or of generalized Grüneisen parameters, which enter the theories of all anharmonic phenomena (13,14). Hydrostatic pressure measurements evaluate only three combinations of the six third-order moduli which occur for high symmetry cubic crystals and therefore our measurements are not sufficient to determine detailed quantities.

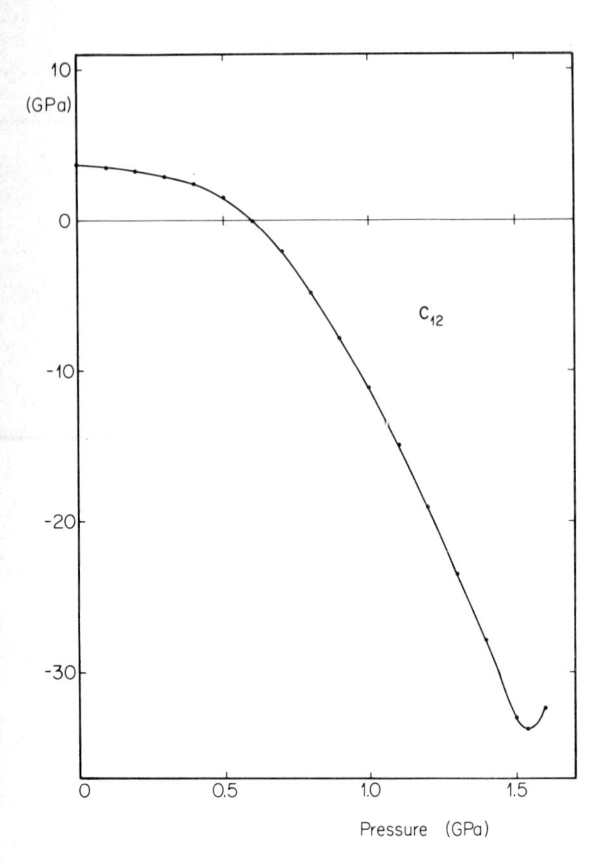

Fig. 3: Pressure dependence of the elastic constant c_{12} of $TmSe_{0.32}Te_{0.68}$

3. DISCUSSION

The most striking feature of phonon dispersion relations of intermediate valent materials is the softening of longitudinal branches compared to the transverse ones. In the [110] and [111] directions the LA branch lies below the TA in in parts of the Brillouin zone (2,15,16). The sound velocities measured in an ultrasonic experiment correspond to the initial slopes of the phonon curves. To show the change of these slopes with pressure we plotted the sound velocities as a function of the direction of the wavevector q at three different pressures (Fig. 4). The three elastic constants c_{11},c_{12} and c_{44}, listed in Table I, were the input data for the calculation of the sound velocities in different q-directions. At ambient pressure (Fig. 4 a) the sound velocities in the symmetry directions reflect a normal behavior. At the pressure of 1.26 GPa c_{12} is equal to minus c_{44}, the consequences of this are $v_L = v_{T_2}$ in [110] and $v_L = v_T$ in [111] directions (Fig. 4 b). A further consequence of the condition $c_{12} = -c_{44}$ is the degeneracy of two sound velocities for all q-vectors in the $[1\bar{1}0]$ plane, indicated by the thick line in Fig. 4 b). At the transition pressure of 1.54 GPa, the same anomalous behavior as in the intermediate valent metallic materials, e.g. $Tm_{0.99}Se$ or $Sm_{0.75}Y_{0.25}S$ can be observed, namely that the longitudinal sound velocity is less than the transverse sound velocity in [110] and [111] directions (Fig. 4c). An interesting point is the change from a pure longitudinal mode in [100] direction to a pure transverse mode in [110] or in [111] direction and vice versa.

Comparing all three plots the main effect is caused by the softening of the $q \to 0$ longitudinal phonon. As discussed above changes in the valence affect the volume and therefore one would expect that the longitudinal and not the transverse phonons would be modified. Transverse deformations to first order in the deformation preserve the volume.

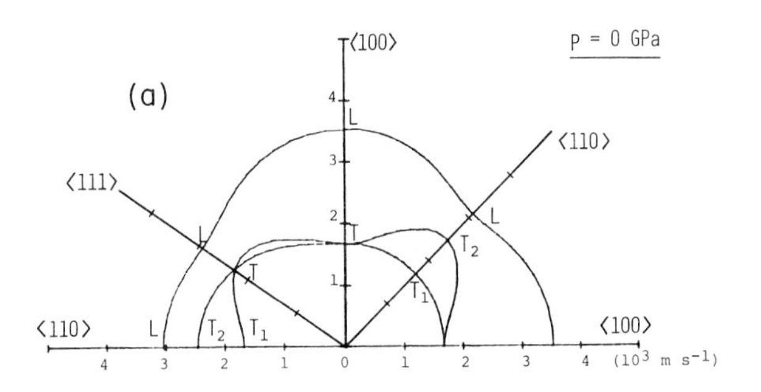

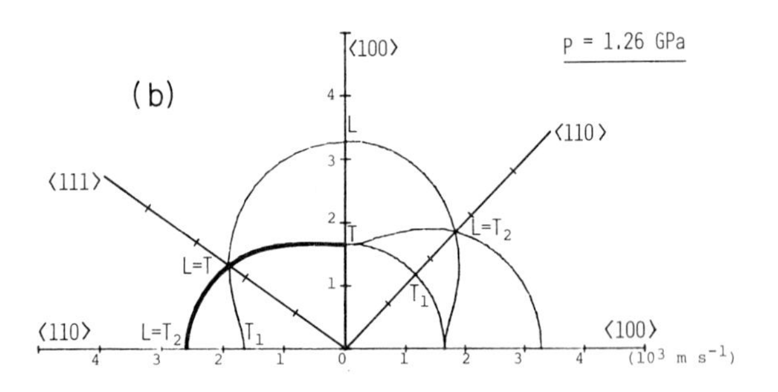

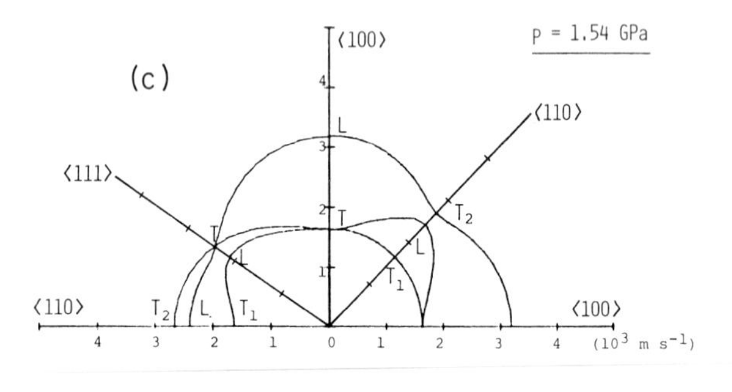

Fig. 4: Sound velocities as a function of the direction of the wavevector q, a) at ambient pressure, b) at 1.26 GPa and c) at the transition pressure of 1.54 GPa.

In a model assuming only central force interactions the so called Cauchy relation $c_{12} = c_{44}$ exists for a cubic crystal. The relation is quite well fulfilled in the alkali halides and in many other materials. A possible failure of the Cauchy relation has been attributed to special ion deformabilities in the alkali halides and to the conduction electrons in the metals. The lattice dynamical behavior of the intermediate valent metallic materials has been treated very successfully by introducing an isotropic deformability called "breathing"-force. According to the "breathing shell model" this implies the new condition $c_{12}-c_{44} < 0$ (17,18). For all our data $c_{12}-c_{44} < 0$ is fulfilled over the whole pressure range.

The anisotropy factor in the cubic crystals is $A = 2c_{44}/(c_{11}-c_{12})$ and is unity when the material is elastically isotropic, and differs from unity otherwise. As one can seen in Table I, A varies from 0.43 at ambient pressure to 0.37 at the transition pressure and is consequently comparable to the anisotropy in the alkali halide series RbF-KCl-KBr (19).

Table I. Semiconducting $TmSe_{0.32}Te_{0.68}$

p (GPa) =	0	1.26	1.54
density (kg m^{-3})	7800	8159	8430
c_{11} (GPa)	97.40	87.80	85.90
c_{12} (GPa)	3.67	-22.2	-33.8
c_{44} (GPa)	21.80	22.2	22.45
anisotropy factor $A = 2c_{44}/(c_{11}-c_{12})$	0.43	0.40	0.37
bulk modulus (GPa)	37	16	6

To summarize our reflections about the variation of the elastic constants through a continuous SMT we start from an ionic solid similar to the alkali halides but already with a considerable degree of breathing deformability. Then around 0.6 GPa c_{12} becomes negative and all measured combinations of elastic constants in the different directions deviate from the originally linear behavior under pressure, because of the softening of the bulk modulus due to the breathing deformability of the rare earth ion. The electronic structure plays a major role in the sense that the admixture of 5d wavefunctions to the localized 4f states is strongly increased by the narrowing of the energy gap. The localized character of the 4f states gets lost and the result is a state which could be called "intermediate valent semiconductor".

In conclusion regarding the elastic properties we find the same feature in the semiconducting range near the transition as it is well known from the intermediate valent metallic compounds.

ACKNOWLEDGEMENT

The authors are very grateful to A. Vonlanthen for the technical assistence and to Dr. R. Monnier for valuable discussions.

REFERENCES

1) Boppart, H., Treindl, A., Wachter, P., and Roth, S., Solid State Commun. 35, 483 (1980)
2) Mook, H.A., Nicklow, R.M., Penney, T., Holtzberg, F., and Shafer, M.W., Phys. Rev. B18, 2925 (1978)
3) Jayaraman, A., in Handbook on the Physics and Chemistry of Rare Earths, Volume 2, 575, ed. by Gschneidner, K.A., and Eyring, L.
4) Boppart, H., Wachter, P., Batlogg, B., and Maines, R.G., Solid State Commun. 38, 75 (1981)
5) Bilz, H., Güntherodt, G., Kleppmann, W., and Kress, W., Phys. Rev. Lett. 43, 1998 (1979)
6) Entel, P., Grewe, N., Sietz, M., and Kowalski, K., Phys. Rev. Lett. 43, 2002 (1979)
7) Batlogg, B., Kaldis, E., and Wachter, P., J. de Phys. 40-C5, 370 (1979)
8) Batlogg, B., and Wachter, P., J. de Phys. 41-C5, 59 (1980)
9) Kaldis, E., Fritzler, B., Jilek, E., and Wisard, A., J. de Phys. 40-C5, 366 (1979)
10) Kaldis, E., Fritzler, B., Spychiger, H., and Jilek, E., these proceedings
11) Lazarus, D., Phys. Rev. 76, 549 (1949)
12) Anderson, O.L., J. de Phys. Chem. Solids 27, 547 (1966)
13) Brugger, K., and Fritz, T.C., Phys. Rev. 157, 524 (1967)
14) Thurston, R.N., and Brugger, K., Phys. Rev. 133, A1604 (1964)
15) Treindl, A., and Wachter, P., Solid State Commun. 36, 901 (1980)
16) Mook, H.A., and Holtzberg, F., in Valence Fluctuations in Solids, Falicov, L.M., Hanke, W., Maple, M.B. (eds), North Holland Publ. Company, 1981
17) Schröder, U., Solid State Commun. 4, 347 (1966)
18) Bilz, H., in "Computational Solid State Physics", ed. by Herman, Dalton and Koehler (Plenum, N.Y., 1972) p. 309
19) Chung, D.H., and Buessem, W.R., J. of Appl. Phys. 38, 2010 (1967)

Valence Instabilities
P. Wachter and H. Boppart (eds.)

NONLINEAR RELATIONSHIP BETWEEN VALENCE AND LATTICE CONSTANT

G.Neumann, R.Pott, J.Röhler, W.Schlabitz, D.Wohlleben and H.Zahel

II. Physikalisches Institut der Universität zu Köln,
Zülpicher Str. 77, D-5000 Köln 41, Fed. Rep. Germany

We show that, in order to avoid a serious systematic error in the determination of the valence from lattice constants, the usual linear interpolation between lattice constants of integral valent compounds a la Vegard should be replaced by a linear interpolation between bulk moduli. The resulting nonlinear relationship between the "true" valence from lattice constants and the valence a la Vegard is nearly universal for Sm, Eu, Tm and Yb mixed valence compounds.

INTRODUCTION

The reliable measurement of the valence is an urgent problem of rare earth (RE) mixed valence research. The valence can be extracted from measurements of lattice constants, susceptibility, Mößbauer isomer shifts, X-ray absorption, UV and X-ray photoemission and several others. The relative merits of the more common methods were discussed recently /1/. Since the beginning of RE mixed valence research, discrepancies of 10% to 20% between values of the valence extracted from different measurements have been the rule. Today it is clear that some of these discrepancies are outside of the error from measuring statistics, i.e. they must be due to systematic errors. The lattice constant values are more widely accepted than those from any other method. However, this confidence is based on the ubiquitous nature of the lattice constant measurement, rather than on anything more rational. In fact, close inspection even of older data e.g. for SmB_6 and TmSe /1/ clearly shows that the valence values from other methods tend to cluster about some particular value while that from the lattice constant is 10% - 15% larger, suggesting that the lattice constant values are off the mark rather than the others.

Recently the x-ray absorption method for measuring the valence from the structure of the L_{III} edge /2/ has become a strong competitor to the lattice constant method /3-7/. It is just as universally applicable but more direct and easier to handle. The x-ray absorption data have added decisive weight to the suspicion that the valence from lattice constants is more than 10% larger than the "true" valence. In Table I we have collected the evidence.

Of course, like any other method the L_{III} method must be screened for problems with systematic errors as well. Where comparison is possible, the L_{III} method delivers values in better accord with other methods than with lattice constants (see Table I). Nevertheless doubts about the reliability of the simple L_{III} valence extraction procedure used in /2-6/ were voiced by photoemission experts /15/, who pointed out possible falsification by final state effects. These doubts were however recently retracted /7,21/, at least for the RE from Sm to Yb. (There still is a lively debate about the validity of the simple extraction procedure for Ce compounds /22/, in which we shall not get involved here).

On this background, inspection of Table I provides ample motivation to search for a possible systematic error in the determination of the valence from the lattice constant. The lattice constant method gives the valence on the basis of three measurements : a_{n+1}, a_n and $a(\nu)$. Here a_{n+1} and a_n are the lattice constants of integral valent reference compounds with n+1 and n 4f electrons and $a(\nu)$ is that of the mixed valent compound. By assuming the validity of Vegard's law for the relationship between valence and lattice constant one obtains the "Vegard valence" v_V via:

$$\begin{aligned} v_V &= Q_{n+1}+(a_{n+1}-a)/(a_{n+1}-a_n) \\ &= Q_{n+1}+\nu_V \end{aligned} \qquad (1)$$

where Q_{n+1} is the lower integral valence. More generally, and in parti-

Table I: Valencies at 300 K from various measurements

Compound	v_V	XPS	Mößbauer	χ (T)	XAS	Reference
YbCuAl	3.0			2.9	2.86	/8,9,10/
$YbCu_2Si_2$	2.94				2.88	/8,10/
$YbNi_2Ge_2$	2.94				2.88	/8,10/
$YbAl_3$	2.93		2.75			/8,11/
$YbInAu_2$	2.86				2.68	/8,10/
YbInPd	2.5				2.42	/8,10/
$YbAl_2$	2.53		2.4		2.4	/8,13,10/
YbInAu	2.30				2.18	/8,10/
TmSe	2.75	2.6		2.6	2.6	/8,14,15,16/
$EuRh_2$	2.93		2.91			/8,17/
$EuPd_2Si_2$	2.3		2.2		2.2	/8,16,10/
SmB_6	2.77	2.65	2.65		2.65	/8,18,19,20/

cular for noncubic compounds, the lattice constants a_i must be replaced by the volumes V_i of the corresponding unit cells. In this paper we argue that expression (1) should be replaced by a nonlinear relationship between valence and lattice constant.

BULK MODULUS AND VALENCE

It is well known that Vegard's "law" can break down. According to Friedel /23/, this should happen whenever the compressibilities of the components of an alloy are sufficiently different. In the case of a RE mixed valence compound, the "components" are the unit cells of the compounds with e.g. the di- and trivalent RE ions. The corresponding compressibilities can be measured directly on such reference

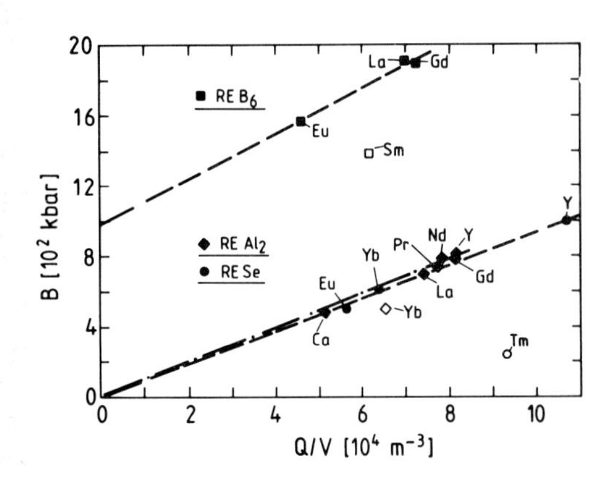

Fig.I
Bulk modulus B as function of Q/V at 300 K (Q=valence, V=molar volume) /24,25,26/.

compounds. Indeed, the compressibilities of intermetallic compounds with stable divalent RE ions are without exception considerably larger than with stable trivalent ones. This is shown in Fig.I for the $REAl_2$ /24/, the RESe /15,25/ and the REB_6 /26/. Following /24/ we have plotted the measured bulk moduli against Q/V, where Q is the RE valence and V the corresponding volume of the unit cell. Closed symbols refer to integral (di- or tri-) valent compounds and open symbols to mixed valent ones. The compressibilities differ by more than a factor two between EuSe (divalent) and YSe (trivalent). All dialuminides and selenides with integral valence fall on a straight line wich goes through the origin. The points for the harder hexaborides may also be connected by a straight line which however intercepts the ordinate at a finite value B_0. All mixed valent compounds fall below the lines defined by those with integral valence.
For integral valence one may write empirically:

$$B = \alpha Q/V + B_o \qquad (2)$$

with the constants α and B_0 characterizing the RE series of compounds. For nonintegral valence this relationship obviously does not hold (Fig.I), but one may formally write by analogy:

$$B(\nu) = \alpha Q(\nu)/V(\nu) + B_o \qquad (3)$$

where ν is the fractional occupation of $4f^n$, and $Q(\nu)=Q_{n+1}+\nu = v$ is the valence. Note that this expression would yield Vegard's law only if $B(\nu)$=const., i.e. if compounds of different integral valence had the same bulk modulus, which is manifestly not true.

Note also, that the bulk moduli on the straight lines of Fig.I were measured at constant valence. This is again obviously not true for the mixed valence compounds (open symbols of Fig.I). For these one must instead write:

$$B_M(\nu) = -V(\nu)\left\{\frac{\partial V}{\partial p}\Big|_{\nu,T} + \frac{\partial V}{\partial \nu}\Big|_{p,T}\frac{d\nu}{dp}\Big|_T\right\}^{-1} = \kappa_M^{-1}$$

$$\kappa_M = \kappa_\nu + \kappa_{\nu c}$$
$$-\kappa_\nu = \frac{\partial \ln V}{\partial p}\Big|_\nu \qquad (4)$$
$$-\kappa_{\nu c} = \frac{\partial \ln V}{\partial \nu}\Big|_{p,T}\frac{d\nu}{dp}\Big|_T$$

In other words, for mixed valence compounds one must distinguish between an isovalent contribution κ_ν to the compressibility and one due to valence change, $\kappa_{\nu c}$. The data on the straight lines of Fig.I and equ.(2) and (3) refer to the inverse isovalent compressibility, κ_ν^{-1}. For mixed valence one needs, besides $\kappa_{\nu c}$, an expression for the ν-dependence of $\kappa_\nu^{-1}=B(\nu)$. In principle this expression is a very complex quantity, and there is no theory. However, any expression that acknowledges the obviously strong ν-dependence of the isovalent bulk modulus is better than the assumption $B(\nu)$=const., which underlies Vegard's law. In the sense of an interpolation we write as simplest choice for the isovalent bulk moduli:

$$B(\nu) = B_{n+1} + \nu(B_n - B_{n+1}) \qquad (5)$$

With this Ansatz in equ.(3) one obtains the following nonlinear relationship between volume of the unit cell and valence :

$$V(\nu) = \frac{(Q_{n+1}+\nu)(B_n - B_{n+1})}{(Q_n V_n^{-1} - Q_{n+1}V_{n+1}^{-1})(B_{n+1} - B_o + \nu(B_n - B_{n+1}))} \qquad (6)$$

DISCUSSION

In Fig.II we have plotted the corresponding relationship for the lattice constants of the cubic compounds $YbAl_2$, TmSe and SmB_6. The compressibilities and lattice parameters of the reference substances used in these curves are collected in Table II. All data refer to 300 K. Also shown is the traditional Vegard's law interpolation and the measured lattice constants of the intermediate valence compounds themselves. One obtains a shift of 0.12, 0.117 and 0.096 towards lower valence for $YbAl_2$, TmSe and SmB_6 respectively when using the nonlinear "sagging" curves rather than Vegard's law to determine the valence from the

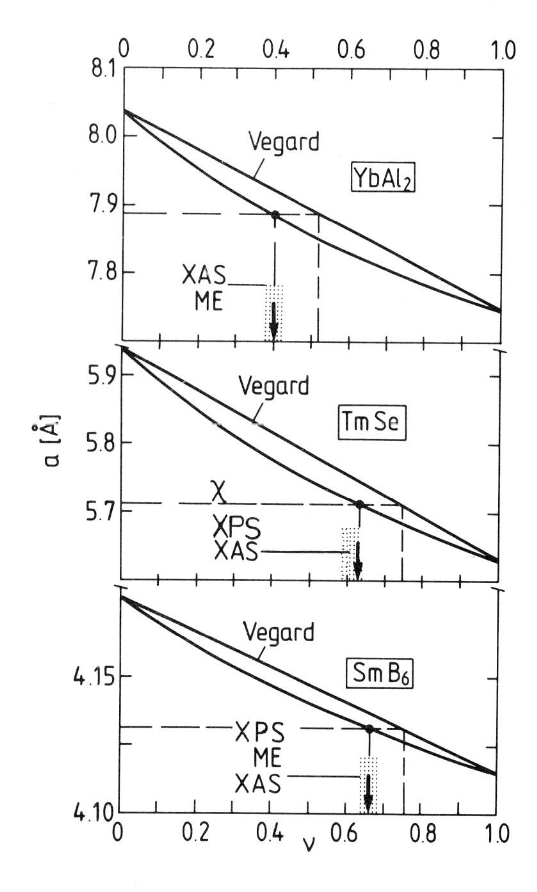

Fig.II
Lattice constant as function of fractional occupation of trivalent state as obtained by linear interpolation of lattice constants (Vegard) and by linear interpolation of bulk moduli. Shaded area gives range of valence values by other methods.

lattice constants. Clearly these shifts bring the valencies from the lattice constants into much better coincidence with values from other methods as indicated by the shaded regions in Fig.II. For these three compounds the mean square deviation of the absolute valencies from different methods are now less than one percent! Further support for one of these values comes from thermodynamics : The high temperature limit of the intermediate valence of Tm should be 2.62, only one percent lower than measured! There is much independent evidence that TmSe is at the high temperature limit at 300K (while $YbAl_2$ is not and SmB_6 has no such limit /1/).

In Fig.III we have plotted the lattice parameters, a, as obtained from equ.(6)

Table II: di and trivalent lattice constants and bulk moduli

Compound	a^{2+}	a^{3+}	B^{2+}	B^{3+}
$YbAl_2$	8.038 Å	7.743 Å	49.5 GPa	83.0 GPa
TmSe	5.940 Å	5.630 Å	59.5 GPa	104.8 GPa
SmB_6	4.178 Å	4.115 Å	150.9 GPa	189.6 GPA

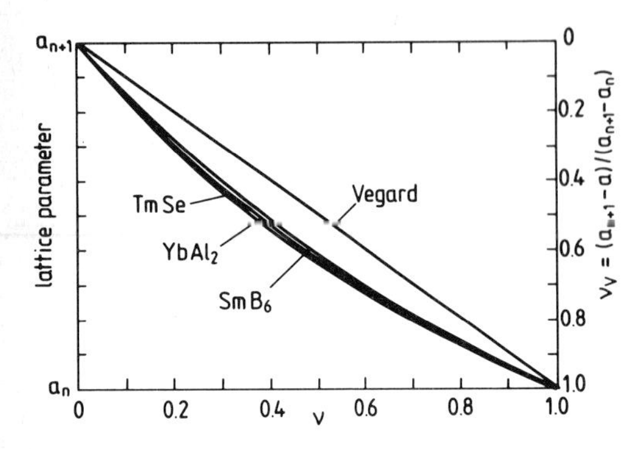

Fig.III
Normalized nonlinear relationship between "true" and Vegard valence. Note the nearly universality..

against ν, the fractional occupation of $4f^n$. The quantity v_V of equ.(1), i.e. the fractional $4f^n$ occupation according to the traditional extraction a la Vegard, is shown on the right hand ordinate. It is remarkable that the sagging curves appear to almost coincide in this normalized representation, in spite of the large difference in bulk moduli of the hexaborides with respect to the others. This is due to the fact that the difference in compressibilities of the compounds is a local ionic effect, i.e. a reflection of the difference of the local compressibilities of the RE ions, while the background compressibility of the matrix of e.g. the REB_6 remains unaffected by the RE valence. This behaviour indeed follows from a simple elastic model which uses only compressibilities and atomic volumes from the periodic table /27/. The accuracy of the normal lattice constant measurement is insufficient to resolve the small differences between the three curves of Fig.III. Therefore an average of such curves may serve as a universal calibration curve to correct Vegard valences from the literature to something much closer to the true valence, without any need to measure the bulk moduli of the reference compounds. Of course, such corrected valence values must still be regarded as approximations, because of the ad hoc nature of the interpolation of the isovalent bulk moduli (equ.5). Due to the complexity of the elastic phenomena in the systems of interest, each situation requires in principle special thought. An example is shown in Fig.IV which reproduces the results of

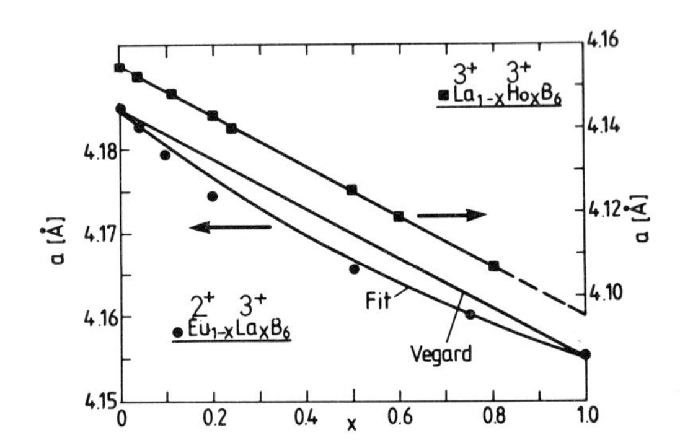

Fig.IV
Lattice constant as function of concentration of inhomogeneous alloys with two trivalent ions (La,Ho) /28/, and one divalent and one trivalent (Eu,La) /29/. Also shown is a fit obtained by equ.(6).

a study of lattice constants as function of concentration in REB_6 with RE ions of the same valence ($La_{1-x}Ho_x B_6$ /28/) and of different valencies ($Eu_{1-x}La_x B_6$ /29/). The first type of alloy follows Vegard's law. The second, which can be regarded as an inhomogenously mixed valent solid, traces a curve which is qualitatively similar to our fit from equ.(6) but deviates quantitatively outside of experimental scatter. We believe that the isovalent B(ν) must take different forms in homogeneously and inhomogeneously mixed valence systems, and that the form of equ.(6) is more appropriate to homogeneous mixtures /27/.
It is obvious that a nonlinear relationship like equ.(6) or some equivalent must be used instead of equ.(1) when extracting temperature or pressure driven valence changes from thermal expansion anomalies or experimental p-V relationships. This will be discussed elsewhere /30/.

Our sucessfull correction of valence values from lattice constants via the nonlinear relationship of equ.(6) cannot be applied to mixed valent Ce compounds at this time. The first reason is uncertanty about the validity of the tetravalent ionic radius /3/, and the second the lack of compressibility data on either end of the scale, for both $4f^1$ and $4f^0$.

This work was supported by the Deutsche Forschungsgemeinschaft, SFB 125.

REFERENCES

/1/ D.K.Wohlleben in "Valence Fluctuations in Solids", Ed. L.M. Falicov, W.Hanke, M.B.Maple, North Holland Publishing Comp., Amsterdam, New York, Oxford (81)

/2/ E.E.Vainsthein, S.M.Blokhin and Yu.B.Paderno, Sov. Phys. Sol. St. 6,231 (65)

/3/ K.R.Bauchspieß, W.Boksch, E.Holland-Moritz, H.Launois, R.Pott and D.Wohlleben in Ref./1/,p.417

/4/ H.Launois, M.Rawiso, E.Holland-Moritz, R.Pott and D.Wohlleben, Phys. Rev. Let. 44, 1271 (80)

/5/ R.M.Martin, J.B.Boyce, J.W.Allen and F.Holtzberg, Phys. Rev. Let. 44, 1275 (80)

/6/ T.K.Hatwar, R.M.Nayak, B.D.Padalia,M.N.Ghatikar, E.V.Sampathkumaran,L.C.Gupta and Vijayaraghavan, Sol. St. Com. 34, 617 (80)

/7/ A.Bianconi, S.Modesti, M.Campagna, K.Fischer and S.Stizza, J.Phys. C: Sol. St. Phys. 14, 4737 (81)

/8/ A recent critical appraisal of Vegard valences can be found in R.Pott, PhD-Thesis, Cologne (82)

/9/ W.C.M.Mattens, PhD-Thesis, Amsterdam (80)

/10/ K.R.Bauchspieß, Diplom Thesis, Cologne (82), and to be published

/11/ J.W.Ross and E.Tronc, J. Phys. F: Met. Phys. 8, 983 (78)

/12/ K.H.J.Buschow, Rep. Prog. Phys. 42, 1393 (79)

/13/ P.Bonville, P.Imbert, G.Jehanno and F.Gonzales-Jiminez, J. Phys. Chem. Sol. 39, 1273 (78)

/14/ B.Battlog, E.Kaldis, A.Schlegel and P.Wachter, Phys. Rev. B14, 5503 (76)

/15/ G.K.Wertheim, W.Eib, E.Kaldis and M.Campagna, Phys. Rev. B22, 6240 (80)

/16/ E.V.Sampathkumaran et al. in Ref./1/, p.193

/17/ E.R.Bauminger, I.Felner, D.Levron, I.Nowik and S.Ofer, Sol. St. Com. 18, 1073 (76)

/18/ J.N.Chazalviel, M.Campagna, G.K.Wertheim, and P.H.Schmidt, Rev. Lett. B14, 4586 (76)

/19/ R.L.Cohen, M.Eibschutz and K.W.West, Phys. Rev. Lett. 24, 383 (70)

/20/ J.M.Tarascon, Y.Isikawa, B.Chevalier, J.Etourneau, P.Hagenmuller and M.Kasaya, J.Physique 41, 1141 (80)

/21/ F.U.Hillebrecht and J.C.Fuggle in press in Phys. Rev. B (82)

/22/ "Meeting on high energy spectroscopy of Ce", KFA Jülich, Fed. Rep. of Germ., (Jan. 82)

/23/ J.Friedel, Adv. in Phys. 3, 446 (54)

/24/ T.Penney, B.Barbara, R.L.Melcher, T.S.Plasket, H.E.King Jr. and S.J.La Placa in Ref./1/, p.341

/25/ A.Jayaraman in "Handbook on the Physics and Chemistry of Rare Earths", Ed. K.A.Gschneidner Jr. and L.R.Eyring, North Holland Publishing Comp., Amsterdam, New York, Oxford (79)

/26/ H.E.King Jr., S.J.La Placa, T.Penney and Z.Fisk in Ref./1/, p.333

/27/ D.Wohlleben et al. to be published

/28/ A.Berrada, J.P.Mercurio, J.Etourneau and P.Hagenmuller, Mat. Res. Bull. 11, 947, (76)

/29/ J.P.Mercurio, J.Etourneau, R.Naslain, P.Hagenmuller and J.B.Goodenough, J. of Sol. St. Chem. 9, 37 (74)

/30/ R.Pott et al. to be published

Valence Instabilities
P. Wachter and H. Boppart (eds.)

ELASTIC PROPERTIES OF THE INTERMEDIATE VALENCE COMPOUND $YbAl_2$ AND THE KONDO COMPOUND $CeAl_2$

B. Barbara,* T. Penney, T.S. Plaskett, H.E. King,+ Jr. and S.J. La Placa
IBM T.J. Watson Research Center, Yorktown Heights, NY 10598

The strongly intermediate valence (IV) rare earth chalcogenide compounds, SmS, $Sm_{1-x}Y_xS$, and TmSe have previously been shown to be soft. That is, they have small bulk moduli and anomalous phonon dispersion compared to similar integral valence materials.[1-5] In order to determine if these effects are characteristic of IV more generally, we have studied the ultrasonic and pressure volume behavior of some intermetallic rare earth aluminum systems.

From lattice constant, magnetic and photoemmission measurements,[6-9] $YbAl_2$ is strongly intermediate valence (v=2.4) while $CeAl_2$ is a Kondo compound with integral valence, v=3.0.[10] Our results show that $YbAl_2$ has a smaller bulk modulus (20%) and smaller Poisson ratio (40%) than one would expect from comparison with divalent $EuAl_2$ and trivalent $LaAl_2$.[11] The size of the effect is similar to that observed in SmS under pressure,[1] but smaller than that found in $Sm_{1-x}Y_xS$ [2] and TmSe.[4]

We conclude that a soft bulk modulus is characteristic of IV, due to the near degeneracy of two electronic configurations f^n and $f^{n-1}d$, which allows the f occupation to change under pressure or ultrasonic strain. Here d is any non f electron at the Fermi energy.

This mechanism will not apply to Kondo compounds because the number of f electrons does not change under stress. We have measured the bulk modulus and Poisson ratio of $CeAl_2$ and find them similar to the values for $LaAl_2$. In contrast, the ultrasonic measurements of Luthi and Lingner[12] have shown the $CeAl_2$ bulk modulus to be about 15% smaller than that of $LaAl_2$. Since both groups have measured five phonon velocities to compute three elastic constants and measured two samples each, we conclude that the bulk modulus is sensitive to sample composition or defects. There may be a softening mechanism, other than the f-d mechanism above, related to the γ - α transition which is manifest at ambient pressure under some conditions.

* Permanent address, Lab. L. Neel, CNRS, 38042 Grenoble, France.
+ Permanent address, Exxon Corporate Research, Linden, NJ.

1. R. Keller, G. Guntherodt, H. Holzapfel, W.B. Dietrich and F. Holtzberg, Solid State Comm. 29, 753 (1979).
2. R.L. Melcher, G. Guntherodt. T. Penney and F. Holtzberg, 1975 Ultrasonics Symposium, Proc. IEEE Cat. #75 CHO 994-4SU, p. 16. T. Penney and R.L. Melcher, J. Physique Colloq. 37, C4-273 (1976).
3. H.A. Mook, R.M. Nicklow, T. Penney, F. Holtzberg, M.W. Shafer, Phys. Rev. B18, 2925 (1978). H. Mook, R. Nicklow, Phys. Rev., B20, 1665 (1979).
4. H. Boppart, A. Treindl, P. Wachter and S. Roth, Solid State Comm., 35, 483 (1980).
5. A. Iandelli, A. Palenzona, Handbook on Physics and Chemistry of Rare Earths, Vol 2, K.A. Gschneidner, L.R. Eyring eds., North Holland, (1980).
6. T. Penney, B. Barbara, R.L. Melcher, T.S. Plaskett, H.E. King Jr. and S.J. LaPlaca, Valence Fluctions is Solids, L.M. Falicov, W. Hanke, M.B. Maple, eds., North Holland, (1981), p. 341.
7. A. Iandelli and A. Palenzona, J. Less Common Metals 29, 293 (1972)
8. J.C.P. Klasse, W.C.M. Mattens, F.R. deBoer and P.F. deChatel, Physica 86-88B , 234 (1977).
9. G. Kaindl, B. Reihl, D.E. Eastman, R.A. Pollak, N. Martensson, B. Barbara, T. Penney and T.S. Plaskett, Solid State Comm. 41 Jan (1982).
10. B. Barbara, M. Rossignol, J.X. Boucherle, J. Schweizer and J.L. Buevos, J. Appl. Phys., 50, 2300 (1979)
11. R.J. Schiltz and J.F. Smith, J. Appl. Phys. 45, 4681 (1974).
12. B. Luthi and C. Lingner, Z. Phys. B34, 157 (1979).

Valence Instabilities
P. Wachter and H. Boppart (eds.)

PHONON-INDUCED INSTABILITIES IN MIXED-VALENCE SYSTEMS

Juan Giner* and François Brouers**

*Institut de Physique, Université de Liège
B-4000 Sart Tilman, Belgique
**Institut für Theoretische Physik, Freie Universität
Berlin, Arnimallee 3, D-1000 Berlin 33.

1. INTRODUCTION

In a recent paper (1) we have reconsidered carefully the results of Entel et al. (2) who have studied the effect of the coupling between 4f electrons and longitudinal optical phonons, and its influence on phase transitions in intermediate valence compounds, within the framework of the periodic Anderson model. The phonon-induced 4f-5d interband transitions participates to the renormalization of the hybridization energy between 4f and 5d conduction electrons. The main characteristic of this model lies in the fact that the renormalized hybridization $\tilde{V}$ can become zero and this is according to these authors one of the requirement for the discontinuous change of the average number of f electrons per site with the position of the f level in the d band.

By reformulating the problem, expressing all quantities in terms of three independent variables, namely the number of f electrons per site $n^f_{i\uparrow}$, $n^f_{i\downarrow}$ and the sum of the one-particle energies of f electrons, one is able to express the renormalized hybridization $\tilde{V}$ as the roots of a second-order equation. One then obtains six solutions, para-, ferro-, and antiferromagnetic with positive and negative $\tilde{V}$. This allowed us to show that the abrupt transition observed in the Entel et al model occurs for nonzero renormalized hybridization and is related to the change of sign of $\tilde{V}$ at a given position of the f level. Moreover, we showed that the stabilizing effect of the hybridization gap is an important contribution to the energy of each Hartree-Fock solution and that the transition occurs between solutions with large absolute value of the effective hybridization $\tilde{V}$.

It has been shown (3) that the d electron-phonon coupling plays an important role in the transition. As in (1) we wanted to discuss Entel et al results, we did not include them. This is the purpose of this paper to investigate the effect of both f and d electron-phonon coupling on this transition.

2. MODEL

We consider as in (1) and (2), the model Hamiltonian describing conduction electrons hybridized to a system of periodically arranged interacting 4f electrons (the first three terms); these f and d electrons are coupled to longitudinal optical phonons (the last 2 terms):

$$H=\sum_{k,\sigma}\varepsilon_k d^+_{k\sigma}d_{k\sigma}+\sum_{i,\sigma}(\varepsilon^f_o+\frac{U}{2}f^+_{i\sigma}f_{i\sigma}f^+_{i-\sigma}f_{i-\sigma})$$
$$+\sum_{i,\sigma}V(f^+_{i\sigma}d_{i\sigma}+c.c.)$$
$$-\sum_{i,\sigma}\left[g_1 f^+_{i\sigma}f_{i\sigma}+g_2(f^+_{i\sigma}d_{i\sigma}+c.c.)+g_3 d^+_{i\sigma}d_{i\sigma}\right]\cdot$$
$$\cdot(b^+_i+b_i)+\sum_i \hbar\omega\,(b^+_i b_i+\frac{1}{2}) \qquad (1)$$

where i represents the site index. We assume the d band to be centered at the zero of energy; the first term in Eq.(1) is the Fourier transform of the tight binding Hamiltonian describing the d electrons:

$$\sum_{k,\sigma}\varepsilon_k d^+_{k\sigma}d_{k\sigma}=\sum_{i\neq j,\sigma}h_{ij}\,d^+_{i\sigma}d_{j\sigma} \qquad (2)$$

where h_{ij} is the hopping integral different from zero only when i and j are first neighbours.

Eliminating the electron-phonon interaction to first-order one obtains within the Hartree-Fock approximation an electron Hamiltonian with the following renormalized quantities :

$$\tilde{\varepsilon}^f_{i\sigma}=\varepsilon^f_o+Un^f_{i-\sigma}-(G_1n^f_i+\sqrt{G_1G_3}\,n^d_i+\sqrt{G_1G_2}\,a_i) \quad (3)$$

$$\tilde{\varepsilon}^d_{i\sigma}=-(\sqrt{G_1G_3}\,n^f_i+G_3n^d_i+\sqrt{G_2G_3}\,a_i) \quad (4)$$

$$\tilde{V}=V-(\sqrt{G_2G_3}\,n^d_i+\sqrt{G_1G_2}\,n^f_i+G_2a_i) \quad (5)$$

where n^f_i and n^d_i are the total number of f and d electrons at site i respectively and where :

$$a_i = \sum_\sigma \langle f^+_{i\sigma} d_{i\sigma} + c.c. \rangle \qquad (6)$$

$$G_\gamma = \frac{2g_\gamma^2}{\hbar\omega}\ ,\ \gamma = 1,2,3 \qquad (7)$$

The expression of the total energy E_t at zero temperature is the following:

$$E_t = \sum_{i,\sigma} \int_{-\infty}^{E_F} dE\ E\left[\rho^f_{i\sigma}(E) + \rho^d_{i\sigma}(E)\right] - \frac{U}{2}\sum_{i,\sigma} n^f_{i\sigma} n^f_{i-\sigma}$$
$$+\frac{1}{2}\sum_i (\sqrt{G_1}\ n^f_i + \sqrt{G_2}\ a_i + \sqrt{G_3}\ n^d_i)^2 \qquad (8)$$

where E_F is the Fermi energy and where $\rho^f_{i\sigma}$ and $\rho^d_{i\sigma}$ are the local density of states at site i corresponding to f and d electrons respectively. In the Hartree-Fock decoupling we have included as in (2) the "excitonic term" a_i. Following the arguments of (1), it is easy to see that $\tilde{V}$ is solution of a second-order equation

$$\tilde{V} = \frac{1}{2}\{V - \sqrt{G_1G_2}\ n^f_i - \sqrt{G_2G_3}\ n^d_i \qquad (9)$$
$$\pm\left[(V - \sqrt{G_1G_2}n^f_i - \sqrt{G_2G_3}n^d_i)^2 - 8G_2\Delta U^f\right]^{\frac{1}{2}}\}$$

with

$$\Delta U^f = \sum_\sigma \left[\int_{-\infty}^{E_F} dE E\rho^f_{i\sigma}(E) - n^f_{i\sigma}\tilde{\varepsilon}^f_{i\sigma}\right] \qquad (10)$$

3. RESULTS

When the position of the f level ε^f_o is moved in the d band (this can be achieved by the effect of pressure, temperature or alloying) an abrupt change of the number of f electrons per site n^f_i can be produced for appropriate values of the parameters U, V, G_1, G_2 and G_3. This abrupt change correspond to a jump between a solution with positive or negative hybridization to a solution corresponding to an effective hybridization of the opposite sign. In this paper, we shall restrict ourselves to the case of a total number of one electron per site. As in (1) we choose V=0.075 and U=0.25 , all energies being in unit of half the bandwith of d electrons before hybridization. The three figures exhibit three typical situations met in the present model.

a) For $G_1=0.3$, $G_2=0.1$, $G_3=0.0$: we obtain a jump from an antiferromagnetic solution with negative renormalized hybridization $\tilde{V}$ to a paramagnetic solution with positive $\tilde{V}$ (fig.1a)

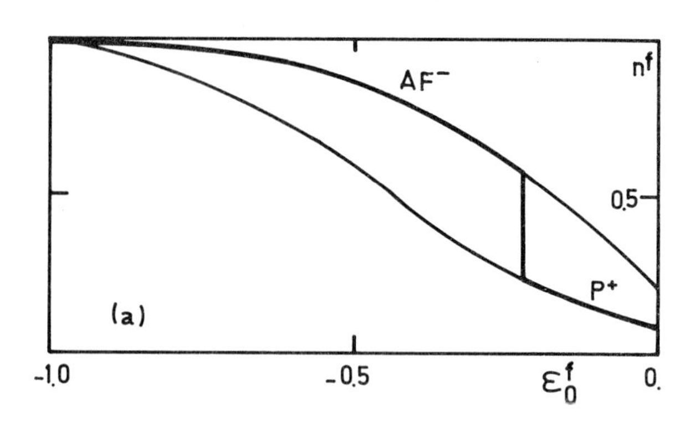

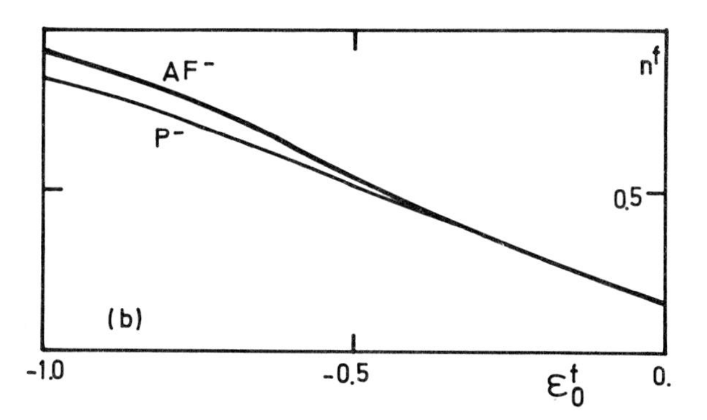

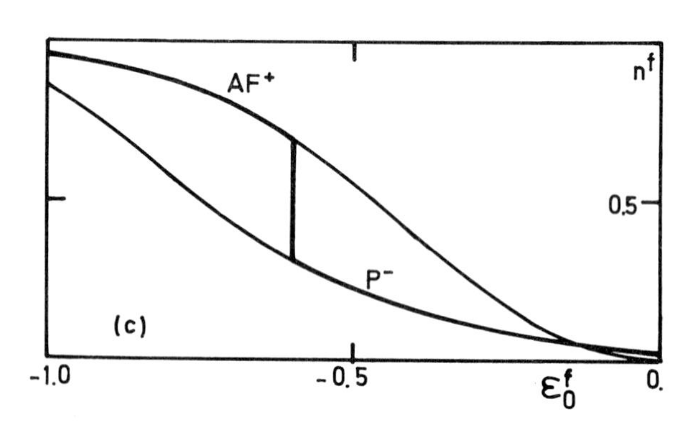

Fig.1: Total occupation n^f of the f level as a function of its position ε^f_o for three typical situations: a) $G_1=0.3$, $G_3=0.0$; b) $G_1=G_3=0.3$; c) $G_1=0.0$, $G_3=0.3$. The other parameter values are V=0.075, U=0.25 and $G_2=0.1$. The heavy lines represent the value of n^f in the state of lowest energy.

b) When both G_1 and G_3 are nonzero the dependence of $\tilde{V}$ upon n_i^f is weaker and this can supress any discontinuity in n_i^f In particular if $G_1=G_3$, Eq.(5) writes:

$$\tilde{V} = V-\sqrt{G_2 G_3} - G_2 a_i \qquad (11)$$

and $\tilde{V}$ depends on n_i^f only indirectly through a_i. For $G_1=0.3$, $G_2=0.1$, $G_3=0.3$ we obtain a continuous transition from an antiferromagnetic solution with negative $\tilde{V}$ to a paramagnetic solution with negative $\tilde{V}$ (fig.1b).

c) For $G_1=0$, $G_2=0.1$, $G_3=0.3$: we obtain a jump from an antiferromagnetic solution with positive $\tilde{V}$ to a paramagnetic solution with negative $\tilde{V}$ (fig.1c)

It is important to point out that : i) this type of first order transition is not of the same type as the one obtained by Falicov and Kimball (4); ii) one can show that the f-d electron-phonon coupling and the f-d electron-electron interaction can lead to similar formal expressions if one consider excitonic correlations as done by Khomskii, and Kocharjan (5). It would be therefore interesting to study the role of the two nonzero solutions given by Eq. (9) in the phase diagram obtained in (5).

4. CONCLUSION

The results of this paper is an illustration of the fact that in this problem, i.e. a d-f Hamiltonian with or without electron-electron and electron-phonon interactions, the nature of the variation of n_i^f with the physical parameters depends more on the approximation chosen than on the nature of the interaction considered.

REFERENCES

(1) Giner, J. and Brouers, F., to be published in Phys. Rev. B (1982).
(2) Entel, P., Leder, J.H. and Grewe, N., Z. Phys. B 30 (1978) 277
(3) Brouers,F. and de Menezes, O.L.T., Phys. Stat. Solidi 104 (1981) 541.
(4) Ramirez, R., Falicov, L.M. and Kimball, J.C., Phys. Rev. B 2 (1970) 3383.
(5) Khomskii, D.I. and Kocharjan, A.N., Solid State Commun. 18 (1976) 985.

Valence Instabilities
P. Wachter and H. Boppart (eds.)

BRILLOUIN SCATTERING IN INTERMEDIATE VALENCE COMPOUNDS

Michael Barth , Gernot Güntherodt

II. Physikalisches Institut , Universität zu Köln

D - 5000 Köln 41 , Fed. Rep. of Germany

Brillouin scattering in intermediate valence compounds has been performed using a high contrast multipass Fabry -Perot interferometer. From the measured sound velocities (bulk, surface) elastic constants could be derived.
The negative c_{12} of TmSe has been confirmed, as well as the anomalous elastic properties of $CeAl_2$ and $YbAl_2$. The absence of long wavelength acoustic phonon anomalies in $CePd_3$ (at 300 K) is consistent with previous measurements.[13]
For intermediate valence $CeCu_2Si_2$, $EuCu_2Si_2$, and $YbCu_2Si_2$ the velocity of the surface acoustic wave (Rayleigh wave) is anomalously small, thus leading to a "soft" bulk modulus. However, no anomalies are observed for intermediate valence $CeNi_2Si_2$, $CeNi_2Ge_2$, and $YbNi_2Ge_2$.

1. INTRODUCTION

Phonon anomalies occuring in intermediate valence systems are due to a strong coupling between the phonon system and the fluctuating 4f -electrons of the rare earth ions [1,2].
The investigation of elastic properties of such systems yields information on the low frequency limit of the electron - phonon coupling, which is closely related to the dependence of the valence on lattice parameters and pressure [3,4].
Measuring elastic constants by ultrasonic techniques or by inelastic neutron scattering is restricted to rather large single crystalline samples or demands elaborate experimental equipment.
Brillouin scattering in intermediate valence compounds proves to be a new, versatile tool for the fast routine - type determination of elastic properties. This inelastic light scattering technique (scattering from thermally activated acoustic phonons) may be performed on single crystalline or polycrystalline samples, their size only restricted by the laser focus diameter.
Here we present our results of the first Brillouin scattering experiments in intermediate valence compounds and some of their stable valence reference systems : TmSe , RE-Al_2 (RE = rare earth), RE-Pd_3 , RE-Cu_2Si_2 , RE-Ni_2Si_2 , RE-Ni_2Ge_2 .

2. EXPERIMENT

Brillouin scattering is the inelastic scattering of visible light from low energy (μeV - meV) excitations, e.g. acoustic phonons. Since the momentum transfer, determined by the scattering geometry, is of the order of the photon momentum, wavelengths of the investigated phonons are of the order of 5000 Å .
In metallic samples the penetration depth of visible light is small (e.g. 100 Å) and thus only the surface projection $q_{\|}$ of the phonon wavevector is conserved. For opaque samples light scattering experiments are performed preferably in 180° - backscattering geometry, with $q_{\|} = 2 q_{Photon} * \sin \Theta$, where Θ is the angle of incidence.
In Brillouin scattering experiments the frequencies of the scattering phonons are of the order of 10 GHz . A piezo - scanned multipass Fabry - Perot interferometer [5] is used to analyze such frequency shifts in the scattered light with a resolution of about 100 MHz (1 %) .
Sound velocities of Rayleigh waves and / or surface components of bulk waves (depending on the penetration depth of the light) are determined from the frequency shifts and momentum transfers. From these velocities elastic constants can be derived.
Acoustic phonons (surface or bulk) measured by Brillouin scattering are related to bulk elastic properties, since the penetration depth of these phonons into the sample is at least of the order of their wavelength, i.e. 10^3 atomic layers.

All results presented here have been obtained at room temperature.

3. RESULTS / DISCUSSION

TmSe

Brillouin scattering has been performed on cleaved (1,0,0) surfaces of $Tm_{1.0}Se_{1.0}$ single crystals.
Since the penetration depth of the visible

light into these samples is of the order of 3000 Å, sound velocities of bulk acoustic waves could be measured. From the velocities of the longitudinal (1,0,0) and (1,1,0) and the transverse (1,1,0) phonons all three elastic constants of the cubic crystal structure can be derived. The values are shown together with values from previous measurements :

TmSe	c_{11}	c_{12}	c_{44}
Brillouin scatt.	16.6 ±0.7	-6.8 ±1.2	3.2 ±0.7
ultrasonic[6]	17.9 ±0.1	-5.7 ±0.1	2.7 ±0.1
neutron scatt.[7]	18.5 ± 1	-6.5 ±0.5	2.6 ±0.5

(elastic constants in 10^{10} Pa)

The negative c_{12} is due to the anomalously strong "breathing" of the Tm ion related to valence fluctuations $4f^{13} \leftrightarrow 4f^{12} + e^-$, which are also evident from polarized Raman scattering experiments [8].

RE - Al_2 (RE = rare earth)

Experiments have been performed on polished polycrystalline samples. Fig. 1 shows the elastic moduli c_R derived from the Rayleigh wave velocities via $c_R = \rho v_R^2$ (ρ : density).

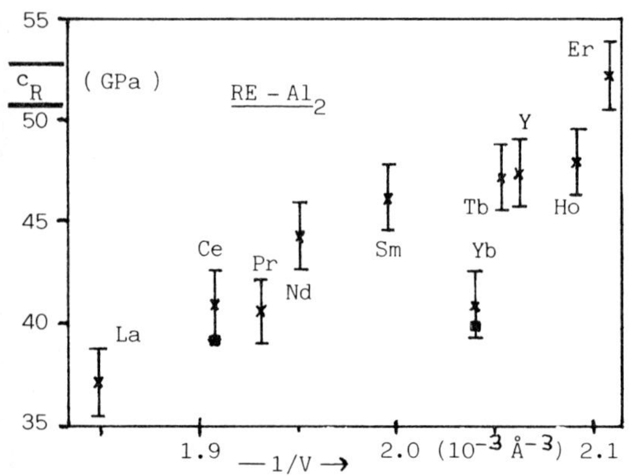

■ : calc. from single cryst. values
$CeAl_2$: ref. 9 , $YbAl_2$: ref. 10

Fig. 1 : Rayleigh wave RE-Al_2

Values of c_R for polycrystalline $CeAl_2$ and $YbAl_2$ were calculated from elastic constants of single crystals [9,10] using an averaging method (Voigt [11]) and the relationship between the Rayleigh wave and the bulk elastic constants [12]. The measured and calculated values agree within experimental error of ±4 %. This supports the statement that the bulk moduli of the Kondo compound $CeAl_2$ and the intermediate valence compound $YbAl_2$ are anomalously "soft" compared to those of integral valence RE-Al_2 systems [9,10].

RE - Pd_3

The Rayleigh wave velocities of polished polycrystalline samples of $CePd_3$, $PrPd_3$, $GdPd_3$, and YPd_3 have been measured. Within experimental error the elastic modulus c_R of $CePd_3$ shows no deviation from the values of stable valence RE-Pd_3 compounds :

	v_R (m/s)	c_R (Pa)	
$CePd_3$	2070±1,5 %	4.63 * 10^{10}	±3 %
$PrPd_3$	2070 ± 2 %	4.62 "	±4 %
$GdPd_3$	2000 ± 2 %	4,62 "	±4 %
YPd_3	2200 ± 2 %	4.94 "	±4 %

For $CePd_3$ the single crystal elastic constants are known from ultrasonic measurements [9]. The c_R calculated from these constants is 4,88 * 10^{10} Pa, which is within 5 % of the value obtained from Brillouin scattering.

The fact that no anomalies are found for the Rayleigh wave of polycrystalline $CePd_3$ at 300 K is in agreement with previous measurements of elastic properties and with theoretical considerations :

- Ultrasonic measurements in $CePd_3$ and $LaPd_3$ single crystals [13] yield the same (within 4 %) bulk moduli for both systems.
- Theoretical work on lattice dynamics [14] taking into account the renormalization of the $CePd_3$ phonon spectrum due to a strong coupling of the phonons to the fluctuating charge density of the Ce ion shows that there are no anomalies to be expected in the elastic regime.

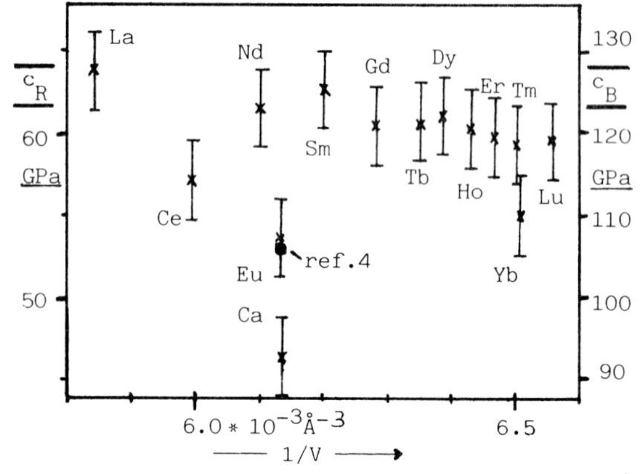

Fig. 2 : Rayleigh wave / Bulk Moduli RE-Cu_2Si_2

Fig. 2 shows the elastic modulus c_R of the RE-Cu_2Si_2 compounds measured with Brillouin scattering from polished polycrystalline samples and the bulk modulus c_B related to c_R via $c_B = 2 * c_R$.

$RE - Cu_2Si_2$

This relation (exactly : $c_B = 1.97\ c_R$) holds for $v_{LA} = \sqrt{3}\ v_{TA}$ ($v_{LA/TA}$: longitudinal/transverse acoustic sound velocity). This is the Cauchy relationship for quasi-isotropic crystals, assuming that the binding forces are independent of direction. The validity of this relationship is proved by Brillouin scattering measurements of the LA bulk phonon velocities in $EuCu_2Si_2$, $YbCu_2Si_2$, and $LuCu_2Si_2$, and is assumed for the other $RE\text{-}Cu_2Si_2$ samples as well.

The bulk modulus for $EuCu_2Si_2$ derived from Brillouin scattering measurements agrees within an experimental error of ±4 % with the value obtained from x-ray diffractometry under pressure[4].

Fig. 2 shows the following results :

- The bulk moduli of trivalent $RE\text{-}Cu_2Si_2$ compounds fall off slightly (- 5 %) over the lanthanide series.
- The bulk modulus of divalent $CaCu_2Si_2$ is about 25 % smaller than the bulk moduli of the trivalent compounds. This is expected, since the "stiffness" of a crystal should be proportional to the valence of its ions.
- The bulk moduli of intermediate valence $CeCu_2Si_2$, $EuCu_2Si_2$, and $YbCu_2Si_2$ lie between the values for divalent and trivalent $RE\text{-}Cu_2Si_2$ compounds, but (at least for $CeCu_2Si_2$ and $YbCu_2Si_2$) they do not scale linearly with the valence.

The linearly interpolated "isovalent" bulk moduli[3] c_B^{iso} for the intermediate valence systems are shown in the following table :

	valence[15]	c_B	c_B^{iso}
$CeCu_2Si_2$	3.05	$11.4 * 10^{10}$ Pa	$12.8 * 10^{10}$ Pa
$EuCu_2Si_2$	2.5	10.8 "	10.8 "
$YbCu_2Si_2$	2.9	11.0 "	12.0 "

(values for c_B : ± 4 %)

The differences in compressibilities $1/c_B - 1/c_B^{iso}$ are related to the pressure dependence of the valence, as is shown by the following simplified equation :

$$1/c_B = \frac{1}{V}\frac{dV(\nu,p)}{dp} = \frac{1}{V}\left.\frac{\partial V}{\partial p}\right|_\nu + \frac{1}{V}\left.\frac{\partial V}{\partial \nu}\right|_p \frac{d\nu}{dp}$$

v : volume p : pressure ν : valence

The first term on the right hand side is the "isovalent" compressibility $1/c_B^{iso}$.
With the c_B^{iso} taken from above and with $\left.\frac{\partial V}{\partial \nu}\right|_p$ taken from lattice parameters (in first approximation by linear interpolation), the pressure dependence of the valence can be roughly estimated, giving e.g. for

$YbCu_2Si_2$: $d\nu/dp = 1.0 * 10^{-11}\ Pa^{-1}$ ($10^{-3}\ kbar^{-1}$),
and for comparison , for
TmSe : $d\nu/dp = 5 * 10^{-10}\ Pa^{-1}$ ($10^{-2}\ kbar^{-1}$).

This simplified estimate does not take into account any non-linear effects . For more detailed calculations see refs. 3, 4 .

$RE - Ni_2Si_2$, $RE - Ni_2Ge_2$

Brillouin scattering measurements of the Rayleigh surface waves do not show any anomalies for the intermediate valence[15] compounds $CeNi_2Si_2$, $CeNi_2Ge_2$, and $YbNi_2Ge_2$ in comparison with stable valence reference systems.
The bulk moduli ($c_B = 2\ c_R$) are constant over the lanthanide series :

$RE\text{-}Ni_2Si_2$: $c_B = 10.6 * 10^{10}$ Pa
$RE\text{-}Ni_2Ge_2$: $c_B = 8.6$ "
($CaNi_2Ge_2$: $c_B = 6.3$ ")

This is quite surprising, since the crystal structure and the lattice parameters of these systems are similar to those of the $RE\text{-}Cu_2Si_2$ compounds. The same refers to the valences of the intermediate valence compounds.
About the absence of long wavelength acoustic phonon anomalies in intermediate valence $CeNi_2Si_2$, $CeNi_2Ge_2$, and $YbNi_2Ge_2$ may only be speculated at the present time.

4. CONCLUSIONS

Brillouin scattering proves to be a new useful experimental tool for the investigation of intermediate valence compounds. It is a method which can give first information (even for poor quality samples) on elastic properties and long wavelength phonon anomalies. Furthermore Brillouin scattering can provide supplementary or confirmative results to other experimental methods investigating phonon and lattice properties.
The question why phonon anomalies occur or are absent in intermediate valence systems needs further considerations, as the results for $RE\text{-}Cu_2Si_2$, $RE\text{-}Ni_2Si_2$, and $RE\text{-}Ni_2Ge_2$ compounds show. To solve these problems, more detailed informations on the phonon dispersion curves of these systems are necessary.

5. ACKNOWLEDGEMENTS

We would like to thank W.Bocksch, E.Cattaneo, P.Entel, K.Fischer, J.C.Fuggle, A.Jayaraman, J.Kahlenborn, G.Kaindl, D.Müller, G.Neumann, R.Pott, J.Röhler, W.Schlabitz, H.Schneider, and D.Wohlleben for providing us with the samples and / or for stimulating discussions , and N.Stüßer and B.Hillebrands for technical assistance.

This work has been supported by the Deutsche Forschungsgemeinschaft / SFB 125 .

6. REFERENCES

1. Entel,P., Grewe,N., Sietz,M., and Kowalski, K., Phys.Rev.Lett. 43 (1979) 2002
2. Bilz,H., Güntherodt,G., Kleppmann,W., and Kress,W., Phys.Rev.Lett. 43 (1979) 1998
3. Neumann,G., Pott,R., Röhler,J., Schlabitz, W., Wohlleben,D., Zahel,H., this conference
4. Röhler,J., Wohlleben,D., Balster, Kaindl,G., to be published
5. Sandercock,J., in Balkanski,M. (ed.), Proc. 2nd Intern. Conference on Light Scattering in Solids (Flammarion, Paris, 1971)
6. Boppart,H., Treindl,A., Wachter,P., and Roth,S., Solid State Commun. 39 (1981) 249
7. Mook,H.A., and Holtzberg,F., in Falicov, L.M., Hanke,W., Maple M.B. (eds.), Valence Fluctuations in Solids (North-Holland , Amsterdam, 1981)
8. Stüßer,N., Barth,M., Güntherodt,G., and Jayaraman,A., Solid State Comm. 39(1981)965
9. Lüthi,B., Takke,R., Assmus,W., Niksch,M., in Crow,J.E. (ed.), Crystalline Electric Fields and Structural Effects in f - Electron Systems (Plenum Press, New York,1980)
10. Penney,T., Barbara,B., Melcher,R.L., Plaskett,T.S., King Jr.,H.E., and LaPlaca, S.J., in Falicov, L.M., Hanke,W., Maple, M.B. (eds.), Valence Fluctuations in Solids (North-Holland, Amsterdam, 1981)
11. Voigt,W., Lehrbuch der Kristallphysik (Teubner, Leipzig, 1928)
12. Landau,L.D., Lifshitz,E.M., Theory of Elasticity (Pergamon, Oxford, 1970)
13. Takke,R., Niksch,M., Assmus,W., Lüthi,B., Pott,R., Schefzyk,R., and Wohlleben,D., Z. Physik B 44 (1981) 33
14. Entel,P., Sietz,M., Solid State Commun. 39 (1981) 249
15. Launois,H., Bauchspieß,K.R., et al., in Falicov,L.M., Hanke,W., Maple,M.B. (eds.), Valence Fluctuations in Solids (North-Holland, Amsterdam, 1981)

Valence Instabilities
P. Wachter and H. Boppart (eds.)

LATTICE DYNAMICS OF $Tm_{0.87}Se$ AND $Tm_{0.99}Se$ BY NEUTRON SCATTERING

Willi Bührer *, Albert Furrer * and Peter Wachter **

* Institut für Reaktortechnik, ETH Zürich
CH-5303 Würenlingen, Switzerland

** Laboratorium für Festkörperphysik, ETH Zürich
CH-8093 Zürich, Switzerland

Inelastic neutron scattering experiments have been performed to study phonon frequencies in trivalent $Tm_{0.87}Se$ and phonon frequencies and linewidths in intermediate valent $Tm_{0.99}Se$. The phonon densities of states have been computed with a fitted Born-von Karman model. The neutron results are compared with ultrasonic and Raman scattering data and discrepancies are discussed.

1. INTRODUCTION

Thulium Selenide is one of the most interesting intermediate valent compounds because the degree of valence mixing can be adjusted with composition: nearly 3^+ for $Tm_{0.87}Se$ and 2.8^+ for $Tm_{0.99}Se$ (1). The physical properties are greatly affected by the valence state. Lattice dynamical anomalies observed in the isostructural compound $Sm_{0.75}Y_{0.25}S$ were assigned to valence changes (2), and in $Tm_{0.99}Se$ a similar behavior is reported in Raman (1,3), ultrasonic (1), and neutron (2) experiments. The present work compares experimental neutron data of integer valent $Tm_{0.87}Se$ with data of intermediate valent $Tm_{0.99}Se$ and discusses the surprising similarity of the dispersion curves.

2. EXPERIMENTAL

The inelastic neutron scattering experiments were performed on a triple-axis spectrometer at the reactor Saphir, Würenlingen. Constant-$\vec{Q}$ and constant-ω modes of operation were used and only phonon creation processes were observed. The analyser energy was kept fixed at 14.9 meV in conjunction with a pyrolithic graphite filter to remove higher order contamination. Since the single crystals were small and thulium has a rather high absorption cross section, the spectrometer was operated in its focusing mode (4), i.e. doubly bent graphite monochromator for the acoustic modes, and in addition horizontally bent graphite analyser for the optic modes. The crystals $Tm_{0.99}Se$ (I) and $Tm_{0.87}Se$ have also been used for the ultrasonic measurements (1) and for the magnetic experiments (5).

Table 1: Sample crystals

	$Tm_{0.87}Se$	$Tm_{0.99}Se$		
a_o (Å)	5.63	5.70		
valence (1)	3 +	2.8 +		
crystal		I	II	III
size (mm^3)	30	30	35	75
orientation	(110)	(110)	(110)	(110)
mosaic (°)	>1.5	>1.5	>1.5	1.2

3. RESULTS

3.1. $Tm_{0.99}Se$

Dispersion curves measured at 295 K are displayed in Fig. 1. The experiments were initiated with crystal I. Along the direction Λ we observed ω(LA) > ω(TA) for intermediate $\vec{q}$-vectors, in disagreement with the results of Mook et al. (2) and the extrapolated velocities of sound (1). Zone boundary frequencies at X agree within experimental error, but at L Mook et al. report remarkably higher values for the acoustic modes. Measurements along Λ were repeated with crystals II and III, they confirmed our results ω(LA) > ω(TA), and we have no explanation for the difference. A line width analysis has been performed for the acoustic phonons (crystal III). The one-phonon peak position and halfwidth, as measured by const.-ω technique, have been corrected for resolution and mosaic spread effects of the sample. Results for directions Σ and Λ are shown in Fig. 2. No line broadening has been observed for Σ(TA), Δ(TA) and Δ(LA) modes. Line shapes of phonons with energies > 6 meV could not be analysed either by const.-ω or const.-$\vec{Q}$ technique, in the former method the crossing point dispersion versus resolution is badly defined, whereas in the latter method an energy dependent background due to incoherent scattering of the thulium ions (see Sect. 4.2 and Fig. 3) maskes the coherent peaks.
The measured intensity distribution for longitudinal modes at the point L is shown in Fig. 3. The dashed lines represent the incoherent and coherent elastic contribution (guide to the eye), the incoherent inelastic contribution (computed according to Sect. 4.2), and tne coherent inelastic one-phonon peaks (calculated by convoluting the scattering cross-section with the spectrometer resolution function and

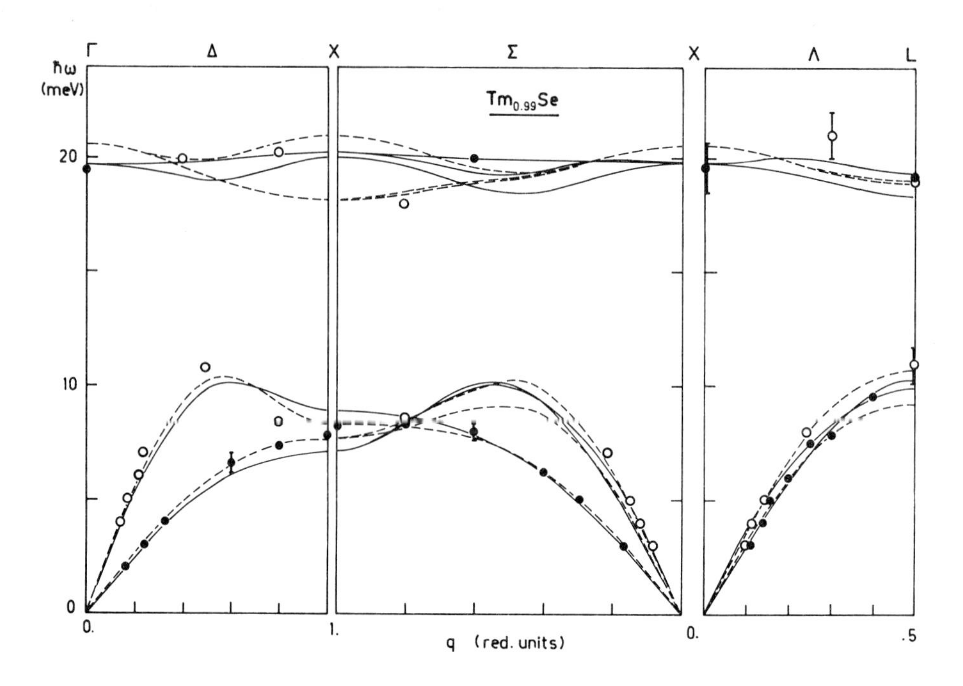

Figure 1: Phonon dispersion in $Tm_{0.99}Se$ at 295 K. • transv. O longitud. ---- Born-von Karman model ---- Screened rigid-ion model.

scaled to the peak maximum of the optic mode), respectively. The LO-peak is considerably broadened, but this effect has been observed for all optic phonons.

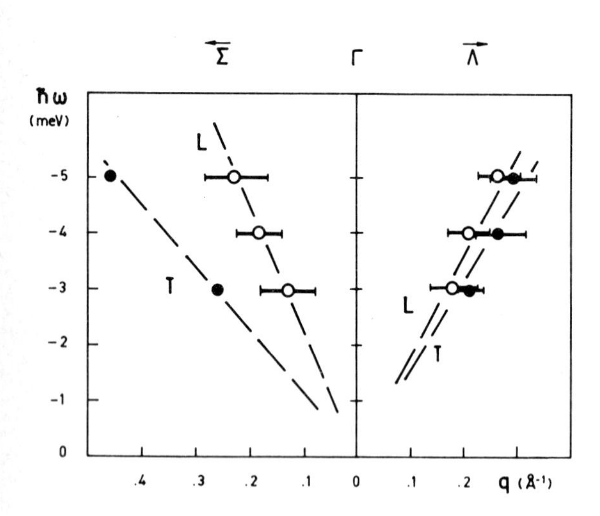

Figure 2: Long wavelength phonon dispersion in $Tm_{0.99}Se$. The bars denote the experimental linewidth (corrected for resolution and mosaic spread.

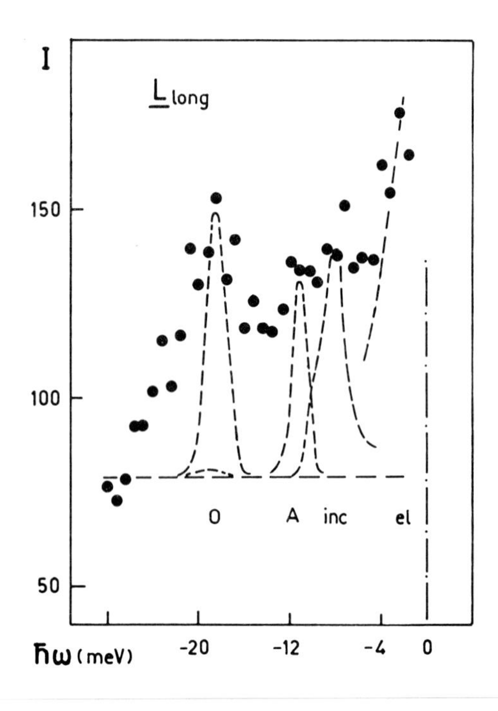

Figure 3: Energy spectrum of neutrons scattered at the L point ($Tm_{0.99}Se$, longitudinal). Dashed lines: Computed coherent and incoherent scattering as described in the text.

3.2. $Tm_{0.87}Se$

The measured dispersion curves are shown in Fig. 4. As compared to $Tm_{0.99}Se$, acoustic branches have slightly higher frequencies, especially those modes where only the Tm-sublattice is involved, i.e. near the zone boundary along Λ , and along the directions Δ and Σ near the middle of the Brillouin-zone; this effect is probably a consequence of the reduced mass due to the Tm-deficiency in the metal sublattice. The energetic separation between the modes Λ(LA) and Λ(TA) is slightly larger than in the stoichiometric compound whereas at the point X the experimental frequencies of both $Tm_{0.99}Se$ and $Tm_{0.87}Se$ agree within experimental error.

4. MODEL CALCULATIONS

4.1. Born- von Karman model and density of states

The experimentally determined frequencies of $Tm_{0.99}Se$ and $Tm_{0.87}Se$ give the necessary information for a computation of the densities of states g(ω) . We used a Born-von Karman model because it provides a convenient set of functions with tensor symmetry to interpolate/extrapolate from experiments in symmetry directions to the whole Brillouin-zone. In addition the Born-von Karman model can be used for integer and intermediate valent Tm_xSe and is not biased by model developments tailored for $Sm_{0.75}Y_{0.25}S$. Least-squares technique has been used to fit 14 force constants (4th neighbour interaction) to the experimental frequencies. The calculated curves are represented by the dashed lines in Figs. 1 and 4. The extrapolation method (6) has been used for the computation of g(ω) and the resulting densities of states are displayed in Fig. 5. The first maxima appear near 8.5 meV (70 cm^{-1}) and are due to the flat acoustic modes around the point X. Both distributions for integer and intermediate valent Tm_xSe show a similar shape; the gap beween the acoustic and the optic modes is much larger and the optic mode band is much narrower than interpreted from Raman experiments (1,3).

4.2. Incoherent Scattering

Phonon frequency and line width measurements were hampered by a strong energy dependent background (Fig. 3), and a possible explanation is incoherent scattering of the thulium ions. The cross section of thulium is not known, however rare earth elements show in general strong incoherent scattering (Er: 7.4 b, Nd 9 b).
We computed the incoherent contribution to the neutron cross section from the eigenvector weighted density of states of the Tm sublattice, multiplied by the population factors and convoluted with the spectrometer resolution function. The result, represented by the broad peak centered near 8.0 meV in Fig. 3, can explain the excess intensity which experimentally is observed in this region. The small peak near 19 meV is due to the thulium contribution in the optic mode region.

4.3. Screened rigid-ion model

The present investigation showed no extraordinary behavior of phonon modes, and because

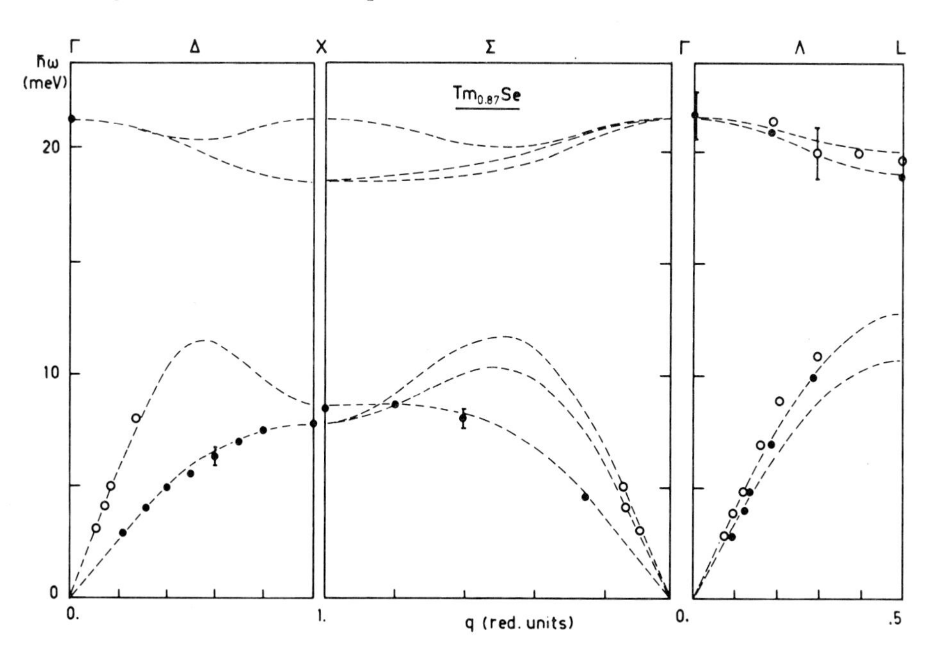

Figure 4: Phonon dispersion in $Tm_{0.87}Se$ at 295 K. • transv. O longitud.
---- Born-von Karman model.

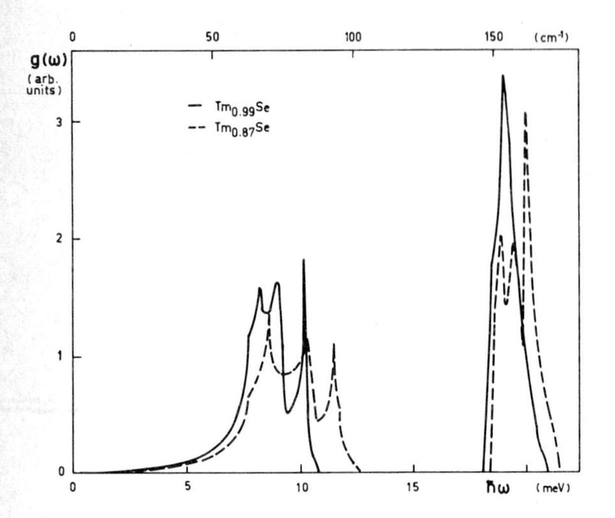

Figure 5: Phonon densities of states of $Tm_{0.87}Se$ and $Tm_{0.99}Se$, computed with the Born-von Karman model.

the amount of data does not warrant the use of very sophisticated models, a simple screened rigid-ion model (7) was tried to analyse the dispersion curves. The result of an 8 parameter fit to $Tm_{0.99}Se$ is shown in Fig. 1 by the solid lines. With the screening distance set to $K_s \cdot a_o = 2$, the effective charge on Tm turned out to be nearly zero. For $Tm_{0.87}Se$ a similar agreement was obtained with slightly different parameters.

5. DISCUSSION

The elastic constant c_{12} of $Tm_{0.99}Se$ determined by static methods and by ultrasonic measurements along Δ and Σ has a negative value. The Cauchy theory of elasticity predicts that the frequency of the longitudinal acoustic mode along Λ should be lower than the transversal one. This is not observed in our neutron data. However in an anharmonic crystal, the elastic constants are frequency dependent: ultrasonic and neutrons probe the first sound and zero sound regime, respectively (8), and at 295 K rather large differences can be expected. Another important point is that the theory of elasticity, mainly due to Cauchy, is not strictly general. The symmetry assumptions for the tensor quantities are not necessarily valid if the crystal is anisotropic and there are noncentral contributions to the force of interaction. The results obtained on $Tm_{0.99}Se$ might support the theory due to Laval (9) in which there are 45 independent elastic constants.

Defect induced first-order Raman-scattering in Tm_xSe showed a strong peak near 60 cm^{-1} (7.4 meV) whose intensity gradually increased with increasing x (1). Different explanations for this peak were given: i) density of states effect of Λ(LA) phonons (1), ii) enhanced Raman intensity due to Γ_1 breathing mode contribution in Λ(LA) (3), and iii) density of states effect due to phonons near X (10).
The computed true density of states of $Tm_{0.99}Se$ and $Tm_{0.87}Se$ call in question all three possibilities because the distributions are very similar and show no peak at 60 cm^{-1}. Inelastic neutron experiments on intermediate valent $Sm_{0.75}Y_{0.25}S$ showed phonon anomalies in the mode Λ(LA) which are presently not observed in $Tm_{0.99}Se$. Both crystals have Rocksalt structure, but SmYS is a mixed crystal with a considerable mass defect in the metal sublattice. The isostructural mixed system $K_xRb_{1-x}Br$ shows strongest linewidth anomalies exactly at the same q-vectors along Λ as SmYS does (11). Therefore we conclude that in SmYS intermediate valence effects are superimposed on a mixed crystal behaviour, and that the lattice dynamics of SmYS can not be considered as typical of an intermediate valent system.
The authors are grateful to E. Kaldis for the preparation and characterization of the crystals.

REFERENCES

(1) Boppart, H., Treindl, A. and Wachter, P., in Falicov, L.M., Hanke, W. and Maple, M.B. (eds.), Valence Fluctuations in Solids (North Holland, Amsterdam 1981) pp. 103-111.

(2) Mook, H.A. and Holtzberg, F., ibid. pp 113-118.

(3) Güntherodt, G., Jayaraman, A., Bilz, H. and Kress, W., ibid. pp. 121-128.

(4) Bührer, W., Bührer, R., Isacson, A., Koch, M. and Thut, R., Nucl. Instr. and Meth. 179 (1981) 259-263.

(5) Furrer, A., Bührer, W. and Wachter, P., Solid State Commun. 40, (1981) 1011-1014.

(6) Gilat, G. and Raubenheimer, L.J., Phys. Rev. 144, (1966) 390-395.

(7) Wakabayashi, N. and Furrer, A., Phys. Rev. B13, (1976) 4343-4347.

(8) Cowley, R.A., Proc. Roy. Soc. 90.(1967) 1127-1141.

(9) Laval, J., C.R. Acad. Sci. (Paris), 238 (1954) 1773-1775.

(10) Celio, M., Monnier, R. and Wachter, P., in Bron, W.E. (ed.), International Conference on Phonon Physics (Journal de Physique 1981) pp. 11-13.

(11) Buyers, W.J.L. and Cowley, R.A., in Proceedings of a Symposium on Inelastic Neutron Scattering (IAEA Vienna 1968) pp. 43-46.

Valence Instabilities
P. Wachter and H. Boppart (eds.)

LATTICE DYNAMICS AND PHASE TRANSITIONS IN INTERMEDIATE VALENCE COMPOUNDS

P.Entel, N.Grewe and M.Sietz

Institut für Theoretische Physik der Universität zu Köln,
Zülpicher Str. 77, D-5000 Köln 41, FRG

A detailed theory of interaction effects between electronic and lattice degrees of freedom for IV-compounds is outlined. It is based on an expansion of the electronic parameters in the periodic Anderson model in terms of ionic displacements and thus is able to aim directly at the pronounced lattice anomalies found in IV-systems. Essential features of phase diagrams and phonon spectra of Ce-, Sm- and Yb-compounds are briefly discussed. In particular, new theoretical results about the lattice dynamics of $Sm_{1-x}Y_xS$, $CePd_3$ and $CeSn_3$ are presented and compared, which allow for a quantitative understanding of measured anomalous phonon spectra. Calculations of the density of vibrational states in Sm(Y)S lead to predictions for the temperature behaviour of the Debye-Waller-factor.

I. THEORETICAL FOUNDATIONS

The existence of strong electron-lattice coupling effects in IV-systems is a fact well established by various experimental results /1/. Among these, particular theoretical attention has been devoted to the phase transition behaviour /2,3,4/, which usually shows large volume changes, and to the pronounced phonon anomalies found in some compounds /5,6,7/. The most direct connection to a microscopic picture of the mixed valent state /8/ uses the Anderson model for a concentrated system,

$$H_{el} = \sum_{\underline{k}\sigma} \epsilon_{\underline{k}\sigma} d^+_{\underline{k}\sigma} d_{\underline{k}\sigma} + \sum_{\nu\sigma}\left(E_\sigma + \frac{U}{2} f^+_{\nu-\sigma} f_{\nu-\sigma}\right) f^+_{\nu\sigma} f_{\nu\sigma} + \frac{1}{\sqrt{N}} \sum_{\nu\underline{k}\sigma} V_{\underline{k}} \left(e^{i\underline{k}\underline{R}_\nu} d^+_{\underline{k}\sigma} f_{\nu\sigma} + h.c.\right), \quad (1)$$

and introduces specific electron-lattice couplings through an expansion of the electronic parameters in the local ionic displacements:

$$E_\sigma = E_{0\sigma} - \sum_n g_1^{(n)} \underline{\xi}_{n\nu} \underline{e}_{\nu n} + \sum_{n,n'} \underline{\xi}_{n\nu}\, \underline{\underline{g}}_1^{(n,n')} \underline{\xi}_{n'\nu} + \ldots ,$$

$$V_{\underline{k}} = V_{0\underline{k}} + \sum_n g_{2\underline{k}}^{(n)} \underline{\xi}_{n\nu} \underline{e}_{\nu n} + \ldots \quad (2)$$

Here E_σ and V are considered at a Rare Earth site ν, $\underline{e}_{\nu n}$ are the unit vectors to the neighbouring ions n, and the relative displacement vectors $\underline{\xi}_{n\nu}$ of these ions are considered as dynamical quantities and expanded like

$$\underline{\xi}_{n\nu} = -\varphi \underline{e}_{\nu n} + (\underline{\xi}_n - \underline{\xi}_\nu), \quad \underline{\xi}_j = \frac{1}{\sqrt{N}} \sum_{q\varkappa} \underline{\eta}_{q\varkappa j} e^{-iq\underline{R}_j}(b^+_{-q\varkappa} + b_{q\varkappa}) \quad (3)$$

into a uniform radial contraction φ of all the cells and a fluctuating distortion due to phonons from different branches $\varkappa$ with corresponding polarization vectors $\underline{\eta}_{q\varkappa j}$.

The general idea behind this kind of description is to concentrate on the particular interaction effects due to charge fluctuations (virtual and real), which are characteristic for the IV state. It is well established that as a consequence of the small valence excitation energy the charge concentrated in the ionic 4f-shell fluctuates and that therefore the mean radii of the valence (5d, 6s)-states and their degree of hybridization with 4f-states of neighbouring ions change considerably. The corresponding couplings, contained in Eq. (2), are an additional feature not present in a "normal" material, and the resulting effects may therefore be calculated using a suitable material in the stable moment regime as a reference system. Insofar, our method is profoundly different from other approaches using the framework of conventional phonon theory /7/. It has to be emphazised that this idea is not restricted to the particular model (1) or any particular method of treating the electronic degrees of freedom. One may as well use an ionic description/9/, which takes the detailed dynamics of 4f-shells more accurately into account including large degeneracies and the infinite U limit, introduce additional Coulomb interactions like electron-hole attraction /10/, or go in any of these models beyond the Hartree-Fock approximation in the presence of the new interactions /11/. A further possible extension, which might be important for the isostructural phase transitions, consists in including a dependence of band parameters and unperturbed phonon energies on the lattice variable φ. Any of these improvements can be incorporated in a straightforward way and its relevance has to be checked in each single case on the basis of experimental information. The interaction constants g also have to be extracted from experiment /8/ or from model calculations. Presumably, they are of the order of characteristic electronic energies like V_k (some tenth of an eV), $g_1^{(n)}$ being positive and $g_{2\underline{k}}^{(n)}$ so that sign changes of the hybridization in the accessible range are possible.

If Eq.(2) and (3) are inserted into the Hamiltonian (1), the f-electrons written in terms of Bloch states, the Hubbard term treated in a Hartree-Fock like manner, and an unperturbed phonon term added, the following model is obtained:

$$H = H_{el}(\varphi) + H_{ph}(\varphi) + H_{el-ph} \quad ,$$

$$H_{el-ph} = -\frac{1}{\sqrt{N}} \sum_{\underline{k}\sigma q\varkappa} [T_{\underline{k}q\varkappa} \tilde{g}_1(\varphi) f^+_{\underline{k}\sigma} f_{\underline{k}-q\sigma} - \tilde{g}_2(\varphi)(T'_{\underline{k}q\varkappa} f^+_{\underline{k}\sigma} d_{\underline{k}-q\sigma} + h.c.)] + \text{higher order terms.} \quad (4)$$

The pure electronic part (1) and the pure lattice Hamiltonian, which includes a static elastic potential, now bear a dependence on the average lattice constant in their parameters via the compression φ. For example:

$$\tilde{E}_\sigma(\varphi) = E_{o\sigma} + U\langle f^+_{\nu-\sigma} f_{\nu-\sigma}\rangle - \sum_n g_1^{(n)} \varphi + \ldots \quad . (5)$$

Likewise, the couplings g_1 and g_2 have acquired a φ-dependency. In Eq.(4) all the important information about the interactions of particular charge density- and lattice-waves are contained in the form factors $T^{(i)}_{\underline{k}q\varkappa}$, which reduce to simple functions of $q,\varkappa$ in case of a dispersionless hybridization $V_{\underline{k}} = V$. They determine how a specific mode $q,\varkappa$ takes part in the relative displacements, Eq.(3), governing the change in electronic parameters via Eq.'s (2). The second essential ingredience in the interaction (4) is the electronic band structure. In the frame of the Hartree-Fock treatment it leads to an essential simplification if the electronic part H_{el} is diagonalized by a canonical transformation. This leads to the typical picture of a broad band hybridized with a dispersionless level exhibiting a (indirect) gap and corresponding flat band pieces at its edges with a strong f-character. The interaction H_{el-ph}, formulated with the new band states, then describes scattering due to phonons, which may become very strong if the Fermi-level lies near the gap and certain resonance conditions are met. This effect is governed by certain phase space factors derived from the new band structure /6/.

On the basis of the Hamiltonian (4) phase transition properties as well as phonon anomalies can be discussed within a unified picture. Equilibrium properties are given through the appropriate thermodynamic potential

$$\mathcal{G} = \langle H\rangle - TS + Np v \quad , \quad (6)$$

which after a mean field like elimination of the phonon coordinates leading to explicitely temperature-dependent effective quantities $E(\varphi,T)$, $V(\varphi,T)$ becomes a function of pressure p, temperature T and lattice constant φ. Variation with respect to φ determines the true equilibrium potential $\mathcal{G}(p,T)$ which contains all information about phase transitions. For a determination of the lattice vibration spectrum one studies the phonon propagator, whose self-energy can be calculated in various degrees of sophistication /6/. Using only the simplest bubble-diagram gives the following renormalization and widths for the phonon-frequencies $(2V_o > \hbar\Omega_{q\varkappa})$:

$$(\hbar\omega_{q\varkappa})^2 = \frac{1}{2}\{(\hbar\Omega_{q\varkappa})^2 + 4V_o^2 - [((\hbar\Omega_{q\varkappa})^2 - 4V_o^2)^2 + 8V_o\hbar\Omega_{q\varkappa} F(\chi_q \tilde{g}_1^2 |T_{q\varkappa}|^2 + \ldots)]^{1/2}\} \quad ,$$

$$\Gamma_{q\varkappa} = \frac{\pi}{4}(\chi'_q \tilde{g}_1^2 |T_{q\varkappa}|^2 + \ldots) \quad , \quad (7)$$

where the quantities F and χ_q, χ'_q incorporate occupation statistics and phase space volume. Additionally, there appears a nearly dispersionless mode with strong electronic weight, which is connected with phonon induced (indirect) transitions over the band gap.

II. LATTICE EFFECTS IN IV-COMPOUNDS

There are essentially two types of experimental phenomena, which show a clear participation of lattice degrees of freedom in the IV-state, namely static measurements of the lattice constant as a function of temperature and pressure or alloying concentration and scattering experiments with light or neutrons. An interesting example for the former type is the phase diagram of Sm(Y)S which shows a first order boundary with a high temperature critical point between a low pressure, semiconducting and essentially integral valent phase and a high pressure $(p \gtrsim 7\,\text{kbar})$ metallic IV-phase. Of particular interest is the shape of this boundary in the pressure-temperature plane. Whereas $dT/dp < 0$ at low temperatures, this line bends back at about $T \approx 100K$ and has a positive slope $dT/dp > 0$ until the critical point is reached /12/. Since $dp/dT = \Delta S/\Delta V|_\perp$ (Clausius-Clapeyron) and ΔV definitely has the opposite sign of Δp across the phase boundary, this slope is directly related to a change ΔS in entropy when entering the IV-phase. There exist some general arguments regarding different contributions to this change in entropy which should properly be incorporated in any good theory. At low temperatures certainly the electronic contributions are dominant, of which there are essentially two. One of them is connected with the increase in band electron entropy $(\sim T)$, when going from an insulating to a metallic state. The other comes from the fact that the ionic multiplet which is additionally admixed in the IV-phase may be more or less degenerate than the ionic groundstate of the phase with stable valency. In the first case, which occurs for Sm, both effects together with $\Delta V < 0$ cause a negative slope of $T_c(p)$, which (absolutely) decreases for increasing T. In the case of Ce-metal for example the ionic mixing entropy favours a positive slope as actually observed. In the calculation outlined above based on the periodic Anderson-model, this different behaviour can be simulated by making use of the electron hole symmetry $E \rightarrow E+U$, which reverses the degeneracy of the state with one electron less relative to the other /4/. For higher temperatures the lattice contribution to the entropy is certainly important which increases (decreases) over the phase boundary when a proper average of the phonon density of states becomes higher (lower). This can give, together with band structure effects, a positive slope of $T_c(p)$ for SmS at higher temperatures /13/. The theory also predicts the possibility of an interesting structure in this phase boundary, including more than one critical points for a general situation /4/.

The next topic which our theory can make a contribution to, concerns the mechanism of the phase transition itself. There exist controversial points of view about the interactions taking part, and in particular it has been questioned if an electronic mechanism alone can account for the first order nature of the transition /14/. In the present model this comes about by a combination of bandstructure changes and the stabilizing effect of the lattice potential when applying pressure or temperature changes to a system with nearly degenerate ionic configurations. Of special importance seems to be the oscillating behaviour of hybridization matrix elements between nearest neighbours with changes φ of the lattice constant. With this properly taken into account rather realistic theoretical results have been obtained for the phase transition behaviour of IV-compounds /8/. This also includes quantities like the coefficient of thermal expansion, which shows quite an interesting dependence on temperature on both sides of the phase boundary /4/. The behaviour of the lattice constant thus obtained also constitutes an important input for the calculation of other elastic quantities like the phonon spectrum or the velocities of sound, as is obvious from the φ-dependence in Eq.(4).

A calculation of the phonon spectrum of IV-compounds with the method outlined consists of two seperate steps. At first, an unrenormalized spectrum is calculated with a Born-von Karman model involving force constants for nearest and next nearest neighbours. This constitutes a good approximation for an integral valent reference substance, since the band structure of the compound in the situation considered is metallic and Coulomb forces are screened. At this stage one needs to fit the force constants at some fixed temperature by relating to some characteristic wave vectors in a measured phonon spectrum. In a second step, one adds the renormalizations of phonon frequencies, which are due to the specific IV-character, according to Eq.(7) and considers the widths $\Gamma_{q\kappa}$. Now an additional characteristic temperature dependence comes in via the behaviour of coupling constants and band structure under changes of φ, which can be derived from the detailed information about the phase diagram discussed before. This method gives for example a quantitative explanation of neutron scattering data on $Sm_{1-x}Y_xS$, and further applications will be discussed in the next section.

There are two essential features which determine if a particular IV-material can exhibit pronounced phonon anomalies. One of these concerns the time scales, on which charge relaxation processes in the 4f-shells and the propagation of lattice waves take place. If both scales are similar, a cooperative motion can come about promoting frequency-shifts. Relaxation rates for a realistic model of various 4f-shell-configurations have recently been calculated /15/ showing that such a coincidence is likely in several IV-compounds. Within our model the corresponding resonance condition involves the average width of the (indirect) gap, the effective coupling constants $gT_{q\kappa}$ measured in units of the unperturbed frequencies Ω, and a phase space factor /6/. This condition is also influenced considerably by the second characteristic feature, namely the steric proportions of the crystal structure under consideration, which enter via the form factor $T_{q\kappa}$. According to the definition

$$T_{q\kappa} = \sum_{\substack{\text{neighbours } n \\ \text{to RE-site } \nu}} \underline{\eta}_{q\kappa n}\, \underline{e}_{\nu n}\, e^{-iq\underline{R}_n} \quad , \qquad (8)$$

the form factors can be calculated using the polarization vectors $\underline{\eta}_{q\kappa n}$ of the unperturbed phonon modes and the normalized vectors $\underline{e}_{\nu n}$ given by the structure of the unit cell. One can obtain a rough picture for the functional dependence of $T_{q\kappa}$ on the wave vector for particular branches κ in the main symmetry directions of the crystal by considering the corresponding changes in volume around the RE ions. In the case of Sm(Y)S with a NaCl-structure these are maximal for zone boundary LO phonons and mid-zone LA phonons in [111]-direction, where the strongest experimentally observed anomalies are situated /16/. In the next section the case of Ce-compounds is discussed, where the steric conditions lead to different predictions.

III. PHONON ANOMALIES IN Ce- AND Sm-COMPOUNDS

An interesting set of quantities derived from the phonon spectrum, which can give direct information about the phase transition behaviour of an IV-compound, are the mean square displacements $\langle u_n^2\rangle$ of different ions n in the unit cell. It has been predicted that in case of SmS the S-ions should display considerable fluctuations in their position, which are maximal near the isostructural first order transition line due to charge fluctuations on neighbouring Sm-sites /5/. We have begun with calculations of $\langle u_n^2\rangle$ and the corresponding Debye-Waller-factor $\exp(-2W)$ on the basis of the renormalized phonon spectrum for Sm(Y)S /17/. Preliminary results are shown in Fig.1. Although these first calculations concentrate on the IV-phase far from the phase transition region a considerable enhancement of the displacements due to phonon softening can be recognized.

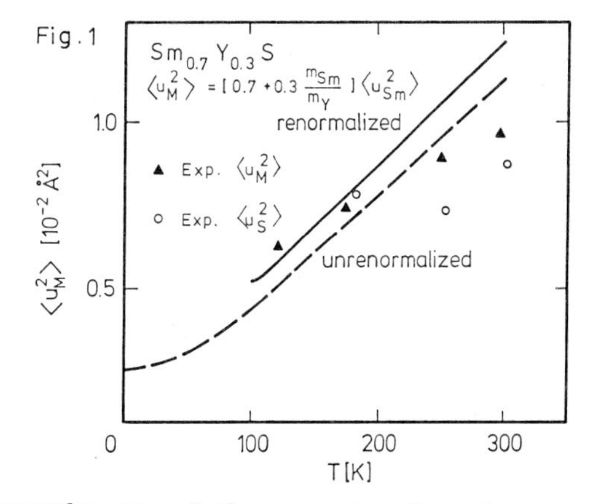

Recently, two of the present authors have reported on current calculations of the phonon spectra of $CeSn_3$ and $CePd_3$ /18/. In contrast to Sm(Y)S these two compounds crystallize in the Cu_3Au-structure suggesting a different interaction of valence fluctuations and lattice modes. The first point to note is that there is more room for the Ce-ions in these compounds than for Sm in the cholcogenide. Therefore the volume effect should be less pronounced, also implying that the change of valence with temperature is less significant. Secondly, the most drastic deformations of the Sn- or Pd-cage around each Ce-ion is not obtained at the same places of the Brilloin-zone as in the NaCl-structure. In the Cu_3Au-structure these are to be expected for LO phonons halfway between the points Γ and R in [111] -direction, where a vibration pattern similar to that at the L-point in Sm(Y)S is found, and at the zone boundary (X-point) in [100] -direction, and for LA phonons generally near the center of the Brilloin-zone in all main symmetry directions /18/. This is directly reflected in the form factors for $CePd_3$ shown in Fig.2.

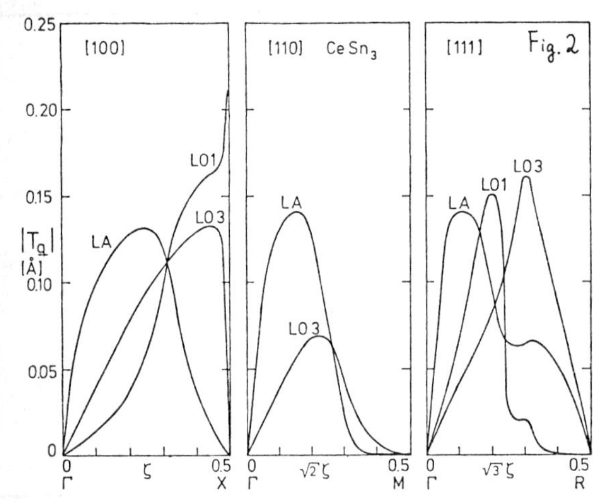

The corresponding phonon dispersions are displayed in Fig.3a. The full lines represent the result of a Born-von-Karman force constant model for the integral valent reference system, which is fitted to optical frequencies obtained with infrared spectroscopy /19/. The dashed lines are renormalized spectra calculated with two different values of the electron phonon coupling constant, g_1 = .05 eV/Å (thin line) and g_1 = .1 eV/Å (thick line)

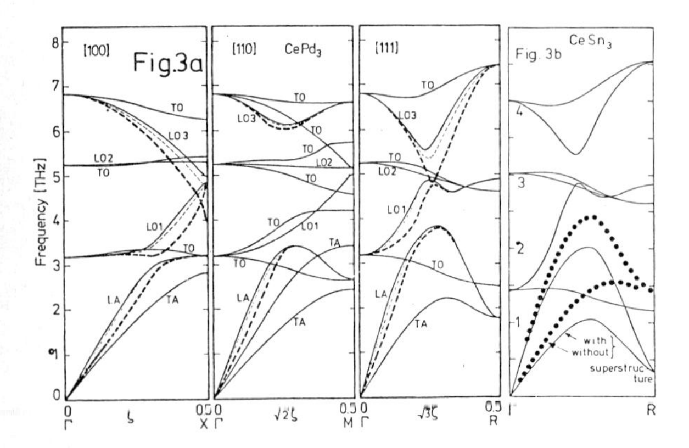

No measurements of the phonon dispersions of $CePd_3$ have been reported so far. The results are qualitatively similar to those of $CeSn_3$, where however the renormalizations come out to be very small due to the near integral valency of about 3.1 (or even smaller), which in turn may be connected with the weakness of the volume effect mentioned above. Since the lattice constant of $CePd_3$ is considerably smaller (4.13 Å compared to 4.77 Å) and the valency much more intermediate (around 3.4) we expect that phonon anomalies in $CePd_3$ might be stronger than in $CeSn_3$ and suggest that corresponding measurements are carried out.

As a last point in this discussion we want to comment on an experimental discrepancy found for the LA and TA modes at the R-point in $CeSn_3$ for which considerably different frequencies have been reported /20,21/. It has been proposed that this crystal is likely to form superstructures and that these might have a significant influence on such zone boundary phonons /22/. We have performed a model calculation by simulating such a superstructure via a substitutional disorder at Sn-places in the unit cell. As a preliminary result a lowering of LA(R) and TA(R) phonons with increasing substitution is found which can be strong enough to explain the differing experimental results, see Fig.3b and compare with Ref./18/.

ACKNOWLEDGEMENTS

The authors are grateful to W.Wiethege whose advice in numerical computations was of considerable help. They also acknowledge a useful collaboration with K.Kowalski.

References:

/1/ Many of the relevant experimental publications are listed in the following two review articles: J.M. Robinson: Physics Reports 51, 1 (1979); N. Grewe, H.J. Leder and P. Entel, in: Festkörperprobleme XX, J. Treusch, ed., (Vieweg, Braunschweig, 1980)

/2/ L.L.Hirst, J. Phys. Chem. Solids 35, 1285 (1975)

/3/ J.H. Jefferson, J. Phys. C 9, 269 (1976)

/4/ P. Entel and N. Grewe, Z. Physik B 34, 229 (1979)

/5/ S. K. Ghatak and K.H. Bennemann, J. Phys. F 8, 57 (1978)

/6/ N. Grewe, P. Entel and H.J. Leder, Z. Physik, B 30, 393 (1978)

/7/ H . Bilz, G. Güntherodt, W. Kleppmann and W. Kress, Phys. Rev. Lett. 43, 1998 (1978)

/8/ N. Grewe and P. Entel, Z. Physik B 33, 331 (1979)

/9/ N. Grewe and H. Keiter, Phys. Rev. B 24, 4420 (1981)

/10/ R. Ramirez, L.M. Falicov and J.C. Kimball, Phys. Rev. B 2, 3383 (1970)

/11/ H. Keiter and N. Grewe, in: Valence Fluctuations in Solids, L.M. Falicov, W. Hanke and M.B. Maple eds. (North-Holland, Amsterdam, 1981)

/12/ D .B. McWhan, S.M. Shapiro, J. Eckert, H.A. Mook and R.J. Birgenau, Phys. Rev. B. 18, 3623 (1978)

/13/ P. Entel and H.J. Leder, Z. Physik B 32, 93 (1978)

/14/ For a discussion of this point see the second review in Ref. /1/

/15/ Y . Kuramoto and E. Müller-Hartmann, in: Valence Fluctuations in Solids, L.M. Falicov, W. Hanke and M.B. Maple eds. (North-Holland, Amsterdam, 1981)

/16/ H.A. Mook, R.M. Nicklow, T. Penney, F. Holtzberg and M.W. Shafer, Phys. Rev. B 18, 2925 (1978)

/17/ P. Entel, N. Grewe, M. Sietz and K. Kowalski, Phys. Rev. Lett. 43, 2002 (1979)

/18/ P . Entel and M. Sietz, Solid State Commun. 39, 249 (1981)

/19/ B. Hillebrands, G. Güntherodt, G. Pott, R. König and A. Breitschwerdt, to be published

/20/ I . Pintschovius, E. Holland-Moritz, D. Wohlleben S. Stähr and J. Liebertz, Solid State Commun. 34, 953 (1980)

/21/ C. Stassis, C.-K. Loong, J. Zarestky, O.D. McMasters and R.M. Nicklow, Solid State Commun. 36, 667 (1980)

/22/ D. Wohlleben, private communication

VALENCE CHANGES

Valence Instabilities
P. Wachter and H. Boppart (eds.)

NEW DATA ON THE SEMICONDUCTOR TO METAL PHASE TRANSITION IN SAMARIUM MONOSULPHIDE AND ITS SOLID SOLUTIONS

I.A.Smirnov[+], I.P.Akimchenko[++], T.T.Dedegkaev[+], A.V.Golubkov[+], E.V.Goncharova[+], N.N.Efremova[+++], T.B.Zhukova[+], V.S. Oskotskii[+], L.D.Finkelstein[+++], S.G.Shulman[+], N.Stepanov[+], N.F.Kartenko[+]

+ A.F.Ioffe Physical-Technical Institute Academy of Sciences of the USSR, Leningrad, I9402I, USSR.

++ P.N.Lebedev Physical Institute Academy of Sciences of the USSR, Moscow, II7924, USSR.

+++ Physics Metals Institute Academy of Sciences of the USSR, Sverdlovsk, 6202I9, USSR.

New data are given on the electrical, galvanomagnetic and X-Ray properties of some new SmS-based solid solutions (SmSBSS). Preliminary results are given on the semiconductor-to-metal phase transition (SMPT) in SmS thin films at the implantation of phosphor ions.

I. INTRODUCTION

During last years the model of the SMPT in $Sm_{I-x}Ln_x^{+3}S$ and $Sm_{I-x}Ln_x^{+2}(M_x^{+2})S$ (Ln-rare earth, M-metal) has got complicated. It is the aim of this work to obtain some new experimental and theoretical data concerning the influence of I) a chemical compression (CC) of the Sm-sublattice due to the substitution of Sm^{+2} by ions with smaller dimensions (P_{CC}), 2) a Sm dilution (SD) and 3) a pressure of the free carriers (PFC) - P_{PFC} on SMPT in SmSBSS. It must be emphasized that SMPT is essentially influenced by the hybridization between nearest Sm-ions depending on all three factors discussed above (CC, PFC and SD). First two factors are closely connected one with another and with the doped cation radius. First of them "squeezes out" extra electron out of the Sm^{+2}- ion and facilitates SMPT, second one diminishes hybridization and prevents SMPT.

For the investigation we choose two systems: $Sm_{I-x}Pr_xS_{I-x}Se_x$ (SmPrSSe) and $Sm_{0.8}Gd_{0.2}S_{I-y}Se_y$ (SmGdSSe).

In the first system a) the CC is negligible (a_{SmS}= 5.97Å , a_{PrSe}=5.95Å) b) the free carrier concentration (n) increases with Pr^{+3}Se doping (n_{PrSe}= I.9 10^{22} cm^{-3}). Thus we expected to find out in SmPrSSe the pure effect of the PFC influence on the SMPT.

The second system is prepared on the base of the metal composition $Sm_{0.8}Gd_{0.2}S$ of the $Sm_{I-x}Gd_xS$. In SmGdSSe with increasing y a) the lattice constant (a) increases (the CC of Sm-sublattice diminishes); b) the degree of the SD remains constant; c) n varies insignificantly (the number of Gd ions keeps constant). In this solid solution we have hoped to separate the influence of the SD and CC on SMPT.

2. EXPERIMENTAL RESULTS

2.I Materials and experiments

The SmPrSSe (x=0 ÷ I) and SmGdSSe (y=0 ÷ 0.3) polycrystals used in this investigation were prepared by the method [I] .
We measured at 300K (R.T.) : a (with the DRON-2, Cu K_α), the electrical resistivity (ρ), the Hall constant (R), the thermoelectric power (α) , ρ under hydrostatic pressure (P) up to I0 kbar, the Sm-ions valency (V_{eff}) (by the L_{III} abrorbtion spectroscopy method)
In SmPrSSe ρ at 4.2 ÷ 300K also was measured.

2.2 SmPrSSe

In the SmPrSSe there is a continuous row of solid solutions with the NaCl-structure (Fig.I). In this system there is no discontinuous SMPT (Fig. I ÷ 4) but there is a continuous SMPT at x=0.2 ÷ 0.3 (see Table I)

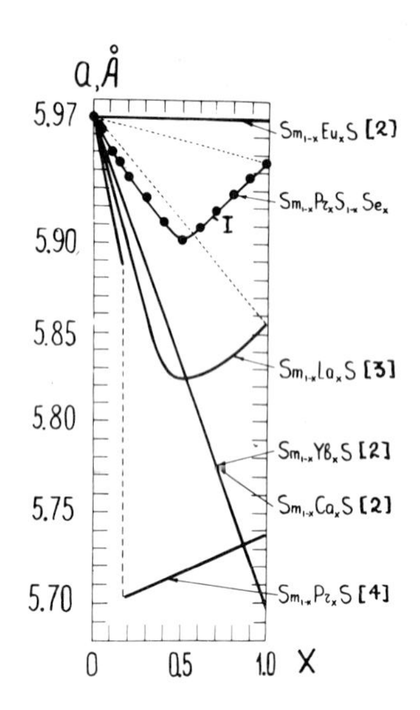

Figure I : a vs. x for SmSBSS at R.T.
I -our data.

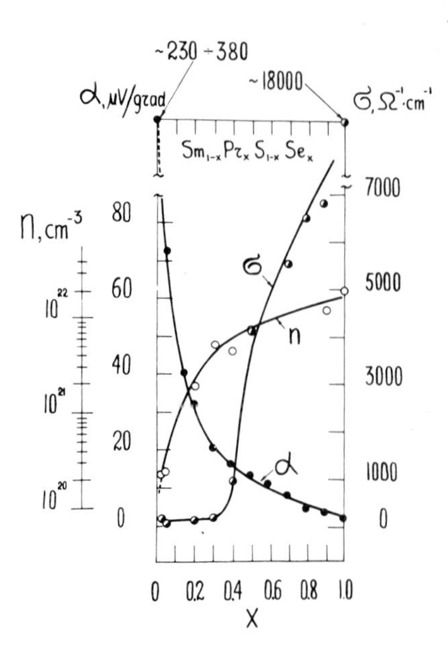

Figure 2: α , n , the electrical conductivity σ (σ = I/ρ) vs. x for SmPrSSe.

Table I.

Physical properties	
$0 \leqslant x \leqslant 0.2 \div 0.3$	$0.2 \div 0.3 \leqslant x \leqslant I$
I) The samples are black.	I) The samples are crimson.
2) The colour changes from black to golden by polishing.	2) The colour does not changes by polishing.
3) ρ(T) has a semiconducting behaviour (Fig.3)	3) ρ(T) has a metallic behaviour (Fig.3).
4) The effective valency of Sm increases from ~+2 to +2.I (Fig.4).	4) The effective valency of Sm is constant (~+2.I) (Fig.4).
5) The electron mobility decreases sharply (from 20 ÷ 80 to 0.03 $cm^2/V\cdot s$.	5) The electron mobility increases from 0.03 to 4 $cm^2/V\cdot s$.
6) σ changes weakly with x (Fig.2).	6) σ increases sharply with x.
7) The discontinuous SMPT is observed under hydrostatic pressure. P_{cr} is 8, 7 and 8.4 kbar for x=0.05, 0.I5 and 0.2 correspondingly (according to the data of ρ vs.P) (Fig.5, 6).	7) The continuous changing of the ρ is observed under hydrostatic pressure (up to I0 kbar).

2.3 SmGdSSe.

Fig.7a represents literature data for the a and Fig.7b gives our data for V_{eff} of Sm in $Sm_{I-x}Gd_xS$. We have plotted (by special choice of the y-scale) in this figures a and V_{eff} for the SmGdSSe. The $Sm_{0.8}Gd_{0.2}S$ is taken for y=0. Some interesting experimental facts must be noted here. It is possible to realize only by increasing the a of the metallic $Sm_{0.8}Gd_{0.2}S$ substituting S for Se (with the constant SD): a) the metal-semiconductor phase transition, b) the nearly exact repetition of the a(x) and $V_{eff}(x)$ dependences of the $Sm_{I-x}Gd_xS$ (for $x < x_{cr}$) An increase of a results in decreasing P_{CC} and diminishing a hybridization between nearest Sm-ions, what makes the total pressure $P_{CC}+P_{PTC}$ insufficient for SMPT.
We are measured ρ vs. P for two compositions with y=0.I and 0.3. In the compounds with y=0.I the pressure-induced phase transition is observed at P_{cr}= 4.8 kbar. In the case of y=0.3 there is no phase transition in the investigated pressure range. Thus SmGdSSe is not exactly analogous to

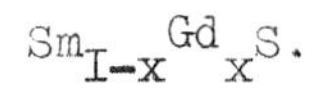
$Sm_{I-x}Gd_xS$.

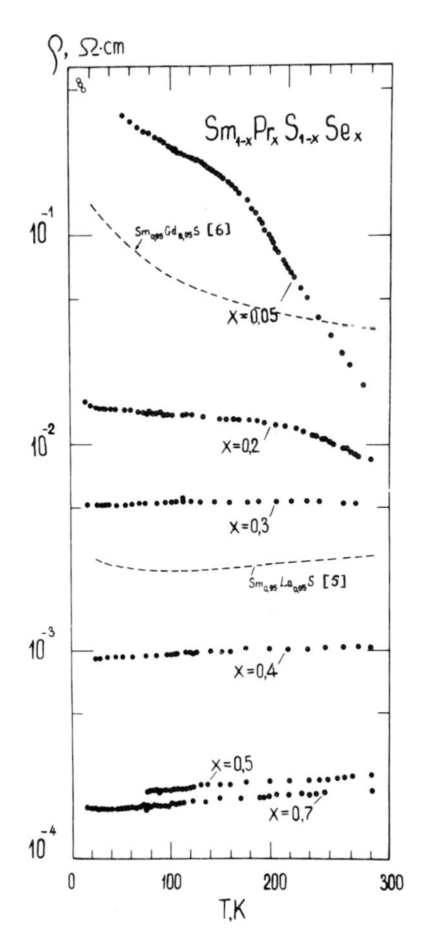

Figure 3: ρ vs. T for SmPrSSe.

3. INTERPRETATION OF RESULTS

Let us examine the inset on the SMPT of the SmSBSS from CC,PFC and SD,more entering into details of PFC (since this effect has not been discussed yet in the literature).

In $Sm_{I-x}Ln^{+3}_xS$ (in SmPrSSe as well) there is a great amount of free electrons producing an additional pressure on Sm-sublattice due to a Coulomb energy gain (P_{PFC}). This pressure can cause a valency change of Sm-ions. To evaluate a P_{PFC} let us consider that this solutions consist of : a) a cation sublattice with the average charge 2+x ; b) uniformly distributed free electrons with the average charge x per cell; c) an anion sublattice with charge -2. Ewald-type calculations give an energy gain:

$$U= - e^2/a\ (I0x + 4x^2\ ) \quad ......(I)$$

to the ionic two-charge lattice energy (for example SmS). In this case an additional pressure is

$$P_{Sm-S}= -\frac{\partial U}{\partial V} = \frac{4}{3}\frac{e^2}{a^4}(I0x + 4x^2\ ) =$$

$$= 237\ x\ (\ I + 0.4x\ )\ \text{kbar} \quad .... (2)$$

P_{Sm-S} compresses the bonds Sm-S in the $Sm_{I-x}Ln^{+3}_xS$. In regard to the metal crystal $Ln^{+3}S$ this pressure is

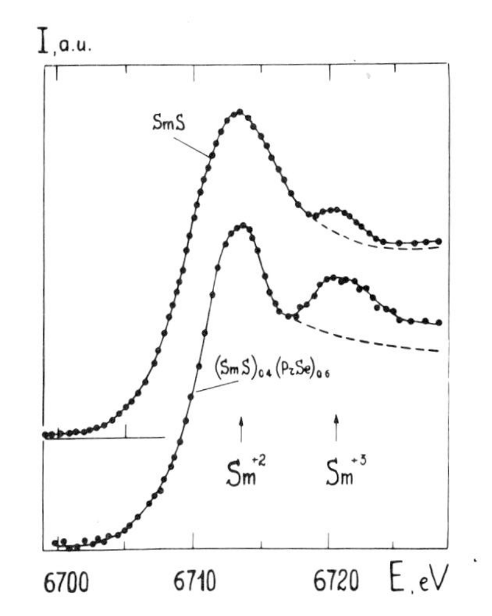

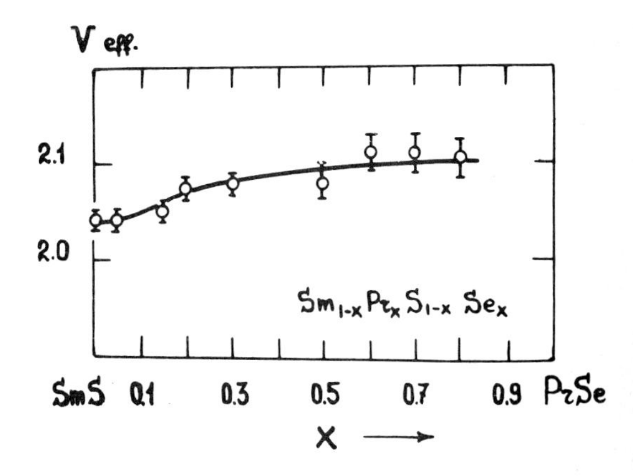

Figure 4: The effective Sm-valency in SmPrSSe calculated from L_{III} -spectra a) for x=0.6 ; b) x=0 ÷ I

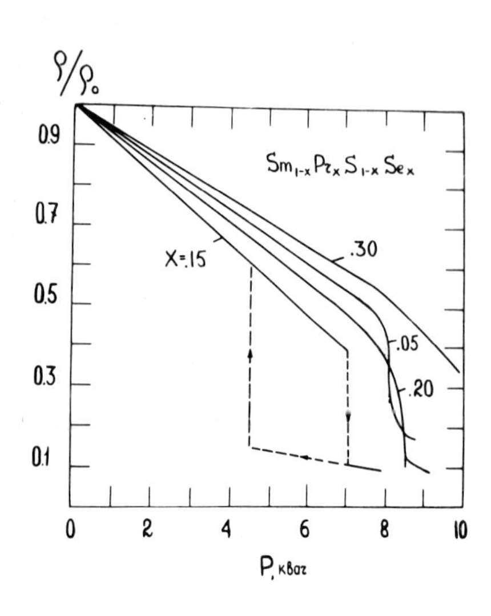

Figure 5: ρ/ρ_o vs. hydrostatic pressure P for some compositions of SmPrSSe at R.T.

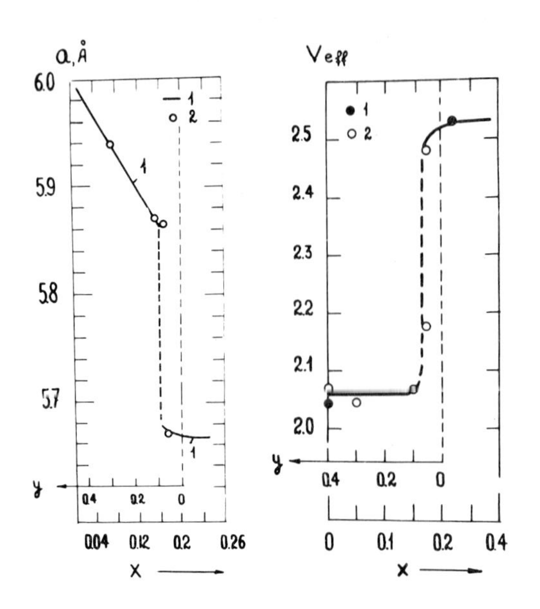

Figure 7: a (Fig.7a) and V_{eff}(Fig. 7b) vs. x (or y) for $Sm_{I-x}Gd_xS$ (I) and SmGdSSe (2) at R.T. Fig.7a: solid curve I - data from [6].

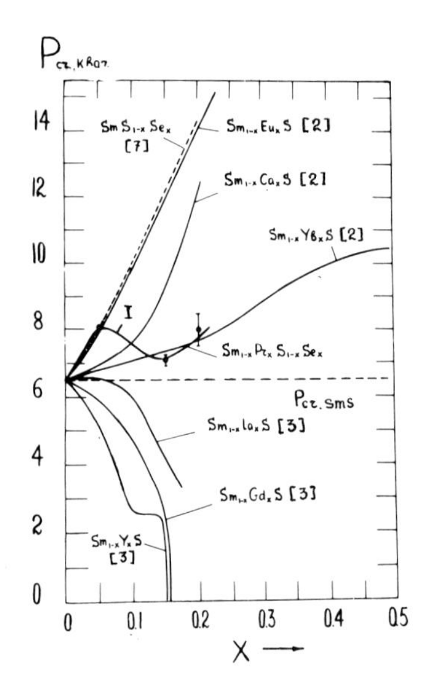

Figure 6: P_{cr} vs. x for SmSBSS at R.T. I - our data (P_{cr}- critical pressure corresponding to discontinuous SMPT).

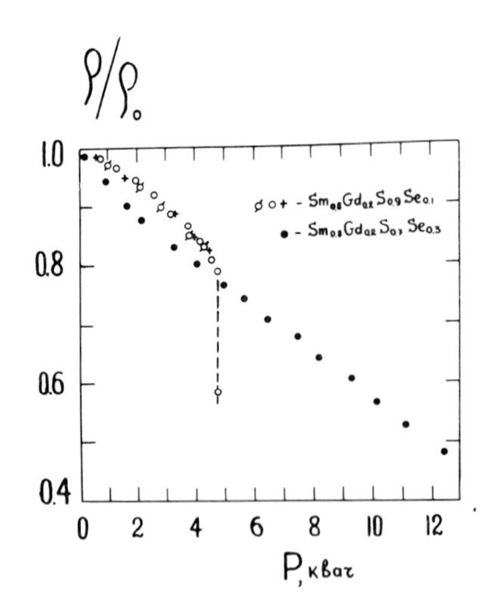

Figure 8 : ρ/ρ_o vs. P for SmGdSSe at R.T.

negative owing to decreasing the average electron charge per cation in a solid solution. It stretches the bonds Ln^{+3}-S in $Sm_{I-x}Ln_x^{+3}S$ and is equal to

$$P_{Ln^{+3}-S} = -237(I-x)\left[I+0.4(I-x)\right] = -237(I-x)(I.4-0.4x) \text{ kbar} \quad ...(3)$$

Eq.(3) is derived from Eq.(2) by substituting (I-x) for x and changing the sign. Under P_{Sm-S} and $P_{Ln^{+3}-S}$ a of $Sm_{I-x}Ln_x^{+3}S$ deviates from the Vegard-law. We take this deviation to be an average change of the a of the solid solution components:

$$\Delta = -a_{SmS}\beta_{SmS}P_{SmS}(I-x) - a_{Ln^{+3}S}\beta_{Ln^{+3}S}P_{Ln^{+3}S}\cdot x \quad ...(4)$$

β - linear compressibility.
If the a_{SmS} and $a_{Ln^{+3}S}$ are nearly equal (as in the case of SmPrSSe), they can be put outside the brackets. Then

$$\Delta = -a\,x(I-x)\cdot 237\left[\beta_{SmS}(I+0.4) - \beta_{Ln^{+3}S}(I.4-0.4x)\right] \quad .....(5)$$

Usually $\beta_{Ln^{+3}S} \approx 0.5\beta_{SmS}$, so

$$\Delta = -237a\,x(I-x)\beta_{SmS}(0.3+0.6x) \quad ..(6)$$

This decrease in the parameter Δ corresponds to the average pressure

$$P_{av} = \Delta/\bar{\beta}\cdot a = 237(I-x)\beta_{SmS}/\bar{\beta}\cdot(0.3+0.6x) \quad ..... (7)$$

where $\bar{\beta} = \beta_{SmS}(I-x) + \beta_{Ln^{+3}S}\cdot x$

P_{av} is essentially smaller then P_{Sm-S} which can reach according to Eq.(2) ~I00 kbar.
P_{Sm-S} compresses Sm-S bonds and can "squeeze out" "extra" electrons from Sm. This pressure appears to be the cause of the intermediate valency of Sm and of an increase of Sm^{+3} concentration in the $Sm_{I-x}Ln_x^{+3}S$. P_{av} is connected with the average distance between the nearest Sm-ions and so influences the hybridization of their electron states and decreases the P_{cr} for pressure-induced SMPT in $Sm_{I-x}Ln^{+3}S$. According to Eq.(7) at small x :

$$P_{av} = 70x\cdot(I+I.5x-I.25x^2) \text{ kbar} \quad ..(8)$$

To elucidate the role of the separate factors (CC,PFC and DS) in SMPT let us examine the dependence of P_{cr} on the x (Fig.6). Fig.6 represents all known data on P_{cr} for various SmSBSS ($P_{cr} < P_{cr\ SmS}$ for $Sm_{I-x}Ln_x^{+3}S$ and $P_{cr} > P_{cr\ SmS}$ for $Sm_{I-x}Ln_x^{+2}(M_x^{+2})S$; $P_{cr\ SmS} = 6.5$ kbar).
In the case of a dilution SmS with divalent cations P_{cr} increases monotonically, that is the effect of the chemical dilution Sm, for example by Ca^{+2} and Yb^{+2}, prevails the effect of a chemical compression. In the case of trivalent cations with smaller radius the compression by the free carriers adds to P_{CC} and decreases P_{cr} by P_{av} according Eq.(8). One can see the role of separate factors in SMPT more clear from the Fig.9 (P_{cr} vs. a of the second solution components : $Ln^{+3}S$, $Ln^{+2}S$, $M^{+2}S$ at x=0.I25).

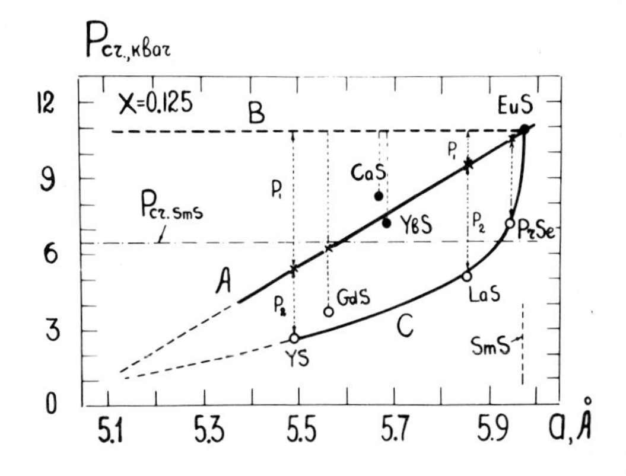

Figure 9: P_{cr} vs. a for SmSBSS at R.T. (x= 0.I25)

For divalent cations Eu, Ca, Yb P_{cr} decreases with decreasing a. Let us extrapolate this dependence by the straigt line. For EuS (a is nearly the same as for SmS) P_{cr} is determined only by dilution (CC and PFC are absent) The horizontal line B ($P_{cr} \approx$ II kbar) describes the influence of dilution on SMPT at x=0.I25. Difference between B and A (P_I) regards to decreasing P_{cr} due to CC. P_{cr} -values for solutions with trivalent cations lie on the curve C . Difference between A and C

(P_2) regards to decreasing of P_{cr} due to additional P_{PFC} (and must be equal P_{av}, Eq.(8)). According to Eq.(8) P_{av} (x=0.I25) $\approx$ I0 kbar. According Fig.9 $P_2=P_{PFC} \approx 3 \div 4$ kbar. This agreement is good enough for such a rough model. As can be seen on Fig.9 P_{PFC} decreases with decreasing the cation radius. This effect can be explained by increasing the localization degree of conducting d-electrons resulting in decreasing the effective density of free electrons. The collapse does not take place for SmPrSSe at any x because a of PrSe is too large and therefore the P_{CC} is too small for the hybridization of the nearest Sm-ions. The additional pressure P_{PFC} is not sufficient to SMPT. Thus there is some critical pressure P^o_{cr} which is necessary for the hybridization between nearest Sm-ions sufficient for appearance of SMPT.

If total pressure $\bar{P} = P_{CC}+P_{PFC} < P^o_{cr}$, then SMPT does not take place. But this pressure may be sufficient for appearence in SmSBSS some concentration of local Sm^{+3}-ions. This does take place in SmPrSSe and, probably, in $Sm_{I-x}La_xS$. Moreover the higher is $\bar{P}$ the higher Sm^{3+} concentration (in SmPrSSe $\bar{P} \simeq P_{PFC}$ and $V_{eff} \sim 2.I$, in $Sm_{I-x}La_xS$ $\bar{P} = P_{CC}+P_{PFC}$ and $V_{eff} \sim 2.3$)

Thus one can speak about two types of phase transitions in the SmSBSS:

I) The discontinuous SMPT are observed in the most $Sm_{I-x}Ln^{+3}_xS$ (in the metallic phase Sm is in intermediate valence state).

2) The continuous SMPT are observed in SmPrSSe and , apparently, in $Sm_{I-x}La_xS$ (in the metallic phase, probably, there is a mixture of Sm^{+2} and Sm^{3+} ions).

Next conclusions can be given:

I) An additional pressure due to conducting electrons P_{PFC} is important for SMPT. However its presence is not sufficient condition for SMPT.

2) P_{PFC} itself is sufficient only to create some concentration of localized Sm^{+3}-ions.

3) Large chemical compression itself is also insufficient for SMPT.

4) SMPT takes place only in the simultaneous presence of sufficiently large P_{CC} and P_{PFC}.

4. COMMENTS ON THE INFLUENCE OF THE DILUTION OF $Sm_{I-x}Ln^{+3}_xS$ ON SMPT.

The degree of a Sm^{+2} dilution by Ln^{+3} (D) is an important parameter for the existence of the SMPT in $Sm_{I-x}Ln^{+3}_xS$. There is some critical value D_{cr}. If $D > D_{cr}$ the collapse in $Sm_{I-x}Ln^{+3}_xS$ is impossible, because there is no sufficient number of neighboring Sm-ions with the superposed wavefunctions. It is very important to evaluate D_{cr} at least approximately. a (x) for all investigated up to now $Sm_{I-x}Ln^{+3}_xS$ can be divided in two parts: I) on the base of semiconductor SmS ($x < x_{cr}$) and 2) on the base of $Ln^{+3}S$ and $Sm^{+3}S$ ($x > x_{cr}$) (a at $x > x_{cr}$ aspires to ~ 5.7Å by extrapolation from $a_{Ln^{+3}S}$ to $a_{x=0}$; 5.7Å is close to 5.62Å for the hypothetical metallic $Sm^{+3}S$).

In the second part the volume concentration of the Sm^{+3}-ions (n_{Sm}) increases at the variation of the solid solutions from $Ln^{+3}S$ to $Sm^{+3}S$. n_{Sm} can be calculated by the Vegard-law:

$$n_{Sm} = \frac{4(I-x)}{\left[a_{Sm^{+3}S} + (a_{Ln^{+3}S} - a_{Sm^{+3}S})x\right]^3} \qquad ....(9)$$

Fig.I0 shows n_{Sm} calculated by Eq.(9) for the $Sm^{+3}_{I-x}La_xS$ (2) and $Sm^{+3}_{I-x}Sc_xS$ (3). LaS and ScS have respectively the maximum and minimum value of a in the series of the LnS . n_{SmS} for the rest of $Sm^{+3}_{I-x}Ln^{+3}_xS$ lie between the curves 2 and 3.

A.V.Golubkov has derived empirically the next rule for D_{cr}: the collapse in $Sm_{I-x}Ln^{+3}_xS$ takes place, when $n_{Sm} = n^o_{Sm}$ (n^o_{Sm} is the Sm-ion concentration in the semiconductor SmS : $n^o_{Sm} = 4 / a^3_{SmS} = I.88 \cdot 10^{22}$ cm^{-3}).

Table 2 shows that the calculated x_{cr}: a) are close to the experimental ones and b) increase with the increasing the atomic number of the

Ln^{+3} (it coincides with the experiment [4]). It gives a hope that the criterion for D : $n_{Sm} = n^{o}_{Sm}$ is close to the reality. It is interesting to note that x_{cr} is determined, mainly, by the value D (not CC or PFC).

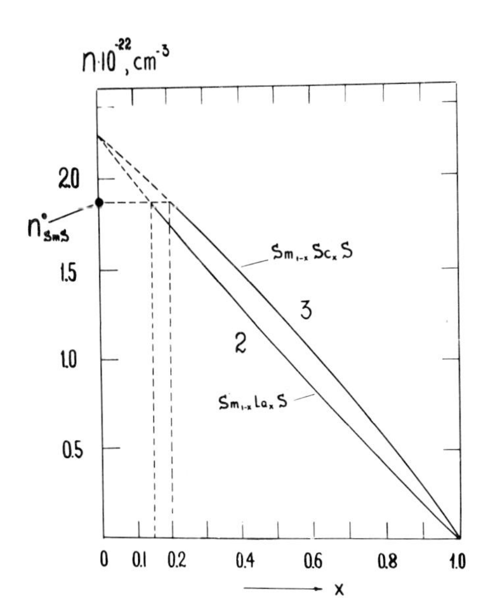

Figure I0: n_{Sm} calculated according to the Eg.(9).

5. SMPT IN $SmS_{I-x}P_x$ FILMS, OBTAINED BY THE IONIC IMPLANTATION OF P INTO SmS .

In the bulk $SmS_{I-x}P_x$ at x_{cr}=0.03 ÷0.06 the discontinuous SMPT takes place [9] (an abrupt decrease of a from 5.94 to 5.7Å occurs, the black color turnes into golden, the Sm-valency increases from +2 to +2.8).
Below preliminary data are cited on an SMPT observed by us in thin SmS films under the P-ions implantation.
Thin films of SmS were prepared by the flash evaporation method. The energy of the implanted ions was 30 keV and the doses (Φ) varied from $2 \cdot 10^{15}$ to $\sim 10^{16}$ cm^{-2}. After implantation a and the colour of the Sm-films changed. The film 40(3) (the thickness d= 0.07μ) turned golden on both the substrate and the free sides. The more thick films 39(3) (d=0.15μ) and 42(3) (d= 0.2μ) became golden only on the free side. According to

Table 2. The experimental and calculated values of x_{cr} for $Sm_{I-x}Ln^{+3}_xS$

Ln in $Sm_{I-x}Ln^{+3}_xS$	x_{cr}	
	calculated (Eq.9)	literature data
Ce	0.I55	0.I6- our data
Pr	0.I57	0.I5 ± 0.02 [4]
Nd	0.I60	0.I5 [2], 0.I25 [8]
Gd	0.I70	0.I4±0.0I5 [4] 0.I6 [2] ,0.I5 [6]
Tb	0.I74	0.2±0.0I5 [4]
Dy	0.I76	0.2±0.02 [4]
Ho	0.I78	0.22±0.0I5 [4]
Er	0.I8I	-
Lu	0.I88	-
Y	0.I76	0.I5 [3]

X- Ray analysis, the film 40(3) was a single-phase with a = 5.72Å . The films 39(3) and 42(3) consist of two phases with a =5.94 and 5.72Å (it is connected with a large ratio of the whole films thickness d to the implanted one d_p. Thus we believe that at the implantation the doping of the SmS by P takes place forming the $SmS_{I-x}P_x$ solid solution.
The P free length in SmS is unknown, so to evaluate d_p we use the values of the free length R_p and the mean square deviation ΔR_p for Al with 30 keV, when it is implanted in W [I0] (the masses of Al and W are nearly equal to ones of P and SmS). We have calculated d_p by Eq.(I0) [II]

$$d_p \sim R_p + 2\Delta R_p \quad \ldots\ldots (I0)$$

According to Eq.(I0) d_p is equal to ~0.04. The P-ions concentration in the implanted layer (N_p) is determined by Eq.(II) [II]

$$N_p = \Phi / 2.5\Delta R_p \quad \ldots\ldots\ldots (II)$$

According to Eq.(II) N_p is equal to $0.57 \cdot 10^{2I}$ (x=0.03), $I.43\ 10^{2I}$ (x=

0.075) and 2.86 $10^{21} cm^{-3}$(x=0.156) for the films 39(3), 40(3) and 42(3) respectively. It must be noted, that evaluation of d_p and N_p is made very roughly.
We have tried to estimate N_p experimentally in the single-phase film 40(3). We have used the X-Ray microanalyser MS-46 and the method described in [12]. The analysis has been made with E=30 keV using K_α -P for the best line contrast. The P-amount implanted into the film appeares to be $x \sim 0.01$, which is 6-times less than x obtained by Eq.(11). Additional experiments are in progress in order to explain this discrepancy.

6. REFERENCES

[1] Golubkov, A.V., Sergeeva, V.M., Physics and chemistry of the Rare-Earth Semiconductors, Preprint (Academy of Sciences of the USSR, 1977).

[2] Jayaraman, A. and Maines, R.G., Study of the valence transition in Eu-Yb-and Ca-substituted SmS under high pressure and some comments on other substitutions, Phys.Rev., 19B (1979) 4154-4161.

[3] Pohl, D.W., Valence transition of $Sm_{1-x}La_xS$ and related compounds, Phys. Rev., 15B (1977) 3855-3862.

[4] Gronau, M., Halbleiter-Metall-Ubergange und Zwischenvalenzen in $Sm_{1-x}SE^{3+}_xS$, Ph.D. Thesis, Dep. Phys. and Astron., Ruhr-Univ. (1979).

[5] Chouteau, G , Pena, O., Holtzberg, F., Penny, T., Tournier, R., von Molnar, S., The influence of Sm^{3+} impurities on properties of pure SmS and its alloys with LaS , J. de Phys., Suppl. 37 (1976) C4-283 - C4-288.

[6] Ohashi, M., Kaneko, T., Yoshida, H. and Abe, S., Insulator-metal transition in the mixed compounds $Sm_{1-x}Gd_xS$, Physica, 86-88B (1977) 224-226.

[7] Bucher, E. and Maines, R.G., Semiconductor-metal transition in SmS-SmSe Mixed Crystals, Sol.St.Comm., 11 (1972) 1441-1445.

[8] Grushko, A.I., Egorov, A.I., Krutov G.A., Mezentseva, T.B., Petrovich, E.V., Smirnov, Yu.P., Sumbaev, O.I., Investigation of the semiconductor-metal transition in $Nd_xSm_{1-x}Se$ and $Nd_xSm_{1-x}S$ by the X-Ray K-line shift, Zh. Exsp. Teor. Fiz., 68 (1975) 1894-1898.

[9] Morillo, J., de Novion, C.H., Senateur, J.P., J.de Phys.C5 , Suppl., 40 (1979) C348-C349.

[10] Tables of Parameters of spatial Distribution of Implanted Impurities (Minsk, 1980).

[11] Maier, J., Eriksohn, L., Devis, J., Ionic doping of the semiconductors (germanium, silicon) (MIR, Moscow, 1973).

[12] Dedegkaev, T.T. and Moshnikova, V.A., Investigation of the inhomogeneity in the semiconductors by electron probe microanalysis, Izv. LETI, N263 (1980) 100-105.

C. GODART: In the case of $Sm_{1-x}Eu_xS$, some people reported that it exists a transition to a metallic state (with about 50% of Eu) or as you said it exists no effect of pressure by lattice (lattice parameters are the same in SmS and EuS) and no effect of free carriers. How do you explain this ? What do you think about the possibility of a stoichiometry effect in EuS in which case the effect of free carriers would not be negligible ?

I.A. SMIRNOV: Yes, I agree with you. The situation with $Sm_{1-x}Eu_xS$ is very complicated. It is very difficult to explain this situation in the scope of a short answer. But your doubt doesn't work in our case, because we consider on Fig. 9 (P_{cr} vs. lattice constant of the second solution components: $Ln^{3+}S$, Ln^{2+} and $M^{2+}S$) the compositions of this solutions only for $x \sim 0.1$.

Valence Instabilities
P. Wachter and H. Boppart (eds.)

THE PHASE DIAGRAM FOR A Eu COMPOUND UNDERGOING CONFIGURATIONAL CROSSOVER: MÖSSBAUER EFFECT MEASUREMENTS

M. Croft, C. U. Segre, J. A. Hodges, and A. Krishnan

Rutgers University, Serin Physics Laboratory
Piscataway, NJ 08854 USA

V. Murgai, L. C. Gupta, and R. D. Parks

Polytechnic Institute of New York
Department of Physics
Brooklyn, NY 11201 USA

Temperature dependent Mössbauer effect measurements on $EuPd_2Si_2$, $Eu_{0.01}La_{0.99}Pd_2Si_2$, and the series $Eu(Pd_{1-x}Au_x)_2Si_2$ are reported. Comparison of the thermally induced valence change in the first two compounds emphasizes the role of intersite valence fluctuation interactions in the $EuPd_2Si_2$ valence transition. Increasing x in the Au substituted compounds is shown to: 1) depress the critical temperature of the valence transition 2) drive the valence transition first order and 3) eventually stabilize an antiferromagnetically ordered ground state. Finally, the global diagram for configurational crossover in this Eu system is discussed.

Introduction

The central result of this paper is to demonstrate a comprehensive (albeit somewhat blurred by material constraints) phase diagram for a Eu compound as it undergoes configurational crossover. This demonstration is useful because such a phase diagram can serve as a "road map" to the physics underlying the crossover between the highly magnetic (J=7/2), Eu^{2+}, $4f^7$ state and the nonmagnetic (J=0) Eu^{3+}, $4f^6$ state. It is significant because, although many other Ce, Pr, Sm, Eu and Yb compounds undergo similar electronic and/or magnetic-nonmagnetic instabilities, no such global phase diagram has here-to-fore been demonstrated.[1]

The starting point for our phase diagram study is the compound $EuPd_2Si_2$. The observation of a precipitous (but continuous) thermally induced valence change near 140K has focused a great deal of attention on this compound over the past year.[2,3,4] Magnetic susceptibility, Mössbauer effect and x-ray diffraction studies on this system have indicated that a valence change (of about $0.5e^-$ in a 50K temperature interval) is accompanied by a strong contraction in the tetragonal compound's a-axis lattice parameter.[2] High pressure, through a stabilization of the Eu^{3+} state, has been shown to enhance the temperature and broaden the extent of the valence transition.[5] After reviewing recent Mössbauer effect results on the pure compound we will discuss the $Eu(Pd_{1-x}Au_x)_2Si_2$ system. Here, by stabilizing the Eu^{2+} state, Au substitution sharpens and depresses the critical temperature of the valence transition. This eventually gives access to an antiferromagnetically ordered ground state, and thereby to the global phase diagram for configurational crossover in this system.

Mössbauer Effect and Experimental Details

As emphasized many times in the past, the wide variation between the $^{151}Eu^{2+}$ and $^{151}Eu^{3+}$ isomer shifts along with the rapid valence fluctuation time scale makes ^{151}Eu Mössbauer spectroscopy a sensitive microscopic probe of valence.[6,7] In the fast valence fluctuation regime, the isomer shift is $I=I_2P_2+I_3P_3$ and the valence is $v=2P_2+3P_3$.[6,7] Here I_2, P_2 and I_3, P_3 are the isomer shift and the occupation probability of the 2+ and 3+ valence states respectively.

In order to obtain an esitmate of nominal valence from the isomer shift of $Eu(Pd_{1-x}Au_x)_2Si_2$ compounds we will use I_3=-.5 mm/s and I_2=-11 mm/s. These values were chosen close to the isomer shifts of 3+ $EuFe_2Si_2$ (-.7 mm/s) and 2+ $EuAg_2Si_2$ (-11.4 mm/s), which are isostructural with $Eu(Pd_{1-x}Au_x)_2Si_2$.[8] A $^{151}Sm_2O_3$ source was used for both our experiments and the determination of the $EuFe_2Si_2$ and $EuAg_2Si_2$ isomer shifts.

The low temperature and Mössbauer effect measurement techniques employed are standard in this field. Sample preparation and characterization are described elsewhere.[2,9]

The Mössbauer absorption spectra were fit using a least squares method. A superposition of one, two, or three Lorentzians were used. The linewidths resulting from the fits were always much larger than the natural linewidth and resulted from a combination of effects, i.e. unresolved quadrupole splittings, relaxation and the continuous distribution of local environments. The inhomogeneity broadening both of the lines and of the valence transitions complicates the fitting of the spectra. The fitting procedures found to yield the most systematic thermal variation in the fit parameters involved constraining the linewidths of all components in a given spectrum to be the same. The linewidth was allowed to vary with temperature. Although alternative procedures are justifiable, this systematic method led to a very understandable interpretation of the spectra despite the inhomogeneity effects.

$EuPd_2Si_2$ Electronic Instability

In figure 1, we present the ^{151}Eu Mössbauer absorption spectra of $EuPd_2Si_2$ at selected temperatures spanning its valence transition.[10] The main line in these spectra (discussed at length below) moves in a strongly nonlinear fashion with temperature from -8.18± 0.01 mm/s at 300K to -1.79± 0.08 mm/s at 13K. The solid lines through the spectra are the results of least squares fits to the data.

In addition to the main line in the experimental spectra, there are also additional weaker, partially or fully resolved components which together account for no more than 20% of the total absorption area. The least squares fits to the spectra indicate that the absorption intensity lying outside the main line shows some systematic variation with temperature. At 80K the subsidiary components in the spectra are best resolved from the main line and lie in the -7 to -8 mm/s isomer shift range. Upon lowering temperature to 12K, approximately 5% of the total absorption area remains in a line in the -7 to -8 mm/s range while a second subsidiary line (containing about 15% of the total area) moves continuously into the -4 mm/s range.

The intensity in the -7 to -8 mm/s range represents Eu sites in the material which retain a valence in the 2.2-2.3 range, similar to the valence of all the Eu sites at 300K. In the isostructural compound $EuCu_2Si_2$, which also exhibits a large (though more gradual) temperature-induced valence change, analogous Eu sites with valences in the 2.0 to 2.3 range have also been observed.[6,8] Previous authors have proposed that Si defect substitution onto the transition metal sublattice gives rise to these Eu sites.[6,8] In interpreting the 15% of the total absorption area which shifts continuously between 80K and 12K it is useful to recall the $\gamma\leftrightarrow\alpha$ volume collapse in Ce and its alloys.[11-13] In particular, the failure of portions of the sample to transform at the bulk transition temperature and to transform only gradually as one gets further from the transition temperature is well known.[11-13] There, as here, strain field and grain boundary effects are presumably important.

The temperature dependence of the isomer shift of the main Mössbauer line in $EuPd_2Si_2$ is shown in figure 2. The observed 7 mm/s variation in isomer shift (corresponding to about 0.6 e^- decrease in 4f occupation number) between 200K and 100K is both the largest and the most rapid temperature-induced isomer shift change in any homogeneous mixed valent Eu compound. It is

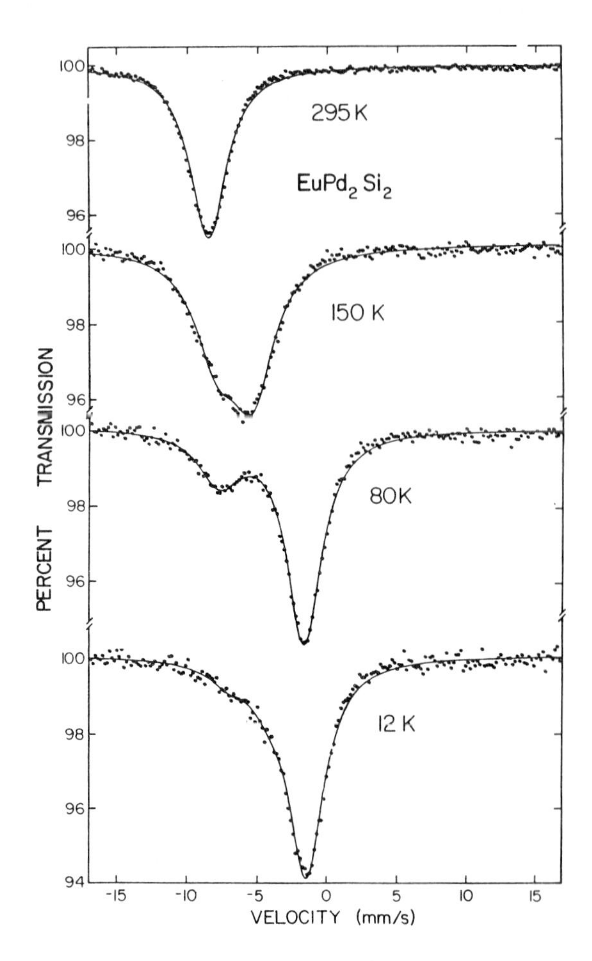

Figure 1: Mössbauer absorption spectra for $EuPd_2Si_2$. The solid lines in this figure and all subsequent spectra figures are the results of least squares fits to the data.

natural to ascribe the precipitous valence change in this system to intersite-charge-fluctuation interactions of the sort that drive first order electronic transitions in SmS or elemental Ce.[1]

In order to verify the role of intersite-valence-fluctuation interactions in $EuPd_2Si_2$ we present in figure 2 the isomer shift versus temperature results of Mössbauer effect measurements on the dilute compound $Eu_{0.01}La_{0.99}Pd_2Si_2$. Here we have proceeded in analogy to the classic method used in magnetic systems to separate single atom effects from intersite magnetic exchange effects, i.e. investigating the dilute analog of a concentrated system.[14] (See footnotes 14 and 15.) It is clear (from figure 2) that the dilute system shows a much smaller and much more gradual temperature- induced change in isomer shift (valence) than observed in the concentrated system. *This provides direct empirical evidence for an intersite interaction mechanism driving the rapid valence change in $EuPd_2Si_2$.*

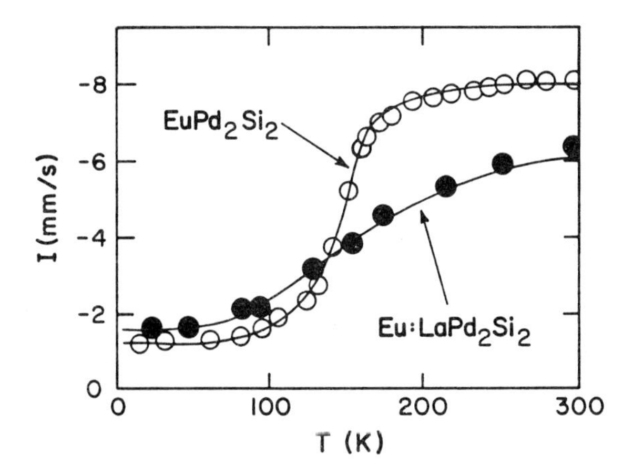

Figure 2: Isomer shift versus temperature for the main line in $EuPd_2Si_2$ and for $Eu_{0.01}La_{0.99}Pd_2Si_2$.

$Eu(Pd_{1-x}Au_x)_2Si_2$ System

Both by analogy with the temperature-pressure phase diagrams of fcc Ce and SmS[1] and by virtue of model calculations[10] it might be anticipated that the valence transition in $EuPd_2Si_2$ could be driven first order by alloy stabilization of the Eu^{2+} state. With this motivation we have investigated a series of compounds in the Eu $(Pd_{1-x}Au_x)_2Si_2$ series. $EuAu_2Pd_2$ supports Eu very close to its 2+ valence state and has the same crystal structure as the palladium compound[9]

Initial magnetic susceptibility measurements on this series (see figure 3) indicated that the Au substitution stabilized the Eu^{2+} state and depressed the valence transition temperature. Here, the signature of the transition is taken to be the inflection point on the low temperature side of the susceptibility maximum[9,11] Two points regarding the magnetic susceptibility results should be noted. First, although the valence transition appeared to sharpen with increasing Au substitution, no evidence for a discontinuous valence transition is apparent. Second, although the peak structure in susceptibility curves is always present, the qualitative shape of this peak appears to change between x=0.175 and x=0.20.

First Order Valence Transition

Inspection of the Mössbauer spectra for $Eu(Pd_{0.85}Au_{0.15})_2Si_2$ shown in figure 4 reveals that a valence transition occurs but that the character of the transition is qualitatively different from that of the pure system. The spectra above 60K (for the x=0.15 sample) are well fit by a single line near -8.5 mm/s (∿2.2 valent). Below 60K a second, well-resolved line near -2 to -3 mm/s (∿2.7 valent) is required in addition to the ∿2.2 valent line. Upon lowering the temperature these distinct lines persist; however, a systematic "dumping" of intensity from the ∿2.2 valent to the ∿2.8 valent line occurs with decreasing temperature. Like the pure system a third intermediate line, with an area less than 20% of the total absorption area, near -6 mm/s (2.5 valence) is also required to adequately fit the spectra below about 60K.

The sharp contrast between the x=0 (pure) and the x=0.15 valence transitions is further illustrated in figure 5a where the isomer shift data for the two systems is plotted versus temperature. In figure 5a (as throughout this paper) in order to avoid placing undue emphasis on minority sites, we plot only the isomer shifts of fitted lines accounting for greater than 20% of the total absorption area. *Figure 5a suggests the immediate interpretation that the valence transition in the x=0.15 sample is a first order phase transition as opposed to the continuous transition in the pure system.* It must be emphasized that the data for the x=0.15 sample in figure 5 do not represent an equilibrium hysteresis loop typical of a first order transition. This point is underscored in figure 5b where a plot of the fractional area of the two lines shows a continuous transfer

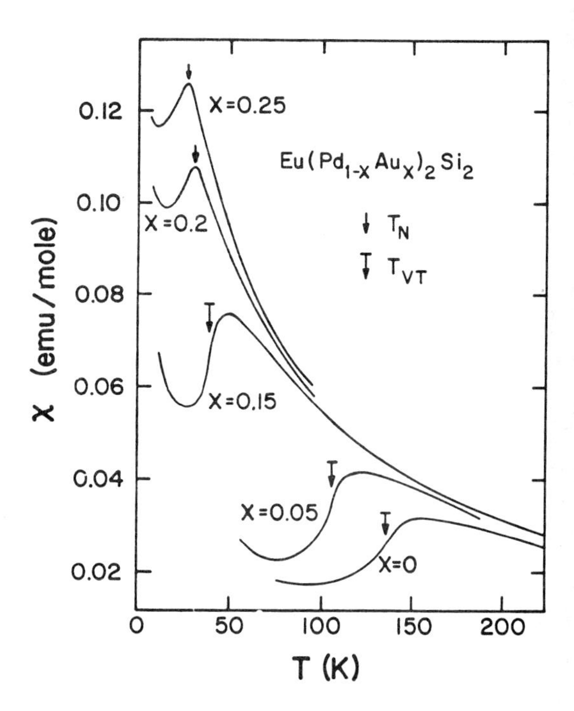

Figure 3: Magnetic susceptibility versus temperature for selected samples in the $Eu(Pd_{1-x}Au_x)_2Si_2$ system. The arrows (↓) indicate Neel temperatures (T_N) and the arrows (↧) indicate valence transition temperatures (T_{VT}).

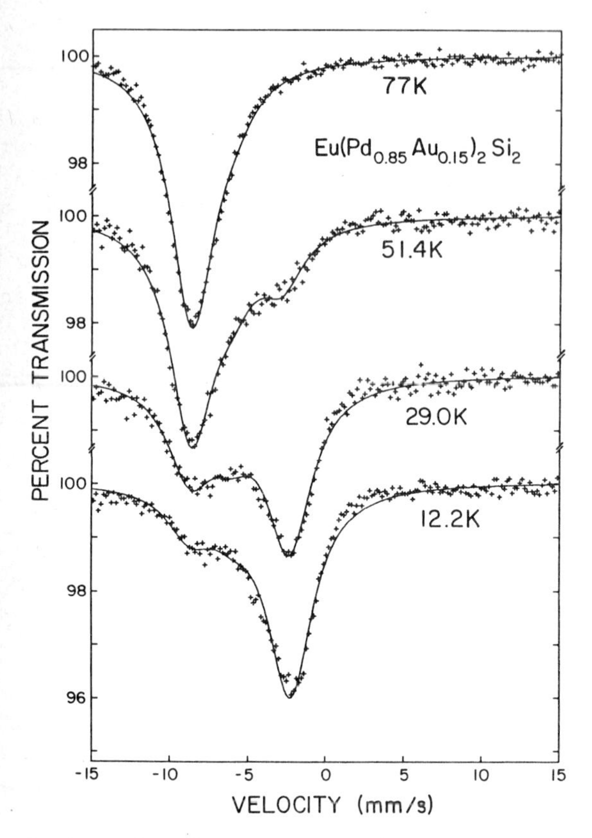

Figure 4: Mössbauer absorption spectra of an x=0.15 sample in the $E(Pd_{1-x}Au_x)_2Si_2$ system.

from the ∿2.2 valent to the ∿2.8 valent absorption line with decreasing temperature. *We interpret this behavior in terms of a first order transition smeared by sample inhomogenitites so that different regions of the sample transform discontinuously at a distribution of transition temperatures.* Note that the area of intermediate line near 6 mm/s has also been included in figure 5b.

The most obvious source for sample inhomogenities in this system is Au-Pd composition fluctuations. The rate of depression of the valence transition line with increasing x would dictate roughly a 6K change in transition temperature per percent fluctuation in Au concentration. Even without composition fluctuations, however, a smearing of the first order valence transition in this system, similar to that observed in the $\gamma \leftrightarrow \alpha$ transition in Ce, is to be expected.[11-13] Indeed, in early x-ray diffraction studies (in Ce) the relative fraction of γ-Ce and α-Ce varies quite similarly to the results in figure 5b (and later 7b).[12]

Approach to the Critical Point

Since the continuous valence transition in $EuPd_2Si_2$ is driven first order by 15% Au substitution there must logically be a critical point (occurring at a T_c and x_c) between x=0 and x=0.15 in the $Eu(Pd_{1-x}Au_x)_2Si_2$ system. As indicated above, Au-Pd composition fluctuations are almost certainly present in this system. Hence, near the critical concentration (x_c), regions of the sample with $x>x_c$ and $x<x_c$ should transform dis continuously and continuously respectively. Approaching the x_c from the $x>x_c$ side, one should see an increasing fraction of the sample transform continuously. A region of a sample which transforms continuously will show a line whose isomer shift

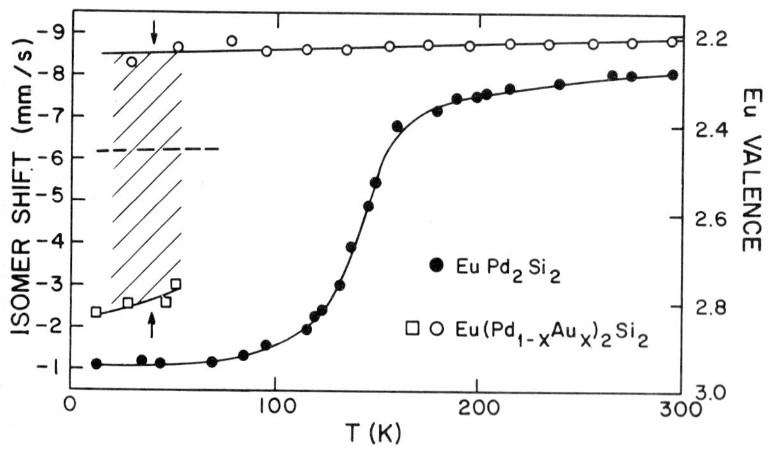

Figure 5a: Thermal variation of the isomer shifts for an x=0.15 sample (and for an x=0.0 sample for comparison). The open circles and open squares are respectively the isomer shifts of the $Eu^{2.2+}$ and $Eu^{2.8+}$ lines in the 0.15 sample. All lines are guides to the eye. The dashed lines indicate the locus of isomer shift results for lines which contain less than 20% of the total absorption area. The shaded region represents the 25%-75% limits of the smeared first order transition. The arrow indicates the transition temperature as determined by magnetic susceptibility.

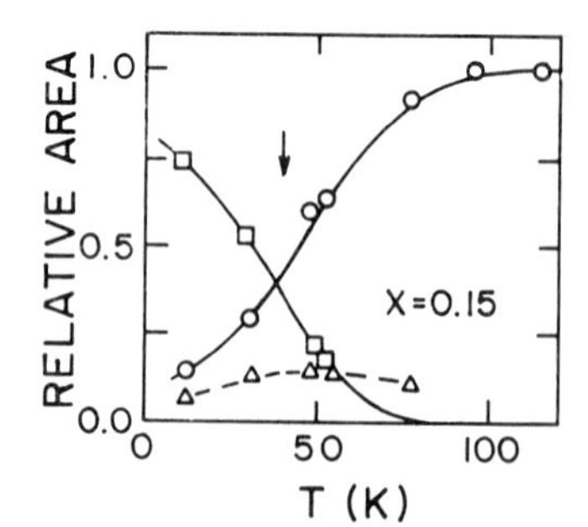

Figure 5b: The temperature dependence of the fractional absorption area in the $\sim Eu^{2.2+}$ line (circles), the $\sim Eu^{2.8+}$ line (squares) and the intermediate $\sim Eu^{2.5+}$ line (triangles) for an x=0.15 sample. The arrow indicates the valence transition temperature as in figure 5a.

moves rapidly (with temperature) but continuously through the intermediate -7 mm/s to -3 mm/s range. Therefore, the approach to x_c from $x>x_c$ should be characterized by an increasing intensity in the intermediate isomer shift range.

This effect is illustrated in figures 6 and 7. In figure 6 the spectra for an x=0.05 sample at T=117K is shown along with the three separate component lines obtained by fitting the spectra. The temperature dependence and fractional absorption areas of the three component lines for the x=0.1 sample are shown in figures 7a and 7b. The results in figure 7a and 7b along with similar results of a detailed study on an x=0.05 sample support the conclusions discussed below.

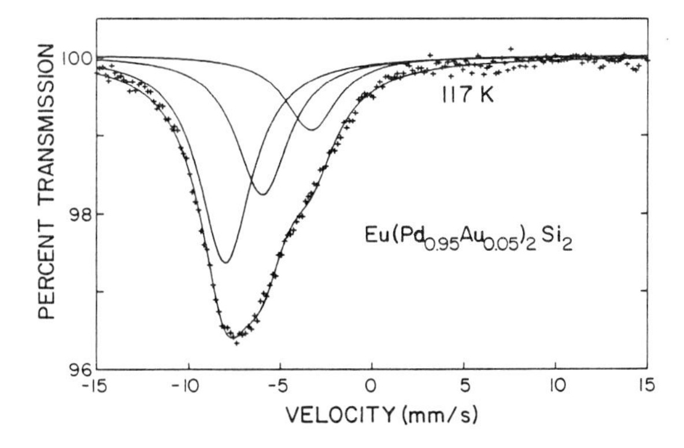

Figure 6: The experimental Mössbauer absorption spectrum for an x=0.05 sample at T=117K. The three individual component lines used to fit the spectrum are shown. The approach to the critical point near $x_c \simeq 0.04$ for $x>x_c$ is characterized by increasing intensity in the central line.

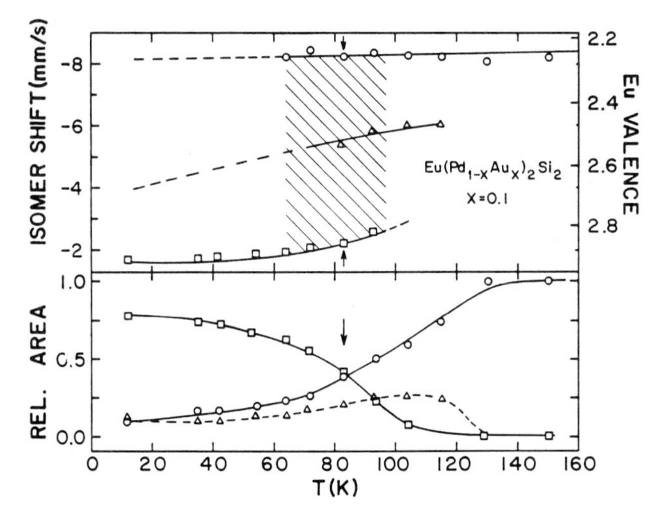

Figure 7a: The isomer shift variation versus temperature for the $\sim Eu^{2.2+}$ line (circles) and the $\sim Eu^{2.9+}$ line (squares) in the x=0.10 sample. The other symbols are as in figure 5a.

Figure 7b: The thermal variation of the fractional absorption area in the $\sim Eu^{2.2+}$ line (circles), the $\sim Eu^{2.8+}$ line (squares) and the intermediate $\sim Eu^{2.5+}$ line (triangles) for the x=0.10 sample. The arrow indicates the valence transition temperature as determined from magnetic susceptibility.

First, the valence transition at x=0.1 is still dominantly first order. However, the intermediate line reaches a maximum of 26% relative area (compared to ∿10% at the valence transition in the x=0.15 sample) and remains above the 20% level over a significant temperature range. The intermediate line in an x=0.05 sample is even stronger, reaching a maximum relative area near 36%. *Based on these results, we qualitatively estimate the critical point in the $Eu(Pd_{1-x}Au_x)_2Si_2$ systems to be in the range $x_c \sim 0.4 \pm 0.015$ and $T_c \sim 115 \pm 9K$.*

Second, at about 85K the Mössbauer effect results indicate that in the x=0.1 sample is proceeding most rapidly and is about half complete. This temperature agrees closely with the inflection point (the point of highest positive slope) in the magnetic susceptibility versus temperature plot (see figure 3). Similar agreement between the Mössbauer effect and magnetic susceptibility results were found for the x=0.05 and x=0.15 samples. *Consequently, we will use the much finer grid magnetic susceptibility results to locate the center of the smeared first order transitions.*

Antiferromagnetic Order

Before summarizing our results in a phase diagram, it is crucial to consider the Mössbauer effect results on the x=0.2 and x=0.25 samples in the $Eu(Pd_{1-x}Au_x)_2Si_2$ system. As mentioned earlier, the temperature dependent magnetic susceptibility for these two samples exhibits a peak which is only subtly different from those observed in the lower concentration samples. Mössbauer spectra for x=0.20 and 0.25 samples at temperatures (T) just above where the peak in the magnetic susceptibility occurs are typical of paramagnetic Eu with a valence near 2.2+. (See top spectra of figure 1 and figure 4 for examples). Mössbauer spectra of these samples for T less than where the peak of the susceptibility occurs show a strong magnetic hyperfine splitting, characteristic of a magnetically ordered Eu compound (see figure 8).

The magnetic ordering temperature can be estimated using the broadening and the intensity decrease of the paramagnetic line when approaching the ordering temperature from above, and the collapse of the magnetic splitting in the spectrum when approaching the ordering temperature from below. *The Mössbauer effect plus magnetic susceptibility results unambiguously indicate that antiferromagnetic order sets in at Neel temperature T_N very close to the peak in the magnetic susceptibility of the x=0.2 and x=0.25 samples. Further, there is no evidence that either of these samples also undergoes a valence transition.*

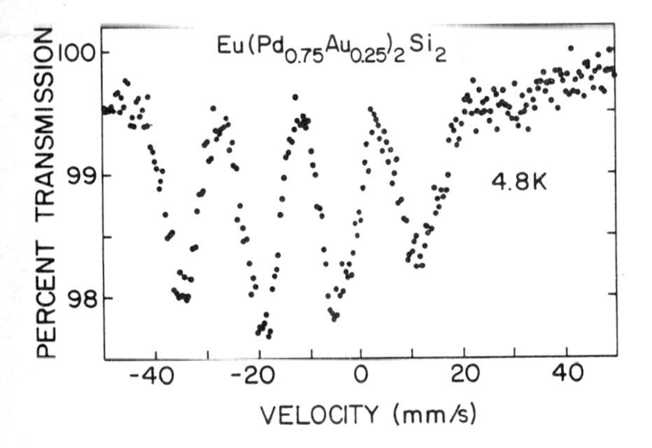

Figure 8: The magnetically split spectrum for an x=0.25 sample at T=4.8K.

Mössbauer effect measurements on an x=0.175 sample show a smeared first order valence transition which begins at about 50K. As described previously, the transformation proceeds by building intensity in an ∿2.8 valent line with decreasing temperature. At temperatures near 32K (the T_N value for the x=0.2 sample) this intensity transfer halts and the more paramagnetic line rapidly broadens and attenuates with decreasing T. This behavior indicates that the untransformed regions of the sample are ordering antiferromagnetically. The relative fraction of the sample in the more 3+ line is about 35% at 27K hence, in the vicinity of the magnetic ordering temperature, about 35% of the sample has transformed into the ∿2.8 valent line. The x=0.175 sample lies very close to a first order coexistence line (broadened by inhomogeneities) between an ordered antiferromagnetic, nominally $Eu^{2.2+}$ phase and a collapsed-volume, nominally $Eu^{2.8+}$ phase.

Phase Diagram and Conclusions

Our results for the $Eu(Pd_{1-x}Au_x)_2Si_2$ system are summarized in the phase diagram in figure 9. The continuous valence transition in $EuPd_2Si_2$ is depressed (dashed line in figure) and sharpened until it is driven first order (solid line in figure) for $x>x_c$ where x_c∿0.04 ± 0.015, T_c≃115 ± 9K is the locus of the critical point. The line with arrows above the critical point indicates an estimate of the sample inhomogeneity-induced uncertainty in x_c. The locus of the first order transition line has been assigned by the positive inflection points in the magnetic susceptibility curves (solid squares in figure) in concert with the Mössbauer analysis. The error bars on three of the solid squares are set where analysis of the Mössbauer spectra indicates the smeared first order transition is 25% and 75% complete.

The first order valence transition line meets a line of second order antiferromagnetic (AF) transitions in the vicinity of T≃32K and x∿0.175. The arrows above x∿0.175 represent the upper limits on inhomogeneity smearing in this portion of the phase diagram. The AF ordering transition appears relatively insensitive to sample inhomogeneities (i.e. the transition is sharp) and the errors in the AF transition temperatures should lie within the solid triangles used to represent them.

This is the first Eu system in which a line of first order valence transitions terminated by a high temperature critical point has been observed. Until now only in the very high pressure phase diagrams of EuO (near 300 kbars) and of elemental Eu (near 150 kbars) have there been suggestions of Eu valence transitions.[16] (The charge ordering transition in Eu_3S_4 falls in a different transition class and the concentration-induced valence change in the $EuPt_{2-x}Rh_x$ system is apparently overwhelmed by local environment effects.[6,7]) Indeed, only in elemental Ce and in SmS have 4f shell electronic transitions previously been shown to exhibit a first order to continuous crossover.[1] It is interesting to note that the first order transition in semiconducting SmS (under pressure) persists to the lowest temperatures whereas the electronic transitions in the metallic alloys of SmS and Ce

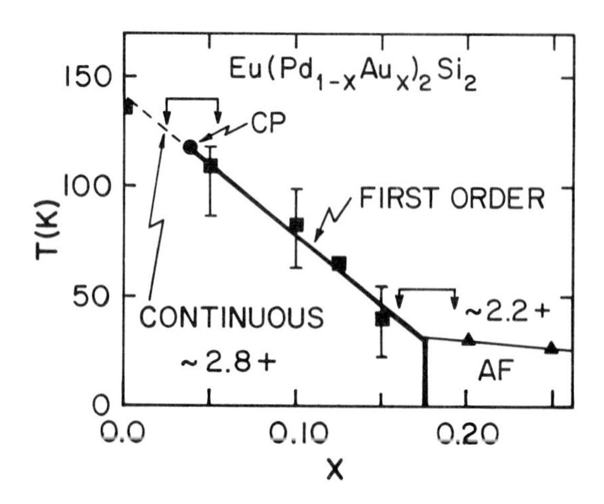

Figure 9: Phase diagram for the $Eu(Pd_{1-x}Au_x)_2Si_2$ system (see text for discussion).

both become continuous at sufficiently low temperatures.[1] *Therefore, the valence transition in the $Eu(Pd_{1-x}Au_x)Si_2$ system also appears to be the sole example of an electronic 4f shell transition in a metallic material which is first order between the two ground state phases.* The extremely large degeneracy of the magnetic Eu^{2+} (J=S=7/2) and its immunity to crystalline electric field degeneracy reduction could play an important role in preserving the first order transition at low temperatures.

In addition to being highly degenerate, Eu^{2+} is of course also highly magnetic. In concentrated systems this leads to a large magnetic ordering energy, a high magnetic ordering temperature, and hence to the potential of a direct competition between an ordered magnetic and a mixed-valent-nonmagnetic ground state. In previous studies of Eu systems, only the hint of a potential valence transition-magnetic ordering transition competition could be gleaned. In EuO, the magnetic transition temperature increases substantially between P=0 kbars and P=50 kbars however, the only known point for the valence transition line is still remote.[16,17] In elemental Eu the ordering temperature is pressure insensitive to about P=150 kbars where predictions have proposed the occurrence of a valence transition.[16,18] Experimental results do appear to confirm a confluence of transition lines near 150 kbars in Eu; however, the character of the various transitions is at this time not at not at all understood.[18] *Our results for $Eu(Pd_{1-x}Au_x)_2Si_2$ make a clear statement on this point. We observe that the magnetic ordering transition increases weakly as the valence transition is approached; that the magnetic ordering transition terminates at the first order valence transition at a critical end point; and that the ordered magnetic (roughly $Eu^{2.2+}$) and nonmagnetic (roughly $Eu^{2.8+}$) ground states are separated by a first order phase transition.*

It is instructive to contrast the magnetic portion of the Eu $(Pd_{1-x}Au_x)_2Si_2$ phase diagram with what is known about the analagous magnetic " γ-like" phase in Ce systems. (The collapsed volume-nonmagnetic phase in Ce compounds is often referred to as the " α-like" phase after the α phase of elemental fcc Ce. Spin fluctuations, due to the Kondo effect, have been shown to play an important role in " γ-like" Ce systems.[1] In particular, both theory and experiment indicate that the degree of quenching of the magnetic moments in " γ-like" Ce systems increases rapidly as the electronic transition into the "α-like" phase is approached.[1,19] This quenching process induces a strong depression of the magnetic ordering temperature in γ-like Ce systems precursive to the electronic transition into the α-like phase.[19,20]

The absence of an analogous precursive depression of T_N in $Eu(Pd_{1-x}Au_x)_2Si_2$ as its valence instability is approached strongly suggests that Kondo effect type spin fluctuations do not play an important role in this Eu compound. The generalization of this observation to include all Eu systems near configurational crossover should be considered closely in future work.

It would also be interesting to compare the magnetic phase diagram of figure 9 to that of a Sm compound near its valence instability. Recent resistivity measurements on the more magnetic phase in SmS do show structure at low temperatures; however, this structure is not believed to be due to magnetic order.[21,22] *Thus far, therefore, $Eu(Pd_{1-x}Au_x)_2Si_2$ appears to be unique in supporting a first order coexistence between weakly mixed valent-ordered-magnetic and a strongly mixed valent-nonmagnetic-ground state.* Again, verifying if and under what circumstances this property generalizes to other Eu systems is an important question.

By virtue of a remarkable good fortune, the $Eu(Pd_{1-x}Au_x)_2Si_2$ has provided a global picture (be it ever so blurred) of configurational crossover for Eu. The fortune lies in the accessibility of the global phase diagram without the use of extremely high pressures or of alloying techniques so drastic that local environmental effects totally dominate. In doing so, this system emphasizes the lack of a global picture for electronic transitions in other Eu compounds and in all the other electronically unstable rare earth compounds. The gradual emergence of more systems for which the general phase diagrams can be determined should elucidate the roles of parameters like conduction band population, 4f level degeneracy, crystal structure, etc. in the electronic instability of these interesting materials.

Acknowledgements: This work was supported in part by the NSF and by the Research Corporation.

References

1) See J. Lawrence, P. Riseborough, and R.D. Parks, Rep. on Prog. in Phys. 44, no.1,sec. 2.1 and 2.2. (1981)

2) See E. Sampathkumaran, R. Vijayaraghavan, K. Gopalakrishnan, R. Pilley, H. Debare, L.C. Gupta, B. Post, and R.D. Parks in "Valence Fluctuations in Solids" L. Falicov, W. Hanke, M.B. Maple editors (North Holland, New York, 1981) p. 193 and references therein.

3) See J.W. Wilkens , ibid 2, p. 495.

4) See several articles in these proceedings.

5) B. Batlogg A. Jayaraman, V. Murgai, L.C. Gupta, R.D. Parks, and M. Croft these proceedings

6) See. Nowik in "Valence Instabilities and Related Narrow Band Phenomena", R. Parks Editor (Plenum, New York, 1977), p. 261 and references therein.

7) See J. Coey and O. Massenet, ibid 2., p. 211.

8) E. Bauminger, D. Freindlich, I. Nowik, S. Ofer, I. Felner, and I. Mayer, Phys. Rev. Lett. 30, 1053 (1973).

9) L.C. Gupta, V. Murgai, Y. Yeshurun, and R.D. Parks these proceedings.

10) M. Croft, J.A. Hodges, E. Kemly, A. Krishnan, V. Murgai, and L.C. Gupta, Phys. Rev. Lett. 48, 826 (1982)
M.Croft, J.A. Hodges, E. Kemly, A. Krishnan, V. Murgai, L.C. Gupta, and R.D. Parks in "Physics of Solids Under High Pressure" J. Schilling, R. Shelton ed. (North Holland Amsterdam, 1981)p. 341.

11) J. Lawrence, M. Croft, and R.D. Parks, in "Valence Instabilities and Related Narrow Band Phenomena", (Plenum, New York, 1977) p. 35.

12) C. McHarque and Y. Yankel Jr., Acta Met. 8, 673 (1960).

13) P. Burgardt, K. Gschneidner, D. Koskenmaki, D. Finnemore, J. Moorman, S. Legvold, C. Stassis and T. Vyrostek, Phys. Rev. B, 14, 2995 (1976)

14) For long standing examples dealing with the separation of crystalline electric field from magnetic exchange effects see W. Penny and R. Schlapp, Phys. Rev. 41, 194 (1932); W. Wallace "Rare Earth Intermetallics "Acad. Press, New York, 1973) p. 49, P. 114 and references therein.

15) For recent examples dealing with the competition between single atom Kondo and inter-site magnetic exchange effects see F. Steglich, J. Aarts, C. Bredl, W. Lieke, D. Menschede, W. Franz, and H. Schafer, J. Mag. and Mag. Mat. 15-18 (1980) and references therein.

16) See A. Jayaraman in "Handbook on the Physics and Chemistry of Rare Earths" K.A. Gschneidner, Jr. and L. Eyring editors (North Holland, New York 1978) p. 707 and references therein.

17) U. Klein, G. Wortmann and G. Kalvins, J. Magn. Mater 3 50 (1976).

18) F. Bundy and K. Dunn in "Physics of Solids Under High Pressure", J. Schitting and R. Shelton editors (North Holland, Amsterdam, 1981) p. 401.

19) See for examples M. Croft, R. Guertin, L. Kupferberg, and R.D. Parks 20, 2073(1979). M. Croft and H.H. Levine J. Appl. Phys. 53 (3), 2122 (1982).

20) See J. Lawrence, J. Appl. Phys. 53(3),2117 (1982) and references therein.

21) J. Wittig ibid 2, p.43.

22) F. Lapierre, M. Ribault, F. Holtzberg and J. Flouquet. ibid 2, p.353.

D. RIEGEL: How can you claim that the Kondo effect plays a relatively unimportant role in Eu compounds by Mössbauer measurements ?

M. CROFT: This is only a conjecture. In γ-like Ce systems the Kondo effect plays an increasingly important role as the collapse into the α-like phase is approached. One of the evidences for this is the depression of the magnetic ordering temperature in Ce systems precursive to the volume collapse into the α-like phase. No such precursive depression in T_N occurs in this Eu system hence the Kondo effect appears not to the same dominant role it plays in Ce systems.

F. HOLTZBERG: Do you consider the region where the isomer shift is between Eu^{2+} and Eu^{3+} to be the homogeneous intermediate valence state as described by Varma ? If so how do you consider the line broadening ?

G. KAINDL: Did you take saturation effects due to finite absorber thickness into account when evaluating the total area of the main Mössbauer absorption line ?

COMMENT: G. KAINDL: A system like $Eu(Pd_{1-x}Au_x)_2Si_2$ with e.g. x=0.15 cannot be expected to be as a whole in a homogeneously mixed-valent state. This is obvious from your Mössbauer spectra, where the three absorption lines observed clearly show that there are different Eu ions with rather different valency behavior. Taking this into account will have some consequences with respect to the proposed phase diagram.

M. CROFT: In any mixed valent alloy there will be excursions from the average valence both on a microscopic scale (due to the distribution of local environments) and on a larger scale (due to alloy composition fluctuations, grain boundary effects etc.). In the high and low temperature states of the $Eu(Pd_{1-x}Au_x)_2Si_2$ system the excursions about the mean valence are small (being reflected by line broadening). Near the temperature of the valence instability a small excursion in the valence can destabilize the state of a region in the sample therefore a strongly bimodal distribution of valences occurs. Similar chemically induced smearing effects are well known near the electronic transitions in both the Ce based and SmS based alloy systems. Indeed in the latter systems the distribution of valence transition temperatures can cause the samples to explode. At present we do not believe that application of the term inhomogeneous mixed valence or modification of the phase diagram should apply to the $Eu(Pd_{1-x}Au_x)_2Si_2$ system any more than to the Ce or SmS based alloys. It should be noted that this Eu system differs significantly from the $Eu(Pt_{1-x}Rh_x)_2$ system where large local environmental effects appear to totally dominate the valence change.

J. ROEHLER: What makes you sure that at low temperatures the valence is 2.9 or 2.95 for $EuPd_2Si_2$? Other work (susceptibility) has shown that it is more close to 2.6.

M. CROFT: I wouldn't put too much weight on exact numbers.

COMMENT: J.A. HODGES: If you take the low temperature valence to be 2.6, then you would have to put the high temperature valence to 1.8.

A. JAYARAMAN: Substituting gold is obviously equivalent to negative pressure. Is this a size effect or electron to atom ratio effect ?

M. CROFT: I think it is not a size-effect, it is the charge transfer which dominates.

Valence Instabilities
P. Wachter and H. Boppart (eds.)

THERMODYNAMIC INSTABILITY OF MIXED-VALENT COMPOUNDS AND SOLID SOLUTIONS

E. Kaldis, B. Fritzler, H. Spychiger, E. Jilek

Laboratorium für Festkörperphysik, ETH Zürich
8093 Zürich, Switzerland

Total energy measurements (calorimetry) and phase relationships are reported for a series of mixed valent compounds and ternary solid solutions with chemically induced metal to semiconductor changes. These investigations show that valence instabilities induce dramatic thermodynamic instabilities. The up to now existing results indicate that the valence changes by discrete steps with rather narrow composition range. Measurements of the enthalpy of formation as a function of nonstoichiometry are a very sensitiv tool for the investigation of valence anomalies. Calorimetric measurements are presented for the mixed-valent systems CeN, TmSe and Sm_3S_4-Sm_2S_3. Miscibility gaps have been measured in the homogeneity ranges of these compounds and of the ternary solid solutions $TmSe_{1-x}Te_x$, $Tm_{1-x}Eu_xSe$ and $TmTe_{1-x}P_x$. The gaps appear due to the existence of repulsive forces when mixing various valence states.

1. INTRODUCTION

Spectroscopies have been among the most frequently used methods of investigations of mixed-valent compounds. Although very informativ, such investigations show two drawbacks:

1) The highly energetic beam may dissociate the initial fluctuating valence state, so that a final state with different properties is actually observed (1). This is in contrast to low energy methods like e.g. compressibility or the lattice constant. The latter, although reflecting also other minor influences like defect concentration, enables a rough but trustworthy estimation of the mixed valence (2,3,16)
2) Only energy differences are measured, which although very important for the study of various excitations, do not give any information about the stability of the compounds.

On the other hand, it is possible that the small changes of the band structure leading to valence instabilities (4,5), may destabilize the chemical bonding and cause appreciable changes of the thermodynamic stability of the comound. If this were true, measurements of the thermodynamic stability would give a sensitiv method for detecting valence anomalies. Further, they would give information about the energy differences between the various valence configurations of the same compound. As it has been shown in the past, the latter can be chemically produced through variation of the metal-metal distance (crystal field splitting of the 5d-band) and of the electron concentration by means of

a) nonstoichiometry (2,3,6,7)
b) mixed crystal formation (8,9,10) or,
c) doping (11)

Last but not least, information about the thermodynamic stability of mixed valent phases would clarify many important material aspects (e.g. phase boundaries, homogeneity) and enable the correct interpretation of many physical measurements.

2. TOTAL ENERGY MEASUREMENTS AS A FUNCTION OF CHEMICAL COMPOSITION

The best experimental method to investigate the thermodynamic stability of mixed valence compounds is the direct calorimetric measurement of the enthalpy of formation ΔH_f^{298} as a function of the chemical composition. The dependence on the chemical composition is of great importance as it allows the controlled change of the valence state (2,3,5,6).

Unfortunately, enthalpy measurements of compounds as a function of nonstoichiometry are almost not existing in the literature. Therefore, meaningful extrapolations about the order of magnitude of the effects expected in mixed valent compounds were not possible.

In the last five years we have performed a series of investigations on TmSe (12,13), CeN (14) and Sm_3S_4-Sm_2S_3 (15) using solution and reaction calorimetry in 4N HCl. A review of the main results about the Tm-Se system are given in this volume (16) and in a review paper (3); about CeN in (11); and about Sm_3S_4-Sm_2S_3 in this volume (17) and in (19-20). A Raman scattering study of the latter material is given in (18).

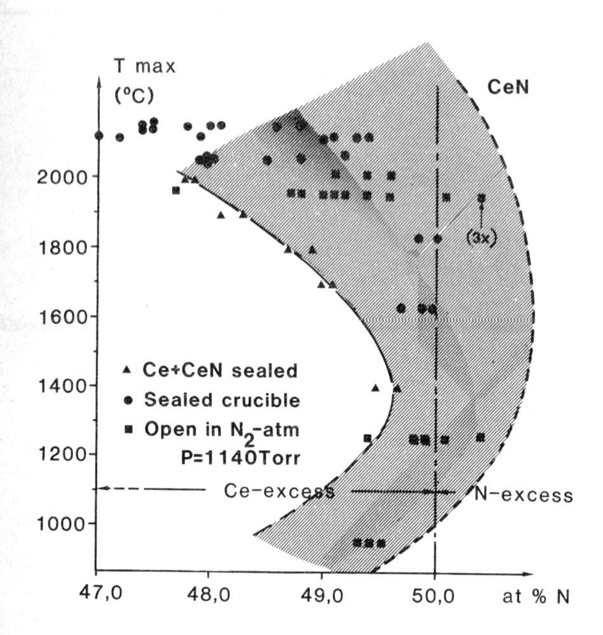

Fig. 1. T-x phase diagram of the Ce-N system, showing the homogeneity range of CeN

2.1. Giant changes of the total energy of CeN upon doping

CeN is the first compound, in which anomalous properties due to mixed valence (unknown at that time) were observed in the 30's. Its abnormally small lattice constant led to the assumption of a high Ce^{4+}-concentration (21). The development of special methods for the crystal growth of rare earth nitrides (22) made spectroscopial investigations possible. XPS-measurements on these crystals support the existence of mixed valence, (23) and photoemission measurements with synchrotron radiation (24,25) indicate a surface binding energy shift of 1.0 eV. The lattice constant of CeN does not react sensitively on the nonstoichiometry. Due to this reason it has been assumed in the past that CeN does not show a homogeneity range (26). Investigation of the T-x phase diagram of the Ce-N system (Fig. 1) shows the existence of a rather narrow nonstoichiometric range with a width between 1-3 at%N in the temperature range 700-2200 C. The curved form of the phase boundaries can be explained with the increasing diffusion rate of nitrogen in the lattice, and the formation of nitrogen vacancies at high temperatures. The interplay of these two opposing effects leads to nitrogen-rich samples for $P_{N2} \simeq 1at$, in the intermediate temperature range. The dependence of the lattice constant on composition (11) shows a slight decrease with increasing nitrogen concentration. This is in contrast to other nitrides (e.g. HoN) which show strong dependence of the opposite sign. Decisive for our investigation has been the discovery of a small but clear dependence of the lattice constant of CeN on the oxygen concentration (11). The lattice constant of pure and stoichiometric CeN is 5.018 Å, that of oxygen doped samples lies in the range $5.022<a<5.025$ Å. The latter samples show dramatic differences of the enthalpy of formation.

Figure 2 shows the enthalpies of formation calculated from measurements of the enthalpies of solution (11) of 75 CeN samples. Chemical analysis and lattice constant determination have been performed in the same sample used for the calorimetric measurements. To reach optimal consistence of these three measurements, the single crystals were powdered (in a train of glove-boxes with an atmosphere of argon containing approx. 1pp oxygen and H_2O) and aliquot parts were used for each measurement. Chemical analysis was performed for both nitrogen and cerium. The micro-Kjeldahl determination of nitrogen is very accurate, ± 0.1%at. Cerium was determined complexometrically with an accuracy of $\pm$ 0.25%at. For the oxygen determination the difference of the sum Ce+N from 100 % was used. The analytical error, therefore, would be 0.7%at for oxygen. However, the precautions necessary to decrease contamination during handling (e.g. weighing in closed tubes containing argon atmosphere) decreased the accuracy. We can claim, therefore, for the determination of oxygen an accuracy better than 1%at. The accuracy of the solution calorimetry (with an LKB mierocalorimeter) is excellent, approx. 1‰. However, for the above mentioned reasons and due to the unavoidable contamination, the reproducibility of samples from the same powdered batch was $\pm$ 1.8%.

We attribute to pure and stoichiometric CeN the sevenfold reproduced value of $\Delta H_f^{298} = -76 \pm 2$ Kcal/mol. These samples have a lattice constant of 5.018 Å. 60% of the samples have values between $-80 < \Delta H_f^{298} < -70$ Kcal/mol. They can be represented by parabolic curves characterizing three thermodynamic phases with different nitrogen concentrations. Between these phases narrow ranges of instabilities may exist. Characteristic for measurements showed in Fig. 2 is the existence of samples with the same nitrogen content but different enthalpies. The differences appearing in this range of enthalpies (80-70) are not dramatic, and may be due to the existence of parallel parabolic curves i.e. different total energy states at the same nitrogen concentration, stabilized by defects. For about 30% of the samples these differences become dramatic. These samples lie on two parabolic curves corresponding:

a) to a phase with $\Delta H_f^{298} \simeq -107$ Kcal/mole and

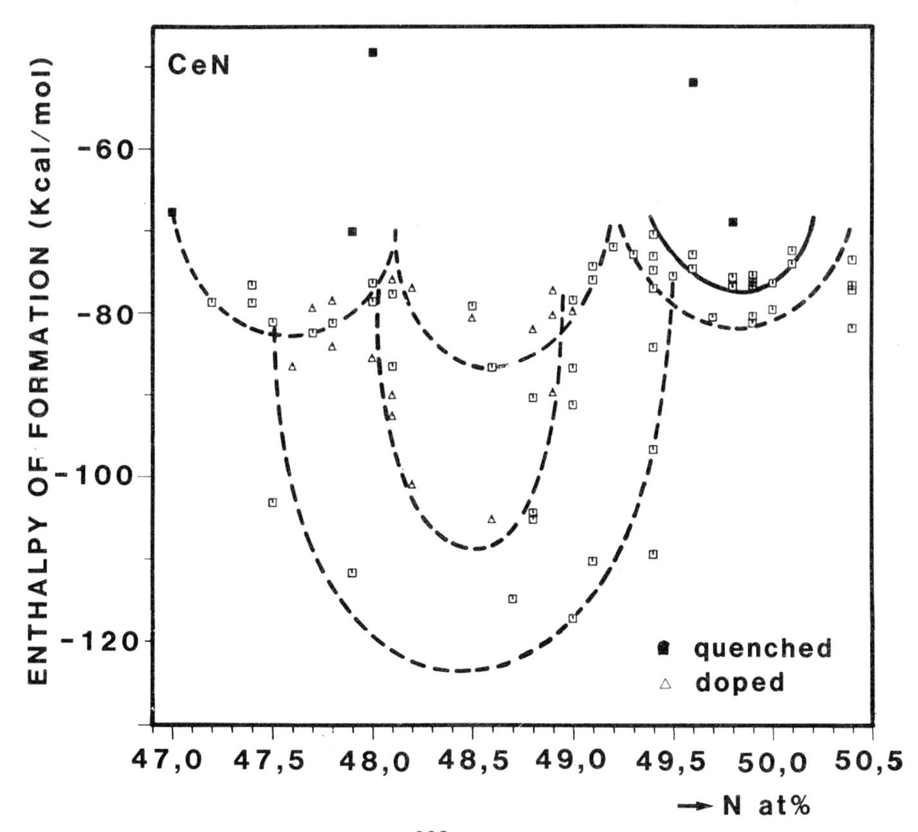

Fig. 2. Enthalpies of formation ΔH_f^{298} of CeN as a function of nitrogen content

~48.5at% N, i.e. 30% more stable than pure CeN, and

b) to a phase with $\Delta H_f^{298} \simeq -117$ Kcal/mole and the same N-concentration i.e. 54% more stable than pure, stoichiometric CeN.

Most of the samples (a) belong to CeN batches which had been doped with oxygen (by addition of CeO_2) and contain according to the chemical analysis not more than 1at% oxygen. In agreement to that they show higher lattice constants.

The samples (b) have not beeing doped on purpose. Some of them, show larger lattice constants indicating contamination. The sum of the chemical analysis of Ce+N, however, shows clearly that this contamination must be smaller than 1at%.

In view of this low contamination level, any thermochemical arguments to explain this huge increase in stability are out of question. Also structural changes do not seem probable. Due to the enormous reactivity of CeN single crystal investigations have not been possible up to now. Powder diagrams, however, do not show any additional reflexions to those of the NaCl-structure.

Changes of the electronic structure remain as the probable explanation. If the pure stoichiometric compound has mixed valence, incorporation of oxygen to specific sites of the lattice could destroy the delicate balance of fluctuating valencies and stabilize the compound. Obviously several stable states (a,b) can exist.

Already in the 60's, it was found that CeN shows an anomalous thermal expansion (27) which has been explained by increasing Ce^{3+}-concentration at higher temperatures (27,28). Temperatures of the order of 1000 C have been extrapolated for the existence of a mainly trivalent CeN. Optical reflectivity measurements as a function of temperature (29) show a decrease of the carriers concentration, which is compatible with an increase of the Ce^{3+}-contribution under the assumption that the effective mass does not change appreciably. On the contrary, magnetic measurements do not give intelligible results (28) as befitting a mixed valent compound.

In terms of a classical redox-reaction doping with oxygen would be expected to increase the Ce^{3+}-concentration. In order to receive Ce^{3+}-rich samples and compare their enthalpies with

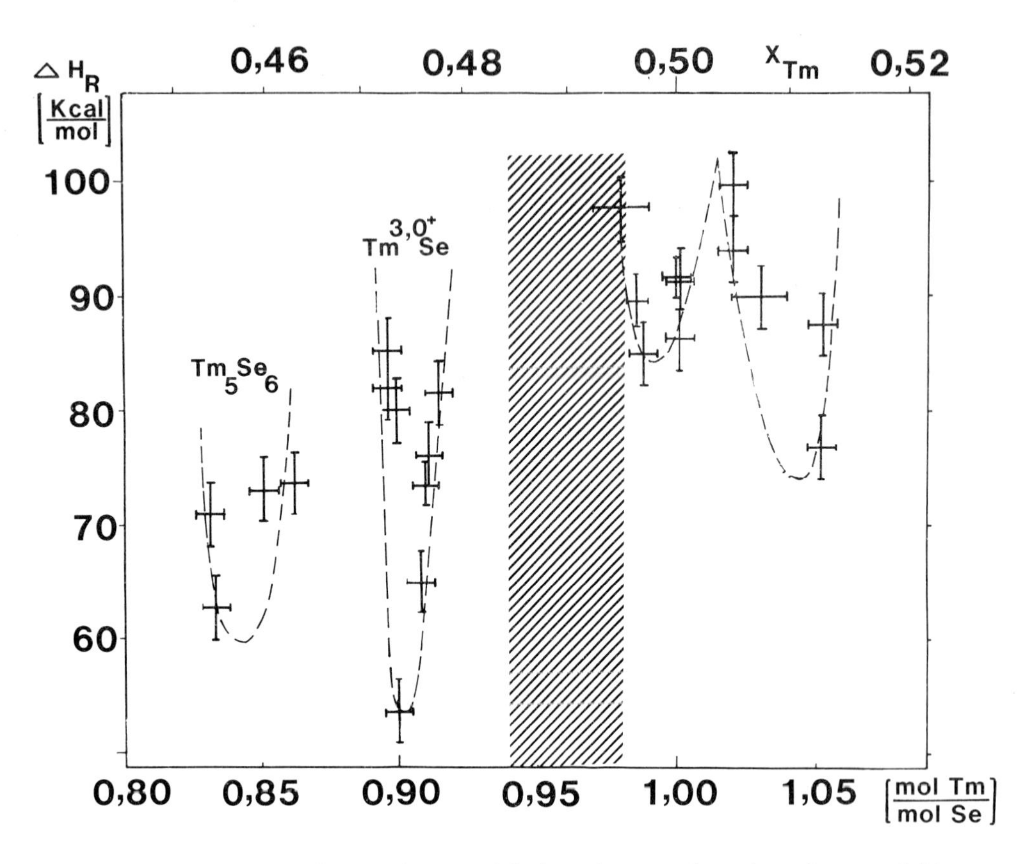

Fig. 3. Reaction enthalpies of TmSe with 4N HCl, as a function of nonstoichiometry

the samples (a) and (b) which we assume to be oxygen doped, we have tried to quench some crucibles with CeN from 2200 C. Due to the large mass, quenching is very difficult and no tendency to highly negativ enthalpy-values could be found. On the contrary some samples showed much less negativ values i.e. they were much more unstable than pure, stoichiometric CeN (up to 30% destabilization). We may, therefore, conclude that the quenched high-temperature state is different than the state resulting from oxygen doping.

2.2. Thermodynamic instability of the fluctuating TmSe

Total energy measurements in TmSe (3,9,12,13,30) show also that valence instabilities induce very strong thermodynamic instabilites. Up to now only reaction calorimetry (with 4N HCl) has been performed in TmSe because this compound, like most chalcogenides and pnictides, cannot be dissolved in acids. Enthalpies of reaction as a function of nonstoichiometry give information about the differences in total energy of TmSe in various valence states (3,13,30). Recently, fluorine combustion calorimetry has been developed in our laboratory (13) in order to measure absolute values of the total energy. Figure 3 shows the reaction enthalpies ΔH_r^{298}. Low values of the enthalpy of reaction correspond to high (negative) values of the enthalpy of formation i.e. to high stability. It is, therefore, clear from the figure that across the miscibility gap (16) the stability of TmSe changes dramatically. Comparison of a hypothetical stoichiometric trivalent thulium selenide $Tm^{3.0+}Se$ with the actual stoichiometric $Tm^{2.8+}Se$ shows a destabilization of approx.30% (13,16). This decrease of thermodynamic stability in the Tm-rich range ($0.98 < Tm/Se < 1.05$) coincides with a strong change of the physical properties of TmSe (5,6) which indicates the onset of valence fluctuation. We can, therefore, conclude that as a result of the fluctuating valence the thermodynamic stability of TmSe decreases strongly. The change of the valence seems to be the reason for the miscibility gap, the repulsive forces developing between the two different valence states.

The calorimetric measurements of Fig. 3. show the appearence of four phases in the nonstoichiometric range of TmSe which originally has been considered as homogeneous. The Tm_5Se_6, a NaCl-superstructure (with double lattice constant) and the $Tm^{3.0+}Se$ are the most stable

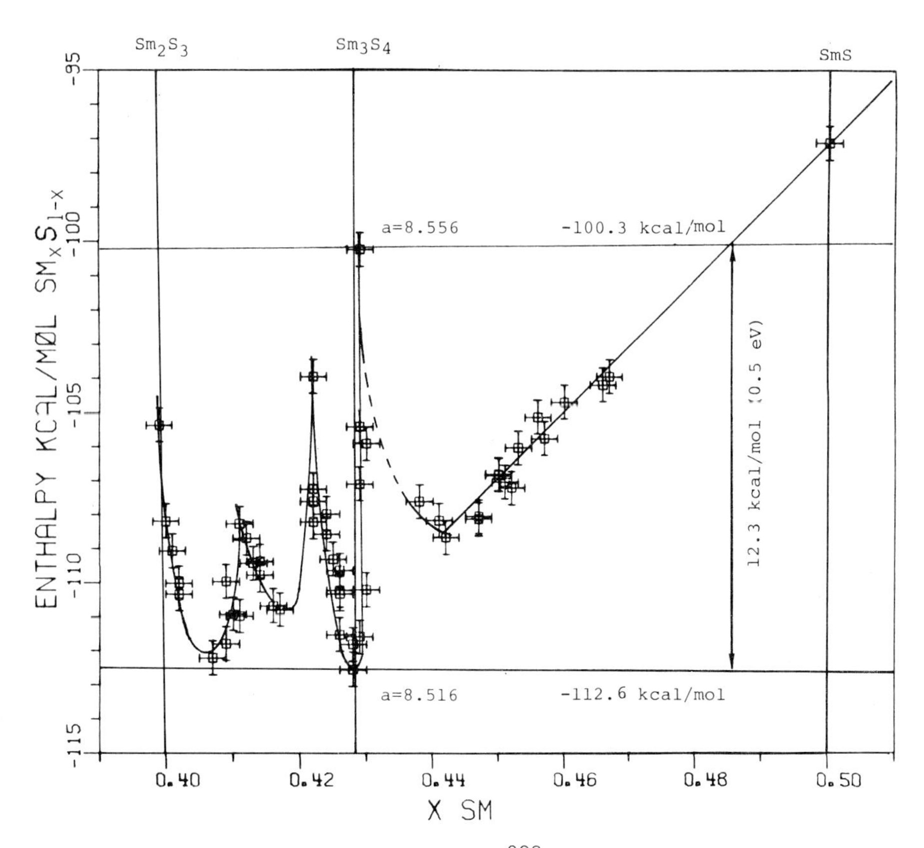

Fig. 4. Enthalpies of formation ΔH_f^{298} of solid solutions Sm_3S_4-Sm_2S_3 as a function of the samarium molefraction x_{Sm}.

phases as expected for Tm^{3+}-compounds. Their trivalent character is clearly shown from the susceptibility measurements and their transport properties (5). In the Tm-rich range (0.99<Tm/Se<1.05) where the TmSe is mixed-valent two different phases appear. The instability at their boundary (Tm/Se $\simeq$ 1.02) indicates that in this range two mixed-valent states may exist. Unfortunately, physical measurements in this composition range are existing only for very few compositions so that such effect could not be traced up to now.

2.3. Thermodynamic instabilities in the Sm_3S_4- Sm_2S_3 mixed crystal range

Some structural properties and phase relationships are discussed in (17), the Raman scattering in (18). Here we discuss only the total energy measurements as a function of nonstoichiometry.

The anomaly of the lattice constants of Sm_3S_4 recently discovered (19,20), triggered the question of the thermodynamic stabilities of the samples with large lattice constant a = 8.549 Å (31) and small lattice constant a = 8.5198 Å (19,32). Further, this mixed crystal series offers an excellent possibility to vary continuously the valence of Sm from 2.66^+ (Sm_3S_4) to 3.0^+ (Sm_2S_3) without perturbations from a third component (17). The question arrises if this change of valence takes place continuously or discontinuous transitions between different valence states appear, as in the case of TmSe (sect.2.2).

Solution calorimetry in 4N HCl gives for this system quantitative information about the enthalpy of formation as a function of the samarium concentration. Figure 4 shows the experimental results. Stoichiometric Sm_3S_4 with the smaller lattice constant is 12.3 Kcal/mole more stable than compositions with high lattice constant which are overstoichiometric. The mixed crystal series Sm_3S_4-Sm_2S_3 is at room temperature not homogeneous but it splitts up into three narrow phases with appreciable instabilities at the phase boundaries. These instabilities indicate the existence of very narrow miscibility gaps which may be outside the normal resolution of structural and chemical character-

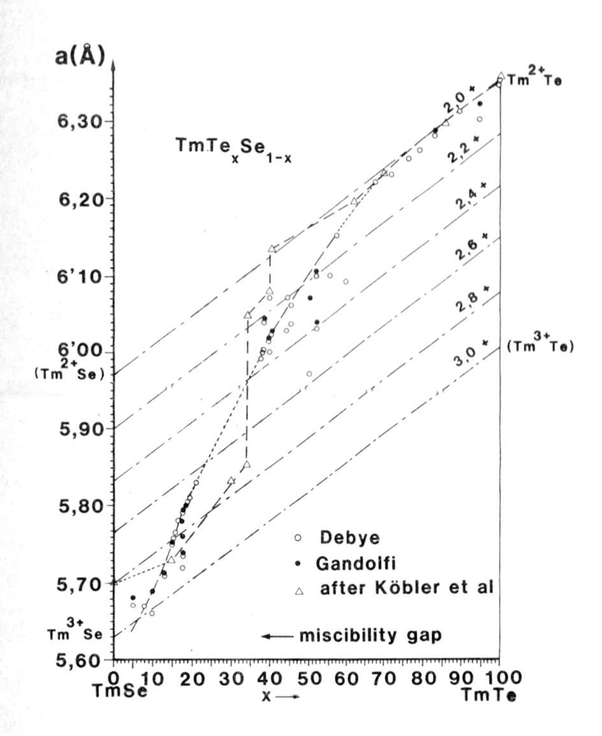

Fig. 5. Lattice constants vs composition in the $TmTe_xSe_{1-x}$ system

ization methods. The complexometric analysis of samarium has an absolute accuracy of $\pm$0.2%at, the high reproducibility of the enthalpy measurements indicates, however, a much better reproducibility for the chemical analysis.

According to the phase diagram Sm-S (17), the two-phase region Sm_3S_4-SmS starts at $x_{Sm}>0.43$. The linear dependence of the enthalpy of formation on composition in the range $0.44<x_{Sm}<0.50$ is in agreement with the existence of two solid phases. However, the slight increase of ΔH_f^{298} at slightly smaller x_{Sm} cannot be understood at the moment. The phase diagram (17) shows two phase transitions at approx. 1550 C and 1650 C. It is possible that hysteresis of one of these phases is the source of this anomaly. A structure refinement (33) is presently performed to clarify the problems of structural homogeneity.

3. MISCIBILITY GAPS IN SOLID SOLUTIONS BETWEEN COMPOUNDS WITH LOCALIZED AND DELOCALIZED ELECTRONS

Miscibility gaps appear as the result of repulsive forces between the phases (or some of their components) representing their phase boundaries. These forces can be due to geometrical factors (ionic mismatch) or to differences in the electronic structure. The first case is well known in many common alloys where due to the isotropic metallic bond the size effect is the source of anisotropic interaction. The second factor becomes important in solid solutions between compounds with localized and itinerant electrons.

The formation of such solid solutions can be used also to change the valence, see e.g. ref. (8,9).

3.1. Chemical substitution to change the internal lattice pressure

In the past we have proposed several solid solutions of TmSe in order to change the valence of Tm from 3.0^+ to 2.0^+ (7,9). The idea has been to alloy

a) with larger anions (Te^{2-}) or cations (Eu^{2+}, Sm^{2+}, Ca^{2+}, Sr^{2+}, Ba^{2+}) in order to decrease the crystal-field splitting of the 5d-band and open a gap between 4f and 5d, and
b) with semiconductors (TmTe, EuSe) in order to decrease the carriers concentration so that the divalent semiconducting state can be reached. The example of e.g. Ca-substitution in SmS (34) has shown clearly that internal lattice pressure alone cannot change the valence if a suitable change of the carriers concentration does not take place.

Figure 5 shows new data in the $TmTe_xSe_{1-x}$ system where a gap clearly exists between 18 and 38% TmTe. This gap divides the metallic from the semiconducting state (35) and it is rather difficult to find, indicating an appreciable thermodynamic metastability. In fact, in another laboratory (36) samples of two compositions in the miscibility gap have been prepared (x=30,34%). From these few measurements the impression is given that the physical properties do not show irregularities at the boundaries of the miscibility gap. Using special synthesis conditions, we have started preparing metastable samples in the miscibility gap in order to find out if a change of the magnetic coupling takes place (37). The existing results indicate also an increase of the Curie temperature without discontinuity at the phase boundary, thus supporting the above mentioned metastability.

For the determination of the phase boundaries of the miscibility gaps, we have used X-ray powder diffraction (chemical analysis is not possible due to the coexistence of several phases). Using this criterion, without considering reaction kinetics, the transition range metal to semiconductor becomes very complicated for the system $Tm_{1-x}Eu_xSe$. Figure 6 shows a tentative dependence of the lattice constant on composition with three miscibility

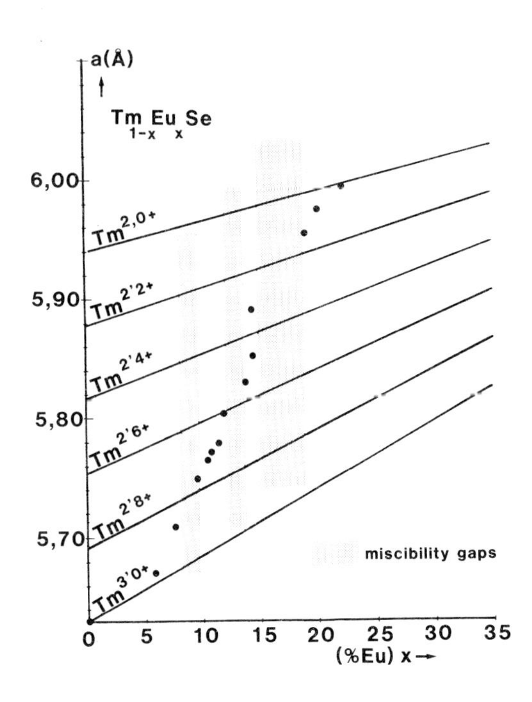

Fig. 6. Lattice constants vs. composition in the $Tm_{1-x}Eu_xSe$ system

gaps.

The above systems aimed to the change of valence from 3.0^+ to 2.0^+ and are coupled to a composition induced metal-semiconductor transition. Recently we have tried the opposite change ($2.0^+ \rightarrow 3.0^+$) starting from semiconductors.

Due to the large misfit of the lattice constants ($\Delta a/a \simeq 12\%$) the $Tm^{2+}Te$-$Tm^{3+}P$ system shows only solubilities near the endcompositions: ~ 0.4% of TmP in TmTe and 15% of TmTe in TmP. On the contrary we could expect a large solid solution range due to the very small ionic mismatch between $Eu^{2+}S$ (a=5.97Å) and a hypothetical $Eu^{3+}As$ with an interpolated lattice constant a ≃ 5.9 Å. Crystals of the nominal compositions $EuS_{0.5}As_{0.5}$ have a = 5.967 Å with the EuS color and are now characterized.

3.2. Isopiestic Substitution

The above systems combine internal lattice pressure effects (substitution with ions having appreciable differences of ionic radii) with variation of the carriers concentration, for the valence change. Therefore, the appearence of the miscibility gap can be attributed to geometrical factors. It would be interesting to find out if a miscibility gap appears also when the valence change is triggered only by the change of the carriers concentration. An ideal system for such "isopiestic" substitution appears in an exchange of divalent Te (r=2.22Å) with trivalent As (r=2.22Å). Figure 7 shows some first results in the $TmTe_{1-x}As_x$ system. A miscibility gap seems to exist in the composition range 12-27% As. Samples cooled abruptly from high-temperatures have X-ray powder diagrams with a whole range of lattice constants $5.870 < a < 6.025$Å. Further investigations will show the properties of this ternary isopiestic system.

4. CONCLUSIONS

The main results emanating from these investigations are summarized in the following:

1. Valence instabilities induce thermodynamic phase instabilities which in some cases (TmSe, CeN) reach dramatic dimensions (30-50%). Total energy measurements (calorimetry) as a function of the chemical composition are a very sensitiv tool for the investigations of valence anomalies.

2. Such instabilities appear not only in the chemically (composition) induced metal to semiconductor changes coupled to the valence change, but also when the valence state changes in the range of the same conduction type e.g. metallic TmSe, semiconducting Sm_3S_4-Sm_2S_3.

3. The experimental results existing up to now, indicate that in mixed-valent compounds the valence changes by several discrete steps which extend only over narrow composition ranges. The reason lies probably in the repulsive forces appearing by the mixing of localized and delocalized states. In compounds with integer valence similar reasons lead to miscibility gaps in alloys of semiconductors with metals.

4. These effects of mixed-valence can be studied in "isopiestic" systems where no substitution with ions of appreciable size mismatch takes place. Such systems are
   a) homogeneity ranges like that of TmSe. The existence of a miscibility gap in the homogeneity range can be attributed only to repulsive forces due to changes of the electronic structure.
   b) Solid solutions like Sm_3S_4-Sm_2S_3. The three narrow instability ranges seem to be the result of the repulsive interaction of different valence states. Vacancy correlation effects possibly coupled with the valence changes must be investigated.
   c) Probably ternary isopiestic systems like $TmTe_{1-x}As_x$, where changes of valence are only produced due to band structure effects. Future investigations will show the properties of such systems.

5. A most unexpected result is the enormous metastability of the unstable compositions.

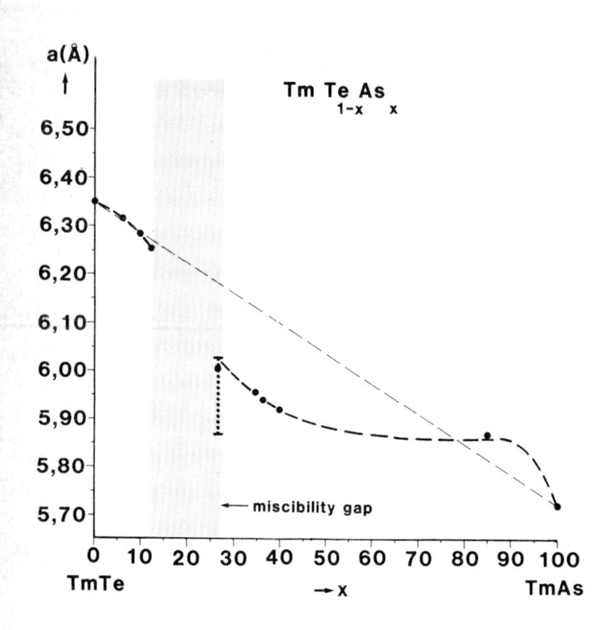

Fig. 7. Lattice constant vs. composition in the $TmTe_{1-x}As_x$ "isopiestic" system.

Samples with these compositions can be synthesized and handled without exploding, although they have enthalpy differences of up to 50 Kcal/mole, to the stable compositions. This must be compared with differences up to 5-10 Kcal/mole which can exist in systems with integer valence.

Obviously, the reasons for this metastability must be sought in the unusual electronic structure of these compounds. Defects of any kind (including impurities → CeN) probably stabilize kinetically these thermodynamically extremely unstable compositions.

In conclusion, the above results show that the interests of solid state physics are moving to a field of materials with complicated but fascinating thermodynamic properties. Coupled investigation of these and the various physical properties as a function of chemical composition will improve our understanding of mixed valence.

REFERENCES

1) Wertheim, G.K., Eib, W., Kaldis, E., Campagna, M., Phys. Rev. B. 22 (1980) 6240
2) Fritzler, B., Kaldis, E., Steinmann, B., Jilek, E. and Wisard A., in "The Rare Earths in Modern Science and Technology" Vol. 3, Mc Carthy et al eds. Plenum Press, (1982)
3) Kaldis, E. and Fritzler, B., "Valence and Phase Instabilities in TmSe Crystals" in "Progress in Solid State Chemistry", Rosenblatt, G.M., Worrell, W.L., eds., Pergamon Press (in press)
4) Batlogg, B., Kaldis, E., Schlegel, A. and Wachter, P., Phys. Rev. B 14 (1976) 5503
5) Batlogg, B., Ott, H.R., Kaldis, E., Thöni, W. and Wachter, P., Phys. Rev. B19, (1979) 247
6) Batlogg, B., Kaldis, E. and Ott, H.R., Phys. Letters 62A (1977) 270
7) Kaldis, E., Fritzler, B., Jilek, E. and Wisard, A., J. de Physique 40, (1979) C5-366
8) Jayaraman, A., Dernier, P.D. and Longinotti, L.D., Phys. Rev. B11, (1975) 2783
9) Kaldis, E., Fritzler, B., J. de Physique 41 (1980) C5-135
10) Batlogg, B., Kaldis, E. and Wachter, P., J. de Physique 40, (1979), C5-370
11) Kaldis, E., Steinmann, B., Fritzler, B., Jilek, E., Wisard, A., in "The Rare Earths in Modern Science and Technology" Vol.3 Mc Carthy, et al. eds., Plenum Press 1982
12) Fritzler, B., Diploma Thesis, Laboratorium für Festkörperphysik ETHZ, 1978
13) Fritzler,B., PhD Thesis, Laboratorium für Festkörperphysik ETHZ, 1982
14) Steinmann, B., Diploma Thesis, Laboratorium für Festkörperphysik ETHZ, 1980
15) H. Spychiger, Diploma Thesis, Laboratorium für Festkörperphysik ETHZ, 1981
16) Fritzler, B., Kaldis, E., this proceedings
17) Spychiger, H., Kaldis, E., this proceedings
18) Mörke, I., Wachter, P., this proceedings
19) Kaldis, E., J. Less Common Metals 76 (1980) 163
20) Kaldis, E., Spychiger, H., Fritzler, B., Jilek, E., in "The Rare Earths in Modern Science and Technology" Vol.3., Mc Carthy et al eds., Plenum Press 1982
21) Iandelli, A., Bolti, E., Rend. Accad. Naz. Lincci 25 (1937) 139
22) Kaldis, E., Zürcher, Chr., Proceedings of the 12th Rare Earth Research Conference, Vol. II, p.915, Lundin, C.E., Ed., Denver Research Institute (1976)
23) Baer, Y., Zürcher, Ch., Phys. Rev. Lett. 39 (1977) 956
24) Gudat, W., this proceedings
25) see e.g. Gudat, W., Rosei, R., Weaver, J.H. Kaldis, E. and Hulliger, F., Solid State Communications 41 (1982) 37 and therein contained references.
26) Brown, R.C., Clark, N.J., Mat. Res. Bull. 9 (1974) 1007
27) Essen, V., Klemm, W., z. Anorg. Allg. Chem. B17 (1962) 16
28) Danan, J., de Novion, D., Lallement, R., Solid State Commun. 7 (1969) 1103
29) Schlegel, A., Kaldis, E., Wachter, P., Chr. Zürcher, Phys. Letters 66A (1978) 125

30) Fritzler, B., Kaldis, E., Jilek, E., in "The Rare Earths in Modern Science and Technology" Vol.3, Mc Carthy, G., et al. eds., Plenum Press 1982
31) Coey, J.M., Cornut, B., Holtzberg, F., and von Molnar, S., J. Appl. Phys. 50 (1979) 1973
32) Batlogg, B., Kaldis, E., Schlegel, A., von Schulthess, G. and Wachter, P., Solid State Commun. 19, (1976) 673
33) Bärnighausen, H., et al. (to be published)
34) Jayaraman, A., Maines, R.G., Phys. Rev. B, 19 (1979) 4154
35) Batlogg, B., Kaldis, E. and Wachter, P., J. Physique 40 (1979) C5-370
36) Köbler, U., Fischer, K., Bickman, K., Lustfeld, H., Magn. and Magn. Mat. 24, (1981) 34
37) Fischer, P., Hälg, N., Schobinger, P., Boppart, H., Kaldis, E. and Wachter, P., these proceedings
38) Varma, C.M., Solid State Commun. 30 (1979) 537

J.C. FUGGLE: Do you have any ideas concerning the mechanism by which ∿1 at% of oxygen stabilizes CeN to such a large extent.

E. KALDIS: This is answered to the best of our present knowledge in the text. To save space we avoid duplication of the answer here.

F. HULLIGER: Is the color of your quenched samples silvery?

E. KALDIS: No. We were looking for such change. The only color change which we have found is in the homogeneity range: orange-gold for N-rich samples; yellow-gold for Ce-rich ones. There might be a change of the position of the plasma resonance frequency.

R. SURYANARAYANAN: What is the contamination due to hydrogen in your crystals?

E. KALDIS: I don't really know, but the temperature at which the samples are grown (2200^{0}C) is rather prohibitive for the incorporation of hydrogen. We also don't expect a contamination during the cooling process.

A. JAYARAMAN: In CeN, what is the valence state of Ce? If Ce is 3+, it is charge compensated. Where does the conductivity and golden color from?

E. KALDIS: XPS measurement showed the presence of both trivalent and tetravalent Ce. What is not in the literature is the percentage of tetravalent Ce because one has no clear cut crystal-chemical method to determine it. The ionic-radius method gives you 84% tetravalent Ce. Another approach was used by Gschneidner and Smoluchowski in their discussion of the $\gamma - \alpha$ transition of Ce and they extrapolated a metallic radius for Ce and if you use that you come down to 40% tetravalent Ce in CeN. Anyway, all existing results indicate the presence of Ce^{4+} and therefore you must have electrons.

A. JAYARAMAN: What is the nature of the miscibility gap in TmSe? Is there a segregation into Tm^{2+} and Tm^{3+} regions?

E. KALDIS: All our investigations show, that the range of composition where valence states can be mixed is very narrow. At some composition, repulsive forces set in and give way to the miscibility gap. In the case of TmSe the repulsion would be between formal valences of 2.88 and 2.84 (12 and 16% of Tm^{2+}).

Valence Instabilities
P. Wachter and H. Boppart (eds.)

RECENT PROGRESS IN THERMODYNAMICS OF VALENCE FLUCTUATIONS

(Extended Abstract)

D. Wohlleben, J. Röhler, R. Pott, G. Neumann and E. Holland-Moritz

II. Physikalisches Institut
der Universität zu Köln
Zülpicherstr. 77
D-5000 Köln 41, FRG

With the aid of phenomenological thermodynamics it is possible to achieve semi-quantitative understanding of the temperature and pressure dependence of the valence, susceptibility, specific heat, thermopower etc. of real valence fluctuation systems by guessing a spectrum and fitting it to the available data. The power of the method lies in the fact that the input parameters are to a large extent fixed because most of them are either well known or measurable properties of stable reference systems (Hund's rule ground states and excited multiplet states with their degeneracies and splittings in the $4f^{n+1}$ and the $4f^n$ configurations on one hand and bulk modulus, volume of the unit cell, linear specific heat coefficient etc. on the other). The bona fide mixed valence parameters which remain adjustable are primarily E_x, the interconfigurational excitation energy, and $\Gamma \equiv k_B T_f$, which has various names such as interconfigurational fluctuation energy or mixing energy or lifetime width, and which is also phenomenologically hard to distinguish from the Kondo or spin fluctuation energies defined in some theoretical approaches. Adjustable fit parameters of secondary importance are the parameters of the crystal field (CEF) splittings, which can demonstrably survive the valence fluctuations e.g. in the $4f^1$ and $4f^{13}$ configurations of Ce and Yb VF compounds. If the number of measured thermodynamic averages equals the number of the adjustable parameters, the spectrum is fixed within limits given by the experimental accuracy, and from then on the method becomes predictive. Not only can other macroscopic thermodynamic averages be anticipated, but also the results of microscopic spectroscopies. The method has gained importance in the last few years with the increasing number and accuracy of data which have become available on any given system.

The first attempts to understand VF data via phenomenological thermodynamics were fits of the temperature dependent Moessbauer isomershift (equivalent to the temperature dependent valence) of $EuCu_2Si_2$ /1/ and of the magnetic susceptibility of a series of VF Yb compounds /2/. The apparently successful introduction of a lifetime width Γ of the $4f^{13}$ Zeeman levels for the VF Yb ions in /2/ led to a search for direct spectroscopic evidence for this quantity by inelastic magnetic neutron scattering and to its detection in the abnormally large width of the quasielastic line /3,4/. Predictions were also made for the temperature dependence of the valence in /2/, e.g. for $YbCu_2Si_2$. They were based on fits to a single quantity, namely $\chi(T)$, and on the assumptions that E_x and Γ are independent of temperature and that there are no surviving CEF splittings. In the meantime a second quantity, the temperature dependence of the lattice constant anomaly, was measured for $YbCu_2Si_2$ by capacitive thermal expansion /5,6/. While a valence change of more than 30% was predicted between 300 and 4 K in /2/, the thermal expansion anomaly gave only about 3%. Most of the discrepancy could be removed by introducing a CEF spectrum in $4f^{13}$ with an overall splitting of about 30 meV. This finding led to an extensive search for CEF splittings in $YbCu_2Si_2$ by neutron scattering, with not only one /4/ but several incoming neutron energies, small and large compared to the suspected CEF splittings. Broadened CEF spectra of the anticipated kind were finally found /7,8/ and it now appears that magnetic line widths in the literature are overestimates /3,4,7,9/ if they were extracted from neutron data with the assumption of zero CEF splittings.

The existence of residual CEF and spin orbit (SO) splittings in VF compounds implies that there is not only one but a series of interconfigurational excitation energies $E_{x,i,j} \equiv E_{n+1,i} - (E_{n,j} + \varepsilon_F)$. The residual CEF splittings do not seem to depend much on temperature /7/. If one however fits the susceptibility /2/ and the thermal expansion anomaly /5/ simultaneously for $YbCu_2Si_2$, using the temperature independent CEF splittings as detected in /7/ with a slightly temperature dependent linewidth as also seen in /7/, one finds that the $E_{x,i,j}$ have a (common) temperature dependence. At $T = 0$ one finds $E_{x,o,o} = E_{14,o} - (E_{13,o} + \varepsilon_F) = 1.5$ meV, and that the dominant population p is in the levels $E_{14,o}$ ($p_{14,o} \simeq 0.2$) and $E_{13,o}$ ($p_{13,o} \simeq 0.8$). Precisely at 1.5 meV we found on the other hand strong singularities of the nonlinear current voltage characteristics of metallic point contacts of molybdenum with $YbCu_2Si_2$ /10/. This coincidence led us to the realization that it is possible to measure the interconfigurational excitation energies $E_{x,i,j}$ directly by metallic point contact spectroscopy (= MPCS), ($E_{x,i,j}$ cannot be "seen" by inelastic magnetic neutron scattering in Ce and Yb VF compounds because the configurational groundstates of $4f^o$ and $4f^{14}$ are nonmagnetic singlets). By point contact spectroscopy we also found the lifetime width Γ_c of the conduction electrons to be close to the lifetime width Γ_m of the $4f^{13}$ Zeeman levels used in the thermodynamic fits, and that there is a very strong asymmetry of the renormalized conduction electron density of states about the Fermi energy in $YbCu_2Si_2$ (and in other Yb and Ce VF compounds). This asymmetry can again be explained semiquantitatively with phenomenological thermodynamics. It is a consequence of the unequal populations p of $E_{14,o}$ and $E_{13,o}$ in thermal equilibrium. Such unequal populations result basically from the existence of residual CEF and SO splittings which force all E_{xij} to be finite in thermal equilibrium in general, and in particular here $E_{x,o,o}$. This is discussed in /6/.

The general temperature dependence of E_x and Γ is expected to be especially clear when the valence changes strongly, (i.e. by an appreciable fraction of one) over the temperature range available in the experiment. The same is true for the pressure dependence of E_x and Γ. We therefore chose $EuCu_2Si_2$ over $YbCu_2Si_2$ for a closer study of $E_x(T,p)$ and of $\Gamma(T,p)$. In $EuCu_2Si_2$ the isomershift indicates a valence change near 0.2 between 4 and 300 k at atmospheric pressure and one near 0.3 between 0 and 53 kbar at 300 K /1, 11, 12/. The configuration $4f^6$ of Eu^{3+} has moreover a rich internal SO energy structure which is fairly insensitive to CEF effects, as the J = 7/2 spin only groundstate of $4f^7$. Therefore two macroscopic averages are sufficient to fix the only adjustable parameters $E_x(T,p)$ and $\Gamma(T,p)$ of VF Eu compounds. We chose the magnetic susceptibility and the Moessbauer isomershift, which happened to be measured over the above mentioned ranges of T and p simultaneously on a single well characterized sample /12,13/. We found a strong variation of both parameters and a strongly nonlinear relationship between them, $\Gamma(E_x)$, which is independent of T and p within the accuracy of the data. This finding implies that there is only one independent bona fide mixed valence energy parameter, for which we chose $E_x(T,p)$ and that the relationship $\Gamma(E_x)$ is determined by two intraconfigurational spectra at integral valence. The "independent" parameter E_x is in turn determined self consistently by the distance of the Fermi energy ε_F to the bottom of the conduction band, by the elastic energy F_L and by the mixing energy Γ, which are all functions of the fractional valence ν. We were able to reproduce the observed $E_x(T,p)$ for $EuCu_2Si_2$ by simple functions $\varepsilon_F(\nu)$ and $F_L(\nu)$ and a more complicated but well-defined term $E_{xm}(\nu)$ in the general expression for E_x. E_{xm} represents the effect which a change of mixing energy upon a change of valence, temperature and pressure has on E_x /11,12/. The form of $F_L(\nu)$ is found by taking the nonlinear relationship between valence and volume into account which follows from the general observation that the bulk moduli of compounds with divalent rare earth ions are much smaller than those with trivalent ones. This leads to a strongly nonlinear dependence of the elastic energy F_L on valence /13/. We find that the shifts of the Fermi energy and of the elastic energy due to shifts of valence cancel each other to a large extent, such that the valence is mainly determined by the temperature dependence of the intraionic entropy at high temperatures and by the temperature dependence of the mixing energy at low temperatures. Since both are intraionic properties, i.e. fairly independent of the matrix, we believe to have found the reason for the often stated but so far puzzling observation that mixed valence effects are fairly independent of concentration, but strongly dependent on the intraionic structure of the VF ion /6/. The most interesting term in the energy balance of mixed valence is the above mentioned E_{xm} /11/, which has the effect of pinning two levels of $4f^6$ and $4f^7$ together, or, in a more simplified but less realistic language to pin the f level to the Fermi level. This pinning maximizes the gain of mixing energy, which can be seen directly by a maximum of $\Gamma(E_x)$ where J=7/2 of $4f^7$ is degenerate with J=1 of $4f^6$ and by a reversal of the sign of E_{xm} when these levels cross /11,12/. At low temperatures this term is responsible for the entire temperature dependence of E_x, which is strong there in spite of the fact that the valence is nearly stationary. Such temperature dependence of E_x and Γ due to the temperature dependence of mixing without appreciable valence change has important consequences for the low temperature specific heat, which are under investigation at present.

So far the results of our work with phenomenological thermodynamics all tend to strengthen our confidence in the soundness of this approach and in the ionic model /14/. We therefore have looked into the consequences of this model at thermal energies which are larger than the residual CEF splittings and the lifetime width, i.e. larger than a few 100 K. If one also has $E_x(T) < k_BT$ there, the ionic model predicts saturation of the valence of Ce, Pr, Tm and Yb at 3.14, 3.40, 2.62 and 2.89 respectively, and a more complicated behaviour for Eu and Sm due to their relatively small intraionic SO splittings /6/. Strong indications for such valence saturations are the disappearance of the anomalies of thermal expansion and of specific heat above a few 100 K in YbCuAl /15/ and $YbCu_2Si_2$ /5, 6,16 / and a similar tendency for the thermal expansion anomaly for Ce compounds /17, 18, 19, 6/. Also, the Curie constants of VF Yb and Ce compounds at high temperatures seem to be closer to the fractional values predicted at saturated mixed valence (6/7 and 8/9 times the trivalent constants for Ce and Yb respectively) than to the integral(tri-) valent values although it is difficult to be sure. We have started measurements of the resistivity of Ce and Yb VF compounds above room temperature and found saturation of the large low temperature anomalies above a few 100 K in every case studied so far /20, 21/. We regard these results as the strongest evidence so far for the high temperature saturation of the valence at its entropy limit and suggest that the high temperature behaviour of VF compounds should be studied much more intensively in the future because it seems to be very simple and therefore should shed valuable light from this end on the often so puzzling low temperature VF anomalies, especially of the transport properties.

The high temperature limit of the mixed valence also has bearing on the question of the existence or nonexistence of tetravalent intermetallics of Cerium. In their L_{III} absorption spectra all "tetravalent" Ce compounds show two peaks at the edge, just as mixed valence compounds of Sm, Eu, Tm and Yb /22/. Naive analysis of this double peaked structure leads to valencies which are never larger than 3.3.(Even the insulators CeO_2 and CeF_4 seem to have valencies near only 3.67 and 3.8 respectively /23/). It turns out that one can fit simultaneously these fractional valencies and the susceptibilities by thermodynamics with the ionic model, this time with somewhat larger mixing energies than usual ($\Gamma \sim$ 30-100 meV) which are however still very small compared to the usual d bandwidths and also compared to the mixing widths extracted for these compounds from photoemission. A most important role is again played by E_{xm} /11/,which is found to pin E_x to "zero" much more strongly than in $EuCu_2Si_2$ and is therefore capable of holding the valence near the high temperature limit against very strong lattice pressure, which tends to drive it towards four.

REFERENCES

/1/ E.R. Bauminger, D. Froindlich, I. Nowik, S. Ofer, I. Felner and I. Mayer Phys. Rev. Letters 30, 1053 (1973)

/2/ B.C. Sales, D.K. Wohlleben, Phys. Rev. Letters 35, 1240 (1975)

/3/ E. Holland-Moritz, M. Loewenhaupt, W. Schmatz and D. Wohlleben Phys. Rev. Lett. 38, 983 (1977)

/4/ E. Holland-Moritz, D. Wohlleben and M. Loewenhaupt J. de Physique 39, C6-835 (1978)

/5/ R. Schefzyk, Diplomarbeit, Universität zu Köln (1980) and to be published

/6/ D. Wohlleben p. 1 in "Valence Fluctuations in Solids" L.M. Falicov, W. Hanke and M.B. Maple (eds.), North Holland (1981)

/7/ E. Holland-Moritz, D.K. Wohlleben and M. Loewenhaupt, to be published in Phys. Rev. B (1982)

/8/ E. Holland-Moritz, this conference

/9/ W.C.M. Mattens, F.R. de Boer, A.P. Murani and G.H. Lander, J. Magn. Magn. Mater. 15-18, 973 (1980)

/10/ B. Bussian, I. Frankowski and D. Wohlleben to be published

/11/ J. Röhler, D. Wohlleben, G. Kaindl and H. Balster, Phys. Rev. Lett. July 5, (1982) (in print)

/12/ J. Röhler et al., this conference

/13/ G. Neumann et al., this conference

/14/ L.L. Hirst, Phys. Kond. Mat 11, 255 (1970)

/15/ R. Pott, R. Schefzyk, D. Wohlleben and A. Junod, Z. Phys. B - Condensed Matter 44, 17 (1981)

/16/ R. Kuhlmann et al., this conference

/17/ R. Pott, R. Schefzyk, W. Boksch and D. Wohlleben, pg. 337 in Valence Fluctuations in Solids, L.M. Falicov, W. Hanke and M.B. Maple, editors, North Holland (1981)

/18/ G. Krill, J.P. Kappler, M.F. Ravet, A. Amamou and A. Meyer, J. Phys. F 10, 1031 (1980)

/19/ R. Pott, Ph.D. Thesis, Universität zu Köln (1982)

/20/ S. Hussain, E. Cattaneo, H. Schneider, D. Müller, W. Schlabitz and D. Wohlleben, to be published

/21/ D. Müller et al., this conference

/22/ K.R. Bauchspieß, W. Boksch, E. Holland-Moritz, H. Launois, R. Pott and D. Wohlleben in "Valence Fluctuations in Solids", L.M. Falicov, W. Hanke and M.B. Maple, editors, North-Holland (1981) pg. 417

/23/ K.R. Bauchspieß, Diplomarbeit, Universität zu Köln (1982) and to be published

T. KAPLAN: Is there any insight into why you replace T by $T+T_f$?

D. WOHLLEBEN: If you look at this expression at T=0 you realize that the distribution of the energy eigenstates is always finite also for the excited ones and if you interpret that as a mixture of the to be measured amplitudes (e.g. susceptibility) in the ground state, that seems to be a good description.

T.A. KAPLAN: An alternative way of getting the same is to replace $T+T_f$ by $\alpha(T)T$, which would give you some sort of scaling.

P. FAZEKAS: To some extent, the use of a composite temperature variable $T^+ = \sqrt{T^2 + T_f^2}$ with a quantum-mechanical (T=0) component T_f can be corroborated by referring to a variational treatment of the nondegenerate Anderson model (this volume). This uses a variational parameter which comes into a Boltzmann-factor-like Ansatz as T_f does, and describes the spread of the wavefunction over a number of configurations due to hybridization, just what T_f is supposed to do. The minimalization of the ground state energy then looks like minimizing a free energy at "temperature T_f", and some of the resulting expressions look rather like those derived in your phenomenological thermodynamic. Also, the "hybridization temperature" turns out to depend on valence, as you say about T_f.

M. CROFT: The T_f in your model plays two roles. It gives you a mixed ground state and it also determines the magnetic response, which may not be the same. Second, in your model you have to have a sixfold change in excitation energy below 200 K and a manyfold change in T_f below 200 K to maintain that constant valence.

D. WOHLLEBEN: The mixing energy is a function of the degeneracy of the levels and therefore the excitation energy may change drastically with changing the temperature.

Valence Instabilities
P. Wachter and H. Boppart (eds.)
© North-Holland Publishing Company, 1982

MAGNETIC AND OPTICAL DETERMINATION OF THE DEGREE OF VALENCE MIXING IN TmSe AND CeO_2

P. Wachter

Laboratorium für Festkörperphysik, ETH Zürich, 8093 Zürich, Switzerland

The advantages and disadvantages of different methods to determine the valency of an intermediate valent material are discussed. The susceptibility of TmSe has been measured up to 950 K where the entropy limit has been reached. CeO_2, which has been claimed to be intermediate valent (4), is shown by lattice constant, susceptibility and optical absorption to be tetravalent.

The degree of valence mixing in intermediate valent (IV) compounds is a steady point of discussion (1, 2, 3, 4). Common methods to determine the degree of valence mixing are measurements of the lattice constant, the Curie constant, the Mössbauer effect, X-ray photoelectron spectroscopy (XPS), ultra-violet photoelectron spectroscopy (UPS), core level spectroscopy and X-ray absorption spectroscopy (XAS). Without discussing more sophisticated methods like EXAFS, even the simple methods have obvious shortcomings, which have to be duly considered.

At an early time in the field of intermediate valence a distinction into "fast" and "slow" measurements with respect to a typical time scale of about 10^{-12} sec has been given. This distinction strictly holds only for mixed valence materials (or inhomogeneously mixed), for which integer valent ions, $4f^n$ and $4f^{n-1}$, sit on equivalent lattice sites in a crystal and thermally stimulated hopping of electrons takes place between adjacent ions of different, but always integer valence. Examples are Fe_3O_4, Sm_3S_4 and Eu_3S_4 (5,6). On the other hand it has been shown experimentally (e.g. 7) that in homogeneously mixed or intermediate valent compounds all rare earth ions are equivalent at zero temperature and one has a coherent hybridization of a $4f^n$ and a $4f^{n-1}5d$ configuration on each rare earth atom. If we use a linear superposition of these states we obtain $a\,|4f^n\rangle + b\,|4f^{n-1}5d\rangle$, where the 5d indicates delocalized electrons in a 5d band. The superposition and hybridization of these states lead to a mixing term in the hamiltonian

$$H_{hyb} = \sum_{m,\sigma} V\,(c^{\dagger}_{m\sigma}\Phi + \Phi^{\dagger}c_{m\sigma})$$

with V the hybridization matrix, $c^{\dagger}_{m\sigma}$ creating an f electron in a one particle orbital and $\Phi^{\dagger}$ creating a 5d electron (8). The hybridization and interaction of 4f and 5d states result in a "4f band" width of about 10 meV. It has been shown in addition (9, 10, 11) that compounds like metallic or "gold" SmS, TmSe, SmB_6 and others whose only conduction electrons are those delocalized from a 4f shell, exhibit a hybridization gap, typically a few meV wide, lying within this "4f band" (12).

It now becomes clear that the term "fast" and "slow" measurements becomes obsolete for hybridized, intermediate valent compounds, especially for $T \rightarrow 0$. Instead, we have to speak of high or low energy measurements with respect to typically 10 meV. To be precise, it is the energy transfer to the system which is of importance. Since the hybridization energy can be considered as the binding energy of the intermediate valent state, high energy measurements like XPS, UPS or XAS break up the hybridization at the site of the absorption of energy and what is left are integer valent ions $4f^n$ and $4f^{n-1}$ with concentration a^2 and b^2, as detected by their final state "fingerprint" like spectra $4f^{n-1}$ and $4f^{n-2}$, respectively.

It is thus an inherent problem of all high energy measurements that they cannot distinguish between homogeneously mixed and inhomogeneously mixed ground states. The claim of XPS or UPS to be able to determine the position of the Fermi level E_F and thus decide between metal or semiconductor (13) holds only in a few special cases. The spectral resolution of these measurements is far beyond typical 4f-5d separations, as e.g. in semiconducting SmS (0.2 eV), Sm_3S_4 (0.15 eV), semiconducting $TmSe_{1-x}Te_x$ and $Tm_{1-x}Eu_xSe$ ($50\ meV < \Delta E < 0.3\ eV$) (5, 14). The fact, that even in homogeneously mixed or intermediate valent "gold" SmS, SmB_6 and TmSe E_F lies in a gap, (the hybridization gap) was not observed with XPS or UPS, but with low energy methods (9, 10, 11).

XPS and UPS methods in addition suffer from surface effects (where we don't mean a trivial oxidation). In TmSe angle resolved photoemission (3) and angle integrated, but excitation energy varied photoemission (15) has shown that the degree of valence mixing is different in the surface and the bulk.

Ce compounds with $4f^1$ and $4f^o$ states have demonstrated the necessity of core level spectroscopic methods, such as e.g. XAS. In principle all inner shells such as 5p, 4d, 3d, 3p etc. can be investigated because their resonance energies vary according with the 4f shell occupation. In the case of TmSe XAS has been used to deduce the degree of valence mixing by comparing TmS, TmSe and TmTe (16). However, screening effects due to the simultaneously occurring change in free electron concentration (TmS: 1/f.u.,TmSe:0.7/f.u., TmTe:0/f.u.) have not been taken into account, thus a quantitative statement regarding the degree of valence mixing is questionable.

In addition one can have excitations from core levels into empty 4f states: $4d^{10}\,4f^{12} \rightarrow 4d^9\,4f^{13} \rightarrow 4d^{10}\,4f^{11}$ + electron. These excitations are not independent from each other and certain branching ratios exist. Another example is core hole creation in Ce compounds. Starting from an $4f^o$ initial state the e.g. $2p^5 4f^o 5d$ excited state can relax by s-d screening or by pulling down the empty $4f^1$ state below E_F. The final state contains amplitudes of $2p^5 4f^1 5d$ (17). It is thus possible that in an integer tetravalent compound like CeO_2, $4f^1$ and $4f^o$ spectra are present (see below) and seemingly make this material appear mixed valent (4).

Low energy methods intrinsically seem more adapted to test the IV ground state, because they don't destroy it. A change of the 4f occupation by one unit essentially changes the ionic radius by as much as 10%, which is unique in the periodic table. Simultaneously the outer 5s, 5p and 5d and the inner shells change their radius because of different shielding of the core charge. Since the IV compounds crystallize in the rocksalt structure the sum of the ionic diameters yields the lattice constant, which thus in turn is a simple measure of the degree of valence mixing, using a (debatable) linear interpolation between di - and trivalent reference compounds. This even holds for complex pseudobinary compounds such as $Tm_{1-x}Eu_xSe$ or $TmSe_{1-x}Te_x$ (18).

The isomer shift of the Mössbauer effect has an energy transfer of the order of 10^{-5} eV, much less than typical hybridization energies. The isomer shift measures the s - electron density at the core which depends on the radii of the s - shells which in turn depend on the 4f occupation. It is thus an indirect measure of the total ion radius, where again linear interpolations between reference materials are used. The valence determined by isomer shift and lattice constant thus should agree and indeed this is the case where comparisons are possible, e.g. in SmB_6 (19, 20). The problem is that isomer shifts with regard to different valence practically are restricted to Eu-and Sm compounds (7).

Another standard method is the valence determination by measuring the effective magnetic moment or rather the Curie constant. Here the usual theories tacidly assume a mixture of $4f^n$ and $4f^{n-1}$ ions with their standard S, L and J values. However, this is only correct for thermal energies high above the hybridization bandwidth, where the system performs real temporal fluctuations of the valence, because kT is a large energy transfer. Room temperature measurements certainly are not enough. Most measurements of the Curie constant, however, are per-

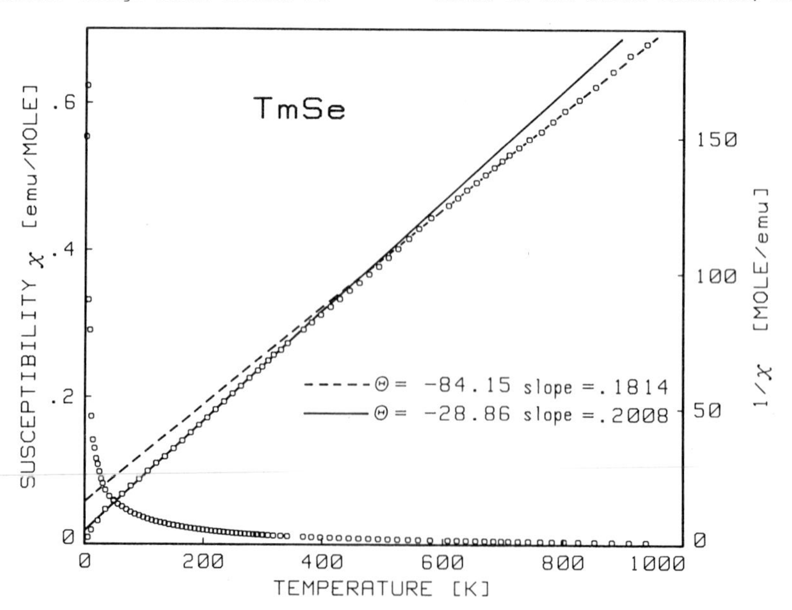

Figure 1 : Susceptibility and reciprocal susceptibility of TmSe

formed at temperatures less than room temperature, where a valence determination cannot be correct. For example in stoichiometric TmSe the valence which can be determined from the Curie constant at temperatures below 300 K is 2.56 (1), contrasted with the valence determined from the lattice constant at room temperature 2.75 (1).

To clear the situation we have measured the susceptibility of TmSe up to 950 K. Such a measurement up to 800 K has been done before (2), but our results are different. The utmost care has been taken to select the sample. It was taken from a batch of stoichiometric TmSe and the lattice constant of the selected crystal at 300 K was 5.708 Å. The sample was sealed under high vacuum in a quartz ampoule. After two reproducible runs to the maximum temperature in a carefully calibrated balance, checked for its linearity with Gd_2O_3, the sample has been transfered to the low temperature apparatus. Afterwards the lattice constant has been checked again and found to be the same as before. Subsequently the crystal has been analyzed for its composition by wet chemical analysis and has been found stoichiometric within the error limits of $\pm$ 0.5%.

Figure 1 shows the results obtained. Below 300 K the measurements of (1) could be reproduced (the numbers in brackets refer to (1)): $\Theta = -28.86$ K (- 29), valence mixing obtained from the Curie constant and linear interpolation between $C_M^{3+} = 7.15$ and $C_M^{2+} = 2.56$ yields 2.53 (2.56). At temperatures above 650 K we obtain $\Theta = -84.1$ K and a valence mixing of 2.64. Thus the valence determined from the lattice constant and from the Curie constant are approaching each other. To be comparable, however, one should use a valence determined from a high temperature lattice constant as well. Such measurements are in progress. It is thus evident, that a valence determined conventionally from lattice constant and Curie constant never agrees. It should also be mentioned that valence determination from high temperature magnetic measurements are only reasonable for Tm and Yb compounds, because only in these rare earth ions the next highest spin orbit split level is with about 0.75 and 1.25 eV, respectively, separated enough from the ground state not to be thermally populated appreciably.

It appears that in TmSe we reach at high enough temperatures the entropy limit, i.e. $kT \gg \Delta E$, where all levels are equally populated. Than the valence mixing should be given by the degeneracy of the J levels: Tm^{2+} $2J + 1 = 8$, Tm^{3+} $2J + 1 = 13$. The valence is then $2 + 13/21 = 2.63$, in excellent agreement with the experiment (21). One can also turn the argument around and say that the spread of energy levels ΔE must be much less than, say 800 K = 66.6 meV, thus we obtain the hybridization width less than $\simeq$ 10 meV.

A detailed theory of the susceptibility of TmSe and other IV compounds should take into account that hybridization gaps exist (9, 10, 11, 12) and one has basically a two level system: because of the high density of f states above and below E_F the d states can be neglected in a first approximation, but as well the occupied as the empty 4f states are hybridized multielectron states which make a theoretical treatment difficult.

Ce, Ce intermetallics and Ce compounds offer a rich field of possible intermediate valence. The problem rests in finding integer valent tri- and tetravalent reference materials. It has been

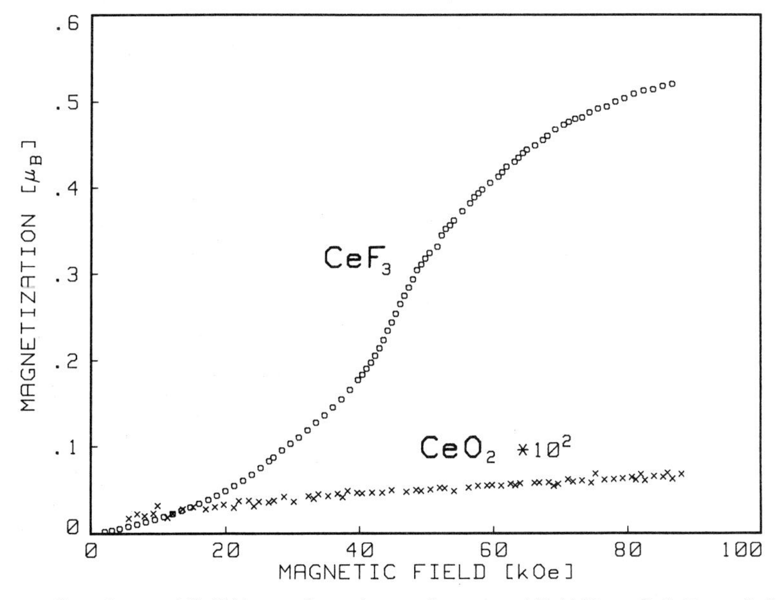

Figure 2 : Susceptibility and reciprocal susceptibility of CeF_3 and CeO_2

claimed (4) that all investigated "obvious" tetravalent compounds show mixed valence, as judged by the L edge XAS. We cannot believe this and thus have looked at CeO_2, a refractory oxide, in use since about 100 years in gas light glowers. CeO_2 is a pale yellowish powder, which we freshly oxidized at 1500 °C to avoid oxygen vacancies which would result in trivalent Ce. The lattice constant of our material was 5.4115 Å and it had the fluorite structure. In agreement with (22) this yields Ce^{4+} ions with ionic radii of 0.94 Å.

In figure 2 we show the susceptibility of CeO_2. Its weak, but paramagnetic value indicates the presence of some spins in the order of 1/1000 per formula unit which can be easily associated with Ce^{3+} impurities induced by oxygen vacancies. For comparison we show also in figure 2 the susceptibility and its reciprocal of commercial, polycrystalline CeF_3. The susceptibility at 4.2 K is about a factor 1000 larger than in CeO_2 and the Curie-Weiss plot yields about the Curie constant expected for $Ce^{3+}(^2F_{5/2})$. CeF_3 seems to be an antiferromagnet with $\Theta = -44$ K.

In figure 3 we show the magnetization of CeO_2 and again for comparison the one of CeF_3. At 90 kOe the magnetization of CeO_2 is about a factor 1000 less than in CeF_3 which again is indicative of about 10^{-3} Ce^{3+} impurity spins in CeO_2. CeF_3 when being single crystalline would be highly anisotropic and exhibit spin flop fields around 40 kOe and above 90 kOe. In the polycrystalline material these transitions are washed out. The magnetic measurements thus indicate that CeO_2 intrinsically is diamagnetic and tetravalent and only about one permille spins are present, probably caused by residual oxygen deficiencies.

CeO_2 is ideally suited to make an optical test of its valency. As a yellowish white powder its absorption edge lies in the ultra-violet. This is verified in figure 4, left hand side, where the powder reemission versus wavelength is plotted, again compared with CeF_3. The reemission is qualitatively comparable with a transmission measurement on single crystals and the poorer transmission of CeO_2 compared with CeF_3 in the visible part of the spectrum is responsible for the pale yellow of CeO_2. Possibly this absorption is invoked by the oxygen vacancies. If the material contains significant Ce^{3+} ions one should see in a transmission measurement the intra 4f transition $^2F_{5/2} \rightarrow {}^2F_{7/2}$ at about 4.3 μ or 0.27 eV. This energy is much smaller than usual $4f^1$ - 5d energies in trivalent Ce compounds which are around 3 eV (13). Indeed also in CeF_3 this $4f^1$ - 5d transition is beyond 3 eV (0.4 μ) (the structures in figure 4 in the near infra-red for CeF_3 have only an oscillator strength of about 10^{-3} and thus are probably extrinsic). Thus with 0.27 eV no excitations into the 5d band are possible. On the other hand 0.27 eV is much larger than any possible hybridization energy (meV) if CeO_2 would be considered intermediate valent, thus with the argumentation above, this energy transfer would break the hybridization at the site of the absorption and one would observe intra 4f transitions of trivalent Ce. We thus have continued our measurements (with another spectrometer) into this energy range, and have used mixtures of CeO_2 or CeF_3 in KBr pills. It is clear from figure 4 right hand side, that the $^2F_{5/2} \rightarrow {}^2F_{7/2}$ transition in CeF_3 is very obvious, experimentally agreeing with the free ion value shown in the lower part of the figure. In CeO_2 no such tran-

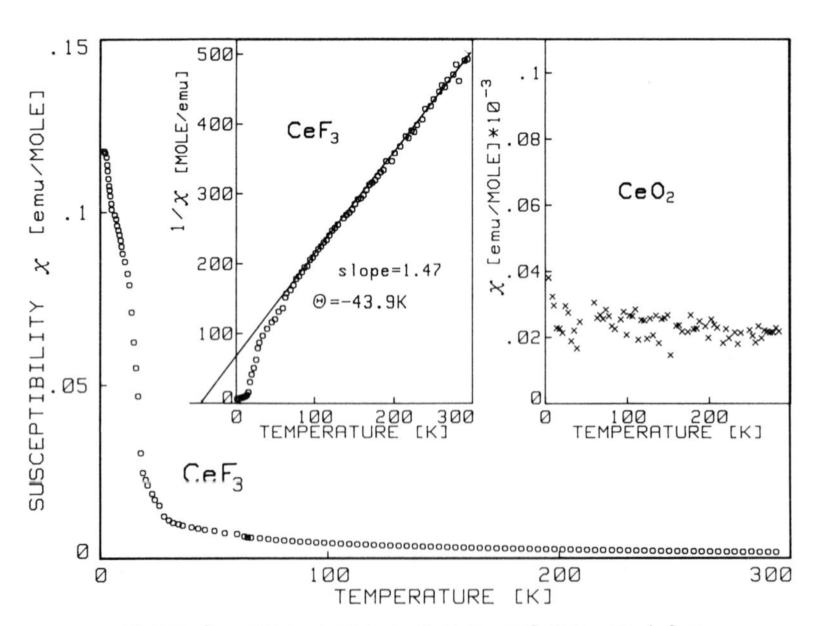

Figure 3 : Magnetization of CeF_3 and CeO_2 at 4.2 K

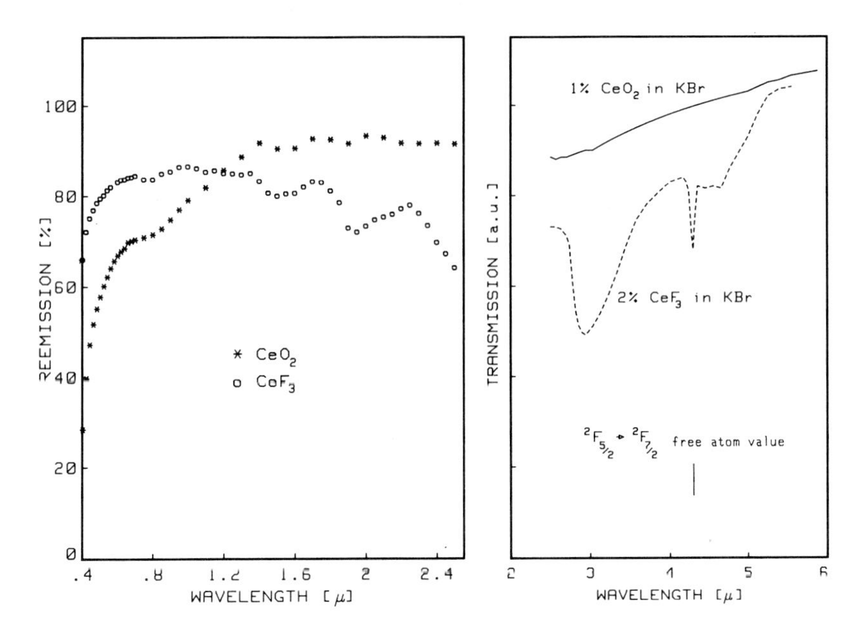

Figure 4 : Optical reemission and transmission of CeO_2 and CeF_3 at 300 K.

sition is visible, not even for 10 % CeO_2 in KBr. (The curves shown in the right hand part of Fig.4 are recorder traces). Considering the limit of detectability we can conclude from these results that the presence of Ce^{3+} or the degree of valence mixing is much less than 5 %.

In conclusion we have shown that conventional valence determination from room temperature lattice constant measurements and low temperature Curie constant can never agree but that also between high temperature Curie constant and room temperature lattice constant discrepancies will remain. A high temperature Curie constant can only be obtained for Tm - and Yb compounds. High energy measurements destroy the hybridization at the site of absorption. The many possible core level excitations and screening effects are not taken into account and no calibration exists. It has been shown by four different methods that CeO_2 is tetravalent. The fact that L edge XAS shows two valencies is evidence of the argumentation above. It definitely shows, even when one has a genuine intermediate valent material, that the degree of valence mixing taken from these measurements is questionable.

At the end it is worth remarking that it is not very important if the degree of valence mixing of any material is 2.6 or 2.7 as long as it is intermediate valent and thus belongs into this fascinating group of materials. It is more important what happens when one changes the valence in a systematic way and investigates physical observables which change concommitantly.

ACKNOWLEDGEMENT

The author would like to thank Dr. E. Kaldis for the single crystals of TmSe and the preparation of CeO_2.and Dr. L. Schlapbach for the high temperature susceptibility measurements. The author is also grateful to E. Jilek, K. Mattenberger, J. Müller and H.P. Staub for technical assistence.

REFERENCES

1) Batlogg,B., Ott, H.R., Kaldis, E., Thöni, W. Wachter, P., Phys. Rev. B 19, (1979) 247
2) Holtzberg, F., Penney, T., Tournier, R., J. Physique 40, (1979) C5-314
3) Wertheim, G.K., Eib. W., Kaldis, E., Campagna M., Phys. Rev. B 22, (1980) 6240
4) Wohlleben,D: Valence Fluctuations in Solids, Falikov, L.M., Hanke, W., Maple, M.B., (eds), North-Holland, Amsterdam (publ.),pg.1 (1981)
5) Batlogg, B., Kaldis, E., Schlegel, A., von Schulthess, G., Wachter, P., Solid State Commun 19, (1976) 673
6) Wachter, P.: Valence Instabilities and Related Narrow Band Phenomena, Parks, R.D.(ed), Plenum New York,(publ.), pg. 337 (1977)
7) Coey, J.M.D., Massenet, O., in 6) p.211
8) Goncalves da Silva, C.E.T., Solid State Commun 30, (1979), 283
9) Allen, J.W., Batlogg, B., Wachter, P., Phys. Rev. B 20, (1979) 4807
10) Frankowski, I., Wachter, P., Solid State Commun 41, (1982) 577
11) Güntherodt, G., Thompson, W.A., Holtzberg, F., in print (1982)
12) Mott, N.F., Phil. Mag. 30, (1974),403
13) Campagna, M., Wertheim, G.K., Baer, Y., Topics in Applied Physics 27, (1977), 217

14) Boppart, H., Wachter, P.,:Physics of Solids under High Pressure, Schilling, J.S., Shelton, R.N., (eds.), North-Holland, Amsterdam, (publ.) pg. 301 (1981)
15) Kaindl, G., Laubschat, C., Reihl, B., Pollak, R.A., Martenson, N., Holtzberg, F., Eastman, D.E., this conference (1982)
16) Launois, H., Raviso, M.,Holland-Moritz, E., Pott, R., Wohlleben, D.K., Phys. Rev. Lett. 44, (1980) 1271
17) Bianconi, A., Campagna, M., Stizza, S. Davoli, I.,Phys. Rev. B (1982) in print
18) Kaldis, E., Fritzler, B., J. Physique, 41, (1980) C5-135
19) Cohen, R.I., Eibschütz, M., West, K.W., Phys. Rev. Lett. 24, (1970) 383
20) Tarason, J.M., Tsikawa, Y., Chevallier, B., Etourneau, J., Hagenmüller, P., Kasaya, M., J. Physique, 41, (1980),1141
21) The author is most grateful to Dr. P. de Chatel for pointing this out to him.
22) Eyring, L.: Handbook on the Physics and Chemistry of Rare Earths, Gschneidner, K.A., Eyring, L. (eds.), North-Holland, Amsterdam (publ.), pg. 337 (1979), Vol. 3

Valence Instabilities
P. Wachter and H. Boppart (eds.)

LOW TEMPERATURE RE-ENTRANT (P,T) PHASE DIAGRAM OF SmS

J. MORILLO
S.E.S.I., Centre d'Etudes Nucléaires, B.P. 6, 92260 Fontenay-aux-Roses, France

M. KONCZYKOWSKI
High Pressure Research Center Unipress, Polish Academy of Sciences,
01-142 Warszawa, Sokolowska 29, Poland

J.P. SENATEUR
ER 155, B.P. 46, 38402 St-Martin d'Hères, France.

For the first time the low temperature (P,T) phase diagram for the semiconductor (SC) to intermediate valence (IV) state transition in SmS is experimentally determined. The "re-entrant" shape of the (P_c, T_c) transition line at low T is clearly established and the existence of a B' phase is refuted. The phase diagram has been determined by resistivity measurements on single crystals and in perfectly hydrostatic conditions. The influence of defects is investigated and the results are analysed with the help of previous results on the transport properties of SmS in both the SC and IV phases.

Pour la première fois le diagramme de phase (P,T) à basse température de la transition semiconducteur (SC) - métal de valence intermédiaire (IV) de SmS a été déterminé expérimentalement. Le signe négatif de dT_c/dP_c à basse température est clairement démontré et l'existence d'une phase B' réfutée. Le diagramme de phase a été déterminé par des mesures de résistivité électrique sur monocristaux dans des conditions de pression parfaitement hydrostatique. L'influence des défauts est étudiée et les résultats sont analysés à l'aide de résultats antérieurs sur les propriétés de transport de SmS dans les deux phases.

1. INTRODUCTION

One of the most peculiar properties of SmS is its pressure induced first order transition (at ≃ 6.5 kbar at 300 K) from a semiconducting (SC) to an intermediate valence state (IV) (1). Up to now the (P,T) phase diagram for this transition has been determined only for temperatures greater than 200 K (2, 3, 4) : the T_c (P_c) transition lines end at a critical point at high T and the slope dT_c/dP_c is positive (figure 1). From the early Bader's measurement of P_c at 4 K (5) and from the (x,T) (Gd_{1-x}, Sm_x) S phase diagram (11) it was stated that at low temperature dT_c/dP_c would change sign and become negative (P_c (4K) was estimated to be about 10 kbar). However since there is a large uncertaincy on the 4K result of Bader (5) and that the 300 K values for P_c range between 5 and 9.9 kbar (2, 3, 4, 5, 6), presently no clear experimental evidence has been given for this "re-entrant" shape of the SmS phase diagram. These large discrepancies (see also figure 1) can be due mainly to the use of samples of different quality and non hydrostatic pressures.We present here the first determination of the low temperature (P, T) phase diagram of SmS showing clearly the re-entrant shape of the (P_c, T_c) transition lines at low temperature.

2. EXPERIMENTAL PROCEDURE

The phase diagram has been determined by resistivity measurements with increasing and decreasing pressure at different constant temperatures T between 360 K and 60 K. All the measurements have been performed on single crystals of nominally stoichiometric composition (typical size 1 x 0.1 x 0.2 mm^3), coming all from the same batch. The preparation method, the caracterization of the samples and their purity have been described elsewhere (7, 8, 9). The high pressure apparatus was UNIPRESS GCA gas compressor coupled with low temperature measuring chamber. The pressure transmitting medium was helium gas, so providing perfectly hydrostatic pressure. We must emphasize that with this apparatus in the explored (P, T) range it was possible to vary independently P and T (isothermal and isobaric runs). The pressure was measured with a manganin sensor at room temperature in the last compressing stage with a ± 50 bar relative precision and a 2 % absolute error. In the vicinity of the transition the pressure was increased by steps less than 50 bar. Since the crystal is destroyed at the reverse transition, the single crystal was changed after each pressure cycle at a given temperature.

3. EXPERIMENTAL RESULTS

3.1 "Pure" SmS

The transition has always been found first order, sharp and well defined at increasing P ($|\Delta P_c|$ < 0.1 kbar) and spread over 1 to 2 kbar at decreasing P. The obtained transition points (P_c , T_c) for the direct and reverse transitions are shown on figure 1 together with previously published data for T > 200 K.

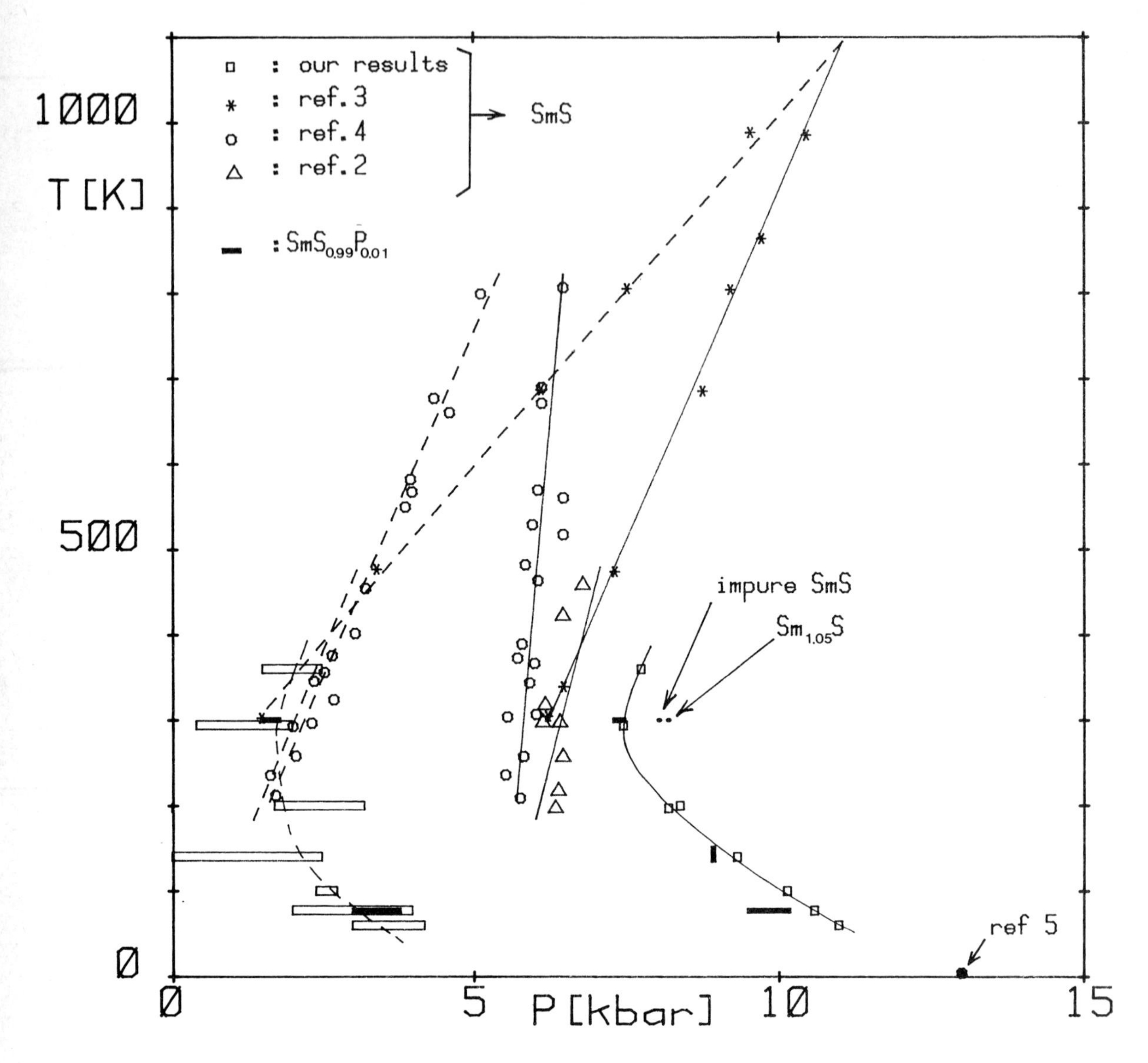

Figure 1 : SmS and $SmS_{0.99}P_{0.01}$ phase diagram (—— : P↑, ---- : P↓)

It appears that dT_c/dP_c is positive for T $\gtrsim$ 300 K and unambiguously negative for T $\lesssim$ 300 K for both the direct and reverse transitions : ($- dT_c/dP_c$) is about 100 K/kbar at ≃ 250 K and decreases down to 44 K/kbar at ≃ 68 K for the direct transition.

This negative character of dT_c/dP_c was also established by performing the following P, T cycle on one of the samples (figure 2) : Initially we applied pressure reversely up to 10 kbar at ≃ 79 K and no transition was observed (① on figure 2). The sample was then warmed up at zero pressure and the pressure was applied at 200 K, the transition occured at 8.4 kbar (② on figure 2).

We tested also the reproducibility of the results by performing several transitions at 300 K on different single crystals : the transition was always found at 7.5 ± 0.1 kbar.

3.2. Influence of defects : $Sm\,S_{1-x}P_x$, $Sm_{1+x}S$

We reported also on figure 1 the results for the transition obtained on doped and "impure" samarium sulphide single crystals.

In the case of $Sm\,S_{0.99}P_{0.01}$ the results are quite similar to those of pure SmS, but as expected with slightly lower transition pressures. However the transitions are not so well defined, indicating either some macroscopic

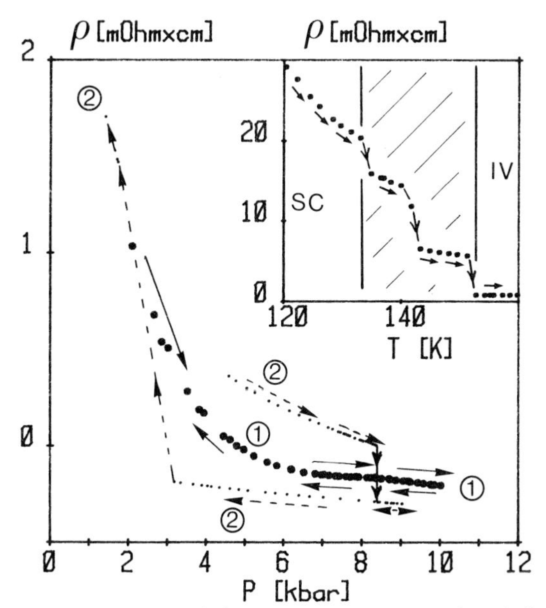

Figure 2 : Resistivity versus pressure for SmS : ① - first cycle, T = 79 K, P ↗ and then P ↘. ② - second cycle, T = 200 K, P_c = 8.39 kbar. Insert : Resistivity versus temperature for $SmS_{0.99}P_{0.01}$ et 9.2 kbar, T ↗ : note the steps corresponding to the transition between 134 K and 152 K.

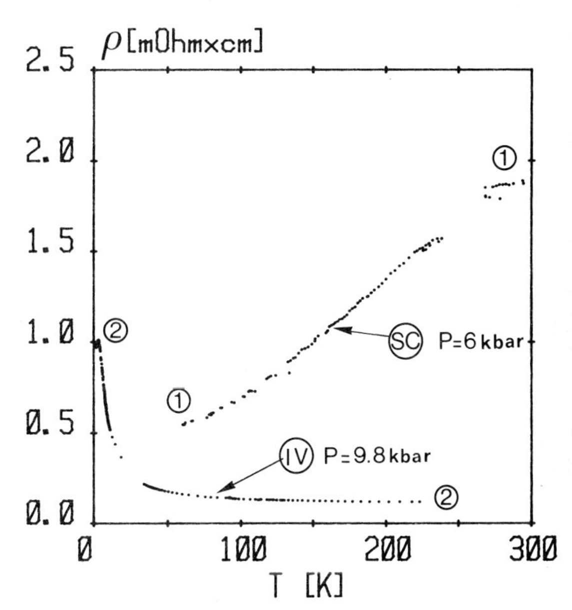

Figure 3 : Resistivity versus temperature for SmS : ① - at P = 6 kbar in the SC phase. ② - at P = 9.8 kbar in the IV phase.

Inhomogeneity and/or the inhomogeneous mixed valence state of $SmS_{0.99}P_{0.01}$ (see fig. 1). This is particularly visible on the insert of fig. 2, where the transition has been performed at constant P and increasing T which implies a negative dT_c/dP_c.

In the case of hyper stoichiometric sample $Sm_{1.05}S$ the transition occurs at room temperature at a slightly higher pressure than for pure SmS (8.20 ± 0.05 kbar) in qualitative agreement with (10). This is also the case for an impure SmS single crystal (P_c = 8.05 ± 0.05 kbar) with a much higher conductivity than pure SmS ($\rho_{20K}/\rho_{300K} \simeq 40$ compared to $\rho_{20K}/\rho_{300K} \simeq 10^8$).

4. DISCUSSION

The use of perfectly hydrostatic pressure and single crystals of constant quality enabled us to establish definitively the negative sign of dT_c/dP_c for 300 K ≳ T ≳ 60 K. For lower temperatures, if we refer to the early estimation of P_c (4 K) ≃ 10 kbar (5) we must expect that dT_c/dP_c will come again positive since P_c (60K) = 11 ± 0.1 kbar. However a more detailed analysis of the transition curve of Bader (fig. 2 in ref. 5) in the most hydrostatic condition (n-pentane-isoamyl alcohol) seems to indicate that the 4K transition occurs more probably at about 13 kbar, giving reasonable agreement with our results as shown on figure 1.

The resistivity variations (figure 2) as well as Hall coefficient variations (8, 9) with pressure at low temperature did not present any discontinuity up to the transition indicating clearly that the B' phase suggested in the (x, T) plane in $Sm_{1-x}Gd_xS$ (11) is absent in SmS.

From the Clapeyron formula ($dP/dT = \Delta S/\Delta V$) and the negative sign of ΔV ($\Delta V \simeq -4.1\ cm^3.mole^{-1}$) at the transition it appears that $\Delta S = S_{IV} - S_{SC}$ is positive at low T (ΔS (60 K) ≃ R) and negative at high T (ΔS (360 K) ≃ −.6 R). Usually the positive sign of ΔS at low T is attributed to the metallic character of the IV phase ($S_{electronic} \propto T$) which dominates at low T the lower lattice contribution to the entropy ($S_{lattice} \propto T^3$) in the SC phase. However from our results of transport properties (on the same single crystals) in both the IV and SC phases (8, 9, 12, 13, 14) it appears that the transition is from a metallic degenerate semiconductor ($n \simeq 10^{20}\ cm^{-3}$) to a narrow gap semiconductor (figure 3) at low T in the region where dT_c/dP_c is negative. The analysis of the transport properties in the SC phase (8, 9, 14) led us to the conclusion that the transport properties are extrinsic for T < 300 K and related to an impurity level, $N_i \simeq 10^{20}\ cm^{-3}$, that enters the conduction band between 2 and 3 kbar giving this degenerate semiconductor behaviour. It appeared also that the 4f intrinsic level position ε_f with respect to the conduction band minimum varied greatly with temperature (the absolute value of the gap increasing with T, $d|\varepsilon_f|/dT \simeq 0.6$ meV/K) and that this gap was far from being closed at 300 K (ε_g (P_c) > 100 meV at 300 K).

Moreover in the frame of the model developed in ref. 8, 9, the leading parameter for the transition is not the gap but the relative position of the 4f level and Fermi energy μ that is almost constant (in temperature units), $(\varepsilon_f - \mu) \simeq 5$ to 6 kT at the onset of the transition and corresponds to about 1 to 2 % ionized samarium ions just before the transition.

One of the questions which arises now is what is the influence of this impurity level on the transition itself. As we have seen, when we introduce defects which increase the conductivity ("impure" SmS, $Sm_{1+x}S$) the transition pressure increases. This is in agreement with the theoretical model developed by Entel et al. (15). In their model an excess free electron concentration shifts the transition line towards higher pressures and the re-entrant shape is accentuated. So we could expect that for a pure and perfectly stoichiometric SmS the transition line would be at lower pressure and with less re-entrant character as we found.

In the case of Sm $S_{0.99}$ $P_{0.01}$ the shape of the transition line is almost the same as for SmS, the transition pressures being, less than .3, .5 and 1 kbar lower than in pure SmS, respectively at 300 K, 150 K and 77 K. This is surprising since each phosphorus ion induces six Sm^{3+} ions in its vicinity with the associated volume collapse (7), giving an equivalent chemical pressure of 1.3 kbar/at % P (with x_c = 5 % and P_c (300 K, SmS) = 6.5 kbar).

5. CONCLUSION

The phase diagram of SmS has been determined for 60 K < T < 360 K showing clearly the negative sign of dT_c/dP_c for T $\lesssim$ 300 K. However some questions remain about the influence of defects on this phase diagram. As a conclusion we would like to recall some points that we have put forward in this paper and that have not been sufficiently taken into account in previous theoretical descriptions of the transition, in particular :

- the large temperature variation of ε_f
- the energy gap doesn't vanish at the transition whereas $(\varepsilon_f - \mu)/kT$ is constant at the onset of the transition and corresponds to about 1 to 2 % ionized samarium ions.
- the exact role played by impurities and defects in the mechanism of the transition.

We wish to thank Dr. C.H. de Novion for many usefull discussions and T. Mazur for its efficient technical assistance.

REFERENCES

(1) Jayaraman, A., Narayanamurti, V., Bucher, E. and Maines, R.G., Phys. Rev. Letters 25 (1970) 1430.
(2) Jayaraman, A., Dernier, P. and Longinotti, L.D., Phys. Rev. B 11 (1978) 2783.
(3) Shubha, V., Ramesh, T.G. and Ramaseshan, S., Solid State Com. 26 (1978) 173.
(4) Tonkov, E. YU. and Aptekar, I.L., Sov. Phys. Solid State 16 (1974) 972.
(5) Bader, S.D., Phillips, N.E. and McWhan, D.B., Phys. Rev. B 7 (1973) 4886.
(6) Keller, R., Guntherodt, G., Holzappel, W.B., Dietrich, M. and Holtzberg, F., Solid State Com. 29 (1979) 753.
(7) Morillo, J., de Novion, C.H. and Senateur J.P., Physics of Metallic Rare-Earths; Journal de Physique C5-40 (1979) 348.
(8) Morillo, J., Konczykowski, M. and Senateur, J.P., Solid State Com. 35 (1980) 931.
(9) Morillo, J., Thèse, Université d'Orsay, France (1981).
(10) Bzhalava, T.L., Shubnikov, M.L., Shul'Man, S.G., Golubkov, A.V. and Smirnov, I.A., Sov. Phys. Solid State 18 (1976) 1838.
(11) Jayaraman, A., Dernier, P. and Longinotti, L.D., Phys. Rev. B 11 (1975) 2783.
(12) Konczzykowski, M., Morillo, J. and Senateur, J.P., in Valence Fluctuations in Solids, Falicov, L.M., Hanke, W. and Maple, M.B. (eds.) North-Holland Pub. Com. (1981) 287.
(13) Konczykowski, M., Morillo, J. and Senateur, J.P., Solid State Com. 40 (1981) 517.
(14) Konczykowski, M., Morillo, J., Portal, J.C., Galibert, J. and Senateur, J.P., this same conference.
(15) Entel, P. and Grewe, N., Z. Physic B 34 (1979) 229.

Valence Instabilities
P. Wachter and H. Boppart (eds.)

NEW INFORMATIONS OBTAINED FROM EXAFS EXPERIMENTS ON INTERMEDIATE VALENT SYSTEMS

G. Krill[+§], J.P. Kappler[+§], J. Röhler[‡§], M.F. Ravet[+], J.M. Léger[°] & F.Gautier[+]

+ L.M.S.E.S. (LA 306) Univ. L. Pasteur - 4, rue B.Pascal 67070 Strasbourg France.

‡ II Physikalisches Inst. Univ. zu Köln, Zülpiche Str. 77, 5000 Köln 41 W.Germ.

° L.E.M.T.A. (ERA 211) 1, Place A. Briand 92190 Meudon Bellevue, France.

§ Laboratoire pour l'Utilisation du Rayonnement Electromagnétique du C.N.R.S., associé à l'Univ. Paris Sud (LURE) Bat. 109 C, 91405 Orsay Cedex, France.

We present the results of an EXAFS study performed on several $SmS_{1-x}O_x$ compounds. For x>0.05, after the collapse of the lattice parameter the samples are in a homogneous intermediate valence (I.V.) state. From our data the EXAFS parameters,i.e. the mean distances (R_i) between the central atom (Sm) and the scattering atom (S(O) or Sm),the number of scattering atoms (N_i) and the Debye-Waller factor (σ_{Ti}) are deduced both for the first (Sulphur or Oxygen) and the second (Sm) shells. The new results is the observance of a strong damping of the EXAFS contribution associated with the Sm-Sm nearest neighbours (nn) in the I.V. state. We show that the best way to explain this damping is to consider an interference process between two Sm-Sm distances separated by $\Delta R \sim 0.25$ Å. We compare these results with recently received data from high-pressure EXAFS experiments on SmS.

I. INTRODUCTION

Mixed valent systems present strong phonons anomalies which are due to the large difference in ionic radii associated with two valence states of the rare-earth (RE) ion[1]. As the rate of the valence transition is supposed to be comparable to the phonon frequencies ($\sim 10^{-2}$ eV) dynamic electron-phonon coupling can take place. Extended X-ray Absorption Fine Structure (EXAFS) experiments, because of their short characteristic time ($\tau \sim 10^{-16}$ sec) give informations about the instantaneous distribution of radial distances to the neighbours of the unstable 4f ions. Therefore this technique seems to be particularly well suited to the study of electron-lattice coupling in Intermediate Valent materials(I.V.). However, recent EXAFS experiments performed on $TmSe$[2] and $Sm_{0.75}Y_{0.25}S$[3] have shown that the surrounding atoms of a given RE ion adopt an average position which coincides with the distance deduced from lattice constants. This paradoxical result may be explained if we consider that for a given value of the valency (V) there is balance between the hybridization, elastic and conduction electron energies[4]. In the I.V. state the hybridization energy can be very large as compared to the other terms and then it strongly reduces the position dependence of the ions which surround the RE[5]. For several reasons an enhancement of the effect on the RE-RE shells may be considered indeed in this case the maximum value for δR is greater, the photoelectrons wave functions are scattered by RE ions which are in different valence states and the force constants are different. Reliable informations on RE shells can only be obtained for systems with small alloy disorder, therefore the study of mixed compounds such as $Sm_{1-x}Ln_xS$ is avoided. The $SmS_{1-x}O_x$ mixed compound where the I.V. state is induced by susbtituting only 5% Sulphur by Oxygen without changing the Sm sublattice is a good case. Moreover, as Oxygen is isoelectronic from Sulphur, it is no extra charge transfer between the RE and the Oxygen atoms as it is the case e.g. for mixed $SmS_{1-x}P_x$ compounds. The experimental situation is thus very similar to that expected for SmS under pressure.

II. EXPERIMENTAL AND DATA ANALYSIS

II.1. Preparation and physical properties of the $SmS_{1-x}O_x$ compounds

$SmS_{1-x}O_x$ ($x \leq 0.13$) compounds have been prepared by high-pressure synthesis (P = 65 kbar, T = 1600°C) starting from pure SmS and SmO compounds[6-7]. When Oxygen is substituted to Sulphur the collapse of the lattice parameter occurs for $x \simeq 0.05$ (Fig.1). Above this concentration the typical physical properties of the I.V. SmS collapsed phase are observed[6] (non-magnetic behaviour at low temperature, semi-metallic character,...). The valence of Samarium in the I.V. state deduced from lattice constant ($V \sim 2.8$) is the same as found for SmS under pressure (P = 7.5 kbar). After synthesis the samples have been checked by X-ray diffraction in order to ensure the absence of parasitic phase. After powdering (10-20 μ), under Argon atmosphere, the samples have been prepared for EXAFS measurement optimizing the thickness in order to obtain the better signal/noise ratio.

II.2. X-ray absorption and EXAFS experiments.

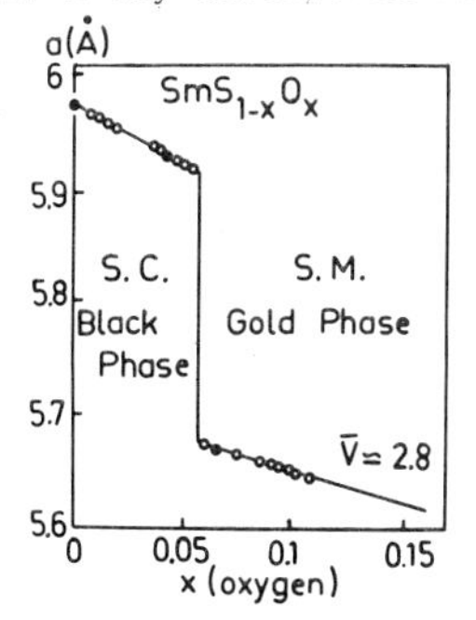

Fig.1 : Lattice constant versus x.

Experiments have been performed at LURE using the X-ray beam delivered by the DCI storage ring. Absorption spectra have been measured on the Samarium L_{III} edge ($\sim$ 6710 eV). For EXAFS, photon energies were swept over 600 eV (up to the L_{II} edge) in 1 eV steps. The DCI storage ring was operated at about 200mA and 1.72 GeV.

II.3. Data analysis
II.3.1. X-ray absorption

The well known double edge character of I.V. Samarium compounds[3] has been used to deduce the valence V. Each component of the edge has been determined by convoluting an asymmetric Lorentzian shape function with a Gaussian which takes into account the spectrometer function. All the parameters, including the energy separation between the Sm^{2+} and Sm^{3+} edges, have been adjusted to obtain the best fit to the experimental data. The absolute error on the determination of V does not exceed ± 0.05.

II.3.2. EXAFS data analysis

The microscopic origin of EXAFS has been discussed at length in the literature (e.g. P.A. Lee at al[8]). The oscillating part of the absorption coefficient (μ) for excitation from a p state normalized to the structureless background (μ_o) is given by *(1)* for systems with small disorder :

$$\chi(k) = (k)^{-1} \sum_j (N_j/R_j)|f_j(\pi,k)|e^{-2\sigma_{Tj}k^2}.e^{-2R_j/\lambda(k)} \times \sin(2kR_j+2\delta_2(k) + \arg(f_j(\pi,k))) \qquad (1)$$

with $k = \frac{1}{\hbar}\sqrt{2m(E-E_o)}$

Here, N_j is the number of neighbour atoms located at a radial distance R_j, $e^{-2\sigma_{Tj}k^2}$ is an attenuation factor, analog to a Debye-Waller term, $e^{-2R_j/\lambda(k)}$ accounts for the fact that electrons which have suffered inelastic losses cannot contribute to EXAFS, $\delta_2(k)$ is the central atom phase shift, $f_j(\pi,k)$ and $\arg(f_j(\pi,k)$ the back - scattering amplitude and back-scattered phase shift,respectively ; E_o is the threshold energy.

The phase shifts we have used are theoretical ones which have been calculated from first principle by Lee[8]. As they give good results for number of divalent and trivalent Samarium compounds, we have real confidence for their use even in the I.V. regime. The mathematical procedure we use to analyze the data is exactly the same as that reported by Raoux et al[9]. As EXAFS takes place on a time scale (10^{-16} sec) much shorter than that of atomic motion we measure an instantaneous snapshot of the atomic configuration.

III. EXPERIMENTAL RESULTS
III.1. Absorption edges

The L_{III} absorption edges of Samarium at T = 10 K are reported on Fig. 2. The I.V. character of $Sm_{0.93}O_{0.07}$ is well illustrated, the valence we found $V \sim 2.55$ is clearly different to that deduced from lattice constant ($V \sim 2.8$) as previously observed on other I.V. systems[2-3]. For SmO we confirm that this compound is still not in a single valence state [7].
In all cases, the energy separation between Sm^{2+} and Sm^{3+} is δ=7.1 eV. For SmO the width of the Lorentzian (correlated both with the whole lifetime (2p) and the 5d bandwidth) has to be increased up to 60 % as compared to SmS or $Sm_{0.93}O_{0.07}$; this traduces the existence of strong changes in the electronic structure between the I.V. state and the nearly trivalent state of SmO. Similar behaviour is observed for SmS at high pressure (P ∿ 80 kbar, Röhler et al, this conference).

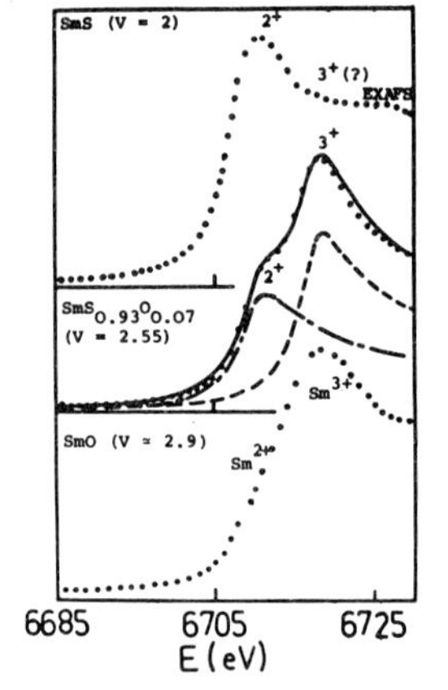

Fig. 2 : Sm L_{III} absorption edges at T=10 K.

III.2. EXAFS results

At low temperature the thermal vibration of the ions is small and the damping factor (σ_T) of EXAFS is reduced. This is the reason why we focus ourselves to the low temperature data (T = 10 K). On Fig. 3, we report the magnitude of the Fourier transform P(R) for $k^3\chi(k)$. The magnitude of the peak centered near 4 Å, due to the Sm-Sm nn, is significantly smaller for the I.V. compound. On the contrary the first peak centered near 2.8 Å which reflects the first shell distances is similar for SmS and the I.V. compounds (the small contribution associated to the Sm-O distance is due to the low Z of Oxygen (Z = 8), thus the $k^3\chi(k)$ procedure we use smeared out their contribution to P(R)[8]).

III.2.1. First shell Sm-S (O) contribution

The partial EXAFS spectrum obtained by backtransforming in k space the first peak in P(R)

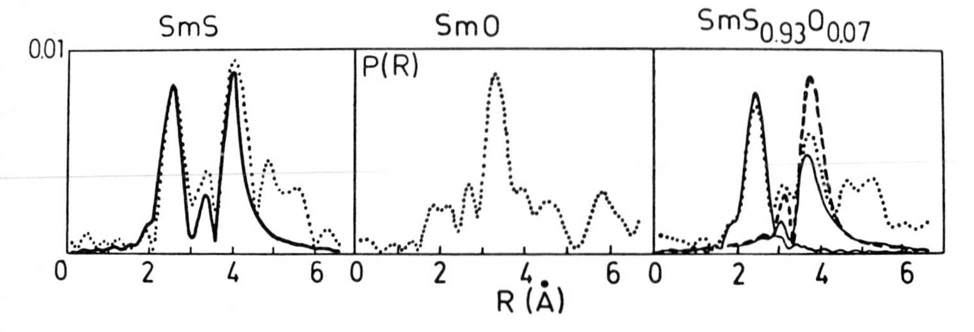

Fig. 3 : F.T. of $k^3\chi(k)$.
... experiment
—— result of our fit.
--- simulation assuming the same EXAFS parameters as for SmS and SmO (except the distances).

(1.5 → 3 Å) for the I.V. compound is shown on Fig. 4. It is well reproduced using the general formula *(1)*. The collapse of the lattice constant is observed but no other effect is detected, particularly the σ_T factor is the same as for SmS indicating that the softening of optical modes observed for I.V. systems does not influence the EXAFS σ_T. The possible existence of two distances Sm^{3+}-S(O) and Sm^{2+}-S(O) which should produce a strong beat of the EXAFS amplitude near 300 eV is ruled out. In the same way the occurrence of a strong disorder which should be followed either by an increase in σ_T or by an impossibility to fit the data by the general formula *(1)* must be discarded.

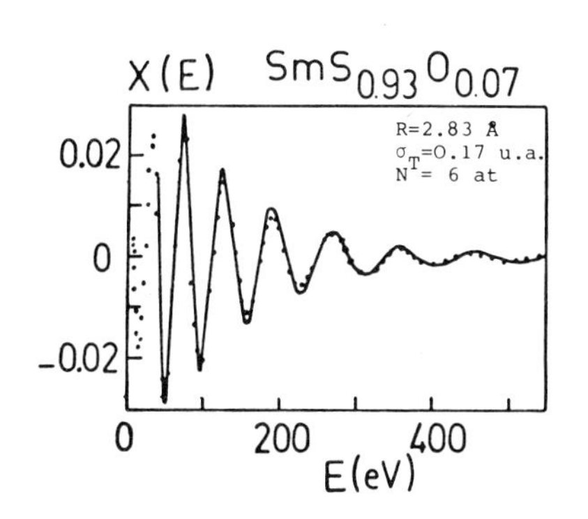

Fig. 4 : Partial EXAFS spectrum of the first shell in I.V. compound (the full line is the fit using (1) with the parameters given in the figure).

In summary, our results for this first shell agree with the earlier experiments performed on I.V. $Sm_{0.75}Y_{0.25}S$ [3] and also with the conclusion that only one simple threshold energy has to be considered for the origin of EXAFS in these I.V. materials. This result is rather surprising considering the 7.1 eV energy difference between the Sm^{2+} and Sm^{3+} absorption edges. However, it is not a general case, indeed if we try to fit the data for a compound in the semiconducting phase (e.g. x = 0.05) then, as expected, two different threshold energies and two distances have to be considered.

III.2.2. Second shell contribution (Sm-Sm)

At evidence, the hypothesis that the Sm-Sm radial distances are identical to those deduced from lattice constant ($\bar{R} = a/\sqrt{2}$), which gives satisfactory results for SmS and SmO, cannot be applied to the I.V. compound (Fig. 5). The damping of the EXAFS near E ≃ 150 eV (k = 6.3 $Å^{-1}$) can be the result of an interference effect between two distances separated by ΔR ≃ 0.25 Å and it is shown on Fig. 6.a. that such an hypothesis yields a good fit to the data. The two distances we found are roughly equal to those expected for Sm^{2+}-Sm^{2+} and Sm^{3+}-Sm^{3+} pairs and it is no evidence for the existence of Sm^{2+}-Sm^{3+} pairs although they should contribute up to 50 % to the EXAFS signal for V ≃ 2.5 (if such pairs are considered it is impossible to reproduce the interference and damping effect). The simplest, but puzzling, way to explain this would be to assume that the Sm^{2+}-Sm^{3+} and Sm^{3+}-Sm^{3+} distances are equal and it is interesting to underline that this hypothesis could explain why lattice constant and L_{III} absorption edges measurements (or Mössbauer, ...) yield to different determination of the valence. Rather than the existence of several definite distances a strong disorder on the Samarium subshells may occur. It can be simulated in EXAFS, either by increasing the σ_T factor (6.b) or by using the general EXAFS formula established for highly disordered systems[8] where a disorder damping factor σ_D is

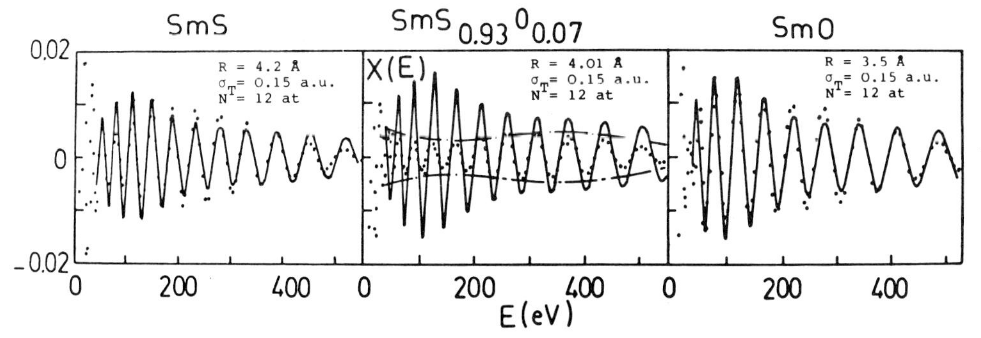

Fig. 5 : Parital EXAFS spectrum of the second shell (assuming that R is given by lattice constant).

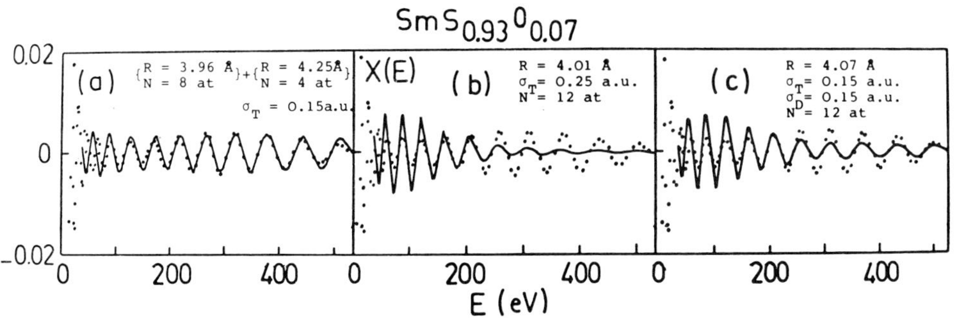

Fig. 6 : Partial EXAFS spectrum of the second shell for the I.V. compound a) contribution of 2 distances b,c) influence of disorder (see text).

introduced independently from the usual σ_T factor (6.c). As shown on Fig. 6.b, 6.c the agreement with the data is rather poor, particularly at high energy E > 300 eV.

III.2.3. Comparison with high pressure EXAFS experiments on SmS

In order to ascertain the existence of the damping effect described above, we recently performed high-pressure EXAFS experiments on SmS up to 80 kbar (Röhler et al, this conference). The magnitude of the F.T. of $k^3\chi(k)$ at T = 300 K for P = 0, 15 and 80 kbar are reported on Fig. 7. At P = 15 kbar, in the I.V. regime (V ≃ 2.6), a decrease in the magnitude

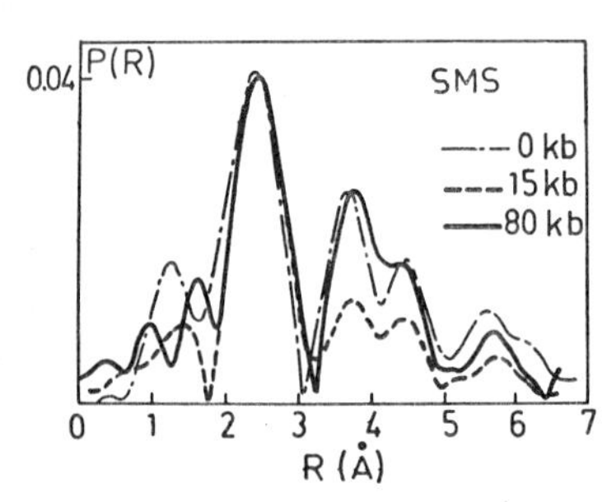

Fig. 7 : F.T. of $k^3\chi(k)$ for SmS as a function of pressure.

of the peak centered near 4 Å occurs ; this effect is quite similar to that observed for I.V. $SmS_{0.93}O_{0.07}$. When the valence approaches 3 (V = 2.9 identical to that of SmO), in the high pressure range, we notice that the intensities of the peak in P(R) become comparable to those observed for P = 0. Preliminary mathematical analysis confirm completely the results obtained for $SmS_{1-x}O_x$ and therefore we believe that the existence of a damping effect on the RE-RE shells is characteristic for the existence of an I.V. ground state.

IV. CONCLUSION

We have shown that EXAFS exhibits particular features characteristic of the I.V. state. However, the interpretation of our results is not yet achieved and still not easy. The reason why the change in the ionic radii of the RE ion does not affect the EXAFS of the first shell itself remains particularly unclear. As stressed recently by Kohn et al[5] it may be due to the large hybridization energy (E_{hy}) which, in the I.V. state, tends to strongly reduce the position dependence of the S^{2-} ions around the RE. If $|E_{hy}| >> E_{elastic}$, δR could then be reduced to values which approach the accuracy limit of EXAFS experiments. As discussed in papers[4-5], for each valency V, the relative importance of E_{hyb}, $E_{elastic}$ and $E_{cond.el.}$ is modified and therefore EXAFS experiments performed on systems where the valency changes significantly with temperature and/or pressure, using the system as its own reference, should be interesting. In our EXAFS study of I.V. $SmS_{0.93}O_{0.07}$ we only explore one particular point in the P-T plane, we must extend it to systems like SmS under pressure at low temperature or $EuPd_2Si_2$ ($EuCu_5Si_2$) where the valency depends strongly from the thermodynamical parameters in order to investigate the possible existence of an effect on all interatomic distances. Let us mention here that the preliminar results we obtained for SmS at T = 300 K in the 20 kb → 50 kb region, where the valence changes from 2.5 to 2.85, seem to indicate the existence of an effect on the Sm-S first shell. However, because the statistic of the experimental data is rather poor, it is premature to conclude.

ACKNOWLEDGEMENTS

We are grateful to Drs P. Lagarde and D.Raoux for their help during the EXAFS experiments and for stimulating discussions about the interpretation of our results. It is also a pleasure to thank Dr Marin and his collaborators for the runs dedicated by the "Laboratoire de l'Accélérateur Linéaire" to LURE. We are indebted to Dr C. Godart and Dr J.P. Sénateur for providing the SmS compounds. We thank Prof. D. Wohleben for his constant interest in our experiments.

REFERENCES

[1] Valence Fluctuations in Solids, International Conference of Santa Barbara, North Holland Pub. Company, Amsterdam (1981).

[2] Launois, H., Raviso, M., Holland-Moritz, E., Pott, R. and Wohlleben, D., Phys. Rev. Lett. 44, 1271 (1980).

[3] Martin, R.M., Boyce, J.B., Allen, J.W. and Holtzberg, F., Phys. Rev. Lett. 44, 1275 (1980) and ibid [1] p. 427.

[4] Röhler, J., Wohlleben, D., Kaindl, G. and Bielster, H. (preprint), submitted for publication in Phys. Rev. Lett.

[5] Kohn, W., Lee, T.K. and Lin-Liu, Y.R. (preprint) to be published.

[6] Abadli, L., Thesis, Univ. Louis Pasteur, France (1981) and to be published.

[7] Krill, G., Ravet, M.F., Kappler, J.P., Abadli, L. and Léger, J.M., Solid State Comm. 33, 351 (1980).

[8] Lee, P.A., Citrin, P.H., Eisenberger, P. and Kincaid, B.M., Rev. Mod. Phys. 53, 769 (1981).

[9] Raoux, D. et al, Rev. Appl. Phys. 15, 1079 (1980).

Valence Instabilities
P. Wachter and H. Boppart (eds.)

PRESSURE-INDUCED VALENCE TRANSITION IN Yb METAL STUDIED BY L_{III}-EDGE SPECTROSCOPY

G. Wortmann, K. Syassen[+], K. H. Frank, J. Feldhaus, and G. Kaindl

Institut für Atom- und Festkörperphysik, Freie Universität Berlin
D-1000 Berlin 33, Fed. Rep. Germany

[+]Physikalisches Institut III, Universität Düsseldorf,
D-4000 Düsseldorf, Fed. Rep. Germany

The L_{III}-absorption edge of Yb metal is investigated at room temperature up to a maximum pressure of 340 kbar. In the fcc-phase (0 - 40 kbar) and in the hcp-phase (above 300 kbar) x-ray absorption edge spectra typical for divalent and trivalent Yb, respectively, are observed. Yb is mixed valent in the high pressure bcc-phase (40 - 300 kbar) with the mean valence increasing rapidly from 2.0 to 2.5 between 40 and 100 kbar.

1. INTRODUCTION

Ytterbium and europium are the only divalent rare earth metals under ambient conditions /1/. Associated with the divalent nature of Eu and Yb is a significantly lower cohesive energy and a larger molar volume compared to the regular trivalent rare earth elements (see Fig. 1).

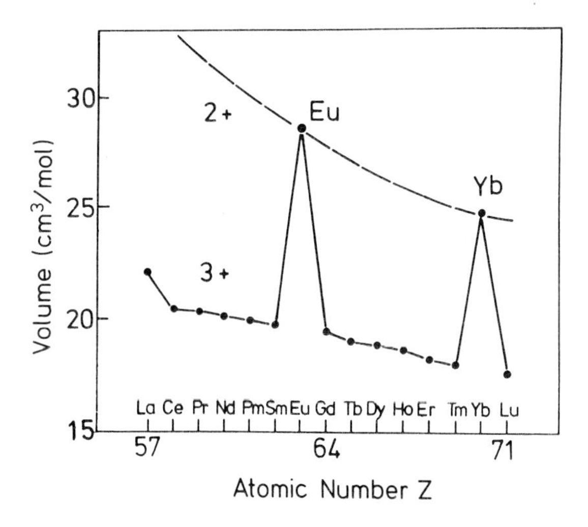

Figure 1: Molar volume of the rare earth elements at ambient conditions.

To our knowledge, the idea of a pressure induced valence change in Yb metal from the $4f^{14}(6s5d)^2$ towards the $4f^{13}(6s5d)^3$ electronic configuration was first put forward by Hall et al. /2/, when they identified a transition (with 3% volume change) from fcc to bcc structure at 40 kbar and 300 K. However, in subsequent high pressure studies /1, 3/ it was concluded, that the 4f electrons are not involved to a major extend in the fcc to bcc transition. Furthermore, the temperature dependence of the resistivity in the bcc phase was taken as indication for Yb to be essentially divalent up to about 160 kbar /3/.

From cohesive energy calculations, Johansson and Rosengren predicted the onset of an electronic transition towards the trivalent state at around 130 kbar /4, 5/. A recent x-ray diffraction study gave strong evidence for the valence change to be almost completed at 300 kbar /6/. As shown in Fig. 2, the molar volume of Yb at 300 kbar is close to the extrapolated pressure-volume isotherms for the neighbouring trivalent elements in the periodic table. At about 300 kbar Yb undergoes a further structural phase transition from bcc to hcp /7/, which is the stable structure of the heavy trivalent rare earth metals under normal conditions /1/. Throughout the bcc-phase, the lattice constant exhibits an anomalous strong dependence on pressure, which is in general an indication for a valence transition in rare earth systems. However, from the

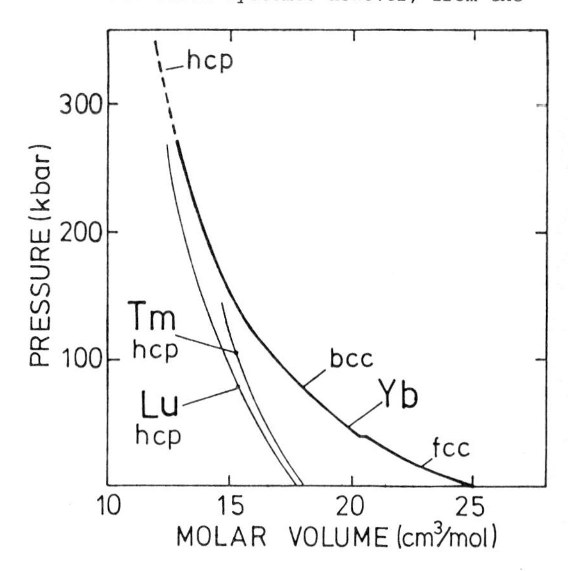

Figure 2: Pressure-volume isotherm (T = 300 K) of Yb metal /1, 6, 7/. For comparison, the compression curves of Tm and Lu metal are included /1/.

pressure-volume data it is difficult to extract any quantitative information about the course of valence transition as a function of pressure or volume because for the elemental rare earths Eu and Yb in their various metallic phases no measure of lattice constants or molar volume in terms of "divalent" and "trivalent" exists.

In order to clarify the question of a valence transition in the high-pressure phases of Yb metal, we have studied the L_{III} x-ray absorption edge at pressures up to 340 kbar. In recent years, L_{III}-edge absorption spectrosopy has been established as an experimental tool to investigate the fractional valence in mixed-valent rare earth systems of Sm, Eu, Tm, and Yb /8 - 12/. As already shown by Vainshtein et al. /8/, information on the mean valence is obtained from position and structure of the L edges in x-ray absorption spectra /13/. Due to an increase in the nuclear potential the L edge is shifted by approximately 7 eV to higher energies /14/ when a 4f electron is promoted into the conduction band. The present L_{III}-edge spectra of Yb metal under pressure clearly identify a continuous valence change in the bcc phase (40 - 300 kbar). Yb is trivalent in the hcp phase above 300 kbar /15/.

2. EXPERIMENT

So far, most of the x-ray absorption studies under high pressure made use of high intensity synchrotron radiation in order to cope with the enormous losses of the photon flux within the high pressure cells. The present experiment has been performed with a conventional laboratory x-ray absorption spectrometer. A Rowland type bent-crystal x-ray monochromator in combination with a 2 kW molybdenum tube was used as a tunable source of monochromatic x-rays (ΔE = 4.5 eV at 8.9 keV). Details of the spectrometer are presented elsewhere /16/. External pressures up to 340 kbar at the Yb metal foil (5 µm thickness, 99,9% purity) were generated in a gasketed diamond anvil cell of the Syassen-Holzapfel type /17/. Silicon oil served as pressure transmitting medium. The pressure was determined using the well known ruby fluorescence method. At the highest pressure achieved the pressure variation over the 0.05 mm^2 sample area was less than ± 5% of the mean pressure. Due to the small sample size and the large photoabsorption in the diamond anvils (more than 99% absorption at 8.9 keV), the photon flux was reduced to about 10 counts per second, which is four orders of magnitude lower compared to normal transmission experiments at ambient pressure.

3. RESULTS

L_{III}-edge absorption spectra of Yb metal at room temperature and at various pressures are shown in Fig. 3. The spectra exhibit a peaked structure ("white line"), which is characteristic for L_{II} and L_{III} edges of 4f systems. The L_{III}-edge

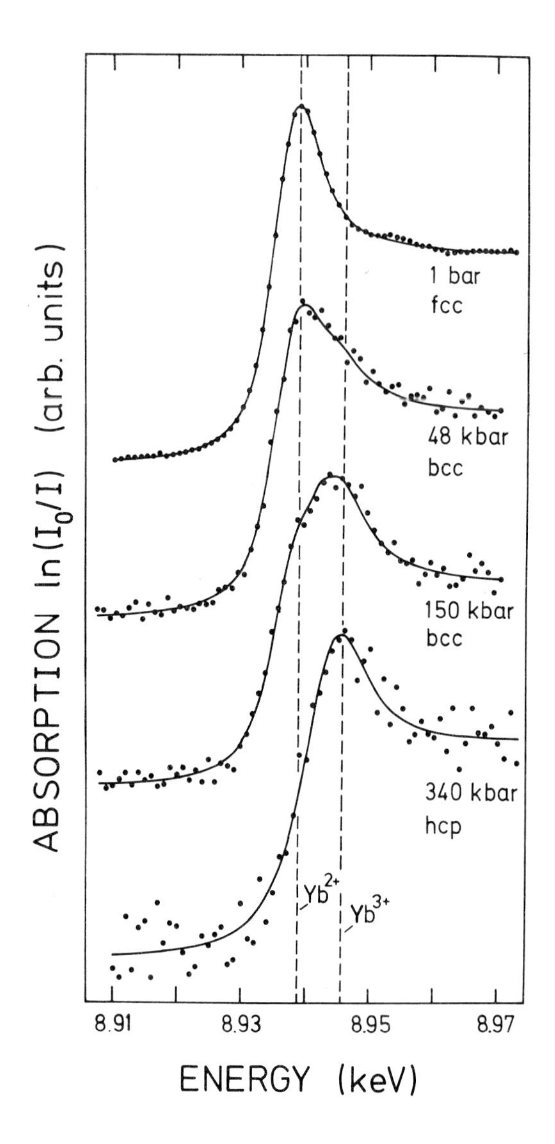

Figure 3: L_{III} edge absorption spectra at room temperature and at various pressures. The solid lines represent the results of least-squares fit analyses (see text). The dashed vertical lines indicate characteristic positions of the L_{III} edge peak for divalent and trivalent metallic Yb systems, respectively.

white line arises from atomic-like transitions from $2p_{3/2}$ core states to unoccupied 5d states above the Fermi level /11/. The energy of the Yb absorption edge at 1 bar is typical for metallic systems with purely divalent Yb /10/. At 340 kbar the L_{III} absorption peak is shifted by 6.5 eV to higher energy. This characteristic energy difference indicates that Yb is very close to the trivlaent state at 340 kbar /10/. The L_{III}-edge spectra of Yb in its bcc phase

(Fig. 3) exhibit features characteristic for a mixed valent state. In this case the edge structure is a superposition of two peaked edges displaced by about 7 eV and caused by divalent and trivalent Yb ions, respectively. The relative intensities of both edge structures are indicative for the mean valence of the mixed valent system /8 - 12/.

The solid lines in Fig. 3 represent the results of least-squares fits to the experimental data points using a superposition of two subspectra for divalent and trivalent Yb, respectively. Each subspectrum is described by an arctan like step function corresponding to the L edge and a Lorentzian corresponding to the white line. The spectral parameters for Yb^{2+} were adjusted from the Yb metal spectrum at ambient pressure, those for Yb^{3+} from additional spectra for compounds containing Yb in the trivalent state /16/. The experimental width of the white line was found to be ~ 8 eV in the divalent state and ~ 10 eV in the trivalent state. The spectral shape of the L_{III} edge of hcp Yb is very similar to the L_{III} edge of Tm^{3+} in (hcp) Tm metal /11/, when the different resolution of the spectrometers is taken into account.

The high pressure L_{III}-edge spectra of Yb were fitted with fixed parameters for the Yb^{2+} and Yb^{3+} subspectra. The spectra at 30 kbar and 340 kbar could be equally well fitted with only one spectrum corresponding to a pure Yb^{2+} and Yb^{3+} state, respectively. The mean valence $\bar{v}$ derived from spectra taken at 7 different pressures are plotted in Fig. 4. One measurement at 100 kbar was taken with decreasing pressure in order to check the reversibility of the pressure induced changes of the L_{III}-edge structure. The dash-dotted line in Fig. 4 is simply a guide to the eye.

4. DISCUSSION

Before discussing the results shown in Fig. 4, some remarks are necessary about the reliability of the mean valence values obtained from L-edge spectroscopy. In all previous applications of this method to mixed-valent rare earth systems of Sm, Eu, Tm, and Yb (excluding the special case of Ce), there was agreement (within 0,15) with valencies determined by other methods, as e.g. Mössbauer isomer-shift and lattice constant measurements /9, 10/. As pointed out recently /9/, each of the methods used up to now for valence determination seems to have its inherent difficulty in giving the right scaling. It has been supposed that final state effects, which are known to play an important role in 3d and 4d x-ray photoemission studies /18, 19/, might masque the actual (initial-state) valence in L_{III}-edge spectroscopy /9, 11 /. However, for all heavy rare earth systems such effects have not been observed up to now. In any case, final-state effects are much less pronounced in L_{III}-edge spectra as compared to XPS spectra because of the localized nature of the quasi-bound final

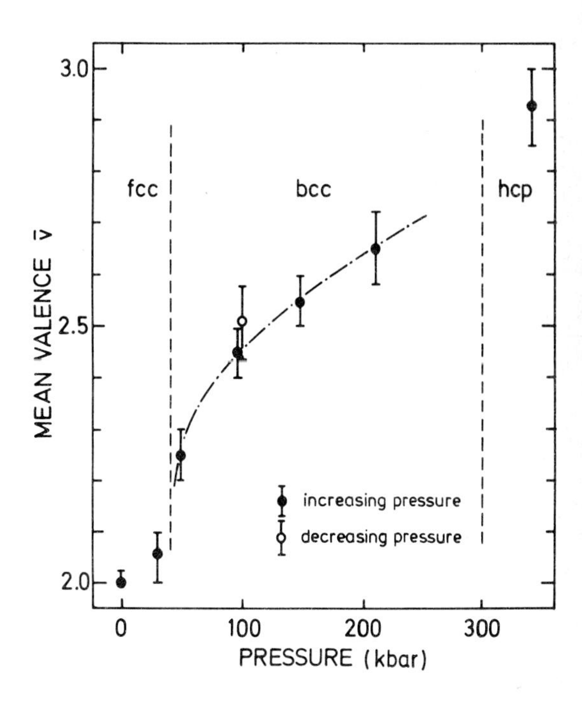

Figure 4: Pressure dependence of the mean valence of Yb metal in its various phases at room temperature.

state of the L absorption process /11, 18/. Therefore, we regard the valences derived from the Yb L_{III}-edge spectra as reliable within their statistical errors and admit a systematic error of 0.1, when comparison is made with other methods.

The results presented in Fig. 4 provide a quantitative picture about the pressure-induced valence transition in Yb metal. Within the accuracy of the present data and in accordance with a variety of previous investigations /1,3/ Yb remains divalent in its fcc phase up to the fcc → bcc transition at 40 kbar. The bcc phase exhibits a mixed valent state over its entire pressure range, which develops rapidly just after the fcc → bcc phase transition and reaches the medium valence 2.5 slightly above 100 kbar. Apparently, the valence transition is not completed within the bcc phase. In the hcp phase, the single data point indicates a trivalent (or nearly trivalent) state of Yb.

Previous high pressure studies of transport properties /3/ were interpreted as evidence for a pure divalent state of Yb in its bcc phase up to 160 kbar. This is apparently in contradiction to the present findings. The discrepancy may be resolved, however, if the temperature dependence of the mean valence in other mixed-valent systems of Yb is considered. Due to the high multiplicity of the $4f^{13}$ configuration, mixed-valent Yb systems like $YbAl_2$, YbCuAl and $YbCu_2Si_2$ show an increase in valence with temperature /9, 20/. Thus, in Yb systems valence transitions towards a mixed-valent or (magnetic) trivalent state are driven by both (increasing) pressure and temperature. The temperature behaviour in Yb systems is opposite to e.g. mixed-valent Eu systems /21/.

Therefore, we emphasize that the mean valence given in Fig. 4 is representative for room temperature only. At lower temperature, the values for the bcc-phase should be shifted towards the divalent state. This may readily explain why resistivity measurements did not detect (strong) transport anomalies. They were performed at relatively low temperatures and pressures. Whether at low temperature Yb is already mixed valent for pressures slightly above the fcc-bcc phase transition is still an open question. The results of a high pressure Mössbauer isomer shift study of Yb metal at 70 kbar and 4.2 K can be interpreted as evidence for a mixed valence of about 2.3 /22/.

As shown in Fig. 5, the mean valence of bcc Yb is (within the accuracy of the data) linearly dependent on volume. For comparison, we include in Fig. 5 the results of a pseudo-potential model calculation for Eu metal by Rosengren and Johansson /5/, which similarly exhibits an almost linear relationship between $\bar{v}$ and volume. Within the Rosengren-Johansson model the difference in slope between the Yb data and the Eu calculation may in part be attributed to the larger bulk moduli of divalent and trivalent Yb

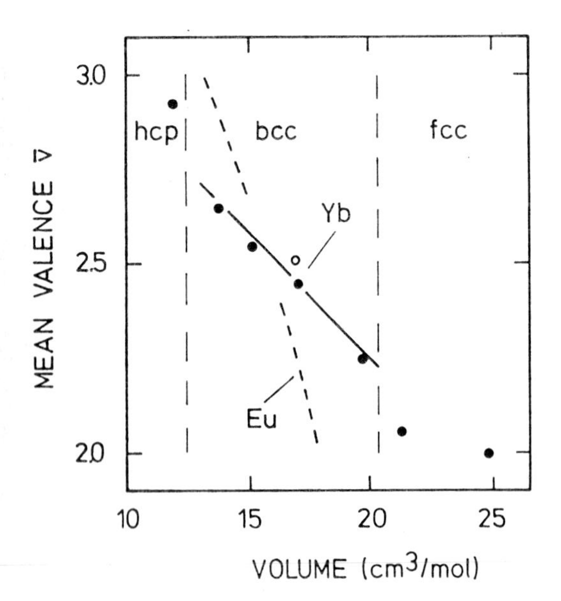

Figure 5: Volume dependence of the mean valence of Yb metal. The dotted line represents the results of a T = 0 model calculation for Eu metal /5/.

as compared to Eu. Also, the effect of temperature on the slope of the curves should be considered. We expect, that in analogy to the Eu results a quantitative description of the pressure induced valence change in Yb can be given, which is (at least for T = 0 K) primarily based on the appropriate pseudo-potential calculation of the cohesive properties of purely divalent and trivalent Yb.

In summary, we have shown that the combination of very high pressure techniques and L-edge x-ray absorption spectroscopy is a widely applicable tool for studying pressure-induced valence transitions in heavy rare earth systems. Yb metal is found to be mixed valent in its high pressure bcc-phase. At room temperature the valence transition is continuous over the entire pressure range corresponding to the bcc-phase. In the hcp-phase above 300 kbar, Yb is trivalent, ar almost trivalent.

Part of this work was supported by the Bundesministerium für Forschung und Technologie, Contract No. 05 115 KA.

REFERENCES

/1/ Jayaraman, A., in Handbook on the Physics and Chemistry of Rare Earths, Gschneidner, K. A. Jr.,Eyring,L., (eds.) (North Holland, Amsterdam 1978), Vol. 1, p. 707; and references therein.
/2/ Hall, H. T., Barnett, J. D., Merrill, L., Science 139, 111 (1963)
/3/ Katzmann, H. and Mydosh, J. A., Z. Physik 256, 380 (1972)
/4/ Johansson, B. and Rosengren, A., Phys. Rev. B 11, 2836 (1975); B 14, 361 (1976)
/5/ Rosengren, A. and Johansson, B., Phys. Rev. B 13, 1468 (1976)
/6/ Syassen, K. and Holzapfel, W. B., in High Pressure Science and Technology, Timmerhaus, K. D. and Barber, M. S. (eds.), (Plenum, New York 1979) Vol. 1, p. 223
/7/ Holzapfel W. B., Ramesh, T. G. and Syassen, K., J. Physique Colloq. 40, C5-390 (1979)
/8/ Vainshtein, E. E., Blokhin, S. M. and Paderno, Y. B., Sov. Phys. - Sol. State 6, 2318 (1965)
/9/ Wohlleben, D. K., in Valence Fluctuations in Solids, Falicov, L. M., Hanke, W. and Maple, M. B., (eds.) (North Holland, Amsterdam, 1981) p. 1
/10/ Bauchspiess, K. R., Boksch, W., Holland-Moritz, E., Launois, H., Pott, R. and Wohlleben, D., see Ref. 9, p. 417
/11/ Bianconi, A., Modesti, S., Campagna, M., Fischer, K. and Stizza, S., J. Phys. C: Solid State Physics 14, 4737 (1981)

/12/ Frank, K. H., Feldhaus, J., Kaindl, G., Wortmann, G., Krom, W., Materlin, G., and Bach, H., Contribution to this volume. Röhler, J., Krill, G., Kappler, J. P., Ravet, M. F. and Wohlleben, D., contribution to this volume.

/13/ This structure extends over an energy range of ~ 30 eV beginning from the edge and has previously been named KOSSEL-structure (Kossel, W., Z. Phys. 1, 119 (1920)). Recently, some authors are using the abbreviation XANES.

/14/ Johansson, B. and Martensson, N., Phys. Rev. B 21, 4427 (1980)

/15/ Syassen, K., Wortmann, G., Feldhaus, J., Frank, K. H. and Kaindl, G., submitted for publication..

/16/ Feldhaus, J., Ph.D-Thesis, Freie Universität Berlin (1982), unpublished.

/17/ Huber, G., Syassen, K., Holzapfel, W. B., Phys. Rev. B 15, 5123 (1977)

/18/ Fuggle, J. C., Campagna, M., Zolnierek, Z., Lässer, R. and Platau, A., Phys. Rev. Letters 45, 1597 (1980)

/19/ Schneider, W. D., Laubschat, C., Nowik, I., and Kaindl, G., Phys. Rev. B 24, 5422 (1981)

/20/ Penney, T., Barbara, B., Melcher, R. L., Blaskett, T. S., King, Jr., H. E. and La Placa, S. J., see Ref. 9, p. 341.

/21/ Röhler, J., Wohlleben, D., Kaindl, G. and Balster, H., submitted for publication.

/22/ Boehm, H.-G., Ph. D. Thesis, Technische Universität München (1970), unpublished

Valence Instabilities
P. Wachter and H. Boppart (eds.)

THE VALENCE VARIATION IN THE $(Ce_{1-x}Y_x)Pd_3$ SYSTEM

M.J. Besnus, J.P. Kappler, G. Krill, M.F. Ravet, N. Hamdaoui and A. Meyer

L.M.S.E.S., Institut de Physique, 3, rue de l'Université, F 67084 STRASBOURG

Measurements of the variation, as a function of concentration, of lattice parameter, resistivity, susceptibility, specific heat, all lead to the conclusion that the intermediate valence state of $CePd_3$ changes continuously to a non magnetic valence state for $x \gtrsim 0.5$.

In binary systems of an Intermediate Valence (IV) compound with a non magnetic isomorphous substitute, different kinds of behaviours have been observed. In some systems, the valency (or n_f occupation number) of the IV atom remains unchanged and the magnetic properties of the system are dilution like, as it has been found to occur for $(Ce-La)Sn_3$ (1) or (Yb-Y)CuAl (2). In others, there is a change in the 4f occupation number as a function of concentration and correlatively a change in the magnetic, electric and other properties of the R.E. atom which outmost may change from an intermediate to an integer valence state. This has been shown to occur for example in $Ce(Pd-Rh)_3$-$Ce(Pd-Ag)_3$ (3) where n_f sweeps continuously from 0 to 1, or in $(Ce-La)Be_{13}$ (4) where it goes from 0.7 to 1. In $(Ce_{1-x}Sc_x)Pd_3$ (5) or $(Ce_{1-x}Y_x)Pd_3$ (6) the lattice parameter variation indicates that n_f goes from 0.5 to 0 when x goes from 0 to about 0.4, as has been confirmed by susceptibility, resistivity and RPE measurements (6)(7).

We present here the result of a further study of the $(Ce_{1-x}Y_x)Pd_3$ system and we will show that lattice parameter, susceptibility, electrical resistivity and specific heat measurements, all lead to the conclusion that the typical physical static manifestations on the IV state have disappeared for $x \gtrsim 0.5$. However our photoemission results (7) do not scale with all the other data : the same spectra were observed on both sides of the critical concentration where the IV state is thought to disappear.

Measurements. Samples : The alloys were prepared from 99.99 pure metals, by arc melting under 99.99 pure argon atmosphere. Most samples were annealed 48 hours at 700°C in evacuated quartz tubes, despite all the investigated properties showed no significant difference if measured before and after annealing. Also X-rays investigation showed no traces of foreign phases.

Lattice parameters. They were measured from 4 to 300 K. Fig. 1 shows their variation as a function of concentration for the 10, 200 and 300 K isotherms and the inset of fig. 1 shows the 0 K mean 4f occupation number n_f deduced from these data and from the assumed temperature dependence a(T) of the virtual $Ce^{4+}Pd_3$ and $Ce^{3+}Pd_3$, identified with that of YPd_3, which was found to be exactly the same as that of $LaPd_3$ and $PrPd_3$. The results recognize a critical concentration near $x \simeq 0.5$ above which the lattice parameter indicates that Ce in these alloys has no longer a mean 4f occupation. Fig. 2 shows the thermal variation of n_f in the IV concentration range : n_f is seen to increase with increasing temperature, similarly for different values of x in the IV concentration range.

Resistivities. The temperature variation of the resistivity has been studied for the whole concentration range (fig. 3). In the Ce rich concentration range our data confirm the previous

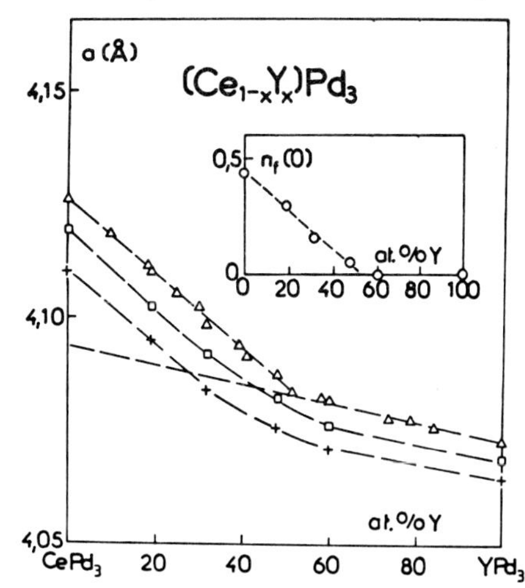

Figure 1 : Variation of the lattice parameter with T and x. Δ : 300 K, : 200 K, + : 10 K. Inset : 0 K mean 4f occupation number.

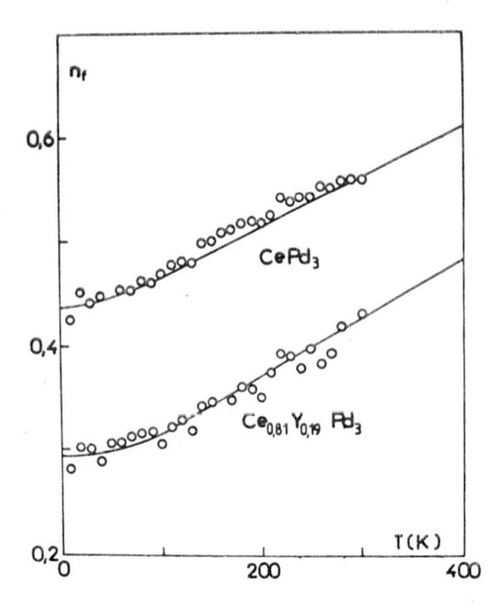

Figure 2 : Variation with temperature of the mean occupation number n_f : o measured, —— calculated.

one from Schneider (8) : the typical thermal variation of $CePd_3$ is strongly modified by small YPd_3 addition leading to an anomalous large increase of the residual resistivity. It decreases for larger x, and takes a near to normal behaviour for $x \gtrsim 0.5$ for a disordered two component alloy system. These results agree with the lattice parameter data : the IV behaviour as reflected by the resistivity is lost for $x \gtrsim 0.5$.

Susceptibilities. At low temperatures, the measured susceptibilities χ_m show an impurity contribution which can be accounted for by a C/T term in a field and temperature range without paramagnetic saturation (H/T small). Plots of $\chi_m T - \chi_o T + C$, allow to extract the moment C of the impurities and χ_o, the initial susceptibility, which appears to stay constant in a temperature intervall ranging from 0 to about 0.3 T_{max}. T_{max} is the temperature of the susceptibility maximum which is seen to increase with increasing x, the maximum becoming less pronounced and being wiped out for $x \gtrsim 0.5$. Fig. 5 and table I show $\chi_o(x)$ corrected for core diamagnetism by Selwoods values (9) : it decreases rapidly with increasing x, becoming very small for $x \gtrsim 0.5$, which here again appears as the limiting concentration for the IV behaviour.

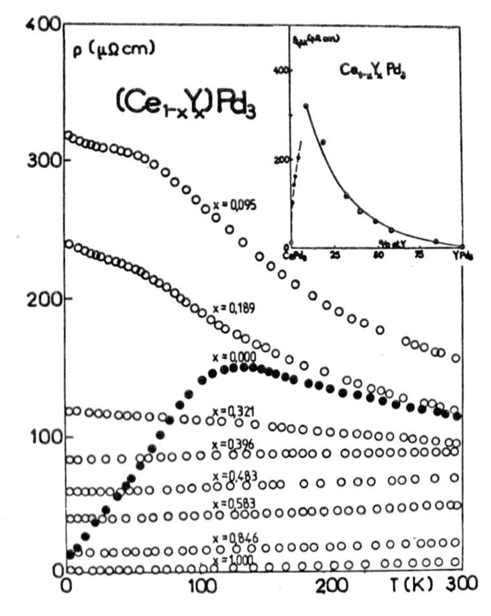

Figure 3 : Resistivity vs T. Inset : Residual Resistivity, data for x<0.1 from (8)

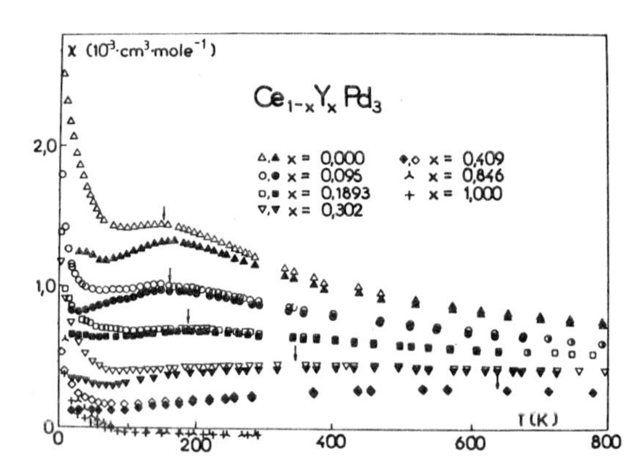

Figure 4 : Susceptibility vs T, open symbols χ as measured, full symbols, χ corrected for impurity contribution

Specific heats. They have been measured in the 1.5-20 K (fig. 6) range. $CePd_3$ as cast or annealed shows a λ anomaly near 6 K already observed by others (3), which can correspond by its entropy to a near 2 at % magnetic Ce^{3+}, Γ_7 impurity, which we tentatively ascribe to cubic Ce_2O_3, as hexagonal Ce_2O_3 is known to present

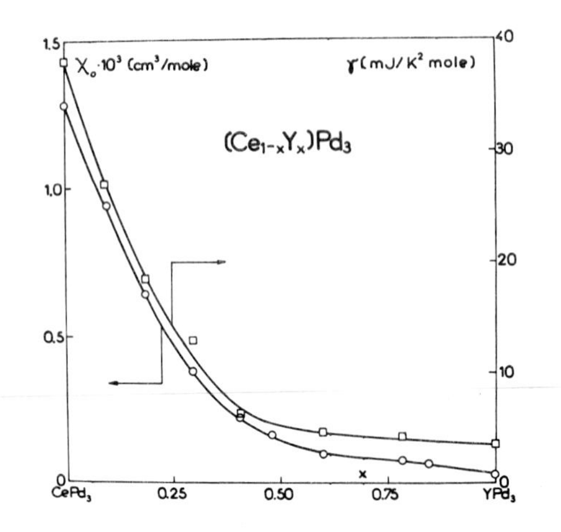

Figure 5 : χ_o (circles) and γ (squares) vs x

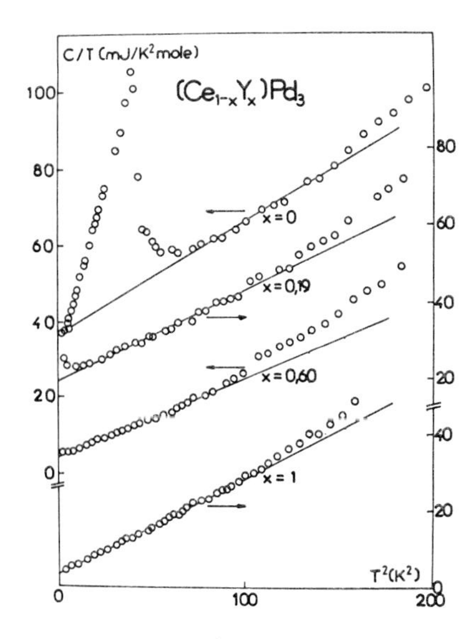

Figure 6 : C/T vs T^2 for several alloys of the $(Ce_{1-x}Y_x)Pd_3$ series

a cooperative anomaly at 8.5 K (10) This anomaly is much reduced and apparently sprayed over a larger temperature range in the $(Ce-Y)Pd_3$ alloys, thus leading to an uncertainty of their measured Debye temperatures. The electronic coefficient γ and Debye temperature are given in table I. Fig. 5 shows that γ decreases rapidly with increasing x like χ_o, while for $x \gtrsim 0.5$ it is almost constant near to the value of YPd_3 : χ_o and γ reflect in the same way the progressive valence change occuring in this system.

Recent theories of Bringer and Lustfeld (11), Newns and Hewson (12) learn that the ratio of the 0 K susceptibility χ_o to the electronic specific heat coefficient γ is characteristic for an atom in an IV state with a value $\chi_o/\gamma = Rg^2\mu_B^2 J(J+1)/\pi^2 k_B^2$ where R should be unity, independently of the value of the 4f occupation number $n_f(o)$. In fig. 7 our χ_o values (corrected for core diamagnetism) are plotted versus γ with the concentration x as an implicit parameter. The plotted values of χ_o and γ are both corrected for a spd back ground, on the basis of

Table 1

at%Ce	100	90.5	81.1	69.8	67.9	59.1	51.7	40.0	21.7	15.4	0
$n_f(0)$	0.44		0.30		0.15		0.05	0			0
$\chi(0)$ (10^{-3} emu/mol)	1.282	0.941	0.641	0.425	0.410	0.219	0.164	0.097	0.076	0.060	0.025
T_{max}(K)	145	165	190	330		640					
γ (mJ/K² mol)	38.2	27.0	18.4	12.45		6.35		4.46	4.19		3.48
θ_D(K)	300	321	317	345		324		341	352		312
R	1.15	1.19	1.19	1.14		1.19		0.75	0.62		0.24
Δ (meV)	6.15	5.4	4.6	2.5		1.1					
$\epsilon_f-\epsilon_F$ (meV)	26.2	28.0	29.9	25.1		24.0					

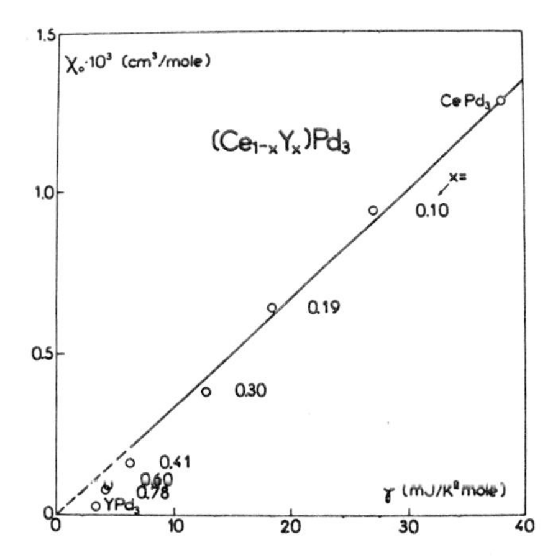

Figure 7 : χ_o vs γ with x as implicit parameter o experimental, —— R = 1.17

the γ value of $LaPd_3$ which we found to be 0.28 mJ/mol K^2. Possible phonon and susceptibility enhancement have been ignored. It is seen that in the IV concentration range a linear relation is obeyed with a proportionality factor R=1.17. The ratio R, as a function of x is given in Table I : it is constant up to $x \approx 0.5$, but at higher x drops toward 0.31, the value for conduction electrons which is $2\mu_B^2/\pi^2 k_B^2$ if taking account of Landau diamagnetism, which is nearly attained for YPd_3. Thus in the IV concentration range of these alloys the χ_o/γ ratio characteristic for Ce atoms, appears as effectively independent of the 4f occupation number $n_f(o)$ which we have shown to vary with concentration, as long as $n_f > 0$; but the ratio changes for $x \gtrsim 0.5$ going toward a value characteristic of conduction electrons.

The Newns and Hewson local Fermi liquid theory

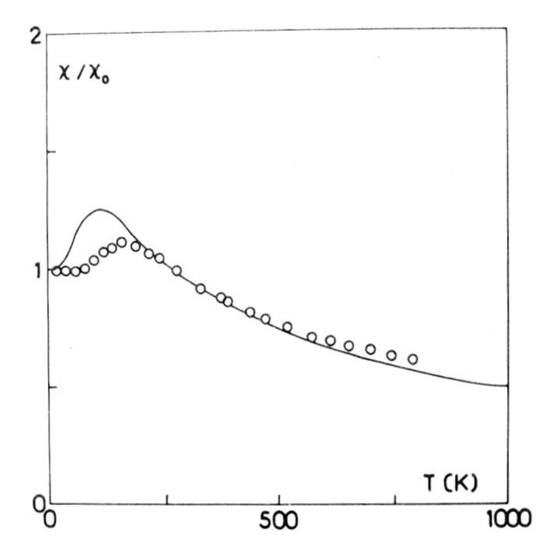

Figure 8 : Theoretical curve $\chi(T)/\chi_o$ (full line) for $CePd_3$ and experimental data

considers impurity f states in a 2J+1 spin-orbit degenerate resonant lorentzian level.The formalism predicts the above mentioned χ_o/γ ratio, but also allows to calculate the temperature variation of n_f, χ and C_v/T. The parameters of the calculations are Δ the half width of the resonant f level and $\varepsilon_f-\varepsilon_F$, its separation from the Fermi energy, which can be deduced from either χ_o or γ, knowing $n_f(o)$. These values are also given in Table I. The calculated n_f variations with T using a conduction band density of 3 eV^{-1} are shown in fig. 2 for two concentrations in the IV state showing agreement with the experimental data. The calculated temperature dependence of susceptibility for $CePd_3$ agrees only quantitatively with the experimental data at high temperature as seen from fig. 8, whereas it has been found to fit satisfactorily the data for $CeSn_3$ and YbCuAl (12). The calculation stresses the fact that the high temperature susceptibility does not follow a Curie Weiss law, as long as $n_f<1$, that is for all concentrations at temperatures below 1000 K. The N-H theory predicts also an anomaly for the electron specific heat which traducts itself by an anomalous low Debye temperature as compared to a non magnetic isomorphous reference. This effect has been observed in YbCuAl, but appears too small to be significative when n_f is small as it is in our case.

All the above mentioned experimental data which agree also with that obtained by EPR (6) are significative for a continuous valence variation in the (Ce-Y)Pd_3 system : the mean n_f occupation number appears to decrease to zero at $x\simeq 0.5$ and above that concentration all investigated properties are that of usual non magnetic RE compounds : the IV state changes to a non magnetic Ce^{4+} state. This conclusion is however not sustained by our photoemission experiments (5) which, at first sight demonstrate the persistence of two valence states in $(Ce_{0.25}Y_{0.75})Pd_3$, a result up to now without clear explanation. However a study presently in course of X-rays, L_{III} absorption edge seems to give results in agreement with the valence change model.

REFERENCES

(1) Dijkman W.H., de Boer F.R. and de Chatel P.F., Physica 98 B (1980) 271.
(2) Mattens W.C.M., de Chatel P.F., Moleman A.C. and de Boer F.R., Physica 96 B (1979) 138.
(3) Mihalisin T., Scoboria P. and Ward J.A., Phys. Rev. Lett. 46 (1981) 862.
(4) Kappler J.P., Krill G., Ravet M.F., Besnus M.J. and Meyer A., Valence Fluctuations in Solids (1981) 271, North-Holland.
(5) Gambke T., Elschner B. and Schaafhausen J., Phys. Lett. 78 A (1980) 413.
(6) Gambke T., Elschner B., Schaafhausen J. and Schaeffer H., Valence Fluctuations in Solids (1981) 447, North-Holland.
(7) Kappler J.P., Krill G. Besnus M.J., Ravet M.F., Hamdaoui N. and Meyer A., 3M Conference, Atlanta (1981).
(8) Schneider H. and Wohlleben D., Z. Physik B 44 (1981) 193.
(9) Selwood P.W., Magnetochemistry, Interscience Publishers, New-York (1956).
(10) Justice B.H. and Westrum E.F., J. Phys. Chem. 73 (1969) 1959.
(11) Lustfeld H. and Bringer A., Sol. St. Com. 28 (1978) 119.
(12) Newns D.M. and Hewson A.C., J. Phys. F 10 (1980) 2429.

Valence Instabilities
P. Wachter and H. Boppart (eds.)

VALENCE CHANGES AT RARE EARTH METAL SURFACES MONITORED BY ELS

G. Strasser, E. Bertel[+], J.A.D. Matthew[++] and F.P. Netzer

Institut für Physikalische Chemie, Universität
Innsbruck, A-6020 Innsbruck, Austria

Electron energy loss spectroscopy in reflection geometry has been used to investigate valence changes at rare earth surfaces. In particular, plasmon energies and "giant resonance" features were found to be of diagnostic value for assessing the valence state of the system. Examples are presented for divalent surface character of Sm metal and for valence changes during oxidation of Yb and Eu.

In electron energy loss spectroscopy (ELS) a primary electron transfers energy and momentum to the substrate in an inelastic scattering process, thereby stimulating excitations in the solid. If the experiment is performed in reflection geometry, the inelastic process has to be preceded or followed by an elastic scattering event, so that the "loss electron" can reach the detector. At energies of the primary electron 100 - 1000 eV the electron mean free path is of the order 5 - 20 Å; ELS in that energy region is therefore surface sensitive, and the probing depth can be tuned by choosing appropriate primary energies. In ELS of rare earth systems it is useful to consider three main processes of electronic excitation:

i) collective plasmon excitations of the quasi-free valence electron gas
ii) low-energy one electron excitations of an interband type of valence electrons to empty states above the Fermi level
iii) inner-core excitations of 4d electrons to empty 4f states corresponding to the "giant resonance" features seen in X-ray absorption (1) and resonant photoemission experiments (2).

Low-energy interband excitations of the 5d6s conduction electrons occur at a similar loss energy 2-4 eV in most rare earth metals; the intensity of corresponding loss features decreases if the conduction band is depleted, e.g. by compound formation, but relatively small differences are observed between systems of different valency. The plasmon energy reflects the number of quasi-free valence electrons in the solid, and significant differences in plasmon energy are observed between divalent (7-9 eV) and trivalent ($\sim$12 eV) rare earth metals. The "giant 4d$\rightarrow$4f resonances" in rare earth materials are localized, atomic-like transitions of the rare earth atoms, and show little environmental dependence during chemical reactions provided there is no change in 4f occupation number. However, the occupation of the 4f shell changes if the rare earth system undergoes a valence change; the structure and energy position of giant resonance features in ELS are therefore sensitive to the valence state of the system.

In the following, several examples will be discussed to demonstrate the usefulness of plasmon and giant resonance spectroscopy for assessing the valence state on rare earth surfaces.

Electron energy loss spectra were recorded in a stainless steel ultrahigh vacuum system with a concentric hemispherical analyser in nearly specular reflection geometry (energy resolution ΔE = const. $\simeq$ 0.25 eV). Details of the experimental set-up are given elsewhere (3). Rare earth samples were in the form of in-situ evaporated metal films, and surface cleanliness and composition was monitored by Auger spectroscopy.

Surface valence change of Sm metal

A surface valence change from trivalent bulk to divalent surface character has been reported recently by several authors using conventional XPS and synchrotron photoemission (e.g. 4,5). In these studies divalency at the surface has been monitored by studying the binding energies of the atomiclike 4f states and 3d or 4d core levels. We have used a different approach, and have studied directly the conduction band changes from the $(sd)^3$ to the $(sd)^2$ configuration via plasmon excitation using the different sampling depths of 1000 eV and 100 eV electrons (6). A plasmon feature is seen for E_p = 1000 eV at around 12 eV (consistent with typical trivalent rare earth metal plasmon energies), but for E_p = 100 eV the plasmon is observed at

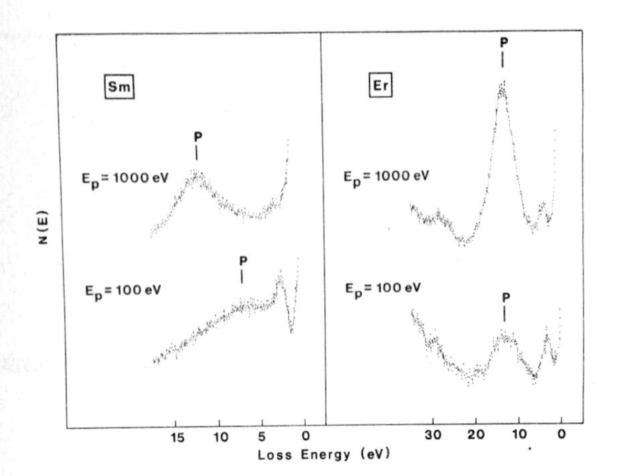

Fig. 1: Comparison of electron loss spectra of Sm and Er. Note the absence of a low-energy-plasmon feature below 10 eV in the case of Er (P = plasmon)

7.5 eV (Fig. 1). It is tempting to associate the lower-energy plasmon with divalent surface character, but the possibility of the dominance of a conventional surface plasmon at the lower primary energy has to be investigated. It is useful to compare the Sm spectrum with the corresponding spectrum of a rare earth where it is certain that no valence change takes place. In Fig. 1 the electron loss spectra of Sm are compared with those of Er. In the case of Er no new feature appears in the spectrum for low incident energy. The 7.5 eV loss in Sm cannot therefore be regarded as a conventional surface plasmon and, given the complementary photoemission evidence, should be associated with divalency at the surface.

Valence change in Yb upon oxidation

Upon oxidation divalent Yb atoms in the metal change to trivalent Yb in the oxide phase. The creation of the hole in the filled metal 4f shell allows the "turning on" of a 4d⟶4f resonance, as seen in resonant photoemission by Johansson et al. (2). The appearance of the giant resonance due to $Yb^{2+} \longrightarrow Yb^{3+}$ during oxidation can also be observed in ELS, and this is shown in Fig. 2. The shape and peak energy of the resonance agree well with the photoemission observations (2). The resonance provides an excellent monitor of the oxidation process on the metal side, and its growth with oxygen exposure can be compared with the build-up of the O KLL Auger signal as discussed by Bertel et al. (3). The formation of trivalent Yb oxide is therefore demonstrated.

Electron loss spectra of clean Yb and Yb exposed to various amounts of oxygen are shown in Fig. 3(a). For comparison, the corresponding spectra for Gd are displayed in Fig. 3(b). Concentrating here on the plasmon behavior, we observe the plasmon at 8.5 eV on clean Yb metal, consistent with a divalent metal value. During oxidation an oxide plasmon feature grows at constant energy ~15 eV thus reflecting the increased electron density in the oxide valence bands.

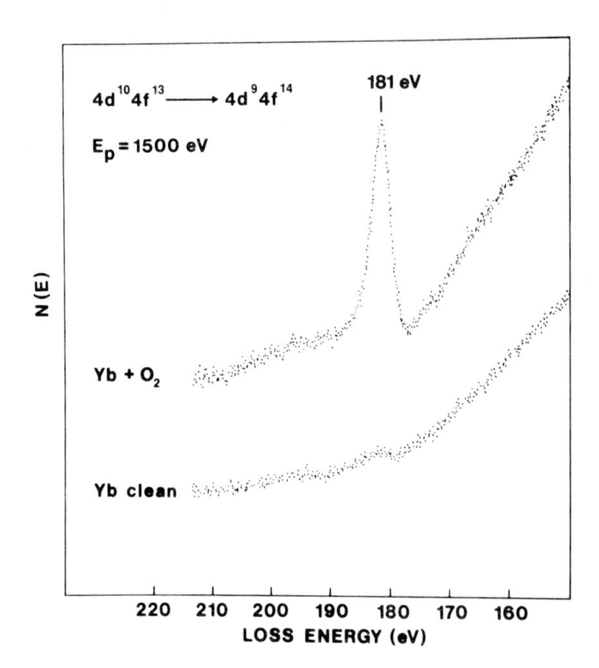

Fig. 2: Giant resonance feature in ELS of clean and oxidized Yb. A linear background has been subtracted from both spectra.

The observed oxide plasmon energy is significantly less than would be estimated assuming 18 free electrons per Yb_2O_3 unit, but is in agreement with typical trivalent rare earth oxide plasmon energies (7). We note that the metal and oxide plasmons are coexisting at intermediate oxygen exposures, and that the oxide plasmon shows no significant energy shift during progressive oxidation. This has been discussed recently in terms of an island growth mechanism of the oxide phase (3). In contrast, the metal plasmon of trivalent Gd at 12 eV shifts continuously to ~14 eV during oxygen exposure (Fig. 3(b)), suggesting a more gradual transformation into a phase with increased mobile electron density. A very similar behavior was observed upon exposure of Er to oxygen (8), and seems to be typical for trivalent rare earth metal oxidation.

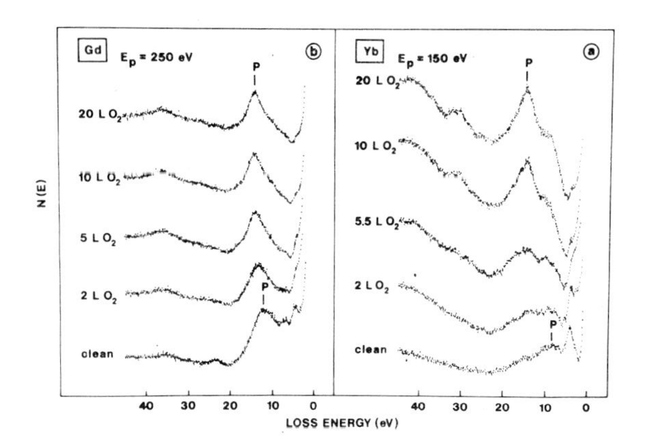

Fig. 3: Electron loss spectra of Yb (a) and Gd (b) as a function of O_2 exposure (1 L = 10^{-6} torr·sec)

Valence change in Eu?

For Eu two different stable oxide stoichiometries have been reported and firmly established under normal temperature and pressure conditions (9): trivalent sesquioxide and divalent monooxide. Eu metal occurs in divalent form, and it is of interest to see whether a valence change can be observed during surface oxidation at room temperature under low pressure conditions. Electron loss spectra of clean Eu metal and Eu exposed to various amounts of oxygen at $\sim 10^{-7}$ torr are shown in Fig. 4. Similarly to Yb we note a divalent metal plasmon at ~ 7 eV and an oxide plasmon at 13.5 - 14 eV. The oxide plasmon energy, however, is not in itself conclusive evidence of trivalent oxide character, because the oxide plasmon in stoichiometric EuO has also been reported at around 14 eV (10). Additional information is therefore needed, and may be obtained from giant resonance data.

In Fig. 5 giant resonance loss curves are compared for clean and oxygen exposed Eu and Gd. The giant resonance losses of Gd are in excellent agreement with recent synchrotron photoemission data of Gerken et al. (11). Upon oxidation the giant resonances of Gd remain virtually unchanged as expected from their insensitivity to chemical environment. In Eu, however, significant changes are observed between clean and oxidized surfaces: the main feature shifts by 2.5 eV to higher loss energy and considerable change in structure is evident. A more detailed discussion of these effects will be given elsewhere (12), but a valence change of Eu atoms during oxidation is suggestive. It appears, therefore, that a surface layer of trivalent Eu oxide or an oxide phase of mixed valency is built up during these experiments.

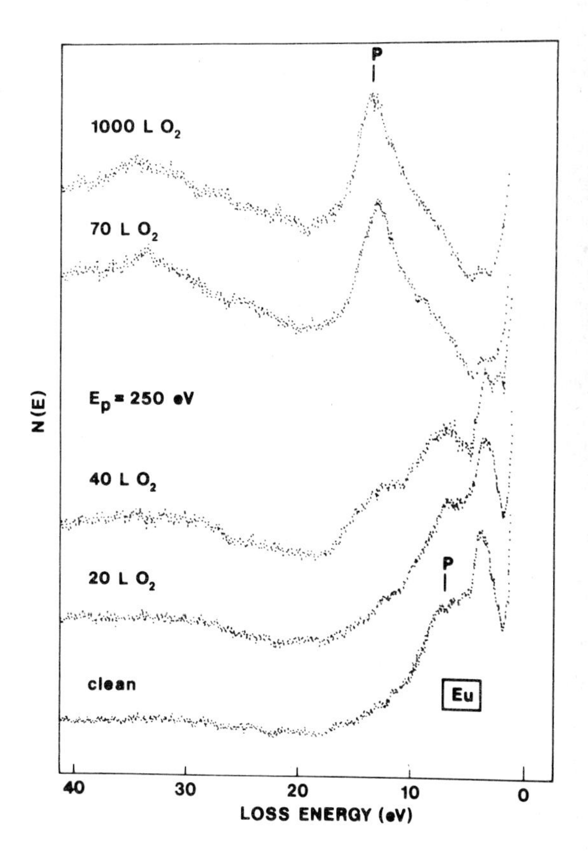

Fig. 4: Electron loss spectra of Eu as a function of O_2 exposure.

Acknowledgements

This work has been supported by the Austrian Fonds zur Förderung der Wissenschaftliche Forschung

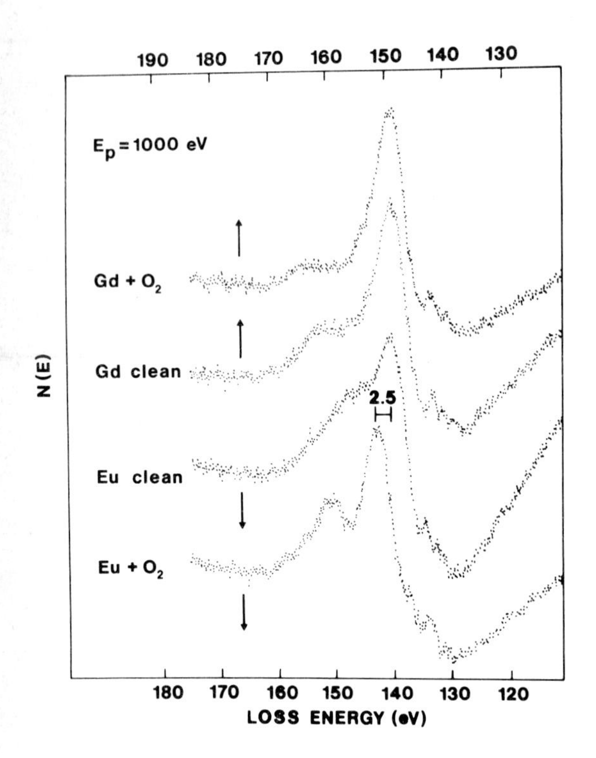

Fig. 5: Giant resonance losses of clean and oxidized Eu and Gd. Upper energy scale: Gd; lower energy scale: Eu.

References:

1) Zimkina, J.M., Formichev, V.A., Gribovskii, S.A. and Zhukova, I.I., Soviet Phys.-Solid State 9 (1967) 1128, 1163

2) Johansson, L.I., Allen, J.W., Lindau, I., Hecht, M.H. and Hagström, S.B.M., Phys. Rev. B 21 (1980) 1408

3) Bertel, E., Strasser, G., Netzer F.P. and Matthew, J.A.D., Surface Sci. 1982, in press

4) Lang, J.K. and Baer, Y., Solid State Commun. 31 (1979) 945

5) Allen, J.W., Johansson, L.I., Lindau, I. and Hagström, S.B.M., Phys. Rev. B 21 (1980) 1335

6) Bertel, E., Strasser, G., Netzer, F.P. and Matthew, J.A.D., Phys. Rev. B 25 (1982)

7) Colliex, C., Gasgnier, M. and Trebbia, P., J. Physique 37 (1976) 397

8) Bertel, E., Netzer, F.P. and Matthew, J.A.D., Surface Sci. 103 (1981) 1

9) Gasgnier, M., phys. stat. sol. (a) 57 (1980) 11

10) Nigavekar, A. and Matthew, J.A.D., Surface Sci. 95 (1980) 207

11) Gerken, F., Barth, J. and Kunz, C., Phys. Rev. Lett. 47 (1981) 993

12) Netzer, F.P., Strasser, G., and Matthew, J.A.D., to be published

+ Present address: Surface Science Division, National Bureau of Standards, Washington, D.C. 20234, USA

++ Permanent address: Department of Physics, University of York, Heslington, York YO1 5DD, England

Valence Instabilities
P. Wachter and H. Boppart (eds.)

PHASE RELATIONSHIPS AND STRUCTURAL INVESTIGATIONS IN TmSe

B. Fritzler and E. Kaldis

Laboratorium für Festkörperphysik, ETH Zürich
8093 Zürich, Switzerland

The gross nonstoichiometric range of TmSe (NaCl-structure) extends on both sides of the stoichiometry (0.87 - 1.05 mol Tm/Se). A miscibility gap separates primarily trivalent TmSe from mixed valent TmSe and indicates a repulsion of their respective electronic structures. The defect structure is studied and a model is presented, showing the existence of Tm clusters in the Tm-rich range. The valence change is determined from the lattice constants using a new and consistent set of ionic radii.

1. PHASE WIDTH OF NONSTOICHIOMETRIC TmSe (NaCl-STRUCTURE)

In this paper only some highlights of our investigations are reported, more detailed discussions are given in recent publications (1,2) and in a review (3).

X-ray and T-x phase diagram investigations, chemical analyses and many other investigations of more than 40 samples of nonstoichiometric TmSe (NaCl-structure) indicate that the homogeneity range extends on both sides of the stoichiometric composition (0.87 < mol Tm/ mol Se < 1.05), as was reported originally (4,5). These investigations also indicate the existence of a miscibility gap, 0.94 < Tm/Se < 0.98 (see below).

Figure 1 shows the dependence of the lattice constants on composition. Three ranges are clearly seen:

- stoichiometric and Tm-rich TmSe (0.98 - 1.05, least square fit line of this range: a(Å) = 0.3955 x + 5.2943; x = mol Tm/mol Se)
- Se-rich TmSe (0.87-0.94, a(Å) = 0.3956 x + 5.2905)
- Tm_5Se_6 superstructure (6) (0.81 - 0.87, a (Å) = 0.2357 x + 5.4297).

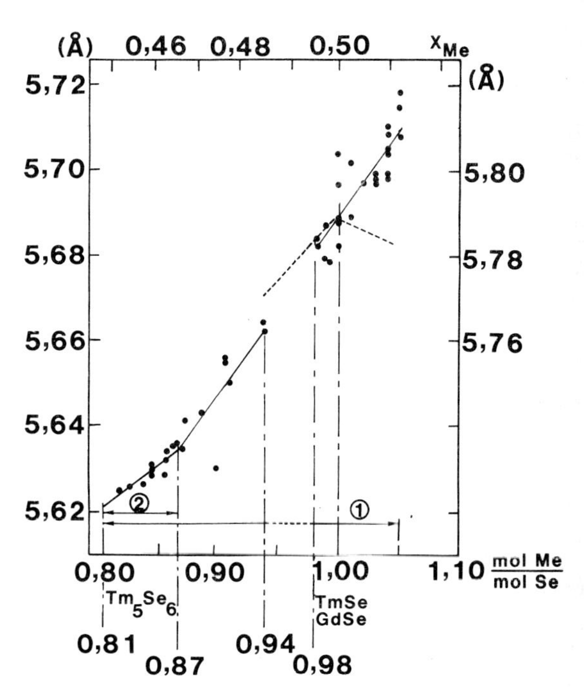

Fig. 1: Lattice constants vs. composition of TmSe (points and full lines, values left scale) and GdSe (dashed line, values right scale, shown for comparison). ① denotes the gross homogeneity range of the NaCl-structure, which extends on both sides of stoichiometric TmSe. ② gives the range of the Tm_5Se_6 superstructure. Between 0.94 and 0.98 mol Tm/Se a miscibility gap occurs.

Figure 2 shows the T-x phase diagram, based on the investigation of over 40 compositions and more than 300 DTA-runs in the temperature range 400 - 2400°C. These investigations were carried out in closed W-capsules, experimental details are given in (7). In the nonstoichiometric range the liquidus and solidus form a maximum melting-point diagram with congruently melting $Tm_{0.96}Se$ at 2065°C (3).

At approximately 1700 and 1100°C two phase transitions have been found in the nonstoichiometric range. These transitions show a large scattering of the DTA peaks (± 50°C), but no structure of these transition could be determined. Comparable phase transitions have been found in $TmSe_{1-x}Te_x$ at about 1600°C which leads to a miscibility gap (8) and in Sm_3S_4 (9). In order to receive more information about these transitions, an investigation with electron diffraction on quenched samples has started.

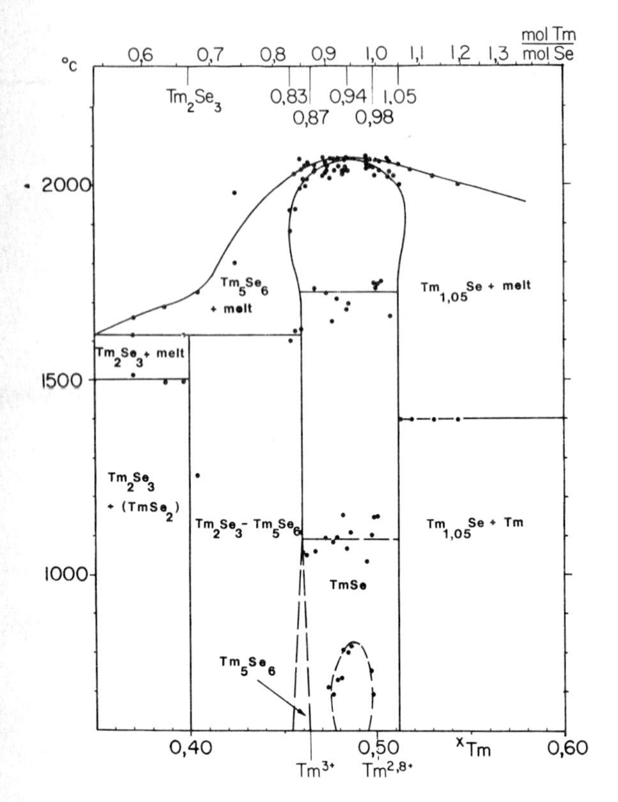

Fig. 2: T-x phase diagram of Se-Tm system. In the nonstoichiometric range two phase transitions at about 1700 and 1100°C and a miscibility gap below 860°C occur. The DTA points of the phase transitions show large scattering but do not reveal a structure, therefore they are given as horizontal lines.

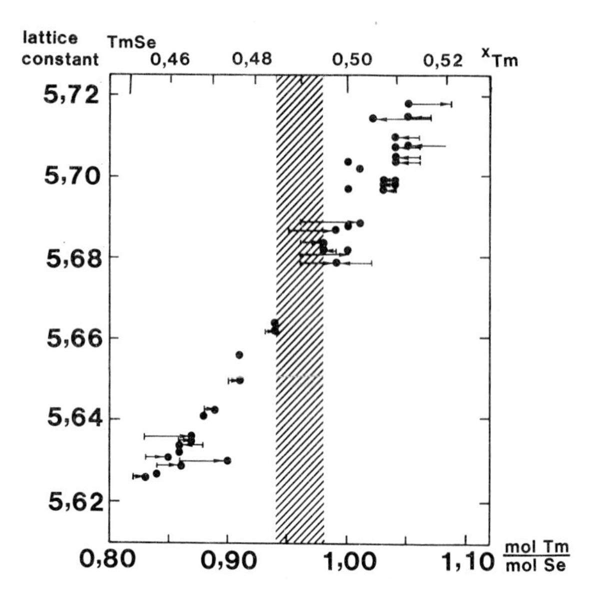

Fig. 3: Drifts of the chemical composition during crystalisation. Vertical lines give the composition of the starting material, points give the final composition of the crystal, both determined by chemical analyses. Large changes occur from the miscibility gap (0.94 - 0.98 mol Tm/Se) and at the phaseboundaries, both can be explained with the form of the T-x phase diagram.

Also high temperature neutron diffraction can give important information, provided that TmSe is handled in closed thin Ta-crucibles and that no side reactions with Ta take place.

An important feature of this phase diagram is the occurence of the curved phase boundary below 860°C in the nonstoichiometric range, which is attributed to a miscibility gap. Miscibility gaps in the low temperature range of solid solutions are the result of a decreasing entropy stabilisation of the solution. The repulsion of the constituents (adjacent phases of the miscibility gap) becomes predominant leading to thermodynamic instability of the solution. Nonstoichiometric TmSe is an "isopiestic" binary system (10), where no substitution with a third component of different size has been made. For this reason repulsive forces between the boundary compositions (0.94 and 0.98 mol Tm/Se) cannot be ascribed to size effects but only to the electronic structure (10). It can be concluded therefore, that a mixing (solution) of essentially $Tm^{3+}Se$ and mixed valent $Tm^{2.8+}Se$ is not possible at room temperature or below.

The low temperatures slow down the decomposition reaction in the miscibility gap, which makes the determination of the phase boundary by DTA difficult. Consequently rather a shift of the DTA base line, indicating a change of the specific heat, than a peak is observed. This signal is not seen in all compositions and in all runs but a statistics of DTA peaks vs. temperature and composition clearly reveals the miscibility gap shown in figure 2 (11). Measurements of the specific heat up to 900°C are therefore desirable.

A strong support for the existence of the miscibility gap is given by the fact, that up to now we did not succeed in growing crystals with compositions between 0.94 and 0.98 mol Tm/Se. Figure 3 shows shifts in composition of the starting material of the crystal growth experiment to the final composition of the crystals, both determined by chemical analyses (3).

2. DEFECT STRUCTURE

Figure 4 shows the densities of TmSe as a function of composition, curves ①, ② and ③

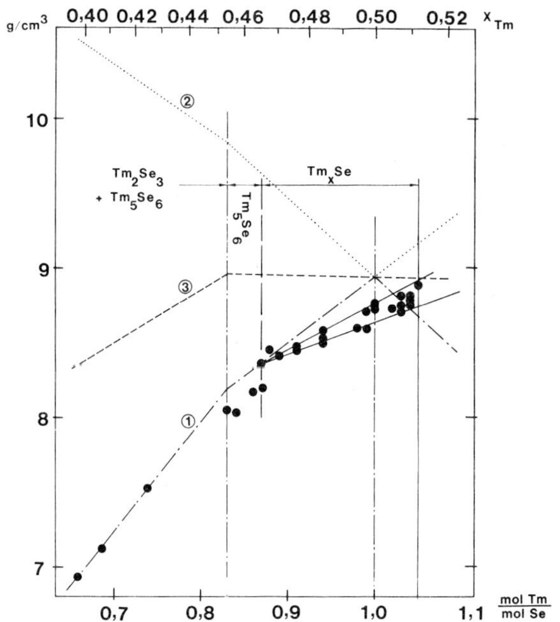

Fig. 4: Experimental densities (points and full lines) and calculated densities of TmSe: ① vacancy, ② interstitial and ③ antisite model for nonstoichiometric defects. Note that the density does not show a maximum at the stoichiometric composition but that it becomes larger than the vacancy model in the Tm-excess range. This indicates the formation of Tm clusters (see text). The strong scattering in the Tm-excess range is attributed to a decreasing reproducibility of the defect structure at low defect concentrations.

give the calculated densities for a vacancy, interstitial or antisite defect model of nonstoichiometric defects. All measured densities lie under the value of the X-ray density of stoichiometric TmSe, stoichiometric TmSe has therefore 2-3% Schottky vacancies (pairs) (1,3). Comparison with the vacancy model ① shows that in nonstoichiometric Se-rich TmSe the concentration of the Schottky-pairs decreases and vanishes at the Se-rich phase boundary. On the other hand an increasing amount of nonstoichiometric vacancies is created, so that the total defect concentration is 6-8% i.e. about 1 vacancy per 4 unit cells.

In the Tm-rich range the important feature is that the density does not show a maximum but a monotonous increase. It becomes larger than the density calculated for the vacancy model and reaches the density of a fully occupied lattice with nonstoichiometric antisite-defects at the Tm-rich phase boundary. The scatter of the experimental points is attributed to less reproducible defect structures at low defect concentrations. It has to be concluded therefore, that in the Tm-rich range excess Tm-ions are filling both cationic and anionic Schottky vacancies, so that no vacancies remain at the phase boundary. This also gives a reasonable explanation of the Tm-rich phase boundary of TmSe to the eutectic with Tm-metal (Fig. 2).

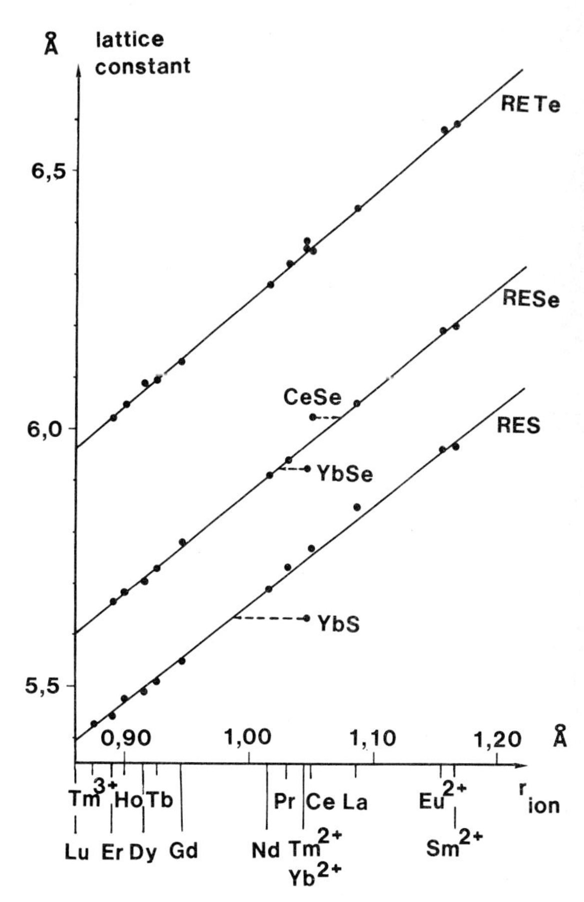

Fig. 5: Effective ionic radii of the trivalent and divalent rare earth ions in monosulfides, -selenides and - tellurides as a function of the experimental lattice constants. Together with the new determined radii of S^{2-} Se^{2-} and Te^{2-}, the lattice constants of hypothetical $Tm^{2+}Se$ and $Tm^{3+}Se$ could be calculated.

A direct consequence of this model is the occurence of octahedral Tm-clusters which are formed when a Tm-ion occupies a Se-vacancy. The maximum concentration of these clusters is 2.5% (1 cluster per 10 unit cells), i.e. a cluster-cluster distance of a little more than 2 lattice constants. These clusters allow a strong Tm-Tm interaction as the Tm-Tm distance (∿ 2.9 Å) is smaller than in Tm-metal (∿ 3.5 Å). Two further consequences are a) the larger space of the Se-vacancies opens the possibility to reduce the

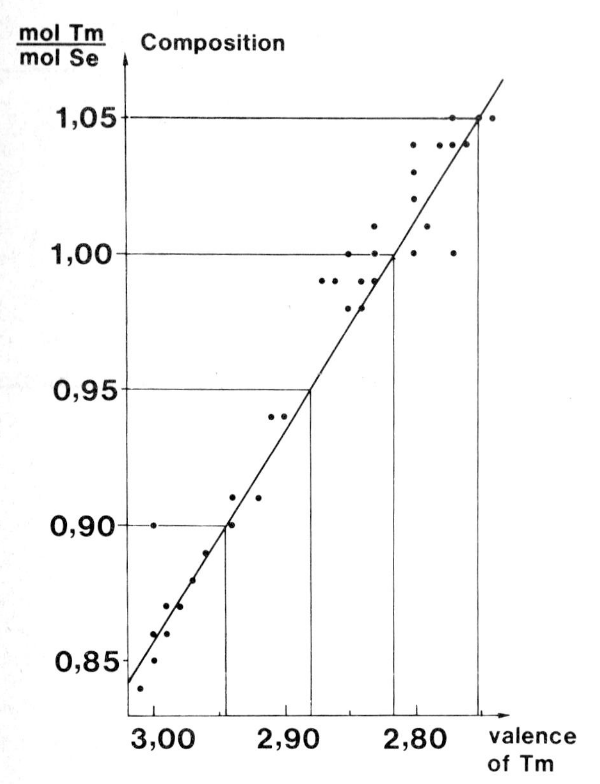

Fig. 6: Valence of Tm in nonstoichiometric TmSe as calculated under the assumption of Vegard behaviour. (see text)

Tm-valency in order to decrease repulsion in the clusters, and b) the high enthalpy of formation of these antisite defects decreases to some extend the stability. It is noteworthy, that this simple model deduced from the density measurements is in agreement with two important properties of the fluctuating TmSe:

- increase of Tm-Tm interaction
- destabilisation of the lattice

Further the cluster concentration is the only chemical parameter varying strongly inside the composition range of the valence fluctuation. How and to what extend these Tm-clusters influence the valence fluctuation must be shown by other measurements.

3. ESTIMATION OF THE VALENCE CHANGE FROM THE LATTICE CONSTANTS

In view of the difficulties arising from various methods that determine the value of mixed valence (3,12), the extrapolation from the lattice constants is a straight-forward method that makes no inherent assumptions on the nature of the mixed valent state. In the following this method is discussed in some detail.

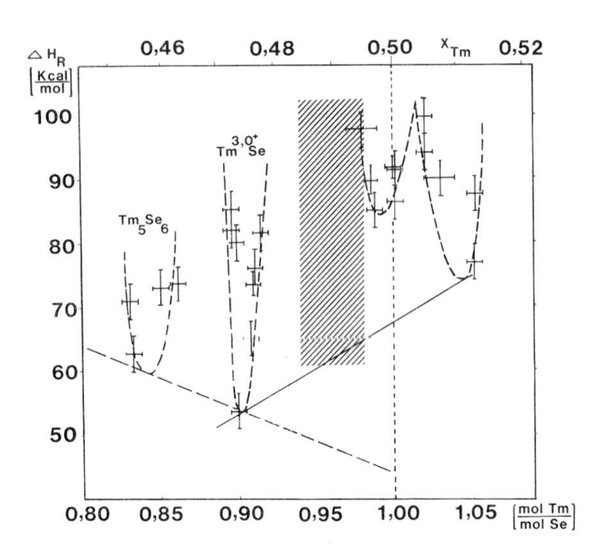

Fig. 7: Enthalpy of reaction of TmSe with 4N HCl as a function of nonstoichiometry. The gross nonstoichiometric range splits up into three stability ranges (most stable compositions: 0.90, 0.99 and 1.04 mol Tm/Se) and the Tm_5Se_6 range. The miscibility gap (shadowed) is clearly shown. The stoichiometric phase is metastable with respect to a two phase mixture of 0.90 and 1.04 mol Tm/Se (full tangent). The dashed tangent defines the stability of a hypothetical trivalent stoichiometric TmSe. The difference in energy is 42 kcal/mol (1.8 eV/Tm) which corresponds to a 29% destabilisation of the mixed valent state.

First the determination of the mixed valence of stoichiometric TmSe from the lattice constants is shown.Since stoichiometric TmSe is mixed valent, lattice constants of hypothetical di- and trivalent stoichiometric TmSe must be determined by summation of the ionic radii. A review (17) shows, that there is no consistent set of ionic radii,which contains both,the rare earth cationic radii and the chalcogen anionic radii. The anionic radii used in all valence determinations up to now are the 50 year old radii of Pauling (18) or of Goldschmidt (19).

Unfortunately these values differ by 0.07 to 0.1 Å, which (a) leads to an arbitrary choice of the cationic radii and (b) corresponds to a valence change of 0.2 - 0.3^+. Therefore a new determination of the anionic radii was carried out using (a) the systematics of the lattice constants of the rare earth chalcogenides and (b) the best and newest available ionic radii

(13,14,17) of the trivalent heavy rare earths and of divalent Eu as determined from oxides and fluorides (Fig.5).

Metallic trivalent chalcogenides of the heavy rare earths and divalent semiconducting Eu-chalcogenides give the same anionic radii which are shown in Table I:

	this work	Gold-Schmidt (19)	Pauling (18)
S^{2-}	1.83(5) ± 0.002	1.74	1.84
Se^{2}	1.94 ± 0.003	1.91	1.98
Te^{2-}	2.13(5) ± 0.004	2.11	2.21

Comparison of the new determined anionic radii in rare earth chalcogenides with the values used up to now.
From this fact two conclusions can be drawn:

- The anionic radii are not affected by the existence of conduction electrons.
- If one calculates from these anionic radii the radii of the metallic light rare earths (experimental lattice constant minus anion diameter) and compares these radii with the ones obtained from oxides and fluorides, they should show a difference due to change of the chemical bonding from ionic to ionic plus metallic. This effect should be clearly seen since the size of the light rare earths is more sensitive against changes of the bonding than that of the heavy rare earths. The comparison (11) however shows no change: The radii calculated from the lattice constants of the metallic rare earth chalcogenides lie between the two values given from oxides and fluorides in (13,14). Therefore the cationic radii change predominately with the ionic charge, independently from the nature of the liberated electrons, - valence or conduction.

In this way the lattice constants of hypothetical, metallic, trivalent and stoichiometric TmSe (5.63 Å) and of hypothetical, nonmetallic, divalent and stoichiometric TmSe (5.97 Å) can be determined by summation of the ionic radii.

The determination of the mixed valence is done by a comparison of the experimental lattice constant with the linear interpolation between the hypothetical lattice constants of $Tm^{2+}Se$ and $Tm^{3+}Se$. This linear interpolation assumes a Vegard behaviour, i.e. the radius of the cation changes linearly with the ion charge whereas the anion radius remaines unchanged (as shown above). This is a zeroth approximation of a principally hyperbolic curve. This hyperbolic curve cannot be determined without additional assumptions, (a) either one takes into account the Tm^{+} radius and changes therefore the chemical bonding to metallic or (b) one determines lattice constant and valence of a third point between 2+ and 3+ with an independent method which necessarily includes assumptions on the nature of the mixed valent state. The mixed valence determined by linear interpolation for stoichiometric TmSe, 2.82 ± 0.02 Å, is therefore a little too large, on the other hand the shift due to the nonlinearity cannot be determined unambiguously.

In the following the determination of the valence of nonstoichiometric TmSe is discussed. It is well known, that the concentration of - nonstoichiometric - defects can change the lattice constant, as shown for nonstoichiometric trivalent GdSe (dashed curve, Fig. 1)

Therefore the lattice constant of purely trivalent nonstoichiometric TmSe is not necessarily identical with the value given above for hypothetical stoichiometric trivalent TmSe. For this reason the lattice constant of purely trivalent nonstoichiometric TmSe has to be determined by an independent method. This can easily be done since the Curie-constants of the trivalent state have a well defined physical meaning. Trivalent TmS (6.46 Oerst. cm^2 K/ (g mol), 15) and $Tm_{0.87}Se$ (6.65 Oerst. cm^2K/ (g mol), 16) have about the same Curie constants, therefore $Tm_{0.87}Se$ is the trivalent nonstoichiometric composition. It is interesting, that this composition has exactly the lattice constant determined for hypothetical stoichiometric TmSe. Having these two points, the valence of each composition can be determined via its lattice constant (Fig. 6).

4. STABILITY RANGES

The relative stabilities have been determined with reaction calorimetry of nonstoichiometric TmSe with 4N HCl. Four phases can be clearly distinguished in figure 7: the Tm_5Se_6 superstructure, an essentially trivalent phase with a most stable composition of 0.90 mol Tm/Se and two phases (0.99 and 1.04) in the stoichiometric and Tm-excess range. The most important features in figure 7 are:
- the miscibility gap (shadowed) is clearly shown;
- the stoichiometric phase is metastable with respect to a mixture of 0.90 and 1.04 mol Tm/Se (full tangent), it is stabilised by the miscibility gap which inhibits the separation into the more stable two-phase mixture (11). This instability of the stoichiometric composition is shown by the lattice instability and the maximum compressibility (16).

The dashed tangent defines the stability of a hypothetical trivalent state. This state has about the same defect concentration and lattice constant as the mixed valent state and it containes no energy term from defect interaction (ordering) if the following assumptions are made, which have been verified above or which are correct within a small error limit:

1) The gain in energy between the short range order over 50-100 Å in $Tm_{0.90}Se$ (more than 1000 unit cells) (16) and long range order in Tm_5Se_6 is negligible, the differences in enthalpy between Tm_5Se_6 and $Tm_{0.90}Se$ are ascribed to the generation of additional defects only.
2) The number of defects varies linearly from Tm_5Se_6 to stoichiometric TmSe (Fig. 4).
3) The lattice constant varies linearly from Tm_5Se_6 to TmSe, which is correct only within an error of 0.01 Å.
4) The energy gained from defect ordering (short range or long range order) is proportional to the number of ordered vacancies.

The following conclusions can be made from the extrapolation mentioned above:

- the hypothetical trivalent stoichiometric TmSe is more stable and
- the difference in energy is 42 kcal/mol (1.8 eV per Tm).

Compared with a preliminary heat of formation of $Tm_{0.99}Se$ determined by fluorine combustion calorimetry (103± 7 kcal/mol) (11) this difference in energy is equivalent to a destabilisation of 29% of the mixed valent state with respect to the hypothetical stoichiometric trivalent state.

5. CONCLUSIONS

The gross homogeneity range of nonstoichiometric TmSe (NaCl-structure) clearly extends on both sides of the stoichiometric composition (0.87 - 1.05 mol Tm/Se). T-x phase diagram and crystal growth experiments show the existence of a miscibility gap (0.94 - 0.98 mol Tm/Se). Since no size effects can be responsible for this miscibility gap in nonstoichiometric TmSe, it has to be concluded that the electronic structures of predominantly trivalent and of mixed valent TmSe are repulsing each other.

Density measurements show a change of the defect structure from the Se-rich stability range (Tm vacancies) to the Tm-excess range (Tm also on Se sites, Tm clusters). The small Tm-Tm distance (∿ 2.9 Å) in these Tm clusters makes a strong Tm-Tm interaction possible. The high enthalpy of formation of these clusters leads to a destabilisation. Both are characteristics of the mixed valent state.

Calorimetric measurements confirm these results and show additionally that the Tm-rich range spilts up into two subphases with the most stable compositions 0.99 and 1.04 mol Tm/Se, the former being metastable. The extrapolation of the most stable compositions to the two primarily trivalent phases to the stoichiometric composition defines a hypothetical trivalent stoichiometric TmSe, which is 42 kcal/mol (1.8 eV/Tm ≙ 29%) more stable than mixed valent TmSe.

REFERENCES

1) Fritzler, B., Kaldis, E., Steinmann, B., Jilek, E.,in "The rare earths in modern science and technology" Vol.3, McCarthy et al. eds. Plenum,(1982)
2) Fritzler, B., Kaldis, E., Jilek, E., ibid
3) Kaldis, E., Fritzler, B., "Valence and phase instabilities in TmSe crystals" in "Progress in solid state chemistry" Rosenblatt, G.M. Worell W.L., eds, Pergamon (1982)
4) Kaldis, E., Fritzler, B., Jilek, E., Wisard, A., J. Physique 40 (1979) C5-366
5) Kaldis, E., Fritzler, B., Peteler, W., Z. Naturforschung 34a (1979) 55
6) Hodges, J.A., Jehanno, G., Debray D., Holtzberg, F., Löwenhaupt, M., J. Physique to be published
7) Kaldis, E., Peteler, W., in "Thermal analysis" Proceedings ICTA 1980, Birkhäuser, (1980) 67
8) Kaldis, E., Fritzler, B., J. Physique 41 (1980) C5-135
9) Spychiger, H., Kaldis, E., Jilek, E., this conference
10) Kaldis, E., Fritzler, B., Spychiger, H., Jilek, E., this conference
11) Fritzler, B. PhD thesis ETH (1982)
12) Wachter, P.,this conference
13) Ahrens, L.H., Geochim. Cosmochim. Acta 2 (1952) 155
14) Templeton, D.H., Dauben, C.H., J. Am. Chem. Soc. 76 (1954) 5237
15) Bucher, E., Andres, K., di Salvo, F.J., Maita, J.P., Gossard, A.C., Cooper, A.S., Phys. Rev. B11 (1975) 500
16) Batlogg, B., Ott, H.R., Kaldis, E., Thöni, W., Wachter, P., Phys. Rev. B19 (1979) 247
17) Shannon, R.D., Prewitt, C.T., Acta Cryst. B25 (1969) 925
18) Pauling, L., J. Am. Chem. Soc. 49 (1927) 765
19) Goldschmidt, V.M. in "Landolt Börnstein" 6.Aufl., 4. Teil, (1955) 521

Valence Instabilities
P. Wachter and H. Boppart (eds.)

INTEGRAL VALENCE SEGREGATION IN NONSTOICHIOMETRIC TmSe : A ^{77}Se NMR STUDY

P. Panissod*, M. Benakki*, and D. Debray†

*LMSES (LA CNRS 306), 4 rue B. Pascal, 67070 Strasbourg, France
†Laboratoire Léon Brillouin, CEN Saclay, 91191 Gif-sur-Yvette, France

^{77}Se NMR measurements on a stoichiometric and a nonstoichiometric sample of the intermediate valence compound TmSe show that the intermediate valence character of the nonstoichiometric sample results from a random static distribution of integral valent Tm^{2+} and Tm^{3+} ions. On the contrary, the results for the stoichiometric sample are consistent with equivalent Tm sites and temporal valence fluctuations on each site.

The physical properties of the system Tm_xSe are known to be highly sensitive to stoichiometry[1-3]. Thus the mean valence state of Tm is found to change from ~ 2.6+ for the stoichiometric sample to ~ 3+ for $x \simeq 0.8$. The intermediate valence state of Tm in stoichiometric TmSe results from single-site charge fluctuations between $4f^{13}Tm^{2+}$ and $4f^{12}Tm^{3+}$ configurations. However, for nonstoichiometric Tm_xSe ($0.8 < x < 1$) there has been so far no experimental microscopic evidence to suggest that valence fluctuation is responsible for the observed mean non-integral valence of Tm in these materials. In this work we present the first microscopic experimental evidence which indicates that the intermediate valence character of Tm in this stoichiometry range results from a segregation of the valences. The present work also gives microscopic evidence to confirm temporal valence fluctuation in stoichiometric TmSe.

We have performed ^{77}Se NMR measurements on a nonstoichiometric (NS) and a stoichiometric (S) powder sample of the intermediate valence system Tm_xSe using a phase-coherent spin-echo spectrometer in the frequency range 6-15 MHz over the temperature range 4.2-400K. The S and NS samples have lattice parameters equal to 5.709(4) and 5.683(4)Å, respectively. The corresponding values of x are, therefore, ~ 1.00 and ~ 0.97 [4,5]. The results of our measurements are summarized below.

In the NS sample, the observed NMR spectra are asymmetrical and exhibit badly resolved structures which are characteristic of environment fluctuations. The asymmetry and structure become more pronounced as the temperature decreases. The spectra of the S sample, on the other hand, are nearly symmetrical and much narrower, especially at low temperatures. This is clearly seen in Fig.1. For the NS sample, the Knight shift K and the linewidth δ, as defined by the centroid and the square root of the second moment, respectively, of the spectra, exhibit Curie-Weiss thermal variation with quite different θ ($\theta_K \simeq -45K$; $\theta_\delta \simeq -30K$). This difference can not be attributed to dynamic hyperfine field fluctuations since dynamic broadening, as measured by the transverse relaxation time T_2, is an order of magnitude smaller than the observed linewidth and field dependence shows the inhomogeneous character of the broadening. Demagnetizing field distribution in the powder sample contributes first to this inhomogeneous broadening but the observed structures, the asymmetry of the spectra and the significantly larger linewidth with respect to the S sample clearly show the existence of a static hyperfine field distribution. This can result only from fluctuations of the local magnetic environment of the Se atoms.

The observed unusual characteristics of the spectra of the NS sample can, however, be explained if two magnetic, i.e., valence states

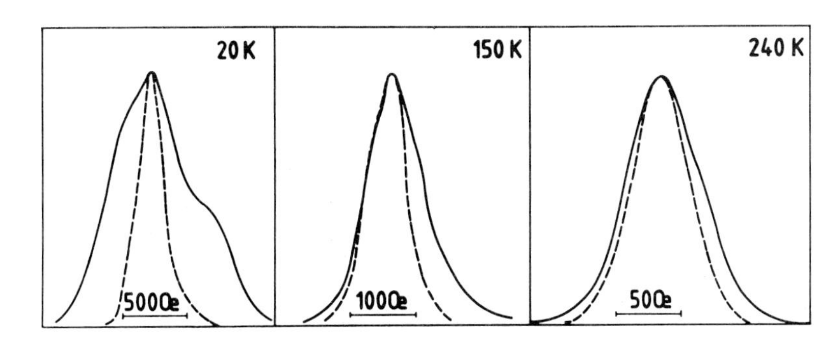

Fig.1. ^{77}Se NMR spectra for the NS (full line) and S (dashed line) samples as a function of temperature.

of Tm atoms with different exchange interactions coexist in the sample. The K and δ data together with the bulk susceptibility χ were, consequently, analysed in a selfconsistent way in the framework of an environment model, assuming a random static distribution of Tm^{2+} and Tm^{3+} ions among the 6 nearest neighbors (nn) of Se atoms. The analysis was restricted to temperatures higher than 70K, where the bulk magnetic susceptibility shows a definite Curie-Weiss behavior, to avoid complications arising from short-range magnetic order observed[6] in this sample and/or possible crystal-field effects. The shape of the spectrum then results from the sum of seven lines (from 6 Tm^{3+} nn to 6 Tm^{2+} nn) with different positions depending on the average surrounding moment (and consequently on the number and nature of the nn's). This model reproduces very well the asymmetry and structure of the observed spectra as shown in Fig.2. It also predicts

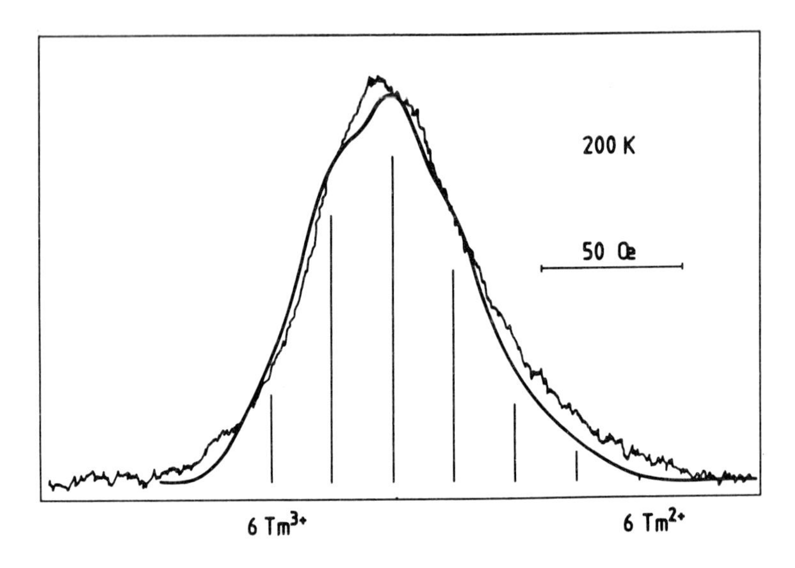

Fig.2. Observed and computed (smooth line) ^{77}Se NMR spectra of NS sample. Vertical bars indicate positions and intensities of the 7 lines

correctly the observed temperature dependence of χ, K, and δ for an average Tm valence of $\sim 2.7+$. At lower temperatures, the model proposed above is still valid ; however, a quantitative analysis is not possible without an exact knowledge of the dynamic hyperfine field fluctuations and crystal-field effects. In a cubic crystal field the ground state of Tm^{3+} is a non-magnetic singlet while that for Tm^{2+} is a magnetic doublet or quadruplet[3]. On the basis of this picture, it is possible to have a qualitative understanding of the enhanced asymmetry of the spectra observed at very low temperatures, e.g., at 20K (Fig.1).

It is worth noting that a random static distribution of Tm^{2+} and Tm^{3+} ions should lead to structural distortion on a microscopic level. Indeed recent Mössbauer measurements[7] on the same sample reveal a distribution of quadrupole splittings related to a distribution of noncubic environments.

We have also carried out a quantitative analysis of the spectra for the S sample. This analysis of δ shows that the broadening originated only from demagnetizing field distribution in the powder sample and not from local environment fluctuation. Thus all Tm atoms are equivalent in this sample. The time-averaged valence of Tm at each site is found to be $\sim 2.5+$, in agreement with literature value.

Finally, we would like to mention the interesting observation that as temperature increases the NMR spectra of the nonstoichiometric and the stoichiometric sample approach each other and become almost identical at room temperature.

REFERENCES

1. Valence Instabilities and Related Narrow Band Phenomena, edited by R.D.Parks (Plenum, New York, 1977).
2. B.Batlogg, H.R.Ott, E.Kaldis, W.Thöni and P.Wachter, Phys. Rev. B19, 247 (1979).
3. O.Peña, Thesis, Docteur ès-Sciences, 1979, Université Scientifique et Médicale, Grenoble, France.
4. E.Kaldis, B.Fritzler, E.Jilek and A.Wisard, J. Physique C5, 366 (1979).
5. F.Holtzberg, T.Penney and R.Tournier, J. Physique C5, 314 (1979).
6. D.Debray (unpublished).
7. J.A.Hodges, G.Jéhanno, D.Debray, F.Holtzberg and M.Loewenhaupt, J. Physique (in press).

Valence Instabilities
P. Wachter and H. Boppart (eds.)

VALENCE TRANSITION DUE TO AN OXYGEN IMPURITY OR A SAMARIUM SELF-INTERSTITIAL IN SmS COMPOUNDS

C. Demangeat, Y. Hammoud, M.A. Khan and J.C. Parlebas

L.M.S.E.S. (LA au CNRS n° 306), Université Louis Pasteur - 4, rue Blaise Pascal
67070 STRASBOURG CEDEX, France

Using a recent band structure model for pure SmS we propose to study the effect of impurities in relation with experimental data. The position of the Oxygen and Sm self-interstitial induced bound states are discussed in terms of local charge neutrality and Friedel's rule. The filling of these bound states are obtained through a careful integration in the complex plane.

1. INTRODUCTION

From magnetization and lattice parameter measurements it is now established that the change of valence in Sm monosulfides can be chemically induced by anion substitution (As, P, O...) and has been the subject of extensive experimental studies (1-4). For example, the conductivity resulting from the presence of phosphorous impurity can be explained in terms of a $Sm^{2+} \rightarrow Sm^{3+}$ transition on each of the six neighbouring atoms to the impurity (1). However, they do not contribute to conduction states as long as the impurity concentration is below the percolation threshold. The pecularity associated with Oxygen is its isoelectronic character as compared to sulfur. Indeed, for dilute concentration, lattice parameter measurements (4) seems to indicate that each Oxygen induces only one $Sm^{2+} \rightarrow Sm^{3+}$ transition in contrast with the case of SmSP (and also SmSAs) dilute systems. However, from susceptibility measurements, the moment at saturation is found to be about 1.2 μ_B per Oxygen atom (4) in agreement with analogous results for P and As impurities. Finally let us point out that recent EXAFS experiments on dilute $SmS_{1-x}O_x$ can be explained through a strong variation of the nearest neighbour-impurity Sm-Sm distance (5).

The high value of the negative magnetoresistivity in nominally stoichiometric SmS coumpounds (from a chemical analysis) has been explained by the presence of an impurity band in the SmS gap (6). These impurities can be Oxygen or sulfur vacancies. Also in the case of irradiated SmS compounds the resistivity measurements have been interpreted by the presence of Sm self-interstitial at tetrahedral position (6). This Sm self-interstitial induces a transition to Sm^{3+} due to new hopping integrals between it and its nearest neighbours. Moreover,it seems that not only the interstitial atom will present this transition but also its nearest neighbouring Samarium ions (7). These assumptions are in agreement with a recent model which display $Sm^{2+} \rightarrow Sm^{3+}$ transition on each of the nearest neighbour to the interstitial impurity (8).

Recent calculations have shown that bound states may appear in the gap of SmS when alloyed with oxygen (9) or in the presence of Sm self-interstitial (8, 10). The present purpose is to discuss these two types of point defects in a more quantitative way by using for the host band structure a three center Slater-Koster fit to the KKR band structure of Davis (11) for 3p (S^{--}) and 6s and 5d (Sm^{++}) states. As usual the periodic Anderson model is applied for the $4f^6$ configuration (represented by a spinless s orbital) hybridized with 5d conduction band states). The SmS partial densities of states (12) display quite a narrow p band roughly 2 eV below the bottom of a broad conduction band and contains a few d electrons through pd mixing processes. More interesting for our purpose is the fact that, even in pure SmS compounds, there is a small d contribution to the f band (and vice-versa). Using Dyson equation the variation of the density of states of the impure system can be related to the density of states of the pure system and the matrix elements of the perturbing potential. These matrix elements are fully determined by the use of Friedel's screening rule and local neutrality condition. The outline of the paper is as follows. Section 2 presents results for Oxygen impurities whereas section 3 is devoted to Sm self-interstitial. Our results are summarized in section 4.

2. ELECTRONIC STRUCTURE OF DILUTE OXYGEN IN SmS

After a brief description of the model used for the description of pure SmS we will discuss our integration procedure in the complex plane. This procedure has been introduced in order to take into account more carefully the filling of the impurity bound states. Also, determination of generalized phase shift and relation with Friedel's rule are briefly discussed.

The Hamiltonian of SmS is written in the Bloch representation from a generalization of the Anderson Hamiltonian to a lattice of 'f impurities' :

$$H^o = \sum_{\alpha k} |\alpha k\rangle \varepsilon^{\alpha} \langle \alpha k| + \sum_{\alpha,\alpha',k} |\alpha k\rangle h_k^{\alpha\alpha'} \langle \alpha' k| + \varepsilon^f \sum_k |fk\rangle\langle fk| + \sum_{mk} (|mk\rangle\, h_k^{mf} \langle fk| + h.c.) \qquad (2.1)$$

The first and second terms of eq. *(2.1)* represent the 3p valence band and the 5d-6s conduction bands (noted C.B. hereafter) formed from the 'spd' tight-binding orbitals $|\alpha\lambda\rangle$ centered on a Samarium site when $\alpha = \{s\ ;\ m=xy,\ yz,\ zx,\ x^2-y^2,\ 3z^2-y^2)$ symmetries (m is labelled from 1 to 5 hereafter) and on a sulfur site when α = x, y, z symmetries. The third term of eq. *(2.1)* describes a zero width band at energy ε^f : this ε^f orbital energy, when counted from the bottom of the C.B., represents the difference of the many-body ground states between the Sm^{3+} configuration : $4f^5(5d6s)^1$ and the Sm^{2+} configuration : $4f^6(5d6s)^0$. Since only one of the f eletrons can fluctuate between C.B. and 'f' level-($4f^4(5d6s)^2$ configuration is energetically far away from $4f^6(5d6s)^0$ configuration- we will represent the valence transition by a non degenerate level able to hold one electron maximum per cation. Finally the last term of eq. *(2.1)* written as a one-electron term, is the simplest contribution to the mixing between the 'f' level and the 'd' states. The 'df' hybridization is of course supposed very small in SmS semi-conducting compounds ; nevertheless it is responsible for a resulting non-zero 'f band' width.

Let us consider now SmS containing one single Oxygen impurity atom substituted at anion site A. We study the effect of the impurity site A on the electronic structure of its six nearest Samarium neighbours. The Oxygen impurity is essentially presumed to (i) increase the absolute value of the transfer integrals $\beta_{RR'}^{mm'}$ between $|mR\rangle$ and $|m'R'\rangle$ 'd' orbitals around A (non diagonal disorder) ; (ii) shift the energy of the corresponding d orbitals (diagonal disorder). As we are dealing with an isoelectronic impurity we will only consider the substitution size effet (i), also assuming that the '3p' band is not directly perturbed. Thus the corresponding alloy Hamitlonian is given (per spin direction) by :

$$H = H^o + \delta V \qquad (2.2)$$

$$\text{with } \delta V = \sum_{\substack{mm' \\ \langle R,R' \rangle}} |mR\rangle\, \delta\beta_{RR'}^{mm'} \langle m'R'| \qquad (2.3)$$

where the notation $\langle R,R' \rangle$ indicates that the summation of eq. *(2.3)* is only over the nearest neighbouring sites among the six surrounding A ; $\delta\beta_{RR'}^{mm'}$ being the dd change of the corresponding transfer integral. For simplicity we will only retain orbital diagonal $\delta\beta_{RR'}^{mm}$ matrix elements. The total number of states for one spin direction displaced by the presence of the Oxygen impurity for energies E' < E is given by :

$$\delta Z(E) = -\pi^{-1} \text{ Arg Det } [1 - \delta V\, G^o(E)] \qquad (2.4)$$

where $G^o(E)$ is the Green function of pure SmS. It is possible to cast $\delta Z(E)$ in a compact form which has exactly the same poles as in the d local DOS called $N^m(E)$ in reference 9. The d and f local DOS relative to a site R, nearest neighbour to the Oxygen impurity, are integrated up to the Fermi level to give the d and local number of electrons. Due to the fact that bound states are present in the SmS gap, uncertainties arise with such integration method. In the present paper we have replaced this integration on the real axis by an integration in the complex plane in order to take into account more carefully the filling of the bound states. The variation of the number of electrons below a given energy E_g is given by integration along the $E_g + iy$ line. To do that we have to compute :

$$G_{\lambda\lambda'}^{o\alpha f}(E_g+iy) = \frac{1}{\sqrt{N}} \sum_{\substack{n \\ k\in B.Z.}} \frac{a_{\alpha}^{n*}(k)\, a_f^n(k)\, e^{ik(\lambda-\lambda')}}{E_g - E_{nk} + iy} \qquad (2.4)$$

which is an intersite matrix element of the host Green function G^o ; E_{nk} and $a_m^n(k)$ are respectively eigenvalue and eigenvector of Hamiltonian H_o. n is the band index and N the number of k points in the Brillouin zone. We have chosen E_g in regions whereas the density of states is zero.

The results obtained (12, 13) show that only two electrons shift locally from the f shells of the Sm_6O molecule into the band structure formed from the $3z^2-r^2$ orbitals of the molecule. If we suppress the host hybridization we completely supress the phenomenon because the empty f impurity level is no more present. Our model underlines the essential role played by fd hybridization and size effects (through $\delta\beta_{RR'}^{dd}$ variation) in SmSO type of dilute alloys. A more complete calculation should relate the dipole force tensor P of the system to the variation of $\delta\beta_{RR'}^{dd}$, as discussed for Hydrogen in Palladium (14). Once this has been done the elastic part of the total energy of the system becomes available if we have a model for the lattice Green function (15).

3. ELECTRONIC STRUCTURE OF Sm SELF-INTERSTITIAL IN SmS COMPOUNDS

From experimental results (6), the Sm interstitial atom is supposed to occupy a tetrahedral λ with respect to the fcc Sm sublattice of SmS : so, before the interactions of the problem are switched on, it is surrounded by 4 Sm^{2+} cations. Also the interstitial atom is assumed to be a charged Sm^{2+} cation because, for example, a neutral Sm atom is far larger as compared to the allowed space at a tetrahedral position. In our simplified picture for 'f' states, the Sm interstitial is considered to bring one extra electron corresponding to a $(4f)^1\ (5d6s)^0$ configuration. It introduces (i) new external states $|\ell\lambda\rangle$ of energy $\varepsilon_{\lambda}^{\ell}$ with ℓ = f, 1...5 ; (ii) new hopping integrals $\beta_{\lambda R}^{\ell\ell'}$ between interstitial states $|\ell\lambda\rangle$ and surrounding states $|\ell' R\rangle$; (iii) diagonal 'f' disorder v_R^f upon each of the four neighbours R of the interstitial site λ. Considering the cluster formed by the first shell 's' of neighbours of λ the simplest Hamiltonian H which describes our point defect problem is deduced from perfect Hamiltonian H^o in the following terms :

$$H = H^o + \sum_{\ell} |\ell\lambda> \varepsilon_{\lambda}^{\ell} <\ell\lambda| + \sum_{R\varepsilon s} |fR> v_R^f <fR|$$

$$+ \sum_{\ell\ell'} \sum_{R\varepsilon s} (|\ell\lambda> \beta_{\lambda R}^{\ell\ell'} <\ell' R| + h.c.) \qquad (3.1)$$

As a consequence, the total number Z(E) of new and perturbed states of energy $E' \leq E$ and the local density of states $n_{\lambda}^{\ell}(E)$ of symmetry $\ell = f, 1...5$ at λ are respectively given by :

$$Z(E) = -\pi^{-1}\{\sum_{\ell} [\mathrm{Arg}\ \Sigma_{\lambda}^{\ell}(E) - \pi] + \sum_{R} \mathrm{Arg}(1 - v_R^f G_R^{of}(E))\} \qquad (3.2)$$

$$n_{\lambda}^{\ell}(E) = -\pi^{-1} \mathrm{Im}\{\Sigma_{\lambda}^{\ell}(E)\}^{-1} \ ; \ \Sigma_{\lambda}^{\ell}(E) = E - \varepsilon_{\lambda}^{\ell} - \Delta_{\lambda}^{\ell}(E) \qquad (3.3)$$

$$\text{with } \Delta_{\lambda}^{\ell}(E) = \sum_{\substack{\ell' R' \\ \ell'' R''}} \beta_{\lambda R'}^{\ell\ell'} g_{R'R''}^{\ell'\ell''}(E)\, \beta_{R''\lambda}^{\ell''\ell}$$

where

$$g_{R'R''}^{\ell'\ell''}(E) = G_{R'R''}^{o\ell'\ell''}(E) + v_R^f [1 - v_R^f G_R^{of}(E)]^{-1} \times$$
$$\times \sum_{R} G_{R'R}^{o\ell f}(E)\, G_{RR''}^{of\ell'}(E)$$

The d and f local DOS relative to interstitial site λ can be integrated up to Fermi level to give the d and f local number of states :

$$N_{\lambda}^{d} = -\pi^{-1}\, \mathrm{Im} \int^{E_F} \sum_{m=d} 2G_{\lambda\lambda}^{mm}(E)\, dE \qquad (3.4)$$

$$N_{\lambda}^{f} = -\pi^{-1}\, \mathrm{Im} \int^{E_F} G_{\lambda\lambda}^{ff}(E)\, dE \qquad (3.5)$$

where

$$G_{\lambda\lambda}^{\ell\ell}(E) = [\Sigma_{\lambda}^{\ell}(E)]^{-1} \qquad (3.6)$$

In the same way we can define the variation of the number of d and f local number of states relative to a site R nearest neighbour to λ :

$$\delta N_R^d = -\pi^{-1}\, \mathrm{Im} \int^{E_F} 2\sum_{m=d} dE(G_{RR}^{mm}(E) - G_{RR}^{omm}(E)) \qquad (3.7)$$

$$\delta N_R^f = -\pi^{-1}\, \mathrm{Im} \int^{E_F} dE(G_{RR}^{ff}(E) - G_{RR}^{off}(E)) \qquad (3.8)$$

where

$$G_{RR}^{\ell\ell}(E) = g_{RR}^{\ell\ell}(E) + \sum_{\ell'} D_{R\lambda}^{\ell\ell'} G_{\lambda\lambda}^{\ell'\ell'}(E)\, D_{\lambda R}^{\ell'\ell} \qquad (3.9)$$

$$\text{with } D_{R\lambda}^{\ell\ell'} = \sum_{\ell'' R'} g_{RR'}^{\ell\ell''}(E)\, \beta_{R'\lambda}^{\ell''\ell'} \qquad (3.10)$$

From symmetry considerations it is relatively easy to classify the $\beta_{\lambda R}^{\ell\ell'}$ integrals between λ and its first nearest neighbours in terms of five independent quantities ; these quantities are themselves proportional to linear combination of a few transfer integrals of the perfect SmS lattice.

Friedel's rule states that the number of electrons initially brought by the interstitial Sm^{2+} (in the $4f^1(5d6s)^0$ configuration) is equal to Z(E) at E_F :

$$Z(E_F) = 1 \qquad (3.11)$$

We have found that the Sm self-interstitial is in its initial Sm^{2+} state whereas the four nearest neighbours are in the Sm^{3+} configuration. By integrating in the complex plane as we did for SmSO alloys, it is possible to calculate the local filling of the different bound states. Within our choice of parameters the filling of the f bound state at site λ is 0.8 electrons whereas the filling of the f bound state on the neighbours is about 0.05 electrons. This result is in qualitative agreement with the model proposed by Morillo (6) in order to explain his transport measurements.

4. CONCLUSION

The present tight-binding scheme has shown to be convenient to explain semi-quantitatively the presence of Sm^{3+} ions in the neighbourhood of substitutional anionic impurities (like Oxygen) or Sm self-interstitials. However, this model relies for the description of the f state to a phenomenological spinless state. Our present objective is to replace it locally by a highly correlated state with a reasonable value of U, the intrasite Coulomb correlation term. This has been tentatively made in the case of Sm self-interstitial (16).

ACHNOWLEDGEMENTS

The authors are grateful to E. Daniel, C. Koenig and R. Riedinger for clarifications in the complex plane integration scheme. Also Y. Hammoud is grateful to Professor F. Gautier for numerous discussions.

REFERENCES

[1] Pena, O., D. Sc. Thesis, Grenoble Univ. (April 1979).
[2] Henry, D.C., Sisson, K.J., Savage, W.R., Schweitzer, J.W. and Cater, E.D., Phys. Rev. B 20 (1979) 1985.
[3] Becken, R.B., Schweitzer, J.W., Phys. Rev. B 23 (1981) 3620.
[4] Abadli, L., Thesis, Strasbourg Univ. (November 1981).
[5] Krill, G., Kappler, J.P., Röhler, J., Ravet, M.F. and Léger, J.M., this conference.
[6] Morillo, J., D. Sc. Thesis, Orsay Univ. (December 1981).
[7] Morillo, J., private communication.
[8] Hammoud, Y., Khan, M.A., Parlebas, J.C. and Demangeat, C., in Chimie et Physique des Sulfures, Séléniures et Tellurures à l'Etat Solide, Paris (1981).
[9] Khan, M.A., Krill, G., Demangeat, C. and Parlebas, J.C., in Valence Fluctuations in

Solids, L.M. Falicov, W. Hanke and M.B. Maple, eds. (North Holland, 1981).

[10] Parlebas, J.C., J.M.M.M. 15-18 (1980) 953.

[11] Davis, H.L., 9th R.E. Reasearch Conf., Field Editor, Vol. 1 (1971) 3.

[12] Hammoud, Y., Khan, M.A., Demangeat, C. and Parlebas, J.C., REAC 82, Durham, to be published by J.M.M.M.

[13] Hammoud, Y., Thesis, Strasbourg Univ. (1982).

[14] Khalifeh, J., Moraitis G. and Demangeat,C., in Hydrogen in Metals, Richmond (1982) to be published by Plenum Press.

[15] Kohn, W., Lee , T.K. and Lin-Liu, Y.R., to be published.

[16] Hammoud, Y., Khan, M.A., Parlebas, J.C. and Demangeat, C., This conference.

Valence Instabilities
P. Wachter and H. Boppart (eds.)

CORRELATION EFFECT IN A LOCAL DESCRIPTION OF IMPURITIES IN SmS COMPOUNDS

Y. Hammoud, M.A. Khan, J.C. Parlebas and C. Demangeat

L.M.S.E.S. (LA au CNRS n° 306) Université Louis Pasteur
4, rue Blaise Pascal 67070 Strasbourg Cedex , France.

The impurity induced bound states due to Sm self interstitial in SmS compounds are described by a model Hamiltonian restricted to the Sm self interstitial and its nearest Sm neighbours. Intrasite Coulomb correlation for f states and intersite hybridization between f and d states are explicitely taken into account. The ground state of the system is found to be singlet and non-magnetic.

1. INTRODUCTION

A recent model for the description of Sm self interstitial in SmS has shown that the defect introduces new external states of energy ε_i^ℓ with ℓ = d, f on interstitial site i (1). In this model the d and f states are approximated by spinless s like state. The position of the impurity induced bound states are adjusted in order to be in agreement with Friedel's screening rule which states that the number of electrons brought by the impurity has to be equal to the phase shift Z(E) at the Fermi level E_F. This tight binding scheme of point defect can explain the presence of Sm^{3+} atoms occupying interstitial position (2). We have recently developed a more realistic band structure for the description of pure semi-conducting SmS compounds (3, 4). We have used a three center Slater-Koster fit to the KKR band structure of Davis (5) for 3p (S^{--}) and 6s and 5d (Sm^{++}) states whereas a s like spinless state is used for the f states. The use of a s spinless orbital for the description of the $4f^6$ state is not entirely satisfying in the sense that the hybridization with a spin dependent and degenerate d state is not trivially understandable. Moreover, we have recently shown (3) that the degeneracy of the d conduction states leads to Γ'_{25} and Γ_{12} bonding states located in the neighbourhood of the SmS gap. In this case, it is impossible to satisfy Friedel's rule as explained by the following simple argument : if Γ'_{25} bound states are present, this means $Z(E_F) \geq +6$ whereas if Γ_{12} bound states are present, we have $Z(E_F) \geq +4$. Moreover, it has recently been shown (6) that in the case of interstitials, extended potential is necessary to satisfy Friedel's rule. From experimental results (2), the Sm interstitial atom is supposed to occupy a tetrahedral position i with respect to the fcc Sm sublattice of SmS : so, before the interaction of the problem is switched on, it is surrounded by 4 Sm^{2+} cations. Also the interstitial atom is assumed to be a charged Sm^{2+} cation. It introduces (i) new external states $|\ell i\rangle$ of energy ε_i^ℓ with ℓ = f, xy, yz, zx, x^2-y^2, $3z^2-r^2$; (ii) new hopping integrals $\beta_{iR}^{\ell\ell'}$ between interstitial states $|\ell i\rangle$ and surrounding states $|\ell' R\rangle$; (iii) diagonal f disorder v_R^f upon each of the four neighbours R of the interstitial site i.

Friedel's rule states that the number of electrons initially brought by the insterstitial Sm^{2+} (in the $4f^1(5d6s)^0$ configuration) is equal to Z(E) at E_F. We have found that the Sm self interstitial is in its initial Sm^{2+} state whereas the four nearest neighbours are in the Sm^{3+} configuration. This result has been obtained by introducing a phenomonological hybridization between the spinless f state and the spin dependent d states.

The present objective is to replace this phenomenological fd hybridization by a spin dependent term. This can be done by introducing an U dependence in the f term and consequently to replace $Z(E_F) = 1$ by $Z(E_F) = 2$ in order to account for the spin. In this case, the following Hamiltonian will describe a Sm self interstitial (at tetrahedral position i) in SmS :

$$H_i = \varepsilon_i^f \sum_\sigma f_{i\sigma}^+ f_{i\sigma} + U^{ff} n_{fi\uparrow} n_{fi\downarrow} + \sum_{c,\sigma} \varepsilon_i^c a_{ci\sigma}^+ a_{ci\sigma} + \sum_{c,j,\sigma} (V_{ij}^{fc} f_{i\sigma}^+ a_{cj\sigma} + V_{ji}^{cf} a_{cj\sigma}^+ f_{i\sigma}) \quad (1.2)$$

where $f_{i\sigma}^+$ and $f_{i\sigma}$ are the creation and annihilation operators of f electrons of spin σ at the site i. The $a_{cj\sigma}^+$ and $a_{cj\sigma}$ create and annihilite conduction electrons 'c' of spin σ at site j. V_{ij}^{fc} is the hybridization term between the f state on impurity site i and the conduction states on nearest neighbouring Samarium atoms j. It is difficult to solve Hamiltonian H_i in the presence of d degeneracy so we have restricted our present task to a model where the d orbitals are replaced by a s-like state (as in the early model (1)). Section 2 presents the diagonalization of this simplified version of the Hamiltonian whereas section 3 is devoted to the conclusion.

2. THE GROUND STATE OF THE HAMILTONIAN H_i

If we use the simplified version (1) for the description of Sm self interstitial in SmS compounds, the inclusion of a Coulomb correlation term leads to the following Hamiltonian :

$$H_i = \varepsilon_i^f \sum_\sigma f_{i\sigma}^+ f_{i\sigma} + U n_{fi\uparrow} n_{fi\downarrow} + \varepsilon_i^c \sum_\sigma c_{i\sigma}^+ c_{i\sigma} + \sum_{j,\sigma} \{ V_{ij}^{fc} f_{i\sigma}^+ c_{j\sigma} + V_{ji}^{cf} c_{j\sigma}^+ f_{i\sigma} \} \qquad (2.1)$$

If we restrict the V_{ij}^{fc} term to intrasite hybridization (which is generally zero from symmetry considerations) the Hamiltonien *(2.1)* reduces to a purely local Hamiltonian which has been resolved analytically (7). The ground state, found by diagonalization of the singlet state for two electrons, is mixed valent in agreement with the case where U is taken to be infinite (8).

To solve Hamitlonian*(2.1)*, the following transformation is made :

$$V = \sqrt{z}\, V_{ij}^{fc} \qquad (2.2)$$

where z is the coordination number. We further introduce the following operator :

$$d_{i\sigma} = \frac{1}{\sqrt{z}} \sum_{j \neq i} c_{j\sigma} \qquad (2.3)$$

We drop then the irrelevant subscript i, so the Hamiltonian reads :

$$H = \varepsilon^f \sum_\sigma f_\sigma^+ f_\sigma + U n_{f\uparrow} n_{f\downarrow} + \varepsilon^c \sum_\sigma c_\sigma^+ c_\sigma + \sum_\sigma V (f_\sigma^+ d_\sigma + d_\sigma^+ f_\sigma) \qquad (2.4)$$

It is not trivial to diagonalize this Hamiltonian because, as compared with the resolution reported in reference (7), the number of basic states is now equal to 48 (4). If we choose ε^c as the origin of the energies (i.e. $\varepsilon^c \equiv 0$) the ground state of Hamiltonian *(2.4)* is found by diagonalization of the singlet state of the two electrons case (4). In the other cases, the problem can be easily solved and the U dependent eigenvalues are reported in reference (4).

Let us now discuss in detail the determination of the eigenstates of the singlet for two electrons. The basis functions of the singlet state are made of :

$$\begin{aligned}
|\varphi_1\rangle &= c_\downarrow^+ c_\uparrow^+ |0\rangle \\
|\varphi_2\rangle &= \frac{1}{\sqrt{2}} (c_\downarrow^+ f_\uparrow^+ - c_\uparrow^+ f_\downarrow^+) |0\rangle \\
|\varphi_3\rangle &= \frac{1}{\sqrt{2}} (c_\downarrow^+ d_\uparrow^+ - c_\uparrow^+ d_\downarrow^+) |0\rangle \\
|\varphi_4\rangle &= d_\downarrow^+ d_\uparrow^+ |0\rangle \\
|\varphi_5\rangle &= \frac{1}{\sqrt{2}} (f_\downarrow^+ d_\uparrow^+ - f_\uparrow^+ d_\downarrow^+) |0\rangle \\
|\varphi_6\rangle &= f_\downarrow^+ f_\uparrow^+ |0\rangle
\end{aligned} \qquad (2.5)$$

Eigenvalues and eigenvectors can be obtained through a diagonalization of the following determinant :

$$\begin{vmatrix}
2\varepsilon^c - E & 0 & 0 & 0 & 0 & 0 \\
0 & \varepsilon^c + \varepsilon^f - E & V & 0 & 0 & 0 \\
0 & V & \varepsilon^c - E & 0 & 0 & 0 \\
0 & 0 & 0 & -E & V\sqrt{2} & 0 \\
0 & 0 & 0 & V\sqrt{2} & \varepsilon^f - E & V\sqrt{2} \\
0 & 0 & 0 & 0 & V\sqrt{2} & 2\varepsilon^f + U - E
\end{vmatrix} \qquad (2.6)$$

In order to solve analytically this 3×3 determinant, we have used the following unitary transformation :

$$\begin{pmatrix} |\psi_4\rangle \\ |\psi_5\rangle \\ |\psi_6\rangle \end{pmatrix} = \alpha \begin{pmatrix} |\varphi_4\rangle \\ |\varphi_5\rangle \\ |\varphi_6\rangle \end{pmatrix} \qquad (2.7)$$

with

$$\alpha = \begin{pmatrix} \frac{U}{\sqrt{1+U^2}} & 0 & \frac{1}{\sqrt{1+U^2}} \\ 0 & 1 & 0 \\ \frac{1}{\sqrt{1+U^2}} & 0 & \frac{-U}{\sqrt{1+U^2}} \end{pmatrix} \qquad (2.8)$$

The eigenvectors of this 3×3 determinant are given by :

$$\phi_n = \sum_i a_i^n |\psi_i\rangle \text{ with } \sum_i |a_i^n|^2 = 1 \qquad (2.9)$$

A numerical estimation of this 3×3 determinant is made for reasonable values of U (9) and for $U \to \infty$. It appears that the eigenvalue which is principally concerned with $|\psi_6\rangle$ is strongly U dependent and is located much above the Fermi level. On the other hand, the eigenvalues concerning $|\psi_4\rangle$ and $|\psi_5\rangle$ have practically no U dependence for $U/V > 15$ (Fig. 1). Moreover, if we solve analytically in the $|\psi_4\rangle$, $|\psi_5\rangle$ basis we obtain the eigenvalues $|4\rangle$ and $|5\rangle$ (Table 1) which are not much different to the numerical solutions of the 3×3 determinant. The ground state labelled $|4\rangle$ is found to be mixed valent.

We report in Fig. 1 the most stable eigenvalues for one, two, three and four electrons in terms of U, for $\varepsilon^c = 0$ and for $\Delta/V = -4$ ($\Delta = \varepsilon^c - \varepsilon^f$). The occupation numbers of the f and conduction level of the ground state are given by :

$$\langle n^f \rangle = 1 - \frac{U^2-1}{U^2+1} \cos^2\alpha \qquad (2.10)$$

$$\langle n^d \rangle = 1 + \frac{U^2-1}{U^2+1} \cos^2\alpha \qquad (2.11)$$

with

$$\mathrm{tg}\alpha = \frac{2\sqrt{2}\, V(1+U)\sqrt{1+U^2}}{\varepsilon^f (U^2-1) - U + \sqrt{[(U^2-1)\varepsilon^f - U]^2 + 8V^2(1+U^2)(1+U)^2}}$$

These occupation numbers are reported on Figure 2.

Table 1

Eigenstates	Energies
$\|1\rangle = c^+_\downarrow c^+_\uparrow \|0\rangle$	$2\,\varepsilon^c$
$\|2\rangle = \frac{1}{\sqrt{2}}\left[\cos\beta\,(c^+_\downarrow f^+_\uparrow - c^+_\uparrow f^+_\downarrow)\,\|0\rangle - \sin\beta\,(c^+_\downarrow d^+_\uparrow - c^+_\uparrow d^+_\downarrow)\,\|0\rangle\right]$	$\frac{\varepsilon^f + 2\varepsilon^c}{2} - \frac{1}{2}\sqrt{(\varepsilon^f)^2 + 4V^2}$
$\|3\rangle = \frac{1}{\sqrt{2}}\left[\sin\beta(c^+_\downarrow f^+_\uparrow - c^+_\uparrow f^+_\downarrow)\,\|0\rangle + \cos\beta(c^+_\downarrow d^+_\uparrow - c^+_\uparrow d^+_\downarrow)\,\|0\rangle\right]$	$\frac{\varepsilon^f + 2\varepsilon^c}{2} + \frac{1}{2}\sqrt{(\varepsilon^f)^2 + 4V^2}$
$\|4\rangle = \frac{\cos\alpha}{\sqrt{1+U^2}}\left[U\,d^+_\downarrow d_\uparrow\|0\rangle + f^+_\downarrow f^+_\uparrow\,\|0\rangle\right] - \frac{\sin\alpha}{\sqrt{2}}\left[f^+_\downarrow d^+_\uparrow\,\|0\rangle - f^+_\uparrow d^+_\downarrow\,\|0\rangle\right]$	$\tau\left[(3+U^2)\varepsilon^f + U\right] - \tau\sqrt{\left[(1-U^2)\varepsilon^f + U\right]^2 + 8V^2\,(1+U^2)\,(1+U)^2}$
$\|5\rangle = \frac{\sin\alpha}{\sqrt{1+U^2}}\left[U\,d^+_\downarrow d^+_\uparrow\,\|0\rangle + f^+_\downarrow f^+_\uparrow\,\|0\rangle\right] + \frac{\cos\alpha}{\sqrt{2}}\left[f^+_\downarrow d^+_\uparrow\,\|0\rangle - f^+_\uparrow d^+_\downarrow\,\|0\rangle\right]$	$\tau\left[(3+U^2)\,\varepsilon^f + U\right] + \tau\sqrt{\left[(1-U^2)\varepsilon^f + U\right]^2 + 8V^2(1+U^2)(1+U)^2}$
$\|6\rangle = \frac{1}{\sqrt{1+U^2}}\,(d^+_\downarrow d^+_\uparrow\,\|0\rangle - U\,f^+_\downarrow f^+_\uparrow\,\|0\rangle)$	$2\tau\left[U^2(U + 2\varepsilon^f)\right]$

Table 1 shows the eigenstates and eigenvalues of the singlet state for two electrons. $\tau = 1/[2(1+U^2)]$; $tg\beta = -2V/[\varepsilon^f - \sqrt{(\varepsilon^f)^2+V^2}]$.

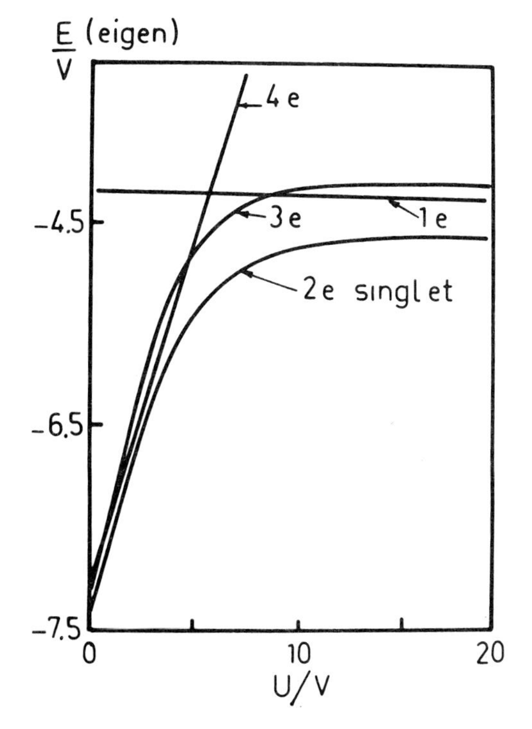

Figure 1 : Ground states energies for one, two, three and four particles as a function of U, for $\varepsilon^c = 0$ and $\Delta/V = -4$.

3. CONCLUSION

In the present paper we have discussed the ground state of a model Hamiltonian which may

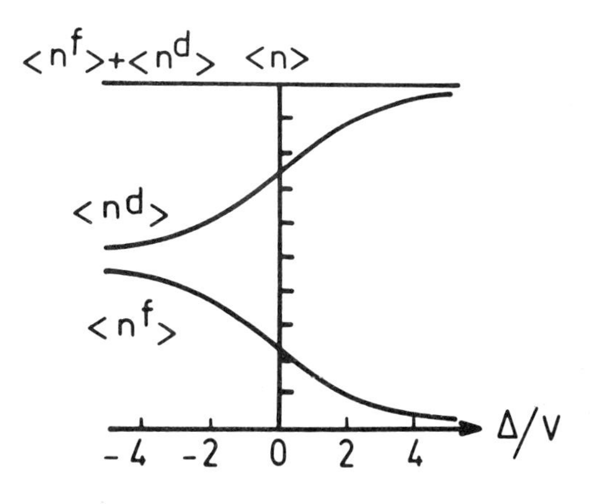

Figure 2 : f and d occupation numbers in terms of Δ/V, for $\varepsilon^c = 0$ and $U/V = 40$.

be representative of the physical situation of an Sm self interstitial in SmS compounds. This is the case if the f and d levels are represented by non-degenerate s like state and if the perturbing potentiel is restricted to diagonal matrix elements on the impurity site and to hopping integrals between the impurity and its nearest neighbouring Sm atoms. This is roughly the simplified model of reference (1) if we restrict to spinless state. If the spin is included and if intersite hybridization takes place we are able to display a mixed valent ground state instead of a $Sm^{2+} \rightarrow Sm^{3+}$ transition. However, if we take into account the extension of the perturbation on the neighbouring sites of the interstitial (3, 10) and also the degeneracy of the d states, the diagonalization of the Hamiltonian is less trivial.

ACKNOWLEDGEMENTS

Y. Hammoud is most grateful to Professor F. Gautier for his many pertinent remarks.

REFERENCES

[1] Parlebas, J.C., J.M.M.M. 15-18 (1980) 953.

[2] Morillo, J., D. Sc. Thesis, Orsay University (December 1981).

[3] Hammoud, Y., Khan, M.A., Demangeat, C. and Parlebas, J.C., REAC 82, Durham, to be published by J.M.M.M.

[4] Hammoud, Y., Thesis, Strasbourg University (1982).

[5] Davis, H.L., 9th R.E. Research Conf., Field Editor, Vol. 1 (1971) 3.

[6] Khan, M.A., Parlebas, J.C. and Demangeat, C., Phil. Mag. B 42 (1980) 111.

[7] Hammoud, Y., Khan,M.A. and Demangeat, C., REAC 82, Durham, to be published by J.M.M.M.

[8] Alascio, B.R., Allub, R. and Aligia, A., J. Phys. C13 (1980) 2869.

[9] Lang, J.K., Baer, Y. and Cox, P.A., J. Phys. F11 (1981) 121.

[10] Demangeat, C., Hammoud, Y., Khan, M.A. and Parlebas, J.C., this Conference.

Valence Instabilities
P. Wachter and H. Boppart (eds.)

X-RAY ABSORPTION STUDY OF THE PRESSURE-INDUCED VALENCE TRANSITION OF SmS

K.H. Frank, G. Kaindl, J. Feldhaus, G. Wortmann, W. Krone

Institut für Atom- und Festkörperphysik
Freie Universität Berlin, D-1000 Berlin 33, Germany

G. Materlik

Hamburger Synchrotronstrahlungslabor am Deutschen Elektronensynchrotron
D-2000 Hamburg, Germany

H. Bach

Institut für Experimentalphysik IV
Ruhr-Universität Bochum, D-4630 Bochum, Germany

L_{III}-absorption spectra were measured for SmS under truly hydrostatic pressure up to 7.50 kbar at room temperature. Values for the mean valence were derived from the double-peaked L_{III}-edge, and nn Sm-S and Sm-Sm distances were obtained from an analysis of the EXAFS structure, both as a function of pressure. The latter scale well with lattice constants from previous high-pressure X-ray diffraction work. A slight pressure-induced change of the mean valence was already observed in the black semiconducting phase, and the semiconductor-metal transition - which is not completed for the studied sample at 7.50 kbar - was found to be smeared over a pressure range of ≅2 kbar. The analysis of the EXAFS spectra supports one threshold energy and one nn Sm-S distance even in the mixed-valent regime. A critical damping of the EXAFS amplitude was observed in the region of the phase transition.

1. INTRODUCTION

SmS has long been considered a prototype of rare-earth systems with an unstable valence, since the Sm-valence can easily be changed both by external pressure and by chemical substitution /1/. When exposed to a pressure of about 7 kbar black semiconducting SmS (NaCl structure) undergoes a supposedly first-order electronic transition to an isostructural golden mixed-valent phase accompanied by a volume collapse of ≅14%. Information on the mean Sm valence in the collapsed phase had previously been obtained from lattice constant measurements /1,2/ and Sm-149 Mössbauer isomer shift studies with relatively low resolution /3/. Recently, a further second-order phase transition to a metallic regime was postulated from a study of electric transport properties /4,5/.

A detailed study of the mean Sm valence as a function of pressure is highly desirable in order to clarify the high-pressure phase diagram of SmS. We have therefore studied L_{III}-absorption spectra of SmS as a function of hydrostatic pressure up to 7.50 kbar. The mean valence was obtained from the structure of the double-peaked L_{III}-edge, and the distances from the central atom to the nn-S and nn-Sm atoms were derived from an analysis of the EXAFS oscillations. Thus both mean valence and lattice parameter can be simultaneously measured for a given sample, allowing a unique check on the validity of Vegard's law, which has been used extensively in deriving mean valences from lattice constant data /6/.

Mixed-valent systems are known to possess strong phonon anomalies due to the large difference in ionic radii associated with the two valence states as well as due to comparable magnitudes of the valence-fluctuation and phonon frequencies /7,8/. X-ray absorption experiments - due to their short characteristic measuring time of the order of 10^{-16} sec - reflect the instantaneous situation concerning valence as well as distances from the absorbing atom to the neighbouring shells. Strong effects of the mentioned phonon anomalies in mixed-valent materials are therefore expected in the EXAFS spectra, which, however, have not been observed in recent experiments on TmSe /9/ and $Sm_{0.75}Y_{0.25}S$ /10/. It may be expected that a study of SmS in the black semiconducting and the high-pressure golden mixed-valent phase at the *same temperature* provides a much clearer check on this question.

2. EXPERIMENTAL DETAILS

The X-ray absorption experiments were performed in transmission geometry at the spectrometer ROEMO at HASYLAB (DESY Hamburg) using a Si(111) double-crystal monochromator optimized for small harmonic content /11/, which provided a tunable X-ray beam with a total width of ∿1.6 eV at 6.7 keV. During the experiments the electron-storage ring DORIS was operated at 2.2 GeV and about 40 mA electron current. The incident photon flux was recorded with a He-filled ionization chamber, while the transmitted X-ray intensity was determined by single-photon counting using an argon-filled proportional counter for minimizing contributions from higher harmonics. The studied absorber was prepared from a single crystal of SmS, which was finely powdered under argon atmosphere and subsequently pressed into paraffin for better homogeneity. The high-pressure experiments were performed at room temperature under hydrostatic conditions employing an oil-pressure cell equipped with two Be-windows. Ambient-pressure data were also taken with the absorber cooled to liquid-nitrogen temperature.

3. RESULTS AND DATA ANALYSIS

Figure 1 shows some typical L_{III}-edge absorption spectra of SmS at room temperature and various hydrostatic pressures. The 1.11-kbar spectrum as well as the ambient-pressure spectrum (not shown here) exhibit prominent peaks at the L_{III}-threshold ("white line"), which arise from transitions from atomic $2p_{3/2}$ core levels into empty 5d states with a high density of states at the Fermi level /12,13/. The energetic position of the prominent peak in the 1.11-kbar spectrum (and also in the ambient-pressure spectrum), is characteristic for divalent samarium /14,15/. In addition, a weak peak shifted by 7.2±0.02 eV to higher binding energy is already observed in the ambient-pressure and the 1.11-kbar spectrum, which arises from trivalent Sm in the sample. This 7-eV shift of the peaked L-edge is due to an increase of the $2p_{3/2}$-core electron binding energy when the Sm^{2+}-$4f^6$ configuration changes to Sm^{3+}-$4f^5$ /16/. It cannot be clarified presently whether the weak Sm^{3+} contribution, which has been observed in all SmS samples studied so far by L-egde spectroscopy /15,10/ is due to bulk Sm^{3+} impurities in the sample, or to the grinding procedure, or to an intrinsic mixed-valent behaviour of SmS.

The dramatic effects of hydrostatic pressure on the electronic properties of SmS in the region of the semiconductor-metal transition are clearly visible in the spectra of Fig. 1. Starting at a pressure slightly above 6 kbar, the Sm^{3+}/Sm^{2+} intensity ratio increases with pressure, which means that an increasing proportion of Sm atoms is found in the Sm^{3+}-$4f^5$ configuration during the 10^{-16} sec exposure time of the L_{III}-edge measurement. On the other hand, no changes in the positions of the Sm^{2+} and Sm^{3+} white lines are observed as a function of pressure within the present accuracy.

In order to derive values for the mean valence $\bar{v}$ a superposition of two identical L_{III}-edge profiles, set equal to the sum of a Breit-Wigner-Fano function plus an arctan function, was least-squares fitted to the experimental data points /17/. The line-shape parameters were determined from the 1.11-kbar spectrum and kept constant in the fits of the other spectra. Almost identical values for the intensities of the Sm^{2+} and Sm^{3+} subspectra resulted from least-squares fits with Lorentzian profiles for the white lines. The solid lines in Fig. 1 represent the results of the Breit-Wigner-Fano profile analyses, and the obtained values for the mean valence $\bar{v}$ are plotted in Fig. 3a as a function of pressure. We observe that the mean valence of the studied SmS sample increases from 2.12±0.05 at ambient pressure to a value of 2.42±0.05 at the highest reached pressure of 7.50 kbar. Previous studies of lattice constants /2/ and Mössbauer isomer shifts of SmS /3/ under pressure indicated values for v around 2.7 at pressures slightly above the semiconductor metal transition. This suggests that the transition is not completed in the studied SmS sample at the highest pressure applied in the present work.

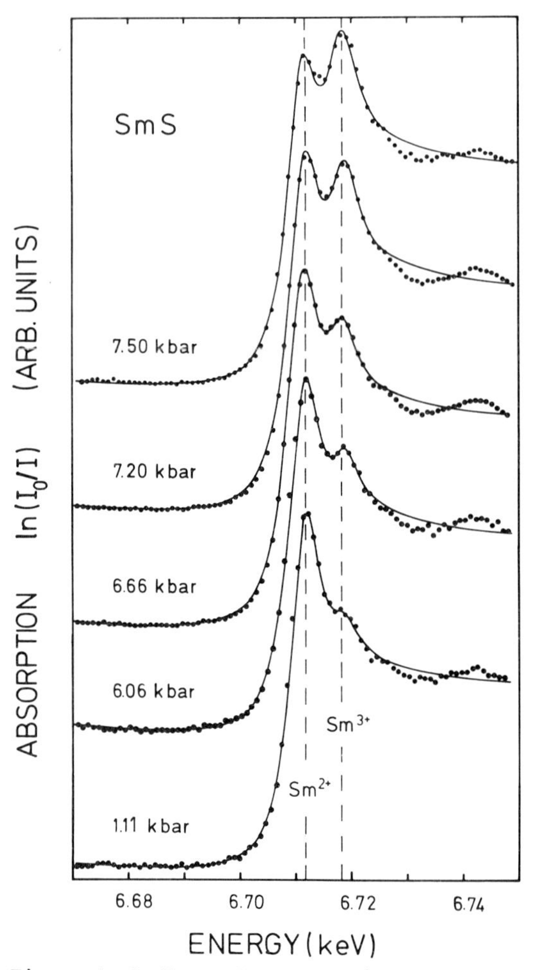

Figure 1: Sm L_{III}-edge absorption as a function of photon energy for SmS at various hydrostatic pressures. The solid lines represent the results of least-squares fit analyses as described in the text.

We have also investigated the EXAFS structure of the X-ray absorption coefficient above the Sm-L_{III} edge for SmS as a function of pressure. Representative room temperature results are displayed in Fig. 2 for two pressures, namely

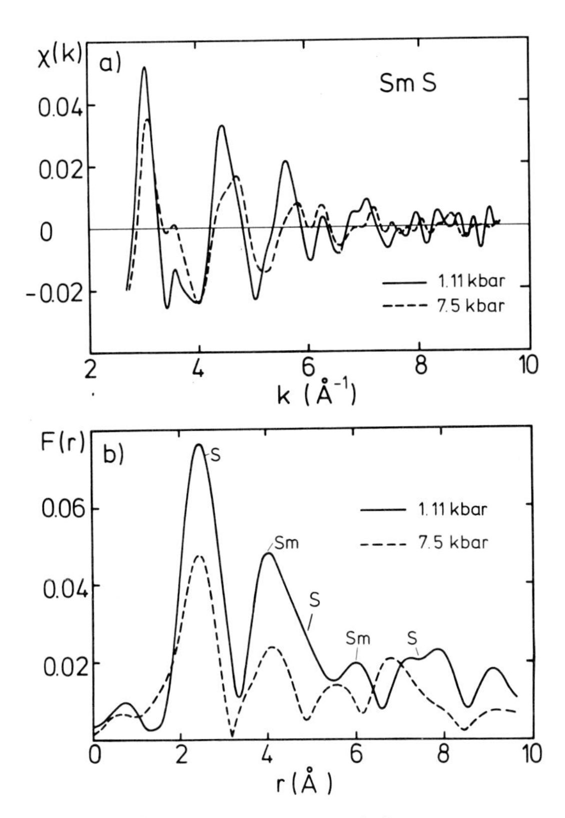

Figure 2: (a) EXAFS-spectra $\chi(k)$ of SmS after background removal at pressures of 1.11 kbar (solid curve) and 7.50 kbar (dashed curve), respectively. (b) Fourier transforms F(r) in distance space of $k^3\chi(k)$ for the two cases.

1.11 kbar (solid lines) and 7.50 kbar (dashed lines). The observed EXAFS spectra $\chi(k)$ clearly show an increase in the lengths of the periods in k-space corresponding to a reduction of the interatomic distances. The magnitude F(r) of the Fourier transform of $k^3\chi(k)$ is shown for the two pressures in Fig. 2b. As indicated in the figure, the nearest five neighbouring sulfur and samarium shells, known from the rocksalt structure of SmS, can be clearly assigned to the observed peaks in F(r). With increasing pressure the peaks in F(r) shift to lower distances. We also observe a quite drastic decrease of the magnitude of F(r) with pressure (to be discussed later).

The EXAFS spectra were analyzed in the usual way by Fourier filtering techniques as described e.g. in Ref. /18/. Because of the relatively low statistical accuracy of the high-pressure EXAFS data, the small Debye-Waller factors at room temperature, and a possible influence of multiscattering processes, only the nn and nnn shells were analyzed. The k-dependent Sm-S scattering phases were obtained from an analysis of the 77-K EXAFS spectrum of SmS at ambient pressure and the known nearest neighbour Sm-S distance under these conditions /1/. From the argument of chemical transferability these scattering phases are expected to be independent of pressure and valence. They were therefore used to derive the pressure dependence of the nn Sm-S distance shown in Fig. 3b. The agreement with the results of the previous X-ray diffraction measurements of Keller et al. /2/ is very good (within ±0.02 Å). This comparison again shows that the semiconductor-metal transition is not fully completed in our SmS sample at 7.50 kbar.

The EXAFS spectra could be evaluated best with a single zero of kinetic energy for the photoelectrons, even though the observed L edge structures suggest 7-eV differences for photoelectrons originating from Sm^{2+} and Sm^{3+}. This finding is in agreement with the results of other EXAFS studies of mixed-valent compounds /9,10/. We also could not detect two different nn Sm-S distances in the observed EXAFS spectra, which would lead to a beating of the Fourier-filtered nn-shell EXAFS oscillations /19/. This again is in accord with the previous EXAFS studies of mixed-valent systems /9,10/. The statistics of the present high-pressure data is not good enough to decide on this question also for the nn Sm shell.

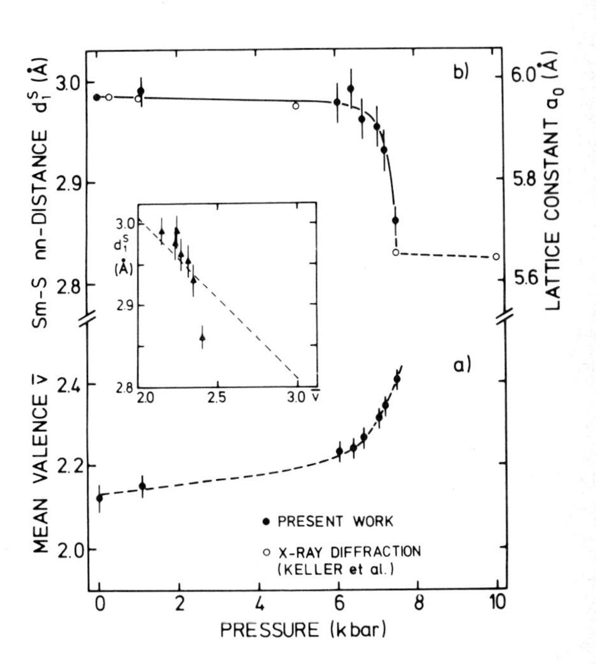

Figure 3: Pressure dependence of (a) the mean valence $\bar{v}$ of SmS derived from an analysis of the L_{III}-edge spectra and (b) of the Sm-S nn-distance d_1^S obtained from the EXAFS spectra. Also shown in (b) are values derived from the X-ray diffraction work of Ref. /2/. In the insert d_1^S is plotted as a function of the mean valence $\bar{v}$. The dashed line represents Vegard's law.

A clear damping of the magnitude of the Fourier transform F(R) is observed at 7.50 kbar as compared to 1.11 kbar (see Fig. 2b). This may be described by an attenuation factor, which is usually accounted for by a Debye-Waller-type term $\exp(-2\sigma_j^2k^2)$ in the EXAFS expression for $\chi(k)$ /18/. The mean-square average of the difference of displacements σ_j^2 between the atoms in the j-th coordination shell and the central atom may have contributions from thermal vibrations and static disorder. We have analyzed the attenuation factor for the nn-S shell as a function of pressure by comparison with the 77-K ambient-pressure EXAFS spectrum using the procedure of Ref. /20/. This leads to the difference $\Delta\sigma^2=\sigma_{300}^2(P)-\sigma_{77}^2(0)$ of the mean-square displacements of the nn-S distances between the room temperature high-pressure and the 77-K ambient-pressure data. The results of this analysis are presented in Fig. 4. In the semiconducting phase the disorder clearly gets smaller up to a pressure of ≅6 kbar, which may be expected from a stiffening of the lattice. At the phase transition a pronounced increase of $\Delta\sigma^2$ is observed, with a peak at ≅7.20 kbar. At the highest pressure studied (7.50 kbar) $\Delta\sigma^2$ has again decreased. It cannot be decided from the present data whether this increase in $\Delta\sigma^2$ at the phase transition is due to the known phonon anomalies in mixed-valent systems or simply due to some static disorder. The observed decrease of $\Delta\sigma^2$ at the highest pressure point indicates a dominant contribution from static disorder at the phase transition. Such an effect has to be expected if the sample undergoes a first-order phase transition smeared out by inhomogeneity effects in the sample. A clarification of this question has to wait for an extension of the hydrostatic pressure range to higher pressures.

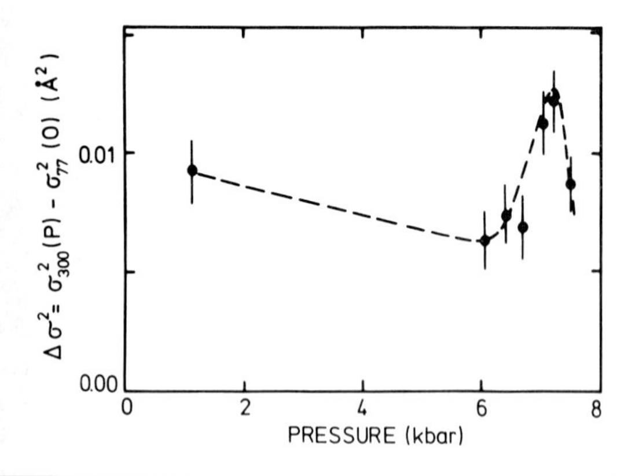

Figure 4: Difference $\Delta\sigma^2$ of nn Sm-S mean-square displacements between the 300-K high-pressure and 77-K ambient pressure data.

4. DISCUSSION

The L_{III}-edge spectra reveal a small Sm^{3+}-derived structure even at ambient conditions as can be seen e.g. in the 1.11-kbar spectrum presented in Fig. 1. The intensity of this trivalent peak was found to vary only slightly for the different samples studied. The lowest Sm^{3+} content was found to be ≅12% corresponding to a mean valence of $\bar{v}$≅2.11 at ambient conditions. Presently it cannot be decided whether this is due to Sm^{3+} impurities in SmS (either intrinsic or created by grinding) or if it reflects a mixed-valent behaviour of SmS already at ambient conditions. The impurity hypothesis is supported by the fact that sample-dependent Sm^{3+}-impurity phases have been identified in various SmS crystals by other methods /21/. On the other hand, recent high-resolution VUV-photoemission experiments on SmS(100) surfaces failed in detecting any Sm^{3+}features in the observed valence bands /22/. In these latter measurements, however, a bulk mixed-valent state of Sm close to divalency would hardly be seen due to the high surface sensitivity and an expected surface-valence transition /22,23/. In order to clarify the addressed question, L-edge measurements on thin single-crystalline discs of SmS should be performed and are in fact planned in the near future.

The results for $\bar{v}$ displayed in Fig. 3a also show that a small change of the 4f occupation number occurs already in the black phase before the semiconductor-metal transition takes place. This is in agreement with a recent theoretical description of the semiconductor-metal transition of SmS in terms of f-d hybridization in the semiconducting phase /24/. A slight increase of the mean Sm valence with pressure in the black phase had previously also been suggested on the basis of the results of high-pressure Mössbauer experiments /3/.

The semiconductor-metal transition of the studied SmS sample seems to be smeared over a pressure range of more than 2 kbar (see Fig. 3). This may be due to an inhomogeneous distribution of transition pressures for the first-order phase transition in our powdered sample. Such an interpretation is supported by the fact that a variation of the transition pressure had already previously been found in X-ray diffraction studies of several SmS samples under hydrostatic pressure /2/. It should be possible to clarify the true nature of the transition by an extension of the present pressure range keeping the truly hydrostatic conditions.

The present X-ray absorption experiments allow a simultaneous determination of the mean valence (from the L-edge structure) and the nn Sm-S distance d_1^S (from EXAFS) as a function of pressure. Since we have found $a_o=2d_1^S$ to hold for all pressure values (see Fig. 3b), a plot of d_1^S versus $\bar{v}$ allows a unique check on the validity of Vegard's law. The data plotted in the insert of Fig. 3 despite the limited accuracy exhibit a considerable deviation from a linear relationship and show that a linear interpolation of a measured lattice constant would lead to a too large value for $\bar{v}$. The dashed line in the insert of Fig. 3 represents Vegard's law assuming d_1^S=2.81 Å for a fictitious $Sm^{3+}S$ and $\bar{v}$=2.11 for SmS under ambient conditions (d_1^S=2.98 Å). The above statement would also hold if the deviation of $\bar{v}$ from 2.0 for SmS under ambient conditions is due to Sm^{3+} impurities. Then the dashed line has to end at d_1^S =3.01 Å for $\bar{v}$=2.0, and all experimental $\bar{v}$ values would have to be shifted by 0.11. The observed deviation from Vegard's law is in qualitative agreement with other findings and has recently been interpreted on the basis of the bulk modulus in the intermediate-valent regime /6/.

Finally, a few words should be made about the reliability and accuracy of the $\bar{v}$ values derived from L-edge spectra of mixed-valent systems. The central question is to what extent the observed double-edge structure of the L-edge represents the initial state valence mixture. There is strong support from a variety of experimental data that in heavy rare-earth systems, where the 4f-states are well localized, a possible influence of final-state screening effects is negligible. This is qualitatively understandable, since in an L_{III}-edge absorption process the electron is excited into a rather localized 5d-state so that the positive core-hole is well screened with respect to the more itinerant valence and conduction electrons, thus making a major redistribution of these electrons unprobable. The situation is quite different in deep core-hole XPS measurements, where the initially unscreened positive core-hole has been found to cause both shake-up and shake-down processes, which may completely masque the initial-state valence /25,26/. It should be noted that such final-state effects have also not been identified up to now in valence-band photoemission spectra of heavy rare-earth systems /27/. We would like to emphasize that these arguments may not apply to the case of Ce systems /28/.

It is therefore reasonable to assume that the observed double structure of the L_{III}-edge reflects the initial-state valence. The *absolute accuracy* of the derived mean valence, however is subject to some uncertainty, since it depends to some extent on the line-shape analysis, particularly due to the broad features of the L-edge white lines (caused mainly by lifetime effects). We have employed the same lineshape parameters for the Sm^{2+}- and the Sm^{3+}-derived edge structures even though a lower intensity of the white line may be expected for the Sm^{3+} ion due to a locally partly-filled d state. If this should be the case the values for $\bar{v}$ have to be considered as lower limits.

ACKNOWLEDGEMENT - This work was supported in part by the Bundesministerium für Forschung und Technologie under contract No. 05 256 KA.

REFERENCES

/1/ See e.g. Jayaraman, A., Handbook of the Physics and Chemistry of Rare Earths, ed. by Gschneidner, Jr., K.A. and Eyring, L.R., (North-Holland, Amsterdam, 1979), Vol. 2, p. 575.
/2/ Keller, R., Günterhodt, G., Holzapfel, W.B., Dietrich, M., and Holtzberg, F., Solid Sate Commun. 29 (1979) 753.
/3/ Coey, J.M.D., Ghatak, S.K., Avignon, M., and Holtzberg, F., Phys. Rev. B14 (1976) 3744,
/4/ Konczykowski, M., Morillo, J., and Senateur, J.P., Solid State Commun. 40 (1981) 517.
/5/ Lapierre, F., Ribault, M., Holtzberg, F., and Flouquet, J., Solid Sate Commun. 40 (1981) 347.
/6/ See e.g. Neumann, G., Pott, R., Röhler, J., Schlabitz, W., Wohlleben, D., and Zahel, H., contribution to this conference.
/7/ See e.g. Bilz, H., Güntherodt, G., Kleppmann, W., and Kress, W., Phys. Rev. Lett. 43 (1979) 1998.
/8/ Mook, H.A. and Nicklow, R.M., Phys. Rev. B20 (1979) 1656.
/9/ Launois, H., Raviso, M., Holland-Moritz, E., Pott, R., and Wohlleben, D., Phys. Rev. Lett 44 (1980) 1271.
/10/ Martin, M., Boyce, J.B., Allen, J.W., and Holtzberg, F., Phys. Rev. Lett. 44 (1980) 1275.
/11/ Materlik, G. and Kostroun, V.O., Rev. Sci. Instr. 51 (1980) 86.
/12/ Materlik, G., Müller, J.E., and Wilkins, J.W., preprint (1982).
/13/ Bianconi, A., Modesti, S., Campagna, M., Fischer, K., and Stizza, S., J. Phys. C14 (1981) 4737.
/14/ Vainshtein, E.E., Blokhin, S.M., and Paderno, Y.B., Sov. Phys. Solid State 6 (1965) 2318.
/15/ Ravot, D., Godart, C., Achard, J.C., and Lagarde, P., Valence Fluctuations in Solids, ed. by Falicov, L.M., Hanke, W., and Maple, M.B., (North-Holland, Amsterdam, 1981), p. 423.
/16/ Johansson, B. and Mårtensson, N., Phys. Rev. B24 (1981) 4484.
/17/ Wei, P.S.P. and Lytle, F.W., Phys. Rev. B19 (1979) 679.
/18/ Lee, P.A., Citrin, P.H., Eisenberger, P., and Kincaid, B.M., Rev. Mod. Physics 53 (1981) 769.
/19/ Martens, G., Rabe, P., Schwentner, N., and Werner, A., Phys. Rev. Lett. 39 (1977) 1411.
/20/ Stern, E.A., Phys. Rev. B21 (1980) 5521.
/21/ Golubkov, A.V, Kartenko, N.F., Sergeeva, V.M., and Smirnov, I.A., Sov. Phys. Solid State 20 (1978) 126.
/22/ Reihl, B., Holtzberg, F., Hollinger, G., Kaindl, G., Mårtensson, N., and Pollak, R.A., contribution to this conference.
/23/ Kaindl, G., Reihl, B., Eastman, D.E.,

Pollak, R.A., Mårtensson, N., Barbara, B., Penney, T., and Plaskett, T.S., Solid State Commun. 41 (1982) 157.
/24/ de Menezes, O.L.T. and Troper, A., Solid State Commun. 38 (1981) 903.
/25/ Fuggle, J.C., Campagna, M., Zolnierek, Z., Lässer, R., and Platau, A., Phys. Rev. Lett. 45 (1980) 1597.
/26/ Schneider, W.D., Laubschat, C., Nowik, I., and Kaindl, G., Phys. Rev. B24 (1981) 5422.
/27/ Schneider, W.D., Laubschat, C., Kaindl, G., Reihl, B., and Mårtensson, N., contribution to this conference.
/28/ Lengeler, B., Materlik, G., and Müller, J.E., DESY report, No. 82/02.

Valence Instabilities
P. Wachter and H. Boppart (eds.)

^{31}P NMR STUDY OF THE $SmS_{1-x}P_x$ INTERMEDIATE VALENT SYSTEM

P. Panissod[+], M. Benakki[+] and J.P. Sénateur[*]

+ L.M.S.E.S. (LA 306) Univ. L. Pasteur 4, rue B. Pascal 67070 Strasbourg France

* E.R. 155 I.E.G. Domaine Universitaire 38042 Saint Martin d'Hère France

^{31}P NMR data have been obtained for $SmS_{1-x}P_x$ (x = 0.1, 0.2) compounds in the collapsed metallic phase. These results are rather controversial against bulk susceptibility data that suggested a homogeneous mixed valence state for Sm in this compound : both static (Knight shift) and dynamic (relaxation rates) local susceptibilities are shown to diverge at low temperature. Knight shift data are consistent with conduction electron polarization due to Sm^{3+} spin moment only and compare well with $Sm^{3+}P$ results. Low temperature relaxation rates can be analyzed in the Fradin's scheme and suggest the occurrence of some magnetic ordering.

1. INTRODUCTION

$SmS_{1-x}P_x$ alloys are known to exhibit similar properties when substituting P to S as pure SmS under pressure. Namely, in the collapsed phase, lattice parameter, XPS and X-rays absorption, high temperature susceptibility suggest a mixture of Sm^{3+} and Sm^{2+} valence state while low temperature susceptibility fails to exhibit the expected magnetic behaviour of Sm^{3+} ions (1-4). This is assumed to be characteristic of a homogeneous mixed valent state for the RE element.

In order to shed some light on the mechanism responsible for the non magnetic behaviour of the Sm ions in $SmS_{1-x}P_x$ ($x \geq 0.1$) we have analyzed the local magnetic properties of this system through ^{31}P NMR. The results of this study, though they do not exclude the possibility of a homogeneous IV state (all Sm sites equivalent) rather favour a scheme with inhomogeneously distributed Sm^{2+} and Sm^{3+} ions, the latter being mostly on the nearest neighbour sites of the P atoms. In any case they also suggest the occurrence of some magnetic ordering at low temperature though this has not been observed yet by other techniques.

2. RESULTS AND DISCUSSION

^{31}P NMR spectra were recorded by spin echo technique at 12 MHz in the 10 K - 400 K temperature range. Relaxation times were also measured in the same range. Below 10 K the NMR Signal could not be observed due to diverging linewidth and relaxation rates as will be seen in the followings. As very little differences in the results were found for x = 0.1 and x = 0.2 we will focus our attention on the $SmS_{0.8}P_{0.2}$ compound.

2.1. Static properties

The measured bulk susceptibility χ (Fig. 1) is in total agreement with Cater's results (1) for the same composition ; the high temperature values T > 100 K are consistent with a mixture

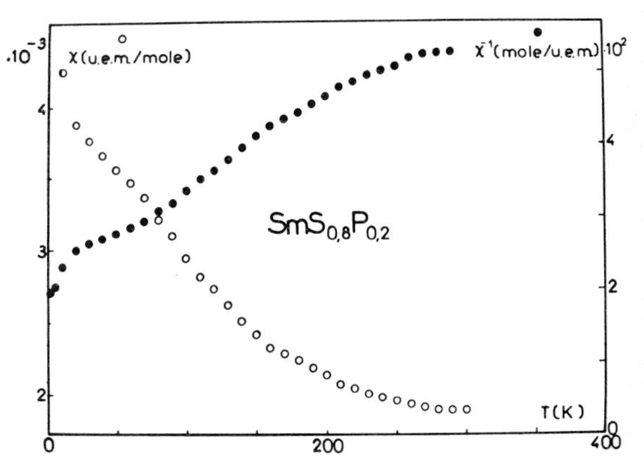

Figure 1 : Temperature dependence of the bulk susceptibility (and reciprocal) in $SmS_{0.8}P_{0.2}$

of 80 % at Sm^{3+} and 20 % at Sm^{2+} ions (for x = 0.2) in general agreement with lattice parameter (2), XPS (3) determination of the valence. But the low temperature susceptibility does not diverge as expected from the Γ_7 doublet state of Sm^{3+} ions.

On the contrary the Phosphorus Knight shift (Fig. 2) does present a divergence at low temperature and changes sign close to room temperature as in pure $Sm^{3+}P$ (5). The knight shift results can be changed in a conduction electron polarization (CEP) scheme where K is the sum of a temperature independent term K_o and a term K_f proportional to the spin component of the RE moment $\langle S_z \rangle$:

$$K = K_o - \frac{H_{hf}}{H_o} \langle S_z \rangle$$

In our case $\langle S_z \rangle$ is hard to evaluate the configuration of Sm atoms being unknown. However we have assumed that $\langle S_z \rangle$ could be written as the weighted average :
(1) $\langle S_z \rangle = (1-c)\langle S_z^{3+} \rangle + c\langle S_z^{2+} \rangle$ at high temperatures as observed for the bulk susceptibility ;

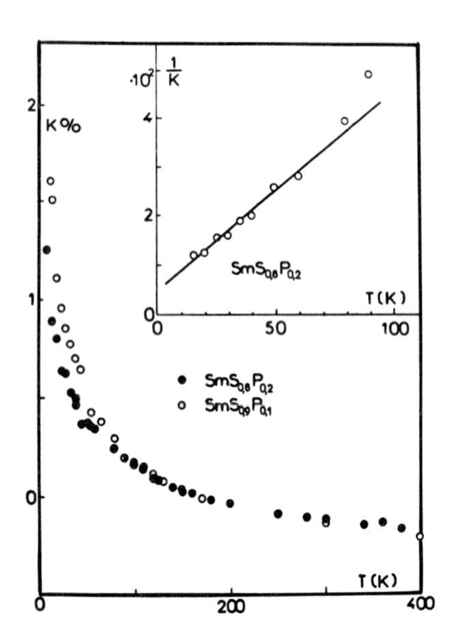

Figure 2 : Temperature dependence of the ^{31}P Knight shift in $SmS_{1-x}P_x$ compounds.

(2) $\chi = (1-c)\ \chi^{3+} + c\ \chi^{2+}$ where c = 0.2 and $<S_Z^{3+}>$, $<S_Z^{2+}>$ χ^{3+}, χ^{2+} are the ionic values including spin orbit and CEF effects. The analysis of the data yields then an unusually large K_o (0.46 %) as compared with K_o = 0.06% in pure SmP and a value for the hyperfine field H_{hf} that is nearly twice higher (- 30 kOe) than in SmP (- 18 kOe). The high value for K_o can be tentatively related to the high density of states at the Fermi level as usually observed in IV systems while the higher hyperfine field indicates an enhanced CEP by the Sm moment with respect to normal compounds.

Since in this analysis we assume a configuration mixing no parallel thermal variation is expected for K and χ. However, numerical calculations according to (1) and (2) show that the deviation from linearity in K vs χ plot is lower than the experimental errors and a linear plot is actually found down to 50 K (Fig. 3.a). Below that temperature an abrupt change of the slope is observed pointing out the very different behaviours of K and χ in the low T's range. Such different temperature dependences have also been observed in other IV compounds (6) and have been tentatively related to either modifications of the transferred hyperfine field between RE and nuclei or a new Pauli like susceptibility component due to the hybridized ground state, although in our case the strong Knight shift varation exclude the latter explanation.

However as already mentioned the ^{31}P Knight shift resembles closely that observed in pure SmP and the thermal variation of K is actually found to parallel that of the susceptibility χ^{3+} of the integral valence state Sm^{3+} (Fig.3.b) down to the lowest temperature. From this

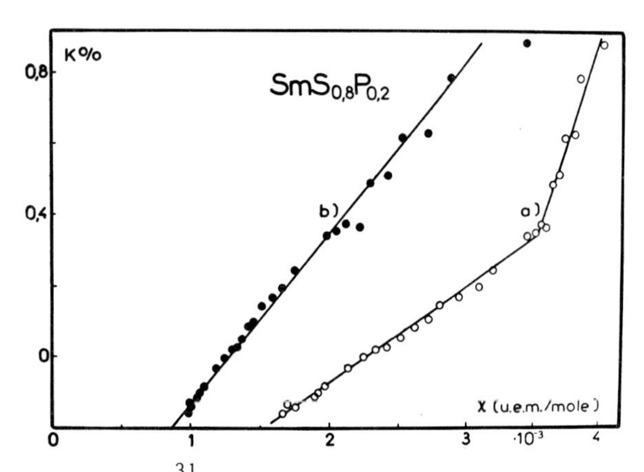

Figure 3 : ^{31}P Knight shift in $SmS_{0.8}P_{0.2}$ versus (a) bulk susceptibility (b) ionic susceptibility of Sm^{3+} (χ^{3+}) (T implicit).

K vs χ^{3+} plot and a least square analysis of K data according to :

$$K = K_o = \frac{H_{hf}}{H} <S_Z^{3+}>$$ with the same spin orbit

and CEF splitting as in pure SmP (according to (5)) we find a much lower K_o (-0.1 %) than in the first analysis and a value for H_{hf} (-14 kOe) quite close to that found in pure SmP. Such a result would indicate that the neighbours of P atoms are in a 3^+ valence state. This in turn can be compared with the analysis of Pena et al (2) of this system in the low concentration range (insulating phase) which showed that substituting P to S induces the valence change from Sm^{2+} to Sm^{3+} of the 6 Sm nearest neighbours of a P atom. However, a fortiori, the non-diverging χ is still unexplained unless some antiferromagnetic ordering occurs at low temperatures which was not observed till now but that NMR dynamic data we analyze now strongly suggest.

2.2. Dynamic data

As shown on Fig. 4 and 5, the linewidth and the relaxation rate $1/T_1$ both diverge below T $\sim$ 10 K. The linewidth is mostly due to homogeneous broadening and fully consistent with transverse relaxation time T_2 whose values are close to T_1's. This homogeneous character is, however, not unconsistent with the assumption of spatial fluctuation of Sm^{2+} and Sm^{3+} states since K data show that in such a case P atoms are surrounded by Sm^{3+} only.

The high temperatures T_1 results are hard to analyze since both crystal field and spin orbit effects are to be taken into account for Sm atoms. However, at low temperatures, the Sm atoms are in their ground state and the results can be analyzed in Fradin's scheme (7):

$$1/T_1T \simeq 1/(T_1T)_K \left[W/(T-\theta(Q)\right]$$

with $w = 4n(ex)\ |J(Q)|^2\ \frac{(g_J-1)^2 J.(J+1)}{3K_B}$

J(Q) is the Q dependent sf exchange interaction and $(T_{1K}\ T^{-1})$ is the temperature independent conduction electron relaxation term.

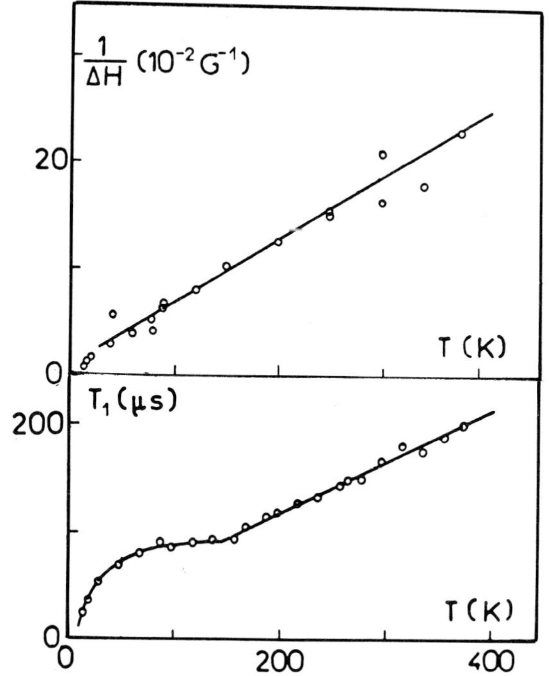

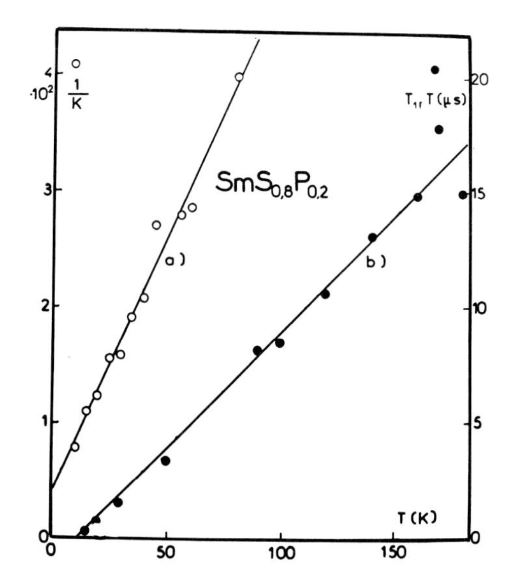

Figures 4 and 5 : Longitudinal relaxation time and reciprocal linewidth temperature dependences in $SmS_{0.8}P_{0.2}$.

Figure 6 : Temperature dependence of the reciprocal relaxation rate T_1T and reciprocal Knight shift K^{-1} in $SmS_{0.8}P_{0.2}$.

As shown on Fig. 6.a, the temperature dependence of T_1T is linear in the 10 K - 150 K range and a critical temperature $\theta(Q) = +\ 10$ K is deduced. As shown by Fradin in a number of 5f systems $\theta(Q)$ is closely related to the magnetic ordering temperature (for both ferro- and antiferromagnetic systems). This is explained by the fact that magnetic ordering and relaxation phenomena are both driven by the first magnetic instability to occur in some direction of reciprocal lattice.

Thus the divergence of both linewidth relaxation rates at low temperature could be explained if the compound magnetically orders below 10 K.

3. CONCLUSION

The analysis of the Phosphorus NMR data in the collapsed metallic phase of the $SmS_{1-x}P_x$ (x = 0.1, 0.2) system has revealed a quite different behaviour of the local magnetic properties as compared with bulk magnetic susceptibility. Particularly while the susceptibility is nearly constant at low temperature the ^{31}P Knigth shift still increases as in SmP where Sm is purely trivalent. This strongly suggests that the IV state of Sm results from spatial distribution of the integral Sm^{3+} and Sm^{2+} states, P atoms being surrounded mostly by Sm^{3+} as in the low P concentration range. However temporal valence fluctuation cannot be excluded ; then a mechanism has to be found that explains the strong temperature dependence of the conduction electron polarization at the P sites together with nearly temperature independent susceptibility. In any case, from dynamic NMR data some magnetic ordering seems to occur at low temperatures that careful neutron diffraction study could possibly reveal.

Bibliography

[1] Cater, E.D., Henry, D.C., Sisson, K.J., Savage, W.R. and Schweitzer, J.W., Phys. Rev. B 20 (1979) 1985-90.

[2] Pena, O., Ph. D. Thesis, Grenoble (1979).

[3] Krill, G., Sénateur, J.P. and Amamou, A., J. Phys. F 10 (1980) 1889-97.

[4] Morillo, J., de Novion, C. and Sénateur, J.P., J. Physique Col. 40 C5 (1979) 348-9.

[5] Jones, E.D., Phys. Rev. 180 (1969) 455-74.

[6] Mac Laughlin, D.E., Valence Fluctuations in Solids, Falicov, L.M., Hanke, W. and Maple, M.B. (ed.) (North-Holland,Amsterdam) (1981) 4321-4.

[7] Fradin, F.Y., J. Phys. Chem. Sol. 31 (1970) 1715.

Valence Instabilities
P. Wachter and H. Boppart (eds.)

SEMICONDUCTOR - METAL TRANSITION IN $Yb_{1-x}Tm_xS$

R.Suryanarayanan, D.Jeanniot[+], G.Brun and G.Zribi

Laboratoire de Physique des Solides, CNRS
92190 Bellevue, France

The lattice constant, optical absorption and the resistivity of $Yb_{1-x}Tm_xS$ films are reported. For $0 \leqslant x \leqslant 0.18$, the samples are semiconducting and the results support the absence of a pure Tm^{3+} ground state. A semiconductor - metal transition takes place for $x \sim 0.2$ associated with a change in valence of Tm from 2 to 3.

1. INTRODUCTION

Thulium monochalcogenides and their solid solutions have been examined recently in order to study the influence of the valence of Tm on different physical properties(1). Whereas TmTe-TmSe alloys(2a-e) represent a concentrated system to study Tm valence effects, TmSe-EuSe(2c) and TmSe-YSe(2g) represent a dilute system to study such effects in different conditions. In this work, we wish to report on the preparation, lattice constant(a_o), optical and electrical measurements of $Yb_{1-x}Tm_xS$ films. YbS is a semiconductor ($f \rightarrow d$ gap $\sim$ 1eV) with a well defined Yb^{2+} ground state(3). It is interesting to note that both Yb^{2+} $f^{14} \rightarrow d$ and Yb^{3+} $f \rightarrow f$ optical transitions occur in a narrow spectral range(1 and 2 eV) easily accessible in a conventional spectrometer. TmS, on the other hand, is an anti-ferro magnetic metal with a Tm valence of three(4). Our aim was to probe the semi-conductor -metal transition (SMT) as a function of x and try to relate it to changes in the valence of Tm.

2. SAMPLE PREPARATION AND CHARACTERISATION

$Yb_{1-x}Tm_xS$ films were prepared by co-evaporating Yb,Tm and S in a vacuum of 1×10^{-6} Torr on glass and CaF_2 substrates heated to 400C in an apparatus described elsewhere(5). This technique allows us to control the stoichiometry by simply controlling the rate of evaporation of Yb,Tm and S independently. The films thus obtaiend are polycrystalline but often show a preffered orientation. For example, films deposited on CaF_2 which were oriented in a X-ray diffractometer to get a maximum (111) intensity showed only Bragg peaks corresponding to (111),(222),(333) and (444) planes indicative of excellent epitaxy. The samples reported here all crystallise in NaCl structure. a_o is calculated either from (111) or (222) peak and the precision varies from ± 0.004 to ± 0.006 A. The composition x is obtained with the aid of a Camebax electron probe analyser with a precision of 2%. The X-ray images of Yb,Tm and S obtained on the samples reported here($5mm^2$) show a perfect homogeneity of the distribution of elements with a spatial resolution better than 2μ. We do not observe any second phase precipitates as observed sometimes in the single crystals of ternary compounds(2b,c). Whereas the slow cooling in the growth of single crystals may favourise segregation and precipitate formation, we believe the film growth is too rapid for favourising such precipitate formation. The depth profiles of these films using SIMS showed constant ratio of Tm/Yb for depths examined upto 500 Å. Many of our samples revealed the presence of oxygen but the ratio O/S decreased rapidly as a function of sputtering time in SIMS indicative of the surface contaminated by oxygen.

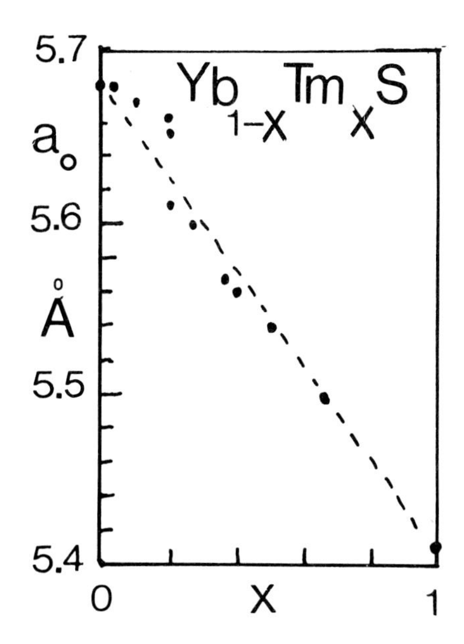

Fig.1. Lattice constant a_o of $Yb_{1-x}Tm_xS$ as a function of x. Experimental points •

3. RESULTS AND DISCUSSION

a. Lattice constant. The variation of a_o with x is shown in Fig.1. The sample with x=0.02 has a_o very close to that of YbS. Since the presence of either Tm^{2+} or Tm^{3+} with such a low concentration is not expected to produce a noticeable change in lattice constant, nothing can be said on the valence of Tm for this sample. If we assume a linear variation in a_o between YbS(5.68 Å) and $Tm^{3+}S$(5.41 A), one expects a_o = 5.630 A for x=0.18. However, samples with x=0.18 had a value of 5.66 Å. This could possibly indicate a deviation from the valence of 3 for Tm since Yb is shown to be divalent from optical studies(see below).The sample BL8 with x=0.20 had a value of 5.61 Å indicating a valence very close to 3 for Tm.Though we have not explored in detail the region 0.18< x <0.35, the lattice constant seems to vary abruptly for x close to 0.20 and hence would indicate a valence change taking place for Tm in this range of composition.The lattice constant for x >0.4 indicate a valence of 3 for Tm. While the samples with $5.65 \leqslant a_o \leqslant 5.68$ Å were green by reflection, BL8 was violet to purple.

b.Optical absorption. The optical density of YbS film at 10K as a function of photon energy is shown in fig.2. Among 10 samples examined, only two of them showed a well defined structure near the two main absorption bands. This structure not observed by others(3) reflects the high quality of these samples(fig2b).The sample with x=0.02 also revealed this structure(fig2a,c). Following an earlier interpretation of YbTe and YbSe spectra (6) and noting $E_2-E_1 \simeq 1.25$ eV $\simeq$ spin-orbit coupling of $4f^{13}$ we assign these bands to Yb^{2+} $4f^{14} \rightarrow 4f^{13}$ $(^2F_{7/2})5d\ t_{2g}$ and $4f^{13}(^2F_{5/2})$ $5d\ t_{2g}$ transitions respectively. Then E_3 is assigned to $4f^{14} \rightarrow 4f^{13}$ $(^2F_{7/2})$ $5d\ e_g$ giving 10 Dq $\simeq$ 1.45 eV. Using the point charge model and taking 10 Dq=2.2 eV(ref7), we estimate 1.5 eV for YbS which is quite close to the value obtained here. The structure seen at E_1 and E_2 is explained due to the interaction of $H_{s.o}(d) \simeq 0.12$ eV which is similar to that obtained in the case of YbTe and YbSe(6). A broad maximum between 4 and 4.3 eV can be attributed to $4f^{14} \rightarrow 4f^{13}(^2F_{5/2})$ $5d\ e_g$. One of our YbS samples showed a very weak shoulder around 1.27 eV (fig. 2d) which can be interpreted as $4f^{13}$ f→f transition indicative of the presence of Yb^{3+} ions in O_h symmetry. This is attributed to deviations from stoichiometry in the cubic NaCl phase.If another phase like Yb_3S_4 were present, then we expect 3 peaks due to orthorhombic symmetry. Indeed, in our samples of composition close to this phase, we do observe 3 peaks between 1.33 and 1.24 eV. This work thus illustrates the possibility of conveniently checking the presence of trivalent Yb optically.

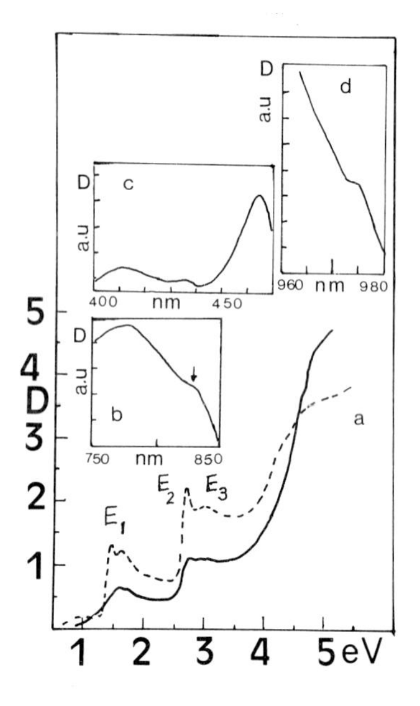

Fig.2 Optical density D at 10K as a function of photon energy.a,b— YbS 0.3μ ; a --- and c x= 0.02, 0.6μ on CaF_2 ; d YbS. a.u=arbitrary units

The optical density for x=0.02,0.18, 0.2 and 0.65 is shown in fig.2(---) and 3. For x=0.02, the absorption bands due to Yb^{2+} transitions are clearly seen. The maximum near 0.93 eV is attributed to an interference fringe(i.f) by the following argument.The refractive index n is calculated from the i.f. of other samples. This value of n is used to calculate i.f. maxima assuming the maximum near 0.93 eV is due to i.f. These are expected at 2.9 and 17μ. The optical density(not shown in fig2) does show a broad maximum around 3 and shows none till 10μ the limit of the measurement imposed by the substrate. Thus there is no evidence of free carrier or any other absorption process till 10μ. Now we assume Tm to be present with valence=3 in this sample.Each Tm atom would then contribute one electron to the conduction band resulting in a plasma resonance at around 0.43 eV(after correcting for the dielectric constant) which is not seen experimentally. Thus we conclude Tm^{3+} is not present in this sample. Tm^{2+} $4f^{13} \rightarrow$ 5d transitions can be expected to occur in the far infrared beyond the limit of our measurement For x=0.18, the principal absorption bands due to Yb^{2+} are clearly seen. The region between 0.5 and 1 eV is dominated by i.f. In the infrared we expect them at 2.7, 3.4, 4.6 and 6.9 μ. Experimentally we observe maxima around 2.9, 3.5, 5.2, 6.6 and 7.1μ.Whereas the first three are due to i.f., the origin of the other two is not understood.Again for this concentration of Tm, the presence of trivalent Tm would lead to a free carrier absorption near 1.35 eV which is clearly absent. Hence, it is very tempting to

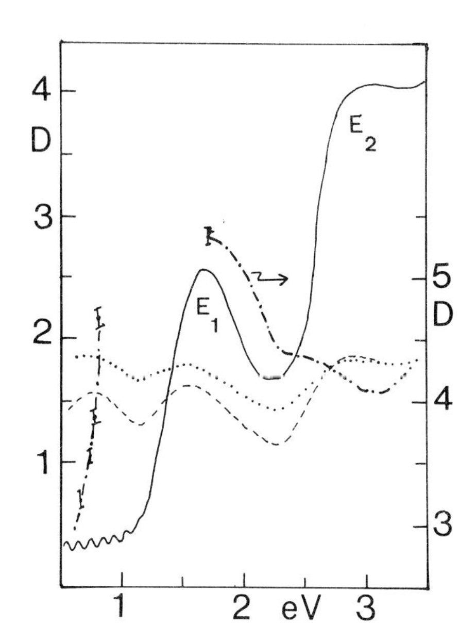

Fig.3 Optical density D at 10K of $Yb_{1-x}Tm_xS$ as a function of photon energy. — x=0.18, 3.2μ, 5.66Å ....BL12 0.44μ, 5.61Å; - - -BL8 0.2, 0.62μ, 5.61Å --- BK20 0.65, 1.8μ, 5.51Å.

postulate the absence of Tm^{3+} ions for $0 < x \leq 0.18$ Two samples having a lattice constant=5.61Å show higher absorption compared to those discussed above. One of them had x=0.20. In spite of the high absorption, Yb^{2+} transitions are clearly identified. No Yb^{3+} f → f transitions are detected. The absorption band at 1.5 eV in particular has broadened compared to that of YbS reflecting the broadening of the 5d band. This also indicates the setting in of a transition to a metallic state. The minimum at 1.15 eV is interpreted as partly due to a free carrier absorption. In $SmS_{.5}As_{.5}$, the free carrier absorption shows a much steeper rise and was thus interpreted as due to the presence of Sm^{3+} (8). The absence of such a steep rise in the present case is perhaps due to an incomplete conversion of Tm into Tm^{3+} ions. It is observed that the positions of E_1 and E_2 are hardly affected. The f→d transitions which are at the origin of these bands are conditioned by the immediate environment of Yb ions. It appears as though Yb-S separation does not seem to change considearably.

The optical density of BK20(x=0.65) reveals a broad shoulder near 2.6 eV which can be attributed to E_2 band. A minimum around 2.4eV is followed by a steep raise. The measurements are uncertain between 1 and 1.7eV due to very high absorption. However, there is an indication of a decrease between 0.8 and 0.5eV. A concentration of 65 at% of Tm^{3+} would lead to a plasma edge around 2.5 eV in close agreement with the observation. The E_1 band is masked by the free carrier absorption in that spectral region. When the metallic(or quasi metallic) phase is approached, more and more free(or nearly free) electrons are available in the d-band for screening. It was proposed(9) that the crucial effect of the screening would be to reduce the electron-hole interactions. Any excitonic effect would then be eliminated and no structure due to f→d transitions will be seen. However, we do see f→d transitions in the metallic phase and hence rule out this explanation. Further, photo conductivity in this spectral region(in the sc state) do not seem to suggest the presence of f→d excitons(10).

c.Resistivity. The resistivity ρ of some of the samples have been measured(fig.4) using 4 point van der Pauw methd for $10 < T < 300K$.In general two activation energies are observed ΔE_1 for $T > 100$ and ΔE_2 for $T < 100$. The ρ of two YbS samples range between 2 and 7 KΩ-cm wheras $\Delta E_1 \sim 0.1$eV which is not very different from that obtained for YbTe and YbSe films(6) and YbTe single crystals(11). This represents an unidentified impurity level.Because of high ρ for $T < 100K$, ΔE_2 could not be determined.BJ-20 (x=0.02) has ρ = 44 Ω -cm with ΔE_1=63meV and ΔE_2 =25meV. Assuming a mobility of 1 $cm^2/V/sec$ for polycrystalline films and all Tm are trivalent we estimate ρ = 9mΩ-cm which is far too small compared to the measured value. This fact combined with the earlier mentioned optical absorption for the same sample indicates the absence of Tm^{3+} in the conduction band.However,

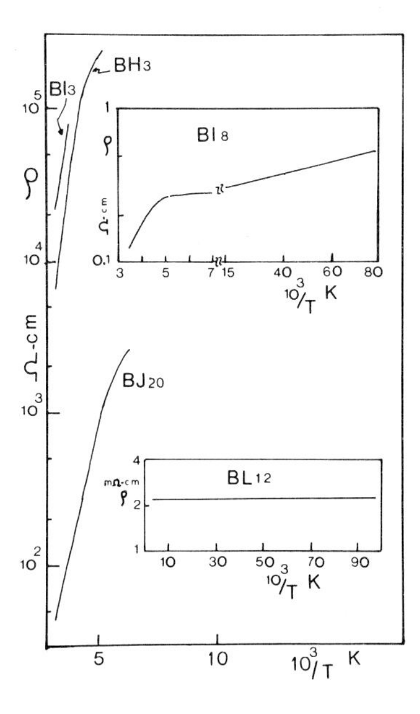

Fig.4. Resistivity ρ as a function of 1/T. BH3 and BI3 are YbS, rest see text.

the low ρ value(compared to YbS) should be related to Tm^{2+} levels lying very close to the conduction band. The ρ of BI10(x=0.18) was not measured. Instead BI8(x=0.10, a_o=5.66A) was measured which had ρ =0.1Ω-cm with E_1=0.9mev and E_2=0.2meV. The estimated value of ρ = 0.002Ω-cm a value smaller compared to the observed value leading us to similar conclusions as above. However, BL8 has ρ =70mΩ-cm and BL12 has ρ =2 mΩ-cm with both ΔE_1 and ΔE_2 = 0. The estimated value is also very close to the observed values and thus it is quite reasonable to propose that at a_o approaching 5.61Å there is a SC-MT in close agreement with the optical data. For $0.4 < x \leq 0.65$, our samples do not show further decrease in ρ but ΔE_1 and ΔE_2 are close to zero. The resistivity measurements on polycrystalline films are subject to impurities, grain boundries defects and deviations from stoichiometry. Hence we do not wish to extract more information than what it contains. However, meaningful qualitative conclusions can still be drawn as is shown by a sudden drop in ρ as x increases and ΔE_1 and ΔE_2 going to zero indicating an overall change in behaviour from a semiconductor to metallic like conduction.

4. Concluding remarks. It is proposed that Tm does not prefer a pure $4f^{12}$ state in YbS for Tm concentrations upto 0.18. In $Yb_{1-y}Sm_yS$, it was shown that an increase in Yb concentration results in shifting the 5d band to higher energies thus favouring Sm^{2+} ground state for all y (8,12). In our case the increase in Yb concentration shifts the 5d band in such a way to to push down the $4f^{13}$ (Tm^{2+}) just below the bottom of the conduction band. On the other hand, in $Sm_{1-q}Tm_qS$, for very low q=0.18, Tm succeeds in bringing down the d band resulting in the configuration cross over of Sm (12,13). It is interesting to note that these results were qualitatively predicted earlier though the author did not mention the dopant concentration at which the cross over would occur (15). Our results while verifying this prediction indicates the dopant concentration for $Yb_{1-x}Tm_xS$. A SC-MT thus takes place at x∽0.2. A simplified electronic structure for this system is proposed based on the results obtained here (fig.5). This work, we hope, will stimulate further interest in studying particularly the magnetic properties, especially for $0 < x \leq 0.2$, the case of the diluted Tm system to look at the influence of the valence of Tm complimenting the results obtained earlier on $Tm_{1-s}Y_sSe$ (2g).

We would like to thank Mrs M.Rommeluere and MrR.Moreau for the microprobe and SIMS analyses and Dr.G.Güntherodt for some helpful comments.

+Present address: CII Honeywell Bull 78340 Les Clayes sous Bois, France.

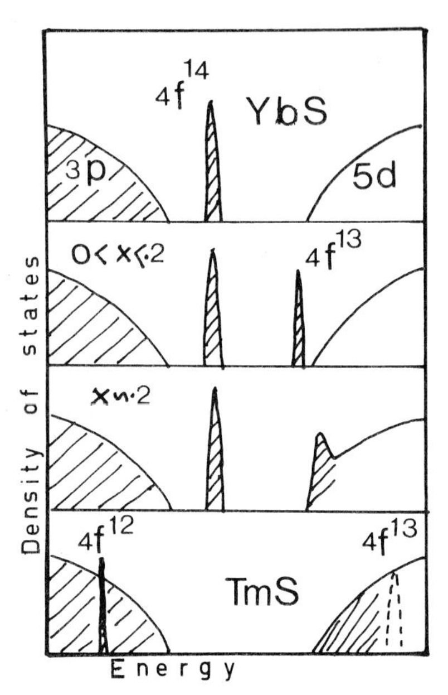

Fig.5. Proposed electronic structure of $Yb_{1-x}Tm_xS$ LHS 3p valence band; RHS 5d-6s conduction band. For Tms see ref.14.

REFERENCES

1. Extensive references can be found in a. Valence Instabilities EdR.D.Parks(Plenum Press 1977); b. Journal de Physique 40,C-5(1979) c. Journal de Physique 41,C-5 (1980); d. Valence fluctuations in Solids Ed.L.M.Falicov et al (North Holland,1981).
2. a.R.suryanarayanan ref1ap541; b.K.G.Barraclough J.Cry.Growth 41(1977)321; c.E.Kaldis et al ref1b p366 and 1cp135; d.B.Batlogg et al. ref1b,p370; ref1c,p59; e.P.Haen et al ref 1d,p313; f.U.Köbler et al. J.Mag.Mag.Mat.24(1981)34; g.F.Holtzberg et al.ref1b,p314
3. V.Narayanamurti et al. Phys.Rev.B9(1974)2521
4. E.Bucher et al. Phys.Rev. B11(1975)500
5. R.Suryanarayanan et al Thin Solid Fil 35(1976) 263; 6. idem Thesis Univ.Paris(1973) and Phys Rev.B9(1974)554.
7. P.Wachter Crit.Rev.Sol.State. 3(1972)189
8. R.Suryanarayanan Phys.Stat.Sol b85(1978)9
9. C.Mariani et al Solid State Comm 38(1981)833
10. P.Wachter and R.Suryanarayanan Unpublished
11. D.Ravot et al ref1c,p357
12. A.Jayaraman et al Phys. Rev.B19(1979)4154*
13. F.Holtzberg and R.Suryanarayanan Unpublished
14. B.Batlogg Phys.Rev. B23(1981)1827; K.Andres et al. Solid State Comm 27(1978)825
15. J.A.Wilson Structure and Bonding vol.32,p58 Ed.J.D.Dunitz et al.(Springer Verlag 1977)
*see also M. Cronau Thesis Univ. Bochum Ruhr 1978

Valence Instabilities
P. Wachter and H. Boppart (eds.)

VERY-HIGH-PRESSURE RESISTIVITY OF $CeIn_3$ AND YbCuAl

J.M. Mignot[+] and J. Wittig

Institut für Festkörperforschung der Kernforschungsanlage Jülich,
Postfach 1913, D-5170 Jülich 1, W.-Germany

The low-temperature electrical resistance of $CeIn_3$ and YbCuAl was measured as a function of pressure up to 200 kbar. In $CeIn_3$, a gradual crossover is observed from γCe-like to αCe-like R-T characteristics. Comparison is made to the related "concentrated Kondo system" $CeAl_2$ and the occurrence of a valence transition in $CeIn_3$ is inferred. In YbCuAl, pressure induces a rapid decrease of the spin-fluctuation temperature up to 80 kbar. The results are consistent with previous specific-heat and thermal-expansion data in the framework of a simple single-energy-scale Fermi-liquid model. At higher pressures, the system approaches the integral-valent Yb^{3+} state,and a new,possibly magnetically ordered,ground state is indicated.

1. INTRODUCTION

Mixed-valence (MV) materials are characterized by a small energy difference between two electronic configurations of the rare-earth element having different f counts n and n+1. Due to the screening of the nuclear potential by the 4f electrons, the ionic radius associated with the outer shells strongly depends on the 4f occupancy. In a MV compound, an unusually strong coupling is therefore to be expected between electronic and lattice properties. Of particular interest to the experimentalist is the case when the valence state of a system can be destabilized by the application of an external pressure. The most famous example is of course the black-to-gold transition of SmS. Another fascinating case is the γ-α transition of elemental cerium. Whereas the leading mechanism of this transition is still a matter of controversy, (1) the classification of many cerium compounds as γ-like or α-like appears experimentally meaningful. For those systems of the first class that are close to the transition, pathological properties can arise as a result of Kondo-type spin fluctuations. (2) In some cases, a gradual change from γ-like to α-like behavior could be achieved by use of a "chemical pressure". (3) However, since alloying can always produce undesired stoichiometry and local-environment effects, the application of an external -ideally hydrostatic- pressure appears preferable. In the case of $CeAl_2$, an electronic transition near 70 kbar (4) with a ∿ 4 % volume collapse (5) was recently reported. The present work describes very-high-pressure resistivity measurements on $CeIn_3$, another "concentrated Kondo compound" with a magnetic ground state. Similar experiments on YbCuAl are also presented and interesting correspondences between the two systems are emphasized in the discussion.

2. EXPERIMENTS

The experimental set-up (6) consisted of a low-temperature Cu-Be press with sintered-diamond Bridgman anvils. Steatite, surrounded by a pyrophyllite gasket was used as pressure-transmitting medium. Due to the brittleness of the substances investigated, it was not possible to prepare bulk samples of the desired dimension ($1\times0.1\times0.01\ mm^3$) from the available material. We therefore used powder samples prepared by crushing small fragments between WC anvils after carefully removing the corroded surface layer. No attempt was made to extract the specific resistance of the material from the experimental data since the geometrical factor of the sample could not be determined. The load was increased in steps at room temperature, then the cell was cooled nearly isobarically to liquid-helium temperature. During each low-temperature run, the pressure was determined from the superconducting transition of a strip of lead foil connected in series with the sample. The detailed procedure was described in a previous paper. (7) The experimental data were taken on the pressure-increasing cycle, where the pressure inhomogeneity, $\Delta P/P$, read from the lead monometer did not exceed 10 %. The only exception was YbCuAl below 100 kbar where $\Delta P/P$ reached 13 %.

3. RESULTS

3.1 $CeIn_3$

In this cubic $AuCu_3$-type compound, cerium occurs in an essentially trivalent state. (8,9) However, the large paramagnetic Curie temperature ($\theta \cong 60$ K) (9), the large linear heat capacity at low temperature ($\gamma \cong 130\ mJ/K^2\times mole$) (10) and the anomalous temperature dependence of the electrical resistivity (11) emphasize the fundamental role played by spin fluctuations in this system. From susceptibility (9) as well as neutron scattering data (12), a spin-fluctuation temperature T_{sf} of the order of 50 to 100 K was estimated. At 10.2 K, the system orders in a simple antiferromagnetic structure. (12) These properties have been interpreted in terms of a periodic Kondo effect, similar to the case of $CeAl_2$. (2)

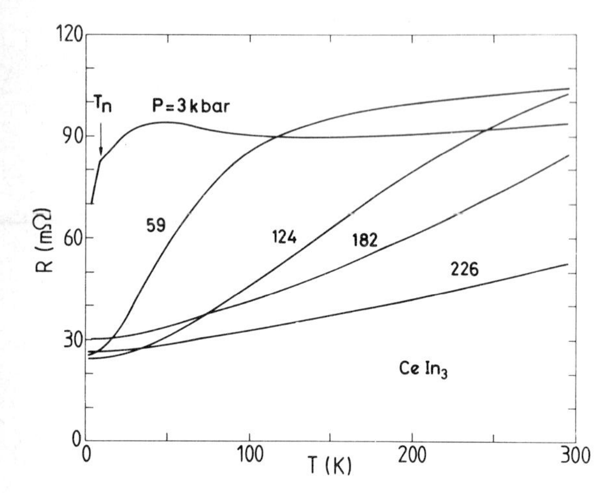

Figure 1 : Temperature dependence of the resistance of $CeIn_3$ at various pressures.

At low pressure (P = 3 kbar) our resistivity data, displayed in figure 1, are comparable to the results of previous ambient-pressure studies (11) : the negative temperature coefficient observed below 150 K is a widespread feature in systems with Kondo-type spin fluctuations. The kink near 10 K on the low-temperature side of the resistivity maximum corresponds to the onset of the AF ordering. Surprisingly enough, the resistivity ratio R(295)/R(1.5) is rather small as compared to previous results for bulk samples, but greatly increases at high pressure.

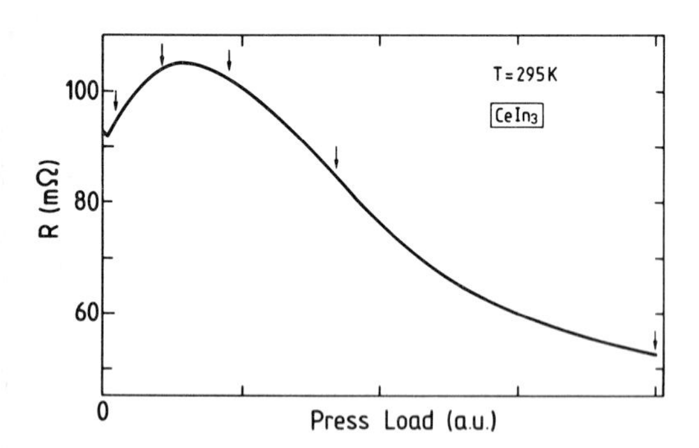

Figure 2 : Room-temperature resistance of $CeIn_3$ vs. press load (data taken during an increase of the pressure). The highest pressure is 226 kbar. The pressure at which the maximum occurs is about 85 kbar. The arrows indicate the values at which the curves of Fig. 1 have been normalized.

Under increasing pressure, the room-temperature resistance (figure 2) initially increases, goes through a maximum at an estimated pressure of 85 kbar, and then steadily decreases up to the highest pressure of 226 kbar. The present data are not corrected for geometrical effects which may contribute a substancial increase of the measured resistance at high pressure. However, a systematic comparison of several Ce compounds in the range 0-40 kbar (13) suggests that, in $CeIn_3$, the resistivity itself increases under pressure. On the other hand, the maximum in the pressure dependence of R is undoubtedly a genuine property of the material. This conclusion is reinforced by the fact that a similar feature was observed in the related system $CeAl_2$ (4) but not in the strongly MV compound $CePd_3$ (14). In the former case the anomaly is clearly connected with the aforementioned $\gamma \rightarrow \alpha$-type transition near 70 kbar. A crude comparison between the two systems suggests that, in $CeIn_3$, a similar phenomenon takes place in about the same pressure range.

In figure 1, the low-temperature resistance is reported for five different pressures. Since not all the curves were taken during the same high-pressure excursion, spurious variations of the geometrical factor had to be corrected for. This was done by normalizing each set of data to the room-temperature value obtained in one particular cycle (the one shown in figure 2) at the same applied load. Within the accuracy of the measurements, all the curves were found to be reversible upon cooling and warming. The results indicate a gradual crossover from the anomalous "Kondo lattice" regime to a rather normal ($LaIn_3$-type) behavior near 200 kbar. The whole set of curves bears close resemblance to that reported previously for $CeAl_2$. (4) It should be noted here that the Ce-Ce spacing at ambient pressure in $CeIn_3$ (d = 4.68 Å) is considerably larger than in $CeAl_2$ (d = 3.49 Å). The fact that both systems undergo a $\gamma \rightarrow \alpha$-like transition in the same pressure range clearly demonstrates that direct 4f-4f overlap (15) cannot be the leading mechanism for the corresponding delocalization of the 4f electron (14).

At 59 kbar, the R-T curve no longer shows any anomaly attributable to a magnetic-ordering transition. The whole shape of the curve is reminiscent of anomalous cerium compounds with a non-magnetic ground state. This observation qualitatively supports the view (3) that in such γ-like compounds, the magnetic ordering can be suppressed with increasing pressure prior to the valence instability. In $CeIn_3$, specific-heat measurements under pressure (16) actually showed a decrease of T_N with $[dT_N/dP] = -0.13$ K/kbar at 12 kbar. More experiments are needed to see if a precursive increase of the spin-fluctuation temperature can be associated with an Anderson-lattice-type behavior of the kind reported for CeAg. (17) Finally, we note that our data follow a T^2 law

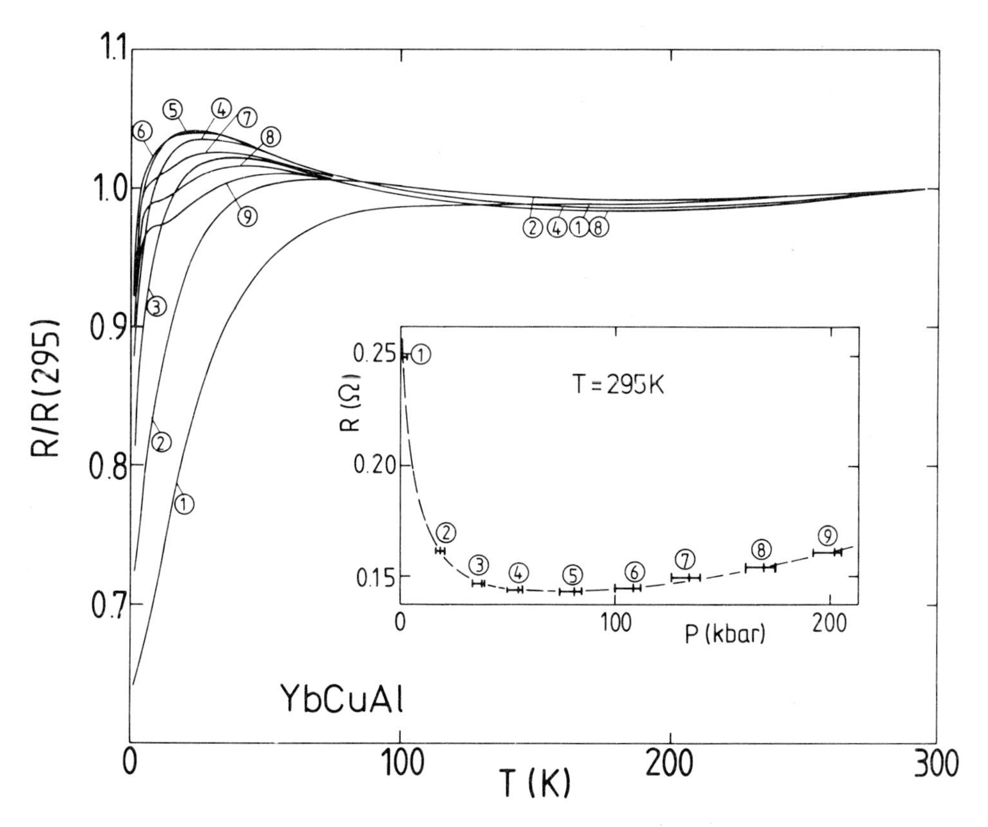

Figure 3 : Temperature dependence of the resistance of YbCuAl at various pressures (from 1 to 9) : 1.5, 19, 38, 55, 81, 108, 134,169 and 202 kbar. Inset : Room-Temperature resistance vs. pressure.

up to 18 K at 59 kbar and 35 K at 124 kbar, pointing to a Fermi-liquid picture of the low-temperature ground state.

3.2 YbCuAl

YbCuAl crystallizes in the hexagonal Fe_2P structure. Extensive lattice-parameter and susceptibility measurements (18,19) have demonstrated that the Yb valence is very close to 3 at room temperature and varies only by a few percent down to liquid-helium temperature. However, the low-temperature properties are typical of a MV compound : i) absence of magnetic ordering, ii) finite susceptibility at T = 0 ($\chi(0) = 3.3\times10^{-7}$ m^3/mole), and iii) large electronic specific heat (γ = 255 mJ/mole). The temperature of the susceptibility maximum is T_{max}=28 K. This temperature can be taken to represent the characteristic energy of the spin fluctuations (20). The two neighboring 4f configurations involved in the valence fluctuations of ytterbium compounds are $4f^{13}$ (Yb^{3+},J=7/2) and $4f^{14}$ (Yb^{2+},J=0). Therefore, in contrast to cerium, ytterbium is magnetic in its higher valence state. Previous susceptibility measurements under moderate pressures (18,21) support the simple view that ytterbium can be made more magnetic by squeezing out one of the 14 f electrons. Due to the relatively low value of T_{max} already at ambient pressure, the possibility exists that the valence transition can be completed within our experimental pressure range.

Figure 3 shows the temperature dependence of the resistance of YbCuAl up to 200 kbar. Since most of the pressure dependence of R(295 K) could be attributed to geometrical effects, the different curves were normalized to the same room-temperature value.

At low pressure, our results are in agreement with previous measurements: the salient features are a quasi-constant resistance down to 100 K, followed by a fall off at low-temperatures which becomes very pronounced below 50 K. Under increasing pressure a shallow minimum appears near 150 K and the resistivity goes through a rounded maximum below 50 K. This maximum shifts rapidely to lower temperatures and reaches 22 K at 81 kbar. Curves 1 and 2 can be roughly scaled as a function of a reduced variable (T/T*) with the characteristic temperature T* decreasing by ≅ 60% between 1.5 and 19 kbar ($(1/T^*)dT^*/dP \cong -34\ 10^{-6}\,bar^{-1}$). With the initial value of T_{max}=28 K derived from the susceptibility maximum, a pressure coefficient of ≅ -1 K/kbar is estimated, in surprisingly good agreement with the low-pressure results of Mattens (18).

In this simple single-energy-scale picture, the logarithmic pressure derivative of T* can be expressed as a function of the molar electronic specific heat C_p, the electronic part of the linear thermal expansion α and the molar volume Ω (22): $(1/T^*)dT^*/dP = \pm 3\alpha\Omega/C_p$. For YbCuAl, Pott et al.(19) find $-29\ 10^{-6}\ bar^{-1}$ at the lowest temperatures. This estimate also agrees with the pressure dependence of the electronic specific heat (23) : $(1/\gamma)d\gamma/dP = +\ 50\ 10^{-6}\ bar^{-1}$. At 80 kbar, a rough analysis of the R-T curve yields a T* value of order 1 K. The system is therefore very close to a magnetic state.

Above 80 kbar, the resistivity maximum shifts back to higher temperatures (≅ 50 K at 202 kbar) and a shoulder appears on the low-temperature side of the maximum. In view of the striking analogy with the ambient-pressure (Kondo-lattice-like ?) resistance of $CeIn_3$, we suggest that these features may reflect the appearance of a magnetic ground state of Yb associated with the Yb^{3+} configuration (see figure 4).
If the low-temperature shoulder is indeed attributable to magnetic ordering, the temperature where it occurs is consistent with the estimates based on de Gennes factors for trivalent RCuAl compounds (24).

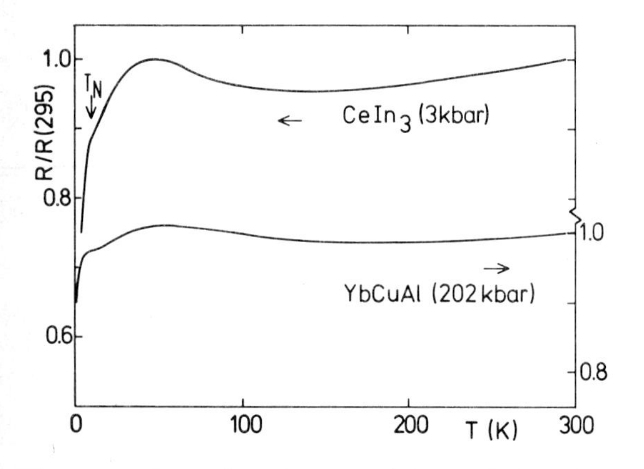

Figure 4. Comparison between $CeIn_3$ and YbCuAl

4. CONCLUSION

The work reported in this paper shows the potential of high-pressure resistivity measurements for the study of MV compounds. The main conclusions are the existence of a $\gamma \rightarrow \alpha$ -like transition in $CeIn_3$ and the appearance of a new (probably magnetic) ground state in YbCuAl above 80 kbar. These results should be supplemented by use of high-pressure X-ray-diffraction and magnetic-susceptibility measurements. Finally, the present method should be extended to a larger number of ambivalent RE elements.

ACKNOWLEDGEMENTS

We are indebted to J. Palleau and W.C.M Mattens for providing samples of $CeIn_3$ and YbCuAl, respectively.

REFERENCES

+ Permanent address: Centre de Recherches sur les Très Basses Températures, CNRS, Grenoble, France.

(1) Koskenmaki, D.C. and Gschneider, K.A.,Jr., in Gschneider, K.A.,Jr., and LeRoy Eyring (eds.) Handbook on Physics and Chemistry of Rare Earths, Vol. 1 (North-Holland, Amsterdam, 1978).
(2) Benoît, A., Boucherle, J.X., Flouquet, J. Holtzberg, F., Schweitzer, J. and Vettier, C. in Falicov, L.M., Hanke, W. and Maple, M. B. (eds.),Valence Fluctuations in Solids (North-Holland, Amsterdam,1981) p 197.
(3) Levine, H.H. and Croft, M.,Ibid.2 p 279 Aarts, J., de Boer, F.R., Horn, S., Steglich, F. and Meschede, D., Ibid. p 301
(4) Probst, C. and Wittig, J., J. Magn. Magn. Mat. 9 (1978) 62.
(5) Croft, M. and Jayaraman, A. Solid State Commun. 29 (1979) 9.
(6) Wittig, J.and Probst, C., in Chu, C.W. and Woollam, J.A. (eds.), High Pressure and Low Temperature Physics (Plenum, New York,1978), p 433.
(7) Wittig, J., Z. Phys. B 38 (1980) 11.
(8) Harris, I.R. and Raynor, G.V., J. Less-Common Metals 9 (1965) 7.
(9) Lawrence, J.M.,Phys. Rev. B 20 (1979) 3770.
(10) Van Diepen,A.M., Craig, R.S. and Wallace, W. E., J. Phys. Chem. Solids 32 (1971) 1867.
(11) Van Daal, H.J.and Buschow, K.H.J., Phys. Stat. Sol. (a) 3 (1970) 853; Elenbaas, R.A., Schinkel, C.J. and van Deudekom, C.J.M., J. Magn. Magn. Mat. 15-18 (1980) 979.
(12) Lawrence, J.M. and Shapiro, S.M., Phys. Rev. B 22 (1980) 4379 .
(13) Croft, M., Levine, H.H. and Neifeld, R. in Schilling, J.S. and Shelton, R.N. (eds.), Physics of Solids under High Pressure (North-Holland, Amsterdam, 1981) p 335.
(14) Mignot, J.M. and Wittig, J., Ibid. 13, p 311.
(15) Hill, H.H. in Miner, W.N. (eds.),Plutonium 1970 and other Actinides (The Metallurgical Society of the AIME , New York, 1970), p 2.
(16) Peyrard, J. Doctorat d'Etat Thesis, Grenoble Univ. (June 1980).
(17) Eiling, A. and Schilling, J.S., Phys. Rev. Lett. 46 (1981) 364.
(18) Mattens, W.C.M., Thesis, Amsterdam Univ. (1980)
(19) Pott, R.,Schefzyk, R. and Wohlleben, D., Z. Phys. B 44 (1981) 17.
(20) Lawrence,J. and Beal-Monod, M-T.,Ibid.2 p 53
(21) Zell, W., Pott, R., Roden, B. and Wohlleben, D. , Pressure and Temperature Dependence of the Magnetic Susceptibility of some Ytterbium Compounds with Intermediate Valence (to be published).
(22) Benoît,A., Berton, A., Chaussy, J., Flouquet, J., Lasjaunias, J.-C., Odin, J., Palleau, J., Peyrard, J. and Ribault, M., Ibid.2 p 283.
(23) Bleckwedel, A. and Eichler, A., Ibid. 13 p 323
(24) Mattens, W.C.M.,private communication.

Valence Instabilities
P. Wachter and H. Boppart (eds.)

EFFECT OF ENVIRONMENT ON THE VALENCE OF Ce

J.G. Sereni

Centro Atómico Bariloche* - Instituto Balseiro#
8400 Bariloche - Argentina

By means of a semi-empirical analysis, the Electrostatic Energy of a charge immersed in the screened Coulomb Potential of the total charge distribution of the metal (E_ψ), was selected among other parameters as the most proper one to correlate different Ce environments. The choice was made by looking at the smallest dispersion in a plot of some physico-chemical parameters versus the rate of depresssion of the superconductive transition temperature of Th-based alloys with Ce impurities.
The analysis was extended to some AB_j compounds (A = elements of the II through V column including Ce and Th, and B_j=N, Sn_3, Tl_3, Rh_3, Co_2, Ir_2, B_6 and Be_{13}). An empirical relationship between E_ψ and the Volume contraction due to the compound formation ($\Delta V/V$) was found.

1. INTRODUCTION

The valence of Ce (Z^*), in different alloys or compounds depends mainly on such parameters as pressure, temperature and the electronic and structural environment. The external variables, pressure and temperature, are able to shift Z^* over some tenths of its total possible range ($3 \leq Z^* \leq 4$), except in those cases in which the system undergoes a first order transition as in pure Ce. Varying the environment through alloying can lead Ce ion to all its possible valence values, in which it retains, on the average, anywhere from one to no electron in its 4f shell [1].

We can compare different environments by looking at the elements able to accede to them. In fig.1, we have ordered some families of isostructural AB_j compounds as a function of the Ce magnetic contribution in the CeB_j phase, normalized with the expected contribution from a trivalent Ce ion at 300 K. The A elements along the Y-axis are ordered following a parameter log E_ψ(A) which will be discussed later. The vertical line for each family extends over the range of A elements for which the AB_j forms. The upper part of the figure shows the valence regime and the magnetic behaviour of Ce in the phases considered below.

The parameters used for comparing different environments are: Electronegativity (ε), introduced by Pauling to denote the tendency of atoms to attract electrons to themselves, the Work Function (Φ), the Electron Density in the Wigner-Seitz cell (n_{WS}) and the Screening Length (λ^{-1}).

Another parameter which we want to take into account is the energy of a charge (q) immerse in the screened Coulomb potential of the total charge distribution of the metal, $E_\psi = q\psi$ [2].

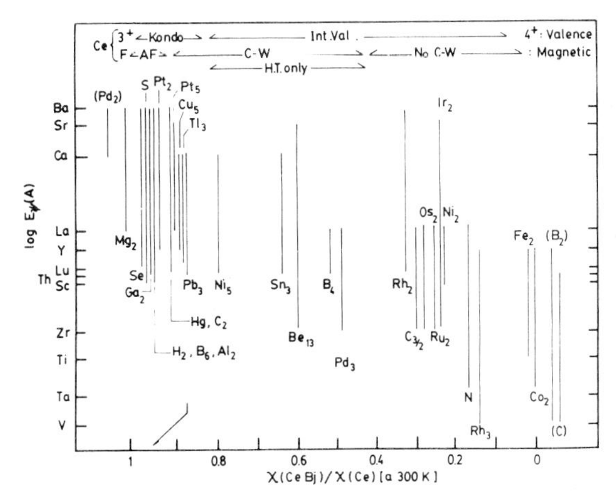

Figure 1: Families of isostructural AB_j compounds ordered as a function of the normalized Ce magnetic contribution in the CeB_j phases.

A simple approximation, useful for our correlation work, is the usual screened potential:

$$\psi \propto \frac{Ze}{r} \exp(-\lambda r) \qquad (1)$$

where: Z = valence of the metal, e = electron charge, r = Goldschmit 12-fold coordination atomic radius and $\exp(-\lambda r)$ = the Thomas-Fermi screening factor. The found correlations justify the use of this approximation.

We now need a physically measurable property, sensitive to the chemical environment, which allows us to select the most appropriate parameter from those proposed above. Such a property is provided by the valence of an element (like Ce, Yb, U, etc) that precisely depends on the chemical environment.

2. ALLOYS

When a paramagnetic impurity is introduced into a superconducting matrix, the transition temperature (T_{co}) is particularly affected by the magnetic moment of the impurity [3]. We can use that property by taking Ce impurities (n_i) as a probe for comparing the chemical environment in different matrices. These matrices are superconducting Th-based alloys in the system Th_iX_j, with i+j=1, and X=La, Y, Sc, Lu, Zr or Hf [4]. The magnetic moment of Ce impurities in these matrices is weakened because of the strong admixture between the localized and conduction electron states. In this case the depression of reduced T_c can be fitted by a modified exponential function: $T_c/T_{co}=\exp[-\alpha n_i/(1-\delta n_i)]$ first proposed by Kaiser [5].

In fig. 2 we show the relationship between the logarithm of the initial rate of depression of reduced T_c: $\alpha = 1/n_i \cdot \mathrm{Ln}(T_c/T_{co})$ (with $n_i \to 0$) and E_ψ. Of all the parameters mentioned above, we find that a E_ψ plotted against the measured $\alpha^{-1/2}$ shows the smallest dispersion.

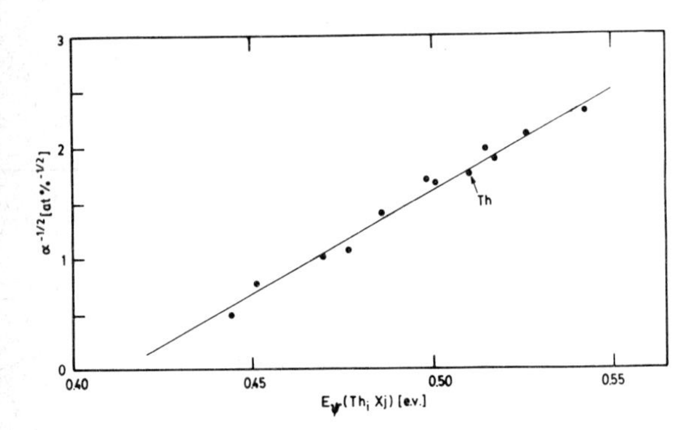

Figure 2: Relationship between "α" and "E_ψ" calculated for the Th_iX_j matrices from eq(2) (X_j = La_j, Y_j, Lu_j, Sc_j, Zr_j, Hf_j and pure Th). "α" is taken from ref.4.

The value of E_ψ for any Th_iX_j alloy is estimated as the weighted average of the values for the components

$$E_\psi(Th_iX_j) = i\,E_\psi(Th) + j\,E_\psi(X) \qquad (2)$$

The valence of Ce in the Th_iX_j alloys can be deduced by assuming that a Ce atom in the matrix will adapt its valence (Z^*) and volume such that its energy in the electrostatic field ($E_\psi(Ce) = Z^*\psi$) equals that of an average matrix atom (M), i.e. $E_\psi(Ce)=E_\psi(M)$. Concerning the evaluation of ψ using eq.(1), Z and λ are taken to have the matrix values because these depend on the valence and the electron density of the entire alloy. The interatomic distance (d) may, however, be locally distorted because the Ce volume in that environment may differ from that average for a matrix atom. We must use the value $d=r^*(Ce)+r(M)$, where r(M) is the average atomic radius of the matrix and $r^*(Ce)$ is a function of Z^* that can be evaluated by using the Gschneidner relationship $Z^* = 13,61 - 5,75\, r^*(Ce)$ [6]. The equation for Z^* results in:

$$Z^* = \frac{E_\psi(M)}{14.4} \cdot \frac{d}{Ze^2} \exp(\lambda d/2) \qquad (3)$$

in the case of ThCe alloy we obtain $Z^* = 3.6$, which is in agreement with the magnetic moment of Ce in that alloy at room temperature, taken from ref.7.

We emphasize that the assumed equilibrium condition used here is the invariance of E_ψ when a matrix atom is replaced by a Ce one.

3. COMPOUNDS

Once we have verified E_ψ as the most appropriate parameter to be used for comparing the chemical environment of Ce in alloys, we can attempt to extrapolate this analysis to a family of compounds that includes a Ce compound. As a family of compounds (AB_j) we consider all the compounds which form with A being an element of the II through V column including also Ce and Th, while the B partner is fixed.

Except in a few cases, it is not possible to know precisely the valence which the partner B assumes in the compound, then we cannot estimate E_ψ as expressed in eq.(1). Nonetheless, while absolute values for $E_\psi(AB_j)$ may not be available, we can compare the $E_\psi(Ce)$ value with the values of $E_\psi(A)$ for the other A elements in the same family of compounds. We are considering the unknown value of $E_\psi(B_j)$ to be constant for a single family of compound. In terms of eq.(2) we can express this as follows $E_\psi(A_1B_j)-E_\psi(A_2B_j) = E_\psi(A_1)-E_\psi(A_2)$. Different partners will simply shift the scale of interest.

For a given AB_j family, we can calculate $E_\psi(Ce)$ if we suppose Ce in the tri or tetravalent state (where we know Z^*), but it remains unknown if Ce is in an intermediate valent state. To estimate $E_\psi(Ce)$ in those cases we have to look for a measurable parameter $\beta(AB_j)$ which, once related with $E_\psi(A)$, allows us to estimate $E_\psi(Ce)$ from the measured $\beta(CeB_j)$ value. We first check this procedure with β = standard heat of formation (ΔH_f) of the nitride family (AN).

In fig. 3a we display $\Delta H_f(AN)$ vs log $E_\psi(A)$, taken from ref. 8 and we propose a full curve that represents the relationship between these parameters, i.e. $\Delta H_f(AN)=f[E_\psi(A)]$. By using that relationship we can estimate $E_\psi(Ce)$ in

CeN from the point where the $\Delta H_f(AN)$ curve reaches the $\Delta H_f(CeN)$ value.

In the upper part of fig.4a we display the valence of Ce, $Z^*(Ce)$, that corresponds to the log E_ψ values of the x-axis.

Because ΔH_f values are lacking for other families of compounds, we must replace it by a parameter that can be estimated for all compounds under study. Such a parameter is the percentage volume contraction ($\Delta V/V$) due to the compound formation. It is defined:

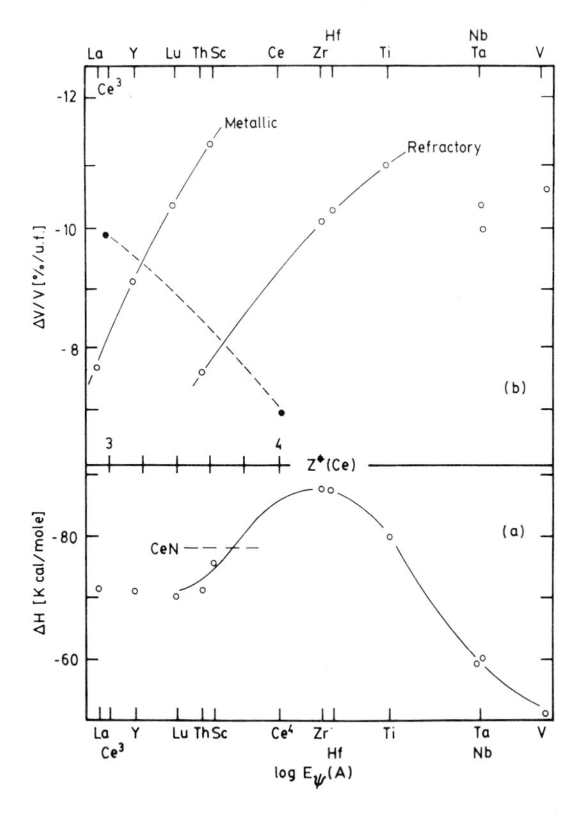

Figure 3a: Standard Heat of formation, H_f, of the nitride family (AN) vs.log $E_\psi(A)$. (o) are the values for each AN compound. $Z^*(Ce)$ is the Ce valence corresponding to the values of log.$E_\psi(A)$.

Figure 3b: Volume contraction, $\Delta V/V$, at the compound formation vs.log. $E_\psi(A)$. The solid lines are the trend of $\Delta V/V$ for the compound, and the deshed line are the values calculated for CeN as a function of $Z^*(Ce)$.

$$B = \frac{\Delta V(AB_j)}{V} (\%) = 100 \left[\left(\frac{V_{AB}}{V_A + V_B} \right) - 1 \right] \quad (4)$$

and it is related to the magnitude of ΔH_f [9].

In fig.3b, we apply this procedure to the nitride family. We note that the volume contraction $\Delta V/V$ versus $E_\psi(A)$ reaches a maximum at $E_\psi(Sc)$, then at greater $E_\psi(A)$ values it appears another branch that corresponds to the refractory nitrides [10].

In the figure we plot $\Delta V/V$ for CeN versus log $E_\psi(Ce)$ as the loci of points for which the coordinates are varied continuously between their values for tri and tetravalent Ce. Then we take the point where that locus crosses the log $E_\psi(AN)$ versus $\Delta V/V$ curve to extract an effective $E_\psi^*(Ce)$ value. In the case of AN family the two branches give two possible $E_\psi^*(Ce)$ values, we choose $E^*_\psi(Ce)$ that corresponds with $Z^* = 3.67$ which is closer to $Z^* = 3.70$ obtained from ΔH_f. The Ce valence estimated from magnetic measurements is 3.65 if we consider E_{ex}=2100 K and Δ~ 1000K in the model by Alascio et al.[11]. We have to note that the E_ψ difference between the two crossing points measured in electron volts units is ~0.16 ev (~1800K).

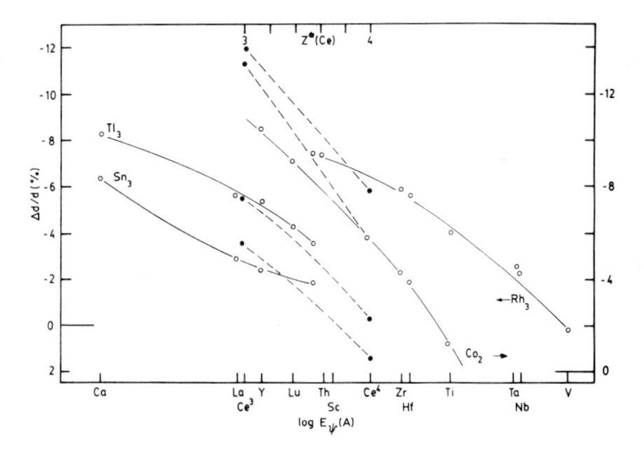

Figure 4: Interatomic distance contraction, $\Delta d/d$, vs.log $E_\psi(A)$ of some cubic compounds. The dashed lines are the values of $\Delta d/d$ calculated as function of the Ce valence, $Z^*(Ce)$. The crossing points with the solid line indicate the Ce valence in each compound.

In fig. 4 we repeat these procedures, now considering the atomic nearing ($\Delta d/d$) versus log $E_\psi(A)$ in some cubic compounds like: ASn_3, ATl_3, ARh_3 and ACo_2. We use, in this case, $\Delta d/d$ instead of $\Delta V/V$ because those values show a significative smaller dispersion in their relationship with log $E_\psi(A)$. The Ce valence in those compounds is estimated to be: 3.3, 3.0, 3.8 and 4.0 respectively, in agreement with, the values given in the literature. We note that the percentual contraction of d is never greater than 10%, as expected.

In fig. 5 we analyze the systems:AB_6, AIr_2 and ABe_{13}, in which the Ce valence results to be

3.0, 3.5 and 3.4 respectively. In that figure the $\Delta V/V$ versus log $E_\psi(A)$ plot shows two branches in the case of the ABe_{13} family, one for the elements A' = La and Th and the other for A" = Y, Gd, Lu, Sc and Zr. Such a striking fact suggests a difference between the electronic band structure of the $A'Be_{13}$ group with respect to the $A''Be_{13}$ one. This fact is in agreement with the difference observed by Kappler [12] in the Curie temperature (θ) of the systems $(Ce_xM_{1-x})Be_{13}$. In the limit of dilute Ce in non magnetic matrices (M = La,Th, Lu and Y) their observed values are: $\theta \sim -140K$ for A"= Lu and Y, and $\theta \sim -70K$ for A'= La and Th.

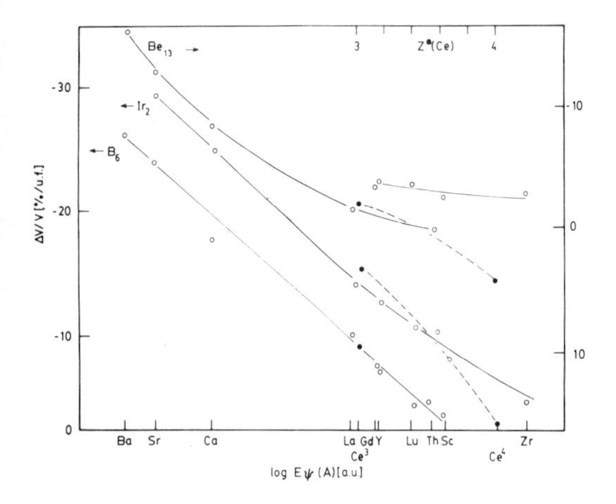

Figure 5: Volume contraction, $\Delta V/V$, vs.log $E_\psi(A)$ of some cubic compounds. The dashes lines are the values of $\Delta V/V$ calculated as function of the Ce valence, $Z^*(Ce)$. The crossing points with the solid line indicate the Ce valence in each compound.

4. CONCLUSIONS

By means of a semi-empirical analysis we have shown that there is a parameter, E_ψ,able to be estimated quantitatively, that allows to correlate the elements of the II through V column of the periodic table, including Ce and Th. Such a parameter is a simple approximation for the "Electrostatic Energy" of a charge inmerse in the screened Coulomb potential of the total charge distribution of the metal. That parameter, which interpretation in terms of microscopic parameters remains an open question, was selected among others because it shows the smallest dispersion with respect to a functional dependence on a measurable property which depends strongly on the environment. This property is the initial rate of depression of the superconductive transition temperature, α.

It was also shown that the volume (distance) contraction, $\Delta V/V(\Delta d/d)$, is a proper parameter to relate the energies involved in the formation of compounds belonging to the same family. The different, positive and negative,curvatures of $\Delta V/V(\Delta d/d)$ vs. log $E_\psi(A)$ relationship cannot be explained from our empirical systematic. The valence and environment relationship can be considered in both senses: one to analyse the valence changes of the unstable valent elements, and the other to compare different environments.

The present study, that introduces the "alloying" as the primary parameter in the analysis, can be extended to all Ce compounds that belongs to a family of compounds with enough known members.

FOOTNOTES

* Comisión Nacional de Energía Atómica.
# Comisión Nacional de Energía Atómica y Universidad Nacional de Cuyo.

REFERENCES

[1] See for example: J.M. Lawrence, P.S. Riseborough and R.O. Parks, Rep.Prog. Phys. 44, 1 (1981).
[2] J.P. Abriata and M.M. Tarchitzky, Scripta Metall. 9, 1003 (1975).
[3] M.B. Maple, Applied Phys. 9, 179 (1976).
[4] J.G Huber and M.B.Maple, "Low Temperature Physics 13", Ed. by K.O. Timmerhaus, W.J. O'Sullivan and E.F. Hammel (Plenum Pub.Corp.) 2, 579 (1972).
[5] A.B. Kaiser, J. Phys. C3, 409 (1970).
[6] K.A.Gschneidner, Jr. and R. Smolnshowski, J.Less.Comm.Metals, 5, 374 (1963).
[7] J.G.Huber, J.Brooks, D.Wohlleben and M.B. Maple, AIP Conference Proc., Ed. by C.D. Graham Jr., G.H.Landes and J.J.Rhyne, N° 24, p. 475 (1974).
[8] J.Kordis and K.A. Gingerich, J.Nucl.Mat., 66, 197 (1977) and A.R. Miedema, J.Less Comm.Met. 46, 67 (1976).
[9] O.Kubaschewski, E.LL.Evans and C.B. Alcock, "Metallurgical Thermochemistry", (Pergamon Press) 1967, p. 228.
[10] "CRC-Handbook of Chemistry and Physics", Ed. by R.C. West, 60th Edition, p.D-51.
[11] B.Alascio, A.Lopez and C.Olmedo, J.Phys. F3, 1324 (1973).
[12] J.P.Kappler, Ph.D. Thesis, University of Strasburg, Dec. 1980, unpublished.

Valence Instabilities
P. Wachter and H. Boppart (eds.)

CHEMICAL PRESSURE EFFECTS IN Sc-SUBSTITUTED YbCuAl

W.C.M.Mattens, J.Aarts, A.C.Moleman, I.Rachman and F.R. de Boer

Natuurkundig Laboratorium der Universiteit van Amsterdam,
Valckenierstraat 65, 1018 XE Amsterdam, The Netherlands

Implications of Sc-substitution for nearly trivalent Yb in YbCuAl will be discussed. Magnetic properties, thermal expansion and specific heat of compounds with Sc-concentrations upto 20 percent will be presented. A possible occurrence of spin-fluctuations growing with increasing Sc-concentration is suggested.

INTRODUCTION

In a number of experiments over the past few years it has been established that dilution of intermediate valence (IV) compounds with non-magnetic atoms only produces a proportional reduction of the bulk properties of the IV state |1,2|. In other words, valence fluctuations at different ions seem to be incoherent and might be described as a single ion effect. Since, therefore, dilution does not disturb any coherency, the same experimental technique might be used to study the effect of chemical pressure upon the IV state. A very suitable system for this purpose is YbCuAl. It is intermediate valent and dilution with Y has been studied extensively |1|. Furthermore, Sc substitution may provide the sought-for chemical pressure effects since Sc is chemically identical to the rare earths, but has a significantly smaller ionic radius than trivalent Yb. Although the compound ScCuAl does not exist, we have found in the present work that single phase $Yb_{1-x}Sc_xCuAl$ samples can be produced up to x = 0.22.

EXPERIMENTAL RESULTS

A detailed description of the sample preparation procedure can be found in ref. |3|. In the $Yb_{1-x}Sc_xCuAl$ system compounds with x = 0.00, 0.05, 0.075, 0.10, 0.15, 0.20 and 0.22 have been prepared. The compounds crystallize in the hexagonal Fe_2P-type structure, in which every Yb atom has the same local environment. The room temperature lattice parameters are given in fig.1.

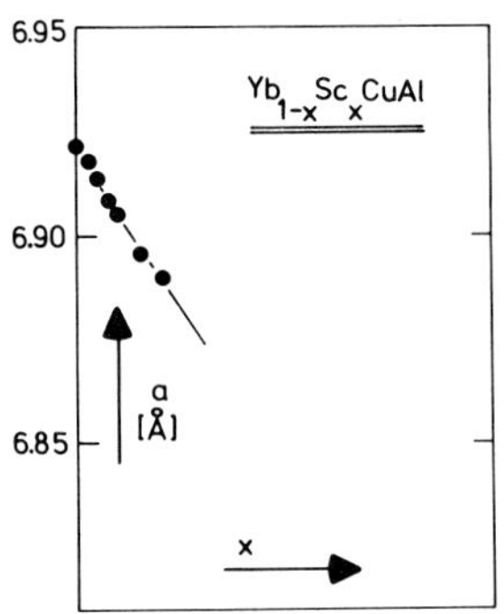

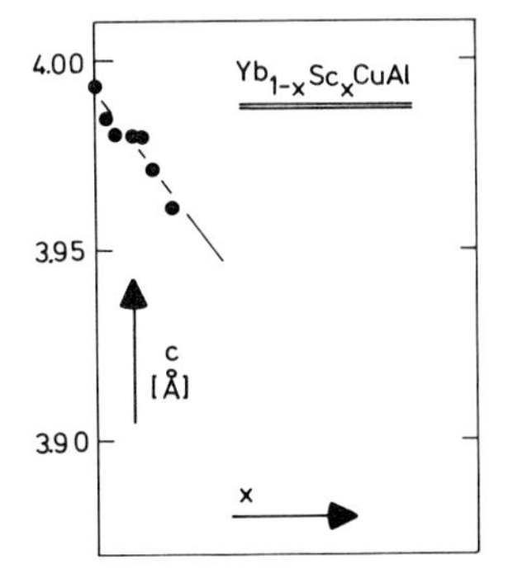

Fig.1.

Thermal expansion data by X-ray diffraction at low temperatures for various Sc-concentrations

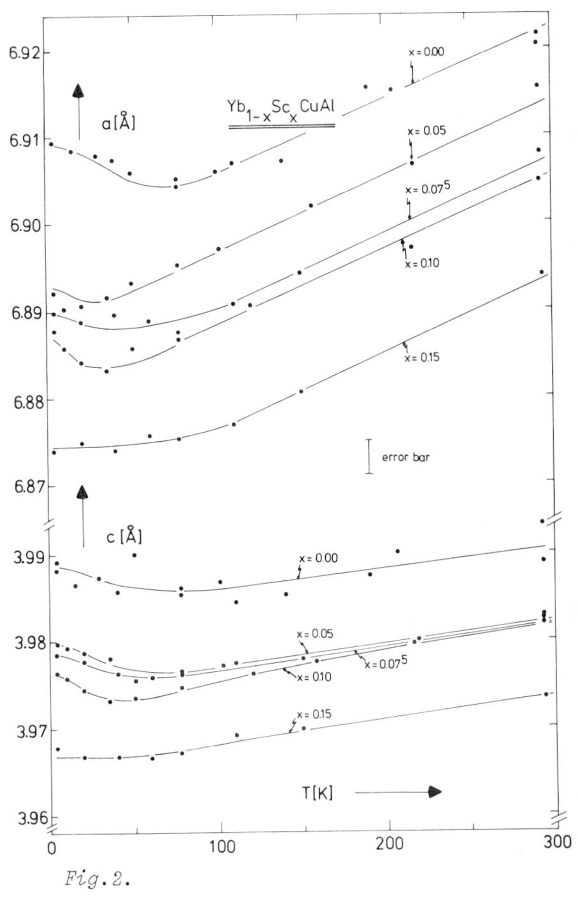

Fig.2.

and also for pure YbCuAl are given in fig.2. The Sc-substituted compounds reveal a behaviour, similar to what earlier has been found for pure YbCuAl |4|: a smaller but clear anomaly is observed, shifting towards lower temperatures upon increasing Sc-substitution. In YCuAl, a non-magnetic dummy, a normal thermal expansion has been observed. At room temperature we found the same slope as for YbCuAl, while below liquid nitrogen temperature the lattice parameters become temperature independent.

In fig.3 as-measured susceptibility curves are given. The susceptibilities given in fig.4 have been corrected for impurity contributions (of the order of about 1% trivalent Yb). It can be seen that the characteristic temperature T_{max}, the temperature of the well known susceptibility maximum, shifts towards lower tempera-

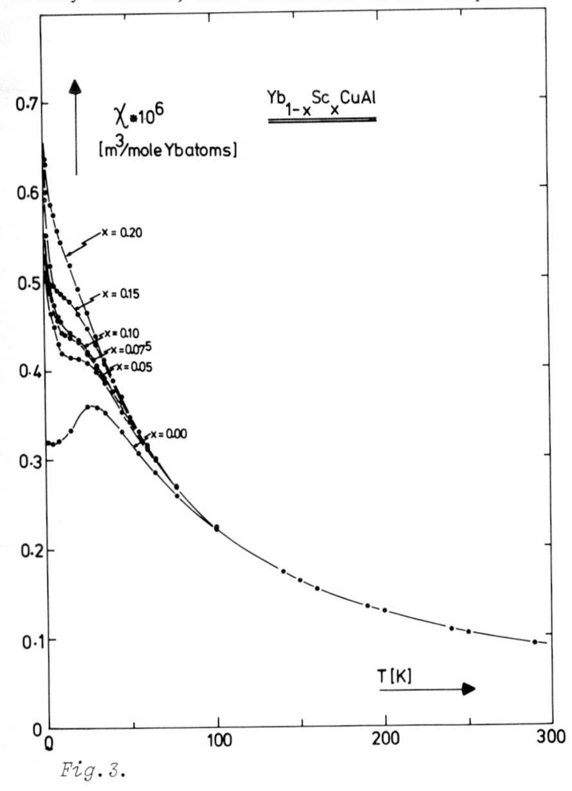

Fig. 3.

tures and the low-temperature susceptibility increases gradually. At high temperatures (up to 1000 K) we find Curie-Weiss behaviour with an effective moment of 4.30 ± 0.05 μ_B/Yb atom and a paramagnetic Curie temperature of 33 ± 5K for all Sc-concentrations, the same as for pure YbCuAl and Y- and Gd-substituted YbCuAl |1|.

The magnetic isotherms at liquid helium temperatures up to 35 tesla are given in fig.5. In fig.6 we again present the magnetization, now corrected for the impurity contributions, which is saturated at 5 T. For x = 0.05 the magnetization is still similar to that of pure YbCuAl. For the higher Sc-concentrations, however, the characteristic shape (upward curvature in the high-field part) vanishes and at the highest Sc-concentrations the curves show a gradually saturating behaviour. If we identify the slope of the low-field part of the magnetization with the initial susceptibility, we find good agreement with the low-temperature susceptibility

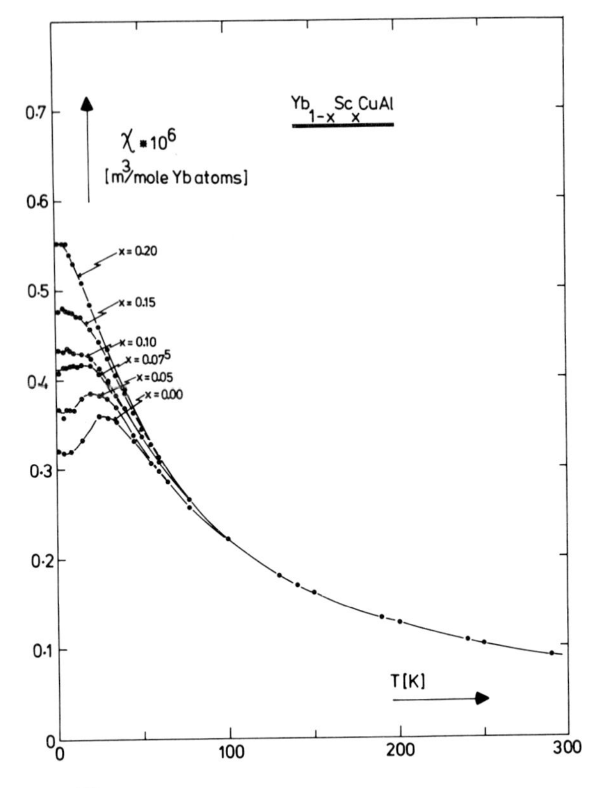

Fig. 4.

values presented in fig.4, which justifies the applied correction procedure.

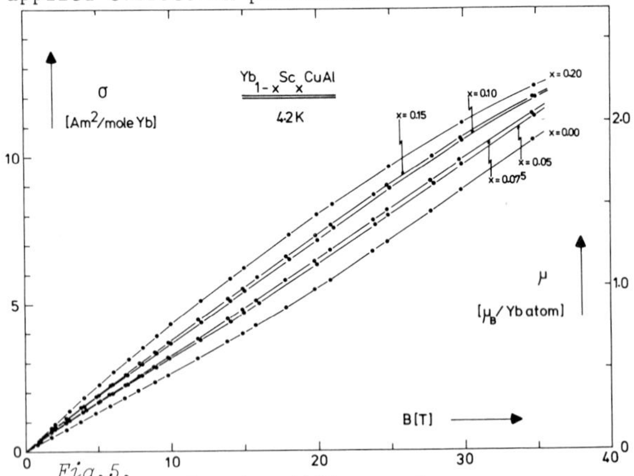

Fig. 5.

The heat capacity for different Sc-concentrations is given in fig.7. It can be seen that the properties have changed drastically with respect to those of YbCuAl. For the x = 0.10 compound the coefficient of the linear term, γ, can be estimated to be about 370 mJ/K^2 mole Yb atoms, compared to 255 mJ/K^2 mole Yb atoms for pure YbCuAl. This relative change is similar to that in the low-temperature susceptibility. For the higher Sc-concentrations γ cannot be determined in a simple way. Furthermore the heat capacities are found to be field-independent at least up to 5 T.

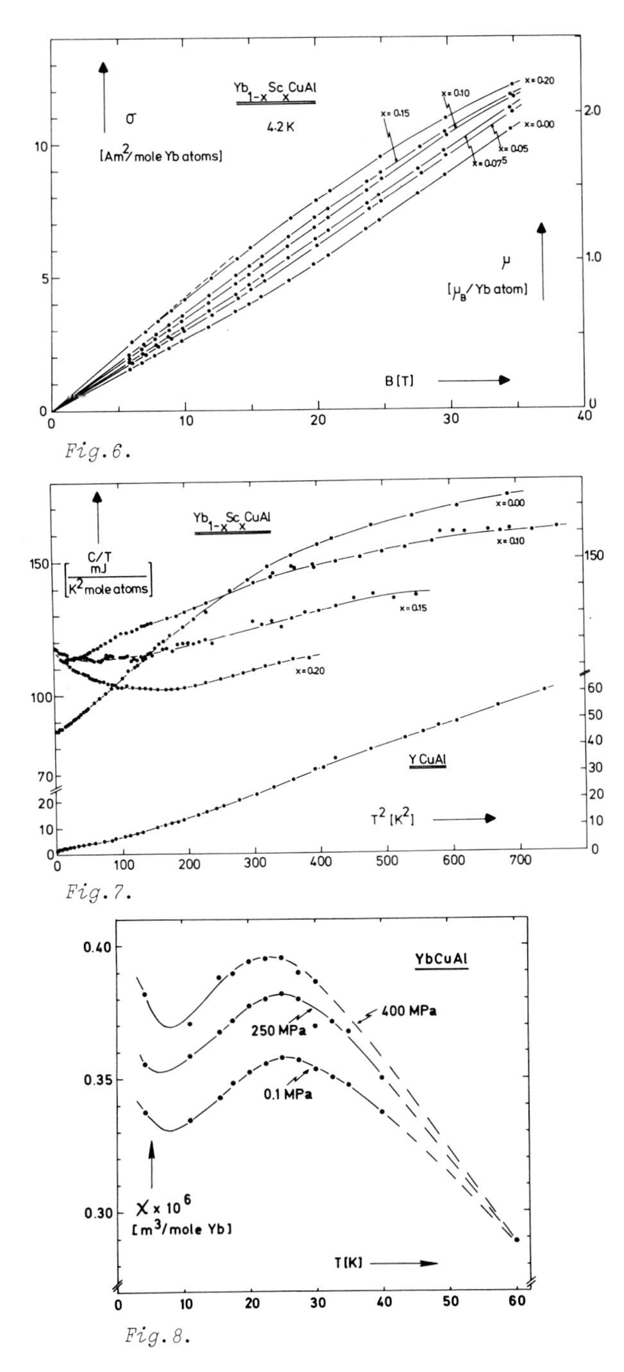

Fig.6.

Fig.7.

Fig.8.

DISCUSSION

The most direct indication of pressure effects due to Sc-substitution would be lattice parameters deviating from Vegard's law. In this case it is not possible to establish this due to the absence of compounds with $x > 0.22$ (see fig.1). On the other hand we can compare the observed magnetic properties with the results of measurements of the magnetic susceptibility of pure YbCuAl when external pressure is applied. In fig.8 these measurements are given. The results show that T_{max} shifts towards lower temperatures when pressure is applied, while the value of the constant low-temperature susceptibility increases. From the susceptibility curves in fig.3 we can conclude that substitution of Sc has a similar effect as application of external pressure. One may object against the use of the impurity-correction procedure in the case of Sc-substituted compounds. It may be that here, due to shift of the maximum towards lower temperatures, the low-temperature susceptibility is no longer temperature independent below 8 K. However, since the χT vs T plots are still linear up to this temperature, it can be assumed that the susceptibility is fairly constant.

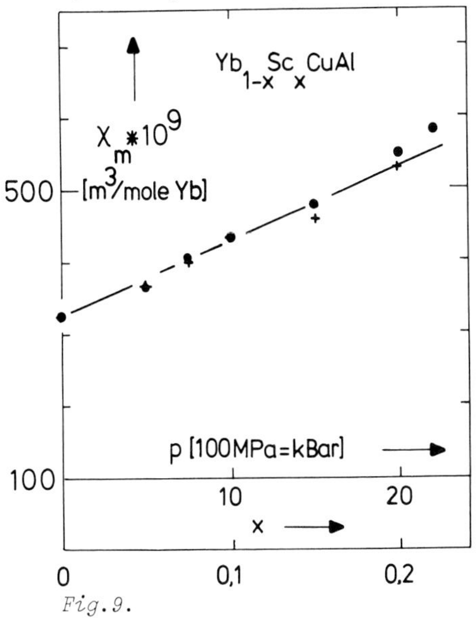

Fig.9.

It has been found (fig.8) that the low-temperature susceptibility of pure YbCuAl increases with 3 % per 100 MPa (1 kBar) |4|. Assuming that this value is independent of pressure, which is probably a very crude approximation, we can make an estimate of the pressures involved with Sc-substitution in YbCuAl. The results of such a procedure are given in fig.9. The low-temperature susceptibility depends essentially linearly upon the Sc-concentration. We can compare the pressures in fig.9 with the pressure required to accomplish a complete valence transition from divalent to trivalent Yb metal. This pressure can be estimated by identifying the transformation energy of 38 kJ/mole |4,5| with the $\int p dV$ term in the free energy. The volume of divalent Yb is 24.9 cm^3/mole. For trivalent Yb metal the molar volume can be approximated by taking the average of the values for Tm(18.1 cm^3/mole) and Lu(17.8 cm^3/mole); with a thus derived value for ΔV of 6.9 cm^3/mole we estimate a pressure of the order of 10 GPa (100 kBar) necessary to achieve the valence transition in pure Yb. Before discussing, however, the consequences of the pressure effect upon magnetization and specific heat, it is worthwhile to mention that substitution of a smaller atom does not always lead to this effect. Majewski et al. |7| have investigated the magnetic properties of various compounds in the (Yb,Sc)Al_3 system. In contrast to what one would expect upon substitution of the smaller Sc-atom, the effect is that $YbAl_3$ becomes less magnetic with increasing Sc-concentration, which implies a shift towards more divalent character. This can be understood when it is remembered that the magnetic behaviour depends upon the valence state which, in its turn, is connected with the volume: a smaller volume is associated with the trivalent magnetic state, a larger volume with the divalent non-magnetic state. Therefore, in dilution experiments the sign of the deviation of the lattice parameter from Vegard's law is indicative for the magnetic behaviour. From ref. |7| it is known that the deviation in (Yb,Sc)Al_3 is positive (fig.10), which is likely to favour the non-magnetic state. At first sight one would say that the magnetic behaviour of mixed-valent Yb in YbCuAl is only gradually changing upon Sc-substitution. On the other hand, the specific heat behaviour is influenced rather drastically. The appearance of a growing upturn below 10 K with increasing Sc-concentration, which is hardly influenced by a magnetic field of about 5 T, betrayes that somehow the system is on the onset of critical behaviour. A more detailed comparison with for instance compounds like UCo_2 and UAl_2 |8,9| reveals a very similar behaviour. In these actinide compounds too, a peak in the specific heat is observed, which does not change when a magnetic field is applied. Therefore the upturn can not be ascribed to local magnetic moments. The specific heat of UAl_2 and UCo_2 has been analysed succesfully in terms of spin-fluctuation theory |10,11|, which adds a $T^3 \ln T$ term to the linear and cubic contributions. In the case of Sc-substituted YbCuAl a detailed, quantitative analysis of the results in the framework of spin-fluctuation theory is hampered by the fact that, besides a spin-fluctuation part, still an appreciable mixed-valent part is contributing to the specific heat. This is underlined by the thermal expansion behaviour: the volume anomaly, characteristic for mixed-valency, remains qualitatively present.

When it comes to the magnetic properties of the $Yb_{1-x}Sc_xCuAl$ compounds in the light of spin-fluctuation behaviour, it is clear that they fit into this framework. Two slopes in the high-field magnetic isotherms, corresponding with the values of the magnetic susceptibility at zero temperature and at a higher characteristic temperature, are also observed in systems like UAl_2 and UCo_2 |8,9|. However, in pure YbCuAl the occurrence of spin-fluctuations should be excluded on the basis of the absence of an upturn in the specific heat.

In conclusion we find that pressure, induced by alloying with Sc, gradually drives the mixed-valent Yb in YbCuAl into a region where critical behaviour appears next to mixed valency. The spin-fluctuation contribution to the physical properties increases with pressure.

REFERENCES

|1| W.C.M.Mattens, P.F. de Châtel, A.C.Moleman and F.R. de Boer, Physica 69B (1979) 138.

|2| W.H.Dijkman, F.R. de Boer and P.F. de Châtel, Physica 98B (1980) 271.

|3| W.C.M.Mattens, thesis, University of Amsterdam (1980).

|4| W.C.M.Mattens, H.Hölscher, G.J.M.Tuin, A.C.Moleman and F.R. de Boer, J.Magn.Magn.Mat. 15-18 (1980) 982.

|5| K.A.Gschneidner, J.Less-Comm.Met. 17 (1969) 13.

|6| F.R. de Boer, W.H.Dijkman, W.C.M.Mattens and A.R.Miedema, J.Less-Comm.Met. 64 (1979) 241.

|7| R.E.Majewski, A.S.Edelstein, A.T.Aldred and A.E.Dwight, J.App.Phys. 49 (1978) 2096.

|8| J.J.M.Franse, P.H.Frings, F.R. de Boer and A.Menovsky, Proc.Conf."Physics under high Pressure", ed. J.S.Schilling and R.N.Shelton, Bad Honnef, North-Holland Publ.Co. (1981).

|9| R.J.Trainor, M.B.Brodsky and L.L.Isaacs, A.I.P.Proc. 24 (1974) 220.

|10| M.T.Béal-Monod, Shang Keng Ma and D.R.Fredkin, Phys.Rev.Lett. 20 (1968) 929.

|11| M.B.Brodsky and R.J.Trainor, Physica 91B (1977) 271.

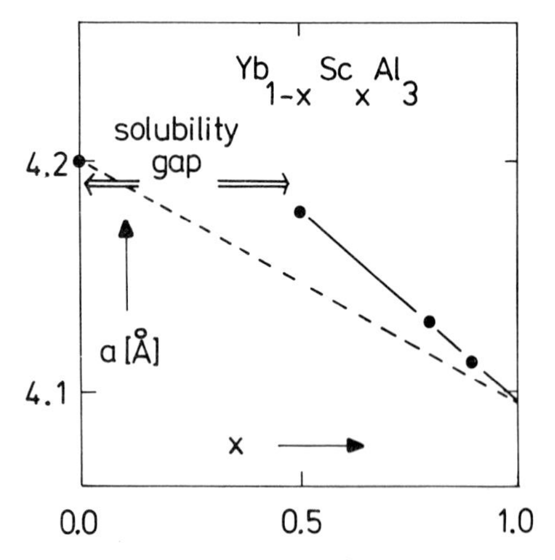

Fig.10. (Data for this plot have been taken from table 1, ref.7).

Valence Instabilities
P. Wachter and H. Boppart (eds.)

PRESSURE DEPENDENCE OF VALENCE OF SmS, $EuPd_2Si_2$, TmSe AND $YbAl_2$ DERIVED FROM L_{III}-ABSORPTION EDGES

J. Röhler[a], G. Krill[b], J.P. Kappler[b], M.F. Ravet[b] and D. Wohlleben[a]

Laboratoire pour l'Utilisation du Rayonnement Electromagnêtique (LURE),
CNRS et Univ. Paris Sud, Bat. 109 C, 91405 Orsay, Cedex, France

[a] II. Physikalisches Institut, Universität zu Köln, Zülpicher Str. 77, 5000 Köln 41, West Germany
[b] LMSES (LA CNRS 306) 4, rue de B. Pascal 670770 Strasbourg, Cedex, France

Pressure induced valence shifts at ambient temperature up to 72 kbar are reported on SmS, $EuPd_2Si_2$, $YbAl_2$, TmSe from L_{III}-absorption edge experiments. A strongly non-linear pressure dependence of the valence is observed in all these compounds. In the case of $EuPd_2Si_2$ the pressure dependence of the interconfigurational excitation energy E_x is calculated.

INTRODUCTION

Pressure induced valence shifts in mixed valence materials are a well known fact. However, there is a great lack of quantitative experimental data describing the valence-pressure relation. It was possible to measure the valence of $EuCu_2Si_2$/1/, SmS /2/ via the Mössbauer isomer shift up to 50 and 12 kbar respectively, with a resolution of the relative valence change of about 1% and 8%. For γ-αCe /4/, $CeNi_5$, $CeCo_2$ /3/ and SmS /4/ some valence data under pressure are also available from K-fluorescence. The L_{III}-method is more generally applicable than the Mössbauer isomer shift, since the structure of the L_{III}-edge can be equally well resolved in all potentially unstable Rare Earth (RE) ions from Ce to Yb. Moreover, the extraction of the valence from the data is simpler and less problematic than by any other method. At least for the RE between Sm and Yb the valence can be extracted by a very simple analysis of the observed double structure with a relative resolution near 1% (see below) without significant falsification by final state effects and without need of data from reference compounds. We have measured the pressure driven valence shift of SmS, $EuPd_2Si_2$, TmSe and $YbAl_2$. All these systems have been previously studied also at high pressure. Therefore our results can be related to a relatively large amount of basic properties, e.g. the pressure dependence of the lattice parameter, susceptibility, resistivity and thermopower.

EXPERIMENTAL AND RESULTS

The L_{III}-absorption studies were performed at L.U.R.E., Orsay, France, using the synchrotron radiation of DCI. The storage ring was usually operated at 1.72 GeV and 250 mA. We used the facilities of the unfocused beam line of EXAFS I (channel-cut Si 220 monochromator), collimating the beam to the entrance window of the high-pressure cell (2 x 0.3 or 4 x 0.3 mm).

The measurements were performed in an opposed Bridgeman-anvil device, using a transmission geometry perpendicular to the axis of the anvils. The gaskets were machined from high-purity Beryllium. The absorber with its typical thickness of 10-15 μ stuck in a slit of 100 μ width inside the gasket, sandwiched in an epoxyresin pressure transmitter. Details of the cell will be described elsewhere /5/. The pressure was determined externally following a load-pressure relation, which has been determined up to 20 kbar by ruby-fluorescence technique using saphire anvils. The distribution of pressure across the sample is estimated to be smaller than 10% and the average pressure is believed to be accurate within 10% of the nominal pres sure. Each spectrum was registered several times in order to improve statistics and to ensure reproducibility. The inserts in Fig. 1-4 show the typical spectra at different pressures, indicating also the characteristic energy positions of the corresponding 2^+ and 3^+ peaks. Drawn out lines connecting the data points are fits resulting from a superposition of two asymmetric Lorentzian lineshapes, taking into account a Gaussian broadening. The two subspectra illustrate the fitting procedure. Valences are obtained by the intensity ratio of 2^+- and 3^+-subspectra. Error bars indicate typical relative errors. The systematic error, mainly due to uncertainties in the choice of the phenomenological asymmetric lineshapes is estimated to be smaller than 5%. All spectra show a pressure dependent shift of the centres of gravity of the white lines. Also a considerable pressure dependent broadening is observed and will be analysed elsewhere/6/.

SmS

Fig. 1 shows the pressure dependence of the valence of SmS up to 72 kbar. Open and closed circles indicate data from different runs in different cell geometries. Triangles mark decreasing pressure. At zero pressure the average

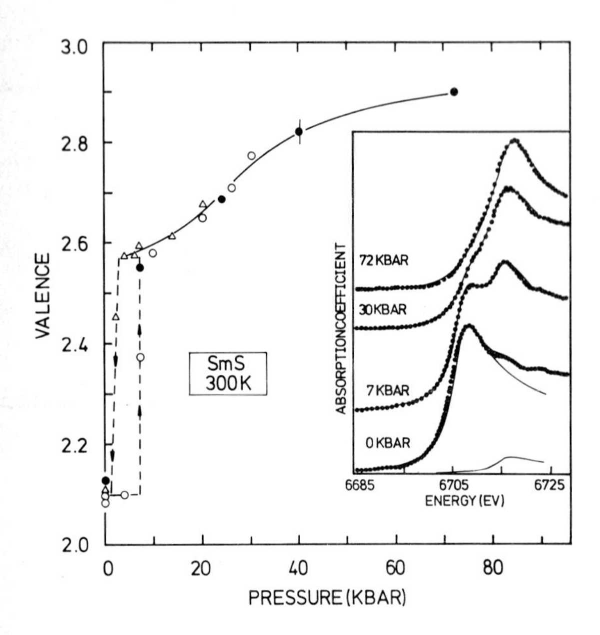

Fig. 1 Pressure dependence of the valence of SmS and typical L_{III}-spectra at different pressures.

valence of the sample is not integral. There is 10% 3^+-contribution in the spectrum.

This observation has been confirmed by the measurements of different samples at ambient temperature and pressure inside and outside the pressure cell, even after eliminating all yellow grains which arose from powdering the sample.

The x-ray diffraction patterns are clean of the Sm_2O_3 or Sm_3S_4 phases down to the detection limit of 5%. The contribution arising from thermal excitation across the semi conducting gap should be smaller than 2% ($E_{gap} > 0.1$ eV) /7/. We do not know the origin of the 10% 3^+ contribution.

The well known first order volume collapse at 6.5 kbar is associated with a drastic valence shift as was established previously by various techniques, e.g. lattice parameters /7/, susceptibility/8/, Mössbauer effect /2/, K-fluorescence /4/. Directly above the transition we find a valence v= 2.57. For comparison K-fluorescence gives 2.62(3) /4/, Mössbauer isomer shift 2.62(8) /2/ and lattice constant 2.77(6) /8/. Thus we have another case of a strong deviation of the valence obtained from lattice constant à la Vegard from the values obtained from other methods /9/. The value from the lattice constants seems to be about 15% too large.

The 3^+ integral valence is not reached, even at 72 kbar.

Note also the remarkable nonlinearities of the pressure dependence of the valence in the mixed valence phase with a point of inflection near 22 kbar. This correlates with an abnormal pressure derivative of the bulk modulus in this range /10/.

$YbAl_2$

Fig. 2 shows the pressure dependence of the valence of $YbAl_2$. The valence increases monotonically by 31% at 30 kbar, but again in a nonlinear fashion. The low resolution of the spectra is due to the large width of the spectrometer function at 9 keV. The difference of the valence at ambient pressure reported here and that reported elsewhere /9/ is believed to be due to different sample qualities, used in the different experiments.

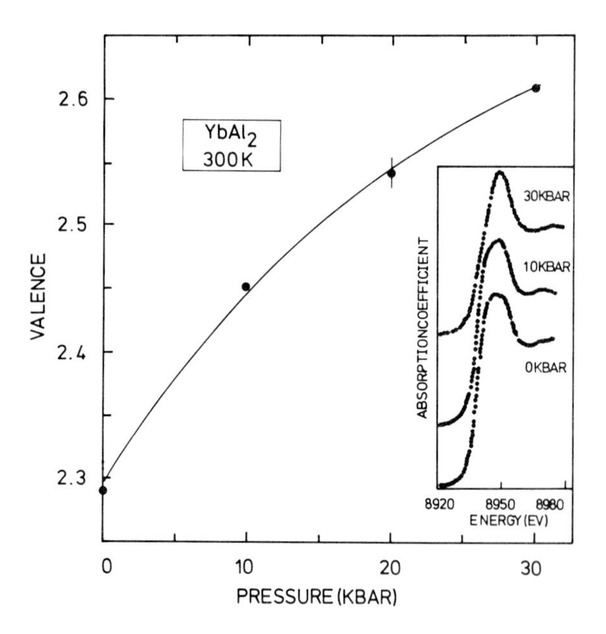

Fig. 2 Pressure dependence of the valence of $YbAl_2$ and typical L_{III}-spectra at different pressures.

TmSe

Fig. 3 shows the pressure dependence of the valence of TmSe. The absorber was prepared from a single crystal, kindly provided and characterized by Dr. K. Fischer, Kernforschungsanlage Jülich. Again the valence change is clearly nonlinear and the stable trivalent state is not reached at 30 kbar.

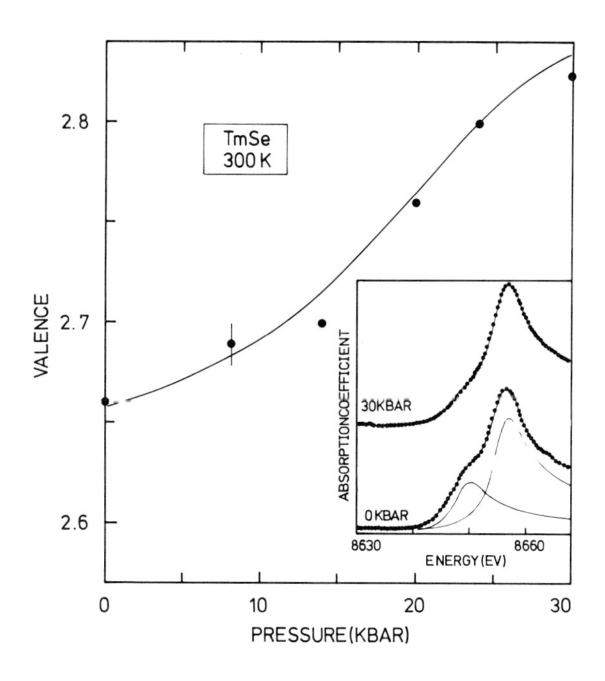

Fig. 3 Pressure dependence of the valence of TmSe and typical L_{III}-spectra at different pressures

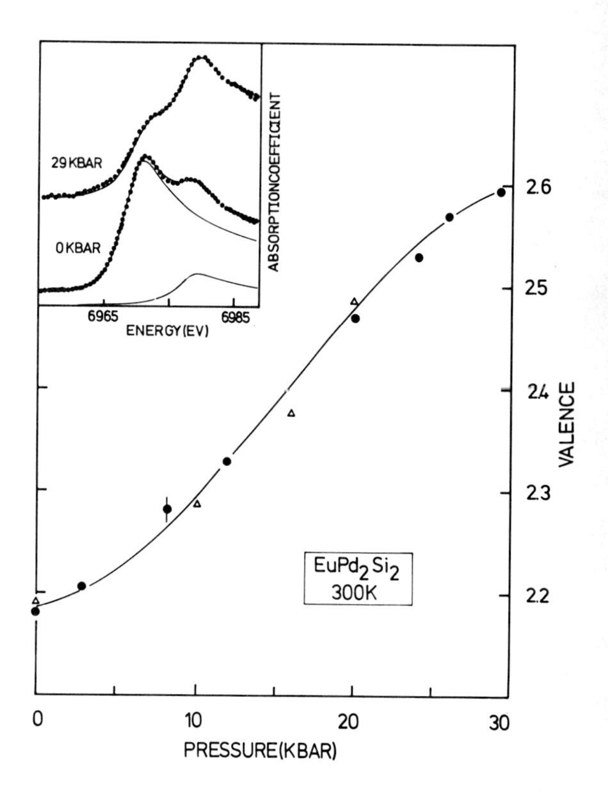

Fig. 4 Pressure dependence of the valence of $EuPd_2Si_2$ and typical L_{III}-spectra at different pressures.

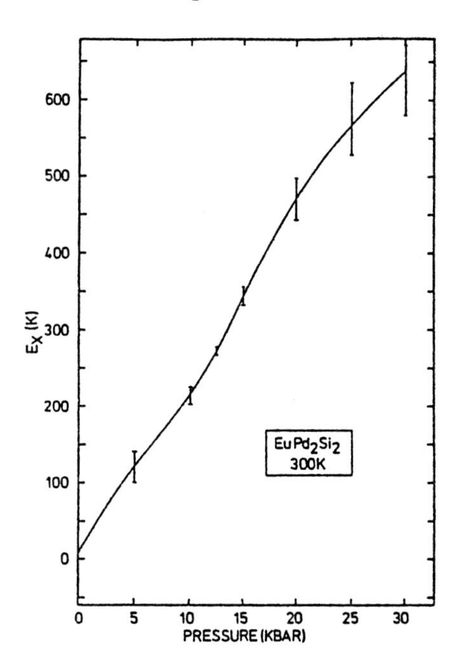

Fig. 5 Interconfigurational excitation energy E_x of $EuPd_2Si_2$ as a function of pressure.

$EuPd_2Si_2$

The $EuPd_2Si_2$ sample was checked by Mössbauer spectroscopy ($^{151}SmF_3$ source). A broadened single line with W=3.8 mm/s was observed at ambient temperatures with S = -7.6 mm/s.

Fig. 4 shows the pressure dependence of the valence of $EuPd_2Si_2$ at room temperature. Filled circles and open triangles indicate increasing and decreasing pressure, respectively. The valence change amounts to 41% at 29 kbar and is reversible within experimental error. This pressure driven shift of the valence corresponds to the amount of the temperature driven valence shift between 300 K and 4.2 K as derived from the Mössbauer isomer shift /11, 12, 13/ and L_{III}-edges /14/ (45%). Both temperature and pressure driven valence changes show a nonlinear (S-shaped) behaviour. The "step" appears however much steeper in the temperature than on the pressure scale. In Fig. 5 we have plotted the pressure dependence of the interconfigurational excitation energy $E_x(p) = E_{7,0} - (E_{6,0} + \varepsilon_F)$ /15/. This quantity was extracted from the pressure dependence of the valence (Fig. 1) via $E_x \equiv kT^+ \ln((1-\nu)\, \zeta_7'/\zeta_6'\nu))$. Here the mixing energy or the interconfigurational fluctuation temperature $T_f = (T^{+2} - T^2)^{1/2}$ has negligable influence on the results: error bars indicate what happens when T_f varies between 0 K and 240 K. Comparison of susceptibility and Mössbauer isomer shift indicates $T_f \simeq 70$ K at ambient pressure and temperature. For $\Delta p = 30$ kbar we find $\Delta E_x \simeq 600$ K. We compare this value with that obtained by calculating the mechanical work done from the outside via the relation $\Delta E_x = (V_{n+1} - V_n)\Delta p$ /16/. Assumig the volume difference between the unit cells of stable 2^+ and 3^+ $REPd_2Si_2$ reference compounds as $(V_7 - V_6) \simeq 8$ Å^3 we find $\Delta E_x \simeq 1700$ K. We conclude that an energy of $\Delta E_x \simeq 1100$ K is absorbed in changes of the elastic-, Fermi- and mixing-energies /17/.

Acknowledgements

We are indebted to K. Fischer, C. Godart, R. Pott and E. Holland-Moritz for letting us use their samples, M. Abd-El Meguid for kindly taking the Moessbauer spectra of $EuPd_2Si_2$. We especially thank P. Hansmann and G. Hagemeier for carefully machining the high-pressure device and K. Syassen and W.B. Holzapfel for valuable discussions. It's also a pleasure to thank Dr. Marin and his co-workers at the Laboratoire de l'Accélérateur Linéaire in Orsay. This work was supported by the Deutsche Forschungsgemeinschaft through Sonderforschungsbereich 125.

REFERENCES

/1/ J. Röhler, D. Wohlleben, G. Kaindl, this conference

/2/ J.M.D. Coey, S.K. Ghatak, AIP Conference (1974)

/3/ W.A. Shaburov, A.E. Sovestnov, O.I. Sumbaev, Phys. Lett., 49A, 83 (1974

/4/ A.E. Sovestnov, V.A. Shaburov, I.A. Markova, E.M. Savitski, O.D. Chistyakov, T.M.Shkatova, Sov. Solid State Phys., 23, 2827 (1981)

/5/ J. Röhler, J.P. Kappler, G. Krill, to be published

/6/ J. Röhler, J.P. Kappler, G. Krill, D. Wohlleben, to be published

/7/ A. Jayaraman, V. Narayanamurti, E. Bucher, R.G. Maines, Phys. Rev. Lett. 25, 368 (1970)

/8/ M.B. Maple, D. Wohlleben, Phys. Rev. Lett., 27, 511 (1971)

/9/ G. Neumann, R. Pott, J. Röhler, W. Schlabitz, D. Wohlleben, H. Zahel, this conference

/10/ R. Keller, G. Güntherodt, W.B. Holzapfel, M. Dieterich, F. Holtzberg, Solid State Comm., 29, 753 (1979)

/11/ E.V. Sampathkumaran, L.C. Gupta, R. Vijayaraghavan, K.V. Gopalakrishnan, R.G. Pillay, H.G. Devare, J. Phys. C, 14, L 237-L 241 (1981)

/12/ M. Croft, J.A. Hodges, E. Kemly, A. Krishnan, V. Murgai, L. Gupta, R. Parks in "Physics of Solids under High Pressure," J.S. Schilling, R.N. Shelton (editors), North Holland, 1981, p. 341

/13/ G. Schmiester, B. Perscheid, G. Kaindl, J. Zukrowsky, this conference

/14/ R. Nagarajan, E.V. Sampathkumaran, L.C. Gupta, R. Vijayaraghavan, B. Haktdarshan, B.D. Padalia, Phys. Rev. Lett. 81A, (1981), 397

/15/ D. Wohlleben in "Valence Fluctuations in Solids", L.M. Falicov, W. Hanke and M.B. Maple (editors), North Holland (1981), p.1

/16/ B. Johansson, A. Rosengren, Phys. Rev. B 14, 361 (1976)

/17/ J. Röhler, D. Wohlleben, G. Kaindl, H. Balster, Phys. Rev. Lett., to be published

Valence Instabilities
P. Wachter and H. Boppart (eds.)

EFFECTS OF PRESSURE AND TEMPERATURE ON THE MEAN VALENCE OF $EuPd_2Si_2$

G. Schmiester, B. Perscheid, G. Kaindl, and J. Zukrowsky*

Institut für Atom- und Festkörperphysik
Freie Universität Berlin, D-1000 Berlin 33, Germany

The pressure and temperature dependence of the ^{151}Eu Mössbauer isomer shift of mixed-valent $EuPd_2Si_2$ is reported for the temperature range 4.2 to 296 K and for pressures up to 53 kbar, and values for the mean valence of Eu are derived. Using the known susceptibility of $EuPd_2Si_2$, the present data yield values for the interconfigurational excitation energy E_x and the mixing width T_f as a function of temperature. It is found that E_x as well as the transition temperature and width of the continuous thermally-induced valence transition strongly increase with pressure. A T-p phase diagram is proposed.

1. INTRODUCTION

Metallic mixed-valent systems of Eu are characterized by strong and continuous changes of the mean valence with both temperature and pressure /1-5/. The best studied system so far is $EuCu_2Si_2$ for which both Mössbauer isomer shift and magnetic susceptibility (as well as several other physical properties) have been investigated as a function of pressure and temperature /1,4-6/. From these rather complete data values for the interconfigurational excitation energy E_x and the mixing width $\Gamma=k_BT_f$, both as functions of temperature and pressure, were recently extracted, and it was shown that the mixing width is at a maximum at configurational crossover /4,5/. In addition, an energy balance was derived for this system, and the pressure and temperature dependences of the different contributions to E_x (Fermi energy, elastic energy, and mixing energy) could be separated out /4,5/.

Recently, $EuPd_2Si_2$ was also characterized as a homogeneously mixed-valent system with a continuous but rather precipitous thermally-induced valence change near 150 K /2,3/. This compound, which is isostructural with $EuCu_2Si_2$ /7/, has quickly attracted great interest because of the narrow temperature range in which the mean valence changes appreciably. Besides the original L-edge X-ray absorption and Mössbauer work /2,3/ the magnetic susceptibility was studied as a function of temperature /8/, and measurements of electrical-transport properties indicated a pressure-induced valence change /9/. We report here on a Mössbauer isomer shift study of $EuPd_2Si_2$ at quasi-hydrostatic pressures up to 53 kbar in the temperature range $4.2 \leq T < 296$ K.

2. EXPERIMENTAL

Polycrystalline $EuPd_2Si_2$ samples were prepared from the elements by cold-crucible induction melting in purified argon atmosphere and subsequent annealing at 800°C in high vacuum. For the high-pressure Mössbauer experiments an opposed-anvil device with sintered B_4C anvils was employed. The pressure was measured in-situ by two different methods: In the liquid-helium temperature range from the superconducting transition temperature of Pb /10/ and at room temperature from the isomer shift of the ^{119}Sn gamma-resonance in β-Sn. For the isomer-shift-pressure relation we used a value of $d\delta/dp(^{119}Sn) = -2.7\times10^{-3}$ mm/s kbar^{-1} /11,12/. $^{151}SmF_3$ and $Ca^{119}SnO_3$ sources were employed for the ^{151}Eu and ^{119}Sn measurements, respectively.

3. RESULTS AND DATA ANALYSIS

Typical ^{151}Eu(21.6-keV) Mössbauer absorption spectra of $EuPd_2Si_2$ at ambient pressure and at three selected temperatures are shown in Fig. 1. The spectra consist of a main component (long-dashed line) moving from $\delta=-(7.61\pm0.05)$ mm/s at 296 K to $\delta=-(0.53\pm0.05)$ mm/s at 4.2 K, plus a relatively weak satellite (short-dashed line) with a much less temperature-dependent position. Such satellite lines were also reported in a previous Mössbauer study of $EuPd_2Si_2$ at ambient pressure /13/ and have in fact been observed in all Mössbauer studies of $EuPd_2Si_2$ /3,14/ and of $EuCu_2Si_2$ /1,4,5/ with varying intensities. A superposition of an electric-quadrupole broadened main component plus a single Lorentzian for the satellite line was least-squares fitted to the data points (solid lines in Fig. 1). The fits consistently resulted in a positive electric-field gradient at the Eu site ($V_{zz}=(2.6\pm0.8)\times10^{17}$ V/cm^2 at 296 K) in close similarity to previous results for isostructural $EuCu_2Si_2$ /5,15/. This electric-quadrupole interaction, which may be understood on the basis of the $ThCr_2Si_2$ structure of $EuPd_2Si_2$ /7/ will not be discussed further here /16/. The results for the isomer shift as a function of temperature are plotted in Fig. 3 for both the main component (solid circles) and the satellite line (open circles). The isomer shifts of the main component are in excellent agreement with previous ambient-pressure data /3/. It is obvious from Fig. 3 that

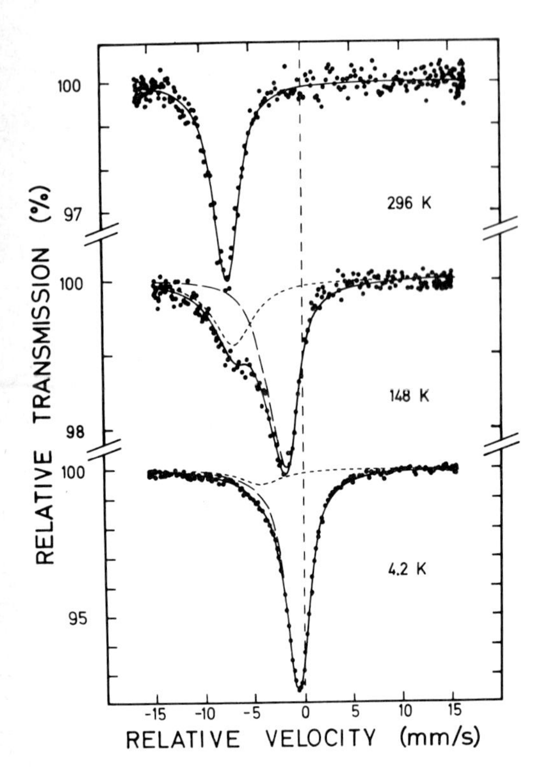

Figure 1: Mössbauer absorption spectra of $EuPd_2Si_2$ as a function of temperature. The solid lines represent the results of least-squares fits of a superposition of two sub-spectra to the data (see text).

the satellite line is also due to Eu in a mixed-valent state with, however, a much lower transition temperature. It is probably connected with Eu ions with some kind of disorder in the nearest-neighbour shell /1/.

The effects of pressure on the gamma-resonance spectrum of $EuPd_2Si_2$ at room temperature are demonstrated by the three representative spectra shown in Fig. 2. Both the main component and the satellite line shift with increasing pressure towards the Eu^{3+}-range of line positions. We have also taken data as a function of pressure at 4.2 K, 78 K, and 198 K in order to obtain information on the T-p phase diagram of $EuPd_2Si_2$.

The isomer shifts resulting from the least-squares fit analysis of these high-pressure spectra are summarized in Fig. 4. At room temperature both the Eu ions associated with the main component and those giving rise to the satellite line experience a pressure-induced valence-transition. The effects of pressure on the isomer shift, however, are much less dramatic for the satellite line than for the main component. This agrees well with the effect of decreasing temperature at ambient pressure

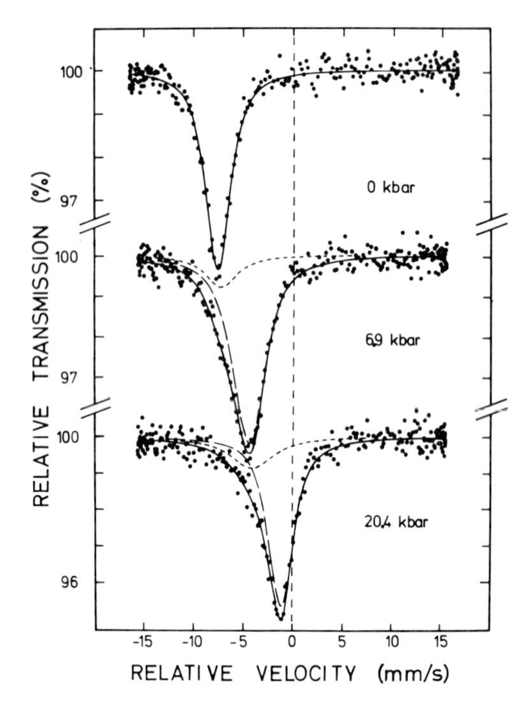

Figure 2: Mössbauer spectra of $EuPd_2Si_2$ at room temperature as a function of pressure.

(Fig. 3), where we also found that the Eu ions associated with the satellite line are much more pinned to a valence close to 2^+ than the main fraction. In the following we will only discuss the results for the main component /16/.

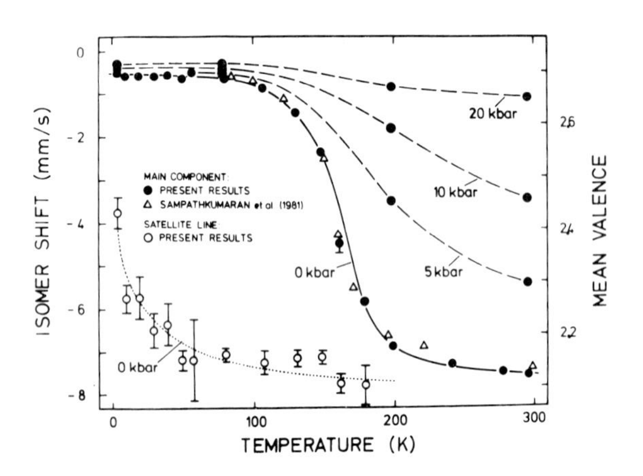

Figure 3: Temperature dependence of the isomer shift of the main component of $EuPd_2Si_2$ at ambient pressure (solid curve) and at three external pressures (dashed curves). For the ambient-pressure case the isomer shift of the satellite line is also given (dotted line).

It is quite obvious from Fig. 4 that the effects of pressure on the isomer shift are drastically different for temperatures above (296 K and 198 K) and below (78 K and 4.2 K) the temperature-induced valence transition of $EuPd_2Si_2$ (see Fig. 3). In the low-temperature region, when the valence is close to 3^+ even at ambient pressure, an approximately linear increase of δ with pressure is observed, with a slope quite similar to that observed for stable-valent Eu compounds /17/. On the other hand, δ changes in a strongly non-linear way with pressure at temperatures above the temperature-induced valence transition. At room temperature this pressure-induced valence change is practically completed at pressures of only about 30 kbar.

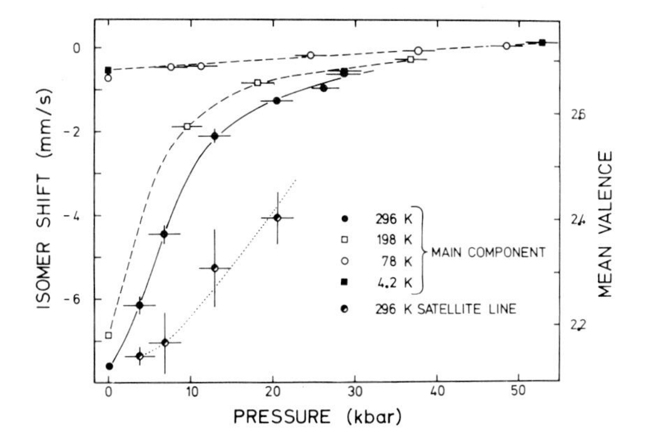

Figure 4: Pressure dependence of the isomer shift of the main component in the gamma-resonance spectrum of $EuPd_2Si_2$ at 296 K, 198 K, 78 K, and 4.2 K, and of the satellite line at 296 K.

We have also plotted the high-pressure data in Fig. 3 as a function of temperature in order to display the effects of pressure on the temperature-induced valence change of $EuPd_2Si_2$. It is obvious that with increasing pressure the effects of temperature on $\bar{v}$ become less and less pronounced: the valence change shifts to higher temperatures and is smeared over a wider temperature range.

4. DISCUSSION

4.1 Mean valence $\bar{v}$

As discussed previously, the characteristic time scale of a Mössbauer measurement is much longer than the inverse valence fluctuation rate in a homogeneously mixed-valent system /1/. Therefore, a motionally-narrowed single absorption line is observed at a time-averaged isomer-shift position

$$\delta(T,V)=\delta_2(T,V)+(\delta_3-\delta_2)\cdot\nu(T,V). \qquad (1)$$

Here $\nu(T,V)$ stands for the fractional occupation of the Eu^{3+}-$4f^6$ state (mean valence $\bar{v}=2+\nu$), and $\delta_2(\delta_3)$ are the isomer shifts of integral-valent Eu^{2+} (Eu^{3+}), respectively, in the same environment. For $\delta_3-\delta_2$ a value of 12 mm/s was assumed, slightly larger than the result found for the semiconductor Eu_3S_4 /18/, and δ_2 was set equal to -9.0 mm/s. This corresponds to the mean value of the isomer shifts observed for the divalent metallic systems $EuNi_2Si_2$ and $EuSi_2$ (δ=-9.6 mm/s) and $EuPd_2$ (δ=-8.5 mm/s) /19,3/. We estimate the uncertainty in δ_2 as ±0.7 mm/s, corresponding to $\Delta\bar{v}\cong\pm0.05$. When deriving values for $\bar{v}$ from the measured isomer shifts, we have neglected the thermal redshift, but corrected $\delta_2(T,V)$ for volume change due to thermal expansion or isothermal compression using $\partial\delta_2/\partial \ln V \cong -6$ mm/s /17/. The thermal volume change was taken from Ref. /3/, while the pressure-volume relation was obtained from X-ray diffraction measurements under hydrostatic conditions /20/. These corrections are small (always <0.3 mm/s) and their uncertainty is totally negligible in the present context.

The resulting mean valences are presented in Fig. 3 and Fig. 4 through the valence scales on the right-hand ordinates. We find a mean valence of $\bar{v}$=2.12±0.05 at ambient conditions, which, upon cooling to 4.2 K changes to 2.71±0.05. A similar variation of $\bar{v}$ to 2.65±0.05 is caused by an external pressure of 28 kbar at room temperature.

4.2 Interconfigurational Fluctuation Model

Next we shall analyze the isobars and isotherms obtained for $\bar{v}$ within the framework of the ionic interconfigurational fluctuation model /21/. In this model the ionic spectra of the $4f^n$ and $4f^{n+1}$ configurations are assumed to remain essentially unaffected by the solid. The configurations are separated by the interconfigurational excitation energy E_x and all ionic levels have a finite mixing width $\Gamma=k_B\cdot T_f$. The mean valence $\bar{v}=2+\nu$ is then obtained by statistical thermodynamics (with n=6 in the case of Eu)

$$\frac{1-\nu}{\nu} = \frac{\xi_7'}{\xi_6'} \exp(-E_x/k_B\cdot T^*). \qquad (2)$$

Here, $\xi_6'(\xi_7')$ are the partition functions of $4f^6(4f^7)$ with energies counted from the lowest interconfigurational levels $E_{6,0}$ and $E_{7,0}$, respectively, and $E_x=E_{7,0}-E_{6,0}$. The effective temperature $T^*=(T^2+T_f^2)^{1/2}$ takes into account the interconfigurational mixing through quantum and thermal fluctuations /22/. We have used the following intraionic spectra: in $4f^7$ one level at 0 K (J=7/2); in $4f^6$ levels at 0 K (J=0), 480 K (J=1), 1330 K (J=2), and 2600 K (J=3).

The mixing width Γ is expected to depend strongly on E_x and the intraionic spectra: Γ should tend to zero for $E_x\to\pm\infty$ and should display maxima near configurational crossover /4,5/. As shown previously for the case of $EuCu_2Si_2$ the two functions $E_x(T,p)$ and $T_f(T,p)$ may be determined from a simultaneous evaluation of two physical observables, e.g. the Mössbauer isomer shift $\delta(T,p)$ and the static magnetic susceptibility $\chi(T,p)$ /4,5/ using equs. 1 and 2 and the following equation for $\chi(T,p)$ /22/:

$$\chi(T,p)=(1-\nu)\chi_7(T^*,p)+\nu\chi_6(T^*,p). \quad (3)$$

χ_7 and χ_6 are the Curie and Van-Vleck susceptibilities of $4f^7$ and $4f^6$.

The static susceptibility of $EuPd_2Si_2$ has been measured as a function of temperature at ambient pressure /3/, but not yet as a function of pressure. Using the $\chi(T,0)$ data and the present results for $\delta(T,0)$ pairs of values for $E_x(T,0)$ and $T_f(T,0)$ are obtained, which are plotted in Fig. 5a as a function of temperature. The solid curves, which serve only as a guide for the eyes, were obtained with δ_2=-9.0 mm/s. The dashed (dotted) curves represent the variations of E_x and T_f for δ_2=-9.7 (-8.3) mm/s.

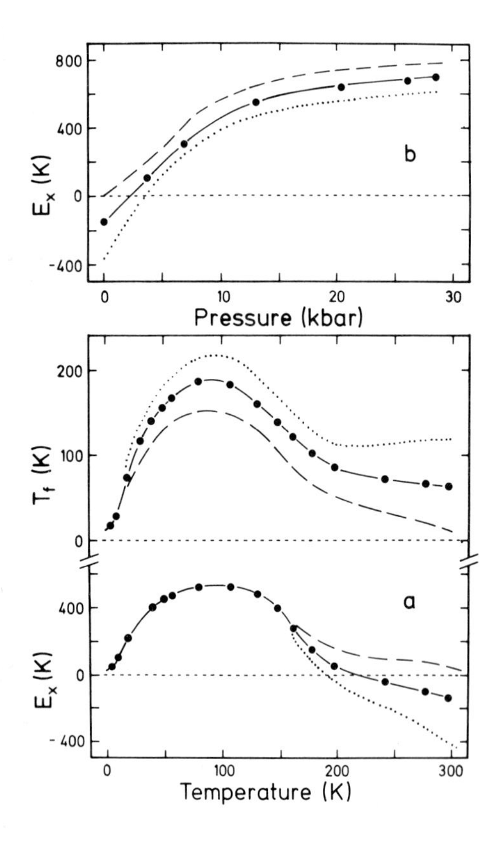

Figure 5: Summary of results for E_x and T_f of $EuPd_2Si_2$: (a) $E_x(T,0)$ and $T_f(T,0)$ as obtained from a simultaneous evaluation of $\delta(T,0)$ and $\chi(T,0)$; (b) $E_x(p)$ obtained from $\delta(296,p)$ with T_f=60 K. The solid curves were obtained with δ_2=-9.0 mm/s, the dashed (dotted) ones with δ_2=-9.7 (-8.3) mm/s.

We find a broad maximum in the $E_x(T,0)$ curve with $E_x(T\cong 100\text{ K})\cong 500$ K, i.e. the $4f^7$ level $E_{7,0}$ seems to be pinned to the first excited level $E_{6,1}$ of $4f^6$ at 480 K in the beginning of the thermally-induced valence transition. During the valence transition E_x decreases strongly with temperature and seems even to become negative around room temperature and ambient pressure. The mixing width T_f shows a strong maximum in the temperature range where E_x is close to 480 K, i.e. at configurational crossover, when the mixing interaction is expected to be strongest. A similar observation was originally made in the case of $EuCu_2Si_2$ /4,5/.

Despite the lack of high-pressure susceptibility data for $EuPd_2Si_2$ we may derive $E_x(296,p)$ from our high-pressure isomer shift results, since at room temperature the influence of quantum fluctuations is rather small on the total interconfigurational mixing ($T_f\cong 60$ K at 296 K and ambient pressure). With this insensitivity to a modest variation of T_f in mind, we have derived $E_x(296,p)$ values from $\delta(296,p)$. The results are plotted in Fig. 5b and show that E_x increases quite drastically with pressure. The variation of T_f with pressure can of course not be derived without e.g. additional susceptibility data, but in analogy with $EuCu_2Si_2$ /4,5/, it may be anticipated that T_f increases with pressure supposedly with a maximum around E_x=480 K.

In the case of $EuCu_2Si_2$ it was recently shown /4,5/ that the temperature- and pressure-induced variation of E_x can be decomposed into three main contributions: an electronic term from a shift of the Fermi level, which - to some extent - is compensated by an elastic energy, plus a contribution from configurational mixing (mixing energy E_{xm}). Without presenting a quantitative analysis of $E_x(T,p)$ of $EuPd_2Si_2$ here, some qualitative statements on the data of Fig. 5 can be made. In the temperature range below about 80 K the strong upward motion of E_x with increasing temperature must be almost entirely due to the mixing term E_{xm}, since - due to a constant valence in this temperature range - the electronic and elastic contributions are negligible. In this temperature region E_{xm} stabilizes E_x at the configurational crossover of $E_{7,0}$ with $E_{6,1}$, which also leads to a maximum of the mixing width T_f (see Fig. 5).

4.3 p-T Phase Diagram of $EuPd_2Si_2$

The pressure-temperature data summarized in Fig. 3 may be used to construct a phase diagram of $EuPd_2Si_2$ in the p-T plane. To this purpose we have plotted in Fig. 6 the transition temperature (solid points) and the width of the continuous phase transition (dashed vertical bars) as a function of pressure. The width of the transition was arbitrarily chosen as the temperature range from 10% to 90% of the full

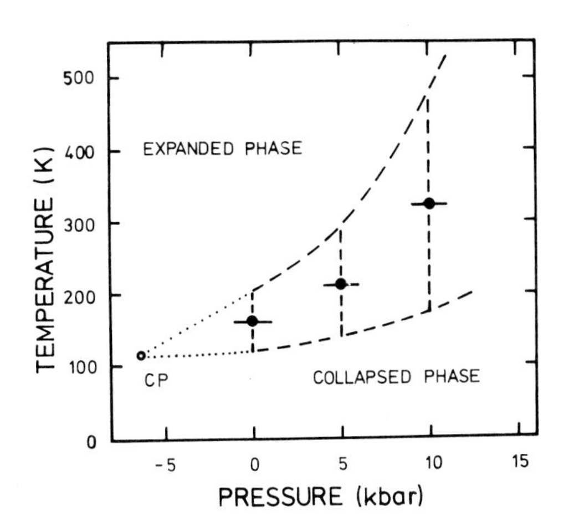

Figure 6: p-T phase diagram of $EuPd_2Si_2$. The width of the continuous valence transition is indicated by dashed lines.

range of thermally-induced valence change. It is obvious from Fig. 6 that with increasing pressure (and E_X) the transition temperature increases and the transition gets smeared over a larger temperature range. When the region of continuous phase transitions is extrapolated to negative pressures (dotted lines in Fig. 6) the width gets narrower and converges into a point, that may be interpreted as the critical point CP of a fictitious first-order valence-change boundary at still higher negative pressures. A similar conclusion has been drawn from very recent high-pressure resistivity measurements on $EuPd_2Si_2$ /23/. It is well known that negative pressures may be simulated by chemical substitution, and there is recent experimental evidence that a first-order phase boundary does indeed exist in the Au-substituted systems $Eu(Pd_{1-x}Au_x)_2Si_2$ /24/.

ACKNOWLEDGEMENT - The authors acknowledge valuable discussions with J. Röhler. This work was supported by the Sonderforschungsbereich-161 of the Deutsche Forschungsgemeinschaft.

* Permanent address: Academy of Mining and Metallurgy, Cracow, Poland.

REFERENCES

/1/ Bauminger, E.R., Froindlich, D., Nowik, I., Ofer, S., Felner, I., Mayer, I., Phys. Rev. Lett 30 (1973) 1053.

/2/ Nagarajan, R., Sampathkumaran, E.V., Gupta, L.C., Vijayaraghavan, R., Bhaktdarshan, Padalia, B.D., Phys. Lett. 81A (1981) 397.

/3/ Sampathkumaran, E.V., Gupta, L.C., Vijayaraghavan, R., Gopalakrishnan, K.V., Pillay, R.G., Devare, H.G., J. Phys. C: Solid State Phys. 14 (1981) L237.

/4/ Röhler, J., Wohlleben, D., Kaindl, G., Balster, H., submitted to Phys. Rev. Lett. (1982).

/5/ Röhler, J., Wohlleben, D., Kaindl, G., contribution to this conference.

/6/ Sales, B.C., Viswanathan, R., J. Low Temp. Phys. 23 (1976) 449.

/7/ Rossi, D., Marazza, R., Ferro, R., J. Less-Comm. Metals 66 (1979) P17.

/8/ Sampathkumaran, E.V., Vijayaraghavan, R., Gopalakrishnan, K.V., Pilley, R.G., Devare, H.G., Gupta, L.C., Post, B., Parks, R.D., in Valence Fluctuations in Solids, edited by Falicov, L.M., Hanke, W., Maple, M.B., North-Holland (1981) p. 193.

/9/ Vijayakumar, V., Vaidyna, S.N., Sampathkumaran, E.V., Gupta, L.C., Vijayaraghavan, R., Phys. Lett. 83A (1981) 469.

/10/ Eiling, A., Schilling, J.S., J. Phys. F: Metal Phys. 11 (1981) 623.

/11/ Möller, H.S., Zeitschr. f. Physik 212 (1968); and references therein.

/12/ A recent recalibration of the SnI-SnII pressure-induced phase transition was taken into account: Decker, D.L., in High-Pressure Science and Technology, edited by Vodar, B., Marteau, Ph., Pergamon Press (1980), Vol. 1, p. 259.

/13/ Croft, M., Hodges, J.A., Kemly, E., Krishnan, A., Murgain, V., Gupta, L., Parks, R., in Physics of Solids Under High Pressure, edited by Schilling, J.S., Shelton, R.N., North-Holland (1981) p. 341.

/14/ Croft, M., Hodges, J.A., Kemly, E., Krishnan, A., Murgai, V., Gupta, L.C., Phys. Rev. Lett. 48 (1982) 826.

/15/ Sampathkumaran, E.V., Gupta, L.C., Vijayaraghavan, R., Phys. Rev. Lett. 43 (1979) 1189.

/16/ A full account of the present work will be published elsewhere.

/17/ Kalvius, G.M., Klein, U.F., Wortmann, G., J. Physique Colloq. 35 (1974) C6-136.

/18/ Röhler, J., Kaindl, G., Sol. State Commun. 36 (1981) 1055.

/19/ Bauminger, E.R., Kalvius, G.M., Nowik, I., in Mössbauer Isomer Shifts, edited by Wagner, F.E., Shenoy, G.K., North-Holland (1978) p. 561.

/20/ Takemura, K., Syassen, K., Kaindl, G., Zukrowsky, J., unpublished results.

/21/ Hirst, L.L., Phys. Kondens. Mat. 11 (1970) 255.

/22/ Sales, B.C., Wohlleben, D.K., Phys. Rev. Lett. 35 (1975) 1240.

/23/ Batlogg, B., Jayaraman, A., Murgai, V., Gupta, L.C., Parks, R.D., Croft, M., contribution to this conference.

/24/ Croft. M., Segre, C.U., Hodges, J.A., Krishnan, A., Murgai, V., Gupta, L.C., Parks, R.D., contrib. to this conference.

Valence Instabilities
P. Wachter and H. Boppart (eds.)

OCCURRENCE OF MIXED VALENCE AND ANTIFERROMAGNETISM IN THE SERIES $Eu(Pd_{1-x}Au_x)_2Si_2$

L.C. Gupta,* V. Murgai, Y. Yeshurun** and R. D. Parks

Department of Physics
Polytechnic Institute of New York
333 Jay Street, Brooklyn, New York 11201

Temperature-dependent susceptibility and resistivity measurements and high field magnetization studies (at 4.2 K) have been made on the series $Eu(Pd_{1-x}Au_x)_2Si_2$ with $0 \leq x \leq 0.25$. For $0 \leq x \leq 0.15$ the system undergoes a valence transition with temperature leading to a non-magnetic ground state, whereas for $x \gtrsim 0.20$ the system undergoes an antiferromagnetic transition. High field studies (up to 21 Tesla) at 4.2 K have established that the low temperature susceptibility tails observed are not intrinsic to the mixed valent state as thought earlier, but, rather, arise from impurities (e.g., untransformed Eu^{2+} ions). A roughly linear relationship is observed between the inverse zero temperature susceptibility and the valence transition temperature, as observed previously for Ce-based alloys.

1. INTRODUCTION

The system $EuPd_2Si_2$ is unique among the known Eu-based mixed valent systems in that it undergoes a cooperative valence transition with temperature.(1) Mössbauer isomer shift measurements indicate that the system is homogeneously mixed valent at all temperatures, with the valence varying from ~ 2.2 above the transition (which is continuous, although nearly second order, with its midpoint at ~ 135 K) to ~ 2.9 below.(1) A frequently exploited technique for modifying the mixed valent nature of a given system is by alloying it suitably (see, e.g., Ref.2).During the course of our investigations of the effects of various additives, e.g., Ag, Au, Ru, Rh and Lu, we observed that alloying $EuPd_2Si_2$ with gold leads to very interesting results. With increasing Au content in the system $Eu(Pd_{1-x}Au_x)_2Si_2$, the valence transition is depressed in temperature, becoming first order for $0.05 \lesssim x \lesssim 0.1$; and, for $x \gtrsim 0.2$, the system undergoes an antiferromagnetic rather than valence transition. Through magnetic susceptibility, resistivity and high field magnetization studies, we have attempted to characterize both the ground states and the phase diagram (mixed valence versus magnetic ordering) for this interesting system .

2. EXPERIMENTAL PROCEDURE

The samples for the various measurements were prepared by arc-melting together, under argon, Eu (99,99%), Pd (99,95%), Au (99.999%) and Si (99.999%) on a water-cooled copper hearth. An excess of Eu (2-3 at. %) was used to compensate for the loss of Eu due to its high vapor pressure. The buttons were flipped and remelted 5-6 times and then vacuum-annealed at 800 C for 4-5 days. X-ray diffraction studies indicated that the samples were single phase (to within the resolution of the technique, viz. $\sim 5\%$) with the $ThCr_2Si_2$ structure. All of the compositions reported below are nominal.

3. RESULTS

3.1 Magnetic Susceptibility

Figures 1 and 2 show the susceptibility and its inverse for three particular compositions chosen to illustrate different regimes. As can be seen clearly in Fig. 2 all three compositions exhibit Curie-Weiss-like behavior [viz., $\chi^{-1} \propto (T+\theta)$] at the higher temperatures; the effective moment per Eu atom being close to that of a free Eu^{2+} ion. The smaller value of the susceptibility at the higher temperatures for x = 0, relative to that for x = 0.125 or x = 0.2, is consistent with the results of L_{III} absorption edge measurements (3), which indicate a larger valence for x = 0 (i.e., further from the magnetic moment-bearing divalent state) than for $x \gtrsim 0.125$ at room temperature. The low temperature Curie-like tails observed for all three compositions arise from impurities consisting mostly of untransformed Eu^{2+} ions, as demonstrated in Section 3.2.

We focus now on the transition regions. We know from Mössbauer experiments (4) that the pure sample (x = 0) undergoes a continuous valence transition, whereas for x = 0.125 the valence transition is first order. The apparent smearing of the first order transition (for x = 0.125) in Figs. 1 and 2 reflects the fact that strain effects prevent the Eu atoms from transforming coherently.(4) Moreover, the detailed shape of the susceptibility curve below the transition (e.g., $10 \lesssim T \lesssim 100$ K for x = 0) varies from sample to sample, which reflects the delayed transformation of a fraction of the Eu atoms. For x = 0.2 the system undergoes an antiferromagnetic rather than valence transition, as confirmed by the appearance of magnetic hyperfine splittings in the Mössbauer spectrum. Note that the shape of the susceptibility anomaly near the antiferromagnetic transition is distinctly different than that near the valence transitions, for in the latter cases there are clear pre-transitional effects, whereas they are absent for the x= 0.2 sample. The temperature of the valence transition is taken as the inflection point and the Neel temperature as the peak in the susceptibility curves of Fig. 1.

A phase diagram based on the valence and anti-ferromagnetic (AF) transitions determined in the above manner is shown in Fig. 3. The dashed lines merely represent visual extrapolations of the two lines of transitions. We estimate that the critical point on the line of valence transitions is contained in the elongated rectangular box, based on our analysis of the Mössbauer (4) and susceptibility results. This places the critical point in approximately the same region of temperature as suggested by pressure experiments on $EuPd_2Si_2$.(5) The broadening of the valence transitions due to the temperature-induced strain effects indigenous to polycrystalline samples of these materials render difficult the exact localization of the critical point on the line of valence transitions and the determination of the exact manner in which the two transition lines (AF and mixed valence) meet.

3.2 High Field Studies

Magnetization measurements have been made at 4.2 K in fields up to 21 Tesla. The results are shown in Fig. 4 for four representative samples (x = 0, 0.10, 0.15 and 0.25), the first three having a mixed valent ground state and the fourth an AF ground state. A significant feature of these results is the complete absence of saturation effects even at the highest available fields (∿21 Tesla). Note that the magnetic energy μH for a free divalent Eu ion in a field of 21 Tesla is ∿ 100 K. This result is explicable in terms of a two level description for $EuPd_2Si_2$, which places the Eu^{2+} level at an energy ∿ 600 K above the Eu^{3+} level.(6) Alternatively, it is also reconcilable in the Varma-Yafet (7) description of the mixed valent ground state if the 4f-(5d,6s) hybridization energy is large compared to 100 K.

The curves for the mixed valent ground state samples can be decomposed into an initial rapidly rising portion contributed by the impurity spins (e.g., Eu^{2+} ions, located probably at grain boundaries), which saturate by 4 Tesla, followed by a linear (Fermi-liquid-like) region extending to the highest field. The values for the zero temperature susceptibility χ(0) determined from the linear regions are close to the values corresponding to the minima in the χ(T) plots (see, e.g., Fig. 1). This proves that the large Curie-like tails which dominate the low temperature susceptibility of all of the samples with $0 \lesssim x \lesssim 0.15$ originate from impurities rather than being an intrinsic property of the ground state as previously suggested.(1)

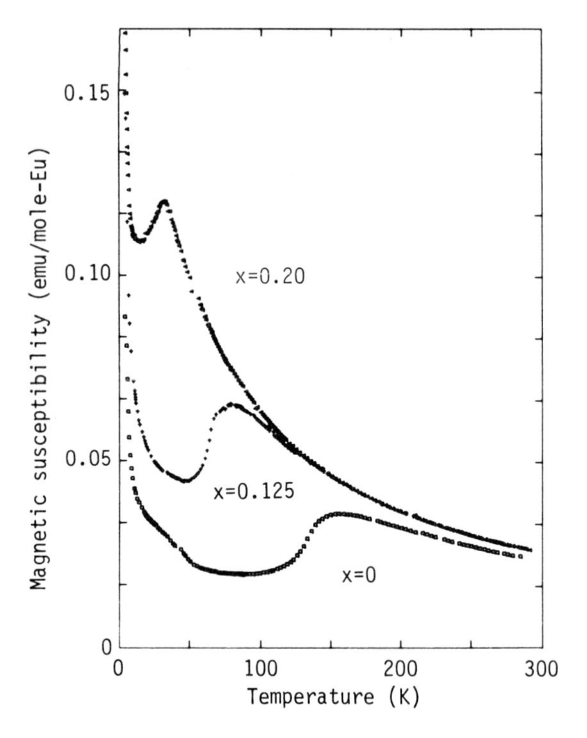

Fig. 1. Magnetic susceptibility of $Eu(Pd_{1-x}Au_x)_2Si_2$ for various values of x.

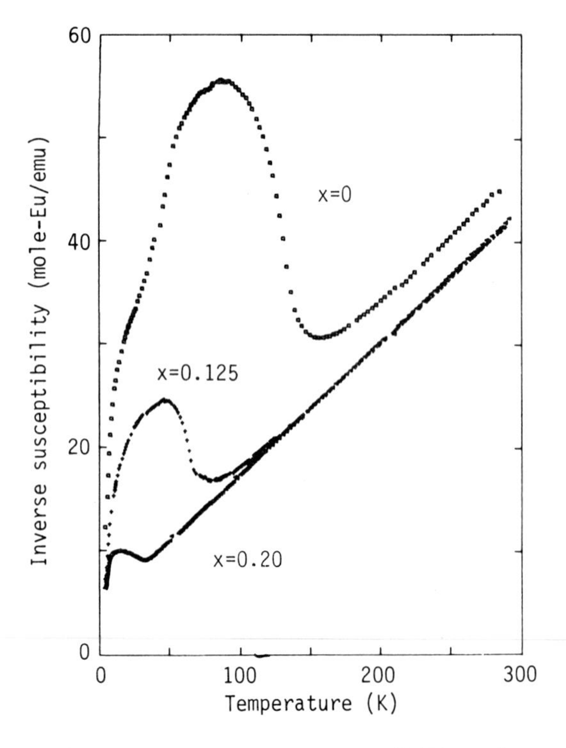

Fig. 2. Inverse magnetic susceptibility of $Eu(Pd_{1-x}Au_x)_2Si_2$ for various values of x.

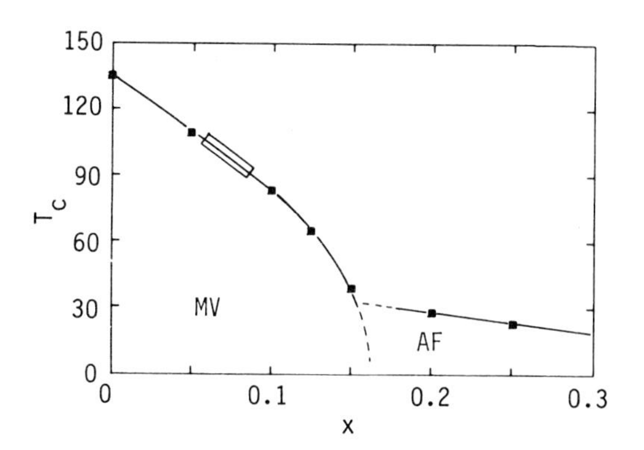

Fig. 3. Phase diagram showing antiferromagnetic AF and mixed valent MV ground states. Rectangular box gives approximate location of critical point on line of valence transitions (see text).

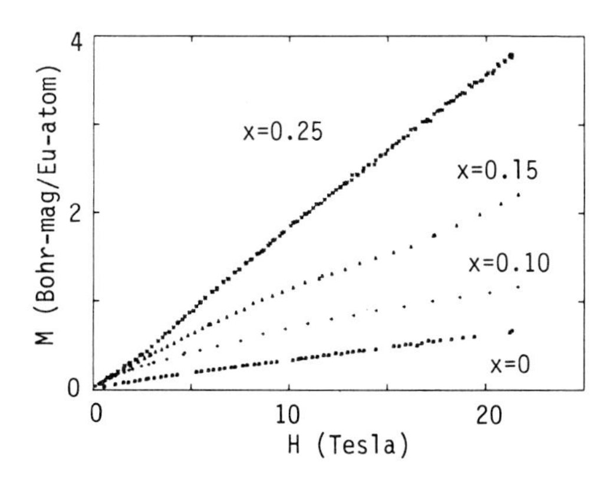

Fig. 4. Magnetization at 4.2 K for three compositions in mixed valent MV ground state and one composition (x=0.25) in AF ground state.

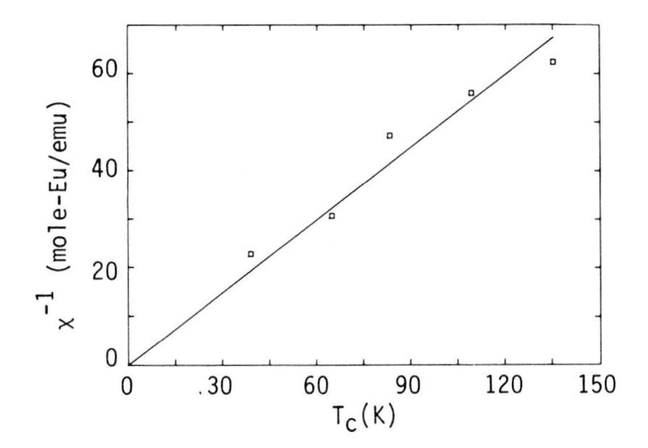

Fig. 5. Inverse susceptibility versus valence transition temperature for various compositions (see Fig. 3).

As seen in Fig.5, an approximately linear relationship is found between $\chi(0)^{-1}$ and the valence transition temperature T_c. This is observed as well in Ce-based alloys (8); and, to our knowledge it has not been satisfactorily explained.

The shape of the magnetization curve for the AF sample (x=0.25), see Fig. 4, is qualitatively different than that for the three mixed valent samples in that the curvature in the field range $0 \leq H \lesssim 6$ Tesla is positive rather than negative, suggesting a field-induced alteration (or destruction) of the AF order parameter. The large linear response which dominates the susceptibility for fields above 7 Tesla suggests either the presence of valence fluctuations or a remnant component of the AF order parameter for which the anisotropy energy is large compared to the ordering energy.

3.3 Electrical Resistivity

During the resistivity measurements large thermal cycling effects were observed for all of the systems exhibiting valence transitions. This is demonstrated for the x=0.10 concentration in Fig. 6, which shows the complete first cycle for a virgin sample and the warming segment of the third cycle. Repeated cyclings led to approximately reproducible behavior. Similar effects, except on a smaller scale, have been observed in the isostructural system $CeCu_2Si_2$ and attributed to the formation of micro-cracks upon thermal cycling.(9) Such cracks, occurring likely at grain boundaries, are caused presumably by the anisotropy both of the volume renormalization accompanying the valence transition (1) and ordinary thermal expansion. The characteristic and reproducible temperature dependence of the resistivity ρ of all of the compositions exhibiting valence transitions is that exhibited by the warming cycles shown in Fig. 6. The qualitative features are a maximum immediately above T_c, a maximum in $d\rho/dT$ at T_c and a rapid decrease in ρ below T_c. The latter two features characterize the behavior of ρ in Ce-based alloys which exhibit valence transitions;(10) however, in the latter case $\rho(T)$ is monotonic in a large temperature interval about T_c.

The peak in $\rho(T)$ just above T_c observed in the $Eu(Pd_{1-x}Au_x)_2Si_2$ systems is inexplicable for a symmetric transition such as the valence transition. For in the latter case, as in a liquid-gas transition, the temperature represents a field variable which couples directly to the order parameter, the latter being the population of 4f-electrons n_{4f} (or the volume V since the two are approximately linearly coupled). Hence, V(T,x) versus T is s-shaped [as is M(H) for a ferromagnet] with a maximum in dV/dT at T_c (the maximum taking the form of a delta function if the transition is first order).(2) In the case of Ce-based alloys $\rho(T)$ is linearly coupled to the volume near T_c and hence mimics the shape of V(T).(10) We can only speculate that the

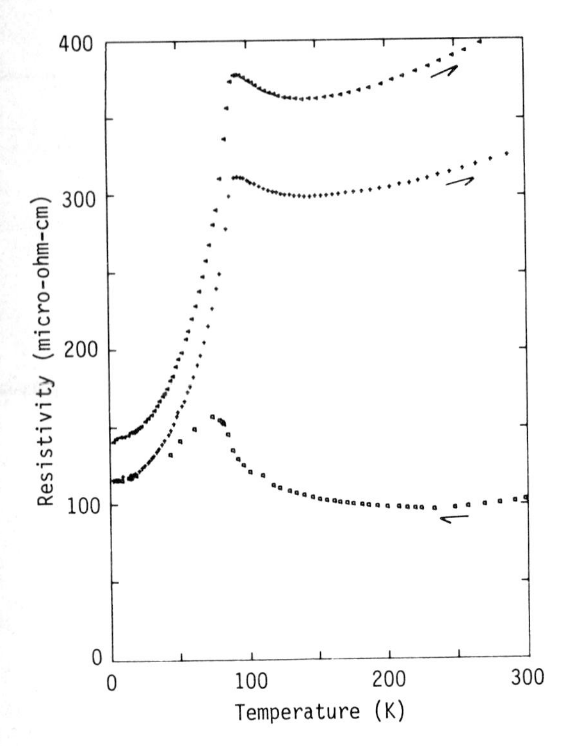

Fig. 6. Resistivity of $Eu(Pd_{0.9}Au_{0.1})_2Si_2$. Two lower traces: cycle 1; uppermost trace: warming segment of cycle 3 (see text).

peak observed in ρ(T) in the present instance arises from some alien mechanism; e.g., perhaps it originates from a subset of micro-cracks which behave in a reversible manner, viz., forming on cooling and annealing out on warming.

4. DISCUSSION

It is clear that $Eu(Pd_{1-x}Au_x)_2Si_2$ is an archetypal system for studies of the competition between mixed valence and magnetic ordering; and, to our knowledge, the first such system exhibited. Factors contributing to the uniqueness of the system, which makes it suitable for this purpose, are: (a) the occurrence of a valence transition with temperature, (b) the large spin of Eu^{2+} which leads to relatively large magnetic ordering temperatures and (3) the facility of being able to vary the stability of the Eu^{2+}-state by varying the Au-concentration.

The effectiveness of the additive Au in stabilizing the Eu^{2+} moment results probably from two effects. One is the increase in the lattice parameter, which simulates a negative pressure, and the second may involve the high efficiency of the Pd 4d-electrons in screening the core hole associated with the Eu^{3+} state (see, e.g., Ref. 11). In $EuPd_2Si_2$ the Pd-derived 4d-band is relatively narrow (∿ 3-4 eV) and situated just below E_F as evidenced by valence band photoemission measurements.(12) A consequence of replacing Pd by Au is to reduce the density of states in the 4d-band. This would reduce the efficiency of the screening, which, in turn, would enhance the stability of the Eu^{2+}-state.

ACKNOWLEDGMENTS

We wish to thank Mark Croft for a number of very useful discussions. This work was supported by the National Science Foundation. The high field studies were performed at the Francis Bitter National Magnet Laboratory, Massachusetts Institute of Technology, which is supported by the National Science Foundation. We are particularly grateful for the helpful assistance given us by the staff of the Magnet Lab.

REFERENCES

* Permanent address: Tata Institute of Fundamental Research, Bombay, 400005, India.

** Present address: Department of Physics, Bar Ilan University, Ramat Gan, Isreal.

1. E. V. Sampathkumaran, R. Vijayaraghavan, K. V. Gopalakrishnan, R. J. Pillay, H. G. Devare, L. C. Gupta, B. Post and R. D. Parks, Valence Fluctuations in Solids, edited by L. M. Falicov, W. Hanke, and M. P. Maple (North-Holland, Amsterdam, 1981)

2. J. M. Lawrence, P. S. Riseborough, and R. D. Parks, Rep. Prog. Phys. 44, 1 (1981).

3. R. D. Parks, V. Murgai, M. denBoer, S. Raaen and L. C. Gupta, to be published.

4. M. C. Croft, J. A. Hodges, E. Kemly, A. Krishnan, C. Segre, V. Murgai, L. C. Gupta and R. D. Parks, these proceedings.

5. B. Batlogg, A. Jayaraman, V. Murgai, L. C. Gupta, R. D. Parks and M. C. Croft, these proceedings.

6. M. C. Croft, J. A. Hodges, E. Kemly, A. Krishnan, V. Murgai and L. C. Gupta, Phys. Rev. Lett. 48, 826 (1982).

7. C. M. Varma and Y. Yafet, Phys. Rev. B 13, 2950 (1976).

8. B. H. Grier, R. D. Parks, S. M. Shapiro and C. F. Majkrzak, Phys. Rev. B 24, 6242 (1981).

9. W. Franz, A. Griessel, F. Steglich and D. Wohlleben, Z. Physik B 31, 7 (1978).

10. J. M. Lawrence and R. D. Parks, J. Physique 37, C4-249 (1976).

11. O. L. T. de Menezes, A. Troper, P. Lederer and A. A. Gomes, Phys. Rev. B 17, 1997 (1978).

12. N. Mårtensson, B. Reihl, W. -D. Schneider, V. Murgai, L. C. Gupta and R. D. Parks, Phys. Rev. B 25, 1446 (1982).

Valence Instabilities
P. Wachter and H. Boppart (eds.)

PRESSURE-TEMPERATURE STUDIES AND THE P-T DIAGRAM OF $EuPd_2Si_2$

B. Batlogg and A. Jayaraman
Bell Laboratories
Murray Hill, New Jersey

V. Murgai, L. C. Gupta and R. D. Parks
Polytechnic Institute of New York
Brooklyn, New York

M. Croft
Rutgers University
Piscataway, New Jersey

The pressure and temperature dependence of the resistivity of $EuPd_2Si_2$ and the linear compressibility as well as thermal expansion at ambient pressure have been determined. By comparison with one atmosphere and high pressure data, the peaks in resistivity, compressibility and thermal expansion are interpreted in terms of valence change in Eu. The peaks occur when the average valence has a value close to 2.5. A P-T diagram is proposed, featuring a first-order transition boundary terminating at a critical point at negative pressures. The slope of the phase boundary is in agreement with the slope calculated from the entropy difference between the $4f^6$ (J=0) and $4f^7$ (J=7/2) ionic configurations

INTRODUCTION

It is now well established through Mössbauer and lattice parameter measurements, (1,2) that the intermetallic compound $EuPd_2Si_2$ undergoes a rather sharp but continuous valence transition in the vicinity of 150K at ambient pressure. The observation of one Mössbauer line and its continuous shift toward the Eu^{3+} state have confirmed the mixed valent nature of Eu. The average valence of Eu increases from 2.2 at 300K to 2.9 at 4.3K.

It is of interest in this connection to investigate the effect of pressure on $EuPd_2Si_2$, for pressure always favors the higher valence state and may be expected to cause the same transition. Indeed the effect of high pressure on the resistivity and thermoelectric power of $EuPd_2Si_2$ was recently investigated(3) using a Bridgman anvil and compacted samples. Although the data obtained were interpreted in terms of a valence change, the measurements, particularly the initial resistivity variation, was masked by the large changes associated with compaction and contact problems. We felt, four probe resistivity measurements on good crystals, performed under hydrostatic pressure would reveal the true nature of the $\rho(P)$ behavior. In this connection it was also felt that a continuous length change measurement under hydrostatic pressure, using a sensitive strain gauge technique, could provide valuable information regarding the compressibility behavior of $EuPd_2Si_2$, in the region of valence collapse. We have performed these measurements. In addition, we have also measured the resistivity of $EuPd_2Si_2$ as a function of temperature at different pressures, and the resistivity and linear thermal expansion as a function of temperature at ambient pressure. The resistivity and linear compressibility along the a-axis exhibit a peak at about 12 kbar. Likewise, the linear thermal expansion and resistivity exhibit a peak centered around 150K. From the pressure-temperature data, a P-T diagram for $EuPd_2Si_2$ is proposed. These results will be presented and discussed in this paper.

EXPERIMENTS AND RESULTS

Hydrostatic pressure was generated in a piston-cylinder device using the well known Teflon Cell technique. Isoamylalcohol which remains fluid up to 37 kbar was used as pressure medium. For resistivity measurements, four leads were pressed on to indium contacts which were previously cold-welded on to the sample. The as-grown material had a layered structure and hence could be shaped roughly into a bar-like sample of about 7-8 mm in length. For low temperature resistance measurements at high pressure, the pressure plate was cooled by immersing into liquid N_2. Length change measurements were made with a very sensitive strain gauge which was attached to a flat sample of about 1 mm in thickness and 4-5 mm across, using the proper adhesive. The sensitive direction of the strain gauge had to be oriented along the a-direction of the layered crystals, which turned out to be along the edges of the platelets. The thermal expansion and resistivity measurements at 1 bar were done on the same samples which were previously exposed to high pressure.

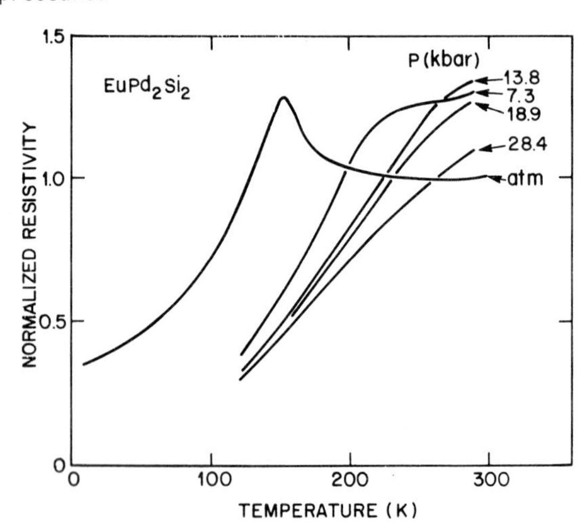

Fig. 1 Temperature dependence of the resistance (normalized to the resistance at RT and 1 atmosphere) of $EuPd_2Si_2$ at different pressures including 1 atmosphere. The prominent peak near 150K is associated with the valence change. Note the broadening and eventual disappearance of the peak at high pressures (see text for explanation).

In Fig. 1 the normalized resistivity vs temperature obtained at different pressures is shown. The peak in the R vs T curve for 1 atmosphere is very striking. With pressure, the peak gets wiped out and the resistivity monotonically decreases. In Fig. 2 the R vs P data are shown. Again, the occurrence of a peak centered around 12 kbar is clear. The dotted curves represent R vs P data at different temperatures, extracted from the primary data shown in Fig. 1.

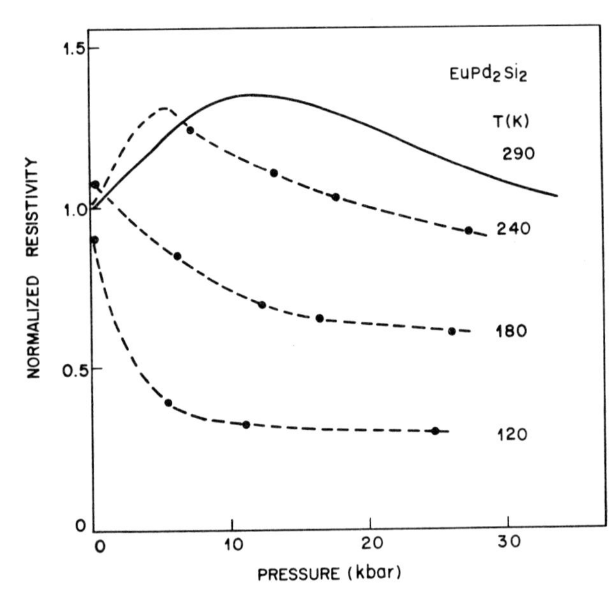

Fig. 2 Pressure dependence of the resistance at a number of temperatures. Note the maximum in the resistance in the 290K data at about 12-15 kbar. This feature is due to pressure-induced valence change and the maximum roughly coincides with the attainment of an average valence of $\sim$ 2.5. The data appearing in the dotted curves were obtained from the primary data in Fig. 1.

It is well established from Mössbauer measurements at ambient pressure that the valence of Eu in $EuPd_2Si_2$ rapidly changes around 150K.(1,2) It is highly significant that the resistance vs T data at atmospheric pressure shows a peak at about 150K. From this we conclude that the resistance anamoly is related to the valence change. Further, the width of the peaks in R(T) and R(P) indicate that the valence change is spread over a large pressure range.

Again, the R vs P data at 290K shown in Fig. 2 exhibits a maximum in resistivity at about 12 kbar. Taking the result of the Mössbauer study(1) and the presently obtained R vs T data, it is to be concluded that the resistance maximum occurs when the valence of Eu attains a value close to 2.5. This is supported by the recent Mössbauer studies under pressure on $EuPd_2Si_2$.(4) If this average valence is already induced through increase in pressure or decrease in temperature, these maxima are not seen in the R(T) and R(P) curves (see Fig. 1 for pressures above 7.3 kbar).

In Fig. 3 the linear compressibility variation with pressure obtained from the length change measurement is shown. In the upper part of the figure the R vs P data at 300K is displayed again for comparison. The maximum in the compressibility corresponds to the maximum in the resistivity from which we conclude that the former maximum would correspond to an average valence of 2.5, and beyond this the lattice gets stiffer with pressure.

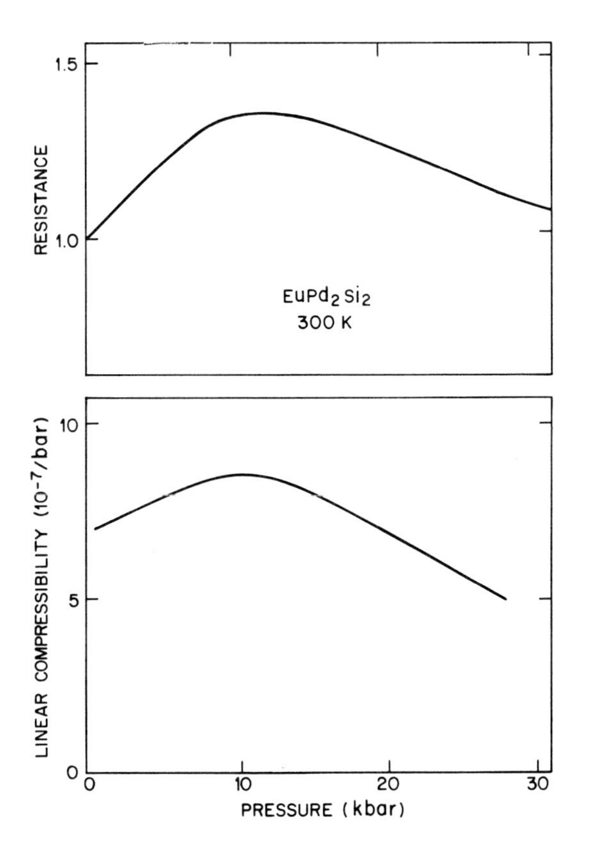

Fig. 3 Pressure dependence of the linear compressibility obtained from length change measurements is compared with the ρ vs P data. The maximum in the two curves occurs at about the same pressure (12-15 kbar). This anomalous behavior in compressibility is attributed to valence change and the maximum to average valence of ∿ 2.5.

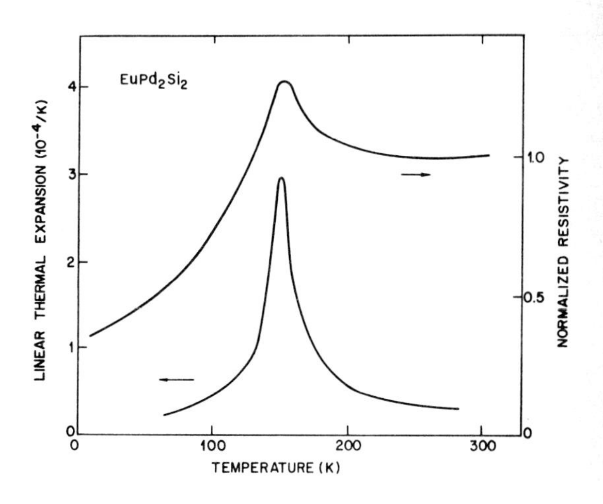

Fig. 4 Comparison of linear thermal expansion and resistance as a function of temperature. The peaks are related to the attainment of an average valence of ∿ 2.5, supported by Mössbauer data.

Fig. 4 is a plot of normalized resistivity as well as the linear thermal expansion α against T, taken at atmospheric pressure. The peaks in the two measured quantities occur in the same temperature region. The small difference in the peak position (4-8K) is significant and will be discussed elsewhere. Obviously the thermal expansion and the compressibility reflect the underlying mechanism, namely the strong coupling between the valence change and volume.

DISCUSSIONS

From the results presented the following scenario can be envisaged for $EuPd_2Si_2$. With decreasing temperature or increasing pressure, the $4f^7$ states of Eu sweep through the Fermi level and is emptied half way at about

150K or ∼ 12-15 kbar. The initial increase in resistivity may be explained in terms of coulomb and spin scattering of electrons involving the 4f states. However in the compressed phase the increase in the number of electrons due to promotion, the normal temperature or pressure effect on electron-phonon scattering, and the decrease in the spin disorder scattering as the valence change progresses, all conspire to decrease the resistivity. With all these contributions operating it is difficult to analyze the resistivity quantitatively, but we believe the above is qualitatively true. From our results it appears that the later terms dominate the resistivity behavior when the average valence is higher than ∼ 2.5.

For the compressibility behavior the explanation may be more involved. We have to keep in mind that the actual course of the phase transition is determined by the interplay between electronic and elastic energies. While the former is related to both the f and d-band density of states, the elastic energy term involves nonlinear interaction terms in stress-strain.

Using our pressure-temperature data, a phase diagram for $EuPd_2Si_2$ in the P-T plane is proposed in Fig. 5. An arbitrary criterion has been chosen and it is obvious that the transition pressure increases with temperature and widens as pressure is increased. This behavior would be consistent if the phase transition terminates at a critical point at negative pressures and low temperatures, for we know that the transition is already continuous at atmospheric pressure. Below the critical point it is first-order, and above it is a continuous transition, as is observed for the γ-α phase transition in Cerium. The trajectory of the first-order transition has been drawn to be consistent with the two limits for the slope of the phase boundary obtained from the present data. The steep intersection with the pressure axis is a thermodynamic necessity, dictated by the vanishing of the entropy at absolute zero. The positive slope of the phase boundary is consistent with

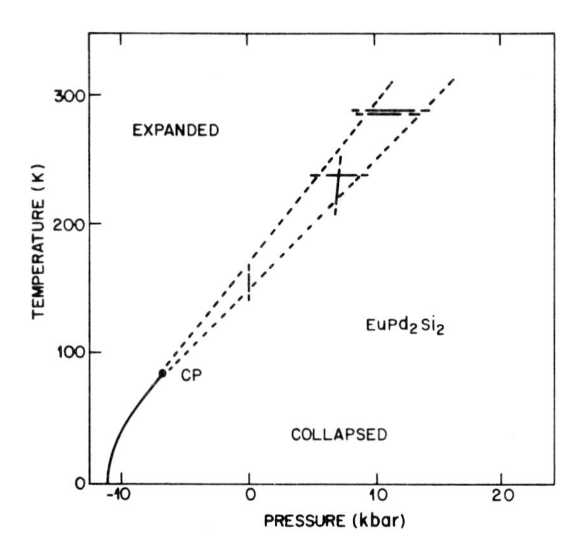

Fig. 5 P-T diagram for $EuPd_2Si_2$. The transition is continuous in the region of positive pressure. At negative pressures and at lower temperatures a first order transition boundary terminating at a critical point is proposed. This appears to be consistent with the observation of a first-order valence change in Au substituted $EuPd_2Si_2$.

the entropy difference between the two phases, arising out of the different degeneracies of the $4f^7$ and the $4f^6$ configuration. The expanded phase is magnetic (J=7/2) while the fully collapsed phase is nonmagnetic. A calculation of the slope for the phase boundary taking into account only the magnetic contribution to the entropy in the two phases yields a dT/dP of 10 degrees/kbar. The volume difference of 1.7 cm^3/mole was taken from the lattice parameter systematics and our direct compressibility and thermal expansion data. The experimentally obtained slope is about 10 degrees/kbar. Even as we have neglected all other contributions to the entropy, the calculated value agrees remarkably well with the experiment. This is meant to show the dominance of the magnetic contribution to the entropy. This T=0 calculation can be improved by considering the next exciting multiplets of the 4f configurations, with a resulting increase of the slope of the phase boundary for higher temperatures.

It is well known that negative pressures can be simulated through chemical substitution.(5) Indeed there are clear indications that with the substitution of gold for Pd in $Eu(Pd_{1-x}Au_x)_2Si_2$, the transition becomes first order for Au concentration exceeding x=0.10. This gives substantial support to the proposed P-T diagram. A simple relationship between the Au concentration and the negative pressure appears to exist.

ACKNOWLEDGMENTS

The authors are thankful to R. G. Maines for assistance in performing high pressure measurements.

REFERENCES

[1] Sampathkumaran, E.V., Vijayaraghavan, R., Gopalakrishnan, K.V., Pilley, R.G., Devare, H.G., Gupta, L.C., Post, B. and Parks, R.D. in "Valence Fluctuations in Solids", Editors L. M. Falicov, W. Hanke and M. B. Maple (North-Holland, Amsterdam, p. 193, 1981).

[2] Croft, M., Hodges, J.A., Kemly, E., Krishnan, A., Murgai, V. and Gupta, L.C., Phys. Rev. Lett. 48, 826 (1982).

[3] Vijayakumar, V., Vaidya, S.N., Sampathkumaran, E.V., Gupta, L.C., Vijayaraghavan, R., Phys. Lett., 83A, (1981) 469.

[4] Schmiester, G., Perscheid, B., Kaindl, G. and Zukrowsky, J., these proceedings.

[5] McWhan, D.B. and Remeika, J.P., Phys. Rev. B2 (1970) 3734.

[6] Gupta, L.C., Murgai, V., Yeshurun, Y. and Parks, R.D., these proceedings.

Valence Instabilities
P. Wachter and H. Boppart (eds.)

SUPPRESSION OF INTERMEDIATE VALENCY IN AMORPHOUS $EuPd_2Si_2$

J.A. Hodges and G. Jéhanno

DPh-G/PSRM - Centre d'Etudes Nucléaires de Saclay
91191 Gif-sur-Yvette Cedex - France

^{151}Eu Mössbauer measurements on amorphous $EuPd_2Si_2$ show that the Eu^{2+} state is stable at all temperatures and that magnetic ordering occurs below 8K. Recrystallisation of the sample by annealing reintroduces the intermediate valency state.

1. INTRODUCTION

Many examples are now known of homogeneous intermediate valence (IV) occurring on dilute 4f impurities in metallic hosts (1)(2). Lattice periodicity of the rare earth atoms is thus not a prerequisite for the existence of IV. All that is required is a suitable interplay between the conduction electrons and the localised 4f electrons of the rare earth atom. Here we examined whether this interplay could be evidenced in an amorphous matrix. The system examined $EuPd_2Si_2$, was chosen as it shows well established IV behaviour in the crystalline state (2)(3) and also because Eu shows IV behaviour when diluted in the various RPd_2Si_2 (2) (4) where the bulk lattice volume changes by about 8% across the series (5). This suggests that the presence of IV in this system is not particularly sensitive to changes in lattice volume. Although we find that in $EuPd_2Si_2$ the transformation from the crystalline to the amorphous state suppresses the IV behaviour, this does not preclude the possibility of establishing IV in other analogous amorphous systems.

2. EXPERIMENTAL RESULTS

The parent crystalline sample was prepared by direct melting in an induction furnace. It was the common starting point for the four samples whose results are briefly presented here.

For the first sample, the parent crystalline sample was annealed at 820C for 2 days. The ^{151}Eu Mössbauer absorption results were of the usual form (2)(3) with an isomer shift (relative to EuF_3) IS=-7.4mm/s (-130 MHz) at 295K and -0.5mm/s (-8 MHz) at 4.2K with a very rapid change in IS (or in the average valence) near 150K. The Europium centres having these characteristics account for 80% of the total Mössbauer absorption area. As in other samples of $EuPd_2Si_2$ (2)(3) the minority component (20% of the absorption area in the present case) which has the same IS value as the main component at room temperature, shows IS changes which occur at temperatures well below 150K.

A 12μ thick amorphous sample was obtained from the parent crystalline sample by sputtering. In the measurement range 295 to 1.4K the IS was temperature independent at -10.0mm/s. This value is characteristic of Eu^{2+} in a metallic environment and is close to the value previously adopted for the Eu^{2+} configuration in crystalline $EuPd_2Si_2$ (2). Below about 8K a magnetic hyperfine structure becomes visible with the average hyperfine field increasing to 240 kOe at 1.4K. As in other amorphous compounds, the thermal variation of the hyperfine field (and of the electronic magnetic moment) is much flatter than that of the associated Brillouin function. The (Mössbauer derived) Debye temperature is 175K which is significantly lower than that observed (215K) in the high temperature region of crystalline $EuPd_2Si_2$ (2).

The third sample was obtained by annealing the amorphous sample at 820C for 2 days. Room temperature X-ray measurements showed that it had recrystallised with the same lattice parameters as for the parent crystalline sample. Mössbauer measurements at 295K show that a small percentage (5%) of the Europium now has an associated IS of -13mm/s characteristic of the presence of a non-metallic Eu^{2+} state and that the dominant contribution (95%) has an IS near -7.6mm/s clearly characteristic of the same IV state as in the parent crystalline sample. As the temperature is lowered a sharp valence change towards the state with IS near -1.0mm/s again occurs near 150C. However this now only concerns ~ 25-30% the total absorption area compared to ~ 80% in the annealed sample. As the temperature is lowered a progressively greater proportion of the surface area is centered at an IS near -1.0mm/s such that at 4.2K, 40% of the total area corresponds to this state. The remaining 60% of the total area is contained in a very broad line which is very roughly centred towards the room temperature IS value.

The parent crystalline sample when annealed and the recrystallised sample thus both show the same IV behaviour at room temperature but somewhat different IV behaviour at lower temperatures. To examine the influence of sample history on these differences, we examined a fourth sample ("quenched" sample) obtained by letting the molten parent sample (held in a levitation furnace) drop into a copper bowl. The sample so obtained again has the same room temperature lattice parameters and IS value as the other crystalline samples. The sharp valence change occurring near 150K now involves ~60% of the total Mössbauer area that is, it involves a percentage which is roughly intermediate between those observed in the annealed and recrystallised samples.

It is possible that the progressive differences in the low temperature results observed in going from the annealed to the quenched and then to the recrystallised sample are due to the presence of varying amounts of disorder among the Eu neighbours for the three samples (6). It is also possible however that these changes are due to the influence of the lattice strains on the coupling between the localised and conduction band electrons.

Whereas the different thermal histories of the three crystalline samples are seen to modify the details of the IV behaviour at low temperatures, the change from the crystalline to the amorphous state is seen to quench out completely the IV state. In going from the crystalline to the amorphous state the relative change occurring in the positions of the localised 4f levels and the Fermi level results in the 2+ level being stabilised well below the Fermi level as it is in $EuPd_2$, $EuSi_2$ and in the surface layers of crystalline $EuPd_2Si_2$ (7).

ACKNOWLEDGEMENTS

We are greatly indebted to Dr. B. Boucher and R. Tourbot for the preparation of the amorphous sample.

REFERENCES

1. D. Wohlleben p.1 in "Valence Fluctuations in Solids" edited by L. Falicov et al. (North-Holland, Amsterdam, 1981).
2. M. Croft et al., Phys. Rev. Lett. 48 (1982) 826.
3. E.V. Sampathkumaran et al., J. Phys. C 14 (1981) L237.
4. R.G. Pillay et al., Int. Conf. on Mössbauer Spectroscopy, Jaipur, India (1981).
5. D. Rossi et al., J. Less Com. Metals 66 (1979) P17.
6. E.R. Bauminger et al., Phys. Rev. Lett. 30 (1973) 1053.
7. N. Martensson et al., this conference.

PHOTOEMISSION

Valence Instabilities
P. Wachter and H. Boppart (eds.)

NATURE OF THE 4f STATES OF CERIUM AS STUDIED BY PHOTOEMISSION

R.D. Parks

Department of Physics
Polytechnic Institute of New York
333 Jay Street, Brooklyn, New York 11201

N. Mårtensson* and B. Reihl**

IBM T.J. Watson Research Center
Yorktown Heights, New York 10598

Photoemission experiments using synchrotron radiation have been performed on $Ce_{0.9}Th_{0.1}$ (Th being present to stabilize the fcc phase) as a function of temperature as the γ-α transition is traversed. A bimodal spectrum is observed for both the γ- and α-states, which consists of a feature near E_F and one displaced by about 2eV. The relative strengths of the two features, but neither their locations (in energy) nor their combined intensity, change at the transition. A comparative study of the photoemission spectra of γ-$Ce_{0.9}Th_{0.1}$, $La_{0.9}Th_{0.1}$ and $Pr_{0.9}Th_{0.1}$ is employed to establish the f-character of the feature near E_F in cerium, thought previously to be principally a 5d-emission. Line shape analyses of the spectra address the question of a possible bulk-to-surface shift in the f-emission of cerium.

1. INTRODUCTION

In the promotional model of homogeneous mixed valence, the two states f^nc^m and $f^{n-1}c^{m+1}$ are isoenergetic, where f represents a localized electron, c a band electron, and n and m are integers (see, e.g., Ref. 1). Valence band photoemission has proved a powerful tool for the study of mixed valence because of the formal identity between the final hole state $f^{n-1}c^{m+1}$ (resulting from emission from the initial state f^nc^m) and the isoenergetic partner of f^nc^m in the mixed valence ground state. This results from the complete local screening of the excited f-hole in metallic systems by one conduction electron which moves into the Wigner-Seitz cell containing the f-hole. This formal correspondence implies that f-emission from the f^nc^m state, since it requires no energy, must occur at E_F. In the presence of multiplet structure in the f-emission, it is that feature with the smallest binding energy which must occur at E_F. This picture works perfectly in the context of valence photoemission studies of Sm-, Eu-, Tm- and Yb-based homogeneous mixed valent systems (see, e.g., Refs. 1-3); whereas, it fails in zeroth order in the case of Ce-based systems.

The photoemission spectra of various Ce-based systems thought previously to be homogeneously mixed valent fail to exhibit the single feature at E_F required by the promotional model. In all cases there is a feature in the f-emission which occurs at a binding energy in the range 1.4 - 3 eV (see, e.g., Refs. 4,5). Failure of the promotional model in describing the 4f instability in cerium is found as well in other contexts. Johansson (6) provided an early challenge by pointing out that the difference in the heat of formation between Ce^{3+} and "Ce^{4+}" is about half the value expected in the simple promotional model, which specifies that the electronic configurations of the two states are $4f^15d^16s^2$ and $4f^05d^26s^2$, respectively. Various experiments [e.g.: positron annihilation (7), Compton scattering (8), neutron scattering (9), and photoemission (10)] suggest that there is no change in the number of 4f-electrons at the γ-α transition in cerium; whereas the promotional model requires a change of about 50%.(1) Equally inexplicable in the promotional model scheme is the observation of a bimodal feature in the L_{III} x-ray absorption edge spectra for Ce-compounds thought to be tetravalent.(11)

The purpose of the present investigation was twofold: (1) to follow the valence band photoemission spectrum of cerium through the γ-α by doing a variable temperature study and (2) to make a comparative study of Ce, La and Pr in order to isolate the 4f-emission from the valence band spectrum. For both segments of the study the use of variable photon energy is essential, since by tuning the incident energy it is possible to vary the ratio of 4f- to 5d-emission. Studies of the γ-α transition in cerium (without the use of external pressure) are complicated by temperature-dependent contamination by the (hexagonal) β-phase, which spoils the transition. One method for circumventing this problem is to use two samples; viz., a conventional bulk or thin film sample for the study of the γ-state at room temperature and a thin film deposited on a low temperature substrate (which enhances the α-phase) for the study of the α-state (see, e.g., Ref.12). A second method, which was employed in the present study, is to stabilize the fcc phase of cerium with a chemical additive, in our case with 10% Th, which has virtually no effect on the γ-α transition.(13) The advantage of the latter method, in the context of the present study, and in view of the fact that the transition is first order, is that it allows the two different states to be sampled in a short time interval, during which all of the parameters in the measurement remain constant and there is negligible surface contamination. This has two consequences: (1) it increases the confidence limit that any observed differences in the γ- and α- spectra reflect only changes produced by the γ-α transition and (2) it allows the monitoring of the 4f-population through the transition.

2. EXPERIMENTAL PROCEDURE

Samples of $La_{0.9}Th_{0.1}$, $Ce_{0.9}Th_{0.1}$ and $Pr_{0.9}Th_{0.1}$ were prepared by arc melting the pure materials on a water-cooled copper hearth in an argon atmosphere. All of the starting materials were nominally 99.9+% pure with respect to metals. The raw materials were melted together and the resulting boule was flipped over and remelted. The latter step was repeated until the samples had been melted 6-8 times. Each sample was then wrapped in tantalum foil, sealed in an evacuated quartz tube, and annealed overnight between 750 and 800 C. From the resulting boules, and for use in the photoemission experiments, samples were extracted and shaped into rectangular parallelpipeds with dimensions ∿ 4mm x 4mm x 6mm.

The photoemission studies were made with a display-type analyzer (14) using synchrotron radiation at the Synchrotron Radiation Center of the University of Wisconsin-Madison. This system was operated in an angle-integrated mode with a total (electrons and photons) energy resolution of ∿ 0.15 at 50 eV. The sample was first introduced into a preparation chamber where the surface was cleaned by machining it with a diamond file. Following this the sample was immediately transferred to the measuring position, where the vacuum was better than 10^{-10} Torr. The surface cleanliness, with respect to all important contaminants, was confirmed by Auger and UPS.

3. CHANGE IN VALENCE BAND SPECTRUM AT THE γ-α TRANSITION

After the $Ce_{0.9}Th_{0.1}$ sample was cleaned and transferred to the measuring position, and during the course of the impending experiment (several hours), Auger and UPS measurements established that surface contaminants comprised less than 0.1 monolayer. Repeated spectra with intervals of only a few minutes, were recorded while the sample was cooled. Between two such recordings there occurred an abrupt change in the spectrum [corresponding to the γ-α transition, which occurs at T ∿ 140 K (13)]. Further cooling down to ∿ 50 K produced no additional changes in the spectrum.

We first view (Fig. 1) the spectrum recorded at hν = 122 eV, where there is a giant resonance in the f-emission, which was first observed by Johansson et al. (15). The spectrum for γ-$Ce_{0.9}Th_{0.1}$ is similar to that reported for γ-Ce previously. (4, 15) The strong feature at 2.1 eV has previously been identified as the 4f position corresponding to a localized $4f^1 \rightarrow 4f^0$ transition. (15, 16) If the promotional model were applicable to cerium, one would expect the 2 eV peak to move close to E_F in the "collapsed" α-phase.(1) Clearly, no such shift is observed. While there is no significant energy shift of any of the spectral features at the transition, there is a definite decrease in the intensity of the 2 eV peak, while the feature near E_F becomes more pronounced. The resonance in the f-emission has been identified as resulting from the transition $4d^{10}4f^1(5d6s)^3 \rightarrow 4d^9 4f^2(5d6s)^3$, followed by the decay process $4d^9 4f^2(5d6s)^3 \rightarrow 4d^{10}4f^0(5d6s)^3 + e$, the latter process being a so-called "Super Coster-Kronig" transition.(15, 17) However, it is not just the f-emission which resonates near 122 eV, as seen for example in the spectra of Ref. 15, which show enhancements of all features within ∿ 5 eV of the f-emission for hν ∿ 120 - 122 eV. This reflects the importance of the auxillary decay channel, $4d^9 4f^2(5d6s)^3 \rightarrow 4d^{10}4f^1(5d6s)^2 + e$, wherein the final state is a band-hole instead of a localized 4f-hole.(15) The presence of this second process complicates the problem of deconvoluting f-features from band-features in spectra taken near hν = 122 eV.

In Fig. 2 spectra are shown for hν = 50 eV. At this phonon energy the change in the spectrum at the transition appears more dramatic than at 122 eV, which results, as is shown in Section 4, from a relative maximum in the intensity of the feature near E_F at hν ∿ 50 eV. As discussed in Section 1, the conditions of the experiment allow a meaningful comparison of the intensities in the γ- and α-state spectra. We find no change in the total photoemission intensity in the valence region (0 - 3 eV) at the phase transition, as monitored by both the 50 eV and 122 eV spectra. Since the ratio of f- to (5d,6s)-emission is different for the two photon frequencies, this result implies that there is a negligible change in the f-count at the transition.

4. COMPARISON OF VALENCE BAND SPECTRA OF γ-$Ce_{0.9}Th_{0.1}$, $Pr_{0.9}Th_{0.1}$ and $La_{0.1}Th_{0.1}$

The motivation for the study reported in this section is to attempt to understand the origin of the feature near E_F in the spectra of Figs. 1 and 2; in particular, to determine to what extent, if any, it has f-content. The approach is to compare the spectra of the adjacent elements La, Ce and Pr: La has no f-electron, Ce has one "unstable" f-electron and Pr has two stable f-electrons. As in the case of Ce,

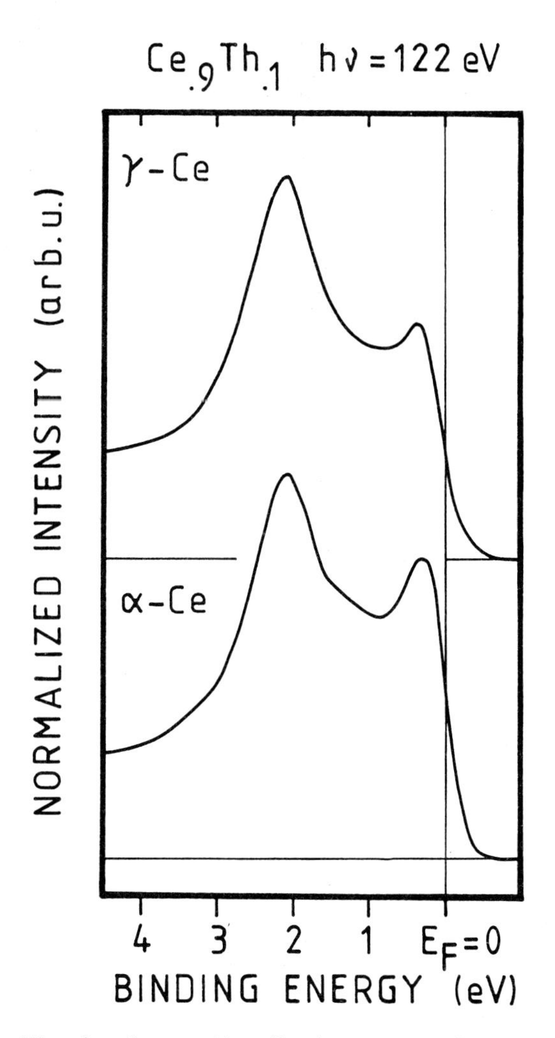

Fig. 1. Energy distribution curves of photoelectrons from $Ce_{0.9}Th_{0.1}$ at hν = 122 eV.

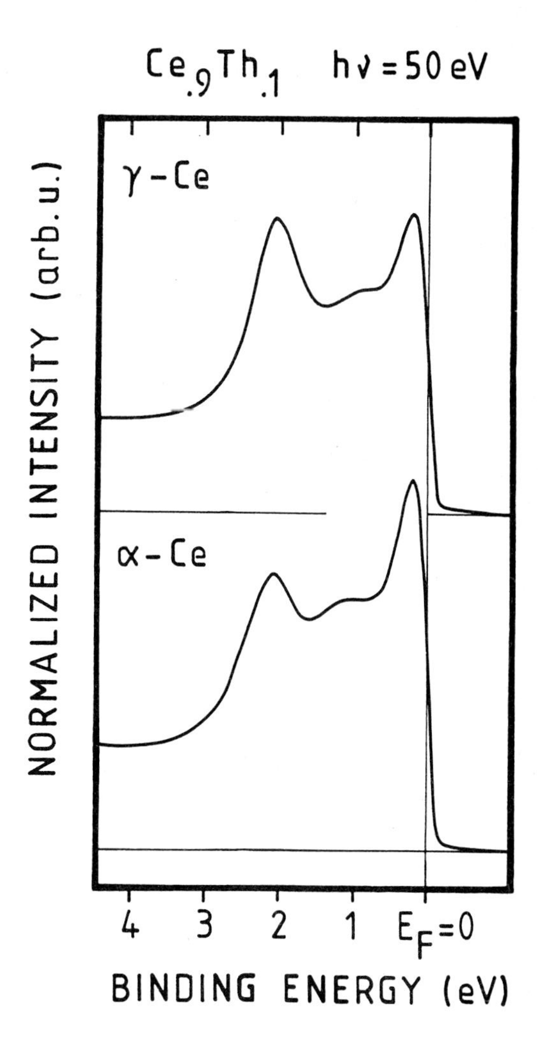

Fig. 2. Energy distribution curves as in Fig. 1 at hν = 50 eV.

10% Th was added in order to stabilize the fcc structures of La and Pr, thereby allowing any comparisons to be meaningful. All of the spectra reported in this section were recorded at room temperature.

Figure 3 reveals how the Pr spectrum evolves as the photon energy is increased from hν = 100 eV (the spectra are not normalized relative to one another). The dominance of the 5d-emission over the 4f-emission at the lowest energy is a consequence of the delayed onset of the 4f-photo-cross-section due to the centrifugal barrier (see, e.g., Ref. 18). At 50 eV this barrier has been essentially overcome, at which point the 4f-emission dominates the spectrum. As the photon energy is further increased the relative strength of the 5d-emission continues to decrease.

In Figs. 4, 5 and 6 we compare the spectra for $La_{0.9}Th_{0.1}$, $Ce_{0.9}Th_{0.1}$ and $Pr_{0.9}Th_{0.1}$ at photon energies of 30, 50 and 70 eV respectively. At 30 eV the 4f-emission is only a few percent of the valence band (VB) emission (which we refer to as 5d VB in the figures because of the dominance of 5d-emission in the relatively narrow emission near E_F, although for the Ce-spectra such a labeling is tentative, see discussion below). Hence, the 30 eV spectra serve as the reference spectra for comparing the VB structure of the three systems; indeed, the three spectra appear very similar. However, in the 50 eV spectra, where the 4f intensity is roughly 70% of the VB intensity, the VB features for the three systems are disimilar. In the Ce spectrum the peak at the Fermi level which we shall refer to as peak A below, is now more prominent and more distinct than in either La or Pr. Note, however, that also in the La spectrum this peak is enhanced and more extended than in Pr. Turning now to the 70 eV spectrum (Fig. 6), the

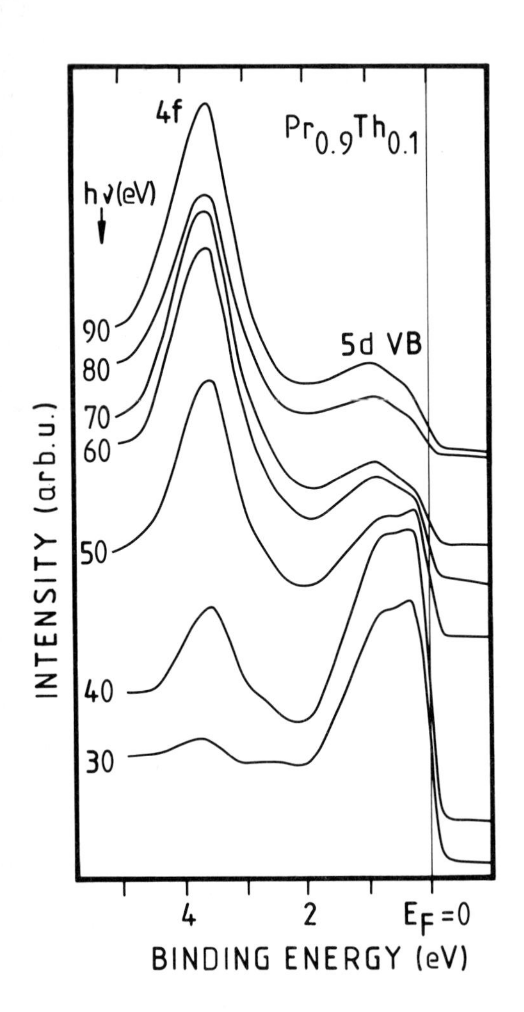

Fig. 3. Energy distribution curves as in Fig.1 for $Pr_{0.9}Th_{0.1}$ at photon energies ranging from 30 to 90 eV.

differences in the VB spectra are even more dramatic. It is now only in the Ce spectrum that the peak at E_F dominates the VB region. In La and Pr the less well defined maximum is found instead at ∿ 0.9 eV from the Fermi level. The evolution of the VB feature in Ce, relative to that of La or Pr as one increases the photon energy, and in turn the 4f dominance of the spectrum, provides clear evidence that the peak near E_F in Ce has considerable 4f-character.

An alternative method for demonstrating the above conclusion is to attempt to construct different spectra in a manner that contrasts Ce to La and Pr. This we attempt in Fig. 7, where we compare the 4f and VB (extended "5d") emissions on an absolute scale, taking into account the variations in the monochromator efficiency, synchrotron beam current, etc. In determining the VB intensities for Pr and La, and the 4f intensity for Pr, we integrated the corresponding spectral features. The procedure for Ce was necessarily different because of the convolution of the 4f and VB portions of the spectrum. The procedure we followed was to use for the VB intensity the integrated intensity from zero to 0.75 eV (calling this the intensity of peak A), thereby avoiding the region of the main 4f emission. For the 4f intensity, we used the intensity at the maximum of the 2eV peak. In Fig. 7 the 5d intensity at hν = 30 eV has arbitrarily been set to 100, and the determined 4f intensities have been divided by two to account for the fact that there are two 4f electrons in Pr. Implicit in all of the above discussions of the VB spectra is the neglect of contributions from the sp-bands, which is justified because of the smallness of the s and p cross-sections at these photon energies. The presence of 10% Th in the three systems should lead to no serious difficulty, since while the Th 6d-emission is expected to contribute to the VB intensity, it should approximately mimic the rare earth 5d-emission at these photon energies. Furthermore, and most important, the contribution is expected to be the same in the spectra of all three systems.

From Fig. 7 a clear review can be obtained of the hν dependence of the 4f and 5d intensities (where throughout we use the terms '5d' and 'VB' interchangeably, but with reservation in the case of Ce as discussed above). The 5d intensity drops off rapidly as a function of photon energy with its values at 50 eV and 70 eV being only ∿ 16% and ∿ 5% respectively of its value at 30 eV. On the other hand the 4f emission is negligibily small at 30 eV because of the centrifugal barrier discussed above. With increasing photon energy this barrier is overcome

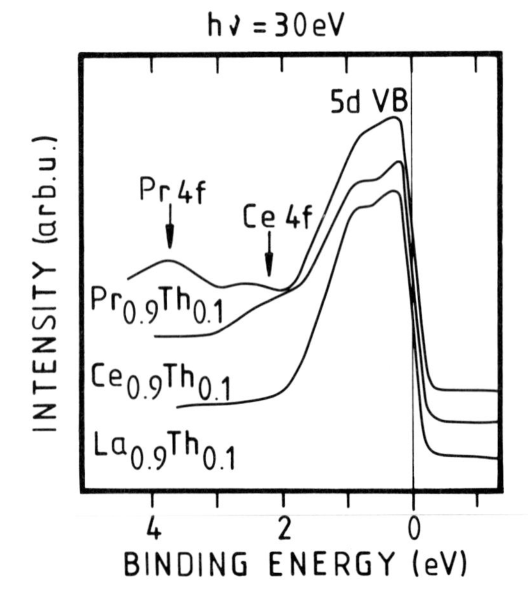

Fig. 4. Energy distribution curves for $La_{0.9}Th_{0.1}$, $Ce_{0.9}Th_{0.1}$ and $Pr_{0.9}Th_{0.1}$ for hν = 30 eV.

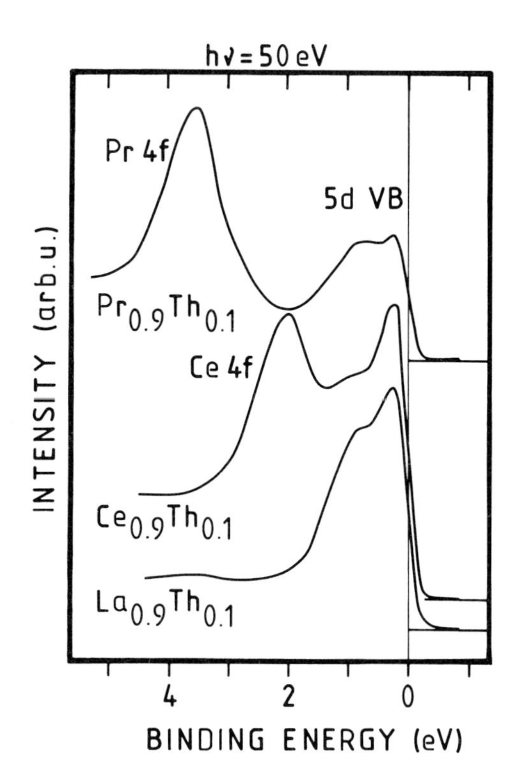

Fig. 5. Energy distribution curves for $La_{0.9}Th_{0.1}$, $Ce_{0.9}Th_{0.1}$ and $Pr_{0.9}Th_{0.1}$ at hν = 50 eV.

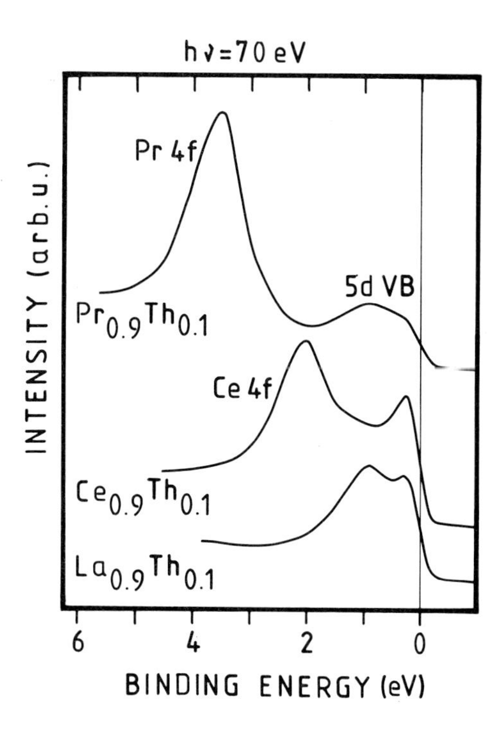

Fig. 6. Energy distribution curves as in Fig. 5 at hν = 70 eV.

and eventually the 4f emission decreases. The net result is a distinctive signature for 4f emission, viz., a well pronounced maximum near 50 eV. Not apparent from the figure is that the 4f/5d intensity ratio reaches its maximum value (∿ 2.6) at a higher photon energy (∿ 80 eV). The fact that the 4f-emission of Ce tracks so well that of Pr is gratifying; since, for Ce it was necessary to resort to the less desirable procedure of plotting the intensity of the 4f-emission peak rather than the integrated 4f-emission (as for Pr).

Turning back to the 5d-emission spectra, we see that the La 5d-curve follows closely the Pr 5d-curve, although the Pr curve is slightly higher. While we cannot rule out the possibility that the difference results from artifacts in the background subtraction procedure, the prospect that the difference is real is suggested by Fig. 6 which reveals qualitative differences in the 5d spectra of La and Pr at hν = 70 eV. The 5d-emission curve for Ce, while following the general photon energy dependence of the Pr and La 5d-curves, is, in detail, different, exhibiting a significantly less steep decrease with increasing photon energy. First of all it is clear that the 5d character of peak A (which is plotted as the "5d-emission") of Ce mainly determines its hν-dependence. In order to attempt to understand the departure from the 5d-behaviour, we plotted in the lower part of Fig.7 the difference curve between the curve for peak A in Ce and the 5d-emission curve for Pr. As can be seen this difference curve has a shape which closely resembles the Ce and Pr 4f-emission curves. This result corroborates our above conclusion based on the study of Figs. 4-6 that peak A in Ce has 4f- as well as 5d-character.

We might ask if the energy dependence of the Ce "A" feature exhibited in Fig. 7 could have arisen solely from a tailing of the main 2 eV emission into the "A" region of the spectrum. This possibility was explored by determining the photon energy dependence of the spectral intensity of the adjacent region (which we label "B") which lies in the energy range 0.75-1.50 eV from E_F. The resulting curve for the "B" region (not shown in Fig. 7) lies close to but slightly below that for the "A" feature (referring to the upper portion of Fig. 7). This confirms that the photon energy dependence of the "A" feature does not originate from the tail of the major f-emission, since the Lorentzian or Gaussian-shaped tail would provide considerably more intensity in the "B" than in the "A" region.

5. LINE SHAPE ANALYSES: POSSIBLE SURFACE SHIFT

In this section we attempt various subtraction procedures and lineshape analyses in order to (1) isolate the f-emission from the measured spectra and (2) explore the question of possible

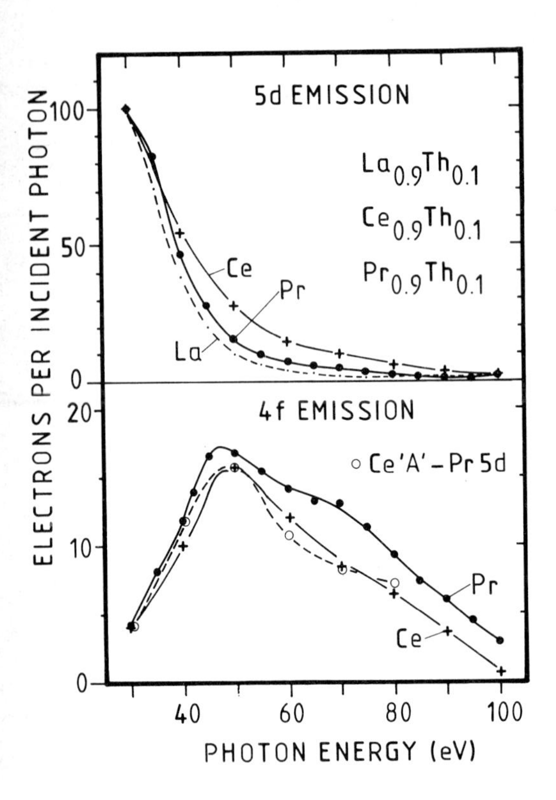

Fig. 7. Upper half: intensity of photoemission from 5d-bands of $La_{0.9}Th_{0.1}$ and $Pr_{0.9}Th_{0.1}$ and from peak A of $Ce_{0.9}Th_{0.1}$ (see text) versus photon energy. Lower half: intensity of photoemission from main f-emission peak in $Ce_{0.9}Th_{0.1}$ and $Pr_{0.9}Th_{0.1}$, and difference in intensity from peak A of $Ce_{0.9}Th_{0.1}$ and 5d-band of $Pr_{0.9}Th_{0.1}$ versus photon energy.

surface shifts in the f-emission.

Recently, surface shifts of the trivalent 4f-emission have been determined for a number of the heavier rare earth elements.(19) This is made possible by the relatively narrow line widths of the 4f-emission in these elements compared to the surface shifts (∿ 0.5 eV). Such shifts reflect the decrease in the extra bonding energy provided by the screening electron (associated with the 4f hole) at the surface, relative to the bulk, which in turn reflects the smaller number of bonds for surface atoms (see, e.g., Ref. 20). Although the large linewidth of the 4f-emission in cerium (also in Pr and Nd) obscures any surface shift (21), it is expected that surface shifts should be present and not markedly different in cerium than in the heavier rare earths, if the bonding in both cases is performed predominately by the $5d^1 6s^2$ conduction electrons. One purpose of the analysis presented below is to establish an upper bound for the surface shift in cerium.

Since f-emission is negligible at an incident photon energy of 30 eV (as established in Section 4). Subtraction of the 30 eV spectrum from the 50 eV spectrum should yield, with reasonable accuracy, the 4f-emission spectrum. The result of such a subtraction for γ-$Ce_{0.9}Th_{0.1}$ is shown in Fig. 8; where, in addition, a (relatively small) background contribution, taken as proportional to the integrated intensity at higher binding energies, has also been subtracted. In the subtraction procedure the relative intensities of the 30 eV and 50 eV spectra are adjusted to give the best fit to the assumed functional form of the spectrum. This leads to slightly different resulting spectra in Fig. 8 and Fig. 9, reflecting differences in the fitting functions used. As seen in Fig. 8, the difference spectrum is well reproduced by fitting each peak with a Doniach-Šunjić (DS) function (22), folded with a 0.15 eV spectrometer function, the latter taken as a Gaussian. For the major peak, the fit corresponds to a Lorentzian width Γ of 0.8 eV (the same value as reported in Ref. 4) and an asymmetry parameter α of 0.1. The "symmetric" 4f position, which is shifted in the spectrum due to the asymmetry, occurs at ∿1.95 eV.

It is shown in Fig. 9 that the difference spectrum for γ-$Ce_{0.9}Th_{0.1}$ can be equally well reproduced by using two DS functions, one representing the surface-shifted emission, to fit the 2 eV peak. The ratio of the intensities of the bulk and surface emission is assumed to be the same as that measured for the heavier rare earths,(19) taking into account the variation of electron escape depth with incident photon energy. In order to minimize the number of fitting parameters, and thus make the analysis meaningful, we have assumed that the spectral widths of the surface and bulk emissions are identical. The fit in Fig. 9 corresponds to $I_{s(urface)}/I_{t(otal)}$=0.4 (from Ref.19), Γ=0.6 eV, α=0.1 and a surface shift Δ_s of 0.3 eV. The (symmetric) 4f position occurs at ∿ 1.85 eV.

In order to demonstrate that the results obtained above are not an artifact of the 50 eV spectrum, we have analyzed also the spectrum for γ-$Ce_{0.9}Th_{0.1}$ at hν = 122 eV. In this case the difference spectrum is obtained by subtracting the $La_{0.9}Th_{0.1}$(122 eV) spectrum from the γ-$Ce_{0.9}Th_{0.1}$(122 eV) spectrum. At this energy the spectrometer function is broader (0.4 eV instead of 0.15 eV) and the surface/bulk emission intensity is smaller, I_s/I_t= 0.3 (19) than for hν=50 eV. The resulting spectrum (Fig.10) can be fitted with parameters nearly identical to those used in Fig. 9; viz., Γ = 0.7 eV, α = 0.1 and Δ_s = 0.3. The (symmetric) 4f position occurs again at ∿1.85 eV.

The corresponding difference spectra for α-$Ce_{0.9}Th_{0.1}$ are shown in Figs. 11 and Fig. 12. There is no reason to demonstrate again that the spectrum can be decomposed into bulk and surface contributions. Such an analysis would have less quantitative significance for the α-phase because we do not know what intensity ratios to

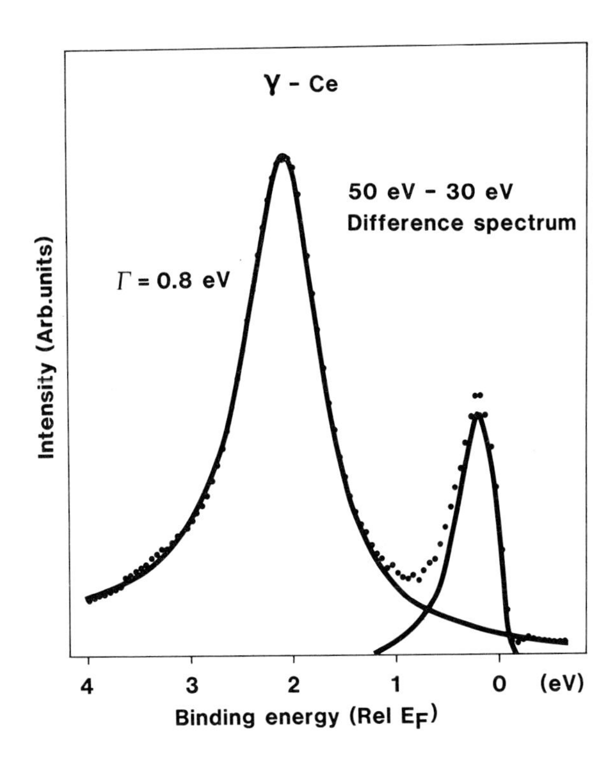

Fig. 8. 4f-emission-dominated difference spectrum for γ-$Ce_{0.9}Th_{0.1}$. Solid line: theoretical fit. See text for fitting procedure.

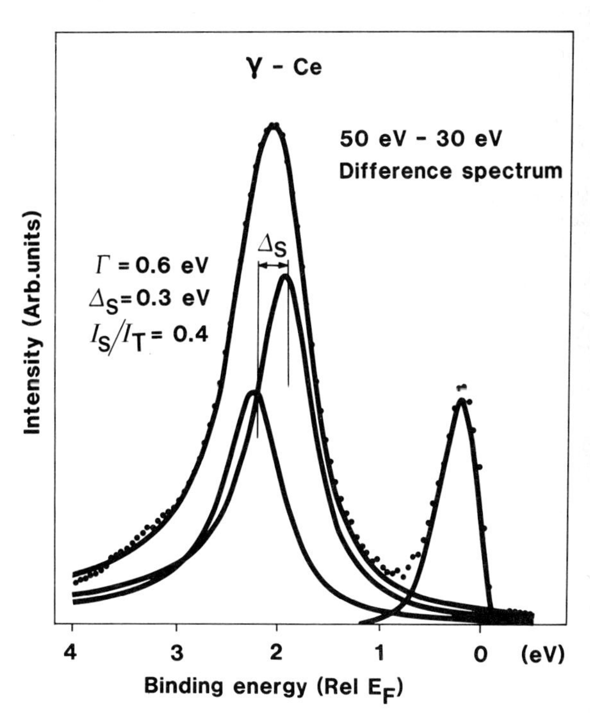

Fig. 9. Same spectrum as for Fig. 8, except fitting routine for major peak now includes surface as well as bulk contribution (see text).

use. If there is a layer of "stable" trivalent Ce on top of the α-Ce crystal, we should expect the ratio I_s/I_b to be different than that for the corresponding situation in a γ-Ce crystal, since the 4f state in γ-Ce is much closer to the stable state than in α-Ce. In Fig. 11 the fitting parameters are α = 0.14 and Γ = 0.9 eV and the 4f position is at $\sim$ 1.95 eV; in Fig. 12 they are α = 0.12, Γ = 0.85 eV and, again, the 4f position is at $\sim$ 1.95 eV.

We have avoided a quantitative analysis of the peak near E_F because its shape and width are more sensitive to the subtraction procedures than are those of the 2 eV peak. However, it is clear that the peak near E_F is much narrower than the 2 eV peak, their uncorrected linewidths being $\sim$ 0.5 eV and $\sim$ 1.0 eV respectively. It is also apparent that the relative narrowness of the peak near E_F precludes the possibility of deconvoluting surface and bulk contributions, unless the surface shift is extremely small, viz., $\lesssim$ 0.1 eV.

The major conclusions generated by the above analysis are: (1) there is no measurable shift in the position of the major 4f peak ($\sim$ 1.9 eV) at the γ-α transition, although there appears to be a slight broadening and increase in asymmetry of the peak in the α-phase and (2) the spectra taken at both $h\nu$ = 50 eV and $h\nu$ = 122 eV can be fitted by assuming there is

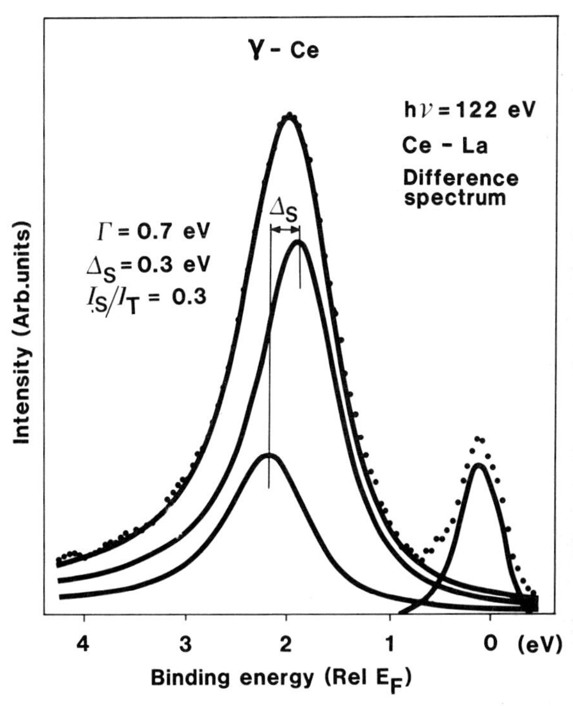

Fig. 10. Difference spectrum obtained by subtracting spectrum of $La_{0.9}Th_{0.1}$ from that of γ-$Ce_{0.9}Th_{0.1}$. Fitting routine same as that used in Fig. 9.

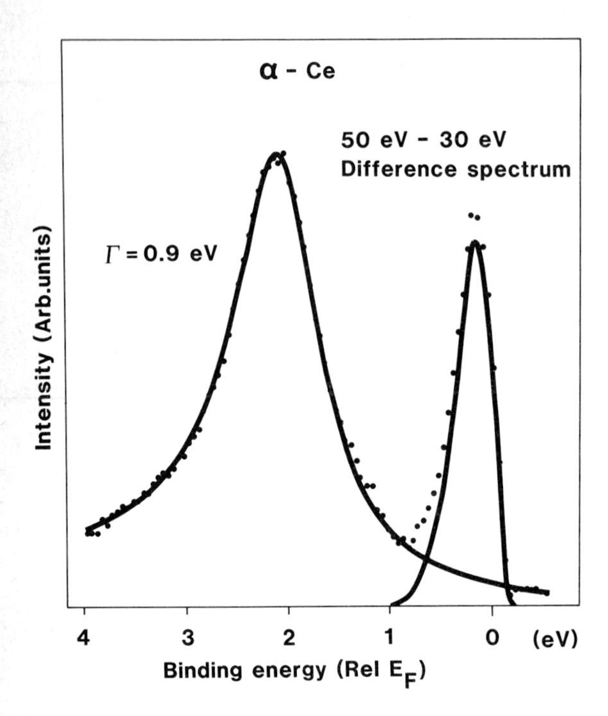

Fig. 11. Difference spectrum for α-$Ce_{0.9}Th_{0.1}$; same subtraction and fitting procedures as used in Fig. 8.

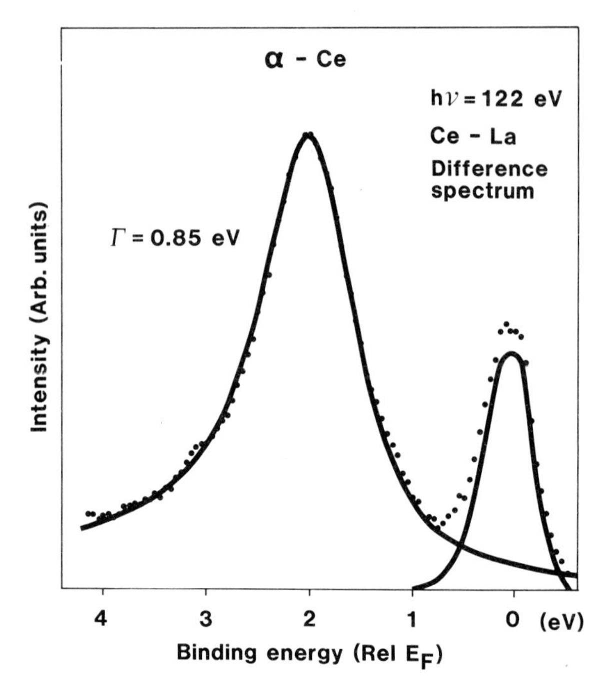

Fig. 12. Difference spectrum for α-$Ce_{0.9}Th_{0.1}$ obtained by same subtraction procedure used in Fig. 10 and same fitting procedure as used in Fig. 8.

a surface shift of 0.3 eV in the major 4f-emission. Allowing the linewidths of the surface and bulk emissions to differ, which is not unrealistic, should lead to a slight redressing of the 0.3 eV value for the surface shift. With some degree of confidence we can set the upper limit for the surface shift in cerium at 0.4 eV, which is slightly smaller than the value measured for the heavier rare earths, e.g., $\Delta_s \sim 0.5$ eV for Gd, Tb and Ho (19).

6. EPILOGUE

The major task of the theory is to explain the bimodal f-emission spectrum observed in Ce-based systems with unstable 4f-electrons, but not observed in any other rare earth systems; and to explain the intensity-trading between the two features which accompanies the volume collapse at the γ-α transition in cerium. In this context, we draw the reader's attention to two theoretical lines of approach presently under pursuit, both of which are in the preliminary stage. One was suggested by an early atomic wave function calculation by Goeppert-Mayer (23); who, using the Thomas-Fermi (statistical) method, obtained a double well potential for the 4f-state, one well being narrow and at the usual 4f position and the second well being broad and shallow and lying near or beyond the radius of the 5d distribution. Of the rare earth elements, only in La, Ce and Pr are the lowest energy levels of the two wells sufficently close to one another to render the second well of any possible significance in the ground state. Schlüter and Varma have re-examined this question, using modern day techniques.(24) Using self-consistent density functional theory, they have demonstrated that for cerium (in the solid state) there are two stationary states with 4f-symmetry, accompanied by different radial distributions of core charge. They have reinterpreted these results in terms of a 4f Green's function which has two peaks separated by a few volts. They suggest that the bimodal property of the 4f Green's function can be mapped onto the bimodal signature of the f-emission spectrum.

A different approach is being pursued by Martin and Allen, which they call the Kondo volume collapse model.(25) Applying this picture to cerium metal, they interpret the γ-α transition as a volume collapse which increases the binding energy of the Kondo lattice ground state sufficiently to offset the increased compressional energy. The constancy of the 4f population through the transition is a natural consequence of the model. They interpret the f-emission feature near E_F as resulting from the Kondo resonance at E_F.

7. ACKNOWLEDGEMENTS

The authors are grateful to J. W. Allen, R. M. Martin, M. Schlüter and C. M. Varma for interesting discussions, and to D. E. Eastman for his generous support. Acknowledgment is also given the staff of the Synchrotron Radiation Center, University of Wisconsin, for their support and cooperation. The work performed by Polytechnic Institute of New York was sponsored in part by the Department of Energy under Contract No. DE-AC02-81ER10862; and that by IBM was sponsored in part by the Air Force Office of Scientific Research under Contract No. F49620-81-C0089.

REFERENCES

* Present address: Linköping University, Dept. Physics and Measurement Technology, S-581 83 Linköping, Sweden.

** Present address: IBM Research Laboratory, 8803-Rüschlikon, Switzerland.

1. J.M. Lawrence, P.S. Riseborough, and R.D. Parks, Rep. Prog. Phys. 44, 1 (1981).

2. N. Mårtensson, B. Reihl, W.-D. Schneider, V. Murgai, L.C. Gupta and R.D. Parks, Phys. Rev. B 25, 1446 (1982).

3. G. Kaindl, B. Reihl, D.E. Eastman, R.A. Pollak, N. Mårtensson, B. Barbara, T. Penney and T.S. Plaskett, Solid State Commun. 41, 157 (1982).

4. J.W. Allen, S.-J. Oh, I. Lindau, J.M. Lawrence, L.I. Johansson and S.B. Hagstrom, Phys. Rev. Lett. 46, 1100 (1981).

5. W. Gudat, M. Campagna, R. Rosei, J.H. Weaver W. Eberhardt, F. Hulliger and F. Kaldis, J. Appl. Phys. 52, 2123 (1981).

6. B. Johansson, Phil. Mag. 30, 469 (1974).

7. D.R. Gustafson, J.O. McNutt, and L.O. Roellig, Phys. Rev. 183, 435 (1969).

8. U. Kornstädt, R. Lässer, and B. Lengeler, Phys. Rev. B 21, 1898 (1980).

9. R.M. Moon and W. C. Koehler, J. Appl. Phys. 50, 2089 (1979).

10. N. Mårtensson, B. Reihl and R.D. Parks, Solid State Commun. 41, 573 (1982).

11. K.R. Bauchspiess, W. Boksch, E. Holland-Moritz, H. Launois, R. Pott, and D. Wohlleben, in Valence Fluctuations in Solids, edited by L.M. Falicov, W. Hanke, and M.B. Maple (North-Holland, 1981) p. 417.

12. Y. Baer and Ch. Zürcher, Phys. Rev. Lett. 39, 956 (1977).

13. J.M. Lawrence and R.D. Parks, J. Physique 37, C4-249 (1976).

14. D.E. Eastman, J.J. Donelon, N.C. Hien, and F.J. Himpsel, Nucl. Instr. and Meth. 172, 327 (1980).

15. L.I. Johansson, J.W. Allen, T. Gustafsson, I. Lindau, and S.B. Hagström, Solid State Commun. 28, 53 (1978).

16. A. Platau and S.-E. Karlson, Phys. Rev. B 18, 3820 (1978).

17. W. Gudat, S.F. Alvarado and M. Campagna, Solid State Commun. 28, 943 (1978).

18. S.M. Goldberg, et al., Electron Spec. 21, 285 (1981).

19. F. Gerken, J. Barth, R. Kammerer, C. Kunz, L.I. Johansson and A. Flodström, to be published.

20. B. Johansson and N. Mårtensson, Phys. Rev. B 21, 4427 (1980).

21. The (large) surface shift reported for CeN (Ref. 5) reflects, in our opinion, a misidentification of the relevant spectral features in Ref.5; see, e.g., discussion in N. Mårtensson, B. Reihl and F. Holtzberg, these proceedings.

22. S. Doniach and M. Šunjić, J. Phys. C 3, 285 (1970).

23. M. Goeppert-Mayer, Phys. Rev. 60, 184 (1941)

24. M. Schlüter and C.M. Varma, private communication; see also these proceedings.

25. R.M. Martin and J.W. Allen, private communication; see also Bull. Amer. Phys. Soc. 27, 275 (1982).

COMMENT: M. CAMPAGNA: Did you consider also surface shifts for the 4f feature near E_F ?

R. PARKS: Yes, the narrowness of the peak precludes a surface shift larger than $\sim$ 0.1 eV.

D. RIEGEL: You have shown in your difference spectra that the linewidth for the feature close to E_F is smaller compared to the feature at 2 eV. Do you have any explanation for this finding ?

R. PARKS: The mere existence of the feature near E_F is puzzling enough. Until its origin is understood, we cannot discuss intelligently its width. What is clear is, that its linewidth is too narrow to be deconvoluted into a bulk and surface contribution unless the surface shift is $\lesssim$ 0.1 eV. This in itself would seem to exclude the possibility that the final state is a deep core hole screened in the usual manner by a 5d electron, since the latter configuration leads to measured surface shifts $\sim$ 0.5 eV.

L. JOHANSSON: Have you tried to disentangle the (5d6s) and 4f contributions by oxidation experiments. It has been shown in earlier reports that the intensity of emission close to E_F is strongly suppressed upon oxygen exposure while the peak around -2 eV is nearly unaffected. Even if you have not done this would you please comment on why you observe a 4f contribution close to E_F in Ce metal but not for a trivalent oxide (where you, at resonance, only see the -2 eV peak).

ANSWER: B. REIHL: The α- and γ- $Ce_{0.9}Th_{0.1}$ spectra are extremely sensitive to contamination, affecting both the 2 eV as well as the E_F peak. No controlled oxydation study has been performed.

P. WACHTER: You have shown that the 4f state in Ce is about 2 eV below E_F. On the other hand the peak at E_F has considerable f-character. In the γ-α transition you observe a transfer of f-intensity from 2 eV below E_F towards E_F. Can't you explain this as a valence transition (not necessarily from $3^+ \to 4^+$)?

R. PARKS: The constancy of the 4f population through the transition implies there is no valence transition per se; however, there may be 4f-electron transfer from a deep position in the atom to a position closer to the Wigner-Seitz boundary, which is possible, in principle, in the 4f double state model (see, e.g., Schlüter and Varma, these proceedings).

G. KAINDL: 1. In the UPS-photoemission experiment described, you are studying a thin surface layer of $Ce_{1-x}Th_x$. Nevertheless, you assume that the phase diagram known for the $Th_{1-x}Ce_x$ alloys in the bulk does also apply for the surface layer. I wonder if this is the case ?

2. In all mixed-valent systems of heavier rare-earths studied so far ($YbAl_2$, $Sm_{1-x}Y_xS$, $EuPd_2Si_2$ TmSe etc.) a valence change has been observed for the outermost surface layer. How would such a situation influence the interpretation of the presented P.E. spectra for $Ce_{1-x}Th_x$?

R. PARKS: We have not assumed that the phase diagram for the bulk modulus applies to the surface layer. As in studies of mixed-valent systems of heavier rare earths, photoemission, at the photon energies employed in the present study, samples both the bulk and surface states, the latter being confined usually to the top monolayer.
Whether or not the surface layer transforms along with the bulk will determine presumably the amount of intensity transfer between the two spectral features at the transition. It is not possible on the basis of the present results to determine whether or not the surface layer transforms.

C.M. VARMA: What sort of measurement gives the valences for the α- and the γ-phases of Ce that you quoted ? Since different measurements do not always agree, which ones should one believe ?

R. PARKS: The spin lifetimes deduced from quasielastic neutron scattering and magnetic susceptibility experiments are a measure of the departure from "stable" 4f behavior. Comparing these lifetimes to those observed in proper mixed valent systems (e.g. Yb-based) yields an effective valence of $\sim$3.15 for γ-Ce and $\sim$3.6-3.7 for α-Ce.

Valence Instabilities
P. Wachter and H. Boppart (eds.)

ELECTRONIC STRUCTURE OF Ce PNICTIDES AND CHALCOGENIDES BY PHOTOELECTRON SPECTROSCOPY WITH SYNCHROTRON RADIATION

[+]W. Gudat , [*]M. Iwan , [‡]R. Pinchaux , and [∞]F. Hulliger

[+]Institut für Festkörperforschung der KFA Jülich
D-5170 Jülich, W. Germany
[*]present address: Fritz Haber-Institut der MPG Berlin
D-1000 Berlin, W. Germany
[‡]LURE-C.N.R.S. Université de Paris-Sud
F-91405 Orsay, France
[∞]Laboratorium für Festkörperphysik, ETH Zürich
CH-8093 Zürich, Switzerland

In this paper we discuss results of resonant and angular-resolved photoemission experiments on formally "trivalent" Ce-pnictides and -chalcogenides. The aim is to provide an understanding of 4f-photoemission features of Ce mixed-valent materials (e.g. CeN, $CePd_3$, CeTh alloys, etc.). Concepts of resonant photoemission spectroscopy are being briefly reviewed and some problems of this technique are pointed out. In the Ce-pnictides two 4f-related emission features are observed. One feature is located close to the Fermi level ($E_B \approx -0.5$ eV) and another one at about 3 eV binding energy. Angular-resolved photoemission data suggest the 3 eV feature to be due to d-screened localized $4f^o$ final hole states. The origin of the low binding energy feature at about 0.5 eV has been previously suggested as due to pf mixing. Alternative explanations for the two 4f-peak structures are also discussed. Systematic trends are observed in the emission intensities which are explained by the particular role of the 5d electrons in bonding. It is concluded that, despite the effort done so far, further investigations are needed to exactly establish the nature of the 4f states not only in "mixed-valent" but also in "stable" Ce-compounds.

I. INTRODUCTION

In the last 2-3 years synchrotron radiation photoemission spectroscopies have been applied with increasing efforts to the study of bulk and surface electronic structure of rare earth (RE) metals and RE compounds with stable and "mixed" 4f valence configuration. In particular the technique of resonant photoemission (RP) /1-3/ has been used with the aim of identifying 4f electron emission in Ce compounds and to locate the emission features with respect to a reference level. This aim cannot in fact be reached by x-ray photoelectron spectroscopy (XPS) on Ce because of unfavourable excitation cross-sections /4/.

The position of the 4f levels with respect to the Fermi-level ($\Delta\_$) as well as their width (Γ) are considered fundamental parameters for a quantitative microscopic understanding of Ce mixed-valent behaviour (e. g. in CeN /5/ $CePd_3$ /6/). These parameters $\Delta\_$ and Γ are equally important for many Ce intermetallic compounds behaving like Kondo systems (e. g. CeB_6 /6/, $CeAl_2$ /7/) as well as for Ce compounds exhibiting a pressure induced isostructural volume collapse (e. g. CeP /8/, CeS /9/) or the well-known $\gamma \longrightarrow \alpha$ phase transition of Ce metal.

Within the renormalized-atom approach of Herbst et al. /10/ under the assumption of a "completely screened" localized final state the 4f photoionisation process /11/ can be related to a transition $4f^n \rightarrow 4f^{n-1}(sd)^1$ which assumes that the 4f-hole state in the final state is screened by s,d electrons. In this picture the experimentally determined 4f binding energy with respect to the Fermi level corresponds to the promotion of a 4f electron to the Fermi level itself /12/. The central question is then how far 4f transitions in Ce can be described by the renormalized atom scheme using localized orbitals.

One traditional way of viewing mixed-valence in Ce-compounds is that of a resonance of at least two distinct configurations $4f^1 \longleftrightarrow 4f^o 5d^1$ in the ground state, giving rise to the mixed-valent state $\psi = a(4f^1) + b(4f^o(5ds)^1)$ with $/a/^2 + /b/^2 = 1$ /13/. This implies a location of the 4f levels near the Fermi energy, i. e. $\Delta\_ \approx o$. In a conceptually identical way, but different wording, the mixed-valence phenomenon is obtained on the basis of hybridization of the 4f levels with extended states /13/. In any case at least one of the configurations contributing to Ψ contains occupied 4f states which should be directly measurable by high energy spectroscopy. XPS core level spectroscopy (and x-ray absorption spectroscopy as well /14/) gives only an indirect information on the ground state through screened core hole final state /15-20/. Therefore the assessment of an intermediate valence state by core - level spectroscopy alone needs a detailed understanding of the screening mechanism involving sd and f states which, in the case of Ce, is not yet available /18,20/.

Early synchrotron radiation 4f photoemission experiments on stable Ce-compounds found the 4f levels to lie in an energy range with $\Delta_- \approx$ up to 2 eV, contradicting current views. Because of these findings we decided to study the 4f emission of Ce compounds in a more systematic way /21, 22/. We were interested in a class of compounds offering the possibility of systematically varying the 5d itinerant electron density and which contained a mixed-valent system as well. This is offered by the rocksalt-type Ce monopnictides and Ce monochalcogenides, where Ce is believed to be formally "trivalent" (i. e. $4f^1$) and mixed-valent in CeN. Formally, the Ce pnictides have been considered as f^1d^0 semi-metals, while the chalcogenides as f^1d^1 metals.

4f derived spectral emission features have been found in a wide energy range from near E_F down to more than 3 eV below E_F with characteristic trends. These investigations showed that in the pnictide series two 4f-derived features can be identified for each compound, in agreement with the findings of Franciosi et al. /23/. One feature is observed at Δ_-=0.5 eV and the other one at Δ_- = 3 eV.

In addition we have performed first angular-resolved photoemission measurements on Ce pnictides allowing a first comparison with bandstructure calculations. They also allow to more firmly identify the 4f emission features in the Ce pnictides.

II. Localized versus extended states

To our knowledge there are no band structure calculations available for Ce pnictides and Ce chalcogenides, but for La- and Gd-pnictides such calculations exist /25,26/. The latter compounds differ from the Ce compounds in their 4f occupation. We will take these calculations as a first approximation for the Ce pnictides, at least as far as the band states are concerned /24/. In Fig. 1 we show as an example the band structure of LaAs /25/. The occupied valence-bands, which are mainly made up of ligand 4p states, have a width of ~3. 5 eV and

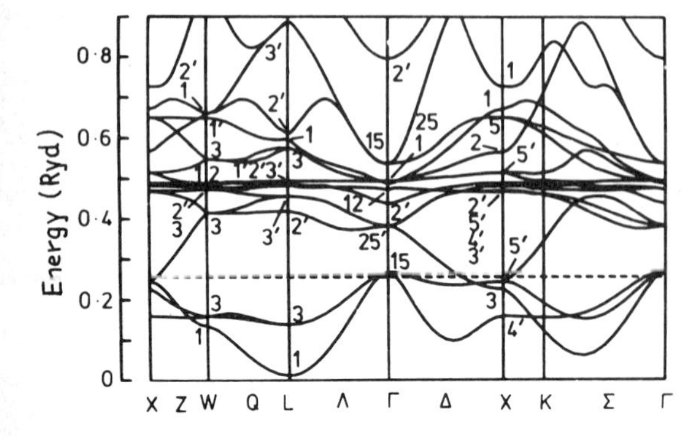

Fig. 1: Energy bands of LaAs for some symmetry directions (from Hasegawa /25/)

show a considerable dispersion. The 5d-like conduction bands slightly overlap the valence bands near the X-point, making LaAs (as well as LaP and LaSb) a semi-metal. The 4f bands are located about 3 eV above the Fermi level. Within this itinerant model the 4f electrons of CeAs are expected to form partially occupied narrow bands crossing E_F.

From an atomic point of view, one would expect for the 4f levels highly correlated core-like levels to be located below E_F. Angular-resolved photoemission spectroscopy (ARPES) has the capability of probing $E(\underline{k})$ band dispersions /27/. This is based on the established facts that in single crystals photoexcitations are direct transitions and that the electron momentum component $\underline{k}_{\parallel}$ parallel to the surface of the sample is conserved in the photoemission process. This is why this technique can be helpful in identifying and classifying 4f emission from Ce-compounds.

We present here as illustration angle-resolved photoemission data of CeAs and CeSb single crystals obtained with monochromatized synchrotron radiation of the ACO storage ring at Orsay. The angular resolution was about 1°. Total instrumental energy resolution of 0. 15 to 0. 6 eV has been used. Single crystals were cleaved in situ at a base pressure of $1x10^{-10}$ mbar.

The electron energy distribution curves (EDC) of CeAs shown in Fig. 2 were measured with mixed sp-polarization of the light to excite band states without polarization selection rules at various polar angles in a (010) emission plane. At $h\nu$ = 30 eV the excitation cross-section of 4f electrons is known to be approximately one to two orders of magnitude smaller than of s-, p-, d-like valence electrons. Thus the spectra of Fig. 2 represent the emission of the itinerant (ligand) valence states and in agreement with the band - structure calculation strong dispersion effects are observed; i. e. strong energy shifts in peak positions upon variation of the polar angle. The total band width is in fair agreement with the one expected from the calculations (see Fig. 1). The onset of the EDC's at E_F indicates a very small density of states, not incompatible with semi-metal behaviour. We note that peaks (at 2.2 eV and Θ = 10-15°) with full-width-at-half-maximum (FWHM) as small as 0. 3 eV are being measured with an instrumental resolution of 0. 2 eV, indicating that intrinsic lifetime broadening is small. Increasing $h\nu$ results in a strong increase in 4f emission. The data of Fig. 3, obtained in normal emission geometry, show the growth of pronounced peaks at -0.5 eV and -3.0 eV (at $h\nu$ = 80 eV) while the structures in between disperses because of $E(k)$ band dispersion. In this work we do not present a full discussion of the valence-band dispersions /28/; we concentrate on discussing the nature of the 4f states as seen by photoemission.

In Fig. 4 we show a set of EDC's for CeSb (100), which were obtained at different polar emission angles at a photon energy of 80 eV. We assign the peak at -3 eV to 4f emission (FWHM ~0.7 eV at an instrumental resolution of 0. 27 eV) because of its $h\nu$ dependence. Upon variation of the polar angle the peak moves only very little (~ 0. 2 eV). This apparent dispersion, to some extent, is probably due to the changes in background slope. In contrast, the structures at smaller binding energies (the situation near E_F is uncertain) do

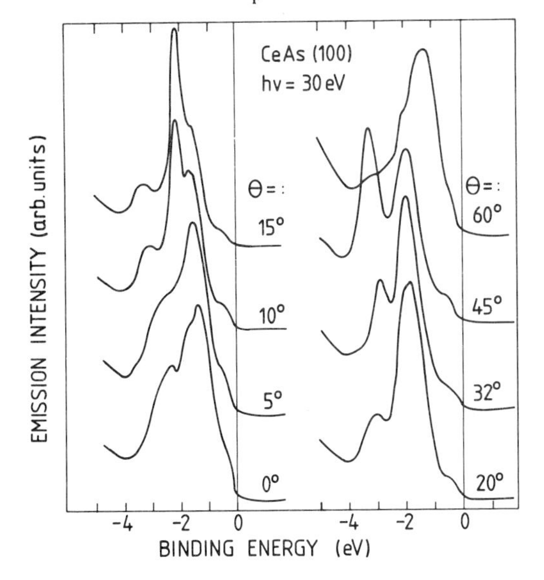

Fig. 2: Angular-resolved photoemission energy distribution curves for increasing take-off angles with respect to the surface normal show that interband transition give rise to the structures observed at low photon energy for CeAs (100) cleavage face. The occupied valence band width is in agreement with the band calculation shown in Fig. 1.

show pronounced dispersion effects and they can thus be ascribed to interband transitions The behaviour of the -3 eV peak is similar to that of a core-level paeak. Thus it is very likely due to emission from a localized state, which we believe is screened to some extent by d-electrons. We identify it as the d-screened final-4f-hole state. In this interpretation we cannot, however, exclude very narrow itinerant bands of widths considerably smaller than the observed line width because of lifetime broadening effects. Because of the bandstructure of LaSb /25/ and particularly because of the trends we have observed in the Ce pnictides and chalcogenides we can exclude for the stationary structure at -3 eV of Fig. 4 a surface state (which should also disperse with $k_{\parallel}$) and a surface resonance as another explanation.

Surface chemical shifts of the localized 4f levels in heavy RE and RE compounds have previously been identified /29/. At a photon energy $h\nu$ = 80 eV the surface sensitivity of ARPES is already very high for valence electron emission. In addition, the sensitivity is further increased for large electron take-off angles ($\Theta \geq 45^{\circ}$, see. Fig. 4).

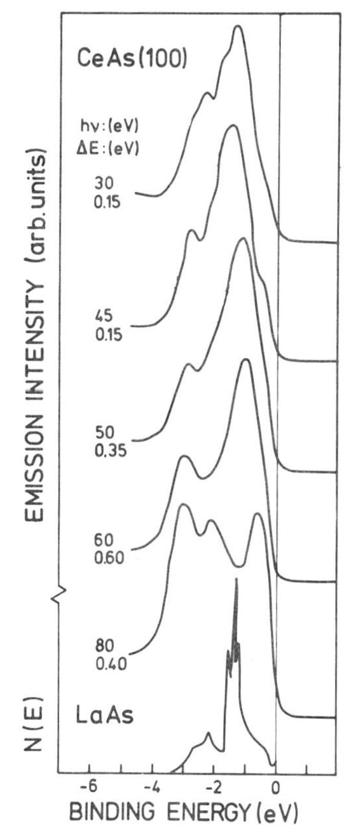

Fig. 3: For increasing photon energy, ARPES data on CeAs show the growth of two pieces of structure close to E_F and near -3 eV which is related to 4f emission. For comparison the occupied part of the density of states is shown as computed by Hasegawa /25/.

The slight changes observed for the lineshape of the 3 eV peak could then be an indication for a 4f surface chemical shift in CeSb. If so, the shift should be small and should give only a small contribution to the observed line width of ~0.7 eV. We note that the observation of surface chemical shifts would also be in agreement with localized 4f state emission.

Comparing the spectra of CeAs (100) and CeSb at $h\nu$=80 (Figs. 3 and 4) we note that in CeAs a prominent 4f feature is observed close to E_F, while in CeSb the most prominent feature is located at -3 eV. We will compare these 4f emission feature in more detail in section IV.

III. Resonant Photoemission

In this technique the emission intensity of the 4f levels can be strongly enhanced by taking EDC's at photon energies near the 4d→4f photoabsorption threshold /1-3,30-34/.

Resonant Photoemission can be considered as a fairly general phenomenon which occurs in atomic systems due to intershell interactions /35/. In a simple step model the emission enhancement is due to two circumstances which are particularly favourable in RE compounds having incomplete 4f shells, but which can also occur , for instance, in 3d- /36/ and 4d-transition metals and in 5f-electron

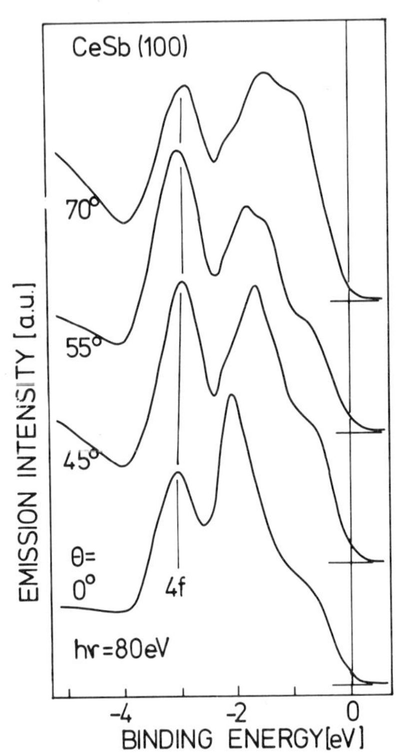

Fig. 4: In CeSb the peak at a binding energy of -3 eV is attributed to the localized final 4f hole state since it is seen to hardly change its maximum position for different polar angles θ. The structures at lower energy ($-3 \leq E_B \leq 1$ eV) are non-stationary because of their itinerant nature.

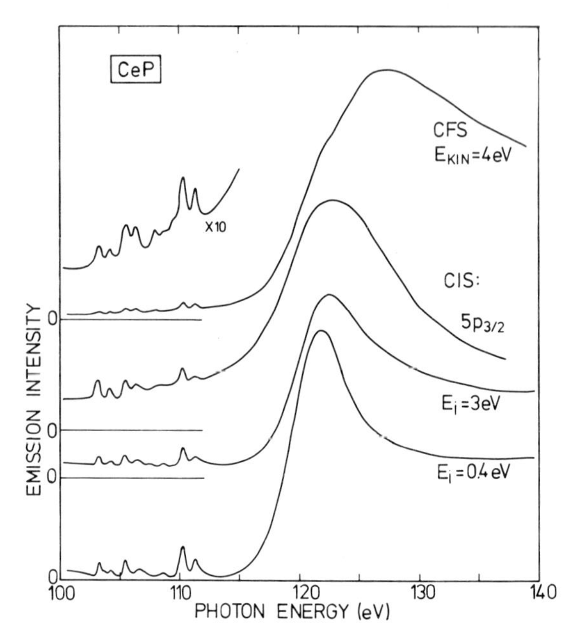

Fig. 5: Approximate (see text) total (CFS) and partial (CIS) cross-section data for CeP in the spectral range of 4d-4f transitions. Emission enhancement is observed for different initial states (Ce5p, 4f derived states, -0.4 eV and -3 eV compare Fig. 6) demonstrating the resonant intershell coupling to the 4d core excitations.

systems /37/. These circumstances are: firstly, a strong increase in photoabsorption in a small spectral range due to transitions into localized or quasi-localized intermediate states and secondly, a decay of these intermediate states resulting in emission of outer core-like or valence electrons because of the released energy.

In RE compounds a so-called "giant absorption resonance" occurs for transitions $4d^{10}4f^N \rightarrow 4d^9 4f^{N+1}$. This is illustrated in Fig. 5 for CeP by the constant-final state (CFS) spectrum. (It is by now well known that CFS spectra, under suitable measurement conditions, are proportional to the spectral dependence of the total excitation cross-section /38/) . The transitions give rise to the discrete multiplet structure of the $4d^9 4f^{N+1}$ excited state configuration and to the giant absorption maximum /1-3/. The exciton-like intermediate state can decay via autoionization and the emission of 4f electrons $4d^9 4f^2 \rightarrow 4d^{10}4f^0 + \varepsilon f$.

Thus the 4f emission is coupled to the strong 4d excitation. Because the initial excitation energy is transferred to the emitted electron, this cannot be distinguished from another directly emitted 4f electron. The spectral dependence of this excitation process can thus be described by Fano's lineshape formula /2,3,30/.

In Fig. 6 we show two EDC s which were obtained at photon energies near the onset of the absorption band (hν = 114 eV) and close to the absorption maximum (hν=121 eV). The EDC s, which are normalized at a binding energy of - 2 eV, show strikingly different structure in the valence band region, while the peaks due to Ce5p electron emission are similar. We attribute the difference of the two spectra in the valence region to resonantly enhanced 4f emission, which implies that two 4f features are observed.

According to Fig. 6 the intensity for particular initial states in the valence band changes drastically in a spectral range of only 7 eV. Detailed information on the partial excitation cross section can be obtained by means of so-called constant-initial-state spectroscopy (CIS /38/) . While tuning the photon energy, the electron spectrometer is manipulated to maintain a selected initial state. With this experimental procedure the absolute partial cross-section cannot be measured in general, because of background emission intensity.
However, it greatly facilitates to follow strong spectral changes, even in very narrow energy intervals as seen in the CIS spectra of Fig. 5. The important message obtained from these spectra is that the emission from both 4f related features at E_B=-0.4 eV and E_B=-3 eV as well as the emission from the 5p core levels is resonantly enhanced. We note that a CIS spectrum for E_B = -2 eV which corresponds mainly to ligand 3p states, shows

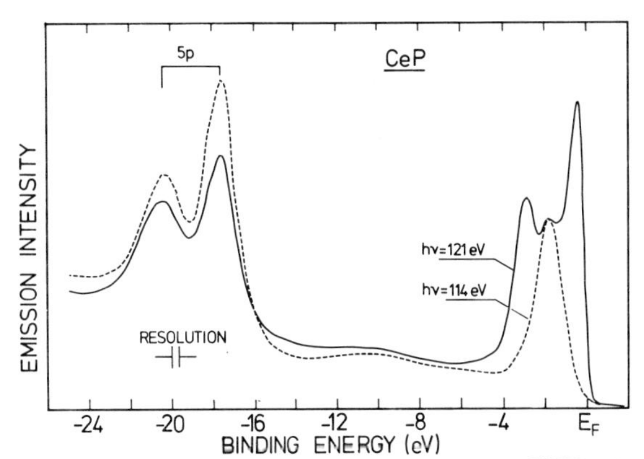

Fig. 6: At the onset of 4d-4f core excitations the valence band photoemission in CeP (100) changes strongly due to resonantly enhanced 4f emission, while the core-like 5p emission line shapes do not change. The variation of the absolute 5p emission intensity (see Fig. 5) is not seen here due to normalization of the angle-integrated EDC s near -2 eV.

a very similar shape as the CIS for the Ce5p state. In the case of CeP and other Ce compounds we did not observe large differences in the peak positions of the various CIS (~1 eV) and it is difficult to establish a trend in the peak positions. The width and the height of the various CIS are, however different, indicating different coupling matrix elements for the partitioning of the 4d excitations into various channels. Further investigations both experimental and theoretical are, however, needed for reaching a better understanding of partial cross-section of Ce (5s, 5p, 4d and 4f) near the 4d-4f transition threshold. It is clear that different peak positions of the partial cross-section spectra would greatly facilitate an identification of the orbital momentum character of degenerate initial states in the EDC (e. g. 4f and 5d states) . A careful modeling of the relevant CIS on the basis of Fano lineshapes could be helpful, if the energy separation of the partial cross-section maxima is not too small.

The CIS spectra of emission intensity depending on the initial state. One can define an enhancement ratio to rationalize these differences and to use them for diagnostic purposes. The simplest way is certainly to use the ratio of emission intensities "on" and "off" resonance, i.e. for Ce-compounds at 112-115 and 121-123 eV. This is what we will use in the following, but other enhancement ratios can be defined and have actually been used /33,39,41/.

Finally we note that the lineshapes of the discrete multiplet peaks (103-112 eV) do look different in total and partial cross-section spectra, just in the same way as the wide maxima do. This observation is only possible by directly measuring the CFS and CIS.

IV. Energy position of the 4f levels

The angular-resolved photoemission data for CeAs (Fig. 3) showed two structures at higher photon energies which we related to 4f emission. For CeSb a high binding energy feature was clearly seen with characteristics of localized 4f emission. A low binding energy peak was also indicated. The "on" and "off" resonance photoemision spectra of CeP (Fig. 6) , CeSb and of CeAs and CeBi previously obtained /23, 40/ showed also very clearly enhanced emission from two peaks in the valence band region. To obtain a "common basis" for a more general comparison we have looked at "on" and "off" resonance spectra for Ce-pnictides and chalcogenides /6,21,22,42/. A collection of pnictide spectra is shown in Fig. 7, where we included to our data previous results on CeBi /23/. Because of the similarity of the spectra for the chalcogenides CeS, CeSe and CeTe we show as an example only "on" and "off" resonance data for CeTe /21/ (Fig. 8). The dashed area between the curves which are normalized at E_B=4. 5 eV /43/ is due to resonant enhanced emission. We expect the enhancement to be strongest because of the strong intershell interactions in the n=4 shell. This is corroborated by investigations of the spectral dependence of the partial photoionization cross-section in Ce chalcogenides /21. 33/. Consequently the peak at 2.6 eV with a measured width of about 1 eV is attributed to 4f emission. In the Ce chalcogenides with formal f^1d^1 configurations the 5d like conduction bands are partially occupied and are expected to be energetically well separated from the ligand p-state derived bands /44/. Thus we are inclined to attribute the small intensity difference near E_F to resonantly enhanced 5d emission. However, we note that this difference becomes smaller in the progression CeS-CeSe-CeTe apparently contrary to what one would expect on the basis of 5d population.

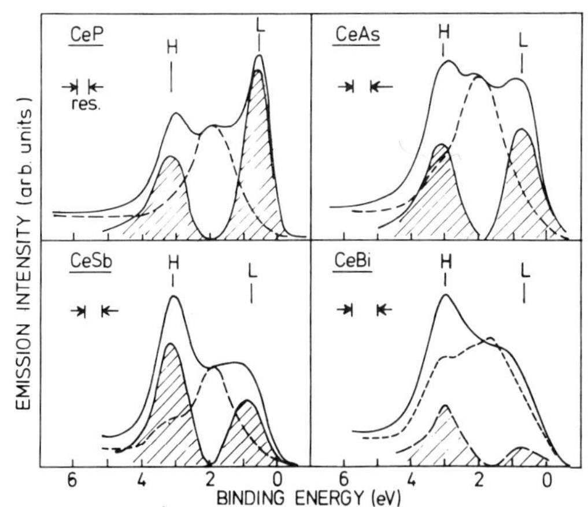

Fig. 7: Two 4f related photoemission features (shaded areas) are clearly observed in difference spectra of EDC's taken "on resonance" (hν=121-122 eV) and "off resonance" (hν =112-115 eV) . The intensity of the low binding energy feature (L) decreases with increasing atomic number of the ligand, while the high binding energy feature increases. The data for CeBi are from ref. 23.

The pnictide spectra of Fig. 7 are normalized near 2 eV binding energy. From the "on" and "off" resonance spectra difference curves are extracted showing directly the resonantly enhanced emission intensity to fall into structures near -0.5 eV and -3 eV. The intensity of the low–energy peak (L) is biggest for CeP and decreases systematically towards CeBi. Evidently the ligand–dependent intensity behaviour of the high–energy feature (H) is exactly opposite, while the total intensity of both features seems to be unchanged when compared in the Ce5p emission /22/. The position of peaks L and H does not change much (~0.3 e V) in the pnictide series. Consequently the energy separation is more or less constant. The low photon energy data (up to 80 eV) suggested already that both peaks are related to 4f emission because of their hν - dependence. One arrives at the same conclusion from the resonant photoemission data by noting that i) the observed trend in emission strength of peak L is opposite to the one of the 5d-electron density, ii) the chalcogenides (f^1d^1 systems) do not show the strong enhancement iii) enhancement ratios for ligand p-valence electrons are much smaller than observed here.

A summary of 4f related emission features of Ce pnictides and chalcogenides is presented in Fig. 9 The heavy horizontal bars give the energy position with respect to the Fermi level, while the adjacent thin lines indicate the observed spectral linewidth. We have included data on CeN wich were also obtained from "off" and "on" resonance spectra /21, 22/ as well as data on γ -Ce /1, 39,) . The dashed lines used for CeN and Ce chalcogenides indicate that the corresponding spectral features are weak. The high binding energy features in the compounds (except CeN) were recently explained as due to d-screened 4f final-hole-states /23/ in agreement with our findings based on ARPES work. The strong intensity variation of this structure in the pnictide serie is then due to limited availability of 5d screening charge /22, 23/. This would be an immediate consequence of the change in bonding due to ionicity. This can also be inferred from bandstructure calculations as shown in Fig. 1. In CeBi, the least ionic compound, the 3 eV feature is strongest and it becomes weaker towards CeP as ionicity increases. The low-energy features have been suggested to be due to p-f hybridization resulting in 4f holes partially delocalized on ligand p-states /22/. The change in emission intensity of this feature in going from pnictides to chalcogenides is in this picture plausible, because of the considerably lower lying p-bands in the chalcogenides. The trend in emission intensity observed with the pnictides is, however, more difficult to understand, since the band structure changes only weakly from CeP to CeBi /25,26/.

What are the consequences, if we extrapolate on the basis of these interpretations to CeN which is considered to be a mixed-valent compound? If we related the small structure, corresponding to the peak

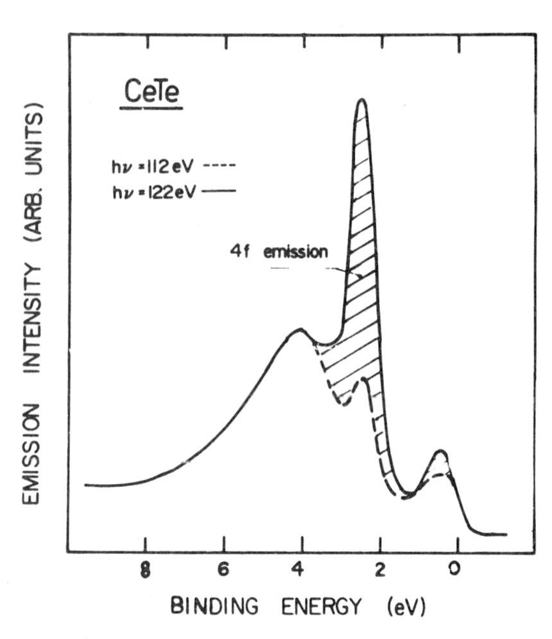

Fig. 8: The dashed area between EDC s of CeTe taken "on" and "off" resonance emphasizes 4f photoemission and allows to identify the localized screened final-4f-hole state at 2. 6 eV.

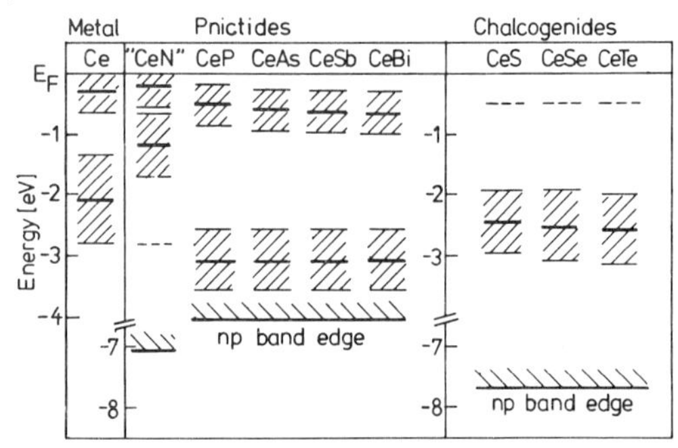

Fig. 9: Summary of 4f related emission features of Ce pnictides, chalcogenides and of γ -Ce as obtained from "on" and "off" resonance photoemission spectra.

position slightly below -3 eV, to the d-screened final 4f-hole state, this would then in our opinion be in contrast to mixed-valence with f-levels pinned at E_F. This would also be the case, if we used the position E_B=-1. 3 eV which belongs to a peak so far attributed to a surface feature /21/. It would further imply that the p-f mixing ionization channel gives a much stronger peak than in CeBi, although the N derived p-bands lie deeper in energy than in CeBi. It thus appears that CeN does not fit to this picture.

If, on the other hand, it were to consider only the energy position of the strongest 4f emission feature for each compound (as determined by its resonance behaviour) then one finds that the 4f-level positions change strongly in the pnictides (CeP: 0. 5 eV to CeBi: 3. 1 eV) while there is only a small change in the chalcogenides (CeS: 2. 4 eV to CeTe: 2. 6 eV). This behaviour would fit with the 5d-electron bonding properties in the compounds as expressed by their ionicity, lattice constant etc. /22/. But we would then have to explain the observed energy and intensity dependence of the second peak for each compound. At present there is therefore no satisfactory explanation for the two 4f features at the same time.

V. Screening and the two-peak structure in Ce

Screening of the excited photoemission final state is obviously of crucial importance. We already related changes in d-screening to drastic photoemission intensity changes of the -3 eV feature in the pnictides and chalcogenides. This means, in other words, that 5d screening is the predominant screening mechanism, but not necessarily the most efficient in the sense of leading to the final state with lowest energy. In their work on excited state properties of 4f systems Herbst et al. /10/ assumed a "completely sd screened" final state. Johansson /11/ has, however, shown that there are serious problems to such an approach.

Recently it has been suggested /45/ that the two peak structure, observed in the photoemission spectrum of γ-Ce, might be due to itinerant d-screening (the 2 eV peak), as well as to screening by f-electrons (the 0. 3 eV peak) with the implication that the 2 eV feature corresponds to the photoemission final 4f hole state, whereas the 0.3 eV peak corresponds to some kind of band-like feature. We could adopt this model to all the Ce pnictides and chalcogenides by assuming that f-screening is responsible for the localized screening. The trends observed in the pnictides would then imply that f-screening (i. e. band-like behaviour) is predominant in CeP and CeN. We have to admit that these considerations are all still highly speculative, but we think that they are worth for serious detailed investigations.

No matter which model we consider the picture which evolves from the presently available photoemission results on the Ce pnictides and chalcogenides appears not to be totally consistent, when one includes the "mixed-valent" compound CeN.

VI. Summary

At present the electronic structure of Ce-compounds with "mixed" and even with "stable" valence cannot be considered as understood. In this paper we have illustrated that angular resolved photoemission provides a tool to detailed investigation of the itinerant states of Ce pnictides and that it can be helpful in identifying and classifying 4f-emission. In the Ce pnictides two 4f related emission features were observed, one close to the Fermi level (E_B=0. 5 eV) and one at about 3 eV binding energy. The high binding energy feature has been tentatively identified as due to d-screened localized $4f^0$-final-hole states. By means of resonant photoemission spectroscopy, striking trends in the intensity of the two 4f features, as already observed previously, have been established for all the Ce pnictides. It is argued that explanations of the trends by means of d-screening and pf hybridization, although apparently applicable for the pnictides, cannot be used for understanding "mixed-valent" CeN. Alternative explanations involving itinerant and localized screening have been discussed.

Acknowledgement

One of us (W. Gudat) thanks A. Bringer for stimulating discussions and M. Campagna for many valuable suggestions. We acknowledge the hospitality and the support of the LURE staff at Orsay and of the HASYLAB staff at Hamburg where the experiments have been performed.

VII. References

/1/ L.I. Johansson, J.W. Allen, T. Gustafsson, I. Lindau, and S. B. M. Hagstrom, Solid State Commun. 28, 53 (1978)

/2/ W. Lenth, F. Lutz, J. Barth, G. Kalkoffen, and C. Kunz, Phys. Rev. Lett. 41, 1185 (1978)

/3/ W. Gudat, S.F. Alvarado, and M. Campagna, Solid State Commun. 28, 943 (1978)

/4/ Y. Baer, H.R. Ott, J.C. Fuggle, and L.E. De Long, Phys. Rev. B24, 5384 (1981)

/5/ Y. Baer, and Ch. Zürcher, Phys. Rev. Lett. 39, 956 (1977)

/6/ H. Sugawara, A. Kakizaki, I. Nagakura, T. Ishii, T. Komatsubara, and T. Kasuya, J. Phys. Soc. Japan in print

/7/ M. Croft, J.H. Weaver, D.J. Peterman, and A. Franciosi, Phys. Rev. Lett. 46, 1104 (1981)

/8/ A. Jayaraman, W. Lawe, L.D. Longinotti and E. Bucher, Phys. Rev. Lett. 36, 366 (1976)

/9/ M. Croft, and A Jayaraman, Solid State Commun. 35, 203 (1980)

/10/ J.F. Herbst, D.N. Lowy, and R.E. Watson, Phys. Rev. B6, 1913 (1972) J.F. Herbst R.E. Watson, and J.W. Wilkins, Phys. Rev. B13, 1439 (1976) ibid, B17, 3089 (1978)

/11/ B. Johansson (Phys. Rev. B 20, 1315 (1979)) has pointed out that the "complete screening picture" /10/ is incomplete because of the neglect of the impurity term, i. e. the interaction of a tetravalent impurity in a trivalent host matrix.

/12/ The quantity $\Delta_+ = U_{eff} - \Delta_-$ is measured in Bremsstrahlung Isochromat-spectroscopy (BIS) (J.K. Lang, Y. Baer, and P. A. Cox, Phys. Rev. Lett. 42, 76 (1976)

/13/ Recently Fuggle et al. /20/ have suggested that the Ce mixed valent ground state might contain a mixture of $4f^0$, $4f^1$ and $4f^2$ configurations.

/14/ J.M. Lawrence, P.S. Riseborough and R.D. Parks, Rep. Prog. Phys. 44, 1 (1981)

/15/ M. Campagna, G.K. Wertheim, and Y. Baer, Photoemission in Solids II, ed. L. Ley and M. Cardona (Springer-Verlag, Berlin, Heidelberg, and New York 1979) Chap. 4, p. 217

/16/ J.C. Fuggle, M. Campagna, Z. Zolnierek R. Lässer, and A. Platau, Phys. Rev. Lett. 45, 1597 (1980)
/17/ G. Crecelius, G.K. Wertheim, and D.N.E. Buchanan, Phys. Rev. B22, 531 (1980)
/18/ B. Johansson and N. Mårtensson, Phys. Rev. B24, 4484 (1981)
/19/ W.D. Schneider, C. Laubschat, I. Nowik, and G. Kaindl, Phys. Rev. B 24, 5422 (1981)
/20/ J.C. Fuggle, F.U. Hillebrecht, Z. Zolnierek, R. Lässer, Ch. Freiburg, and M. Campagna, this conference proceedings
/21/ W. Gudat, M. Campagna, R. Rosei, J.H. Weaver, W. Eberhardt, F. Hulliger, and E. Kaldis, J. Appl. Phys. 52, 2123 (1981)
/22/ W. Gudat, R. Rosei, J.H. Weaver, E. Kaldis, and F. Hulliger, Solid State Commun. 41, 37 (1982)
/23/ A. Franciosi, J.H. Weaver, N. Martensson, and M. Croft, Phys. Rev. B24, 3651 (1981)
/24/ A. Bringer, in Valence Fluctuations in Solids, L. M. Falicov, W. Hanke, M. B. Maple (eds), North-Holland Publ. Co., 1981
/25/ A. Hasegawa, J. Phys. C 13, 6147 (1980)
/26/ A. Hasegawa, and A. Yanase, J. Phys. Soc. Jpn 42, 492 (1977)
/27/ see e.g. F.J. Himpsel, Appl. Opt. 19, 3964 (1980), Y. Petroff, and P. Thiry, Appl. Opt. 19, 3957 (1980)
/28/ A full account of our ARPES work (W. Gudat, R. Pinchaux) will be published elswhere.
/29/ see e.g. the various contributions in this volume
/30/ W. Gudat, S.F. Alvarado, M. Campagna, and Y. Petroff, J. Physique C5 suppl. 6-14 (1980) C5-1.
/31/ A. Zangwill and P. Soven, Phys. Rev. Lett. 45, 204 (1980)
/32/ M. Aono, T.-C. Chiang, J.A. Knapp, T. Tanaka and D. E. Eastman, Phys. Rev. B21, 2661 (1980)
/33/ M. Croft, A. Franciosi, J.H. Weaver, and A. Jayaraman, Phys. Rev. B 24, 544 (1981)
/34/ W.F. Egelhoff jr., G.G. Tibbets, M.H. Hecht and I. Lindau, Phys. Rev. Lett. 46, 1071 (1981)
/35/ G. Wendin: Vacuum Ultraviolet Radiation Physics, ed. E. Koch, C. Kunz, and R. Heansel (Pergamon-Vieweg, Braunschweig, 1974) p. 225.
/36/ C. Guillot, Y. Ballu, J. Paigné, J. Lecante, K. P. Jain, P. Thiry, R. Pinchaux, Y. Pétroff, and L. M. Falicov, Phys. Rev. Lett. 39, 1632 (1977)
/37/ R. Baptist, M. Belakhovsky, M.S.S. Brooks, R. Pinchaux, Y. Baer, and O. Vogt, Physica 102 B 63 (1980)
/38/ W. Gudat, and D.E. Eastman, in Photo emission and the Electronic Properties of Surfaces, B. Fitton, B. Feuerbacher, and R. F. Willis (eds), J. Wiley and sons, New York (1978)
/39/ J.W. Allen, S.-J. Oh, I. Lindau, J.M. Lawrence, L. I. Johansson, and S. B. Hagström, Phys. Rev. Lett. 46, 1100 (1981)
/40/ M. Iwan, W. Gudat, to be published
/41/ R.D. Parks, N. Mårtensson, B. Reihl this volume
/42/ M. Campagna, W. Gudat, W. Eberhardt, R. Rosei, J. H. Weaver, E. Kaldis, and F. Hulliger, Physica 102B, 367 (1980)
/43/ the normalization of the two spectra near $E_B=-4,5$ eV, an energy range dominated by ligand p electron emission, has the effect of visually diminishing the absolute resonance enhancement in inverse proportion to the enhancement ratio of the extended states.
/44/ F. Hulliger, in Handbook on Physics and Chemistry of Rare Earths, K. A. Gschneidner jr., and L. R. Eyring (eds), Vol. 4, North Holland, Amsterdam 1979
/45/ S. Hüfner, and P. Steiner, private communication and contribution in this volume.

M. CROFT: I'm obliged to my co-workers to point out that the results and interpretations regarding CeP, CeAs and CeBi that you discussed here were published by A. Franciosi, J. Weaver, N. Martensson and M. Croft in Phys. Rev. B 24, 3651 (1981). Of course we haven't done any angular resolved work.

W. GUDAT: This is correct. I mentioned your results and conclusions in my talk.

B. REIHL: The conclusion that the absence of an angular dependence of a photoemission peak indicates that it comes from a localized level cannot be made. At these high photon energies you get $\vec{k}$-broadening effects that wash out any band dispersion.

W. GUDAT: I agree that momentum broadening occurs in $\vec{k}_\perp$. In our spectra we looked at both $\vec{k}_\parallel$ and $\vec{k}_\perp$. Nevertheless, detailed band structures have been obtained at the same high energies for various metals (see e.g. F.J. Himpsel et al., Phys. Rev. B 22, 4604 (1980)).

G. KRILL: You show that the structure at 2 eV below E_F does not disperse. Is this not a usual feature of a surface state ?

W. GUDAT: A surface state (band) does not disperse with $\vec{k}_\perp$, but it does with $\vec{k}_\parallel$. However, from the trends which we observed in Ce-pnictides and chalcogenides and from the photon energy dependence of the structure near 3 eV we exclude a surface state interpretation.

P. WACHTER: You stated, as many others, that CeN is a mixed valent material. However, we have just heard all the controversial views on Ce and one should remember, that in CeN, the Ce-Ce distance is less than in γ-Ce. Therefore, if the γ - α transition can be described in terms of the Johansson-model, you will have direct f-f overlap in CeN.

W. GUDAT: Yes.

N. MARTENSSON: Your spectra suggest that CeN is not mixed valent but analogous to α-Ce. You find the 5d-screened $4f^0$ state in CeP at 3 eV. This seems to contradict that the pressure induced phase transition in CeP and, in analogy, the ground state of CeN is due to a 4f-promotion. The two peak structure in CeN is then not due to a surface shift but reflects the same situation as in α-Ce.

W. GUDAT: If the 3 eV peak in CeP is the 4f final hole state screened by 5d electrons and if the same formal transition $4f^1d^n \rightarrow 4f^0d^{n+1}$ applies to the pressure induced transition and to the photoemission, then the observation of peak at 3 eV in CeP is in contrast to the high pressure data. Because of the different energy separations involved in the two peak structures (CeN ∿ 1 eV, γ/α Ce ∿ 2 eV, Ce-pnictides $\lesssim$ 2.5 eV). I still believe that surface effects have to be considered for CeN.

T. KASUYA: It is wrong to say that the core hole charge is well screened by the conduction electrons. Screening is mostly made by the inter-band transition and rather poorly. When the screening is complete, the f^0 final state level in CeN should be at the Fermi level and that of CeBi about 1 eV below the Fermi energy.

W. GUDAT: We have two possible explanations for the 4f-derived two peak structure clearly seen in the pnictides. For the high binding energy feature we consider a d-screened final hole state in the sense of screening by itinerant states. It should not imply that we are dealing with a (energetically) well screened peak.

M. CROFT: Two comments: First I think it is not necessary to debate whether f electron delocalization or 4f promotion occurs in every different Ce system. The transition is probably always the same. Second I think the ideas about direct f-f overlap should be dropped. There is a lot of work (see e.g. St. Barbara proceedings) that the nonmagnetic state can be stabilized in dilute systems on isolated Ce atoms and may be regarded as a single ion state.

J.C. FUGGLE: Comment: I do not agree with the chairman's previous statement that we all agree that the peak at the Fermi level is a $4f^1$ final state. I do not.

Valence Instabilities
P. Wachter and H. Boppart (eds.)

Bistable 4f-States in Rare-Earth Solids

M. Schlüter and C. M. Varma

Bell Laboratories
600 Mountain Avenue
Murray Hill, New Jersey 07974

We show with self-consistent one electron theory that in rare-earth solids, particularly those of Cerium, two stationary states with different radial extent but both labeled 4f in terms of one-electron quantum numbers can exist. These results are interpreted in terms of a pole in the self-energy of the 4f states due to Coulomb interactions with the core-states and the higher lying states. The anomalous photoemission and X-ray absorption edge spectra of Ce are explained on this basis.

The 4f-wavefunctions for atoms with nuclear charge less than Barium are hydrogenic and lie well outside the 5d and 6s wavefunctions. For the larger nuclear charge of the rare-earth atoms the 4f wavefunctions are pulled in and, unlike the outer s, p, d wavefunctions, having no orthogonality requirements, they collapse close to the nucleus and lie well inside of even the "core-like" 5s and 5p wavefunctions. The elements La, Ce, Pr lie near the border-line of this transition.[1]

In recent years several anomalous experimental results[2–8] have been reported in experiments on Ce and its compounds, raising questions about the nature of the ground state and the excitation spectra of the border-line elements.

X-ray absorption edge experiments[5] show that essentially all compounds with Ce^{4+} ions have two energy thresholds differing by about 8 eV. This is in contrast to experiments in compounds with Ce^{3+} ions and indeed any other rare-earth with stable valence where, as expected, a single threshold is observed. Ultra-violet photoemission experiments[4,6,7] on a number of Cerium compounds with either one f electron or of a mixed-valence character reveal two peaks associated with f state emission. Especially intriguing are the results[6,7] on $Ce_{0.9}Th_{0.1}$, and Ce which as a function of decreasing temperature transform from the γ-phase to the α-phase. The photoemission experiments in the γ-phase reveal a narrow f-like peak, as expected, about 2.2 eV below the Fermi-energy. There is *also*, however, a fairly narrow peak of lesser weight near the Fermi-energy. In the α-phase, there are again two peaks separated by about the same energy, but the weight of the peak near the Fermi-energy is now higher than the other. Sharp satellite structures due to *shake-up* and *shake-down* processes utilizing resonances to which the initial or the final states are coupled are often seen in photoemission spectra.[9] But, for Ce no additional resonance which might explain the observations is known; the nearest *known* resonance is the f^2 level which isochromat spectra[8] reveal to be about 5 eV above the Fermi-level.

These experimental findings have led us to enquire whether in Ce (and possibly in elements close to it) there exists more than one sharp feature in the spectral distribution associated with the 4f-states. Unlike the Fermi-liquid theory of metals, no one to one correspondence need exist between the excitation spectra of an atom with electron-electron interactions and the states obtained in one-electron theory. If, due to many-body effects, additional features exist in the spectral distribution of the 4f-states, they ought to be revealed in a strong energy dependence of the self-energy of the 4f-states. Although a correct description of the experimental results necessitates such a many-body description, we have found that self-consistent field one-electron theory can quantitatively establish the existence of the sharp resonance in the 4f-self-energy. A first principles many-body calculation for this inhomogeneous problem would be extremely difficult to carry out.

The eigenstates and eigenvalues of a self-consistent field one-electron approximation like the density-functional method (or the Hartree-Fock method) are obtained from the solution of *non-linear* equations which in principle allow the existence of different stationary 4f-states (with the same occupation). We have found that a bistable situation with two different 4f-like states exists in Ce because of the competition between the nuclear potential and the valence and the core charge redistribution. We then reinterpret these results in a many-body language and discuss the experiments in terms of it.

We begin with an all-electron self-consistent local density functional calculation including exchange, correlation and self-interaction corrections. The eigenstates and eigenvalues of all states except the 4f-state are calculated in the standard way by solving the self-consistent Schrödinger equation while an explicit variational procedure is used for the 4f-states. We assume for the radial part of the 4f-state a simple variational function of the form

$$\phi_f(r) = r^3 \exp(-r/r_0) \left[\frac{(2/r_0)^9}{8!}\right]^{1/2} \tag{1}$$

and calculate the *total energy* of the atomic system as a function of the variational parameter r_0. For a given r_0 self-consistency of all other states is achieved to an accuracy of 10^{-5} a.u. The solid line in the lower part of Fig. (1) is the result of this calculation for the free Ce atom as a function of $r_{max} = 3r_0$, which is the position of the maximum of the f-charge. The well confined f-state of Ce at $r_{max} \approx 0.6$ a.u. is, of course, in

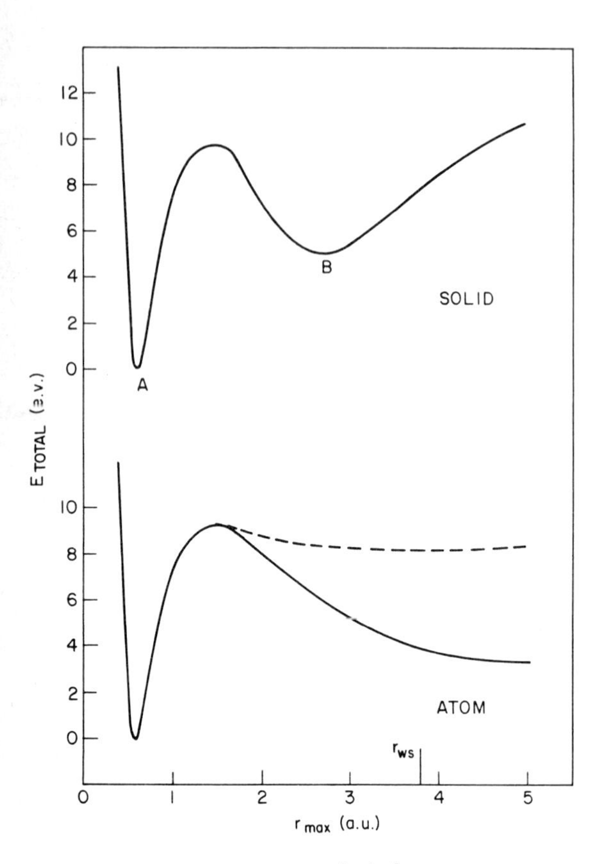

Fig. 1. Calculated total energy of $4f^1 5d^1 6s^2$ atomic Ce as a function of the parametrized 4f wavefunction using the self-consistent field local density approximations (r_{max} is the radial position of the maximum of the trial wavefunction). For the free atom, results for an unrelaxed core (dashed curve) and fully relaxed core (full curve) are shown for comparison. The addition of "embedding" contributions in the solid (see text) yields two stationary states.

evidence. The most remarkable feature of this curve, however, is the "bump" around 1.5 a.u. To understand the origin of this feature we recalculated the total energy as a function of r_0, but with the charge configuration of the remaining electrons "frozen" at the distribution obtained for the stationary state at $r_{max} \approx 0.6$ a.u. . The dashed curve then results. This result as well as an examination of the eigenvalues and wavefunctions of the core states lead to the conclusion that the "bump" arises due to the *relaxation towards the nucleus of the core-electrons* (primarily 5s and 5p) whose charge is concentrated near $r \approx 1.5$ a.u. as the variational 4f-charge moves past this region. This breathing of the core-charge may be considered in a many-body description as the condensation of a core-plasmon.

When we consider the solid, a few changes need to be made. First, the tail of the 4f-state, Eq. (1) must be orthogonalized to the core-states in the Wigner-Seitz cells of the surrounding atoms. To calculate this correction, we approximate the tail to be mostly s-like with respect to surrounding atom and calculate the orthogonality potential[10] due to a neighbor to be:

$$\hat{V} \approx \sum_n \epsilon_{ns} \langle \phi_{ns} | 4f \rangle | \phi_{ns} \rangle \tag{2}$$

where ϕ_{ns} are the s-like core-states of the neighbor with eigenvalue ϵ_{ns}. This potential is then suitably spherically averaged and integrated over the range of the tail of the 4f wavefunction. Second, Wigner-Seitz boundary conditions are used for the 5d and 6s conduction electrons. The total energy per atom in the solid calculated with this procedure is shown in the top part of Fig. (1). Two stationary states are obtained; the one labelled A being the usual "collapsed" rare-earth 4f-state, the one labelled B is about 5 eV higher than A and has a maximum of its charge at about 2.8 a.u., well inside the Wigner Seitz cell ($R_{\omega s} \approx 3.8$ a.u.). The localization of the B state is likely to increase for trial wavefunctions with more variational freedom. It is worth emphasizing that the orthogonality etc., and therefore the upwards turn of the "solid" results for large r-values in Fig. (1) are normal and customary solid-state results; the interesting new feature, i.e., the stationary state B arises from the occurrence of the "bump" which is due to core charge relaxation and already present in the results for the free atom. We have repeated the calculation with the atomic volume reduced to that of the α-phase ($r_{ws} \approx 3.6$ a.u.). We find that the energy difference of the A and B state remains unaffected, while the total energy/atom goes down by about 3 eV.

It is important to note that the states A and B are *not* orthogonal to each other. This arises because the self-consistent field Hamiltonians of the two states are different due to the different core-charge distributions.

From our numerical calculations and the above considerations, we can proceed to a proper many-body description of the problem. Let us start with the usual lowest (orthogonal) eigenstates of a one-electron calculation. Since the effective one-electron Hamiltonian is based on a self-consistent field it is clear that the normal 4f state (A-like) is coupled by Coulomb interactions to the core-charge and the higher one-electron eigen states, (and to "external fields" like the nuclear charge and the core of surrounding atoms). This coupling is represented in the self-energy diagram for the A state, Fig. (2).

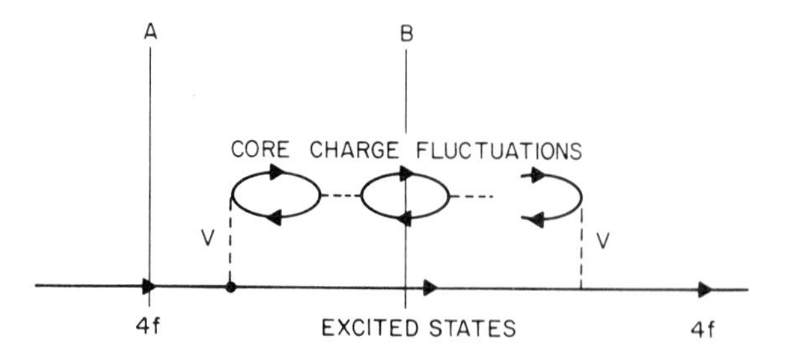

Fig. 2. A self-energy diagram for the 4f Green's function. The A-like state is admixed to higher excited states through Coulomb coupling with the core-charge fluctuations giving rise to a B-like resonance.

The existence of the stationary B state in our variational calculation then means that the self-energy, Fig. (2) has a pole whose residue has the same angular symmetry (f) as the A state:

$$\Sigma_f(\omega) = V^2/(\omega - \epsilon_B' + i\delta) \tag{3}$$

The spectral distribution for f-electrons is given by the one-electron Green's function

$$G_f(\omega) = (\omega-\epsilon_A'-\Sigma(\omega)+i\delta)^{-1}$$

$$= Z_A/(\omega-\epsilon_A+i\delta) + Z_B/(\omega-\epsilon_B+i\delta) \quad . \qquad (4)$$

In (3), V is as shown in Fig. (2) the matrix element of the many electron Hamiltonian admixing the "one electron" 4f and other states due to charge fluctuations of the core. Its magnitude is in the range of a few eV but it is hard to be precise about it. In (4) we approximately identify ϵ_A and ϵ_B as the energies of the stationary states A and B of Fig. (1). The weights Z_A, Z_B depend upon $V/(\epsilon_B-\epsilon_A)$. $G_f(\omega)$ has a representation in real space, $|\{r\}\rangle G_f(\omega)\langle\{r'\}|$; so do Z_A and Z_B. We identify $Z_{A,B}(\{r\},\{r'\}) \sim \Psi^*_{A,B}(\{r\})\Psi_{A,B}(\{r'\})$ where $\Psi_{A,B}(\{r\})$ refer to the core and f-wavefunctions corresponding to A and B states respectively.

In terms of the two-peak structure found in the one-electron Green's function, Eq. (4), the qualitative features of the photoemission results can be interpreted. The admixture of A or B characters to the ground state depends on the position of the chemical potential with respect to ϵ_A and ϵ_B. We regard the γ-phase as one in which the Fermi-level is pinned by the B resonance. This makes the total 4f occupation slighlty less than 1 in accordance with the estimates based on the lattice constants and yields a ground state with considerable A, B admixture. In the photoemission, the peak near the zero energy loss (at E_F) is then due to the excitation of the B resonance while the one below E_F is due to the excitation of the A resonance. The relative weight is determined by the spectral weight of the A and B resonances.

The α-state has an atomic volume about 15% less than the γ-phase which lowers the s-d band and the chemical potential. One possibility is that the chemical potential is pinned in the α phase by the A-resonance. The ground state charge with B character is, therefore, zero, however, the amount of A-like f-charge is in first order unchanged from the γ-phase. The total f-occupation in the α-phase is then less than in the γ-phase consistent with the atomic volume. In the photoemission results the zero-energy loss peak is identified as due to the A resonance. The B resonance is, however, available for a strongly coupled shake-up process leading to an energy loss peak. This picture is consistent with the difference in the energy of the two peaks remaining the same in the α and the γ-phase and with the reversal in the intensity of the two peaks. To agree with experiments, we require $|V/(\epsilon_A-\epsilon_B)|^2 \sim 0.4$. Alternative explanations, of the α-phase results, given Eq. (4) are certainly possible.

Quantitatively, the calculated energy difference between A and B states is 5 eV for Ce which is about twice the experimental value. This discrepancy results from the use of a 4f wavefunction with restricted variational freedom (Eq. (1)), from the neglect of details in the metallic screening process of the various final state 4f-holes as well as from some systematic errors due to the use of the local density approximation, and perhaps most importantly, from the neglect of the detailed 4f-core electron correlations.

We have also investigated some of the other near-by metals. We find no evidence for any bistable behavior in Cs, a slight non-monotonicity of the dE/dr_0 vs. r_0 in Ba and definite evidence for bistability in La. For Pr and Nd, and we suspect for all the higher rare-earths, pronounced "bumps" in E vs. r_0 exist for the free-atom. We infer, however, from our calculations that the B states are placed far above the Fermi-level. The weight of the B structure in the 4f-Green's function is therefore much smaller than for Ce. Closest to Ce is an excited neutral Pr with one 4f electron. This models the fully relaxed final state in X-ray absorption experiments on Ce^{4+} which, as already mentioned shows edge structures separated by about 8 eV. We have estimated the total energy of this final state and find it bistable with an A-B energy difference of about 12 eV. We identify the two edge structures as due to relaxation in the final state into the A and the B resonances respectively.

We are grateful to R. D. Parks for an extensive discussion of the experimental results and to P. W. Anderson, E. I. Blount, D. R. Hamman and G. K. Wertheim for stimulating remarks.

References

1. M. Goeppert-Mayer, Phys. Rev. *60*, 184 (1941).
2. B. Johansson, Phil. Mag. *30*, 469 (1974); A. Platau and S. E. Karlsson, Phys. Rev. *B18*, 3820 (1978).
3. C. Stassis, C.-K. Loong, B. N. Harmon, S. H. Liu, and R. M. Moon, J. Appl. Phys. *50*, 7567 (1979); B. Barbara, M. F. Rossignol, J. X. Boucherle, J. Schweizer, and J. L. Buevoz, J. Appl. Phys. *50*, 2300 (1979).
4. J. W. Allen, S.-J. Oh, I. Lindau, L. I. Johansson and S. B. Hagström, Phys. Rev. Letts. *46*, 1100 (1981); W. Gudat, M. Campagna, R. Rosei, J. H. Weaver, W. Eberhardt, F. Hulliger and F. Kaldis, J. Appl. Phys. *52*, 2123 (1981).
5. K. R. Bauchspiess, W. Boksch, E. Holland-Moritz, H. Launois, R. Pott and D. Wohlleben, in *Valence Fluctuations in Solids*, edited by L. M. Falicov, W. Hanke, and M. B. Maple (North-Holland, 1981) p. 417.
6. N. Martensson, B. Reihl and R. D. Parks, Solid State Communications (to be published).
7. D. Wieliczka, J. H. Weaver, D. W. Lynch and C. G. Olson (Preprint).
8. Y. Baer, H. R. Ott, J. C. Fuggle and L. E. DeLong, Phys. Rev.
9. For a review, see G. K. Wertheim in *Valence Fluctuations in Solids*, edited by L. M. Falicov, W. Hanke and M. B. Maple, (North-Holland, 1981).
10. P. W. Anderson, Phys. Rev. Letts. *21*, 13 (1968).

Valence Instabilities
P. Wachter and H. Boppart (eds.)

THE ELECTRONIC STRUCTURE OF γ-Ce AND α-Ce

Stefan Hüfner and Paul Steiner

Fachbereich Physik, Universität des Saarlandes, Saarbrücken,
Germany

The experimental spectroscopic data on the electronic structure of γ-Ce and α-Ce are inspected. It is concluded that in both modifications the f-level is close to the Fermi energy in the groundstate. Structure observed in photoemission at 2 eV below the Fermi energy is ascribed to a poorly screened final state. The interpretation of the photoemission spectra of Ce is similar to those of Ni and U.

Cerium is plagued by the problem, that the data from thermodynamic measurements and those obtained by socalled high energy probes (e.g. photoemission spectroscopy - PES, X-ray absorption spectroscopy - XAS) are not easy to reconcile with each other. PES experiments give e.g. positions of the 4f level 0.8 eV below the Fermi energy (E_F) for $CePd_3$ and 2.4 eV below E_F for γ-Ce [1], whereas in the thermodynamic limit for these two materials a position of the 4f level close to E_F is expected [2]. The problem is most apparent in CeO_2 commonly assumed to be a fourvalent material which shows L_{III} XAS data indicating that it is an intermediate valence (IV) compound with a valence of 3.32 (4f occupancy 0.68) [3]. In order to come to an understanding of the PES data of Ce metal [1],[4]-[14] it is useful to compare them with those of materials which seem to show similar effects [15]. This is the case for Ni and U and the discussion here will be restricted to Ni. The core lines of Ni show a satellite [16], as do the spectra of Ce metal [6]. In Ni the satellite is ascribed to a $3d^9(4sp)^2$ final state configuration (assuming that $3d^9(4sp)^1$ is the groundstate in Ni metal) and the main line is ascribed to a $3d^{10}(4sp)^1$ screening configuration [17]-[19]. Accordingly the main and satellite core line in Ce metal are ascribed to a $4f^1(5d6s)^4$ and a $4f^2(5d6s)^3$ screening configuration [6]. Note that in the literature the line which is called in Ni the satellite is called the main line in Ce and vice versa.

The effect of the two screening configurations on the valence band spectra of Ni and Ce is schematically shown in fig. 1 [17]-[19]. PES of a valence band 4d electron in Ni can lead to two final states. In one of them the screening is produced by the (4sp)-band leading to a $3d^8(4sp)^2$ final state (satellite). In the second final state the screening charge is produced by the 3d band giving $3d^9(4sp)^1$ which leads to what is generally refered to as the main peak [16]. This interpretation for Ni is substantiated by the resonance photoemission measurements of Guillot et al. [20]. These authors find that the 6 eV satellite in Ni resonates in intensity if the photon energy is swept through the 3p energy (67 eV). The explaination of this phenomenon is given as follows:

$$3p^6 3d^9 (4sp)^1 + h\omega(67\ eV) \rightarrow 3p^5 3d^{10} (4sp)^1 \xrightarrow{\text{Auger}} 3p^6 3d^8 (4sp)^2 + e \qquad (1)$$

The final state of this process is thus identical to the one found in direct photoemission of the valence band and (4sp) screening, a fact which is responsible for the enhancement.
We note that a very similar process has been found in U metal [21].
It is suggested that the reasoning given just for Ni can also be applied to Ce [15], which in the PES data shows a broad $(5d5s)^3$ band with

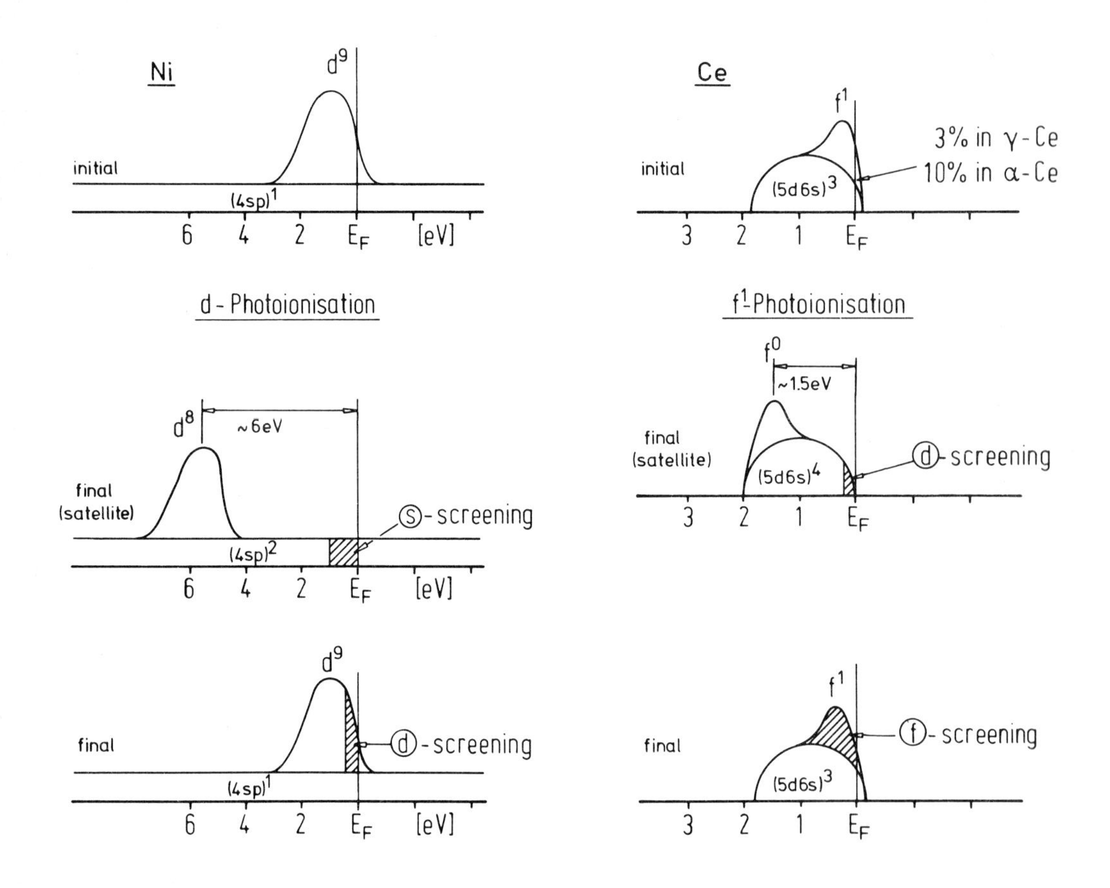

Figure 1 : Schematic drawing of the density of states of Ni and Ce and of the two possible final states after valence band photoemission [16]-[19]

narrow structures at 2 eV and 0.2 eV below E_F (fig. 2). It is assumed that the 2 eV peak is a $4f^o(5d6s)^4$ final state and the 0.2 eV peak a $4f^1(5d6s)^3$ final state. In agreement with this assignement, the 2 eV peak is found to resonante in a way similar to that observed for the 6 eV satellite [5] in Ni and therefore one can assigne this peak to a poorly screened [22] signal $(f^o(5d6s)^4)$. An equation in analogy to equ.[1] yields:

$$4d^{10}4f^1(5d6s)^3 + h\omega(120\text{ eV}) \rightarrow 4d^9 4f^2(5d6s)^3$$

$$\xrightarrow{\text{Auger}} 4d^{10}4f^o(5d6s)^4 \qquad (2)$$

One may argue that a $4f^o(5d6s)^4$ configuration at 2 eV below E_F precludes any itineracy or IV behaviour of the f electrons in α-Ce. Note however that PES produces a very localiced excitation which certainly has a higher binding energy than the groundstate. A case in point is again Ni, where the $3d^8(4sp)^2$ configuration in the PES data is found at ~ 6 eV below E_F, which agrees with an atomic calculation [23].
The present understanding of the d-band properties in Ni assumes however $d^8(4sp)^2$ charge fluctuation states also in the d-band [24], meaning near E_F. Therefore in Ce a $4f^o(5d6s)^4$ localized excitation can be observed in the PES data while such a charge fluctuation is of course near E_F.

Having established an interpretation of the PES data of γ-Ce (α-Ce) we now turn to the question

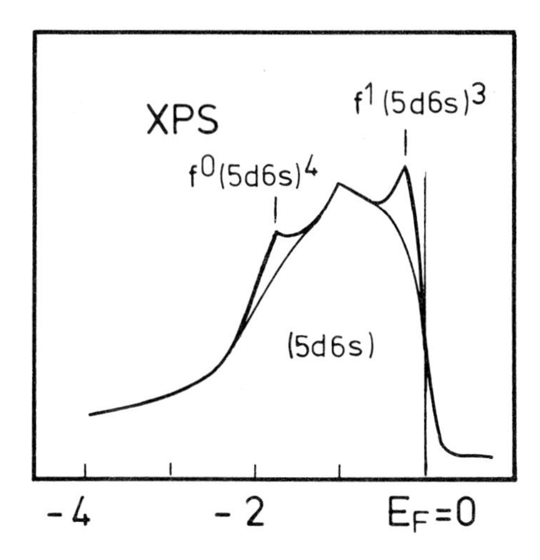

Figure 2 : Valence band photoemission spectrum of γ-Ce [10]

of the difference in electronic structure of γ-Ce and α-Ce. To that purpose we show in fig.3 the PES data of the valence band of γ-Ce [9], $CePd_3$[9]and CeN [14], the BIS data of γ-Ce [9] and $CePd_3$ [9] and the reflectivity data of CeN [25][27]. (Note that the valence band PES of γ-Ce is very similar to that of α-Ce [8]). The BIS spectrum of γ-Ce shows the f^2 configuration at 5 eV above E_F. The BIS spectrum of $CePd_3$ shows also this f^2 configuration. In addition however a structure at 2 eV above E_F can be seen. If $CePd_3$ is assumed to be a intermediate valence compound with a wavefunction $4f^1(5d6s)^3 + (5d6s)^4$ the structure at 2 eV must be due to a $4f^1(5d6s)^4$ final state. In order to find out whether similar structure can be found in CeN we have plotted the reflectivity of this material [25][26], which shows two narrow lines at 2 eV and 4 eV above E_F. Although different interpretations are possible we tentatively ascribe these peaks to $4f^1(5d6s)^4$ and a $4f^2(5d6s)^3$ structure. The 4 d and 3 d spectra of α-Ce [8][14], $CePd_3$ [27] and CeN [14] all shows a $4d^5 4f^o(5d6s)^5$ component in the spectra, indicating that their initial state contains a $4f^o(5d6s)^4$ component.
At this point it cannot be decided from the high energy probe measurements wether α-Ce is

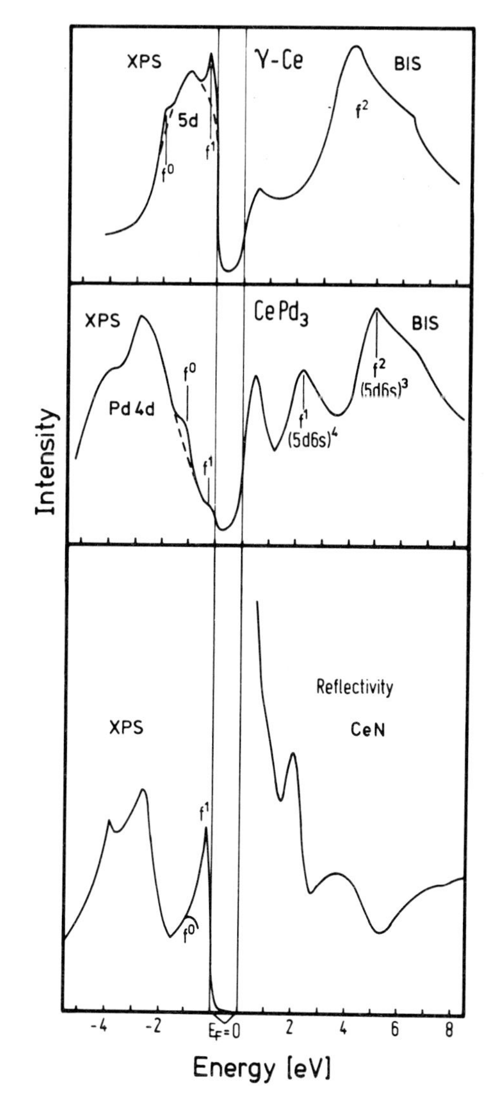

Figure 3 : XPS valence bands of γ-Ce, $CePd_3$ and CeN [9], [14], BIS spectra of γ-Ce and $CePd_3$ [9] and reflectivity data for CeN [25], [26]

a f-band metal [28] or an IV compound. We note that μ-SR data [29], core line systematics [30] and the large $f^o - f^1$ separation relative to $CePd_3$ and CeN may point to an IV behaviour.

In conclusion it has been demonstratet that also for Ce in experiments with high energy probes a highly excited final state is observed. The correlation of a spectrum measured by these techniques with the groundstate electronic structure is very difficult and can only be performed with extreme reservations.

Acknowledgement: We thank Y. Baer, M. Campagna, A.J. Freeman, J.C. Fuggle, B. Johansson, F. Steglich, D. Wohlleben and P. Wachter for very informative discussions.

This work has been supported by the Deutsche Forschungsgemeinschaft and by a Nato Research Grant (221.80).

References

[1] J.W.Allen, S.J. Oh, I.Lindau, J.M.Lawrence, L.I.Johansson, S.B.Hagström: Phys.Rev.Lett. 46, 1100 (1981)

[2] D.M.Newns and A.C. Hewson: Valence Fluctuations in Solids, L.M.Falicov, W.Hanke, M.B.Maple (eds), North Holland Publishing Company (1981), p.27

[3] K.R.Bauchspiess, W.Boksch, E.Holland-Moritz, H.Launois, R.Potts and D.Wohlleben: Valence Fluctuations in Solids, L.M.Falicov, W.Hanke, M.B.Maple (eds), North Holland Publishing Company (1981), p.417

[4] A.Platau, S.E.Karlson: Phys.Rev. B 18, 3820 (1978)

[5] L.I.Johansson, J.W.Allen, T.Gustafsson, I. Lindau, S.B.Hagström: Solid State Commun. 28, 53 (1978)

[6] G.Crecelius, G.K.Wertheim, D.N.E.Buchanan: Phys.Rev. B. 18, 6519 (1978)

[7] Y.Baer, G.Busch: J.Electron.Spectrosc. 5, 611 (1974)

[8] R.A.Pollak, S.P.Kowalczyk, R.W.Johansson: Valence Instabilities and Related Narrow Band Phenomena. R.D.Parks (ed), pl 463, New York, Plenum Press 1977. N.Martensson, B.Reihl, R.D.Parks: Solid State Commun. 41, 573 (1982)

[9] Y.Baer, H.P.Ott, J.C.Fuggle, L.E.DeLong: Phys.Rev. B. 24, 5384 (1981)

[10] J.K.Lang,Y.Baer and P.A.Cox: J.Phys. F; Metal Phys. 11, 121 (1981)

[11] P.Steiner, H.Höchst, S.Hüfner: J.Phys.F 7, L 145 (1977)

[12] Y.Baer, G.Busch:Phys.Rev.Lett 31, 35(1973)

[13] C.R.Helms, W.E.Spicer: Appl.Phys.Lett. 21, 237 (1972)

[14] Y.Baer, Ch.Zürcher: Phys.Rev.Lett 39, 956 (1977)

[15] S.Hüfner, P.Steiner: Z.Physik B 46,37 (1982)

[16] S.Hüfner and G.K.Wertheim: Phys.Lett 51A, 301 (1975)

[17] A.Kotani,Y.Toyozawa: J.Phys.Soc.Jpn. 35, 1082 (1973)

[18] K.Schönhammer, O.Gunnarsson: Z.Phys.B Condensed Matter 30, 297 (1978)

[19] L.C.Davis and L.A.Feldkamp: J.Appl.Phys. 50, 1944 (1979); Phys.Rev. B 22, 3644 1980

[20] C.Guillot, Y.Baller, J.Paigné, J.Lecante, K.P.Jain, P.Thiry, R.Pinchaux, Y.Petroff, L.M.Falicov: Phys.Rev.Lett. 39, 1632 (1977)

[21] M.Iwan, E.E.Koch, F.J.Himpsel: Phys.Rev. B 24, 613 (1981)

[22] J.C.Fuggle, M.Campagna, Z.Zolnierek, R. Lässer, A.Platau: Phys.Rev.Lett 45, 1957 (1980)

[23] N.Martensson, B.Johansson: Phys.Rev.Lett 45, 482 (1980)

[24] C.Herring, Magnetism IV, G.T.Rado and H. Suhl (eds), Academic Press (1966)

[25] J.Schoenes: Physica 102 B, 45 (1980)

[26] A.Schlegel, E.Kaldis, P.Wachter and Ch. Zürcher: Phys. Lett. 66A, 125 (1978)

[27] R.Lässer, J.C.Fuggle, M.Beyss, M.Campagna, F.Steglich, F.Hullinger: Physica 102B, 360 (1980)

[28] B.Johansson:Philos.Mag. 30, 469 (1974); Phys.Rev. B20, 1315 (1979); J.Phys. F6, L 169 (1974)

[29] H.Wehr, K.Knorr, F.N.Gygax, A.Schuck and W.Studer: Phys.Rev. B 24, 4041 (1981)

[30] J.C.Fuggle, Private Communication

Valence Instabilities
P. Wachter and H. Boppart (eds.)

CORE LEVEL SPECTROSCOPIES OF Ce AND ITS COMPOUNDS

John C. Fuggle, F. Ulrich Hillebrecht, Zygmunt Zołnierek*, Ch. Freiburg≠
and Maurice Campagna

Institut für Festkörperforschung, der KFA Jülich,
D-5170 Jülich, W. Germany.

* Permanent address, Institute for Low Temperatures, Wroclaw, Poland.

≠ Zentralinstitut für Chemische Analyse, der KFA Jülich.

Screening of core holes by delocalized and localized valence electrons is discussed in the light of a scheme involving configuration interaction in both initial and final states. The similarities of this formulation to the models of Friedel, Toyozawa and Kotani, and Schönhammer and Gunnarsson is pointed out and used to reach a qualitative understanding of Ce intermetallic compounds. Evidence is presented for the importance of f-level hybridization and a surprisingly large f-level width of the order of 0·5 to 1·5 eV. Reasons are given why it is not possible to deduce ground state f-counts directly from, for instance, peak intensities in core level spectroscopies. Our discussions imply, however, a need for revision of current views of the electronic structure of Ce and its compounds.

The understanding of photoemission and x-ray absorption or emission experiments involving Ce is, at present, not fully quantitative because of the rôle of the 4f states. Despite the large effort of the last years not much quantitative information is yet available about their degree of localization in Ce. We feel, on the other hand that recent discussion about high energy experiments on Ce(hv≥30 eV) has not taken sufficient account of the fundamental physics of photoemission and x-ray processes that is at present understood. In this paper we attempt to improve this situation. We will restrict ourselves to core level spectroscopies and will argue that they can already contribute to the understanding of the ground state structure of Ce. However, we do still support the arguments of J. Wilkins [1] in his opening lecture pointing out that theoretical developments are needed in order to make progress towards quantitative interpretation and we will imply that some conclusions in the literature on Ce have questionable validity.

In both XPS and XAS we consider the photoemitted (photoexcited) electron separately from the N-1 electrons. The suddenly created core hole is screened by these N-1 electrons. The spectrum of the core hole left behind (e.g. in XPS the distribution of kinetic energies of the photoelectron) will therefore contain the structure of the possible final states, ψ_i^* (N-1) reached by the N-1 electrons. In the sudden approximation [2-7] one says that the ground state wave function, ψ (N-1) is expanded in terms of the available final states, ψ_i^* (N-1). The transition probabilities, P_i, to a given final state ψ_i^* (N-1) are thus given by [2-6]

$$P_i = |\langle \psi_i^* (N-1) | \psi (N-1) \rangle|^2$$

The physical situation can be described by saying that after core photoexcitation the N-1 electrons are not in a wave function of the final state Hamiltonian, but in a mixture of states. The probability that photoexcitation will result in a given final state is given by equation (1). The larger the perturbation caused by core photoexcitation, the larger will be the spreading amongst different final states in the spectrum. In XAS the photoexcited electron helps screen the core hole from the N-1 electrons, but the perturbation is still large and one expects to find similar effects in XAS and XPS[7]. However there have, so far, been more detailed studies of the effects relating to equation 1 in XPS than XAS.

As is the case for charged impurities in metals, [8,9] a large part of the screening of a deep core hole is done by the valence electrons so that the form of screening and the consequences from the XPS and XAS spectra depend strongly on the chemical environment of the atom containing the core hole. In free-electron metals the valence electrons react collectively to the core hole potential. Here the XPS spectra show a main peak due to excitation to the core-hole state of lowest energy and in addition there are higher binding energy peaks due to simultaneous excitation of a core hole and plasmons. In materials with more localized valence electrons the attractive core hole potential in the XPS or XAS final states can pull down a localized state from the unoccupied bands[8-12]. In this case two lines are expected to final states in which the "Friedel exciton" or "screening orbital" is left empty or full[10-15]. If the screening orbital is derived from highly localized orbitals it is decoupled from the other valence electrons and the probability of it being filled during photoemission is small, even though the final state in which it is occupied may be the core ionized state of lowest energy The 4f levels in La and Ce represent an intermediate case where, in the example of the 3d XPS lines, two peaks are seen due to final states screened by the 4f and the 5d6s electrons, respectively [10-15]. In these cases the relative weights of the peaks are sensitive to chemical environment. In some of the heavy rare earths the 4f levels seem to be so strongly localized that no peaks due to 4f screening are noticeable (see e.g. ref. 15). At one time it was thought that core level spectra of mixed valence compounds with f^n and f^{n-1} configurations would always contain two sets of peaks due to f^n and f^{n-1} final states. It was also thought that the spectral weight of the f^n and f^{n-1} peaks would give the amplitudes of the f^n and f^{n-1} configurations in the ground state. It is now recognized that this is not true because of screening effects in La, Ce, Pr[12-17] and possibly even in Eu[18] compounds with integral f-count, but to hold approximately for Sm and Yb compounds[19-20]. It is one of the advances made recently that in Ce and its compounds (as well as for La and to a minor degree also for Pr) one has realized that the f level width W_f[21] due to hybridization with the other valence electrons is significant. We know of no theoretical reason why XAS should differ from XPS in this respect, as is sometimes suggested. We note also that it is mistaken to believe that this width is the only factor in determining the peak intensities in core level spectroscopies, or that the peak intensities in core level spectroscopy bear no relation to the ground state properties and f-counts.

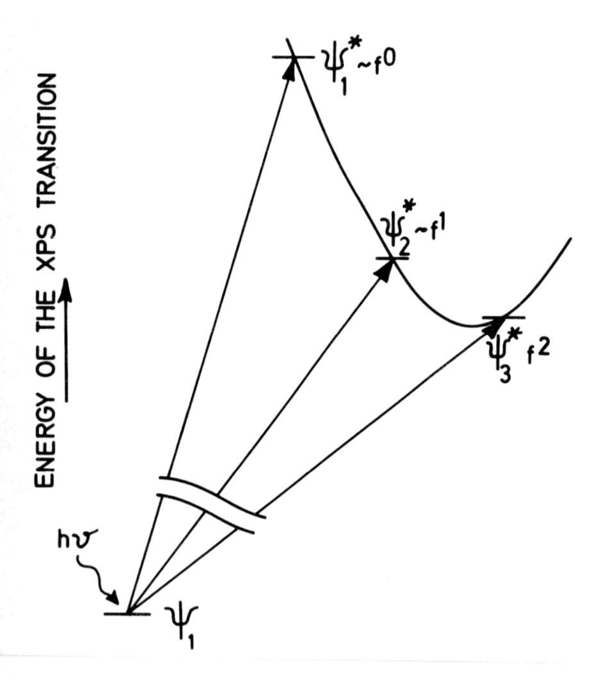

Figure 1: Schematic diagram of transitions made as a result of 3d core hole creation in XPS of Ce and its compounds.

In previous papers[12,14,15] we have concentrated on relating our core level photoemission data to models which can describe the "dynamics" of the photoemission process and rationalize the relative weights in the different peaks. Here we describe the interpretation of the results in slightly different terminology in the hope that it will widen the understanding of the problems and implications of such spectra. We still stand by the validity of our previous treatments.

Core level photoemission from Ce and its solid compounds can be schematically represented as in figure 1, with transitions from a single initial state to three different final states. In the spirit of configurational interaction, the initial state, ψ_1, can be written as a combination of three n-particle determinants, $\phi(f^n)$

$$\psi_1=a_1\phi(f^o) + b_1\phi(f^1) + c_1\phi(f^2) \quad ..... (2)$$

where $\phi(f^n)$ describes the configuration

$$Ce \ ....3d^{10}.......4f^n(5d6s)^y$$

In these configurations described by the ϕf^n, (n+y)=4 for gas phase Ce. Without some perturbation of the spherical symmetry the ϕf^n are orthogonal so that in the gas phase mixing of the $\phi(f^n)$ is forbidden. These restrictions are relaxed in the solid where the 5d and 6s electron are band-like. Charge-transfer <u>from</u> and <u>to</u> the partner element in compounds may lead to non-integral values of (n+y).

There are, in addition, two other states formed by mixing of the $\phi(f^n)$ in the absence of a core hole which may be written as

$$\psi_2=a_2\phi(f^o) + b_2\phi(f^1) + c_2\phi(f^2)$$
$$\psi_3=a_3\phi(f^o) + b_3\phi(f^1) + c_3\phi(f^2) \quad .... (3)$$

We will proceed in the assumption that the coefficients b_1, a_2, and c_3 are approximately 1 and a_1, c_1, b_2, c_2, a_3 and b_3 are all approximately zero.[22] Thus ψ_1, which is taken to have the lowest energy in the <u>absence</u> of a core hole, will be close to a pure f^1 configuration. ψ_2 and ψ_3 are then to be viewed as two possible excited states (e.g. the valence band f-photoionized final state of PS and the final state reached in Bremsstrahlung Isochromate Spectroscopy, BIS[23,24] of nearly pure ϕ_1 systems). As discussed above core level photoemission occurs to more than one final state with probabilities given by the overlap between the initial and final states. In our notation the final states can be written as

$$\psi_1^*=d_1\phi^*(f^o) + e_1\phi^*(f^1) + g_1\phi^*(f^2)$$
$$\psi_2^*=d_2\phi^*(f^o) + e_2\phi^*(f^1) + g_2\phi^*(f^2) \quad ....(4)$$
$$\psi_3^*=d_3\phi^*(f^o) + e_3\phi^*(f^1) + g_3\phi^*(f^2)$$

where now the $\phi^*(f^n)$ is an N-1 particle determinant describing the configuration

$$Ce \ldots\ldots\ldots 3d^9 \ldots\ldots 4f^{n'}(5d6s)^{y'}$$

In these configurations n'+y' is one larger than in the ground states because an electron is taken from the extended valence states to screen the core hole. The final state, or more precisely the energy of the XPS transition, involving an increase in f-count is lowest in energy. Again, as in the ground state, the $\Psi^*_{1,2,3}$ all contain contributions from the $\emptyset^*(f^0)$, $\emptyset^*(f^1)$ and $\emptyset^*(f^2)$ so that in order to obtain the spectral function it is necessary to know the twelve coefficients, a_1, b_1, c_1, $d_{1,2,3}$ $e_{1,2,3}$ and $g_{1,2,3}$ [22]. Calculation of these twelve coefficients is probably not possible with present computational techniques although model calculations with approximate 4f wave functions would be useful to get an order of magnitude estimate for the mixing of the $\emptyset$ and $\emptyset^*$. In this sense a quantitative understanding of the photoelectron spectra is not possible. (The same is true for x-ray absorption). Nevertheless some conclusions are inescapable when we compare the 3d XPS data from a series of alloys. To illustrate typical trends in core level Ce XPS spectra we choose the Ce 3d spectra of Ce-Ni compounds shown in figure 2. These spectra represent part of a much larger body of data we have collected on La and Ce alloys [25].

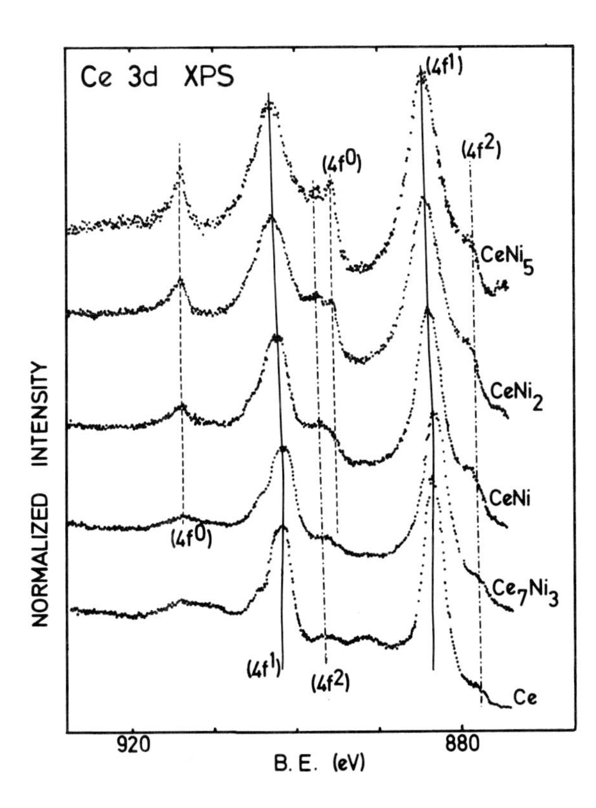

Figure 2: Ce 3d XPS peaks from Ce-Ni intermetallic compounds.

The spin-orbit splitting of the Ce $3d_{5/2-3/2}$ core holes is about 17 eV so that all features of the spectrum are seen doubled. We have labelled the salient XPS peaks as f^0, f^1 and f^2, according to the approximate numbers of f electrons present in the final states of the XPS process ψ_1^*, ψ_2^*, and ψ_3^* in our notation. We believe that observation of these multiple peaks in the XPS spectra has important implications for the mixing of configurations in not only the final states, but also in the initial state. As we see both f^0, and f^2-like final states in the XPS spectra there must be mixing of f^0 and f^2 configurations in either the initial, or final states, or both. In order to get some idea about this mixing, i.e. about the coefficients of equations (1) - (4) we note that mixing is reduced by $\sim 1/\Delta E$ if the configurations involved are separated by an energy ΔE. From the data of figures 2 and 3 we see that in the f^0-like final state, ψ_1^* to be nearly pure $\emptyset^*(f^0)$, i.e. $d_1 \sim 1$ and e_1 and g_1 are nearly zero. Thus observation of an f^0 XPS peak, as for instance in $CeNi_2$, or $CeNi_5$, indicates that the amplitude, a_2, of $\emptyset(f^0)$ in the ground state is definitly not

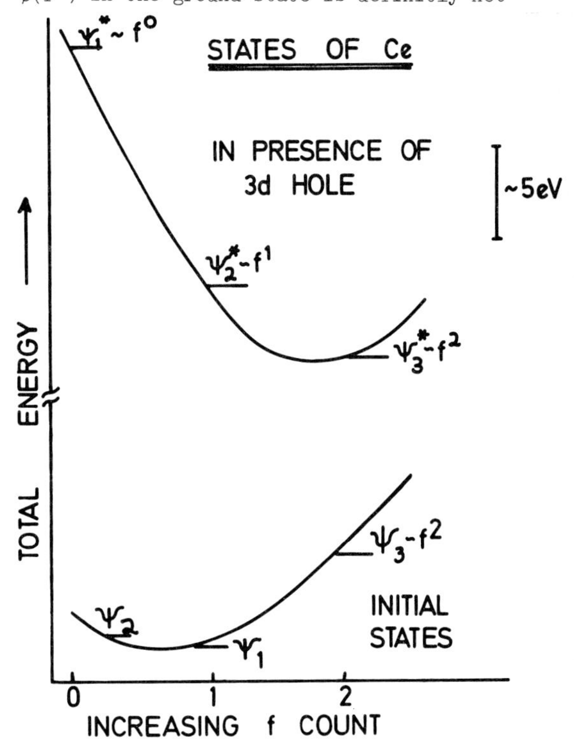

Figure 3: Schematic diagram of the relative energies of the N-particle states in Ce and its compounds with, and without a 3d core hole. As shown symbolically the f-counts in these states need not have precisely values should be deduced from this diagram. Widths of the states due to hybridization are not shown. The approximate energies of the states may be derived from PS[23,24] and BIS[23,24] results, as described in more detail elsewhere.

zero. We cannot derive the coefficient a_1 for the $\phi(f^o)$ contribution to the ground state unless we can evaluate the transition probabilities to ψ_2* and ψ_3* as well and for this b, c, $e_{2,3}$ and $g_{2,3}$ must be calculated. Concerning the f^2 peak in the spectra of Ce alloys, we must ask the question if it indicates a $\phi(f^2)$ contribution in the ground state. Within the above context we note that as the separation of ψ_2 and ψ_3(f^1 and f^2-like states) is similar with and without a core hole, f^1-f^2 mixing must be handled on an equal footing and must be present in both initial and final states. Returning to the problem of the ground state f-count, we cannot give an accurate value without accurate estimates of the f^o and f^2 contributions. We note, however, that there is a general tendency for screening to increase spectral weight in the f^1 and f^2-like final states with respect to the ground states. This shift in spectral weight may partially account for the differences between Ce "valencies" derived from, for instance, lattice constants and the relative intensities of the core level Ce f^o and f^1 peaks. We have to point out that there are, however, discrepancies between observed XPS spectra and conventional ideas about the 4f levels in Ce compounds. For instance so-called "mixed valent" Ce compounds such as $CePd_3$ were once considered to have a f level straddling the Fermi level with a width of 10-100 meV and an occupation of ~0·6. Such a structure would yield an XPS spectrum with only $3d^9f^o$ and $3d^9f^1$ final state peaks in the intensity ratio of 4:6. What is observed is $3d^9f^o$, $3d^9f^1$ and $3d^9f^2$ peaks with the approximate ratio 0·8:10:3[14]. As another example, $CeRu_2$ has long been thought to have a zero f count because of its superconductivity and magnetic properties. If this were true the XPS spectrum should show f^o:f^1:f^2 peaks in the ratio 1:0:0 whilst the observed ratio is ~0·6:10:3[25]. Now as pointed out by Oh and Doniach[17] the core level XPS results could be brought into agreement with an f count of ~0·6 if the perturbation of the f levels due to hybridization, V_{fc}, is of the order of 1·5 eV. Note that such a width implies that the f^2 levels, centred ~5 eV above E_F in the ground state[23,24] play a significant rôle in the ground state. One cannot reconcile the $CeRu_2$ XPS spectrum with an f-count of zero in the ground state.

The widely accepted picture for the electronic structure of the Ce compounds is that they have a narrow band of f levels: These may be located below E_F, as in so-called f^1, three-valent materials; straddling E_F and only partially occupied, as in the so-called mixed valence materials like CeN or $CePd_3$; or above E_F as in the so-called f^o, four-valent materials such as CeO_2 or CeF_4. When we consider the core-level XPS and x-ray absorption spectra of a range of Ce compounds such as $CeNi_x$, $CePd_x$, CeO_2, etc. we come to the conclusion that many aspects of the electronic structure of Ce and its compounds must clearly be reconsidered. First of all we have by now experimental evidence that hybridization between the Ce 4f and other valence levels is important. A quantitative estimate of Oh and Doniach[17] gives an f level hybridization, V_{fc}, of about 1·5 eV. Doniach and Oh admitted that their model has limited validity at such large f band widths. The value is, however, of the right order of magnitude. Secondly we note that if the 4f levels are strongly hybridized with other levels many of the arguments for changes in the f-count in Ce compounds between 0 and 1 inferred from "static" measurements, such as susceptibility, do not hold any more.

Concluding Remarks

We wish to stress the following points:

1) Whilst further theoretical work is required for quantitative understanding of Ce core level spectra, qualitative understanding can be achieved by reference to the advances in understanding of photoelectron and x-ray spectroscopy over the last fifteen years.

2) There is no theoretical evidence or justification for deducing numerical values Ce f-counts directly from core level spectra of Ce compounds if the f levels form broad resonances due to hybridization with the other valence levels.

3) XPS (and XAS) now give indirect evidence that the Ce f levels have widths of at least an order of magnitude larger (i.e. ~1 eV) than believed so far. This is due to hybridization with other valence levels in Ce compounds.

4) Whilst no quantitative estimates of f-counts in Ce compounds can be obtained from core level spectroscopy we presented evidence that the f-count is never identical to zero in intermetallics of Ce.

Acknowledgements

We thank Y. Baer, A. Bringer, O. Gunnarsson, R.O. Jones, B. Lengeler, S.-J. Oh, K. Schönhammer and D.K. Wohlleben for stimulating discussions as well as M. Beyss, J. Keppels and J.-M. Welter for technical assistance.

References

1. J. Wilkins, first paper in this conference.
2. The term "sudden approximation" may sound dubious to the reader who does not specialize in spectroscopy. It is, in fact, a tried and tested approximation used in photoemission for well over a decade with good results and only small deviations in PS with low energy photons. See e.g. references 3-6.
3. L. Hedin and A. Johansson, J. Phys. B. 2 (1969) 1336.
4. R. Manne and T. Åberg, Chem. Phys. Lett. 7 (1970) 282.
5. C.S. Fadley p1 in C.R. Brundle and A.D. Baker (eds.) Electron Spectroscopy, (Academic Press, London 1978).
6. J.C. Fuggle, R. Lässer, O. Gunnarsson and K.

Schönhammer, Phys. Rev. Lett. 44 (1980) 1090.
7. L. Hedin Rp 226 in L. Azároff (ed.) X-ray Spectroscopy (Mc Graw-Hill, New York 1974).
8. J. Friedel, Comments Sol. State. Phys. 2 (1969) 21. and references therein.
9. M. Combescot and P. Nozière, J. Phys. (Paris) 32 (1971) 913.
10. A. Kotani and Y. Toyozawa, Jpn. J. Phys. 35 (1973) 1073, 1082 and 37 (1974) 912.
11. K. Schönhammer and O. Gunnarsson, Solid State Comm. 23 (1977) 691 and 26 (1978) 147, 399.
12. J.C. Fuggle, M. Campagna, Z. Zolnierek, R. Lässer, and A. Platau, Phys. Rev. Lett. 45 (1980) 1597.
13. G. Crecelius, G.K. Wertheim and D.N.E. Buchanan, Phys. Rew. B. 18 (1978) 6519.
14. R. Lässer, J.C. Fuggle, M. Beyss, M. Campagna, F. Steglich and F. Hulliger Physica 102 B (1980) 360.
15. F.U. Hillebrecht and J.C. Fuggle, Phys. Rev. B (1982) in press
16. G. Krill, J.P. Kappler, A. Meyer, L. Abadli and M.F. Ravet, J. Phys. F. 11 (1981) 1713 and references therein.
17. S.-J. Oh and S. Doniach, in press
18. W.D. Schneider, C. Laubschat, I. Nowik and G. Kaindl, Phys. Rev. B. 24 (1981) 5422.
19. G. Krill, J.P. Senateur and A. Amamou, J. Phys. F. 10 (1980) 1889.
20. K.R. Bauchspiess, W. Baksch, E. Holland-Moritz, H. Launois, R. Pott and D.K. Wohlleben, p417 in L.M. Falicov, W. Hanke and M.P. Maple (Rds.) Valence Fluctuations in Solids, (N. Holland Amsterdam, 1981)
21. Note that we have reservations about representing the broadening of the f levels by a single number, W_f, representing the width. As the f levels become broader due to hybridization the f-character density may begin to show structure several eV away from the main peak. Note also that even a Lorentzian peak has 50% of its weight outside the limits defined by 2Γ.
22. In this discussion we ignore the consequences of differences between the $\emptyset(f^n)$ and $\emptyset^*(f^n)$. We hope to discuss this approximation at a later date, as well as the consequences of d^9-f^n multiplet splitting.
23. J.K. Lang, Y. Baer and P.A. Cox, J. Phys. F 11 (1981) 121
24. Y. Baer, H.R. Ott, J.C. Fuggle and L.E. de Long, Phys. Rev. B. 24 (1981) 5384.
25. F.C. Fuggle, F.U. Hillebrecht, Z. Zolnierek, R. Lässer and H. Hillebrand, to be published.

Valence Instabilities
P. Wachter and H. Boppart (eds.)

SURFACE VALENCE STATE OF SAMARIUM AND THULIUM METALS

Anders Rosengren

Institute of Theoretical Physics
University of Stockholm
S-113 46 Stockholm, Sweden

and

Börje Johansson

Institute of Physics
University of Aarhus
DK-8000 Aarhus C, Denmark

The stability of the surface of the samarium and thulium metals against a divalent state is investigated within a pair-bonding model. We find that a divalent surface is favoured for samarium but not for thulium, although it is close to become stable also for the latter. The position for the surface divalent samarium f^6-level relative to the Fermi energy is calculated and compared with experimental data. As a test of the pair-bonding approximation employed we calculate surface core-level shifts and compare with recent experiments on europium and ytterbium metals. Finally some interesting possible physical consequences of the divalent samarium surface layer are pointed out.

I. INTRODUCTION AND MODEL

The intermediate valence (IV) state has now been observed in many rare earth systems of which SmS probably is the most well-known example [1]. One popular way to probe this IV-state has been to use X-ray photo-electron spectroscopy (XPS). The spectra obtained show peaks originating from both divalent (f^{n+1}) and trivalent (f^n) initial configurations and the intensity of these peaks are often used to derive a value of the degree of valence mixing for the material in question. When the 4d spectrum of Sm metal showed peaks attributable to both Sm^{2+} and Sm^{3+} [2] claims were made that Sm metal was in an IV state. However the question remained whether the observed divalent signal originated from bulk or surface atoms [2]. In view of the facts that the trivalent metallic state is favoured energetically relative to the divalent state by about 6 kcal/mol [3,4], and that the bulk properties of Sm closely follow those of the other trivalent rare earth metals and intra rare earth alloys [3], the interpretation that bulk samarium metal is in a mixed valence state seemed unlikely, and it was suggested that the presence of the divalent peak in the spectrum was exclusively a surface effect [5]. New surface sensitive XPS-measurements by Wertheim and Crecelius [6] strongly supports this conjecture. However the question then arose whether the divalent samarium signal was an initial state surface feature or a final state, charge transfer satellite. This discrimination can be made on basis of the 4f spectra, since they are not accompanied by satellites [2]. The presence of the $4f^6V^2$ initial state near E_F (the Fermi energy) was in Ref. 6 only detected by a uniform increase of the intensity in the valence band region relative to the +3 intensity with increasing take-off angle. A somewhat safer indication of the presence of divalent samarium atoms at the surface was made by Allen et al [7,8] who compared partial yield spectra of different surface sensitivity to a bulk absorption spectrum. A direct XPS-observation of the $4f^6 \rightarrow 4f^5$ transition at the surface was then reported by Lang and Baer [9], who found the binding energy of the $4f^6$ level relative to E_F to be 0.77 ± 0.25 eV in good agreement with the value 0.65 eV measured by Allen et al [8]. Lang and Baer also reported BIS (Bremsstrahlung Isochromat Spectroscopy) experiments for the reverse transition $f^5 \rightarrow f^6$, for bulk atoms, and obtained the unoccupied f^6 level 0.46 ± 0.2 eV above E_F. From the measured angular dependence of the relative ratio between the intensities of the divalent and trivalent $3d_{5/2}$ levels, Wertheim and Crecelius [6] concluded that only about 40% of the surface atoms were in a divalent state. This together with the large value found for the binding energy of the divalent f^6 component makes the interpretation that the samarium surface atoms were in an IV-state unlikely, and would instead imply an inhomogeneous valence mixture at the surface. However, it should be noted that divalent Sm atoms are larger than trivalent ones. Thus the number of divalent Sm atoms needed to cover the same surface area as covered by trivalent ones is quite significantly reduced (≈25%) which to some extent might explain the observed relatively low intensity of the divalent signal. Another, and possibly more important, explanation to this low intensity is surface roughness.

The main purpose of the present work, besides demonstrating the stability of the divalent Sm surface, is to investigate how the phase change at the surface influences on the energy position of the divalent f^6 level. We also study the corresponding valence problem for thulium.

There are two essential assumptions underlying the theoretical treatment performed. The first one is that if a f-electron is photoemitted, the final state will be completely screened. This means e.g. that in case the f-electron is emitted from a Sm^{2+} ion we can treat the final state as a fully screened Sm^{3+} ion (impurity picture of the final state).

The other assumption, or rather approximation, we have used is that the total configurational energy of our system can be written as a sum of interaction energies between nearest-neighbour atom pairs only. In our calculations we will deal with two kinds of atoms, Sm^{2+} and Sm^{3+} (called A and B, respectively, for brevity in the equations). We consider a semi-infinite system, the surface of which is assumed to be atomistically plane. This system is divided into layers parallell to the surface, numbered with the index λ ($\lambda=0$ is the surface layer). If we by $N_{ij}^{\lambda\mu}$ denote the total number of nearest-neighbour (N.N.) pairs in which an atom of type i lies in the λ'th layer and an atom of type j in the μ'th layer and by ε_{AA}, ε_{BB}, and ε_{AB} the bond enthalpies for N.N. pairs of A and/or B atoms, the total configurational energy can be written as [10]

$$U=\sum_{\lambda=0}[\varepsilon_{AA}(N_{AA}^{\lambda\lambda}+N_{AA}^{\lambda\lambda+1})+\varepsilon_{AB}(N_{AB}^{\lambda\lambda}+N_{AB}^{\lambda\lambda+1}+N_{BA}^{\lambda\lambda+1})+\varepsilon_{BB}(N_{BB}^{\lambda\lambda}+N_{BB}^{\lambda\lambda+1})]. \quad (1)$$

Numerical values for the bond enthalpies ε_{AA} and ε_{BB} are obtained from cohesive energies. The cohesive energy of hypothetical divalent Sm metal can with high accuracy be interpolated from the cohesive energies of Ba, Eu, and Yb [11] to be 43.0 kcal/mol. In the pair-bonding scheme the bond enthalpy for trivalent Sm has the trivalent atomic state as reference level and the corresponding appropriate cohesive energy is then 105$\pm$1 kcal/mol [12]. Thus simply $Z\varepsilon_{AA}/2=-43.$ and $Z\varepsilon_{BB}/2=-105.$, where Z is the bulk coordination number. For the energy difference between bulk hypothetical divalent and actual trivalent Sm metal, $\Delta H_{II,III}$, we find using the value 55.7 kcal/mol for the excitation energy [12] of the divalent (A) atom to the appropriate trivalent atomic configuration (B), f^5ds^2, the value

$$\Delta H_{II,III}=43.+55.7-105.\approx-6.3 \text{ kcal/mol}. \quad (2)$$

I.e the trivalent state is stable by 6.3 kcal/mol relative to the divalent metal.

To determine ε_{AB} we have used the BIS experiment. In this experiment for bulk Sm one starts from the B metal and from the complete screening picture of the excitation process one ends up with a final state where one B atom (f^5) has been transformed into an A atom (f^6). This gives immediately (using 1 eV/atom =23.05 kcal/mol) the following relation

$$-Z\varepsilon_{BB}-55.7+Z\varepsilon_{AB}=0.46\cdot 23.05, \quad (3)$$

where 0.46$\pm$0.2 eV is the value for the BIS excitation measured by Lang and Baer [9]. The left-hand side of eq.(3) can be rewritten in a more transparent form as

$$-Z\varepsilon_{BB}-55.7+Z\varepsilon_{AB}=$$
$$-Z\varepsilon_{BB}/2+Z\varepsilon_{AA}/2+Z[\varepsilon_{AB}-(\varepsilon_{AA}+\varepsilon_{BB})/2]-55.7=$$
$$E_{coh}(B)-E_{coh}(A)+E_A^{imp}(B)-55.7, \quad (4)$$

where $E_{coh}(B)$, $E_{coh}(A)$, and $E_A^{imp}(B)$ are the cohesive energy of trivalent Sm metal (relative to the trivalent atomic configuration), the cohesive energy of hypothetical divalent Sm metal, and the energy of dissolving a divalent (A) Sm metallic impurity in the trivalent (B) host respectively. We find $E_A^{imp}(B)$ to be positive (4.3 kcal/mol), which is gratifying since it is known that the divalent rare earth metals Eu and Yb do not form solid solutions or alloys with the trivalent rare earths.

With the model and its parameters now defined we will in the next section consider various physical situations of interest in the present context of the Sm surface. At the same time we will also investigate the corresponding surface problem for Tm metal.

II. CALCULATIONS

a) Stability of a divalent Sm surface

We now first turn to the question if a completely trivalent Sm crystal is energetically stable against a crystal state where the surface layer consists of divalent samarium and the rest of the system remains trivalent. As already pointed out divalent Sm ions are considerably larger than trivalent ones, which means that the number of surface atoms for a divalent surface will be less than it is for a trivalent one. Since an energy comparison of the two states defined above has to be made with the total number of atoms fixed, this means for the case of a divalent surface layer that the "missing" surface atoms have to be transported to the bulk. If we by N_2 and N_3 denote the number of atoms in the surface layer and in a bulk layer respectively, their ratio is (assuming the divalent surface layer density is the same as for hypothetical pure divalent Sm metal) $\alpha=N_2/N_3=(19.95/30.0)^{2/3}\approx0.76$, where 30.0 cm^3/mol and 19.95 cm^3/mol are the atomic volumes for divalent Sm (interpolated) and trivalent Sm metal respectively.

Related to this difference in atomic size is the question how to define the number of bonds between the divalent surface layer and the first trivalent layer, N_{AB}^{01}. The remaining $N_{ij}^{\lambda\mu}$:s of eq.(1) can be expressed in terms of N_2, N_3, Z_{LL}, and Z_{IL}, where Z_{LL} and Z_{IL} denote the number of nearest neighbours of an atom in the same layer and in the underlying layer respectively (e.g. N_{AA}^{00} for the divalent surface is equal to $N_2Z_{LL}/2$). In the commensurate situation N_{AB}^{01} is simply equal to N_3Z_{IL}. In order to get an estimate of the value for N_{AB}^{01} in the incommensurate or non-register situation we have represented the point densities in the planes $\lambda=0$ and $\lambda=1$ by "smeared out" functions. By performing the "overlap" integral for these functions, we find that N_{AB}^{01} is reduced in the incommensurate situation by the factor $\beta=(4/5)\sqrt{N_2/N_3}$ relative to its value for the commensurate case, $N_3 Z_{IL}$ [13].

Now having defined N_{AB}^{01} we return to the question we first addressed ourselves namely if a pure trivalent Sm crystal is energetically stable against a Sm crystal where the top layer is formed by divalent Sm atoms. The energy of the completely trivalent crystal will be called U_0 and the energy of the Sm crystal with the divalent top layer will be denoted U. From eq.(1) the energy difference can be written as

$$\begin{aligned}U-U_0&=\varepsilon_{AA}N_{AA}^{00}-\varepsilon_{BB}N_{BB}^{00}+\varepsilon_{AB}^{IL}N_{AB}^{01}-\varepsilon_{BB}N_{BB}^{01}-\\&-N_2\cdot 55.7+(N_3-N_2)\varepsilon_{BB}(Z/2)=\\&=\varepsilon_{AA}N_2(Z_{LL}/2)-\varepsilon_{BB}N_3(Z_{LL}/2)+\varepsilon_{AB}^{IL}\beta N_3 Z_{IL}-\\&-\varepsilon_{BB}N_3 Z_{IL}-N_2\cdot 55.7+(N_3-N_2)\varepsilon_{BB}(Z/2).\end{aligned}\tag{5}$$

Equation (5) can be understood as follows: if we start from the completely trivalent crystal, the second and fourth terms describe the energy it will cost to take away the topmost layer of metallic B (trivalent) atoms and separate it into free B atoms with an appropriate trivalent atomic configuration ($4f^5 5d^1 6s^2$). The term preceding the last term accounts for the gain in binding energy when we let N_2 of these N_3 free B atoms change their atomic configuration to $4f^6 6s^2$. The so obtained N_2 A (divalent) atoms are then brought to crystallize on top of the crystal, the gain in binding energy of this process being described by the first and third term. The last term describes the gain in energy when the remaining N_3-N_2 free B atoms crystallize as B metallic bulk atoms.

Since the number of vertical bonds are reduced for the divalent surface by the factor β, there will be a tendency to somewhat compensate for this by making the remaining bonds stronger. Therefore we renormalize the vertical AB bonds in this case by $\varepsilon_{AB}^{IL}=\varepsilon_{AB}(1+\delta_{AB}^{IL})$, where $\delta_{AB}^{IL}=0.1$ should be a reasonable choice. The last entities to be specified in eq.(5) are Z_{LL}, Z_{IL}, and Z. We perform here the calculations for the most densely packed surface of the fcc structure, the (111) surface, where Z_{LL}, Z_{IL}, and Z are 6,3, and 12 respectively. In reality Sm crystallizes in a hexagonal type structure [14] but we note here that for the hcp (0001) surface with an ideal axial ratio the different Z numbers will be the same as for the fcc (111) plane. Since for the Sm crystal structure the effective axial ratio is 1.61 this is a most reasonable substitution. With the bond parameters defined in the preceding section we find that $(U-U_0)/N_3$ equals −0.28 eV/atom. This means that a divalent surface on top of a trivalent Sm crystal is stable against the completely trivalent Sm crystal.

b) Position of the f^6 level

We now turn to the question of the position of the f^6 level, i.e. the energy of the $4f^6V^2 \rightarrow 4f^5V^3$ transition at the surface. The initial state, with energy U, is the trivalent bulk crystal covered with a divalent layer. The final state, with energy U', is the same except for that one of the divalent atoms in the surface layer is replaced by a trivalent atom. Straightforward application of eq.(1) gives that U'−U equals 0.65 eV, in good agreement with the experimental values 0.77±0.25 eV [9] and 0.65 eV [8]. The main point here is that the reconstruction at the surface to a divalent layer brings the f^6 level far down below E_F.

In order to see this more clearly we consider the hypothetical case with "isolated" divalent Sm atoms in the surface plane, i.e. the divalent atoms are surrounded by trivalent atoms only. The position of the f^6 level is in this case found to be 0.26 eV below E_F. This illustrates the importance of the surface reconstruction for obtaining a high binding energy of the divalent f^6 surface level.

c) Surface core-level shift for a completely divalent samarium metal

As an independent test of the present pair-bonding approximation the surface core-level shift of the $f^6 \rightarrow f^5$ transition for a divalent Sm metal can be used. Although divalent Sm does not exist, we can directly compare our calculated value with experiment on Eu, the divalent neighbour element of Sm in the Periodic Table. The initial state is the fully divalent Sm metal with total energy denoted U_0^{II}. The final state is a trivalent bulk Sm impurity and a trivalent Sm surface impurity, respectively. If the total energies of these two states are called U_B and U_S respectively, the surface core-level shift, Δ_C^S, is given by

$$\Delta_C^S=U_S-U_0^{II}-(U_B-U_0^{II})=(\varepsilon_{AA}-\varepsilon_{AB})Z_{IL}.\tag{6}$$

The calculated shift is 0.63 eV in nice agreement with the experimental value for Eu metal, 0.63±0.02 eV [15].

d) Stability of the thulium surface

For Tm the appropriate excitation energy from the divalent atom to the trivalent one is significantly lower, 47.2 kcal/mol [12] compared to the value 55.7 kcal/mol for Sm. The other parameters in the model will only be slightly modified in Tm as compared to Sm. Thus the trivalent cohesive energy is 103 kcal/mol [12] and the divalent one (interpolated) is 37.8 kcal/mol. From the BIS experiment [16], giving that the f^{13} level is situated 1.10±0.2 eV above E_F, we obtain, using eq.(3), that ε_{AB}=−11.kcal/mol. Finally for Tm the parameter $\alpha(=N_2/N_3)$ is 0.81.

Using these values for the parameters we calculate the energy difference $(U-U_0)/N_3$ between the Tm metal with a divalent top layer and the completely trivalent metal to be 0.15 eV/atom. Thus we find that the trivalent surface layer is stable in Tm, which is in agreement with experiment [7,17]. Still the calculation shows that the divalent state is close to become stable at the thulium surface.

For the hypothetical divalent Tm surface we find the energy position of the f^{13} level to be 0.22 eV below E_F. Thus from the point of view of an one-atom excitation process the divalent surface state should be stable! This points to the collective nature of the surface transition within our model.

Finally we find the surface core-level shift of the f^{13} level for the completely divalent Tm metal to be 0.63 eV, in nice agreement with the experimental value 0.63±0.03 eV for Yb [18,19].

III. SUMMARY AND DISCUSSION

From the given analysis it seems clear that a trivalent Sm surface is unstable against a valence change, a result strongly supported by the experimentally observed features in Refs. 6-9. In the present work we have simplified the problem and treated the surface layer as a rigid divalent fcc (111) surface on top of a fcc (111) trivalent layer and applied a pair-bonding model. We then faced the problem of how to define the number of bonds between two non-commensurate layers. This problem was treated by deriving an average number, but admittedly, the method applied is by no means unique. Still, the most serious approximation in our treatment is probably the neglect of the possibility of various deformations, like e.g. induced buckling of the surface layer.

From the above mentioned difficulties it is clear that what is presently most needed is an experimental surface structure determination of samarium. Also the experimental indication that only 40% of the surface atoms are divalent needs further clarification.

Besides demonstrating the plausibility of a divalent samarium surface, the main task of the present work was to account for the experimentally observed high binding energy of the divalent f^6 level at the surface. We have argued that the reason for this is a surface reconstruction, such that at the surface there will be a macroscopic number of divalent atoms. As a partial check on the general validity of our treatment we also calculated surface core-level shifts and found good agreement with experiments on Eu and Yb metals. Also the valence state of the Tm metal surface was investigated and we found, in agreement with experimental indications, that the trivalent state is stable although the divalent state was found to be close in energy.

Finally we want to stress that for an elemental metal a unique physical situation is met with in Sm, namely that it has a surface with very different characteristics compared to those of its interior parts. This should affect several physical properties. E.g., one might expect that the divalent surface should melt at a considerably lower temperature than the bulk. One might then speculate about the possibility that this early surface melting could have some influence on the bulk melting temperature. We notice that the experimental melting temperatures for the lanthanides [20] do indicate a somewhat anomalously low melting temperature for Sm.

We have already mentioned that the Tm surface seems to be stable against a valence change. However, at high temperatures the situation is less clear. Whether the slightly anomalous melting temperature of Tm has anything to do with the possibility of a valence change of the Tm surface at elevated temperatures can for the moment only remain speculations, but it is certainly remarkable that the two lanthanide metals which are most likely to have a divalent surface both show anomalous melting temperatures.

References

[1] Jayaraman, A., Singh, A.K., Chatterjee, A. and Usha Devi, S., Phys. Rev. B9 (1974) 2513.

[2] Wertheim, G.K. and Campagna, M., Chem. Phys. Letters 47 (1977) 182

[3] Johansson, B. and Rosengren, A., Phys. Rev. B11 (1975) 2836

[4] Johansson, B. and Rosengren, A., Phys. Rev. B11 (1975) 1367

[5] Johansson, B., Inst. Phys. Conf. Ser. 37, (1978) Chap. 3, p39

[6] Wertheim, G.K. and Crecelius, G., Phys. Rev. Lett. 40 (1978) 813

[7] Allen, J.W., Johansson, L.I., Bauer, R.S., Lindau, I. and Hagström, S.B.M., Phys. Rev. Lett. 41 (1978) 1499

[8] Allen, J.W., Johansson, L.I., Lindau, I. and Hagström, S.B.M., Phys. Rev. B21 (1980) 1335

[9] Lang, J.K. and Baer, Y., Sol. St. Commun. 31 (1979) 945

[10] Kumar, V., Kumar, D. and Joshi, S.K., Phys. Rev. B19 (1979) 1954

[11] Brewer, L., LBL Rep. no. LBL3720 (unpubl.)

[12] Johansson, B., (unpubl.)

[13] Rosengren, A. and Johansson, B., Phys. Rev. B (accepted for publication)

[14] Donohue, J., The Structure of the Elements (Wiley, New York, 1974)

[15] Kammerer, R., Barth, J., Gerken, F., Flodström, A. and Johansson, L.I., (unpubl.)

[16] Lang, J.K., Baer, Y. and Cox, P.A., J. Phys. F. 11 (1981) 121

[17] Johansson, L.I., Allen, J.W. and Lindau, I. (unpubl.)

[18] Hecht, M., Johansson, L.I., Allen, J.W., Oh, S.J. and Lindau, I., Proc. VI VUV-Conf (Charlottesville, 1980) p. I-64

[19] Johansson, L.I., Flodström, A., Hörnström, S.E., Johansson, B., Barth, J. and Gerken, F. (unpubl.)

[20] Spedding, F.H., Handbook of Chemistry and Physics (Chemical Rubber Co, Cleveland, Ohio, 1970) p. B-253

Valence Instabilities
P. Wachter and H. Boppart (eds.)

VALENCE BAND SATELLITES IN THE 4f-PHOTOEMISSION SPECTRA OF Eu- AND Yb-METAL

W.D. Schneider, C. Laubschat, and G. Kaindl

Institut für Atom- und Festkörperphysik
Freie Universität Berlin, D-1000 Berlin 33, Germany

B. Reihl and N. Mårtensson
IBM Thomas J. Watson Research Center
Yorktown Heights, New York 10598, U.S.A.

Photoemission spectra of the valence bands of Eu and Yb metal, taken at VUV (20 eV $\leq h\nu \leq 100$ eV) and X-ray energies (AlK_α and MgK_α) exhibit characteristic satellite structures at higher binding energies. In contrast to a previous suggestion these structures could not be identified as two-hole final states in the 4f shells. The experimental observations strongly support an interpretation of these satellites in terms of collective conduction electron excitations, i.e. bulk and surface plasmon losses.

Core-level photoemission spectra of many rare earth systems exhibit characteristic satellite structures, which are usually caused by changes in the 4f-occupation numbers during the photoemission process (1). Particularly in the case of mixed-valent solids it is difficult to discriminate on the basis of core-level spectra alone between such final-state effects and a mixed-valent initial state (2,3). Therefore, the concept of "replicate" core-level spectra must be considered as rather problematic when applied for determining absolute values for the mean valence. In this context the question arises whether valence-band photoemission spectra of the 4f-levels themselves are subject to similar final-state satellites, particularly, whether there is a possibility for two localized 4f-holes being created in the photoemission process. Effects of this kind are well known for Ni-metal (4) as well as for U-compounds (5), where in the final state of the photoemission process two localized holes are created in the Ni-3d and the U-5f shell, respectively. For valence-band spectra of 4f-systems, however, the formation of such two-hole final states has only marginally been considered (6) but not studied experimentally yet.

In this communication we report on a photoemission study of the valence bands of Eu- and Yb-metal at both VUV- and X-ray energies. The observed spectra reveal characteristic satellite structures, which could in principle be caused by two possible mechanisms: (A) Eu-$4f^5$ and Yb-$4f^{12}$ final states, respectively, due to 4f→5d shake-up induced changes in the 4f-occupation numbers and (B) multielectron effects due to conduction-electron excitations. From comparisons with loss functions for both metals known from reflectivity measurements (19) and with similar satellite structures reported for the photoemission spectra of Ba and Sr metal (16) it is concluded that mechanism B is responsible for the observed satellite structures.

The XPS-measurements were performed with a VG-ESCA-3 spectrometer with a total-system resolution of ≅1 eV caused mainly by the linewidth of the unmonochromatized AlK_α and MgK_α X-rays. The samples of Eu and Yb metal were evaporated in-situ using standard UHV-techniques. Furthermore, solid polycrystalline samples of Eu and Yb metal were studied. In these cases the surfaces were mechanically cleaned in a vacuum of $4x10^{-10}$ Torr, and this process was frequently repeated in order to keep the oxygen contamination of the surfaces below the limit of detectability (by monitoring the O 1s signal). Both sample preparation techniques gave essentially the same results. The photoemission measurements in the photon-energy range up to 100 eV were performed at the Synchrotron Radiation Center of the University of Wisconsin-Madison, using a display-type photoelectron analyzer |7|. This system was operated in an angle-integrated mode with a full 86°-acceptance cone of emission angles and with an energy resolution of 0.15 eV. Angle-integrated energy distribution curves (EDC's) were recorded from evaporated Eu and Yb samples in a vacuum of $4x10^{-11}$ Torr. The surface conditions were then monitored by Auger spectroscopy.

Fig. 1 shows the valence-band spectra of Eu- and Yb-metals excited with synchrotron radiation at $h\nu$=70 eV (solid lines) and with MgK_α and AlK_αX-rays (dashed-lines). Both X-ray energies yielded essentially identical results. Apart from surface core-level shifts in the 4f-region, which are only resolved in the VUV-spectra (8), both sets of spectra reveal two intense and rather broad satellite structures at higher binding energies (see Table I, columns 4 and 5). These structures are temperature independent in the temperature range 77 K≤T≤300 K, but are more intense at $h\nu$=70 eV as compared to

X-ray excitation (see Fig. 1). This intensity increase is most prominent for the spectral features around 7 (8) eV binding energy in Yb (Eu). A similar double structure has previously been observed in the valence-band spectra of Yb-metal by Lang et al. (9), Brodén et al. (10), Baer and Busch (11) and by Alvarado et al. (12). These authors associated the satellite structure with surface contaminations. Under the experimental conditions of the present measurements, however, oxygen contamination can be dismissed as a possible cause for the satellites.

Let us first consider mechanism A as a possible cause for the observed satellites. Then two-hole final state satellites would be expected at the $4f^5$- and $4f^{12}$-final-state multiplet positions (13), which are indicated in Fig. 1 by the dashed-bar diagrams. In the case of Yb shape and position of the observed structure could be assigned to the final-state multiplet of the trivalent rare earth ion. Its broad shape could in principle be understood by lifetime-broadening effects. Thus, in the case of Yb metal an interpretation of the satellite structures in terms of two-hole final states in the 4f-shell cannot be excluded. In the case of Eu, however, it is evident from Fig. 1 that the observed satellite structure is not compatible with shape and position of the $4f^5$-final state multiplet. Contributions from excited final-state multiplets, as recently found for Gd and Eu metals with resonant photoemission within a narrow photon energy range (15), can also be excluded for the following two reasons: (i) from the photon energies used in the present experiments and (ii) from the absence of fine structure in the satellite spectra.

As will be outlined in the following, the similarities in position and shape of the observed satellite structures for Eu and Yb metal suggest an interpretation on the basis of mechanism B. Similar structures as observed here were previously reported for the photoemission spectra of Ba and Sr metal (1,16), which are isostructural and isoelectronic (apart from the 4f-shell) to Eu and Yb metal, respectively. In these cases the observed structure was interpreted as due to extrinsic and predominantly intrinsic plasmon excitations on the basis of calculated plasmon energies (16). Furthermore, similar structures have been reported by Kunz (17) for Ba and Sr metal and by Colliex et al. (18) for Yb metal in high-energy electron-loss experiments. Endriz and Spicer (19) derived surface and volume loss functions from a reflectance study of Ba, Sr, Eu, and Yb metal in the energy range from 1 eV to 11 eV. These volume loss functions are in excellent agreement with the high-energy electron-loss data of Refs. 17 an 18. We therefore use the volume and surface loss functions of Ref. 19 for an interpretation of the observed satellite structures. For a comparison the relevant energies are summarized in Table I. It is obvious that excellent agreement exists between the positions of the satellite features in the photoemission spectra and the volume- and surface-loss energies of Ref. 19. We therefore interpret the satellite features at 5 eV (9 eV) below the main peak as due to surface-plasmon (bulk-plasmon) excitation. This interpretation is supported by the variation of the intensity of the structure around 5 eV below the main line with photon energy. As expected from the much higher surface sensitivity of the 70-eV spectra as compared to the XPS spectra a higher intensity of the surface-plasmon satellite is observed. The appreciable broadening of the loss peaks is characteristic for materials with a relatively large density of interband transitions near their collective oscillation frequencies (20).

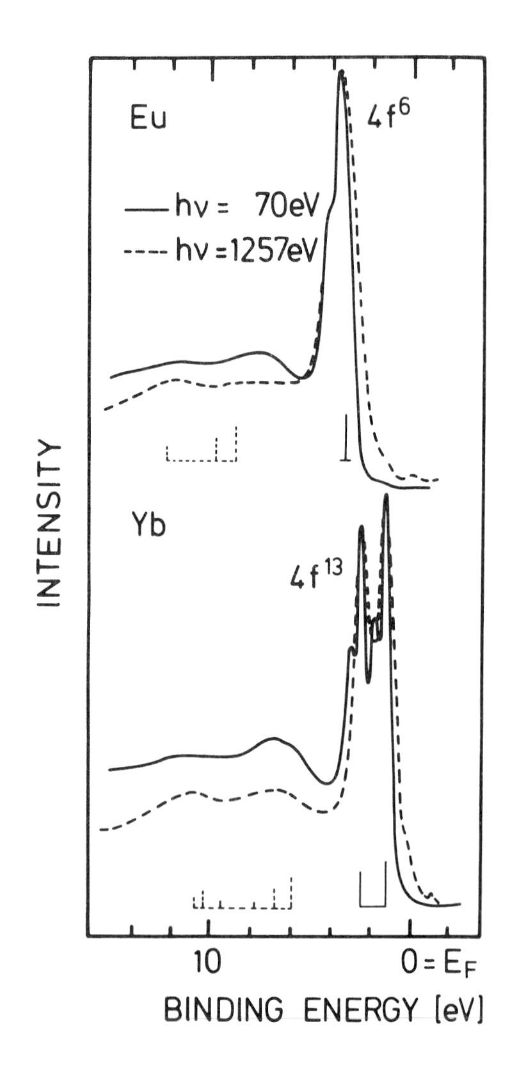

Fig. 1: Photoelectron spectra of the valence band region of Eu and Yb metal

Table I. Comparison of experimentally observed volume-loss-function and surface-loss-function peak positions (19), high-energy electron-loss peak positions (18) and photoemission satellite peak positions with respect to the center of gravity of the main 4f lines for Eu and Yb metal.

	volume loss (eV)[a)]	surface loss (eV)[a)]	volume loss (eV)[b)]	photoemission 4f-satellites (eV)[c)]	
Eu	8.66	3.9±0.2	-	8.5±0.2	3.9±0.2
Yb	9.57	4.8±0.2	9.7	9.5±0.2	5.0±0.2

a) Ref. 19
b) Ref. 18
c) This work

The interpretation of the observed satellite structures in Eu and Yb metal as due to plasmon excitations is also supported by the fact that neither in $EuAl_2$ (21) nor in $YbZn_2$ (22) loss features at the above energies are observed in the XPS-spectra. Due to changes in the conduction electron densities of these intermetallic compounds as compared to the pure metals different plasmon frequencies and therefore different loss features are expected.

To summarize, photoemission spectra of Eu and Yb metal reveal characteristic satellite structures in the valence bands. Comparisons with loss functions from reflectivity measurements and with similar photoemission loss features in Ba and Sr metal strongly support an interpretation of these valence-band satellites in terms of collective conduction electron excitations, i.c. bulk and surface plasmon losses. Thus photoemission valence-band spectra of heavy rare-earth systems seem not to be influenced by final-state induced changes in the 4f-occupation numbers as observed in deep core-level spectroscopy.

Acknowledgement - This work was partly supported by the Bundesministerium für Forschung und Technologie, Contract No. 05 241 KAP, and by the Air Force Office of Scientific Research under Contract No. F49620-81-C0089.

REFERENCES

|1| Wertheim, G.K., Valence fluctuations in solids, in Falicov, L.M., Hanke, W., Maple, M.B., (eds.), (North-Holland, Amsterdam 1981), p. 67.
|2| Fuggle, C., Campagna, M., Zolnierek, Z., Lässer, R., and Platau, A., Phys. Rev. Lett. 45 (1980) 1597.
|3| Schneider, W.D., Laubschat, C., Nowik, I., and Kaindl, G., Phys. Rev. B24 (1981) 5422.
|4| Hüfner, S. and Wertheim, G.K., Phys. Lett. 51A (1975) 299.
|5| Schneider, W.D. and Laubschat, C., Phys. Rev. Lett. 46 (1981) 1023.
|6| Johansson, B. and Mårtensson, N., Phys. Rev. B21 (1980) 4427.
|7| Eastman, D.E., Donelon, J.J., Hien, N.C., and Himpsel, F.J., Nucl. Instr. and Meth. 172 (1980) 327.
|8| Schneider, W.D., Reihl, B., Mårtensson, N., and Kaindl, G., to be published.
|9| Lang, W.C., Padalia, B.D., Fabian, D.J., and Watson, L.M., J. Electron Spectrosc. Rel. Phenom. 5 (1974) 207.
|10| Brodén, G., Hagström, S.B.M., and Norris, C., Phys. Rev. Lett. 24 (1970) 1173.
|11| Baer, Y. and Busch, G., J. Electron Spectr. Rel. Phenom. 5 (1974) 611.
|12| Alvarado, S.F., Campagna, M., and Gudat, W., J. Electron Spectr. Rel. Phenom. 18 (1980) 43.
|13| Buschow, K.H.J., Campagna, M., and Wertheim, G.K., Solid State Commun. 24 (1977) 253.
|14| Mårtensson, N., Reihl, B., Schneider, W.D., Murgai, V., Gupta, L.C., and Parks, R.D., Phys. Rev. B25 (1982) 1446.
|15| Gerken, F., Barth, J., and Kunz, C., Phys. Rev. Lett. 47 (1981) 993.
|16| Ley, L., Mårtensson, N., and Azoulay, J., Phys. Rev. Lett. 45 (1980) 1516.
|17| Kunz, C., Ph. D. Thesis, Hamburg (1966), unpublished.
|18| Colliex, C., Gasgnier, M., and Tebbia, P., J. Physique 37 (1976) 397.
|19| Endriz, J.G. and Spicer, W.E., Phys. Rev. B2 (1970) 1466.
|20| Arakawa, E.T., Hamm, R.N., Hanson, W.F., and Jelinék, T.M., Optical Properties and Electronic Structure of Metals and Alloys, ed. F. Abelés (North-Holland, Amsterdam, 1966) p. 374.
|21| Schneider, W.D. and Laubschat, C., Phys. Rev. B20 (1979) 4417.
|22| Kowalczyk, S.P., Ph. D. Thesis (University of California, 1976) (unpublished).

Valence Instabilities
P. Wachter and H. Boppart (eds.)

SURFACE EFFECTS ON CORE LEVEL BINDING ENERGIES AND VALENCE IN THULIUM CHALCOGENIDES

G. Kaindl, C. Laubschat

Institut für Atom- und Festkörperphysik,
Freie Universität Berlin, D-1000 Berlin 33, Germany

B. Reihl, R.A. Pollak, N. Mårtensson, F. Holtzberg, and D.E. Eastman

IBM Thomas J. Watson Research Center,
Yorktown Heights, New York 10598, USA

The (100)-surfaces of mixed-valent TmSe, divalent TmTe, and trivalent TmS were studied by high-resolution photoelectron spectroscopy. Surface shifts of the 4f levels to higher binding energy were found for TmSe (Δ_c=0.32±0.04 eV) and TmTe (Δ_c=0.41±0.05). In both TmSe and TmS the topmost surface layer is divalent. Exposure of the surfaces to ≅1 Langmuir of O_2 completely quenches the surface-derived spectral features. The bulk mean valence of nearly stoichiometric TmSe is determined: $\bar{v}$=2.55±0.05. Values for the electron mean-free path are derived and found to decrease with decreasing electron kinetic energy down to ≅45 eV.

1. INTRODUCTION

Photoemission (PE) studies of rare-earth solids have revealed in a number of cases surface-induced changes of both the mean valence of the rare-earth ion and of 4f-binding energies /1-4/. These effects may be understood as a consequence of the decrease in cohesive energy due to reduced coordination when a rare-earth ion is brought from the bulk to the surface /5-7/. In all cases the 4f levels were found to be shifted to higher binding energy at the surface, in agreement with theoretical expectation /5/. Quite recently it was shown that in mixed-valent systems even more than one surface layer may be involved in a valence transition /3,4/. No surface effects on valence or 4f-binding energy, however, have been reported up to now for Tm systems /8/.

The present work was therefore mainly concerned with a systematic study of such surface effects in Tm systems, especially the Tm monochalcogenides: TmS, TmSe, and TmTe. These isostructural compounds (NaCl structure) possess remarkably different physical properties: metallic TmS is trivalent, semiconducting TmTe is divalent, and TmSe is homogeneously mixed-valent /9/. The Tm monochalcogenides, including mixed-valent TmSe, order antiferromagnetically at low temperatures /9,10/. Both the bulk mean valence and the lattice constant of TmSe are known to vary appreciably with stoichiometry /10,11/. The mean valence $\bar{v}$ of stoichiometric TmSe (a_o≅5.71 Å) has been subject to some controversy, since from lattice constant systematics $\bar{v}$≅2.75 was derived /10,11/, while XPS studies /12/, X-ray absorption L-edge measurements /13,14/, and magnetic measurements /10/ resulted in $\bar{v}$=2.55 to 2.60.

2. EXPERIMENTAL

The angle-integrated PE experiments were performed at the Synchrotron-Radiation Center of the University of Wisconsin-Madison, using a display-type photoelectron spectrometer /15/ behind a toroidal-grating monochromator. Photoelectrons were accepted within a 35°-cone around sample normal. The total system (spectrometer and monochromator) allowed a resolution of ≅0.17 eV at hν=70 eV.

(100)-surfaces were prepared by cleaving single crystals of the studied specimens in ultra-high vacuum (1×10^{-10} Torr in the preparation chamber, 4×10^{-11} Torr in the spectrometer main chamber). These favourable vacuum conditions together with the short sampling times of typically 7 minutes per spectrum allowed studies on very clean surfaces. All measurements were performed with the samples at room temperature.

3. RESULTS AND DATA ANALYSIS

3.1 TmSe(100)

Fig. 1 shows valence-band PE spectra of freshly-cleaved and oxygen-exposed TmSe(100) at various photon energies. The spectra are a superposition of $4f^{12}$ and $4f^{11}$ final-state multiplets separated by a Coulomb correlation energy of ≅6 eV /12/. This is in accord with a mixed-valent groundstate of TmSe, where the divalent ($4f^{13}5d^0$)- and trivalent ($4f^{12}5d^1$)-configurations are energetically degenerate. The clean-surface spectra are characterized by (i) a dominant divalent $4f^{12}$ emission, which indicates a valence change at the surface, and (ii) a $4f^{12}$ final-state multiplet spectrum that has to be described by a superposition of a bulk and a surface multiplet separated by ≅0.3 eV.

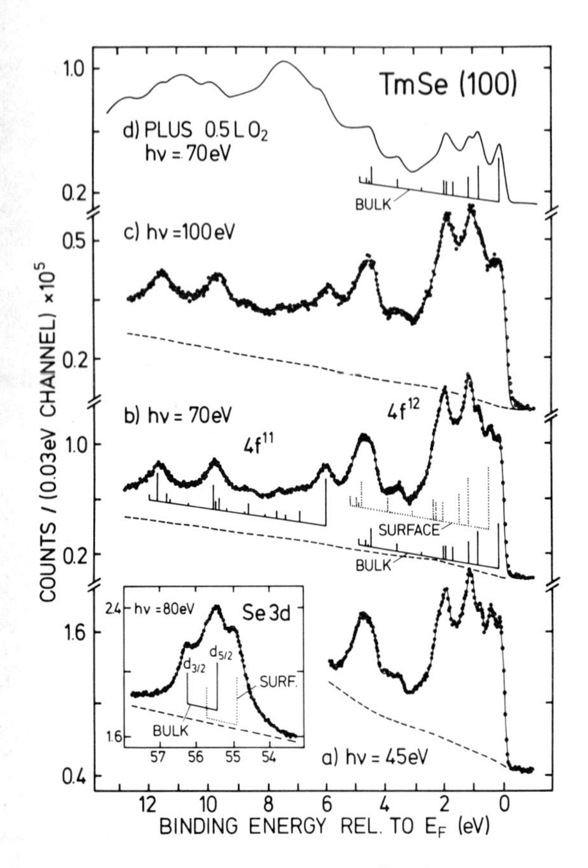

Figure 1: P.E. spectra of TmSe(100) at various photon energies: (a)-(c) freshly cleaved, (d) after exposure to 0.5 L of O_2. The bar diagrams represent the positions and relative intensities of the final-state multiplets originating from the bulk (solid bars) and from a surface layer (dotted bars). The spectrum obtained of the Se-3d core levels is given in the inset.

Emission from divalent Tm at the outermost surface layer is completely quenched by exposure of the surface to ~0.5 L of O_2, whereupon a single $4f^{12}$-multiplet spectrum from bulk Tm^{2+} is sufficient to describe the observed spectrum (Fig. 1d). The oxygen exposure has created an additional $4f^{11}$ final-state feature, which is shifted by about 1.3 eV to higher binding energies relative to the $4f^{11}$ multiplet in the spectrum of clean TmSe. It originates from a trivalent surface layer like e.g. $(TmO)_2Se$ /16/. These observations clearly show that the $4f^{12}$ multiplet spectrum from the outermost surface layer is shifted to higher binding energies. This assignment is supported by the observed photon-energy dependence of the relative intensities of the surface- and bulk-$4f^{12}$ multiplet spectra (see Fig. 1a-1c).

The valence-band spectra of Fig. 1a-1c were least-squares fitted to a superposition of two $4f^{12}$ multiplets (bulk and surface) and one $4f^{11}$ multiplet, with relative energetic positions and relative intensities of the individual multiplet lines taken from UV-absorption measurements /17/ and intermediate-coupling calculations /18/, respectively. While the separations of the $4f^{12}$-multiplet lines could be directly taken from the optical levels of Tm^{3+}, those of the $4f^{11}$ multiplet were obtained from the optical levels of Er^{3+} expanded by a factor of 1.11. This factor was a free parameter in the least-squares fits and describes the increased effective nuclear charge for the 4f shell in Tm^{4+} as compared to Er^{3+}. For some multiplet lines empirical adjustments were made to the theoretical intensities in order to reach optimum least-squares fits /19/. Doniach-Šunić lineshapes /20/ folded with a Gaussian approximating the spectrometer function were employed in the fits. In addition, an extrinsic background proportional to the integrated photoemission above a given energy was added, with two different proportionality constants for bulk and surface emission. The extrinsic background created by emission from the surface layer was consistently found smaller than that from the bulk. This is also the case for the Doniach-Šunić singularity index α. Emission from Se-4p levels was approximated by a Gaussian with 5-eV width (FWHM) and positioned 7.5 eV below the Fermi level /21/.

The results of the fit analysis are represented in Fig. 1a-1c by the solid lines, and it is obvious that the agreement with the measured spectra is good. A mean value of Δ_c=0.32±0.04 eV is obtained for the surface shift of the $4f^{12}$ levels, and the binding-energy difference between the lowest-binding-energy lines of the $4f^{11}$ and $4f^{12}$ multiplets is found to be 5.91± ±0.02 eV. The fractional intensities of the three multiplets are given in Table 1 for photon energies of 70 eV and 100 eV, respectively.

We have also observed a surface shift of the Se-3d core levels to lower binding energy by Δ_c=0.52±0.05 eV (see inset of Fig. 1). The spectrum could be fitted very well by a superposition of two spin-orbit split doublets (Δ_{so} =0.83 eV as in crystalline Se) using the same lineshape functions as described above.

3.2 TmTe(100)

Similar P.E. studies were performed on TmTe(100) and representative spectra are shown in Fig. 2. In agreement with the divalent character of TmTe the valence-band spectra obtained for hν≥50 eV are dominated by 4f emission from $4f^{12}$ final-state multiplets. The weak spectral features seen in the $4f^{11}$ region are probably due to some trivalent surface contamination, but final-state effects cannot be ruled out /22,23/. The complicated structure of the $4f^{12}$ features of the clean surface reveals the presence of a surface-shifted $4f^{12}$ multiplet as in the TmSe case. Again O_2 exposure completely quenches the $4f^{12}$-surface emission (Fig. 2c) and gives rise to an intense $4f^{11}$ multiplet spectrum originat-

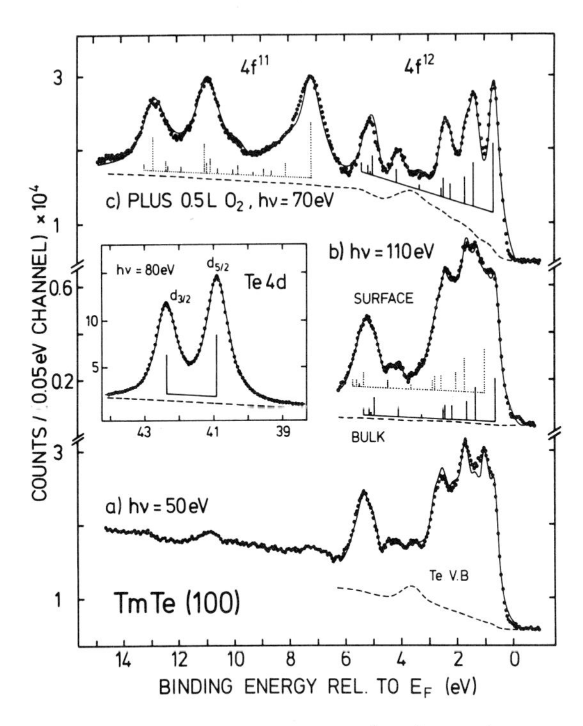

Figure 2: P.E. spectra of TmTe(100) obtained with various photon energies: (a) and (b) freshly cleaved, (c) after exposure to 0.5 L of oxygen. In the inset the spectrum of the Te-4d levels is given.

ing from a trivalent surface oxide, most probably $(TmO)_2Te$ /16/. No surface core-level shift could be resolved for the Te-4d core levels (see inset of Fig. 2). The least-squares fit analysis, performed in a similar fashion as described for the TmSe case, resulted in a weighted mean value of Δ_c=0.41±0.05 eV for the surface-induced shift of the 4f levels. In addition, the following ratios R_1 of integrated $4f^{12}$-surface to $4f^{12}$-bulk multiplet intensities were obtained for four photon energies: R_1=0.40±0.06 for hν=110 eV; R_1=0.45±0.06 for 90 eV; R_1=0.62 ±0.08 for 70 eV; R_1=0.73±0.10 for 50 eV. A clear increase in the surface sensitivity of the PE measurements is observed with decreasing photon energy.

3.3 TmS(100)

We have also studied surface effects on the PE spectra of TmS(100) in the photon-energy range 30 eV≤hν≤90 eV (Fig. 3). The 30-eV spectrum (Fig. 3a) is dominated by emission from the conduction band and the sulfur-derived valence band. At hν=50 eV and higher photon energies emission from Tm-4f states is dominating the spectra (Fig. 3b-3d). Despite the trivalent bulk nature of TmS /9,24/ the observed spectra clearly exhibit multiplet structures from both $4f^{11}(Tm^{3+})$ and $4f^{12}(Tm^{2+})$ final states. This observation is interpreted as due to a surface-

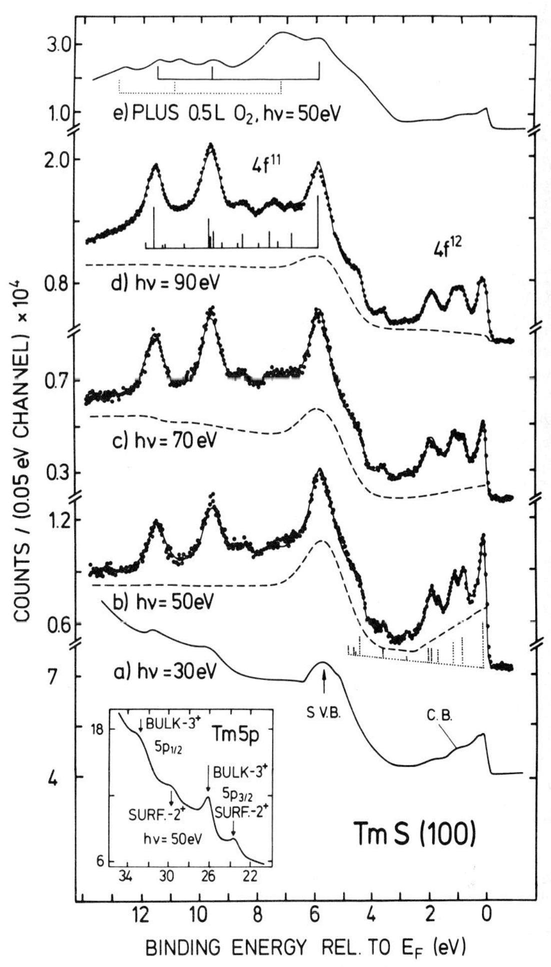

Figure 3: P.E. spectra of TmS(100) at various photon energies.

induced valence change in the initial state of the Tm ions for the following reasons: (i) O_2 exposure of the surface completely quenches emission from $4f^{12}$ final states (see Fig. 3e), giving rise to an additional $4f^{11}$-multiplet from a trivalent surface oxide (e.g. $(TmO)_2S$). Only a single $4f^{12}$ multiplet located at E_F is observed for the clean surface revealing that the divalent Tm ions are limited to the outermost surface layer of TmS(100). If TmS would be mixed-valent in the bulk or if the surface region of divalent Tm would extend deeper a superposition of two $4f^{12}$ final-state multiplets shifted by a surface core-level shift of similar magnitude as observed for TmSe and TmTe would have to be expected. (iii) The ratio R_2 of intensities of $4f^{12}$-surface and $4f^{11}$-bulk spectral features increases with decreasing photon energy: R_2=0.50±0.07 at hν=90 eV, R_2=0.62±0.09 at 70 eV, and R_2=0.66±0.10 at 50 eV. These fit results were obtained with a similar curve-fitting procedure as described above. Emission from the sulfur-derived valence band was simulated by a Gaussian at 5.5 eV below E_F, and from the conduction band by a triangular-shaped function extending 2.5 eV below E_F. The solid lines in Fig. 3b-3d represent the least-squares fit results.

The surface valence transition is also observed via the Tm-5p PE spectrum shown in the inset of Fig. 3. Both the $5p_{1/2}$ and $5p_{3/2}$ final-state levels are split into two peaks originating from trivalent-bulk and divalent-surface TmS, respectively.

4. DISCUSSION

4.1 Bulk-mean valence and surface valence transition in TmSe

The PE spectra of TmSe(100) presented in Fig. 1 exhibit a dominant emission from $4f^{12}$ (Tm^{2+}-derived) final states, which is only in accord with a bulk mean valence $\bar{v}$ in the range from 2.5 to 2.75 /10-14/ if the topmost surface layer is in a divalent or almost divalent state. If no surface valence transition were taken into account, a value of $\bar{v}$=2.36 would be obtained from the fractional intensities $I^{12}=I_B^{12}+I_S^{12}$ and I^{11} listed in Table 1, in obvious disagreement with results from other measurements on stoichiometric TmSe ($a_o \cong 5.71$ Å).

Table 1. Fractional integrated intensities I of the two $4f^{12}$ (bulk and surface) and the $4f^{11}$ multiplets, as obtained from the least-squares fit analyses of the TmSe spectra. The error bars given in parentheses are in units of the last digit.

hν(eV)	I_B^{12}	I_S^{12}	I^{11}
100	0.32(2)	0.31(2)	0.37
70	0.30(2)	0.38(2)	0.32

To obtain an estimate for the thickness Δs of the divalent surface layer of TmSe(100) we use the relation:

$$I_S/I_B=I_S^{12}/(I_B^{12}+13I^{11}/12)=\exp(\Delta s/\ell)-1, \qquad (1)$$

where the observed intensities were nomalized to the same number of 4f electrons in the initial state. Using the fractional intensities in Table 1 we arrive at $\Delta s/\ell$=0.41 and 0.45 for hν =100 eV and 70 eV, respectively. With an estimate of $\ell \cong 6$ Å for the mean-free path of 60-eV electrons /3/ this gives $\Delta s \cong 2.76$ Å, which is very close to the thickness $a_o/2$ of the topmost surface layer of TmSe(100) neglecting a possible surface relaxation ($a_o/2$=2.855 Å). The conclusion that only one surface layer is divalent is plausible in view of the small surface core-level shift observed. This is in contrast to other recently studied mixed-valent systems, like $YbAl_2$ /3/, $EuPd_2Si_2$ /25/, and $Sm_{1-x}Y_xS$ /4/, where much larger surface core-level shifts and surface valence transitions extending over more than one surface layer were observed. There is strong support now for the conclusion that surface valence transitions in at least one surface layer are always present in mixed-valent rare-earth materials.

Due to the separation of surface and bulk emission, the present data allow a determination of the bulk mean valence $\bar{v}$ of TmSe using the following equation:

$$\bar{v} = 2+I^{11}/(I^{11}+12\ I_B^{12}/13). \qquad (2)$$

Using the fractional intensities of Table 1 we arrive at values for $\bar{v}$ of 2.56 (2.53) from the 100-eV (70-eV) results, leading to a weighted mean value of $\bar{v}$=2.55±0.05. This value is again smaller than the Vegard's-law estimates from the lattice constant /10,11/, but fits well in the range of values from other methods /12-14/.

4.2 Surface valence transition in TmS(100)

TmS has metallic properties and its bulk-trivalent character has been proved previously /9,24/. The observed valence change at the surface of TmS(100) represents the first such valence transition in a Tm system. Tm metal does not seem to undergo a valence change at the surface /8,26/, in agreement with theoretical expectation /5/.

4.3 Electron mean-free path ℓ

The present results may be used to derive absolute values for the electron mean-free path ℓ as a function of the kinetic energy of the photoelectrons, if we assume that just the topmost (100)-surface layer (of thickness $a_o/2$) is involved in the surface valence transition (TmSe, TmS) or surface-shifted $4f^{12}$-multiplet (TmSe, TmTe). This assumption is strongly supported by the experimental observations in each case.

In a continuum model we arrive at the following expression for ℓ

$$\ell = \Delta s/\ln(1+R), \qquad (3)$$

where R represents the ratio of emitted 4f intensity from the surface layer to that emitted from all deeper layers. Values for R are derived from the fractional intensities I for TmSe listed in Table 1 and from the $R_1(R_2)$ values given for TmTe (TmS) in the text. All 4f-intensities are normalized to the same number of 4f electrons in the initial state, and we have assumed equal photoemission cross-sections for surface and bulk layers.

The resulting values for ℓ are plotted in Fig. 4 as a function of the kinetic energy of the photoelectrons above E_F for the three Tm com-

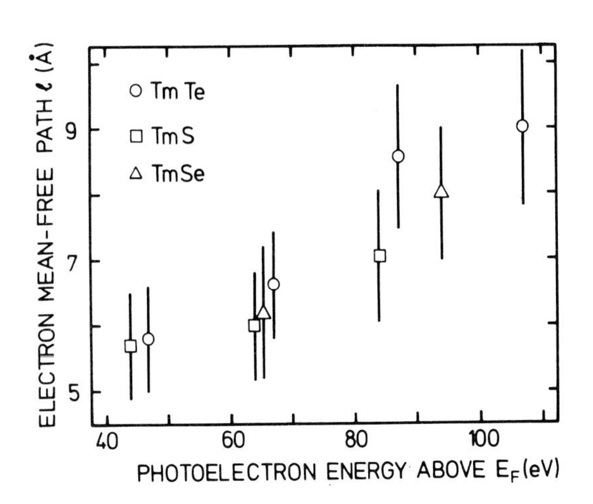

Figure 4: Electron-mean free path ℓ as a function of kinetic energy of the photoelectrons above E_F for TmS, TmSe, and TmTe.

pounds studied. In all three cases ℓ decreases with decreasing kinetic energy even below 50 eV. No minimum is observed in the studied energy range in contrast to the so-called "universal curve" proposed for the energy dependence of ℓ /27/. A quite similar behaviour of ℓ has recently also been observed for several rare-earth metals /28/. Despite the large error bars the ℓ values for metallic TmS and TmSe seem to be systematically smaller than those for semiconducting TmTe.

ACKNOWLEDGEMENTS - The authors would like to thank Prof. P.A. Cox for making available unpublished results of intermediate-coupling calculations for $4f^{11}$-multiplet intensities. The work was partly supported by the Bundesministerium für Forschung und Technologie (contract 05 127 KA) and the Air Force Office for Scientific Research (contract F-49620-81-C0089).

REFERENCES

/1/ Wertheim, G.K., Crecelius, G., Phys. Rev. Letters 40 (1978) 813.
/2/ Alvarado, S.F., Campagna, M., Gudat, W., J. Electron Spectr. Relat. Phenom. 18 (1980) 43.
/3/ Kaindl, G., Reihl, B., Eastman, D.E., Pollak, R.A., Mårtensson, N., Barbara, B., Penney, T., Plaskett, T.S., Sol. State Commun. 41 (1982) 157.
/4/ Reihl, B., Holtzberg, F., Hollinger, G., Kaindl, G., Mårtensson, N., Pollak, R.A., contribution to this conference.
/5/ Johansson, B., Phys. Rev. B19 (1979) 6615.
/6/ Johansson, B., Mårtensson, N., Phys. Rev. B21 (1980) 4427.
/7/ Rosengren, A., Johansson, B., Phys. Rev. B23 (1981) 3582.
/8/ Johansson, L.I., Allen, J.W., Lindau, I., Phys. Letters 86A (1981) 442.
/9/ Bucher, E., Andres, K., di Salvo, F.J., Maita, J.P. Gossard, A.C., Cooper, A.S., Hull Jr., G.W., Phys. Rev. B11 (1975) 500.
/10/ Batlogg, B., Ott, H.R., Kaldis, E., Thöni, W., Wachter, P., Phys. Rev. B19 (1979) 247.
/11/ Holtzberg, F., Penney, T., Tournier, R., J. Physique Colloq. 40 (1979) C5-314.
/12/ Wertheim, G.K., Eib, W., Kaldis, E., Campagna, M., Phys. Rev. B22 (1980) 6240.
/13/ Launois, H., Rawiso, M., Holland-Moritz, E., Pott, R., Wohlleben, D., Phys. Rev. Letters 44 (1980) 1271.
/14/ Bianconi, A., Modesti, S., Campagna, M., Fischer, K., Stizza, S., J. Phys. C: Solid State Phys. 14 (1981) 4737.
/15/ Eastman, D.E., Donelon, J.J., Hien, N.C., Himpsel, F.J., Nucl. Instr. Methods 172 (1980) 327.
/16/ See e.g. Flahaut, J. in Handbook of the Physics and Chemistry of Rare Earths, ed. by Gschneidner Jr., K.A., and Eyring, L.R., North-Holland Publ. Corp. (1979), Vol. 4 p. 1.
/17/ Carnall, W.T., Fields, P.R., Rajnak, K., J. Chem. Phys. 49 (1968) 4424.
/18/ Cox, P.A., Lang, J.K., unpublished results; Lang, J.K., PhD-Thesis, ETH Zürich, unpublished.
/19/ A full account of the present work will be published elsewhere.
/20/ Doniach, S., Šunić, M., J. Phys. C: Solid State Phys. 3 (1970) 285.
/21/ Campagna, M., Rowe, J.E., Christman, S.B., Bucher, E., Sol. State Commun. 25 (1978) 249.
/22/ Schneider, W.D., Laubschat, C., Nowik, I., Kaindl, G., Phys. Rev. B24 (1981) 5422.
/23/ Schneider, W.D., Laubschat, C., Kaindl, G., Reihl, B., Mårtensson, N., contribution to this conference.
/24/ Campagna, M., Wertheim, G.K., in Structure and Bonding, Springer Verlag Berlin (1976), Vol. 30, p. 99.
/25/ Mårtensson, N., Reihl, B., Schneider, W.D., Murgai, V., Gupta, L.C., Parks, R.D., Phys. Rev. B, in print.
/26/ Baer, Y., Busch, G., J. Electron Spectrosc. Relat. Phenom. 5 (1974) 611, and references therein.
/27/ Lindau, I., Spicer, W.E., J. Electron Spectrosc. Relat. Phenom. 3 (1974) 409.
/28/ Gerken, F., Barth, J., Kammerer, R., Johansson, L.I., Flodstrøm, A., Surface Science, in print.

Valence Instabilities
P. Wachter and H. Boppart (eds.)

SURFACE-INDUCED VALENCE CHANGES, CORE-LEVEL SHIFTS, AND PHONON BROADENING IN $Sm_xY_{1-x}S$

B. Reihl

IBM Zurich Research Laboratory,
8803 Rüschlikon, Switzerland

F. Holtzberg, R.A. Pollak, and G. Hollinger

IBM T.J. Watson Research Center,
Yorktown Heights, NY 10598, USA

G. Kaindl

Institut f. Atom- u. Festkörperphysik, Freie Univ.
1000 Berlin 33, West Germany

N. Mårtensson

Dept. of Physics and Measurement Technology
Linköping Univ., 58183 Linköping, Sweden

We have studied (100) cleavage faces of $Sm_xY_{1-x}S$ ($0 \leq x \leq 1$), using high-resolution ($\Delta E < 150$ meV) photoemission with synchrotron radiation. A surface-induced core-level shift of $\Delta E_S = (0.6 \pm 0.05)$ eV is determined for the $Sm^{2+}(4f^5)$ final-state multiplet. It is practically independent of x. No $Sm^{3+}(4f^4)$ signal can be detected for the mixed-valent compounds indicating a possible Sm segregation to the surface. Temperature-dependent studies reveal a stronger phonon broadening for semiconducting SmS than for metallic $Sm_{0.7}Y_{0.3}S$.

1. INTRODUCTION

$Sm_xY_{1-x}S$ exhibits an interesting electronic behavior as a function of x. The system ranges from a divalent (black) semiconductor (x = 1.0), through a mixed-valent (golden) phase ($0.9 > x > 0.2$), to a trivalent metal (x = 0).(1) SmS has been studied extensively in the past by X-ray photoemission spectroscopy (XPS) (2,3) and ultraviolet photoemission spectroscopy (UPS). (4,5) Due to the differences between the more surface-sensitive UPS and the more bulk-sensitive XPS data, Gudat *et al.*(5) suggested a surface-induced shift of the Sm^{2+} multiplet by ΔE_S ~0.3-0.5 eV. Here, we present UPS measurements for the whole concentration range of $Sm_xY_{1-x}S$, including O_2-adsorption studies (to quench the surface signal), low-temperature measurements (to determine the phonon broadening), and a detailed line-shape analysis (to determine ΔE_S more accurately, as well as its possible x-dependence).

2. EXPERIMENTAL

Experiments were performed with the IBM 2D spectrometer (6) using synchrotron radiation in the range $5 < h\nu < 130$ eV from the Synchrotron Radiation Center of the University of Wisconsin — Madison. The overall resolution (electrons and photons) was set to $\Delta E = 150$ meV at $h\nu = 70$ eV. Single crystals of $Sm_xY_{1-x}S$ ($0 < x < 1$) were cleaved along their (100) surfaces in a vacuum of 2×10^{-10} Torr and immediately afterwards transferred to the measurement position (vacuum of 5×10^{-11} Torr). For the cold measurements, SmS and $Sm_{0.7}Y_{0.3}S$ were mounted on a Displex refrigerator, which reached a minimum temperature of (33±3) K. They were cleaved at low temperature and transferred to the measurement position by lowering the whole cryostat.

3. RESULTS AND DISCUSSION

In Figure 1, we present energy distribution curves (EDC's) of photo-electrons from $Sm_xY_{1-x}S$ (100) at room temperature (Curves a, b, d - g) and at 33 K (Curve c). As obvious from Curve b, 1 Langmuir of O_2 adsorbed on a freshly cleaved (100) surface quenches the surface-shifted Sm^{2+} multiplet (see, e.g., Refs. 7 and 8). In addition, bulk *and* surface Sm^{2+} signals are clearly seen in the mixed-valent $Sm_xY_{1-x}S$ samples (Curves d - g) or in SmS at low temperature (Curve c). This observation is consistent with our studies of the surface-induced core-level shifts in the mixed-valent systems $YbAl_2$,(9) $EuPd_2Si_2$,(10) and TmSe.(11) In $Sm_xY_{1-x}S$, the surface core-level shift, ΔE_S, can be obtained direct from the well-separated ^2P multiplet peaks in Figure 1 at ~3.1 eV (bulk) and ~3.7 eV (surface): ΔE_S ~0.5-0.6 eV.

To obtain peak positions more accurately, we performed a detailed line-shape analysis. It turned out that the SmS spectrum of Curve a in Fig. 1 could not be fitted unambiguously with respect to the surface-to-bulk intensity ratio I_S/I_B. Therefore, we first fitted the cold SmS spectrum (Curve c) by using $Sm^{2+}(4f^5)$ final-state multiplets (12,13) which were Gaussian broadened with full widths at half-maximum W_S and W_B for surface and bulk, respectively. The bulk-energy positions obtained and the I_S/I_B ratio were then used as starting parameters to fit the 300 K SmS spectrum, too. The bulk positions were also identical to those obtained by fitting the O_2-covered surface of Curve b. The same procedure was likewise used to fit the cold (not shown) and warm (Curve e) $Sm_{0.7}Y_{0.3}S$ data. The results of this line-shape analysis are summarized in Table I and are now discussed in more detail: i) The Gaussian widths, W_S and W_B, decrease upon cooling of

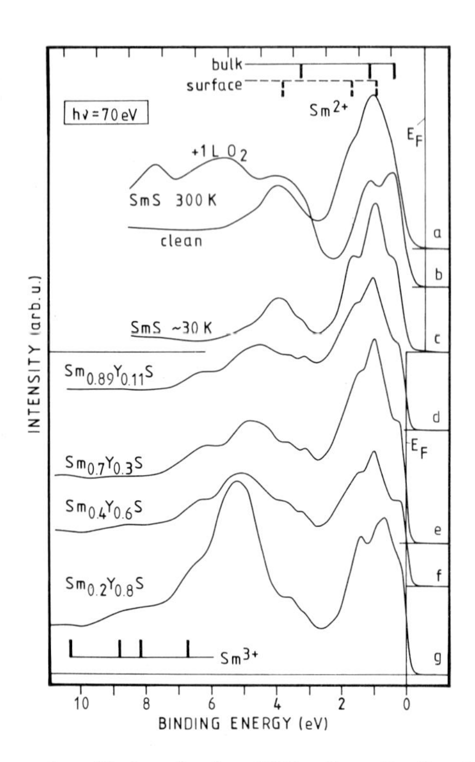

Figure 1 : Photoemission EDC's from $Sm_xY_{1-x}S$ (100) at hν = 70 eV at room temperature (a, b, d - g) and at 33 K (c). The bar diagrams indicate the surface and bulk $Sm^{2+}(4f^5)$ final-state multiplets, as well as the possible (from Ref. 3) position of a $Sm^{3+}(4f^4)$ multiplet.

the sample. This reflects phonon broadening.(14) Using Gaussians and neglecting the vibrational broadening at low temperatures, we define the phonon broadening as $\Delta^{ph} = [W^2(300) - W^2(33)]^{0.5}$, which yields essentially the same values for the surface and the bulk, $\Delta_S^{ph} \sim \Delta_B^{ph}$, but reveals a significant difference between semiconducting SmS and metallic $Sm_{0.7}Y_{0.3}S$. Namely, the conduction-band electrons, present in $Sm_{0.7}Y_{0.3}S$, soften the phonons and therewith reduce the phonon broadening. ii) The surface signal at all temperatures is broader than the bulk signal, $W_S > W_B$, which suggests that phonon broadening for surface atoms is important already at low temperature. iii) The surface core-level shift, ΔE_S, varies between 0.57 eV and 0.61 eV with an estimated uncertainty of ±0.05 eV. It is practically independent of x (see also Curves d - g). iv) The surface-to-bulk intensity ratio decreases in going from SmS to $Sm_{0.7}Y_{0.3}S$. Simple arguments, e.g., a higher surface sensitivity in metallic $Sm_{0.7}Y_{0.3}S$ due to a shorter photoelectron escape depth, or Sm segregating to the surface in $Sm_{0.7}Y_{0.3}S$, would result in a change of I_S/I_B in the opposite direction.

A $Sm^{3+}(4f^4)$ multiplet signal is to be expected in the mixed-valent phase ($x \leq 0.85$) at binding energies $6 < E_B < 10$ eV. A trivalent signal was readily visible with bulk-sensitive XPS.(3) The fact that we detect hardly any Sm^{3+} emission in Curves d - g neither in the cold $Sm_{0.7}Y_{0.3}S$ spectrum (not shown in the figure) is attributed to a 6 Å (escape depth in TmSe at hν = 70 eV, Ref. 11) thick surface layer which is converted to a divalent state. Two reasons can be given: i) The internal pressure due to the smaller Y ions is relaxed at the surface. The lower

Table I (in eV)

	W_S	W_B	Δ_S^{ph}	Δ_B^{ph}	ΔE_S	I_S/I_B
SmS, 300 K	0.99	0.66	0.66	0.56	0.57	4.4
SmS, 33 K	0.74	0.35			0.57	4.7
$Sm_{0.7}Y_{0.3}S$, 300 K	0.75	0.46	0.32	0.33	0.61	3.0
$Sm_{0.7}Y_{0.3}S$, 33 K	0.68	0.32			0.55	3.0

W_S: Gaussian FWHM of surface Sm^{2+} multiplet line

W_B: Gaussian FWHM of bulk Sm^{2+} multiplet line

Δ_S^{ph}, Δ_B^{ph}: $\Delta^{ph} = [W^2(300) - W^2(33)]^{0.5}$ for surface and bulk, respectively

ΔE_S: Energy difference between the surface and bulk Sm^{2+} multiplets

I_S/I_B: Intensity ratio of the surface and bulk Sm^{2+} signals

pressure in the surface region thus brings it back to an SmS-like state, which is divalent. This would result in a phase diagram different for the surface and the bulk. ii) Since Sm^{2+} is more stable at the surface,(15,16) there is the possibility of Sm segregation to the surface, accompanied by a Y-depletion. This would increase the Sm^{2+} signal. In both cases, a possible dependence of ΔE_S on the Sm concentration x is masked and a comparison with the calculated (17) $\Delta E_S(x)$ behavior is not possible.

ACKNOWLEDGEMENTS

We thank D.E. Eastman for his interest and help with the present investigation, and the staff of the Synchrotron Radiation Center for the excellent support. This work was sponsored by the Air Force Office of Scientific Research under Contract-No. F49620-81-C-0089.

REFERENCES:

[1] Tao, L.J., and Holtzberg, F., Phys. Rev. B11 (1975) 3842.

[2] Campagna, M., Bucher, E., Wertheim, G.K., and Longinotti, L.D., Phys. Rev. Lett. 33 (1974) 165.

[3] Pollak, R.A., Holtzberg, F., Freeouf, J.L., and Eastman, D.E., Phys. Rev. Lett. 33 (1974) 820.

[4] Freeouf, J.L., Eastman, D.E., Grobman, W.D., Holtzberg, F., and Torrance, J.B., Phys. Rev. Lett. 33 (1974) 161.

[5] Gudat, W., Campagna, M., Rosei, R., Weaver, J.H., Eberhardt, W., Hulliger, F., and Kaldis, E., J. Appl. Phys. 52 (1981) 2123.

[6] Eastman, D.E., Donelon, J.J., Hien, N.C., and Himpsel, F.J., Nucl. Instrum. and Methods 172 (1980) 327.

[7] Duc, T.M., Guillot, C., Lassailly, Y., Lecante, J., Jugnet, Y., and Vedrine, J.C., Phys. Rev. Lett. 43 (1979) 789.

[8] Mårtensson, N., Reihl, B., and Vogt, O., Phys. Rev. B25 (1982) 824.

[9] Kaindl, G., Reihl, B., Eastman, D.E., Pollak, R.A., Mårtensson, N., Barbara, B., Penney, T., and Plaskett, T.S., Solid State Commun. 41 (1982) 157.

[10] Mårtensson, N., Reihl, B., Schneider, W.D., Murgai, V., Gupta, L.C., and Parks, R.D., Phys. Rev. B25 (1982) 1446; and these Conference Proceedings.

[11] Kaindl, G., Laubschat, C., Reihl, B., Pollak, R.A., Mårtensson, N., Holtzberg, F., and Eastman, D.E., Phys. Rev. B (1982) in press.

[12] Cox, P.A., Structure and Bonding 24 (1975) 59.

[13] Carnall, W.T., Fields, P.R., and Rajnak, K., J. Chem. Phys. 49 (1968) 4424.

[14] Citrin, P.H., Eisenberger, P., and Hamann, D.R., Phys. Rev. Lett. 33 (1974) 965.

[15] Wertheim, G.K., and Crecelius, G., Phys. Rev. Lett. 40 (1978) 813; Allen, J.W., Johansson, L.I., Lindau, I., and Hagstrom, S.B., Phys. Rev. B21 (1980) 1335.

[16] Rosengren, A., and Johanson, B., these Conference Proceedings.

[17] Broksch, H.-J., Tománek, D., and Bennemann, K.H., these Conference Proceedings.

Valence Instabilities
P. Wachter and H. Boppart (eds.)

SURFACE CORE-LEVEL SHIFTS IN MIXED-VALENT SYSTEMS*

H.-J. Brocksch, D. Tománek
Institut für Theoretische Physik, Freie Universität Berlin
and
K.H. Bennemann**
Institut für Theoretische Physik, ETH Zürich

A semiempirical model for surface core-level shifts Δ_c^s in mixed-valent compounds is presented. The interplay of electronic and Madelung-type contributions to Δ_c^s is discussed by including incomplete final state screening. It is shown that Δ_c^s depends sensitively on the surface change in the electronic configuration and screening capacity. In the core-level screening both the dynamic response via the itinerant charge and the dielectric polarizability of the medium are considered. To demonstrate the range of validity of this model, surface 4f-level shifts are calculated for TmSe, TmTe and $YbAl_2$ and for systems with varying binding character such as $Sm_{1-x}Y_xS$ as a function of composition.

1. INTRODUCTION

For many solids surface core-level shifts have been observed. This shift is caused by a change in the electrostatic potential resulting from the local chemical environment, the local electronic configuration and the final state screening at the core-excited site. At transition metal surfaces the surface core-level shift has been directly related to the geometrical arrangement and other interesting thermodynamic properties[1,2]. In this paper it is shown for the first time how this shift reveals also interesting information on the valence state of the solid and how it can be used to study mixed-valent systems. The surface core-level shift depends sensitively on the valence and the screening capacity.

In section 2 we develop a simple model for Δ_c^s. In section 3 this model is applied to mixed-valent systems such as $Sm_{1-x}Y_xS$ [3], TmSe, $YbAl_2$ and to TmTe. In section 4 a brief summary is presented.

2. THEORY

The surface core-level shift $\Delta_c^s(Z)$ of an atom with atomic number Z is given by[4]

$$\Delta_c^s(Z) = E_c^S(Z) - E_c^B(Z) , \qquad (2.1)$$

which is the difference of the core-level binding energies (with respect to the Fermi energy E_F) at the surface and in the bulk. The core-level binding energy can be expressed in terms of total energies E^{tot} of the system in the initial state and in the final state, after a photoelectron is emitted, as

$$E_c^{B(S)}(Z) = E_f^{tot}(\text{core-excited atom } Z^* \text{ in the bulk (surface) of the unperturbed host}) - E_i^{tot}(Z \text{ atom in the bulk (surface) of the unperturbed host}). \qquad (2.2)$$

For pure metal surfaces, where to a good approximation complete final state screening can be assumed, $\Delta_c^s(Z)$ is the difference of surface energies of Z and (Z+1) metal[5]. In the case of semiconductors and insulators $\Delta_c^s(Z)$ is expected to depend also on the final state screening. In solids with partial ionic character Δ_c^s can be decomposed into an electronic part $\Delta_e(Z)$ and a Madelung part $\Delta_M(Z)$. Then,

$$\Delta_c^s(Z) = \Delta_e(Z) + \Delta_M(Z). \tag{2.3}$$

Δ_e and Δ_M can be decomposed as

$$\Delta_e(Z) = \Delta E_e(Z^*) - \Delta E_e(Z) \tag{2.4}$$

and

$$\Delta_M(Z) = \Delta E_M(Z^*) - \Delta E_M(Z) , \tag{2.5}$$

where $\Delta E_{e(M)}$ is the difference between the electronic (Madelung) contribution to the cohesive energy at the surface and in the bulk, and where $Z^*(Z)$ denotes the core-excited (ground state) atom.

First consider the electronic part Δ_e of Δ_c^s in eq. (2.4). In the one-electron picture ΔE_e can be obtained by integration over the local density of states $g^i(E)$ and summation over all occupied bands i, as

$$\Delta E_e(Z) = \sum_i \Big(\int_{-\infty}^{E_F} dE(E-E_S^i) g_S^i(E) - \int_{-\infty}^{E_F} dE(E-E_B^i) g_B^i(E) \Big) . \tag{2.6}$$

Here, E^i denotes the center of gravity of the band i. In the following, the indices S(B) refer to surface (bulk) properties. Note also that only partially filled bands will contribute to ΔE_e. Using rectangular band shapes[6] for $g^i(E)$ and appropriate band parameters for the ground and excited state, $\Delta_e(Z)$ can easily be calculated from eq. (2.4) and eq. (2.6). A formula similar to eq. (2.6) can be obtained fo the final Z^*-state if $g^i(E)$ is replaced by $g^{i,*}(E)$, etc.

Next we turn to the Madelung shift Δ_M. From the definition of the quantities in eq. (2.5) it follows that

$$\Delta E_M(Z^*) = \int d^3r \int d^3r' \frac{(\rho_S^*(r) - \rho_B^*(r))\, \rho(r')}{|r - r'|} , \tag{2.7}$$

which represents the interaction of the charge distribution $\rho_{S,B}^*(r)$ at the core-excited site with the charge distribution of the host. Note that ρ^* depends sensitively on the screening mechanism. The corresponding formula for $\Delta E_M(Z)$ is given by just omitting the asterisks in eq. (2.7). In a simple approximation both $\Delta E_M(Z)$ and $\Delta E_M(Z^*)$ can be obtained from the lattice energy in the adiabatic rigid lattice point-charge model[7]. In evaluating ΔE_M, the appropriate Madelung constants[7] reflecting the crystal symmetries in the bulk and at the surface must be chosen, and one has to take into account the nuclear core-repulsion by using, e.g., a Born-Mayer-like potential.

$\Delta_c^s(Z)$ can in principle be calculated, once the relevant parameters for the ground state of a system are known. The only remaining point to be discussed is the charge distribution $\rho^*(r)$ or the total charge Q^* at the core-excited site in eq. (2.7). Clearly, the charge distribution after the core-ionization must be considered in evaluating both $\Delta E_e(Z^*)$ from eq. (2.6) and $\Delta E_M(Z^*)$ from eq. (2.7).

In the following we briefly discuss the screening capacity of the host in terms of extra-atomic dynamic screening and of dielectric static screening. The latter gets especially important in the case of semiconductors.

Consider now a rare-earth atom Z in the host, where a 4f-photoelectron is being emitted. Screening of the core-hole can be achieved by itinerant conduction band electrons locally available at the core-excited site. Complete screening is obtained if the initial state occupation number N is larger than one. A positive excess charge, however, is expected to remain at the core-excited site in the case where N is smaller than one. A localized state will form in this case due to the new attractive scattering potential. It will accomodate the screening charge, thereby draining the conduction band and hence reducing the electronic binding energy at the Z^*-site. In such a case of incomplete local screening we expect the surrounding electron cloud to relax toward the core-excited site and to produce some charge transfer for additional screening of the excess charge. As there is no simple way to describe this process quantitatively, we introduce an adjustable parameter S to account for this effect. In addition to this dynamic screening we expect the bound electronic charge of the host to get polarized by the positive excess charge. We can account for this static screening by considering the host as an effective medium whose electronic polarizability is described by a high-frequency dielectric constant ϵ_{oo}. Hence, the final state charge Q* of the rare-earth ion can be obtained as

$$Q^*_{B(S)} = Q_{B(S)} + f_{B(S)}(\epsilon_{oo}^{B(S)}) \cdot S_{B(S)} \cdot (1 - N_{B(S)}) \quad (2.8)$$

where Q is the ground state ionic charge and N is the conduction band occupation number. The function $f(\epsilon_{oo})$ and the parameter S describe static and dynamic screening effects, respectively. f is a well-behaved function which - alike S - differs for bulk and surface. Note that eq. (2.8) also reproduces correctly the limiting cases of a metal ($f_{B(S)} = S_{B(S)} = 0$) and that of an insulator ($f_{B(S)} = S_{B(S)} = 1$).

As the Madelung energy of a crystal can be much larger than the electronic binding energy, the final state charge and the screening parameters turn out to be of crucial importance for the surface core-level shift. Due to the sensitivity of Δ_c^s to the screening mechanism, it should be possible to obtain information about the electronic configuration (valence) and the screening of charged impurities in intermediate-valent systems from photoemission measurements.

3. APPLICATIONS

First we use our model to calculate Sm 4f-level shifts[3] at the (100) surface of $Sm_{1-x}Y_xS$ as a function of x. The bulk valence of Sm, $V_B(x)$, has been obtained by Tao and Holtzberg[8] for the whole composition range from lattice parameter measurements. We assume a common d-band for Y and Sm and obtain for the 5d-band occupation number in the bulk

$$N_B(x) = (1-x)(V_B(x)-2) + x. \quad (3.1)$$

At the surface, there is strong evidence[9] for divalent Sm, so that only Y will contribute to the conduction band filling, giving

$$N_S(x) = x \quad (3.2)$$

As $V_B(x)$ is always less than 3, the d-band occupation never reaches one. Hence, we expect incomplete final state screening and zero contribution to Δ_e from the final state. The 5d-bandwidth of $Sm_{1-x}Y_xS$ has been obtained from linear interpolation between those of SmS and YS as done previously[10]. The contribution of the 4f-electrons to the cohesive energy turns out to be negligible due to the small 4f-bandwidth.

In the calculation of the Madelung shift Δ_M the core-repulsion term can be related to the bulk modulus $B(x)$[11]. With increasing Y concentration in $Sm_{1-x}Y_xS$, the system undergoes a semiconductor-metal transition at x= 0.15 at room temperature. At this point B(x) drops to zero and also causes Δ_M to vanish. The structure in the Madelung shift is predominantly due to discontinuities in the $V_B(x)$ and B(x) curves, especially near the transition point.

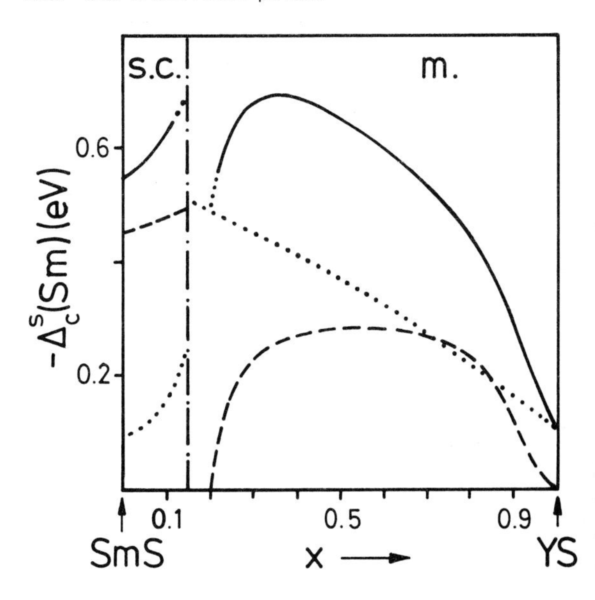

Fig. 1
Surface 4f-level shift Δ_c^s of Sm in $Sm_{1-x}Y_xS$, together with the constituent electronic and Madelung shift Δ_e and Δ_M (dotted and dashed)

Δ_M is expected to be large in the semiconducting phase for small x and to vanish in the limit of metallic YS. In Fig.1 the calculated 4f-level shifts $\Delta_c^s(x)$ of Sm for one set of screening parameters $S_{B(S)}$ is presented together with data for $\Delta_e(x)$ and $\Delta_M(x)$.

It is interesting to see that by applying the simple model presented in this paper and using experimental data for the surface (bulk) valence $V_{S(B)}$, the density of states, the bulk modulus and optical reflectivity one obtains for some other related systems the following results:

	TmSe	TmTe	$YbAl_2$
V_B	2.55	2.0	2.4 [12]
V_S	2.0	2.0	2.2 [12]
Δ_e (eV)	0.6	0	~0.6..0.7
screening	incomplete		full
Δ_M (eV)	~-0.3..-0.2	~0.2..0.6	0
Δ_c^s(theor)(eV)	~ 0.3...0.4	~0.2..0.6	~0.6..0.7
Δ_c^s(exp) (eV)	0.32	0.41	0.92 [12]

4. SUMMARY AND CONCLUSIONS

We have presented a simple model for the calculation of surface 4f-level shifts Δ_c^s in both semiconducting and metallic mixed-valent compounds. In the general case, Δ_c^s consists of an electronic and of a Madelung part. Especially the latter is very sensitive to changes in the valence and the screening capacity between bulk and surface. We have shown that Δ_c^s contains direct information about screening of excited atoms or charged impurities and the electronic configuration.

Acknowledgement

We acknowledge useful discussions with Dr. P. Schlottmann.

* This work is supported by the DFG-Sonderforschungsbereich 6

** On leave of absence from FU Berlin.

References

1) Tománek, D., Kumar, V., Holloway, S. and Bennemann, K.H., Solid State Comm. 41(1982), 273.

2) Kumar, V., Tománek, D. and Bennemann, K.H., Solid State Comm. 39 (1981) 987.

3) Brocksch, H.-J., Tománek, D. and Bennemann, K.H., to be published in Phys. Rev. B.

4) Rosengren, A. and Johansson, B., Phys. Rev. B22 (1980), 3706.

5) Johansson, B. and Mårtensson, N., Phys. Rev. B21 (1980), 4427.

6) Pettifor, D.G., Phys. Rev. Lett. 42 (1979), 846.

7) Tosi, M.P. in Seitz, F. and Turnbull, D. (eds.), Solid State Physics, Vol. 16 (Academic Press, New York and London, 1964).

8) Tao, L.J. and Holtzberg, F., Phys. Rev. B11 (1975), 3842.

9) Lang, J.K. and Baer, Y., Solid State Comm. 31 (1979), 945; Gudat, W., Campagna, M., Rosei, R., Weaver, J.H., Eberhardt, W., Hulliger, F. and Kaldis, E., J. Appl. Phys. 52 (1981), 2123.

10) Bilz, H., Güntherodt, G., Kleppmann, W. and Kress, W., Phys. Rev. Lett. 43 (1979), 1998.

11) Penney, T., Melcher, R.L., Holtzberg, F. and Güntherodt, G. in AIP Conf. Proceedings No. 29 (1976), 392

12) Kaindl, G., Reihl, B., Eastman, D.E., Pollak, R.A., Mårtensson, N., Barbara, B., Penney, T. and Plaskett, T.S., Solid State Comm. 41 (1982), 157; and private communication.

Valence Instabilities
P. Wachter and H. Boppart (eds.)

VALENCE CHANGE AND SURFACE CORE-LEVEL SHIFTS IN MIXED-VALENT $EuPd_2Si_2$

W.-D. Schneider*, N. Mårtensson, and B. Reihl
IBM T.J. Watson Research Center
Yorktown Heights, New York 10598

and

V. Murgai, L.C. Gupta, and R.D. Parks
Department of Physics
Polytechnic Institute of New York
Brooklyn, New York 11201

Synchrotron radiation excited photoelectron spectroscopy is employed to study the bulk and surface states of the mixed valent system $EuPd_2Si_2$, which undergoes a valence transition from $Eu^{2.2+}$ at room temperature to $Eu^{2.9+}$ at low temperature ($\lesssim$ 100 K). The high resolution employed, coupled with a relatively large bulk to surface energy shift (1.2 eV) of the Eu^{2+} emission, allows a definitive identification of the bulk and surface states of Eu^{2+}. The Eu^{3+} emission is not observed at room temperature, but is readily apparent in the low temperature spectra.

Surface-induced changes of both the mean-valence of rare earth ions and of the 4f-binding energy are being reported recently in photoemission studies of mixed and integral valent rare-earth systems. Theoretically, these effects may be understood as a consequence of the decrease in cohesive energy due to the broken bonds at the surface of a solid (1-3). The first surface valence transition was observed for bulk-trivalent Sm metal, where the surface layer was shown to be essentially divalent (4). More recent studies have shown that in mixed-valent $YbAl_2$ more than one surface layer is involved in a valence transition to the divalent state (5), giving rise to predominantly divalent features in the VUV-photoemission spectra. Furthermore, the presence of both a bulk and energy-shifted surface spectral component for the same initial state configuration has been found for, e.g. Sm^{2+} in SmB_6 (6), SmS (6), SmSe (7), for Tm^{2+} in TmSe (8,9), for Yb^{2+} in $YbAl_3$ (10) and $YbAl_2$ (5), or for Ce^{3+} in CeN (7). In all these cases the 4f levels exhibit a shift to higher binding energies at the surface, in agreement with theoretical predictions (1).

In this communication we present a high-resolution photoemission study of the mixed valent intermetallic compound $EuPd_2Si_2$ which was carried out in order to gain more insight into the subtle interplay between surface and volume states in a mixed valent situation. The high resolution employed, coupled with a relatively large bulk to surface energy shift of the Eu^{2+} emission (1.2 eV) allowed a definitive identification of the bulk and surface states of Eu^{2+}. Moreover, by taking advantage of a Cooper minimum for the Pd 4d emission at hν = 120 eV excitation energies, the change of the bulk valence in $EuPd_2Si_2$ is apparent in the low temperature spectra (50 K).

$EuPd_2Si_2$ crystallizes in the tetragonal $ThCr_2Si_2$-type structure (11), wherein the Eu atoms reside on planes well separated by layers composed of the Pd and Si atoms (see Fig. 1). It is unique among the few Eu-based mixed valent systems studied to date (viz., $EuCu_2Si_2$, $EuFe_4Al_8$, and $EuRh_2$) (12) in that it undergoes a strong, albeit continuous, valence transition with temperature. The characterization of the valence transition by Mössbauer, magnetic susceptibility, and lattice constant measurements is reported in Ref. 13. The Mössbauer isomer shift measurements indicate that the system is homogeneously mixed valent at all temperatures,

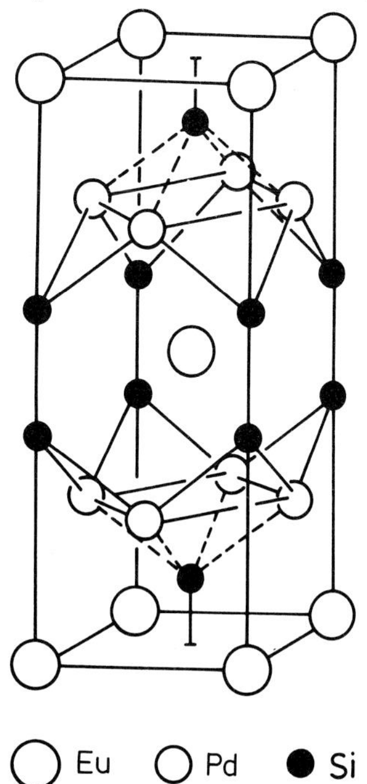

Fig. 1: Crystal structure of $EuPd_2Si_2$

with the valence varying from ∿2.2 above the transition (whose midpoint is at ∿150 K) to ∿2.9 below. At least 90% of the valence change takes place within plus and minus 50 K from the midpoint.

Photoemission data were taken with a display-type analyzer (14) using synchrotron radiation at the University of Wisconsin-Madison. The system was operated in an angle-integrated mode with a total energy resolution of ∿0.15 eV at 70 eV and ∿0.3 eV at 120 eV photon energy. The measurements were performed at 300 K and 50 K on freshly fractured crystals in a vacuum better than 10^{-10} Torr. No signs of oxygen or other contaminants were seen during the course of the measurements.

Fig. 2a shows the spectrum of $EuPd_2Si_2$ recorded at 50 K with a photon energy of hν=70 eV. The spectrum is dominated by a relatively narrow Pd 4d band centered ∿3.5 eV below F_F. Closer to the Fermi level there are two structures separated by 1.1-1.2 eV which we interpret as the bulk and surface Eu 4f emission from Eu^{2+} ions. In the energy range 6-10 eV where we expect the trivalent Eu multiplets to occur, especially in this low-temperature phase which has a valence of 2.9, only a few very weak structures might be discerned. In order to make the Eu 4f emission more dominant in the spectrum, we can utilize the well-known fact that the Pd 4d emission has a Cooper-minimum at photon energies around 120 eV (15). Fig. 3 vizualizes the influence of this effect on the valence band spectra with increasing photon energy at 50 K. The Pd 4d emission is almost completely suppressed at hν=115 eV and the Eu 4f emission dominates the spectrum. Due to the reduction of the Pd 4d emission, the trivalent multiplets in the 50 K spectrum are clearly seen. In Figs. 2b and 2c the spectra at 300 K and 50 K are compared at 120 eV photon energy. With this clearer view of the spectrum the interpretation is now more obvious. Closest to the Fermi level we find the bulk emission from Eu^{2+} ions corresponding to the excitation $f^7(5d6s)^2 \rightarrow f^6(5d6s)^3$, wherein the final state is fully screened. The lowest multiplet line falls in the near vicinity of the Fermi level which is a necessary condition for homogeneous mixed valency, i.e. that the $f^6(5d6s)^3$ and $f^7(5d6s)^2$ states are energetically degenerate. The divalent portion of the spectrum is, however, dominated by a structure at 1.1-1.2 eV higher binding energy which we interpret as a surface-shifted Eu^{2+} emission. When the low temperature phase is entered the bulk divalent spectral contribution is reduced and the characteristic trivalent multiplet pattern appears in the energy range 6-10 eV. The surface 4f emission seems, however, to be unaffected.

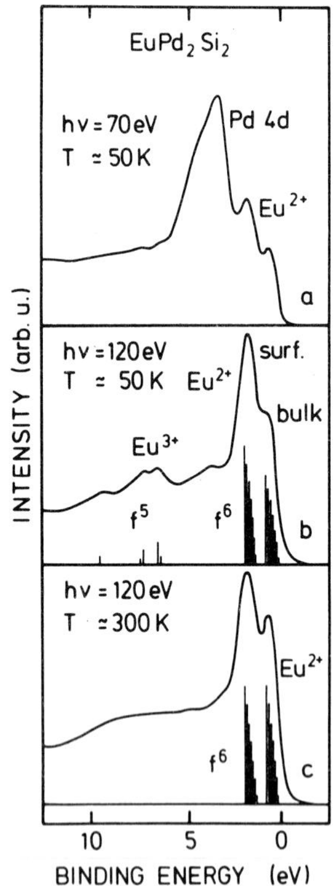

Fig. 2: Angle integrated energy distribution curves (EDC's) of photoelectrons emitted from $EuPd_2Si_2$ at (a) hν=70 eV and T=50 K, (b) hν=120 eV and T=50 K, and (c) hν=120 eV and T=300 K.

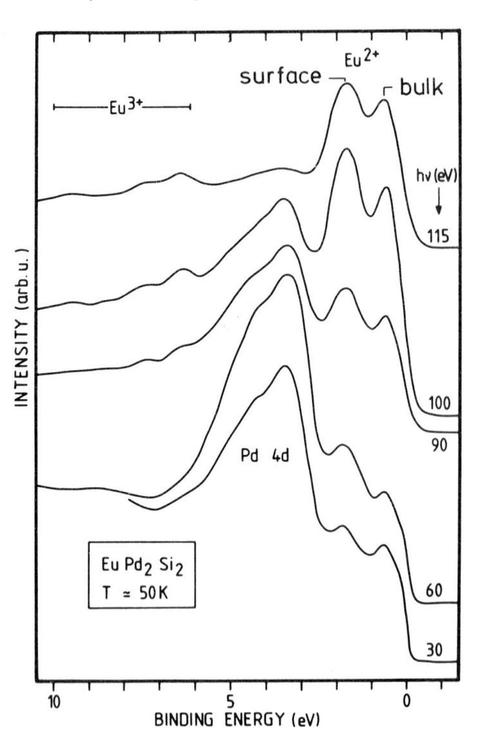

Fig. 3: Angle integrated EDC's of photoelectrons emitted from $EuPd_2Si_2$ at various photon energies to show the effect of the Cooper minimum for the Pd 4d emission near hν=120 eV.

What information can be obtained from the relative intensities of the various 4f features? At $h\nu$=120 eV the electron escape depth λ is ~6Å (16), while in $EuPd_2Si_2$ the Eu-Eu interlayer distance is ~5Å. With these parameters we sample to ~55% the top monolayer of Eu atoms, to ~25% the second monolayer and to ~20% the remaining layers. Hence, in the simple picture where the top monolayer is in a surface-shifted divalent state and all the other atoms exhibit the bulk mixed valence, we would expect the intensity of the "bulk" Eu^{2+} feature to be nearly completely transferred to the Eu^{3+} feature at 50 K, where the bulk valence is 2.9. Clearly this is not the case, since at 50 K ~20% of the total 4f intensity is found in the Eu^{3+} feature and ~35% in the "bulk Eu^{2+} feature. A possible explanation is that the second monolayer of Eu atoms is divalent at all temperatures and not measurably shifted from the deep bulk Eu^{2+} emission. However, it is important to note that the second layer is not a well defined concept since we are not measuring a surface which has been well-characterized topologically. Depending on how the $EuPd_2Si_2$ crystal actually fractures, the second Eu layer referred to above might consist of Eu atoms in certain cleavage planes where they are not surface atoms, but where they are still the topmost Eu atoms, being covered by a 2-4Å thick layer of Pd and Si atoms. While we find this to be the most reasonable interpretation of the observed "second layer", we concede that uncertainties in our estimated intensities, the applied value of λ and the surface topology render as speculative any of our attempted descriptions of the second layer (17).

In Eu metal a surface shift in the Eu^{2+} emission of 0.6 eV has been observed (18,19), whereas it appears to be 1.1-1.2 eV in $EuPd_2Si_2$. We shall attempt to explain this difference with arguments which focus on possible differences in the nature of the bonding between Eu metal and $EuPd_2Si_2$. Recently, it has been shown (2) how the metallic influence on core level binding energies can be calculated within a scheme where the metallic screening of the photoionized hole state is explicitly taken into account. One term ΔE_{coh}, which enters into this scheme, describes how the bonding properties (hence cohesion) of the ionized site changes due to the photoemission of a core electron in the presence of the concomitant metallic screening within the valence shell. If the screening takes place in a bonding orbital this increases the cohesive energy, corresponding to $\Delta E_{coh}>0$; conversely if the screening takes place in an antibonding orbital, $\Delta E_{coh}<0$.

In the case of a surface atom we must consider the fact that it has an incomplete coordination. Due to the broken bonds ΔE_{coh} will therefore be reduced at the surface (2,3). If we write $\Delta E^{surf}_{coh} = \alpha\ \Delta E^{bulk}_{coh}$, then the surface shift will be $(1-\alpha)\ \Delta E^{bulk}_{coh}$. In Ref. 2, the estimate α=0.8 was used, based on earlier empirical studies. Thus, in Eu metal, ΔE^{bulk}_{coh} = 2.4 eV (20) gives a calculated surface shift of 0.5 eV towards higher binding energies, which agrees closely with the experimental result, 0.6 eV. Within this model chemical shifts in metallic compounds can also be described by changes in ΔE_{coh} (2,21). If the chemical shift is such that the magnitude of ΔE_{coh} increases, we expect also the surface shift to increase. In the compound $EuPd_2Si_2$, the bulk 4f emission is shifted by 1.4 eV towards E_F (the direction of increasing ΔE_{coh}) relative to Eu metal; hence $\Delta E^{bulk}_{coh}(EuPd_2Si_2)$=3.8 eV. Since ΔE^{bulk}_{coh}(Eu metal)= =2.4 eV yields a measured surface shift of 0.6 eV, the corresponding surface shift expected for $EuPd_2Si_2$ is (3.8/2.4)0.6≅1.0 eV, which is close to the measured value 1.2 eV reported herein. In other words the surface shift in $EuPd_2Si_2$ relative to that in Eu metal can be well accounted for within this model. Better agreement would be fortuitous in view of the uncertainty in α and its dependence on crystal structure orientation (2) and in the case of a compound its dependence on the specific surface composition.

In summary, we have both identified and accurately measured the surface shift in a mixed valent compound. The facility of being able to obtain spectra for widely different bulk valences (2.2+ at 300 K and 2.9+ at 50 K) has allowed us to more thoroughly characterize the nature of the surface state than otherwhise possible. Our conclusion is that the surface state of $EuPd_2Si_2$ is confined to the top one or two monolayers and it is divalent (rather than mixed valent) irrespective of the valence of the bulk. The relatively small change in the "bulk" divalent 4f emission when the valence is changed from 2.2 to 2.9 suggests that the mixed valent bulk is not simply terminated by one divalent layer of Eu surface atoms. As one possibility it is shown that with a reasonable value for the electron escape depth it is possible to reproduce the spectrum by assuming the existence of a "second layer" of Eu atoms which are in a stable valent, but (almost) unshifted, Eu^{2+} state. The fact that the surface shift of the Eu^{2+} emission (1.1-1.2 eV) is roughly twice that reported for Eu metal (0.6 eV) is reconcilable within the screening model discussed in Ref. 2, if one takes into account the measured differences in the bulk emission spectra of the two systems.

The authors gratefully acknowledge the generous support by D.E. Eastman and the cooperation of the staff of the Synchrotron Radiation Center, University of Wisconsin. The work carried out at Polytechnic Institute of New York was sponsored in part by the Department of Energy and the National Science Foundation; that of IBM was partly supported by the Air Force Office of Scientific Research under contract No. F49620-81-C0089 and by the Deutsche Forschungsgemeinschaft, Sfb-6.

REFERENCES

* Present address: Institut für Atom- und Festkörperphysik, Freie Universität Berlin, 1000 Berlin 33, Germany

|1| Johansson, B., Phys. Rev. B19 (1979) 6615.

|2| Johansson, B., and Mårtensson, N., Phys. Rev. B21 (1980) 4427.

|3| Rosengren, A., and Johansson, B., Phys. Rev. B22 (1980) 3706.

|4| Wertheim, G.K., and Crecelius, G., Phys. Rev. Lett. 40 (1978) 813.

|5| Kaindl, G., Reihl, B., Eastman, D.E., Pollak, R.A., Mårtensson, N., Barbara, B., Penney, T., and Plaskett, T.S., Solid State Commun. 41 (1982) 157.

|6| Allen, J.W., Johansson, L.I., Lindau, I., and Hagström, S.B., Phys. Rev. B21 (1980) 1335.

|7| Gudat, W., Campagna, M., Rosei, R., Weaver, J.H., Eberhardt, W., Hulliger, F., and Kaldis, E., J. Appl. Phys. 52 (1981) 2123.

|8| Wertheim, G.K., Eib, W., Kaldis, E., and Campagna, M., Phys. Rev. B22 (1980) 6240.

|9| Kaindl, G., Laubschat, C., Reihl, B., Pollak, R.A., Mårtensson, N., Holtzberg, F., and Eastman, D.E., Phys. Rev. B, to be published.

|10| Buschow, K.H.J., Campagna, M., Wertheim, G.K., Solid State Commun. 24 (1977) 253. The spectra for $YbAl_3$ and $EuCu_2Si_2$ reported here, when considered in the light of more recent developments, suggest the presence of both bulk and surface divalent states.

|11| Rossi, D., Marazza, B., and Ferro, R., J. Less-Common Metals 66 (1979) 17.

|12| E.g., see Lawrence, J.M., Riseborough, P.S., Parks, R.D., Rep. Prog. Phys. 44 (1981) 1.

|13| Sampathkumaran, F.V. Vijayaraghavan, R., Gopalakrishnan, K.V., Pilley, R.G., Devare, H.G., Gupta, L.C., Post, B., and Parks, R.D., Valence Fluctuations in Solids, in Falicov, L.M., Hanke, W., and Maple, M.B., (eds.), (North-Holland, Amsterdam 1981) p. 193.

|14| Eastman, D.E. Donelon, J.J., Hien, N.C., and Himpsel, F.J., Nucl. Instr. and Meth. 172 (1980) 327.

|15| See, e.g. Goldberg, S.M. Fadley, C.S., and Kono, S., J. Electron Spectrosc. Rel. Phen. 21 (1981) 285.

|16| See, e.g. Feuerbacher, B., and Willis, R.F., J. Phys. C9 (1976) 169.

|17| Neither can we, in principle, exclude the possiblity that the second layer is mixed valent, but that its effective valence is reduced from the true bulk value of 2.9 to 2.3 - 2.4, while still exhibiting the same kind of temperature dependence as the deep bulk.

|18| Kammerer, R., Barth, J., Gerken, F., Flodström, A., and Johansson, L.I., Solid State Commun. 41 (1982) 435.

|19| Schneider, W.D., Mårtensson, N., Reihl, B., and Kaindl, G., to be published.

|20| This value for $Eu(4f^7)$ is obtained by interpolating between the divalent metals Ba $(4f^0)$ and Yb $(4f^{14})$, for which the same values as in Ref. 2 have been used.

|21| Steiner, P., Hüfner, S., Mårtensson, N., and Johansson, B., Solid State Commun. 37, (1981) 73.

Valence Instabilities
P. Wachter and H. Boppart (eds.)

Eu VALENCE AND SURFACE CORE LEVEL SHIFTS IN $EuPd_x$ COMPOUNDS AS STUDIED BY PHOTOELECTRON SPECTROSCOPY

V. Murgai, L.C. Gupta, and R.D. Parks

Polytechnic Institute of New York,
Brooklyn, NY 11201, USA

N. Mårtensson* and B. Reihl+

IBM T.J. Watson Research Center,
Yorktown Heights, NY 10598, USA

For the $EuPd_x$ system (with the addition of $EuPd_2Si_2$), we study how the Eu 4f photoemission spectrum evolves as the Eu valence is brought from a divalent, through a mixed-valent to a trivalent state. In the divalent and mixed-valent compounds, a 4f surface shift is clearly seen. In the trivalent Eu compounds, as well as in mixed-valent $EuPd_2Si_2$, we detect a stable-valent Eu^{2+} surface layer. The trivalent compounds $EuPd_3$ and $EuPd_5$ form a most interesting class of magnetic materials with a non-magnetic Eu^{3+} bulk covered by a few Å thick layer of magnetic Eu^{2+} atoms. The 4f surface peak shows a smaller chemical shift as a function of Pd alloying than the bulk peak. This is explained in terms of the reduced number of Eu-Pd bonds for a surface Eu atom.

I. INTRODUCTION

In its intermetallic compounds, europium occurs in three valence states, namely, Eu^{2+}, Eu^{3+} and mixed-valence state. Eu metal is divalent with a stability of about 1 eV relative to the trivalent configuration. (1,2) In intermetallics with highly electronegative elements which have a high density of states at the Fermi level, the trivalent state of europium is favored, whereas those with a broad s-band, have europium in a divalent state. For instance, binary intermetallics of europium with noble metals and s-p elements always have europium in a 2+ state. However, the (binary) intermetallics with transition metals such as Rh, Ir and Pd exhibit all three valence states, viz., 2^+, 3^+ and mixed-valence.

The series of europium intermetallics, EuPd, $EuPd_2$, $EuPd_3$ and $EuPd_5$ is a particularly interesting one in that there is a transition of the valence state of europium, as one goes from one member to the other, in a very systematic fashion.

Eu is in a truly divalent state in EuPd, (3) it is in a nearly divalent state in $EuPd_2$ (4,5) (this material undergoes a ferromagnetic phase-transition at 80 K, however, an incipient instability of the divalent state towards trivalent has been suggested on the bases of the magnitudes of the hyperfine field and the nuclear quadropole interaction observed in this system) and transforms to a trivalent state in $EuPd_3$.(6) This behavior is linked with the progressively higher concentration of electronegative Pd atoms.

Recently, 4f surface shifts have been observed in pure RE metals (7,8) as well as in several RE compounds. (9-11) In Sm metal and a few other cases, it has also been observed that the surface RE atoms are in a lower valence state than the bulk atoms. (10-14)

In this paper, we study how the bulk and surface Eu4f photoemission spectrum evolves as the europium valence is brought from a divalent, through a mixed-valent to a trivalent state. Eu^{2+} atoms give 4f intensity in the energy region 0-2 eV below the Fermi level, whereas the Eu^{3+} intensity occurs in the energy region 6-9 eV. The europium valence can thus be determined by the presence of the typical Eu^{2+} and Eu^{3+} 4f emission features. Furthermore, the energy position of the Eu^{2+} 4f level gives direct information about the relative stabilities of the divalent and trivalent europium states.

2. EXPERIMENTAL

Photoemission measurements on polycrystalline samples of EuPd, $EuPd_2$, $EuPd_3$ and $EuPd_5$ were made with a display-type analyzer (15) using synchrotron radiation at the University of Wisconsin-Madison. For all the samples except $EuPd_5$, fresh surfaces were prepared by fracturing the crystals *in situ*. $EuPd_5$ was cleaned with a diamond file instead. The presently shown spectra were recorded at a photon energy of 120 eV with a combined monochromator and analyzer resolution of about 0.3 eV.

3. RESULTS AND DISCUSSION

Figure 1 shows the evolution of the Eu 4f spectrum as the europium valence is brought from Eu^{2+} in EuPd through mixed-valence in $EuPd_2Si_s$ to Eu^{3+} in $EuPd_3$ and $EuPd_5$. In Figure 2, we show how the EuPd spectrum can be decomposed into two peaks (each peak consisting of several closely spaced multiplet lines) separated by about 0.8 eV. The peak closest to the Fermi level corresponds to the bulk 4f emission, while the structure at higher binding energies originates from a few Å thick surface layers. Due to the complete screening of the photoionized state, the Eu^{2+} photoemission process corresponds

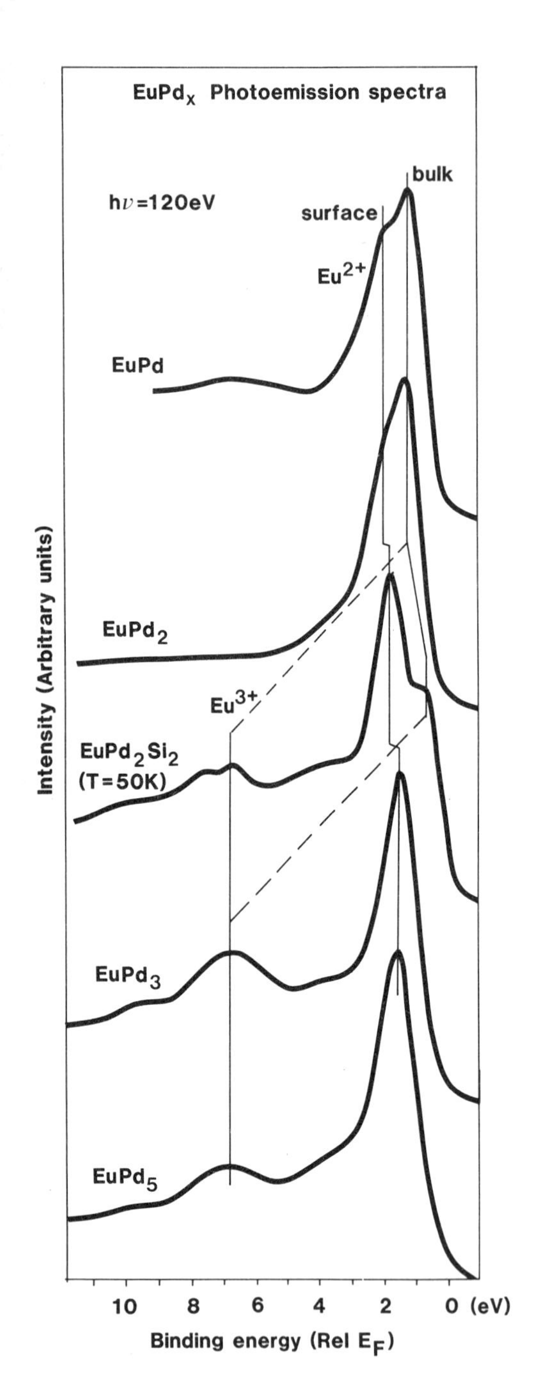

Figure 1 : Eu 4f spectra from several $EuPd_x$ compounds. The spectrum from mixed-valent $EuPd_2Si_2$ (10) is included for completeness. The spectra are almost totally dominated by the Eu 4f emission due to a Cooper-minimum in the Pd 4d intensity.

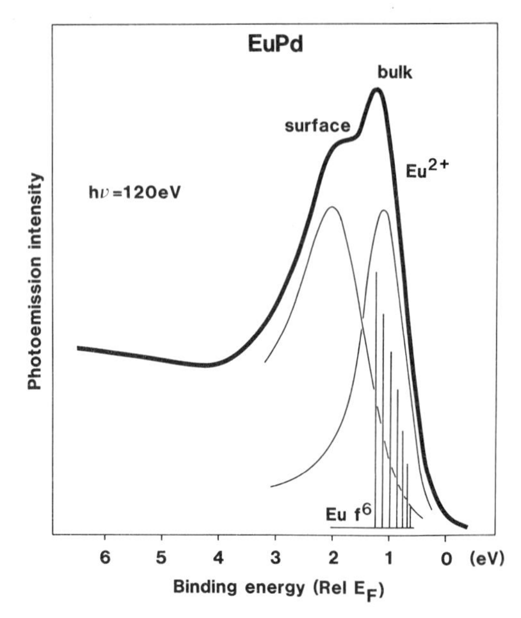

Figure 2 : Bulk and surface 4f emission features in divalent EuPd.

directly to a local $Eu^{2+} \rightarrow Eu^{3+}$ transition. By measuring the 4f position, the photoemission experiment thus probes in a quantitative way the relative stabilities of the Eu^{2+} and Eu^{3+} valence states. The higher binding energy of the surface peak in EuPd and $EuPd_2$ implies that there is an increased stability of the divalent state in the surface layer as compared to the bulk.

In $EuPd_2Si_2$, we observe (10) how the Eu^{2+} bulk and surface emission features have both shifted towards E_F. Analyzing the bulk 4f emission spectrum, one finds that the lowest-lying multiplet line is located in the vicinity of the Fermi level. This implies that the Eu^{2+} and Eu^{3+} configurations are essentially degenerate in energy, and a necessary condition for homogeneous mixed-valence is fulfilled. The mixed-valent character of this system is then established in the spectrum by the simultaneous appearance of Eu^{3+} emission features at around 6 eV binding energies.

By further alloying in $EuPd_3$ and $EuPd_5$, we observe how the bulk 4f intensity has been completely transferred to the Eu^{3+} region; the bulk samples becoming completely trivalent. Although the effects of the alloying are sufficient to bring the bulk to an integral valent Eu^{3+} state, we observe how the surface europium atoms remain divalent.

In Figure 3, we show the spectrum from $EuPd_3$ in greater detail. In this figure, we have also marked the positions of the Eu^{2+} bulk 4f multiplet levels.

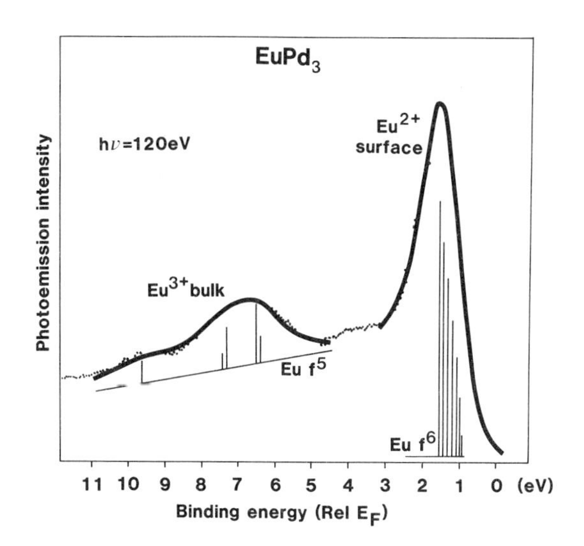

Figure 3 : Bulk and surface 4f emission features in trivalent $EuPd_3$.

Identifying the measured 4f position with the total energy difference between the divalent and trivalent europium states, we obtain an energy of 1.5 eV for the trivalent Eu metal, and around 1.0 eV for $Eu^{3+}Pd$ and $Eu^{3+}Pd_2$, relative to the divalent states. Now we know that the energy required to create trivalent Eu metal is only 1.0 eV. (1,2) The discrepancy between this value and the spectroscopically measured energy originates from the neglect of final-state impurity terms in our simplified interpretation of the photoemission results. When we create a Eu^{3+} final state by photoionization, this trivalent ion will be an isolated Eu^{3+} impurity in an otherwise divalent lattice. This defines an energy which in the present case tends to increase the energy of the photoexcited final state. Furthermore, when we compare the 4f positions in several ordered compounds which all have different crystal structures, we have to consider the possibility of the impurity effect differing from system to system.

For the $EuPd_x$ compounds, we can compare our results to the calculated Eu^{3+} energies by Miedema. (16) He correctly predicts that the $EuPd_x$ systems lie very closely to a valence transition from divalent to trivalent Eu. According to these calculations, however, all the $EuPd_x$ compounds are divalent with $Eu^{2+} \rightarrow Eu^{3+}$ transition energies of 0.3 eV in EuPd and 0.1 eV in the remaining compounds. Our spectroscopic results indicate a stronger concentration dependence of this energy difference. As we noted above it has been suggested that $EuPd_2$ is very close to mixed-valent. It is therefore interesting to compare the $EuPd_2$ spectrum to the spectra from EuPd and $EuPd_2Si_2$ in Figure 1. We find that the spectra from EuPd and $EuPd_2$ are very similar but that they are qualitatively different from the $EuPd_2Si_2$ spectrum. The lowest-lying bulk multiplet line in EuPd lies at ~1.0 eV. Unless the final-state impurity effects are much more important in this compound than in europium metal, the large binding energy of the 4f level gives no indication that $EuPd_2$ is particularly close to a mixed-valent state.

In the Eu 4f spectra, we also note a strong chemical influence on the separation between the bulk and the surface 4f levels (ΔE_s). In Eu metal (cf. ref.8), the bulk peak has its maximum at 2.0 eV, while the surface emission is located at 0.6 eV higher binding energies (ΔE_s = 0.6 eV). As the 4f level is shifted towards the Fermi level, we observe an increase of ΔE_s to ~0.8 eV in EuPd and $EuPd_2$ and to 1.1 eV in $EuPd_2Si_2$. This trend is a direct consequence of the incomplete atomic coordination at the surface. To demonstrate this, in Figure 4 we have plotted the shift of the surface peak (relative to the position in europium metal) against the corresponding shift in the bulk peak. From this figure, we find that the shift of the surface 4f signal is reduced to about 70% of the bulk value. Due to the reduced number of Eu-Pd bonds for a surface Eu atom, we expect any chemical shift effect to be smaller at the surface than for a completely coordinated atom in the bulk. A strict proportionality, however, cannot be expected. The exact size of the surface shift will depend on factors like crystal structure, surface reconstruction, valence state of the bulk (divalent or mixed-valent), etc.

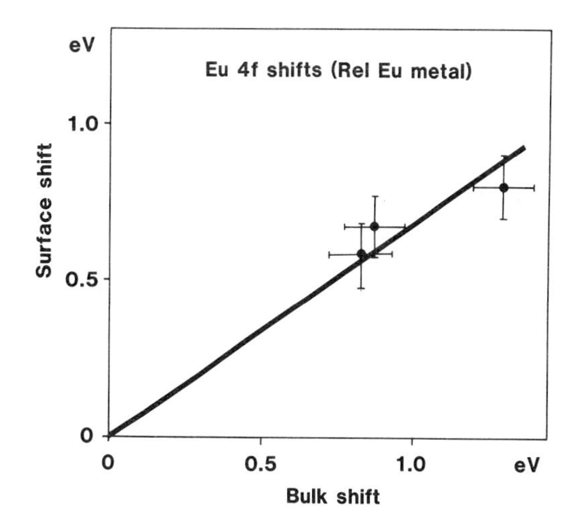

Figure 4 : A comparison of the bulk and surface 4f level shifts in EuPd, $EuPd_2$ and $EuPd_2Si_2$. Zero shift corresponds to Eu metal.

Finally, we want to point out some consequences of the divalent surface layers in $EuPd_3$ and $EuPd_5$. From Figure 3, it is possible to determine the surface Eu^{2+} to bulk Eu^{3+} 4f intensity ratio. With this information, we could then determine the escape depth for the photoelectrons in this compound. There are, however, some inherent problems in such an analysis which are often overlooked. The valence change of a europium atom from Eu^{3+} to Eu^{2+} is accompanied by a 45% increase in the atomic volume. The equilibrium bond distances will therefore be substantially longer at the surface than in the bulk. Due to this mismatch between the bulk and surface equilibrium geometries, we expect drastic surface rearrangements to occur. This situation will be encountered in any system which shows a surface-valence transition.

The existence of a divalent surface layer on top of the trivalent compounds $EuPd_3$ and $EuPd_5$ is also interesting from the point of view of magnetism. The divalent Eu configuration has a J=7/2 magnetic ground-state, while trivalent europium is nonmagnetic. These compounds thus comprise a most interesting class of two-dimensional magnetic systems with a nonmagnetic bulk covered by a possibly magnetic surface layer.

4. SUMMARY

We have compared the photoemission spectra from divalent EuPd and $EuPd_2$, mixed-valent $EuPd_2Si_2$ and from trivalent $EuPd_3$ and $EuPd_5$. In the divalent compounds, a surface shift of about 0.8 eV is observed which increases to 1.1 eV in mixed-valent $EuPd_2Si_2$. The successive increase from its value of 0.6 eV in Eu metal is shown to be a direct consequence of the reduced number of Eu-Pd bonds at the surface. In the trivalent compounds, the surface layer is found to be divalent. It is pointed out that these compounds therefore comprise a most interesting class of magnetic systems with a nonmagnetic bulk covered by a layer of magnetic europium atoms.

ACKNOWLEDGEMENTS

The authors gratefully acknowledge the generous support of D.E. Eastman, and the assistance of J.J. Donelon, A. Marx, and the staff of the Synchroton Radiation Center at the University of Wisconsin. The work carried out at IBM was supported in part by the Air Force Office of Scientific Research under Contract No F49620-81-C-0089; that of Polytechnic Institute of New York was partly sponsored by the Department of Energy and the National Science Foundation.

REFERENCES:

* Present address: Linköping University, Dept. of Physics and Measurement Technology, S-581 83 Linköping, Sweden.

+ Present address: IBM Zurich Research Laboratory, 8803 Rüschlikon, Switzerland.

[1] Gschneidner, K.A., J. Less-Common Met., 17 (1969) 1.

[2] Johansson, B., Phys. Rev. B20 (1979) 1315.

[3] Longworth, G. and Harris, I.R., J. Less-Common Met. 33 (1973) 83.

[4] Kropp, H., Dormann, E. and Buschow, K.H.J., Solid State Commun. 32 (1979) 507.

[5] Wickman, H.H., Wernick, J.H., Sherwood, R.D. and Wagner, C.F., J. Phys. Chem. Solids 29 (1968) 181.

[6] Harris, I.R. and Longworth, G., J. Less-Common Met. 23 (1971) 281.

[7] Alvarado, S.F., Campagna, M. and Gudat, W., J. Electron Spectrosc. Relat. Phen. 18 (1980) 32.

[8] Kammerer, R., Barth, J., Gerken, F., Flodström, A. and Johansson, L.I., Solid State Commun. 41 (1982) 435.

[9] Johansson, L.I., Flodström, A., Hörnström, S.-E., Johansson, B., Barth, J. and Gerken, F., Solid State Commun. 41 (1982) 427.

[10] Mårtensson, N., Reihl, B., Schneider, W.-D., Murgai, V., Gupta, L.C. and Parks, R.D., Phys. Rev. B25 (1982) 1446.

[11] Kaindl, G., Reihl, B., Eastman, D.E., Pollak, R.A., Mårtensson, N., Barbara, B., Penney, T. and Plaskett, T.S., Solid State Commun. 41 (1982) 157.

[12] Gudat, W., Campagna, M., Rosei, R., Weaver, J.H., Eberhardt, W., Hulliger, F. and Kaldis, F., J. Appl. Phys. 52 (1981) 2123.

[13] Wertheim, G.K. and Crecelius, G., Phys. Rev. Lett. 40 (1978) 813.

[14] Wertheim, G.K., Wernick, J.H. and Crecelius, G., Phys. Rev. B18 (1978) 875.

[15] Eastman, D.E., Donelon, J.J., Hien, N.C. and Himpsel F.J., Nucl. Instrum. and Methods 172 (1980) 327.

[16] Miedema, A.R., J. Less-Common Met. 46 (1976) 167.

Valence Instabilities
P. Wachter and H. Boppart (eds.)

SYSTEMATICS OF THE 4f IONIZATION ENERGIES IN THE RARE EARTH PNICTIDES

N. Mårtensson,
Linköping University,
Dept of Physics and Measurement Technology,
581 83 Linköping, Sweden

B. Reihl,
IBM Research Laboratory,
8803-Rüschlikon, Switzerland

F. Holtzberg,
IBM T.J. Watson Research Center,
Yorktown Heights, NY 10598, USA

The 4f binding energies have been measured for several trivalent rare earth antimonides: NdSb, GdSb, ErSb and LuSb. It is shown that the 4f level shift, relative to the pure metals, is essentially a constant over the lanthanide series. The shift for CeSb is extrapolated and used to identify the screened $4f^0$ photoionized final state in the spectra from γ-Ce and CeSb. In a similar way the measured 4f position in GdP is used to show that the Ce^{3+} to Ce^{4+} excitation in CeP requires an energy of about 3 eV. This seems to suggest that the high-pressure phase of CeP, and in analogy the ground state of CeN, should not be described as mixed-valent systems within the framework of the promotional model. Instead we suggest that these systems are analogous to the α-phase of Ce which by f-counting methods is shown to have essentially a $4f^1$ configuration.

1. INTRODUCTION

Photoelectron spectroscopy provides through measurements of the 4f ionization energies direct information on the energy relation between rare earth (RE) atoms in different valence states. The effect of the 4f ionization is to locally increase the valence of the RE atomic site by one. Thereby the measured photoionization energy provides a measure of the total energy difference between the system in its original state and the next higher valence state. The 4f ionization energy can be separated into two terms, one of which is purely atomic, while the other one contains all the solid state effects. The second term is approximately a constant for all RE´s and will only vary smoothly over the lanthanide series (lanthanide contraction). Knowing the atomic contributions to the RE 4f ionization energies it should therefore be sufficient to measure the 4f position for a few RE compounds of a certain kind (e.g. RE antimonides) to establish the 4f energies for the whole series of such compounds.

In the present paper we demonstrate how a systematic treatment of the 4f ionization energies in a system of RE compounds can be made. The 4f level shifts between the metals and the antimonides have been determined for Nd($4f^3$), Gd($4f^7$), Er($4f^{11}$) and Lu($4f^{14}$). We show that the shifts fall closely on an straight line as a function of Z. By extrapolation we can then determine the corresponding shift for Ce. The knowledge of this shift turns out to be a most valuable piece of information for the interpretation of the Ce photoemission spectra and it allows us to identify the $Ce^{3+} \rightarrow Ce^{4+}$ transition in γ-Ce as well as in the Ce pnictides. The interpretation of these spectra are not straight-forward and is presently a matter of great controversy. Namely, recent photoemission studies have revealed the existence of two 4f derived spectral features in Ce pnictides (1) (CeP, CeAs and CeBi) as well as in the α- and γ-phases of Ce-metal.(2) In Ce metal one finds for photon energies at which the 4f emission is enhanced one spectral feature within a few tenths of an eV from the Fermi level and another peak 2 eV below E_F. In CeP, CeAs and CeBi the corresponding peaks are located at binding energies 0.6 eV and 3.0-3.1 eV.(1)

Of fundamental importance for the understanding of the valence state in Ce systems is the proper identification of the $4f^1\,[ds^2] \rightarrow 4f^0\,[d^2s^2]$ process in the spectra. In several investigations of γ-Ce, the peak at 2 eV has been identified with the fully (5d) screened $4f^0$ final state. (2-4) The same conclusion is arrived at when this excitation energy is calculated from atomic data, under consideration of the influence of the metallic environment. (5) This energy position has also been derived from thermodynamical data. For instance it has been demonstrated that the existence of Ce_2O_3 and the nonexistence of $CeCl_4$ is only consistent with $Ce^{3+} \rightarrow Ce^{4+}$ transition energies of larger than 0.4 eV and 0.8 eV, respectively. (6) This identification of the 4f level position has, however, been strongly questioned. Based on XPS (7) and BIS (8) measurements it has been claimed that the 4f level in Ce is located much closer to the Fermi level. Recently Liu et al. (9) have made a theoretical treatment of the screening problem in Ce metal. They claim that the peak at the Fermi level is due to a 5d screened $4f^o$ final state. The 2 eV feature is instead associated with an unscreened $4f^o$ final state. Wieliczka et al. (10) have also interpreted their photoemission results within the framework of this model.

Based on the results for the trivalent RE antimonides we find that it is the 2 eV feature in γ-Ce that has to be identified with the completely 5d-screened $4f^o$ final state. We also show that one by similar considerations can use the measured 4f position in GdP to determine the energy of tetravalent CeP. We find that the Ce^{3+} state lies about 3 eV above the trivalent configuration. The implication of this energy value on the description of the isostructural phase transition in CeP at 100 kBar is discussed.

We also discuss what the results for the compressed phase of CeP suggest for the description of the ground state of CeN. This compound is generally considered to be mixed-valent with a 4f-count of about 0.6/Ce-atom.

2. EXPERIMENTAL

We have used synchrotron radiation to study the photoemission spectra from single crystals of several RE antimonides: NdSb, GdSb, ErSb and LuSb. The spectra were recorded from in situ cleaved (100) surfaces using a two-dimensional display-type spectrometer (11) which was operated in an angle-integrated mode. The combined monochromator and spectrometer resolution was about 0.15 eV at a photon energy of 70 eV.

3. RESULTS AND DISCUSSION

Fig. 1 shows the valence region of the photoemission spectra from NdSb, GdSb, ErSb and LuSb, in which the RE atoms all have a trivalent configuration. The spectra in Fig. 1 are dominated by the rare earth 4f contribution. The 4f emission features are all very sharp (uncorrected FWHM 0.50 eV in LuSb) and no sign of a surface shift is observed. Considering the widths of the observed levels this implies that any surface shift must be smaller than ∿ 0.1 eV.
The multiplet patterns are identical to the ones in the spectra from the pure metals, and the characteristic energy shifts between the antimonides and the elemental metals can be determined with high accuracy.

Chemical shifts are caused by rearangements in the initial and photoionized final state valence electron distributions. Chemically all the lanthanides in the same valence state are equivalent except for a smooth variation over the series (lanthanide contraction). We therefore expect a

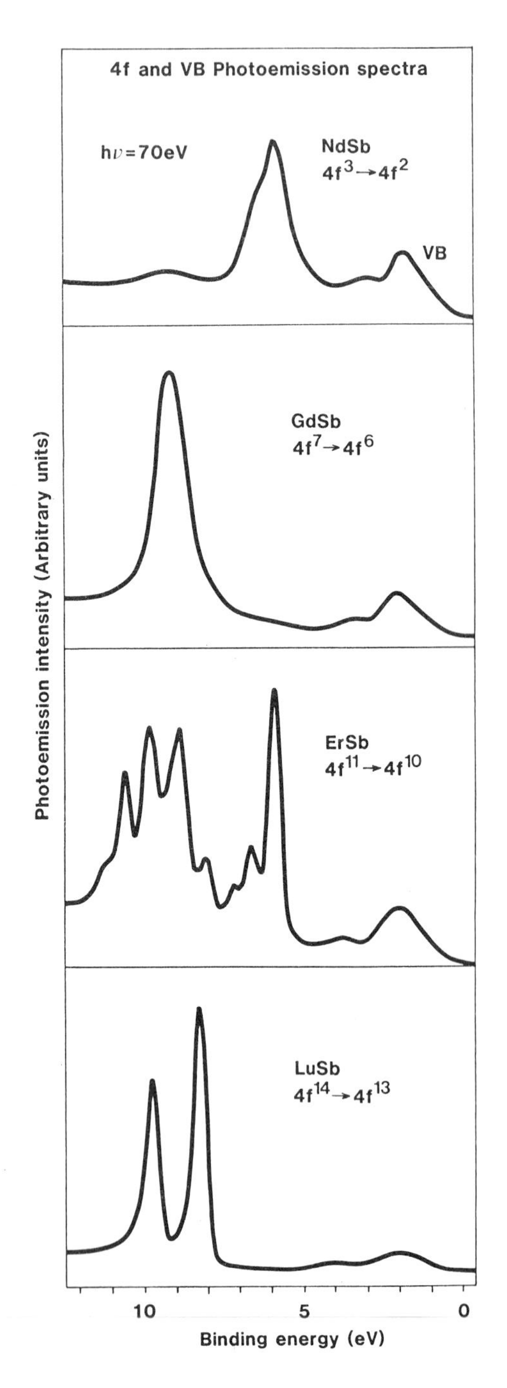

Figure 1: Photoemission spectra of some trivalent rare earth antimonides

corresponding regular behavior with atomic number Z of the chemical shifts in the rare earth systems. The presently determined energy shifts are summarized in Fig. 2. It is clear from this figure that they fall very closely on a straight line, with essentially a constant shift throughout the lanthanide series. Additional shift values can be obtained from previous XPS measurements on rare earth antimonides.(12) Analysing these spectra we find shifts in the energy range 0.9-1.2 eV for another six trivalent antimonides, i.e. close to the values in Fig. 2.

The regular Z-dependence of the 4f binding energy shift in the rare earth antimonides is clearly demonstrated in Fig. 2. If we extrapolate the values in this figure to Ce ($4f^1$) we find a shift of 1.0 eV between Ce and CeSb. In Fig. 3 we investigate how this extrapolated shift compares with the observed Ce 4f shifts. Fig. 3. shows the photoemission spectrum of γ-$Ce_{.9}Th_{.1}$, recorded at hν=50 eV.(2) (The 10% Th is added to stabilize the FCC phase and has no significant influence on the Ce 4f spectrum). The two 4f related features are clearly seen at this photon energy. The pnictide spectra (CeP, CeAs and CeBi) show the same type of two peak structure. (1)
The spectrum from CeSb was not measured but since CeAs and CeBi (As is located above and Bi below Sb in the periodic table) gave identical 4f positions of 0.6 eV and 3.0-3.1 eV, the same positions will certainly be found in CeSb. In Fig. 3 we have marked these energies by vertical lines. In this figure we also show how these energy positions should be modified by the previously extrapolated shift of 1.0 eV. From Fig. 3 it is immediately clear that only the deep features in Ce and the Ce pnictides are related by this energy shift. The 2 eV peak in γ-$Ce_{.9}Th_{.1}$ shows the same characteristic displacement when the antimonide is formed as was consistently found for the 4f levels ($f^n[ds^2] \rightarrow f^{n-1}[d^2s^2]$ transition) in the other RE elements. Based on these findings we conclude that the 2 eV feature in γ-Ce (as well as the 3 eV feature in the pnictide) corresponds to a process, which in analogy with the other RE´s produces a localized d-screened $4f^0$ final state configuration ($Ce^{3+} \rightarrow Ce^{4+}$ transition).

If we, like Liu et al., (9) want to describe the two 4f-peaks in γ-Ce as due to different screening processes, we first of all have to associate the deep feature at 2 eV with what they considered to the well-screened process (5d screening). To explain the feature at E_F it must correspond to a final state which is more effectively screened (in terms of screening energy). This would require a 4f-like screening charge.

Above we have shown that we by extrapolation in the RE series can locate the 4f level in Ce-systems. We will now use this approach to determine the energy of the tetravalent state of CeP. At 100 kBar CeP undergoes an isostructural phase transition, with a large volume contraction of ~8%. This transition has been interpreted as due to a 4f promotion.(13) This interpretation requires a relatively low value for the $Ce^{3+} \rightarrow Ce^{4+}$ transition energy. Previous XPS results for CeP identified a 4f peak at 0.4 eV and gave strong support for the promotional model.(14) Recently, however, it has been shown that there in CeP is another 4f derived spectral feature at ~3 eV below E_F. Before concluding anything

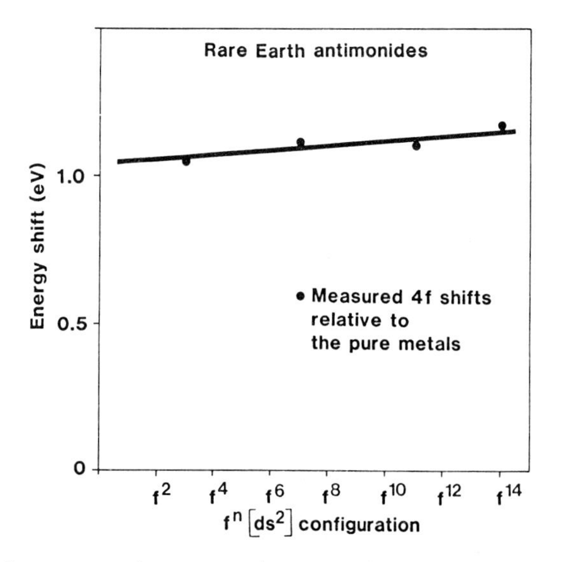

Figure 2: Shift of the 4f level in the rare earth antimonides. The 4f binding energies for the pure metals are taken from ref. 7

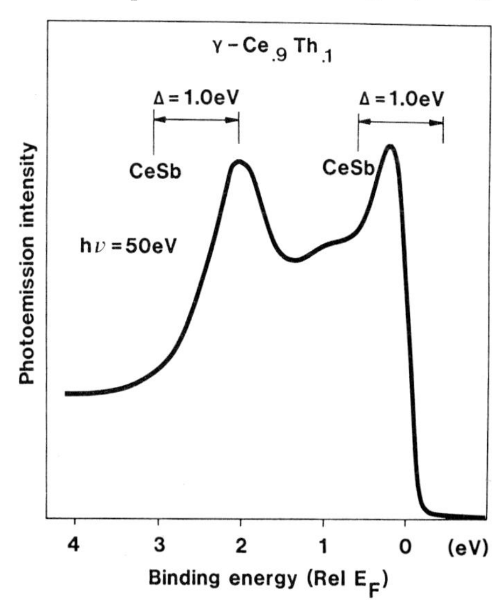

Figure 3: Spectrum of γ-$Ce_{.9}Th_{.1}$ recorded at hν = 50 eV.(2) The energies of the 4f features in CeSb (see text) are marked at 3.05 and 0.6 eV. The arrows show how these spectral features in CeSb would be displaced according to the extrapolated Ce4f shift from Figure 2.

about the valence transition in CeP we have to determine what spectral feature corresponds to the $4f^1[ds^2] \rightarrow 4f^0[d^2s^2]$ transition. Again we utilize the extrapolation method outlined above. Using XPS, Baer et al.(14) identified the 4f peak in GdP at 8.6 eV. The 4f position in GdP is thus significantly shifted towards higher binding energies from its value in Gd metal (peak position 8.0 eV). Above we showed that the $Ce^{3+} \rightarrow Ce^{4+}$ process in γ-Ce requires an energy of about 2 eV. If we make the most reasonable assumption that the shifts in the phosphides show the same Z-dependence as the antimonides in Fig. 3 we have to identify the 3 eV feature in the CeP spectrum with the $Ce^{3+} \rightarrow Ce^{4+}$ transition.

This casts doubts on the interpretation that the pressure induced phase transition in CeP corresponds to a 4f promotion. Instead it seems to suggest that the transition is related to the γ-α transition in Ce-metal, for which no change in the 4f-count is observed.(2,15) This revised view would then apply also to CeN which has traditionally been considered to be mixed-valent with a 4f-count of about 0.6/Ce-atom. Also the CeN photoemission spectrum shows a two peak structure but the deep feature is now located about 1.2 eV below E_F.(16) This two peak structure has been interpreted as due to a surface shift of the 4f emission. Based on the above discussion we instead suggest that the CeN spectrum is analogous to the spectrum from α-Ce, (2) with both peaks being characteristic of the bulk material.

4. CONCLUSIONS

The 4f shift between the RE antimonides and the RE metals form a smooth function over the lanthanide series. This regularity is due to the similarity of the cohesive properties for various RE elements, and will be a general property of chemical shifts in RE compounds. This recognition provides a most valuable method for determining 4f ionization energies by interpolation or extrapolation among the RE elements. The interpretation of the 4f spectrum in γ-Ce is still a matter of great controversy. There are two peaks in the Ce-spectra which show 4f-character, one close to E_F and another peak at 2 eV binding energies. In particular it is important to identify the fully screened $4f^1 \rightarrow 4f^0$ process in the spectrum. We show how the 2 eV feature can be identified with this process through its shift to CeSb. In a similar way we have determined the energy of tetravalent CeP. The 4f level in Gd is shifted to higher binding energies in GdP. With a similar shift in CeP we find that the tetravalent state lies about 3 eV above the trivalent state. This casts doubts on the generally accepted interpretation that the isostructural phase transition in CeP at 100 kBar is due to a 4f promotion. Instead we propose that the high pressure phase in CeP, as well as the ground state in CeN, is analogous to α-Ce, which cannot be described within the promotional model.

ACKNOWLEDGEMENTS

The authors gratefully acknowledge the generous support from D.E. Eastman and the assistance of J.J. Donelon, A. Marx, and the staff of the Synchrotron Radiation Center at the University of Wisconsin, as well as valuable discussions with B. Johansson. This work was supported in part by the Air Force Office of Scientific Research under contract No. F49620-81-C-0089.

REFERENCES:

(1) Franciosi, A., Weaver, J.H., Mårtensson, N. and Croft, M., Phys. Rev. B24 (1981) 3651.

(2) Mårtensson, N., Reihl, B. and Parks, R.D., Solid State Commun. 41 (1982) 573.

(3) Platau, A. and Karlsson, S.-E., Phys. Rev. B18 (1978) 3820.

(4) Johansson, L.I., Allen, J.W., Gustafsson, T., Lindau, I. and Hagström, S.B., Solid State Commun. 28 (1978) 53.

(5) Johansson, B., Phys. Rev. B20 (1979) 1315, and Phil. Mag. 30 (1974) 469.

(6) Johansson, B., J. Phys. Chem. Solids 39 (1978) 467.

(7) Lang, J.K., Baer, Y. and Cox, P.A., J. Phys. F11 (1981) 121.

(8) Baer, Y., Ott, H.R., Fuggle, J.C. and DeLong, L.E., Phys. Rev. B24 (1981) 5384.

(9) Liu, S.H., Davis, L.C. and Ho, K.M., Bull. Am. Phys. Soc., 27 (1982) 276.

(10) Wieliczka, D., Olson, C.G., Lynch, D.W. and Weaver, J.H., Bull. Am. Phys. Soc., 27 (1982) 276.

(11) Eastman, D.E., Donelon, J.J., Hien, N.C. and Himpsel, F.J., Nucl. Instr. and Meth. 172 (1980) 327.

(12) Campagna, M., Wertheim, G.K. and Bucher, E., Structure and Bonding 30 (1976) 99.

(13) Jayaraman, A., Lowe, W., Longinotti, L.D. and Bucher, E., Phys. Rev. Lett. 36 (1976). 366.

(14) Baer, Y., Hauger, R., Zürcher, Ch., Campagna, M. and Wertheim, G.K., Phys. Rev. B18 (1978) 4433.

(15) Kornstädt, U., Lässer, R. and Lengeler, B., Phys. Rev. B21 (1980) 1898.

(16) Gudat, W., Rosei, R., Weaver, J.H., Kaldis, E. and Hulliger, F., Solid State Commun. 41 (1982) 37.

SPECTROSCOPIES

Valence Instabilities
P. Wachter and H. Boppart (eds.)

POINT CONTACT SPECTROSCOPY OF SmB_6, TmSe, "gold" SmS AND $TmSe_{1-x}Te_x$: GAPS

I. Frankowski and P. Wachter

Laboratorium für Festkörperphysik, ETH Zürich
8093 Zürich, Switzerland

We measured the dynamic resistance dU/dI(U) and its derivative $d^2U/dI^2(U)$ of point contacts with intermediate valent SmB_6, "gold"SmS, TmSe and ferromagnetic $TmSe_{0.82}Te_{0.18}$ and their integer valent reference compounds LaB_6, LaS, LaSe and $Tm_{0.87}Se$ at liquid helium temperatures. While dU/dI(U) of point contacts with the integer valent compounds does not show any anomalous structures, SmB_6, "gold"SmS and TmSe (in the antiferromagnetic phase) exhibit a strong resistance peak at zero bias with a characteristic width of 5meV, 6.5meV and 2.5meV, respectively. This is taken as evidence of hybridization gaps. Point contacts with ferromagnetic $TmSe_{0.82}Te_{0.18}$ show below T_C a shape of dU/dI(U) which is similar to that of TmSe in its field induced ferromagnetic phase.

INTRODUCTION

SmB_6, "gold"SmS and TmSe belong to a group of intermediate valent compounds in which the resistivity increases with decreasing temperature suggesting that these materials are insulating at zero temperature. The possibility of a nonmetallic groundstate has been discussed in several theoretical models and has been attributed to the existence of a narrow energy gap due to hybridization (1)-(4).

We investigated SmB_6, "gold"SmS and TmSe by means of point contact spectroscopy (PCS), which has been shown to be a powerful tool for the investigation of scattering processes of electrons with energies close to ε_F; integer valent LaB_6, LaS, LaSe and $Tm_{0.87}Se$ were used as reference compounds. In addition we report on PCS measurements on metallic and ferromagnetically ordering $TmSe_{1-x}Te_x$.

PCS is based on the following phenomenon: If a narrow constriction is created between two metals (this can be realized experimentally either by electric breakdown of a metal-oxide-metal junction (5) or by pressing a sharp metal tip onto a bulk sample (6)), then the current I through the constriction is no longer ohmic when the mean free path of the electrons becomes longer than the radius a of the contact area. The nonlinearities of I as a function of the applied voltage U can be resolved by measurement of the first and second derivative dU/dI(U) and $d^2U/dI^2(U)$, respectively.

I(U) has been calculated by solving the Boltzmann equation for the distribution of the electrons by an iteration with respect to the scattering term (7),(8). Regarding the low temperature limit, the two main results of this calculation should be underlined:

first: the dynamic resistance of a point contact is directly proportional to the energy dependent scattering rate τ^{-1};

second:-in application of the electron phonon interaction, as most of the point contact experiments up to now deal within this field - for a slow variation of dU/dI(U), the second derivative $d^2U/dI^2(U)$ is directly proportional to $\alpha^2F_p(\omega)$ (requiring a constant density of states for energies $eU \ll \varepsilon_F$); α^2 is the squared matrix element of the electron phonon interaction and $F_p(\omega)$ is the phonon density of states which differs from the usual function $F(\omega)$ by an efficiency function $\eta(\theta)$ depending on the angle θ between the electron momenta before and after a scattering event (7),(8).

EXPERIMENTS

Our point contact experiments were performed at a temperature of 1.7 K or 1.8 K. The contacts were established in liquid helium pressing a sharply etched Mo wire (tip radius $\sim 1\mu$) against a cleaved single crystal of the sample with a size of about 3x3x2mm; for a detailed description of the apparatus see ref. (9).

Fig. 1 and 2 show typical PCS spectra of LaB_6 and LaSe, respectively. For both materials the dynamic resistance dU/dI(U) has the usual shape expected when the electron scattering processes are dominated by the electron phonon interaction: a flat bottom over a range of some mV followed by a weak rise at a certain voltage (positive and negative) corresponding to the electron phonon scattering. For LaB_6 the second derivative $d^2U/dI^2(U)$ shows two prominent peaks at about 35 mV and 68 mV and a weak structure

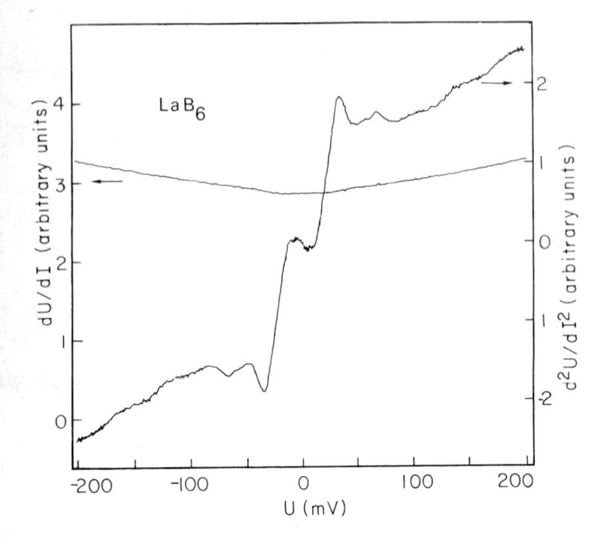

Fig. 1: Dynamic resistance dU/dI(U) and derivative $d^2U/dI^2(U)$ of a LaB_6 point contact (dU/dI(0) = 12Ω); temperature T = 1.8 K.

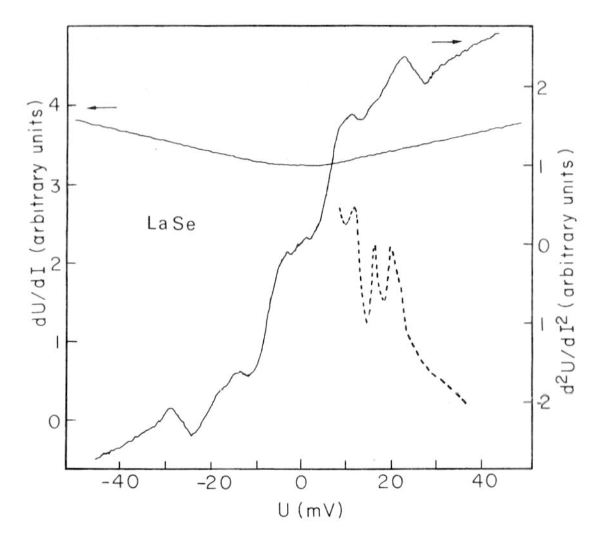

Fig. 2: Dynamic resistance dU/dI(U) and derivative $d^2U/dI^2(U)$ of a LaSe point contact (dU/dI(0) = 2 Ω); temperature T = 1.8 K. Dashed line: Raman scattering intensity (ref.(10)) at T = 300 K.

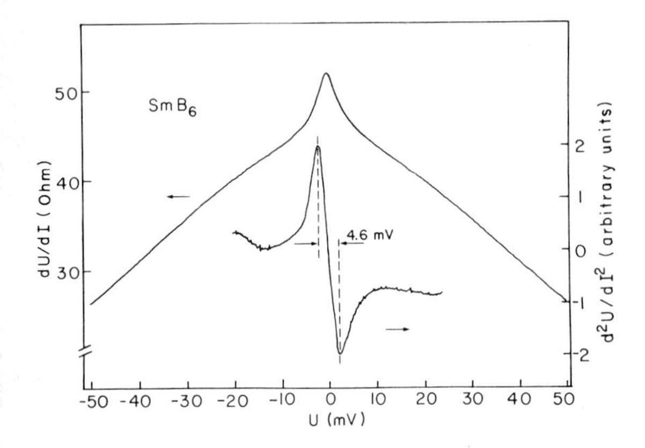

Fig. 3: Dynamic resistance dU/dI(U) and derivative $d^2U/dI^2(U)$ of a SmB_6-point contact; temperature T = 1.8 K.

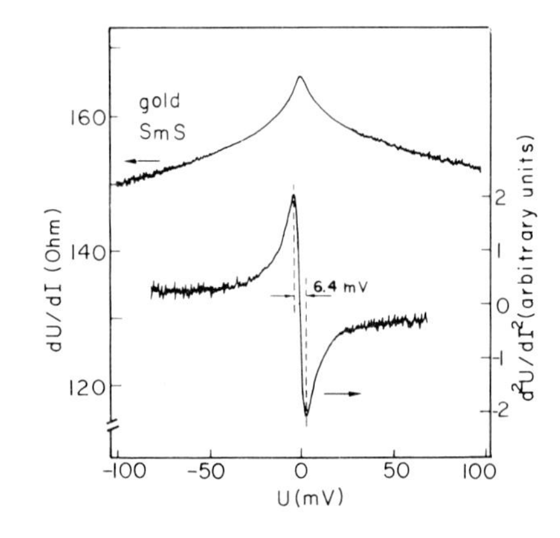

Fig. 4: Dynamic resistance dU/dI(U) and derivative $d^2U/dI^2(U)$ of a SmS point contact (with SmS polished), temperature T = 4.2 K.

around 140 mV. In the case of LaSe a shoulder at 12 mV and a peak at 23 meV is observed in $d^2U/dI^2(U)$. Similar results on LaS were already reported in ref. (9). There we could verify that PCS really measures α^2F_p by comparing it with Raman scattering measurements on the same crystals. In these fcc compounds the first order Raman effect is forbidden by symmetry, but one observes a defect induced second order spectrum which measures α^2F. A good agreement between PCS- and Raman spectra has been obtained. For LaSe (also fcc structure) we found relatively good agreement between the structures in $d^2U/dI^2(U)$ and those of the Raman scattering intensity (dashed line in Fig.2) (10). In LaB_6 (CaB_6-structure), however, first order Raman effect is allowed, thus Raman scattering measures Γ phonons, their energies corresponding only roughly with PCS on LaB_6 (11).

In Fig. 3,4 and 5 representative dU/dI(U) curves of point contacts with the intermediate valent compounds SmB_6, "gold" SmS and TmSe are plotted. Their common- and for PCS up to now totally new- feature is a strong decrease of dU/dI(U) with increasing voltage. For all three compounds this basic shape is superimposed by a peak at zero voltage, less pronounced for "gold"SmS than for

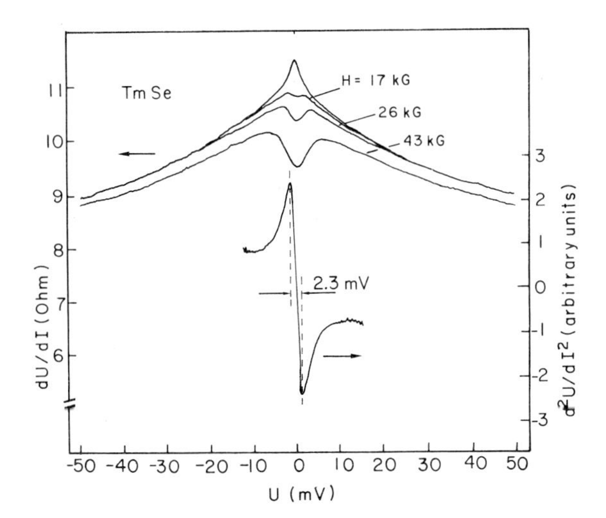

Fig. 5: Magnetic field dependence of the dynamic resistance dU/dI(U) of a TmSe point contact (Tm to Se ratio: 0.98) and derivative $d^2U/dI^2(U)$ (H=0); temperature T - 1.7 K.

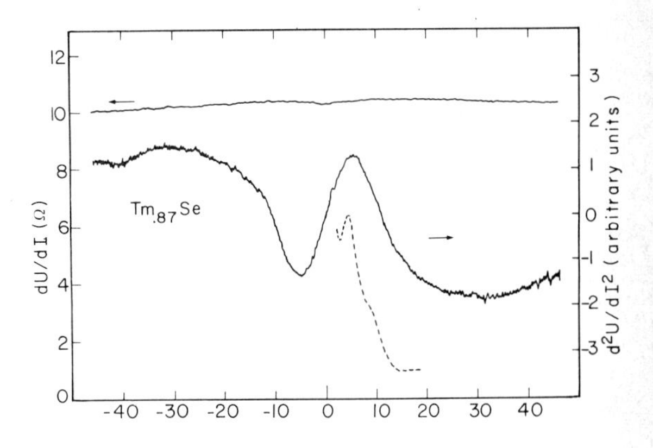

Fig. 6: Dynamic resistance dU/dI(U) and derivative $d^2U/dI^2(U)$ of a $Tm_{0.87}Se$ point contact; temperature T = 1.8 K. Dashed line: neutron scattering intensity at T = 1.5 K (ref. (13)).

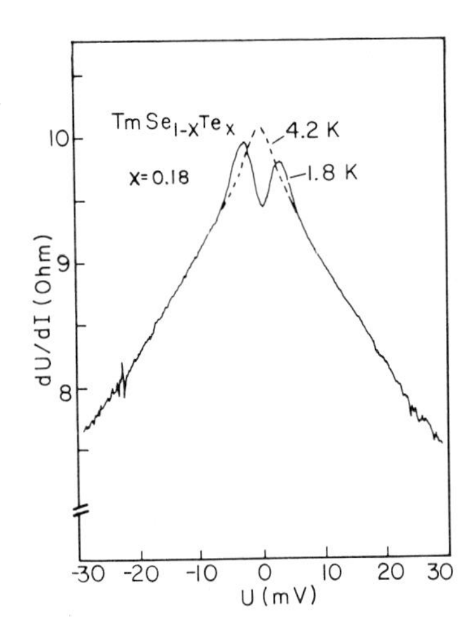

Fig. 7: Dynamic resistance dU/dI(U) of a $TmSe_{0.82}Te_{0.18}$-point contact

SmB_6 and TmSe. By measurement of the second derivative $d^2U/dI^2(U)$, we found characteristic values for the peak width of ∿ 5meV, ∿ 6.5meV and ∿ 2.5meV for SmB_6, "gold"SmS and TmSe, respectively. $d^2U/dI^2(U)$ did not reveal any structures that could be due to electron phonon interaction. Magnetic fields up to H = 50 kG and parallel to the contact area did not influence the SmB_6-and SmS- dU/dI(U) peak but changed it for TmSe (which orders antiferromagnetically below T_N ∿ 3K and becomes ferromagnetic in fields H > 5kG (12)) in a way shown in Fig. 5: the central peak disappears with increasing field and turnes into a relative minimum in fields higher than ∿ 15 kG.

In TmSe the degree of valence mixing can be changed from 3^+ to 2.7^+ by varying the ratio of Tm to Se ions from 0.87 to 1.05 (12). Thus, trivalent $Tm_{0.87}Se$ served us as a further reference compound. The typical shape of dU/dI(U) (Fig. 6) is characterized by a weak increase of dU/dI(U) with increasing voltage followed by a wide and very weak maximum. The second derivative $d^2U/dI^2(U)$ has a peak at about 5meV. Quite recently inelastic neutron scattering has been performed on single crystals of $Tm_{0.87}Se$ (13) (dashed line in Fig. 6); the dominant feature there is a peak at about 4.5meV and a weak shoulder at about 8meV which were explained within a crystal field scheme. Thus, we ascribe the peak at 5meV in PCS to the same excitation, which would mean the first experimental observation of a crystal field excitation in PCS. Further PCS measurements on materials where crystal field excitations are well known appear necessary.

Fig. 7 shows the dynamic resistance of a point contact with metallic $TmSe_{0.82}Te_{0.18}$ which orders ferromagnetically below T_C ∿ 3K (14), (15). The similarity of dU/dI(U) around zero voltage (i.e. the relative minimum) with that of TmSe in the ferromagnetic phase is obvious. Above T_C the minimum turnes into a peak (dashed line).

DISCUSSION

Obviously PCS spectra of SmB_6, "gold"SmS and TmSe are totally different from those taken at metallic integer valent compounds as well as from the spectra which have been observed in point contacts with intermediate valent materials in which the resistivity behaves metallic (e.g. $YbCu_2Si_2$, $CePd_3$,...) (16). But again they follow the general rule that the dynamic resistance dU/dI as a function of U reflects - at least qualitatively - the temperature dependence of the resistivity (12),(17),(18).

As already proposed in ref. (19), we interprete the well pronounced dU/dI peak at zero voltage in SmB_6 as evidence for a small hybridization gap, which has already been inferred from resistivity and Hall effect (17), optical absorption and tunneling (20). The width of the gap which we derived from the half width of the dU/dI(U) peak (distance of the turning points) is in agreement with the results reported there (20).

Within the framework of the point contact model it is generally assumed that N(E) is constant near ε_F. However, it seems reasonable to assume that a strong change in N(E) will heavily influence dU/dI(U). Especially a gap in N(E) (or a deep minimum (pseudo-gap)) will lead to a peak in dU/dI at zero bias. We verified this in PCS on superconductors like for example Pb. Thus, structures in dU/dI(U) which are due to a non-constant N(E) at the Fermi energy are zeroth order effects. This would explain why we could not observe any structures due to electron phonon interaction in these compounds, which is only possible in higher order.

"Gold"SmS and SmB_6 both do not order magnetically, but they have different crystallographic structures namely fcc and bcc, respectively. Theory predicts (2),(3) that only SmB_6 should have a hybridization gap and also TmSe in its antiferromagnetic phase. However, we believe that the observed peak structure in dU/dI(U) of "gold"SmS point contacts is also a sign for a hybridization gap as in SmB_6; the smearing of the structure is probably due to the large number of lattice defects induced by the mechanical polishing of the sample surface.

The observed magnetic field dependence of the resistance peak in TmSe point contacts is in accordance with the drastic decrease of the bulk resistivity in magnetic fields (12) and we propose a vanishing of the gap in the ferromagnetic phase which is in agreement with (2),(4). We think that the gap we associate with the peak in dU/dI is due to hybridization rather than to antiferromagnetism as ε_F for TmSe is predicted to fall into the hybridization gap, but not into the antiferromagnetic gap (4). The results on $TmSe_{0.82}Se_{0.18}$, which below T_C is in a similar magnetic phase as TmSe above its metamagnetic transition (14), support very much the statement of the nonexistence of a gap in the ferromagnetic phase and rather indicate that a gap exists in the paramagnetic phase.

The value of 2.5 meV for a gap in TmSe is at variance with the results of neutron scattering experiments where a 10meV excitation was interpreted as an excitation over the hybridization gap (21); it is in reasonable agreement with the results of tunneling experiments reported in ref. (22). For SmB_6 the value of 5meV is nearly the same as that reported in ref. (20).

In the case of "gold"SmS, our value of 6.5 meV has to be taken with caution because of the uncertainty introduced by polishing of the surface.

ACKNOWLEDGEMENT

We are most grateful to Dr. E. Bas, Dr. F. Hulliger and Dr. E. Kaldis for growing the single crystals.

1) Mott, N.F., Phil. Mag. 30 (1973) 403
2) Martin, R.M. and Allen, J.W., J. Appl. Phys. 50, (1979) 11
3) Brandow, B.H., Solid State Commun. 39 (1981) 1233
4) Alascio, B., Aligia, A.A., Mazzaferro, J. and Balseiro, C.A., this conference
5) Yanson, I.K. and Kulik, I.O., J. Phys. C: Solid State Phys. 6 (1978) 1564
6) Jansen, A.G.M., van Gelder, A.P. and Wyder, P., J. Phys. C: Solid State Phys. 13, (1980) 6073
7) Kulik, I.O., Omel'Yanchuk, A.N. and Shekter, R.I., Sov. J. Low Temp. Phys. 3 (1977) 740
8) van Gelder, A.P., Solid State Commun. 25 (1978) 1097
9) Frankowski, I. and Wachter, P., Solid State Commun. 40 (1981) 885
10) Treindl, A., Dissertation ETH 6873, (1981) Library ETH Zürich, unpublished
11) Mörke, I., Dvorak, V. and Wachter, P., Solid State Commun. 40, (1981) 331
12) Batlogg, B., Ott, H.R., Kaldis, E., Thöni, W. and Wachter, P., Phys. Rev. B19, (1979) 247
13) Furrer, A., Bührer, W. and Wachter, P., Solid State Commun. 40 (1981) 1011
14) Batlogg, B., Ott, H.R. and Wachter, P., Phys. Rev. Lett. 42, (1979) 278
15) Fischer, P., Hälg, W., Schobinger-Papamentellos, P., Boppart, H., Kaldis, E. and Wachter, P., this conference
16) Bussian, B., Frankowski, I. and Wohlleben, D.K., Phys. Rev. Lett. (1982), in press
17) Allen, J.W., Batlogg, B. and Wachter, P., Phys. Rev. B20, (1979) 4812
18) Haen, P., Lapierre, F., Mignot, J.M., Tournier, R. and Holtzberg, F., Phys. Rev. Lett. 43 (1979) 304
19) Frankowski, I. and Wachter, P., Solid State Commun. 41 (1982) 577
20) Batlogg, B., Schmitt, P. and Rowell, J.M., in"Valence Fluctuations in Solids", ed. by Falicov, L.M., Hanke, W. and Maple, M.B., p. 267 (1981)
21) Grier, B.H. and Shapiro, S.M.,(ref.20)p.325
22) Güntherodt, G., Thompson. W.A., Holtzberg, F., Fisk, Z., this conference

Valence Instabilities
P. Wachter and H. Boppart (eds.)

ELECTRON TUNNELING SPECTROSCOPY OF INTERMEDIATE VALENCE MATERIALS

G. Güntherodt[+], W.A. Thompson, F. Holtzberg and Z. Fisk[++]

IBM T.J. Watson Research Center Yorktown Heights, N.Y. 10598

[+]II. Physikalisches Institut, Universität Köln, 5000 Köln 41, FRG

[++]Los Alamos Scientific Laboratory, Los Alamos, N.M. 87545

Electron tunneling spectra of TmSe, SmS, SmB_6 and $CePd_3$ have been measured using the GaAs Schottky barrier probe tunneling method. TmSe shows an energy gap 2Δ (FWHM)=1.2meV only in the antiferromagnetic phase. In situ pressure-transformed metallic SmS exhibits a gap 2Δ=1.7 meV and SmB_6 shows a gap 2Δ=2.7 meV, which is independent of magnetic field. For $CePd_3$ an inelastic excitation is found near $\pm$ 14 meV, which is absent in YPd_3. An interpretation of the tunneling spectra in terms of quasiparticle excitation energies of bound electron-hole pairs is presented.

INTRODUCTION

The interaction of localized 4f states with conduction electrons near the Fermi energy E_F in intermediate valence (IV) materials raises fundamental questions concerning the nature of the low temperature coherent ground state and its low energy excitations. A central issue in recent research concerns the possibility of an insulating singlet ground state in certain IV materials as suggested by their low temperature transport anomalies and proposed by various theoretical models /1/. Although electron tunneling spectroscopy appears to be extremely well suited for probing the low energy excitations in the immediate vicinity of E_F, measurements have only been performed on SmB_6 /2/.

In certain IV materials, such as SmB_6, SmS or TmSe, the low temperature resistivity increase, partly affected by sample purity and stoichiometry, has been subject to various conjectures in terms of gaps due to hybridization or antiferromagnetism, impurity or Kondo scattering, Anderson localization or Wigner lattice formation /1,3,4/. Deviations at the lowest temperatures from an activated resistivity behaviour have been attributed to extrinsic effects /5,6,7/, partly in view of theoretical predictions of gaps in the low energy excitation spectrum of IV states using the Kondo lattice /1,8/ or Anderson lattice /1,9,10/ models. In particular, the formation of a hybridization gap /11/ has been proposed for SmB_6 /12/, but has been questioned later on /13/. On the other hand, the above three materials can be contrasted with the majority of IV materials which show a decrease of the resistivity at low temperatures, like for instance $CePd_3$ /14/.

EXPERIMENTAL RESULTS

To investigate the low energy excited states of the IV materials SmB_6, SmS, TmSe and $CePd_3$, we have applied the GaAs Schottky barrier probe tunneling method /15/, which overcomes inherent difficulties in fabricating oxide barrier tunnel junctions. We have measured the differential resistance R'= dV/dI vs. the applied voltage V up to $\pm$ 100 meV at temperatures between 1.5<T<4.2 K and in magnetic fields up to 20 kOe. Reproducible, symmetric tunneling spectra were obtained only after carefully removing surface contaminants by an in situ (within liquid helium) sputter cleaning method /15/. The preparation of single crystals of SmS, YSe and TmSe /16/, and of SmB_6 /17/ has been described elsewhere. The annealed polycrystalline $CePd_3$ (ρ/ρ_0 = 5) and YPd_3 samples have been prepared by E. Cattaneo, Universität Köln.

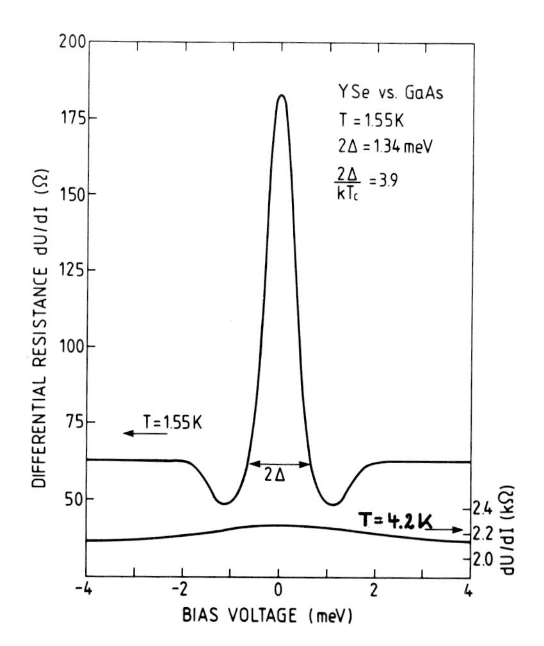

Fig. 1 Tunneling spectrum of a cleaved (100) face of YSe above (4.2 K) and below (1.55 K) T_c = 4.0 K with a BCS-defined gap 2Δ.

The proper handling of the tunneling and sputtering techniques has been tested using appropriate reference materials. In Fig. 1 we show R' vs. V of a cleaved (100) face of superconducting YSe (T_c = 4.0 K /18/). The tunneling spectrum at 1.55 K shows a BCS-defined gap at $R' = R'_{max}(V=0) - 0.9\cdot(R'_{max}(V=0) - R'_{min}(V\neq 0))$ of 2Δ = 1.34 meV, with $2\Delta/kT_c$ = 3.9 exceeding the BCS weak coupling limit 3.52. This result as well as measurements on a variety of other superconducting materials /15,19/ give good confidence in the reliability of this versatile and powerful technique.

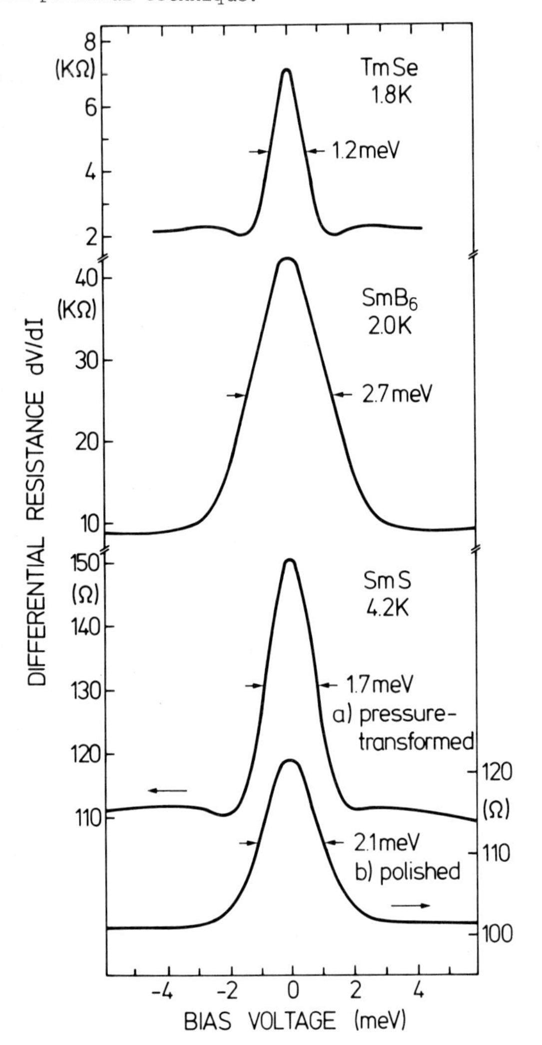

Fig. 2 Tunneling spectra of a cleaved (100) face of TmSe at 1.8 K, single crystalline SmB_6 at 2.0 K and a pressure-transformed (a) and mechanically polished (b) cleaved (100) face of SmS at 4.2 K. The gap 2Δ (FWHM) is indicated.

In Fig. 2 we have assembled the tunneling spectra of all "gap-type" IV materials. A more detailed account of this work, including temperature and magnetic field dependence of the spectra will be given elsewhere /20/. A cleaved (100) face of nominally stoichiometric TmSe (a_0 = 5.705 Å) shows just below the Néel temperature T_N = 3.5 K /16/ a tunneling spectrum (see Fig. 2) similar to YSe, but with R'(V=0, H=0) = 7 kΩ. At 1.8 K we deduce an energy gap 2Δ = 1.2 meV at FWHM, i.e. by replacing 0.9 in the above definition by 0.5. (To stress the quantitative difference between TmSe and superconducting YSe we quote also a BCS-defined gap for TmSe 2Δ = 2.0 meV, which is quite different from that of YSe). The resistance minimum of TmSe near ± 1.4 meV disappears for H = 6.6 kOe and the tunneling spectrum becomes field independent above $H_\perp$ = 13.1 kOe in the same way as the dc resistivity /21/. We conclude that in the ferromagnetic phase of TmSe ($H_\perp \geq$ 13 kOe /21/) the energy gap has disappeared /20/.

From the tunneling spectrum of the SmB_6 single crystal at 2.0 K in Fig. 2 we deduce an energy gap 2Δ(FWHM) = 2.7 meV, which is within ± 0.3 meV practically independent of magnetic field up to 15.3 kOe /20/. (2Δ = 4.9 meV is quoted for comparison). We find that R'(V=0) is strongly reduced by excessively pressurizing the GaAs-SmB_6 contact.

The effective pressure in a GaAs probe point tunneling experiment can actually reach the yield pressure of the sample surface /15/. This unique opportunity has been exploited for tunneling into the pressure-transformed metallic IV-phase of a cleaved (100) face of SmS. Gradually pressurizing the SmS surface by the GaAs tip exhibits abruptly a tunneling spectrum with a well pronounced gap 2Δ(FWHM) = 1.7 meV as shown in Fig. 2(a), with R'(V=0) = 150 Ω. The size of this gap remains unchanged upon further pressurizing SmS, with a change in R' up to only 5%. (2Δ = 2.9 meV is quoted for comparison). A very similar tunneling spectrum as for pressure-transformed SmS is obtained for mechanically polished, metallic ("gold") SmS with 2Δ(FWHM) = 2.1 meV as shown in Fig. 2(b).

The tunneling spectra of both pressure-transformed and polished, metallic SmS show a magnetic field dependence /20/, which saturates for fields above 8.8 kOe for which the above gap definition becomes meaningless. This behaviour is unexpected in view of the field insensitive resistivity up to 18 kG /7/ and because of the field independent gap of SmB_6 /20/. However, since the pressure-transformed IV phase of SmS only occurs in the contact region, magneto-resistance effects in the remaining untransformed divalent SmS /22/ may be responsible for the observed magnetic field dependence.

Contrary to the tunneling spectra of TmSe, SmS and SmB_6, the Schottky barrier probe tunneling into the intermetallic IV material $CePd_3$ exhibits no such pronounced gap-like structure. Instead with increasing |V| we observe a dip near $\Delta' = \pm$ 14 meV in the monotonically decreasing R' which is absent in the reference YPd_3 as shown in Fig. 2 of Ref. 23 and schematically

(R' in arb. units) in Fig. 4. The dip near 14 meV can be correlated with a resonant scattering process of conduction electrons at empty $4f^1$ states 14 meV above E_F as identified in an analysis of far infrared data /23,24/ including those of Ref. 25. The tunneling spectrum of $CePd_3$ is independent of magnetic field up to 15 kOe.

DISCUSSION

Because of the 1.2 meV gap of TmSe for $T<T_N$ = 3.5 K and H=0 we disagree with the interpretation /10,26/ that the 10 eV excitation /26,27/ found in neutron scattering below 50 K results from the excitation of f-electrons across the hybridization gap. No evidence for such a large gap excitation has been found in our experiments.

For SmB_6 a Wigner crystal localized state with a large gap of about 160 K (14 meV) has been invoked /28/, which, however, is not supported by our tunneling results.

Asymmetric tunneling spectra have been obtained using SmB_6-oxide-Pb junctions /2/. A minimum of conducting states 5-8 meV around E_F in SmB_6 has been deduced. A rough estimate of 2Δ(FWHM)$\simeq$ 10 meV (based on our definition) is four times larger than our value. In our experiments when making contact to the SmB_6 surface without sputter cleaning we observe very similar spectra. The asymmetry in our case is attributed to leakage resistance, which is tunneling assisted by intermediate states due to surface contamination. It is possible that the oxide barrier of the SmB_6-oxide-Pb junctions /2/ may not have been continuous throughout the junction.

The characteristic features of our tunneling spectra seem to correlate qualtitatively with the low temperature resistivity behavior. However, the results raise fundamental questions concerning the interpretation of the tunneling spectra and the nature of the gaps in IV materials. Since the electronic density of states of IV materials is generally believed to be asymmetric around E_F, it is somewhat surprising that we observe symmetric tunneling spectra in all materials. Instead of using the one-electron density of states picture, which in any case may not be adequate for tunneling between normal metals /29/, we propose a description of our spectra in terms of a particle representation.

In Fig. 3 we show a schematic representation of the tunneling contact between the IV material under investigation and the p-type GaAs, whose Schottky tunneling barrier is not indicated. In the IV material the E(k) relation of electrons with dE/dk>0 is schematically represented for $E>E_F$, whereas for $E<E_F$ it has been inverted (dashed line) in order to represent the hole states with dE/dk<0. In our tunneling spectra

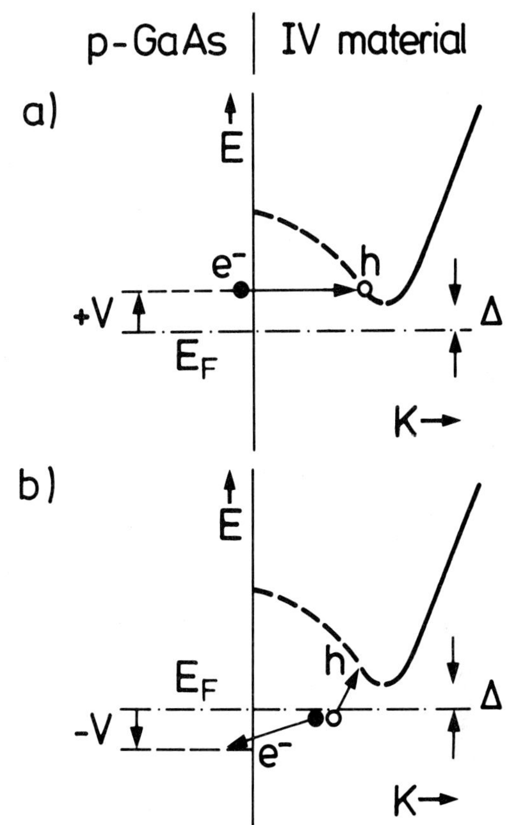

Fig. 3 Schematic particle representation of the tunneling mechanism of electron (e)-hole (h) pairs between p-GaAs and an IV material (see text): a) e-h pair condensation, b) e-h pair breaking.

we consider Δ' or Δ (which corresponds more closely to Δ' than 2Δ(FWHM)/2) as quasiparticle excitation energies of bound electron (e-)-hole ($4f^{n-1}$) pairs of the $4f^n$ configuration. The average excitation energy is indicated by Δ in Fig. 3. For positive bias voltage +V exceeding the quasiparticle excitation energy Δ (Fig. 3a)) electrons can tunnel from p-GaAs into the IV material, resulting in an electron (e^-)-hole ($4f^{n-1}$) pair condensation, with the average electron-hole pair binding energy 2Δ being given up by the system. On the other hand, for negative bias voltage -V exceeding Δ, bound electron-hole pairs can be broken up, since the electron-hole pair binding energy 2Δ can be overcome. This electron-hole pair tunneling mechanism is analogous to that of Cooper pairs in superconductors and should account for the characteristic features of the heavily renormalized Fermi liquid in IV materials.

A unifying description of the tunneling spectra of the IV materials with increasing (SmB_6, SmS) or decreasing ($CePd_3$) low temperature resistivity evolves by assuming that $1/2\Delta$ or $1/2\Delta'$ is a measure of the electron-hole lifetime τ_{eh}. This is suggested by the schematic representa-

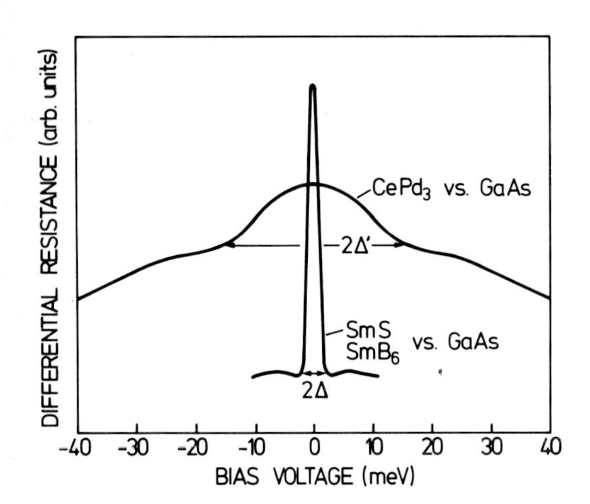

Fig. 4 Schematic representation (R' in arb. units) of the tunneling spectra of IV materials with increasing (SmB_6, SmS) or decreasing ($CePd_3$) low temperature resistivity.

tion of the two types of tunneling spectra in Fig. 4. The "broadened" tunneling spectrum of $CePd_3$ compared to the pronounced one of SmS or SmB_6 is attributed to the shorter $\tau_{eh}=1/2\Delta'$ in $CePd_3$ compared to $1/2\Delta$. On the other hand, τ_{eh} of IV materials is also determined by the charge relaxation rate $\tau_{eh}\sim 1/\Gamma_c$. Indeed, the roughly five times larger $2\Delta'$ of $CePd_3$ compared to 2Δ of IV SmS or SmB_6 is consistent with the approximately five times larger Γ_c of $CePd_3$ in comparison to IV SmS or SmB_6 /30/.

In the case of SmB_6 our interpretation is supported by recent far infrared transmission measurements of a gap of similar magnitude /31/.

The comparison of tunneling into $CePd_3$ with Ga-$CePd_3$ (metal-metal) point contacts, which can also be formed by our method, as well as with point contact spectroscopy of IV materials /32, 33/ in general will be discussed elsewhere /20/. In metal-metal point contact spectroscopy between Mo and SmB_6 (or TmSe) the resistance peak at zero voltage has been interpreted by analogy to tunneling experiments of SmB_6 (Ref. 2) as signature of a narrow gap /32/.

Acknowledgements

We thank S. von Molnar, T. Penney, D. Wohlleben and B. Hillebrands for stimulating discussions, and S.F. Hanrahan and J.M. Rigotty for technical help.

REFERENCES

/1/ Valence Fluctuations in Solids, ed. by L.M. Falicov, W. Hanke and M.B. Maple, North-Holland, Amsterdam, 1981

/2/ B. Batlogg, P.H. Schmidt and J.M. Rowell, in Ref. 1, p. 267.

/3/ Valence Instabilities and Related Narrow Band Phenomena, ed. by R.D. Parks, Plenum Press, New York 1977

/4/ For review see: J.W. Allen and R.M. Martin, J. Physique Coll. C5, 171 (1980)

/5/ M. Ribault, J. Flouquet, P. Haen, F. Lapierre, J.M. Mignot and F. Holtzberg, Phys. Rev. Lett 45 1295 (1980)

/6/ A. Benoit, J.X. Boucherle, J. Flouquet, F. Holtzberg, J. Schweizer and C. Vettier, in Ref. 1, p. 197.

/7/ M. Konczykowski, J. Morillo and J.-P. Senateur, in Ref. 1, p. 287

/8/ R. Jullien, P. Pfeuty, A.K. Bhattacharjee, B. Coqblin, J. de Physique 41, Colloque C5, 331 (1980); R. Jullien, P. Pfeuty, B. Coqblin, in Ref. 1, p. 169

/9/ R.M. Martin and J.W. Allen, J. Appl. Phys. 50, 7561 (1979); R.M. Martin and J.W. Allen, in Ref. 1, p.85

/10/ A.J. Fedro and S.K. Sinha, in Ref. 1, p.329

/11/ B. Coqblin and A. Blandin, Adv. Phys. 17, 281 (1968)

/12/ N.F. Mott, Phil. Mag. 30, 403 (1974)

/13/ P.W. Anderson, in Ref. 1, p. 451

/14/ H. Schneider and D. Wohlleben, Z. Phys. B 44, 193 (1981)

/15/ S. von Molnar, W.A. Thompson and A.S. Edelstein, Appl. Phys. Lett. 11, 163 (1967); W.A. Thompson and S. Von Molnar, J. Appl. Phys. 41, 5218 (1970)

/16/ F. Holtzberg, T. Penney and R. Tournier, J. Physique 40, Coll. C5, 314 (1979)

/17/ Z. Fisk, A.S. Cooper, P.H. Schmidt and R.N. Castellano, Mat. Res. Bull. 7, 285 (1972)

/18/ T. Penney, private commun.

/19/ G. Güntherodt, W.A. Thompson, H.-J. Güntherodt, T.S. Plaskett, Verhandl. DPG (VI) 17, 1031 (1982).

/20/ G. Güntherodt, W.A. Thompson, F. Holtzberg, Z. Fisk, to be publshed

/21/ P. Haen, F. Holtzberg, F. Lapierre, T. Penney and R. Tournier, in Ref. 3, p. 495

/22/ M. Konczykowski, J. Morillo, J.C. Portal, J. Galibert and J.P. Senateur, these Conference Proceedings

/23/ B. Hillebrands, G. Güntherodt and W.A. Thompson, these Conference Proceedings

/24/ B. Hillebrands, G. Güntherodt, R. Pott, W. König and A. Breitschwerdt, to be published

/25/ F.E. Pinkerton, A.J. Sievers, J.W. Wilkins, M.B. Maple and B.C. Sales, Phys. Rev. Lett. 47, 1018 (1981)

/26/ B.H. Grier and S.M. Shapiro, in Ref. 1, p. 325

/27/ M. Loewenhaupt and E. Holland-Moritz, J. Appl. Phys. 50, 7456 (1979)

/28/ T. Kasuya, K. Takegahara, Y. Aoki, K. Hanzawa, M. Kasaya, S. Kunii, T. Fujita, N. Sato, H. Kimura, T. Komatsubara, T. Furuno and J. Rossat-Mignod, in Ref. 1, p. 215.

/29/ J.M. Rowell, in Superconductivity, Vol. 2, ed. by P.R. Wallace, Gordon and Breach, New York, 1969, p. 721

/30/ E. Müller-Hartmann, in Springer Series in Solid-State Sciences, Vol. 29, T. Moriya (ed.), Springer, Heidelberg, 1981, p. 178

/31/ S. von Molnar, T. Theis, private commun; S. von Molnar, T. Theis, J. Flouquet and Z. Fisk, to be published

/32/ I. Frankowski and P. Wachter, Solid State Commun. 41, 577 (1982), and these Conference Proceedings

/33/ B. Bussian, I. Frankowski and D. Wohlleben, to be published

Valence Instabilities
P. Wachter and H. Boppart (eds.)

CRYSTALLINE ELECTRIC FIELD IN THE THULIUM MONOCHALCOGENIDES

Albert Furrer *, Willi Bührer * and Peter Wachter **

* Institut für Reaktortechnik, ETH Zürich
CH-5303 Würenlingen, Switzerland

** Laboratorium für Festkörperphysik, ETH Zürich
CH-8093 Zürich, Switzerland

Inelastic neutron scattering experiments have been carried out to measure the magnetic response of the thulium monochalcogenides. From the excitation spectra observed for trivalent TmS and $Tm_{0.87}Se$ as well as for divalent TmTe we have determined the crystalline electric field level scheme, whereas no crystal-field splittings were observed for intermediate valent $Tm_{0.99}Se$. The results are correlated in terms of reduced crystalline electric field parameters which point to similar charge distributions among the thulium monochalcogenides. Considerable dispersion effects were detected for $Tm_{0.87}Se$. The line broadening effects vary through the thulium monochalcogenide series and are most clearly seen in TmTe where both sharp and broad magnetic response coexist.

1. INTRODUCTION

The thulium monochalcogenides, TmX (X=S,Se,Te), are one of the most unique series among the rare earth monopnictides and monochalcogenides because of their unstable valence. All three compounds crystallize in the NaCl structure and exhibit magnetic phase transitions. The valence instability becomes evident when considering the lattice constants a and the effective magnetic moments μ of the whole series (1). These properties unambiguously demonstrate the transition from the trivalent state of TmS to the divalent state of TmTe, whereas TmSe has an intermediate valence of ∿2.8. The valence state of TmSe can be influenced by varying the chemical composition (2). Some basic properties of the thulium monochalcogenides studied in the present work are summarized in Table I.

Table I: Basic properties of the thulium monochalcogenides. The data are either derived in this work or taken from Refs. (1), (2) and (7).

Compound	a (Å)	$\mu(\mu_B)$	T_N (K)	Valence of the Tm ion
TmS	5.41	7.19	8.9	+3
$Tm_{0.87}Se$	5.63	7.29	4.2	+3
$Tm_{0.99}Se$	5.70	6.39	2.9	+2.8
TmTe	6.35	4.96	0.2	+2

For a quantitative understanding of the magnetic properties of f-electron systems information on the crystalline electric field (CEF) is indispensable. From comparison with inelastic neutron scattering (INS) studies of the thulium monopnictides (3,4) the thulium monochalcogenides are expected to exhibit a non-magnetic CEF ground-state. However, several attempts to measure CEF splittings in the thulium monochalcogenides were unsuccessful (1). This is an inherent difficulty of Tm monochalcogenides and reflects the basic instability of the trivalent state in these compounds. Recently we have been able to determine by INS experiments the CEF level structure of trivalent $Tm_{0.87}Se$ which yields a magnetic triplet ground state (5). This is drastically different from the expectations based on the CEF splittings known for the thulium monopnictides. It is the purpose of the present work to summarize the earlier and some new INS results for TmSe and to extend these measurements to TmS and TmTe in order to arrive at a systematic insight into the CEF level schemes of the thulium monochalcogenides.

2. THE CRYSTALLINE ELECTRIC FIELD

The Tm ions in the thulium monochalcogenides experience a cubic point symmetry. For the polar axis along the body diagonal which is the easy axis of magnetization for trivalent $Tm_{0.87}Se$ (6) and TmS (7) (for TmTe the magnetic structure is unknown), the CEF Hamiltonian is given by

$$\hat{H}_{CEF}=B_4(\hat{O}_4^0-20\sqrt{2}\,\hat{O}_4^3)+B_6(\hat{O}_6^0+\frac{35\sqrt{2}}{4}\hat{O}_6^3+\frac{77}{8}\hat{O}_6^6), \quad (1)$$

where the B_n are the CEF parameters and the $\hat{O}_n^m$ are operator equivalents (8). It is customary to parametrize eq. (1) in the following way (9) :

$$B_4F_4 = W\,x \,, \quad B_6F_6 = W\,(1-|x|). \quad (2)$$

Here W is a measure of the overall CEF splitting and x essentially gives the ratio between the fourth- and sixth-order terms in eq. (1). Eq.(1) has been diagonalized for $-1\leq x\leq 1$, and the eigenvalues E_n and eigenfunctions $|n\rangle$ as well as the numerical factors F_n have been tabulated (9).

In the magnetically ordered state an exchange term has to be added to the CEF Hamiltonian; in molecular-field (MF) theory we have

$$\hat{H} = \hat{H}_{CEF} - (g\mu_B)^2 \lambda \langle\hat{\vec{S}}\rangle \cdot \hat{\vec{S}} \quad , \qquad (3)$$

where λ denotes the MF parameter. The Hamiltonian (3) completely removes the degeneracy of the CEF levels.

For a system of N non-interacting ions the thermal neutron cross-section for CEF or MF transitions in the dipole approximation (10) is given by

$$\frac{d^2\sigma}{d\Omega d\omega} = \frac{N}{Z} \left(\frac{1.91\, e^2}{2\, m\, c^2} g\right)^2 \exp\{-2W\}\, F^2(Q)\, \frac{k_1}{k_0} \sum_{n,m} \exp\{-\frac{E_n}{k_B T}\}\, |\langle m|\hat{\vec{S}}_\perp|n\rangle|^2\, \delta(E_n - E_m - \hbar\omega) \quad , \qquad (4)$$

where Z is the partition function, F(Q) the form factor and $S_\perp$ the component of the total angular momentum perpendicular to the scattering vector $\vec{Q}$. The remaining symbols have their usual meaning.

3. EXPERIMENTAL

The samples used were either single crystals or polycrystalline ingots. A structure analysis by x-ray and neutron diffraction did not indicate any parasitic phases or lattice distortions; the resulting lattice constants are given in Table I. The single crystals were mounted with the $(1\bar{1}0)$ axis vertical. The polycrystalline samples were sealed in cylindrical aluminium containers of 6 mm diameter.

The INS experiments were performed on a triple-axis spectrometer at the reactor Saphir, Würenlingen. The constant-Q measurements were carried out in the neutron energy-loss configuration at various temperatures and momentum transfers. The analyzer energy was kept fixed at 13.7 or 14.9 meV, and a pyrolithic graphite filter was used to reduce higher-order contamination. Since the sample volumes were rather small (0.03 cm^3 for the single crystals, 0.2 cm^3 for the polycrystalline ingots), the experiments were performed with maximum beam focusing, i.e. a doubly bent graphite monochromator and a horizontally bent graphite analyzer (11) was used. For each compound the energy spectra were taken in the range $0 \le \hbar\omega \le 34$ meV. By measuring in different experimental configurations the magnetic origin of the inelastic scattering contributions could be verified. In particular the CEF transitions were identified by means of their characteristic dependence upon both the scattering vector $\vec{Q}$ through the form factor F(Q) and the temperature T through the Boltzmann statistics as seen in eq. (4).

4. RESULTS

4.1. TmSe

INS experiments were performed for single crystals of trivalent $Tm_{0.87}Se$ and intermediate valent $Tm_{0.99}Se$ above and below T_N. Typical energy spectra taken in the paramagnetic state are shown in Fig. 1. For $Tm_{0.87}Se$ there is a broad inelastic line at ∿5 meV which proves the existence of CEF excitations in this system, whereas the analogous spectrum observed for $Tm_{0.99}Se$ is governed by quasielastic scattering, i.e. there is no evidence for CEF transitions. Some aspects of these experiments have already been published (6) and can be summarized as follows: (i) The width of the quasielastic line in $Tm_{0.99}Se$ is a direct measure of the magnetic relaxation rates, i.e. the lifetimes of the 4f levels. A detailed analysis of the data suggests that the quasielastic scattering consists of a superposition of two Lorentzians with very different linewidths (e.g. at 10 K: $\Gamma_1^- = 0.6\pm0.3$ meV, $\Gamma_1^+ = 4\pm1$ meV). In fact the presence of two quasielastic lines has been predicted for intermediate valent Tm compounds (12) and is due to the fact that the Hund's rule ground states of both Tm^{3+} and Tm^{2+} are magnetic. In the ordered state a spin-

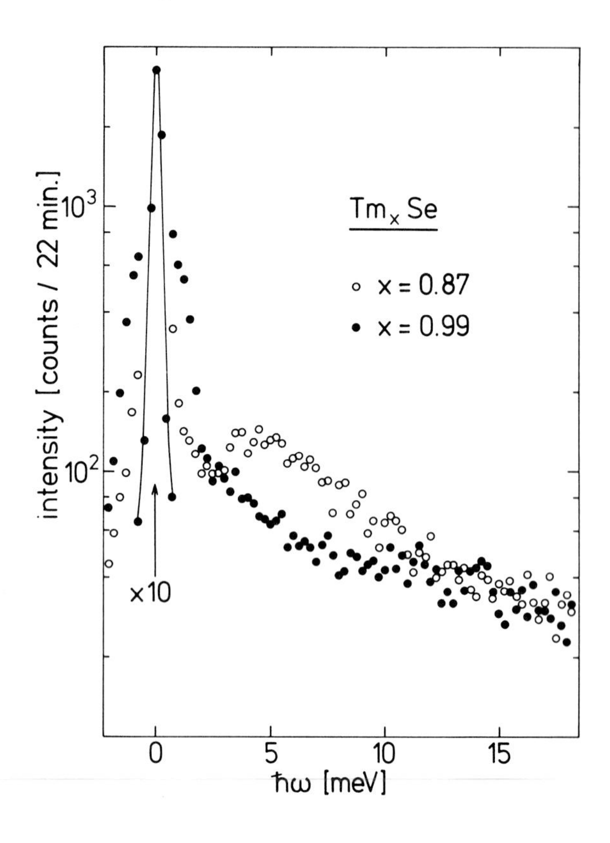

Figure 1: Energy spectra of neutrons scattered from Tm_xSe at $\vec{Q}=2\pi/a(0.97,0.97,0.97)$ at 10 K.

Table II: CEF parameters of the thulium monochalcogenides determined in the present work .

Compound	W (meV)	x	B_4 (10^{-4}meV)	B_6 (10^{-6} meV)	A_4a^5 (10^4 meV Å)	A_6a^7 (10^6 meV Å)
TmS	0.033±0.003	-0.26±0.03	-1.4±0.3	3.2±0.4	-4.1±0.9	-0.64±0.14
$Tm_{0.87}Se$	0.059±0.002	-0.10±0.02	-1.0±0.2	7.0±0.3	-2.9±0.6	-1.40±0.06
TmTe	-0.59±0.02	-0.28±0.03	28 ± 4	-337±26	-7.4±1.1	-1.72±0.13

wave like excitation was observed at around 1 meV, but the measured energy spectra gave no evidence for an inelastic line at 10 meV which was detected in other INS experiments (13,14).
(ii) From the energy spectra observed for $Tm_{0.87}$ Se in the paramagnetic as well as in the ordered state we were able to determine the CEF level structure. The CEF parameters given in Table II were derived by fitting the experimental data to the cross-section formula (4).

We have now extended the measurements for trivalent $Tm_{0.87}Se$ at 10 K to some other points in reciprocal space. It turned out that the inelastic peak at ∿5 meV identified as the $\Gamma_5^1 \to \Gamma_3$ excitation showed considerable dispersion effects as illustrated in Fig. 2. The excitation energy is lowest at the L-point as expected for a type-II antiferromagnet corresponding to an ordering wave-vector $\vec{q}_0=2\pi/a(0.5,0.5,0.5)$. From the dispersion direct information about the exchange interaction can be obtained. In the random phase approximation the magnetic excitation energies of the system are determined by the poles of the dynamic susceptibility (15)

$$\chi(\vec{q},\omega) = \frac{\chi_0(\omega)}{1 - J(\vec{q})\,\chi_0(\omega)} , \qquad (5)$$

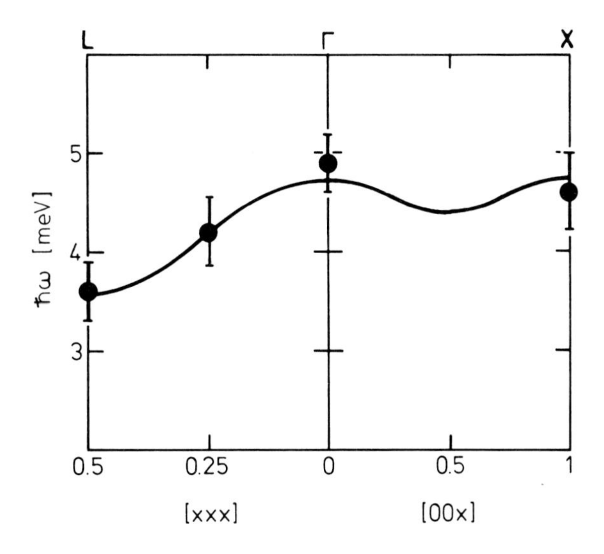

Figure 2: Dispersion of the $\Gamma_5^{(1)} \to \Gamma_3$ CEF excitation of $Tm_{0.87}Se$ at 10 K. The line is the result of a least-squares fit as explained in the text.

where $J(\vec{q})$ is the Fourier transform of the exchange coupling and $\chi_0(\omega)$ the single-ion susceptibility. The calculation becomes particularly simple when only the two lowest CEF states are considered. For T<<Δ where Δ is the $\Gamma_5^1 \to \Gamma_3$ CEF splitting we find

$$\omega(\vec{q}) = \left(\Delta\{\Delta - \frac{2}{3} J(\vec{q})|\langle\Gamma_3|\hat{\vec{S}}_\perp|\Gamma_5^1\rangle|^2\}\right)^{1/2} . \qquad (6)$$

For the actual CEF parameters of $Tm_{0.87}Se$ we have $|\langle\Gamma_3|\hat{\vec{S}}_\perp|\Gamma_5^1\rangle|^2 = 29.6$. For the calculation of $J(\vec{q})$ we only took into account the next-nearest neighbour exchange coupling J_2 which is usually the dominant exchange term in this class of compounds. Fitting the data of Fig. 2 to eq. (6) yields Δ=4.2±0.2 meV and $J_2=(-1.0\pm0.2)\times10^{-2}$ meV.

4.2. TmS

Energy spectra of neutrons scattered from polycrystalline TmS in the paramagnetic state are shown in Fig. 3. At 12 K there is a partially resolved line at ∿2 meV, which becomes less pronounced when the temperature is raised. The spectra are very similar to those observed for $Tm_{0.87}Se$ (5), but the overall CEF splitting is considerably reduced. The CEF parameters resulting from a least-squares fit based on eq. (4) are listed in Table II. The reliability of the CEF parameters is supported by INS measurements in the ordered state where excellent agreement is obtained between the experimental data and the calculations as shown in Fig. 4. The MF parameter λ was determined from T_N.

4.3. TmTe

Fig. 5 shows a measured energy spectrum for TmTe in the paramagnetic state. Besides an intense elastic peak there are two well defined inelastic lines at 8.6 and 17.3 meV. When the temperature is raised the intensities decrease as expected for ground-state CEF transitions. In the data analysis the elastic line was fitted by a Lorentzian, whereas the CEF transitions were fitted to Gaussians. The ground-state multiplet of Tm^{2+} is split by the cubic CEF into two doublets Γ_6 and Γ_7 and a quartet Γ_8. Since the CEF transition $\Gamma_6 \to \Gamma_7$ is not allowed by the selection rules, we can immediately conclude that the quartet Γ_8 is the ground state. Following Ref. (9) we have therefore W<0 and $-0.6 \le x \le 0.7$. Furthermore the line at 17.3 meV can be identified as the $\Gamma_8 \to \Gamma_6$ transition which is more intense than the $\Gamma_8 \to \Gamma_7$ transition. Thus we arrive

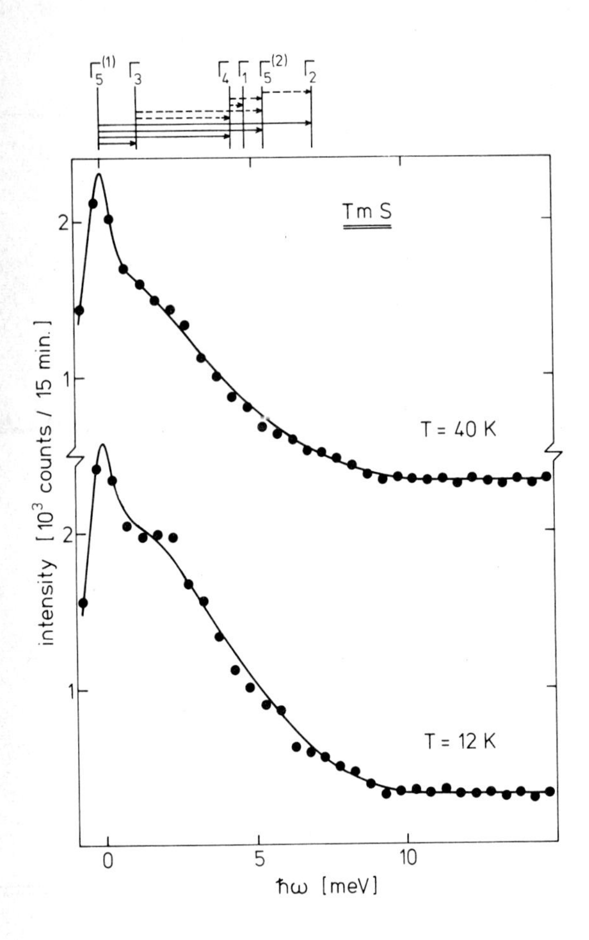

Figure 3: Energy spectra of neutrons scattered from polycrystalline TmS in the paramagnetic state at Q=1.5 $Å^{-1}$. The lines are the results of the least-squares fitting procedure as explained in the text. The top of the figure shows the resulting CEF level scheme. The allowed transitions are denoted by arrows.

at an unequivocal set of CEF parameters for TmTe as listed in Table II. Very puzzling is the presence of the strong quasielastic scattering which will be discussed in the next section. It is of Lorentzian shape with a linewidth of 1.2 meV at 10 K.

5. CONCLUSION

We have presented INS experiments for the thulium monochalcogenides and determined the CEF parameters. No CEF excitations were found for intermediate valent $Tm_{0.99}Se$. The data analysis for the trivalent metals TmS and $Tm_{0.87}Se$ was made difficult by the fact that the widths of the CEF transitions were rather large, thus the CEF parameters were obtained by a profile fit of

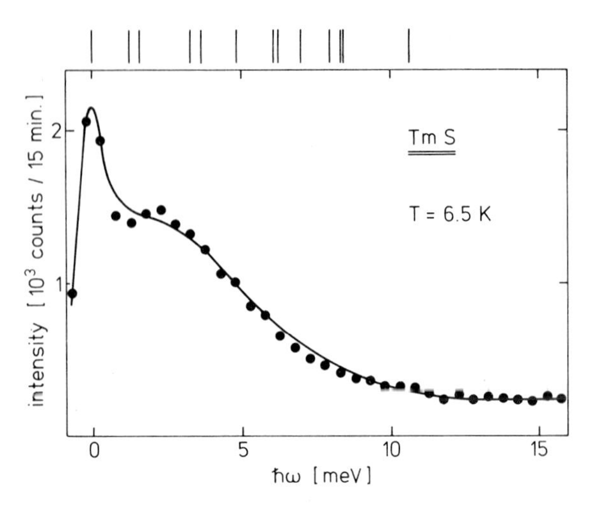

Figure 4: Energy spectrum of neutrons scattered from polycrystalline TmS in the ordered state at Q=1.5 $Å^{-1}$. The line is as in Fig. 3. The top of the figure shows the resulting MF levels.

the cross-section formula (4) to the measured energy spectra. On the other hand direct and clear spectroscopic evidence was obtained for the CEF parameters of the divalent semiconductor TmTe. Therefore it is interesting to correlate our results by calculating the reduced CEF parameters

$$A_n a^{n+1} = \frac{B_n a^{n+1}}{\langle r^n \rangle \chi_n} , \qquad (7)$$

where χ_n is a reduced matrix element (8) and $\langle r^n \rangle$ the n-th moment of the radial distribution of the 4f electrons. For $\langle r^n \rangle$ we took the relativistic values tabulated by Lewis (16). The results are listed in Table II. There is a rather good agreement of the reduced CEF parameters within the whole series of the thulium monochalcogenides indicating that the effective charge distributions are similar. However, the thulium monochalcogenides behave very different in comparison with the corresponding monopnictides where the CEF parameters are consistent with an effective fourth-order point-charge of -2 (3,4) in agreement with simple valency considerations. This is not at all surprising, since we do expect the electronic band structures to be drastically different in electronically stable trivalent rare earth compounds compared to "anomalous" rare earth compounds which are close to the coexistence of two energetically equivalent electronic configurations. The fact that the thulium monochalcogenides belong to the latter category is further supported by the large intrinsic line widths of the CEF transition peaks which are typically 3 meV for TmTe, 5 meV for TmS and 7 meV for $Tm_{0.87}Se$ at T= 10 K. Finally, for $Tm_{0.99}Se$ the broadening of the magnetic response is so large that no CEF

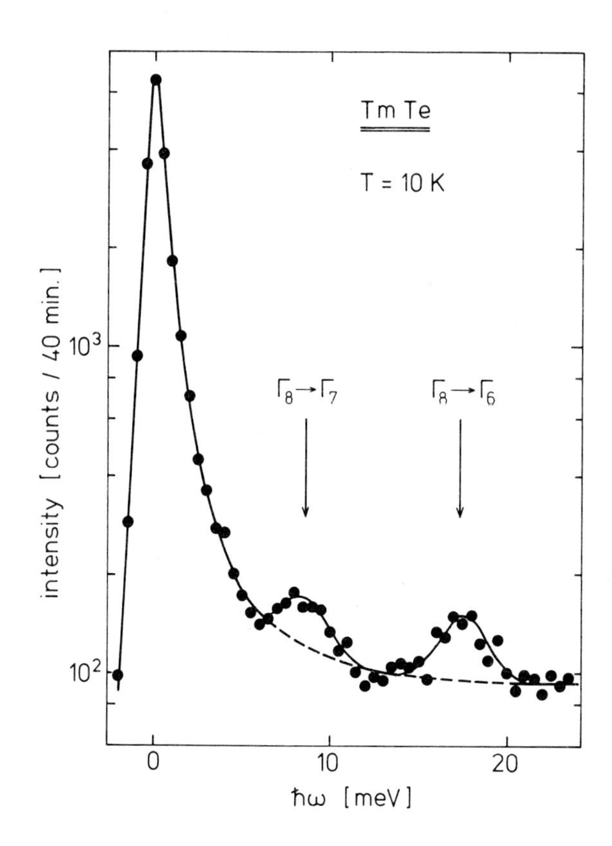

Figure 5: Energy spectrum of neutrons scattered from polycrystalline TmTe at Q=1.8 Å^{-1}. The lines are the results of a least-squares fit as explained in the text.

excitations can be observed, and we have the typical intermediate valence behaviour. Of particular interest are the energy spectra observed for TmTe, where sharp and broad magnetic response coexist. There are two well defined CEF transitions which are indicative of a stable electronic configuration. Besides these sharp features, which point to localized behaviour, there is a broad band of quasielastic scattering typical for intermediate valence behaviour. It cannot be explained by a conventional CEF transition within the Γ_8 ground state, since its intensity exceeds the corresponding quasielastic line strength when scaled to the inelastic CEF transitions by two orders of magnitude. Its origin is not yet understood. Similar phenomena were observed for UTe (17) which also exhibits both sharp and broad excitations.

The autors are grateful to B.R. Cooper for a stimulating discussion and to E. Kaldis for the preparation and characterization of the samples.

REFERENCES

(1) Bucher, E., Andres, K., die Salvo, F.J. Maita, J.P., Gossard, A.C., Cooper, A.S. and Hull, Jr., G.W., Phys. Rev. B 11 (1975) 500-513.

(2) Batlogg, B., Ott, H.R., Kaldis, E., Thöni, W. and Wachter, P., Phys. Rev. B 19 (1979) 247-259.

(3) Birgeneau, R.J., Bucher, E., Passell, L. and Turberfield, K.C., Phys. Rev. B 4 (1971) 718-725.

(4) Davis, H.L. and Mook, H.A., in Graham, Jr., C.D. and Rhyne, J.J. (eds.), Magnetism and Magnetic Materials - 1973 (AIP Conf. Proc. No. 18, New York, 1974) pp 1068-1072.

(5) Furrer, A., Bührer, W. and Wachter, P., Solid State Commun. 40 (1981) 1011-1014.

(6) Vettier, C., Flouquet, J., Holtzberg, F. and Mignot, J.M., J. Magn. Magn. Mater. 15-18 (1980) 987-988.

(7) Koehler, W.C., Moon, R.M. and Holtzberg, F., J. Appl. Phys. 50 (1979) 1975-1977.

(8) Stevens, K.W.H., Proc. Phys. Soc. A 65 (1962) 209-215.

(9) Lea, K.R., Leask, M.J.M. and Wolf, W.P., J. Chem. Phys. Solids 23 (1962) 1381-1405.

(10) Trammell, G.T., Phys. Rev. 92 (1953) 1387-1393.

(11) Bührer, W., Bührer, R., Isacson, A., Koch, M. and Thut, R., Nucl. Instr. and Meth. 179 (1981) 259-263.

(12) Müller-Hartmann, E., in Moriya, T. (ed.), Electron correlation and magnetism in narrow-band systems (Springer, Berlin, 1981) pp 178-195.

(13) Loewenhaupt, M. and Bjerrum-Møller, H., Physica 108 B (1981) 1349-1350.

(14) Grier, B.J. and Shapiro, S.M., in Falicov, L.M., Hanke, W. and Maple, M.B. (eds.), Valence Fluctuations in Solids (North-Holland, Amsterdam, 1981) pp 325-328.

(15) Fulde, P. and Peschel, I., Advances in Physics 21 (1972) 1-67.

(16) Lewis, W.B., in Ursu, I. (ed.), Proc. XVIth Congress Ampere (Publishing House of the Academy of Romania, Bucharest, 1970) pp 717-722.

(17) Buyers, W.J.L., Murray, A.F., Jackman, J.A., Holden, T.M., Du Plessis, P. de V. and Vogt, O., J. Appl. Phys. 52 (1981) 2222-2224.

Valence Instabilities
P. Wachter and H. Boppart (eds.)

SINGLE-ION MIXED VALENCY BEHAVIOUR OF DILUTE Ce IN (La,Th) ALLOYS

T.M. Holden, W.J.L. Buyers and P. Martel

Atomic Energy of Canada Limited
Chalk River Nuclear Laboratories
Chalk River, Ontario, KOJ 1J0, Canada

and

M.B. Maple* and M. Tovar**
Institute for Pure and Applied Physical Sciences
University of California, San Diego
La Jolla, California, 92093 U.S.A

Measurements have been made of the magnetic inelastic neutron scattering from dilute cerium in $Ce_{0.1}(La_xTh_{1-x})_{0.9}$ for x = 0.9, 0.6 and 0.0. The smaller thorium ionic size has the effect of applying chemical pressure to the individual cerium ions. A well-defined crystal field transition is observed at 2.9 ± 0.2 THz at the lanthanum-rich end of the alloy series, whereas no crystal field transition is observable up to 6 THz at the thorium-rich end. The previously measured susceptibility is Curie-Weiss like at the lanthanum-rich end but almost independent of temperature at the thorium-rich end. Our results show that the inverse lifetime of the f^1 configuration increases in step with the spin-fluctuation temperature that characterizes the susceptibility. The transition to mixed valent behaviour in Th rich alloys is largely a single-ion effect.

1. INTRODUCTION

We have measured the neutron inelastic scattering from Ce ions in dilute alloys of the form $Ce_{0.1}(La_xTh_{1-x})_{0.9}$ in order to focus on the magnetic properties of the individual ions and to lessen the importance of interionic exchange. The magnetic character of a Ce ion changes markedly with concentration, x, exhibiting a magnetic moment in the La rich region where the lattice parameter is largest, and becoming demagnetized as Th is added and the lattice parameter decreases. If the Ce ion is in a 3+ valency state with a single 4f-electron localized below the Fermi level, as in the conventional rare-earth ions, the crystal field experienced by the ion splits the J = 5/2 free-ion level into a Γ_7 doublet and a Γ_8 quartet. The dipole matrix element for this single transition is large and the splitting, Δ, between Γ_7 and Γ_8 may readily be observed by conventional neutron scattering methods. If this transition is absent, this indicates directly that the Ce ion is no longer magnetic. In fact, we observe features even at large La concentrations, where magnetic effects are expected, which suggest that the simple crystal field approach is too simple and that the lifetime of the excitation is limited even when a strong peak is observed.

The magnetic susceptibility of Ce dilutely dissolved in La_xTh_{1-x} alloys has been previously studied [1]. For La-rich alloys the Ce incremental susceptibility resembles a Curie-Weiss law suggesting well-developed local Ce moments with a low spin-fluctuation temperature. From a theoretical calculation, [2], employing the s-f exchange Hamiltonian and a crystal field model it was concluded that the susceptibility could be described by a Γ_8 level lying 60 K (1.25 THz) above the ground doublet. Maple, Wittig and Kim [3] showed that Ce ions destroy the superconductivity of La and, interpreting their results in terms of the Anderson model, suggested that the effect of pressure on Ce ions is to increase the coupling between the 4f electrons and the conduction electrons by decreasing the separation between the position of the 4f-level and the Fermi level until the self-consistent conditions for the stability of the magnetic moment can no longer be satisfied. Alloying Th with La is the chemical analog of applying pressure to the Ce. On the other hand, Huber and Maple [4] showed that Ce ions in Th are non-magnetic, although there are still 0.75 4f electrons associated with every Ce ion in a non-magnetic configuration.

Several neutron inelastic scattering experiments on fcc Ce alloys have been performed. In an alloy with a large Ce concentration $Ce_{0.76}Th_{0.24}$, in which a valence transition occurs near 150 K, Shapiro et al.[5] found a broad distribution of scattering, Lorentzian in form, centred on zero frequency and extending beyond 17 THz, with no evidence for crystal field peaks. The inverse width is interpreted as the lifetime of the magnetic moment on the Ce ion [6,7], not as interatomic exhange, as is the similar result for Ce metal [8]. In a further series of experiments [9] in which La was partly substituted for Th, a peak was observed near 3.4 THz with an intrinsic width which was interpreted as the Γ_7-Γ_8 crystal field transition; there was still however a major quasi-elastic component of scattering. In the present work with a low concentration of Ce ions it has been possible to observe a well-defined crystal-field peak and to monitor its lifetime as Th is added.

2. EXPERIMENTS AND RESULTS

The experiments were carried out on polycrystline samples in the form of cast buttons suitably heat treated,[1], and contained in a thin-walled aluminum tube with the $Ce_{0.1}(La_x$-$Th_{1-x})_{0.9}$ sample separated by 1 cm of Cd from a sample of $La_{0.1}(La_xTh_{1-x})_{0.9}$. In this way either sample can be placed in the same position in the neutron beam simply by raising or lowering the variable temperature cryostat containing the samples. The Ce scattering alone can then be obtained by subtraction. The measurements were made with triple-axis crystal spectrometers at the NRU reactor, Chalk River, operated in the constant-$\underset{\sim}{Q}$ mode with fixed scattered frequency E_1. For the experiments with x = 0.9 and x = 0.0 a (113) plane of a Ge crystal was used as monochromator with a pyrolytic graphite analyser set for E_1 = 7.2 THz at an overall frequency resolution of 1.0 THz. For the experiment with x = 0.6, a (111) plane of Si crystal was used as monochromator with a pyrolytic graphite analyser set for E_1 = 4.5 THz which gave an overall resolution at 3.5 THz of 0.65 THz.

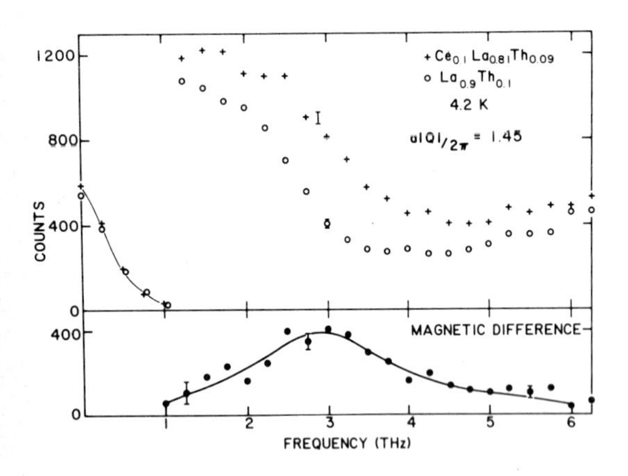

Fig. 1 Inelastic neutron scattering from $Ce_{0.1}(La_{0.9}Th_{0.1})_{0.9}$ and $La_{0.9}Th_{0.1}$ alloys at 4.2 K and the Ce difference scattering.

The results for x = 0.9 and the corresponding blank are shown in Fig. 1 for a reduced wavevector $\zeta(a|Q|/2\pi)$ = 1.45 together with the difference count which represents the Ce scattering. Corrections for the different nuclear scattering lengths of the constituents and the difference in mass of the samples were less than 5%. The Ce scattering has the form of an inelastic transition centred on 2.9 ± 0.2 THz with an intrinsic halfwidth of 0.8 THz. The peak intensity shows a wavevector dependence which is consistent with a $4f^1$ form factor and its shape is independent of wavevector in the range ζ = 0.8 to 1.5 confirming the single ion character of the scattering.

As the strong scattering below 3 THz is of phonon character for accurate results it is essential to derive the magnetic component by subtracting the scattering from the blank rather than relying on its wavevector dependence. The rise in the scattering for frequencies beyond 5 THz comes from an increased component of fast neutron background at small scattering angles and this effect can be subtracted also with the blank sample.

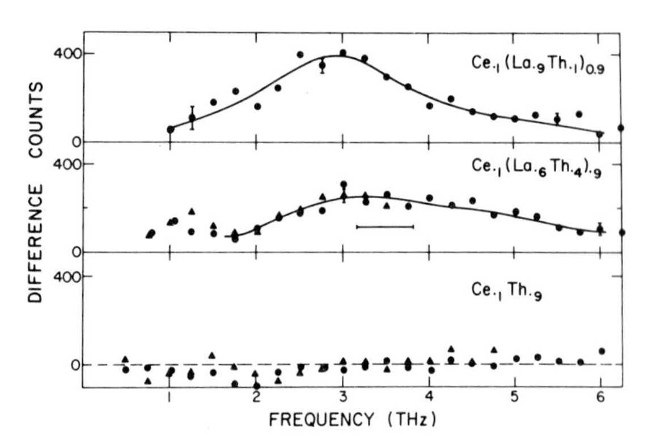

Fig. 2 Cerium difference scattering for La concentrations x = 0.9, 0.6, 0.0 at 4.2 K. Circles and triangles correspond to different wavevectors.

The magnetic difference counts for x = 0.0, 0.6 and x = 0.9 are shown in Fig. 2. The scattering from Ce in Th appears to be absent within the errors; note that if the scattering had merely scaled with concentration from the results for $Ce_{0.76}Th_{0.24}$ [5] it would have been observable in the present experiment. This provides evidence that lifetime effects and even the existence of a magnetic moment are changed by the presence of neighbouring Ce ions which also appear to enhance the quasi-elastic component observed at high Ce concentrations [5],[9]. The scattering for x = 0.6 appears to show a broader peak at a somewhat higher frequency than the peak for x = 0.9. Finally Fig. 3 shows the temperature dependence of the Ce scattering for x = 0.9 for the sum of all the wavevectors studied between ζ = 0.86 and 1.5 to give the highest possible statistical accuracy.

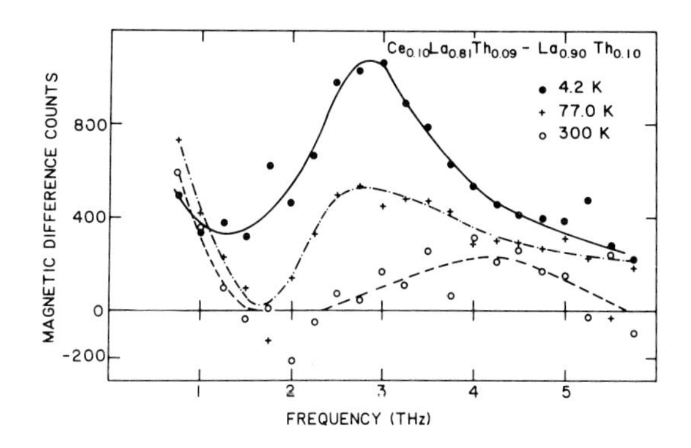

Fig. 3 Temperature dependence of the Ce difference scattering for the $Ce_{0.1}(La_{0.9}Th_{0.1})_{0.9}$ alloy.

The effect of temperature is to depress the scattering rapidly, much more rapidly, as we shall show, than for a crystal field model.

3. ANALYSIS AND DISCUSSION

By unfolding the experimental resolution from the observed scattering the parameters, Δ, characterizing the splitting and Γ, the half-width at half-height of a Lorentzian profile, were obtained and are given in Table 1.

Table 1. Transition frequency ,Δ, intrinsic width, Γ, and apparent Curie-Weiss θ, [ref. 1] for dilute Ce alloys. The fractional La concentration is x.

x	Δ(THz)	Γ(THz)	Θ(THz)
0.9	2.9±0.2	0.8±0.4	0.5
0.6	3.5±0.3	1.6±0.2	1.35

The parameters obtained by Shapiro et al [9] for $Ce_{0.5}La_{0.4}Th_{0.1}$ are similar (Δ = 3.4 THz, Γ = 0.8 THz) suggesting that for the low thorium concentrations the electronic configuration of the Ce ion is remaining the same, presumably close to the integral valence situation since $\Delta \gg \Gamma$. The increase in the splitting as x changes from 0.9 to 0.6 may be consistent with the decrease in the lattice parameter; crystal field theory would predict an increase of 10%. It is interesting to note that the apparent Curie-Weiss parameters Θ derived from the susceptibility measurements [1] have the same numerical value when expressed in THz (Table 1) as the measured intrinsic widths, to within the experimental accuracy.

Table 2. Static susceptibility ($10^{-6}\mu$B oe^{-1}Ce $atom^{-1}$) of Ce in La_xTh_{1-x} alloys calculated from the observed splitting Δ and assuming a full 3+ state on each Ce ion. The figures in brackets give the percentage which the experimental susceptibility [1] is of the expected crystal field susceptibility. The experimental result for x = 0.6 is interpolated from results in ref. [1].

x	77 K		300 K	
	exp	theory	exp	theory
0.9	1.39 (86%)	1.62	0.43 (90%)	0.48
0.6	1.04 (68%)	1.52	0.36 (76%)	0.47

The crystal field susceptibility (Van Vleck plus Curie contributions) has been calculated from Δ using the Ce^{3+} matrix elements and taking Γ_7 lowest, and the results are compared with the experimental susceptibility per Ce ion [1] in Table 2. The crystal field susceptibility is clearly the major contribution to the susceptibility but it is notable that the measured susceptibility falls below the calculated value in each case and that the fractional discrepancy increases with Th content.The results can be interpreted as showing that only a fraction of the full Ce^{3+} moment contributes to the static susceptibility. The fraction falls from ≈ 0.9 for the La-rich alloy to ≈ 0.7 for x = 0.60 and presumably close to zero for pure Th.

The temperature dependence of the strength of the inelastic neutron scattering peak can be calculated from crystal-field theory if Δ is known. For x = 0.9 the calculated intensity at 77 and 300 K was found to be 75% and 44% respectively of its strength at 4.2 K. Fig. 3 shows that the peak intensity falls to ≈ 50% at 77 K and ≈ 20% at 300 K. That is, the intensity associated with the so-called crystal field transition disappears much faster than predicted by crystal field theory. In an experiment [9] on $Ce_{0.4}La_{0.5}Th_{0.1}$ the parameter Δ decreased with rising temperature, whereas in the present case the response appears to shift to higher frequencies.

4. CONCLUSIONS

The results show that the magnetic intensity decreases with Th content and temperature. The former result can be interpreted as evidence for a valence change of the Ce ion in the sense that the effective number of polarized 4f electrons is reduced below unity. As a result of chemical pressure as Th is added, the 4f-conduction band hybridisation becomes larger, the f level moves closer to the Fermi energy, and the effective moment of the Ce ion is reduced. Our interpretation of the static susceptibility of Ce in La_xTh_{1-x} alloys also suggests that the relative weight of the magnetically correlated f^1 state decreases as x decreases. At the same time the intrinsic width of the observed peak, Γ, grows with Th content, as might be expected on the basis of the calculated magnetic relaxation of Kuramoto and Müller-Hartmann [6,7]

Although the idea of the relative weight of the magnetic state can help to explain the static susceptibility, it cannot account for the rapid decrease of the inelastic scattering with increasing temperature. The effect of magnetic relaxation [6,7] would be to produce a broadening of the crystal field peak with temperature rather than the rapid attenuation of the integrated intensity that is observed. It appears that the magnetically correlated state is therefore weakened by temperature as well as by addition of Th. This is consistent with the interpretation of Kondo-like behaviour of the La-rich alloys, as observed in their resistivity [10]. The f-level (half-width 230 K) was found to lie 650 K below the Fermi energy for La. In this regime it seems quite plausible that the state could be rapidly decorrelated as the temperature is raised to 300 K. As Th is added the f-level rises towards the Fermi energy at the rate of 10 K per at. % Th and should make the temperature dependence more rapid for x = 0.6 than for x = 0.9. For x = 0.0 the f-level lies above but still overlaps the Fermi energy and contains 0.75 electrons, but even at low temperatures moment formation does not take place.

REFERENCES

[1] Huber, J.G., Brooks, J. Wohlleben D. and Maple, M.B., A.I.P. Conf. Proc. 24 (1975), 475-6.
[2] DeGennaro, S. and Borchi, E. Phys. Rev. Lett. 30 (1973) 377-80.
[3] Maple, M.B., Wittig, J. and Kim K.S. Phys. Rev. Lett. 23 (1969) 1375-78.
[4] Huber, J.G. and Maple M.B., J. Low Temp. Phys. 3 (1970) 537-44.
[5] Shapiro, S.M., Axe, J.D., Birgeneau, R.L., Lawrence, J.M. and Parks, R.D., Phys. Rev. B 16 (1977) 2225-34.
[6] Kuramoto, Y. and Müller-Hartmann, E. in "Valence Fluctuations in Solids" Eds. Falicov, L.M., Hanke, W. and Maple, M.B. (North-Holland, Amsterdam, 1981) pps. 139-46.
[7] Kuramoto, Y., Z. Phys. B37 (1980) 299-305.
[8] Rainford, B.D., Buras, B. and Lebech, B. Physica 86-88B (1977) 41-2.
[9] Shapiro, S.M., J. Appl. Phys. 52 (1981) 2129-33.
[10] Pena, O. and Meunier, F., S.S. Comm. 14 (1974) 1087-9.

Valence Instabilities
P. Wachter and H. Boppart (eds.)

OPTICAL PROPERTIES OF $CePd_3$ AND $ThPd_3$ AND THEIR RELATION TO PHOTOEMISSION DATA

J. Schoenes

Laboratorium für Festkörperphysik, ETH Zürich,
8093 Zürich, Switzerland

Optical reflectivity measurements on $CePd_3$ and $ThPd_3$ have been performed for photon energies from 0.03 to 12 eV. A strong similarity between the Kramers-Kronig derived optical conductivity and X-ray photoelectron spectra is observed. The 4f binding energy in $CePd_3$ is derived to be <0.3 eV.

1. INTRODUCTION

The binding energy of the $4f^1$ state in intermediate valent $CePd_3$ is the subject of a keen debate. X-ray photoelectron spectroscopy (1) places the $4f^1$ state within 0.4 eV at E_F, while resonant photoemission data (2) have been interpreted as evidence for a binding energy of $\sim$ 1 eV. To clarify this discrepancy (3) we have performed reflectivity measurements from 0.03 to 12 eV on $CePd_3$ and also on $ThPd_3$ as a reference material (4). The advantage of optical spectroscopy compared to photoemission is to probe several hundred Å deep bulk properties, avoiding thus possible misinterpretations due to surface binding energy shifts. Also 4f electrons can be excited with large oscillator strength into 5d states allowing to probe the 4f states with photon energies which are comparable to the 4f binding energy. The reflectivity spectra have been Kramers-Kronig transformed to obtain the complex dielectric function, the optical conductivity, the energy loss function and the effective number of electrons contributing to optical transitions.

2. RESULTS AND DISCUSSION

A remarkable similarity is observed between the optical conductivity above 1.5 eV and the XPS spectra (1,5) if the binding energy of the latter ones is plotted so as to coincide with the photon excitation energy in the optical spectra Fig. 1. This indicates that the final states of the optical transitions are very close to E_F and that matrix element effects play only a minor role.

Comparing the optical conductivity of $CePd_3$, $ThPd_3$ and Pd (6) (Fig. 1, inset) the structures labelled B,C,D and E can be assigned to palladium d→p transitions. Relative to the Pd d-states the Fermi energy is found to increase by 1.2 eV for $CePd_3$ and by 1.7 eV for $ThPd_3$ (4). Below $\sim$ 2eV the optical conductivity of $CePd_3$ on one hand and the corresponding spectra for $ThPd_3$ and Pd on the other hand differ markedly.

In $CePd_3$ we observe two peaks marked A and A' but no peak B. The comparison of the XPS spectra of $CePd_3$ and $ThPd_3$ shows for both compounds a shoulder B and in addition for $CePd_3$ a shoulder A at 0.3-0.4 eV binding energy. It is this structure A which has been assigned to occupied 4f states in $CePd_3$ (1). Similarly, the most straight forward assignments for the structures A and A' in the optical conductivity are f (Ce) → d (Ce) and d (Pd) → p (Pd) transitions, respectively. Thus peak A' in $CePd_3$ corresponds to peak B in $ThPd_3$ and Pd. The apparent shift by nearly 1 eV between A' in $CePd_3$ and B in $ThPd_3$ can be explained by the two following effects of about equal size:

a) A shift of the peak energy due to the superposition in $CePd_3$ of peak A' on a steep negative slope (which originates from peak A) while peak B in $ThPd_3$ is superposed on a steep positive slope. This is an extrinsic effect.
b) An intrinsic shift of the Fermi energy by 0.5 eV to lower energy going from $ThPd_3$ to $CePd_3$.

The assignment of peak A' to a Pd d→p transition is strongly supported by a recent selfconsistent LAPW band structure calculation for $CePd_3$ (7) which shows a large and narrow peak in the density of Pd d states at -1.4eV in addition to the main Pd d peak near -2.8eV.

The only remaining peak A at 0.25eV has then to be assigned to a f→d transition. Because empty cerium 5d states (as well as Pd 4d states) are found at E_F the final states of the f→d transition are expected to be very close to E_F. We can not exclude a reduction of the excitation energy due to electron-hole interaction. However, the excellent agreement between the excitation energy in the optical experiment and the binding energy from the XPS data indicate that this effect is small in $CePd_3$ and also in UPd_3 (4).

It should be mentioned that two more interpretations for the optical conductivity have been given very recently (8,9). In the former reference peak A is assigned to an intra 4f i.e. a

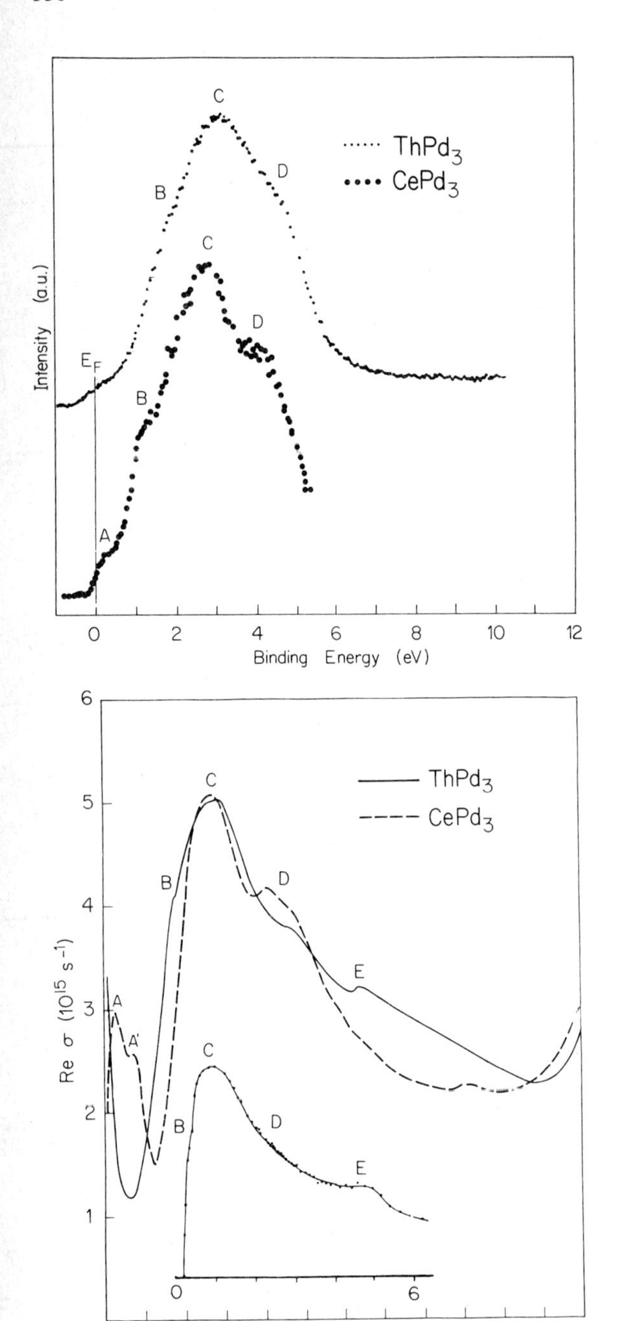

Fig. 1. Comparison between optical conductivity (bottom) and X-ray photoelectron spectra (1,5) (top) for $CePd_3$ and $ThPd_3$. Note the different spectral resolution for the XPS of $CePd_3$ (0.3eV) and $ThPd_3$ (0.7eV). Also shown is the optical conductivity of pure Pd (inset) after subtraction of the intraband part (6).

$^2F_{5/2} \rightarrow {}^2F_{7/2}$ transition. In (9) peak A and A' are assigned to d → f and f → d transitions, respectively. Taking a valence of 3.5 for Ce in $CePd_3$ (10) and assuming that occupied d-states are created solely by f delocalization the product of occupied and available empty states leads to a ratio of ∿ 1:20 between d → f^1 and f^1 → d transitions (11). This estimated ratio is of the order of what we do observe in CeN (12) but it is much smaller than the intensity ratio between peak A and A' in Fig. 1. An intra 4f transition is electric dipole forbidden and its intensity is expected to be several orders of magnitude smaller than the dipole allowed f → d transition. Also if peak A' corresponds to a f → d transition the disappearence of peak B in $CePd_3$ remains to be explained.

3. CONCLUSION

In conclusion we have shown that in $CePd_3$ and $ThPd_3$ a rather exceptional similarity exists between optical conductivity and X-ray photoelectron spectra. This is particularly interesting because the photon excitation energies in the two techniques differ by a factor of thousand. The discrepancies with the resonant photoemission data are yet not fully understood. Due to the large surface sensitivity of the latter technique one may invoke a surface binding energy shift as has been found in many rare earth materials (13). Alternatively a different screening of the final state (13,14) in the various techniques may be the origin of the apparently contradicting binding energies.

ACKNOWLEDGEMENT

The author is much indebted to D.D. Koelling for communicating results of his band structure calculation for $CePd_3$ before publication. I would like to thank J.C. Fuggle and Z. Zolnierek for providing the samples and J. Allen, Y. Baer and P. Wachter for lively discussions. The technical assistance of J. Müller is gratefully acknowledged

REFERENCES

1. Y. Baer, H.R. Ott, J.C. Fuggle and L.E. Long, Phys. Rev. B24, 5384 (1981)
2. J.W. Allen, S.-J. Oh, I. Lindau, J.M. Lawrence, L.I. Johansson and S.B. Hagström, Phys. Rev. Letters 46, 1100 (1981)
3. To this problem, occuring in many Ce systems, see also "Valence Fluctuations in Solids" edited by L.M. Falicov, W. Hanke and M.B. Maple, (North-Holland, Amsterdam 1981)
4. A full version of this paper including UPd_3 will appear under: J. Schoenes and K. Andres, Solid State Commun. (1982)
5. J.C. Fuggle and Z. Zolnierek, Solid State

Commun. 38, 799 (1981)
6. J. Lafait, Proc. "Transition Metals", Inst. Phys. Conf. Ser. No. 39, (1978) p. 130
7. D.D. Koelling, private Communication
8. B. Hillebrands, G. Güntherodt and R. Pott, Verhandl. DPG (VI) 16, 292 (1981) and this conference
9. J.W. Allen, R.J. Nemanich and S.-J. Oh, J. Appl. Phys. 53, 2145 (1982)
10. E. Holland-Moritz, M. Löwenhaupt, W. Schmatz and D. Wohlleben, Phys. Rev. Lett. 38, 983 (1977)
11. For a larger f occupation, as is expected if the binding energy is larger, the intensity of the d → f transition would decrease further compared to the f → d transition.
12. J. Schoenes, E. Kaldis and P. Wachter, to be published
13. See for instance: W. Gudat, R. Rosei, J. H. Weaver, E. Kaldis and F. Hulliger, Solid State Commun. 41, 37 (1982)
14. See for instance: S. Hüfner and many other papers of the Photoemission session of this conference.

Valence Instabilities
P. Wachter and H. Boppart (eds.)

INFRARED RESPONSE AND ELECTRON TUNNELING SPECTROSCOPY OF $CePd_3$, YPd_3 AND $PrPd_3$

B. Hillebrands and G. Güntherodt
II. Physikalisches Institut†, Universität zu Köln, 5000 Köln 41, FRG

and

W. A. Thompson
IBM Research Center, Yorktown Heights, N. Y. 10598, USA

The optical conductivity of intermediate valence $CePd_3$ exhibits at 300 K a maximum near 16 meV, which exceeds the static value and which is absent in the reference materials YPd_3 and $PrPd_3$. This anomaly is attributed to a resonant electron-electron scattering process of conduction electrons at the localized 4f states near the Fermi level E_F. From a fit of our data using a Drude model modified by a memory function ansatz we obtain for the width Γ of the 4f states and the separation Δ between the localized 4f states and E_F, respectively, Γ = 5 meV and $\Delta < k_BT$ = 26 meV. Applying this analysis to the 4.2 K absorptivity data of Pinkerton et al. yields Γ = 8 meV and Δ = 14 meV. In the electron tunneling spectrum of $CePd_3$ using a Schottky barrier tunneling technique an inelastic excitation is observed near the same energy $\Delta = \pm$ 14 meV, which is absent for YPd_3. This excitation is attributed to an electron-hole binding energy 2Δ.

INTRODUCTION

The near degeneracy of the localized 4f states with the Fermi energy E_F in intermediate valence (IV) $CePd_3$ should manifest itself in anomalous effects of transport and related optical properties, such as temperature dependent resistivity (/1/), tunneling phenomena (/2/) or frequency dependent optical conductivity $\sigma(\omega)$ or absorptivity $A(\omega)$ (/3/). Since $\sigma(\omega)$ is a linear measure of energy absorption processes in solids it may reveal resonant scattering processes of conduction electrons at the localized 4f states. $\sigma(\omega)$ has been obtained from a Kramers-Kronig analysis of room temperature reflectivity data of $CePd_3$ and its reference materials YPd_3 and $PrPd_3$. As a reference material without 4f electrons we prefered YPd_3 instead of $LaPd_3$ because of its superior chemical stability. On the other hand, $PrPd_3$ has a stable $4f^2$ configuration. Because of anomalous contributions to $\sigma(\omega)$ of $CePd_3$ at 300 K near 16 meV exceeding the static value $\sigma_o = \sigma(\omega=0)$ we have also performed electron tunneling experiments at 4.2 K using the GaAs Schottky barrier probe tunneling technique (/2/, /4/). In order to further corroborate these studies we have analyzed the optical absorptivity data of Pinkerton et al. (/3/) in the energy range from 4 meV to 40 meV at temperatures between 4.2 K and 150 K.

EXPERIMENTAL RESULTS

The room temperature optical reflectivity of $CePd_3$, YPd_3 and $PrPd_3$ has been measured in the energy range from 5 meV to 5 eV using a Bruker FIR spectrometer, a Nicolet MX1 spectrometer and a Cary 17D spectrometer. The optical conductivity $\sigma(\omega)$ has been derived by a Kramers-Kronig analysis of the reflectivity data and is shown in Fig. 1. To get a good reliability of the transformation over the full measured energy range the experimental data have been extrapolated

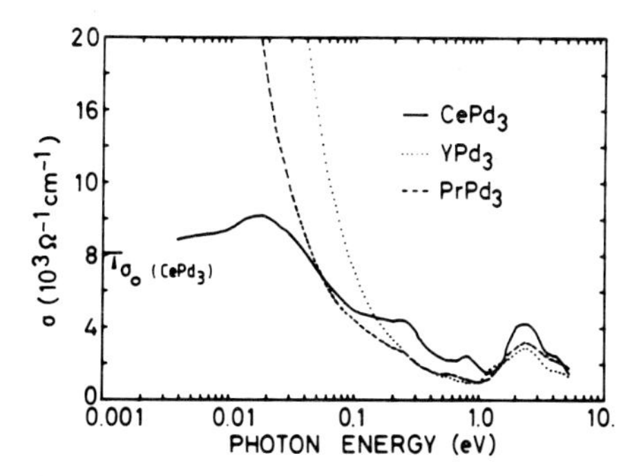

Fig. 1: Optical conductivity of $CePd_3$, YPd_3 and $PrPd_3$ versus energy, obtained by a Kramers-Kronig transformation of the measured reflectivity data.

to $\omega = 0$ using a Hagen-Rubens relation and to $\omega \to \infty$ using a $1/\omega^2$ dependence of the reflectivity for $\omega_{max} < \omega < A\cdot\omega_{max}$ and a $1/\omega^4$ dependence for $A\cdot\omega_{max} < \omega$, where ω_{max} is the highest measured frequency value. The fit parameter A has been chosen in such a way that the optical constants obtained by the Kramers-Kronig transformation coincide with values measured independently at some energies in the visible range.

The electron tunneling spectra of $CePd_3$ and YPd_3 have been measured using the GaAs Schottky barrier probe tunneling technique, which has been described elsewhere (/4/). We have measured the differential resistance R = dV/dI versus the applied bias voltage V up to ± 100 meV at 4.2 K. Reliable symmetric tunneling spectra have been obtained after removing surface contaminants by an in situ sputter cleaning method (/4/). Fig. 2 shows the spectra (/2/).

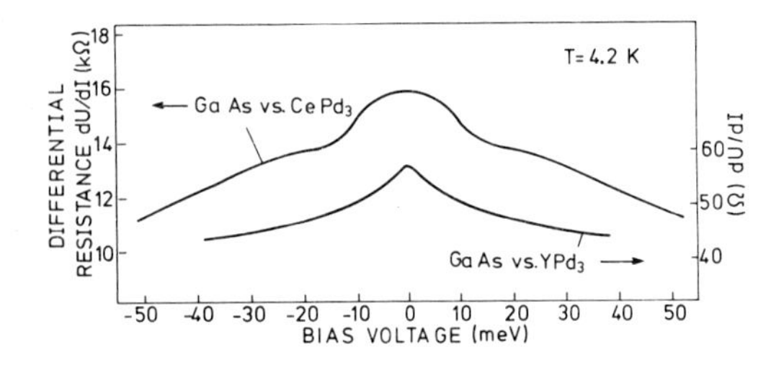

Fig. 2: Tunneling spectrum of $CePd_3$ and YPd_3 versus GaAs at 4.2 K.

DISCUSSION

a. Optical conductivity: resonant electron-electron scattering processes

As shown in Fig. 1 the optical conductivity of the reference materials YPd_3 and $PrPd_3$ behaves Drude-like within the experimental error. In $CePd_3$ $\sigma(\omega)$ differs essentially from Drude behavior. This is illustrated in Fig. 3, where $\sigma(\omega)$ is shown on an expanded scale together with a Drude fit and a fit using a modified Drude model described below. With decreasing frequency $\sigma(\omega)$ increases from small values near 100 meV to a maximum near 16 meV and then it decreases to an extrapolated value $\sigma_o = \sigma(\omega{=}0)$ of $8100\ \Omega^{-1}cm^{-1}$, which agrees well with the independently measured static conductivity $\sigma_o = 8300\ \Omega^{-1}cm^{-1}$ (/1/). This maximum near 16 meV may be due to two possible scattering mechanisms: Electron-phonon scattering or

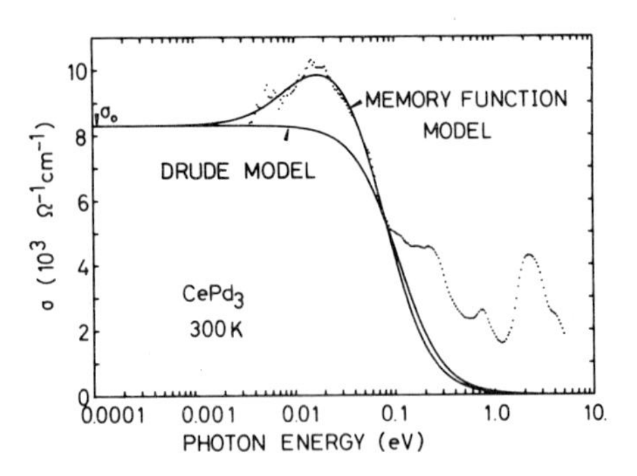

Fig. 3: Optical conductivity of $CePd_3$ versus energy (dotted curve) and fits (solid curves).

scattering of conduction electrons at localized (Ce)4f states near E_F. As discussed elsewhere (/5/) the influence of electron-phonon scattering on the optical conductivity of $CePd_3$, YPd_3 and $PrPd_3$ is negligible because of the high reflectivity below 100 meV and the absence of q=0 optical phonon anomalies in $CePd_3$ (/6/) and, obviously, in YPd_3 and $PrPd_3$. Thus we attribute the maximum near 16 meV in $\sigma(\omega)$ of $CePd_3$ to a resonant electron-electron scattering process at the localized 4f states of Ce. Resonant scattering of the conduction electrons at localized 4f states can take place if the localized states lie in the vicinity of the Fermi energy. If the width Γ of the localized states is small compared to the electronic relaxation rate $1/\tau_o$, which describes all frequency independent scattering processes of the conduction electrons, the scattering time of the conduction electrons can be neglected with respect to the lifetime of the localized state. Consequently, resonant scattering processes give a measure of the width Γ of the localized states. Such resonant scattering processes have been observed in recent studies of SmB_6 (/7/) and $CePd_3$ (/3/). For a numerical analysis we used a modified Drude model proposed by J. W. Allen et al. (/7/) in which the usually frequency independent relaxation rate $1/\tau_o$ is replaced by a frequency dependent complex relaxation rate described by a memory function ansatz $M(\omega)$:

$$\sigma(\omega) = \omega_p^2\ (M(\omega) - i\omega)^{-1}$$

The simplest model of resonant scattering at localized states of width Γ is a description by an exponential decay in the time t of an excitation or perturbation at a previous time t'. Then the memory function M(t-t') becomes

$$M(t-t') = \frac{1}{\tau_o}\,\delta(t-t') + A^2\, e^{-\Gamma(t-t')}\,\Theta(t-t')$$

where $\Theta(t-t')$ is the step function and provides causality. Then the optical conductivity is given by

$$\sigma(\omega) = \omega_p^2 (M(\omega)-i\omega)^{-1} = \omega_p^2 (\frac{1}{\tau_o} + \frac{A^2}{\Gamma - i\omega} - i\omega)^{-1}$$

$M(\omega)$ is the Fourier transform of $M(t-t')$. τ_o is a frequency independent relaxation time and ω_p the plasma frequency of the conduction electrons; A reflects a coupling constant of the resonant f-d scattering process and is of the order of the f-d hybridization energy V_{fd}. We have fitted our data of $\sigma(\omega)$ at 300 K by this model with the fit parameters Γ, A, τ_o and ω_p. The results are Γ = 5 meV, A = 11 meV, τ_o = 8.5 10^{-15} sec and ω_p = 2.15 eV. The value of σ_o = 8100 Ω^{-1} cm^{-1} resulting from our fit coincides with an independently measured value and confirms the reliability of the fit. Because of the simplicity of this model and also because of the sensitivity to experimental errors, these values reflect only the order of magnitude of the parameters. We would like to point out that similar parameters habe been obtained by independent methods like point contact spectroscopy (Γ = 4.5 meV, (/8/)) and temperature dependent resistivity (Γ = 4.2 meV (/1/), A = 10 meV (/9/)).

In view of the anomalous temperature dependence of the resistivity of $CePd_3$ it is particularly interesting to investigate the temperature dependence of the resonant scattering process. Pinkerton et al. (/3/) have measured the absorptivity $A(\omega) = 1-R(\omega)$ of $CePd_3$ and YPd_3 in the far infrared at temperatures between 4.2 K and 150 K. They found a maximum near 19 meV at 4.2 K which is superposed on the Drude absorption. With increasing temperature the maximum vanishes but the remaining broad background at energies above 10 meV still deviates from normal Drude behavior exhibited by

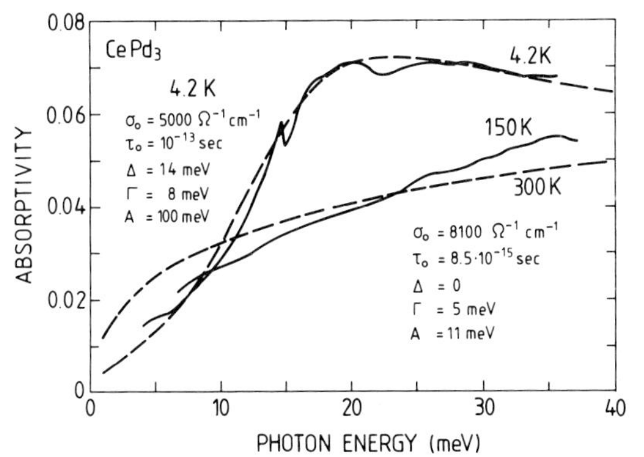

Fig. 4: Optical absorptivity of $CePd_3$ versus energy (from /3/) (solid curve) and the memory function fit (dashed curve).

YPd_3. A good quantitative fit of the maximum in the absorptivity at 4.2 K (Fig. 4) can be obtained with the above described modified Drude model by introducing a separation Δ between the 4f level and E_F, i. e. by adding an imaginary part $i\Delta$ to the width Γ. Thus Δ changes from Δ = 0 at 300 K to Δ = 14 meV at 4.2 K. A value Δ = 17.6 meV has been derived by a model fit of Pinkerton et al.. The width Γ changes slightly from Γ = 5 meV at 300 K to Γ = 8 meV at 4.2 K. In Fig 5 our results of Γ and Δ are summarized. It should be taken into consideration that a separation between the 4f level and E_F can only be observed in the optical conductivity, if Δ is larger than k_BT. Because of that the fitted value of Δ = 0 at 300 K can only be considered as an estimate of Δ < k_BT = 26 meV.

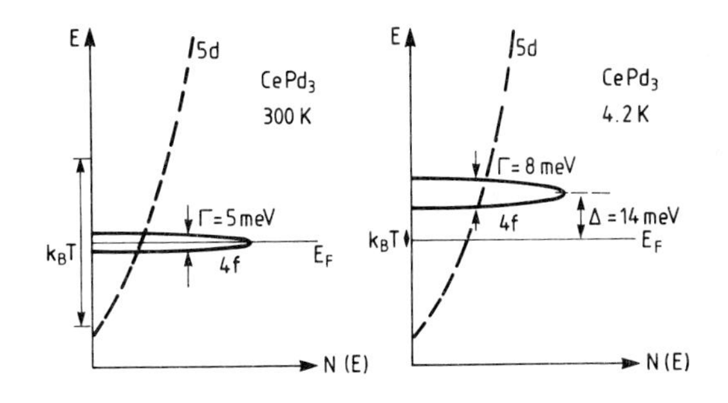

Fig. 5: Quasiparticle density of states of $CePd_3$ near the Fermi energy E_F at 300 K and 4.2 K.

b. Electron tunneling experiments: Evidence for quasiparticle excitations of bound electron-hole pairs.

In the tunneling spectra of $CePd_3$ an inelastic excitation is observed near $\Delta = \pm$ 14 meV, which is absent for YPd_3 as shown in Fig. 2 (/2/). On the basis of the above deduced asymmetric quasiparticle density of states of $CePd_3$ at 4.2 K (Fig. 5) it is somewhat surprising that we observe a symmetric tunneling spectrum. To overcome this difficulty with the single particle density of states scheme we attribute the value Δ to a quasiparticle excitation energy of bound electron-hole pairs, where 2Δ = 28 meV is a measure of the electron-hole binding energy. For these electron-hole pairs a symmetric tunneling spectrum is obtained in analogy to the tunneling process of Cooper pairs in superconductors. The electron-hole pair description is consistent with the independence of the spectrum to magnetic fields up to 15 kOe.

CONCLUSIONS

The optical conductivity of $CePd_3$ exhibits at 300 K a maximum near 16 meV, which is attributed to a resonant scattering process of the conduction electrons at the localized 4f states. An analysis of this process using a Drude model modified by a memory function ansatz yields a relaxation rate $\Gamma = 5$ meV, a f-d hybridization energy A = 11 meV and an estimate of the separation between the 4f state and E_F of $\Delta < k_BT = 26$ meV. Analyzing the $CePd_3$ low temperature far infrared absorptivity data of Pinkerton et al. we have found a separation $\Delta = 14$ meV and a width $\Gamma = 8$ meV. The same energy Δ has been found in the tunneling spectrum of $CePd_3$ as inelastic excitations, which we attribute to quasiparticle excitation energies of bound electron-hole pairs.

We do not know exactly, why our model fit yields values of Γ, which are smaller than those of Δ: In the case of valence fluctuations of electrons between the 4f state and E_F one must assume an overlap between the hybridized 4f state and E_F. However, this can only take place, if Γ is of the order of or greater than 2Δ. This discrepancy may be caused by the simplicity of the underlying model. Because Γ is interpreted as the charge relaxation rate Γ_D of the valence fluctuation we can estimate Γ_D to be of the order of or greater than $2\Delta = 28$ meV. The magnetic relaxation rate $\Gamma_M = 13$ meV of $CePd_3$ has been determined by Holland-Moritz et al. from neutron scattering data (/10/). Then the ratio Γ_D/Γ_M becomes 2.2 or greater and is in agreement with a theoretical prediction of Müller-Hartmann (/11/): Assuming a low temperature valence 3.3 of $CePd_3$ (/12/) these calculations of Γ_D and Γ_M give a ratio of $\Gamma_D/\Gamma_M = 3.2$.

ACKNOWLEDGEMENTS

We would like to thank D. Wohlleben and P. Entel for helpful discussions. We also thank R. Pott, E. Cattaneo and J. C. Fuggle for providing the samples and A. Breitschwerdt, W. König. and L. Laidig for technical help.

+Part of the work was supported by the Deutsche Forschungsgemeinschaft through Sonderforschungsbereich 125, Aachen, Jülich, Köln.

REFERENCES

/ 1/ H. Schneider, D. Wohlleben, Z. Phys. B 44, 193 (1981)

/ 2/ G. Güntherodt, W. A. Thompson, F. Holtzberg, Z. Fisk, to be published

/ 3/ F. E. Pinkerton, A. J. Sievers, J. W. Wilkins, M. B. Maple, B. C. Sales, Phys. Rev. Lett 47, 1018 (1981)

/ 4/ S. von Molnar, W. A. Thompson, A. S. Edelstein, Appl. Phys. Lett 11, 163 (1967); W. A. Thompson, S. von Molnar, J. Appl. Phys. 41, 5218 (1970)

/ 5/ B. Hillebrands, G. Güntherodt, R. Pott, A. Breitschwerdt, W. König, to be published

/ 6/ P. Entel, M. Sietz, Solid State Commun. 39, 249 (1981)

/ 7/ J. W. Allen, R. M. Martin, B. Batlogg, P. Wachter, J. Appl. Phys. 49, 2078 (1978)

/ 8/ B. Bussian, I. Frankowski, D. Wohlleben, to be published

/ 9/ P. Entel, B. Mühlschlegel, Y. Ono, Z. Phys. B 38, 227 (1980)

/10/ E. Holland-Moritz, M. Loewenhaupt, W. Schmatz, D. K. Wohlleben, Phys. Rev. Lett. 38, 983 (1977); E. Holland-Moritz, these conference proceedings

/11/ E. Müller-Hartmann, in: "Springer Series in Solid-State Sciences", Vol. 29, T. Moriya (ed.), Springer, Heidelberg, 1981, p. 178

/12/ I. R. Harris, M. Norman, W. E. Gardner, J. Less. Comm. Materials 29, 299 (1972); K. R. Bauchspiess, W. Boksch, E. Holland-Moritz, H. Launois, R. Pott, D. Wohlleben, in: "Valence Fluctuations in Solids", L. M. Falicov, W. Hanke, M. B. Maple (eds.), North Holland Publishing Company, Amsterdam (1981), p. 417

Valence Instabilities
P. Wachter and H. Boppart (eds.)

ON THE VALENCE OF CERIUM IN $CeRu_2$

S.H. Devare and H.G. Devare

Tata Institute of Fundamental Research
Bombay, India

and

H. de Waard

Laboratorium voor Algemene Natuurkunde
Groningen, The Netherlands

Mossbauer spectra of RRu_2 compounds (R = La, Ce, Pr, Nd, Sm) and $ThRu_2$ are reported. It is shown that $CeRu_2$, unlike the other RRu_2 compounds gives a quadrupole split Mossbauer spectrum. This is attributed to Ce being in 4+ valence state in $CeRu_2$.

The rare-earth intermetallic compound $CeRu_2$ has a cubic Laves phase structure of the type $MgCu_2$ and exhibits several interesting properties. It is seen, for instance, that the lattice constant of trivalent RRu_2 Laves phase compounds varies smoothly from a = 7.702 A° for $LaRu_2$ to a = 7.580 A° for $SMRu_2$ [1]. The value of a = 7.536 A° for $CeRu_2$ however, deviates considerably from this smooth trend. It was suggested [1] that this deviation is due to cerium being in the 4+ valence state in $CeRu_2$ instead of in the 3+ state as in the other cases. This is supported by Tessema et al [2] on the basis of bulk measurements like susceptibility and the observation that $CeRu_2$ has a superconducting transition below 5.8 K. Both Ce and La in these compounds have no localized 4f electrons. Tessema et al have also shown that on hydrogenation of $CeRu_2$, the superconducting transition disappears and a peak in the susceptibility is observed at ∿1.4 K. This was interpreted as cerium changing from the tetravalent to the trivalent state. Our Mossbauer spectroscopy measurements of ^{57}Fe as a dilute impurity in RRu_2 compounds (R = La, Ce, Pr, Nd, Sm) revealed a smooth variation of the quadrupole splitting as a function of the atomic number of R for all RRu_2 compounds except for $CeRu_2$. This deviates from the 3+ trend [3] . However, the quadrupole splitting and isomer shift in the case of a probe like ^{57}Fe may be dominated by local contributions and therefore, it may not be very sensitive to cerium valence. For this reason, we also performed Mossbauer measurements of ^{99}Ru in RRu_2 compounds. These clearly demonstrate that this technique can be used profitably to determine the valency of cerium.

The RRu_2 compounds and $ThRu_2$ were prepared by argon arc melting stoichiometric amount of Ru and rare-earth metals. The alloys were annealed at 800°C for about a week. The X-ray characterization of the samples revealed > 95% single RRu_2 phase and ∿5% of unreacted Ru metal except in $CeRu_2$ which had ∿10% free Ru metal. The samples were powdered and absorbers of about 700 mg/cm^2 thickness were used. The source was ^{99}Rh in Ru matrix, produced by the Ru(p,xn) reaction in Ru metal with 45 MeV protons. The irradiated metal containing the ^{99}Rh activity was used as the Mossbauer source without any chemical separation or heat treatment. It was seen that the linewidth of Ru metal absorber was 0.33 mm/s which is about twice the natural linewidth, but comparable to the experimental linewidths reported so far [4]. The Mossbauer spectra were recorded with a constant acceleration drive with both the source and the absorber at 4.2 K. The 90 keV γ-rays were detected with a Ge(Li) detector to minimize the contribution from other radiations. Fig. 1A shows the spectrum for $CeRu_2$ along with that for $ThRu_2$ which is a compound of Th- an isostructural member of the 5f series. The spectra of $NdRu_2$ a typical trivalent compound and of Ru metal are shown in Fig. 1B. It is seen that the spectra corresponding to trivalent rare-earth compounds have a single absorption line

Table 1

Sample	Lattice Parameter a(Å)	Line Width mm/s	Q.S (mm/s)	I.S (mm/s) w.r.t. Ru metal
$LaRu_2$	7·702	0·33(4)	≤ 0·10	0·035(13)
$CeRu_2$	7·536		0·35(4)	0·140(12)
$PrRu_2$	7·624	0·40(3)	≤ 0·10	0·046(8)
$NdRu_2$	7·614	0·35(2)	≤ 0·10	0·039(8)
$SmRu_2$	7·580	0·35(2)	≤ 0·10	0·020(8)
$ThRu_2$	7·649		0·26(2)	0·022(12)

Summary of the results for RRu_2 and $ThRu_2$ compounds.

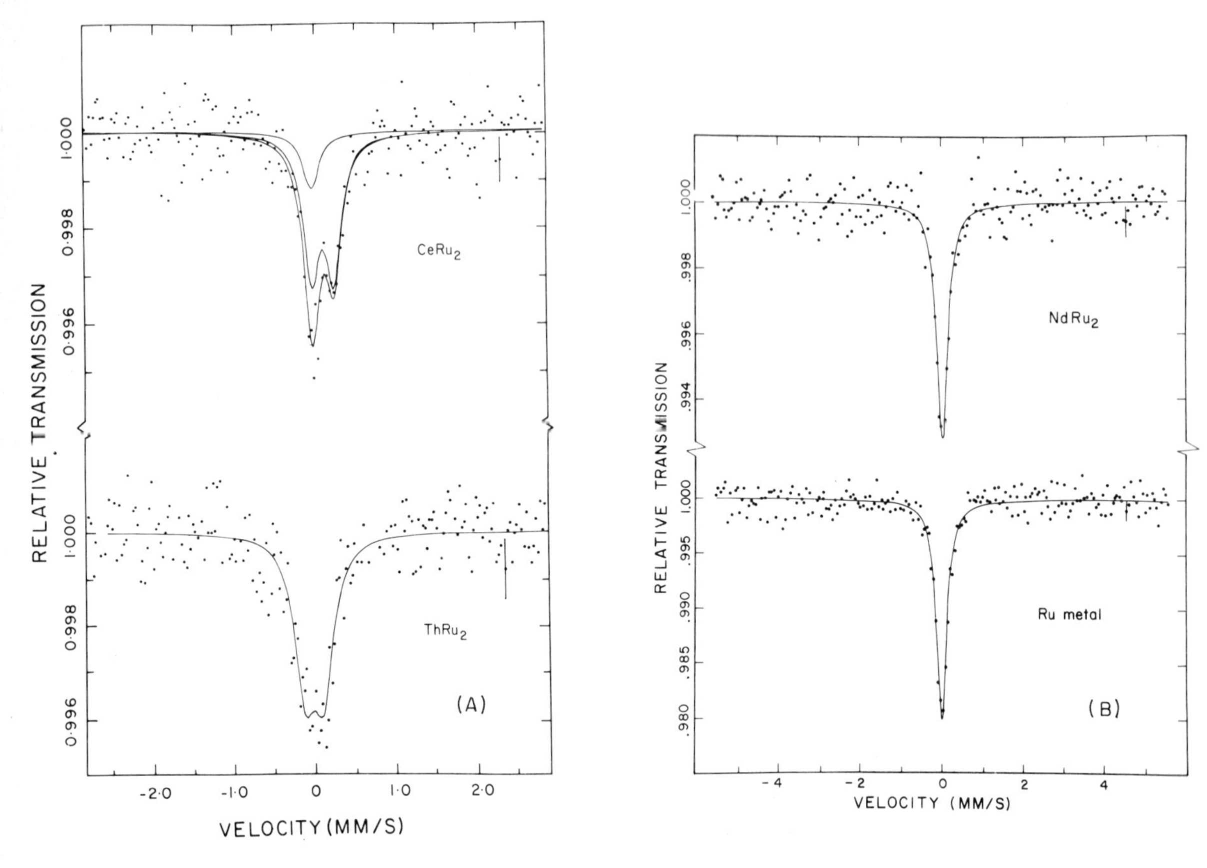

Fig.1. Mossbauer spectra of (A) $CeRu_2$ and $ThRu_2$ and (B) $NdRu_2$ and Ru metal all at 4.2 K along with least squares fitted curves. The asymmetry of the doublet in the case of $CeRu_2$ is due to the single line contribution from free **Ru** metal which has been taken into account in the fit.

while doublets are observed for $CeRu_2$ and $ThRu_2$. It may be noted that even though the magnetic ordering temperatures of the trivalent rare-earth compounds are well above 4.2 K and the site symmetry at Ru is non cubic, no magnetic hyperfine or electric quadrupole splitting is observed and the single absorption line has a width very close to the experimental line-width observed for Ru metal absorber. Furthermore, the isomer shift for these compounds with respect to Ru metal is close to zero. The Mossbauer spectrum of $CeRu_2$ differs drastically from that of the other RRu_2 compounds and exhibits a quadrupole splitting ΔE_Q = 0.35(4) mm/s and an isomer shifts S = 0.14(1) mm/s. These values of ΔE_Q and S measured by us agree well with those reported by Wagner [5]. The large electric field gradient can be interpreted as due to the 4+ state of cerium. This is further supported by the fact that the isostructural compound of $ThRu_2$ in which Th has a 4+ valence shows a similar quadrupole splitting ΔE_Q = 0.26(2) mm/s. Thus, these measurements clearly demonstrate that the valence of Ce in $CeRu_2$ can be directly monitored by the Mossbauer spectrum of ^{99}Ru and any change in cerium valency due to hydrogenation or any other reason should be clearly observable by this technique.

The large difference in the electric field gradient at the Ru site for 3+ and 4+ valency of the rare-earth can be understood in terms of a simple point charge model if the charge on Ru is assumed to be between 1.4e to 1.5e. Using this range of Ru charge in $CeRu_2$ and expressing the observed electric field gradient as $eq_{expt} = eq_{latt}\ (1-\gamma_\infty)\ (1-K)$, the value of the parameter (1-K) lies between 2 and 3 just as in the case of pure metals [6]. The observed positive isomer shift in the case of $CeRu_2$ corresponds to an increase in the charge density at the Ru nucleus as the change of nuclear radius $\Delta R/R$ is positive for ^{99}Ru [5]. Apparently, the extra electron contributed by the cerium to the conduction band leads to an increase in the S-electron density at the Ru site. A similar increase in charge density has been observed [3] using ^{57}Fe as the Mossbauer probe in $CeRu_2$. In this case a negative isomer shift is found as $\Delta R/R$ is negative in the case of ^{57}Fe.

The authors are thankful to F. ten Brek and F. Pleiter for their help in the production and handling of the activity. They wish to thank R.G. Pillay for performing the lattice sum calculations and for some useful suggestions.

Two of the autors (S.H.D. and H.G.D.) would like to acknowledge with thanks the hospitality of the the Hyperfine Interactions group of the Laboratorium voor Algemene Natuurkunde, Groningen. They are also grateful for the financial support from FOM (H.G.D.) and the Rijksuniversiteit te Groningen (S.H.D.). This work was performed as a part of the research programme of the "Stichting voor Fundamenteel Onderzoek der Materie" (FOM) subsidized through the "Nederlandse Organisatie voor Zuiver Wetenschappelijk Onderzoek" (ZWO).

References:

1. Gschneider, K.A., Rare Earth Alloys (D. van Nostrand Company Inc., Princeton, 1961) p.380.

2. Tessema, G.X., Peyrard, J., Nemoz, A., Senateur, J.P., Rouault, A. and Fruchart, R., Z. fur Phys. Chem. Neue Folge, 116 (1979) 793-798.

3. Devare, S.H., Pillay, R.G. and Devare, H.G., Proceedings of the International Conference on the Applications of the Mossbauer Effect, Jaipur, 1982 (To be published).

4. Good, Mary L., A review of the Mossbauer spectroscopy of Ru-99 and Ru-101, in Stevens, John G. and Stevens, Virginia E. (eds.), Mossbauer Effect Data Index (Plenum, New York, 1972) 51-70.

5. Wagner, F.E. and Wagner, U., Mossbauer isomer shifts in 4d and 5d transition elements, in Shenoy, G.K. and Wagner, F.E. (eds.), Mossbauer Isomer Shifts (North-Holland, Amsterdam, 1978) p.497.

6. Raghavan, R.S., Raghavan, P. and Kaufmann, E.N., Phys. Rev. Lett. 34 (1975) 1280.

Valence Instabilities
P. Wachter and H. Boppart (eds.)

THE ENERGY BALANCE OF MIXED VALENCE IN $EuCu_2Si_2$

J. Röhler (a), D. Wohlleben (a) and G. Kaindl (b)

(a) II. Physikalisches Institut der Universität zu Köln
D 5ooo Köln 41, Germany
(b) Institut für Atom- und Festkörperphysik
Freie Universität
D 1ooo Berlin, Germany

The pressure dependence of the ^{151}Eu Mössbauer isomer shift (and of the volume) of $EuCu_2Si_2$ is reported up to 53 kbar at 3oo K. It is analyzed together with the pressure dependence of the susceptibility and with the temperature dependence of isomer shift and susceptibility at ambient pressure to yield unambiguous values of the interconfigurational excitation and mixing energies. From these data we derive an energy balance of the system and trace the pressure and temperature dependence of the Fermi energy, the elastic energy and the mixing energy.

INTRODUCTION

Strong temperature dependence of the valence (from 2.8 - 2.9 near T=O to 2.4 - 2.2 near 700 K) accompanies the valence fluctuations of Europium in intermetallic compounds (1,2) or in concentrated (3,4) or dilute (5) alloys. Such large valence changes imply e.g. shifts of order eV of the conduction electron Fermi energy (ε_F) in concentrated systems. This energy is not available thermally. The observed valence changes can therefore occur only in the course of a much smaller shift of some energy balance in which ε_F is just one component, whose shifts are largely compensated by shifts of some combination of other components like the averaged configurational ("spectroscopic") energy ($\Delta_{n+1,n}$), the elastic energy (F_L), the mixing energy and the "bound" (entropic) energy TS. Since external pressure can also induce valence changes of order one, either continuously as in SmSe and SmTe (6) or discontinuously as in SmS (6), mechanical work done from the outside should be included in the balance.

In this paper we describe an attempt to properly identify and to trace quantitatively the most important components of this energy balance as function of temperature and pressure in $EuCu_2Si_2$ (7). The temperature and pressure dependence of the elastic energy F_L and of the Fermi energy ε_F can be calculated directly if one knows the valence changes, the bulk moduli and the volumes of the integral valent reference compounds, and the density of states at the Fermi energy, $\rho(\varepsilon_F)$, which one obtains from the linear specific heat coefficient of the reference. The spectroscopic energy $\Delta_{n+1,n}$ is an ionic constant. The main task is the proper measurement of the energy balance itself and of the mixing energy.

A measure of the energy balance is the well known interconfigurational excitation energy E_x, which can be extracted from thermal averages like Mössbauer (ME) isomer shift, susceptibility etc. by statistical thermodynamics. This can be done quite reliably in the case of Eu, because of some peculiarities of the spectra of that ion in its di- and trivalent configurations $4f^7$ and $4f^6$, which remain unaffected by the solid. (There are e.g. no CEF effects in their Hund's rule ground states $E_{7,o}$ (J=S=7/2) and $E_{6,o}$ (J=O)). A spectrum constructed by simple superposition of the ionic $4f^7$ and $4f^6$ spectra at mutual distance $E_x \equiv E_{7,o}-E_{6,o} > 0$, describes the observed Mössbauer isomer shifts reasonably well above 200 K for fixed E_x (1-5). At lower temperatures a lifetime width Γ of all levels (8,5) or some equivalent (1-4) must be introduced into this spectrum as an essential additional feature in order to account for the unequivocal evidence from ME and L_{III} (16) that the ion does not condense into $E_{6,o}$ at T=O, i.e. does not become completely trivalent. In the following we shall take Γ as a measure of the mixing energy.

Both E_x and Γ depend on temperature and pressure in principle. They are not independent of each other, since the mixing energy must be at maximum at configurational crossover and must go to zero for $|E_x| \to \infty$, at stable, integral valence. However, their functional dependence is not known a priori, and therefore they must be treated as independent variables initially. Temperature dependence of the energy balance (E_x) was noted early in more refined analyses of the isomer shift at high temperatures (1,3,4) and no pair of temperature independent values of E_x and Γ can be found to fit the Mössbauer data within experimental error at low temperatures either (8,5). The evidence for temperature dependence becomes stronger when one tries to fit two measurements (e.g. isomer shift δ and susceptibility χ) simultaneously(4,5,8,9). Indeed, if Γ (T) and E_x(T) are regarded as independent variables, the proper way to determine them from statistical thermodynamics should be fits to two independent observables. We have in the following determined E_x(T,p) and Γ(T,p) from simultaneous fits to measurements of δ(T,p) and χ(T,p) along the isobar p=0 between 4 and 300 K and along the isotherm at 300 K between 0 and 50 kbar. For this we use data from the literature (1,9,8) and measurements of the isomer shift and of the volume under pressure (10) which we describe first in the following. We then discuss E_x(T,p) and Γ(T,p) and the function $\Gamma(E_x)$ which seems nearly independent of p and T. We finally show a decomposition of the energy balance into its components. For a discussion of the thermodynamic equations used in the analysis we refer to (7,11) and (12).

EXPERIMENTAL PROCEDURE AND RESULTS

The $EuCu_2Si_2$ sample used in the high pressure Mössbauer experiment (10) was the same as used in the measurements of the susceptibility as function of temperature at ambient pressure and as function of pressure at 300 K ((8,9) and Fig. 2). An opposed anvil device with B_4C anvils was employed. The pressure was determined in situ by two different methods: At low temperature from the superconducting transition temperature of Pb and at room temperature from the isomer shift of the ^{119}Sn γ-resonance in β-Sn (13). The compressibility data were taken with a diamond cell using Mo-K_α X-rays. Details of the experimental technique are described in (10). Typical Mössbauer spectra of $EuCu_2Si_2$ under pressure are presented in Fig.1. The solid lines represent the results of least-squares fits of the spectra with a superposition of electric quadrupole split Lorentzian. lines. This reveals a small electric quadrupole splitting of eqQ(7/2) = +0.55mm/s at ambient pressure, which decreases slightly with pressure (by about 20% at 53 kbar). In addition, a weak satellite line at - 7.5 mm/s (at ambient pressure) with a relative intensity of only 6.5% was taken into account.

In Fig. 2a the isomer shift is plotted as a function of pressure together with the pressure dependence of the relative volume at 300 K. A combination of both measurements yields a linear relationship between isomer shift and relative volume at 300 K with a slope of $d\delta/d(\ln V) = -(80\pm10)$mm/s, which is very large compared with the average for stable valent Eu solids ($d\delta/d(\ln V)|_{av} = -(6\pm2)$mm/s (14)).

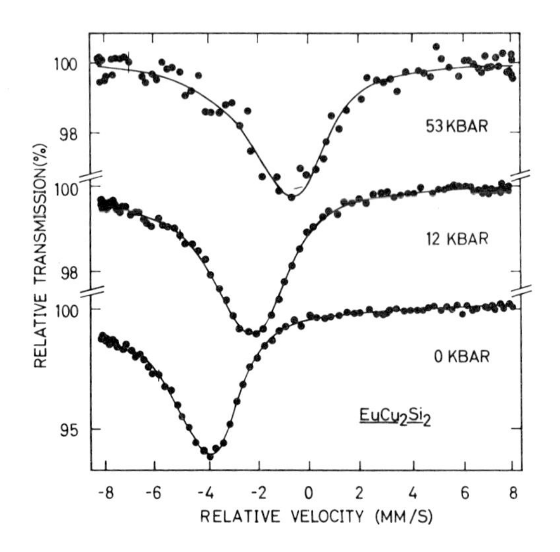

Fig. 1 ^{151}Eu(21.7-keV)-Mössbauer absorption spectra of $EuCu_2Si_2$ at 300K and at the indicated pressures.

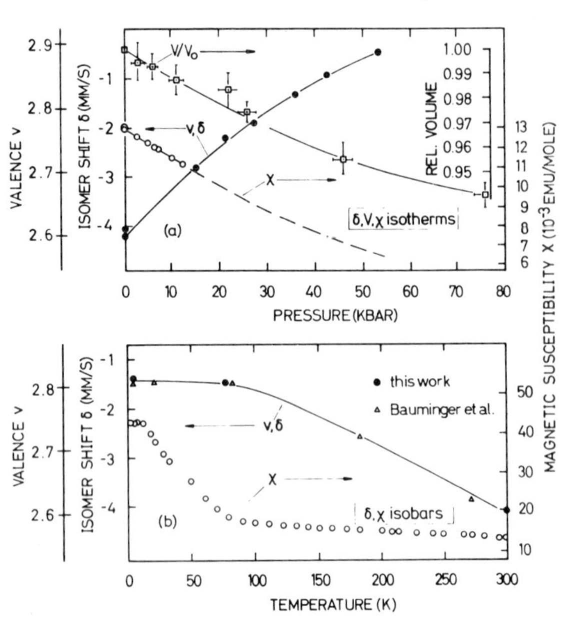

Fig. 2 (a) Pressure and (b) temperature dependence of the ^{151}Eu isomer shift δ and of the magnetic susceptibility χ for $EuCu_2Si_2$. In (a) also the relative volume at 300 K is given as a function of pressure. Next to the δ ordinate a scale is added for the valence.

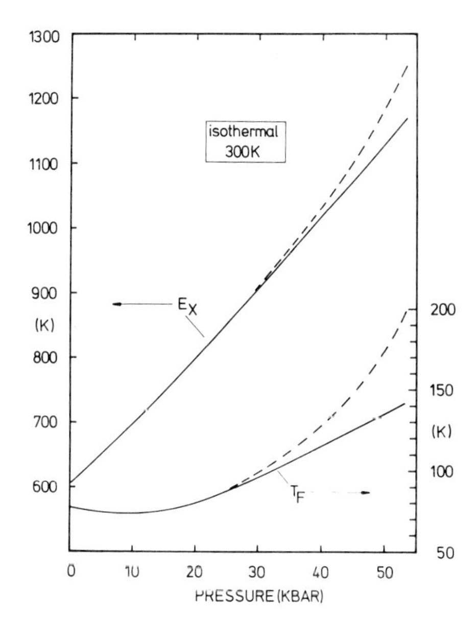

Fig. 3 Pressure dependence of the interconfigurational excitation and fluctuation energies, E_x and k_BT_f, of $EuCu_2Si_2$ at 300 K.

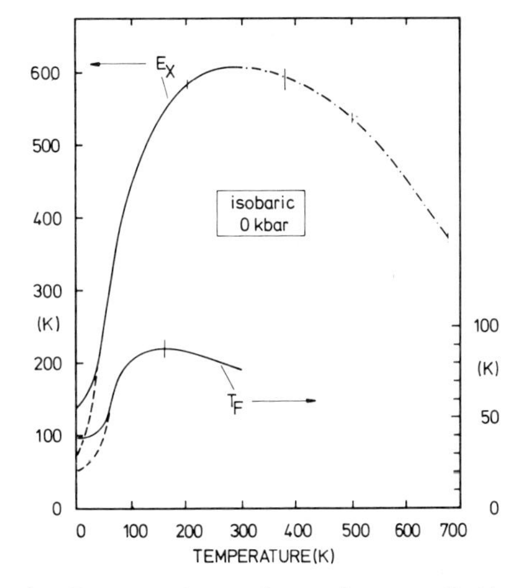

Fig. 4 Temperature dependence of the interconfigurational excitation and fluctuation energies, E_x and k_BT_f, of $EuCu_2Si_2$ at atmospheric pressure.

In Fig. 2b the variation of δ for our sample of $EuCu_2Si_2$ is plotted as a function of temperature at ambient pressure together with the results of Bauminger et al.(1) Both measurements agree well with each other, if the data of (1) are shifted by - 0.4 mm/s. This shift is correlated with intensity and position of a satellite line (1o), which was much stronger in the sample of (1) and in our own samples than in that of B. Sales used here.

The isomer shift measures the valence v= 2+ν via

$$(1) \quad \delta(T,V) = \delta_2(T,V) + (\delta_3-\delta_2)\nu(T,V)$$

Here ν is the fractional occupation of $Eu^{3+}(4f^6)$. δ_3 and δ_2 are the isomer shifts of stable Eu^{3+} and Eu^{2+} ions. In deriving $\nu(T,V)$ from equ.(1) small corrections were applied which take the volume dependence of δ_2 into account: In the high pressure experiment $\delta_2(V)$ was obtained from $d\delta/d(\ln V)_{av}$ = -6mm/s (14) using $d(\ln V)/dp$ as measured here (Fig. 2a). In the zero pressure experiment the thermal redshift contribution to δ_2 as well as the temperature dependence caused by the known thermal expansion of $EuCu_2Si_2$ (9) were included. For $\delta_3-\delta_2$ a value of 13±1 mm/s was assumed, which is slightly larger than the corresponding difference determined for the semiconductor Eu_3S_4 (11.8 mm/s) (15). (In a metallic IV system we expect $\delta_3-\delta_2$ to be slightly larger). δ_2 in equ.(1) was set equal to -12 mm/s at ambient pressure and room temperature. The valence scale obtained in this way is indicated next to the isomer shift in Fig. 2.

From the pressure and temperature dependence of valence ν and susceptibility χ we extract $E_x(T,p)$ and $\Gamma(T,p)$ by use of the following expressions (1,11)

$$(2) \quad \frac{1-\nu}{\nu} = \frac{\zeta_7'}{\zeta_6'} \exp-\left[\frac{E_x(T,p)}{k_BT^+(T,p)}\right]$$

$$(3) \quad \chi(T,p) = (1-\nu)\chi_7(T^+,p) + \nu\chi_6(T^+,p)$$

Here the ζ_j' are the ionic partition functions with energies measured from the lowest interconfigurational levels $E_{j,o}$. χ_7 and χ_6 are the Curie and Van Vleck susceptibilities of $4f^7$ and $4f^6$. In the ζ_j''s and χ_j's we have used the usual intraionic spectra (4,5). In all ζ_j' and χ_j the temperature is replaced by an effective temperature $T^+=(T^2+T_f^2(T,p))^{1/2}$ (11). We identify $k_BT_f(T,p)\equiv\Gamma(T,p)$ as the mixing energy.

Pairs of values for E_x and T_f are shown for the isothermal pressure variation in Fig. 3 and for the isobaric temperature variation in Fig. 4, where we also show E_x beyond 300 K out to 700 K, as extracted from the data of Bauminger et al.(1). Our $E_x(T)$ varies from theirs, because we have included more multiplet states of $4f^6$ and because of a more realsitic choice of δ_3 (4,5). It is not necessary to know $\chi(T)$ here, because at these high

temperatures T_f has no significant influence on δ and χ: bars indicate changes of E_x if T_f is varied between 0 and 200 K. We can of course not give any values of T_f in this region. At the lowest temperatures we have used both high and low field susceptibilities of (9) for the determination of E_x and T_f (drawn out lines at H= 8 T, dashed lines at H=0, data shown in Fig. 2b are 8 T values only). We emphasize that we are now confident, on the basis of much independent evidence, that most of the rise of $\chi(T)$ below 80 K in Fig. 2b is an intrinsic property of the mixed valence state and not due to residual divalent Eu in the sample.
The dashed lines in Fig. 3 indicate our uncertainty of E_x and T_f arising from the extrapolation of $\chi(p)$ in Fig. 2a (dashed line) beyond the measurement (9).

DISCUSSION

If one plots E_x and T_f of Figs 3 and 4 against each other (7), one obtains a curve in which the isobaric and the isothermal data join smoothly, without break in slope. The curve $T_f(E_x)$ has values for 100 K < E_x < 1100 K, a maximum near E_x = 500 K, and a minimum near $E_x \approx$ 700 K. It suggests that $T_f(E_x)$ is fairly independent of T and p and has maxima at configurational crossover, e.g. at E_x = 480 K, where the $4f^7$ level $E_{7,o}$ crosses the J=1 level $E_{6,1}$ of $4f^6$. One may speculate that there are maxima at E_x=0 and E_x= 1330, the crossings of $E_{7,o}$ with $E_{6,o}$ and $E_{6,2}$. These maxima are just beyond the range of the measured E_x, but within this range the curve is consistent with a ratio of 1:3:5 for T_f at E_x=0, 480 and 1330, as expected for the behavior of the mixing energy as function of E_x at a sequence of crossovers of one level ($E_{7,o}$) with three others, with degeneracy ratio 1:3:5.

If there is a function $T_f(E_x)$ which depends on the spectra of the involved configurations but not on T and p, then there is just one variable, whose temperature and pressure dependence needs explanation. For this we chose $E_x(T,p)$. From the condition of thermal equilibrium with respect to valence ν, we obtain from the Gibbs free energy

$$(4)\quad E_x \equiv k_BT^+\ln(\zeta'_{n+1}\nu/\zeta'_n(1-\nu)) - (\pi k_BT^+)^2/6\varepsilon_F(\nu)$$
$$= \Delta_{n+1,n} - \varepsilon_F(\nu) + \partial F_L/\partial\nu + p\partial V/\partial\nu \qquad + E_{xm}$$

V is the cell volume. The partial derivatives are taken at constant T and p. For Eu, n=6. E_{xm} is a well defined but complicated function of the intraionic and conduction electron free energies and entropies and of E_x and dT_f/dE_x which arises from the dependence of the mixing energy on valence, i.e. from the fact that $\partial T_f/\partial\nu \neq 0$, (7), (12).

Fig. 5 shows a decomposition of $E_x(T,p)$ (7) as "measured" (first line of equ. 4 and Figs. 3,4) into components (second line of equ. 4). The term labelled "conduction electrons" was calculated with the valence changes of Fig.2 with

$$(5)\quad \Delta\varepsilon_F(T,p) = (\nu(T,p)-\nu_o)/\rho(\varepsilon_F)$$

using $\rho(\varepsilon_F)$ = 1.65 (states /eV cell) from the linear specific heat coefficient of $LuCu_2Si_2$ (17) and ν_o = 0.8 at T=0 and p=0. The term labelled "elastic" was derived from an Ansatz

$$(6)\quad F_L = C(V(T,p,\nu)-V_o)^2B(\nu)/2V_o$$

Here V_o is a constant volume and

$$(7)\quad B(\nu) \equiv V_o(\nu B_6/V_6 + (1-\nu)B_7/V_7)$$

is an isovalent bulk modulus as interpolated linearly between those of the integral valent reference compounds. We use B_6 = 1240 kbar from the high pressure end of V(p) in Fig. 2 and B_7=930

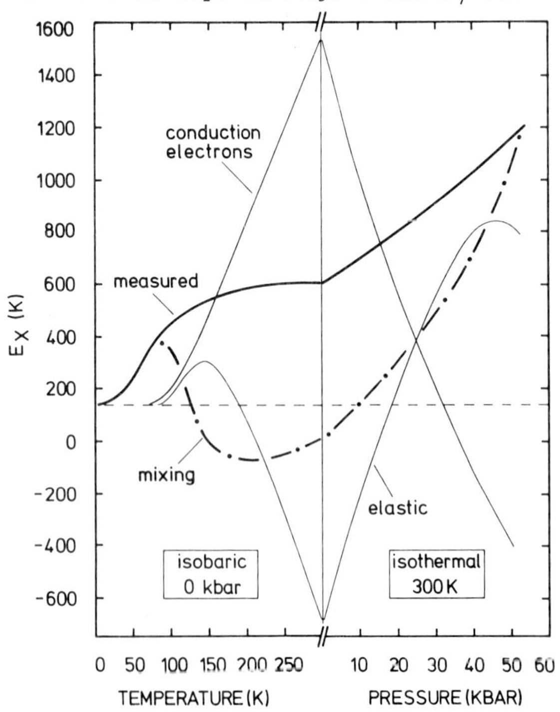

Fig. 5 Temperature and pressure dependence of the excitation energy E_x and its decomposition into various contributions.

kbar from a measurement of $CaCu_2Si_2$ (18). For $V(\nu,T,p)$ we take into account the deviations from Vegard's law discussed elsewhere at this conference (19). We find $v_o = 41$ Å^3 and c=2.

Note that the shift of the Fermi energy, which gives a very large contribution to E_x, is nearly compensated by an opposite shift of the elastic energy, in both the isobaric and the isothermal experiments. This is probably the reason why the valence changes in mixed valent compounds and impurities are so similar (11): They are primarily controlled by entropy and by the mixing term.

The component E_{xm} labelled "mixing" was calculated from the "measured" $E_x(T,p)$, $T_f(T,p)$, dT_f/dE_x, T^+ and the $\zeta_j^!(T^+)$. This term, which would be zero in the absence of mixing, has a tendency to pin E_x to the nearest configurational crossover, because there the system can take best advantage of the mixing energy. The dashed line is the zero of E_{xm}; E_{xm}=0 occurs whenever dT_f/dE_x=0, i.e. at the maxima and minima of $T_f(E_x)$. This term is responsible for the entire dramatic temperature dependence of $E_x(T)$ below 80 K, where the valence is nearly stationary, so that conduction electron - and elastic energies do not change. The term shows how heating leads into mixed valence and is probably at the root of those anomalies which are often observed at low temperatures and labelled Kondo anomalies.

This work was partially supported by SFB 161 and SFB 125 of the Deutsche Forschungsgemeinschaft. We thank I. Nowik, W.B. Holzapfel and J.Schilling for discussions and N.v.Randow for her efficient typesetting.

REFERENCES

(1) E.R. Bauminger, D. Froindlich, I. Nowik, S. Ofer, I. Felner, I. Mayer, Phys. Rev. Letters 30, 1o53 (1973)

(2) E.V. Sampathkumaran, L.C. Gupta, R. Vijayaraghavan, K.V. Gopalakrishnan, R.G. Pillay and H.G. Deware, J. Phys. C 14, L 237 (1981)

(3) E.R. Bauminger, I. Felner, D. Froindlich, D. Levron, I. Nowik, S. Ofer and R. Yanovsky J. Physique Coll. 35, C 6-61 (1974)

(4) I. Nowik, pg.261 in "Valence Instabilities and Related Narrow Band Phenomena", R.D. Parks, editor, Plenum Press, New York, London (1977)

(5) W. Franz, F. Steglich, W. Zell, D. Wohlleben and F. Pobell, Phys. Rev. Lett. 45, 64 (198o)

(6) A. Jayaraman, V. Narayanamurti, E. Bucher and R.G. Maines, Phys. Rev. Lett. 25, 368 and 143o (197o)

(7) J. Röhler, D. Wohlleben, G. Kaindl and H. Balster, to be published

(8) B.C. Sales, D.K. Wohlleben, Phys. Rev. Lett. 35, 124o (1975)

(9) B.C. Sales, R. Viswanathan, J. Low. Temp. Phys. 23, 449 (1976)

(1o) J. Röhler, Diplom-Thesis, Ruhr-Universität Bochum (1979)

(11) D. Wohlleben, pg. 1 in "Valence Fluctuations in Solids", L.M. Falicov, W. Hanke and M.B. Maple, editors, North Holland, Amsterdam (1981)

(12) D. Wohlleben, J. Röhler, R. Pott, G. Neumann and E. Holland-Moritz, this conference.

(13) H.S. Möller, Z. für Physik 212, 107 (1968); V.N. Panyushkin, Sov. Phys. Sol. State 1o, 1515 (1968)

(14) G.M. Kalvius, U.F. Klein, G. Wortmann, J. Physique Colloqu. 35, C 6-136 (1974); U.F. Klein, Ph-D Thesis, Techn. Univ. München (1976)

(15) O. Berkooz, M. Malamud, S. Shtrikman, Solid State Comm. 6, 185 (1968) J. Röhler, G. Kaindl, Solid State Comm. 36, 1055 (1981)

(16) T.K.Hatwar, R.M. Nayak, B.D.Padalia, M.N. Ghatikar, E.V.Sampathkumaran, L.C. Gupta and R. Vijayaraghavan, Solid State Comm. 34, 617 (198o)

(17) R. Kuhlmann, H.-J. Schwann, R. Pott, W. Boksch and D. Wohlleben, this conference

(18) M. Barth, Diplom-Thesis, Universität zu Köln (1981)

(19) G. Neumann, R. Pott, J. Röhler, W. Schlabitz, D. Wohlleben and H. Zahel, this conference.

Valence Instabilities
P. Wachter and H. Boppart (eds.)

NEUTRON STUDY OF CRYSTAL FIELD EFFECTS IN INTERMEDIATE VALENCE COMPOUNDS

E. Holland-Moritz

II. Physikalisches Institut
Universität zu Köln
Zülpicherstr. 77
5000 Köln 41
W. Germany

It was established several years ago by inelastic neutron scattering that the Zeeman levels of Ce and Yb intermediate valence (IV) compounds have a structureless magnetic spectrum with a width of order 10 meV even near T=0 K. This width is of the order of the usual crystal field splittings of stable valent Ce- and Yb-compounds and therefore the question arises, whether residual crystal field splittings might be hidden in the above spectra which are usually fitted by a single Lorentzian with slightly temperature dependent linewidth Γ_S. It is indeed always possible to fit the observed spectra also with a set of crystal field levels and their transition lines; all smeared out by a width $\Gamma_{CF} < \Gamma_S$ /1/. In order to decide whether a single line spectrum or a multiline CF-spectrum comes closer to the truth, it is necessary to perform measurements with incoming neutron energies which are smaller as well as larger than the expected crystal field splittings. Here we describe work of this nature on the IV-compound $YbCu_2Si_2$ with three different incoming neutron energies: 3.5, 12.5 and 51 meV. We found three arguments for the existence of inelastic transition lines:

1. The total magnetic scattering cross-section of a single line fit is two times larger than the cross-section of Yb^{3+}, i.e. impossibly large.
2. With E_o=51 meV the scattering intensity in the energy range 20-30 meV increases with decreasing temperature as expected for a magnetic excitation from the ground state.
3. The most convincing argument for the existence of CF-transitions is given in fig. 1. There the quasielastic line width is plotted in dependence of temperature. The results in the upper part are from the analysis with single line fits Γ_S. One sees that the experimental points obtained with different incoming neutron energies do not coincide. This discrepancy can only be resolved by introducing crystal field splittings. Careful fits of all available data lead finally to the crystal field scheme shown in the inset of the lower part of fig. 1. The corresponding temperature dependence of the quasielastic linewidth (Γ_{CF}) is presented below.

We arrive at the conclusion that our more elaborate effort unequivocally points to crystal field

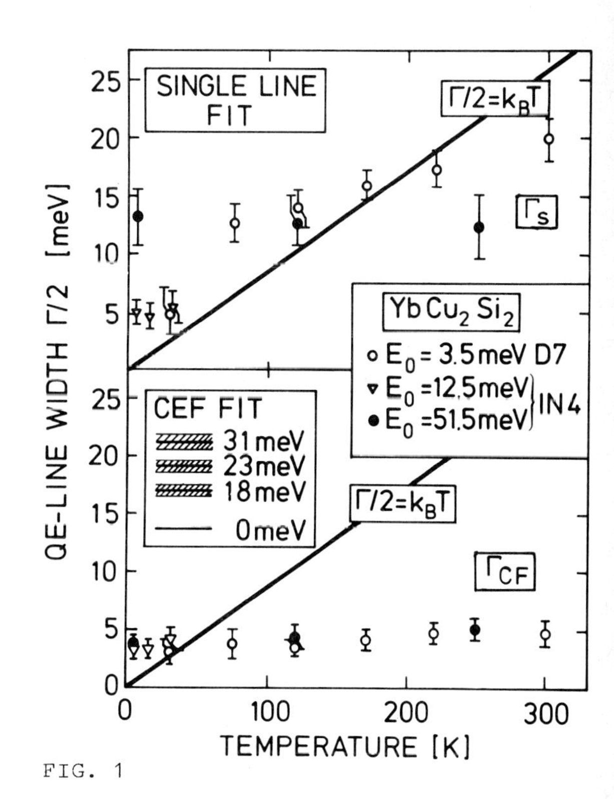

FIG. 1

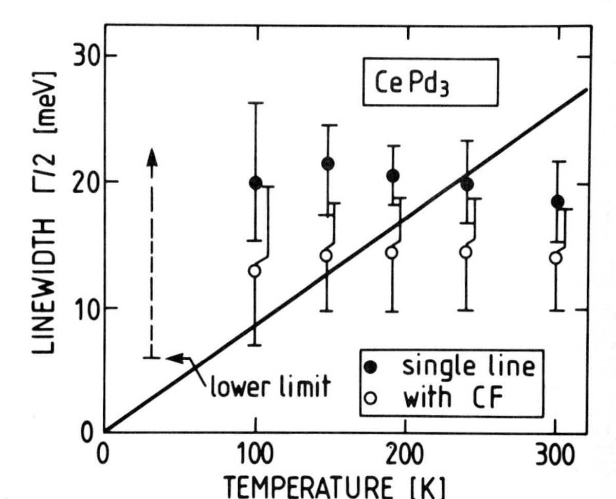

FIG. 2

splittings in $YbCu_2Si_2$ rather than a single line spectrum. Efforts comparable to the one applied to $YbCu_2Si_2$ and described in more detail in ref. /2/ will be needed to decide whether crystal field splittings are also hidden in the spectra of $CePd_3$, $CeSn_3$, $CeBe_{13}$, /3/ and YbCuAl /4/. For the time being the (single) linewidths reported for these compounds must be regarded as upper limits to the true linewidth. Fig. 2 shows the difference between the single line analysis and the CEF-analysis in the case of $CePd_3$ assuming a crystal field splitting energy of 10 meV. The quasielastic linewidth remains temperature independent in a way that $\Gamma_{CF} = 2/3\ \Gamma_S$. One finds similar behaviour for the other IV-Cerium compounds.

References

/1/ E. Holland-Moritz, M. Loewenhaupt, W. Schmatz, D. Wohlleben; Phys. Rev. Lett. 38 983 (1977)

/2/ E. Holland-Moritz, D. Wohlleben, M. Loewenhaupt; to be published in Phys. Rev. B 25 (May or June 1982)

/3/ E. Holland-Moritz; JÜL SPEZ 14 (1978)

/4/ W.C.M. Mattens, F.R. de Boer, A.P. Murani, G.H. Lander; J. Magn. Magn. Mat. 15-18, 973 (1980)

KONDO

Valence Instabilities
P. Wachter and H. Boppart (eds.)

MIXED VALENT AND KONDO SYSTEMS: CONFIGURATIONAL AND FERMI LIQUID APPROACHES

T.V. Ramakrishnan

Department of Physics, Indian Institute of Science, Bangalore 560012, India

The microscopic theory of a mixed valent impurity in the large orbital degeneracy limit ($n_\lambda >> 1$) is briefly reviewed. The description of results in terms of effective ionic configurations or a local Fermi liquid is made precise. The importance of alloying pressure in determining valence is emphasized and a simple model for incorporating its effect is suggested. We then describe the Fermi liquid or resonant level model for the lattice. A semiconducting ground state appears unlikely in this model, and some alternative suggestions are made. Finally, the resistivity of mixed valent metals and alloys is discussed. The most important effect of impurities and lattice vibrations on resistivity is via local strain induced valence change and consequent scattering. From a model for this, the saturation behaviour, maximum metallic resiativity as a function of valence, and low temperature T^2 term are explained.

1. INTRODUCTION

Two phenomenological approaches for the mixed valent impurity have been successfully used recently. One of them, due to Wohlleben and coworkers [1], emphasizes persistence of ionic configurations. The other is a local Fermi liquid model discussed in detail by Newns and Hewson [2]. Using a microscopic theory of the mixed valent impurity developed recently [3,4], we discuss exactly what configurational and Fermi liquid approaches mean, and how their parameters are related to those of a microscopic Hamiltonian. In the mixed valent regime, there are strong cohesive or alloying energy contributions which determine the valence. These are not generally considered; we show a simple way in which they can be included. This discussion sets the stage for a Fermi liquid model for the lattice, which consists simply of a lattice of effective f-level resonances. The f-level position is constrained by valence. This narrow band metal is specially prone to disorder, and for strains induced thermally or by weak disorder can produce local valence fluctuations sufficient to make it behave as an incoherent collection of resonant level scatterers. Using this idea we derive an expression for maximum metallic resistivity for the Kondo regime, and as a function of valence in the intermediate valence regime. There is both qualitative and quantitative agreement with experimental results on $CePd_3$, Ce and their alloys. For the resistivity, the large T^2 terms, the saturation tendency, and the limiting value are all explained in a single model. The large orbital degeneracy of one of the configurations plays a crucial simplifying role in all our considerations. For example, corrections to the theory outlined here are of order $(1/n_\lambda)$ i.e. about 10%.

2. MIXED VALENT IMPURITY

2.1 Configurational Model

The theory of a mixed valent impurity has been discussed recently at length [3,4] following the earlier work of Bringer and Lustfeld [5]. One considers the situation where two different ionic configurations f^n and f^{n-1} with energies ε_λ and ε_o are very close in energy, i.e. $(\varepsilon_\lambda - \mu) - \varepsilon_o = (\tilde{\varepsilon}_\lambda - \varepsilon_o)$ is small, μ being the chemical potential. The former configuration (state $|\lambda\rangle$) is magnetic, with large orbital degeneracy n_λ while the latter (state $|0\rangle$) is nondegenerate and nonmagnetic. The configurations admix via hybridization with the conduction band, and the basic energy scale of the problem is $\Delta = |V_{k\lambda}|^2\rho(\mu)$. The principal results obtained [3,4] are the following.

The large orbital degeneracy n_λ plays an important role in several ways [6]. The mixed valent regime [7] is enlarged by a factor n_λ and covers the range $\infty > (\tilde{\varepsilon}_\lambda - \varepsilon_o) \gtrsim \{-n_\lambda\Delta \ln(D/n_\lambda\Delta)\} = \varepsilon_c$ where 2D is the conduction electron bandwidth. In this regime, the zero temperature valence $n_v = \text{Prob.}(f^n) = \sum_\lambda |\langle \psi_g|\lambda\rangle|^2$ varies from $0(\tilde{\varepsilon}_\lambda = \infty)$ to 0.8 or 0.9 (for $\tilde{\varepsilon}_\lambda \simeq \varepsilon_c$).

In Cerium, for example, this would range from 4^+ to nearly trivalent (valence 3.1 to 3.2). Low order Brillouin-Wigner perturbation theory (see below) is adequate in this regime. For example, when $\tilde{\varepsilon}_\lambda - \varepsilon_o = 0$, the second order term is $[n_\lambda \ln(D/n_\lambda\Delta)]^{-1}$ times the leading term, and is thus very small, since n_λ is typically six or more. The expansion parameter increases as $(\tilde{\varepsilon}_\lambda - \varepsilon_o)$ decreases and is of order 25% at $(\tilde{\varepsilon}_\lambda - \varepsilon_o) \simeq \varepsilon_c$. To summarize, there is a simple $(1/n_\lambda)$ expansion for the mixed valent regime and a quantitative theory is possible for the thermodynamic properties of a mixed valent impurity.

In the lowest order one has the following self-consistent equations for real, sharp effective energy levels $\tilde{E}_\lambda$ and E_o.

$$E_o = \varepsilon_o + \sum_{k\lambda} [|V_{k\lambda}|^2 f^-_k/\{E_o - (\tilde{\varepsilon}_o - \varepsilon_k)\}] \qquad (1a)$$

$$\tilde{E}_\lambda = \tilde{\varepsilon}_\lambda + \sum_k [|V_{k\lambda}|^2 f^+_k/\{\tilde{E}_\lambda - (\varepsilon_o + \varepsilon_k)\}] \qquad (1b)$$

In terms of these, the partition function

$$Z = Z_{cond.}[\exp(-\beta E_o) + \sum_\lambda \exp(-\beta \tilde{E}_\lambda)]$$

i.e. it has the same form as in the absence of configurational admixture. The conduction electron states are integrated over and their effect is contained entirely in E_o and $\tilde{E}_\lambda$. The physics of the mixed valent impurity can thus be discussed in terms of what happens to these levels as a function of temperature T, bare separation $(\tilde{\varepsilon}_\lambda - \varepsilon_o)$, magnetic field etc.

For example, when the configurations are energetically degenerate, i.e. $(\tilde{\varepsilon}_\lambda - \varepsilon_o) = 0$, the lowering of E_o is n_λ times $\tilde{E}_\lambda$ because the state $|0\rangle$ can hybridize with n_λ intermediate states $|\lambda\rangle$ and lower its energy, whereas the state $|\lambda\rangle$ hybridizes with only one state $|0\rangle$[7]. The ground state is thus a hybridization stabilized singlet [8], with a stabilization energy $(E_o - \tilde{E}_\lambda) \simeq -\varepsilon_c$ at $(\varepsilon_o - \tilde{\varepsilon}_\lambda) = 0$. The phase space preference mentioned above depresses E_o relative to $\tilde{E}_\lambda$ till about $(\tilde{\varepsilon}_\lambda - \varepsilon_o) \simeq \varepsilon_c$, where low order perturbation theory begins to break down. One enters the Kondo regime for $(\varepsilon_o - \tilde{\varepsilon}_\lambda) >> -\varepsilon_c$. Here the renormalized $\tilde{E}_\lambda$ lies lower, and the singlet is due to the residual exchange coupling J of the magnetic multiplet to conduction electrons. This regime is not considered here.

In the mixed valent regime, the effective energy levels are temperature dependent on a scale Δ. for example, starting at high temperatures $(T >> \Delta)$ with i.e. $\tilde{E}_\lambda \simeq \varepsilon_\lambda = \varepsilon_o \simeq E_o$ (the entropy regime), on cooling E_o falls below $\tilde{E}_\lambda$, i.e. the separation $|E_o - \tilde{E}_\lambda|$ decreases as temperature increases. This simple fact explains an ubiquitous feature in the susceptibility of mixed valent systems: the positive T^2 slope at low temperatures. The ground state is a polarizeable singlet. Its van Vleck polarizeability depends inversely on the separation $|E_o - \tilde{E}_\lambda|$ between the ground state (energy E_o) and the magnetic configurations (energy $\tilde{E}_\lambda$) and calculation shows that the susceptibility $\chi \sim \Delta^2/|E_o - \tilde{E}_\lambda|$. Since as temperature rises, $|E_o - \tilde{E}_\lambda|$ decreases, χ increases. At higher temperatures, the effective level $\tilde{E}_\lambda$ beings to be thermally occupied, this being also favoured by the degeneracy factor. The susceptibility maximum at $T = T_{max}$ is due to a competition between Curie law increase and rapid thermal depopulation; the singlet contribution to $\chi(T)$ varies relatively slowly with temperature for $T \lesssim T_{max}$ and influences T_{max} weakly as detailed calculations show [4].

In this extreme configurational picture where conduction electron degrees of freedom are completely integrated over, the effective levels are sharp, and are known smooth functions(Eq.(1)) of the bare Hamiltonian parameters $(\tilde{\varepsilon}_\lambda - \varepsilon_o)$, Δ, n_λ, and the temperature T. There is also a weak dependence on bandwidth D. The most detailed phenomenological theory [9] uses crystal field split ionic configurations $|\lambda\rangle$, corresponding level separations $(\tilde{B}_\lambda - B_o)$, and effective level widths Γ. The latter two are functions of temperature; detailed comparison with experiment shows them to be similar functions of temperature so that there is only one temperature dependent quantity. It is easy to incorporate crystal field splitting. The singlet is still favoured because it hybridizes with all the multiplets. In phenomenological analysis Γ is a temperature cutoff, so that near $T = 0$, quantities behave as $(T/\Gamma)^2$ rather than as $\exp(-\alpha T)$. In the Brillouin-Wigner perturbation theory (Eq. (1)) where $\Gamma = 0$, this is achieved by the self-consistency condition, namely the presence of E_o and $\tilde{E}_\lambda$ in the denominators on the right-hand side of the equations for them. Thus, low temperature expansion involves $(T/E_o)^2$. The fact that Γ and $(B_o - \tilde{B}_\lambda)$ are seen to behave similarly with temperature is not therefore surprising; they are seen to be the same thing.

This configurational approach, while precise, is limited to thermodynamic properties of a single impurity. The effective levels E_o and $\tilde{E}_\lambda$ are not to be identified with peaks seen in a spectroscopic experiment, low or high energy. The dynamics of a single impurity in a configurational model [10] requires considerable additional work, though E_o and $\tilde{E}_\lambda$ are important inputs. Similarly, for the lattice, with conduction electrons hopping in and out of sites, it is not enough to know E_o and $\tilde{E}_\lambda$. The full many body theory has been discussed by Grewe and Keiter [11]. An alternative, which is physically deeper and richer is the Fermi liquid approach. We now discuss this, presenting three somewhat plausible 'derivations'.

2.2 Fermi Liquid Theory

Since the ground state is a singlet, conduction electrons are scattered as from a potential. The low energy states of the system are electron hole excitations whose density is modified locally. This local Fermi liquid theory is specially simple if self-interaction effects are weak. A calculation of the T = 0 susceptibility $\chi(0)$ and the linear term γ in the specific heat for a mixed valent impurity shows [2,4] that

$$[\pi^2\chi(0)/\mu^2_{eff}\ \gamma] = 1 - (\Delta/|E_o - \tilde{E}_f|) \qquad (3)$$

In the strongly mixed valent regime, the second term is $<(1/n_\lambda)$, of order 10% or less.

This Wilson ratio is close to unity, the value for a non-interacting Fermi system. At this level of accuracy, i.e. in the large n_λ limit, self interaction effects can be neglected, and the system can be treated as a Fermi gas with local scattering.

de Chatel [13] has proposed that Fermi liquid parameters be obtained in the standard way by differentiation of the ground state energy E_o regarded as a functional of electron occupation number $\{f^-_k\}$. The interaction parameters are small, $O(1/n_\lambda)$ and there is no exchange term. We note however that E_o is expected to be a functional of the local density $\psi^+(r)\psi(r)$ (see e.g. the local Fermi liquid theory of Nozieres and Blandin [14] for the Kondo case), whereas on averaging over free electron states as in Eq.(1), it becomes diagonal in k and a function of $\{f^-_k\}$. The form of the local Fermi liquid theory is now the same as that for a homogeneous Fermi liquid.

Neglecting self interaction effects, the mixed valent impurity is characterized entirely by the t-matrix for scattering of conduction electrons of frequency v_ℓ from state $\vec{k}$ to state $\vec{k}'$. This can be calculated using many body methods to be

$$t^\lambda_{k,k'}(v_\ell) = z(V_{k\lambda}V_{\lambda k'})/\{v_\ell - (\tilde{E}_\lambda \ -E_o)\} \qquad (3)$$

where z is a quasiparticle renormalization constant of order unity. The correct scattering amplitude has a resonance of width Δ at energy $(\tilde{E}_\lambda - E_o)$ rather than a pole; to lowest order in $\Delta/(E_\lambda - \tilde{E}_o)$ i.e. to order $(1/n_\lambda)$, they give the same scattering amplitude near Fermi energy. The pole is a low order artifact. The scattering is that from an effective ionic level located $(\tilde{E}_\lambda - E_o)$ above the Fermi energy, so that the f level is nearly empty.

Perhaps the best insight into the suitability of a Fermi liquid description comes from renormalization group methods [7, 15]. Starting from the above Hamiltonian, i.e. a $U - \infty$ Anderson Hamiltonian with orbital degeneracy n_λ, one traces the evolution of effective Hamiltonians as the bandwidth D is scaled down. In the empty orbital regime $\tilde{\varepsilon}_\lambda - \varepsilon_o > \varepsilon_c = - n_\lambda\Delta\ \ell n(D/n_\lambda\Delta)$ $H_{eff} \simeq H$ with renornalized parameters. In particular, near the fixed point (i.e. $D_{eff} \simeq \tilde{\varepsilon}^*_\lambda$) H_{eff} has parameters $\tilde{\varepsilon}^*_\lambda = (\tilde{E}_\lambda - E_o)$, $\Delta^* = \Delta$.

The shift in $\tilde{\varepsilon}_\lambda$ is due to the level repulsion effect discussed in B-W perturbation theory, Eq. (1). Now the remarkable fact is that in this regime, and for large n_λ, the properties of H_{eff} (a $U = \infty$ model) are described fairly accurately by a U = 0 or resonant level model with the bare parameters $\tilde{\varepsilon}^{RL}_\lambda = \tilde{\varepsilon}^*_\lambda = (\tilde{E}_\lambda - E_o)$ and resonant level width $\Delta^{RL}_\lambda = \Delta$. This is because the two differ by multiple occupancy corrections; in the former case the f level cannot be multiply occupied. However, double occupancy probability is of order (1/2) $(\Delta^*/\tilde{\varepsilon}^*_\lambda)^2$ i.e. of order $\frac{1}{2}(\Delta^*/\varepsilon^*_\lambda)$ relative to single occupancy.

This is about $(1/2\ n_\lambda)$, i.e. less than 10%. Thus at this level of accuracy, in the mixed valent regime a resonant level model is adequate for low energy and low temperature properties, with the above identification of parameters. Such a model has been proposed and used by Newns and Hewson [2]. The discussion here hopefully clarifies its microscopic basis. It can be shown, by comparison with direct calculations, that the impurity resistivity is correctly given by this resonant level model. The valence at T = 0 from microscopic theory is $n_\lambda\Delta/|E_o - \tilde{E}_f|$. This is very nearly the area in the f resonance integrated up to the Fermi level.

2.3 Alloying Pressure Effect

The model Hamiltonian used so far is incomplete. For fixed $\tilde{\varepsilon}_\lambda$, the change in f electron number with say temperature can be quite large. However, as the valence increases(n_v decreases) the conduction electron number increases correspondingly. These electrons stay around the impurity, increasing the local kinetic energy and also the electron -f hole Coulomb attraction. These large energy changes, depending on the impurity and host electronic structure, control valence. These effects are absent in a model with fixed $\tilde{\varepsilon}_\lambda$. The valence changes occuring in such a model are unrealistic. One simple way of including alloying effects is to consider $\tilde{\varepsilon}_\lambda$ to be function of valence, i.e.

$$\tilde{\varepsilon}_\lambda(n_v) = \tilde{\varepsilon}_\lambda + (1 - n_v)\,\eta \qquad (4)$$

The quantity η is the electrochemical alloying potential. It measures the tendency of the impurity to transfer charge to the host. For example, in an alloy of concentration c, the bulk Fermi level may be expected to rise by $[-c\delta n_v/\rho(\mu)]$ for $-c\delta n_v$ electrons added to the Fermi sea. If this were the only effect, one has $\eta = [-c/\rho(\mu)]$, i.e. η is negative. However, if electron-f hole attraction is strong, η can change sign and be positive. In the former case, the f shell is relatively stable, and valence changes are reduced; in the latter case the f shell is unstable and valence charge is promoted.

Alloying pressure has a strong effect on physical properties of mixed valent systems. For example, assuming that $\eta = 0$, one generally finds large valence changes with temperature, e.g. $\Delta n_v \simeq 0.4$ for the case $(\tilde{\varepsilon}_\lambda - \varepsilon_o) = 0$. Similarly, the ratio of maximum susceptibility χ_{max} to that at zero temperature $\chi(0)$, has the ratio $[\chi_{max}/\chi(0)] \simeq 2.5$. Experimentally, most systems show smaller valence changes, and a ratio $[\chi_{max}/\chi(0)] \simeq 1.3$. A modest alloying pressure of $\eta \simeq -20\Delta$ e.g. due to the Fermi level increase with f promotion, reduces the total valence change (from $T = \infty$ to $T = 0$) dramatically to about 0.15 or so, and the susceptibility ratio to ~ 1.3 [4]. The maximum now occurs at much lower temperatures than before; as $\tilde{\varepsilon}_\lambda$ decrease on cooling, the characteristic energy scale of the system decreases). The calculations are done self-consistently; starting at very high temperatures with given bare level positions, as the valence changes (n_v decreases) on cooling, the f level is moved up or down according to Eq. (5) till consistency is achieved.

In some systems, e.g. Eu Pd_2 Si_2 and Yb Al_2 there are large changes in valence and the ratio $[\chi_{max}/\chi(0)]$ is large, so that a negative η is not invariably the case. An extensive study of Ce impurity in a large number of hosts, due to Riegel and co-workers [16], shows correlation of valence with host size and Fermi energy similar qualitatively to that discussed above.

3. MIXED VALENT LATTICE

It is clear from the above that a good first approximation to the mixed valent solid is a lattice of renormalized f levels which resonantly scatter conduction electrons. The many body aspects of the ionic configuration based lattice problem have been analyzed diagrammatically by Grewe and Keiter [11]. A natural description is in terms of time ordered diagrams or processes on site and Feynman diagrams for intersite conduction electron hopping. When the on site state is a renormalized singlet a large well-defined class of diagrams [17] is summed by the multiple scattering series for the resonant level lattice. This series incorporates the 'excluded volume' effect, i.e. the condition that an electron hop to another scatterer before returning to the previous one. It does not include higher order polarization processes e.g. those coupling two impurities via many conduction electron hole pairs. Such intersite many body terms are assumed absent. One of them, i.e. that which couples two renormalized magnetic levels $|\tilde{\lambda}>_i$ and $|\tilde{\lambda}>_j$ could be important. For example, if this coupling J_{ij} is say ferromagnetic and larger than the singlet stabilization energy $|E_o-\tilde{E}_\lambda|$ i.e. $Z\,J_{ij} > |E_o-\tilde{E}_\lambda|$, the system would prefer a ferromagnetic alignment. This is unlikely because a simple estimate of the sort made in Ref. [3] shows J_{ij} to be down by a factor $n_\lambda\{\ln(D/n_\lambda\Delta)\}^2$ or so compared to $|E_o-\tilde{E}_\lambda|$. We assume that the ground state is a normal metal.

The resonant level lattice is characterized by the following:

(i) The scattering is from renormalized ionic levels, i.e. in conduction electron scattering channels with given J and m_J at each site rather than orbital angular momentum channels (ℓ, m_ℓ). This is because intra-atomic spin orbit coupling $\lambda\vec{L}\cdot\vec{S}$ is larger than band crystal field splitting energies characterized by Δ . Typically $\lambda \sim 0.2$ to 0.5 eV whereas $\Delta \sim 0.03$ to 0.1 eV.

(ii) The valence at each site j is fixed by appropriate choice of $\tilde{\varepsilon}_f^j > 0$. The resonant levels of width Δ are nearly empty such that the excitation number or probability of high degeneracy configuration f^n is

$$n_v = (2J+1)\ \delta_J(\mu)/\pi \qquad (5)$$

where δ_J is the phase shift at the Fermi energy, and by definition

$0 < n_v < 1$

(iii) Further alloying pressure constraints on $\tilde{\varepsilon}_f$ are included as necessary.

We briefly outline some largely qualitative consequences of such a model for the lattice.

3.1 Low temperature tails

These are ubiquitous in mixed valent systems and are generally described as impurity effects. However, it is clear that at temperaturs $k_BT < \Delta$, lattice coherence effects will develop. This temperature could be well below that for singlet formation by a factor of order n_λ. Thus there is at low temperatures, a fine structure in the density of states, and this could very well be the origin [6] of the rise in $\chi(T)$, seen at low temperatures in α-Ce, Ce Pd_3, Ce Sn_3 etc. The low temperature tails are in that case a purity rather than an impurity effect; they should become less marked and disappear with disorder, e.g. off stoichiometry and irradiation.

3.2 Mixed valent semiconductor

A hybridization gap has been proposed by Mott as an explanation for the semiconducting ground state of mixed valent systems such as SmB_6 and SmS. The resonant level lattice, at least for transition metals, always leads to a metallic ground state. The reason has been discussed by Heine [18] and others. In a mixed tight binding-plane wave representation, the tight binding bands acquire a width due to admixture with higher energy Bloch wave states. The hybridization gap occurs around different energies at different points in k-space. It seems very unlikely therefore that one will have a semiconducting ground state in this model.

Two candidates for a semiconducting ground state are the following. Consider a conventional semiconducting ground state with the f level below but close to the conduction band edge. Because of the hybridization effects, the ground state wave function will have some f^{n-1} admixture. With a free electron density of states for the conduction band, the admixture is not large (less than about 0.3, this being the value for a gap of order Δ) but excitonic effects known to be strong in semiconducting SmS, increase it considerably. For $(V_{eh}/D) \sim 0.5$ where D is the bandwidth and V_{eh} the electron-f hole attraction, the f^{n-1} fraction is seen to be ~ 0.7 by a calculation similar to that of Section 2.

With application of pressure, such a mixed valent semiconductor will go metallic, as happens in SmS. On the other hand, the simplest expectation is that the hybridization gap increases with pressure when there is no symmetry change.

The second possibility is an excitonic instability of the metallic state, leading to a small gap. Since the f electrons are heavy and screening of electron hole interaction is poor because of low density of conduction electrons (say in SmS where $n_e < 1$ per formula unit), excitonic effects are quite likely. While for a single impurity these effects are well known from the X-ray emission or absorption problem, the corresponding Mahan-Noziéres de Dominicis lattice has not been investigated.

4. RESISTIVITY OF MIXED VALENT METALS AND ALLOYS

The resistivity of mixed valent metals and alloys shows spectacular changes as a function of temperature and alloying, for example. In the best studied example, i.e. Ce Pd_3 and its alloys [19, 20] one finds that thermal resistivity of pure Ce Pd_3 is ≃ 150 μΩ cm at about 150°K, an order of magnitude higher than that of transition metals. For small alloying concentrations, e.g. for x = 0.03 in Ce_{1-x} Gd_x Pd_3, the residual resistivity is very large, ≃ 200 μΩcm. The variation of resistivity with temperature in this limit is the same, whether the disorder is due to alloying by replacement of Ce or Pd or is caused by off stoichiometry. The low temperature resistivity for relatively pure Ce Pd_3 goes as T^2. We present here a simple disordered lattice model which accounts for these features.

It is clear from the size of impurity effect ($\Delta\rho \sim 70$ μΩcm for x ≃ 1%) that direct impurity scattering cannot account for it. For example, an impurity with three sp electrons will lead to a resistivity of about 6-10 μΩcm per %, an order of magnitude less. In this estimate, the impurity is described with respect to the perfect lattice, by a scattering matrix -t (see Eq. 4) in the f channel with phase shift fixed by valence (n_v = 0.55) and (nonunique) phase shifts corresponding to a Friedel sum of 3 in the s-p channels. It is clear that there is an indirect effect which dominates. This is also obvious from the effect of lattice vibrations. The mean squared displacement of ions is the same as in other systems, but the thermal resistivity is much larger.

We propose that the main effect of static or thermal disorder is to produce incoherence so that the phase of the electron fluctuates randomly from site to site as it scatters resonantly. The phase coherence energy, i.e. the width of the f band is very small (of order Δ). Further, there is a large number n_λ of channels and as the electron propagates from one site to another, it changes channels, losing phase memory. It is thus easy to produce enough disorder so that each mixed valent Ce ion scatters nearly independently. This is the saturation limit, to which resistivity will tend. Since resistivity in this limit depends essentially on valence (see below) which is strongly constrained by alloying pressure etc., further disorder will not cause big changes in the resistivity, except for localization effects. This is most clearly seen in the marginal effect static disorder has on resistivity above say $T \sim 150°K$, whereas below 50°K, the same disorder changes $\rho(T)$ spectacularly. For $T > 150°K$, thermal disorder is such that one is in the random phase limit; further disorder has no effect.

We now discuss the maximum residual resistivity expected in the mixed valent regime. Assuming independent scattering from each mixed valent ion, and negligible scattering in angular momentum channels other than $\ell = 3$, we have

$$\rho = (\pi\hbar/e^2k_F)\ (2\ell+1)\ \sin^2 \eta_\ell \qquad (6)$$

Normally, the Mott maximum metallic resistivity is estimated by allowing $\sin^2 \eta$ to take its maximum value, i.e. unity so that one has $\rho_{Mott} = (\pi\hbar/e^2k_F)\ (2\ell+1)$. For $\ell = 3$ and $k_F \sim 10^8\ cm^{-1}$, this gives a value of 900 $\mu\Omega$cm, much higher than the observed number. The fact of course is that η_ℓ is constrained by the valence. Using Eq. (3) for the t matrix and the theoretical expression for valence, i.e. $n_v = (n_\lambda\Delta/|E_0-\tilde{\varepsilon}_\lambda|)$, we find that

$$\rho_{max} = (\pi^2\hbar/e^2k_F)\ \ \{\pi/(2J+1)\}\ n_v^2 \qquad (7)$$

For Ce Pd_3, this gives a valence constrained residual resistivity of about 200 $\mu\Omega$cm, in good agreement with experiment. Eq. (7) has several interesting features. There is a quadratic dependence on valence. There is support for this in the data of Parks et al [2] on $\rho(0)$ of $Ce_{0.9-x}La_xTh_{0.1}$. As x increases, Ce tends towards 3^+, and n_v tends towards unity. $\rho(0)$ increases dramatically, till one gets into the local moment regime and the ground state is probably a spin glass with little scattering from frozen spins. The resistivity then drops abruptly. In the former regime, the rise is nearly quadratic. The results of Section 2 can be used to relate $\rho(0)$ to $\chi(0)$, and one finds for a mixed valent alloy, that

$$\rho(0)\ \chi(0)^{-2} = \text{constant}. \qquad (8)$$

This relation, which again agrees with experimental results on $Ce_{0.9-x}La_xTh_{0.1}$, needs to be checked more extensively also.

In the Kondo regime also the residual resistivity can be calculated, assuming that one is in the dense impurity limit. This assumption is not very good if T_k is small, since the physical size of a singlet is $\xi_k \sim (\hbar v_F/T_k)$ so that there can be important overlap and interference effects. However we neglect these. We use $\eta_J = \pi\{2J/(2J+1)\}$ which corresponds to the strong coupling ground state picture [14] where 2J conduction electrons 'bind' to the impurity having one f electron outside a closed shell, and angular momentum J.

$$\rho_{max}^{Kondo} = (\pi\hbar/e^2k_F)(2J+1)\ \sin^2\{2\pi J/(2J+1)\} \qquad (9)$$

If there are no crystal field effects ($J = 5/2$), this gives for Ce Pd_3, in the Kondo limit, a resistivity $\sim 800\ \mu\Omega$cm. This is close to what Mihalisin et al [22] find in Ce $Pd_{3-x}Ag_x$ alloys. There are other indications, from $\chi(0)$, $\gamma(0)$ etc. that one is in the Kondo regime. However, a common possibility is that due to crystal field splitting $J_{eff} = 1/2$. In this case, Eq. (6) gives (with $\eta = \pi/2$) value of about 1000 $\mu\Omega$cm. We thus see that limiting residual resistivities close to those observed are obtained when the phase shifts consistent with the correct ground state are used.

We now return to the general case. The local scattering depends on the valence, and this is affected strongly by local longitudinal strain. We thus assume that

$$\delta\eta^j = A\ e_{\alpha\alpha}^j \qquad (10)$$

A very rough estimate of A is from the $\gamma - \alpha$ transition in Cerium, where using Eq. (5), we have $A \sim 4$. Static strains due to impurities etc., and thermal strain fluctuations, lead to local scattering as the phase shift fluctuates according to Eq. (10). It is clear that large fluctuations in phase shift whose randomizing effect is multiplied by off diagonal intersite propagation in the large number of scattering channels, is possible.

At low temperatures, for relatively pure systems, there is a large T^2 term. We argue that this could arise from electron phonon (or electron-thermal strain fluctuation) scattering, as follows. A phonon can decay into an electron-hole pair, via electron phonon coupling, so that in thermal resistivity there is a T^2 term due to the electron decaying into another electron and an electron hole pair. In normal metals, this term is very small because it depends on the decay probability of a phonon into an electron hole pair which has a phase space factor $(v_s/v_F) \ll 1$. In mixed valent systems, carriers are very heavy, and Fermi velocities small, so that (v_s/v_F) is not very small. Further, the strain induced fluctuations in phase shift (Eq. 10) are large (A is nearly four) and the resistivity due to the mechanism suggested here goes as A^4. There can thus be a sizeable T^2 contribution to resistivity from this source.

For large disorder, e.g. $x > 2\%$ in $Ce_{1-x} Gd_x Pd_3$, there is an upturn in the resistivity at low temperatures ($T < 10°K$). This could be due to localization effects; the signature is characteristic. Observation of negative magnetoresistance would strengthen this interpretation.

ACKNOWLEDGMENTS

I should like to thank P.W. Anderson for many illuminating discussions on the mixed valence problem. I am thankful to C.M. Varma for pointing out the striking resistivity behaviour of Ce Pd_3, and to H.R. Krishnamurthy and N. Kumar for discussions about this.

REFERENCES

[1] Sales, B.C. and Wohlleben, D.K., Phys. Rev. Lett. 35, 1240 (1975).

[2] Newns, D.M. and Hewson, A.C., J. Phys. F, 10, 2429 (1980).

[3] Ramakrishnan, T.V., in "Valence Fluctuations in Solids", Falicov, L.M., Hanke, W. and Maple, M.B. editors,(North Holland, Amsterdam, 1981).

[4] Ramakrishnan, T.V. and Sur, K, Phys. Rev. B (to be published).

[5] Bringer, A. and Lustfeld, H., Z. Phys. B, 22, 213 (1977).

[6] Anderson, P.W. in "Valence Fluctuations in Solids", Falicov, L.M., Hanke, W. and Maple, M.B., editors (North Holland, Amsterdam, 1981).

[7] This regime was first identified by Haldane (Phys. Rev. Lett. 40, 416 (1977)) using scaling arguments, and as the empty orbital regime by Krishnamurthy, Wilkins and Wilson (Phys. Rev. B 21, 1044 (1980)) using renormalization group methods. The authors considered the asymmetric Anderson model ($n_\lambda = 2$).

[8] Varma, C.M. and Yafet, Y. Phys. Rev. B, 13, 2950 (1976).

[9] Röhler, J., Wohlleben, D. and Kaindl, G., this conference.

[10] Schlottmann, P., this conference. Müller-Hartmann, E,,ibid.

[11] Grewe, N. and Keiter, H., Phys. Rev. B, 24, 4420 (1981).

[12] Lustfeld, H. and Bringer, A., Solid State Comm. 28, 119 (1978).

[13] de Chatel, P.F., this conference.

[14] Noziéres, P. and Blandin, A., J. de Physique 41, 193 (1980).

[15] Krishnamurthy, H.R. (unpublished) has explicitly worked out results for the n_λ fold degenerate Anderson model in the mixed valent regime, using RNG methods.

[16] Riegel, D., this conference.

[17] See for example, Grewe, N., this conference.

[18] Heine, V., Phys. Rev. 153, 673 (1962).

[19] Scoboria, P. Crow, J.L. and Mihalisin, T., J. Appl. Phys. 50, 3 (1979).

[20] Schneider, H. and Wohlleben, D., Z. Phys. B 44, 193 (1981).

[21] Parks, R.D. et al in "Valence Fluctuations in Solids", Falicov, L., Hanke, W. and Maple, M.B., editors (North Holland, 1981).

[22] Mihalisin, T. et al, ibid.

COMMENT: C.M. VARMA: It is true that the effective interaction effects between f-states at different sites are 10% of the interaction between f and conduction electron states at a site. However, to conclude from this that they may be neglected, by using "local" Fermi-liquid theory, is throwing away the most interesting parts of the problem. If you applied similar reasoning to transition metals, you would conclude that Ni has a non-magnetic ground state.

T.V. RAMAKRISHNAN: The ground state is non-magnetic if the on-site stabilization energy of the renormalized singlet with respect to the magnetic multiplet is much larger than the inter-site coupling between the magnetic multiplets. This is what happens for large multiplet degeneracy n_λ in the mixed valent regime. No such system is known to order magnetically. However, when the valence is close to that for the magnetic configuration, the singlet stabilization energy is small, comparable to magnetic interaction energy. Magnetic ordering can occur then and does. These results depend on a configurational description being applicable ($U \simeq \infty$), and n_λ being large. In transition metals, $U \lesssim \pi\Delta$, and because of crystal field splitting, orbital degeneracy is not very large. Many, strongly modified ionic configurations admix. These well known problems about Hund's rule, number of atomic configurations etc. make it unlikely that the simpler results discussed here can be carried over without more thought.

N. GREWE: You will have recognized that many of the compound properties you proposed are in agreement with the renormalized perturbation expansion I have presented, including the resonant-level lattice and the key role played by the impurity T-matrix. The expression for this quantity you presented is very suggestive, but wouldn't you agree that it still needs a lot more than the first Brillouin-Wigner approximation to arrive at really reliable conclusions ?

T.V. RAMAKRISHNAN: The low order Brillouin-Wigner theory is accurate for thermodynamic properties in the mixed valence regime. It gives results for Green functions which have to be interpreted with care, but are quite consistent with results for thermodynamic quantities.

H.R. OTT: What is the conceptional difference between your resonance level model and the model of virtual bound state formation put forward to explain the behavior of $CeAl_3$. The latter model gives a maximum resistivity of about 150 $\mu\Omega$cm, quite in good agreement with experiment.

T.V. RAMAKRISHNAN: I have shown how, starting from an ionic description, one ends up with a resonant-level like model, and have shown, how its parameters are related to those of the bare Hamiltonian. In the mixed valent regime, I show how the resistivity for the dense impurity limit depends on valence. $CeAl_3$ is nearly trivalent and I believe Kondo-like effects modified by its being a lattice, are important. The low energy properties including resistivity of a Kondo impurity with spin 1/2 can also be mimicked by a resonant level model. However, renormalization effects are large and affect different physical quantities differently. Less is known about the relation between a Kondo-impurity ground state for large n_λ and a resonant-level model. In addition there are large intersite or lattice coherence effects in the Kondo regime. So I do not know how much meaning one could read into simple resonant-level model fits for properties of $CeAl_3$.

M.T. BEAL-MONOD: I think it would be fair to remark that what is called the "local Fermi liquid" approach of Newns and Hewson is nothing else than the local paramagnon approach derived long ago by Mills and Lederer, Doniach and Rice.

T.V. RAMAKRISHNAN: I do not believe this is correct. The local paramagnon theory is for the regime $U \lesssim \pi\Delta$, and treats the effect of U in the RPA or similar approximations. What has been shown here is that in a configuration based description, where $U=\infty$, the local Fermi-liquid picture of Newns and Hewson emerges.

M.T. BEAL-MONOD: The interactions involved in the local paramagnon picture can be gathered altogether in a kind of internal field acting on one impurity so that the overall picture is that of a single impurity picture, in the sense of the Friedel virtual bound state.

T.V. RAMAKRISHNAN: This is a very interesting point. For local correlations, as also emphasized by Brinkman and Rice, both the susceptibility χ and the specific heat coefficient γ are enhanced identically, so that one has a Wilson ratio of unity. I don't know whether all physical properties can be described by a suitable parametrized local weakly interacting Fermi liquid or, to specialize, by an effective phase shift $\delta(t)$.

Valence Instabilities
P. Wachter and H. Boppart (eds.)

DENSE KONDO BEHAVIORS IN CeB_6, Ce-MONOPNICTIDES AND THEIR ALLOYS

T. Kasuya, K. Takegahara, Y. Aoki, T. Suzuki, S. Kunii, M. Sera, N. Sato, T. Fujita, T. Goto,† A. Tamaki† and T. Komatsubara*

Department of Physics, Tohoku University, Sendai, Japan
†Research Institute for Scientific Measurement, Sendai, Japan
*Institute of Material Science, Tsukuba University, Ibaraki, Japan

Dense Kondo behaviors in CeB_6, CeSb and CeBi are reviewed. CeB_6 is the most typical dense Kondo state with respect to the transport properties. Other properties are, however, very unusual compared with other well known dense Kondo systems. These unusual properties seem to originate partly in an critical situation of the 4f wave function in CeB_6 and partly in strong Γ_7-Γ_8 mixing type interactions. Well known T^2 dependence of resistivity in the lowest temperature is shown to be due to impurity effect. It is shown in CeSb and in CeBi that the Γ_7 states make dense Kondo state, even though stronger p-f coupled Γ_8 states never show such character but exhibit unusual magnetism. These are the first case in semimetals, and the coherent effect among individual Kondo states seems to be strong.

1. INTRODUCTION

The Kondo state is a typical example of a single site many body problem and has attracted many theoretical and experimental physicists. The most serious problem for the experimental studies, in particular for the 3d Kondo state was to measure the properties of very dilute impurities accurately because mutual interactions destroyed the single site Kondo state very easily. A lot of theoretical works has been published revealing the essential physical picture of the Kondo state, and still now, the recent development of the exact solution is revealing new features of the Kondo state.(1) Furthermore, recent discovery of dense Kondo state or Kondo lattice in rare earth compounds has added an essentially new feature on the Kondo state because the single site many body picture should be enlarged to cover the many site or even periodic system. (2) Already a lot of works has been done both theoretically and experimentally but still we are far from the essential understanding of the physical picture for the dense Kondo state. For example, $CeAl_2$ and $CeAl_3$ may be typical examples of dense Kondo systems studied in detail so far. (3)(4) In the former magnetic orders appear in low temperature competing with the Kondo state while in the latter no magnetic order appears and the ground state is called as coherent Kondo state or Kondo lattice. In any case, T-linear resistivity and T-square specific heat were reported in the lowest temperature region.

In the following, we report detailed properties of CeB_6 and Ce-monopnictides which are new types of dense Kondo system showing a lot of unusual properties, some of which are very unique in these materials. The main purpose of this paper is to identify what anomalous properties are originated from the dense Kondo state and what are from the rather conventional interactions, even though detailed investigations have not yet been done. CeB_6 crystallizes in the CsCl type structure in which a B_6 molecule replaces a Cl atom.(5) Ce atom is considered to be trivalent, in principle, and thus has one 4f electron. This is the most typical dense Kondo system in the sense that the temperature dependence of the resistivity shows the most beautiful Kondo like behavior in a large temperature region as shown in Fig.1.(6) Two transition temperatures, T_1=3.15K and T_2=2.3K for no applied field, are seen in Fig.1 and thus three phases, called as the phase I, II and III in the sequence of decreasing temperature, are identified in the H-T plane as given in Fig.2.(7) In the phase I, incoherent Kondo scattering seems to dominate in the resistivity. However, other properties are very unusual. Phase II is the most mysterious phase. In this phase, occurence of the coherent Kondo state seems to be the key event but the situation is so puzzling to understand the real situation. In phase III, real magnetic orderings similar to that in $CeAl_2$ seem to occur but the effect of coherent and incoherent Kondo states in lowest temperature seems to be more complicated than that considered before. The detail will be shown in section three.

Ce-monopnictides crystallize in the rock-salt type structure.(8) Except CeN, the valency of Ce is considered to be trivalent. Among them, CeBi and CeSb are well known by their very unusual magnetic properties.(9) Recently we could explain these anomalous properties by the p-f mixing model.(10) It was shown that both CeBi and CeSb are well compensated semimetals with carrier numbers of several percent per Ce atom.(11)(12) Unusual magnetic properties come from the strong mixing between the 4f Γ_8 and the valence band at Γ-point.(13) Hall constant measurement shows that the current is carried mostly by the conduction electrons which mix mainly with the 4f Γ_7. The dense Kondo behavior of Γ_7 has been observed in various properties. The detail is shown in section four.

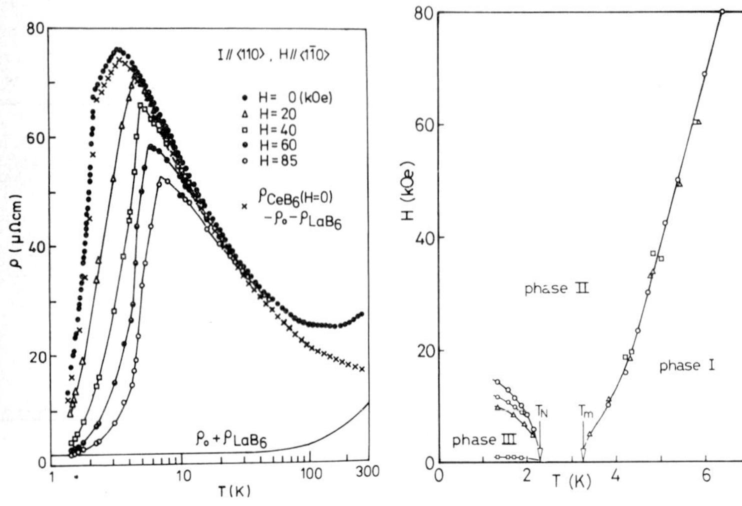

Figure 1 : Eelctrical resistivity of CeB_6 under various magnetic fields is plotted vs logT. The crosses mean the values subtracted by the non-magnetic impurity scatterng ρ_0 and ρ_{LaB_6}.

Figure 2 : Phase diagram of CeB_6 in the H-T plane obtained by the magneto-resistance experiment.

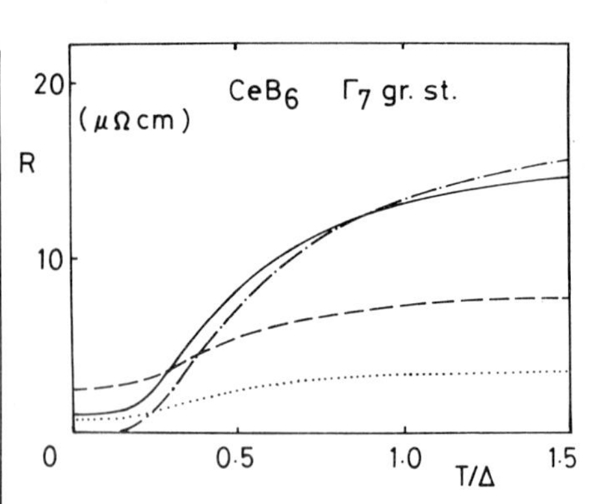

Figure 3 : Calculated value of resistivity for CeB_6 with Γ_7 ground state with the crystal field splitting Δ. The dotted line is due to isotropic exchange scattering, the dashed line for anisotropic exchange, the dot-dashed for multiple and the solid line for the total resistivity.

It should be noted that the dense Kondo behavior in semimetals in which the carrier number is only several percent per Ce is the first case and it should be called as super-dense Kondo state casting a more puzzle on the physical picture of the dense Kondo state.

2. FUNDAMENTAL CONSIDERATIONS

There are two main interactions between 4f electrons and band electrons including both conduction and valence bands electrons. In the following the above interactions are called simply as c-f interaction. In the compounds we are considering in this paper, in general, the bottom of the conduction bands is usually made mostly of the atomic 5d character on the rare earth atom and thus called simply as d-band, while the top of the valence bands is made mostly by the p-character of the anion and thus called as p-band. The most important c-f interaction comes from the intra-atomic d-f Coulomb interaction which is written as (14)

$$H_{df}=-2J_0S_dS_f+\Sigma_{\kappa=1}^{4}\{A_\kappa-B_\kappa(\frac{1}{2}+\frac{1}{S_f}S_dS_f)\}u_d^{(\kappa)}u_f^{(\kappa)}, \quad 1$$

where the first term means the usual isotropic d-f Coulomb-exchange interaction, the A_κ terms mean the d-f multipole-multipole interactions and the B_κ terms the anisotropic d-f exchange interactions, where $u^{(\kappa)}$ is the κ-th unit tensor matrix. These atomic parameter values are known as follows for Ce; J_0=2443, B_1=394, B_2=166, B_3=369, B_4=47, A_2=252 and A_4=30, all in the unit of K.(15) While the d-band state may be written as the linear combination of the atomic 5d state $\psi_{dm}(r-R_n)$ at lattice site R_n as follows,

$$<r|\nu k>=\sum_{n,m}<ndm|\nu k>\ \psi_{dm}(r-R_n)e^{ikR_n}+<r|\nu k>', \quad 2$$

in which $|\nu k>'$ means the remaining part. The parameter $<ndm|\nu k>$ is evaluated from the band calculation and thus the effect of the d-f Coulomb interaction is evaluated numerically. Usually, the first term is considered to be most important and the anisotropic terms have been neglected in many cases. However, the anisotropic terms are much larger than the isotropic term in particular in Ce-compounds.

The most clear indication is seen in the crystal field splitting of the 4f levels. In CeB_6, for example, the usual point charge model, in which Ce and B_6 are assumed as divalent, gives the Γ_8 quartet ground state and the Γ_7 doublet to be situated 150K above Γ_8.(15) However, by considering eq.(1), the sign of the level splitting is reversed in agreement with experiment. The similar situation exists also in PrB_6 and NdB_6.(16)(17) We have tried to fit both the magnitude and temperature dependence of the magnetic resistivity in PrB_6 and NdB_6 by postulating a common reduction factor in the parameters in eq.(1). It became clear that a single reduction factor is not enough to explain the experimental results but reduction factors of 0.7 in A_κ and 0.4 in J_0 and B_κ, namely the exchange terms, give good agreement with experimental results. The same reduction factors are also satisfactory to explain the observed crystal field splitting in CeB_6. This means that $|\nu k>'$ term has also some different contributions on the Coulomb and the exchange terms. Note that the lower part of the conduction band occupied by the conduction electrons is made mostly by the 5d e_g states and thus in Ce the 4f Γ_8 states have strong d-f Coulomb interaction. Both the A_K and B_K terms are large and have opposite signs but the former is a little larger. This difference is still large enough to reverse the result of the point charge model. In PrB_6 and NdB_6, too, the 4f crystal-split levels with the larger d-f interaction are pushed up into higher levels. Concerning the magnetic scattering, A_K and B_K terms contribute much larger than the usually considered J_0 term. An example for CeB_6 is given in Fig.3, in which the reduction factor is taken to be unity, that is, the lower two conduction bands are approximated simply by the linear combination of 5d e_g states and $|\nu k>'$ is put zero. It is seen that the contribution from the isotropic exchange scattering is very small. In PrB_6 and NdB_6 the contribution of J_0 term becomes larger but still is not the dominant term. Note that the magnetic resistivity in CeB_6 is much larger than those in PrB_6 and NdB_6, in which the magnetic resistivities are only around several μΩcm at room temperature. In CeB_6, too, the present Born scattering resistivity is considered to be less than those in PrB_6 and NdB_6. The same situation exists in the RKKY type indirect f-f interaction. For example again in CeB_6, in the approximation taking into account only the first conduction band with the reduction factor unity, the overall energy level splitting for the nearest neighbour pair 4f electrons is evaluated to be about 25K.(15) Again the most important contribution comes from A_K and B_K terms and thus the actual interaction form is of very complicated anisotropic mixed multipole-multipole interaction. It is also remarkable that the usual isotropic exchange interaction and the interaction among Γ_7-Γ_7 multiplets are very small. The present situation is not restricted to the hexaborides but should be common to all light rare earth compounds and alloys and thus we should be very careful to subtract out the anomalous dense Kondo properties from experiments because the non-Kondo part is also not simple but may include various anomalous properties due to the strong anisotropic d-f interaction.

The above-considered d-f Coulomb exchange interaction can induce the dense Kondo state in principle. It is well known that within the ground J-multiplet the isotropic d-f exchange interaction is rewritten as

$$-2(g-1)J_0\, J_f \cdot S_d \qquad 3$$

in which the g-values for Ce, Pr and Nd are 6/7, 4/5 and 8/11, respectively. Therefore, in all of their compounds, J_f and S_d couple antiferromagnetically and thus the Kondo state can be formed. In particular in Ce, within the ground Γ_7 doublet, J_f is replaced by the effective 1/2 spin operator. Even though, the actual treatment of eq.(3) is not so easy because of the multivalley structure of the d-band. Note that the screening d-band spin density has essentially the e_g character with the largest amplitude at the central Ce site. Therefore, this is considered essentially s-wave scattering for d-bands. The situation becomes more complicated for the more dominant anisotropic d-f exchange interaction, but in principle we can expect Kondo state due to the d-f Coulomb interaction. However, it is hard to attribute the dense Kondo behavior in CeB_6 to this interaction because then we should expect the similar dense Kondo behavior in PrB_6 and NdB_6, too, in disagreement with experiment. One possible way to overcome the above difficulty may lie in the special feature of the 4f wave function in CeB_6. It is known in the atomic physicists that due to the strong centrifugal force the double well potential is formed for the 4f atomic state.(18) Ce atom is thought to be in a critical situation where the 4f wave function moves from the outer well to the inner well, while in all other heavier rare earth atoms the 4f state is trapped to the inner well and thus forms the well known tightly shrinked state situated inner of the 5s, 5p closed shells. Therefore, it may be possible that in CeB_6 a substantial part of the 4f state is extended to the outer well. Applicability of this kind of model to other Ce compounds is also beginning to be considered.(19) Because the extension of the 4f wave function trapped to the outer well is nearly equal to that of the 5d state, we can expect a large d-f interaction. If this is true, we should expect various anomalous properties other than the Kondo behavior, for example, anomalously large crystal field splitting. So far, however, such anomalies have not yet been observed and thus we can not expect such a large shift of the 4f wave function to the outer well.

From the preceding considerations, the main origin for the dense Kondo behavior in CeB_6 seems to be due to the d-f mixing mechanism. We have also calculated the d-f mixing matrix and the indirect f-f interaction through the d-f mixing matrix. However, there is big ambiguity in evaluating the d-f mixing matrix and thus the result is also not so convincing. But in any case the question as before, that is, why no Kondo behavior in PrB_6 and NdB_6, should be answered. In the present case, however, the above mentioned shift of 4f electron to the outer well seems to work because even small amount of shift, which does not alter the preceding d-f Coulomb interaction substantially, can cause large change in the d-f mixing matrix because the latter is essentially due to inter-atomic overlapping. Note that the screening spin density in the mixing mechanism is mostly of f and p character with nearly no amplitude at the central Ce site. Therefore it is essentially p-wave scattering for d-band and thus coexists with the d-f Coulomb screening mentioned before. In the simple model, only the s-wave scattering is taken into account both for the Coulomb exchange and the mixing interactions, which thus cause direct competition with each other.

Relative position of the 4f level to the Fermi energy is a very important quantity in the d-f mixing model. Usually, the photo-emission experiment is thought to give the 4f level position fairly correctly in metals. The following values were obtained from the photo-emission experiment, 2.5eV for CeB_6 and 3.5eV for CeSb.(20)(21) Real positions seem, however, to be much smaller. It is known that in the inner core photo-emission, when the energy of emitted electron is sufficiently large, the main peak corresponds to the generalized Frank-Condon situation, that is, the occupied band states do not follow the change in the core potential. Therefore, if nothing others happen, the core potential change is not screened and thus the effective core level is observed to be much deeper than the static one. Actually, however, various kinds of screening occur and the effective shift of the core level is not so large. The same situation occurs also for the 4f level emission. The main screening mechanisms are then considered to be as follows. (i) The intra-atomic and inter-band polarization. This is the most important mechanism. The reduction of matrix element due to this polaron effect is not large. This corresponds to the optical dielectric constant. (ii) The lattice screening, which is much smaller than (i) because the usual Frank-Condon principle is applicable fairly well. The actual calculation is not so easy but a preliminary calculation shows that the actual static 4f level in CeB_6 and CeSb may be one to one and a half eV below from the Fermi level. In any way, a more detailed theoretical calculation is necessary to evaluate the actual 4f level from experiment.

3. DENSE KONDO BEHAVIOR IN CeB_6

As was stated above, there are three phases in CeB_6. In phase I the resistivity and thermoelectric power show typical Kondo like behaviors. Recently, T_1 in the NMR on B^{11} has been measured by Takigawa et al.(22) offering interesting result as shown in Fig.4. The anomalous temperature dependence was interpreted as follows. In low temperature, the life time of the 4f level, τ_f is determined by the f-f interaction. But, as temperature increases, the life time due to the strong d-f Kondo scattering dominates, which causes the increase of τ_f^{-1} as TlnT and thus the decrease of T_1^{-1} in agreement with experiment. Note that the dipole-dipole and the transfered hyperfine interactions are nearly equally important for the 4f-B^{11} nuclear spin interaction. It is, however, a puzzle that T_1 is nearly constant from 4K to 30K in which the f-f interaction seems to be dominating to determine τ_f. The fitting of the temperature dependence of the susceptibility χ indicates that the effective crystal field splitting is about 50K and there are strong antiferromagnetic exchange interactions among the Γ_8 and in between the mixing Γ_7-Γ_8, in which the original anisotropic exchange interaction was simplified in the form of S_nS_m type and the anisotropy was included only in the Γ_7-Γ_8 dependence of the coupling constants.(23) If we accept the above model, we expect fairly large change of T_1 below 30K. Therefore the constancy of T_1 should be attributed to an accidental cancellation. Another interesting result comes from the sound velocity measurement by Goto et al.. From that, three elastic constants were obtained as listed in Table I.(24) For comparison, the values for the typical valence fluctuating materials are also tabulated.(25)(26) It is remarkable that the negative value of C_{12}, which has been considered to be typical character of valence instability, in CeB_6 is larger than those of the typical mixed valence materials. However, the bulk modulus is substantially larger in CeB_6. This may indicate that the

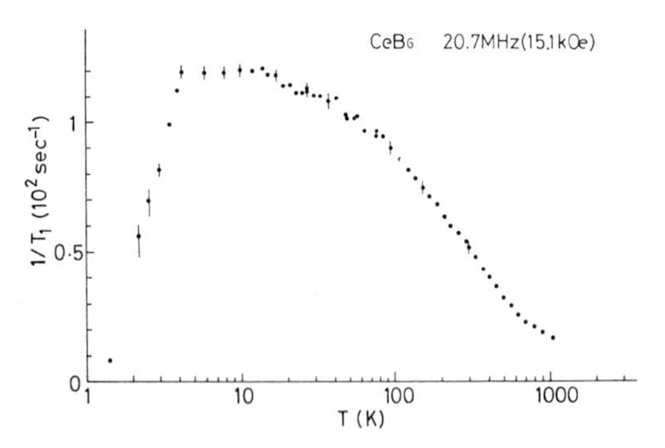

Figure 4 : T_1^{-1} of B^{11} in CeB_6 is plotted as a function of logT. Frequency is 20.7 MH_z but the frequency dependence is small

Table I : Values of elastic constants for CeB_6 and for some mixed valence materials. The unit is 10^{12} erg cm^{-3}

	C_{11}	C_{12}	C_{44}	$(C_{11}+2C_{12})/3$	
$Sm_{.75}Y_{.25}S$	11.3	-4.13	2.9	1.0	(25)
$Tm_{.99}Se$	17.9	-5.7	2.7	2.1	(26)
CeB_6	40.6	-9.3	7.8	7.3	

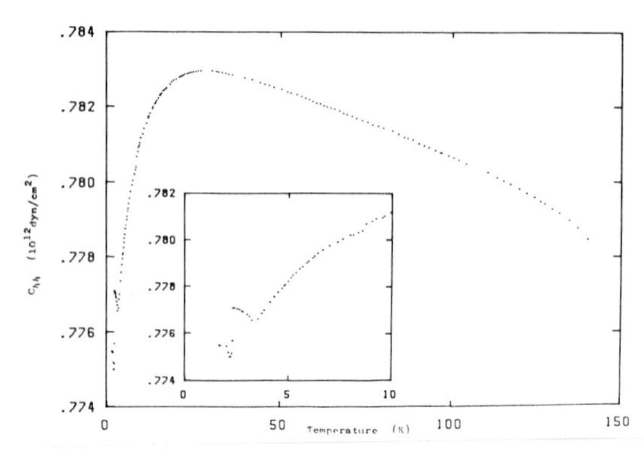

Figure 5 : C_{44} of CeB_6 obtained from the sound velocity is given as a function of temperature T. An inserted figure shows a detail in low temperature region.

shift of the 4f wave function to the outer well is induced by the pressure, which increases the d-f mixing and thus increases the valency of Ce. Note that there is nearly no temperature dependence in C_{12}. Great mystery in this phase is absence of neutron inelastic scattering due to the crystal field splitting which was reported by Steglich et al..(27) More extended experiment has been done by Rossat-Mignod group using our samples but so far the similar result has been observed. The experiment indicates that the crystal field splitting is larger than 400K or the level width is larger than 400K, in disagreement with the model obtained from the susceptibility fitting mentioned before. One possibility to explain the above mystery is the assumption of large non-exchange type multipole multipole interactions in Γ_8 and Γ_7-Γ_8 mixture, which are not directly concerned with the susceptibility. This will be discussed later again. It should be also pointed out that the temperature dependences of the elastic constant also can not be fitted by the simple crystal field splitting of 50K as shown in Fig.5.

The phase II is the most mysterious phase. First of all, the phase boundary I-II is unusual in the sense that the transition temperature $T_1(H)$ increases with increasing magnetic field H. Furthermore, the character of the phase transition is very much different from the usual second order transition.(2) For small H, the transition looks similar to the case of the singlet ground state model.(28) For larger H, the transition becomes clearer and the temperature dependence of the specific heat looks similar to the superconducting transition.(29)(30) However, as seen in Fig.5, there is a sharp minimum in the sound velocity at $T_1(H=0)$. NMR measurement indicates that below $T_1(H\neq0)$ complicated line splittings occur indicating some kind of antiferromagnetic order.(31)(32) The splitting decreases with decreasing field and at H=0 no magnetic order seems to exist. On the other hand, so far, neutron experiment can not detect any magnetic ordering,(32)(28) and sound wave attenuation shows also no structure in the phase II.(33) It should be noted that the entropy just above $T_2(H=0)=2.3K$ is already about ln2. From these features, the following picture seems to be plausible. Similar to the singlet ground state problem, the strong Γ_7-Γ_8 mixing interaction, which should include non-exchange type interactions,causes a large dispersion on the Γ_8 levels and at T_1 an ordering of the mixing type magnetic moment occurs. Because of the Γ_7 doublet ground state, instead of the singlet ground state, there remains two degrees of freedom due to Γ_7. In low field, the Kondo state within Γ_7 destroys the magnetic order of the mixed moment. In high field, the Kondo state is destroyed by the field and unbalance in the Γ_7 doublet is induced. This causes a broader dispersion for an unbalanced mixed moment and thus the increase of T_1. High field experiment offers also interesting information, which is now in progress.(34)

In the phase III, the magnetic order was detected by the neutron experiment,(28)(32) as well as by NMR.(31)(22)(7) Although the detailed ordering structure is not yet established, it is expected to coexist with the Kondo lattice. To check the properties of Kondo lattice, detailed studies of specific heat and resistivity have been done on various samples down to 70mK. Some results are shown in Figs.6 and 7 and in Table II. In various materials, the specific heat shows T-linear dependence in temperature below

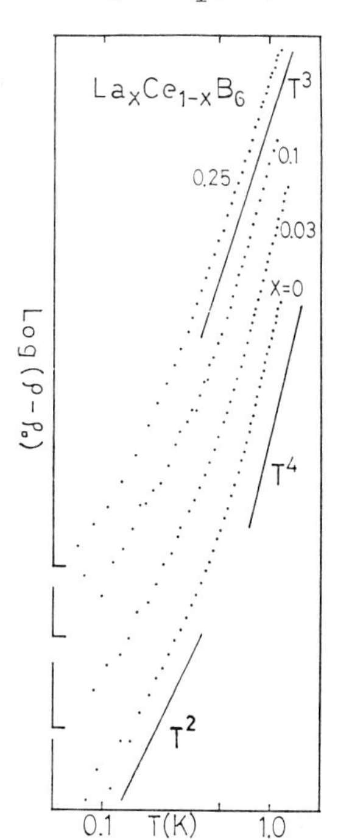

Figure 6 : Resistivities of $La_xCe_{1-x}B_6$ are plotted as functions of temperature T in log-log scale. Solid lines are drawn for references.

Figure 7 : Specific heats of $Ce_{1-x}La_xB_6$ are plotted as functions of temperature T in log-log scale.

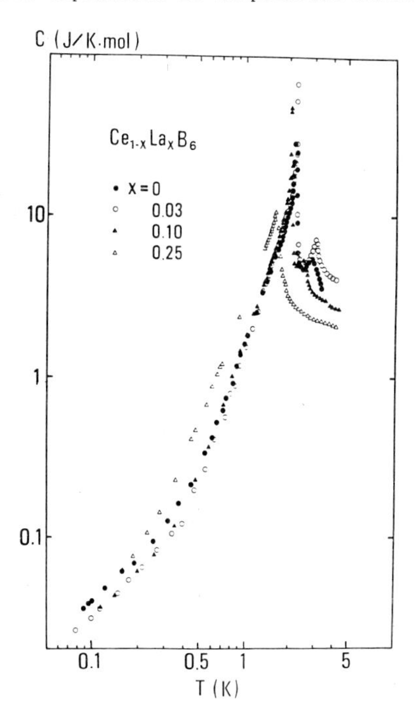

about 0.5K and T-cube dependence above 0.5K.(30) The former is thought to be due to the dense Kondo state and the latter due to the antiferromagnetic spin waves. When the magnetic field is applied the specific heat in the higher temperature region is suppressed as expected in the usual antiferromagnet but no visible change was found in the lower temperature region. Effect of randomness is not observed. In the resistivity, too, T-fourth to cube dependence in the higher temperature and T-square dependence in the lower temperature are observed.(35) The former is thought to be due to antiferromagnetic magnon scattering but the resistivity is too large to be a simple spin scattering. It is thought that the spin fluctuation destroys the coherent Kondo state and thus induces incoherent Kondo scattering. The latter is thought to be due to Kondo scattering but it depends strongly on impurity concentration, type of impurity and in particular the amount of residual resistivity for zero field. This means the following picture. The residual resistivity at H=0 is mostly due to incoherent Kondo scattering induced around defects. Actually, by applying high field, the residual resistivity decreases very much.(36)(6) As temperature increases the region and degree of incoherent Kondo state increases at around each defect. In this sense, T-square dependence is not the intrinsic character of the perfect Kondo lattice but rather depends on more complicated process. Actually, when we tried to fit the whole temperature region by the form $AT^{\alpha}+BT^{\beta}$ with the four common parameters, we found that there is a big ambiguity in α, extending from 1.1 to 1.8. In this sense, T-square dependence in low temperature is not so clear now and thus the results reported so far on other materials should be checked carefully. We also tried to obtain the intrinsic character by extrapolating the results to the zero-residual resistance but there is too much ambiguity to get plausible result.

4. CeBi AND CeSb. SUPER DENSE KONDO STATE

In CeBi and CeSb the strong p-f(Γ_8) mixing induces anomalous magnetic properties, in which a strong nonlinear p-f mixing effect is the main mechanism to push the Γ_7 in lower temperature.

Table II : Resistivities for various samples are fitted by $\rho_{0L}+AT^2$ in low temperature and by $\rho_{0H}+BT^m$ in high temperature and these parameters are listed. Note that $BT_N{}^m$ is nearly constant in all samples.

	ρ_{0L}	A	ρ_{0H}	B	n	$\rho_{0L}-\rho_{0L}(CeB_6)$	A/B
CeB_6 No.1	0.71	0.98	0.71	3.7	4	0	0.26
CeB_6 No.2	0.63	0.91	0.59	3.6	4	-0.08	0.25
La 3%	2.48	1.34	2.41	4.6	3.7	1.77	0.29
La 10%	6.92	2.60	6.91	8.2	3.5	6.21	0.32
La 25%	13.8	5.67	13.6	19.9	3	13.1	0.28
Y 10%	1.52	1.03	1.53	4.5	4	0.81	0.23
Y 25%	10.1	3.16	10.1	11.6	3.2	9.4	0.27
Ca 10%	6.46	1.53	6.19	7.2	3.3	5.75	0.21

It is interesting to notice that the strong p-f mixing did not induce any Kondo behavior in the Γ_8 state at all. This may be due to the semi-metallic character of the valence band. The Fermi energies of the valence and the conduction bands are estimated to be 0.5 and 0.3eV, respectively, while the p-f mixing matrix (pfσ) is evaluated to be about 0.5eV.(10)(11)(12) Therefore the Fermi energy is smaller than the mixing energy. On the other hand the Γ_7 state mixes with the lower part of the 5d(t_{2g}) conduction bands, which have three minima at the X-points. This mixing energy is thought to be about several hundreds degree and thus substantially lower than the Fermi energy. Therefore, an interesting question of whether the Γ_7 doublet shows Kondo behavior arises. Already resistivity measurement on some diluted samples by Hessel Andersen et al. shows some hints on the present problem but their results are not conclusive and they rather denied the Kondo state. (37) In the following our more detailed investigations are shown. Due to the limitted space, mostly CeSb alloys are considered.

Resistivities of various $Ce_{1-x}La_xSb$ alloys are shown in Fig.8. As x increases, the first order transition temperature T_{N1} to the FP phase decreases and for x=0.2, T_{N1} is extrapolated to be nearly zero. Note that in the FP phase the ferromagnetic layers made mostly by the ordered Γ_8 state and the paramagnetic layers stack in various ways making various FP phases. In Fig.9 the resistivities subtracted by that of LaSb are plotted by logT scale. The result is very similar to that of a typical dense Kondo system such as $CeAl_3$.(4) The same plot is shown in Fig.10 for $Ce_{1-x}La_xBi$. A sharp peak at T_N is considered to be due to critical scattering. In $Ce_{1-x}La_xBi$ no paramagnetic layers appear. It is remarkable that below T_N the resistivity decreases very sharply compared with the usual magnetic materials. This is also explained by the rapid decrease of the Γ_7 Kondo scattering. Residual resistivity in the ordered Γ_8 state is also very small.

The main problem is the character of the so called para-plane in $Ce_{1-x}La_xSb$. From the magnetostriction experiment, this is thought to be made mostly by the Γ_7 doublet.(38) However, the present experiment shows no indication of the doublet ordering. Therefore, again, the Γ_7 Kondo picture is favorable. The only obstacle to this picture is the entropy of ln2 in the para-layer.(39) Therefore careful measurements of specific heat have been done both for CeSb and for $Ce_{.8}La_{.2}Sb$. The result is very much different from that reported so far.(39) As shown in Fig.11, the entropy just above T_{N1} obtained from our experiment is only 0.6 of that reported in ref.(39) for CeSb. Therefore, the entropy of the para-layer is thought to be reduced by the nearly the same ratio. This is consistent with the Γ_7 Kondo model. In Fig.11, the entropy calculated from the simple crystal field splitting model with the reported split-

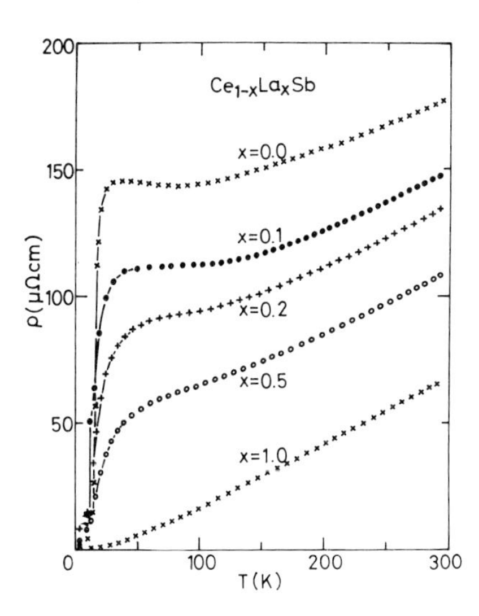

Figure 8 : Resistivities of various $Ce_{1-x}La_xSb$ are plotted as functions of T.

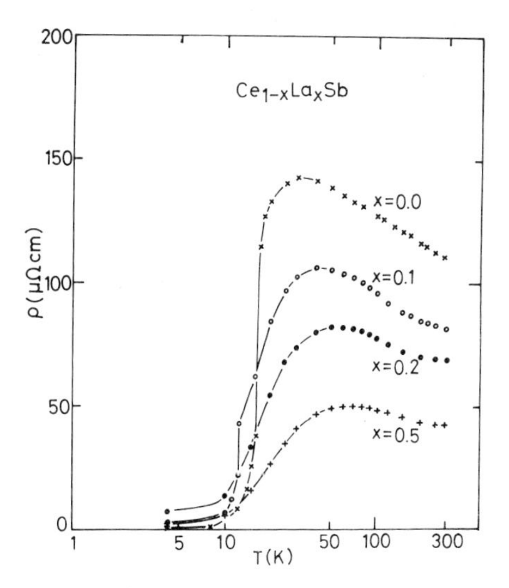

Figure 9 : Resistivities of $Ce_{1-x}La_xSb$ subtracted by that of LaSb are plotted as functions of log T.

ting, 37K,(40) is also shown for comparison. Main part of the large reduction seems to be due to the Γ_7 Kondo effect. It is, however, difficult to estimate the Kondo temperature T_K consistently to all experimental results. Perhaps, interference effect among individual Kondo states seems to be important in such a semimetallic system with a small number of carriers. More experiments are needed to understand this point. Specific heat for $Ce_{.8}La_{.2}Sb$ has a peak at around 4K, which corresponds to the first order transition to FP phase but is broadened by La alloys. Again the entropy just above the peak is smaller than that expected from that calculated by the simple crystal field model of 37K splitting indicating the Γ_7 Kondo state. Extension to the lower and the higher temperature regions is necessary.

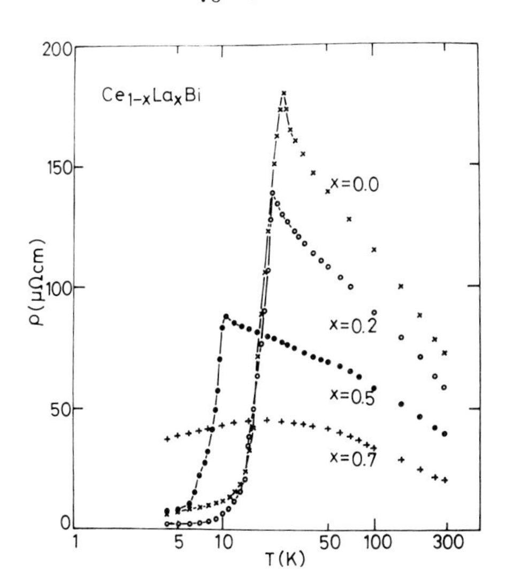

Figure 10 : Resistivities of $Ce_{1-x}La_xBi$ subtracted by that of LaBi are plotted as functions of log T.

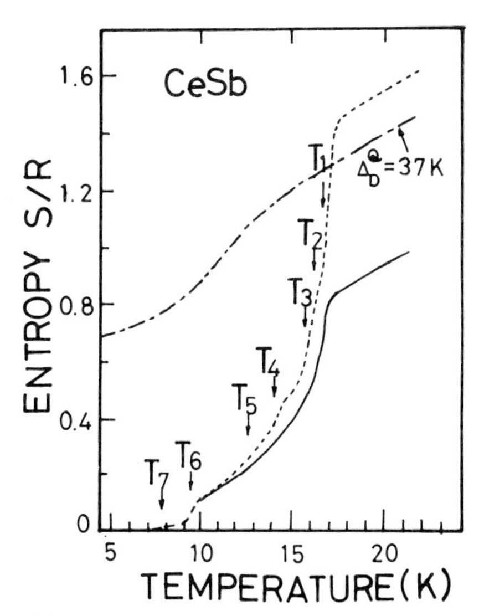

Figure 11 : Magnetic entropy of CeSb obtained by subtracting that of LaSb is shown by solid line. The value of ref.(39) is shown by dotted line for comparison with arrows showing various transition temperatures. Dot-dashed line is the calculated result for Δ=37K.

CONCLUSIONS

Dense Kondo behaviors in CeB_6, CeSb and CeBi are reviewed briefly mostly based on the studies in our group. There are a lot of mysterious behaviors in CeB_6. More elaborated studies are needed. CeSb and CeBi are unique in the sense that they are semimetals. It is particularly convenient to study the interference effect among the individual Kondo states.

Finally, the authors express their cordial thanks to Prof. W. Sasaki, Mr. T. Furuno, Prof. H. Yasuoka, Mr. H. Takigawa, Prof. M. Date, Mr. T. Sakakibara, Prof. A. Yanase, Prof. A. Hasegawa, Prof. M. Kawakami, Prof. T. Fujimura, Dr. T. Nakajima, Prof. S.B. Woods and Dr. J. Rossat-Mignod for their cooperative works on the present subject.

REFERENCES

[1] Wiegmann, P.B., J. Phys. C: Solid State Phys. 14 (1981) 1463-1478.
[2] Kasuya, T., Takegahara, K., Aoki, Y., Hanzawa, K., Kasaya, M., Kunii, S., Fujita, T., Sato, N., Kimura, H., Komatsubara, T., Furuno, T. and Rossat-Mignod, J., in Falicov, L.M., Hanke, W. and Maple, M.B. (eds.), Valence Fluctuations in Solids (North-Holland, N.Y., 1981) pp.215-223.
[3] Barbara, B., Rossignol, M.F., Boucherle, J.X., Schweizer, J. and Buevoz, J.L., J. Appl. Phys. 50 (1979) 2300-2307.
[4] Andres, K., Graebner, J.E. and Ott, H.R., Phys. Rev. Letters 35 (1975) 1779-1782.
[5] Etourneau, J., Mercurio, J.P. and Hagenmuller, P., in Matkovich, V.I. (ed.), Boron and Refractory Borides (Springer-Verlag, Berlin, 1977), pp.115-138.
[6] Takase, A., Kojima, K., Komatsubara, T. and Kasuya, T., Solid State Commun. 36 (1980) 461-464.
[7] Kawakami, M., Kunii, S., Komatsubara, T. and Kasuya T., Solid State Commun. 36 (1980) 435-439.
[8] Hulliger, F. J. Magn. Magn. Mater. 8 (1978) 183-205.
[9] Bossat-Mignod, J., Burlet, P., Villain, J., Bartholin, H., Wang Tcheng-Si, Florence, D. and Vogt, O., Phys. Rev. B16 (1977) 440-461.
[10] Takegahara, K., Takahashi, H., Yanase, A., Kasuya, T., Solid State Commun. 39 (1981) 857-861.
[11] Hasegawa, A., J. Phys. C: Solid State Phys. 13 (1980) 6147-6156.
[12] Suzuki, T., Kitazawa, H., Sera, M., Oguro, I., Shida, H., Yanase, A. and Kasuya, T., in Guertin, R.P., Mulak, J. and Suski, W. (eds.), Crystalline Eelctric Field and Structual Effects in f-Electron Systems (Plenum Press, N.Y. 1982) in press.
[13] Kasuya, T., Takegahara, K., Kasaya, M., Isikawa, Y., Takahashi, H., Sakakibara, T. and Date, M., in Chikazumi, S. and Miura, N. (eds.), Physics in High Magnetic Fields (Springer-Verlag, Berlin, 1981) pp.150-160.
[14] Yanase, A. and Kasuya, T., Progr. Theoret. Phys. 46 (1970) Suppl. No.46 388-410.
[15] Aoki, Y., in preparation.
[16] Fisk, Z. and Johnston, D.C., Solid State Commun. 22 (1977) 359-362.
[17] Fisk, Z., Solid State Commun. 18 (1976) 221-223.
[18] Connerade, J.P. and Mansfield, M.W.D., Phys. Rev. Letters 48 (1982) 131-134.
[19] Bauchspiess, K.R., Boksch, W., Holland-Moritz, E., Launois, H., Pott, R. and Wohlleben, D., in Falicov, L.M., Hanke, W. and Maple, M.B. (eds.), Valence Fluctuations in Solids (North-Holland, N.Y., 1981) pp.417-421.
[20] Sugawara, H., Kakizaki, A., Nagakura, I., Ishii, T., Komatsubara, T. and Kasuya, T., J. Phys. Soc. Japan 51 (1982) 915-921.
[21] Baer, Y., Hauger, R., Zürcher, Ch., Campagna, M. and Wertheim, G.K., Phys. Rev. B18 (1978) 4433-4439.
[22] Takigawa, M. and Yasuoka, H., private communication and to be published.
[23] Aoki, Y. and Kasuya, T., Solid State Commun. 36 (1980) 317-319.
[24] Goto, T. et al. to be published.
[25] Entel, P., Grewe, N., Sietz, M. and Kowalski, K., Phys. Rev. Letters 43 (1979) 2002-2005.
[26] Boppart, H., Treindl, A., Wachter, P. and Roth, S., Solid State Commun. 35 (1980) 483-486.
[27] Horn, S., Steglich, F., Loewenhaupt, M., Scheuer, H., Felsch, W. and Winzer, K., Z. Phys. B42 (1981) 125-134.
[28] Cooper, B.R., in Elliott, R.J. (ed.), Magnetic Properties of Rare Earth Metals (Plenum Press, N.Y., 1972) pp.17-80.
[29] Fujita, T., Suzuki, M., Komatsubara, T., Kunii, S., Kasuya, T. and Ohtsuka, T., Solid State Commun. 35 (1980) 569-572.
[30] Furuno, T., Sasaki, W., private communication and to be published.
[31] Kawakami, M. et al., to be published
[32] Kunii, S., et al., this conference.
[33] Komatsubara, T., Suzuki, T., Kawakami, M., Kunii, S., Fujita, T., Isikawa, Y., Takase, A., Kojima, K., Suzuki, M., Aoki, Y., Takegahara, K. and Kasuya, T., J. Magn. Magn. Mater. 15-18 (1980) 967-968.
[34] Date, M., Hanzawa, K., to be published.
[35] Sato, N. et al., to be published.
[36] Sato, N., Komatsubara, T., Kunii, S., Suzuki, T. and Kasuya, T., in Falicov, L.M., Hanke, W. and Maple, M.B. (eds.), Valence Fluctuations in Solids (North-Holland, N.Y., 1981) pp.259-262.
[37] Hessel Andersen, N., in Crow, J.E., Guertin, R.P. and Mihalisin, T.W. (eds.), Crystalline Electric Field and Structural Effects in f-Electron Systems (Plenum Press, N.Y., 1980) pp.373-387.
[38] Sera, M. et al., this conference.
[39] Rossat-Mignod, J., Burlet, P., Bartholin, H., Vogt, O. and Lagnier, R., J. Phys. C: Solid State Phys. 13 (1980) 6381-6389.
[40] Heer, H., Furrer, A., Hälg, W. and Vogt, O., J. Phys. C: Solid State Phys. 12 (1979) 5207.

This paper is dedicated to Professor S. Methfessel for his 60th anniversary of birthday.

M.T. BEAL-MONOD: About the T^2 term in the low-temperature resistivity: in the Fermi-liquid approach the resistivity is a function of T times $\chi(T)$, say: $f[T_2 \cdot \chi(T)]$, so that in the best cases it is a pure T^2 law, i.e. $f \simeq T^2 . \chi^2(0)$. Depending on the temperature dependence $\chi(T)$ you may have a less well defined law except very close to T=0. So one should not be so pessimistic about not always finding a pure T^2 law.

T. KASUYA: What we tried to check is whether the reported T^2 law of resistivity at low temperatures in the dense Kondo system is really intrinsic or not. We found that the T^2 term depends strongly on the residual resistivity at zero magnetic field. There is too much ambiguity to ascertain the T^2 law in pure materials.

F. STEGLICH: Comments: 1) In contrast to the antiferromagnetically ordered "Kondo lattice" $CeAl_2$, where the specific heat coefficient γ remains almost unaffected by high external fields, we have found that for CeB_6, γ increases by a factor of three upon application of 16 kG, but, again, comes close to the B=0 value at 80kG.
2) We have observed that the "Kondo-type" incremental Ce-resistivity, which you have reported for the Ce-rich (Ce,La)Sb alloys, does not exist for dilute (Ce,La)Sb alloys, with Ce-concentration $\lesssim$10at%.

T. KASUYA: Reply: 1) We measured the specific heat of CeB_6 in a magnetic field of 6 kOe but no visible change was observed in the whole temperature region. For 36 kOe of magnetic field, the specific heat above 0.5 K decreased substantially but no change was observed below 0.5 K showing the same γ value of 300 mJ/K^2 mole. Three La-doped samples have also nearly the same γ values.
2) As shown in the figures, we have measured only up to 50% La alloy in CeSb. We have still clear Kondo-like behavior but the intensity is decreasing rapidly and perhaps you can not see it at x=0.9. This is because for larger x the mobility of conduction electron becomes smaller due to stronger d-f (Γ_7) Kondo scattering but the mobility of valence bands becomes larger due to decreasing population of Γ_8 and you are seeing normal valence band conduction. Actually, the Hall constant measurement shows change of sign to positive for x=0.5. For $Ce_{1-x}La_xBi$, because of smaller crystal field splitting, the sign of Hall constant is still negative for x=0.7 and thus we can see Kondo effect.

Valence Instabilities
P. Wachter and H. Boppart (eds.)

GENERALIZED FERMI LIQUID THEORY FOR MIXED-VALENT SYSTEMS

P.F. de Châtel

Natuurkundig Laboratorium, Universiteit van Amsterdam,
Valckenierstraat 65, 1018 XE Amsterdam, The Netherlands

The perturbation treatment of the asymmetric Anderson model in the $U \to \infty$ limit |1| leads to an excitation spectrum which, within the manifold of states involving the mixed-valent impurity in the singlet (nonmagnetic) state, can be described in terms of Landau's Fermi liquid theory |2|. For these low-energy excitations quasiparticle energies are evaluated and the incremental density of states, due to hybridization, is shown to give rise to a specific heat contribution, which is related to the corresponding susceptibility contribution by

$$\chi(0)T/C = J(J+1)\, g_J^2 \mu_B^2 / \pi^2 k_B^2 \; .$$

This relationship has been experimentally verified, has been derived earlier |3| and follows from a "local Fermi liquid" model neglecting all fermion-fermion interactions ($U = 0$) |4|. In the present theory the absence of a susceptibility enhancement due to exchange interactions is found to be due to the nature of the quasiparticle interaction, which is independent of spin. The derivation of this result gives some insight into the physical origin of the spin-independent interaction.

At higher temperatures, excitations to states involving the mixed-valent impurity in the degenerate (magnetic) state also have to be taken into account. In the absence of hybridization, the occupancy of these states follows a generalized Fermi statisctics |5|,

$$n_f = \frac{1}{\frac{1}{2J+1} e^{(\varepsilon_f - \mu)/k_B T} + 1} ,$$

in which the degeneracy of the magnetic configuration plays a role. Making use of the one-to-one correspondence between perturbed and unperturbed states which is exploited in Landau's Fermi liquid theory, a generalized Fermi liquid theory can be set up to describe the high-temperature behaviour in the presence of hybridization.

|1| T.V. Ramakrishnan, in Valence Fluctuations in Solids (Santa Barbara), p.13.
|2| P.F. de Châtel, Solid State Commun. 41, 855 (1982).
|3| H. Lustfeld and A. Bringer, Solid State Commun. 28, 119 (1978).
|4| D.H. Newns and A.C. Hewson, J. Phys. F 10, 2429 (1980).
|5| R. Ramirez and L.M. Falicov, Phys. Rev. B 3, 2425 (1972).

Valence Instabilities
P. Wachter and H. Boppart (eds.)

RESONANCE AND PAIRING EFFECTS IN THE ANDERSON LATTICE†

A. J. Fedro
Northern Illinois University, DeKalb, Illinois 60115 and Argonne National Laboratory, Argonne, Illinois 60439 U.S.A.
and
S. K. Sinha
Argonne National Laboratory, Argonne, Illinois 60439 U.S.A.

We have calculated the one electron Green's functions for the Anderson lattice beyond Hubbard I by considering the equation of motion for the spin fluctuations arising from the large ($U = \infty$) correlation energy in a manner formally equivalent to that of Nagaoka in the Kondo impurity problem. This leads directly to frequency dependent self energy corrections yielding lifetime effects and, more importantly, a resonance in the f density of states at the Fermi energy, μ, resulting in heavy Fermi liquid behavior for states near μ, regardless of the position of the f level. In addition, at low temperatures, we find the formation of a highly coherent state, driven by the correlations, in which the f and conduction electrons undergo singlet pairing resulting in anomalies in specific heat and susceptibility.

1. Model and the Resonance Effect

A puzzling feature of recent experimental results on Ce-based systems is the fact that photoemission and other measurements indicate that the valence of the f-shell is essentially integral, while other measurements indicate a large density of electron states at the Fermi level, μ, much as one would expect in true intermediate valence (IV) systems.[1-4] It is well-known from the theory of the Kondo effect for dilute magnetic impurities that a many-body resonance exists at μ, but this has never been demonstrated explicitly for the Anderson lattice. Martin has recently proposed such a resonance for the Anderson lattice by appealing to the Friedel sum rule.[5] In this paper, we sketch how such a resonance can be derived, by extending an earlier mean-field model obtained by ourselves and others[6,7,8] to take into account dynamical spin-fluctuation terms. We obtain a theoretical model which should apply both in the "IV" and "Kondo" limits. We also demonstrate that a more general decoupling scheme for the equations of motion of the Green's functions allows a solution yielding a new correlated "exotic-paired" ground state.

As a model which incorporates the essential features of the mixed-valence system, consider a simple (s-like) conduction band described by pure Fermion operators $\{C_{k\sigma}\}$ with momentum k and spin σ and with dispersion ε_k and bandwidth W hybridizing with a set of localized (f-like) electrons. The f electrons have an arbitrarily large, $U \to \infty$, correlation energy restricting the f electron configuration at each site to f^n and f^{n+1}. Previously we have shown how to include arbitrary multiplet structures in the n and n+1 configurations,[6] but, for simplicity and space limitations, we take $n = 0$ and the f^1 configuration to be doubly degenerate (spin $1/2$).

Our Hamiltonian in the paramagnetic region can then be written as follows:

$$H = \sum_{j,\sigma} \varepsilon_f f^+_{j\sigma} f_{j\sigma} + \sum_{k,\sigma} \varepsilon_k C^+_{k\sigma} C_{k\sigma} + \frac{1}{\sqrt{N}} \sum_{k,j,\sigma} [V_k e^{-ikj} C^+_{k\sigma} f_{j\sigma} + \text{h.c.}] \quad (1)$$

where ε_f is the f level energy, V_k is the hybridization matrix element and N the total number of sites. Here the $\{f_{j\sigma}\}$ refer to f-electron state operators which were defined in Ref. 6 and ensure the proper occupancy of the f level at each site j. These operators obey the relations

$$[f_{j\sigma}, f_{j'\sigma'}]_+ = 0 \; ; \; [f_{j\sigma}, f^+_{j'\sigma'}]_+ = \delta_{jj'} [\delta_{\sigma\sigma'} - F^j_{\sigma\sigma'}] \; ,$$

where (2a)

$$F^j_{\sigma\sigma'} = \delta_{\sigma'\sigma} f^+_{j\bar{\sigma}} f_{j\bar{\sigma}} - \delta_{\sigma'\bar{\sigma}} f^+_{j\bar{\sigma}} f_{j\sigma} \quad (2b)$$

and reflects the departure from pure Fermi behavior due to the correlation energy. The equations of motion for the one-electron Green's functions of momentum k, spin σ, and complex frequency $\bar{\omega} \equiv \omega - i\varepsilon$; $\varepsilon = 0^+$ become

$$(\bar{\omega} - \varepsilon_k) G^{c-a}_{k\sigma}(\bar{\omega}) - V_k G^{f-a}_{k\sigma}(\bar{\omega}) = \frac{1}{2\pi} \langle [C_{k\sigma}, a^+_{k\sigma}]_+ \rangle \quad (3a)$$

$$-V^*_k G^{c-a}_{k\sigma}(\bar{\omega}) + (\bar{\omega} - \varepsilon_f) G^{f-a}_{k\sigma}(\bar{\omega}) = \frac{1}{2\pi} \langle [f_{k\sigma}, a^+_{k\sigma}]_+ \rangle + \Gamma^a_{k\sigma}(\bar{\omega}) \quad (3b)$$

Here we employ the notation of Zubarev and define

$$G^{b-a}_{k\sigma}(\bar{\omega}) \equiv \langle\langle b_{k\sigma};a^{+}_{k\sigma}\rangle\rangle \quad (4a)$$

$$\Gamma^{a}_{k\sigma}(\bar{\omega}) \equiv -\frac{1}{N}\sum_{\substack{j\\k_1,\sigma_1}} V^{*}_{k_1}\, e^{-i(k-k_1)j}\langle\langle F^{j}_{\sigma\sigma_1} C_{k_1\sigma_1};a^{+}_{k\sigma}\rangle\rangle \quad (4b)$$

where a,b are appropriately f or c and $a_{k\sigma}$ is the standard Fourier transform of the operator $a_{j\sigma}$. The Green's function, $\Gamma^{a}_{k\sigma}(\bar{\omega})$ is formally identical to that used by Nagaoka in his treatment of the Kondo impurity problem and reflects the spin fluctuations[9]. Lowest order decoupling of $\Gamma^{a}_{k\sigma}(\bar{\omega})$, i.e., in (3b) replace

$$F^{(j)}_{\sigma\sigma_1} \to \langle F^{(j)}_{\sigma\sigma_1}\rangle \equiv \delta_{\sigma,\sigma_1}\langle n_{\bar{\sigma}}\rangle \quad (5a)$$

where

$$\langle n_{\bar{\sigma}}\rangle \equiv \langle f^{+}_{j\bar{\sigma}} f_{j\bar{\sigma}}\rangle \quad (5b)$$

yields results equivalent to Hubbard I.[6,7,8] and merely treat the spin fluctuations in mean field. To go beyond mean field we now calculate the equation of motion for $\Gamma^{a}_{k\sigma}(\bar{\omega})$ and substitute this back in Eqs. (3). We now decouple the Green's function on the R.H.S. of (3) in standard Nagaoka fashion bearing in mind the exact relation $F^{j}_{\sigma\sigma'} f_{j\sigma'} = 0$ from (2). One then finds the following:

$$\vec{G}^{-1}_{k\sigma}(\bar{\omega}) = \begin{pmatrix} \bar{\omega}-\varepsilon_k & -V_k \\ -V^{*}_k & \dfrac{\bar{\omega}-\tilde{\varepsilon}_f(\bar{\omega})}{1-\langle n_{\bar{\sigma}}(\bar{\omega})\rangle} \end{pmatrix} \quad (6)$$

where the element $G^{11}_{k\sigma}(\bar{\omega})$ refers to $G^{c-c}_{k\sigma}(\bar{\omega})$ etc... and where the complex and frequency dependent renormalizations are given by

$$\tilde{\varepsilon}_f(\bar{\omega}) \equiv \varepsilon_f + \langle n_{\bar{\sigma}}\rangle \frac{1}{N}\sum_k \frac{|V_k|^2}{\bar{\omega}-\varepsilon_k} + \frac{1}{N}\sum_k \frac{|V_k|^2[\langle C^{+}_{k\bar{\sigma}}C_{k\bar{\sigma}}\rangle-(\langle f^{+}_{k\bar{\sigma}}f_{k\bar{\sigma}}\rangle-\langle n_{\bar{\sigma}}\rangle)]}{\bar{\omega}-\varepsilon_k} \quad (7a)$$

$$\langle n_{\bar{\sigma}}(\bar{\omega})\rangle \equiv \langle n_{\bar{\sigma}}\rangle + \frac{1}{N}\sum_k \frac{V^{*}_k\langle f^{+}_{k\bar{\sigma}} C_{k\bar{\sigma}}\rangle}{\bar{\omega}-\varepsilon_k} \quad (7b)$$

Equations (3) and (7), together with the usual equations relating $\langle f^{+}_{k\sigma} f_{k\sigma}\rangle$ and $\langle C^{+}_{k\sigma} C_{k\sigma}\rangle$ to the imaginary parts of the appropriate Green's functions, lead to a set of coupled integral equations which must be solved self-consistently to obtain the Fermi energy, μ, $\langle n_\sigma\rangle$ etc. at any temperature.

Here we see that if we neglect all frequency dependent corrections in (7) we again reduce to the mean field expressions derived earlier. Also, if the only renormalization in (7) is that of the second term on the R.H.S. of (7a) we essentially recover the C.P.A. result for this problem as given by Martin and Allen.[10] The origin of this renormalization is the $f^{+}_{j\bar{\sigma}} f_{j\bar{\sigma}}$ term in $F^{(j)}_{\sigma\sigma}$ i.e., the charge fluctuation term. The rest of the renormalization in Eqs. (7) is due to the spin-flip part of $F^{j}_{\sigma\sigma'}$ i.e., the $f^{+}_{j\bar{\sigma}} f_{j\sigma}$ term, and here we can see the appearance of the Kondo-like integral in the last term in (7a) as $\bar{\omega} \to \mu$ where μ is the Fermi energy. Thus the self energy correction exhibits a temperature dependent resonance at μ regardless of whether the f level is "at" the Fermi energy or not and results in heavy Fermi liquid behavior for states near μ. Such a many-body resonance was first proposed in a phenomenological way for the magnetic impurity alloys by Zlatic et al.[11]. It is important to note that the origin of this resonance is the spin flip process which occurs due to the large correlation energy in the problem exhibited by the second term in $F^{(j)}_{\sigma\sigma'}$ of (2b). If we treat the last term on the right hand side of (7a) in lowest order we find

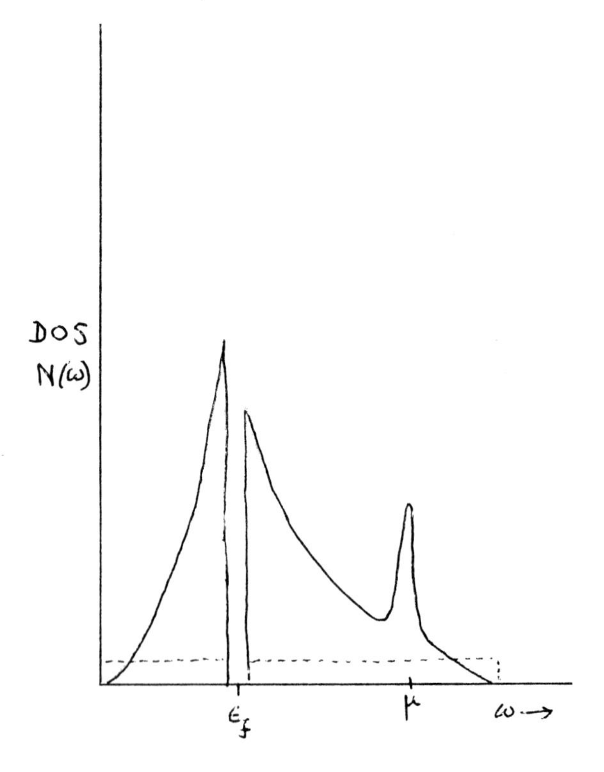

Fig. 1 Effective density of states for an Anderson lattice with ε_f well below the Fermi level. The full line represents the f-like density of states and the dashed line the conduction like density of states. μ denotes the Fermi level.

$$\frac{1}{N}\sum_k \frac{|V_k|^2 \langle C^+_{k\bar\sigma} C_{k\bar\sigma}\rangle_o}{\bar\omega - \varepsilon_k} \approx \frac{|V|^2}{W} \ln\left(\frac{W}{2\pi k_B T}\right) \tag{8}$$

for $\bar\omega = \mu$. Thus the self energy shows a strong temperature dependent resonance at μ. This is reflected in the effective f-like density of states, $\rho^{f-f}(\bar\omega)$, which can be shown to vary as follows:

$$\rho^{f-f}(\omega) \sim [\omega - \tilde\varepsilon_f(\omega)]^{-2} \tag{9}$$

Thus for $\omega=\mu$ we find

$$\mu - \tilde\varepsilon_f(\mu) \approx \text{const} + \frac{|V|^2}{W} \ln\left(\frac{W}{2\pi k_B T}\right) \tag{10}$$

where const is a small correction of order $|V|^2$. Thus, in the temperature range, T_k defined by

$$k_B T_k \sim W \exp\left[\frac{-W(\mu - \tilde\varepsilon_f(\mu))}{|V|^2}\right] \tag{11}$$

there is a large enhancement of the f-like density of states at μ. (See Fig. 1)

2. Pairing

Another feature of our theory is the possibility of formation of a highly coherent state, driven by the correlations, where the f and conduction electrons can undergo singlet pairing at low temperatures much like the Cooper pairing in BCS theory. This can be seen by decoupling the equation of motion for $\Gamma^a_{k\sigma}(\bar\omega)$ assuming anomalous type averages exist as in the Gorkov treatment of BCS and solve for these anomalous averages (order parameters) self-consistently. This leads to frequency dependent order parameters much like those in the Elashberg theory of superconductivity. In this paper, due to space limitations, we do a simpler decoupling which already illustrates the possibility of pair formation. What we do is decouple $\Gamma^a_{k\sigma}(\bar\omega)$ of (4b) directly in all possible ways using (2b). We find

$$-\frac{1}{N}\sum_{k_1,j_1,\sigma_1} V^*_{k_1} e^{-i(k-k_1)j_1} F^{j_1}_{\sigma\sigma_1} C_{k_1\sigma_1} \approx -\langle n_{\bar\sigma}\rangle C_{k\sigma}$$

$$-\frac{1}{N}\sum_{k_1} V^*_{k_1} \langle f^+_{k_1\sigma_1} C_{k_1\sigma_1}\rangle f_{k\sigma} + \gamma^*_\sigma f^+_{-k\bar\sigma} \tag{12}$$

where the singlet order parameter, γ_σ, is defined by

$$\gamma_\sigma \equiv \frac{1}{N}\sum_{k_1} V_{k_1} \langle C^+_{k_1\bar\sigma} f^+_{-k_1\sigma} - C^+_{k_1\sigma} f^+_{-k_1\bar\sigma}\rangle = -\gamma_{\bar\sigma} \tag{13}$$

Use of (12) in (3) leads in a straightforward manner to the following "gap" equation for γ.

$$\gamma = \gamma\, K(T, \gamma) \tag{14}$$

where $\gamma \equiv \{|\gamma_{\bar\sigma}|^2\}^{1/2} = \{|\gamma_\sigma|^2\}^{1/2}$ and

$$K(T,\gamma) \equiv \frac{1}{N}\sum_k \frac{|V_k|^2 \xi_k}{[\Omega^+_k]^2 - [\Omega^-_k]^2} \times \left[\frac{1}{\Omega^+_k}\tanh\left(\frac{\beta\Omega^+_k}{2}\right) - \frac{1}{\Omega^-_k}\tanh\left(\frac{\beta\Omega^-_k}{2}\right)\right] \tag{15}$$

with $\beta \equiv 1/k_B T$. Here

$$\Omega^\pm_k = \{\tfrac{1}{2}[B_k(\gamma) \pm \sqrt{B^2_k(\gamma) - 4D_k(\gamma)}]\}^{1/2} \tag{16}$$

with

$$B_k(\gamma) \equiv (E^+_k)^2 + (E^-_k)^2 + \gamma^2 \tag{17a}$$

and

$$D_k(\gamma) \equiv (E^+_k E^-_k)^2 + \gamma^2 \xi^2_k \tag{17b}$$

and the energies measured relative to the Fermi energy, μ, is given by

$$E^\pm_k \equiv \frac{1}{2}(\xi_k + \xi_f) \pm \frac{1}{2}\sqrt{(\xi_k - \xi_f)^2 + 4|V_k|^2(1 - \langle n_{\bar\sigma}\rangle)} \tag{18a}$$

with

$$\xi_k \equiv \varepsilon_k - \mu \;;\; \xi_f \equiv \varepsilon_f - \frac{1}{N}\sum_k V^*_k \langle f^+_{k\bar\sigma} C_{k\bar\sigma}\rangle - \mu \tag{18b}$$

The physical origin of the singlet pairing between f and conduction states is the negative (attractive) effective exchange between f and conduction electrons induced by correlation effects. In the limit of ε_f being well below μ, this reduces to the usual Schrieffer-Wolff exchange, $J \equiv -|V|^2/|\varepsilon_f - \mu|$, and one can show that

$$T_c \sim \frac{W}{a} \exp\left[-\frac{W}{a|J|}\right] \tag{19a}$$

where

$$a \equiv \frac{|V|^2}{|\varepsilon_f - \mu|^2} \tag{19b}$$

Thus the effect is largest when ε_f is only slightly below μ. The point is that in the usual Kondo regime, the f electrons are localized so that this attractive exchange cannot lead to pairing. In our case, however, the f-electrons have a finite density of states at μ and thus pairing can result. Preliminary calculations indicate T_c can be ~ a few millidegrees K. It would be very interesting to see if experimental evidence for this kind of pairing can be found in these systems.

3. References

1. Lawrence, M. Riseborough, P. S., and Parks, R. D., Rep. Prog. Phys. 44, 379 (1980).
2. Allen, J. W., Oh, S.-J., Lindau, I., Lawrence, J. M., Johansson, L. I., and Hagstrom, S. B., Phys. Rev. Lett. 46, 1100 (1981).
3. Johannsson, W. R., Crabtree, G. W., Edelstein, A. S., and McMasters, O. D., Phys. Rev. Lett. 46, 504 (1981).
4. See papers in Valence Fluctuations in Solids, edited by L. M. Falicov, W. Hanke, and M. P. Maple (North-Holland, Amsterdam, 1981).
5. Martin, R. M. Phys. Rev. Letters 48 (1982) 362.
6. Fedro, A. J. and Sinha, S. K., Valence Fluctuations in Solids, (Edited by L. M. Falicov, W. Hanke and M. B. Maple, North Holland, 1981) p. 329.
7. Varma, C. M. and Yafet, Y., Phys. Rev. B 13 (1976) 2950.
8. Goncalves da Silva, C.E.T. and Falicov, L. M., Solid St. Comm. 17, (1975) 1521.
9. Nagaoka, Y., Phys. Rev. 138, (1965) A1112.
10. Martin, R. M. and Allen, J. W., J. Appl. Phys. 50, (1979(7561.
11. Zlatic, V., Gruner, G. and Rivier, N. Solid St. Comm. 14, (1974), 139.

Valence Instabilities
P. Wachter and H. Boppart (eds.)

TRANSPORT PROPERTIES OF THE KONDO LATTICE

M. LAVAGNA, C. LACROIX, M. CYROT

Laboratoire Louis Néel,
C.N.R.S. 166 X,
38042 Grenoble-Cédex, France

The transport properties of a Kondo lattice are studied in the model introduced by Lacroix and Cyrot (1) taking account of both thermal and spatial fluctuations. At high temperature, the resistivity calculated within the phase shift method, is found to exhibit a logarithmic decrease characteristic of a Kondo effect. At low temperature, the behaviour of the resistivity is either metallic for $n<1$ or semiconductor for $n=1$. The thermoelectric power can be easily deduced within the same model. Qualitative agreement with experiments is found.

1. INTRODUCTION

Among the various Kondo anomalies observed in concentrated alloys of anomalous rare earth, the transport properties are perhaps the most interesting : both the resistivity and the thermoelectric power take specialy large values.

All these compounds exhibit a large logarithmic decrease of the resistivity at high temperature. At low temperature, two behaviours can be observed depending on the compound studied : $CeAl_2$, $CeIn_3$, $CePd_3$ ... exhibit a metallic behaviour, with increasing resistivities with temperature ; a maximum occurs at a temperature of the order of T_K.
SmB_6, TmSe, MSmS ... behave as a semi-conductor with an infinite resistivity at T=0 K followed by a large decrease with temperature. A particularly interesting case is provided by Tm_xSe : out of stoichiometry, the compound exhibits a metallic behaviour at low temperature while the residual T=0 resistivity diverges as stoichiometry is approached.

On the other hand, the thermoelectric power is found to take giant values at low temperature, going through an extremum approximately at the same temperature as the resistivity. Its sign may be as well positive in $CeCu_2Si_2$, $CeAl_2$ ... as negative in $CeAl_3$ ...

Many authors agree that these properties must find an explanation in the Kondo lattice model.

In fact, all the theories developed so far are more interested in the mixed valence regime (2,3,4) or in the crystalline field effect (5,6). The purpose of this paper is to study the transport properties in a Kondo lattice. For that, we use the model of Lacroix and Cyrot (1) which is a generalization to the Kondo lattice of a method of functional integration first introduced by Yoshimori and Sakurai (7) for the single impurity case. In this manner, they transformed the Kondo interaction into a fictitious s-f hybridization. The hamiltonian obtained is :

$$H = \sum_{k\sigma} (\varepsilon_k + \frac{J}{4}) c^+_{k\sigma} c_{k\sigma} + \sum_{i\sigma} (E_o + \frac{Jn}{4}) d^+_{i\sigma} d_{i\sigma}$$

$$+ \frac{J}{2} \sum_i x_i (d^+_{i\uparrow} c_{i\uparrow} + c^+_{i\downarrow} d_{i\downarrow}) + \frac{J}{2} \sum_i y_i (d^+_{i\downarrow} c_{i\downarrow} + c^+_{i\uparrow} d_{i\uparrow}) \qquad (1)$$

where c^+_k, d^+_i, ε_k, E_o have the usual meaning.

J is the Kondo s-f interaction

n is the number of conduction electrons per atom

x_i and y_i represent the fictitious s-f hybridization and can be considered as Kondo order parameters : $<x_i>=<y_i>=0$ above the Kondo temperature T_k.

In this model, a gap of width T_K occurs in the density of states around the impurity level E_o. When n=1, E_F falls in this gap and the system is semi-conductor.

We recall in section 1 the different stages of the resistivity calculation in this model, detailed in a previous paper (8) and show how is important the introduction of thermal and spatial fluctuations in the lattice. Once established the expression of the conductivity $\sigma(\omega)$, the thermoelectric power will be easily deduced in section 3.

2. ELECTRICAL RESISTIVITY

The way to treat the problem depends on the temperature magnitude. At high temperatures, the coherence between impurities has completely disappeared and the system may be treated as a collection of incoherent impurities. At much lower temperatures, coherence develop between impurities and the material behaves as a Kondo lattice.

a) High temperatures limit

The problem is reduced to that of incoherent impurities and can be solved with a single impurity model. In our formalism, this is equivalent to the study of a virtual bound state of width $\Delta=\pi(Jx/2)^2 1/2D$ (where 2D is the width of the conduction band). The calculation is done with the phase shift method.

* Importance of thermal fluctuations

In a mean-field theory, $\langle x\rangle=\langle y\rangle=0$ just above T_K and the resistivity obtained is that of free electrons. This is in contradiction with experimental results which show a large decrease of the resistivity until 100 or 200 K, well above T_K(5 or 10 K). To account for this, we must consider the thermal fluctuations. These one are evaluated from the expression of the free energy at high temperatures :

$$\begin{cases} F = \dfrac{J^2x^2}{4D} \operatorname{Ln} \dfrac{T}{T_K} - kT\operatorname{Ln}2 \\ \text{with } kT_K = 1.14\ De^{2D/J} \end{cases} \tag{2}$$

giving :

$$\begin{cases} \langle x^2\rangle = \dfrac{k_B T}{\dfrac{J^2}{2D} \operatorname{Ln} \dfrac{T}{T_K}} \\ \langle x^4\rangle = 3\left(\dfrac{k_B T}{\dfrac{J^2}{2D} \operatorname{Ln} \dfrac{T}{T_K}}\right)^2 \ \ldots \end{cases} \tag{3}$$

The resistivity is calculated by means of the phase shift $\phi(\omega)$:

$$\begin{cases} \rho(\omega) \sim \sin^2\phi(\omega) \\ \text{with } \phi(\omega) = \operatorname{Arctg} \dfrac{\Delta}{\omega} \end{cases} \tag{4}$$

Then, suming over all the ω :

$$\rho \sim \frac{3\ J^4\ x^4}{16D^2(k_BT)^2} \tag{5}$$

and then over all the x with the statistic distribution defined above :

$$\rho \sim \frac{9}{4\operatorname{Ln}^2 \frac{T}{T_K}} = \frac{9}{4}\left(\frac{J}{2D}\right)^2\left[1 + \frac{J}{D} \operatorname{Ln} \frac{k_BT}{D}\right] \tag{6}$$

We recognize here the usual Kondo expression with a logarithmic decrease in $J \operatorname{Ln}(k_BT/D)$ reported in figure 1.

b) Low temperature limit

If the Kondo parameter x_i is uniform on all the sites, the hamiltonian (1) can easily be diagonalised and nothing is able to scatter electron. We postulate here that the spatial fluctuations of the s-f hybridization is the scattering potential :

$$\begin{cases} V = \sum\limits_i \dfrac{J\delta_i}{2} (d^+_{i\uparrow}c_{i\uparrow} + c^+_{i\downarrow}d_{i\downarrow} + d^+_{i\downarrow}c_{i\downarrow} + c^+_{i\uparrow}d_{i\uparrow}) \\ \text{with } \delta_i = x_i - \langle x_i\rangle \end{cases} \tag{7}$$

The average value $\langle\delta^2\rangle$ is evaluated to $kT/2|J|$ (8) and the resistivity is calculated in a Boltzmann theory where the relaxation time $\tau(\omega)$ is directly related to the T-matrix :

$$\tau_i(\omega)^{-1} = -\operatorname{Im}.T_{ss} = \left(\frac{J\delta_i}{2}\right)^2 \rho_d(\omega) \tag{8}$$

* For n lower than 1 (but close to 1), only the states in the ω_2 neighbauhood are to consider and we find :

$$\sigma(\omega) = \frac{4e^2}{3m^*} \frac{1}{\left(\frac{J\delta}{2}\right)^2\left(\frac{Jx}{2}\right)^2} (\omega_2-\omega)\ (\omega - E_o - \frac{Jn}{4})^2 \tag{9}$$

ω_2 being the lower limit of the gap in the density of states. Suming over all the ω :

$$\rho = \frac{\rho(o)}{1 + \left(\frac{T}{T_1}\right)^2} \tag{10}$$

with $\rho(o) \sim T/(1-n)^{2/3}$

$T_1 \sim 0.367\ T_K(1-n)^{1/3}$

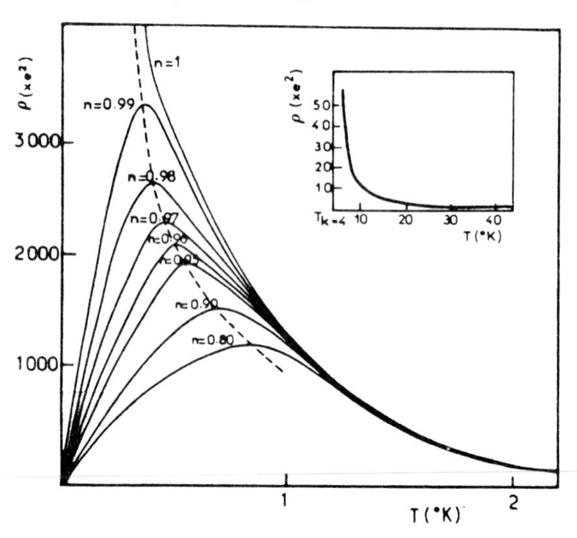

Figure 1 : Thermal variation of the resistivity (for $|J|/D=0.2$)

The corresponding variation of the resistivity is represented in figure 1. It is of metallic type with an increase of the resistivity with temperature. The resistivity passes through a maximum at the temperature $T_1 \sim T_K(1-n)^{1/3}$. The value of the maximum varies then as $1/(1-n)^{2/3}$, becoming infinite for n=1.

* For n=1, the resistivity is characteristic of a semi-conductor with a behaviour in $e^{T_K/T}$ (see figure 1).

3. THERMOELECTRIC POWER

The thermoelectric power is studied in the Kondo lattice model as shown in a previous paper (9). It can be deduced from the expression of the conductivity by the usual formula :

$$S(T) = \frac{1}{eT} \frac{\int \left(-\frac{df}{d\omega}\right) (\omega - E_F)\sigma(\omega)\,d\omega}{\int \left(-\frac{df}{d\omega}\right) \sigma(\omega)\,d\omega} \qquad (11)$$

(where e is the negative electron charge)

At low temperature and for n<1, we can write by use of equation (9) :

$$S(T) = S(0) \frac{1 + \left(\frac{T}{T_2}\right)^2}{1 + \left(\frac{T}{T_1}\right)^2} \qquad (12-$$

with
$$S(0) = -\frac{2\pi^2 k_B}{3e\left[\frac{3\pi}{4}(1-n)\right]^{2/3}} \frac{T}{T_K}$$

$$\begin{cases} T_1 = 0.367\ T_K(1-n)^{1/3} \\ T_2 = 0.269\ T_K \end{cases}$$

As n is very close to 1 in almost all these compounds, the thermoelectric power takes giant positive values at low temperature coming from the presence of $(1-n)^{2/3}$ in the denominator. On the other hand, $T_1 \ll T_2$ generally and the thermoelectric power exhibits a maximum at a temperature slightly higher than T_1, temperature of the maximum of the resistivity. It is easy to transpose this to the case n>1 :

$$S(0) = \frac{2\pi^2 k_B}{3e\left[\frac{3\pi}{4}(n-1)\right]^{2/3}} \frac{T}{T_K} \qquad (13)$$

The thermoelectric power takes again giant values but of negative sign with a peak slightly above $T_1' = 0.367\ T_K(n-1)^{1/3}$. So, the sign of (n-1) determines the nature (maximum or minimum) of the peak. The experimental cases $CeCu_2Si_2$, $CeAl_2$ and $CeAl_3$ probably correspond to each of these two possibilities. The results of Jaccard (10) about the variation of S(300 K) with x in Tm_xSe confirm that a changement in the sign of the temperature occurs near the stoichiometry (i.e. n=1). The thermal variation of the thermoelectric power is shown in figure 2 for various n.

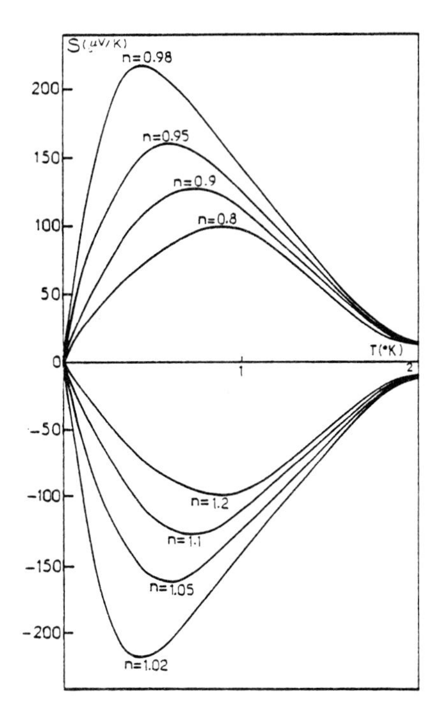

Figure 2 : Thermal variation of the thermoelectric power.

4. CONCLUSION

The Kondo lattice model is a good starting point to discuss the transport properties of these systems. The results appear to agree with experiments : the resistivity exhibit a logarithmic decrease at high temperature, with either a metallic behaviour at low temperature for n≠1 (and a maximum at $T_1 \sim T_K(1-n)^{1/3}$) or semi-conductor for n=1. The thermoelectric power is found to take giant values at low temperature with a peak at approximately the same temperature as the resistivity maximum : its sign can be positive or negative depending on the number of conduction electrons n.

REFERENCES

1. LACROIX C., CYROT M., Phys. Rev. B20, 1969 (1979)
2. GHATAK S.K., AVIGNON M., BENNEMANN K.H., J. Phys. F, 6, 1441 (1976)
3. ENTEL P., MUHLSCHLEGEL B., ONO Y., Z. Phys. B38, 227 (1980)

4. COQBLIN B., BHATTACHARJEE A.K., JULLIEN R., Journal of Magn. Magn. Mat. 15-18, 995 (1980)

5. PESCHEL I., FULDE P., Z.Phys. 238, 99 (1970)

6. BHATTACHARJEE A.K., COQBLIN B., Phys. Rev. B13, 3441 (1976)

7. YOSHIMORI A., SAKURAI A., Prog. Theor. Phys. Suppl. 46, 162 (1970)

8. LAVAGNA M., LACROIX C., CYROT M., J. Phys. F12, 745 (1982)

9. LAVAGNA M., LACROIX C., CYROT M., Phys. Letters (to appear)

10. JACCARD D., SIERRO J., Helvet. Phys. Acta (1982) to appear.

Valence Instabilities
P. Wachter and H. Boppart (eds.)

p-f MIXING MECHANISM IN Ce MONOPNICTIDES

Hiroko Takahashi, Katsuhiko Takegahara, Akira Yanase and Tadao Kasuya

Department of Physics, Tohoku University, Sendai 980, Japan

Anomalously small crystalline field splitting in the paramagnetic region in CeSb and CeBi is explained on the basis of the mixing mechanism between the 4f states and the valence bands. Considering the p-f mixing and the crystalline field by the point charge model, the effective crystalline field splitting is determined to minimize the total free energy of the valence bands, the conduction bands and the 4f states in good agreement with experiment. The formula of the second order transition temperature, at which the population ratio of the 4f Γ_8 states begins to be unbalanced, is also derived. In the case of CeBi, the second order transition is possible to occur, while in the case of CeSb, the second order transition does not occur at least before the first order transition occurs.

1. INTRODUCTION

Various unusual properties of cerium monopnictides with the rock-salt type structure have been attracting much attention recently. In the Ce monopnictides except CeN, Ce is considered to be in the stable trivalent configuration. The magnetic properties originate from the single 4f electron of Ce^{3+} with the free ion $^2F_{5/2}$ multiplet as the ground state. Because of a large spin-orbit interaction, the excited J multiplet $^2F_{7/2}$ has a energy of 0.28eV above the ground level and may be ignored in the following discussion. In an octahedral crystalline field, the J=5/2 multiplet splits into a Γ_7 doublet and a Γ_8 quartet. The point charge model predicts that the ground state is the Γ_7 doublet. Among these compounds, CeP and CeAs behave as normal trivalent magnetic materials with the Γ_7 ground state. But the crystalline field splitting is about a half of that expected from the extrapolation of other rare earth monopnictides.(1) On the other hand, CeSb and CeBi show various anomalous phenomena as follows. The observed crystalline field splittings in CeSb and CeBi are unusually small, 37K in CeSb and 8K in CeBi in the paramagnetic region, compared with those expected from the extrapolated values, 264K and 247K, respectively.(1) Afterwards, the latter values are cited for simplicity as the values for the point charge model, even if they include various complicated mechanisms. In the ferromagnetic state, however, the magnetic anisotropy is extraordinarily large with the easy axis in the [001] direction, contrary to [111] expected from the point charge model. In CeSb, even a magnetic field up to 15T in the [111] direction cannot induce any visible change in the direction of the moments.(2) Such a large anisotropy may suggest a large anisotropic exchange interaction. However, the observed Néel temperatures of CeSb and CeBi are not so high, 16K and 26K, respectively. Note that the former is the first order transition, while the latter is the second order.(3) To explain these unusual magnetic properties, we have proposed the anisotropic p-f mixing model.(4)-(9) A detailed description of the effective crystalline field splitting at room temperature and the ferromagnetic ground state have been already given in a previous paper. In this paper, we treat the temperature dependence of the crystalline field splitting in the paramagnetic region and the magnetic transition in detail.

The trivalent rare earth monopnictides are known as typical indirect narrow gap semiconductors or weakly overlapping semimetals.(10) The APW band calculations exhibit that the top of the valence bands derived mostly by the p states on pnictogens is at the Γ point and the bottoms of the conduction bands derived mostly by the $5dt_{2g}$ states on cerium atoms are at three X points.(11),(12) The band gap becomes narrower for heavier Ce pnictides. CeSb and CeBi are definitely semimetals with the equal number of holes and electrons. The 4f level is expected to be a little below the Fermi level.(13) As mentioned before, the p-f mixing is much larger than the d-f mixing.(9),(14) By the p-f mixing the 4f level is pulled down and the valence bands are pushed up because of the bonding-antibonding effect. When the valence bands are filled no real energy gain is obtained. In CeP and CeAs this effect seems to be not so important because of the small number of holes. However, in CeSb and CeBi the number of holes increases to about a few percents per Ce atom and thus the p-f mixing effect should play an important role.

2. EFFECTIVE CRYSTALLINE FIELD SPLITTING IN THE PARAMAGNETIC REGION

In the paramagnetic state, the 4f state is denoted by the eigenstate of the usual cubic crystalline field, $|Qi\rangle$ ($i=\kappa,\lambda,\mu,\nu$) for the Γ_8 quartet and $|Di'\rangle$ ($i'=\alpha'',\beta''$) for the Γ_7 doublet. We assume that the Bloch states of the valence bands are made by the linear combinations of the atomic p states on each pnictogen.

Because the number of the valence band holes is small, we use the simplified model to replace all the existing hole states by the six states at the Γ point. It is convenient to introduce the eigenstate with the spin-orbit coupling for the basic p states, $|Pi\Gamma\rangle$ ($i=\kappa,\lambda,\mu,\nu$) for $j=3/2$ and $|P'i'\Gamma\rangle$ ($i'=\alpha',\beta'$) for $j=1/2$. Only the nearest neighbour mixings are taken into account. Then it is derived that the nonvanishing matrix elements with a common value are in between $|Qi\rangle$ and $|Pi\Gamma\rangle$

$$\langle Qi|V|Pi\Gamma\rangle=\sqrt{18/7}[(pf\sigma)-\sqrt{3}/2(pf\pi)], \quad (1)$$

where $(pf\sigma)$ and $(pf\pi)$ represent the p-f two center integrals. The following values, $(pf\sigma)=0.5$eV and $(pf\pi)=-0.5(pf\sigma)$, are estimated from the actual band calculation of LaSb. We consider the simplified system which contains the valence bands, the conduction bands and the 4f states. We calculate the total free energy as a function of the population ratio of the Γ_8 state X(Qi).

$$F(X(Qi))=U_{vb}+U_{cb}+U_{4f}-TS_{4f}\equiv U_{tot}-TS_{4f}, \quad (2)$$

where U_{vb}, U_{cb}, U_{4f} and U_{tot} are the energies for the valence bands, the conduction bands, the crystalline field energy of the point charge model and the sum of them, respectively, and S_{4f} is the entropy of the 4f states. The entropy by the band electrons are neglected because it has no essential effect in the present problem. The energy shift of Pi state due to the p-f mixing energy may be given by

$$\Delta E(Pi\Gamma)=(\Delta E_f/2)(\sqrt{a}-1) \quad (3)$$

in the mean field theory, where

$$a=1+(4MX(Qi)/\Delta E_f), \quad (3')$$

$$M=|\langle Qi|V|Pi\Gamma\rangle|^2/\Delta E_f \quad (3'')$$

and ΔE_f is the 4f level energy relative to the top of the valence band without the p-f mixing effect and is chosen to be positive. The Fermi level E_F is determined so as to give equal number of holes and electrons. We assume that in the paramagnetic region the degeneracies within the doublet and the quartet exist and thus X(Qi) and X(Di') are written as X_Q and X_D, respectively, in which $4X_Q+2X_D=1$. The equilibrium value X_{Q0} is given by requiring the minimum of $F(X_Q)$ with respect to X_Q. We define the effective crystalline field splitting Δ_{QD} as follows,

$$\Delta_{QD}=(1/4N_R)(\partial U_{tot}/\partial X_Q)_{X_Q=X_{Q0}}, \quad (4)$$

where Δ_{QD} is positive for the Γ_7 lower and N_R means the number of Ce atoms. Then we obtain

$$\Delta_{QD}=\Delta_{pc}-n(Pi)M/\sqrt{a}, \quad (5)$$

where n(Pi) represents the number of Pi state holes and Δ_{pc} is the splitting without the p-f mixing effect, or the value for the point charge model in our conventional definition. The value for X_{Q0} calculated by this effective crystalline field splitting is consistent with that calculated before from the minimum condition for $F(X_Q)$. In the following numerical calculations, the lowest part of the conduction bands is treated by the nearly free electron model with the effective mass of 0.25m according to the APW band calculations, the valence bands by the simple tight binding model and the number of the valence band holes are chosen to be 0.05/Ce in CeSb and 0.055/Ce in CeBi. An example is shown in Fig. 1 in which the values of Δ_{QD} at room temperature are chosen to be consistent with the observed values. Δ_{QD} begins to increase appreciably for temperatures below 50K in CeSb and below 20K in CeBi. However, in lower temperatures the degeneracy in the Γ_8 quartet may be solved and the long range order may be established. This is treated in the following section.

3. SECOND ORDER TRANSITION TEMPERATURE

In this section, we study the second order transition temperature T_c in which the degeneracy in the Γ_8 quartet is lifted. F(X(Qi)) is now expanded at around X_{Q0} up to the second order of $(X(Qi)-X_{Q0})$,

$$F_2=\{1-(M^2X_{Q0}/akT)[D_p(E_F)-(2n(Pi)/\Delta E_f\sqrt{a})]\} (kT/2X_{Q0})\Sigma_i(X(Qi)-X_{Q0})^2, \quad (6)$$

where $D_p(E_F)$ means the density of states of the p band at the Fermi level. Then T_c is determined as the temperature at which the second order term vanishes, that is,

$$kT_c=(M^2X_{Q0}/a)[D_p(E_F)-(2n(Pi)/\Delta E_f\sqrt{a})]. \quad (7)$$

To solve eq.(7), the function f(T) given in the double bracket { } in eq.(6) is considered. Then the Curie temperature T_c is given from the zero point of f(T). If there is no zero point the second order transition does not exist. We examine how f(T) changes as the parameters M, Δ_{pc}, ΔE_f etc. are changed. One example is shown in Fig.2 in which M and Δ_{pc} are changed so as to keep the value of Δ_{QD} at room temperature in agreement with the experimental value of CeSb. While other parameters are fixed with the values shown above. It is seemed that for a small value of M there is no second order transition and the system is always in paramagnetic state with the Γ_7 ground doublet. In the real material there should be a sort of exchange interaction among the Γ_7 and finally at a certain low temperature some kind of the Γ_7 ordering should occur. When the value of M becomes larger than 1.77ev, the second order transition occurs. However, there are two transition temperatures, T_{c1} and T_{c2}, which are caused by the following reason. As the temperature decreases in the second term of f(T), at first the denominator kT decreases more rapidly than the numerator X_{Q0} and then the ordering in the excited Γ_8 quartet occurs at T_{c1}. In the present case, however, the ground state for the paramagnetic state is Γ_7, so in lower temperatures the exponential decrease of X_{Q0} overcomes the decrease in the denominator. Therefore the Γ_8 ordering turns out to disappear again at T_{c2}. In other words, when we approach from lower temperature of the Γ_7 ground state, Γ_8 ordered appears at T_{c2}. This situation is changed by the higher order terms. If the real ground state is not the Γ_7 but the ordered Γ_8 including some amount of Γ_7, T_{c2} does not exist except a special case in which three transition temperatures exist. In the real CeSb, the following set of the values for M and Δ_{pc}, 1.67eV and 228K, seems to be most reasonable from the band calculation and the extrapolation from other compounds as mentioned above. Then, as is seen in Fig.2, the second order transition does not occur. However, as was shown already in previous papers, the real ground state is the ordered Γ_8 state. Therefore, in CeSb, the first order transition should occur in agreement with experiment. Then it is remarkable in Fig.1 that, at T_c, Δ_{QD} increase about 70% of that at room temperature. This point will be discussed later.

The same treatment has been done for CeBi and obtained results are shown as function of M in Fig.3. At first, T_c in CeSb is always lower than that in CeBi with respect to the same value of M. This is due to the larger value of X_{Q0} in CeBi as shown in eq.(7) and Fig.1. It is remarkable that in CeBi the second order transition can occur for the value of M as low as 0.8eV at which T_{c1} is 10K. This is entirely due to a smaller value of Δ_{QD} in CeBi. For the most plausible value of M=1.67eV, the second order transition exist in CeBi in agreement with the experiment. The theoretical value T_{c1}=60K, however, is substantially larger than the experimental value 26K. This point will be discussed in the following section.

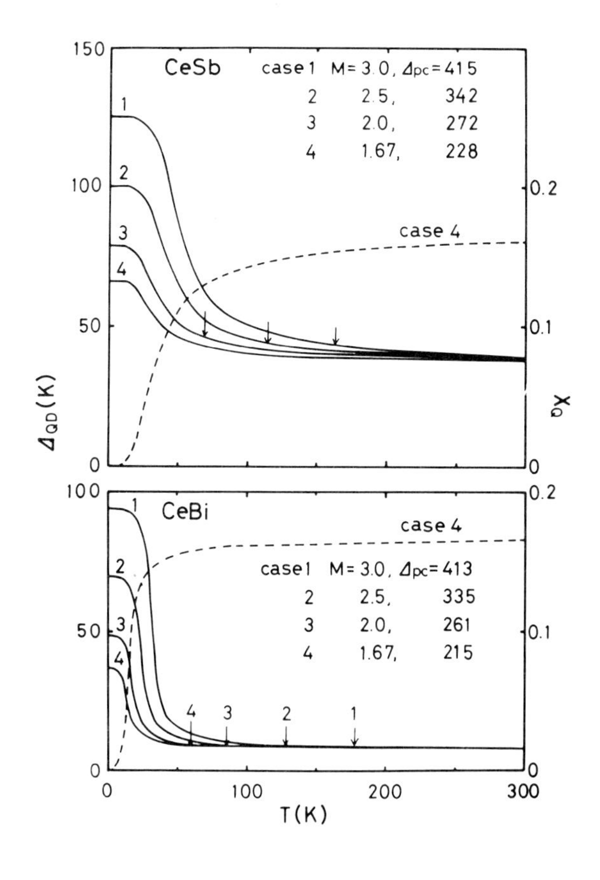

Figure 1 : Temperature dependence of the effective crystalline field splitting for some sets of M and Δ_{pc} are shown by solid lines. The units of M and Δ_{pc} are eV and K, respectively. Broken lines show the population of the Γ_8 states X_Q for case 4. The arrow shows the higher second order transition temperature in each case.

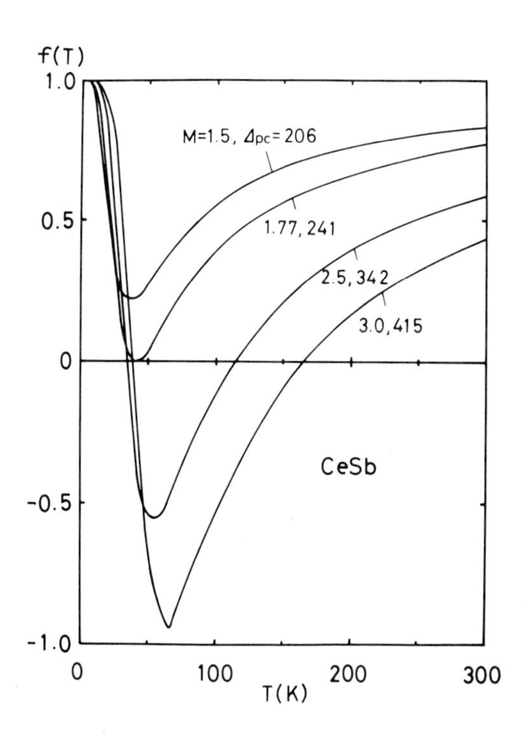

Figure 2 : f(T) is shown as a function of T for varing values of the parameters M and Δ_{pc} in which the crystalline field splitting at 300K is fitted with the experimental value of CeSb. The units of M and Δ_{pc} are eV and K, respectively.

4. DISCUSSIONS AND CONCLUTIONS

Our present calculations based of the p-f mixing model could predict the second order transition in CeBi and the first order transition in CeBi practically without any adjustable parameters in consistent with a previous calculation. Our present results are, however, based on the molecular field approximation and thus the following two effects should be taken into account to improve the theory and to obtain better agreement with experimental results. The first is the short range order effect. Disagreement of T_c in CeBi is expected to be mostly due to this effect. Note, however, that in Sera's paper in this conference the deviation of the thermal expansion in CeBi from the molecular field theory begins at about 60K in good agreement with the present calculation. It is also noted that in CeSb, too, a large deviation is observed in Sera's paper just only at around 40K but not at all for La-diluted CeSb. This is also consistent with the present calculation, because just in CeSb the second order transition is begining to occur at 40K where a large short range order fluctuation in Γ_8 is expected. Another deficiency of the molecular field approximation that it becomes poor for dilute cases. This means both the real diluted alloys and the case of small X(Qi) in dense system. Large increase of Δ_{QD} in CeSb mentioned in the preceding section will be reduced substantially by considering the dilute situation more in detail. These calculations are now in progress.

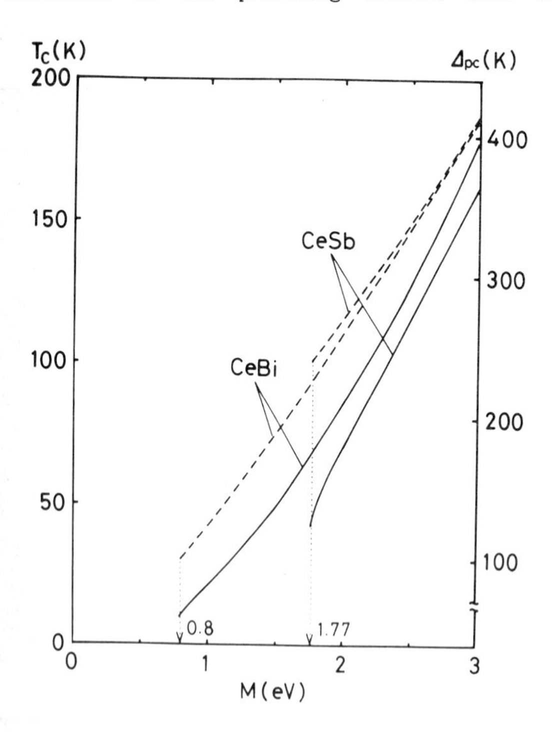

Figure 3 : Second order transition temperatures T_{c1} are shown by solid lines as functions of p-f mixing parameter M. The point charge crystalline field splittings Δ_{pc} are shown by broken lines.

The authors greatly appreciate valuable commonts and discussions with Professors A. Hasegawa and T. Suzuki.

REFERENCES

(1) Birgeneau,R.J.,Bucher,E.,Maita,J.P., Passell,L. and Turberfield,K.C., Phys.Rev. B8(1973)5345-5347.

(2) Burlet,P.,Rossat-Mignod,J.,Bartholin,H. and Vogt,O., J.de Phys. 40(1979)47-50

(3) Hulliger,F.,Landolt,M.,Ott,H.R. and Schmelczer,R., J.Low Temp.Phys. 20(1975)269-284.

(4) Takegahara,K.,Yanase,A. and Kasuya,T., J.de Phys. 41(1980)C5 327-329., Erratum J.de Phys. 41(1980)1231.

(5) Takegahara,K.,Takahashi,H.,Yanase,A. and Kasuya,T., J.Phys.C:Solid St.Phys. 14(1981)737-743.

(6) Kasuya,T.,Takegahara,K.,Kasaya,M., Isikawa,Y.,Takahashi,H.,Sakakibara,T. and Date,M.,in Chikazumi,S. and Miura,N.(eds.), Physics in High Magnetic Fields (Springer-Verlag,Berlin,1981) p150.

(7) Takegahara,K.,Takahashi,H.,Yanase,A. and Kasuya,T., Solid State Commun. 39(1981)857-861.

(8) Kasuya,T., in Moriya,T.(ed.), Electron Correlation and Magnetism in Narrow-Band Systems (Springer-Verlag,Berlin,1981)

(9) Takegahara,K.,Takahashi,H.,Yanase,A. and Kasuya,T., in Guertin,R.P.,Mulak,J. and Susuki,W.(eds.), Crystalline Electric Field and Structual Effects in f-Electron Systems (Plenum Pub.,1982).

(10) Guntherodt,G.,Kaldis,E. and Wachter,P., Solid State Commun. 15(1974)1435-1440.

(11) Hasegawa,A., J.Phys.C:Solid St.Phys. 13(1980)6147-6156.

(12) Yanase,A. and Hasegawa,A., in Moriya,T.(ed.) Electron Correlation and Magnetism in Narrow-Band Systems (Springer-Verlag,Berlin,1981).

(13) Baer,T.,Hauger,R.,Zurcher,Ch.,Campagna,M. and Wertheim,G.K., Wertheim,G.K., Phys.Rev. B18(1978)4433-4439.

(14) Kasuya,T., J.de Phys. 37(1976)C4 261-265.

Valence Instabilities
P. Wachter and H. Boppart (eds.)

A REEXAMINATION OF THE THEORETICAL STUDY OF THE KONDO LATTICE : SCHRIEFFER-WOLF TRANSFORMATION AND RENORMALIZATION GROUP STUDY

L. C. Lopes[+], R. Jullien, A. K. Bhattacharjee and B. Coqblin

Laboratoire de Physique des Solides[++], Bât. 510
Université Paris-Sud, 91405 Orsay, France

In the case of a lattice, the Schrieffer-Wolff transformation yields extra terms in addition to the classical s-f exchange Hamiltonian. The study of the Kondo-lattice Hamiltonian including these new terms by the renormalization-group method shows the possibility of a magnetic-nonmagnetic transition versus exchange constant, in good agreement with the experimental situation of cerium compounds.

1. INTRODUCTION

It is well known that the Schrieffer-Wolff transformation (1) yields an equivalence between the Anderson and the Kondo Hamiltonians for one anomalous rare-earth impurity in the so-called "Kondo limit", i. e., the ionic limit with, however, the impurity 4f level close to the Fermi level. Such an equivalence was very useful to study the properties of Kondo alloys, typically LaCe or AuYb alloys.(2)

On the other hand, it was observed that certain Cerium compounds, at very low temperatures, can be either magnetic ($CeAl_2$ or CeB_6) (4)(5) or non magnetic ($CeAl_3$).(6) This situation can be accounted for by an analog of the Kondo-lattice Hamiltonian in one dimension (3) where one finds a magnetic-nonmagnetic transition in the ground state. But on the opposite, there is always a nonmagnetic insulating ground state in the case of the Kondo-lattice Hamiltonian itself with one conduction electron per site, for every value of the Kondo coupling.(7)

In this paper, firstly we reexamine in detail the Schrieffer-Wolff transformation for the lattice of anomalous rare-earths. New extra terms are found in addition to the classical Kondo-lattice Hamiltonian. Secondly, we study the obtained Hamiltonian by use of real-space renormalization group method and a possibility of a transition magnetic-nonmagnetic is then found.

2. THE SCHRIEFFER-WOLFF TRANSFORMATION FOR THE ANDERSON-LATTICE HAMILTONIAN

Let us derive now the Schrieffer-Wolff transformation for the Anderson Hamiltonian written for a lattice of rare-earth atoms. For sake of simplicity, we do not consider here the orbital degeneracy of the localized electrons.

Thus, the Anderson-lattice Hamiltonian can be written (with usual notations) as :

$$H = H_o + H_1 \tag{1}$$

with

$$H_o = \sum_{k,\sigma} \varepsilon_k a^+_{k\sigma} a_{k\sigma} + \sum_{i,\sigma} E_o n_{i\sigma} + \sum_i U n_{i\uparrow} n_{i\downarrow}, \tag{2}$$

and

$$H_1 = \frac{V}{\sqrt{N}} \sum_{i,k,\sigma} (e^{-i\vec{k}.\vec{R}_i} c^+_{i\sigma} a_{k\sigma} + e^{i\vec{k}.\vec{R}_i} a^+_{k\sigma} c_{i\sigma}). \tag{3}$$

We take here an infinitely narrow (zero bandwidth) 4f band and a constant k-f coupling V. We use the notation $a^+_{k\sigma}$ (or $a^+_{i\sigma}$) for conduction electrons and $c^+_{i\sigma}$ for localized electrons.

We use the classical procedure of the Schrieffer-Wolff transformation which yields a new Hamiltonian, by imposing that the term in first order in V vanishes. The k dependence of ε_k is considered through a Fourier transform :

$$\frac{V^2}{E_o - \varepsilon_k} + \frac{V^2}{\varepsilon_k - E_o - U} = - \sum_j J_j e^{i\vec{k}.\vec{R}_i} \tag{4}$$

where only the two first terms J_o and J_1 are retained. The transformed Hamiltonian is :

$$\tilde{H} = \tilde{H}_o + \tilde{H}_1 \tag{5}$$

$$\tilde{H}_o = \sum_i J_o (2s^+_i S^-_i + 2s^-_i S^+_i + s^z_i S^z_i), \tag{6}$$

$$\tilde{H}_1 = \sum_{\substack{ij \\ (j\in NNi)}} J_1 (2s^+_{ij} S^-_i + 2s^-_{ij} S^+_i + s^z_{ij} S^z_i) \tag{7}$$

where the f electrons are sketched by $\frac{1}{2}$ spin S_i. S^z_i, S^+_i and S^-_i are the Pauli matrices.

This Hamiltonian, with obvious notations for $\vec{s}_i$ and $\vec{s}_{ij}$, describes simultaneously spin exchange in the Kondo Hamiltonian itself (6) and conduction electron hopping from a site i to another nearest neighbour j through the extra term.(7)

J_o is the one-impurity exchange integral :

$$J_o = \frac{V^2}{E_F - E_o} + \frac{V^2}{E_o + U - E_F} \tag{8}$$

It is rather difficult to evaluate the terms J_j, since they all diverge for E_o tending to the Fermi energy. However, in the simplest one-dimension example of a half-filled s band treated in tight binding, i. e.,

$$\varepsilon_k = W(1 - \cos ka) \tag{9}$$

we can show that J_o and J_1 have opposite signs, $J_o > 0$ and $J_1 < 0$. This remark will be useful later on where we will look for the ground state of the Hamiltonian (5) by renormalization-group techniques.

3. RENORMALIZATION-GROUP STUDY

To see the influence of the extra terms found in the Schrieffer-Wolff transformation, we have performed a renormalization-group calculation restricted to one dimension and to a half-filled band. We have considered an "extended" Kondo lattice hamiltonian written as :

$$H=\frac{W}{2}\sum_{\ell,\sigma}\left[(1+\alpha\vec{s}_\ell\vec{S}_\ell)a^+_{\ell\sigma}(1+\alpha\vec{s}_{\ell+1}\vec{S}_{\ell+1})a_{\ell+1,\sigma}+\text{h.c.}\right] + J_o\sum_\ell \vec{s}_\ell.\vec{S}_\ell + U\sum_\ell(a^+_{\ell\uparrow}a_{\ell\uparrow}-\frac{1}{2})(a^+_{\ell\downarrow}a_{\ell\downarrow}-\frac{1}{2}) \tag{10}$$

This Hamiltonian considers extra terms compared to the usual Kondo lattice hamiltonian. When expanding in α the modified hopping term, we recover, to first order in α, the extra term described by (7) with :

$$J_1 = \frac{1}{2} W \alpha \tag{11}$$

The other extra term is a Coulomb-like term which is naturally generated by the renormalization group procedure and which is taken to be zero at the first iteration.

We have applied to hamiltonian (10) the blocking method analog to that already used in ref. 7. The method proceeds as follows :

1) The lattice is split into adjacent blocks of n_s sites (here we have taken $n_s = 3$). The Hamiltonian for each block is diagonalized exactly.

2) We select eight states which form a singlet-triplet-quadruplet system, so that the block Hamiltonian can be written as a "renormalized" one-site Hamiltonian with new parameters U' and J_o'. (Note that here we have retained the lowest triplet as in ref. 7, but this is not perhaps the best choice, as discussed in ref. 8).

3) The original interblock hamiltonian is rewritten in terms of the new-block states in order that the interblock part takes the same form as the original intersite interaction but with new parameters W' and α'.

The renormalization-group procedure allows us to evaluate $W^{(n)}$, $J_o^{(n)}$, $\alpha^{(n)}$, $U^{(n)}$ after n iterations and then to obtain their asymptotic values when $n \to \infty$, in order to reach finally the fixed-point Hamiltonian.

In ref. 7, we have restricted the study by taking $\alpha^{(o)} = 0$, $U^{(o)} = 0$ at the beginning of the iteration scheme, in order to describe the term (6) alone of the Kondo-lattice Hamiltonian. In this restricted case, we have found that $J_o^{(n)}$ tends to a finite value $J_o^{(\infty)}$, while $W^{(n)}$ tends to zero, so that the system tends to a Kondo-like fixed-point Hamiltonian with a finite gap $G = 4J_o^{(\infty)}$. This gap has been plotted as a function of J_o/W, as shown by the curve of Fig. 1 corresponding to $\alpha = 0$.

Here we have performed the same calculation but taking $U^{(o)} = 0$ and $\alpha^{(o)} = 2J_1/W$. Figure 1 shows the plot of the gap G versus J_o/W for several α values. The results depend greatly on the sign of α, i.e., on the sign of J_1. When $J_1 > 0$, the gap starts from a finite

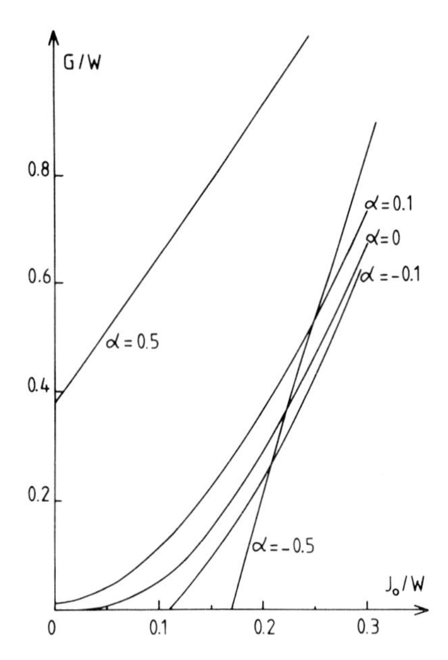

Figure 1 : Renormalization-group results on the 1D Kondo lattice by use of an eight-level method for $n_s = 3$: Plot of the singlet-triplet gap G vs J_o/W for several values of $\alpha = 2J_1/W$.

value for $J_o = 0$ and is continuously increasing with increasing J_o. On the contrary, when $J_1<0$, we find a transition in the ground state for a critical value of J_o/W, above which only the gap becomes nonzero and then remains increasing with increasing J_o. Below the critical value of J_o/W, $J_o^{(n)}$ tends to zero and above it, $J_o^{(n)}$ tends to a finite value $J_o^{(\infty)}$: in this sense, the transition occurring with increasing J_o/W values can be regarded as a magnetic-nonmagnetic transition. Since we have seen that J_o and J_1 have opposite signs, a magnetic-nonmagnetic transition is obtained when the extra terms (7) are considered.

4. CONCLUDING REMARKS

Thus, we have shown that a careful derivation of the Schrieffer-Wolff transformation yields new extra-terms which describe a "delocalized" spin-flip scattering. When studying these terms by real-space renormalization-group in the half-filled one dimensional case, we find that a magnetic-nonmagnetic transition by increasing the Kondo coupling can be recovered. This result obtained in a restrictive situation (one dimension, half-filled case) appears to be in contradiction with finite size calculations on the Anderson lattice which are in favour of a gap for all values of the parameters, including the Kondo limit.(8) Perhaps, the case studied here is too peculiar and we cannot consider too seriously the present conclusions. However, it remains that our study shows clearly that such delocalized Kondo interaction tends to stabilize the magnetic phase. This mechanism could explain why either a magnetic phase (in $CeAl_2$, Ce_3Al_{11}, $CeIn_3$) or a nonmagnetic phase (in $CeAl_3$) are experimentally observed in three dimensional Cerium compounds.(9)

REFERENCES

1. Schrieffer J. R. and Wolff P. A., Phys. Rev. 149, 491 (1966)
2. Coqblin B., Bhattacharjee A. K., Jullien R. and Flouquet J., J. Phys. (Paris) 41, C5-297 (1980) and references therein
3. Jullien R., Fields J. N. and Doniach S., Phys. Rev. Lett. 38, 1500 (1977) ; Phys. Rev. B 16, 4889 (1978)
4. Barbara B., Boucherle J. X., Buevoz J. L., Rossignol M. F. and Schweitzer J., J. Phys. (Paris) 40, C5-321 (1979)
5. Rossat-Mignod J., Burlet P., Kasuya T., Kunii S. and Komatsubara T., Solid State Commun., 39, 471 (1981)
6. Andres K., Graebner J.E. and Ott H., Phys. Rev. Lett. 35, 1779 (1975)
7. Jullien R., Pfeuty P., Bhattacharjee A. K. and Coqblin B., J. Appl. Phys. 50, 7555 (1979)
8. Jullien R., This conference
9. Berton A., Chaussy J., Chouteau G., Cornut B., Flouquet J., Odin J., Palleau J., Peyrard J. and Tournier R., J. Phys. (Paris) 40, C5-326 (1979)

+On leave of absence from Instituto de Fisica da U.F.R.J., Ilha do Fundão, Cidade Universitaria, 21.941 - Rio de Janeiro, RJ, Brazil.
Partially supported by CNPq - Brazil

++Laboratoire associé au C. N. R. S.

TRANSPORT PROPERTIES

Valence Instabilities
P. Wachter and H. Boppart (eds.)

STUDY OF THE ENERGY GAP IN SINGLE CRYSTAL SmB_6

S. von Molnar and T. Theis
IBM Thomas J. Watson Research Center, Yorktown Heights, NY 10598

A. Benoit, A. Briggs, J. Flouquet, and J. Ravex
CRTBT, CNRS, B.P. 166X, 38042 Grenoble Cedex, France

and

Z. Fisk
University of California, Los Alamos Scientific Laboratory, Los Alamos, NM 87545

We review the available experimental data which are consistent with a low energy gap in the electron excitation spectrum of SmB_6 and offer new evidence for such a gap in the form of low temperature specific heat and optical experiments.

INTRODUCTION

The suggestion that the interaction between the extended states of the 5d, 6s conduction bands and the more localized 4f levels in intermediate valence materials might lead to a gap in the density of states near or at E_F,[1] the Fermi energy, has led to considerable theoretical activity.[2,3]

Transport,[4,5] NMR,[6] electron tunneling and far infrared absorption measurements[5] on SmB_6 are consistent with the presence of an energy gap, ΔE, near the Fermi energy at low temperature. However, no low energy optical data on single crystals exist in the literature and recent tunneling experiments[7,8] on single crystals differ from the earlier results. In addition, various large values for γ, the coefficient of the linear term in the specific heat, have been reported.[9,10] If γ is to be interpreted as proportional to the density of electron states at E_F, these results speak against the existence of a gap.

To clarify this situation we have made transport, susceptibility, specific heat and low temperature far infrared transmission measurements on several small single crystals of SmB_6. Both the transport and susceptibility data are in excellent agreement with earlier reports.[4,5,9,10] We find, however, that the low temperature specific heat differs markedly from the published data.[1,13] The simple law $C = \gamma T + \beta T^3$ is not obeyed over any reasonable temperature range. An interpretation in terms of a linear electronic contribution is, therefore questionable. The results are, however, consistent with the existence of a gap.

A direct confirmation of the gap comes from our far infrared transmission results at low temperatures. These show a well defined absorption edge corresponding to a gap of ~3 meV.

In the following section we review past experiments. We also discuss our own results and compare them, wherever possible with already existing data. The final section summarizes all the results which lead to the conclusion that an energy gap has been observed in SmB_6.

EXPERIMENT

A. Transport

We have measured the temperature dependence of the resistivity, ρ and Hall effect, R_H, on a small single crystal of SmB_6 as a function of temperature between 2 and 300 K. Coincidentally, we have also observed that the magnetoresistance in this specimen is very small (< 3% at 4.2 K and 15 kOe). Our data are given in Fig. 1 where we have plotted both ρ and R_H using the same convention as Allen *et al.*[4] Our data are very similar to those of Ref. 4 and yield an exponentially varying ρ between 6 and 14 K. Following Ref. 4, we write $\rho = \rho_o e^{-\Delta E_\rho/kT}$ and obtain $\Delta E_\rho \approx 2.6 \times 10^{-3}$ ev As pointed out in Ref. 4, it is not possible to make a complete analysis of the transport properties in the absence of a detailed model for the electronic structure of SmB_6 and the values for ΔE must therefore be considered order of magnitude. This type of analysis gave, however, the first evidence for ΔE (see, e.g. Ref. 1).

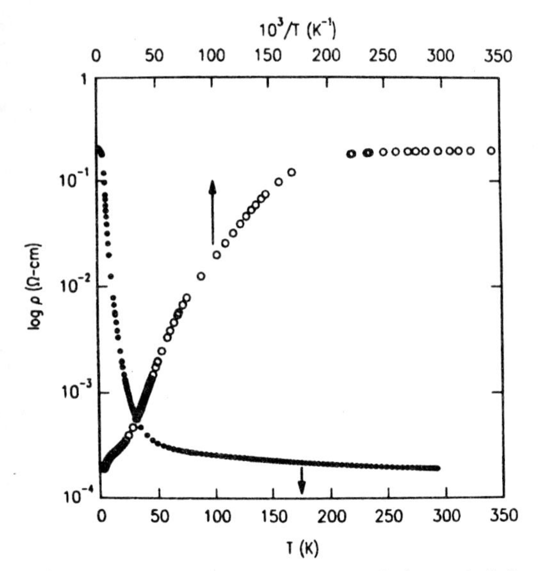

Fig. 1a Temperature dependence of the resistivity of SmB_6.

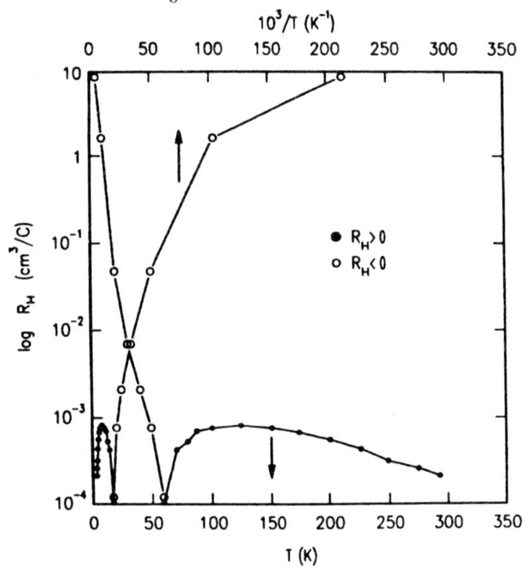

Fig. 1b Temperature dependence of the Hall coefficient, R_H.

B. NMR

In Fig. 2 we present the data of Péna and Mc Loughlin,[6] which show the temperature dependence of the ^{11}B relaxation rate in SmB_6 as a function of temperature. Above 10 K the temperature dependence is exponential with $\Delta E = 5.6\times10^{-3}$ ev. The authors interpret their results as the consequence of fluctuations of the 4f spins. The physical model thus relates the measured line width to the contribution to the hyperfine field from these fluctuations. Since the fluctuations appear to die away exponentially, it is supposed that the 4f energy spectrum contains a gap, $\Delta E_{NMR} = 5.6\times10^{-3}$ ev. The NMR experiment is the only one under discussion in this article in which the results purport to be specific to the 4f spectrum. It is, therefore, noteworthy that ΔE_{NMR} is comparable to the other measurements of ΔE, e.g. transport or tunneling, in which the measurement cannot distinguish between the contribution of different electronic states, but where it is usually assumed that the lighter particles, the conduction 5d6s bands, play the most important role.

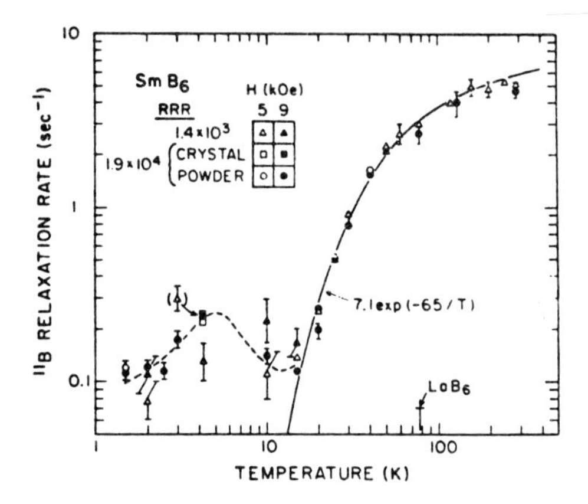

Fig. 2 Temperature dependence of the ^{11}B relaxation rate in SmB_6. Solid curve: high-temperature Arrhenius law describing activated relaxation. Dashed curve: low-temperature anomaly, seen most clearly in the data for residual resistance ratio $\rho = 1.9\times10^4$. An upper bound on $1/T_1$ for LaB_6 at 77 K is also shown; at 4.2 K $T_1(LaB_6) \gtrsim 250$ sec . (From Ref. 6.)

Below 15 K the linewidth behaves in an unpredictable manner, is sample dependent and magnetic field sensitive. Similar effects have been observed independently by Takigawa *et al.*[7]

C. Tunneling and Point Contact Spectroscopies

Battlog *et al.*[5] attempted the first tunneling experiments in SmB_6. They employed both the normal sandwich structure, in which the native oxide on the surface of an SmB_6 sputtered film served as the insulator separating the hexaboride from a Pb counter electrode and oxidized single crystal surfaces. They demonstrated tunneling by observing the well-known dynamic resistance due to superconducting Pb, quenched the Pb signal by applying a magnetic field and found that an additional asymmetric peak in the dynamic resistance remained. From this structure they estimated a gap of between 5 and 8×10^{-3} ev. In the same study,[5] Battlog *et al.* also measured the optical transmission of a thin film of SmB_6 at low temperatures between 2×10^{-3} and 12×10^{-3} ev. Although there was some increase in the transmission as the energy was decreased the effect was weak (presumably because the film contained various defects). It was argued, however, by continuity that a similar experiment on single crystal material would show a more dramatic effect since its resistance at low temperatures is orders of magnitude higher than a comparable film. Their estimate for the optical gap was $\lesssim 10^{-2}$ ev.

More recently Frankowski and Wachter[8] employed a weak link spectroscopy, in which a metal (Mo) point is pressed against a single crystal SmB_6 sample. They also find a small peak, rather more symmetric, in the dynamic resistance. By defining the gap empirically as the distance between inflection points of this peak, they obtain a gap of $\sim 5 \times 10^{-3}$ ev. It should be remarked at this point that these two techniques are very different from one another since in the point contact method the electrons do not tunnel through a barrier.

Very recently Güntherodt *et al.*[9] have used a Schottky barrier point contact method to estimate the energy gap. In this method it is also possible to look at single crystals, but the contact is made with a degenerately doped GaAs semiconducting tip. The Schottky barrier created by the contact serves as a tunnel barrier and the method is, therefore, similar to the thin film tunneling method described earlier. The spectrum in this case also shows a dynamic resistance peak at, but very symmetric about, zero bias. Estimates of the energy gap vary between 2.7 and 4.9×10^{-3} ev depending on where the width of the resistance peak is measured.

D. Optical

Clearly, although structure of order 5×10^{-3} ev had been seen in all of the spectroscopies discussed in the last section, there are important shape differences depending on the technique used. We have, therefore, made optical measurements on the same small single crystal which was used for the Schottky barrier point contact tunneling work.[9] Our motivation was based in part on complementing the thin film measurements by Battlog *et al.*[5] described briefly in the last section but its primary purpose was to see if an energy gap comparable to that determined from the tunneling measurements could be seen optically. Details of both method and analysis will be described elsewhere.[10] The measuring instrument is a Fourier transform infrared spectrometer in which the optics were optimized to examine small surface areas. For example, the dimensions of the single crystal of SmB_6 under discussion are approximately $.0122 \times .32 \times .24$ cm^3. Our transmission data in the far infrared are shown in Fig. 3. The

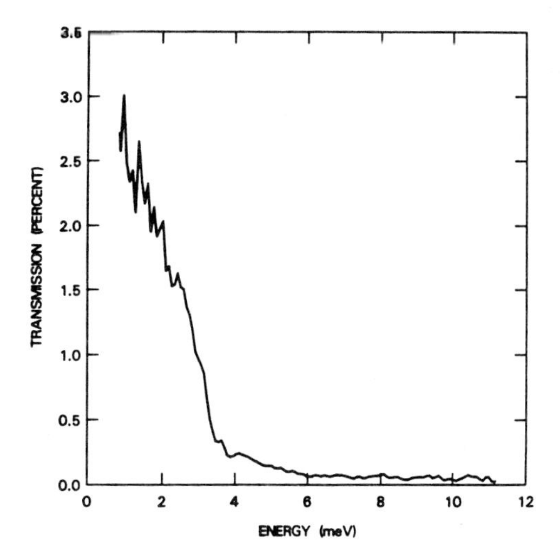

Fig. 3 Transmission of SmB_6 at 4.2 K as a function of energy.

most obvious feature of the data is the sharp increase in transmission below approximately 3.5×10^{-3} ev indicating the presence of a gap or sharp decrease in the density of states of the energy spectrum governing the optical excitations. The peaks near and below $\sim 2.5 \times 10^{-3}$ are interference fringes from which the low energy index of refraction, n = 24.5, was determined. With this information it is possible to calculate the frequency dependence of the conductivity through the relation $\sigma(\omega) = cn\alpha/4\pi$, where c is the velocity of light, α is the transmission coefficient and n is assumed to be constant. $\sigma(\omega)$ is plotted as a function of energy in Fig. 4. The conductivity is limited at low energies to approximately 11 Ω^{-1}cm^{-1}, presumably due to possible defects in the boron lattice and impurities.[5,11] The values of $\sigma(\omega)$ above $\sim 5 \times 10^{-3}$ ev must be regarded as a lower limit. This is because a small amount of light may have "leaked" around the nearly opaque sample, shifting the zero transmission baseline slightly downward in

Fig. 3 This possibility is being currently investigated, and its correction could significantly increase the values of $\sigma(\omega)$ at the highest measured energies.

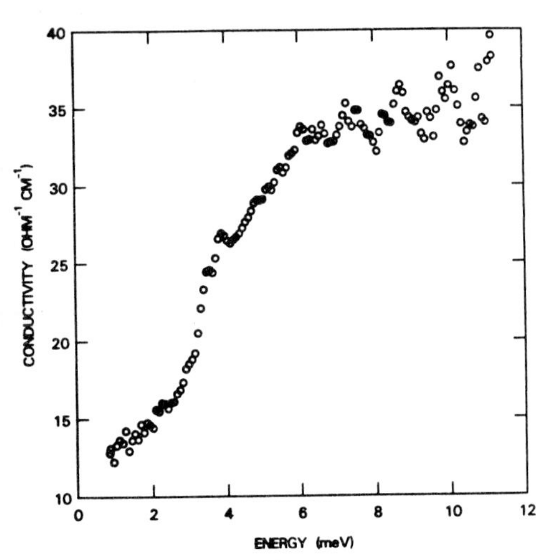

Fig. 4 Real part of the far infrared conductivity as a function of energy.

In our opinion the low temperature optical transmission data on single crystals represents the most direct evidence so far for a gap in the energy spectrum of SmB_6.[12] It is also gratifying that the value $\Delta E_{opt} \approx 3\times10^{-3}$ ev is in good agreement with the values derived from tunneling spectroscopy on the same sample[9] implying that the dynamic resistance peak also reflects gap structure.

E. Specific Heat

We have measured both the susceptibility and specific heat of the various single crystals discussed in the previous sections. The susceptibility values are in good agreement with published data[1] and only weakly dependent on sample with a scatter of approximately 10% at 4.2 K. Details of these measurements will be given elsewhere.[10]

The high temperature ($12\ K \leq T \leq 80\ K$) specific heat measurements are in excellent agreement with one another and earlier work by Kasuya *et al.*[13] In order to analyze our data we have assumed that the total specific heat is given by three terms, i.e.

$$C_T = C^i_{el} + C_{IV} + C_L = \gamma T + C_{IV} + \beta T^3 \quad (1)$$

where C^i_{el} is an electronic term, presumably due to impurities, C_{IV} is a term characteristic of the intermediate valence state, and C_L is the usual lattice term. Since C^i_{el} contributes only a small part to the specific heat in the high temperature regime, we shall ignore it for the present. The problem is then reduced to finding the proper form for C_L. We assume here that $C_L(SmB_6)$ may be estimated by multiplying the known lattice specific heat of LaB_6,[10] $C_L(LaB_6)$, by a scaling factor determined from the compressibilities of the compounds.[14]

This rational expresses the fact that $C_L \propto (\Theta_D)^{-3}$, where Θ_D is the Debye temperature, and $\Theta_D \propto (K_o)^{1/2}$, where K_o is the compressibility.[15] Therefore $\Delta C_L/C_L \sim -3/2\ \Delta K_o/K_o$, and with $K_o(SmB_6) = 139$ GPa and $K_o(LaB_6) = 191$ GPa,[13] one obtains

$$C_L(SmB_6) = C_L(LaB_6)\times 1.27. \quad (2)$$

Of course, Eq. (2) is only a rough approximation since the relations between C_L, Θ_D and K_o are valid strictly only at low temperatures, i.e., $T<<\Theta_D$. Nonetheless, the method affords an estimate of the contribution to the specific heat from the intermediate valence character of the compound. This contribution, C_{IV}, is shown in Fig. 5.

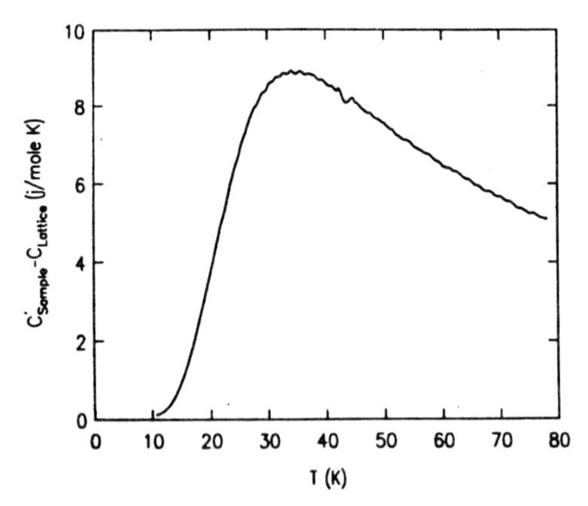

Fig. 5 Specific heat of SmB_6 without an estimated lattice term as a function of temperature.

In the absence of a theory for the temperature dependence of C_{IV}, we present, in Fig. 6, the temperature variation of the entropy, $\int_0^T (C_{IV}/T)dT$. The total entropy of the I-V state is known to be the sum of two terms, one magnetic, the other the entropy of mixing of the two integer valence configurations.[15] Following Kasuya *et al.*,[13] we define C_3 as the fraction of Sm^{3+} and C_2 as the fraction of Sm^{2+} and write the total entropy with the help of Stirling's approximation as

$$S = R\{C_3 \ln(2J+1) - C_3 \ln C_3 - C_2 \ln C_2\} \quad (3)$$

where J = 5/2 for SmB$_6$ and R = 8.314 Joules/moleK. For a valence of 2.6, 2.7 or 2.8, Eq. 3 gives S = 14.5, 15.5, and 16.1 Joules/moleK respectively. The experimental value (see Fig. 6) approaches these calculated numbers and lends credence both to the lattice correction and to the accuracy of the model for the total entropy of the intermediate valence state. It should also be noted that the results appear to rule out important crystal field effects. Splittings would lower the effective J value and, therefore, reduce the calculated entropy.

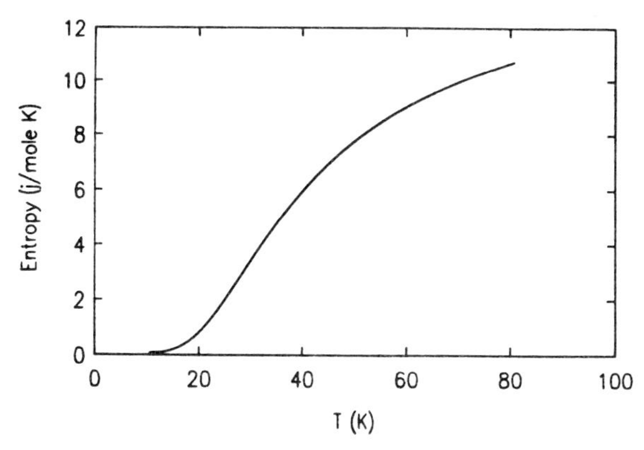

Fig. 6 Temperature variation of the entropy, $S(T) = \int_0^T C_{IV}/T dT$, for SmB$_6$.

The specific heat below ~ 15 K is much more difficult to analyze since it is extremely sample dependent. This is shown in Fig. 7 where we have collected all of the specific heat data, including preliminary data taken on a large (1 gm) polycrystalline sample between 50 mK and 2 K. The disparity between the results on various samples below 12 K is dramatically evident. We assume, therefore, that the low temperature specific heat (at least between .5 K and 15 K) is impurity dominated and concentrate our attention on the sample with the lowest specific heat, i.e., Sample No. 6344. A more convenient plot of this data is given in Fig. 8 where we plot C_T/T vs. T^2 for both SmB$_6$ and a single crystal of LaB$_6$[10] Whereas the LaB$_6$ curve is linear with $\gamma = 2.3\times10^{-3}$ j/moleK^2 and $\beta = 3.05\times10^{-5}$ j/moleK^4, the SmB$_6$ curve can be considered linear only in a very limited temperature range (in the present case 5 K ≤ T ≤ 9 K) with a $\gamma \sim 2.1\times10^{-3}$ j/moleK^2. In addition, although both the low temperature anomaly and a linear term in C/T vs. T^2 were found for various samples, the values varied by ~ 200% in γ and ~ 30% in β. We believe, therefore, that extrapolations of these curves to obtain γ values and, therefore, values for the density of states at E_F are unreliable. The anomaly below 5 K, which has also been observed in many other I-V systems, is not understood.

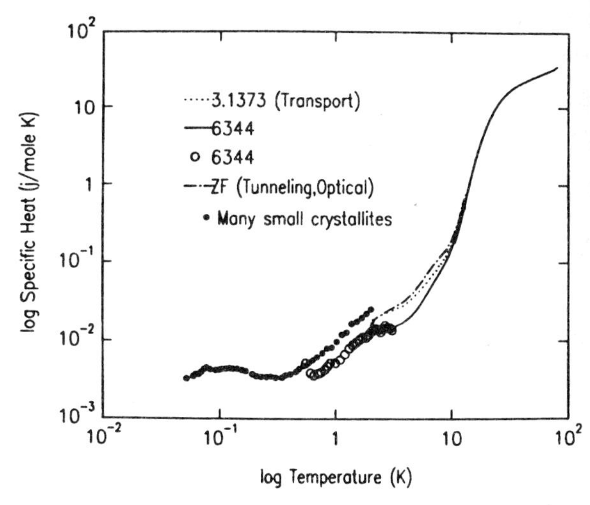

Fig. 7 Summary of the temperature dependence of specific heat for various samples.

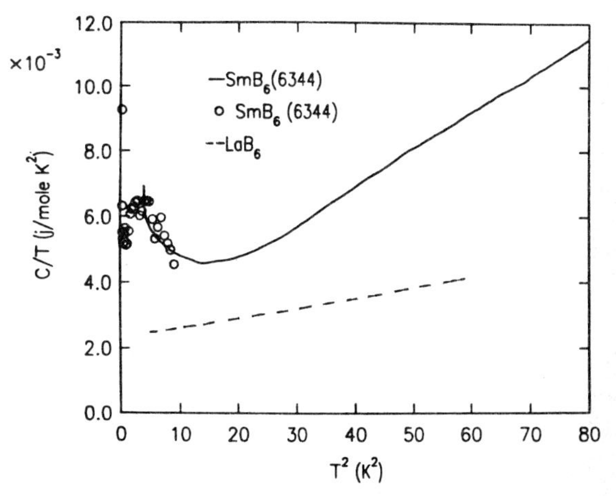

Fig. 8 C_{Total}/T vs. T^2 for both SmB$_6$ and LaB$_6$. The dots are preliminary low temperature data on another sample from the same batch.

In an attempt to analyze these data we have taken another approach. We assume that the "linear" term in C is impurity dominated. Using Eq. (1), we write $C_{IV} = C_T - "\gamma"T - C_L$, where $"\gamma" = 2.1 \times 10^{-3}$ j/moleK and C_L is defined by Eq. (2). Fig. 9 shows a plot of log C_{IV} versus T^{-1}. The dashed line is a guide to the eye and indicates the temperature range over which the specific heat has the approximate functional form $Ae^{-\Delta E/kt}$ with $\Delta E \approx 1.8 \times 10^{-3}$ ev. The deviation for $T^{-1} > .2$ (i.e. $T < 5$ K) is expected because the low temperature anomaly was not subtracted in our determination of C_{IV}. Deviations for $T^{-1} \lesssim 0.1$ ($T \gtrsim 10$ K) may be due to the beginning of the magnet and mixing entropy terms.

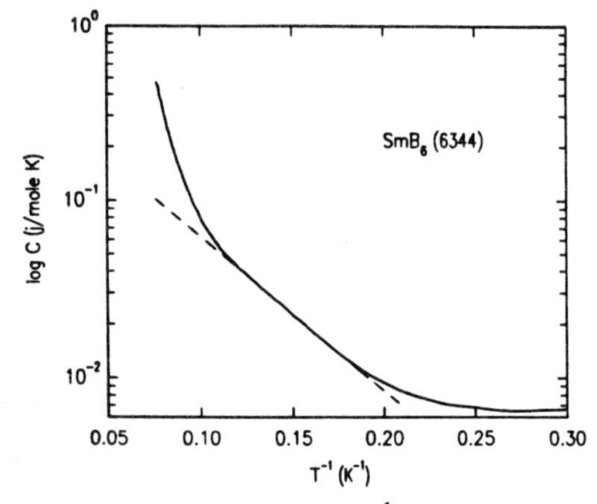

Fig. 9 Log C_{IV} vs. log T^{-1}. The dashed line helps to indicate the temperature over which the law $C = Ae^{-\Delta E/kT}$ is approximately obeyed.

We speculate that the exponential dependence[17] is another indirect piece of evidence for the gap and suggest that purer samples may obey this or a similar law for lower temperatures.

CONCLUSIONS

We have reviewed existing transport, tunneling optical and specific heat data on SmB_6. We have found that all of data are consistent with sharp structure in the density of states near E_F. Our far infrared transmission results at 4.2 K on a small single crystal give direct evidence for a gap in the excitation spectrum of order 3×10^{-3} ev. It should be noted, however, that none of the experiments carried out so far can distinguish between various models for the gap. Neither can they determine if the structure is truly a gap (with no allowed states) or a sharp decrease in the density of states.

In addition we have demonstrated that the low temperature specific heat varies from sample to sample and appears to be impurity and defect dominated. A simple law $C = \gamma T + \beta T^3$ is not obeyed over any reasonable temperature range. An interpretation in terms of a linear electronic contribution is, therefore, probably erroneous.

References

1. J. C. Nickerson, R. M. White, K. N. Lee, R. Bachmann, T. H. Geballe, and G. W. Hull, Jr., Phys. Rev. **B3**, 2030.
2. See the section on ground state properties in this volume.
3. T. Kasuya, K. Takegehara, Y. Aoki, K. Hanzawa, M. Kasaya, S. Kunii, T. Fujita, N. Sato, H. Kimura, T. Komatsubara, T. Furuno, and J. Rossat-Mignod, in "Valence Fluctuations in Solids" L. M. Falicov, W. Hanke, and M. B. Maple (eds.), (North Holland Publishing Company, 1981) p. 215.
4. J. W. Allen, B. Battlog, and P. Wachter, Phys. Rev. **B20**, 4807 (1979).
5. B. Battlog, P. H. Schmidt, and R. M. Rowell, in "Valence Fluctuation in Solids," L. M. Falicov, W. Hanke, and M. B. Maple (eds.), (North Holland Publishing Company, 1981) p. 267.
6. O. Pena, M. Lysak, and D. E. MacLoughlin, Solid State Comm. **40**, 539 (1981).
7. M. Takigawa, H. Yasuoka, Y. Kitaoka, T. Tanaka, H. Nozaki, and Y. Ishizawa, J. Phys. Soc. Japan **50**, 2525(1981).
8. I. Frankowski and P. Wachter, this conference, contribution S1.
9. G. Güntherodt, W. A. Thompson, F. Holtzberg, and Z. Fisk, this conference, contribution S2.
10. von Molnár *et al.* to be published.
11. The susceptibility of this SmB_6 sample increases by approximately 6% as the temperature is lowered from 20 K to 4.2 K, indicating a small "Curie" tail.
12. In a post deadline paper at this conference P. Wachter reports reflectivity data on single crystal SmB_6 at room temperature which also shows low energy structure.

13. T. Kasuya, K. Takegehara, T. Fujita, T. Tanaka, and E. Bannai, J. de Physique **40**, C5-308 (1979).
14. H. E. King, Jr., S. J. La Placa, T. Penney, and Z. Fisk, in "Valence Fluctuations in Solids"L. M. Falicov, W. Hanke, and M. B. Maple (eds.), (North Holland Publishing Company, 1981) p. 333.
15. See, e.g., E. S. R. Gopal, "Specific Heats at Low Temperatures," (Plenum Press, New York, 1966).
16. J. H. Jefferson and K. W. H. Stevens, J. Phys. C: Solid State Phys. **9**, 2151 (1976).
17. A low temperature Schottky form for the specific heat, i.e., $C = A(\Delta E/kT)^2 e^{-\Delta E/kT}$ with $\Delta E \approx 3\times10^{-3}$ev, also fits the data approximately over the same temperature range.

D. WOHLLEBEN: From the collected evidence, can you say whether the structure around E_F (gap or not) is symmetric or asymmetric ?

S. VON MOLNAR: The only data reflecting on the shape of the density of states is the tunneling work, which in the case of the GaAs pointcontact spectroscopy shows symmetry about zero bias. The other tunneling spectroscopies (metal point contact and sandwich configuration) appear to be asymmetric. I have no reason to doubt the GaAs work considering my own experience using the Schottky barrier tunneling technique on superconductors.

C. BUSCH: What about the determination of an energy gap from transport-measurements ? If one does not know the temperature dependence of the mobility, the gap values might all be wrong.

S. VON MOLNAR: In this talk I have quoted the results of Allen, Batlogg and Wachter (Phys. Rev. B<u>20</u>, 4812 (1979)). These authors quote activation energies from log ρ and log R_H vs 1/T. This number may well be wrong because any temperature dependence of the mobility is excluded. The analysis does serve to demonstrate the presence of a gap.

B. COQBLIN: Do you know if there are presently experiments with pressure or with departure from stoichiometry in the case of SmB_6, as in the case of TmSe, in order to see what happens to the hybridization gap ?

S. VON MOLNAR: I don't know of any pressure work but I have found out during this conference that radiation studies on SmB_6 are being made by J. Morillo.

G. GUENTHERODT: Comment: The gap structure seen in GaAs probe tunneling into SmB_6 disappears upon further pressurizing the SmB_6 sample by the GaAs. A high yield pressure of the sample surface can be reached considering the extreme hardness of SmB_6.

T.A. KAPLAN: Did you actually try to account for the large linear term in the specific heat by impurities and, or lack of stoichiometry ?

S. VON MOLNAR: No, we have no quantitative model nor do we know the lack of stoichiometry well enough.

G. CZYCHOLL: Can you explain the finite zero temperature static susceptibility, when there is a gap ?

S. VON MOLNAR: I have not thought about this in detail, but it appears to be a problem.

Valence Instabilities
P. Wachter and H. Boppart (eds.)

MIXED VALENCE COMPOUNDS AT LOW TEMPERATURES: EVIDENCE AGAINST A MINIMUM METALLIC CONDUCTIVITY?

N.F. Mott

Cavendish Laboratory, Cambridge

The compound SmB_6 shows a resistivity which rises sharply as the temperature falls below 5K and then flattens out [1], [2], [3], [4]. The hypothesis that the rise is the result of a gap, and that the low temperature behaviour is to be explained by impurity conduction, has been criticised because the low-temperature metallic conductivity in some specimens seems unacceptably small, and for other reasons [2], [5]. With this in view, a comparison is made in this paper with recent measurements by Rosenbaum et al [6] on the metal-insulator transition in uncompensated Si:P. These, it is claimed, can be explained only if the specimens contain long-range fluctuations of composition, producing barriers too thick for tunnelling, and so classical percolation channels. It seems probable that the same occurs for SmB_6, and that the hypothesis of impurity conduction is correct. The high resistivity of TmS, apparently not a mixed valency material, is also described. We give also a discussion of the origin of the gap, including the 4f - 5d hybridisation model and that of Kasuya [2], [7] which envisages the production of a gap by a charge density wave.

A gap forms only, as far as we know, in those rare earth compounds in which the conduction electrons are of 5d type, so that a metal-insulator transition must occur if the bands separate (as SmS); it does not occur for instance for $CeAl_3$ where s-p conduction electrons are involved. Some discussion is given of the latter type. In the absence of a gap, or above the temperature at which the gap is effective, resistivity is the result of transitions of conduction electrons to the 4f band where the density of states is very high. For such transitions we expect:

(1) A very large residual resistance due to impurities.

(2) A rapid rise of resistivity with T, followed by a fall which may show the Kondo-like logarithmic behaviour.

1. INTRODUCTION

The resistivity of SmB_6, metallic at room temperature, increases as the temperature falls and then flattens out [1], [2], [3], [4], to a constant value below about 5K. The increase has been widely interpreted as a consequence of a small gap which forms at low temperatures separating occupied from empty states. There is now direct experimental evidence for this gap from electron tunnelling (Guntherodt et al [8]) and from point contact spectroscopy (Frankowski and Wachter [9]). On the origin of the gap, hybridisation between the 4f and 5d electrons was suggested by the present author and others, [8], [9], [10], [11], [12], [13], [14], though strong criticisms of the model have been advanced by Kasuya et al [2], [7], and Anderson [15]. Kasuya believes that an incommensurate charge density wave forms and produces the gap at low temperatures.

Whatever the origin of the gap, we have to explain the flattening-off at low temperatures, and the main purpose of this paper is to show that the observations are compatible with the hypothesis that the effect is due to impurity conduction caused by non-stoechiometry. First, however, we make a few remarks about intermediate valence (IV) in general.

The gap phenomenon is peculiar to IV materials in which the conduction electrons are of 5d type. We believe that in all IV materials, there is an overlap between a conduction band and the 4f band. The mobile entities in the 4f band are of course quasiparticles in which spin-orbit interaction is not broken down; none the less hybridisation can occur [13]. The Fermi surface, therefore contains areas where the density of states N(E) is very high and the wave functions 4f-like, and others where N(E) is normal. The latter carry the current, and resistivity is due to transitions to the 4f region of the surface where N(E) is high, either due to phonons or, at low T, to electron-electron collisions or scattering by spin fluctuations. In this they resemble the model first put forward by the author [16] in 1936 for transition metals. The transition probability is by Fermi's golden rule

$$(2\pi/\hbar)\,|M|^2\,N(E_F),$$

where M is the matrix element of the scattering potential and where $N(E_F)$ is the density of the states after the collision. The scattering cross section by impurities can, under such circumstances be much larger than the size of the scatterer, in agreement with observation (Schneider and Wohlleben [17]), a point on which Anderson [15] comments. Similar behaviour is found in V_2O_3 (McWhan et al [18]); according to Mott [12, p. 184] this again is due to a high value of $N(E_F)$, this quantity being enhanced by correlation as predicted by Brinkman and Rice [19]. For such materials, if pure and in the absence of a gap, we expect ρ to rise (first as T^2) and saturate at about

$$\sigma_{IR} \sim \frac{1}{3} e^2/\hbar a$$

(the Ioffe-Regel value), and then drop. The reason for the drop is that, if kT becomes larger than the energy interval for which N(E) is large, not all the electrons at the Fermi energy will be strongly scattered. Such a drop of course also occurs in Kondo scattering by isolated impurities, $\Delta\rho$ beginning as T^2 followed by logarithmic behaviour. We do not know of an investigation of scattering into 4f at temperatures where for the 4f part of the Fermi surface $N(E_F)$ kT is no longer small, but a logarithmic behaviour seems likely. If so, materials which show it could either be of normal IV type, or a true Kondo lattice as described by Jullien et al [20], in which the Kondo temperature for each rare earth ion is greater than the RKKY interaction.

2. LOW TEMPERATURE RESISTIVITY OF SmB_6

We return now to SmB_6, and consider that a small gap, due either to hybridisation or a charge density wave, separates filled from empty states and that excess metal produces a small number of donor states, in sufficient concentration to give metallic conduction. Our understanding of metallic impurity conduction and of the metal-insulator transition has been considerably changed by some recent work on Si:P. In the first place the criterion for metallic (non-activated) conduction, first given by the present author [21] in 1949, is very successful [22], [23] in giving the concentration n_c at which the transition takes place, for a wide variety of materials. The equation is

$$n_c^{1/3} a_H \simeq 0.26 \tag{1}$$

where a_H is the hydrogen radius

$$a_H = \hbar^2 \kappa / m_{eff} e^2 .$$

For SmB_6, with its very small gap, the dielectric constant κ should be large; for a conduction band formed by hybridisation with 4f, the effective mass should be large too, so it does not seem possible at present to estimate the concentration of excess metal at which metallic conduction should begin.

The present author now believes [24], [25] that there is a significant difference between the mechanisms for the transition in heavily ($\sim 50\%$) compensated and uncompensated materials. In the former, near the transition conduction is in an impurity band, the transition at zero temperature, induced by change of concentration, magnetic field or stress, being from variable-range hopping ($\sigma = A \exp(-B/T^{1/4})$ to metallic behaviour. B should go continuously to zero; at the transition the density of electronic states and specific heat should show no discontinuity, but $\sigma(T=0)$ should jump discontinuously from zero to a value σ_{min} given by [26], [27]

$$\sigma_{min} = C e^2/\hbar a \tag{2}$$

Here a is the mean distance between donors and C lies in the range 0.025 to 0.05. There is considerable evidence for the correctness of (2) for a wide range of semiconductors, and as far as we know no evidence against it. It has however been widely criticised on different grounds for instance by Götze [28] and by Abrahams et al [29]. These criticisms are discussed by Mott and Kaveh [30] and by Mott [31]. In the second paper the present author shows that if σ is not finite at the transition, it must rise extremely rapidly with concentration, as $(n - n_c)^{0.06}$, but a discontinuous transition still seems to us the most likely.

For uncompensated Si:P the work of Rosenbaum et al [6], Sasaki [32], Berggren [33], Mott and Davies [34] now give the following picture. For concentration below the transition conduction is in an impurity band, and is either by hopping or excitation to a mobility edge with excitation energy ε_2; ε_2 seems to tend to zero as the transition is approached. But at the transition there is a discontinuous (first order) change, the Fermi energy jumping discontinuously from the impurity band into the conduction band. For metallic conduction, the electrons are not in an impurity band as was previously thought [27], but in the conduction band. The evidence is that the specific heat γT does not differ greatly from the formula for free electrons in the conduction band, and that the many-valley nature of the band describes the effects of magnetic field and stress up to quite near the transition.

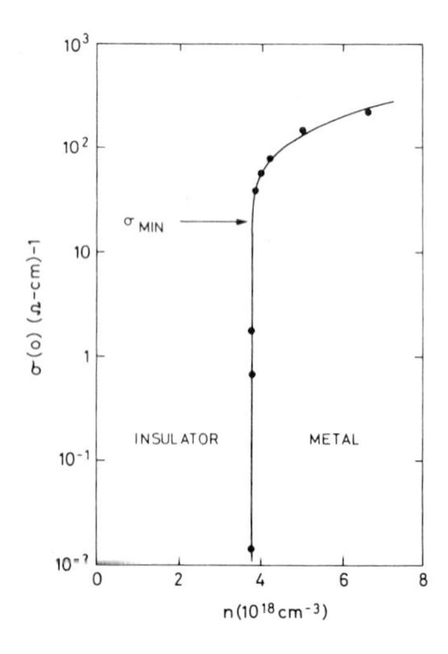

Figure 1: Conductivity in the limit as $T \to 0$ of Si:P as a function of phosphorus concentration

Rosenbaum et al [6] have measured σ extrapolated to zero temperature from 2 mK, and their results are shown in figure 1. The conductivity drops well below the Ioffe-Regel value and this is accounted for by Berggren [33] in terms of the same correction which gives the ln T term in two-dimensional problems; Berggren gives (i)

$$\sigma = \left[1 - \frac{3}{(k_F\ell)^2}\left\{1 - \frac{\ell}{L_i}\right\}\right]\sigma_{IR}\,\ell/a \tag{3}$$

where ℓ is the mean free path and L_i the inelastic diffusion length. The minimum conductivity will be that at which the discontinuous transition occurs - and nothing to do with the author's σ_{min}.

There is widespread experimental evidence [36], [37], however, that for uncompensated semiconductors the lowest metallic conductivity is near to the value (2), for reasons which are not clear. In view of this, and since according to our arguments a discontinuous drop in σ(T = 0) to zero must occur, we consider it highly likely that the lowest values of σ(T=0) observed by Rosenbaum et al are due to long-range fluctuations in concentration leading to classical percolation channels. Figure 2 shows our expectations.

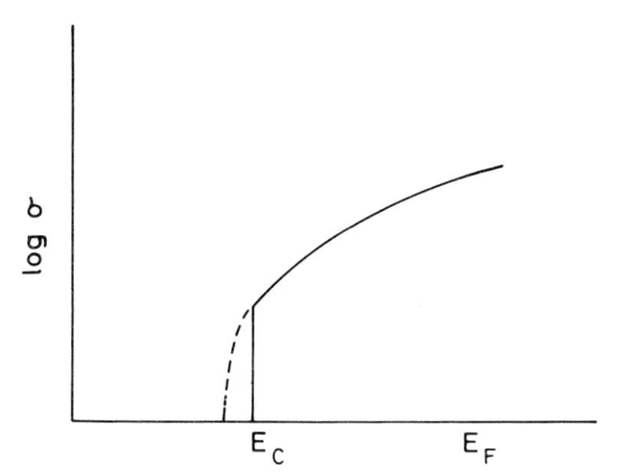

Figure 2 - Conductivity at T = 0; - if σ_{min} exists - - - if σ_{min} does not exist.

For SmB_6 the resistivity-temperature curves have been obtained by Nickerson et al [1], Kasuya et al [2], Lapierre [3], and Allen et al [4]. The resistivity in the low-temperature regime depends on the preparation of the specimen, and the latest work gives the highest values. In Lapierre's work the conductivity is almost independent of T below 5K and equal to about 2 $\Omega^{-1}cm^{-1}$. Kasuya et al find that in the low temperature regime the conductivity can be expressed as

$$\sigma = A + 29.4\exp\left(-32/T^{\frac{1}{4}}\right) \tag{4}$$

with the constant term A varying between 0.1 and 10 $\Omega^{-1}cm^{-1}$. Allen et al find a conductivity 0.3$\Omega^{-1}cm^{-1}$ in a specimen for which the Hall effect gives 5 x 10^{17} cm^{-3} carriers at 4K. With this number of carriers, equation (2) gives σ_{min} in the range 1.5 - 3 $\Omega^{-1}cm^{-1}$, 5 to 10 times that observed.

If SmB_6 as a consequence of excess metal is an uncompensated semiconductor, our discussion of Si:P shows that the lowest metallic conductivity for a homogeneous specimen cannot lie below some value, for which we have no theoretical estimate but which from extensive observation should be of order (2). On the other hand we can only interpret the observations of Rosenbaum et al on the assumption that long-range fluctuations in composition do exist, leading to classical percolation channels and low metallic conductivities. We think it highly likely that such channels exist in SmB_6, accounting both for the low value of σ in the work of Allen et al, and also the form (4) in the results of Kasuya et al, suggesting hopping and metallic regions in series with each other.

Perhaps the results of Berger [38] et al, and of Lapierre on TmS give further evidence of long range fluctuations. These are reproduced in figure 3, and the remarkable result is that, with no obvious mechanism for a residual resistance, the conductivity at low T is about 4000 $\Omega^{-1}cm^{-1}$, about what one expects in the Ioffe-Regel regime ($k_F\ell \sim 1$), if scattering is inelastic (due to phonons). This material does not have mixed valence, all ions being in the state Tu^{3+}, and the moments are ordered. However, the energy to bring an electron from the Fermi level E_F to form a Tu^{2+} state is small, and composition-dependent. If regions exist, where on account of concentration fluctuations, the Tu^{2+} is at E_F, very strong resonance scattering is expected, which might account for the high residual resistance.

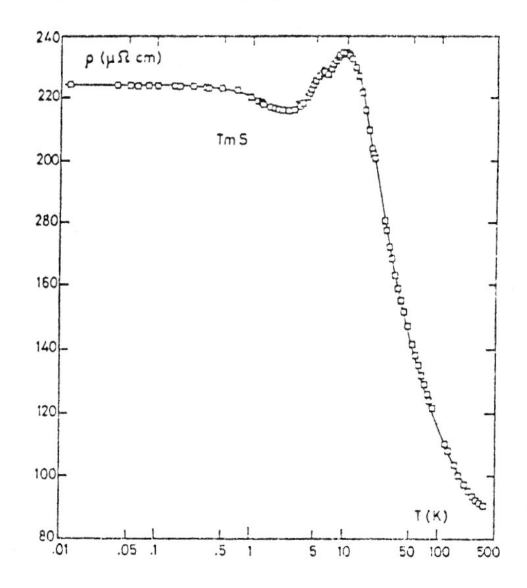

Figure 3

Kasuya et al [7], as already stated, argue against the gap in SmB_6 being due to hybridisation and propose some kind of Wigner crystallisation or Wigner glass formation (ii) to account for the rise in the resistivity below 5K. We think this model may well be correct; but it cannot explain the constant value of at low T. In our view, such a model, if the material remained metallic, should lead to equation (2) for the minimum unactivated conductivity with a the scale in space of the fluctuations, a few Angstroms and thus to much too high a value for the conductivity. Of course, with this model, as with any other, a small conductivity can be explained by long-range fluctuations of potential.

FOOTNOTES

(i) The term $1 - \ell/L_i$ is added by Kaveh and Mott [35] and tends to zero as the temperature rises, or if the collisions determining ℓ are inelastic, as in interaction with phonons.

(ii) For a discussion of this concept applied to Fe_3O_4 see reference 39.

REFERENCES

[1] J.C. Nickerson, R.M. White, K.N. Lee, R. Bachmann, T.H. Geballe, and G.W. Hall, Phys. Rev. B3 (1971), 2030.

[2] T. Kasuya, K. Takegahara, T. Fujita, T. Tanaka and E. Bannai, Journal de Physique 40, (1979), C5-, 308.

[3] F. Lapierre, Thèse de l'Institut National Polytechnique de Grenoble, 1980.

[4] J.W. Allen, B. Batlogg and P. Wachter, Phys. Rev. B20, (1979), 4807.

[5] C.M. Varma, Rev. Mod. Phys. 48, (1976), 219.

[6] K.F. Rosenbaum, K. Andres, G.A. Thomas and R.V. Bhatt, Phys. Rev. Lett. 43 (1980), 1723.

[7] T. Kasuya, K. Takegahara, Y. Aoki, K. Hanzawa, M. Kasuya, S. Kunii, T. Fujita, N. Sata, M. Kimura, T. Komatsubara, T. Furano and J. Rossat-Mignod, Proc. Int. Conf. on Valence Fluctuations in Solids, eds. L.M. Falikov et al., North Holland, 1981.

[8] G. Gunterodt, W.A. Thompson, F. Holtzberg and Z. Fisk (in press).

[9] I. Frankowski and P. Wachter, Solid State Commun. 40 (1981), 885.

[10] B. Coqblin and A. Blandin, Adv. Phys. 17, (1968), 281.

[11] N.F. Mott, Phil. Mag. 30, (1974), 403.

[12] N.F. Mott, Metal-Insulator Transitions, Taylor & Francis, London, (1974).

[13] N.F. Mott, Journal de Physique, 41, (1980), C5-57.

[14] J.W. Allen and R.M. Martin, J. de Physique, 41, (1980), C5-17.

[15] P.W. Anderson, in Valence Fluctuations in Solids, ed. L.M. Falicov et al, North Holland, 1981, p.451.

[16] N.F. Mott, Proc. Roy. Soc. A 153, (1936), 699.

[17] H. Schneider and D. Wohlleben, Z. Phys. B.44, (1981), 193.

[18] D.B. McWhan, A. Menth, J.P. Remeika, W.F. Brinkman and T.M. Rice, Phys. Rev. B7, (1973), 1920.

[19] W.F. Brinkman and T.M. Rice, Phys. Rev. 2, (1971), 4302.

[20] R. Jullien, P. Pfeuty and B. Coqblin, Valence Fluctuations in Solids, ed. L.M. Falicov et al, North Holland, 1981, p. 169.

[21] N.F. Mott, Proc. phys. Soc. A 62, (1949), 416.

[22] P.P. Edwards and M.J. Sienko, Phys. Rev. B17, (1978), 2573.

[23] P.P. Edwards and M.J. Sienko, J. Am. Chem Soc. 103, (1981), 2967.

[24] N.F. Mott and M. Kaveh, Phil. Mag. B (1982) in press.

[25] N.F. Mott, Proc. R. Soc. A (1982) in press.

[26] N.F. Mott, Phil. Mag. 26 (1972), 1015.

[27] N.F. Mott and E.A. Davis, Electronic Processes in Non-Crystalline Materials, 2nd ed., Clarendon Press, Oxford, 1979.

[28] W. Götze, J. Phys. C. 12, (1979), 1279; Phil. Mag. B43, (1981), 219.

[29] E. Abrahams, P.W. Anderson, D.C. Licciardello and T.W. Ramakrishnan, Phys. Rev. Lett. 42, (1979), 693.

[30] N.F. Mott and M. Kaveh, J. Phys. C (Solid State Phys.) 14, (1981) L659.

[31] N.F. Mott, Phil. Mag. B44, (1981), 265.

[32] W. Sasaki, Phil. Mag. B42, (1980), 723.

[33] K.F. Berggren, J. Phys. C. 1982(in press)

[34] N.F. Mott and J. H. Davies, Phil. Mag. B42 (1980), 845.

[35] M. Kaveh and N.F. Mott, J. Phys. C (in press).

[36] H. Fritzsche, The Metal non-Metal Transition in Disordered Systems, 19th Scottish Univ. Summer School in Physics, ed. L.R. Friedman and D.P. Tunstall,

1978, p.193.

[37] M. Pepper, J. non-crystalline solids, 32, (1979), 161.

[38] A. Berger, E. Bucher, P. Haen, F. Holtzberg, F. Lapierre, T. Penney, and R. Tournier, in Valence Instabilities and Narrow-band Phenomena, ed. R.D. Parks, Plenum Press, N.Y., (1977), p.491.

[39] N.F. Mott, Festkörperprobleme, 19, (1979) 331.

D. WOHLLEBEN: At high temperature, when kT is larger than the width of the narrow band, experimental evidence exists for saturation of the strong resistivity anomaly, after subtraction of the phonon scattering. You seem to state, on the other hand that $\Delta\rho$ should drop, like T^{-1}. Can you comment ?

N.F. MOTT: I think that the fraction of the electrons within a range kT of the Fermi energy which can be scattered is $\Delta E/kT$, where ΔE is the width of the resonance. But the matrix element (squared) of the interaction also comes in, denoted by $|M|^2$. If the scattering is due to phonons, $|M|^2$ should be proportional to kT, so that T cancels. Scattering can also be due to electron-electron collisions, spin fluctuations and impurities, and for these cases kT should not cancel.

Valence Instabilities
P. Wachter and H. Boppart (eds.)

THE FERMI SURFACE AND THE ENERGY GAP IN MIXED VALENCE MATERIALS

N.F. Mott

Cavendish Laboratory, Cambridge

Anyone who attended the Zürich meeting must have been struck by the wide variety of theoretical approaches to the problem of intermediate valency (IV) and the incomplete understanding of the connections between them. This note seeks to defend the model put forward by the author on previous occasions (1), (2) namely that IV materials are similar in principle to transition metals, a 4f rather than a 3d band being hybridised with a conduction band. I shall also discuss the "gap" in SmB_6, now shown experimentally to exist, and the criticism of the hybridisation model given by P.W. Anderson at Santa Barbara (3), and emphasized as an outstanding problem by Professor Wachter in his opening remarks at the Zürich meeting.

Starting with normal rare earth metals, the outer electron in the 4f shell lies at a large energy E below the Fermi energy E_F of the conduction electrons; the moments are ordered through the RKKY interaction. For materials where E is fairly small, the Kondo energy

$$\Delta \exp(-E/\Delta),$$

where Δ is a self energy, can however be greater than the RKKY interaction. There is then no ordering of the moments, and each is doing a "spin flip" independent of the others. Such a material can be called a Kondo metal. The resistivity ρ is zero for a perfect crystal at T = 0, and should rise to a value of the order $3\hbar a/e^2$ in a temperature range T_K given by

$$kT_K \sim \Delta \exp(-E/\Delta)$$

and then decrease. The reason why it decreases is the same as for materials containing isolated Kondo impurities, namely that, if $T > T_K$, only a fraction T_K/T of the electrons within a range kT of the Fermi energy can be scattered into the region of a high density of states. Thus if the scattering is due to impurities or any other mechanism that does not increase with T, we expect a decrease in ρ.

A Kondo metal does in a sense have intermediate valence, because a fraction exp $(-E/\Delta)$ of the rare earth ions will have a valency different by $\pm$ 1 from the majority. But in the typical intermediate valency material we believe that E = 0, and that this is not an "accident". The narrow 4f band acts as a "sink" so that the conduction electrons can take up the number which will minimise the energy of the crystal. The 4f electrons must hybridise with the conduction electrons, giving a Fermi surface, as in Ni or Pd, which has regions of high effective mass and of low effective mass. The one IV material, $CeSn_3$, for which a Fermi surface has been determined (4) experimentally, has just this property.

We believe, then, that for all IV materials for which there is no "gap", we must suppose that a Fermi surface of this kind exists. In the parts of the surface with heavy mass, the quasiparticles ($4f^1$ in cerium compounds, $4f^5$ moving through the non-magnetic $4f^6$ in Sm compounds) are such that the spin-orbit coupling is <u>not</u> broken down, and the effective mass may be enhanced by polaron formation. Nevertheless each quasiparticle has, as $T \to 0$, a definite k-value. As emphasized by Brandow at this conference, the Hubbard U for the f-like region is virtually infinite; this in our view should lead to strong electron-electron scattering and a resistivity varying as T^2, but no major effect on the bandwidth. As the temperature is raised, electron-electron scattering will become strong, $\Delta k/k$ will tend to unity in the regions of high effective mass, and the model of a Fermi liquid goes over into that of independent moments, as illustrated in figure 1 of Anderson's paper (3).

In the low temperature region, the current will be carried by that part of the Fermi surface where m_{eff} is low, and scattering is to that part where m_{eff} is high and which therefore makes a large contribution to $N(E_F)$. Even in the latter part there may be large variations of m_{eff} over the Fermi surface. The scattering should be given by Fermi's golden rule:

$$(2\pi/\hbar)\ |M|^2\ N(E_F)$$

where M is the scattering matrix and $N(E_F)$ the density of states in the final state. We see no reason why very high values of $N(E_F)$ should not exist on some parts of the Fermi surface, as supposed by D. Jaccard and J. Sierro at this meeting, to explain some results on the thermopower, but that to render the <u>whole</u> high mass region non-degenerate a higher temperature than $1/kN(E_F)_{max}$ is required. Moreover, as far as we can see, scattering by impurities according to Fermi's golden rule may give a cross section much greater than the size of the scatterer, a point commented on by Anderson (3) and discussed by P. Entel and P. Gies at this conference.

If kT is greater than the width of the 4f band; or rather greater than $\Delta E = 1/N(E_F)$ for the high mass region of the Fermi surface, we may expect that, as for Kondo metals, the scattering behaves like

$$|M|^2/kT$$

If $|M|^2$ is due to phonons, so that $|M|^2 \propto kT$, the scattering into the region with high density of states may be independent of T. Thus Schneider and Wohlleben (5) find that the excess resistivity of $GdPd_3$ above that of YPd_3, presumably due to spd → 4f scattering, is little dependent on T above 75 K. Similar results for $YbCu_2Si_2$ and $CeCu_2Si_2$ are found by Hussain et al (6).

Finally we look at the nature of the hybridisation and the possibility that it may be responsible for a gap in SmB_6, as first proposed by Coqblin and Blandin (7). Anderson (3) criticised the model, because the spin degeneracy of the conduction electron is 2, while that of the $4f^5$ state is 9. He supposed that only two of the 9 4f-bands can hybridise, leaving 5 unhybridised 4f bands. But it seems to us that all 4f bands will behave in the same way; into whichever a quasiparticle is excited, it can move *via* the conduction band. So if s is the number of Sm ions per atom in the $4f^5$ state and n its degeneracy, then each will be filled up to only s/n states. The E-k plots for each of the n quasiparticles will then be identical - and the degeneracy of the $4f^5$ state does not affect the gap, if indeed one exists in which the Fermi energy lies, as maintained by Allen and Martin (8) for SmB_6.

Kasuya (9) argues against the hybridisation gap mainly on the grounds of observations of the specific heat anomaly, which extends to higher temperatures than would be expected if due simply to the gap deduced from the conductivity and other measurements; he supposes that some kind of charge density wave is set up. If so, we feel it should enhance the hybridisation gap; it must have the wave-length for which the 4f and 5d bands cross, appropriate to s/n particles in the f band.

Tarascon et al (10) showed two years ago that the degree of oxidation of Sm in SmB_6 changed from 2.60 at 300 K to 2.53 at 4.2 K. Further results of this kind are discussed by Wohlleben and co-workers in this volume (see also Bauchspiess et al (11)). This means that as the temperature is raised there is a decrease of 17.5% in the number of Sm^{2+} ions (that is the non-magnetic $4f^6$). The proportion of magnetic ions increases. We follow Wohlleben in supposing:

a) At 300 K the f-like part of the Fermi surface is non-degenerate so that we can consider each Sm ion separately.
b) At 300 K we have at least 6 states from J = 5/2, and perhaps more due to crystal field splitting.

If the number of states is λ, we expect a free energy per $4f^5$ state of kT ln λ. If then n is the proportion of $4f^5$ Sm ions in this state and n_o the value at T = 0, the total free energy should be of the form

$$A\,(n-n_o)^2 - nkT \ln \lambda.$$

The equilibrium value of n is thus

$$n = n_o + kT \ln \lambda/2A.$$

Since the right hand side is observed to be of order 15%, then assuming $\ln\lambda \sim 2$ and $n_o \sim 1/2$ we see that at 300 K

$$A \sim kT \ln \lambda/\ 0.15$$

$$\sim 0.33\ \text{eV}.$$

A should depend on the energies of the conduction electrons, and the value 0.33 eV seems reasonable, though we might perhaps expect it to be rather larger.

REFERENCES

1) Mott, N.F., Phil. Mag. 30, 403 (1974)
2) Mott, N.F., J. de Physique, C5, suppt. to no 6, 41, C5-51 (1980)
3) Anderson, P.W., in Valence Fluctuations in Solids (Santa Barbara Conference), Falicov, L.M., Hanke, W. and Maple, M.B., (eds), North Holland, (publ.) p. 451 (1981)
4) Crabtree, G.W., Johanson, W.R., Edelstein, A.S. and McMasters, O.D., in Ref. (3) p.93 (1981)
5) Schneider, H., and Wohlleben, D., Z. Phys. B - Condensed Matter, 44, 193 (1981)
6) Hussain, S., Cattaneo, E., Schneider, H., Schlabitz, W., and Wohlleben, D., (in press), (1982)
7) Coqblin, B., and Blandin, A., Adv. Phys. 17, 281 (1968)
8) Martin, R.M., and Allen, J.W., Ref. (3) p. 85 (1981)
9) Kasuya, T., Takegahara, K. , Aoki, Y., Hanzawa, K., Kasaya, M., Kunii, S., Fujita, T., Sato, N., Kimura, H., Komatsubara, T., Furuno, T., and Rossat-Mignod, J., Ref. (3) p. 215 (1981)
10) Tarascon, J.M., Isikawa, Y., Chevalier, B., Etourneau, J., Hagenmuller, P., and Kasaya M., J. de Physique 41, 1141 (1980)
11) Bauchspiess, K.R., Boksch, W., Holland-Moritz, E., Launois, H., Pott, R., and Wohlleben, D., Ref. (3) P. 417 (1981)

Valence Instabilities
P. Wachter and H. Boppart (eds.)

ELECTRICAL RESISTANCE IN INTERMEDIATE VALENCE COMPOUNDS ARISING FROM IMPURITY SCATTERING

P. Entel, W. Wiethege and P. Gies

Institut für Theoretische Physik der Universität zu Köln
Zülpicher Str. 77, D-5000 Köln 41, FRG

The temperature dependent impurity resistivity in mixed valence compounds is calculated. For simplicity the mixed valence compounds are described within the framework of the periodic Anderson model. The normal effect of impurities consists of random potential scattering. The resulting resistivity can be considerably enhanced by the presence of localized 4f-states around the Fermi surface giving rise to resonant scattering. A similar result was obtained in Ref. /1/. However, the huge increase of the residual resistivity in $CePd_3$ due to a few impurities cannot be explained by resonance scattering alone.
Therefore, we have enlarged the model Hamiltonian and considered the effect of lattice distortion caused by impurities. The static distortion will change the energy of the 4f-levels of a large number of rare earth ions neighbouring the impurity site. This gives rise to a non local scattering potential which cannot be treated by ordinary impurity scattering theory. The resulting effect consists of a considerable enhancement of the electrical resistivity.

The electrical resistivity in mixed valence compounds seems to be unequivocally connected with the mixed valence phenomenon. In these compounds the existence of characteristic charge (and spin) relaxation rates is obvious from measurements of typical response functions such as the magnetic susceptibility or the conductivity /2/. Thus, pronounced maxima in magnetic susceptibility and resistivity curves can approximately be related to the fluctuating valence via a characteristic temperature $k_B \Gamma_f \propto \pi N_F V^2$. Here, V denotes the hybridization energy of localized 4f states and conduction band states and N_F denotes the density of states of conduction electrons. Therefore, in any theoretical description of transport properties the detailed nature of mixed valency should be taken into account. This is an unsolved problem.

Here, we face the problem of the influence of impurity scattering on the residual resistivity in mixed valence compounds. As is well known the huge increase of the residual resistivity due to a few impurities as observed in CePd3 is a unique phenomenon so far not observed in other mixed valence compounds /3/. Our aim is to give an estimate for the order of magnitude for the resistivity arising from impurity scattering. Therefore, typical mixed valent aspects are only taken into account in very approximative form. The model consists of the interacting system of localized, correlated 4f electrons and non localized band electrons (Anderson lattice) in the presence of impurities:

$$\mathcal{H} = \sum_{\underline{k},\sigma} \varepsilon_{\underline{k}} d^+_{\underline{k}\sigma} d_{\underline{k}\sigma} + E_0 \sum_{\underline{k}\sigma} f^+_{\underline{k}\sigma} f_{\underline{k}\sigma} + V \sum_{\underline{k}\sigma} [d^+_{\underline{k}\sigma} f_{\underline{k}\sigma} + h.c.] + U \sum_i f^+_{i\uparrow} f_{i\uparrow} f^+_{i\downarrow} f_{i\downarrow} + \sum_{\underline{k},\underline{k}',\sigma,\{\underline{R}_i\}} e^{i\underline{R}_i(\underline{k}-\underline{k}')} \{ U_0 d^+_{\underline{k}\sigma} d_{\underline{k}'\sigma} + U_1 [d^+_{\underline{k}\sigma} f_{\underline{k}'\sigma} + h.c.] + U_2 f^+_{\underline{k}\sigma} f_{\underline{k}'\sigma} \}. \qquad (1)$$

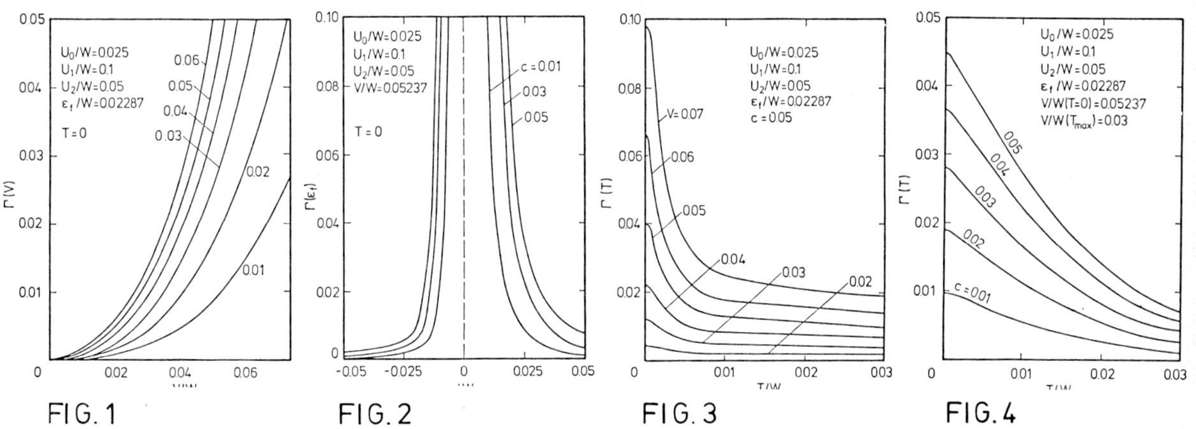

FIG. 1 FIG. 2 FIG. 3 FIG. 4

As a first approximation we replace E_0 by $\varepsilon_f = E_0 + U\langle n^f_\downarrow \rangle$ (Hartree-Fock approximation) which leads to

$$\mathcal{H}_{HF} = \sum_{\underline{k},\underline{k}',\sigma} (d^+_{\underline{k}\sigma}, f_{\underline{k}\sigma}) \underline{\underline{H}}_{\underline{k},\underline{k}'} \begin{pmatrix} d_{\underline{k}'\sigma} \\ f_{\underline{k}'\sigma} \end{pmatrix}, \tag{2}$$

$$\underline{\underline{H}}_{\underline{k},\underline{k}'} = \underbrace{\begin{pmatrix} \varepsilon_{\underline{k}} & V \\ V & \varepsilon_f \end{pmatrix}}_{\equiv \underline{\underline{H}}^0_{\underline{k}}} \delta_{\underline{k},\underline{k}'} + \sum_{\{R_i\}} e^{iR_i(\underline{k}-\underline{k}')} \cdot \underbrace{\begin{pmatrix} U_0 & U_1 \\ U_1 & U_2 \end{pmatrix}}_{\equiv \underline{\underline{U}}}. \tag{3}$$

The matrix Green's function is calculated in second order Born approximation:

$$\underline{\underline{G}}(\underline{k},z) = \left(z - \underline{\underline{H}}^0_{\underline{k}} - \mathrm{Re}\,\underline{\underline{\Sigma}}(\underline{k},z) + i\,\mathrm{sgn}(\mathrm{Im}\,z) \cdot \underline{\underline{\Gamma}}(\underline{k},\omega)\right)^{-1} \tag{4}$$

with

$$\underline{\underline{\Gamma}}(\underline{k},\omega) = c\pi N_F \sum_i \int \frac{d\varepsilon_q}{\varepsilon_1(q) - \varepsilon_2(q)} \int d\Omega_q \cdot$$

$$\cdot\, \underline{\underline{U}}(\Theta) \cdot \delta(\omega - \varepsilon_i(q)) \begin{pmatrix} [\varepsilon_i(q) - \varepsilon_f] & V \\ V & [\varepsilon_i(q) - \varepsilon_q] \end{pmatrix} \underline{\underline{U}}(\Theta), \tag{5}$$

where c is the concentration of impurities and ε_i (i=1,2) stands for the hybridized band energies. In the metallic limit the real part of the self-energy vanishes and one obtains for the band electron Green's function the following expression:

$$G_{dd}(\underline{k},z) = \left(z - \varepsilon_{\underline{k}} + i\Gamma_{11}(\omega)\,\mathrm{sgn}(\mathrm{Im}\,z) - \frac{V^2 + \Gamma^2_{12}(\omega)}{z - \varepsilon_f + i\Gamma_{22}(\omega)\,\mathrm{sgn}(\mathrm{Im}\,z)}\right)^{-1} \tag{6}$$

with

$$\underline{\underline{\Gamma}}(\omega) = c\pi N_F \underline{\underline{U}} \begin{pmatrix} 1 & \frac{V}{|\varepsilon_f - \omega|} \\ \frac{V}{|\varepsilon_f - \omega|} & \left(\frac{V}{\varepsilon_f - \omega}\right)^2 \end{pmatrix} \underline{\underline{U}} . \tag{7}$$

Equations (5) and (7) clearly show the possibility of resonance scattering of conduction electrons on the resonance level: The imaginary part of the self-energy and hence the resulting resistivity will be considerably enhanced for small values of $(\varepsilon_f - \omega)$. For simplicity $\underline{\underline{\Gamma}}(\omega)$ has been evaluated with the help of $\underline{\underline{H}}^0_{\underline{k}}$ and not with the full Green's function.

Since we are dealing with extended and highly localized states we may (in a very good approximation) use the following simple expression for the current operator /4,5/:

$$j_\alpha = \frac{1}{\hbar} \sum_{\underline{k}\sigma} \frac{\partial \varepsilon_{\underline{k}}}{\partial k_\alpha} d^+_{\underline{k}\sigma} d_{\underline{k}\sigma} \tag{8}$$

For the electrical conductivity

$$\sigma(z) = -\frac{ie^2}{z}\chi(z) + \frac{i\omega_p^2}{4\pi z}, \tag{9}$$

$$\chi(z)\,\delta_{\alpha,\beta} = -\langle\langle j_\alpha ; j_\beta \rangle\rangle_z \tag{10}$$

we then obtain in lowest order of the ladder approximation the final expression

$$\sigma(T,\omega=0) = \frac{e^2 n_d}{m_d} \frac{\hbar}{2} \int_{-\infty}^{+\infty} dx \left(-\frac{\partial f}{\partial x}\right) \cdot \frac{1}{\Gamma_{11} + (V^2 + \Gamma^2_{12}) \frac{\Gamma_{22}}{(x-\varepsilon_f)^2 + \Gamma^2_{22}}} . \tag{11}$$

We would like to note that (11) is qualitatively different from the result obtained in Ref. /1/ where essential contributions to the current operator have been neglected.

It is easy to obtain from (11) the following enhancement factors:

1. For vanishing hybridization the electrical resistivity at T=0 K becomes:

$$\rho_0 = \frac{m_d}{e^2 n_d} \frac{2}{\hbar} \Gamma_{eff}, \quad \Gamma_{eff} = 2c\pi N_F U_0^2 \tag{12}$$

   where the factor 2 in the expression for Γ_{eff} is the only enhancement factor.

2. In the limit of large values for $V/|\varepsilon_f - \omega| \propto 1/V$ the effective scattering rate becomes:

$$\Gamma_{eff} = 2c\pi N_F U_1^2 \left(\frac{V}{\varepsilon_f - \omega}\right)^2 \tag{13}$$

Using typical values for $CePd_3$ ($V \approx 0.05237$ eV from $T_f \approx 130$ K, $|\varepsilon_f - \omega| \approx 0.02287$ eV from $\gamma = 36$ mJ/(mole·K)) one gets $\Gamma_{eff} \approx c$ [eV] which for c=5 at% corresponds to an effective transport relaxation rate of $\tau \approx 1.45\ 10^{-14}$ sec. This obviously does not describe the huge increase of the residual resistivity found in CePd .

The behaviour of Γ for a typical set of parameter values can be seen in Fig. 1-4.

Next, we consider the influence of the local lattice distortions caused by each impurity ion on the electrical resistivity. It is well known that such distortions give rise to extra terms in the resistivity which are proportional to T and T^2, respectively /6/. The occurence of these terms is caused by additional inelastic electron scattering in the regions of static distortions. In mixed valence compounds static distortions cause a shift of the ionic 4f-levels at sites neighbouring the impurity site. This gives rise to an effective interference term of impurity scattering at the impurity site and resonance scattering at neighbouring sites which may considerably enhance the resistivity.

The evaluation and handling of the effective nonlocal scattering matrix element is rather cumbersome. One has to include in the Hamiltonian the displacement variable φ at every lattice site:

$$\mathcal{H}_\varphi = -g_1 \sum_{i\sigma} f^+_{i\sigma} f_{i\sigma}\,\varphi - g_2 \sum_{i\sigma} [d^+_{i\sigma} f_{i\sigma} + h.c.]\,\varphi$$

$$- \tilde{g}_1 \sum_{\sigma,\{R_i\}} f^+_{i\sigma} f_{i\sigma} (\tilde{\varphi}_i - \varphi)$$

$$- \tilde{g}_2 \sum_{\sigma,\{R_i\}} [d^+_{i\sigma} f_{i\sigma} + h.c.] (\tilde{\varphi}_i - \varphi) \qquad (14)$$

whereby $\tilde{\varphi}_i$ refers to the local distortion introduced by the impurity at the impurity site i and φ denotes the displacement variable of the regular lattice.

If we replace $\mathcal{H} + \mathcal{H}_\varphi$ by the best Ersatz-Hamiltonian an effective scattering-term is obtained in which the impurity site is coupled to the whole crystal. In a first rough approximation we may then replace this non local scattering term by a local one times an effective concentration. (For example, in CePd 1 at% impurity concentration corresponds to an effective concentration of 50 at%.)

A last remark concerns preliminary results about an improved treatment of the mixed valence phenomenon itself. We have evaluated the d-band Green's function in an improved perturbation expansion up to order V^2. In the presence of impurities this leads to rather complicated self-consistency equations for the frequency dependence of the real and imaginary part of the self-energy. An approximative treatment shows that the addition of impurities should not lead to a drastic shift of the imaginary part of the self-energy at the Fermi level. It therefore remains unclear whether an improved treatment of the mixed valence phenomenon itself can lead to an understanding of the extreme sensitivity of the residual resistance to stoichiometry and to impurities.

REFERENCES

/1/ Riseborough, P.S., Solid State Commun. 38, 79 (1981).

/2/ Kuramoto, Y. and Müller-Hartmann, E., in: "Valence Fluctuations in Solids", ed. L.M. Falicov, W. Hanke and M.B. Maple (North-Holland, Amsterdam, 1981).

/3/ For recent work and further references see:
Schneider, H. and Wohlleben, D., Z. Physik 44, 193 (1981).

/4/ Entel, P., Mühlschlegel, B. and Ono, Y., Z. Physik B 38, 227 (1980).

/5/ Czycholl, G. and Leder, H.J., Z. Physik B 44, 59 (1981).

/6/ Zhernov, A.P., Sov. Phys. Solid. State 22, 332 (1980).

Valence Instabilities
P. Wachter and H. Boppart (eds.)

THERMOELECTRIC POWER OF SOME INTERMEDIATE VALENCE COMPOUNDS

D. Jaccard and J. Sierro

Département de Physique de la Matière Condensée
32 Bd d'Yvoy, 1211 Geneva 4, Switzerland

The thermopower of intermediate valence compounds (IVCs) shows typical features correlated with the valence of the rare earth ion. Using the Hirst picture of an IVC, the low temperature thermopower can be evaluated with a two band model. Our measurements strongly suggest that $YbCu_{3.5}$, $YbCu_{4.5}$, $YbSi_{2-x}$, Yb_3Si_5, YbSi, Yb_2CuSi_3 and $YbNi_2Si_2$ are new IVCs.

I. INTRODUCTION

Within the frame of the study of IVCs , relatively few works have been devoted to the thermopower of these materials. This situation is due partly to experimental difficulties and partly to the still very qualitative interpretation of the anomalous transport properties of IVCs. This paper reports for the first time, on a systematic experimental study of the thermopower of IVCs in order to stimulate a theoretical effort about this subject. The samples preparation and the measuring technique have been described elsewhere [1,2].

2. RESULTS

2.1. Ytterbium compounds

Figure 1 shows the temperature dependence of the thermopower of some Yb IVCs. Results about $YbA\ell_2$ and $YbA\ell_3$ [3] have been included for completeness. In order to show some characteristic features correlated with the valence of the Yb ion we shall present these data by going from divalent to trivalent compounds. Valence has been roughly estimated either from the lattice constant or from the room temperature susceptibility.

$YbCu_2$ is the more divalent of these fifteen compounds. Its thermopower shows above 200K a negative deviation from the usual linearity which reflects a slight tendency of Yb towards intermediate valence. $YbA\ell_2$ and Yb_4Bi_3 which are both still rather divalent exhibit a quite similar and large thermopower while their resistivity differs by more than an order of magnitude [3,4]. Following our list we note that $YbA\ell_3$ presents at 230K an enormous negative contribution of about $k/e=-86\mu V/K$. More trivalent compounds show similar giant negative peak at a temperature which roughly decreases with increasing valence. Such curves may be seen in fig.1 for well known IVCs like $YbCu_2Si_2$ and $YbCuA\ell$ and also for new compounds like $YbCu_{3.5}$ and $YbCu_{4.5}$. Nearly trivalent YbSi and $YbNi_2Si_2$ behave differently. At low temperature their thermopower has a positive peak and then changes to negative value at higher temperature.

The following points are noteworthy about these results: i) No drastic impurity, stoichiometric or annealing effects have been observed. Negative thermopower is found for $YbCu_2Si_2$ and Yb_4Bi_3 contrary to reference 5 and 6.
ii) In order to extract the anomalous contribution to the thermopower of IVCs we have measured isotypic compounds with Lu or with another rare earth and found that the normal thermopower can be considered as negligible.
iii) At low temperature, a linear variation of the thermopower is generally observed except for Yb_4Sb_3, $YbCu_{3.5}$ and problably for YbSi below 1K. This linearity extends up to 50K for compounds like $YbCu_2Si_2$.
iv) In the silicon rich side of the Yb-Si system, two phases appear at very close compostion [7]. Our sample $Yb_{37.5}Si_{62.5}$, called $YbSi_{2-x}$ in ref.8 crystallizes with a defect disordered $A\ell B_2$-type structure, while $Yb_{39}Si_{61}$ is a two phases sample. It crystallizes partly with the $A\ell B_2$-type and partly (about 40%) with Th_3Pd_5-type structure. Single phase sample with this last structure type has not been obtained. The interesting point is the very similar thermopower of both samples which suggests that no correlation exists in these compounds between crystalline disorder and intermediate valence phenomena. On the other side, the residual resistivity differs by a factor five. [1]

The homogeneous nature of the intermediate valence of most of these Yb compounds is still not well established. However the fact that the Yb atom occupies an unique crystalline site for all compounds of known structure favors this possibility. An homogeneous valence is also suggested by the similitude of the anomalous thermopowers of the well known and the new IVCs.

Specific heat and electrical resistivity have also been measured in order to check if their behavior is compatible with intermediate valence. The linear term of the low temperature specific

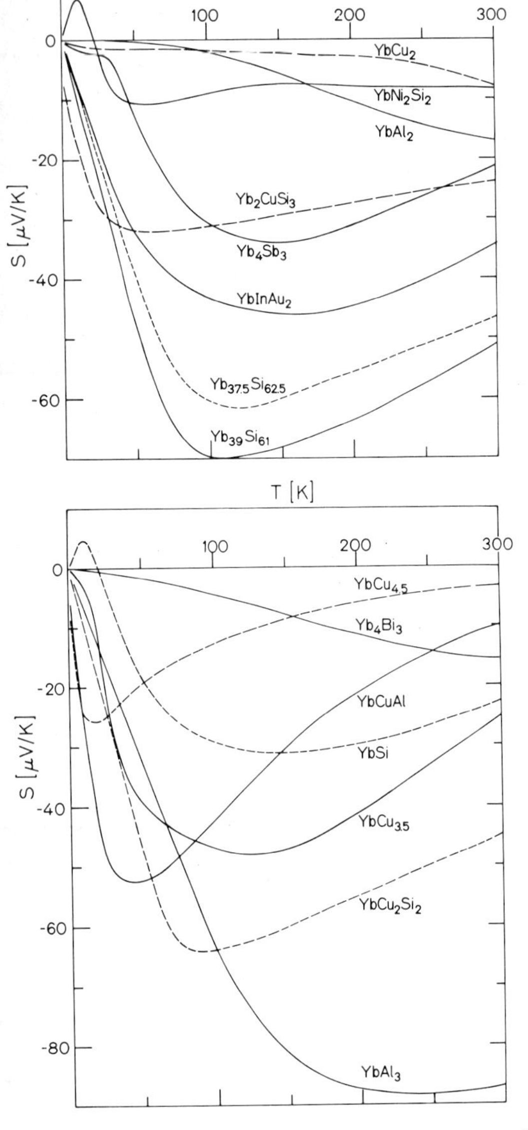

Fig. 1. Absolute thermopower vs temperature of some ytterbium compounds.

heat [8] gives the following results:
$\gamma \simeq 600 mJ/K^2$ mole for $YbCu_{4.5}$,
$\gamma \simeq 90 mJ/K^2$ mole for $YbCu_{3.5}$ and
$\gamma > 100 mJ/K^2$ mole for YbSi. Deviation from linearity at low temperature and presence of traces (<0.5%) of magnetic Yb_2O_3 affect the accuracy of the γ values.

Figure 2 shows some typical temperature dependence of the resistivity of Yb IVCs. The measures have been performed from 0.07 to 300K for YbSi and from 1.4 to 300K for the other compounds. Like the termopower, the electrical resistivity

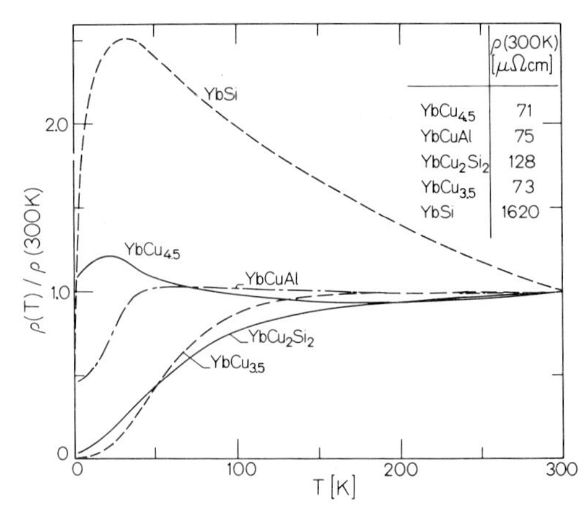

Fig. 2. Resistivity vs temperature of some ytterbium compounds.

shows features correlated with the valence of the rare earth. The resistivity of $YbCu_{3.5}$ is strongly temperature dependent ($\rho(300K)/\rho(1.4K) \simeq 100$) whereas a relatively high and constant resistivity is always observed for $YbCu_{4.5}$. We note a gradual increase of the residual resistivity with increasing valence. For $YbCu_{4.5}$ the partial loss of the coherent electronic propagation suggests that each Yb ion mainly behaves like an isolated impurity. Nevertheless, at sufficiently low temperature the resistivity of all compounds follows approximately a T^2 law, $\rho(T) - \rho(T{=}0) = \alpha T^2$ but with very different values for α. The result for nearly trivalent YbSi is very surprising. Quite similar behaviors have been measured on three samples of different batches. The temperature dependence seems typical of a dense Kondo system with a still decreasing resistivity at 300K. However very high values are observed at all temperatures. We find a T^2 law under 1.5K with $\rho(T{=}0) \simeq 800 \mu\Omega cm$ and $\alpha \simeq 400 \mu\Omega cm K^{-2}$. In contrast to YbSi, the isotypic compounds LuSi, TmSi and ErSi show a normal metallic resistivity with values at room temperature of 77, 122 and 130 μΩcm repectively.

In the binary and ternary systems Yb-Cu-Si we know today eight compounds which show typical anomalous properties of a valence instability of the Yb ion. Such compounds are formed in the ~25at.% Yb region of the phase diagram. It is striking that similar situations occur for the systems Yb-Cu-Aℓ or Yb-In-Au.

2.2. Cerium compounds

As for the Yb compounds, the thermopower of Ce IVCs also presents a valence dependence but with opposite sign as a result of the electron/hole symmetry. For example, it may be seen in figure 3 that nearly trivalent $CeCu_2Si_2$, $CeAℓ_2$ and

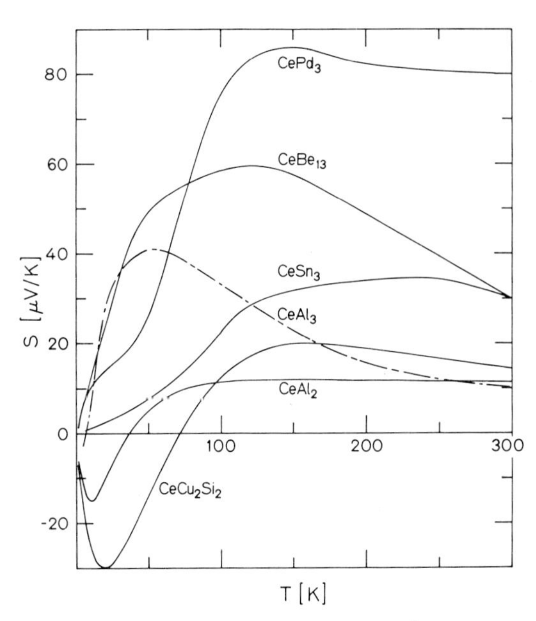

Fig. 3. Absolute thermopower of cerium compounds. Results for $CeAl_3$, $CePd_3$, $CeBe_{13}$ and $CeSn_3$ are from ref.9,10 and 11.

$CeAl_3$ (0.9-1 4f electron) qualitatively behave as the nearly trivalent Yb compounds (0.9-1 4f hole) but with opposite signs. Their thermopower shows a negative peak at low temperature followed by a positive contribution at higher temperature. $CeBe_{13}$, $CePd_3$ and $CeSn_3$ are typical examples of the other behavior with only one positive maximum. Contrary to $YbCu_2Si_2$, the magnitude (but not the characteristic temperature) of the transport anomalies of $CeCu_2Si_2$ might change drastically from sample to sample. This can be due to a partial disorder in the occupation of the Cu and Si cystalline site [1,12].

2.3. Europium and samarium compounds

In opposition to what is presented in reference 5 the thermopower of $EuCu_2Si_2$ exhibits a well-defined positive peak as shown in figure 4. However in a certain range, the position of the peak and the low temperature thermopower contribution are samples dependent probably for the same reason as for $CeCu_2Si_2$. The drop of the thermopower of $EuCu_2Si_2$ above 200K and the change to negative value originate from the well known decrease with temperature of the valence of the Eu ion. By comparison, the isotypic compound $EuPd_2Si_2$ which valence is also strongly temperature dependent has a similar thermopower (not shown in fig.4) but shifted to lower temperature with a change to negative value at about 150K. In both compounds, the zero thermopower occurs for the same 2.6 valence [13]. These typical behaviors of Eu IVCs have also to be compared with the quite normal transport properties of $EuNi_2Si_2$ which is an inhomogeneous IVC [1,12]. In this sense, the result shown in fig.4 for Sm_4Bi_3 is consistent with an at least partly homogeneous intermediate valence of the Sm ion.

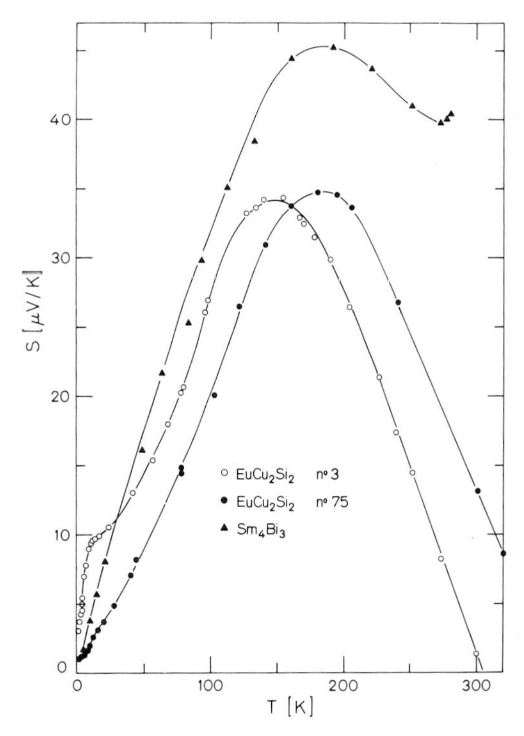

Fig. 4. Absolute thermopower of $EuCu_2Si_2$ and Sm_4Bi_3.

2.4. Thermopower under pressure

The consistence of our classification of the thermopower behavior versus the valence of a rare earth ion can be tested at room temperature by the measure of the pressure dependence of the transport properties. Some typical results for different samples are shown in figure 5. For nearly divalent $YbCu_2$ the pressure favors the trivalent state and consequently the thermopower decreases towards giant negative values, while the resistivity rises by a factor of four at 70 kbar [2]. From the measure of $YbCu_2Si_2$ which is a more trivalent compound, we can predict that the thermopower of $YbCu_2$ might have a minimum at roughly 100 kbar followed by an increase towards smaller negative value at still higher pressure. The just symmetrical case of $YbCu_2$ from an electron/hole point of view is $CeCu_2Si_2$. The strong increase of its thermopower under pressure is an evidence of the non integer valence of the Ce ion. With $EuCu_2Si_2$, hydrostatic pressure and temperature have a reverse effect on the valence of the Eu ion and therefore we observe

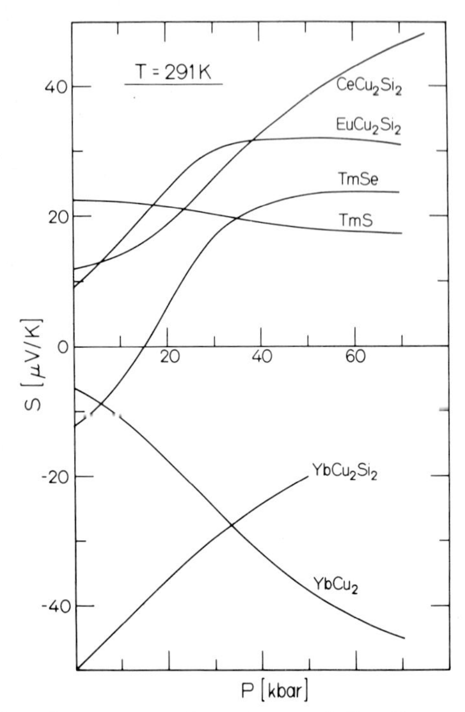

Fig. 5. Pressure dependence of the thermopower of some IVCs and TmS.

an increase of the thermopower under pressure. The last example showing that the thermopower is a good parameter to follow a pressure induced valence transition is that of TmSe. In this case the anomalous negative contribution vanishes between 30-40kbar and at higher pressure TmSe behaves like TmS which thermopower is representative of the pressure independent normal contribution. Similar thermopower results have been obtained on non-stoichiometric Tm_xSe [14] for which the valence transition is induced by chemical pressure.

3. DISCUSSION

According to the Hirst picture [15] of an ICV, the integrated density of states of the 4f excitation band expected near the Fermi level can be put in the form

$$Y = Q^-(1-z) + Q^+z$$

where Q^- and Q^+ are constants for a given rare earth ion and have been calculated if an LSJ spin-orbit ground level is taken as the interconfiguration fluctuation manifold, z and 1-z are the probalities of the ion being in a $4f^{n-1}$ or $4f^n$ state respectively and $Q^-(1-z)$ is the number of occupied states below the Fermi level.

Since the f density of states $N_f(E)$ is much larger than the s or d at the Fermi level, s,d-f scattering dominates and the relaxation time τ can be evaluated with a two band model [16]

$$\frac{1}{\tau} \simeq \frac{1}{\tau_{s,d-f}} \quad \alpha \; [N_f(E)]_{E=E_F}$$

and the thermopower S is given by

$$S = -\frac{\pi^2 k_B^2 T}{3|e|}\left[\frac{3}{2E_F} - \frac{1}{N_f(E)}\frac{\partial N_f(E)}{\partial E}\right]_{E=E_f}$$

At low temperature the first term in S is small and S depends then mainly on the logarithmic derivative of $N_f(E)$ at E_F. If now the spread of ionic energies included in the interconfiguration fluctuation manifolds is small compared to the width W of the 4f band, a Lorentzian density of states centered at an energy E_o can be assumed. The knowledge of the valence and the electronic specific heat $C_{e\ell}=N_f(E_F)(\frac{1}{3}\pi^2)k^2T$ of an IVC allows us to evaluate W and E_o-E_F and for $k_BT \ll W$ one obtains :

$$S = -566T \frac{T_o}{T_o^2 + (\frac{1}{2}T_w)^2} \quad \mu V/K \qquad (1)$$

where $T_o=(E_F-E_o)/k_B$ and $T_w=W/k_B$.

Some experimental and calculated values of the low temperature linear coefficient of the thermopower are given in Table 1. Very good fit can be obtained for $YbCu_{4.5}$ and $YbCuA\ell$ but we must keep in mind that eq.(1) depends strongly

Table 1

Compound	Structure type	valence	$N_f(E_F)$ [eV•RE at.]$^{-1}$	T_w [K]	T_o [K]	$(\Delta S/\Delta T)_{cal}$ [$\mu V/K^2$]	$(\Delta S/\Delta T)_{exp}$ [$\mu V/K^2$]
YbSi	CrB	2.95	>40			>0	+0.6
$YbCu_{4.5}$	unknown	2.84	255	55	9.3	-6.2	-6
$YbCuA\ell$	Fe_2P	2.8	110	120	35	-4.1	-4
$YbCu_2Si_2$	$ThCr_2Si_2$	2.8	57	233	67	-2.1	-1
$YbCu_{3.5}$	unknown	2.75	38	305	132	-1.8	-1
$YbA\ell_3$	Cu_3Au	2.7	19	511	297	-1.1	-0.6
$CeCu_2Si_2$	$ThCr_2Si_2$	3.12	420	27	2.2	-6.3	-5.4

on the assumed band shape and is valid in principle for elastic scattering only. The thermoelectric peak is observed at a temperature lying between $T_W/2$ and $T_W/3$. It corresponds to a degeneracy effect of the 4f band. For the other compounds of lower valence the fit is rather bad. However, in these cases the scattering s,d-f is not predominant and then according to the Nordheim-Gorter rule, S is reduced by a factor ρ_f/ρ where ρ_f is the s,d-f scattering contribution to the total resistivity ρ.

From the values of Q^- and Q^+ we deduce that $S \gtrless 0$ if the valence $v \lessgtr 3.14$ for Ce compounds or $v \gtrless 2.89$ for Yb compounds. It is striking that the low temperature thermopowers of the nearly trivalent Ce and Yb compounds follow these signs predictions. However eq.(1) is qualitatively unable to describe the high temperature behavior.

AKNOWLEDGMENTS

We wish to thank R. Cartoni for his helpful technical assistance. This work was partly supported by the Swiss National Science Foundation.

REFERENCES

1. D.Jaccard, Thesis, University of Geneva (1981) unpublished
2. D.Jaccard, F.Haenssler and J.Sierro, Helv.Phys.Acta 53,590(1980)
3. H.J.Van Daal, P.B.Van Aken and K.H.J.Buschow, Phys.Lett. 49A,246(1974)
4. E.Bucher, A.S.Copper,D.Jaccard and J.Sierro, in Valence Instabilities and Related Narrow Band Phenomena, edited by R.D.Parks(Plenum, New York, 1977) p.529.
5. B.C.Sales and R.Viswanathan, J.Low.Temp.Phys. 23,449(1976)
6. V.Vijayakumar, S.N.Vaidya, E.V.Sampathkumaran, L.C.Gupta and R.Vijayaraghavan,J.Phys.Soc. Japan 50,19(1981)
7. A.Iandelli, A.Palenzona and G.L.Olcese, J.Less-Common Met. 64,213(1979)
8. D.Jaccard,A.Junod and J.Sierro, Helv.phys. Acta 53,583(1980)
9. P.B.Van Aken, H.J.Van Daal and K.H.J.Bushow, Phys.Lett. 49A,201(1974)
10. H.Sthioul,D.Jaccard and J.Sierro, this issue
11. J.R.Cooper, C.Rizzuto and G.Olcese, J.Phys. C1-32,1136(1971)
12. I.Mayer and I.Felner, J.Phys.Chem.Sol. 38 , 1031 (1977)
13. E.V.Sampathkumaran et al., in Valence Fluctuations in Solids, L.M.Falicov,W.Hanke, M.B.Maple(eds.) North-Holland, Publishing Company(1981)p.193
14. D.Jaccard and J.Sierro, Helv.Phys.Acta, (to be published)
15. L.L.Hirst, Phys.Rev. B15,1(1977)
16. R.D.Barnard,Thermoelectricity in Metals and Alloys,p.149.Taylor and Francis,London(1972).

Valence Instabilities
P. Wachter and H. Boppart (eds.)

THEORY OF SPIN FLUCTUATIONS AND THE DE HAAS-VAN ALPHEN EFFECT

Peter S. Riseborough

Department of Physics
Polytechnic Institute of New York
333 Jay Street
Brooklyn, New York 11201

A calculation of the effect of spin fluctuations on the De Hass-Van Alphen oscillations is presented. The spin fluctuations have two main effects. That is they renormalize the spin splitting factor by an enhancement associated with the static susceptibility. The amplitude of the oscillations is governed mainly by the enhancement of the specific heat. However, for systems with a large amplitude of spin fluctuations, other terms can contribute to the amplitude of the oscillations in the magnetization. Such effects may have been observed in De Hass-Van Alphen measurements on Pd.

1. INTRODUCTION

De Hass-Van Alphen measurements have been made on $CeSn_3$[1], a material which is known to be in a mixed valence phase. The valence of $CeSn_3$, as deduced from lattice constant measurements [2], has been found to vary from 3.3 at T 0°K to 3.1 at room temperature. The system is non-magnetic, the static susceptibility[3] shows a broad maxima near 140°K and follows a Curie-Weiss law at higher temperatures. The low temperature non-magnetic state is assumed to be caused by the fluctuation of the 4f spins which occur when the electrons hop between the 4f orbitals and the conduction band states. Inelastic neutron scattering experiments[4] shows a quasi-elastic peak with a temperature independent width of 22 MeV. This is in accord with the picture that the 4f orbitals hybridize with the conduction band wavefunctions leading to a significant amplitude of spin fluctuations of the 4f electron wavefunctions.[5] These spin fluctuations are expected to effect the eletrical resistivity by scattering the conduction electrons.[6] The resistivity of $CeSn_3$ shows a T^2 term[7] that can be interpreted as being due to scattering from spin fluctuations, the size of the coefficient of the T^2 term is compatable with the estimates of the spin fluctuation temperature deduced from the magnetic susceptibility. The above data suggests that it might be appropriate to describe $CeSn_3$ as a coherent, fermi liquid.

The De Haas-Van Alphen measurements of Crabtree et al[1b] have reinforced this picture of $CeSn_3$ as having a spatially coherent ground state. The experimental results not only indicate that the conduction electrons hybridize with the 4f orbitals, as evidenced by the high effective masses, but they also show that the hybridization is coherent. The extent of the spatial correlation can be estimated from the size of the cyclotron orbitals, this leads to a correlation length which has to be greater than 7,000 Å. Since an electron in a 4f shell interacts strongly with the other electrons in the 4f shell, it is natural to ask what the effect of these interactions are on the De Hass-Van Alphen oscillations. In this paper we shall study the effect of the spin fluctuations on the De Haas-Van Alphen effect. As in prior theoretical studies[8a,b] we find that the manybody effects renormalize the De Haas-Van Alphen oscillations in two manners. First, the spin splitting factor contains an enhancement of the band mass. This enhancement is not the mass enhancement associated with the specific heat,[9,11] but it is enhamced by a factor associated with the static susceptibility[10] instead. The second manybody effect enters into the amplitude of the De Haas-Van Alphen oscillations. The amplitudes A_l are found to be governed by the specific heat enhancement, as in previous calculations[8] which considered the effect of electron-phonon interactions. However, we find that there are also other contributions to the amplitudes which can become important where the susceptibility specific heat ratio indicates large exchange enhancements. From this we are lead to expect that if the De Haas-Van Alphen measurements can be made on a variety of Mixed valent systems, there should be a discrepancy between the effective masses deduced from the De Haas-Van Alphen effect and the effective masses deduced from measurements of the specific heats in applied fields. The discrepancy should increase with increasing susceptibility enhancements.

Such discrepancies have already been found in studies of transition metal systems[12]. In Pd Crabtree et al have found an effective mass Meff which is 15% less than that found from specific heat experiments. This is to be contrasted to Pt and Nb where the agreements is of the order of 1%. The present calculation shows a possible source of these discrepancies. In the next section we shall outline a simple theoretical model that describes a system, in a finite magnetic field, that exhibits spin fluctuations. We shall formulate the oscillating part of the magnetization in terms of the electron self energy. We shall calculate the self energy in the Random Phase Approximation which takes into processes whereby spin fluctuations are emitted and absorbed. In section III we shall present and analyze our results.

2. THE DE HAAS-VAN ALPHEN EFFECT

In the presence of a magnetic field H, a system of non-interacting electrons will traverse Landau orbitals. The electronic spectrum is described by

$$\mathcal{E}_\sigma(k_z,\ell) = \frac{k_z^2}{2m} + (\ell+\tfrac{1}{2})\omega_H + g\mu_B\sigma H \tag{1.1}$$

in which ω_H is the cyclotron frequency

$$\omega_H = \frac{eH}{m^*c}$$

The degeneracy of the level with energy $\mathcal{E}_\sigma(k_z,\ell)$ is proportional to $\frac{m^*\omega_H}{2\pi}$. Thus as H increases the degeneracy of the Landau levels and their spacing increases. This can result in a Landau level crossing the fermi surface as H increases. Each crossing leads to oscillations in the thermodynamic potential and the magnetization.

We shall incorporate the electron-electron interactions via a short ranged Hubbard interaction. The Coulomb interaction is assumed to be so highly screened that it only acts between the electrons in the same atomic orbital. Since the band is non-degenerate, the Pauli exclusion principle restricts the interaction to occur between pairs of electrons with opposite spins. Thus the Coulomb interaction can be written as

$$\hat{H}_I = \frac{U}{2}\sum_{i\sigma} a^+_{i\sigma} a^+_{i-\sigma} a_{i-\sigma} a_{i\sigma} \tag{1.2}$$

where $a^+_{i\sigma}$ and $a_{i\sigma}$ respectively create and destroy an electron of spin σ in the atomic orbital at site i. We shall proceed in the following manner, first we shall express the oscillating part of the thermodynamic potential in terms of the electron self energy. Then we shall evaluate the self energy due to the electron - electron interactions described by equation 1.2, in the Random Phase Approximation. The thermodynamic potential Ω can be expressed as

$$\Omega = \int\frac{dz}{2\pi} f(z) \sum_{k_z \ell \sigma} \log_e\left\{\mathcal{E}_\sigma(k_z,\ell) + \Sigma'_\sigma(z) - z\right\} \frac{m\omega_H}{2\pi} \tag{1.3}$$

in which $\Sigma_\sigma(z)$ represents the self energy of an electron with spin. We may transform the integration over z into a sum over the Matsubara frequencies ω_n,

$$\omega_n = \frac{2\pi}{\beta}(n+\tfrac{1}{2})$$

by using the Poisson summation formulae. We then separate out the term in the self energy which gives a manybody contribution to the splitting between the up and downspin electron bands via

$$\Sigma'_\sigma(z) = \Sigma(z) + \sigma g\mu_B H\alpha$$

On performing the trace over the z component of the momentum k_z, we find that the oscillating part of Ω can be written as

$$\Omega_{osc} = -\left(\frac{m^*\omega_H}{4\pi}\right)^{3/2} \sum_{\ell=1}^{\infty} \frac{(-1)^\ell}{\ell^{3/2}} \cos\left\{\frac{\pi\ell g m^*}{m}(1+\alpha)\right\} \times$$
$$\times \frac{1}{\beta}\sum_n \exp\left[-\frac{2\pi\ell}{\omega_H}(\omega_n + i\Sigma(i\omega_n))\right] \cos\left(\frac{2\pi\ell\mu}{\omega_H} - \frac{\pi}{4}\right) \tag{1.4}$$

The spin splitting factor contains an enhancement $S = 1 + \alpha$, which is associated with the static susceptibility, and that the amplitude of the oscillations is governed by second factor.

The self energy is written as the sum of two terms, the average potential due to all electrons with spin $-\sigma$, and a term which represents the repeated interaction between the electron and a hole with opposite spin. These terms are depicted diagramatically in figure 1. The terms in figure 1b can be summed up to yield

$$\frac{1}{\beta}\sum_{q\omega_n} \frac{U\Gamma(q,\omega_n)}{1 - U\Gamma(q,\omega_n)} G_{k-q}(z - i\omega_n)$$

which can be reinterpreted as a process whereby an electron emits a spin fluctuation and later reabsorbs it. The factor $\Gamma(q,\omega_n)$ is the non-interacting susceptibility defined by

$$\Gamma(q,\omega_n) = \frac{1}{\beta}\sum_{n'k} G_{k+q}(\omega_{n'}+\omega_n) G_k(\omega_{n'})$$

The self energy can then be written as

$$\Sigma'_\sigma(z) = \Sigma(z) + \frac{U\Gamma(0,0)}{1 - U\Gamma(0,0)}\sigma\mu_B H \tag{1.5}$$

The field dependent part is recognized as being proportional to the Hartree Fock static susceptibility. We shall analytically continue $\Sigma(z)$ on to the imaginary z axis. We find

$$i\Sigma(i\omega_n) = \frac{iU}{N}\sum_{k_z,\ell} f(k_z,\ell) +$$
$$+ \frac{3}{2}U^2 N(\mu)\int_0^\infty d\omega \int dq\, q\, \frac{Im}{\omega}\left\{\frac{\Gamma(q,\omega)}{1 - U\Gamma(q,\omega)}\right\} \times$$
$$\times \left\{\omega_n - \frac{2\pi}{\beta}\sum_{m=1}^{n} \frac{m^2}{m^2 + (\frac{\omega\beta}{2\pi})^2}\right\} \tag{1.6}$$

In this expression the factor

$$\frac{3}{2}U^2 N(\mu)\int dq\, q\, Im\left\{\frac{\Gamma(q,\omega)}{1 - U\Gamma(q,\omega)}\right\} \tag{1.7}$$

is interpreted as the product of an electron-spin fluctuation coupling strength times the spin fluctuation density states. The correspondence between the multiple electron-hole scattering and the emission of spin fluctuations is depicted in figure 2. The factor $\frac{3}{2}$ incorporates both the longitudinal and transverse spin fluctuations. We shall reabsorb the constant Hartree Fock term into the definition of the fermi energy μ . If we use the T = 0, small q and ω expansion of $\Gamma(q, \omega)$ we obtain

$$i\Sigma(i\omega_n) = \frac{9}{2}\omega_n \log_e\left\{1+\frac{S}{3}\right\} +$$

$$-\frac{3\pi}{64}\left\{\frac{\omega_n^2-\frac{\pi^2}{\beta^2}}{\mu}\right\}\left[S^{3/2}\, 3\tan^{-1}\sqrt{\frac{S}{3}} + \frac{S^2}{1+S/3}\right] \tag{1.8}$$

$$+ \dots\dots$$

where we have introduced a phenomenological cut off for $q = 2k_f$, and $S = [1-UN(\mu)]^{-1}$ is the Stoner Wohlfarth enhancement factor. The first term gives the logarithmic enhancement of the specific heat[9]. For higher temperatures $T \sim T_{sf}$, the same approximation yields

$$i\Sigma(i\omega_n) = \frac{9}{2}\frac{\pi}{\beta}\log_e\left(1+\frac{S}{3}\right) +$$

$$+\frac{\mu}{\pi}\left\{C+\log_e n+\frac{1}{2n}-\frac{1}{12n(n+1)}+\dots\right\}+\dots \tag{1.9}$$

in which C is Eulers constant.

In the next section we shall present the results for the oscillating magnetization and discuss their implications.

3. RESULTS AND DISCUSSION

The De Hass-Van Alphen oscillations in the magnetization are given by

$$M_{osc} = -\frac{m^*}{\pi^2}\left(\frac{2m_e\mu_B}{H}\right)^{\frac{1}{2}}\sum_{\ell}\frac{(-1)^\ell}{\ell^{1/2}}\cos\left[\frac{\pi\ell g m^* S}{2m_e}\right]$$

$$\times k_B T\sum_{n=0}^{\infty}\exp\left[-\left(\frac{2\pi\ell}{\omega_H}\right)(\omega_n+i\Sigma(i\omega_n))\right]\sin\left(\frac{2\pi\ell\mu}{\omega_H}-\frac{\pi}{4}\right) \tag{2.1}$$

In the limit $T \rightarrow 0$, this simplifies

$$M_{osc} = -\frac{m^*}{\pi^2}\left(\frac{2m\mu_B}{H}\right)^{\frac{1}{2}}\omega_H \times$$

$$\times\sum_{\ell}\frac{(-1)^\ell}{\ell^{3/2}}\cos\left(\frac{\pi\ell g m^* S}{2m}\right)\sin\left(\frac{2\pi\ell\mu}{\omega_H}-\frac{\pi}{4}\right)A_\ell \tag{2.2}$$

and the amplitude of the oscillations is given by

$$A_\ell = \frac{m^*}{m_{eff}}\left\{1+\frac{3}{128}\left(\frac{\omega_H}{\mu\ell}\right)\left[\sqrt{3}\, S^{\frac{3}{2}}\tan^{-1}\sqrt{\frac{S}{3}} + S^2/\left(1+\frac{S}{3}\right)\right]+\dots\dots\right\} \tag{2.3}$$

In this expression

$$m_{eff}/m^* = 1+\frac{9}{2}\log_e\left(1+\frac{S}{3}\right)$$

is the spin fluctuation enhancement that occurs in the specific heat. We see that although the amplitude A_1 is dominated by the specific heat effective mass there are other terms present. These terms can be important for systems with large susceptibility enhancements S. These terms should provide a discrepancy between the effective masses deduced from De Haas-Van Alphen measurements and specific heat measurements. These discrepancies should remain, even if one were to perform the specific heat measurements in the same field [11] as used in the De Hass-Van Alphen measurements. Such discrepancies have already been observed in transition metals.[12] Pd exhibits a 15% discrepancy between the effective masses, while in Pt and Nb the agreement is about 1%. The field dependent terms could provide an explanation of these discrepancies.

At higher temperatures $T > T_{sf}$, when the number of thermally activated spin fluctuations is of the order of K_BT, the electron spin fluctuation scattering increases, giving a concomittant decrease in the amplitude of the oscillations. We find that at these temperatures the amplitude of the oscillations is

$$A_\ell = \frac{2\pi\ell}{\beta\omega_H}\left(1+\frac{S}{3}\right)^{-\frac{9\pi^2\ell}{\beta\omega_H}}\exp\left[\frac{-2\pi^2\ell}{\beta\omega_H}\right] \tag{2.4}$$

This should be compared with the free particle formulae in which the mass in $h\omega_H$ is the spin fluctuation enhanced mass M_{eff}.

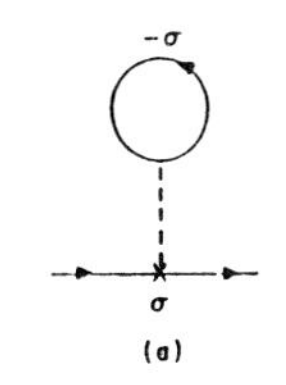

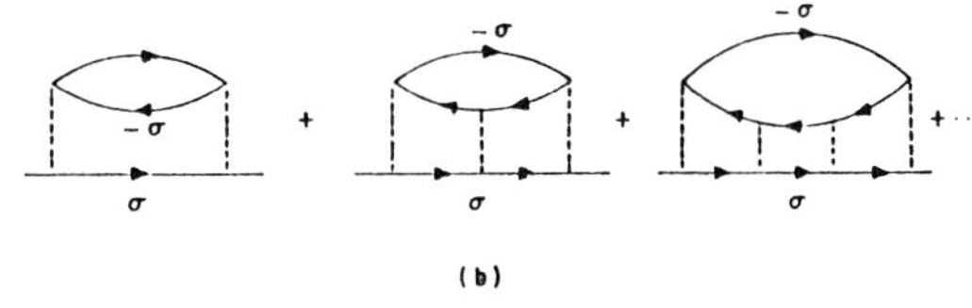

Figure 2

The electronic selfenergy.

Figure 1

Spinfluctuations in terms of multiple scattering.

In the conclusion, we have shown that the spin fluctuations are expected to renormalize the De Haas-Van Alphen oscillations. At low temperatures, systems with large susceptibility enhancements , S , should show discrepancies between effective masses deduced from specific heat and De Haas-Van Alphen oscillations. Such discrepancies have been observed in transition metals and should also be present in some mixed valent compounds

REFERENCES

1a. W.R. Johanson, G.W. Crabtree, D.D. Koeling A.S. Edelstein and O.D. McMasters, J. App. Phys. (1980)

1b. G.W. Crabtree, W.R. Johanson, A.S. Edelstein and O.D. McMasters, Valence Fluctuations in Solids, L.M. Falicov, W. Hanke and M.P.Maple editors, North Holland, New York 1981

2. E. Umlauf, P. Sutsch and E. Hess, Crystalline Electric Field and Structural Effects in F Electron Systems, J.E.Crow, R.P. Guertin and T.W. Mihalasin editors, Plenum New York 1980.

3. J.M. Lawrence, Phys. Rev. B 9, 3770 (1979).

4. M. Lowenhaupt and E. Holland-Moritz, J.App. Phys. 50, 7456 (1979).

5. H.J. Leder and B. Muhlschlegel,Z. Phys B 29, 341 (1978).

6. P.S. Riseborough and D.L. Mills, Valence Fluctuations in Solids, L.M. Falicov, W. Hanke and M.P. Maple, North Holland Publishing Co., New York 1981.

7. B. Stalinski, Z. Kletowski and Z. Henkie, Phys. Stat. Solidi. A 19 , K 165 (1973).

8a. S. Engelsberg and G. Simpson, Phys. Rev. B 2, 1657 (1970).

8b. M. Fowler and R. Prange, Physics 1, 315 (1965).

9a. S. Doniach and S. Engelsberg, Phys.Rev.Lett., 17, 750 (1966).

9b. N.F. Berk and J.R. Schrieffer, Phys. Rev. Lett. 17, 433 (1966).

10. Izuyama D.J. Kim and R. Kubo, J. Phys.Soc., Japan 18, 1025 (1963).

11. W.F. Brinkman and S. Engelsberg, Phys. Rev. B 169, 417 (1968).

12a. G.W. Crabtree, D.H. Dye, S.A. Campbell,J.B. Ketterson, J.J. Vuillemin and N.B.Sandesara, Inst. Phys. Conf. Ser. No. 55 (1981).

12b. D.H. Dye, S.A. Campbell, G.W. Crabtree,J.B. Ketterson, N.B. Sandesara and J.J.Vuillemin, Phys. Rev. b 23, 462 (1981).

Valence Instabilities
P. Wachter and H. Boppart (eds.)

UPPER CRITICAL MAGNETIC FIELD OF THE EXOTIC SUPERCONDUCTOR $CeCu_2Si_2$

J.Aarts[§], C.D.Bredl[+], W.Lieke[+], U.Rauchschwalbe[+], F.Steglich[+], B.Batlogg[*] and D.J.Bishop[*]

[§]Universiteit van Amsterdam, NL-1018 XE Amsterdam, The Netherlands

[+]Technische Hochschule Darmstadt, SFB 65, D-6100 Darmstadt, F.R.G.

[*]Bell Laboratories, Murray Hill, N.J.07974, U.S.A.

Previous results on both the normal and superconducting states of nearly trivalent Ce Cu_2Si_2 are reviewed and discussed along with new results of the upper critical field, $B_{c2}(T)$. We conclude that $CeCu_2Si_2$ behaves as a "heavy-fermion superconductor", but does not show "triplet pairing". Key parameters of the low-temperature system of "heavy fermions" are derived.

UNUSUAL NORMAL-STATE BEHAVIOR

In the tetragonal intermetallic compound $CeCu_2Si_2$ the Ce ions are nearly trivalent [1], presumably not far from a valence instability [2]. Above the "spin-fluctuation temperature" $T^* \cong 10K$ the Ce ions in $CeCu_2Si_2$ exhibit a magnetic moment which is dominated by the crystal-field (CF) splitting of the $J=5/2$ state of Ce^{3+} [3]. The CF-level scheme deduced from the magnetic neutron-scattering experiments consists of three Kramers doublets with excitation energies corresponding to 140K and 364K. The magnetic relaxation rate of the Ce ions, as given by the half-width of the quasielastic neutron line, is very similar to that of archetypical "Kondo lattices" like $CeAl_2$ and CeB_6 [4], but differs strongly from the relaxation rate in typical mixed-valence compounds like $CePd_3$ and $CeSn_3$ [5].

When $CeCu_2Si_2$ is cooled below its "spin-fluctuation temperature", a new type of collective state is entered: a) Local magnetic degeneracies become removed, i.e. the effective Ce moment disappears, as is shown both by a pronounced specific-heat anomaly at $T_{max} \cong T^*/3$, the entropy change connected with it being $\cong k_B \ln 2$ per Ce ion [6], and by a transition of the susceptibility from a high-temperature Curie-Weiss behavior to a temperature-independent ($T \lesssim 6K$) "Pauli-like" susceptibility χ_o [7].
b) In addition to the latter, which is strongly enhanced compared with χ_o of simple metals, the low-temperature ($T \lesssim 1K$) variation of the specific heat is found to approach a γT behavior with an extremely large coefficient $\gamma (\cong 1 Jmole^{-1}K^{-2})$ [6], and the resistivity is found to follow an AT^2 dependence, again with a giant coefficient [8].

All these observations show that $CeCu_2Si_2$ behaves nonmagnetically below a few K and exhibits the characteristic features of a strongly interacting Fermi system [9]. The existence of "heavy fermions" appears to be related to the existence of the 4f electrons at the Ce sites, since the d-band reference compound $LaCu_2Si_2$, lacking 4f electrons, does not show such "heavy-fermion" properties at low temperature (e.g.: the specific-heat coefficient $\gamma \cong 4mJ mole^{-1}K^{-2}$ in this case [10]). Similar simple-power laws in the low-temperature variation of the aforementioned quantities, with even larger coefficients, are already known for $CeAl_3$ [11]. However, while $CeAl_3$ remains in the enhanced Pauli paramagnetic state down to ultralow temperatures, it was surprising to find that $CeCu_2Si_2$ assumes a superconducting state below $T_c \cong 0.6$ K [6].

SUPERCONDUCTING PROPERTIES: MEISSNER EFFECT, SPECIFIC HEAT AND UPPER CRITICAL FIELD

While for previous samples spurious reflections were observed in the X-ray patterns [6], it was recently demonstrated that by annealing at 1100°C $CeCu_2Si_2$ can be obtained in the proper $ThCr_2Si_2$ structure [7]. Such single-phase samples were found to show the highest values of the static Meissner signal [7], as measured by DC magnetization, namely by cooling the samples in a fixed external B-field through T_c. Up to 60% of the volume of powdered samples was observed to exhibit the Meissner

effect. This has to be considered as a lower bound of the superconducting volume fraction [7].
A large superconducting volume fraction was also concluded [6] from the size of the reduced specific-heat jump at T_c, $\Delta C/\gamma T_c$, which, from sample to sample, ranges from 0.8 to 1.4. The latter number is close to the BCS value $\Delta C/\gamma T_c$=1.43.

It has been argued [6] that the observation of a jump height ΔC comparable to the gigantic normal-state specific heat γT_c can only be understood by assuming that the Cooper pairs in $CeCu_2Si_2$ are formed by the "heavy fermions". In fact, the isostructural intermetallic $LaCu_2Si_2$, which does not show those "heavy-fermion" effects, also does not become superconducting [6].

In order to check this conclusion from specific heat, we have recently measured the upper critical field $B_{c2}(T)$ [12]. We expected B_{c2} to be very high since the pair-breaking effect of an external magnetic field on the orbital motion of the quasiparticles forming the Cooper pairs should be rather weak in the case of a small Fermi velocity. In fact, the absolute slope of $B_{c2}(T)$ near T_c, B'_{c2}, is proportional to $\gamma^2 T_c$ in the "pure limit" (quasi-particle-mean-free path $\ell \gg$ BCS coherence length ξ_0), while an additional term to B'_{c2}, $\sim \gamma \ell^{-1}$, enters in the "dirty limit" ($\ell \ll \xi_0$) [13].

In fig. 1 are shown the $B_{c2}(T)$ results of three different $CeCu_2Si_2$ samples, of which two (#7, #4) have been subject to a recent investigation [12]: Sample #7 was a rather pure one, annealed at 1100°C [7] and showing a residual ($T\to 0$) resistivity in the normal state, ρ_0, as small as 3.5μΩcm [8]. The upper critical field for this sample is ≅ 1.7 T (as $T\to 0$), which is a very high value considering the low transition temperature T_c=0.69 K [7]. The initial slope, B'_{c2} = 5.8 T/K, is comparable to the highest values found so far for the Chevrel-phase superconductors ($B'_{c2} \lesssim 7.5$ T/K; $T_c \lesssim 15$K [14]). This demonstrates the dominance of the giant γ on the slope of $B_{c2}(T)$ in $CeCu_2Si_2$. In addition, on going to the less pure sample #4 (annealed at only 900°C and showing a residual resistivity $\rho_0 \lesssim 40$μΩ, i.e. a mean free path (mfp) about one order of magnitude smaller than that

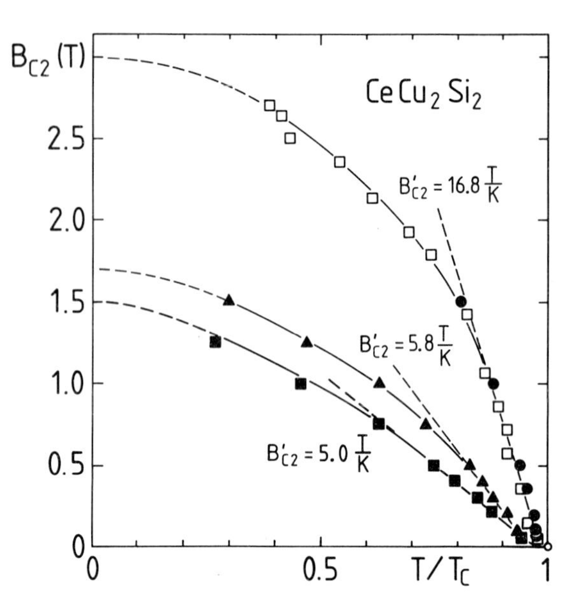

Fig. 1. Upper critical field as function of reduced temperature for $CeCu_2Si_2$. Data were obtained from AC susceptibility [□, #4, T_c=0.66K; ▲, #7, T_c=0.69K; ■, #10, T_c=0.45K] and specific heat [●, #4, T_c=0.56K].

of sample #7) the expected increase is found in $B_{c2}(T)$: $B_{c2}(0)$ amounts to about 3.0 Tesla and B'_{c2} to 16.8 T/K, an extremely large value. $B_{c2}(T)$ slopes of similar magnitude were also reported for $CeCu_2Si_2$ at this conference by Ishikawa et al. [15].

In addition to the data of samples #4 and #7 discussed before, we present in fig. 1 $B_{c2}(T)$ results of a new sample (#10), for which neither the resistivity nor the specific heat has been measured so far. Preliminary susceptibility results, however, have revealed a low-temperature contribution of "magnetic impurities" smaller by about 25% compared with what was found for sample #7 [see ref. 7]. As for this sample, we could not find any spurious reflections in the X-ray pattern of the new one. In accordance with these observations, the $B_{c2}(T)$ data points lie slightly below those of sample #7, hinting at a slightly larger ratio ℓ/ξ_0 in sample #10.

PARAMETERS OF THE "HEAVY-FERMION SUPERCONDUCTOR"

A semi-quantitative analysis can be performed [12], using the following experimental input data:
1) B'_{c2} [$=\alpha T_c v_F^{-2} + \beta (v_F \ell)^{-1}$] and T_c, along with

low-temperature results, in the normal state, for 2) the resistivity $\rho[\sim(k_F \ell)^{-1}]$ and the specific-heat coefficient $\gamma[\sim k_F^2 v_F^{-1}]$. Assuming a spherical Fermi surface, one can estimate key parameters for both the system of "heavy fermions" and the novel superconducting state. For sample #7 we found [12] for the Fermi wave number $k_F \cong 1.7 Å^{-1}$, which is of the same order as the Fermi wave number of the system of (non-renormalized) conduction electrons as estimated from the high-temperature resistivity [16]. Next, the effective mass was found to be two orders of magnitude larger ($m^* = 220 m_o$), and both the Fermi velocity ($v_F = 8.7 \cdot 10^5 cm\, s^{-1}$) and Fermi temperature ($T_F = 540K$) two orders of magnitude smaller, than the corresponding free-electron values. It is interesting to note that the Fermi temperature T_F is considerably larger than the "spin-fluctuation temperature", $T^* \cong 10$ K [3]. The mfp was found to be $\ell \cong 120Å$, i.e. not too much smaller than the BCS coherence length, $\xi_o \cong 190Å$. The London penetration depth is unusually large, because of the large m^*: $\lambda_L(0) \cong 2000$ Å. Finally, the Ginzburg-Landau parameter was estimated to be $\kappa = 22$ for sample #7 and $\kappa \cong 10$ in the "pure limit". This clearly proves $CeCu_2Si_2$ to be a type II superconductor. In fact, the lower critical field was experimentally found to be $B_{c1}(0) \cong 2.3$ mT [12], i.e. much smaller than the upper critical field, $B_{c2}(0) \cong 1.7T$.

The consistency of the above analysis could be checked by using measured B_{c2} and T_c data along with the estimated κ value and calculating certain quantities which can be either directly measured (specific-heat-jump height, ΔC) or calculated from the results of independent measurements (thermodynamic critical field, $B_{cth}(0)$). Surprisingly close agreement has been obtained in both cases [12].

The measurements of $B_{c2}(T)$ have, therefore, confirmed that superconductivity in $CeCu_2Si_2$ is created by the system of "heavy fermions". However, while this material appears as the first example of a "heavy-fermion superconductor", it is very likely not a "p-wave superconductor", for the latter should react sensitively to mfp reduction [17], whereas an order-of-magnitude change in the mfp was found to leave essential superconducting parameters (T_c, ΔC) almost unchanged.

OUTLOOK

In order to get a better understanding of the two unresolved problems, which are proably connected intimately with each other, i.e. the formation of the "heavy fermions" and the attractive fermion-fermion interaction causing the novel superconducting state, new experiments have been initiated. For example, we have found that the application of an external pressure of 11kbar to $CeCu_2Si_2$ results in an increase of T_c by ≅6%. Furthermore, a study of the influence of both "internal" pressure and paramagnetic impurities on the superconducting properties of $CeCu_2Si_2$ is presently underway.

REFERENCES

[1] R.Lässer, F.C.Fuggle, M.Beyss, M.Campagna, F.Steglich and F.Hulliger, Physica 102B, 360 (1980); R.D.Parks, V.Murgai, B.Reihl, N.Mårtensson and F.Steglich, to be published.

[2] E.Umlauf and E.Hess, Physica 108B, 1347 (1981).

[3] S.Horn, M.Loewenhaupt, E.Holland-Moritz, F.Steglich, H.Scheuer, A.Benoit and J.Flouquet, Phys.Rev.B 23, 3771 (1981).

[4] S.Horn, F.Steglich, M.Loewenhaupt and E.Holland-Moritz, Physica 107B, 103 (1981).

[5] M.Loewenhaupt and E.Holland-Moritz, J.Appl. Phys. 50, 7456 (1979).

[6] F.Steglich, J.Aarts, C.D.Bredl, W.Lieke, D. Meschede, W.Franz and H.Schäfer, Phys.Rev. Lett. 43, 1892 (1979).

[7] W.Lieke, U.Rauchschwalbe, C.D.Bredl, F.Steglich, J.Aarts und F.R.de Boer, J.Appl.Phys. 53, 2111 (1982).

[8] F.Steglich, K.H.Wienand, S.Horn, W.Klämke and W.Lieke, Proc.Int.Conf. on "Crystalline Electric Fields and Structural Effects in f-Electron Systems", Wrocław, 1981, forthcoming.

[9] L.D.Landau, Sov. Phys. JETP 3, 920 (1956).

[10] D.Meschede, Diploma Thesis, University of Cologne, 1979 (unpublished).

[11] K.Andres, J.E.Graebner and H.R.Ott, Phys. Rev.Lett. 35, 1779 (1975).

[12] U.Rauchschwalbe, W.Lieke, C.D.Bredl, F.Steglich, J.Aarts, K.M.Martini and A.C.Mota, to be published.

[13] T.P.Orlando, E.J.McNiff, Jr., S.Foner and M.R.Beasley, Phys.Rev.B 19, 4545 (1979).

[14] Ø.Fisher, Appl.Phys. 16, 1 (1978).

[15] M.Ishikawa, D.Jaccard and J.L.Jorda, this conference.

[16] W.Franz, A.Griessel, F.Steglich and D. Wohlleben, Z.Phys.B 31, 7 (1978).

[17] R.Balian and N.R.Werthamer, Phys.Rev. 131, 1553 (1963).

Valence Instabilities
P. Wachter and H. Boppart (eds.)

KONDO-TYPE RESISTIVITY AND MAGNETORESISTANCE OF MIXED-VALENT THULIUM IMPURITIES IN DILUTE (Y,Tm)Se ALLOYS

P. Haen, O. Laborde, F. Lapierre and J.M. Mignot
CRTBT-CNRS, BP 166 X, 38042 Grenoble-Cedex, France

and

F. Holtzberg and T. Penney
IBM T.J. Watson Research Center, Yorktown Heights N.Y. 10598, USA

In this paper, we report systematic electrical-resistivity measurements on dilute $Y_{1-x}Tm_xSe$ alloys (0.01 ⩽ x ⩽ 0.2) at very low temperatures (T ⩾ 20 mK) and in high magnetic fields (H ⩽ 76 kOe). In this system, the valence of thulium remains strongly intermediate (v ≅ 2.6–2.7) for all concentrations. Strong evidence is found for a Kondo-like scattering from the thulium impurities, and a rather low characteristic temperature ($T_K \lesssim 0.3$ K) is estimated. The results are compared to those for example Kondo alloys with 3d or 4f impurities. The possibility for Kondo effect and valence fluctuations to coexist in the present system is probably a consequence of the unique magnetic ground state of MV thulium ions.

1. INTRODUCTION

Valence fluctuations were originally discovered in periodic rare-earth (RE) materials. However, typical intermediate-valence (IV) properties have now been reported for a number of dilute systems in which a large fraction of the RE atoms is substituted by various transition-metal atoms (1). Among all ambivalent RE elements, thulium is unique because of the property that both Tm^{2+} and Tm^{3+} configurations are magnetic. This is generally regarded as the reason why the homogeneously mixed-valent compound TmSe (v ≅ 2.6–2.7) exibits a Curie-Weiss-like susceptibility over a wide range of temperature and orders in an AF, fcc type I, structure below T_N ≅ 3.5 K. In a previous work (2), an IV state was reported for dilute Tm impurities in $Y_{1-x}Tm_xSe$ alloys down to 1 % thulium concentration. In this system, both YSe and TmSe have almost identical lattice parameters (a_o=5.71 Å), so that undesired chemical-pressure effects are minimized. In the solid solutions, the same lattice parameter is found for the whole range of concentrations (figure 1). Since there is a large difference in the atomic volumes of divalent and trivalent RE atoms, the observed constancy of the lattice parameter indicates that the Y substitution does not introduce any substantial change in the Tm valence. For the more dilute compounds, the lattice parameter is no longer a sensitive measure of the valence but the molar Curie constant clearly remains intermediate between the Tm^{2+} and Tm^{3+} free-ion values for all concentrations (x ⩾ 0.01). Recent X-ray-absorption (3) and photoelectron-spectroscopy (4) experiments also confirm this conclusion.

At low temperature, the magnetic susceptibility of $Y_{1-x}Tm_xSe$ alloys shows a strong Curie-like temperature dependence, indicating that, as in pure TmSe, the ground state of thulium is magnetic. Recently, low-frequency ac measurements on dilute alloys (x ⩽ 0.2) (5) below 1 K revealed sharp susceptibility peaks which were ascribed to a spin-glass freezing of the thulium moments. The same alloys exhibit a resistivity minimum around 10-15 K, followed by a region with a negative temperature coefficient of resistivity (2,6). This is indeed a typical Kondo-like behavior.

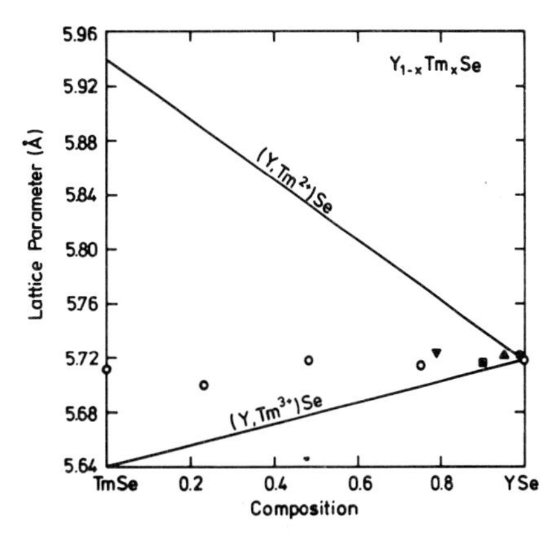

Figure 1 : Lattice constant *vs.* composition for the $Y_{1-x}Tm_xSe$ system. The TmSe value shown corresponds to a nearly stoichiometric sample.

In order to better characterize this anomalous scattering from the thulium impurities we have performed systematic electrical-resistivity measurements down to very low temperatures (T ⩾ 20 mK) and in high magnetic fields (H ⩽ 76 kOe).

2. EXPERIMENTS

The systems $Y_{1-x}Tm_xSe$ all condense in the rock-salt structure. Single crystals with concentrations x = 0.01, 0.052, 0.1 and 0.2 were prepared and their lattice parameters were determined by X-ray diffraction (solid symbols in figure 1). The samples were cleaved parallel to the (100) planes to the shape of small, nearly square platelets, 1 to 2 mm in side and ∿ 0.5 mm thick. Indium contacts were made to the four corners to allow measurements by the Van der Pauw method. Down to liquid-helium temperature, both dc and ac measurements were performed, and the results were found to agree within experimental accuracy. At lower temperatures, the ac technique was adopted, as the best compromise between maximum sensitivity and lowest power dissipation in the specimen. The experimental uncertainty on the measured resistance was generally less than 1 % and never exceeded 3 % even at the lowest temperatures. On the other hand, the absolute value of the resistivity is not defined to better than 10 % due to imprecision on the sample thickness.

The magnetoresistivity was measured up to 76 kOe in a superconducting solenoid with the (100) plane of the crystal perpendicular to the field (the Van der Pauw geometry allows only $\rho(H_\perp)$ measurements). Between 1.7 and 4.2 K, the sample was immersed directly in a regulated helium bath. Above 4.2 K it was placed inside a thermally insulated bulb filled with ^{4}He gas. The temperature of this bulb was first measured in zero field with a carbon resistor and then monitored by the helium pressure acting on a capacitive sensor at room temperature (7). This setup allows a temperature stability better than 10^{-4} during the application of the field.

3. RESULTS

3.1 Resistivity in zero field

The resistivities of the four samples and of pure YSe between room temperature and liquid-helium temperature are shown in figure 2. The irregular concentration dependence of the resistivity both at room temperature and at 4.2 K is attributed to imprecision in the absolute value of ρ (see § 2) and in the total concentration of magnetic impurities (most serious for x=0.01). Whereas the resistivity of YSe exhibits a normal metallic temperature dependence and saturates at a residual value of 10 μΩcm at low temperatures*, the Kondo-like resistivity minimum near 10-15 K is clearly seen in the alloys. In figure 3, we have plotted our low-temperature data in a logarithmic temperature scale between 0.025 and 100 K. We emphasize that the matrix term is essentially constant up to 12 K, so that the low-temperature part of the curves reflects the thermal variations of the impurity resistivity only. The linear dependence of ρ *vs.* lnT is unambiguous, especially for x = 0.01 where it extends over two decades in temperature, from T = 0.05 to 5 K. At higher concentrations, the range of validity of the lnT law is reduced by negative deviations occurring at low temperatures (∿ 1 decade for x = 0.10). The normalized slope $\frac{1}{x}|d\rho/d\ln T|$ is 0.25 μΩcm/% Tm in the most dilute sample x = 0.01, then it seems to gradually decrease in the more concentrated ones (see table I), although this may be an artefact of the reduced temperature range in which the lnT law is valid. It should be noted that whereas the present values are comparable to those

* YSe actually undergoes a superconducting transition below 4 K, in agreement with previous observations (8).

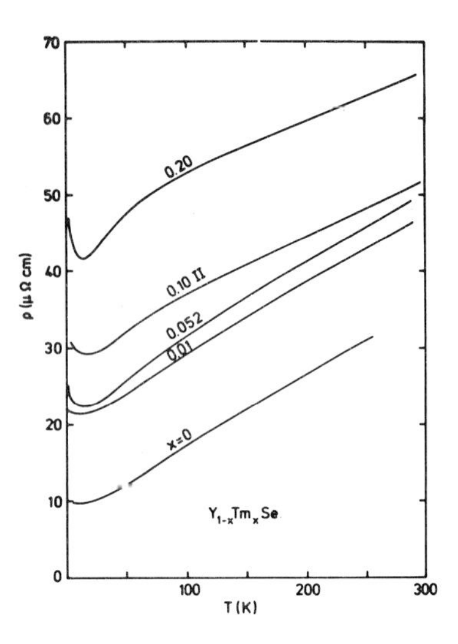

Figure 2 : Resistivity *vs.* temperature for the $Y_{1-x}Tm_xSe$ system.

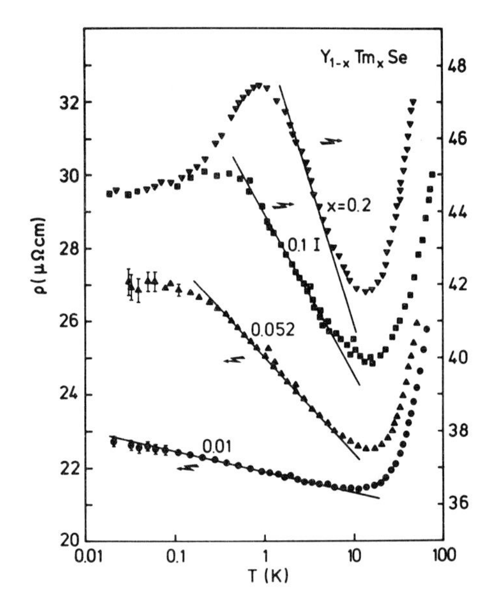

Figure 3 : Very-low-temperature resistivity *vs.* lnT of $Y_{1-x}Tm_xSe$.

Table I

x	0.01	0.052	0.10	0.20
a_o (Å)	5.72(2)	5.719	5.716	5.722
$-d\rho/d\ln T$ (μΩcm)	0.25	1.16	1.92	3.5
T_{max} (mK)	—	—	32±5	90±5
T_o (mK)	27	85	180	380

obtained for Kondo alloys with cerium impurities (≅ 0.15 μΩcm/% Ce in the same matrix), they are about 18 times smaller than in pure TmSe above T_N (≅ 4.5 μΩcm/% Tm). This result clearly indicates that cooperative effects are essential in the latter case. On the other hand, the lnT slope in trivalent TmS (67 μΩcm (9)) is comparable to that extrapolated from the dilute limit (≅ 0.4 μΩcm/% Tm (10)).

The low-temperature deviations from the lnT slope are strongly concentration-dependent : For x = 0.01 and 0.052, the resistivity levels off below 0.1 K. A weak maximum occurs near T_{max} = 0.3 K for x = 0.1, and it shifts to T_{max} = 0.9 K while becoming much more pronounced for x = 0.2. This behavior reflects the onset of a presumably mictomagnetic ordering. For the last two samples the temperature T_o of the ac susceptibility maximum (5) is substantially smaller than T_{max} (table I), as it is generally observed in spin glasses (11). A gradual change from a maximum to a plateau with decreasing impurity concentration occurs in the resistivity of many Kondo alloys with transition metal impurities. In those systems, the observation of a maximum at a given concentration generally indicates that $T_K < T_{max}$. Applying the same criterion to the -admittedly much more concentrated- $Y_{1-x}Tm_xSe$ alloys yields $T_K < 0.3$ K. It is therefore not surprising that the usual plateau corresponding to the strong-coupling limit of the Kondo effect for a single impurity is not observed in our temperature range.

3.2 Magnetoresistance

In order to confirm our interpretation of the low-temperature resistivity anomaly in dilute (Y,Tm)Se, we have investigated the magnetoresistance of these alloys in high magnetic fields. In the case of the Kondo effect, the resonant scattering of the conduction electrons is gradually quenched in an external field, leading to a negative magnetoresistance. The physical origin of this effect is twofold (12) : In low magnetic fields ($g\mu_BH < k_BT$) and for $T > T_K$, the magnetoresistance is associated with the freezing out of spin-flip scattering due to the aligning of the impurity spins by the magnetic field. The variation of the conduction-electron scattering amplitude -which determines the T-dependence of ρ in zero field- then becomes important when the first effect reaches saturation.

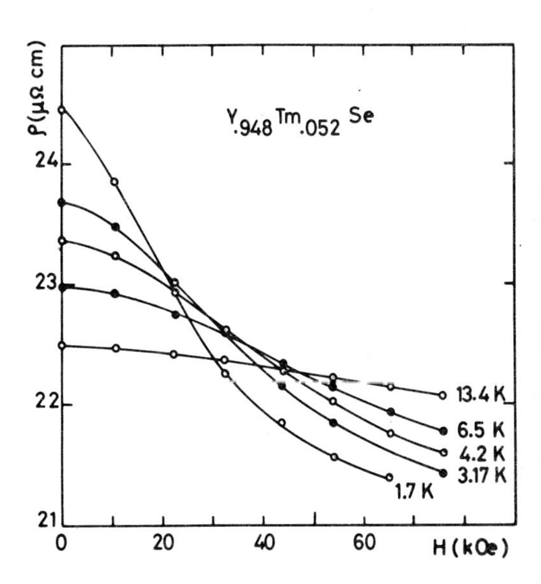

Figure 4 : Magnetoresistance *vs.* magnetic field of $Y_{0.948}Tm_{0.052}Se$ for different temperatures.

Figure 4 shows the field dependence of ρ at different temperatures for the sample containing 5 % thulium. A negative magnetoresistance is observed at all temperatures, but even at 1.7 K, a saturation is not reached in 76 kOe. The main effect is a competition between the magnetic field and thermal fluctuations. Indeed, $\Delta\rho(H,T)$ scales with H/T for all temperatures T > 1.7 K, as it can be expected if $T \gg T_K$. Another interesting feature is the crossing of the various curves leading to a positive temperature coefficient of ρ in high fields. Furthermore, in intermediate fields, ρ varies non-monotonically with temperature. This is better seen on figures 5 and 6 which show ρ-T plots at fixed

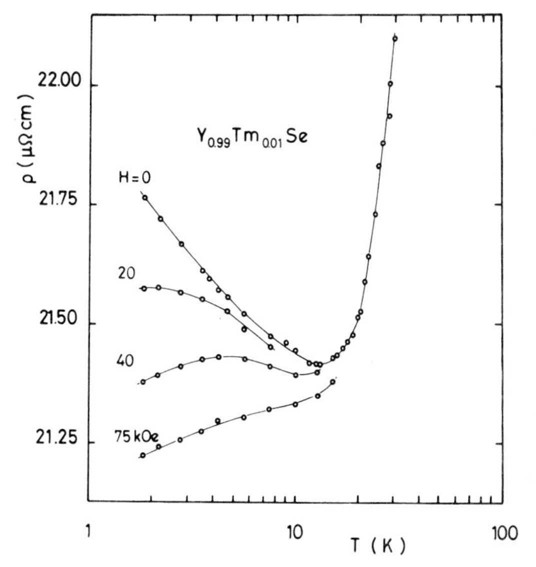

Figure 5 : Resistivity *vs.* lnT of $Y_{0.99}Tm_{0.01}Se$ in different magnetic fields.

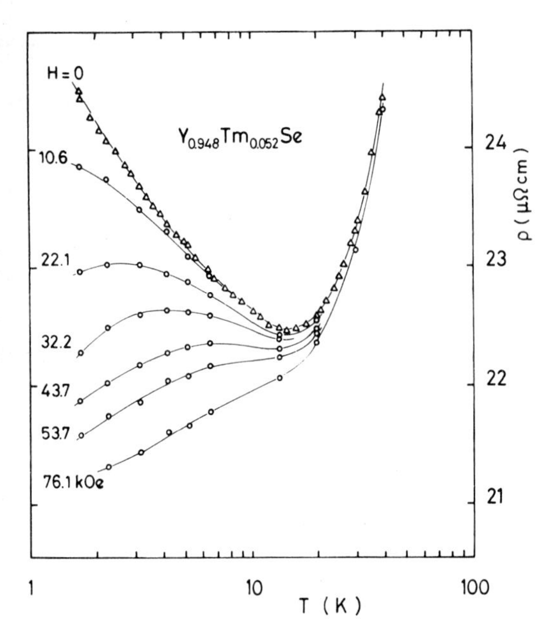

Figure 6 : Resistivity *vs* lnT of $Y_{0.948}Tm_{0.052}Se$ in different magnetic fields.

H for two samples x = 0.01 and 0.052 : A broad maximum appears for H > 20 kOe and shifts gradually to higher temperatures with increasing magnetic field. The same effect exists for x = 0.1 and 0.2. This type of behavior has been observed in a number of Kondo alloys with 3d (<u>Cu</u>Mn (13)) or 4f ((<u>La</u>Ce)Al_2 (14),(<u>La</u>Ce)B_6 (15)) impurities, and basically reflects the more rapid aligning of the impurity spin in the external field at lower temperatures. Interestingly enough, it was noted by Samwer and Winzer (15) that a maximum is observed experimentally only for $H > H_K = (k_B T_K / g\mu_B)$. Assuming $g \cong 3.8$ as proposed recently for pure TmSe (16), and $T_K \lesssim 0.3$ K, a maximum should exist down to $H \cong 2$ kOe at temperatures as low as 0.1 K, provided that x is chosen small enough for interaction effects to be negligible.

4. CONCLUSION

To summarize, the present work establishes the first observation of Kondo effect due to dilute thulium impurities. The resulting transport properties are quite comparable to those of classical 3d or 4f systems with similar Kondo temperatures ($T_K < 0.3$ K). In the case of 4f impurities however, Kondo effect was always observed so far in alloys where the valence of the rare-earth (generally Ce or Yb) is very nearly integral, whereas the present system corresponds to a strongly mixed valent situation. This difference probably reflects the unique magnetic nature of the low-temperature thulium ground state. Whereas some simple models (17) can account qualitatively for the latter property, the problem of treating the residual coupling of this local degree of freedom with the conduction band is presently unsolved.

<u>Acknowledgements</u> : The authors are indebted to P.G. Lockwood and L. Nesterovskaya for sample preparation and X-ray data, and to J. Souletie and K. Matho for extensive discussions.

REFERENCES

(1) For references, see : Wilkins, J.W., Falicov, L.M., Hanke, W. and Maple, M.P. (eds.), in Valence fluctuations in solids, (North-Holland, Amsterdam, 1981), p. 459.
(2) Holtzberg, F., Penney, T. and Tournier, R., J. Physique Colloq. 40 (1979) C5-314.
(3) Launois, H., Rawiso, M., and Mignot, J.M., unpublished.
(4) Mårtenson, N., Reihl, B., Pollak, R.A., Holtzberg, F., Kaindl, G. and Eastman, D.E., to be published.
(5) Genicon, J.L., Haen, P., Holtzberg, F., Lapierre, F. and Mignot, J.M., Physica 108B (1981) 1355.
(6) Berger, A., Haen, P., Holtzberg, F., Lapierre, F., Mignot, J.M., Penney, T., Peña, O. and Tournier, R., J. Physique Colloq. 40 (1979) C5-364.
(7) Zotos, X., Thesis, University of Grenoble (1979), unpublished.
(8) Hulliger, F. and Hull, G.W., Jr., Solid State Commun. 8 (1970) 1379.
(9) Lapierre, F., Haen, P., Coqblin, B., Ribault, M. and Holtzberg, F., Physica 108B (1981) 1351.
(10) Haen, P., Lapierre, F., Mignot, J.M., Flouquet, J., Holtzberg, F. and Penney, T., ICM 82 Conference, to be published.
(11) Rammal, R. and Souletie, J., in Cyrot, M., (ed), Magnetism in metals and alloys (North-Holland, Amsterdam, 1982) p. 379
(12) Béal-Monod, M.-T. and Weiner, R.A., Phys. Rev. 170 (1968) 552.
(13) Monod, P., Phys. Rev. Lett. 19 (1967) 1113.
(14) Felsch, W. and Winzer, K., Solid State Commun. 13 (1973) 569.
(15) Samwer, K. and Winzer, K., Z. Phys. B 25 (1976) 269.
(16) Loewenhaupt, M. and Bjerrum Møller, H., Physica 108B (1981) 1349.
(17) Aligia, A.A., Mazzaferro, J., Balseiro, C.A. and Alascio, B., this conference.

Valence Instabilities
P. Wachter and H. Boppart (eds.)

ANOMALOUS PRESSURE DEPENDENCE OF THE SUPERCONDUCTING T_C IN YPr ALLOYS: NEW EVIDENCE FOR THE VALENCE INSTABILITY OF PRASEODYMIUM

J. Wittig

Institut für Festkörperforschung, KFA Jülich, D-5170 Jülich, W. Germany

The depression of the superconducting transition temperature T_C of yttrium by substitutional praseodymium impurities shows a pronounced maximum as a function of pressure at $\simeq$260 kbar. At the maximum, the specific depression $\Delta T_C/x$ amounts to 17 K/at/o vs. 13 K/at/o in the analogous LaPr system. We suggest that the localized $4f^2$ shell of single Pr impurity atoms starts to delocalize under such high pressure and begins to demagnetize into a nonmagnetic $4f^2$ virtual bound state (to be attained at pressures beyond the present limit). Alternatively, a pressure-induced change of the 4f shell configuration, $4f^2$ to $4f^1$, via a mixed valence regime, may take place.

1. INTRODUCTION

It was recently discovered that praseodymium undergoes a valence transition under high pressure which may prove to be the counterpart of the cerium γ/α transition. Heretofore the following three pieces of evidence existed. A phase transition occurs in pure Pr metal at 212 kbar (at 300 K) which is accompanied by (i) a volume collapse of 19% [1] and by (ii) strong resistance anomalies resembling those at the γ/α transition in Ce [2,3]. It was also found that (iii) the depression of the superconducting transition temperature of La by substitutional Pr impurities, $\Delta T_C(P) = T_{C,La} - T_{C,LaPr}$, shows a pronounced maximum at 230 kbar [4]. The specific depression $\Delta T_C(P)/x$ (with x being the Pr concentration) amounts to $\simeq$ 0.5, 13 and 7 K/at/o at pressures of 120, 230 and 270 kbar, respectively. The impetus for that investigation came from the classical experiment by Maple et al. [5] on the LaCe system which revealed a maximum in ΔT_C under pressure and demonstrated that the valence transition of bulk Ce metal (regardless of its precise nature) has a counterpart in this dilute Ce system. In an effort to investigate the valence instability of Pr further we have now studied the pressure dependence of T_C for YPr alloys up to approximately 300 kbar. Yttrrium was chosen as the matrix metal since it is a pressure-induced d-band superconductor with T_C as high as 9 K (see below) in contrast to La which has an additional non-negligible 4f character in the author's opinion [6]. By employing such a clearcut d-band host metal it should be possible to investigate the interesting question of whether the presence of a 4f occupation on the surrounding matrix atoms (as in LaPr) plays a decisive role in triggering the valence instability of the single Pr atom, or not. Y appeared to us also particularly suitable because it is trivalent like Pr and both element's metallic radii are quite close. It is thus hoped that Pr is dissolvable in Y as a single impurity for the present concentrations up to 1 at/o. Experimental details are deferred to another publication.

2. RESULTS

Figure 1 shows the pressure dependence of T_C for pure Y (upper curve) and for a YPr alloy containing 0.99 at/o Pr (lower curve). Y is known to undergo a rather sluggish phase transition, Y I$\rightarrow$II, between $\simeq$130 and $\simeq$160 kbar [7]. The sluggishness of the transition to Y II is reflected in the rather gradual decrease of T_C from 2.7 K at 137 kbar to 2.0 K at 167 kbar. Within this 30 kbar - pressure range the state of the sample changes from being predominantly Y I to exclusively Y II. Intermediate T_C's can perhaps be explained by the superconducting proximity effect. This would require that both phases are in an intimate contact with each other on a fairly small scale. The quantity $\Delta T_C = T_{C,Y} - T_{C,YPr}$ equals to 0.9 K at 167 kbar in phase II just above the transition. With further pressure, the two $T_C(P)$ curves diverge. Quantitatively ΔT_C increases by a factor of 4 to 3.6 K at 211 kbar. In marked contrast, T_C amounts to only a few tenths of a degree in the pure phase Y I for $P < 120$ kbar.

Fig. 2(a) shows T_C vs. pressure for pure Y and a more dilute YPr alloy containing 0.35 at/o Pr. Here the pressure range extends from 180 to almost 300 kbar. The strong upward trend of T_C with P for Y continues. With T_C=9 K, Y becomes eventually an equally "good" superconductor as niobium [8]. The $T_C(P)$ dependence of the alloy (triangles) passes through a maximum which is followed by a minimum. The relevant quantity, ΔT_C, is plotted in Figure 2(b). ΔT_C culminates at $\simeq$260 kbar with a maximum specific depression $\Delta T_C/x$=17 K/at/o. Finally we note that these data are in good agreement with our previous $T_C(P)$-results for Y [7] and also with preliminary data for the YPr system [10].

3. DISCUSSION

The purpose of this experiment was to investigate whether single Pr impurities give rise to a similarly strong maximum of the pair breaking power under pressure when they are dissolved in the d-band host superconductor Y rather than in La which is probably not a pure sd band metal. The results show that this is in fact the case at quite comparable pressures (260 vs. 230 kbar for LaPr). We note that the specific depression $\Delta T_C/x$ at the maximum is even slightly larger

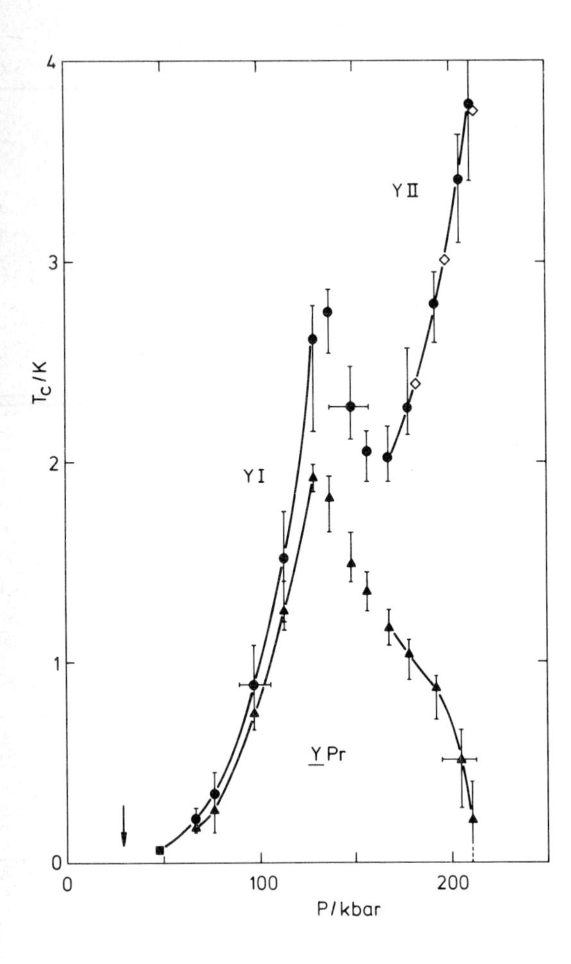

Fig. 1 : Superconducting transition temperature T_C vs. pressure for Y and a YPr alloy (triangles) containing 0.99 at/o Pr. The curves strongly diverge in the high pressure phase Y II. Bars indicate experimental pressure inhomogeneity and superconducting transition width.

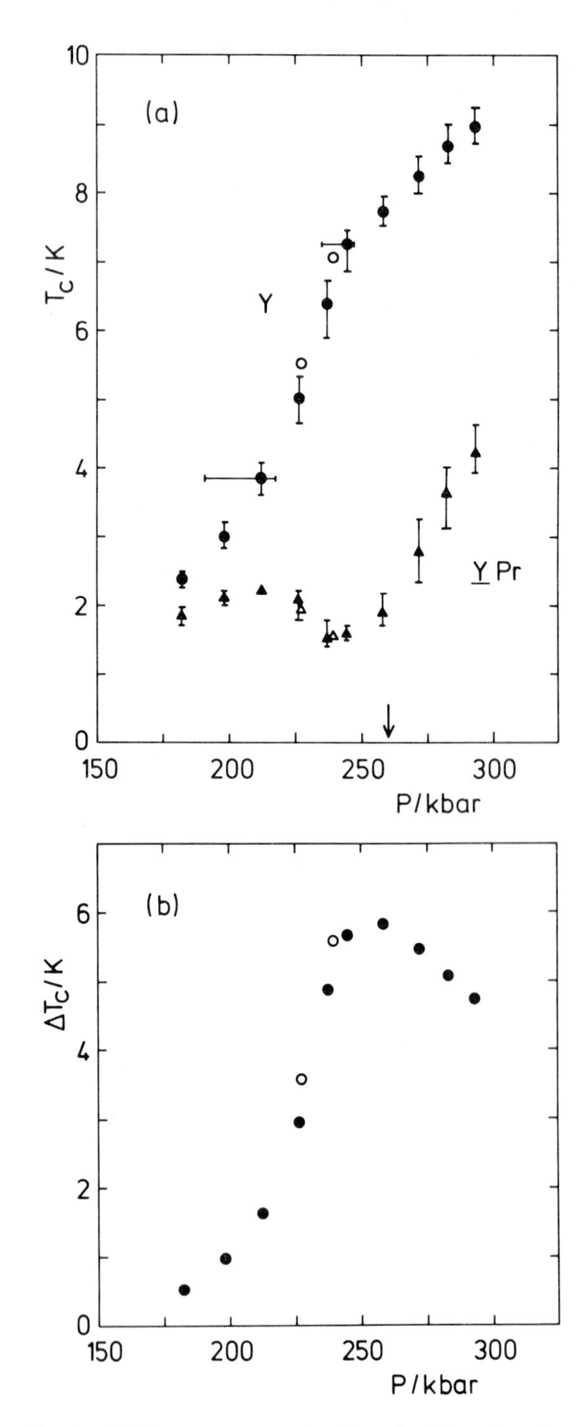

Fig. 2 : (a) T_C vs. pressure for Y and a YPr alloy doped with 0.35 at/o Pr. Arrow indicates absence of superconductivity down to 0.1 K in a 0.59 at/o alloy.
(b) The difference ΔT_C vs. pressure.

than in the LaPr alloys (17 K/at/o vs. 13 K/at/o, respectively). This quantity is in both instances approximately three times larger than for Gd impurities in La at normal pressure. Gd possesses the largest de Gennes factor $(g-1)^2\ J(J+1)$ and shows therefore normally the strongest pair breaking among the rare earth metals. As mentioned in the Introduction, we associate the phenomenon with a valence instability of Pr's $4f^2$ shell. The present findings imply that this valence instability occurs in isolated Pr impurity atoms irrespective of the electronic nature of the host metal. Since the Y matrix atoms have no atomic-like 4f electrons, our experiment proves rather unambiguously that direct 4f-4f overlap is unimportant for triggering the valence instability. This clearcut conclusion will probably also hold for the "electronic phase transition" of pure Pr metal contrary to what we had suggested in an earlier publication [2]. We think in the case of Pr metal it is now very likely that the anticipated delocalization of the $4f^2$ shells into itinerant $4f^2$ band states [3,10] proceeds via energy resonance, i.e. hybridization with the existing $(sd)^3$ Bloch states rather than by 4f-band formation through direct $4f^2$-$4f^2$ overlap. We would like to suggest a model where the electrons of the $4f^2$ shell may tunnel through the angular momentum barrier into the sd conduction sea with greatly increasing tunneling probability at high pressure. Similar ideas may be appropriate for Ce metal and for dilute Ce alloys likewise [11].

In conclusion we can say that we have identified the valence instability of Pr (regardless of its precise nature) as a single impurity or as an atomic property. From this point of view it is, in retrospect, not at all surprising that a similar "electronic transition" occurs in bulk Pr metal at approximately the same pressure as in the dilute alloys.

Concerning the nature of the valence transition we favor the model of a gradual delocalization of the $4f^2$ shell into an, eventually, nonmagnetic $4f^2$ virtual bound state [3,4,10]. The still very strong depression of T_c at 300 kbar (Fig. 2) however shows that much higher pressures are necessary in order to arrive at the anticipated nonmagnetic limit. Alternatively, the maximum of the pair breaking could be associated with a pressure-induced change of the configuration, $4f^2 \rightarrow 4f^1$, with one 4f electron being promoted to the sd band. Riegel has recently argued in favor of this possibility [12]. In this model, Pr adopts the magnetic Ce configuration at very high pressure. A study of the pair breaking power of Pr impurities at much higher pressures than in the present work may therefore help to find out which of the two different models is appropriate.

REFERENCES

1 Mao, H.K., Hazen, R.M., Bell, P.M. and Wittig, J., J. Appl. Phys. 52, 4572 (1981).

2 Wittig, J., Z. Physik B 38, 11 (1980).

3 Wittig, J., in Falicov, L.M., Hanke, W. and Maple, M.B., (eds.) Valence Fluctuations in Solids, p.43 (North-Holland, Amsterdam, 1981).

4 Wittig, J., Phys. Rev. Lett. 46, 1431 (1981)

5 Maple, M.B., Wittig, J. and Kim, K.S., Phys. Rev. Lett. 23, 1375 (1969).

6 Wittig, J., Comments Solid State Physics VI, 13 (1974).

7 Wittig, J., Verhandl. DPG (VI) 14, 412 (1979) and to be published. See also: Vohra, Y.K., Olijnik, H., Grosshans, W. and Holzapfel, W.B., Phys. Rev. Lett. 47, 1065 (1981).

8 The increase of T_c by more than two orders of magnitude between 50 and 300 kbar is related to the pressure-induced $s \rightarrow d$ transfer. This and other features as e.g. the kink-like anomaly in $T_c(P)$ at $\simeq 240$ kbar will be discussed elsewhere.

9 Wittig, J., Phys. Rev. Lett. 24, 812 (1970).

10 Wittig, J., in Shelton, R.N. and Schilling, J.S., (eds.) Physics of Solids under High Pressure, p.283 (North-Holland, Amsterdam 1981).

11 The first evidence, that an isolated 4f shell, imbedded in a d-band host, can be delocalized under P, has been published for the YCe system by Maple, M.B. and Wittig, J., Solid State Comm. 9, 1611 (1971).

12 Riegel, D., Phys. Rev. Lett. 48, 516 (1982).

Valence Instabilities
P. Wachter and H. Boppart (eds.)

LOCAL DENSITY OF STATES OF THE ANDERSON CHAIN IN THE ALLOY APPROXIMATION[a]

P. Schlottmann[x,b] and C.E.T. Gonçalves da Silva[xx,c]

(x) Institut für Theoretische Physik, Freie Universität Berlin, Germany

(xx) Instituto de Física "Gleb Wataghin", Universidade Estadual de Campinas, Brasil

We employ a novel real space renormalization technique to study the Anderson chain within Hubbard's alloy approximation and evaluate local densities of states for d- and f-like states. The effect of intra-site and inter-site f-d mixing, direct f-f hopping, orbital degeneracy of the f level and antiferromagnetic long-range order on the densities of states is discussed.

The exciting properties of intermediate valence compounds[1-3] arise from the presence of two types of electronic states near the Fermi level: atomic-like highly correlated 4f-states and band-like weakly correlated states derived from s, p and d states. Although many different approximations have been applied to this problem, still no reliable quantitative solution exists.

Recently, the techniques of the renormalization group were employed to study simplified hamiltonians, either analytically[4,5] or numerically[6,7]. The coherent potential approximation has also been used to study intermediate valence compounds[8,9] within Hubbard's alloy approximation[10]. In a recent communication[11] we combined the alloy approximation and a novel renormalization method[12,13] to study the local density of states of a model intermediate valence compound. The purpose of this paper is to present some extensions of our previous work[11] and to summarize our results.

In ref. 11 the starting point of our calculation is the standard Anderson chain hamiltonian of orbitally nondegenerate spin 1/2 fermions. We take into account the Coulomb interaction only within the f-states manifold, but we consider the possibility of intra-site and inter-site f-d mixing and also direct f-f hopping. We evaluate the local densities of states for d- and f-like states in the absence of magnetic or charge long-range order.

We now extend the calculation in the following two aspects:

(1) we include the orbital degeneracy of the f-electrons for the case of intra-site f-d mixing, and
(2) antiferromagnetic long-range order is incorporated for the case of the symmetric Anderson model with on-site hybridization only.

(1) The orbitally degenerate case

The hamiltonian can be written as $H=\sum_\sigma (H_{o\sigma}+H_{h\sigma})$,

where

$$H_{o\sigma}=\sum_{im}\Big\{\Big(\varepsilon_f+U\sum_{m'\sigma'\neq m\sigma} f^+_{im'\sigma'}f_{im'\sigma'}\Big)f^+_{im\sigma}f_{im\sigma} + V^*_{im}f^+_{im\sigma}d_{i\sigma}+V_{im}d^+_{i\sigma}f_{im\sigma}\Big\} \qquad (1)$$

$$H_{h\sigma}=t\sum_{\langle ij\rangle} d^+_{i\sigma}d_{j\sigma} \qquad (2)$$

The operators $f^+_{im\sigma}$ and $d^+_{im\sigma}$ create electrons at the site i, with spin and orbital momentum components σ and m, of the type f and d, respectively. Here, ε_f is the f-level energy, U the Coulomb repulsion, V denotes the hybridization matrix element and $H_{h\sigma}$ describes a wide uncorrelated band. The sites i form a one-dimensional lattice, with $\langle ij\rangle$ indicating nearest neighbor sites in the chain. The above hamiltonian is the standard Anderson chain, where spin orbit coupling and exchange are neglected. Note that the total spin of the system is conserved, while the orbital momentum z-component is not a good quantum number.

Two approximations are made to solve the problem. The first is the so-called alloy approximation[10] by which an f-electron in the state m at the site i sees all other states in a frozen in configuration. All the dynamic effects upon the motion of one electron due to electrons in other states is hereby neglected. Due to the spin conservation up and down spins can be treated separately. The problem is most conveniently formulated in terms of (2l+2)x(2l+2) matrices in analogy to ref. 11, where l is the orbital momentum. The second approximation is the real space renormalization approach introduced by Gonçalves da Silva and Koiller[12,13]. This decimation procedure yields recursion relations for the renormalized hopping matrix element and the renormalized electron selfenergies. In the limit of large U and $|\varepsilon_f|<2t$ we obtain that

(i) the f-density of states (DOS) is neglegible if there are one or more f-electrons at the site and
(ii) the d-DOS is almost independent on the number of f-electrons at the site under

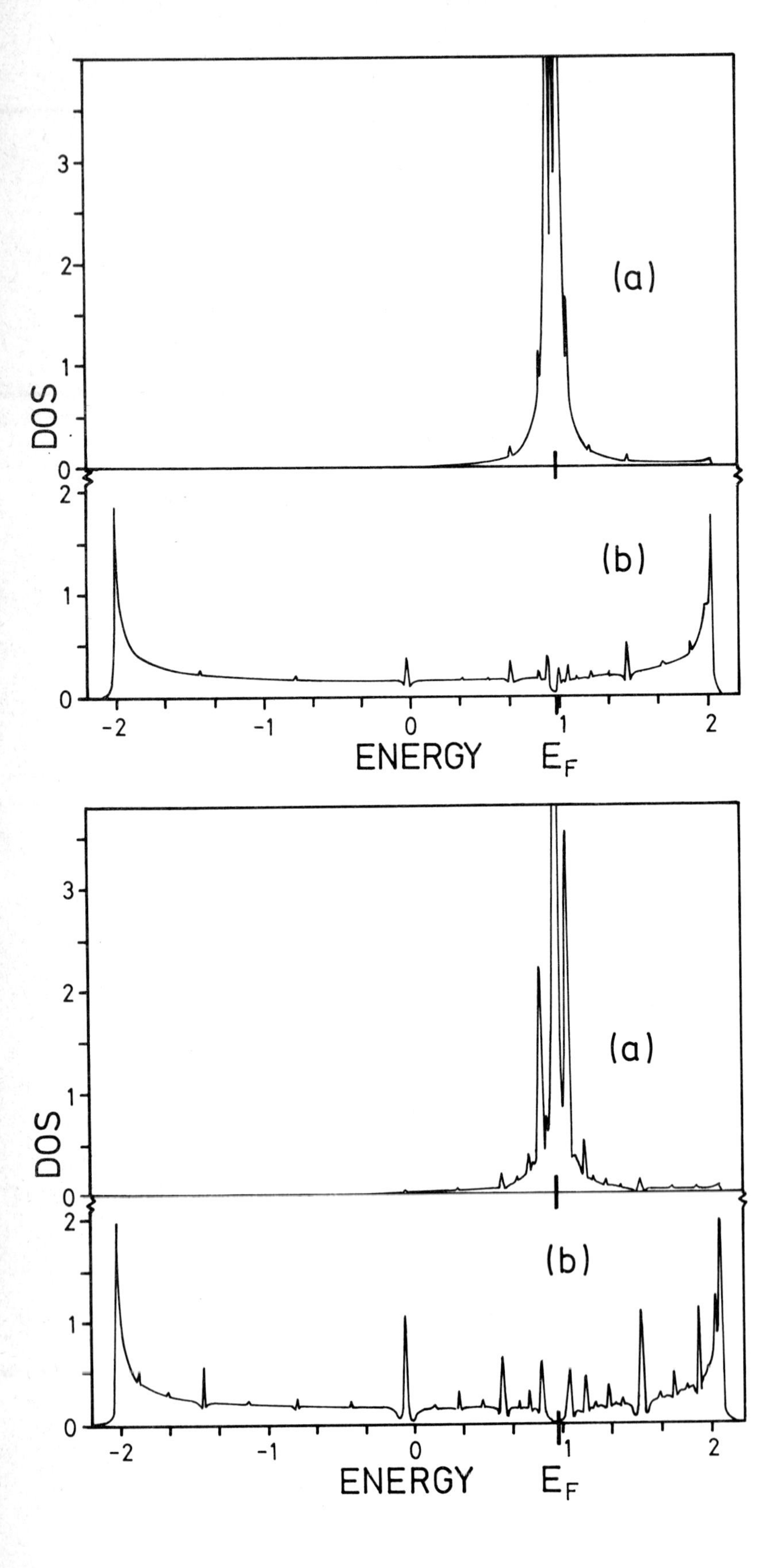

Fig. 1: The local density of states for on-site hybridization V = 0.2 and no orbital degeneracy, l = 0. (a) f-like states at a site unoccupied by an opposite spin f-electron. (b) The same for a d-like electron. The f-level is at E_f=0.97 with the Fermi level slightly above it at E_F=0.99. This corresponds to a total occupancy of 2 electrons per site, 1/3 being f-like and 2/3 d-like. The imaginary part of the energy is 0.005 and the hopping matrix element is t=1.

Fig. 2: Same as Fig. 1, but with l=1. The degeneracy of the f-level is 6, which resembles the J = 5/2 Hund's rule ground multiplet of Ce.

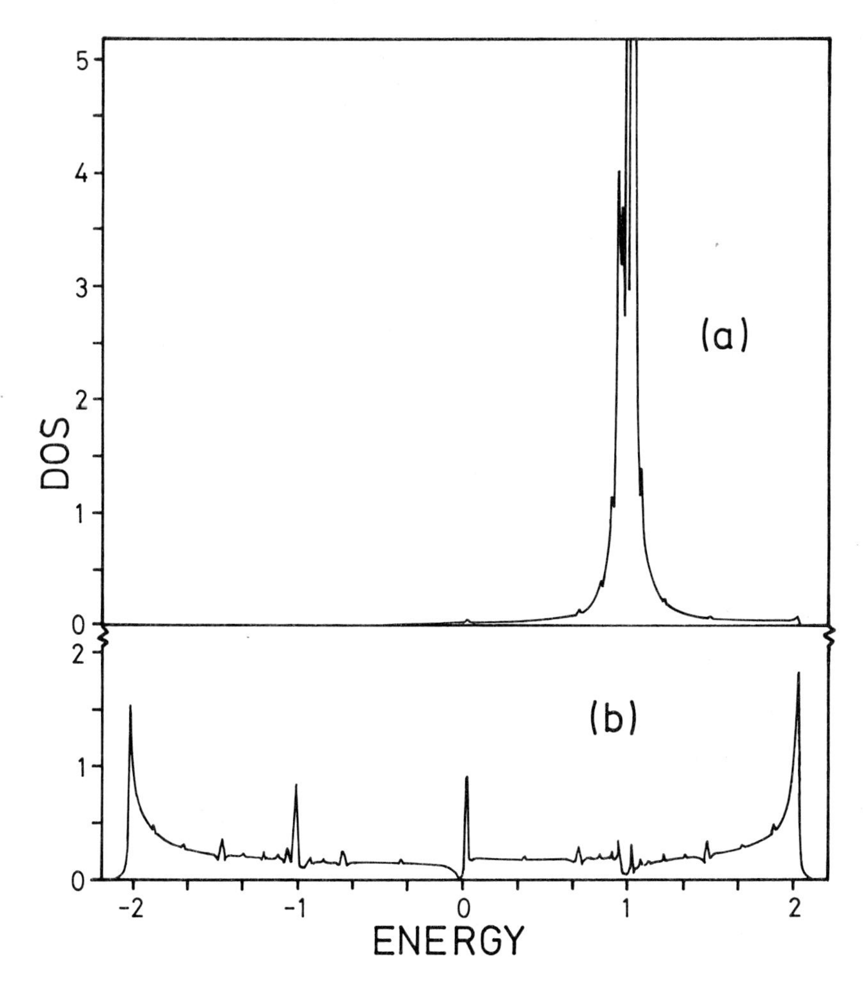

Fig. 3: The local up-spin DOS for on-site hybridization, V=0.2, no orbital degeneracy, ℓ=0, and antiferromagnetic long-range order m=0.6. (a) f-like states at a site occupied by an opposite spin f-electron. (b) The same for a d-like electron. We have chosen $\varepsilon_f=-1$, U=2, t=1 and the imaginary part of the energy is 0.005. The Fermi level is at $E_F=0$, corresponding to a total occupancy of 2 electrons per site, one being f-like and one d-like. The DOS at a site unoccupied by an opposite spin f-electron has its main resonance at ε_f.

consideration.
The recursion relations can then be simplified to

$$t_n = t_{n-1}^2 \left\{ \frac{1-X}{z-\Sigma_{n-1}^{dd}} + \frac{X}{z-\Sigma_{n-1}^{dd} - \frac{(2\ell+1)V^2}{z-\varepsilon_f}} \right\}, \quad (3)$$

$$\Sigma_n^{dd} = \Sigma_{n-1}^{dd} + 2t_n \ , \quad (4)$$

with the initial conditions $\Sigma_0^{dd}=0$, $t_0=t$. Here Σ^{dd} is the d-electron selfenergy and $X=(1-\tilde{n}_f)^{4\ell+1}$, $\tilde{n}_f$ being the average f-occupation per state, is the probability that all other f-states at i are empty. Note that since only the d-electrons hop, Σ^{dd} is the only selfenergy that is renormalized. The f- and d-DOS are finally given by

$$\rho_f = -\frac{1}{\pi} \mathrm{Im} \left\{ \frac{1}{z-\varepsilon_f} \left[1 + \frac{V^2}{(z-\varepsilon_f)(z-\Sigma_\infty^{dd}-(2\ell+1)V^2)} \right] \right\}, \quad (5)$$

$$\rho_d = -\frac{1}{\pi} \mathrm{Im} \left\{ 1 \Big/ \left[z - \Sigma_\infty^{dd} - \frac{(2\ell+1)V^2}{z-\varepsilon_f} \right] \right\} . \quad (6)$$

The above expressions are valid in the paramagnetic phase with local charge neutrality and under the assumption of a random alloy.

In Fig. 1, we present the (a) f- and (b) d-local DOS for l=0 at a site with no opposite spin f-electron. Delta function singularities in these densities are broadened by using a finite imaginary part of the energy. The f-DOS, with the resolution above, shows three main peaks around the Fermi level. The d-DOS at the same site shows, in addition to the van Hove singularities at the band edges, several resonances throughout the band and one pronounced anti-resonance just below the Fermi level.

The localized states, i.e. delta functions, are suppressed if we choose a small imaginary part of the energy. The DOS then shows only extended states. We found[11] a sharp structure of the f-DOS and absence of extended states just below the Fermi level.

Fig. 2 shows the local DOS as above for the case of a sixfold degenerate f-band (l=1). Compared with Fig. 1 the three peaks of the f-DOS are now clearly separated. The resonances of the d- and f-DOS are considerably more pronounced than in Fig. 1, as a consequence of the enhanced scattering due to the larger number of alloy components. The anti-resonance at the Fermi level is also larger, indicating that the gap is due to the hybridization mechanism.

The effect of f-f-hopping and inter-site hybridization has been discussed in ref. 11.

(2) The case of antiferromagnetic long-range order

We restrict ourselves to the symmetric Anderson model ($\varepsilon_f = -U/2$) without orbital degeneracy and two electrons per site (half-filled d-band) and introduce two equivalent sublattices. The hamiltonian is given by Eqs. (1) and (2). If m is the f-magnetization of the sublattice of even sites, the f-occupation numbers are given by

$$n_{ef\sigma} = \frac{1}{2}(1+\sigma m) \,, \quad n_{of\sigma} = \frac{1}{2}(1-\sigma m) \,, \tag{7}$$

where $\sigma = \pm 1$. The first step in the decimation eliminates one of the sublattices, e.g. the odd-site sublattice, yielding

$$t_{1\sigma}(e) = t^2 \left\{ \frac{n_{of\bar{\sigma}}}{z - V^2/(z-\varepsilon_f-U)} + \frac{1-n_{of\bar{\sigma}}}{z - V^2/(z-\varepsilon_f)} \right\} \,, \tag{8}$$

$$\Sigma^{dd}_{1\sigma}(e) = 2\,t_{1\sigma}(e) \,. \tag{9}$$

The recursion relations for $n > 1$ are given by

$$t_{n\sigma}(e) = t^2_{n-1\sigma}(e) \left\{ n_{ef\bar{\sigma}} \Big/ \left[z - \Sigma^{dd}_{n-1\sigma}(e) - \frac{V^2}{z-\varepsilon_f-U} \right] + (1-n_{ef\bar{\sigma}}) \Big/ \left[z - \Sigma^{dd}_{n-1\sigma}(e) - \frac{V^2}{z-\varepsilon_f} \right] \right\} \,, \tag{10}$$

$$\Sigma^{dd}_{n\sigma}(e) = \Sigma^{dd}_{n-1\sigma}(e) + 2\,t_{n\sigma}(e) \,, \tag{11}$$

and the corresponding DOS by

$$\rho^k_{f\sigma}(e) = -\frac{1}{\pi} \mathrm{Im} \left\{ \left(z - \Sigma^{dd}_{\infty\sigma}(e)\right) \Big/ \left[(z - \Sigma^{dd}_{\infty\sigma})(z-\varepsilon_f-Uk) - V^2 \right] \right\} \,, \tag{12}$$

$$\rho^k_{d\sigma}(e) = -\frac{1}{\pi} \mathrm{Im} \left\{ (z-\varepsilon_f-Uk) \Big/ \left[(z - \Sigma^{dd}_{\infty\sigma})(z-\varepsilon_f-Uk) - V^2 \right] \right\} \,, \tag{13}$$

where k=0,1 indicates whether the site is occupied by an opposite spin f-electron.

In Fig. 3 we show the up-spin f- and d-DOS at a site occupied by an opposite spin f-electron for $\varepsilon_f = -U/2 = -t$, m=0.6 and an average total occupancy of two electrons per site. Beside the f-resonance and d-antiresonance at $\varepsilon_f + U$, the d-DOS shows a peak at ε_f that arises from the f-resonance when no opposite spin f-electron is present at the site. In addition a gap below and a peak above the Fermi level, $E_F = 0$, is obtained. The gap is due to the f-d-hybridization while the asymmetry is caused by the antiferromagnetic order.

Our analysis in ref. 11 and the present extensions lead us to the following conclusions:

(a) At least for the local density of states, the cases of only intra-site hybridization, only inter-site hybridization, or intra-site and direct f-f hopping are not drastically different. These results justify the standard Anderson lattice with on-site hybridization only, neglecting the f-f hopping, as well as the inter-site hybridization.

(b) Our calculation has the limitations of the alloy approximation, i.e. it is a static potential approximation. As such, it neglects dynamic effects which involve energies of the order of $\pi\rho V^2$. This may modify the fine structure of the density of states, in particular the coherence of the f-d hybridized state tends to reinforce the gaps close to the f-level energy.

(c) The treatment of the disordered static potential through the renormalization technique is superior to usual CPA calculations[8,9], particularly in this one-dimensional case. While CPA is the best single site alloy approximation, it does not incorporate compositional fluctuations that give rise to the rich structure in the density of states, which, on the other hand, are included by the renormalization procedure.

References

(a) Work supported in part by the CNPq (Brasil)-DAAD (Germany) exchange program
(b) On a Heisenberg-fellowship of the DFG (Germany)
(c) CNPq Senior Research Fellow

1. Valence Fluctuations in Solids
Falicov, L.M., Hanke, W. and Maple, M.B., eds, (North-Holland, 1981).
2. Robinson, J.M., Phys. Repts. 51, 1 (1979).
3. Valence Instabilities and Related Narrow-Band Phenomena
Parks, R.D., ed., (Plenum, New York 1977).
4. Schlottmann, P., Phys. Rev. B22, 613 and 622 (1980).
5. Schlottmann, P., in ref. 1, p. 159.
6. Hirsch, J. and Hanke, W., in ref. 1, p. 353.
7. Jullien, R. and Martin, R.M., private communication.
8. Martin, R.M. and Allen, J.W., J. Appl. Phys. 50, 7561 (1979).
9. Leder, H.J. and Czycholl, G., Z. Phys. B35, 7 (1979).
10. Hubbard, J., Proc. Roy. Soc. A281, 401 (1964)
11. Gonçalves da Silva, C.E.T. and Schlottmann, P., Solid State Commun., in press.
12. Gonçalves da Silva, C.E.T. and Koiller, B., Solid State Commun. 40, 215 (1981).
13. Koiller, B., Davidovich, M.A. and Gonçalves da Silva, C.E.T., preprint.

Valence Instabilities
P. Wachter and H. Boppart (eds.)

THE THERMAL EXPANSION AND MAGNETOSTRICTION IN $Ce_{1-x}La_xSb$ AND $Ce_{1-x}La_xBi$

M. Sera, T. Fujita, T. Suzuki and T. Kasuya

Department of Physics, Tohoku University, Sendai 980, Japan

The thermal expansion and magnetostriction of $Ce_{1-x}La_xSb$ and $Ce_{1-x}La_xBi$ single crystals have been measured. The phenomena that the thermal properties of $Ce_{1-x}La_xSb$ are very similar to those of $Ce_{1-x'}La_{x'}Bi$ for some concentration x<x' have been discovered. These phenomena can be explained naturally by the p-f mixing model. We have measured also the specific heat of $Ce_{1-x}La_xSb$. These results show that Γ_7 in the paramagnetic region and the para plane in the FP phase consists not only of doublet but also of singlet. This singlet state should cause a Kondo like behaviour as observed in our resistivity measurement. Thus, the anomalous magnetic properties are due to the p-f(Γ_8) mixing and the transport properties due to the d-f(Γ_7) mixing.

1. INTRODUCTION

Ce monopnictides crystalize in the rock-salt type structure. Among them, CeSb and CeBi are particurarly interesting because of their anomalous magnetic properties.(1) Recently, our theoretical group have proposed the p-f mixing model, which can explain those anomalous magnetic properties consistently.(2) The APW band calculation on La monopnictides(3) exhibits that the valence band consists mainly of the p-states on pnictogen with the top at the Γ point and the conduction band consists mainly of the 5d (t_{2g}) states on La with the bottom at the three X points. LaN is a narrow gap semiconductor, but, as the pnictogen becomes heavier, a narrow gap semiconductor changes into a semimetal. In CeSb and CeBi, the 4f level is expected to be about 1eV bellow the Fermi level.(4) Then, the p-f(Γ_8) mixing effect plays a more important role than the d-f(Γ_7) mixing effect because the distance between the nearest anion and the cation is smaller than that between the nearest Ce atoms. The p-f mixing effect is due to the interaction between the holes of the valence band and the 4f electrons, so as the number of the holes is larger, the strength of the p-f mixing effect is expected to be larger. Therefore, among the pure Ce monopnictides, the p-f mixing effect is expected to be largest in CeBi. In CeAs, the number of holes becomes very small and thus, the anomalous magnetic properties becomes not to be excellent so as CeSb and CeBi. $Ce_{1-x}La_xSb$ and $Ce_{1-x}La_xBi$ are the very interesting systems to investigate the p-f mixing mechanism which can be changed smoothly by changing x. This effect has been already reported by H. Bartholin et al. on the magnetic measurements on $Ce_x(La_{0.76}Y_{0.24})_{1-x}Sb$,(5) or the same time as the very similar effect of an applied pressure on CeSb. In our case, we expect that the p-f mixing interaction of $Ce_{1-x}La_xSb$ becomes comparable in its strength to that of $Ce_{1-x'}La_{x'}Bi$ for some concentration of x<x'. So, if the number of the 4f electrons is decreased by the substitution of La for Ce, the similar phenomena are expected to appear in both materials. Our first purpose is to check this point. The fact that a negative thermal expansion exists in CeSb,(6) but does not exist in CeBi,(7) was known. We had already explained these difference between CeSb and CeBi by the position of the Γ_8 level relative to the Γ_7 level on the paramagnetic region.(8) To study the character of Γ_7 not only on the paramagnetic region but also on the para-planes in the FP phases, we have also measured carefully the specific heat of $Ce_{1-x}La_xSb$.

2. EXPERIMENTAL RESULTS AND DISCUSSIONS

2.1 The thermal expansion and the magneto striction

We have obtained good single crystals of $Ce_{1-x}La_xSb$ and $Ce_{1-x}La_xBi$, and have measured the thermal expansion and the magnetostriction by the three terminals capacitance bridge method. The thermal expansion of $Ce_{1-x}La_xSb$ and $Ce_{1-x}La_xBi$ along c-axis are shown in Fig.1 and Fig.2. Note that the c-axis means one of x,y, and z directions where the magnetic field is applied. The direction of the magnetic moment is parallel to that of the applied field except the (+-) phase of CeBi and its alloys in which the moment is perpendicular to the applied field.(9) In CeSb, a fairly sharp minimum of the lattice parameter exists at about 40K; the lattice parameter increases fairly rapidly with decreasing temperature below about 40K but shrinks very rapidly at T_N. In x=0.2, the ordered state does not appear down to 4.2K, but the minimum still exists even though it is not so sharp as in CeSb. The temperature for the minimum increases gradually with increasing the La concentration x. On the other hand, in $Ce_{1-x}La_xBi$ no minimum exists in the pure compound but it appears with increasing the La concentration x. In $Ce_{0.5}La_{0.5}Bi$, the minimum is very shallow and

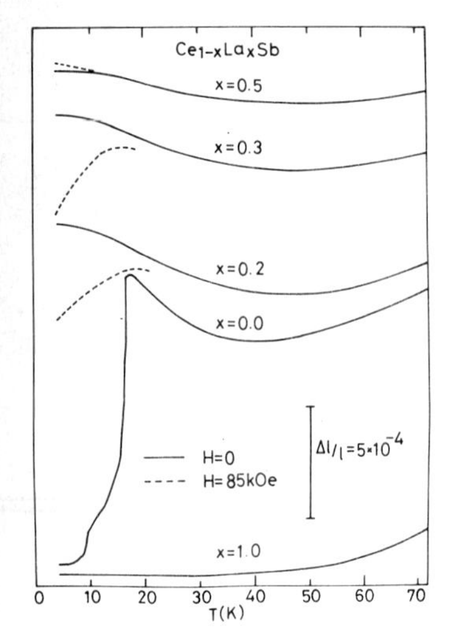

Fig.1. The thermal expansion of c-axis in $Ce_{1-x}La_xSb$. The solid line shows the result under H=0 and the broken line under H=85.4 kOe.

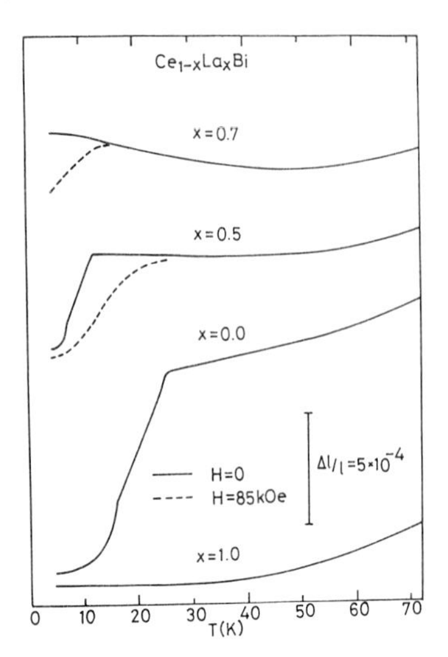

Fig.2. The thermal expansion of c-axis in $Ce_{1-x}La_xBi$. The solid line shows the result under H=0 and the broken line under H=85.4 kOe.

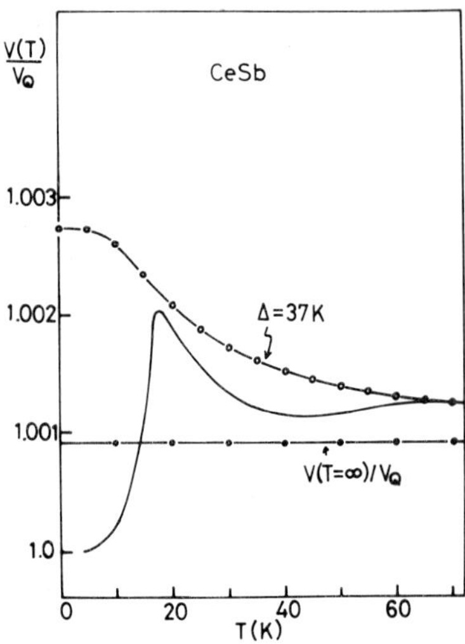

Fig.3. The magnetic part of the thermal expansion of CeSb. The solid line shows the experimental result and -∘- the calculated result by the simple crystal field splitting model with Δ_D^Q =37K.

take place at about 30K, in $Ce_{0.3}La_{0.7}Bi$ it moves at about 50K and becomes clearer. The behaviour of $Ce_{0.3}La_{0.7}Bi$ is similar to those of $Ce_{0.8}La_{0.2}Sb$ and $Ce_{0.7}La_{0.3}Sb$ both in H=0 and H=85.4 kOe. Therefore, as mentioned in the introduction, these phenomena can be explained by the p-f mixing model. Without the p-f mixing effect the splitting between the Γ_7 and Γ_8 levels, Δ_D^Q, is large(10) as is extrapolated from other monopnictides, about 230K for CeSb and 215K for CeBi. A strong p-f(Γ_8) mixing pushes the Γ_8 down close to the Γ_7 level. The ordering of a special type of Γ_8, say Γ_{8i}, pushes the Γ_{8i} level further down below the Γ_7 level making it the ground state. This effect is stronger in CeBi than in CeSb because of the larger number of holes and the smaller value for Δ_D^Q without the p-f mixing effect in CeBi. Therefore, a larger La dilution in CeBi corresppnds to the smaller dilution in CeSb. Note that the crystal shrinks for more population of Γ_8 because of the p-f(Γ_8) mixing energy gain for a smaller anion-cation distance. But the simple p-f mixing model can not explain some difference between the behaviour of $Ce_{1-x}La_xBi$ and $Ce_{1-x}La_xSb$. Fig.1. and Fig.2. show that a discontinuity of the lattice parameter exists for $Ce_{1-x}La_xSb$ but not for $Ce_{1-x}La_xBi$. On another hand, FP phases do not appear in $Ce_{1-x}La_xBi$. These seem to correlate with the dense Kondo behaviour of Γ_7 which mixes mostly with the conduction bands. Note that T_K of CeSb seems larger than that of CeBi. Some results are shown in ref.(11)

Fig.3. and Fig.4. show the magnetic part of the thermal expansion of CeSb and CeBi, respectively. The solid line shows the experimental result and -∘- the result obtained by the simple mean field approximation with assuming the crystal field splitting Δ_D^Q =37K for CeSb and Δ_D^Q =8K for CeBi. In CeSb, a large difference exists between the experimental result and the calculated one at 50K. This difference is explained consistently with the result calculated by H.Takahashi et. al.(12); i.e. by their theory, the magnetic order has a strong tendency to appear with the second order transition just around 50K. Then, the intense two dimensional short range order occurs in the same temperature region and causes the increase of the Γ_8 population just around 50K. When it is diluted by La, such critical situation disappears and thus the temperature dependence of $Ce_{0.8}La_{0.2}Sb$ is well fitted by the same simple calculation. In CeBi, the molecular field calculation by H.Takahashi et.al.(12) predicts the 2nd order transition at 60K. Due to a strong two dimentional short range order fluctuation, however, the real transition temperature is suppressed to 26K. But, we can see the short range order effect clearly above 26K up to 50K in close agreement with the prediction of the Takahashi's calculation. The same behaviour is observed up to x=0.5 where still the 2nd order transition exists. For x=0.7 no transition appears and the Γ_7 is the real ground state.

As shown in Fig.2, $Ce_{0.3}La_{0.7}Bi$ presents in low temperature a negative thermal expansion which can be accounted for assuming Γ_7 the ground state

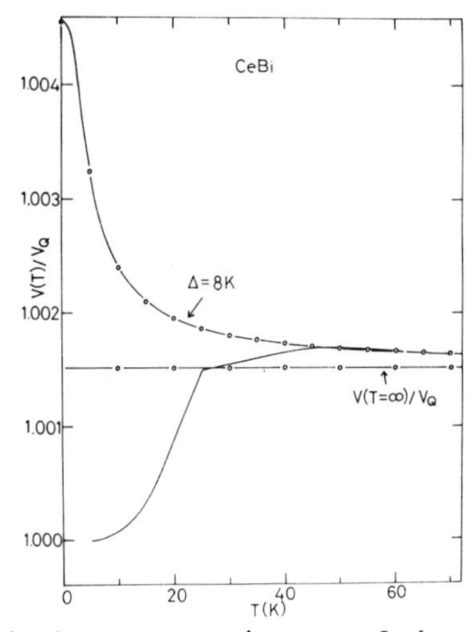

Fig.4. The magnetic part of the thermal expansion of CeBi. The solid line shows the experimental result and -o- the calculated result by the simple crystal field splitting model with Δ_D^Q =8K.

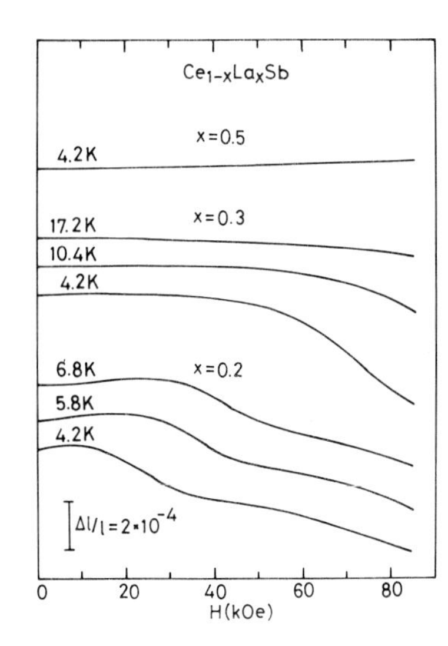

Fig.5. The longitudinal magnetostriction of $Ce_{1-x}La_xSb$ for various fixed temperature.

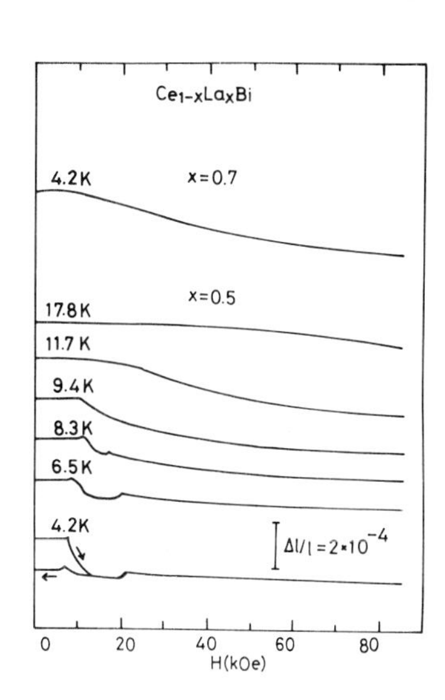

Fig.6. The longitudinal magnetostriction of $Ce_{1-x}La_xBi$ for various fixed temperature.

On the other hand, under a magnetic field H=85.4 kOe at 4.2K, the thermal expansion becomes positive and the magnetic moment reaches a value of $1.57\mu_B$/Ce at 4.2K. The ground state under the magnetic field of 85.4kOe is supposed to be nearly Γ_8. Thus, with increasing the magnetic field, the ground state changes from Γ_7 to Γ_8 by the p-f mixing effect, which is still strong enough to cause this change even if 70% of Ce are replaced by La.

Some examples of more detailed magnetostriction measurement are shown in Fig.5. and in Fig.6. The behaviours of $Ce_{0.8}La_{0.2}Sb$ and $Ce_{0.7}La_{0.3}Sb$ are similar, showing the transitions from the Γ_7 ground state to the FP phase and furthermore to the ferro phase, the last transition is, however, not observed for x=0.3 sample at higher temperature. The phase diagrams thus obtained are essentially consistent with those reported previously in $Ce_{1-x}(La_{0.76}Y_{0.24})_xSb$.(13) For $Ce_{0.5}La_{0.5}Sb$ the lattice parameter increases rather gradually with increase of applied field up to 85kOe showing no transition to the FP phase within this region. The increase of lattice parameter with increasing field in the Γ_7 ground state region is observed also in the x=0.2 and 0.3 samples. This increase seems also to originate from the Γ_7 Kondo state because the Γ_7 singlet Kondo state is destroyed by the increase of the applied magnetic field. Then, the lattice distance is expected to increase because the d-f inter-atomic mixing, which is responsible to the Γ_7 Kondo state and increases with decreasing lattice distance, becomes needless. In $Ce_{1-x}La_x$ Bi, the behaviour of magnetostriction is very similar to that of CeBi up to the x=0.5 sample. The phase diagram is simply scaled down with increasing of x value. This is again consistent with the previous results of the phase diagram for $Ce_{1-x}(La_{0.76}Y_{0.24})_xBi$ obtained from the magnetic measurement.(14) For x=0.7, the lattice shrink begins to occur from a weak magnetic field This seems to indicate the fact that the ground state is mainly Γ_7 but perhaps mixed with domains of short range ordered Γ_8 clusters because still at x=0.7 these two energies are close each other. This seems also to correspond to a large residual resistivity of this sample.(11)

2.2 The specific heat

Electrical resistivity of $Ce_{1-x}La_xSb$ shows a Kondo like behaviour. It is found from our measurement of the Hall effect at high temperature that the current is carried mostly by the conduction electron. The conduction electron made by the $5d(t_{2g})$ state on Ce mixes with Γ_7 of the nearest Ce atom and the Kondo like behaviour appears. Thus, Γ_7 is considered to become singlet. On the other hand, it is concluded from our measurements of the thermal expansion and magnetostriction that the lower level in the paramagnetic region in Γ_7 and the para-planes in FP and AFP phases are also made by Γ_7. The only evidence that Γ_7 remains to be doublet in the para-plane in FP phase is the entropy of ln2 which is obtained from a detailed analysis of the specific heat measurement. Therefore, we have measured the specific heat of CeSb and $Ce_{0.8}La_{0.2}$ Sb carefully by the adiabatic heat pulse method using large samples. As shown in Fig.7., our values of specific heat in CeSb are substantially smaller than the values reported before.(15) Therefore, the entropy at T_N obtained from our measurement is about 40% smaller than that in ref.(15) and about 30% smaller than the value

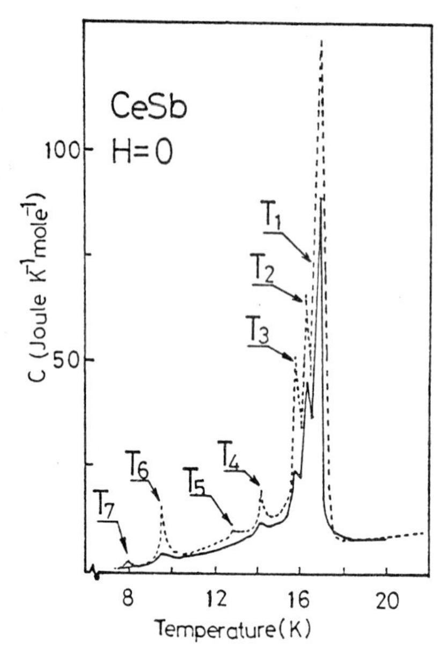

Fig.7. The specific heat of CeSb. The solid line shows our result, for comparision, the result of ref.(14) is shown by the dotted line.

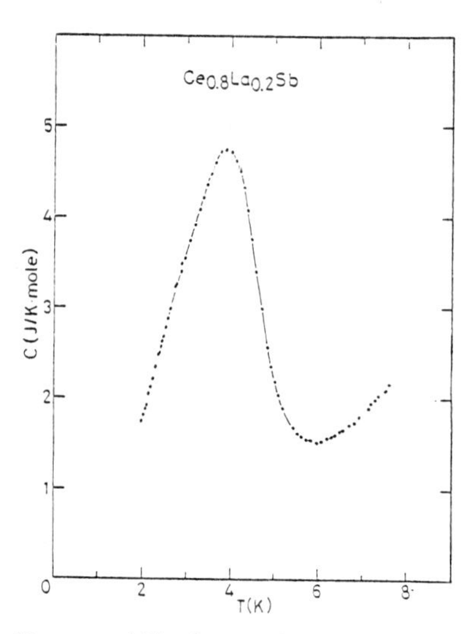

Fig.8. The specific heat of $Ce_{0.8}La_{0.2}Sb$ as a function of temperature T.

obtained by the simple crystal field splitting model with Δ_D^Q =37K. This reduction shows that a singlet Kondo state should be the ground state of Γ_7 doublet in agreement with other experiments reported so far. This is also consistent with the fact that the induced moment in the para-planes is much smaller than that expected from the simple para state of Γ_7 doublet. This situation is clearer in $Ce_{0.8}La_{0.2}Sb$. In this material, a broad peak of the specific heat exists around 3.9K, as shown in Fig.8. This peak can be assigned to the transition from the Para phase to the FP phase. Because we used a large sample with many small single crystals, distribution of T_N seems to cause broadening of the transition. As shown in ref.(11) for resistivity, a sharp first order transition is observed in a single crystal.

If the para-planes in FP phase are made by Γ_7 doublet, they must order at lower temperature. However, no such ordering seems to occur even though our measured temperature region may not be sufficiently low enough. Furthermore, the estimated entropy at 5K is substantially lower than the simple calculation mentioned before and the value of the specific heat above 5K is also larger. Thus, the para-planes in FP phases are considered to be made mostly by Γ_7 singlet and the Kondo effect seems to occur in $Ce_{1-x}La_xSb$ which is semimetal. More experiments are needed to understand the Kondo effect in this system.

REFERENCES:

(1) Busch, G. and Bogt, O., Phys.Lett. 25A(1967) 449
Birgeneau, R. J., Bucher, E.,Maita, J.P., Pussel, L. and Turberfield., Phys.Rev. B8 (1973) 5345
Rossat-Mignod, J., Burlet, P., Bartolin,H., Vogt,O. and Lagnier,R., J/Phys. C13(1980)
Bartolin,H., Burlet,P/, Quezel,S., Rissat-Mignod,J. and Vogt,O., J.de Phys. 40(1979) C5-130
(2) Takegahara,K., Takahashi,H., Yanase,A. and Kasuya,T., IV International Conference on Crystal and Structual Effects in f-Electron Systems. Sept. 1981. Wroclaw, Poland
(3) Hasegawa,A., J.Phys.C:Solid State Phys. 13 (1980) 6147
(4) Bear,Y., Hauger,R., Zucher,Ch, Compagna,M., and Wertheim,G.K., Phys.Rev. B18(1978) 4433
(5) Bartolin,H., Effantin,J.M., Burlet,P., Rossat-Mignod,J., abd Vogt,O., International Conference on Crystal Field Structure and Structural Effects in f-Electrons Systems Wroclaw (Poland) Sept. 1981.
(6) Levy,F., Physik Kondens Materie 10(1969) 85
(7) Hulliger,F., Landolt,M., Ott,H.R., and Schmelczer,R., J.Low.Temp.Phys. 20(1975) 269
(8) Nakajima,T., Suzuki,T., Sera,M. and Kasuya,T IV International Conference on Crystal Field and Structual Effects in f-Electron Systems. Sept. 22-25, 1981. Wroclaw, Poland.
(9) Rossat-Mignod,J., in private communication.
(10) Birgeneau,R.J., Bucher,E., Maita,J.P., Passel,L. and Turberfield,K.C., Phys.Rev. B8(1973)5345
(11) Kasuya,T., et al., in this conference
(12) Takahashi,H., et al., in this conference
(13) Bartholin,H., Florence,D., and Vogt,O., phys.stat.sol.(a)52,(1979)647
(14) Bartholin,H., and Vogt,O., phys.stat.sol. (a)52(1979)315
(15) Rossat-Mignod,J., Burlet,P., Bartholine,H., Vogt,O., and Lagnier,R., J. Phys. C: Solid State Phys.,13(1980) 6381

Valence Instabilities
P. Wachter and H. Boppart (eds.)

EXTRAORDINARY SUPERCONDUCTOR WITH NEARLY TRIVALENT CERIUM, $CeCu_2Si_2$

M. Ishikawa, D. Jaccard and J.-L. Jorda

Département de Physique de la Matière Condensée ,
Université de Genève, 32,Bd d'Yvoy, 1211 Genève 4
Switzerland

Concentrating on the ternary phase diagram, we have performed a complementary investigation on $CeCu_2Si_2$ and convinced ourselves that the compound containing nearly trivalent cerium ions is a new type of superconductor with T_c around 0.5K. The analyses of the upper critical field curve support the description of the compound by heavy fermion quasiparticles. Other particular features of this compound are also presented.

$CeCu_2Si_2$ is one of the unusual compounds of cerium which show anomalous transport properties at high temperatures [1-3]. Recent experiments indicate that the valence state of Ce in this compound is slightly greater than three [4,5] and that there is no evidence for long range magnetic order down to 20 mK [6]. These experimental facts are consistent with a description of the unstable 4f character of Ce. The compound has received further attention with the discovery of superconductivity by Steglich et al. [7]. They found a very large specific heat coefficient, γ, of about 1 J/mole·K^2 and characterized the compound as a system of heavy fermion quasiparticles. However, there remained several unanswered important questions [8] such as why its physical properties including the T_c vary so much from sample to sample and why other workers have been unable to find bulk superconductivity in similar samples [9]. Moreover, the crystal structure ($ThCr_2Si_2$ type [10]) of this compound, which is very common for transition metal silicides, does not appear to be very favorable for superconductivity [11].

As pointed out by Steglich et al. [7], the metallurgy of this compound is indeed very complicated. In the present work, we have, first of all, attempted to clarify the metallurgical problems with various techniques available in our laboratory.Using three different lots of Ce metal from two suppliers (Rare Earth Products Ltd. and Leico Industries, Inc.), the samples were prepared by either arc-melting or induction-melting and subjected to different heat treatments. They were analyzed for the lattice constants and impurity phases with the X-ray Guinier powder method. A tetragonal structure of the $ThCr_2Si_2$ type [10]without any appreciable deformation was confirmed for all samples. The lattice constants are listed in Table 1 with other physical parameters. Ce_2CuSi_3 and "Cu_3Si" were identified as impurity phases in some of our samples. By differential thermal analysis (DTA) we found the melting point at 1540 ± 15°C and no other phase transformations down to 600°C. Samples were accordingly annealed between 700 and 1500°C. One arcmelted sample was annealed at 1500°C for one hour under four atm. of Ar and very slowly cooled down. No second phase was detected with X-ray analysis but a considerable amount of microcracks was present. This sample is denoted as No 101 in the Table. The sample was then remelted in a levitation coil and cast in a 4mm diameter cylinder. A metallographic examination of this sample showed that the second phase (Ce_2CuSi_3) of about 5% in volume reappeared but the sample was extremely dense and free of microcracks. So, this sample (denoted as No 100) was considered appropriate for resistivity measurements and used for the determination of the upper critical field curve (H_{c2} vs T) (see Fig. 2). Annealed samples contained very often a lot of microcracks, but we do not know at the moment why such microcracks are formed during annealing. It is also noted here that some samples disintegrated into small pieces in about one year at room temperature but X-ray analyses showed no modification of the crystal structure and lattice constants. All of our samples were superconducting with T_c ranging from 0.3 to 0.7K, as summarized in Table 1. Typical ac susceptibility (χ_{ac} vs T) curves are shown in Fig. 1 with those of annealed Cd and Zn of the same volume (within 3%) and about the same mass (within 10%) as the sample No 103 for comparison.

We also checked for superconductivity several other ternary and binary compounds such as $CeCuSi_2$, $CeCu_{1.6}Si_{1.4}$, Ce_2CuSi_3, $CeCu_5$ and Cu_3Si, which surround the compound of interest in the phase diagram of Bodak et al. [12] . The crystal structures reported for these compounds by Bodak et al. were confirmed except for Cu_3Si. The latter bianry compound is here simply denoted as "Cu_3Si" with the nominal concentration. Only this binary among these neighboring compounds was found to be superconducting above 70 mK(T_c ~ 0.3K). In order to confirm that the superconductivity found in our $CeCu_2Si_2$ samples is not due to this superconducting second phase of "Cu_3Si", we made a sample, $CeCu_{1.7}Si_{1.7}$ (at + in the phase diagram). After annealing at 850°C

Table 1 : Experimental results of $CeCu_2Si_2$

Sample number	T_c (K)		$\rho_{1.5\ K}$ ($\mu\Omega \cdot cm$)	$[dH_{c2}/dT]_{T_c}$ (KOe/K)	Lattice constants (Å)		Fe content (at. ppm)
	From ρ	From χ_{ac}			a	c	
103	0.45± .08	0.29±.05	110 ± 5	50	4.102	9.927	6000
133	0.70± .05	0.45±.07	49±5	29	4.102	9.921	2000
134	0.59± 0.07	0.43±.07	77±5	31	4.101	9.923	3000
100	0.49 ±0.06		50±3	182	4.103	9.926	< 200
101		0.27 ±.1			4.102	9.925	< 200
120		0.35±.1			4.105	9.923	2000

for 6 days, the sample was three-phase ($CeCu_2Si_2$, Ce_2CuSi_3 and $CeCu_{1.6}Si_{1.4}$), but free of "Cu_3Si", as expected from the phase diagram. For this sample we found bulk superconductivity as for the $CeCu_2Si_2$ samples, i.e. T_c=0.495±0.05 and $(\frac{dH_{c2}}{dT})_{T_c}$ = 31 kOe/K.
Accordingly, this preliminary investigation convinced us to believe that $CeCu_2Si_2$ is an intrinsic superconductor with T_c around 0.5K. It is, of course, consistent with the observation of the bulk nature of the superconducting transition in the specific heat measurements by Steglich et al. [7]. However, we do not know yet why the value of T_c among other parameters varies so much from sample to sample. We tried to find a correlation between T_c and other parameters such as the lattice constants, the normal resistivity, the estimated iron content and heat treatments etc. As can be seen from the Table, there is no convincing simple correlation. It is tempting, however, to assume that a slight atomic disorder due to either heat treatments or a change in concentration resulting from a homogeneity range may influence the physical properties of $CeCu_2Si_2$ as found. It is also noteworthy in this respect that an additional step in χ_{ac} vs T was found in some of our samples (see, for example, the curves for No 133 and 134 in Fig. 1).

Fig. 2 shows the H_{c2} vs T curve for the sample No 100 deduced from the dc resistance measurements in applied magnetic fields. One should notice the extremely large value of

$$(\frac{dH_{c2}}{dT})_{T_c} \qquad (\sim 182\ kOe/K)$$

which is much higher than for any other "ordinary" superconductors known at present. The value for the other samples estimated from χ_{ac} (T,H) range from 30 to 50 kOe/K, which is still considered to be very high (see Table 1).

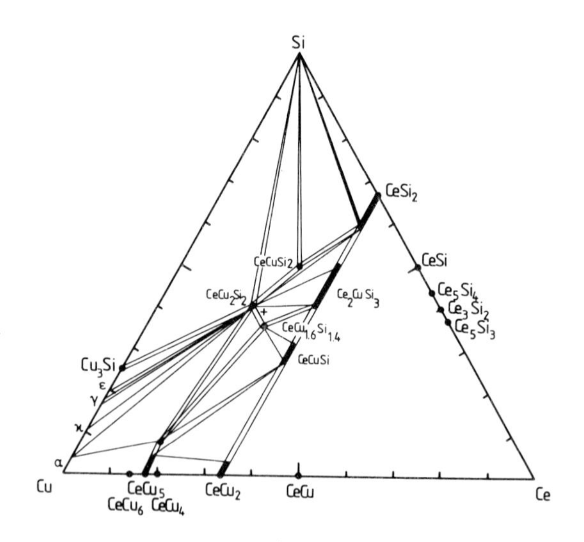

Partial phase diagram of Ce-Cu-Si (Bodak et al. (Ref. 12))

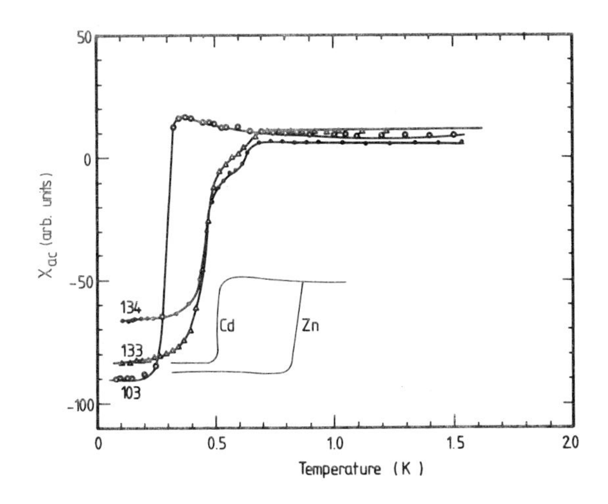

Figure 1 : χ_{ac} vs T curves of $CeCu_2Si_2$ and Cd and Zn of the same volume as No 103.

Using the value of 182 KOe/K, we estimate the critical field at 0K, $H_{c2}(0)$, in the dirty limit by $H_{c2}(0) = -0.686\, T_c (\frac{dH_{c2}}{dT})_{T_c}$ and found $H_{c2}(0) = 61$ kOe. Our experimental curve, however, extrapolates to about 17 kOe. We may attribute this difference to the strong paramagnetic effect with an intermediate value of the spin-orbit parameter, λ_{so}. Taking into account this effect with $\lambda_{so}=2.1$, we found 17.3 kOe for $H_{c2}(0)$ from the formula

$$H_{c2}(0)=\left[1.16\times10^{-3}\,\frac{H_{c2}^{2}(0)}{\lambda_{so}T_c} - 0.686T_c\right](\frac{dH_{c2}}{dT})_{T_c}\,.$$

Since $H_{c2}(0)$ is inversely proportional to the Fermi velocity of the relevant electrons, the high value of $H_{c2}(0)$ implies a very small Fermi velocity. This is consistent with the heavy fermion description for this new type of superconductor proposed by Steglich et al.[7]. The present experiments, combined with previous evidence, leave very little doubt that superconductivity is an intrinsic property of $CeCu_2Si_2$.

It must be emphasized here that the discovery of this new type of superconductor with such a high value of $(\frac{dH_{c2}}{dT})_{T_c}$ would be of great importance not only from a fundamental point of view but also from a practical point of view.
The next important question is then to know whether related cerium compounds or intermediate valence compounds attain a higher T_c.

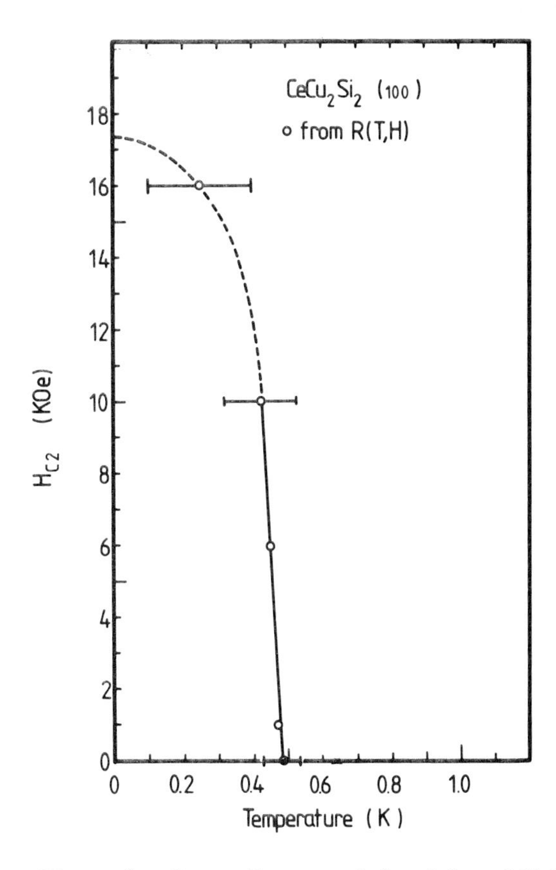

Figure 2 : H_{c2} vs T curve deduced from R(T,H) of sample No 100.

ACKNOWLEDGMENTS

The authors are grateful to Professors J. Muller, J. Sierro and Dr. H. Braun for valuable discussions and comments. They also thank Mr. F. Liniger and A. Naula for their excellent technical assistance.

REFERENCES

[1] Sales, B.C. and Viswanathan, R., J. Low Temp. Phys. 23 (1976) 449.
[2] Jaccard, D., Ph. D. Thesis, Univ. de Genève (1981).
[3] Franz, W., Griessel, A., Steglich, F. and Wohlleben, D., Z. Physik, B31 (1978) 7.
[4] Jaccard, D. and Sierro, J., in the Proceedings of this conference.
[5] Umlauf, E. and Hess, E., Physica 108B (1981) 1347.
[6] Horn, S., Holland-Moritz, E., Loewenhaupt, M., Steglich, F., Scheuer, H., Benoit, A. and Flouquet, J., Phys. Rev. B23 (1981) 3171.

[7] Steglich, F. et al., Phys. Rev. Lett. 43 (1979) 1892 and also in the Proceedings of this conference.
[8] Steglich, F., (private communication, 1981).
[9] Hull, G.W., Wernick, J.H., Geballe, T.H., Waszczak, J.V. and Bernardini, J.E., Phys. Rev. B24 (1981) 6715.
[10] Rieger, W. and Parthé, E., Monat. für Chemie, 100 (1969) 444.
[11] Ishikawa, M., (unpublished results).
[12] Bodak, O.I. et al., Inorg. Materials (USA), 10 (1974) 388.

Valence Instabilities
P. Wachter and H. Boppart (eds.)

THERMOELECTRIC POWER OF $CePd_{3+\varepsilon}$, $Ce(Pd_{1-x}Rh_x)_3$ and $Ce(Pd_{1-y}Ag_y)_3$

H. Sthioul, D. Jaccard and J. Sierro

Département de Physique de la Matière Condensée
32 Bd d'Yvoy, 1211 Genève 4, Switzerland

The anomalous thermopower of $CePd_3$ is roughly independent of a slight stoichiometric deviation, contrary to the electrical resistivity. The introduction of Rh or Ag into $CePd_3$ shifts the maximum of the thermopower towards higher or lower temperature indicating a more tetravalent or trivalent state of the Ce ion, respectively. The systems $Ce(Pd_{1-x}Rh_x)_3$ and $Ce(Pd_{1-y}Ag_y)_3$ allow a unification of the thermopower behaviours observed in very different compounds.

I. INTRODUCTION

Intermediate valence compounds (IVCs) show typical anomalous transport properties like high electrical resistivities and giant thermopowers. $CePd_3$ is such a compound [1,2] with an almost temperature independent valence of about 3.45 as deduced from lattice constant and thermal expansion measurements [3].

$CePd_3$ and $CeRh_3$ crystallize in the $AuCu_3$ structure and solid solutions $Ce(Pd_{1-x}Rh_x)_3$ exist in the whole range, $0 < x < 1$ [4]. Although the compound $CeAg_3$ does not exist, samples of $Ce(Pd_{1-y}Ag_y)_3$ are single phased for $0 < y < 0.4$. The Ce valence increases with x up to 4 when $x > 0.2$ and decreases with y to reach 3 when $y > 0.13$ [5]. It is then possible to study in this system the correlation of the transport properties with the Ce valence and to compare the results with other rare earth IVCs [6].

2. SAMPLE PREPARATION

Samples were prepared either by arc melting the pure elements under 500 Torr of high purity Argon or by induction melting in a levitation furnace under 4 atm. of Argon. Weight losses observed during melting or heat treatment were lower than 0.5% for all the samples, so that nominal compositions might be accepted. X-ray and metallographic analysis showed that $CePd_3$ is homogeneous and almost single phased. Although the X-ray analysis on $Ce(Pd_{1-x}Rh_x)_3$ and $Ce(Pd_{1-y}Ag_y)_3$ did exhibit only the $AuCu_3$ phase with lattice constants in good agreement with previous work [5], the metallographic examinations of these alloys revealed about 10% of a secondary phase probably rich in Ag. Annealing for 72 hours at 700 or 800°C improved the homogeneity of the samples but did not suppress the secondary phase.

3. RESULTS

3.1. $CePd_{3+\varepsilon}$

Figure 1 shows the thermopower S versus temperature for as cast and annealed $CePd_3$ and for non-stoichiometric samples $CePd_{3+\varepsilon}$ with $\varepsilon = \pm 0.08$. The thermopower of $CePd_3$ is very large (about 80 µV/K) at room temperature, shows a broad maximum around 150K and starts decreasing below T ~ 100K. This behaviour is typical for a Ce compound with an intermediate valence of about 3.5. The low temperature shoulder may be due to a small variation of the valence [3]. A slight deviation from the stoichiometry did not modify this characteristic temperature dependence, especially the peak temperature. This fact may suggest that all these samples have the same valence.

A different situation occurs for the low temperature resistivity as shown in figure 2. A slight excess of Pd strongly increases the residual value whereas no large effect is obser-

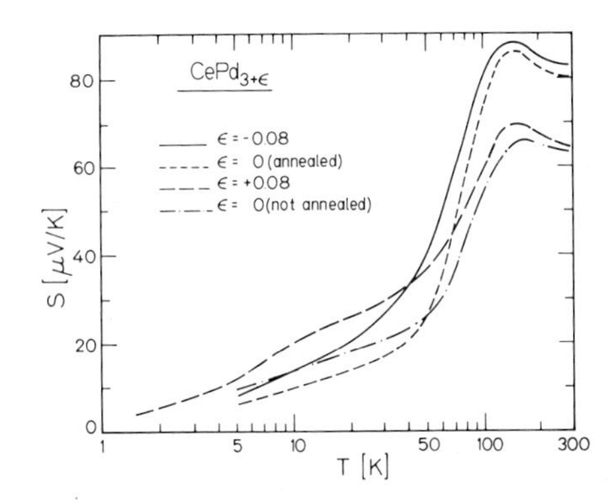

Fig. 1. Absolute thermopower of different $CePd_{3+\varepsilon}$.

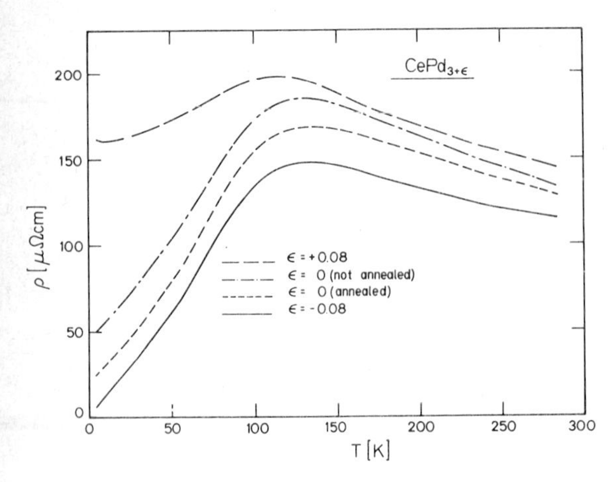

Fig. 2. Resistivity of the samples of fig.1.

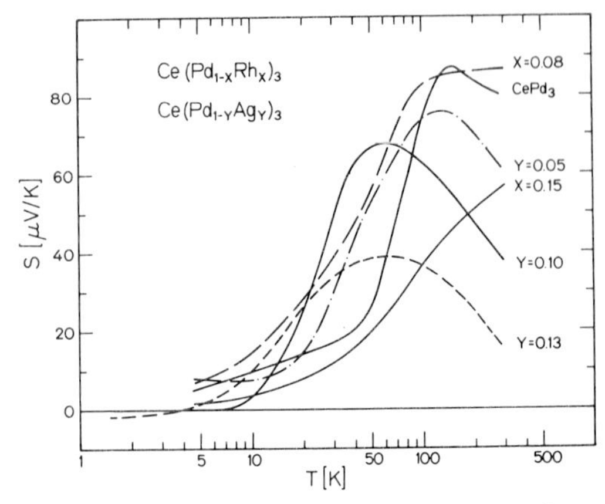

Fig. 3. Thermopower versus temperature of $Ce(Pd_{1-x}Rh_x)_3$ and $Ce(Pd_{1-y}Ag_y)_3$.

ved for the sample with an excess of Ce, contrary to what was reported by Scoboria et al [2]. As $CePd_3$ exists over a narrow composition range of approximately 75-76 at % Pd [7], we note that high resistivities are observed for Pd rich compounds. Moreover, the annealing effects found by Schneider and Wohlleben [8] correspond to our results for compounds with an excess of Pd.

3.2. $Ce(Pd_{1-x}Rh_x)_3$ and $Ce(Pd_{1-y}Ag_y)_3$

Figure 3 shows the temperature dependence of the thermopower of $Ce(Pd_{1-x}Rh_x)_3$ and $Ce(Pd_{1-y}Ag_y)_3$ with different values of x or y. The main effect of the introduction of Rh or Ag into $CePd_3$ is to shift the maximum of the thermopower towards higher or lower temperature, indicating a more tetravalent or trivalent state respectively. If we interprete the variation of the lattice constant with the Rh or Ag concentration as a continuous variation of valence, a correlation can be established between the temperature of the maximum and the Ce valence. For nearly trivalent compounds ($y \simeq 0.13$) a small negative contribution appears at $T \sim 2K$ well below the main positive maximum. This change of sign of S at a low temperature has been observed for all nearly trivalent Ce compounds [6]. Our results of thermopower are qualitatively identical to those of the following systems :

x = 0.08	$CeSn_3$	[9]
y = 0.08	$CeBe_{13}$	[9]
y = 0.10	$CePb_3$	[9]
y = 0.13	$CeAl_3$	[10]

The series $Ce(Pd_{1-x}Rh_x)_3$ and $Ce(Pd_{1-y}Ag_y)_3$ allow a unification of the thermopower behaviours observed for very different IVCs.

The temperature dependence of the resistivity is shown for several samples in figure 4. The results are in qualitative agreement with reference 5, but our values of resistivities are about a factor two lower. It should be noted that, for the alloys with Ag, the shift of the maximum towards low temperature corresponds to the similar shift of the maximum in S. The sample (y=0.11) with the very high resistivity shows the negative thermopower at low temperatures (fig.3).

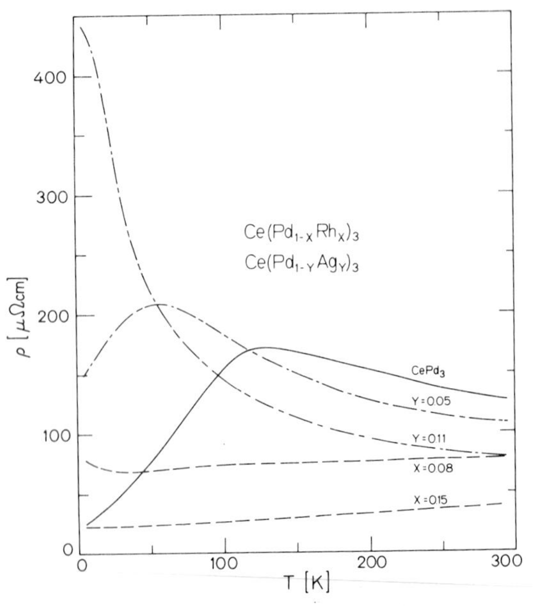

Fig. 4. Resistivity of $Ce(Pd_{1-x}Rh_x)_3$ and $Ce(Pd_{1-y}Ag_y)_3$.

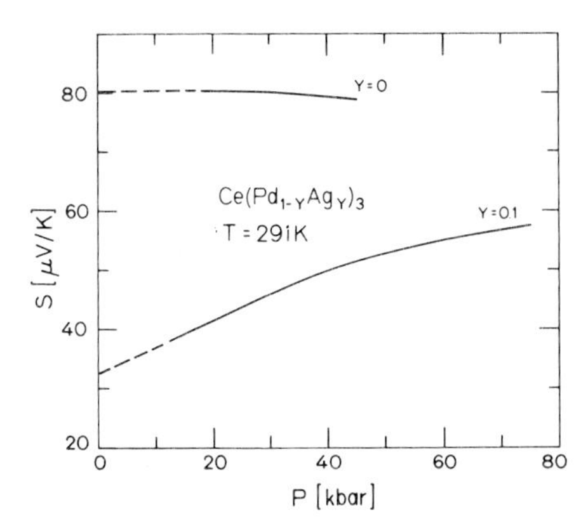

Fig. 5. Thermopower versus pressure of $Ce(Pd_{1-y}Ag_y)_3$.

In order to verify the correlation with the Ce valence, we also measured the pressure dependence of the thermopower at room temperature. The results for the two samples (y=0 and y=0.1) are shown in fig.5. Since hydrostatic pressure always favors the tetravalent state of the Ce ion,one expects the thermopower of $Ce(Pd_{0.9}Ag_{0.1})_3$ to increase towards that of $CePd_3$. In fact, we estimate that the valence of this sample increases from about 3.2 at normal pressure to about 3.3 at 80 kbar. For y=0 a nearly constant value is observed because the thermopower versus pressure just passes through its maximum.

4. DISCUSSION

Using the Hirst picture of an ICV [11], the low temperature linear thermopower can be evaluated by a two band model if we know the valence and the electronic specific heat of the compound. Using the results of Mihalisin et al. [5], we obtain for $CePd_3$: $S/T \simeq 1.2\ \mu V/K^2$. This ratio slightly increases with the Ag concentration up to y ≃ 7% (v=3.25) and then drops abruptly to a negative value for y ≃ 10% (v=3.14). Consequently our results are qualitatively explained by this model.

$CeA\ell_3$ has a thermoelectric behaviour similar to $Ce(Pd_{0.87}Ag_{0.13})_3$ with the small negative peak at low temperature. Bhattacharjee and Coqblin [12] have interpreted the large positive contribution by a Kondo scattering on the crystalline field levels of the 4 f^1 cerium configuration. However the negative peak at low temperature is not explained by their model.

Since all the nearly trivalent compounds with very different structure type have the same behaviour, we think that both negative and positive thermoelectric peaks are intermediate valence effects.

ACKNOWLEDGMENTS

We wish to thank R.Cartoni for his helpful technical assistance. This work was partly supported by the Swiss National Science Foundation.

REFERENCES

[1] R.J. Gambino, W.D. Grobman and A.M.Toxen, Appl.Phys.Lett. 22, 506 (1973)
[2] P. Scoboria, J.E. Crow and T. Mihalisin, J.Appl.Phys. 50, 1895 (1979)
[3] J.M. Lawrence, P.S. Riseborough and R.D. Parks, Rep. Progress Phys. 44, 1(1981)
[4] I.R. Harris, M. Norman and W.E. Gardner, J. Less-Common Metals 29, 299 (1972)
[5] T. Mihalisin, P. Scoboria and J.A. Ward, in Valence Fluctuations in Solids, L.M. Falicov, W. Hanke, M.B. Maple (eds), North Holland Publishing Company (1981) p.61
[6] D. Jaccard and J. Sierro, in these Proceedings
[7] J.R. Thomson, J.Less-Common Metals 13,307, (1967)
[8] H. Schneider and D. Wohlleben, Z.Phys. B44, 193 (1981)
[9] J.R. Cooper, C. Rizzuto and G. Olcese, J. Phys. C1 32, 1136 (1971)
[10] P.B. van Aken, H.J. van Daal and K.H.J. Buschow, Phys.Lett. 49A, 201 (1974)
[11] L.L. Hirst, Phys.Rev. B15, 1 (1977)
[12] A.K. Bhattacharjee and B. Coqblin, Phys.Rev. B13, 3441 (1976)

Valence Instabilities
P. Wachter and H. Boppart (eds.)

GIANT NEGATIVE MAGNETORESISTANCE IN SEMICONDUCTING SmS

M. Konczykowski, High Pressure Research Center "UNIPRESS",
Polish Academy of Sciences, 01-142 Warszawa
Sokolowska 29, Poland

J. Morillo, DECPu/SESI - C.E.N., 92260 Fontenay-aux-Roses, France

J.C. Portal and J. Galibert, INSA, Av. de Rangeuil, Toulouse
France

J.P. Senateur, CNRS, ER-155, St-Martin d'Hères, France

The negative magnetoresistance in nominally stoichiometric SmS samples was measured in pulsed magnetic fields up to 350 KΘe at temperatures from 60°K to 280°K. The resistivity decrease exceeding 99 % was observed (64°K, above 200 KΘe). The Hall coefficient measurements at 77°K up to 120 kΘe, indicate that negative magnetoresistance is due to an increase of free electron concentration under magnetic field. A qualitative model, involving a magnetic cluster or impurity is presented in order to explain these results.

1. INTRODUCTION

The low pressure phase of nominally stoichiometric SmS behaves as a small gap semiconductor. The resistivity, ρ, and Hall coefficient, R_H, are thermally activated in wide temperature range with apparent activation energy, E_A, ranging from 0.03 to 0.3 eV |ex. 1,2|. The band structure calculations predict that bottom of the conduction band of 5d character is located at X point of Brillouin zone and that 4f shells of Sm^{2+} ions act as a valence band |3|. The meaning of observed activation energies for ρ and R_H is not clear, mainly due to the uncertainty on extrinsic or intrinsic origin of conducting electrons. The recent investigations of electronic transport in SmS in the function of temperature and hydrostatic pressure point out the extrinsic origin of conducting electrons |4,5|. In order to clarify this question we focuse our attention on the negative magnetoresistance observed in most stoichiometric SmS samples |4,5,6|.

2. EXPERIMENT

The monocrystaline SmS samples, cleaved from the same ingots as in our previous studies |4,5| were used. The compositions of bulks was nominally stoichiometric. Sample preparation and characterization is described elsewhere |7|. Indium contacts were soldered on the faces of samples for resistivity measurements. The transverse magnetoresistance and Hall coefficient were measured with H parallel to <100> axis. The magnetoresistance measurements in pulsed magnetic fields were done in Toulouse with INSA facilility. Four probe AC method at 100 kHz was used for resistivity measurements. The precision of these measurements was 0.4% due to resolution of the digital storage oscilloscope used for recording of ρ vs. H curves. Above 60°K the results obtained on different samples coincide; below this temperature some erratic ρ vs. H variations was observed. It was probably due to sample inhomogenity or/and surface conduction, which starts to be important at low temperature.

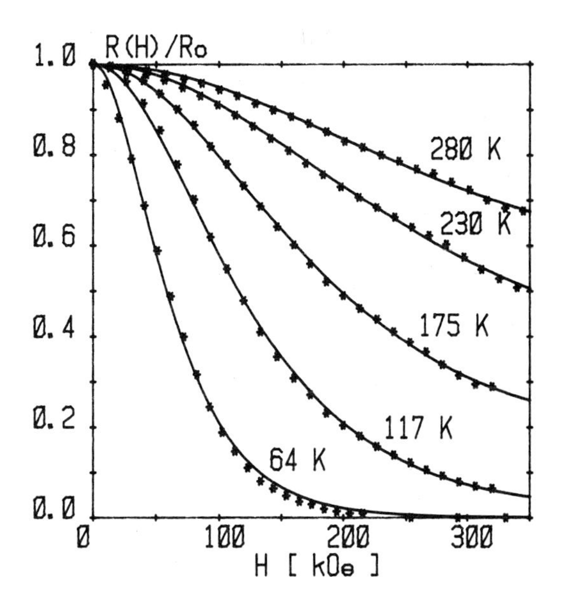

Figure 1. The typical resistivity vs. magnetic field dependencies measured on two samples of pure SmS. Stars represent some of experimental points continuous lines represent calculated variations of ρ(H).

In fig. 1 the typical ρ vs. H curves are presented. At low magnetic fields, the H^2 proportional decrease was observed at all temperature range. The resistivity drop under magnetic field increases as temperature is lowered. The observed decrease of resistivity exceeds two orders of magnitude below 100°K.

In the first paper reporting the observation of this negative magnetoresistance by Goncharova et al. |6|, this effect was attributed to an increase of electron mobility. As it seemed to us unlikely to increase the mobility by more then two orders of magnitude under magnetic field, we checked this hypothesis by measuring the Hall coefficient up to 120 kΘe in static magnetic field at liquid nitrogen temperature.

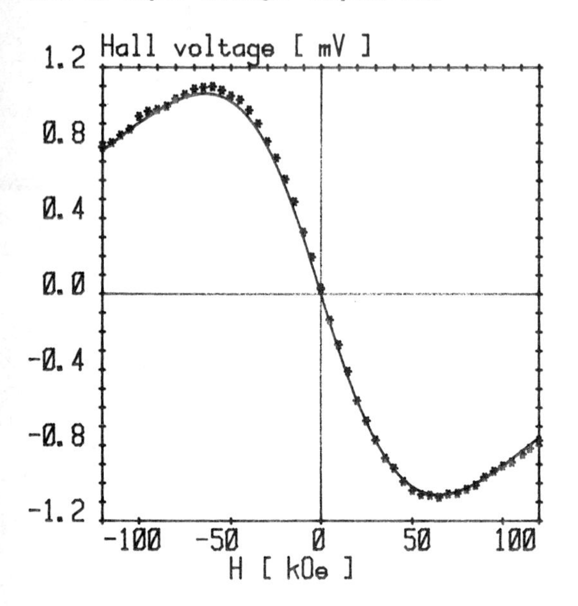

Figure 2. The variation of the Hall voltage vs. magnetic field at 77°K. Stars represent the experimental points; continuous line is the calculated V_H vs. H variation.

This was done at SNCI Grenoble with use of Bitter coil. The standard DC, van der Pauw method was used for measurements of Hall voltage and resistivity vs. magnetic field. In fig. 2 the variation of the Hall voltage, V_H, with magnetic field is presented. The S-like form of V_H vs. H curve means the decrease of R_H with magnetic field. The decrease of ρ, measured simultaneously is nearly the same as decrease of R_H resulting from $V_H(H)$ variation. It means that the Hall mobility, μ_H, is nearly invariant with magnetic field. The appearence of extraordinary Hall effect is unlikely in view of small magnetic susceptibility (see discussion). At low magnetic fields the resistivity decrease proportionaly to H^2, $\rho(H) = \rho_o\ (1-\alpha\ H^2)$. The magnetoresistance coefficient, α, was measured in the function of hydrostatic pressure for several temperatures. This was done at High Pressure Research Center "UNIPRESS" in Warsaw with use of helium gas compressor. Details of these measurements are presented in ref. 4 and 5. In fig. 3 some of magnetoresistance coefficient α, vs. pressure variations are presented. At pressures below 2 kbar α is invariant with pressure ; between 2 and 4 kbar α decreases rapidly and above 4.5 kbar the magnetoresistance disappears, ($|\alpha| < 10^{-6}$ $k\Theta e^{-2}$).

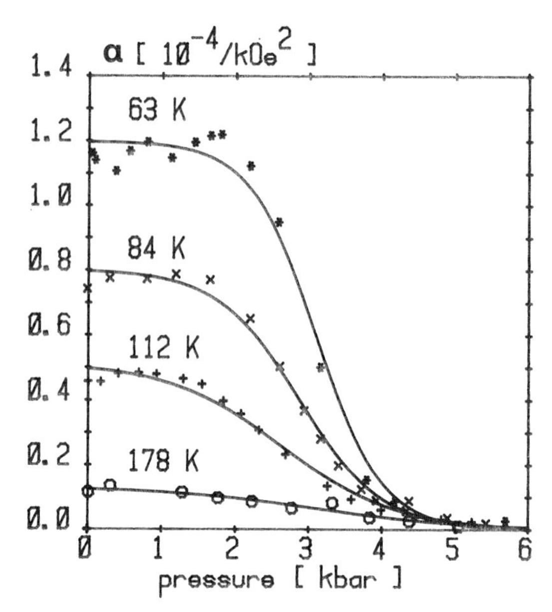

Figure 3. The typical magnetoresistance coefficient α vs. pressure variations. Lines represent the variations calculated in frame of the model presented in text.

3. DISCUSSION

The decrease of R_H with magnetic field indicates that the negative magnetoresistance is due to an increase of free electron concentration. The intrinsic origin of this effect seems unlikely. Indeed, the semiconducting SmS is a weak Van Vleck paramagnet (the ground state of intrinsic Sm^{2+} ion is a nonmagnetic $^7F_0$ singlet, first excited multiplet $^7F_1$ is 35 meV above), the magnetic susceptibility is nearly invariant with temperature and no magnetic order was observed even at very low temperature |2|. The magnetization measurements up to 150 kΘe do not reveal any striking anomaly |9|. The magnetic susceptibility increase slowly under hydrostatic pressure up to 6 kbar |8|, and no drastic changes of bulk magnetic properties corresponding to the disappearence of negative magnetoresistance at 2÷4 kbar was seen. In contrast, the existence of local magnetic anomalies in form of superparamagnetic clusters was detected in SmS : P,As |9|. From the other side the arguments for extrinsic origin of conducting electrons were presented in ref. 4 and 5. An impurity center of complex character was proposed in ref. 5 in order to explain the large pressure coefficient of activation energy for promotion of electrons from this impurity. The existence of the magnetic moment can be expected for ground and excited charge state of such a complex. (by excited charge state we consider this which gives the electron in the conduction band). The magnetic Sm^{3+} ions can be involved in it and large crystal field effect can perturb the initialy nonmagnetic ground state of neighbouring Sm^{2+} ions. Moreover the exchange interactions inside of the complex can lead to an increase of apparent Lande factor g |10|.

The magnetic field induced splitting of excited charge state multiplet facilitates the promotion of electrons from the complex to the conduction band. The splitting of the ground state acts in the opposite sense. To focuss our attention let us consider the impurity with singlet ground state. The concentration of the conducting electrons promoted from the impurity is given by formula 1 :

$$/1/ \quad n_e(H) = \frac{N_d}{1 + \frac{sh(x/2)}{sh((J+1/2)x)} \exp\{\frac{-(E_a + \varepsilon_F)}{kT}\}}$$

where $x = g\mu_B H/kT$, and E_a and ε_F denote respectively activation and Fermi energy at H=0. Assuming Boltzman statistics for conduction electrons, we obtain :

$$/2/ \quad n_e(H) = N_d\{\frac{2}{\sqrt{4C\frac{sh(x/2)}{sh((J+1/2)x)} + 1} + 1}\}; \quad C = \frac{N_d}{N_{ef}} e^{E_a/kT}$$

Supposing that Sm^{3+} ions are involved in the impurity complex we put J = 5/2 as for $^6H_{5/2}$ configuration of $4f^5$ shell. We fitted the variations of ρ vs. H presented in fig. 1 with formula 2. The conduction band parameters and impurity concentration were the same as in ref. 4 and 5. ($N_d \sim 2\times10^{20} cm^{-3}$). The Lande factor g and activation energy E_a were adjustable parameters. The continuous lines in fig. 1 and 2 represent the calculated ρ and V_H vs. H variation. For all temperatures we obtain nearly the same value of g factor (g = 16±2) and activation energy E_a (E_a = 50±15 meV). The accuracy of the fitting is satisfactory, despite that at highest magnetic fields the assumption of Boltzman statistics for conduction electrons is not valid. The change of J value does not affect the quality of fitting, only the values of adjusted g factor modifies roughly following $g J \approx$ const. The magnetoresistance coefficient α, can be calculated from the formula /1/ using Fermi-Dirac statistic for conducting electrons:

$$/3/ \quad \alpha = \frac{\{(2J+1)^2 - 1\} g^2 \mu_B^2}{48(kT)^2}\{1 + (2J+1)\exp\{\frac{-E_a - \varepsilon_F}{kT}\}\}^{-1}$$

The variation of magnetoresistance coefficient α, with pressure was interpreted as a result of decrease of activation energy E_a with pressure. As it was found in ref. 4, E_a is about 50 meV at ambient pressure (at 77°K) and it decrease at the rate ∿20meV/kbar when pressure is applied. The decrease of E_a leads to total depopulation of impurities above 4.5 kbar. We calculated the variations of α vs. pressure using /3/ and E_a vs. pressure variation from ref. 4. The solid lines in fig. 3 represent these dependences for g=13±1.

4. CONCLUSION

The presence of defect complex involving Sm^{3+} ions in pure SmS was detected by EPR measurements |10|. From the analysis of Lattice parameter and density in the function of stoichiometry results that SmS always contains at least 0.5% of sulphur vacancies or/and samarium interstitials |11|. The large perturbation of 4f shell of Sm ions neighbouring such a defect is expected |15|. The ionization energies can also be affected |14|. The formation of superparamagnetic clusters was observed in SmS:As and SmS:P |9,12|. In view of all this facts the presented interpretation of the negative magnetoresistance seems to be well founded. The presented model allowed us to explain quantitatively the behaviour of negative magnetoresistance in the function of magnetic field, temperature and pressure. However the model has only qualitative character. The g factor of the order of 16 is unrealistic for isolated ion. The large value of pressure coefficient $\partial E_a/\partial p \sim 20$ meV/kbar also indicates the complex character of ionization proces.

More detailed study, such as EPR under pressure are necessery to dress the microscopic model. The preliminary EPR measurements revealed the line corresponding to g∿13 in the spectra of highly nonstoichiometric SmS samples |13|.

REFERENCES

(1) Ditchenko, R. Gortsema, F.P., J. Phys. Chem. Sol. 24,(1963) 863
(2) Peña, O., Thèse, Université de Grenoble, France (1979)
(3) Davis, H.L., Rare Earths and Actinides, the Institute of Physics, London 1971, p. 126
Davis, H.L., Proceedings of the 9th Rare Earth Research Conference (Field P.E. edt.) 1971,vol.1
Farberovitch, O.V., Sov. Phys. Solid State, 21,(1979) 1982 and 22,(1980) 393
(4) Morillo, J., Konczykowski, M., Senateur, J.P., Solid State Commun., 35,(1980) 931
(5) Morillo, J., Thèse, Université d'Orsay, France (1981)
(6) Goncharova, E.V., Romanova, M.V., Sereeva, V.M., Smirnov, I.A., Fiz. Tverd. Tela., 19,(1977) 911
(7) Morillo, J., de Novion, C.H., Senateur, J.P., J. de Physique, 40, C5 (1979) 348
(8) Maple, M.B., Wohlleben, D., Phys. Rev. Lett., 27, (1971) 54
(9) Peña, O., Tournier, R., Senateur, J.P., Fruchart, R., J. Mag. Magn. Materials, 15-18, (1980)
Henry, D.C., Sisson, K.J., Savage, W.R., Schweitzer, J.W., Phys. Rev. B20, (1979) 1985
Krill, G., Senateur, J.P., Amamou, A., J. Phys. F, 10 (1980) 1889
(10) Walsh, W.M.Jr., Bucher, E., Rupp, L.W. Jr., Longinotti, L.D.,A.I.P. Conf. Proc., 24, (1975) 34
(11) Golubkov, A.V., Kartenko, N.F., Sergeeva, V.M., Smirnov, I.A., Fiz. Tverd. Tela., 20, (1978) 228
(12) Holtzberg, F., Peña, O., Penney, T., Tournier, R., in Valence Instabilities and Related Narrow Band Phenomena, R.D. Parks edt. (Plenum Press N.Y. 1977) p.507
(13) Beneu, F., private communication.
(14) Kasuya, T., J. de Physique, 10, C4 (1976) 261
(15) Parlabas, J.C., J. Mag. Magn. Materials, 15-18, (1980) 953

Valence Instabilities
P. Wachter and H. Boppart (eds.)

THE HALL EFFECT OF $CePd_3$

E. Cattaneo, U. Häfner and D. Wohlleben

II. Physikalisches Institut
Universität zu Köln
Zülpicher Str. 77
5000 Köln 41
West Germany

The Hall coefficient was measured for mixed valent $CePd_3$ and its normal reference compound YPd_3 as function of temperature. The Hall coefficient of $CePd_3$ is strongly abnormal in magnitude and temperature dependence while YPd_3 behaves like a normal metal. This is the first measurement of the Hall effect of a mixed compound which cannot be suspected of exhibiting a gap at low temperature.

INTRODUCTION

The Hall coefficient has rarely been measured in intermediate valence (IV) compounds.

It is known for SmB_6 /1,2/, for TmSe /3/ and for SmS /4/ in the high pressure collapsed phase. In all these systems the resistivity exhibits a steep increase at low temperatures suggesting the existence of a gap in their electronic energy spectrum. In most IV compounds on the other hand, the resistivity tends to go to zero for $T \to 0$, the more so the purer the samples. To our knowledge the behaviour of the Hall coefficient has never been studied in the latter situation. In other words, the Hall effect has never been studied in a truly metallic IV compound /5/. In this paper we report measurements of the Hall effect in such a compound, namely in $CePd_3$ (and in its reference compound YPd_3) between Nitrogen and Helium temperatures and at room temperature. Other transport properties of this metallic compound like electrical and thermal conductivity and thermopower are well known /6/.

EXPERIMENTAL

The polycrystalline samples of $CePd_3$ and YPd_3 were prepared in an induction furnace by melting stoichiometric amounts of the elements. The samples were annealed in vacuum at 900 C during four days. Portions of both samples were x-rayed and only lines belonging to the cubic Cu_3Au structure were observed above the usual sensitivity limit of 5%. The lattice constant values are in agreement with literature values within 0.4%.

An Indium film was evaporated on the sample through a mask. Copper leads (50 µ) were then attached to it with Indium solder. The samples were rectangular in shape with cross-section $(3.25 \times 0.38)mm^2$ and 6.55 mm length for $CePd_3$ and $(0.43 \times 3.23)mm^2$ with 5 mm length for YPd_3.

The resistivity of both samples was also measured between 300 and 4.2 , with the following results: For $CePd_3$ the resistivity ratio was 5.8 with a resistivity at 4.2 K of ρ_o=21.4 µΩcm. For YPd_3 the corresponding values are 12.6 and ρ_o=0.7 µΩcm.

The Hall effect was measured in the five lead configuration /7/ with an ac current of 117 Hz. The current was 10 mA for $CePd_3$ and 50 mA for YPd_3. The transverse ac voltage was measured with a Lock-In with a preamplifier to enable voltage measurements in the nanovolt range. The Hall voltage was measured in both magnetic field directions; the maximal field, applied with a superconducting magnet was 50 K gauss. The temperature range was 1.6 to 110 K. At room temperature the Hall voltage of $CePd_3$ was measured up to 8 K gauss in ac and in dc in order to obtain the absolute sign of the Hall coefficient. Three different samples were measured at this temperature and two at low temperatures.

No geometry corrections were applied to R_H.

RESULTS

Fig. 1 shows the Hall voltage V_H as function of magnetic field and temperature.

At high temperatures V_H is linear in field whereas a curvature in V_H vs. H sets in below 50 K; the Hall voltage changes sign just above 10 K. This is shown with more resolution in Fig. 2 where the Hall constant $R_H=(V_H \cdot d)/(H \cdot I)$ is shown on some critical isotherms. Clearly ist is necessary to specify not only temperature but also magnetic field for the zero crossing of the Hall coefficient; at low fields R_H crosses zero at a few tenths of a K below 10 K.

Fig. 3 shows R_H as taken at 26 K gauss plotted as a function of temperature below 110 K for $CePd_3$ and YPd_3. The error bars in temperature are due to thermal drift in the low temperature equipment. In Fig. 3 we show also R_H of $CePd_3$

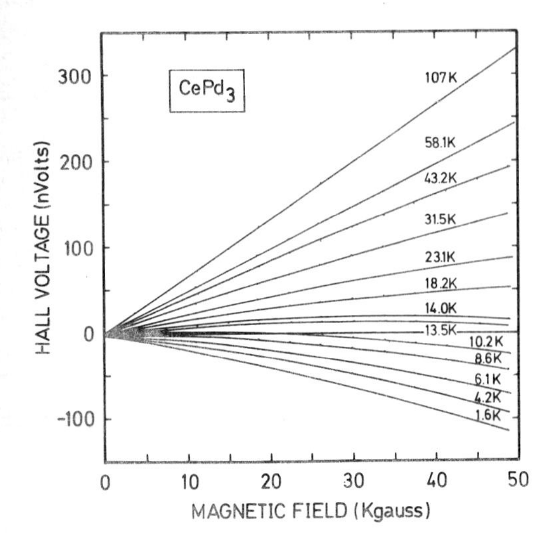

Fig. 1 Magnetic field dependence of the Hall voltage V_H of $CePd_3$ at different temperatures.

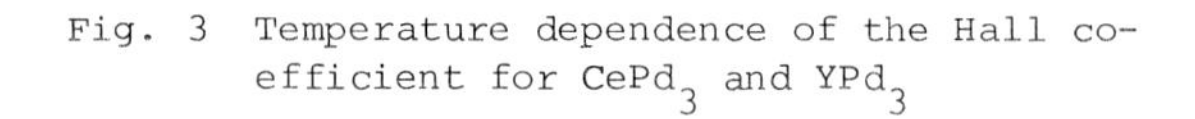

Fig. 3 Temperature dependence of the Hall coefficient for $CePd_3$ and YPd_3

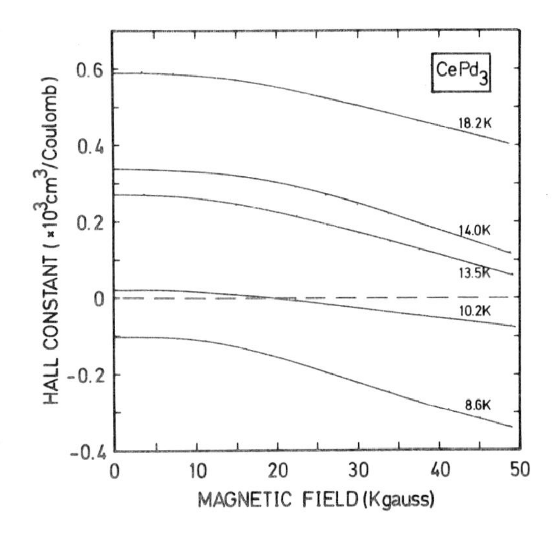

Fig. 2 Magnetic field dependence of the Hall coefficient $R_H=V_H.d/I.H$ at temperatures near the sign reversal (V_H= Hall voltage, d = thickness of the sample, I = current and H = magnetic field)

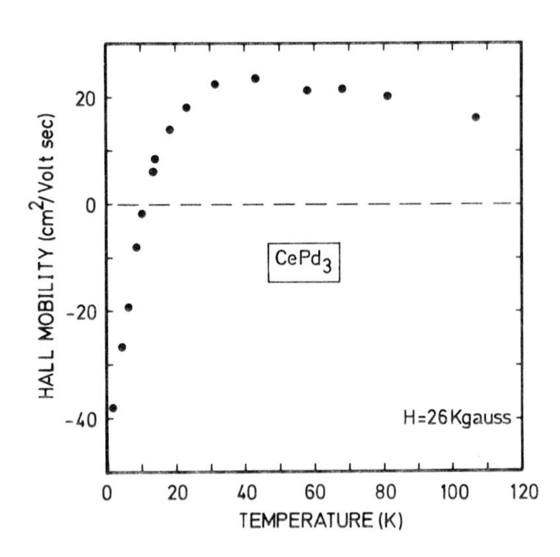

Fig. 4 Temperature dependence of the Hall mobility $\mu_H= R_H/\rho$ for $CePd_3$ (R_H = Hall coefficient, ρ = electrical resistivity)

as measured at 300 K in a different experimental arrangement.

Fig. 4 shows the Hall mobility of $CePd_3$ as defined by $\mu_H = R_H/\rho$. The resistivity values used in this plot were measured simultaneously with the Hall effect.

The Hall constants in Fig. 2 and Fig. 3 have an absolute error of about 15%.

DISCUSSION

The main result of this experiment is shown in Fig. 3. The Hall coefficient of YPd_3 is negative with only weak temperature dependence and with an absolute value around 8×10^{-5} cm^3/cb. This corresponds to a density of carriers of $n=(eR_H)^{-1}=7.8 \times 10^{22}/cm^3$ or about 4 electrons per formula unit of YPd_3. In other words the Hall coefficient of YPd_3 is that of a normal metal.

The Hall coefficient of $CePd_3$ is up to 30 times larger than that of YPd_3. It shows a strong temperature dependence, including a reversal of sign near 10 K and a maximum near 120 K. Above 10 K the sign is positive. All these properties are quite abnormal for a metal. The order of magnitude of the Hall coefficient of $CePd_3$ at high temperatures is comparable to that of the already mentioned other mixed valent compounds (SmB_6, TmSe, SmS in the collapsed phase). At low temperatures however the Hall coefficient of $CePd_3$ remains much smaller in magnitude than that of the others. This was to be expected because of the metallic character of $CePd_3$.

Since $CePd_3$ definitely derives from YPd_3 and must have a fraction of one conduction electron more than this reference compound, the large magnitude of the Hall coefficient of $CePd_3$ cannot be due to a lower conduction electron density. Instead it must be due to strong scattering.

We cannot derive any strong conclusions from these preliminary results on a single metallic IV compound; measurements on $YbCu_2Si_2$ and others are in progress.

Acknowledgements

We thank N. Rüßman, F. Ackermann, J. Kalenborn, A. Alpsancar and B. Schlicht for their advice at different stages of the experiment. We also thank S. Wood and F. Simons for technical assistance.

This work was supported by the Deutsche Forschungsgemeinschaft through SFB 125 and by the Deutscher Akademischer Austauschdienst by a scholarship to E.C.

REFERENCES

/1/ J.C. Nickerson, R.M. White, K.N. Lee, R. Bachman, R.H. Geballe and G.W. Hull, Jr., Phys. Rev. B3, 2030 (1971)

/2/ J.W. Allen, B. Batlog and P. Wachter, Phys. Rev. B 20, 4807 (1979)

/3/ P. Haen, F. Lapierre, J.M. Mignot and R. Tournier, Journal of Magnetism and Magnetic Materials 15-18, 989-990 (1980)

/4/ M. Konczykowski, J. Morillo, J.P. Senateur, in Valence Fluctuations in Solids, L.M. Falicov, W. Hanke and M.B. Maple (eds.) North Holland Publishing Company (1981) p. 287 and references therein.
T.L. Bzhalava, M.L. Shubnikov, S.G. Shul'man, A.V. Golobkov and I.A. Smirnov, Sov. Phys. Solid State 18, 1201 (1976)

/5/ The only exception known to us is $CeCu_2Si_2$ for which a carrier density $n=2 \times 10^{21}/cm^3$ was extracted from the Hall coefficient at 4.2 K by F.G. Aliev, N.B. Brandt, E.M. Levin, V.V. Moshalkov, S.M. Chudinov and R.I. Yasanitski, Solid State Physics (Russian original), 24, 289 (1982)

/6/ P. Scoboria, J.E. Crow and T. Mihalisin, J. Appl. Phys. 50, 1895 (1979)
H. Schneider and D. Wohlleben, Z. Phys. B 44, 193 (1981)
R.J. Gambino, W.D. Grobman and A.M. Toxen, Appl. Phys. Lett. 22, 506 (1973)

/7/ Albert I. Schindler and Emerson M. Pugh, Phys. Rev. 89, 295 (1953) and references therein.

Valence Instabilities
P. Wachter and H. Boppart (eds.)

THE SPECIFIC HEAT OF $YbCu_2Si_2$ AND SOME $Yb_xY_{1-x}Cu_2Si_2$ ALLOYS

R.Kuhlmann , H.-J. Schwann , R. Pott , W. Boksch and D. Wohlleben

II. Physikalisches Institut der Universität zu Köln, Köln, Federal Republic of Germany

The specific heat of the mixed valent compound $YbCu_2Si_2$, of the reference compound $LuCu_2Si_2$ and of the series of alloys $Yb_xY_{1-x}Cu_2Si_2$ (x=1.,.75,.5,.25) was measured between 1.5 and 400 K. The entropy of the specific heat anomaly of $YbCu_2Si_2$ is consistent with Rln9 at 400 K, as expected for the high temperature limit of the valence of Yb. The linear specific heat coefficient per Yb-atom shows a slight concentration dependence.

INTRODUCTION

It is well known, that most intermetallic Rare Earth mixed valence compounds show large linear specific heat coefficients at helium temperature , one to three orders of magnitude larger than in normal metals /1//2/. This implies the absence of a gap between the ground state and the excited states of the electronic spectrum. Instead the Fermi-energy must lie in a sharp peak of this spectrum at a density of states of order 1 to 100 states/eV atom. In the case of $YbCu_2Si_2$ (γ=135 mJ/Mol K^2 ,/3/) we have about 55 states/eV Yb-atom at the Fermi-level, from which one estimates a peak width of order 200 K. By several experimental techniques one finds a characteristic temperature of about T_f=70 K in this system /3/. By varying the temperature between helium and roomtemperature, one should therefore be able to change the thermal energy from smaller to larger than the peak width. From measurements of the specific heat through this temperature range one can then derive the entropy and the number of states per atom contributing to this peak. In the ionic model, according to which the $4f_{13}$-configuration with a J_{13}=7/2 Hund's rule ground state is degenerate with the $4f_{14}$-configuration (J_{14}=0), one expects for the entropy per Yb-atom $S=R\ln 2(J_{13}+J_{14}+1)=R\ln 9$ /4/. This value is expected in the ionic model, independent of the details of the low temperature structure of the spectrum, or its temperature dependence and independent of the concentration of the Yb-atoms. In this paper we present specific heat measurements of $YbCu_2Si_2$, $LuCu_2Si_2$ and of $Yb_xY_{1-x}Cu_2Si_2$ between 1.5 K and 400 K with the aim of measuring the entropy due to the Yb-ions and to extract, if possible, more information about the structure of the excitation spectrum of the Yb-ion , in particular of its concentration dependence.

EXPERIMENTAL PROCEDURE AND RESULTS

The samples of $LuCu_2Si_2$ and YCu_2Si_2 were molten together in an arc furnace from the stochiometric weights of the constituents and remolten 4 to 5 times. Weightlosses were of order 0.3 a|o. $YbCu_2Si_2$ was molten together in an induction furnace under an argon atmosphere from the stochiometric weight of the constituents, however with 5 a|o surplus Yb. The weightloss during the melting procedure was assumed to be due to loss of Yb. The sample was remolten with additional Yb as many times as necessary to reach the stochiometric weight within 2 a|o, at least 4 times. The $Yb_xY_{1-x}Cu_2Si_2$-alloys were molten together in the induction furnace from the appropriate ratio of $YbCu_2Si_2$ and YCu_2Si_2, remolten at least 3 times. Weight losses were less than 1 a|o.

The specific heat measurements were performed between 1.5 K and 400 K, in a calorimeter described elsewhere /5//6/. Between 1.5 K and 40 K the specific heat was measured step by step and between 25 K and 400 K by a continuous method /7/. The absolute error of the specific heat, measured in the step by step method, is 5% and in the continuous method 3%.

Fig.1 shows the low temperature specific heat of $LuCu_2Si_2$. From the insert we extract a linear specific heat coefficient of γ_l=(2.3±.2)mJ/Mol K^2 and a Debye temperature of Θ_1=(376±3)K. In the region between 17 K and 27 K there is a second aproximately linear section, from which one can extract a T^3-coefficient, corresponding to a Debye temperature of Θ_2 =(277±10)K.

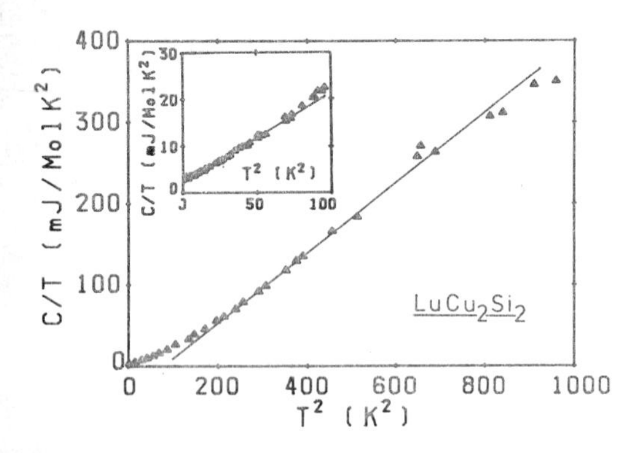

Fig.1: Specific heat of $LuCu_2Si_2$ from 1.5 K to 30 K.

In Fig.2 we show the low temperature specific heat of $YbCu_2Si_2$ between 1.5 K and 30 K and again between 1.5 K and 20 K in the insert. A straight section between 10 K and 14 K in the insert yields $\gamma_1=(131\pm13)$ mJ/Mol K^2 and $\Theta_1=(334\pm15)$K. It is possible to find a second straight section of C/T vs. T^2 between 16 K and 28 K, which gives $\Theta_2=277$ K. The deviation from linearity below 10 K and the strong rise below 3 K is sample dependent. We attribute the low temperature rise to the presence of Yb_2O_3 in the sample.

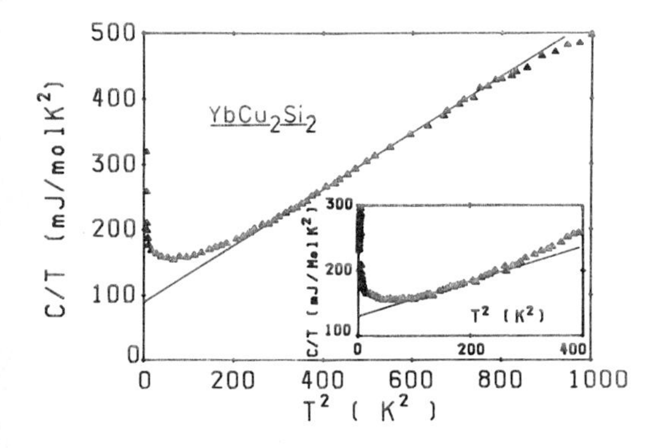

Fig.2: Specific heat of $YbCu_2Si_2$ from 1.5 K to 30 K.

In Fig. 3 we show the specific heat of $LuCu_2Si_2$ and the difference ΔC between the specific heats of $YbCu_2Si_2$ and $LuCu_2Si_2$. At high temperatures the difference is as large as our systematic error (indicated by the bars). The undulations are an artefact of the data analysis (Spline fit). At low temperatures however, where the large specific heat anomaly of $YbCu_2Si_2$ dominates the total specific heat, the relative accuracy of ΔC becomes much better.

Fig.4 shows $\Delta C/T$, as obtained from ΔC in Fig.3, and ΔS , the entropy of the specific heat difference up to 400 K. The absolute error of $\Delta C/T$ becomes smaller relative to that of ΔC at high temperatures. The integration of $\Delta C/T$ yields a value of $\Delta S=(19\pm5)$J/Mol K at 400 K.
Fig.5 is a collection of data, obtained from the alloy series. These data were extracted from the specific heat data in a manner completely analogous to the procedure described above for $YbCu_2Si_2$. However the reference compound for the alloys was YCu_2Si_2 throughout.

DISCUSSION

Our low temperature measurement of the specific heat coefficient of $YbCu_2Si_2$ agrees well with the value given by Sales and Visvanathan /3/, while the Debye temperature $\Theta_2=277$ K, is 20% larger than theirs. The scatter of our data is less severe than in /3/, presumably because our sample size is much larger. In fact our data agree both in γ_1 and in Θ_1 with data of a large sample presented in /8/. The straight section of C/T vs. T^2 between ≃15 K and ≃27 K in both, $YbCu_2Si_2$ and $LuCu_2Si_2$, yields the same value $\Theta_2=(280\pm5)$K. It therefore seems, that this feature is due to phonon specific heat, while the large γ_1 and the deviation of Θ_1 of $YbCu_2Si_2$ from that of $LuCu_2Si_2$ is a consequence of the open 4f-shell of Yb. The change of Debye temperature ($\Theta_1=334$ K for $YbCu_2Si_2$ vs. $\Theta_1=376$ K for $LuCu_2Si_2$) corresponds to an electronic contribution of the T^3-term. As shown

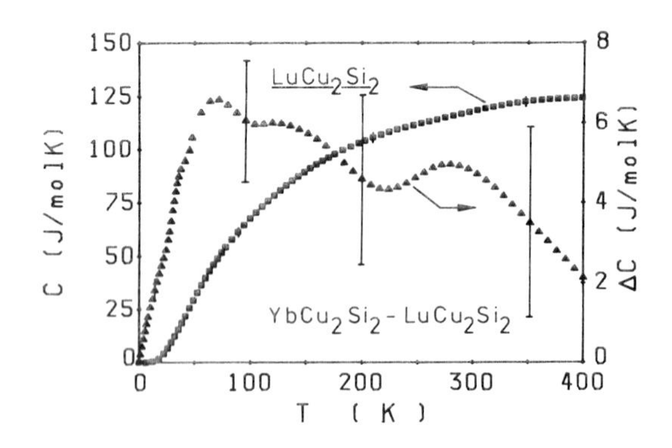

Fig.3: Specific heat C of $LuCu_2Si_2$ and the specific heat anomaly, the difference ΔC between the spec. heats of $YbCu_2Si_2$ and $LuCu_2Si_2$ between 1.5 K and 400 K.

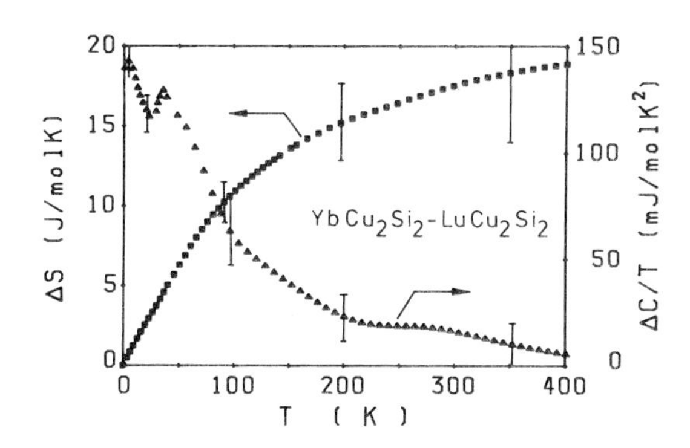

Fig.4: The reduced specific heat anomaly $\Delta C/T$, and the entropy of the anomaly, between 1.5 K and 400 K.

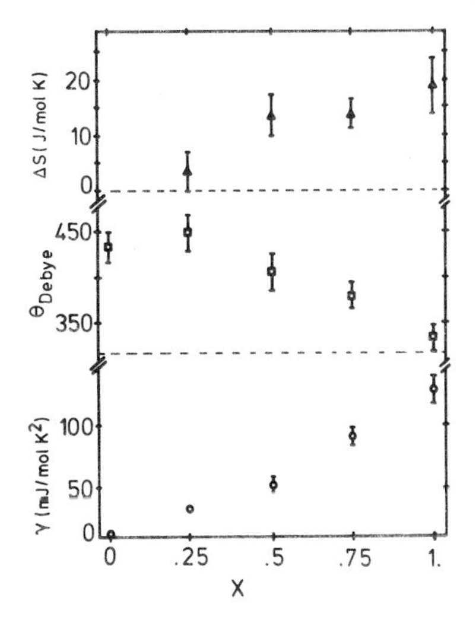

Fig.5: Entropy of specific heat anomaly at 400 K, ΔS, Debye temperature, Θ_D and linear specific coefficient, γ for $Yb_xY_{1-x}Cu_2Si_2$ alloys.

in /4/, it is not due to phonon softening. This contribution causes an increase of the T^3-term of 42% with respect to $LuCu_2Si_2$. This increase should be compared with an increase of 200% of the T^3-term of YbCuAl, compared with that of LuCuAl /4/. The electronic contribution of the T^3-term arises from a combination of the first and second derivatives of the electronic density of states with respect to energy /4/. Apparently the smaller contribution and the smaller linear specific heat coefficient of $YbCu_2Si_2$ are indicating the same thing, namely that the peak in the electronic spectrum is lower and wider in $YbCu_2Si_2$ than in YbCuAl. This can also be seen from the position of the maxima of the specific heat anomalies ($T_{\Delta Cm}$=28 K in YbCuAl and 70 K in $YbCu_2Si_2$). The entropy of the peak, however, $\Delta S=(19\pm5)$J/K mole Yb, is again very close to S=Rln9=18.3 J/K mole Yb, as expected for mixed valent Yb-ions in the high temperature entropy limit /9/ and as found in YbCuAl ((17.5±2)J/mol K) /4/ and in $YbNi_2Ge_2$ ((18.5±5)J/mole K) /10/. The error of the measured entropy is of course large. But it is remarkable that the entropy is almost saturated (Fig.4) where the theoretical value is reached experimentally. This and the fact, that the thermal expansion anomaly has alsc reached zero at 400 K /11//9/, strengthens our confidence in the analysis.

In the alloy measurements of Fig.5 the linear specific heat coefficient is the most reliable number. Clearly this coefficient decreases with concentration somewhat faster than linearly outside of experimental error. While in the compound we have 55 states/eV Yb-atom, at x=.5 and x=.25 we have 35 states/eV Yb-atom. This indicates some slight concentration dependance of the mixed valence phenomena on the Yb-ions in the series of alloys. Nothing much can be said about the entropy of the alloys, except that it is independent of concentration within the considerable experimental error.

This work was supported by SFB 125 of the Deutsche Forschungsgemeinschaft.

REFERENCES

/1/ Valence Instabilities and Related Narrow Band Phenomena, R.D.Parks, ed., Plenum Press,New York (1977)

/2/ Valence Fluctuations in Solids, L.M.Falicov, W.Hanke and M.B.Maple, eds. North Holland, Amsterdam (1981)

/3/ B.C.Sales and R.Visvanathan, J. Low Temp. Phys. 23, 449 (1976)

/4/ R.Pott,R.Schefzyk,D.Wohlleben and A.Junod, Z. Phys. B - Condensed Matter 44, 17 (1981)

/5/ H.-J.Schwann, Diplomarbeit, Universität zu Köln (1982)

/6/ R.Kuhlmann, Diplomarbeit, Universität zu Köln (1981)

/7/ A.Junod, J. Phys. E12, 945 (1979)

/8/ B.Sales, Ph.D Thesis, University of California, San Diego (1974)

/9/ D.Wohlleben, pg 1 in Ref.2

/10/ R.Pott, Ph.D.Thesis , Universität zu Köln (1982)

/11/ R.Schefzyk, Diplomarbeit, Universität zu Köln (1980)

Valence Instabilities
P. Wachter and H. Boppart (eds.)

VALENCE CHANGE IN (Ce,Sc)Al_2 - STUDIED BY ELECTRONIC TRANSPORT PROPERTIES

S.Horn, W.Klämke and F.Steglich

Institut für Festkörperphysik, Technische Hochschule Darmstadt,
and Sonderforschungsbereich 65, D-6100 Darmstadt, Fed. Rep. Germany

Measurements of the resistivity and thermoelectric power of $Ce_{1-x}Sc_xAl_2$, performed in the temperature range 1K to 300K, reveal several qualitative differences between $x \leq 0.4$ and $x \geq 0.5$. They confirm the existence of a valence transition near x=0.5, as concluded from neutron-scattering.

INTRODUCTION

$CeAl_2$ contains (nearly) trivalent Ce ions and is often classified as "Kondo lattice". It shows complex antiferromagnetism below $T_N \cong 3.9K$.[1] When subject to an external pressure of ≅65 kbar (at 300K), $CeAl_2$ undergoes an (isostructural) transition into an intermediate valent (IV) state of the Ce ions.[2] A similar valence transition has been inferred to exist in the quasibinary alloy system $Ce_{1-x}Sc_xAl_2$ near x=0.5.[3-5] In these alloys, $ScAl_2$, which is isostructural to $CeAl_2$, but with smaller lattice parameter, gives rise to what is sometimes called "internal pressure". Since the valence change takes place in a rather narrow concentration range,[5] $Ce_{1-x}Sc_xAl_2$ is well suited to compare in a reliable way physical quantities in the two valence states. Of course, local environment effects may complicate analyses in a disordered alloy system. In fact, such effects were found in previous ESR results on (Ce,Sc)Al_2.[4] In this note we report on the (absolute) resistivity and the thermoelectric power (TEP) in the whole composition range of the $Ce_{1-x}Sc_xAl_2$ system. While the former quantity samples the majority of Ce ions, the latter is sensitive to single-ion effects and may, therefore, reflect different local environments of the Ce ions. Preliminary results have been included in a recent report.[6] Relative resistivities of $Ce_{1-x}Sc_xAl_2$ for $x \lesssim 0.5$ showed a negative temperature coefficient (NTC).[3]

SAMPLE PREPARATION AND CHARACTERIZATION

$Ce_{1-x}Sc_xAl_2$ was prepared in an induction furnace from stoichiometric amounts of Ce, Sc and Al. Weight losses were less than 1‰. X-ray powder diffractometry showed that all samples had the proper $MgCu_2$ structure. Compared to the pure compounds $CeAl_2$ and $ScAl_2$, the X-ray reflections of $Ce_{1-x}Sc_xAl_2$ were found to increase with substitutional disorder, i.e. by 25% for x=0.6 and by 50 % for x=0.4. For the x=0.5 sample the reflections were about twice as broad as for the x=0.4 sample.

RESULTS AND DISCUSSION

<u>$Ce_{1-x}Sc_xAl_2$ alloys with $x \leq 0.4$.</u> The temperature dependence of the resistivity, $\rho(T)$, shown in fig. 1, is qualitatively similar to that reported, for the same concentration range, in ref.3: it is dominated by a Kondo anomaly due to Ce^{3+}, modified by a structure near 70K, which is caused by the Γ_7-Γ_8 crystal-field (CF) splitting of Ce^{3+}. For $x \leq 0.3$, a peak near 5K appears which is probably connected with Ce^{3+}- Ce^{3+} magnetic (short-range) interactions. We observe that the low-temperature resistivity, ρ_{5K}, which is certainly dominated by the action of the Ce^{3+} scatterers, surprisingly increases with <u>decreasing</u> Ce concentration, 1-x (see fig.3). This hints at a strong interference between electron scattering from the Ce^{3+} ions and from non-magnetic scatterers, as was also concluded from earlier results on the related systems (La,Ce) Al_2[7] and $(La_{1-z}Y_z$,Ce)Al_2.[8]

As is shown in fig.2, the thermal variation of the TEP, S(T), of $CeAl_2$ and the Ce-rich alloys confirms the above conclusions from the resistivity: there exists a minimum at $T_s^{(-)}$=5-10K, this temperature being close to the "Kondo" or "spinfluctuation-temperature" T* as derived from neutron scatte-

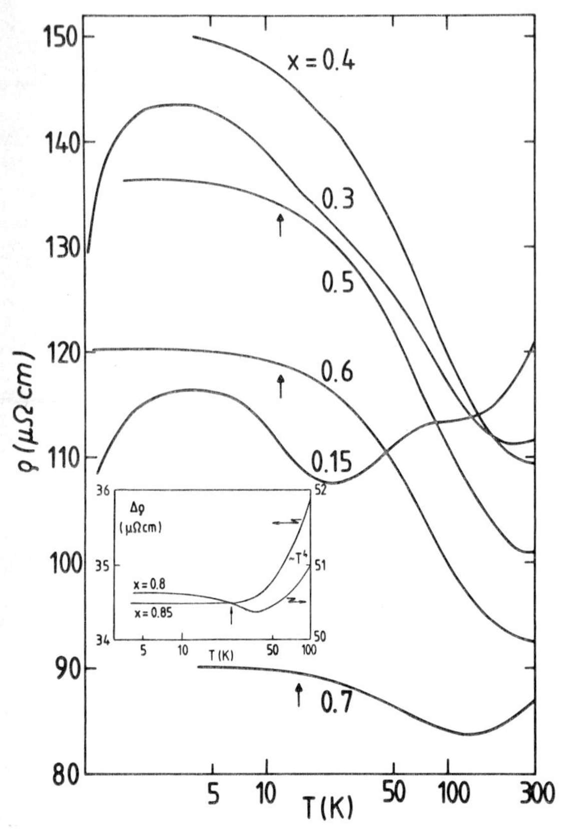

Fig.1 ρ vs T (logarithmic scale) for various $Ce_{1-x}Sc_x Al_2$ alloys. Inset shows $\Delta\rho = \rho_{Ce_{1-x}Sc_xAl_2} - \rho_{ScAl_2}$ for x=0.8 and x=0.85. Arrows mark temperature T_ρ (see text).

ring.[5] In addition, the TEP of $CeAl_2$, when compared to that of $LaAl_2$ (inset) exhibits a maximum near $T_s^{(+)} \cong$ 130K. This feature is probably connected with the CF splitting of Ce^{3+} into a Γ_7 ground state doublet and excited Γ_8 states. The overall shape of the observed S(T) curve is similar to theoretical results by Bhattacharjee and Coqblin.[9] On a quantitative scale, we find $T_s^{(+)}$ to be comparable to the overall CF splitting Δ_{CF},[10] although it should range between 1/3 and 1/6 of Δ_{CF} according to ref.9. There are two possible reasons for this discrepancy: (i) while the theory treats the case of dilute Ce^{3+} ions, the measurements were done with a Ce intermetallic, and (ii) the CF level scheme is complicated for $CeAl_2$ and its quasibinary alloys.[10] As usual, the influence of non-magnetic scattering centers tends to depress the Ce-derived TEP anomalies. This can be seen in the inset, where the results of $CeAl_2$ are compared with those of $Ce_{0.7}Sc_{0.3}Al_2$. It is likely that the change in bandstructure on going

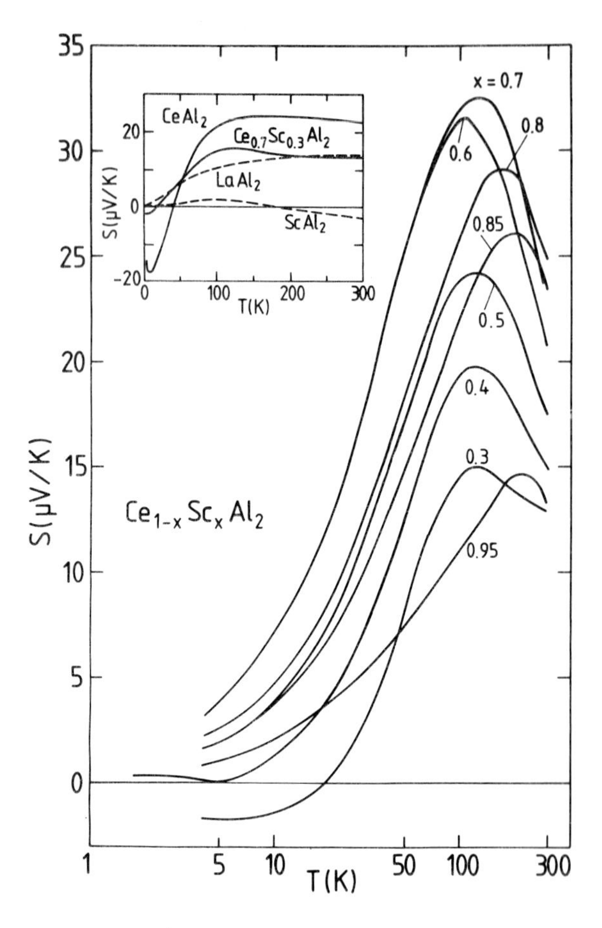

Fig.2 S vs T (logarithmic scale) for various $Ce_{1-x}Sc_x Al_2$ alloys. In the inset are compared data of the x=0.3 sample with data of $CeAl_2$, $LaAl_2$ and $ScAl_2$ [ref.13]. Coefficients of linear low-T TEP, $B=\frac{S}{T}$, are (in μVK^{-2}): 0.40 [x=0.5]; 0.71 [0.6;0.7]; 0.41 [0.8]; 0.39 [0.85] and 0.21 [0.95].

from $CeAl_2$ or $LaAl_2$ to $ScAl_2$ may influence the actual shape of S(T) for the disordered systems. In particular, this could cause the very flat low-T maximum, which almost completely masks the minimum, for the x=0.4 sample. Finally, an increase is found in the height of the high-T maximum when comparing the results for x=0.4 with those of x=0.3. This could originate from local-environment effects, i.e. from a small number of IV-Ce ions, which give rise, as we will see below, to a positive TEP peak at T≧100K. Positions of both the minima and maxima in S(T) are given as function of concentration in fig.3.

$Ce_{1-x}Sc_xAl_2$ alloys with x≧0.5. According to ear-

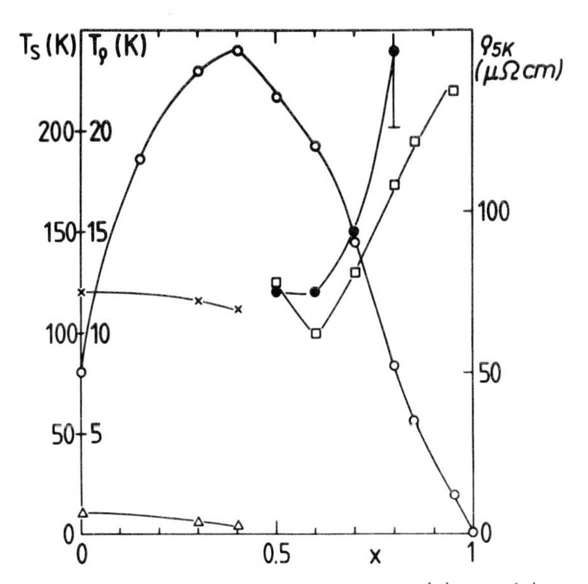

Fig.3 Characteristic temperatures, $T_s^{(-)}$[△], $T_s^{(+)}$[x], $T_{s,max}$[□] and T_ρ [•] (see text), as well as low-T resistivity, ρ_{5K}[o], for $Ce_{1-x}Sc_xAl_2$ as function of x.

lier experiments,[3-5] the Ce ions are expected to be in a homogeneous IV state in this concentration range. It is, therefore, not surprising that the electronic transport properties behave, in part, distinctly differently from those of the trivalent alloys (with x≦0.4): Although the resistivity curves show a "Kondo-like" NTC for concentrations ranging up to x=0.8, no CF-derived structure is resolved. For x=0.5 (with largest substitutional disorder), the low-T resistivity, ρ_{5K}, should be largest, if the Ce ions were to remain trivalent. However, a substantial reduction of ρ_{5K} is found when compared to x=0.4. As is shown in fig.3, the ρ_{5K} vs x dependence becomes increasingly steeper up to x=0.8, and bends off slightly for higher x values. The TEP, S(T), shows only one (positive) peak rather than a positive and a negative peak as found for the Ce-rich samples. As in Kondo systems, the temperature of the TEP peak in IV system usually is of the same order as the "spinfluctuation temperature" T^* from neutron scattering. For x=0.5, we find $T_{s,max}$≅125K to be roughly one half of T^*=(200±50)K.[5] We also find $T_{s,max}$≅100K for x=0.6 and, in addition, a continous increase up to $T_{s,max}$≅220K, when x becomes increased to x=0.95 (see fig.3). The height of the TEP peak, S_{max}, assumes the giant value of ≅33μV/K for the x=0.7 sample. Our transport results, presented so far, confirm the conclusions drawn from the previous neutron-scattering experiments on $Ce_{0.5}Sc_{0.5}Al_2$ that (i) no CF splitting can be resolved and (ii) the "spinfluctuation temperature" is much larger than for $Ce_{0.6}Sc_{0.4}Al_2$.[5]

At sufficiently low temperature, we observe typical "Fermi-liquid effects" in both ρ(T) and S(T): Below a temperature T_ρ (for x=0.5: above ≅5K), the resistivity can be well fitted by $\rho=\rho_{5K}(1-AT^2)$ and below 15-20K, the TEP strictly follows a dependence S=BT. T_ρ is indicated in fig.1 by the arrows and plotted vs x in fig.3. It is found to roughly track the concentration dependence of the peak temperature of the TEP, $T_{s,max}$. The coefficients B of the low-T TEP are collected in the caption of fig.2. Following Jaccard and Sierro,[11] we can correlate the coefficient B through $B\cong -566T\cdot T_o[T_o^2+(T_w/2)^2]^{-1}$ with the energy parameters T_w and T_o, which characterize, in a model by Hirst,[12] the band of low-energy excitations ("fermion quasiparticles") in the low-T phase of an IV system. Here, $W=k_BT_w$ measures the width and $E_F-E_o=k_BT_o$ the separation of this quasiparticle band from the Fermi energy E_F. The ratio T_w/T_o is unequivocally linked to the valence of the Ce ions. If we assume that (i) the "spinfluctuation temperature" $T^*\cong 2T_{s,max}$, inferred for x=0.5, holds for x > 0.5, too, we arrive, from the measured coefficients B, at two different average values of the low-T valence, i.e. v≅3.16-3.17, with $-T_o$≅(10-30)K, and v≅3.75-3.85, with $-T_o$≅(800-2700)K. From the present transport data alone, we cannot decide which of these alternative situations is closer to reality. This must be left to other techniques. To derive a possible x-dependence of the valence, we would need to know the incremental (Ce-derived) parts of the TEP. This would require determining incremental resistivities, namely by subtracting from the data of fig.1 the resistivities of related host alloys, e.g. $La_{1-x}Sc_xAl_2$. Unfortunately, preparation of good-quality samples of this latter system has proven to be impossible up to now. Another point must be clarified by future investigations: as is indicated in the inset of fig.1, the difference Δρ of the resistivities of $Ce_{1-x}Sc_xAl_2$ and $ScAl_2$ shows a temperature dependence with NTC for x=0.8, but is temperature independent below 15K and proportional to T^4 for x=0.85. Such a T-dependence is common to several d-band metals, most notably $LaAl_2$.[7] If this rather rapid change in $\Delta\rho(T)$ could be located to an even narrower concentration range, this would possibly indicate a complete demagnetization of Ce in the very dilute limit of $Ce_{1-x}Sc_xAl_2$. The TEP peak observed for the x=0.95 sample, would, then, have to be explained with local environment effects, i.e. the existence of a few IV-Ce ions in an otherwise non-magnetic system.

CONCLUSION

Earlier experiments[3-5] have indicated that in $Ce_{1-x}Sc_xAl_2$ a transition takes place, in a rather narrow concentration range around x=0.5, from a (nearly) trivalent state to an intermediate-valent state of the Ce ions. This valence transition is well reflected by distinct differences in the temperature dependences of both resistivity and thermopower. In the whole concentration range of IV, the resistivity shows a negative temperature coefficient, with $\rho \sim (1-AT^2)$ at low temperature. Our results constitute a noteworthy difference in the conduction-electron scattering from either IV-Ce ions or IV-Eu ions in the same $ScAl_2$ host: While Ce ions give rise to a negative, Eu ions cause a positive temperature coefficient in the respective incremental resistivity.[13] This difference deserves further consideration. Giant TEP peaks were found in IV-$Ce_{1-x}Sc_xAl_2$. The peak position $T_{s,max}$, which may be considered a good experimental measure of the width of the low-energy fermion quasiparticle band in an IV system, was found to increase by a factor of two upon increasing the Sc concentration. Analysis of the low-T TEP, $S \sim T$, shows that the low-T value of the valency of Ce is, in fact, considerably non-integral.

REFERENCES

[1] Barbara, B., Boucherle, J.X., Buevoz, J.L., Rossignol, M.F. and Schweizer, J., Solid State Commun. 24 (1977) 481.

[2] Croft, M. and Jayaraman, A., Solid State Commun. 29 (1979) 9.

[3] Levine, H.H. and Croft, M., in Falicov, L. M., Hanke, W. and Maple, M.B. (eds.), Valence Fluctuations in Solids (North-Holland, Amsterdam, 1981), p. 279.

[4] Preusse, N., Schäfer, W. and Elschner, B., ibid., p. 317.

[5] Loewenhaupt, M., Horn, S. and Steglich, F., Solid State Commun. 39 (1981) 295.

[6] Steglich, F., Wienand, K.H., Klämke, W., Horn, S. and Lieke, W., in Proc. Int. Conf. on Crystalline Electric Field and Structural Effects in f-Electron Systems, Wroclaw 1981 (forthcoming).

[7] Steglich, F., Z.Phys.B 23 (1976) 331.

[8] Steglich, F., Franz, W., Seuken, W. and Loewenhaupt, M., Physica 86B (1977) 503.

[9] Bhattacharjee, A.K. and Coqblin, B., Phys. Rev. B 13 (1976) 3441.

[10] Loewenhaupt, M., Rainford, B.D. and Steglich, F., Phys.Rev.Lett. 42 (1979) 1709.

[11] Jaccard, D. and Sierro, J., this conference.

[12] Hirst, L.L., Phys.Rev.B 15 (1977) 1.

[13] Franz, W., Steglich, F., Zell, W., Wohlleben, D. and Pobell, F., Phys. Rev. Lett. 45 (1980) 64.

Valence Instabilities
P. Wachter and H. Boppart (eds.)

THE HIGH TEMPERATURE RESISTIVITY OF SOME Ce AND Yb COMPOUNDS WITH UNSTABLE VALENCE

D. Müller, S. Hussain, E. Cattaneo,+ H. Schneider, W. Schlabitz and D. Wohlleben

II. Physikalisches Institut
der Universität zu Köln
5000 Köln 41, West Germany

The resistivity was measured between 300 and about 1000 K on $CeCu_2Si_2$, $CePd_3$, $CeAl_2$, $YbCu_2Si_2$, $YbAl_2$, $YbAl_3$, $YbInAu_2$ and on appropriate reference compounds with stable valence. In all the Ce and Yb compounds (except $YbAl_2$) the complicated low temperature resistivity anomalies saturate at high temperatures at values between 40 and 100 μΩc. We argue that this happens when Ce and Yb ions reach the high temperature entropy limit of their valence. $LaAl_2$ also shows such a large, saturating resistivity anomaly.

INTRODUCTION

Very little attention has so far been paid to the behaviour of mixed valence systems at high temperatures. From several points of view, this lack of interest is regrettable. For one thing phenomenologically the window from helium to room temperature does not usually allow a full view of the anomalies associated with unstable 4f shells. Secondly the various theoretical models in use today predict different behaviours at high temperature and should therefore be distinguishable at least there, if not at the more complicated low temperature end. If for instance the term Kondo compound is still meant to imply a resistivity anomaly $\Delta\rho\sim -\ln T$, then this behaviour must be expected to persist to temperatures of order 20000 K in certain key cerium compounds according to recent interpretations of photo emission data (1,2). On the other hand, virtual bound state and other narrow band theories (e.g.Fermi liquid theories) should all have their own simple predictions in the regime where the thermal energy is large compared to the band parameters, i.e. large compared to a few 100 K.

The ionic model (3) seems particularly well suited for predictions of high temperature properties (4). In this model the solid causes weak perturbations of the energy structures of two more or less degenerate ionic rare earth configurations $4f^{n+1}$ and $4f^n$+ e. The complicated low temperature behaviour is caused by a low energy spectrum dominated by the temperature and pressure dependent interconfigurational mixing and excitation energies, $\Gamma(T,p)$ and $E_x(T,p)$, and characterized by crystal field and intra-ionic spin orbit splittings of the two configurations. While the overall crystal field splitting (W) and the mixing energy (Γ) are a few hundred K or less, the spin orbit (multiplet) splittings (E_λ) are much larger in Ce, Pr, Tm and Yb. Therefore if E_x is smaller than a few hundred K, the electronic excitation spectrum of a mixed valent system with Ce, Pr, Tm and Yb should extend from zero to a limiting energy E_1 of order a few hundred K, above which there should be a gap of the local ionic spectrum, followed by more narrow ionic states at about E_λ/k_B= 3000 K in Ce and Pr and at much higher (thermally inaccessible) energies in Tm and Yb. In the gap between E_1 and E_λ the electronic spectrum has only the low density conduction electron states. At thermal energies larger than E_1 but smaller than E_λ the electronic entropy is then rather accurately given by

$$(1)\quad S_\infty = k_B \ln 2(J_{n+1} + J_n + 1) + \pi^2 k_B^2 \rho(\varepsilon_F) T/3$$

where the Ji are the angular momenta of the two ionic Hund's rule groundstates and $\rho(\varepsilon_F)$ is the (free electron) conduction electron density of states. At temperatures up to 1000 K the second term on the RHS of equ. (1) remains negligible against the first, and therefore the valence of the above ions must approach its so-called entropy limit (4).

$$(2)\quad v(T) \to v_\infty = Q_{n+1} + (1+(2J_{n+1}+1)/(2J_n+1))^{-1}$$

Q_{n+1} is the lower RE valence. The values are v_∞ = 3.14, 3.4, 2.62 and 2.89 for Ce, Pr, Tm and Yb. They are independent of the matrix. The limiting values must be approached whenever an ion shows signs of configurational instability at low temperatures (eg. even a slight lattice constant anomaly etc.) because then with increasing temperature the large mixing entropy quickly forces E_x to become smaller than k_BT (5).

Because of its simplicity and independence of the matrix it seems worthwhile to look for manifestations of this limit experimentally. There are a few high temperature measurements of the susceptibility of Ce (6,7) and Yb (8) compounds which show Curie constants close to but not quite at the trivalent values. Unfortunately it is experimentally hard to distinguish between the Curie constants of trivalent Ce and Yb and of mixed valent Ce and Yb in the high temperature limit, since the latter are only 16 and 11% lower; background susceptibilities and in Ce the relative closeness of the excited spin orbit multiplet J=7/2 prevent a clear cut separation. In the case of YbCuAl (9) and $YbCu_2Si_2$ (10) it was possible to measure nearly the full entropy predicted by equ. (1) and to come close to the end of the thermal expansion anomaly expected at the high temperature limit, when the valence no longer changes. In this paper we describe a search for effects of the high temperature entropy limit of the valence on the resistivity anomaly of Ce and Yb compounds.

EXPERIMENTAL

The elements used for the sample preparation were of nominal purity, 99.999 for Al, In, Au and 99.99 for Lu, Yb, Sc, Pd. The samples of $RECu_2Si_2$ and of $REPd_3$ (all polycrystalline) were prepared as described previously (12,13,14). The samples of $YbAl_3$, $ScAl_2$, $LuInAu_2$ were produced by arc melting, $YbAl_2$ and $YbInAu_2$ in an induction furnace. The other $REAl_2$ were single crystals (15). The strong weight losses which occurred during melting of the Yb samples were compensated for by breaking the samples, adding surplus Yb and remelting. This procedure was repeated several times until the desired stoichiometry was attained according to the weight, assuming each time that the weight loss was due to the preferential evaporation of Yb. $ScAl_2$, $LuInAu_2$ and $YbInAu_2$ were annealed for seven days at 800 C, while $YbAl_2$ and $YbAl_3$ were not annealed because of their high vapour pressure. X-ray analysis was made for all samples with the exception of $LuInAu_2$ and $YbCu_2Si_2$. All crystals were single phased except for $CeCu_2Si_2$ and $YbAl_3$ where very weak reflexes of some foreign phase could be detected.

Below room temperature the resistivity was measured as described in (12,14) and above room temperature as described in (16). The high temperature measurements were performed by an AC four probe technique using a lock-in amplifier as described in (17). The vacuum oven to maintain the sample at the desired temperature is described in (18). The temperature was measured with a thermo-couple Pallaplat. The sample was held in a vacuum of about 10^{-6} Torr on a machined and fired frame of ceramic and contacted by four springloaded molybdenum tips. The high temperature measurements were taken in heating and cooling cycles. One run from 300 K to 1000 K and back took up to 16 hours. The resistivity of most samples showed a slight hysteresis; it was a few percent lower during cooling than during heating, with nearly identical slopes. The resistance of a $YbCu_2Si_2$ sample dropped irreversibly by about 16%, when heating from 550 C to 750 C, while heating below 500 C caused only a small hysteresis. Similar, but less strong effects (about 5%) were observed in $CeCu_2Si_2$ and $LuCu_2Si_2$. We attribute this behaviour to healing of microcracks during heating which presumably opened up during cool down from the melt due to differential anisotropic contraction in the polycrystalline samples of the tetragonal structure of the $ThCr_2Si_2$ type. Taking data on many samples of a given compound and looking for the smallest resistivity in our own data and in the literature we estimate our systematic error in the absolute values of each $RECu_2Si_2$ compound to be less than 10%. No such problems were encountered in the single crystals or in any of the other samples, whose cross-sections appeared thermally stable.

RESULTS AND DISCUSSION

Fig. 1 shows the resistivity of some $RECu_2Si_2$ compounds (11). Note first the nearly linear increase of the phonon resistivities of the stable compounds (RE = Lu, Tb, Tm). For Tb and Tm one also observes magnetic disorder resistivity (13) (appropriately small for the small magnetic ordering temperatures). Ce and Yb show their well known, large and complicated, i.e. strongly temperature dependent low temperature anomalies (12,19), which however become very simple not far above room temperature, where they saturate at slopes close to the phonon slopes of the stable compounds. T_{CEF} indicates the overall CEF splittings as detected by inelastic neutron scattering (20, 21), which also finds the mixing energies at 60 and 130 K for Yb and Ce at room temperature, i.e. much smaller than T_{CEF}. One gets the clear impression that saturation sets in at $T>T_{CEF}$, which is strengthened by Fig. 2 where we have plotted the resistivity increment $\Delta\rho$ against lnT. The straight section with negative slope noted earlier for $CeCu_2Si_2$ (12) starts near 140 K and ends abruptly near 360 K. Both these numbers correspond to crystal field excitations identified by neutron scattering (21). We do not believe that the slight drop of $\Delta\rho$ of all these compounds in Fig. 2 at still higher temperatures should be viewed as new straight sections of $\Delta\rho$ vs lnT. Firstly, our control over the cross-section of the $RECu_2Si_2$ samples is not better

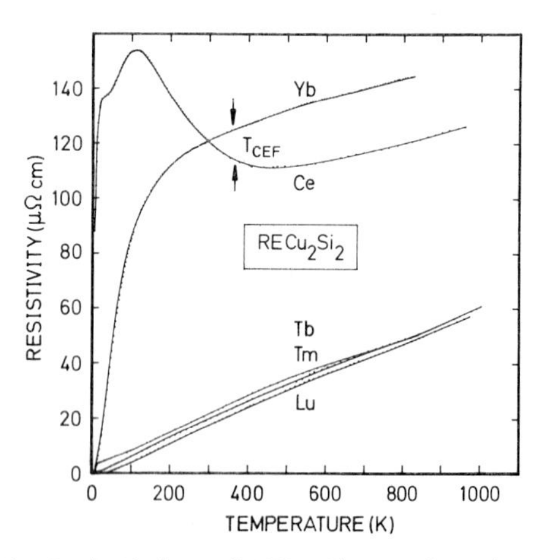

Fig. 1 Resistivity of $RECu_2Si_2$ as function of temperature. Residual resistivities of less than 2.5 μΩcm were subtracted except for Ce. Arrows indicate overall crystal field splittings (20,21).

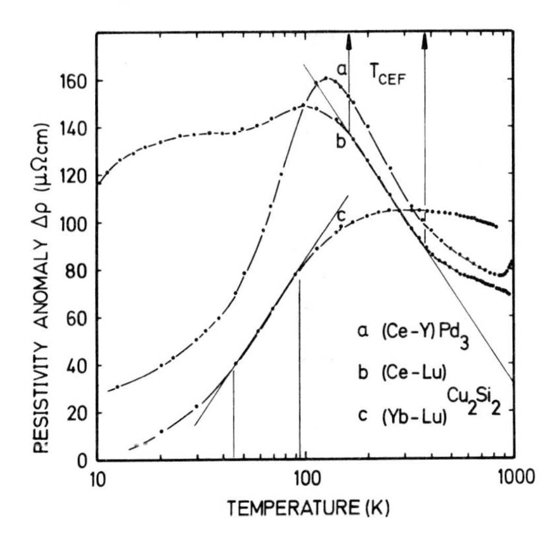

Fig. 2 Resistivity anomaly of (Ce, Yb) Cu_2Si_2 and $CePd_3$ plotted against lnT. Arrows indicate crystal field splittings

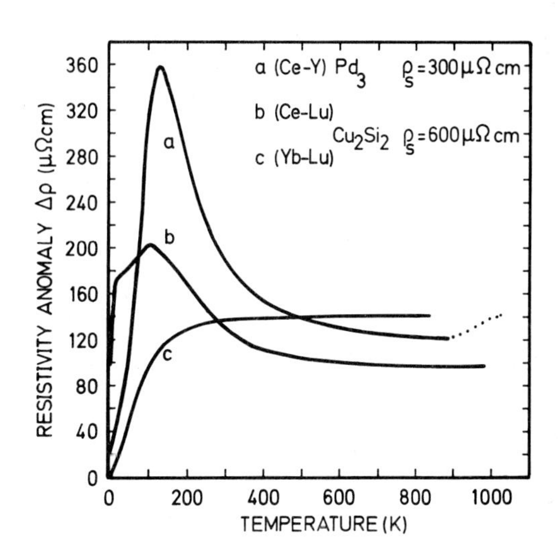

Fig. 3 Resistivity anomaly plotted against temperature after application of common shunt correction of 600 μΩcm to $RECu_2Si_2$ and of 300 μΩcm to $REPd_3$.

than 5-10%. Secondly, since the resistivities of Ce and Yb are near the upper limit of metallic conductivity (22,23), their phonon slope is expected to decrease near this limit, the faster, the higher the resistivity; this would result in negative $\partial\Delta\rho/\partial T$ for all such compounds at high temperatures. We have attempted to account for this effect by the shunt correction (24).

$$\sigma_{id}(T) = \sigma_m(T) - \sigma_s \quad (3)$$

Here σ_m, σ_{id} and σ_s are the measured, the ideal and the shunt conductivity. σ_s is due to a mean free path equal to the conduction electron wavelength at the Fermi energy. Fig. 3 shows what happens to $\Delta\rho(T)$ if we apply ρ_s = 600 μΩc to all $RECu_2Si_2$ and ρ_s = 300 μΩc to the $REPd_3$ (14). Perfect simultaneous saturation can be achieved for the $RECu_2Si_2$, while in $CePd_3$ the picture is spoiled near 900 K by an order-disorder transition. In view of the cross-section problem we do not wish to attribute too much significance to the exact value of ρ_s for $RECu_2Si_2$. However, there is no question that the resistivity anomaly saturates just above T_{CEF} in both compounds where also the valence is expected to saturate if $\Gamma, E_x < k_B T_{CEF}$. That the valence saturates there is also supported by the fact that the electronic entropy was found near $k_B \ln 9$ in $YbCu_2Si_2$ at 400 K (10) and that the thermal expansion anomaly which signals a temperature driven valence change, has gone to zero there (4,25).

Fig. 4 shows the behaviour of two $REInAu_2$ compounds, and Fig. 5 of two $REAl_3$'s. The $YbAl_3$ data on the heavy line are our own. The weak curved line shows $YbAl_3$ as measured by Havinga et al. (26) and $LuAl_3$ is also taken from

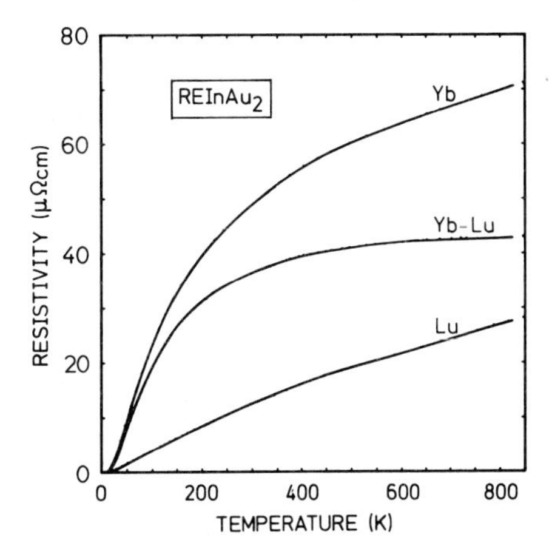

Fig. 4 Resistivity of $YbInAu_2$, $LuInAu_2$ and the resistivity anomaly (residual resistivities of 3.1 μΩcm (Yb) and 4.76 μΩcm (Lu) are subtracted).

that paper (26), with our extrapolation added (dashed line). Obviously both Yb compounds behave very much alike and like $YbCu_2Si_2$, i.e. they saturate at the phonon slope.

The resistivity of five $REAl_2$ compounds are shown in Fig. 6. A sixth, $LuAl_2$, measured from 1.5 to 300 K (26) coincides with Y, Sc and Yb within their mutual scatter and is not shown here. $YbAl_2$ was only measured up to 800 K, where evaporation of Yb prevented further increase of the temperature. At 800 K the susceptibility of $YbAl_2$ is still increasing (8). It has not passed over the maximum into the Curie-Weiß regime yet. $YbAl_2$ is therefore still far from the limiting valence at 800 K since still $E_x>kT$. So we cannot expect resistivity saturation here. $CeAl_2$, on the other hand, saturates again, as the other two Ce compounds, but this time at a lower value ($\Delta\rho \approx 50\mu\Omega c$) and after passing through a quite weak decrease of $\Delta\rho$ à la Kondo.

In Fig. 7 we show $\Delta\rho(T)$ for RE=Ce, La and Yb, using both Sc and Y as phonon references. The corresponding differences give an idea of the reliability of the phonon background correction when there are no problems with sample cross-sections (the cross-sections of the $REAl_2$ remained thermally stable).

It is clear from Fig. 6 and 7 that $LaAl_2$ is not a good reference compound, when compared to the three others with the s^2d^1 configurations (Sc,Y and Lu). While the latter are very close to each other, $LaAl_2$ shows a resistivity anomaly nearly as big as $CeAl_2$! $LaAl_2$ is also a superconductor while the other three s^2d^1 metals are not, and has a strongly temperature dependent susceptibility, reminiscent of a mixed valence compound (27). All this is in line with the old suspicion (28) that there is significant 4f character in the electronic spectrum of some La metals.

In conclusion, we have found a common property of Ce and Yb compounds with unstable valence, namely that the large resistivity anomaly tends to saturate not far above room temperature at values of order 100 μΩc. This saturation cuts off any Kondo like behaviour. It occurs in a temperature regime where the valence of these compounds is expected to saturate at its high temperature entropy limit and probably heralds this fact. An explanation for the occurence of a large, constant resistivity anomaly at the entropy limit of mixed valence, based on the ionic model, has been given elsewhere (11).

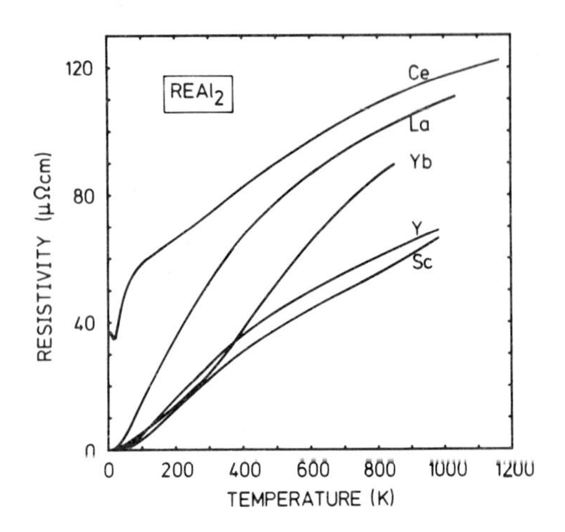

Fig. 6 Resistivity of some $REAl_2$ compounds (residual resistivities subtracted: ρ_o = 2.19, 7.15, 0.7, 0.37, and 1.4 μΩcm for Yb, Y, La, Sc and Ce (29)).

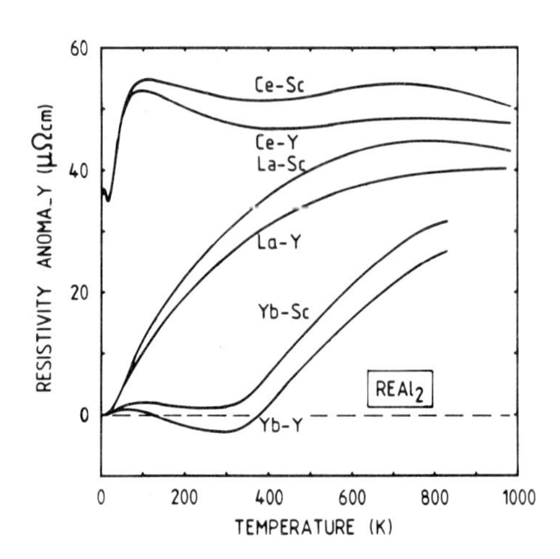

Fig. 7 Resistivity anomaly of $REAl_2$ as obtained by using two different reference compounds.

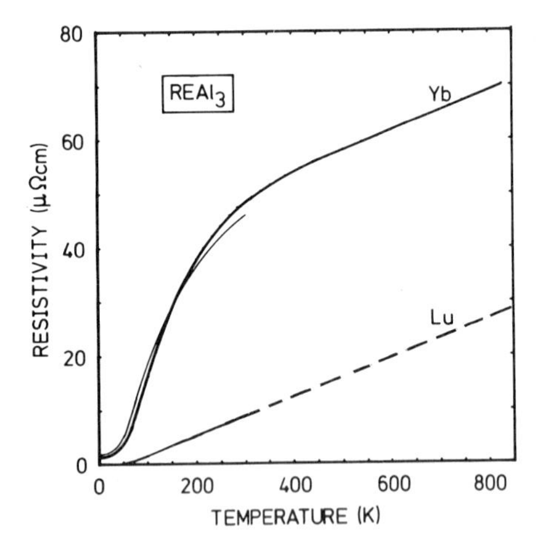

Fig. 5 Resistivity of $YbAl_3$ as measured here (heavy line) and by Havinga et al. (26) (thin line) and of $LuAl_3$ (26) with our extrapolation (dashed line).

This work was supported by Deutsche Forschungsgemeinschaft through SFB 125.

REFERENCES

(1) M. Croft, J.H. Weaver, D.J. Petermann and A. Franciosi; Phys. Rev. Lett. 46, 1104 (1981)

(2) J.W. Allen, S.-J. Oh, I. Lindau, J.M. Lawrence, L.I. Johansson and S.B. Hagström; Phys. Rev. Lett. 46, 1100 (1981)

(3) L.L. Hirst, Phys. Rev. B 15, 1 (1977)

(4) D.K. Wohlleben, Proc. of Valence Fluctuations in Solids, eds. L.M. Falicov, W. Hanke and M.B. Maple (North-Holland Publishing Company (1981) p. 1

(5) D. Wohlleben, J. Röhler, R. Pott, G. Neumann and E. Holland-Moritz; this conference

(6) W.H. Dijkman, F.R. de Boer, P.F. de Châtel and J. Aarts; J. Magn. Magn. Mat. 15-18, 970 (1980)

(7) H.W. Ludwigs, Doktorarbeit, Universität zu Köln, (1981)

(8) J.C.P. Klaase, W.C.M. Mattens, A.H. van Ommen, F.R. de Boer and P.F. de Châtel; AIP Conf. Proc. 34 184 (1976)

(9) R. Pott, R. Schefzyk, D. Wohlleben and A. Junod; Z. Phys. B44, 17 (1981)

(10) R. Kuhlmann, A. Schwann, R. Pott, W. Boksch and D. Wohlleben; this conference

(11) S. Hussain, E. Cattaneo, H. Schneider, D. Müller W. Schlabitz and D. Wohlleben; to be published

(12) W. Franz, A. Grießel, F. Steglich and D. Wohlleben; Z. Phys. B 31, 7 (1978)

(13) E. Cattaneo and D. Wohlleben; J. Magn. Magn. Mat. 24, 197 (1981)

(14) H. Schneider and D. Wohlleben; Z. Phys. B 44, 193 (1981)

(15) M. Beyß, W. Uelhoff and A. Fattah; Berichte der KFA Jülich, Nr. 1416 (1977)

(16) S. Hussain, Diplomarbeit, Universität zu Köln (1981)

(17) A.M. George and I.K. Gopalkrishan; J. Phys. E 8, 13 (1974) and references therein

(18) R. Kohlhaas, Arch. Eisenhüttenwesen 36, 437 (1965)

(19) B.C. Sales and R. Viswanathan; J. Low Temp. Phys. 23, 449 (1976)

(20) E. Holland-Moritz, D. Wohlleben and M. Loewenhaupt; Phys. Rev. B 25, (1982)

(21) S. Horn, E. Holland-Moritz, M. Loewenhaupt, F. Steglich, H. Scheuer, A. Benoit and J. Flouquet; Phys. Rev. B 23, 7 (1981)

(22) N.F. Mott, "Metal-Insulator Transitions" (Taylor and Francis, London (1974)) and A.F. Ioffe and A.R. Tegel; Prog. Semicond 4, 237 (1960)

(23) J. Friedel, Nuovo Cimento Suppl. 7, 287 (1958)

(24) H. Wiesman, M. Gurvitch, H. Lutz, A. Ghosh, B. Schwarz, Myron Strongin, P.B. Allen and J. W. Halley; Phys. Rev. Lett 38, 14 (1977)

(25) R. Schefzyk, Diplomarbeit, Universität zu Köln (1980)

(26) E.E. Havinga, K.H.J. Buschow and H.J. van Daal; Solid State Commun. 13, 621 (1973)

(27) M.B. Maple, Ph.D. Thesis, University of California, San Diego (1969)

(28) B.T. Matthias, private communication

(29) We use $\rho-\rho_0 = 34$ μΩcm at 14 K, (F. Steglich, private communication)

+) DAAD Stipendiat

MAGNETIC PROPERTIES

Valence Instabilities
P. Wachter and H. Boppart (eds.)

DYNAMIC SUSCEPTIBILITY OF INTERMEDIATE VALENCE IMPURITIES

P. Schlottmann[+]

Institut für Theoretische Physik, Freie Universität Berlin,
1000 Berlin 33, Arnimallee 3, Germany

We consider an Anderson impurity with orbital degeneracy in the $U \to \infty$ limit, such that all configurations except $4f^{n+1}$ and $4f^n$ are excluded. Only the Hund's rule ground-multiplets are taken into account and a rotationally invariant hybridization is employed. The f-level Green's function, the dynamic charge and spin susceptibilities and the a.c. resistivity are calculated by means of Mori's technique for all frequencies and temperatures. The f-level Green's function within the leading and next-leading logarithmic order essentially agrees with the expression derived by Grewe and Keiter.

The dynamic susceptibilities show a quasielastic peak with a Lorentzian shape and at low T bumps at energies ω close to the energy separation between the two multiplets. This threshold behavior is also present in the static susceptibilities, the specific heat and the resistivity. The spin-relaxation rate for Ce is finite and weakly temperature dependent, while the linewidth for Tm follows a Korringa law at low T. The depression of the superconductor T_c due to mixed-valent impurities is also discussed.

1. INTRODUCTION

In intermediate-valence systems[1-3] two configurations of the highly correlated f-states of the rare-earth ions have large probability of occupation. Many of the unusual properties of mixed-valence systems are caused by the competing effects of three types of energies: (a) the strong electron-electron repulsion in the f-shell, (b) the delocalization of the conduction electrons and (c) the hybridization mixing the f- and conduction-band states.

The intermediate-valence impurities are usually described in terms of the Anderson model[4]. Due to the large electron-electron repulsion in the f-shell only two ionic configurations, $4f^{n+1}$ and $4f^n$, need to be considered. We employ Hund's rule to find the ground-multiplets of these configurations and denote with J_1 and J_2 respectively their total angular momenta and with $E_{J_1M_1}$ and $E_{J_2M_2}$ their energies. The hybridization of the f-states with the conduction electrons induces charge-fluctuations in the f-electron occupation and mixes the two configurations. The conduction-electron states are expanded in partial waves at the impurity site and only the states with total angular momentum $j = 5/2$ $(7/2)$ if we have a light (heavy) rare earth impurity are assumed to contribute to the hybridization. This simplification corresponds to a particular jj-coupling instead of the usual Russell-Saunders coupling scheme. The Hamiltonian may then be written $H = H_o + H_V$, where

$$H_o = \sum_{km} \varepsilon_k c^+_{km} c_{km} + \sum_{M_1} E_{J_1M_1} B_{J_1M_1} + \sum_{M_2} E_{J_2M_2} B_{J_2M_2}$$

$$H_V = V\sqrt{2J_2+1} \sum_{kmM_1M_2} (J_2M_2jm|J_2jJ_1M_1)\left[A^+_{M_1M_2} c_{km} + c^+_{km} A_{M_1M_2}\right] \qquad (I.1)$$

Here V is the hybridization and C_{km} denotes the destruction of a conduction electron with momentum k, angular momentum j and z-component m. The Clebsch-Gordan coefficient selects $m = M_1 - M_2$. The operators $B_{J_1M_1}$, $B_{J_2M_2}$ and $A^+_{M_1M_2}$ are number operators and a flip operator, respectively, most conveniently expressed in terms of bra and kets

$$A^+_{M_1M_2} = |J_1M_1\rangle\langle J_2M_2| \qquad (I.2)$$

$$B_{J_2M_2} = |J_2M_2\rangle\langle J_2M_2| \; , \; B_{J_1M_1} = |J_1M_1\rangle\langle J_1M_1|$$

The model corresponds to the $U \to \infty$ limit of an Anderson impurity, which excludes states other than the J_1-manifold of the $4f^{n+1}$- and the J_2-manifold of the $4f^n$-configurations.

A mixed valence impurity may be alternatively interpreted either as a system with noninteger valence arising from a linear superposition of the two configurations or as a system with fluctuating valence and having an intermediate valence in time average. Both descriptions are correct; the former is more adequate for static low temperature (ground-state) properties, while the latter is more appropriate at high T and/or when large external energies are involved. The fluctuating picture is the natural description for the magnetic susceptibility. This is clear for Ce, Sm, Eu or Yb impurities (one configuration has a singlet ground-state), where magnetic measurements only probe one configuration (the argument is valid in general unless all Landé factors are equal $g_{J_1} = g_{J_2} = g_e$; i.e. the magnetization is conserved).[2] The charge and spin fluctuations determine the dynamics of the f-electrons. In addition to the static charge and spin susceptibilities a complete description of the problem requires the corresponding relaxation rates which characterize the dynamics of the impurity.

The smallest energy parameter in the model is the hybridization matrix element and hence it is natural to treat it as the perturbation. The dynamic magnetic susceptibility, i.e. the f-electron relaxation rate has been discussed in second order in the hybridization by several authors[5-8]. These results are valid only for temperatures much higher than the resonance width. By using a mode-mode coupling approach within Mori's technique we have extended the range of validity of these results to all temperatures.[9] This approach has been successfully applied previously to the Kondo problem[10] and to the resonant level[11] (our model with $J_1=J_2=0$). The resonant level can be solved exactly and the agreement with the approximation is satisfactory in the mixed valence regime. The theory needs static quantities as input parameters, e.g. the susceptibility and the f-electron occupation. In ref. 9 we determined the static quantities self-consistently from the f-electron propagator.

The purpose of this paper is to present an alternative way to calculate the dynamic susceptibilities. Instead of the mode-mode coupling approach[9] we use the Brillouin-Wigner technique[12] to resum the perturbation theory in the hybridization. Instead of the standard diagrammatic technique[12-15], we use Mori's formulation by introducing a novel "scalar product" between operators, which is appropriate to treat highly correlated f-states (section II). In section III we rederive the f-level Green's function within the leading and next-leading logarithmic order[16], which essentially agrees with the expression obtained by Grewe and Keiter[15]. The properties and drawbacks of the solution are discussed and the thermodynamics in leading logarithmic order is obtained via the fluctuation dissipation theorem. In section IV the dynamic charge and spin susceptibilities are expressed in terms of a relaxation kernel, which is then evaluated within the Brillouin-Wigner method up to next leading logarithmic order. The results are discussed in the context of mixed-valent Ce and Tm ions. In section V the resistivity[17] and the reduction of the superconducting transition temperature[18] due to intermediate valence impurities is given. Conclusions are presented in section VI.

II. INNER PRODUCT AND MORI'S FORMALISM[19-21]

Let us consider the space of operators built up by all possible products of f-electron operators and creation and annihilation operators of conduction electrons. We denote with (A,B) a scalar product defined onto this space, where A and B are two arbitrary vectors of the Hilbert space. Operators on this Hilbert space are denoted by script letters, e.g. the Liouville operator $\mathcal{H}$, defined by $\mathcal{H}A = [H,A]$, the resolvent $\mathcal{R}(z) = (z-\mathcal{H})^{-1}$ and projectors $\mathcal{P}$ and $\mathcal{Q}$. Green's functions and correlation functions can be expressed in terms of resolvent matrix elements if a convenient scalar product is chosen. Let $\{A_\alpha\}$ be the set of N operators that are relevant to the properties under discussion. Mori's projection technique[19-21] then yields the following expression for the resolvent matrix elements between these operators

$$\hat{\phi}(z) = [z\hat{1} - \hat{\Omega} - \hat{M}(z)]^{-1}\hat{\chi}^{\circ} , \qquad (II.1)$$

where $\hat{1}$ is the N x N unit matrix,

$$\phi_{\alpha\beta}(z) = (A_\alpha, \mathcal{R}(z)A_\beta) , \qquad (II.2)$$

$$\chi^{\circ}_{\alpha\beta} = (A_\alpha, A_\beta) , \qquad (II.3)$$

$$\hat{\Omega} = \hat{\omega}\,\hat{\chi}^{\circ -1} , \quad \omega_{\alpha\beta} = (A_\alpha, \mathcal{H}A_\beta) , \qquad (II.4)$$

$$\hat{M}(z) = \hat{m}(z)\hat{\chi}^{\circ -1} , \quad m_{\alpha\beta}(z) = (\mathcal{Q}\mathcal{H}A_\alpha, \mathcal{R}_Q(z)\mathcal{Q}\mathcal{H}A_\beta) . \qquad (II.5)$$

Here Q projects onto the subspace orthogonal to the set $\{A_\alpha\}$ and $R_Q(z)$ is the resolvent within this subspace $R_Q(z) = [zQ - Q\mathcal{H}Q]^{-1}$. Note that $\hat{m}(z)$ can be interpreted as a resolvent matrix element on the reduced subspace. Eqs. (II.1-5) can succesively be applied and a continued fraction is generated in this way. $\hat{\Omega}$ and the real part of $\hat{M}(z)$ play the role of a restoring force, while the imaginary part of $\hat{M}(z)$ yields the life-time. The many-body memory-effects of the correlation function are contained in $\hat{M}(z)$.

The success of the approach depends then on the choice of the scalar product (which is not unique) and the appropriate set $\{A_\alpha\}$. Two scalar products are commonly used

(a) the static susceptibility

$$(A,B) = \chi_{A^+B}(z=0) = i\int_0^\infty dt\, e^{i0^+t}\langle[A^+(t),B]\rangle \qquad (II.6)$$

and (b) the expectation value of the anti-commutator

$$(A,B) = \langle\{A^+,B\}\rangle . \qquad (II.7)$$

The former is appropriate for commutator correlation functions (susceptibilities), while (b) is convenient for fermion propagators.

The f-states, however, are highly correlated and do not obey Fermi statistics. The $[(2J_1+1)+(2J_2+1)]$ f-states are mutually excluding, i.e. only one can be occupied at a time. It is convenient to separate explicitly these states in the trace of the thermal averages

$$\langle X\rangle = \sum_{M_1}\langle X\rangle_{J_1M_1} + \sum_{M_2}\langle X\rangle_{J_2M_2} , \qquad (II.8)$$

where $\langle\cdots\rangle_{JM}$ denotes the trace over matrix elements within the state $|JM\rangle$. We now define the inner operation[16]

$$(A,B)_{JM} = \langle A^+B\rangle_{JM} , \qquad (II.9)$$

which has the following properties:

(i) conjugate bilinear

(ii) $(A,A)_{JM} \geq 0$, positive definite

(iii) $(A,A)_{JM} = 0$ if and only if $A = 0$ or A^+ has no matrix elements with final state $|JM\rangle$.

(iv) hermitian non-symmetric, i.e. in general $(A,A)_{JM} \neq (A^+, A^+)_{JM}$.

Since the inner operation is not hermitian symmetric, it is not an ordinary scalar product and our operator space is not a Hilbert space. We have shown[16] that Mori's expressions, Eqs. (II.1-5), are still valid under these conditions (we have to distinguish between projections from the right and from the left).

III. f-LEVEL GREEN'S FUNCTION AND THERMODYNAMICS

In this section we rederive the f-level Green's function[16] by means of Mori's technique[19-21], which was previously obtained diagrammatically by Grewe and Keiter[15]. First we give the formal expression of the f-level propagator and discuss then its properties and drawbacks. We connect the result with thermodynamics within the leading logarithmic order approximation by making use of the fluctuation-dissipation theorem. Finally we point out some difficulties that arise when the dynamic susceptibilities are directly calculated by this method.

(a) The f-level propagator

The f-level Green's function can be written as the sum of two resolvent matrix elements[16]

$$\langle\langle A_{M_1M_2}; A^+_{M_1M_2}\rangle\rangle_z = \langle A_{M_1M_2} \frac{1}{z-\mathcal{H}} A^+_{M_1M_2}\rangle + \langle A^+_{M_1M_2} \frac{1}{z+\mathcal{H}} A_{M_1M_2}\rangle . \qquad (III.1)$$

We explicitly separate the f-states in the trace as indicated in (II.8) and evaluate the resolvent matrix elements using the inner product (II.9). Since the quantity

$$(A^+_{M_1M_2}, A^+_{M_1M_2})_{Jm} = \langle B_{J_2M_2}\rangle_{Jm} , \qquad (III.2)$$

vanishes unless $J = J_2$ and $m = M_2$, the expression (II.1) can only be used in this case. We first evaluate this contribution in leading and next-leading logarithmic order for the first term in (III.1) in the absence of an external magnetic field and crystal fields. We obtain[16]

$$\langle A_{M_1M_2} \frac{1}{z-\mathcal{H}} A^+_{M_1M_2}\rangle_{J_2M_2} = \frac{\langle B_{J_2M_2}\rangle}{z - E_{J_1} + E_{J_2} + \omega_{J_2} - V^2(2J_2+1)\sum_k \frac{1-f(\varepsilon_k)}{z-\varepsilon_k+\omega_{J_2}-R_{J_2}(z)}} , \qquad (III.3)$$

where ω_{J_2} is the energy shift of the J_2 multiplet given by

$$\langle B_{J_2M_2}\rangle \omega_{J_2} = V\sqrt{2J_2+1}\sum_{kM_1m}(J_2M_2jm|J_2jJ_1M_1)\langle c^+_{km} A_{M_1M_2}\rangle_{J_2M_2} . \qquad (III.4)$$

Note that the remainder $R_{J_2}(z)$ is another resolvent matrix element. Each step in the continued fraction corresponds to one hierarchy in the logarithmic order approximation.

The hybridization mixes the two configurations by forming bonding and antibonding states, having degeneracy min (J_1,J_2) and max (J_1,J_2), respectively.[13,14] The energies of the new states are $E_{J_1} + \omega_{J_1}$ and $E_{J_2} + \omega_{J_2}$, where ω_{J_1} is defined in complete analogy to (III.4). We now introduce the energy difference between the bonding and antibonding states

$$\Delta E = E_{J_1} + \omega_{J_1} - E_{J_2} - \omega_{J_2} \qquad (III.5)$$

and define $N_{J_2}(z)$ by rewriting (III.3) as

$$\langle A_{M_1M_2} \frac{1}{z-\mathcal{H}} A^+_{M_1M_2}\rangle_{J_2M_2} = \frac{\langle B_{J_2M_2}\rangle}{z-\Delta E - N_{J_2}(z)} . \qquad (III.6)$$

In (III.3) we considered only the contributions in which the initial and final state is $|J_2M_2\rangle$. In order to obtain the remaining contributions we apply two times the equation of motion to $\langle A_{M_1M_2}(z-\mathcal{H})^{-1}A^+_{M_1M_2}\rangle$

$$\langle A_{M_1M_2} \mathcal{R}(z) A^+_{M_1M_2}\rangle = \langle B_{J_2M_2}\rangle/(z-\Delta E - N_{J_2}(z)) + (\langle X_{M_1M_2}A^+_{M_1M_2}\rangle + \langle X_{M_1M_2}\mathcal{R}(z)X^+_{M_1M_2}\rangle)/(z-\Delta E-N_{J_2}(z))^2, \qquad (III.7)$$

where

$$X_{M_1M_2} = -(\mathcal{H}_V - \omega_{J_2} + \omega_{J_1} + N_{J_2}(z))A_{M_1M_2} . \qquad (III.8)$$

In (III.7) we keep only the most relevant terms involving $X_{M_1M_2}$. Several terms cancel and after some algebra[16] the last term yields

$$-\langle B_{J_1}\rangle^0 N_{J_1}(-z)/[z-\Delta E - N_{J_2}(z)]^2 , \qquad (III.9)$$

where $\langle B_{J_1}\rangle^0$ denotes that the thermal average is to be evaluated in leading order only. This is necessary in order to be consistent, since (III.6) contains leading and next leading order terms and (III.9) only next leading logarithmic terms. The function $N_{J_1}(z)$ is defined in a similar way as $N_{J_2}(z)$

$$N_{J_1}(z) = V^2(2J_1+1)\sum_k \frac{f(\varepsilon_k)}{z-\varepsilon_k-\omega_{J_1}-R_{J_1}(z)} + \omega_{J_2} , \qquad (III.10)$$

where $R_{J_1}(z)$ is a remainder which can be expressed again as a resolvent matrix element.

The second term on the r.h.s. of (III.1) yields a similar contribution, which can be expressed as

$$\langle A^+_{M_1M_2} \frac{1}{z+\mathcal{H}} A_{M_1M_2}\rangle = \langle B_{J_1}\rangle/[z-\Delta E-N_{J_1}(z)] - \langle B_{J_2}\rangle^0 N_{J_2}(-z)/[z-\Delta E - N_{J_1}(z)]^2 . \qquad (III.11)$$

These results essentially agree with those derived by Grewe and Keiter[15] for the impurity f-level Green's function by resumming Goldstone diagrams. The advantage of the present method is that it requires just the evaluation of commutators, one for each logarithmic order, and traces. Up to next leading order we have only calculated the first and second time derivative of the operator $A^+_{M_1M_2}$.

(b) Selfconsistent energy shifts

The energy shifts ω_{J_1} can be obtained through the resonance condition of the Green's function, i.e. by the zeroes of the real part of the energy denominators. By definition this occurs at

$z = \Delta E$, such that

$$\mathrm{Re}\, N_{J_1}(z=\Delta E) = \mathrm{Re}\, N_{J_2}(z=\Delta E) = 0 \qquad (\mathrm{III}.12)$$

determines the energy shifts. In leading logarithmic order we neglect the remainder $R_1(z)$ and $R_2(z)$ and obtain

$$\omega_{J_1} = -\frac{\Gamma}{\pi}(2J_2+1)\left[\ln\frac{D}{2\pi T} - \mathrm{Re}\,\psi\left(\frac{1}{2} - i\,\frac{E_{J_1}-E_{J_2}+\omega_{J_1}}{2\pi T}\right)\right] \qquad (\mathrm{III}.13)$$

ω_{J_2} is determined by a similar equation. Here $\Gamma = \pi\rho_F V^2$, ψ is the digamma function and a half-filled square density of states has been assumed. As pointed out by other authors the energy shifts are negative[13-15] and proportional to the degeneracy of the other configuration[14,15]. Ramakrishnan[14] concluded that higher order approximations do not change the result considerably if $|J_1 - J_2|$ is large.

(c) Consequences of the approximation

We first discuss the excitation spectrum of the impurity in leading order approximation. As pointed out by Grewe and Keiter[15] the Fermi level as well as the renormalized charge transfer energy $\omega = \Delta E$ lie in a gap. The edges of the gap are at $-|\omega_{J_2}|$ and $|\omega_{J_1}|$. As a consequence the excitation spectrum has a δ-function at $\omega = \Delta E$ with the weight of

$$\frac{\pi\langle B_{J_1}\rangle}{1-\left.\frac{\partial N_{J_1}}{\partial\omega}\right|_{\omega=\Delta E}} + \frac{\pi\langle B_{J_2}\rangle}{1-\left.\frac{\partial N_{J_2}}{\partial\omega}\right|_{\omega=\Delta E}} \qquad (\mathrm{III}.14)$$

and a continuum of excitations beyond the gap edges. This gap is unphysical and just an artifact of the approximation as concluded gy Grewe and Keiter[15]. In next leading order the gap is closed, although the δ-function resonance is still present.

Higher order approximations for (III.3) yield a continued fraction in which each step consists of a Fermi function and an energy denominator. A remarkable property of the continued fraction is that the arguments of the Fermi functions are such that their continuum contribution vanishes for $\omega > 0$ at $T = 0$. The properties of $N_{J_1}(z)$ are similar and the continuum contribution vanishes for $\omega < 0$. The physical interpretation of this result is the following: The propagator describes the transition between the two configurations. Perturbatively the f-states are infinitely-lived, such that the Fermi functions remain sharp step functions and are not smeared (no linewidth). $N_{J_2}(\omega)$ represents the absorbtive part of the propagator while $N_{J_1}(\omega)$ corresponds to the emission.

There are, however, non-analytic contributions to $N_{J_2}(\omega)$ ($N_{J_1}(\omega)$). In order to obtain them one has to calculate at least three (four) steps in the continued fraction. At a certain step the real and imaginary parts (Fermi functions) of an energy denominator may vanish and give rise to a pole. The position of the pole is given by a transcendental equation of the type (assume $\Delta E > 0$)

$$X - \Delta E + \frac{\Gamma}{\pi}(2J_2+1)\ln\left|\frac{\Delta E}{X}\right| = 0\,, \qquad (\mathrm{III}.15)$$

where the continuum vanishes for $X < 0$. This equation has a root for $X < 0$, approximatively given by

$$X_0 \simeq -\Delta E\,\exp\left[-\frac{\pi\Delta E}{(2J_2+1)\Gamma}\right]\,, \qquad (\mathrm{III}.16)$$

which is of the Kondo type. This pole has the important consequence of smearing the step functions. In other words, the poles generate a finite lifetime for the f-states.

If the continued fraction is carried on far enough the functions $N_{J_1}(\omega)$ and $N_{J_2}(\omega)$ may be approximated by

$$N_{J_1}(\omega) = -\frac{\Gamma}{\pi}(2J_1+1)\left\{i\frac{\pi}{2} - \mathrm{Re}\,\psi\left(\frac{1}{2}+\frac{(2J_2+1)\Gamma}{2\pi T} - i\frac{\Delta E}{2\pi T}\right) + \psi\left(\frac{1}{2}+\frac{(2J_2+1)\Gamma}{2\pi T} - i\frac{\omega}{2\pi T}\right)\right\}\,, \qquad (\mathrm{III}.17)$$

$$N_{J_2}(\omega) = -\frac{\Gamma}{\pi}(2J_2+1)\left\{i\frac{\pi}{2} + \mathrm{Re}\,\psi\left(\frac{1}{2}+\frac{(2J_1+1)\Gamma}{2\pi T} - i\frac{\Delta E}{2\pi T}\right) - \psi\left(\frac{1}{2}+\frac{(2J_1+1)\Gamma}{2\pi T} - i\frac{\omega}{2\pi T}\right)\right\}\,. \qquad (\mathrm{III}.18)$$

Note that (III.12) is satisfied. The Γ in the argument of the digamma functions represent the relaxation, which is proportional to the respective degeneracy of the multiplet.

In summary the Green's function has been expanded as a series of continued fractions, of which we kept only the two most important. This approximation, in connection with the separate treatment of the absorption and emission in (III.1), leads to violations of the fluctuation-dissipation theorem in all orders in V higher than the second. The fluctuation-dissipation theorem relates the imaginary parts of the first and second terms of (III.1). This drawback does not seem to affect drastically the thermodynamics, since the occupation probabilities of the f-levels calculated via the relation

$$\langle B_{J_1M_1}\rangle - \langle B_{J_2M_2}\rangle = \int\frac{d\omega}{\pi}\,\mathrm{Tanh}\,\frac{\omega}{2T}\,\mathrm{Im}\,\langle\langle A_{M_1M_2}; A^+_{M_1M_2}\rangle\rangle_\omega \qquad (\mathrm{III}.19)$$

are compatible (in leading logarithmic order) with the partition function[13-15]

$$Z = \sum_{M_1}\exp[-\beta(E_{J_1M_1}+\omega_{J_1M_1})] + \sum_{M_2}\exp[-\beta(E_{J_2M_2}+\omega_{J_2M_2})]. \qquad (\mathrm{III}.20)$$

The drawback, however, seems to be the origin for the large change in the excitation spectrum from leading to next leading order.

Within the next leading order approximation the thermodynamics is still given by (III.20), but with the energy shifts determined selfconsistently from

$$\omega_{J_1M_1} = -\frac{\Gamma}{\pi}\sum_{M_2}\left[\ln\frac{D}{2\pi T} - \mathrm{Re}\,\psi\left(\frac{1}{2} - i\frac{\Delta E_{M_1M_2}}{2\pi T}\right)\right]\,, \qquad (\mathrm{III}.21)$$

where $\Delta E_{M_1M_2} = E_{J_1M_1} + \omega_{J_1M_1} - E_{J_2M_2} - \omega_{J_2M_2}$. Here $\omega_{J_2M_2}$ is determined from a similar expression. This result also holds in the presence of magnetic and crystal fields. Note that in

(III.21) we neglected the linewidth, otherwise the expression (III.20) of the partition function would not be a simple sum of Boltzmann factors.

We attempted to calculate the dynamic susceptibilities by the same method. In analogy to (III.1) the anticommutator correlation function can be obtained, but one arrives at serious drawbacks when one uses the fluctuation-dissipation theorem to get the susceptibility. An alternative way is to express the commutator correlation function as a difference of resolvant matrix elements, which also lead to contradictions related to the violation of the fluctuation-dissipation theorem.

An approach that avoids these difficulties is presented in the next section.

IV. DYNAMIC SUSCEPTIBILITIES

In a recent publication[9] we calculated the dynamic susceptibilities for a mixed-valent ion by using a mode-mode coupling approach within Mori's technique. A similar procedure has been successfully applied to the Kondo problem[10] and the resonant level[11]. The result is valid for all frequencies and temperatures and contains Kuramoto and Müller-Hartmann's[7,8] relaxation rates for the high temperature limit as a special case. In this section we present an alternative calculation of the dynamic susceptibilities, which leads to similar results. We first express the dynamic charge and spin susceptibilities in terms of a relaxation function, which is then evaluated in a similar way as the selfenergies in the previous section. Finally we specialize the results in view of Ce and Tm impurities.

(a) The relaxation kernel

In order to calculate a Green's function by Mori's method it is necessary to choose the relevant set of operators and a convenient scalar product as mentioned in section II. Since susceptibilities are commutator correlation functions, (II.6) is an appropriate scalar product. The dynamic susceptibility can then be expressed in terms of the static susceptibility $\hat{\chi}^{o}$, the restoring force $\hat{\Omega}$ and the memory function $\hat{M}(z)$ as

$$\hat{\chi}(z) = \left[z\hat{1} - \hat{\Omega} - \hat{M}(z)\right]^{-1}\left[-\hat{\Omega} - \hat{M}(z)\right]\hat{\chi}^{o}, \qquad (IV.1)$$

where the quantities are defined by (II.3-6). Note that (IV.1) already satisfies the static limit.

The relevant operator for the charge susceptibility is the f-charge, which may be defined as

$$q = \sum_{M_1} B_{J_1M_1} = 1 - \sum_{M_2} B_{J_2M_2} \qquad (IV.2)$$

$\hat{\chi}(z)$ is therefore a scalar given by $\chi_{ch}(z) = -\langle\langle q;q\rangle\rangle_z$ and Ω vanishes by the symmetry of the problem. We introduce the charge current operator

$$j_{ch} = [H_V, q] = V\sum_{mM_1M_2}\sqrt{2J_2+1}\,(J_2M_2jm|J_2jJ_1M_1) \times\left\{c^{+}_{km}A_{M_1M_2} - A^{+}_{M_1M_2}c_{km}\right\} \qquad (IV.3)$$

and since q and j_{ch} are orthogonal operators within the chosen scalar product we have for the relaxation function

$$m_{ch}(z) = \left(j_{ch}, R_Q(z)\,j_{ch}\right) = +(2J_1+1)(2J_2+1)N(z) \qquad (IV.4)$$

where Q projects onto the space orthogonal to q. The projector Q in $R_Q(z)$ does not contribute in leading and next leading logarithmic order and can be neglected. In the absence of magnetic and crystal fields we approximately have then

$$(2J_1+1)(2J_2+1)N(z) = +V^2(2J_2+1)\sum_{kmM_1M_2}(J_2M_2jm|J_2jJ_1M_1)^2 \left\{\langle\langle c^{+}_{km}A_{M_1M_2}; A^{+}_{M_1M_2}c_{km}\rangle\rangle_z + \langle\langle A^{+}_{M_1M_2}c_{km}; c^{+}_{km}A_{M_1M_2}\rangle\rangle_z - (z=0)\right\}/z \qquad (IV.5)$$

The relevant operators for the spin susceptibility are the total angular momentum operators of the two multiplets

$$S_{J_1} = \sum_{M_1} M_1 B_{J_1M_1}, \quad S_{J_2} = \sum_{M_2} M_2 B_{J_2M_2}. \qquad (IV.6)$$

The susceptibility $\hat{\chi}(z)$ is then a 2x2 matrix and the final spin susceptibility is obtained as

$$\chi_s = (g_{J_1}, g_{J_2})\,\hat{\chi}(z)\begin{pmatrix} g_{J_1} \\ g_{J_2}\end{pmatrix} \qquad (IV.7)$$

where g_{J_1} and g_{J_2} are the Landé factors of the multiplets. $\hat{\Omega}$ vanishes and repeating the arguments used above for the charge susceptibility we obtain in the absence of magnetic and crystal fields

$$m^{s}_{11}(z) = \tfrac{1}{3}(2J_1+1)(2J_2+1)J_1(J_1+1)N(z) \qquad (IV.8a)$$

$$m^{s}_{22}(z) = \tfrac{1}{3}(2J_1+1)(2J_2+1)J_2(J_2+1)N(z) \qquad (IV.8b)$$

$$m^{s}_{12}(z) = m^{s}_{21}(z) = -\tfrac{1}{6}(2J_1+1)(2J_2+1)\left[J_1(J_1+1)+J_2(J_2+1)-j(j+1)\right]N(z) \qquad (IV.8c)$$

Hence, we have expressed the spin and charge susceptibilities in terms of N(z), given by (IV.5).

In order to evaluate N(z) we write the correlation functions in (IV.5) as a difference of resolvent matrix elements, separate the trace of the thermal averages according to (II.8) and use the inner operation (II.9). The procedure is then completely analogous to the calculation of $N_{J_1}(z)$ and $N_{J_2}(z)$ in the previous section. A series of continued fractions is obtained from which only the four most relevant ones (non-vanishing in second order in V) are kept. The Fermi functions again yield vanishing continuum contributions for $\omega > \Delta E$, $\omega < \Delta E$, $\omega > -\Delta E$ and $\omega < -\Delta E$, respectively. Poles analogous to (III.15-16), but now around the edges at $\pm\Delta E$ smear the step functions and we may approximately write

$$N(\omega) = (\Gamma/\pi)\langle B_{J_1}\rangle\left\{2\,\mathrm{Re}\,\psi\left(\tfrac{1}{2} + \tfrac{(2J_1+1)\Gamma}{2\pi T} - i\tfrac{\Delta E}{2\pi T}\right) - \psi\left(\tfrac{1}{2} + \tfrac{(2J_1+1)\Gamma}{2\pi T} - i\tfrac{\omega+\Delta E}{2\pi T}\right) - \psi\left(\tfrac{1}{2} + \tfrac{(2J_1+1)\Gamma}{2\pi T} - i\tfrac{\omega-\Delta E}{2\pi T}\right)\right\}/\omega + (J_2 \leftrightarrow J_1). \qquad (IV.9)$$

The origin of the Γ in the argument of the digamma functions is the finite linewidth of the f-states, which is proportional to the corresponding degeneracies of the multiplets.

The energy shifts, the occupation numbers and the static susceptibilities are the input quantities needed for the dynamic susceptibilities, which are obtained through (III.20-21). If we consider the second order approximation in V for N(z) and insert the bare static quantities (V=0) we recover in the limit $\omega \to 0$ the expressions for the relaxation rate by Kuramoto and Müller-Hartmann[7,8]. Their results, of course, are only valid in the high temperature regime.

(b) The case of Ce impurities

The $4f^0$ and $4f^1$ configurations of Ce yield $J_1=j=5/2$, $J_2=0$ and Landé factors $g_{J1}=6/7$, $g_{J2}=0$. Due to the difference in the degeneracies of the two multiplets, we have $\Delta E>0$ in the mixed valence regime, such that the groundstate is a singlet. The low T susceptibility is then of the van Vleck type and finite. At high T all the levels are equally populated and a Curie-Weiss law is obtained. The results of Brillouin-Wigner theory in next-leading order (Eqs. (II.20-26)) and the equation of motion method (see ref. 9) are compared in Fig. 1 for E=0, $D=100\pi\Gamma$. The reason for the good agreement is that the convergence of the perturbation expansion improves with $2|J_1-J_2|$ as pointed out by Ramakrishnan[14]. At intermediate temperatures, $T\sim\Delta E$ the susceptibility shows a maximum which is caused by the higher lying multiplet in (III.20) getting populated. The dots in Fig. 1 are data by Luszik-Bhadra et al.[22] for CeTh scaled with $\Gamma = 100$ K. The agreement at high T is good, while at low T the theories do not fit the experiment within the mixed valence regime. The inclusion of a crystal field could change the situation.

In Fig. 2 we show the charge and spin relaxation rates defined as the half-width of the quasi-elastic peak. The agreement between the two theories is remarkable. The spin relaxation is almost temperature-independent and shows a small maximum at $T\sim\Delta E$. Since the groundstate is a singlet it is finite a T=0. The charge-relaxation is always larger than the spin-relaxation. The charge fluctuations are dropped by the Boltzmann-factors when the temperature is reduced. At T=0 only the intrinsic charge-relaxation within the singlet remains.

Fig. 3 shows the dynamic spin susceptibility as a function of frequency. At high T its shape is Lorentzian, while at low T a second peak for $\omega>\Delta E$ develops due to photon-induced charge-excitations. Such bumps have been discussed previously by several authors[5,9-11,23-25].

(c) The case of Tm impurities

The two active configurations for Tm are $4f^{12}$ and $4f^{13}$. It is convenient to argue with f-holes instead of f-electrons, such that $J_1=6$, $J_2=j=7/2$ and $g_{J1}=7/6$, $g_{J2}=8/7$. Since both configurations are magnetic the groundstate in the mixed-valent regime is a multiplet of degeneracy $(2J_2+1)$ (Kondo bound-states are neglected). The spin susceptibility is then of the Curie-type and we have a Korringa relaxation. This is seen in Figs. 4 and 5 for E=0, $D=100\pi\Gamma$. The Curie constant for χ_s^o (dashed) shows a bump close to $T\sim\Delta E$. It is caused by the Boltzmann-factors in (III.20). This Schottky behavior is also responsable for the maximum in χ_c^o, the anomalies in the partial spin susceptibilities χ_{ij}^o and the minimum in the charge relaxation rate. A second anomaly appears at a higher temperature (maximum of $1/T_{1c}$, minimum of $\chi_{J_1J_1}^o$, change of slope of $T_{1s}T$) and it arises from derivatives of the energy shifts. The energy shift ΔE corresponds to the binding energy of the mixed-valent resonant state. Note that $\chi_{J_1J_1}^o T\to 8.66$, $\chi_{J_1J_2}^o T\to 0$ and $\chi_{J_2J_2}^o T\to 2$ for $T\to\infty$. The partial susceptibilities can be compared with the results of ref. 26.

The spin relaxation rate shows a crossover from a Korringa law at low T to a saturation at high T. The change in this behaviour is approximately at $T\sim\Delta E$ and it is in qualitative agreement with the linewidth for TmSe observed by neutron scattering.[27] Experimentally an inelastic peak similar to the one shown in Fig. 3 has been found. Our theory yields such an inelastic peak for Tm impurities only for $\Delta E>20\Gamma$ and a fit of the experimental data is not possible. In compounds like TmSe, however, the shift of the chemical potential and the interaction among the Tm are expected to be important.

V. RESISTIVITY AND REDUCTION OF SUPERCONDUCTOR Tc

(a) Resistivity[17]

The resistivity is calculated through Kubo's formula by making use of the procedure developed by Götze and Wölfle.[28] The conductivity is expressed in terms of a relaxation function $\tilde{M}(z)$

$$\sigma(z)=i(e^2N_e/m_e)/(z+\tilde{M}(z)) \quad , \qquad (V.1)$$

where N_e is the number of electrons and e and m_e are their charge and mass, respectively. For a dilute alloy the relaxation function is proportional to the concentration of impurities, c,

$$\tilde{M}(z)=c\,m_e\left(\langle\langle A;A\rangle\rangle_z-\langle\langle A;A\rangle\rangle_{z=0}\right)/z \quad , (V.2)$$

where A is the time derivative of the current operator in z-direction

$$A=\sum_{k\sigma}\frac{k_z}{m_e}\left[c_{k\sigma}^{+}c_{k\sigma},H\right] \quad . \qquad (V.3)$$

Note that for $M(z)=i/\tau$ expression (V.1) leads to Drude's formula.

The error introduced by projecting the momenta onto the Fermi surface is of the order of Γ/D. Making use of the total charge conservation we can express M(z) in terms of the charge susceptibility

$$\tilde{M}(z)=c(k_F^2/3m_e)\chi_c^o/\left[z+(2J_1+1)(2J_2+1)N(z)/\chi_c^o\right] \quad (V.4)$$

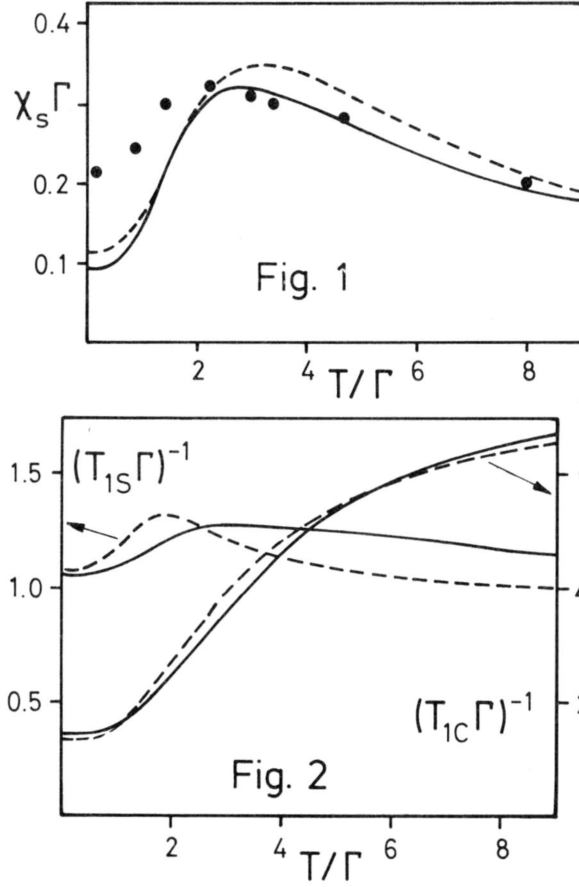

Figs. 1 and 2: Static spin susceptibility and spin and charge relaxation rates for Ce impurities (E=0, D=100$\pi\Gamma$): (a) present theory related to next-leading Brillouin-Wigner (full line), (b) equation of motion method (see ref. 9) (dashed) and (c) data for CeTh from ref. 22 scaled with Γ = 100 K.

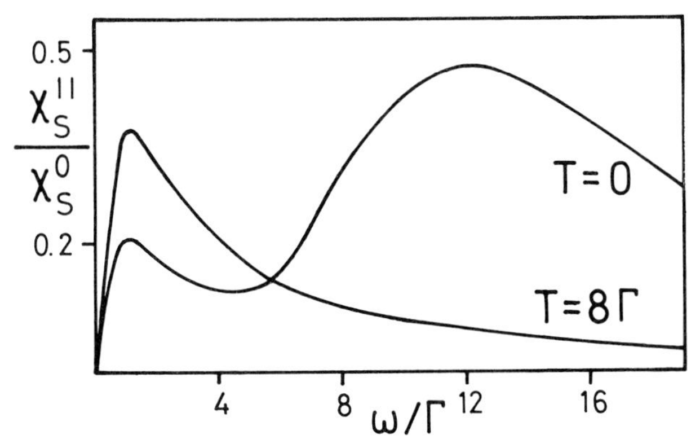

Fig. 3: Dynamic spin susceptibility for Ce impurities as a function of frequency (E=0, D=100$\pi\Gamma$). The second resonance at low T is due to induced transitions with energy $\omega \gg \Delta E$.

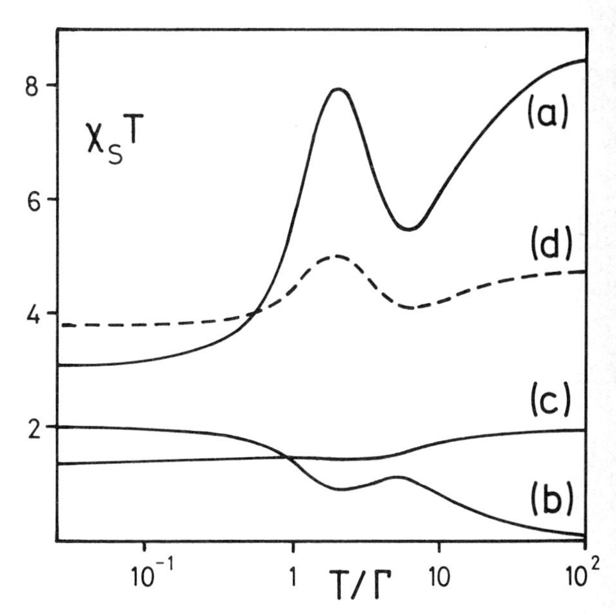

Fig. 4: Static spin susceptibilities (Curie constants) for Tm impurities (E=0, D=100$\pi\Gamma$): (a) $\chi^\circ_{J_1J_1}$, (b) $\chi^\circ_{J_1J_2}$, (c) $\chi^\circ_{J_2J_2}$ (full) and (d) total susceptibility χ°_S times 1/3 (dashed). The partial susceptibilities do not contain the Landé factors.

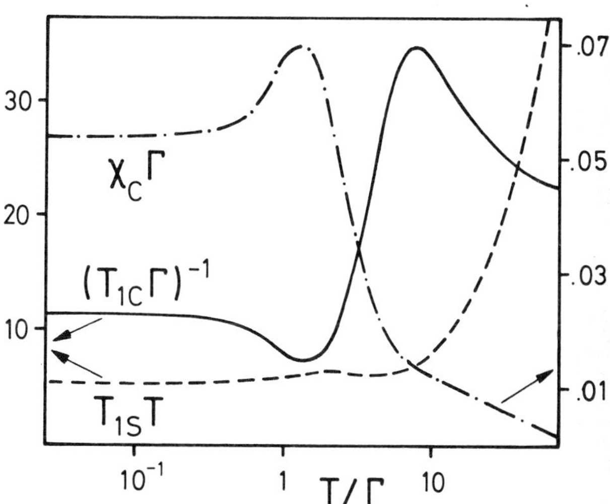

Fig. 5: Charge susceptibility (dash-dotted), spin relaxation time (dashed) and charge relaxation rate (full) as a function of T (E=0, D=100$\pi\Gamma$) for Tm impurities.

The dc-resistivity can then approximately expressed as

$$\rho_o = c\left(k_F^2/3e^2N_e\right) T_{1c}\,\chi^o_c \,. \qquad (V.5)$$

This expression is not valid in the presence of an external magnetic field. The magnetoresistivity has been discussed by Foglio and Schlottmann[17].

(b) Reduction of the transition temperature of a superconductor

The properties of Ce impurities in a superconductor have been derived within the equation of motion technique in ref. 18. A procedure similar to that of section III gives the following approximate expression for Tc:

$$\ln \frac{T_c}{T_{co}} + \left(1-\frac{r}{t}\right)\left[\ln \frac{\omega_D}{2\pi T_c} - \psi\left(\frac{1}{2}+\frac{s/t}{2\pi T_c}\right)\right] = 0, \qquad (V.6)$$

where T_{co} refers to the pure superconductor and ω_D is the Debye energy. The quantities r, $s=s_2-s_1$ and t are given by

$$r = 1 - c\,(\Gamma/\omega_D \lambda\rho)\,Y$$
$$t = 1 + cY \;, \quad s_1 = c\Gamma Y \qquad (V.7)$$
$$s_2 = c\Gamma Y (2J_2+1) \;, \quad Y = \tfrac{1}{2}(2J_1+1)(V/\Delta E)^2 ,$$

where $\lambda\rho$ is the electron-phonon coupling constant. We have assumed that $\Delta E \gg \omega_D$.

s/t is the Cooper pair breaking parameter and (1-r/t) is the pair weakening parameter. In the case of Ce we have $J_2=0$ and s=0. Since the groundstate is a singlet, mixed-valent Ce impurities are nonmagnetic and suppress superconductivity by pair weakening rather than pair breaking. We obtain an exponential dependence of T_c on the concentration

$$T_c/T_{co} = \exp\left[-\frac{1}{\lambda\rho}\left(\frac{t}{r}-1\right)\right] , \qquad (V.8)$$

in agreement with the experimental situation[29]. Tm impurities, on the other hand, are magnetic at low T and one expects a quenching of superconductivity by the pair breaking mechanism. Indeed $s\neq 0$ and T_c essentially follows an Abrikosov-Gorkov dependence on c.

V. CONCLUDING REMARKS

In summary, we have calculated the dynamic charge and spin susceptibilities and the a.c. resistivity for all temperatures for a mixed valence impurity. The perturbation series has been summed up into continued fractions by using Mori's formalism, which is modified in the spirit of the Brillouin-Wigner approximation[16]. The impurity Green's function calculated diagrammatically by Grewe and Keiter[15] has been reproduced. The dynamic susceptibilities are in good agreement with a previous mode-mode coupling approach[9] and reproduce the Kuramoto-Müller-Hartmann results[7,8] for the high temperature limit. Two approaches for the thermodynamics of impurities, namely the Brillouin-Wigner theory[13,14] and the mode-mode coupling method[9] yield very similar results and the reduction of the superconductor T_c due to mixed valence impurities has the expected behavior. We can conclude that most of the known dynamic and static features of mixed-valence impurities are correctly obtained.

REFERENCES

+On a Heisenberg-fellowship of the Deutsche Forschungsgemeinschaft

1 - Varma, C.M., Rev. Mod. Phys. 48,219 (1976)
2 - Valence Instabilities and Related Narrow-Band Phenomena, edited by R.D. Parks (Plenum, New York 1977)
3 - Valence Fluctuations in Solids, edited by L.M. Falicov, W. Hanke and M.B. Maple (North-Holland, Amsterdam 1981)
4 - Anderson, P.W., Phys. Rev. 124, 41 (1961)
5 - Balseiro, C.A. and López, A., Solid State Commun. 17, 1241 (1975)
6 - Foglio, M.E., J. Phys. C 11, 4171 (1978)
7 - Kuramoto, Y., Z. Physik B 37, 299 (1980)
8 - Kuramoto, Y. and Müller-Hartmann, E., in ref. 3, p. 139
9 - Schlottmann, P., Phys. Rev. B 25, ... (1982)
10 - Götze, W. and Schlottmann, P., Solid State Commun. 13, 861 (1973), J. Low Temp. Phys. 16, 87 (1974)
11 - Schlottmann, P., Thesis, T.U. München (1973), unpublished
12 - Keiter, H. and Kimball, J.C., Intern. J. Magnetism 1, 233 (1971)
13 - Bringer, A. and Lustfeld, H., Z. Physik B 28, 213 (1977)
14 - Ramakrishnan, T.V., in ref. 3, p. 13
15 - Grewe , N. and Keiter, H., Phys. Rev. B 24, 4420 (1981)
16 - Schlottmann, P., Phys. Status Sol., in press
17 - Foglio, M.E. and Schlottmann, P., Solid State Commun., in press
18 - Schlottmann, P., J. Low Temp. Phys. 47, 27 (1982)
19 - Mori, H., Prog. Theor. Phys. (Kyoto) 34, 399 (1965)
20 - Zwanzig, R., J. Chem. Phys. 33, 1338 (1960)
21 - Götze, W. and Michel, K.H., in Lattice Dynamics, edited by A.A. Maradudin and G.K. Horton (North-Holland, 1974), p. 500
22 - Luszik-Bhadra, M., Barth, H.J., Brocksch, H.J., Netz, G., Riegel, D. and Bertschad, H., Phys. Rev. Lett. 47, 871 (1981)
23 - Götze, W. and Wölfle, P., J. Low Temp. Phys. 5, 575 (1971)
24 - Schlottmann, P., Phys. Rev. B 22, 622 (1980)
25 - Mazzaferro, J., Balseiro, C.A., Alascio, B., Phys. Rev. Lett. 47, 274 (1981)
26 - Schlottmann, P. and Falicov, L.M., Phys. Stat. Solidi 107, 165 (1981)
27 - Loewenhaupt, M. and Holland-Moritz, E., J. Appl. Phys. 50, 7456 (1979)
28 - Götze , W. and Wölfle, P., Phys. Rev. B 6, 1226 (1972)
29 - Huber, J.G. and Maple, M.B., J. Low Temp. Phys. 3, 537 (1970)

H. KEITER: I want to bring at least one of your statements back onto the right track: I was not facing a regularization problem in my talk, but I presented the solution to it. There are no ghosts anymore, which have to be thrown away (!?), in the final formula (10) of the paper given by Czycholl, Niebur and myself.

P. SCHLOTTMANN: So far I remember in your talk you discussed the partition function and not the Green's function where the problem occurs.

N. GREWE: You stated that the result derived by Keiter and me for the f-Green's function of the IV-impurity is obtained in the most divergent approximation and that we maintain there is a gap around the Fermi-level. In reality in our paper (PRB 24, 4420 (1981)) we wrote down the next divergent approximation explicitely, discuss that the gap is reduced and state, that it "apparently is an artifact of the (low order BW-) approximations". As far as I could see you have assumed that the density of states has a continuous Lorentzian form. That is exactly what I have pointed out in my talk on this conference to be observable as a trend (with some additional structure) from the low order of the renormalized perturbation scheme.

P. SCHLOTTMANN: I agree. I did not say anything that contradicts your statement.

T.M. HOLDEN: What was the experimental system with which you compared your theory for Cerium ions.

P. SCHLOTTMANN: It is CeTh measured by Riegel and coworkers (PRL 47, 871 (1981)) (ref. 22 of the paper).

I. NOWIK: Can you show the temperature dependence of the resistivity ?

P. SCHLOTTMANN: The resistivity is constant for $T \ll \Delta E$, has a bump at $T \sim \Delta E$ due to the Schottky anomaly and decreases at high T. It is closely connected with the charge susceptibility. As a function of frequency it is of the Drude type.

Valence Instabilities
P. Wachter and H. Boppart (eds.)

ANALYSIS OF Ce^{3+} FORM FACTOR IN ANOMALOUS CERIUM COMPOUNDS

J.X. Boucherle

DRF/DN, Centre d'Etudes Nucléaires
85X, 38041 Grenoble Cédex, France

Polarized neutron diffraction allows to measure with precision magnetic form factor and to study magnetization densities. In the case of a Ce^{3+} ion this magnetization density is very anisotropic, and the form factor changes drastically with the physical parameters. It is a favourable case to determine with accuracy the ground state of the ion. We will show how polarized neutron diffraction has improved the understanding of the anomalous compounds $CeAl_2$, CeSb, CeTe.

1. INTRODUCTION

In valence instability phenomena, the study of magnetic behaviour is important in determining the properties of both the electrons localized on the atoms and the conduction electrons delocalized in the whole cell. The measurements of magnetic form factors by polarized neutron diffraction is a powerful method : it allows very accurate measurements of magnetic structure factors which can give the magnetization density by Fourier transformation.

Cerium compounds are favourable cases for such studies. i) The cerium 3+ ion has only one electron on the magnetic 4f shell. The instability of this electron tends to induce a transition to a 4+ state where there is no electron on the 4f shell. The study of the localized part gives the Ce^{3+} contribution alone. ii) The 4f electron is very sensitive to the influence of the environment (especially crystal field effects). The important related anisotropy of the magnetization density allows accurate determination of the ground state of the ion.

Two aims can be achieved by polarized neutron studies. The first is to obtain a good description of the Ce^{3+} ground state which corresponds to the localized contribution. The second is to investigate whether there are other contributions, more delocalized than the 4f one.

2. PROPERTIES OF Ce^{3+} MAGNETIZATION DENSITY

Due to the strength of the spin orbit coupling, the resulting angular momentum is J = 5/2. The sixfold degenerated ground multiplet is split by electrostatic and magnetic field. As for Ce^{3+} ion only one electron is involved, the number of orbitals is small and the angular distribution of the $|J,M\rangle$ states is highly anisotropic [1] (fig.1). For cubic symmetry the six states are a combination of the $|\pm5/2\rangle$, $|\pm3/2\rangle$, $|\pm1/2\rangle$ basis vectors arranged in two levels : a doublet Γ_7 and a quartet Γ_8 with a splitting energy Δ.

The crystal field effects are peculiarly important for the Ce^{3+} ion. First, the reduction of the magnetic moment is considerable ($0.714\mu_B$ for the Γ_7 compared to the saturated ion value $2.14\mu_B$). Secondly, the shape of the magnetization density is very anisotropic and different for the two states. A simulation of form factor

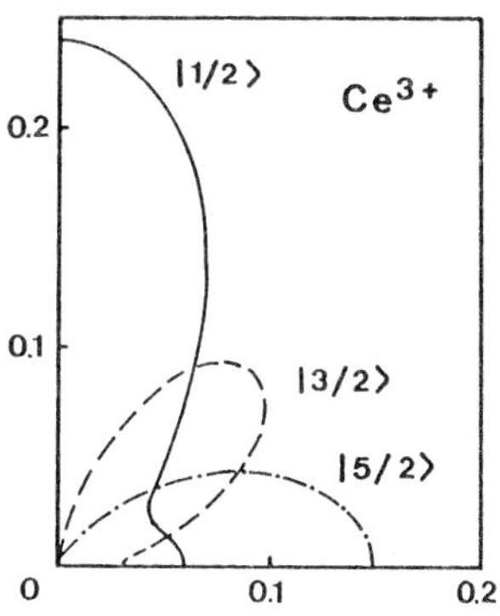

fig.1 : Density of probability $P(\Theta)$ for $|5/2\rangle$, $|3/2\rangle$, $|1/2\rangle$ states of Ce^{3+} in an axial symmetry.

and magnetization density has been done, for $[01\bar{1}]$ projection, in the case of Γ_7 and Γ_8 states (fig.2) with the wave functions (see appendix) :

$$|\phi\rangle_{\Gamma_7} = 0.612\ |5/2\rangle - 0.646\ |1/2\rangle - 0.456\ |-3/2\rangle$$

$$|\phi\rangle_{\Gamma_8} = 0.597\ |5/2\rangle + 0.756\ |1/2\rangle - 0.268\ |-3/2\rangle$$

The calculated points are far from a continuous curve, but dispersed between two curves corresponding to reflections (h00) and (0kk). This huge dispersion is unusual and characteristic of Ce^{3+} ion. Moreover it may be noted that for the two levels the anisotropy is completely different. For the Γ_8 state, the density is more spread out along [100] than along [011] and in the reciprocal space, the form factor of (h00) reflections is more contracted than for (0kk) reflections. For the Γ_7 state, the properties are reversed. This important anisotropy of the form factor leads to a very sensitive determination of the ground state of the Ce^{3+} ion. It may be anticipated that, when a magnetic field mixes the two states Γ_7 and Γ_8, the effects will be drastic, depending on the relative strength of the crystal field and of the exchange or applied field.

3. THE "KONDO" COMPOUND $CeAl_2$

$CeAl_2$ presents very unusual properties at low temperatures. In particular the modulated

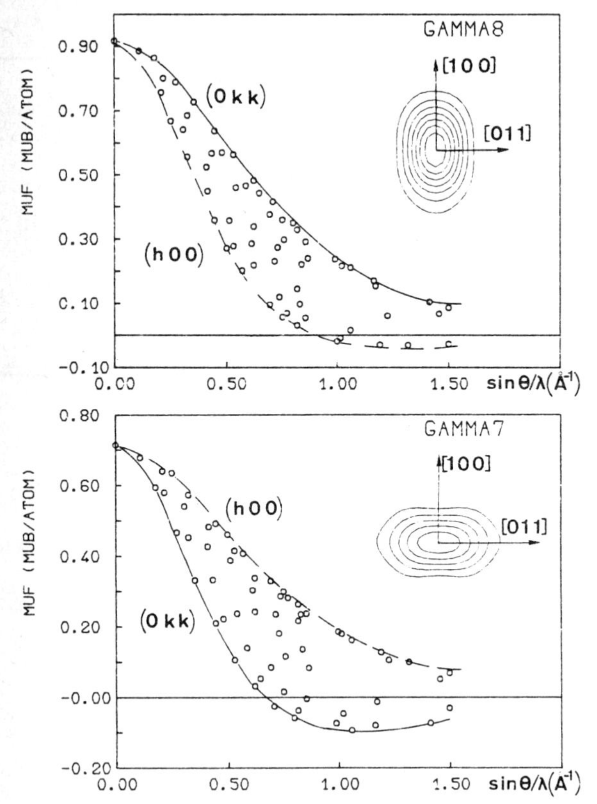

fig.2 : Simulation of form factor and magnetization density, corresponding to a projection along [01$\bar{1}$] axis, for the two states Γ_8 (above) and Γ_7 (below).

magnetic structure [2] has been explained only by invoking anomalous coupling with conduction electrons of Kondo type. To determine the state of the 4f ion a form factor study has been done at T = 1.5 K on the ferromagnetic component induced by an applied field of 4.65 T parallel to [01$\bar{1}$] [3]. $CeAl_2$ is a case of a strong crystal field ($\Delta \sim$ 100 K, Γ_7 ground state) and small exchange. The measurements (fig.3) are close to the calculation performed for the pure Γ_7 state (fig.2). A considerable scattering of the points is observed and the (0kk) reflections have the smallest values. At T = 1.5 K, only the ground state is populated and it is possible to refine its wave function (appendix 2) together with a factor k corresponding to a phenomenological reduction of the 4f moment : k = 0.452 (5)

$|\phi\rangle = 0.691(6)\,|5/2\rangle - 0.627(10)\,|1/2\rangle - 0.358(21)\,|-3/2\rangle$

For $\sin\theta/\lambda > 0.25$ Å^{-1}, the agreement between the experimental (full circles) and the calculated values (open circles) is excellent (fig.3). Thus almost all the magnetization density in $CeAl_2$ can be explained by a Ce^{3+} model. The wave function, slightly different from that of a pure Γ_7 level, can be explained by an exchange field mixing the Γ_7 and Γ_8 states.

However at low $\sin\theta/\lambda$ some discrepancies appear due to a positive extra magnetization. We have shown that this contribution, less localized

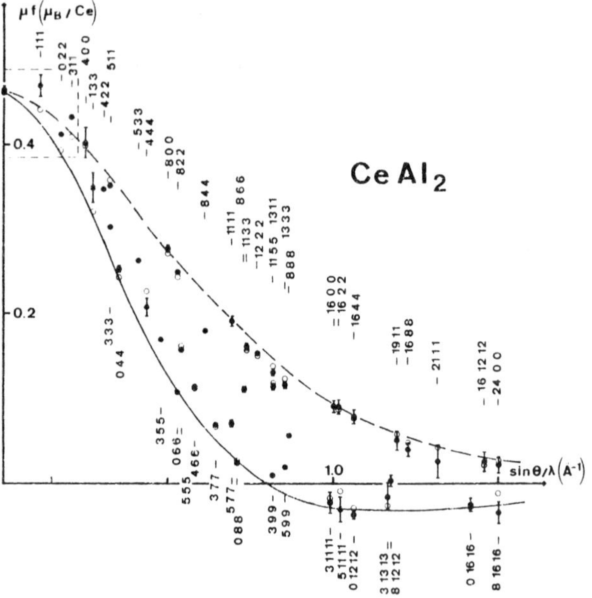

fig.3 : Observed (•) and calculated (o) values of the form factor for $CeAl_2$ at T = 1.5K and H = 4.65 T.

than the 4f one, is very similar to that observed for other rare earth compounds and shows a 5d like localization [4]. But the negative coupling between 4f and conduction electron spins is anomalous. It can be related to Kondo type behaviour of $CeAl_2$. Moreover such a positive 5d contribution has been shown in $CeSn_3$ [5].

4. CeSb : THE FERROMAGNETIC STATE

The magnetic behaviour of CeSb is complex below 16.5 K [6]. At T = 4.2 K and in an applied field of 4.65 T parallel to [001], this compound is ferromagnetic. To know precisely and to understand the value and the origin of the observed

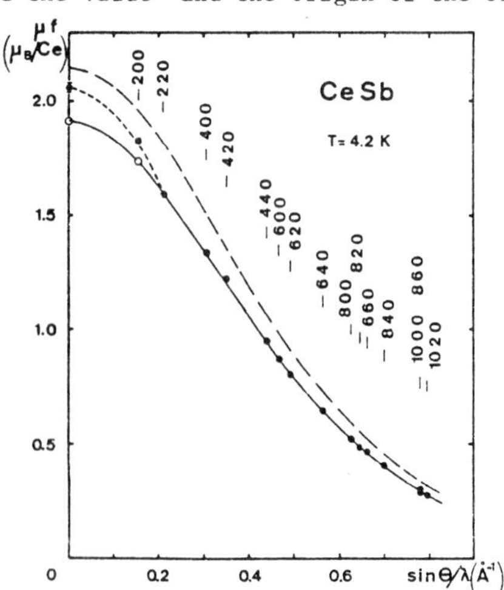

fig.4 : Observed (•) and calculated (o and -) values of the form factor for CeSb in the ferromagnetic region (T = 4.2 K, H = 4.65 T). The values for a saturated ion (---) are also shown.

magnetic moment ($\mu \sim 2\mu_B$), form factor measurements were performed [7]. We are in a case of a small crystal field and important exchange interactions. The experimental form factor (fig.4), which presents no anisotropy, is very different from that observed in $CeAl_2$. It is close to the saturated ion form factor (dashed line fig.4), but with a small reduction of the moment. The results are well described by the following wave function :

$|\phi\rangle = 0.995(1)|5/2\rangle - 0.100(10)|-3/2\rangle$, k=0.92(1)

This study clearly shows that the Ce^{3+} ion is nearly in the saturated state as indicated by its wave function. However this wave function cannot be explained by the usual formalism involving exchange : the exchange interactions would be too large. This result must be related to the f-p mixing which seems to be very important in this compound [8].

5. CeSb : CRYSTAL FIELD STUDY AT HIGH TEMPERATURE

As a consequence of the low crystal field splitting ($\Delta < 40$ K), it is not obvious whether the ground state is Γ_8 or Γ_7. Taking advantage of the very different shape of the form factor for the two states, we have tried to solve this problem. The polarized neutron experiment was performed in the paramagnetic region (T = 35.5 K) with an applied field of 4.65 T. The measured form factor [7] shows only a small anisotropy

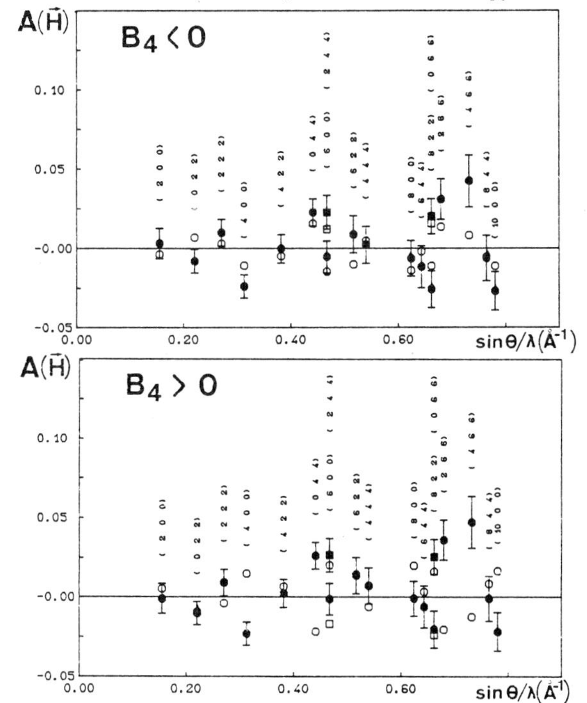

fig.5 : Observed (● and ■) and calculated (o and □) anisotropy (see appendix 1) of the form factor for CeSb in the paramagnetic region (T = 35.5 K, H = 4.65 T).
Above : Γ_8 ground state, B_4 = -16.6 K, Θp = 5.3 (2) K (R = 2.0%).
Below : Γ_7 ground state, B_4 = +16.6 K, Θp = 6.1 (5) K (R = 5.2%).

due to the thermal averaging. Observed (full symbols) and calculated (open symbols) anisotropy (see appendix 1) are compared (fig.5) for two calculations performed for $B_4 < 0$ (Γ_8 ground state) and $B_4 > 0$ (Γ_7 ground state) (see appendix 3). It may be noted that in the first case the agreement is excellent and that in the other case the calculated anisotropy is opposite to the observed one. Thus at T = 35.5 K and H=4.65 T the results are well described by the usual model of exchange field and crystal field with negative value for B_4.

However it is not obvious that this result can be directly extrapolated to the paramagnetic state without field or to the ordered state. More sophisticated theories may be necessary to explain the magnetic field effects in this compound [8].

6. CeTe : THE PROBLEM OF THE MOMENT REDUCTION

Below T = 2 K CeTe presents an antiferromagnetic structure. Its striking property is the low value of the magnetic moment measured by neutron diffraction ($\mu < 0.3\ \mu_B$) [9]. To study the origin of this reduction, we measured accurately the form factor at T = 1.5 K and in an applied field of 4.65 T [10]. An anisotropy is observed (fig.6) but smaller than in $CeAl_2$. The mixing of the Γ_7 and Γ_8 states is important as a result of the small crystal field splitting ($\Delta \sim 20$ K). It gives a more spherical magnetization density than in $CeAl_2$. However the reflections (Okk) have the smallest values : this result, which is characteristic of a Γ_7 ground state is in agreement with measurements by other techniques.

The wave function which best represents the experimental data is :

$|\phi\rangle = 0.77(1)|5/2\rangle - 0.60(2)|1/2\rangle - 0.23(9)|-3/2\rangle$

with a moment reduction k = 0.67 (1). The calculated values (open symbols) are in excellent agreement with the experimental points (full symbols). The origin of the reduction factor is not obvious and will be discussed elsewhere [10]. But it cannot explain the low magnetic moment of

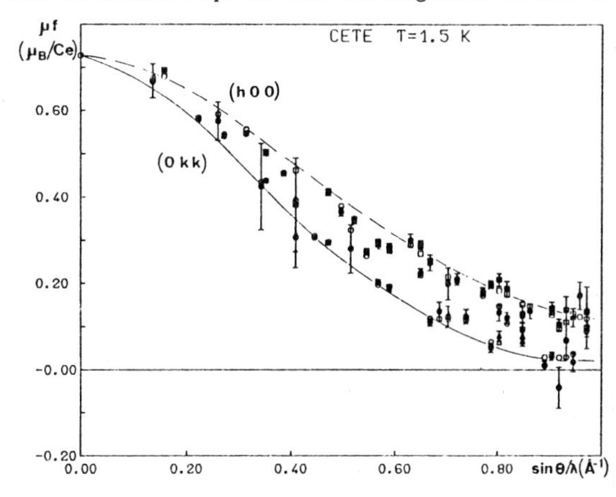

fig.6 : Observed (full symbols) and calculated (open symbols) values of the form factor for CeTe at T = 1.5 K and H = 4.65 T.

the antiferromagnetic structure. One can think that the influence of the applied field is important and that it can screen the zero field properties.

7. CONCLUSION

We have shown how the Ce^{3+} form factor can vary considerably with the different relative importance of the crystal field and magnetic field. The three studies, which have been presented, are good examples of the different possible cases. For $CeAl_2$, the exchange is small compared to the crystal field and the observed form factor is close to that of a pure crystal field state. When the two contributions are of the same order, as for CeTe, the anisotropy is reduced. Finally, for CeSb in ferromagnetic state, the exchange is much higher than the crystal field and an almost saturated ion is observed.

In each case it was possible to obtain directly a good determination of the ground state of the 4f ion. These accurate results are important to improve the understanding of such anomalous compounds and may help to find suitable models to explain their properties.

8. APPENDIX : CALCULATION OF THE FORM FACTOR

Using the tensor operator method [11], it is possible to calculate the magnetic form factor from the 4f wave function :

$$|\phi\rangle = \sum_M a_M \, |J,M\rangle$$

$$\mu f(\vec{H}) = \sum_{K''Q''} Y^{Q''}_{K''}(\hat{H}) \sum_{K'} \langle j_{K'}(H)\rangle \sum_{MM'} a^{*}_{M} a_{M'} C_{K''Q''K'MM'}(H)$$

The radial integrals $\langle j_K \rangle$ are best suited for rare earth atoms when they are obtained from an atomic Dirac-Fock calculation [12]. The spherical harmonics describe the orientation of the scattering vector $\vec{H}$ by means of the polar angles Θ and Φ. The values of $C_{K''Q''K'MM'}$ are calculated using Clebsh-Gordon and tabulated [13] coefficients.

8.1. Case of a projection

If the magnetic structure factors are collected only in a plane, they correspond to the projection of the magnetization density on this plane. In this case $\Theta = \pi/2$ and the form factor depends only on the Φ angle. By isolating the isotropic part independent of Φ, $\mu f_{Ical}(H)$, it is possible to define an anisotropy of the form factor (calculated and observed) :

$$A_{cal}(\vec{H}) = \frac{\mu f_{cal}(H,\Phi) - \mu f_{Ical}(H)}{\mu f_{cal}(0,0,0)}$$

$$A_{obs}(\vec{H}) = \frac{\mu f_{obs}(H,\Phi) - \mu f_{Ical}(H)}{\mu f_{cal}(0,0,0)}$$

8.2. Low temperature case : Refinement of the wave function

At low temperature, only the ground state is populated. The local symmetry imposes the value ΔM between the basis vectors $|J,M\rangle$. The components of the wave function are determined and only the a_M coefficients are unknown. The form factor can be written :

$$\mu f(\vec{H}) = \sum_{MM'} a^{*}_{M} a_{M'} B_{MM'}(\vec{H})$$

It is possible, from a comparison with the experiment, to obtain, by a least square method, the values of the a_M coefficients. Thus the ground state wave function can be refined.

8.3. High temperature case : Crystal field model

At high temperatures, several levels are populated and it is no longer possible to take into account only the ground state. It is now necessary to use a model which gives the wave function and the energy of each level. The usual Hamiltonian (crystal, exchange and applied field) is diagonalized :

$$\mathcal{H} = \mathcal{H}_{CEF}(B_4) + \mathcal{H}_{ex}(\Theta p) + \mathcal{H}_{ap}(Hap)$$

B_4 is the fourth order crystal field parameter and Θp the paramagnetic Curie temperature. The form factor is calculated for each level and a thermal averaging is performed. By a comparison between the experimental and calculated form factors, the parameters of the model are refined.

REFERENCES

[1] Schmitt, D., to be published

[2] Barbara, B., Boucherle, J.X., Buevoz, J.L., Rossignol, M.F., Schweizer, J., Sol. Stat. Comm. 24 (1977) 481

[3] Barbara, B., Boucherle, J.X., Desclaux, J.P., Rossignol, M.F., Schweizer, J., Crystal Field in Metal and Alloys, Ed. Furrer, A., (Plenum) (1977) 168

[4] Boucherle, J.X., Givord, D., Gregory, A., Schweizer, J., 27th M.M.M. Conf. to be published in J. Appl. Phys. (1982)

[5] Stassis, C., Loong, C.K., Harmon, B.M., Liu, S.H., Moon, R.M., J. Appl. Phys. 50 (1979) 7567

[6] Rossat-Mignod, J., Burlet, P., Villain, J., Bartholin, H., Wang Tcheng-Si, Florence, D., Vogt, O., Phys. Rev. B16 (1977) 440

[7] Boucherle, J.X., Delapalme, A., Howard, C.J., Rossat-Mignod, J., Vogt, O., Physica 102B (1980) 253

[8] Takegahara, K., Takahashi, H., Yanase, A., Kasuya, T., IV Int. Conf. Crystal Field and Structural Effects in f-Electron System Wroclaw (1981) to be published

[9] Ravot, D., Burlet, P., Rossat-Mignod, J., Tholence, J.L., J. de Phys. 41 (1980) 1117

[10] Boucherle, J.X., Burlet, P., Ravot, D., Rossat-Mignod, J., Schweizer, J., to be published

[11] Lovesey, S.W., Rimmer, D.E., Rep. Prog. Phys. 32 (1969) 333

[12] Freeman, A.J., Desclaux, J.P., J.M.M.M. 12 (1982) 11

[13] Lander, G.H., Brun, T.O., J. Chem. Phys. 53 (1970) 1387

Valence Instabilities
P. Wachter and H. Boppart (eds.)

MUON KNIGHT SHIFT INVESTIGATIONS IN Ce, Ce-Th AND $CeSn_3$ (*)

H. Wehr, K. Knorr
Institut für Physik, Universität Mainz, D-6500 Mainz, FRG
F.N. Gygax, A. Schenck and W. Studer
LHE/ETH Zürich, c/o SIN, CH-5234 Villigen

Positive muons have been used as local probes of the spin polarization of the conduction electrons in Ce, $Ce_{0.74}Th_{0.26}$ and $CeSn_3$. In the Curie-Weiss type region of the individual systems the muon Knight shift ($K\mu(T)$) reflects the temperature dependence of the bulk susceptibility $\chi_B(T)$ which is explained by an RKKY-type hyperfine field arising from the 4f-electrons. In the α-phase of Ce and Ce-Th the proportionality between $K\mu$ and χ_B remains nearly unchanged suggesting that the hyperfine-coupling to the 4f-state is not much affected by the valence transition. In $CeSn_3$ $K\mu$ decreases much stronger than χ_B below 200 K presumably because of an additional negative hyperfine field arising from 5d-electrons.

INTRODUCTION

The intermediate valence state encountered in many rare earth compounds is expected to influence the spin polarization of the conduction electrons. Due to the 4f-conduction band hybridization the RKKY approach usually applied in systems with stable 4f-state is questionable.
Muon spin rotation experiments offer the chance to investigate the conduction electron spin polarization at an interstitial site. Thus this method promises a somewhat more direct access to the quantity of interest than the other magnetic resonance techniques ESR and NMR.
In the present work we study the muon Knight shift in γ and α Ce, (Ce,Th) and $CeSn_3$.
As is common practice we assume that the Knight shift $K\mu$ can be split up into contributions of different origin. $K\mu$ probes the local spin density, hence the shift due to spin polarized conduction electrons of type say "i" may be given by:

$$K^i\mu = (B_i^{hf}/\mu_B)\ \chi_i$$

where B_i^{hf} is the induced hyperfine field and χ_i is the corresponding spin susceptibility (per atom). The electron spin density at an interstitial muon will in general be different from its value in the unperturbed case since the positive muon charge presents a rather strong perturbation of the electronic system. The understanding of the hyperfine fields at the muon is still at an early stage, nevertheless the following features are commonly accepted:
Conduction electrons in wide bands which have a non zero density at the muon can contribute directly to $K\mu$, B_i^{hf} being then equivalent to the Fermi contact field. In this case the Knight shift should be positive and temperature independent. In systems with a high density of d-electrons at the Fermi level often negative muon Knight shift values are observed (1). These results suggest some analogy to the core polarization effects at the nucleus measured in NMR studies though the positive muon may not possess an electron core in the usual sense. In the presence of stable 4f-moments B_i^{hf} may represent the induced hyperfine field due to the RKKY exchange interaction and χ_i should be identified with the 4f-spin susceptibility. Finally one has to take into account a negative and temperature independent contribution K_{dia} which results from a diamagnetic shielding of the muon charge by the surrounding electrons. The value of K_{dia} is supposed to be about -20 to -30 ppm in most metals (2).

RESULTS AND DISCUSSION

In the present study polycristalline samples have been used which were prepared from the elements by levitation melting after prereaction in an argon arc furnace. The experiments were carried out at the SWISS INSTITUTE FOR NUCLEAR RESEARCH (SIN) in transverse fields of 3.7 kG or 7.4 kG. A stroboscopic method was applied which bases on a coherent superposition of spin polarized muon bursts in the sample (3). For more experimental details see ref.(4).

Ce, $Ce_{0.74}Th_{0.26}$, $La_{0.74}Th_{0.26}$

We first comment on the local muon site in these systems. It is commonly accepted that the muon preferentially occupies octahedral interstitial sites in fcc-metals (5). Since the linewidth results of our stroboscopic signals do not conflict with this assumption (see ref. 4) one might conclude - without being a stringent proof - that the Knight shift data in Fig. 1 are first of all representative for the octahedral sites. Due to the variation of the local environment in the alloy $K\mu$ represents a mean value over the sample.

In the 4f-free reference system La-Th the shift is almost zero and temperature independent (Fig.1).

Presumably the positive Fermi contact field is cancelled by the diamagnetic shielding of the muon charge. In γ - Ce and in Ce-Th positive values of $K\mu$ are observed which follow the Curie-Weiss type behaviour of the 4f-spin susceptibility $\chi_{4f}(T)$. Throughout this article we identify $\chi_{4f}(T)$ with the bulk susceptibility $\chi_B(T)$, corrected for the susceptibility of the corresponding La-sample, as determined experimentally. From the linear relation between $K\mu(T)$ and $\chi_{4f}(T)$ (see Fig. 3) we obtain the induced hyperfine field per $4f^1$ moment in the γ-phase:

$B^{hf}_{4f}/_{\gamma} \sim 0.35$ kG/μ_B and $B^{hf}_{4f}/_{\gamma} \sim 0.46$ kG/μ_B

for Ce-Th and Ce resp. These values are by about two orders of magnitude lower than typical muon Fermi contact fields telling us that the muon experiences the induced moments of the 4f-states rather indirectly i.e. via the RKKY-interaction and not directly by a contact interaction.

Deviations from the linear relation between $K\mu(T)$ and $\chi_{4f}(T)$ are observed around the γ-to-α phase transition temperature T_s in Ce-Th. The spin density at the muon seems to be systematically increased over the spatially averaged values of the bulk measurement (circles in Fig. 3). In the same temperature range the linewidth of the muon signals was anomalously increased possibly because of the influence of density fluctuations but perhaps also because of lattice inhomogeneities induced by the large volume collapse. Unfortunately the present experiments cannot distinguish between true relaxation effects and static field distributions due to inhomogeneities, hence it cannot be unambiguously decided whether the larger values of $K\mu$ result from an increased exchange coupling or from a site change effect of the muon.

One could expect that in the collapsed α-phase the direct Fermi contact field at the muon and hence the hyperfine coupling to the former $4f^1$ electrons is increased due to the delocalization of the 4f-state. The experiments do not provide evidence for this to be the case since the linear relation between $K\mu(T)$ and $\chi_{4f}(T)$ of the high temperature range seems to hold again without change in α-Ce-Th well below T_s (Figs. 1, 3) and the shift in α-Ce is slightly below the extrapolated $K\mu(\chi_{4f})$ value (Fig. 3). One may thus conclude that the admixture of $4f^1$- electrons into wide band states is too small to be experienced by the muon. Later on we will discuss the possible influence of 4f-5d hybridization.

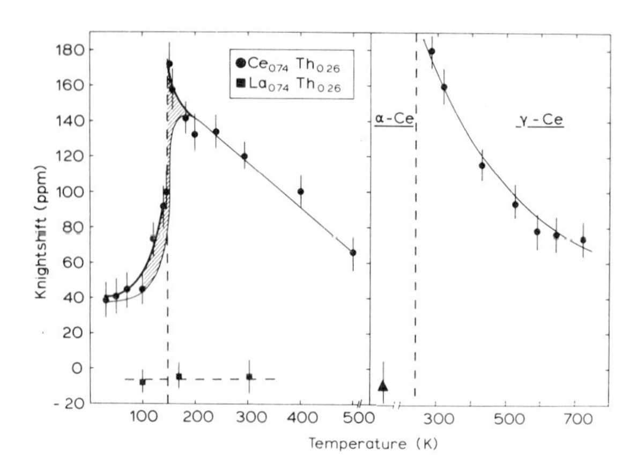

Fig.1 (left side) The muon Knight shift in $Ce_{0.74}Th_{0.26}$ and $La_{0.74}Th_{0.26}$ as a function of temperature. The dotted line and the heavy solid lines are guides to the eye. The light solid line gives the bulk susceptibility taken from ref. (11) and scaled to fit the data at high temperatures.
(right side) The muon Knight shift in γ -Ce and α -Ce as a function of temperature. The light solid line gives the bulk susceptibility taken from ref.(12) after appropriate scaling to fit the Knight shift.

$CeSn_3$, $LaSn_3$, $PrSn_3$

Two non-equivalent octahedral interstitial sites with a relative occurrence of 1:3 exist in (rare earth)-Sn_3 compounds: one with six Sn, the other one with two rare earth ions and four Sn-ions in the next neighbour shell. If both sides are occupied by the muons the stroboscopic signal should in principle consist of two resonance frequensies unless the muon diffuses thus performing a site average over the sample. We could not resolve two precession frequencies in our polycristalline samples presumably because their difference is too small with respect to the width of the signals which is given by the muon lifetime (2.2μ sec) and transverse muon spin relaxation effects. From the symmetry of the local sites it is obvious that considerable inhomogeneous line-broadening can arise from the dipole fields of the induced 4f-moments. Furthermore it is known that sample inhomogeneities often present a delicate problem in these compounds. Hence muons at irregular lattice sites can also contribute to an overall broadening of the signals.
We have fitted the stroboscopic signals with a single lorentzian line, the frequency shift is shown in Fig.2.

We first concentrate on the results found in $LaSn_3$. The rather large shift - when compared to La-Th - suggests that the non-transition metal Sn contributes a considerable proportion of wide band electrons to the muon site. This assumption is supported by our recent Knight

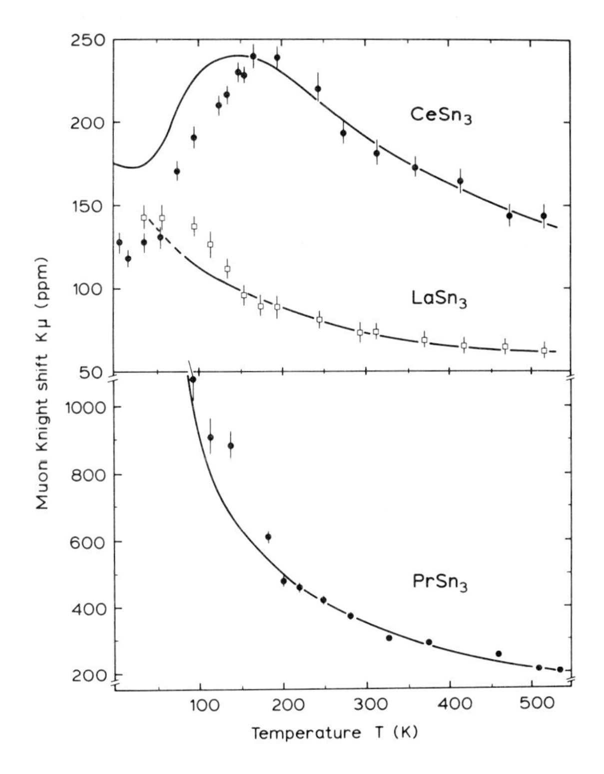

Fig.2 (upper part) The muon Knight shift in $CeSn_3$ and $LaSn_3$ as a function of temperature. The solid lines give the ($CeSn_3$)-bulk susceptibility (corrected for the 4f-free background and magnetic impurity contributions according to refs.(13,14)) and the ($LaSn_3$)-bulk susceptibility (taken from ref.(7)) after appropriate scaling to fit the Knight shift data at higher temperatures. (lower part) The muon Knight shift in $PrSn_3$. The solid line represents the bulk susceptibility taken from ref.(15) and scaled to fit the data.

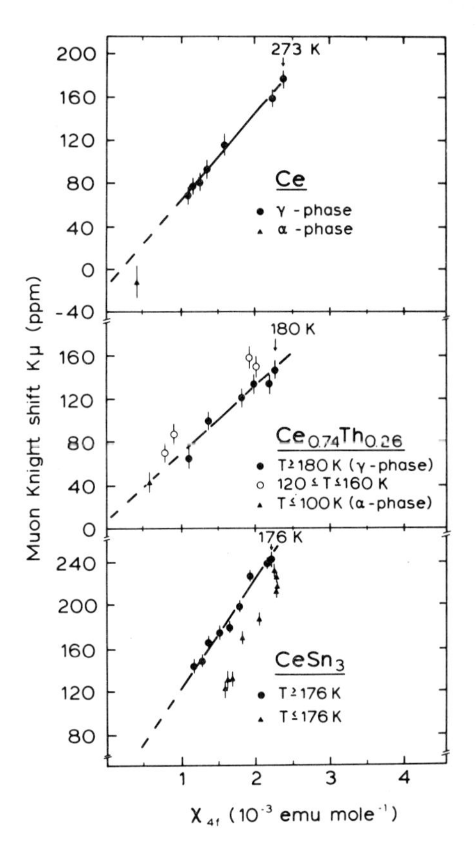

Fig. 3 Dependence of the muon Knight shift on the 4f spin susceptibility χ_{4f} in and Ce, Ce-Th and $CeSn_3$. χ_{4f} is identified with the bulk susceptibility after correction for the 4f-free back ground and magnetic impurity contributions (see refs.(11,12, 16) for Ce, Ce-Th, refs.(13,14) for $CeSn_3$).

shift study in $LaIn_3$ where $K\mu$ is always about 30 ppm lower than in $LaSn_3$ presumably because of the smaller number of outer electrons of In (6). The temperature dependence of $K\mu$ in $LaSn_3$ proposes that the total conduction electron system can not be treated as free electron like. A similar temperature behaviour has been observed in the bulk susceptibility χ_B (though there are noticeable differences between the data of different authors). For a comparison with our data we follow the work of Welsh et al (7) who proposed a Curie-Weiss -type law for $T > 100$ K. The linear fit $K\mu(\chi_B(T))$ yields a temperature independent Knight shift part $K_o \sim 15$ ppm and a positive hyperfine field $B_{hf} \sim 1.65$ kG/μ_B. We believe that the analogy between $K\mu(T)$ and $\chi_B(T)$ is not accidental but rather proposes that the spin density of $LaSn_3$ in fact shows some temperature dependence though the exact origin is yet unclear.

The muon Knight shift curve in the instable valence system $CeSn_3$ shows remarkable features (Fig.2). In the Curie-Weiss regime above 200 K a fairly large hyperfine field B^{hf}_{4f} when compared to the values found in Ce and Ce-Th is deduced from the linear relation between $K\mu(T)$ and $_{4f}(T)$ i.e. ($B^{hf}_{4f} \sim 0.55$ kG/μ_B, $K_o \sim 22$ ppm). (see Fig.3)

Actually the muon Knight shift in $CeSn_3$ at room temperature is about twice as large as in $CeIn_3$ (6) which suggests a very effective coupling between the 4f-electrons and the conduction electrons in $CeSn_3$.

Below 200 K $K\mu$ decreases more strongly than χ_{4f} and crosses the Knight shift curve of $LaSn_3$ below 100 K (see Figs. 2,3). This is a rather large reduction of K in view of the fact that $CeSn_3$ is supposed to undergoe only a minor valence change. The results propose that Knight shift contributions other than those of the high temperature range are involved and from the discussion in Ce and Ce-Th it is clear that these can not result from wide-band electrons. Low temperature anomalies have also been observed in NMR-Knight shift studies at the Sn-site of $CeSn_3$ (8), however the sign of the NMR anomaly (increase of B^{hf}) is opposite to the muon results (see Fig. 3). From neutron form factor measurements in $CeSn_3$ an increased 4f-5d(e_g) hybridization at low temperatures has been deduced (9). Increased 5d-Knight shift contributions could explain the anomalous reduction of the muon Knight shift since d-conduction electrons often contribute via negative hyperfine fields to $K\mu$. This is supported by our recent Knight shift study in $LaAg_{1-x}In_x$ compounds where we found that a population of the 5d(e_g) states systematically shifts $K\mu$ to lower values (10).

To contrast the results in the instable valent $CeSn_3$ we finally present in Fig. 2 the muon Knight shift in $PrSn_3$. In the temperature range which could be studied so far Curie-weiss behaviour is observed which can be fitted with the parameters:

($E^{hf}_{4f} \sim 0.36$ kG/μ_B , $K_o \sim 20$ ppm)

Note the smaller 4f-hyperfine field in $PrSn_3$ when compared to the value found in the Curie-Weiss regime of $CeSn_3$. The increased error bars at low temperatures mainly result from inhomogeneous line broadening due to the induced Pr-moments as pointed out above.

SUMMARY

One conclusion from the muon experiments, which holds for α-Ce, α-Ce-Th and $CeSn_3$ likewise, is the obvious absence or undetectable small admixture of $4f^1$-electrons into delocalized wide-band states which should increase the hyperfine-field with respect to its value in the Curie-regime. From a comparison of $K\mu$ in $LaSn_3$ and $LaIn_3$ i.e.($K\mu(LaSn_3) - K\mu(LaIn_3)) \sim 30$ ppm, whose conduction electron systems differ by about one itinerant electron, one might expect that an increase of perhaps 0.2 electrons in wide bands should be resolvable by muon Knight shift measurements in these systems.
In α-Ce-Th the hyperfine coupling between the muon spin and the 4f-electrons seems to remain unchanged with respect to its value in the high temperature regime since no deviations from the linear $K\mu(\chi_{4f})$ relation are observed.
Evidence for the onset of a new hyperfine mechanism is found at low temperatures in $CeSn_3$ and possibly also - because of a small anomaly - in α-Ce. It is not surprising that the spin density at the muon seems to be influenced differently in the instable valence regime of Ce, Ce-Th and $CeSn_3$ since the electronic environment might be rather different as can be seen from the Knight shift in the corresponding La-samples or from the different hyperfine field values B^{hf}_{4f}.

REFERENCES

(1) see for example: F.N. Gygax et al., Solid State Comm., 30, 1245 (1981)
(2) E. Zaremba, D. Zobin, Phys.Rev.B,22, 5490 (1980)
(3) M. Camani et al.,Phys.Lett. 77B, 326 (1978)
(4) H. Wehr, K.Knorr, F.N. Gygax, A.Schenck and W. Studer, Phys.Rev.B, 24, 4041 (1981)
(5) see for example: M. Camani etal., Phys.Rev. Lett., 39, 836 (1977)
(6) H. Wehr et al., to be published
(7) L.B. Welsh, A.M. Toxen and R.J. Gambino, Phys.Rev.B, 13, 3132 (1976)
(8) D.E. MacLaughlin, in "Valence Fluctuations in Solids", L.M. Falicov, W. Hanke, M.B. Maple (eds.), North-Holland Publishing Company, 1981
(9) C. Stassis, C.K. Loong, O.D. McMasters and R.M. Moon, J.Appl.Phys. 50, 2091 (1979)
(10) H. Wehr, K. Knorr, F.N. Gygax, A.Schenck and W. Studer, "Proc. 4th. Int.Conf. on Crystal Electric Field and Structural Effects in f-Electron Systems, Wroclaw, 22-25 Sept. 1981 (to be published)
(11) J.M. Lawrence and R.D. Parks, J.Phys.(Paris) 37, 64 (1976)
(12) C.R. Burr and S. Ehara, Phys. Rev. 149, 551 (1966)
(13) J. Lawrence, Phys. Rev. B, 20, 3770 (1979)
(14) J. G. Sereni, J.Phys.F,10, 2831 (1980)
(15) P. Lethuillier and J. Chaussy, Phys. Rev.B, 13, 3132 (1976)
(16) R. A. Elenbaas, C.J. Schinkel and E. Swakman, J.Phys.F,9, 1261 (1979)

(*) Work supported in part by the "Bundesminister für Forschung und Technology" F R G .

Valence Instabilities
P. Wachter and H. Boppart (eds.)

STRONG FERROMAGNETIC EXCHANGE AMONG MIXED VALENT Ce IN $CeMn_6Al_6$

I. Felner and I. Nowik

Racah Institute of Physics, The Hebrew University, Jerusalem, Israel

X ray, resistivity, magnetization and Mossbauer studies of dilute probes of Fe^{57} and Gd^{155} in $CeMn_6Al_6$ were performed. All studies indicate that Ce in $CeMn_6Al_6$ is in a mixed valent state. The magnetization studies show that the Ce contribution to the susceptibility is very small at low temperatures, increases to a huge maximum at 100K ($\sim 3.10^{-2}$emu/mole) and follows Curie Weiss law at high temperatures. The experimental observations are interpreted in terms of interconfigurational fluctuations between Ce^{4+}(ground state) and Ce^{3+} ions. The detailed analysis, which considers also coupling with the Mn sublattice, yields for the Ce interconfigurational excitation energy, the value of 223K and for the Ce-Ce exchange coupling 47K.

1. INTRODUCTION

Recent studies of RMn_6Al_6 [1,2] reveal that for R = light rare earth, the systems crystallize in the rhombohedral Th_2Zn_{17} structure. Magnetization studies of RMn_6Al_6 show weak antiferromagnetic exchange in both the Mn and rare earth sublattices, T_N=25K for $SmMn_6Al_6$ and T_N=4K for $LaMn_6Al_6$ [2]. In all systems the effective moment of the Mn ion was $2.6\mu_B$. The lattice dimension measurements [1] have shown that Ce in $CeMn_6Al_6$ is probably in a mixed valent state.

Here we present a detailed study of $CeMn_6Al_6$. Measurements of its resistivity, magnetization and Mossbauer spectra of dilute probes of Fe^{57} and Gd^{155}, were performed. The magnetic susceptibility curves of $CeMn_6Al_6$ and the isostructural compounds $LaMn_6Al_6$, $Ce_2Mn_7Al_{10}$, $La_2Mn_7Al_{10}$, $Ce_2Mn_{8.5}Al_{8.5}$ are shown in fig. 1. In comparison to all the other systems, the susceptibility curve of $CeMn_6Al_6$ looks very peculiar. The Mn sublattice orders at about 10K, fig.2. The Ce contribution to the susceptibility is very low at temperatures below 50K, and increases to a huge maximum at 100K. This increase, $\sim 3.10^{-2}$ emu/mole, is four times larger than the free ion Ce^{3+} susceptibility at this temperature. The magnetic susceptibility of the samples doped with 1% Fe^{57} or Gd^{155} look very similar to that of pure $CeMn_6Al_6$, fig.2.

The rather peculiar magnetic behaviour of $CeMn_6Al_6$ was interpreted in terms of valence fluctuations and strong exchange interactions among the Ce ions. A least square fit procedure of the simple theory to the experimental results yields the interconfigurational excitation energy, 223K, and the Ce-Ce ferromagnetic exchange coupling, 47K.

2. EXPERIMENTAL DETAILS AND RESULTS

The resistivity measurements of a $CeMn_6Al_6$ rod of dimensions 1.1x1.1x5 mm were performed by a standard four point probe method. The results of the temperature dependence of the specific resistivity are displayed in fig.2. One observes a minimum at about 65K. The general shape is typical to a mixed valent state of Ce, similar to that of $CeAl_2$ or $CeAl_3$. The observed rise in resistivity at the very low temperatures is probably due to the magnetic Mn.

The Mossbauer studies of the 14.4 keV transition of dilute Fe^{57} in $CeMn_6Al_6$ and the 86 keV transition of Gd^{155} in $CeMn_6Al_6$ prove that

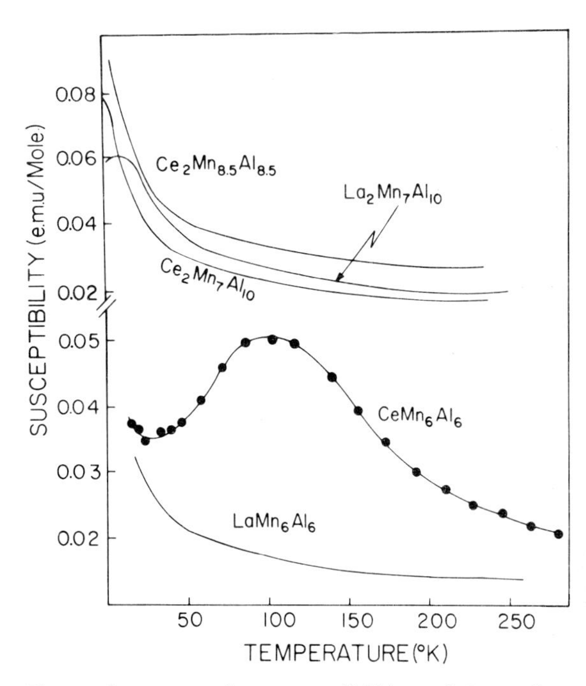

Figure 1 : Magnetic susceptibility of $LaMn_6Al_6$, $CeMn_6Al_6$ and isostructural systems. The solid curve for $CeMn_6Al_6$ is a theoretical fit.

$CeMn_6Al_6$ is magnetically ordered at 4.1K yet not at 77K. As one observes in fig.3, the 4.1K spectrum of Fe^{57} is split by magnetic interactions, whereas the 80K and the 300K spectra are identical, exhibiting only quadrupole structure. The Gd^{155} studies, using a $SmPd_3$ source, show magnetic structure at 4.1K and none at 77K.

The magnetic susceptibility measurements were performed using a PAR 155 sample vibrating magnetometer. The magnetization was linear in magnetic field at all temperatures down to about 10-15K, the range of temperatures where the Mn sublattice orders, fig.2. The higher susceptibility at low temperatures in the Fe and Gd doped samples, fig.2, in comparison to pure $CeMn_6Al_6$, is consistent with the additional contribution of the Fe and Gd, respectively.

Considering the results of the Mossbauer studies it is clear that the large maximum observed in the temperature dependence of the $CeMn_6Al_6$ susceptibility, fig.1, is not due to a magnetic transition. It must arise from the mixed valency properties of the Ce ion. We thus analyzed the $CeMn_6Al_6$ susceptibility curve in terms of a simple model for the magnetic behavior of a mixed valent system [3,4], with the addition of exchange interactions among Mn-Mn, Mn-Ce and Ce-Ce. The analysis yields the interconfiguration excitation energy, its width and the three exchange parameters.

The analysis indicates that a Ce mixed valent system can interact through exchange and may in principle order, as claimed for Tm mixed valent systems [5].

3. THEORETICAL ANALYSIS

The analysis of the susceptibility curve shown in fig.1 was done by the following procedure. It was assumed that the Ce ion has a probability p_3 to be in its trivalent state.

$$p_3 = 6/(6 + e^{E_{exc}/kT}) \qquad (1)$$

Here E_{exc} is the interconfigurational excitation energy, the position of the 4f localized level above the Fermi level. To include a possible fluctuation width to E_{exc} we replace the temperature T by $T^{+}=(T^2 + T_f^2)^{1/2}$ [4]. The Ce Curie constant will be given by C_3p_3 where C_3 is the Ce^{3+} free ion Curie constant. The Curie constant of Mn was assumed to be that found for $LaMn_6Al_6$ and the other RMn_6Al_6 systems [2] ($P_{eff}(Mn)=2.6\mu_B$). If the intrasublattice exchange constants in the Ce and Mn sublattices are J(Ce,Ce) and J(Mn,Mn) respectively and J(Ce,Mn) is the intersublattice exchange, then the magnetic susceptibility as a function of temperature can be expressed in a closed-form formula given in text-books treating ferrimagnetism [6].

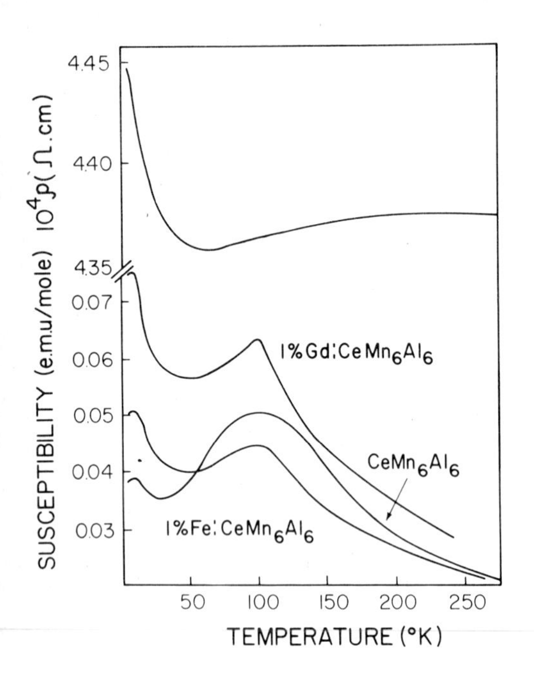

Figure 2 : Magnetic susceptibility of $Fe:CeMn_6Al_6$ and $Gd:CeMn_6Al_6$ and specific resistivity of $CeMn_6Al_6$

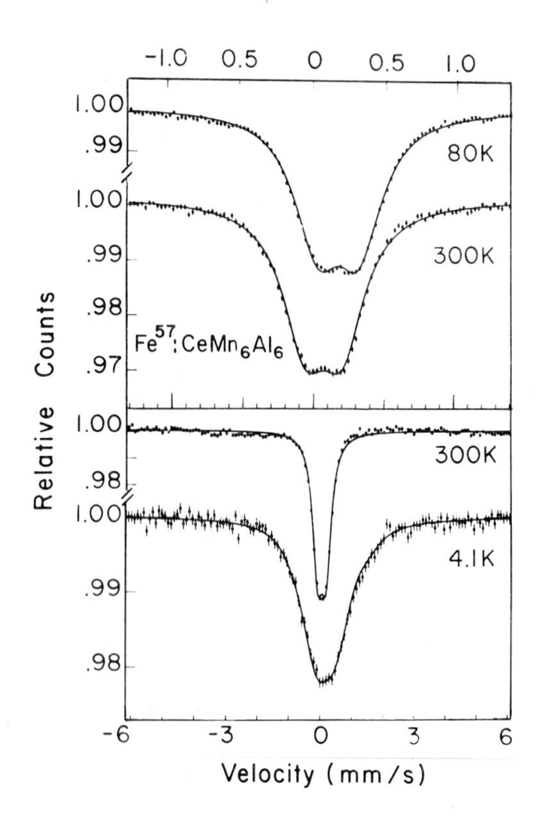

Figure 3 : Mossbauer spectra of Fe^{57} doped in $CeMn_6Al_6$

If the Ce and Mn sublattices are assigned the symbols A and B, then the molecular fields acting on the two sublattices in the presence of an external field H, are given by

$$H_A = \lambda_{AA}M_A + \lambda_{AB}M_B + H$$
$$H_B = \lambda_{AB}M_A + \lambda_{BB}M_B + H \qquad (2)$$

The exchange parameters J_{ij} expressed in ^oK are the molecular field constants λ_{ij} multiplied by μ_B^2/K. The magnetic susceptibility is given by [6]

$$\frac{1}{\chi} = \frac{T^+}{C} - \frac{1}{\chi_0} - \frac{\sigma}{T^+ - \theta} \qquad (3)$$

$$C = C_A + C_B, \quad C_A = C_3 p_3(T^+), \quad C_B = C_{Mn}$$

$$\frac{1}{\chi_0} = \frac{1}{C^2}(C_A^2\lambda_{AA} + C_B^2\lambda_{BB} + 2C_AC_B\lambda_{AB})$$

$$\sigma = \frac{C_AC_B}{C^3}\{C_A^2(\lambda_{AA} - \lambda_{AB})^2 + C_B^2(\lambda_{BB} - \lambda_{AB})^2 - 2C_AC_B[\lambda_{AB}^2 - (\lambda_{AA} + \lambda_{BB})\lambda_{AB} + \lambda_{AA}\lambda_{BB}]\},$$

$$\theta = \frac{C_AC_B}{C}(\lambda_{AA} + \lambda_{BB} - 2\lambda_{AB}).$$

The formula contains five free parameters, E_{exc}, T_f and the three exchange constants. A least square fit procedure of this formula to the experimental curve (fig. 1) yields the parameters E_{exc}= 223(5)K, J(Ce,Ce) = 47(7)K, J(Mn,Mn) = -9(1)K, J(Ce,Mn) = 16(2)K and $T_f \leq$ 5K. The $LaMn_6Al_6$ curve yields J(Mn,Mn) = -13(1)K.

4. CONCLUSIONS

From the susceptibility curves shown in fig.1 it is obvious that the Ce ion in $Ce_2Mn_7Al_{10}$ and in $Ce_2Mn_{8.5}Al_{8.5}$ is tetravalent yet it is in a mixed valent state in $CeMn_6Al_6$, although the lattice dimensions of the last two compounds are almost identical (a = 9.010 A^o, c=13.12 A^o for $CeMn_6Al_6$ and a=9.010 A^o, c=13.09 A^o for $Ce_2Mn_{8.5}Al_{8.5}$). The equality of the lattice dimensions results from the fact that the dimensions are determined mainly by the Mn to Al ratio and very little by the occupancy of the rare earth site[2]. It is here worthwhile mentioning that we found that $CeMn_5Al_5$ has also the same structure and the Ce ion is again in a mixed valent state. The susceptibility curve of $CeMn_5Al_5$ is similar to that of $CeMn_6Al_6$,fig.1.

The experimental observations and the theoretical analysis of the susceptibility curves lead us to the conclusion that the magnetic properties of $CeMn_6Al_6$ and of $CeMn_5Al_5$ can be well explained in terms of;

1) A constant interconfiguration excitation energy of small width and thermodynamic equilibrium between the pure Ce^{3+} excited state and the Ce^{4+} ground state.

2) Ferromagnetic coupling within the Ce sublattice. This ferromagnetic exchange is probably due to the extra conduction electrons at the Fermi level present in $CeMn_6Al_6$ and absent in the other RMn_6Al_6 systems, in which J(R,R) is antiferromagnetic [2].

It seems that this system is a good example for the validity, at least at finite temperature, of the fluctuating valency approach, fluctuations between well defined $4f^n$ states. The ions in such a state may interact with each other and even may order magnetically [7].

REFERENCES

[1] Felner, I., J. Less Comm. Metals 72 (1980) 241.

[2] Felner, I. and Nowik, I., J. Phys. Chem. of Solids (in press, 1982).

[3] Havinga, E.E., Buschow, K.M.J. and van Daal, H.J., Solid State Comm. 13 (1973) 621; de Chatel, P.F., Aarts, J. and Klaasse, J.C.P., Comm. on Physics 2 (1977) 151.

[4] Franz, W., Steglich, F., Zell, W., Wohlleben, D. and Robell, F., Phys. Rev. Letters 45 (1980) 64.

[5] Batlogg, B., Ott, H.R. and Wachter, P., Phys. Rev. Letters 42 (1979) 278.

[6] Morrish, A.H., The Physical Principles of Magnetism (John Wiley, 1965), pp.492.

[7] Goncalves da Silva, C.E.T., Phys. Rev. B 19 (1979) 3656.

Valence Instabilities
P. Wachter and H. Boppart (eds.)

PERIODIC LATTICE MODEL FOR VALENCE FLUCTUATIONS BETWEEN TWO MAGNETIC CONFIGURATIONS

B. Alascio, A.A. Aligia, J. Mazzaferro and C.A. Balseiro

Centro Atómico Bariloche* - Instituto Balseiro#
8400 Bariloche - Argentina

We present a simplified model that describes a lattice of intermediate valence ions fluctuating between two magnetic configurations. The model consists of a lattice of spin one-half ions, on which it is possible to add electrons either in localized states of the ions or in a band of extended states. Localized and extended states are hybridized through a one-particle Hamiltonian. We study the stability of ferro and antiferromagnetic phases and the ensuing metalic or insulating character of the system. For n=1 we find that the most stable phase is insulating antiferromagnetic. Antiferromagnetic phases can be driven ferromagnetic by application of an external magnetic field. We briefly discuss the implication of the model to the understanding of intermediate valence Tm chalcogenides.

1. INTRODUCTION

The magnetic properties of intermediate valence Tm compounds (I.V.Tm) are unique among the large family of intermediate valence rare earth systems. The peculiarity of I.V. Tm ions is that the two accesible consigurations are magnetic. TmSe, with an average 4f shell occupation near 12.5 is the only known I.V. system that orders magnetically, antiferromagnetic order sets in at T_N 3.5K [1].

The magnetic phase diagram drawn from susceptibility, thermal dilation [2] and neutron scattering data [3] shows that TmSe is a metamagnet with an antiferro to ferromagnetic transition at $H_c \sim 7KO_e$ at low temperatures and normal pressure.

The dependence of the magnetic phase diagram boundaries on stoichiometry and pressure, [1,4] as well as on alloying of TmSe with EuSe or TmTe [5,6] supports the idea that the magnetic interactions in TmSe are dominated by the same parameters that determine the valence changes in these substances.

The high electrical resistance of TmSe in the antiferromagnetic phase suggests an insulating state. A dramatic decrease of this resistance with magnetic field saturating at about H_c has been reported [4].

Recently, an investigation of the stability of different magnetically ordered phases for an intermediate valence system fluctuating between two magnetic configurations in terms of V, the hybridization energy and E, the energy difference between the two configurations has been carried out [7]. It allows to conclude that stoichiometric samples order antiferromagnetically and are insulating, but rather unstable against a ferromagnetic and metallic phase for small changes in stoichiometry. However the insulating behaviour is due to the particular band structure with perfect nesting at the Fermi surface, used in these calculations. In addition the validity of the calculations in Ref.[7] is restricted to band-widths W small as compared to $(E^2+V^2)^{1/2}$.

It would be of interest to investigate to what extent the fluctuations between two magnetic configurations can explain the properties of TmSe avoiding if possible these limitations.

We have recently presented an exactly solvable one impurity model for valence fluctuations between two magnetic configurations that shows that many of the qualitative features of the paramagnetic phase of TmSe can be understood simply on the basis that the valence fluctuations take place between two magnetic configurations [8].

In this paper we extend the model to consider a periodic array of sites which are hybridized to a conduction band. The Hamiltonian can be solved exactly for any ordered state of the local spins. We have investigated particularly the stability of the ferro vs the antiferromagnetic phase and the corresponding electronic spectral densities for local and conduction electron. We find that the antiferromagnetic phase is favored at stoichiometry, but a small deviation from stoichiometry can favor the ferromagnetic phase. The density of states shows a gap at the Fermi level for the stoichiometric antiferromagnetic phase, while out of stoichiometry or in the ferromagnetic phase, a metalic ground state is found if the localized levels fall within the continuum of extended states.

2. DESCRIPTION OF THE MODEL

The model is the periodic version of an exactly solvable one-impurity Hamiltonian presented earlier [9]. It can also be regarded as the extreme anisotropic version of the Hamiltonian proposed by Schlottmann and Falicov in Ref. 7.

We consider a lattice of highly correlated states (representing the 4f states in TmSe, for example) hybridized to a band of conduction states. We represent the posible states of the $4f^n$ configuration by $|i,\sigma>$, where $\underline{i}$ indicates the site location in the periodic array and $\sigma=\uparrow$ or $\downarrow$ are degenerate states that can be split by an external magnetic field. Similarly, the $4f^{n+1}$ configuration states are represented by only two states $|i,s>$, where s = + or -.

The hybridization term mixes the $|i,+>$ ($|i,->$) state with the $|i,\uparrow>$ ($|i,\downarrow>$) state by promoting a spin up (down) electron to the conduction band. If we identify the $|i,\sigma>$ states with a spin-one-half configuration, and $|i,s>$ states with a spin one configuration as in references [7] and [9] the present model describes the highly anisotropic limit where the zero z-component of the spin one configuration is projected out of the Hilbert space. These assumptions concerning the structure of the localized states and its consequences on hybridization are not realistic for Tm compounds and invalidate quantitative conclusions. However they retain what we believe is the essential feature of these compounds: hybridization between two magnetic configurations, while allowing the model to be solved exactly at T=0 for any value of the intervening parameters.

In terms of the states defined above, the Hamiltonian reads:

$$H = \sum_{j\sigma} E_\sigma |j,\sigma><j,\sigma| + \sum_{js} E_s |j,s><j,s| + \sum_{k\sigma} \varepsilon_k C^+_{k\sigma} C_{k\sigma} + V \sum_i (|j,+><j,\uparrow| C_{j\uparrow} + |j,-><j,\downarrow| C_{j\downarrow}) + h.c. \qquad (1)$$

where C^+_{js} creates an electron in the conduction band at the Wannier state corresponding to site i (coordinate R_i), and

$$C^+_{k\sigma} = \frac{1}{N^{1/2}} \sum_i e^{ik.R_i} C_{i\sigma} \qquad (2)$$

To include different magnetic moments for the two configurations we take: $E = E_0 \pm \mu_0 B$ and $E_s = E_1 \pm \mu_1 B$ in the presence of an external field B.

For simplicity, we have assumed in (1) on-site hybridization. This assumption, which is not valid for systems with inversion symmetry [10] can be relaxed and its consequences are not too important to the conclusions we will draw, as discussed below.

The highly anisotropic limit we consider, completely inhibits spin-flip scattering of conduction electrons. Consequently the direction (but not the magnitude) of the local spin at each site is conserved. Spin up conduction electrons can go into $|j,\uparrow>$ sites to produce the $|j,+>$ states, but do not "see" the spin down local states. Thus, to calculate the energy of any given spin configuration it is only necessary to calculate the energy distribution of band states.

The model can also be regarded as a lattice of spin one-half sites (playing the role of a multiply degenerate "vacuum") on which we can add electrons either on each site (with spin paralel to that of the site) or on a conduction band. The added electrons can hop from the sites to the band and back to a different site thus providing an interaction between different sites. This way of looking at the model makes clear its similarity to the Anderson-Hasegawa model for double exchange between two sites [11].

3. THE FERROMAGNETIC PHASE

The ferromagnetic states are those for which in all sites only the spin up states of each configuration can be occupied. Thus we consider the subspace of states $|\phi>$ where

$$<\phi|(|j\uparrow><j\uparrow| + |j+><j+|)|\phi> = 1 \text{ for all j's.}$$

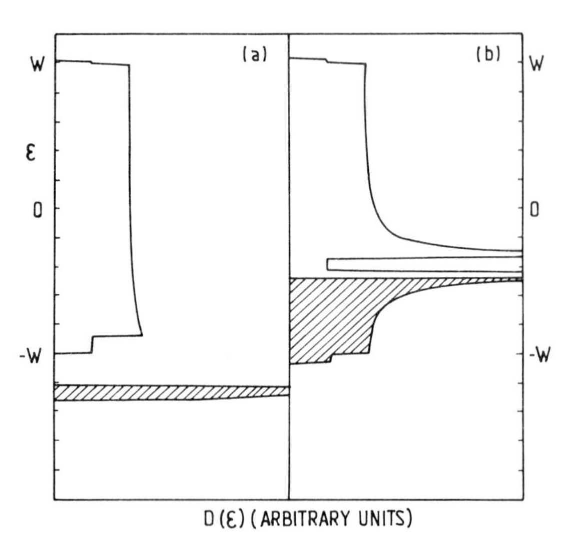

Fig.1: Density of states of the Ferromagnetic phase for V/W = 0.2. a) Δ/W=-1.2, b) Δ/W=-0.4. Shaded areas indicate occupied states for n=1.

In this subspace, the conduction states with spin down are decoupled from the localized ones, while the conduction electrons with spin up are hybridized with the local states at every site. In this case, the Hamiltonian reduces to a one body Hamiltonian.

Defining $\bar{\mu} = \mu_1 - \mu_0$, straightforward calculations give the following dispersion relation for the eigenstates with spin up:

$$\varepsilon_{\substack{\alpha k \\ (\beta)}} = \frac{\Delta - \bar{\mu}B + \varepsilon_{k\uparrow}}{2} \overset{-}{(+)} \sqrt{\left(\frac{\Delta - \bar{\mu}B - \varepsilon_{k\uparrow}}{2}\right)^2 + V^2} - \mu_0 B \quad (3)$$

Assuming for simplicity a constant density of states for conduction electrons, extending from -W to W, the total spectral density of states is shown in figure 1.

The total density of states shows a gap if $-\Delta > W - V^2/2W$ (fig.1(a)), and the lowest band can accomodate one electron per site. The ferromagnetic state is then insulating if there is one added electron per site.

For $\Delta > -W + V^2/2W$ (fig.1(b)), there is no gap in the total density of states and the ferromagnetic state is always a good metal due to spin down conduction.

4. THE ANTIFERROMAGNETIC PHASE

To describe the antiferromagnetic phase we separate the lattice into two sublattices: a and b. We consider then the subspace of states $|\psi\rangle$ where

$$\langle\psi|(|j\uparrow\rangle\langle j\uparrow| + |j+\rangle\langle j+|)|\psi\rangle = \begin{cases} 1 & \text{if } j \text{ in a} \\ 0 & \text{if } j \text{ in b} \end{cases}$$

$$\langle\psi|(|j\downarrow\rangle\langle j\downarrow| + |j-\rangle\langle j-|)|\psi\rangle = \begin{cases} 0 & \text{if } j \text{ in a} \\ 1 & \text{if } j \text{ in b} \end{cases}$$

The conduction states with spin up (down) hybridize only with the localized states of sublattice a(b).

We consider an antiferromagnetic structure characterized by a vector Q, such that the Brillouin zone of the antiferromagnetic structure is half the original Brillouin zone. For each vector k' of the new zone, the problem reduces to a three by three matrix which secular equation is:

$$(\omega-\Delta)(\omega-\varepsilon_{k'})(\omega-\varepsilon_{k'+Q}) - \frac{V^2}{2}(2\omega-\varepsilon_{k}-\varepsilon_{k+Q}) = 0 \quad (4)$$

At this point it is necessary to make some assumption about the band structure. The simplest one is to take

$$\varepsilon_{k'} = -\varepsilon_{k'+Q} \quad (5)$$

the origin of energies is at the center of the band which extends from -W to W with a constant density of states 1/2W. The conclusions we will draw from the model do not depend on this particular band structure.

Fig.2(a) shows the eigenvalues of (4) as a function of $\varepsilon_{k'}(-W \leq \varepsilon_{k'} \leq 0)$. The corresponding density of states is shown in Fig.2(b). Exactly the same result is obtained for the spin down states. Two gaps open up indicated by 1 and 2 in Fig.2(b).

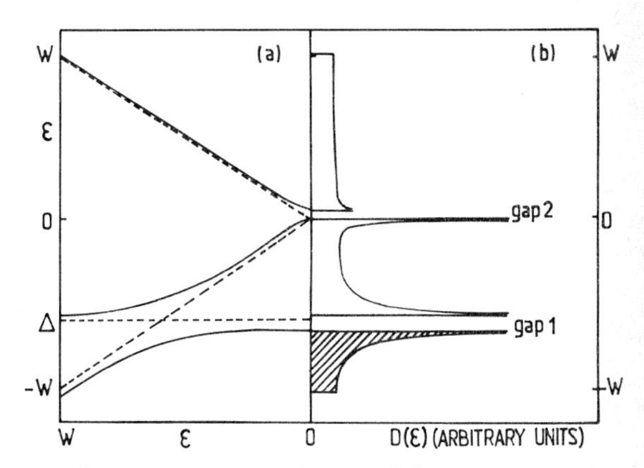

Fig.2: Dispersion relation (a) and density of states (b) for the antiferromagnetic phase for V/W=0.2 and Δ/W=-0.6.

Since k' runs over half the original Brillouin zone, each branch can accomodate one-half electron per site. Counting spin up and down states, the total number of states in the lowest band is one. This means that for one electron per atom (added on the spin one-half lattice "vacuum") the Fermi level falls within gap 1 in Fig.2 and the antiferromagnetic state is insulating. Note, the insulating state appears for one electron per site, not for one conduction electron per site as in the Kondo model [12].

Gap 2 is a consequence of the "perfect nesting" assumption (Eq.5), while gap 1 arises from the hybridization of local and conduction states. Relaxing assumption (5) then closes gap 2, but in general gap 1 remains open. The conclusion that the antiferromagnetic phase is insulating is then not necessarily modified when more realistic band structures are considered.

5. THE PHASE DIAGRAM

Calculations of the energies for different number of electrons per site "n" and different positions of the local level (Δ), leads to the phase diagrams shown in fig.3.

In agreement with ref.7 we find that for one electron per atom added to the spin one half lattice (n=1), the antiferromagnetic phase is the most stable.

To study the stability of the antiferromagnetic against the ferromagnetic phase we have calculated the energies of both phases in the presence of a magnetic field parallel to the quantization axis. The results are shown in figs.3a. The phase space occupied by the AF phase shrinks appreciably when the magnetic field increases, particularly if $n \gtrsim 1$.

6. DISCUSION

To give a qualitative interpretation of some

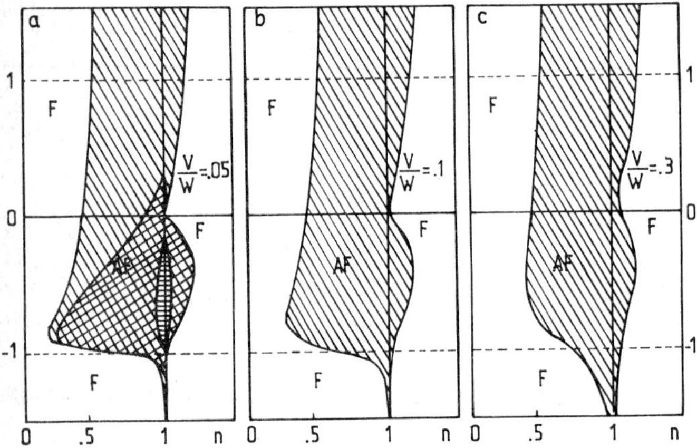

Fig.3: Ferro-antiferromagnetic phase diagram. The shaded area indicates the region where the antiferromagnetic phase is stable. The values of V/W are indicated in each figure. Figure 3a shows the magnetic phase diagram for three different values of an external magnetic field B: B/W=0, 0.5 and 20 kOe/eV (μ_0=4.5 μ_B and μ_1=7.3 μ_B). An increase of the magnetic field produces a shrinkage of the antiferromagnetic region. Figure 3b and 3c are for B=0.
Due to the symmetry of the Hamiltonian the phase diagram for 1.5$\leqslant$n$\leqslant$3 can be obtained by replacing n by 3-n and Δ by $-\Delta$.

properties of Tm chalcogenides and its alloys, we identify the lowest energy states of Tm^{2+} ($4f^{13}$) with states $|+\rangle$ and $|-\rangle$ of our model Hamiltonian. Similarly the lowest states of Tm^{3+} ($4f^{12}$) are represented here by $|\uparrow\rangle$ and $|\downarrow\rangle$. "Stoichiometric" Tm chalcogenides fall on the lines n=1 of fig.3. Non-stoichiometric chalcogenides would fall slightly to the right (excess Tm) or to the left (defect Tm) if they were homogeneous.

Assuming that the hybridization matrix element V does not vary very much from one compound to another, the effect of pressure or of chalcogen substitution can be interpreted as changing the value of Δ. Pressure increases Δ while, assuming homogeneity, the substitution of Se by Te should decrease Δ.

According to ref,[6] Tm in $Se_{1-x}Te$, for $x > 0.2$ is divalent (or almost so) which indicates that Δ falls bellow the bottom of the conduction band. This implies a semiconductor state and Figs.3 shows that the magnetic interactions arising from hybridization favor the ferromagnetic phase except for n exactly equal to one where the ground state is antiferromagnetic, but unstable against small magnetic fields.

For $x \leqslant 0.2$ in $TmSe_{1-x}Te_x$ we expect Δ to be such that $-W<\Delta<0$. Then, for n=1 and below the antiferromagnetic ordering temperature, the Fermi level falls within the hybridization gap making the compound a semiconductor. For TmTe the $4f^{13}$ configuration is dominant ($\Delta < -W$) and all magnetic phases are insulating.

Finally, TmS where Tm is almost triply ionized should correspond to $\Delta > 0$. Here, for n=1 and for temperatures bellow the ordering temperature, the Fermi level should fall in gap 2. But, as pointed out earlier, this gap is a consequence of our simplifying assumption of "perfect nesting". Relaxing this assumption we can conclude that TmS or perhaps TmSe under high pressure should be metallic.

Among the most important conclusions we can draw are the following: a) Intersite magnetic interactions arising from hybridization favor an antiferromagnetic ordering of the magnetic moments of Tm ions in "stoichiometric" samples. b) The antiferromagnetic order looses stability away from "stoichiometry" or in the presence of external magnetic fields. c)"Stoichiometric" antiferromagnetic, intermediate valence, Tm chalcogenides are insulators. Lack of stoichiometry or antiferromagnetic order make them metallic. The insulating state is a consequence of the hybridization gap.

7. REFERENCES

* Comisión Nacional de Energía Atómica.
# Comisión Nacional de Energía Atómica and Universidad Nacional de Cuyo.

1. B.Batlogg, H.R.Ott, E.Kaldis, W.Thoni and P. Wachter; Phys.Rev.B19,247 (1979).
2. H.R.Ott, K.Andres, E.Bucher, in Mag.and Mag.Mat.24, 40 (1974).
3. H.B.Møller, .S.M.Shapiro and R.J.Birgeneau; Phys.Rev.Lett.69, 1021 (1977).
4. P.Haen, F.Hotzberg, F.Lapierre, T.Penney and R.Tournier, "Valence instabilities and related Narrow Band Phenomena," Ed.R.D. Parks (P.P.,N.Y., 1976), pg.495.
5. R.P.Guertin, S.Foner and F.P.Missell; Phys. Rev.Lett. 37, 529 (1976).
6. B.Batlogg; Phys.Rev. B23, 650 (1981).
7. P.Schlottmann and L.M.Falicov; Phys.Rev. B23, 5916 (1981).
8. C.A.Balseiro and B.Alascio, to be published in Phys.Rev.B, March (1982).
9. J.Mazzaferro, C.A.Balseiro and B.Alascio, Phys.Rev.Lett.47, 274 (1981).
10. C.Balseiro, M.Passegi and B.Alascio, Solid State Comm. 16, 737 (1975).
11. P.W.Anderson and H.Hasegawa, Phys.Rev.100, 675 (1955).
12. R.Jullien, P.Pfeuty, J.N.Fields and S. Doniach, J.de Physique 40, C5-293 (1979).

Valence Instabilities
P. Wachter and H. Boppart (eds.)

4f INSTABILITIES OF ISOLATED RARE EARTH IONS IN ELEMENTS

D. Riegel, H.J. Barth, M. Luszik-Bhadra, and G. Netz

Fachbereich Physik, Freie Universität Berlin

D-1000 Berlin 33, Germany

It is demonstrated that the perturbed angular γ-ray distribution method following heavy ion reactions allows microscopic measurements of the local susceptibility and 4f spin dynamics of extremely dilute 4f ions in many simple metals over wide temperature ranges. We present various microscopic investigations of the single-ion behavior in Kondo-like and intermediate valence systems, which considerably exceed the possibilities of the methods applied hitherto. The formation and dynamics of the 4f configuration of isolated light rare earth ions in elements are discussed in detail.

I. INTRODUCTION

The formation of local moments and their interaction with the metallic surroundings are basic problems in the field of magnetism in metals. Numerous experiments have been performed by macroscopic and microscopic methods to study the behavior of metallic systems containing 4f, 3d or 5f ions. In the center of current interest are 4f configurational instabilities, especially for light rare earth systems[1-3]. At present, there exists deep uncertainty in the description of intermediate valence phenomena and their relation to less pronounced 4f instabilities which usually are described in terms of Kondo theories.

We have schematically indicated in Fig.1 the various situations which might occur for the important example: Ce in metals. For magnetic Ce^{3+} systems the $4f^1$ level is located below the Fermi level E_F. The linewidth might arise predominantly from mixing exchange interaction between the 4f levels and conduction electrons. This hybridization is widely regarded as the source of various Kondo anomalies.

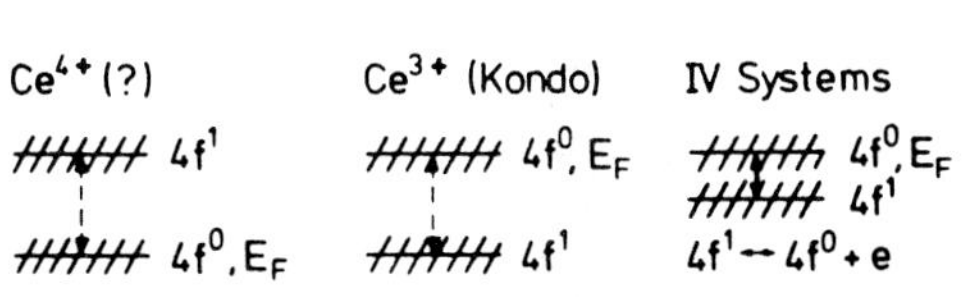

Ce^{4+} (?) Ce^{3+} (Kondo)   IV Systems

$4f^1$ $4f^0, E_F$ $4f^0, E_F$ / $4f^1$

$4f^0, E_F$ $4f^1$ $4f^1 \leftrightarrow 4f^0 + e$

Fig. 1: Pictures for Ce systems in the 4^+, 3^+, and intermediate valence state.

For Ce^{4+} systems the $4f^1$ level might be placed above E_F. As an alternative one can also assume an extremely broadened $4f^1$ level below E_F, so that the $4f^1$ electron delocalizes. It has been suggested[1] that Ce^{4+} does not exist in concentrated Ce systems.

Intermediate valence systems might be characterized by the conditions that (a) the energy difference between the two states $4f^1$ and $4f^0$ is small and (b) due to strong hybridization the linewidths do have a finite overlap.

Albeit simplified, the pictures in Fig.1 partly reflect the fundamental problems in the field of 4f instabilities in metals:

1. Location of the 4f levels
2. Linewidth of the 4f levels
3. Single-ion versus collective contributions

These three parameters control the manyfold phenomena for both intermediate valence and Kondo-like instabilities in dilute and concentrated rare earth systems. However, it is very hard to extract information about these basic quantities from experiment or theory. Of particular importance are investigations in dilute 4f systems, since the very many results in concentrated systems usually depend on all three basic parameters which are hard to disentangle. Microscopic measurements of the single-ion-behavior are highly attractive, but are nearly impossible to perform by the methods known hitherto[1,4].

In the following we present accurate investigations of the single-ion-behavior in many systems. We first discuss basic methodical aspects by which scope and limitations of local 4f moment studies by the perturbed angular γ-ray distribution method in connection with a heavy ion accelerator are determined. The method allows microscopic measurements of the local susceptibility and 4f spin dynamics of extremely dilute 4f ions in many simple metals over wide temperature ranges. Examples will be given that the local susceptibility provide information about the 4f configuration and the valence in all types of systems shown in Fig.1. The formation of local Ce moments in elements will be discussed within a thermodynamic approach. Next we shall demonstrate the importance of measurements of the 4f spin dynamics for isolated 4f ions in both Kondo-like and intermediate valence systems. The various results provide information about all three basic parameters given above.

II. LOCAL 4f MOMENT STUDIES BY TDPAD

The time differential perturbed angular γ-ray distribution (TDPAD) method has been widely used for measurements of nuclear moments of excited states and of nuclear magnetic and quadrupole interactions. General descriptions of the basic grounds and applications of the TDPAD method and closely related techniques can be found in e.g. Refs.5-7. Further we want to note two modifications of this technique: the stroboscopic observation[8] of PAD and the perturbation of the γ-ray distribution by NMR[9] which both might be applied to accurate local moment studies in the future.

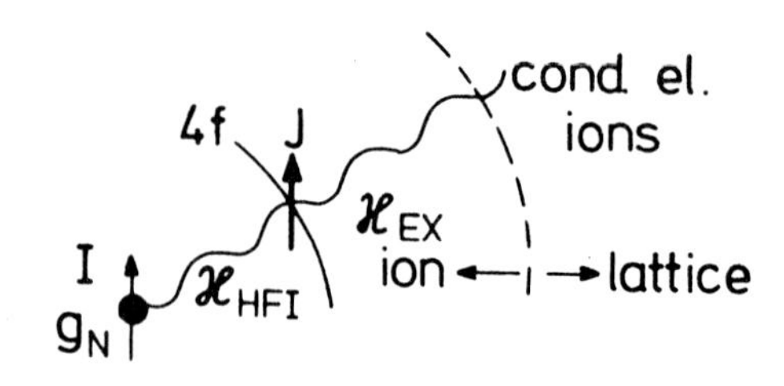

Fig. 2: see text

In this chapter we want to discuss some basic methodical aspects by which scope and limitations of local 4f moment studies by TDPAD are determined. Many methodical aspects are governed by the heavy ion reactions by which the systems are produced. Furthermore, the heavy ion reactions excite and orient the nuclear states which serve as probes for measurements of the static and dynamic fields of its local environment. As indicated in Fig.2 the method aimes to get microscopic information about the 4f configuration (J) and its interactions with the metallic surroundings (e.g. the exchange interaction $\mathcal{H}_{EX}$) via the detection of the magnetic hpyerfine interaction($\mathcal{H}_{HFI}$) at the site of the rare earth nucleus (I, g_N).

II.1 THE PRODUCTION OF 4f SYSTEMS

The systems are produced by heavy ion reactions. To be specific we want to use as an example[10,11] the compound reaction ^{124}Sn(^{16}O,4n)^{136}Ce. By bombarding a Sn target (enriched in ^{124}Sn) with a 90 MeV ^{16}O beam, a highly excited ^{140}Ce nucleus is produced which decays within 10^{-16} s into ^{136}Ce by the emission of four neutrons.

As a consequence this reaction leads to ^{136}Ce ions in a Sn host (Fig.3a). One can estimate that the concentration of the Ce ions produced during a TDPAD measurement is smaller than 1 ppm; thus we really have isolated Ce ions in the host matrix.

Even more important is the possibility of recoil implantation following heavy ion reactions. The recoil energy which is about 10 MeV for

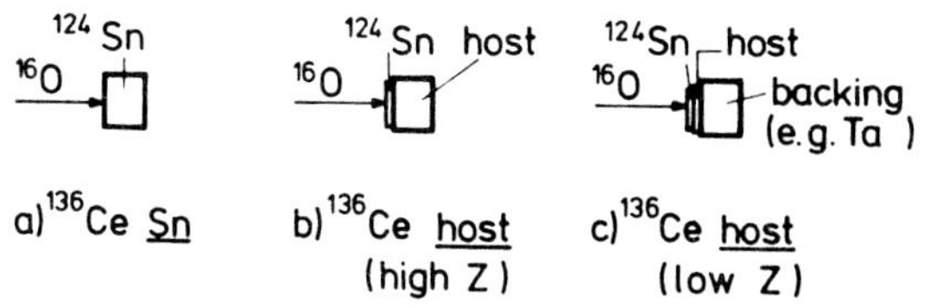

^{124}Sn ^{16}O a) ^{136}Ce Sn

^{124}Sn host ^{16}O b) ^{136}Ce host (high Z)

^{124}Sn host ^{16}O backing (e.g. Ta) c) ^{136}Ce host (low Z)

Fig. 3: Illustration of various implantation techniques.

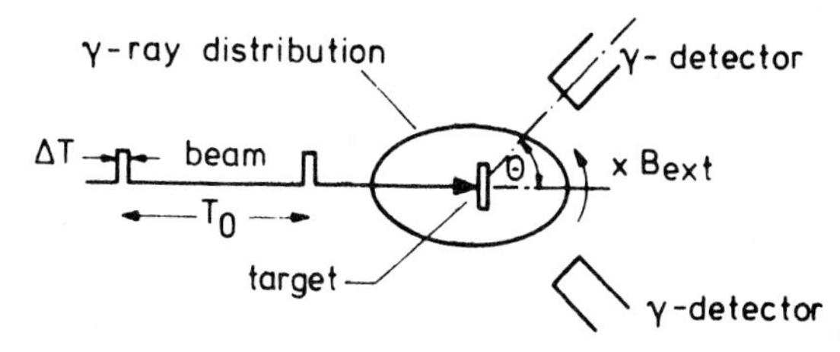

Fig. 4: Principle of a TDPAD experiment

^{136}Ce for the example given can be used to implant the rare earth ions out of a thin target layer (e.g. $\sim$0.8 mg/cm^2 ^{124}Sn) into a host matrix of physical interest (Figs.3b,3c). Fig.3b mainly refers to hosts with high Z, where nuclear reactions in the host are suppressed by the high Coulomb barrier. In order to diminish γ-ray background arising from nuclear reactions in the hosts (especially hosts with low Z) it is often necessary to prepare sandwich targets consisting out of the ^{124}Sn layer, a thin host layer ($\sim$ 1 mg/cm^2) in which all ^{136}Ce recoils are stopped and a backing of high Z material in which the beam is stopped (Fig.3c). These powerful techniques of implantation can be used to produce very many and very simple 4f systems, e.g. isolated Ce ions in elements. Furthermore, implantation allows the production of 4f systems which cannot be produced by metallurgy.

Due to the high recoil energies the rare earth ions are stopped deep within the bulk (10^3-10^4 Å).

II.2 THE TDPAD METHOD AND THE NUCLEAR PROBES

How can one measure the magnetic response of such extremely dilute 4f systems? Simultaneously to the production of the systems the nuclear reaction excites and orients nuclear isomeric states which serve as nuclear probes for the measurement of the magnetic hyperfine interaction. The example ^{124}Sn(^{16}O,4n) leads to the 10^+ isomeric state in ^{136}Ce with a long halflife $T_{1/2}$=2000 ns. Since this state is aligned by the nuclear reaction process (transfer of angular momentum), the γ-rays of the decaying isomers have an anisotropic angular distribution (Fig. 4). An external field B_{ext} applied perpendicular to the beam-γ-detector plane causes a Larmor precession ω_L of the nuclear spins and consequently of the γ-ray distribution (Fig.4). The observation of the γ-ray intensity modulation requires pulsed beam techniques[6-9] e.g. in order to fulfill the timing conditions $\Delta T \omega_L \ll 1$ and $T_o \sim 5\ T_{1/2}$ for the width ΔT (usually $\Delta T \sim 1$ ns) and the repetition time T_o of the beam. Furthermore pulsed beams are necessary in order to suppress the high prompt γ-ray background.

Conveniently the time dependence of the perturbed angular γ-ray distribution is measured in at least two detectors (NaI(Tl) or Ge) placed under $\Delta\Theta=90^o$ which allows to form the normalized ratio[5-7]

$$R(t)=\frac{3A_2\ \exp\{-t/(\tau_N)_2\}}{4+A_2\ \exp\{-t/(\tau_N)_2\}}\ \cos\{2(\omega_L t+\varphi)\} \qquad (1)$$

This formula is already simplified to the case that the tensor degree of the γ-ray distribution is described by k=2, which holds for all experiments described in this paper; we define the nuclear spin relaxation time $(\tau_N)_{k=2} = \tau_N$. The parameter A_2 is the effective γ-ray anisotropy which depends on the nuclear reaction and on parameters involved in the particular γ-transition observed.

A typical R(t) spectrum (or spin rotation pattern) is shown in Fig.5 for our example ^{136}CeSn following the reaction ^{124}Sn(^{16}O,4n). The pattern reflects a clear modulation and a clear exponential damping from which the quan-

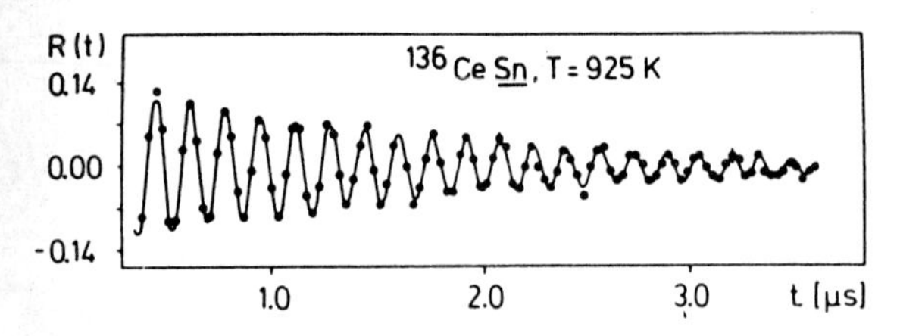

Fig. 5: Spin rotation pattern of the 10^+ state of ^{136}Ce in liquid Sn.

tities ω_L, τ_N, and A_2 can be deduced with high accuracy using Eq.1. As will be shown below the frequency ω_L provides microscopic information about the 4f configuration and the magnetic part of the nuclear relaxation rate τ_N^{-1}(magn) can be correlated to the 4f spin dynamics of the system.

Accurate measurements of ω_L and τ_N by the TDPAD method require long lived nuclear states ($T_{1/2} \gtrsim 5$ ns) and additionally $T_{1/2}$ of the isomer and τ_N must be of comparable order to measure the 4f spin dynamics of the systems. Thus the first decisive step in planning local moment studies by TDPAD is the search for appropriate nuclear probes. One of the powerful features of TDPAD is that many nuclear isomers in nearly all rare earth elements can be reached by heavy ion

Table I: Some nuclear probes in light rare earth isotopes. a) Refs. 10,11,12, b) Ref. 10 and this work, c) Ref. 13, d) Ref. 14, e) Refs. 15,16, f) Ref. 16

Nucleus	I^π	$T_{1/2}$ [ns]	g_N	Reaction
^{136}Ce^a	10^+	2000	-0.180	^{124}Sn(^{16}O,4n)
^{138}Ce^b	10^+	82	-0.176	^{130}Te(^{13}C,5n)
^{139}Ce^b	$19/2^-$	74	+0.420	^{130}Te(^{13}C,4n)
^{139}Pr^c	$11/2^-$	40	+1.3	^{139}La(α,4n)
^{138}Nd^c	10^+	330	-0.175	^{122}Sn(^{20}Ne,4n)
^{142}Sm^d	7^-	170	+0.06	^{126}Te(^{20}Ne,4n)
^{147}Eu^e	$11/2^-$	760	+1.281	^{139}La(^{12}C,4n)
^{148}Eu^f	9^+	230	+0.685	^{139}La(^{12}C,3n)

reactions. This might be documented by the nuclear probes in light rare earth ions listed in Table I. Within only three years of work these isomers have been shown to be ideal probes for various local 4f moment studies by our small group in Berlin. The measured g_N factors are discussed briefly in Ref.14.

II.3 COMMENTS ON THE FATE OF THE IMPLANTED RARE EARTH IONS

Lattice location and ionic states of the stopped rare earth ion have to be known in order to interpret the measured results reliably. As shown in Fig.6 one first has to consider some characteristic times involved in a TDPAD experiment. The nuclear reaction process is finished within 10^{-16} s, and the stopping process of the recoil is finished within a few 10^{-12} s after the nuclear reaction. During the stopping process isomeric ionic states (e.g. J=7/2 of Ce^{3+}) might be populated. However, in metallic systems these states should decay immediately (<< 1 ns) by exchange interactions. Since the measurement of the spin rotation pattern starts ∿ 10 ns after the nuclear reaction (see Fig.6), the implanted rare earth ions are in the ionic ground states.

Next we consider lattice perturbations caused by the recoiling rare earth ions. Due to the microscopic nature of TDPAD measurements lattice perturbation caused by the beam particles can usually be neglected.

In liquid metals lattice perturbations are healed out rapidly (10^{-12} s) and have no influence on the quantities observed.

Concerning the lattice location of implanted non-rare earth ions in solid metals, there exists a broad experimental basis that the dominant fraction of those recoils which contribute to R(t), come to rest at normal lattice sites, if volume and mass of recoil and host atoms are not extremely different. This rule is based on channeling experiments and on a large

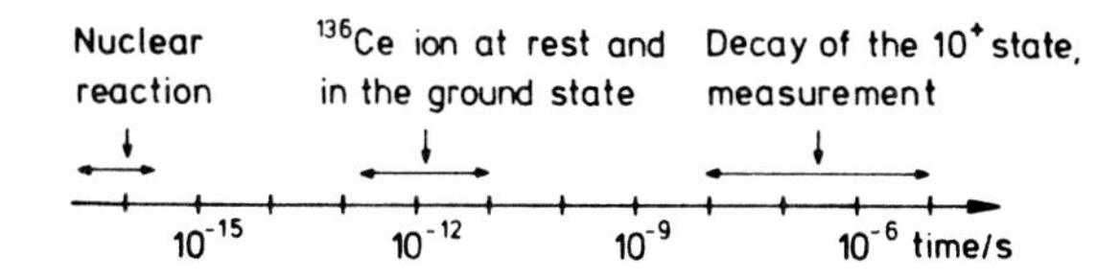

Fig. 6: Some characteristic times involved in a TDPAD experiment.

number of hyperfine investigations (magnetic and quadrupole interaction) at implanted ions performed by the Mössbauer and TDPAD method[17].

These general findings are consistent with the results of our investigations in about fifty solid 4f systems following implantation. Including intermediate valence systems we always have observed only one frequency ω_L, suggesting that all measured events stem from rare-earth ions at normal lattice sites. In several systems e.g. CeW we have found a too small amplitude of R(t) which must be interpreted by assuming that a certain fraction of recoils do not contribute to R(t). Such events might reflect strong lattice perturbations but have no influence on the extracted ω_L and τ_N values. We also have observed in several systems an additional damping of R(t) which arises from interaction of the nuclear quadrupole moment with small electric field gradients produced by radiation damage around the substitutional recoil. However all these experiments do not show any measurable indication that the magnetic response of the systems is influenced by lattice perturbations.

A more exhaustive discussion of these somewhat complex considerations is beyond the scope of this paper and will be given in Ref.18 where the successful implantation of Ce ions into more than forty metallic hosts will be described.

We conclude that the 4f ions which contribute to R(t) are in the ionic ground states and are located predominantly at normal lattice sites (excluding exotic systems like CeC, CeB), being in thermal and chemical equilibrium with the metallic matrix at the starting point of the TDPAD measurements. Further we note that the alignment of the nuclear states usually survives the implantation process since the correlation times of the fluctuating hyperfine fields during the stopping process are extremely short ($\lesssim 10^{-14}$ s).

III. THE LOCAL STATE OF THE 4f CONFIGURATION

The frequency of a R(t) pattern is given by

$$\omega_L(T) = \hbar^{-1} g_N \mu_N \beta(T) B_{ext} \qquad (2)$$

where the paramagnetic enhancement factor β[19] scales the internal magnetic field B_{int} at the site of the nucleus: $\beta B_{ext} = B_{int} + B_{ext}$. The quantity $\beta - 1 = B_{int}/B_{ext}$ is defined analogously to the macroscopic susceptibility, we therefore call $\beta-1$ the local susceptibility. If the atomic Zeeman splitting and the crystal electric field splitting are small compared to the thermal energy at the measurement temperatures, $\beta(T)-1$ follows a Curie law

$$\beta - 1 = \beta' = g_J \mu_B (J+1) B(0)/3k_B T \qquad (3)$$

where the Curie constant contains the magnetic field B(0) at the nucleus at T=0. The influence of an excited ionic state (index 2) lying δ above the ground state (index 1) on β' can be estimated by

$$\beta' = \frac{(2J_1+1)\beta_1' + (2J_2+1)\beta_2' \exp(-\delta/k_B T)}{(2J_1+1) + (2J_2+1)\exp(-\delta/k_B T)} \quad . \qquad (4)$$

For non-S-state rare earth ions B(0) essentially arises from the localized 4f electrons. As discussed in some detail by Bleaney[20] the magnetic fields of the free ions depend on spin algebra factors and on the wave function $\langle r_{4f}^{-3} \rangle$. See Ref. 21 for modern calculations of $\langle r_{4f}^{-3} \rangle$.

In view of the accuracy of TDPAD results we want to discuss minor contributions to B(0) also. Usually B(0) is regarded[22,23] as the sum of the free ion value B_{4f} and B_{ce}. The latter term arises from conduction electron polarization caused by exchange interactions. Hitherto known B_{ce} values are of the order of a few 0.1 MG which are considerably smaller than the B_{4f}

values which range from 2-8 MG. The B_{4f} values contain a small core polarization, B_{core}. For S-state rare earth ions $B_{4f} = B_{core}$, usually B(0) and B_{ce} for metallic S-state systems are of comparable magnitude.

Determinations of β or B(0) require the knowledge of the nuclear g_N factors. These can be extracted from measurements of $\omega_L(T) \propto g_N\, \beta(T)$ over wide temperature ranges. An example is given in Fig.7 for the 10^+ isomer in ^{136}Ce. According to Eqs.(2),(3) a plot of $g_N\, \beta(T)$ versus 1/T yields a straight line. The intersection at $T \to \infty$ determines the $|g_N|$ value and B(0) can be extracted from the slope. We have performed such measurements for various systems yielding the g_N factors listed in Table I. The sign of g_N can be extracted from the phase of the spin rotation. The g_N values for the isomers in Ce, Nd and Eu isotopes (Table I) are among the most accurate ones known for nuclear states in rare earth isotopes.

In summary the measurement of the frequency $\omega_L(T)$ provides information about the local susceptibility, the magnetic fields and nuclear g_N factors. The β(T) and B(0) values are microscopically sensitive to the local 4f configuration and its exchange interaction with conduction electrons.

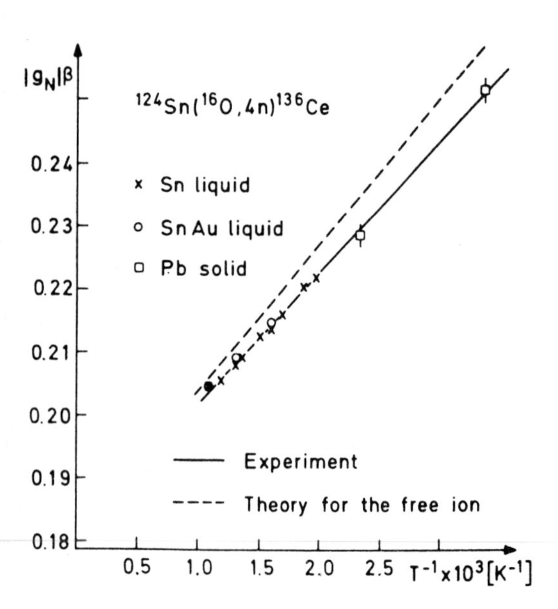

Fig. 7: T-dependence of $|g_N|\beta$ for ^{136}Ce in liquid and solid hosts.

III.1 NEARLY INTEGRAL VALENCE SYSTEMS

The TDPAD method can be applied to deduce the localization of 4f electrons and the valence of isolated rare earth ions in liquid and solid systems. Fig.8 contains some examples of β(T) measurements in the high temperature range (T larger than a possible crystal field splitting). The systems ^{147}Eu<u>La</u>[16], ^{136}Ce<u>Sn</u>[10], and ^{138}Nd<u>Sn</u>[13] were produced by the heavy ion reactions ^{139}La (^{12}C,4n), ^{124}Sn (^{16}O,4n), and ^{122}Sn (^{20}Ne, 4n), respectively. The system ^{139}Ce<u>Pt</u>[18] was produced by recoil implantation of ^{139}Ce into Pt following the reaction ^{130}Te(^{13}C,4n). The β(T) values shown in Fig.8 were deduced from the

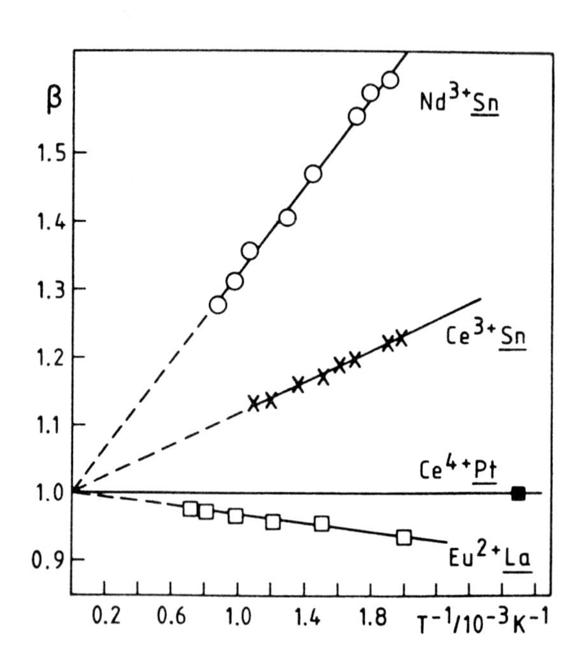

Fig. 8: Local susceptibility for nearly integral valence systems as a function of temperature. The lines represent fits by Eq.3. The extrapolations $T \to \infty$ nicely yield β=1 for all measurements.

T dependence of the frequencies ω_L of various isomers listed in Table I.

The slope of $\beta(T)$ for the system Eu<u>La</u> is negative and small reflecting a small and negative B(0) value. The fit by Eq.(3) yields B(0) = -0.17 MG which can be decomposed[16] into B_{core} = -0.34 MG and B_{ce} = +0.17 MG. Unambiguously the valence of the Eu ions in solid and liquid La is determined to be 2^+. The Curie-like behavior of $\beta(T)$ and the B(0) values extracted, which are B(0) = +1.8 MG for Ce<u>Sn</u> and B(0) = +3.6 MG for Nd<u>Sn</u>, strongly support the picture of nearly stable Ce^{3+} and Nd^{3+} ions in liquid Sn[10,13]. As can be estimated by Eq.(4) the contribution to $\beta(T)$ arising from the excited ionic states (J=7/2 in Ce^{3+}, J=11/2 in Nd^{3+}) are relatively unimportant for T $\lesssim$ 1000 K. A comparison with estimated free ion values yields negative B_{ce} values for both systems Nd^{3+}<u>Sn</u> and Ce^{3+}<u>Sn</u>.

For the system Ce<u>Pt</u> β is found to be 1 and consequently B(0) = 0. It follows in very good approximation that isolated Ce ions in Pt are tetravalent. To illustrate this important finding more clearly, we have plotted in Fig.9 R(t) spectra for ^{139}Ce implanted into Yb, Th and into Pt observed under identical experimental conditions. The obvious differences in the ω_L frequencies reflect the fact[18] that Ce ions in Yb are trivalent, tetravalent in Pt and intermediate valent in Th.

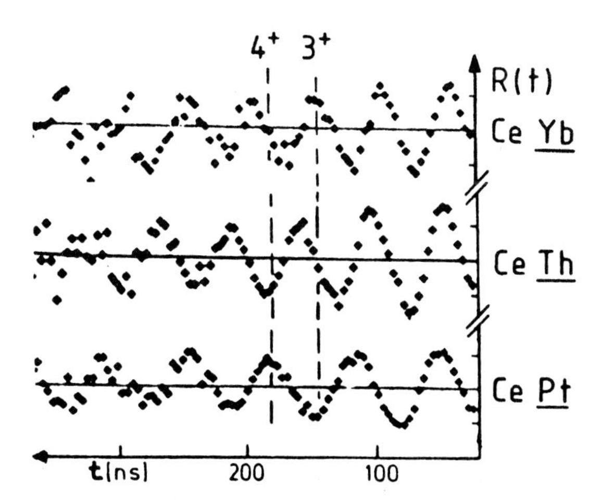

Fig. 9: Spin rotation for isolated ^{139}Ce ions in Yb, Th and Pt at $\sim$ 370 K under identical experimental conditions. The dashed lines are plotted at the third maxima of R(t) for Ce^{3+}<u>Yb</u> and Ce^{4+}<u>Pt</u> respectively. The frequency for ^{139}Ce<u>Th</u> reflects intermediate valence.

The results shown in Fig.8 and Fig.9 document in a nice manner the power of the TDPAD method to determine the valence of extremely dilute 4f systems by frequency measurements.

III.2 INTERMEDIATE VALENCE OF EXTREMELY DILUTE 4f SYSTEMS

As discussed in Refs.1,4,12 microscopic investigations of the single-ion behavior of a mixed-valence impurity are highly desirable. Unfortunately the methods used up to now fail or are very difficult to apply, except Mössbauer isomer shift measurements for dilute Eu ions[4].

Fig.10 represents the results of the first application of TDPAD on a mixed valence impurity. Extremely dilute Ce ions in fcc thorium were produced by recoil implantation following the reaction ^{124}Sn (^{16}O,4n) ^{136}Ce. As shown in Fig.10, the T dependence of the local magnetic susceptibility for Ce<u>Th</u> deviates dramatically from both the Ce^{3+} and the Ce^{4+} behavior. With decreasing temperatures $\beta(T)$ reflects a continuous change from more Ce^{3+} behavior towards more Ce^{4+} behavior. At low temperatures $\beta(T)$ approaches a constant value β=1.12, which is significantly different from β=1 observed for tetravalent Ce systems e.g. for Ce<u>Pt</u>. The dash-dotted line in Fig.10 represents a parametrization[12] of the $\beta(T)$ data within a simple two level approach which has been used by Wohlleben et al.[1,4] to fit the macroscopic susceptibility and isomer shift in intermediate valence systems.

As discussed in more detail in Ref.12 some of the most essential conclusions are:

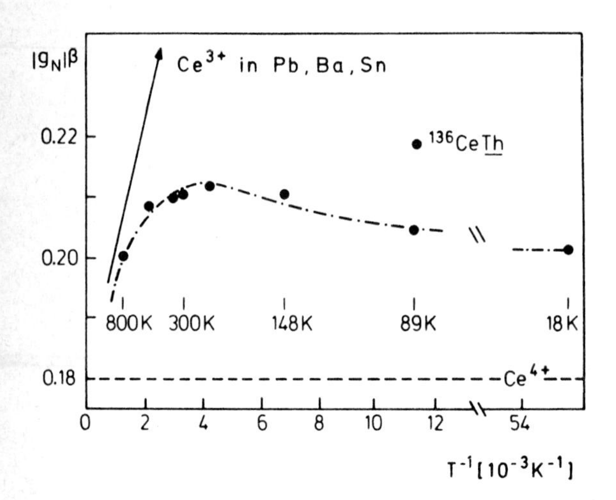

Fig. 10: T-dependence of g_N $\beta(T)$ for isolated ^{136}Ce ions in Th. Solid line: experimental Ce^{3+} behavior observed in Ba, Pb, and Sn. Dashed line: $|g_N|\beta$ for diamagnetic Ce^{4+}.

(i) Only one frequency ω_L is observed in the R(t) spectra, which represents a direct experimental proof of the dynamic nature of the mixed valence phenomena.

(ii) An alternative interpretation of $\beta(T)$ in terms of Kondo theories seems to be not justified; the experiment strongly indicates the occurrence of intermediate valence for isolated Ce ions in Th.

(iii) A comparison to closely related concentrated Ce systems indicates that intermediate valence phenomena in concentrated Ce systems are predominantly governed by the single-ion behavior.

Dynamic aspects of this system will be discussed below.

III.3 FORMATION OF THE 4f CONFIGURATION OF ISOLATED CE IONS IN ELEMENTS

One of the most fundamental problems in the field of magnetism concerns the formation of local moments in metals. For the following reasons the TDPAD method is very well suited to study this central problem: (a) the implantation technique allows the investigation of the matrix dependence of local moments in many simple systems. (b) The microscopic state of the 4f configuration of isolated ions can be accurately probed by observing the frequency ω_L. We have measured[18] the local susceptibility for isolated Ce ions in about 40 elements. Simultaneous implantation of two isomers in ^{138}Ce and ^{139}Ce was achieved by ^{130}Te (^{13}C,xn) reactions (x = 4,5). The measurements were performed around 400 K, so that crystal field effects on β can be neglected. As described above we have analyzed the valence of the Ce ions from the measured β values. The results are summarized in Fig.11. We have found many new systems in which the isolated Ce ions reflect intermediate valence and several hosts in which Ce occurs as a tetravalent ion. These experiments imply important results in many aspects[18], here we only want to comment on the correlation of the magnetic behavior to the heats of solution of Ce^{3+} and Ce^{4+} in the various elements. The difference of the heats of solution between Ce^{3+} and Ce^{4+} in a certain host, ΔH_s, was calculated by use of the semi-empirical theory of Miedema et al.[24]. Within a thermodynamic approach[25] ΔH_s might roughly scale the relative shifts of the $4f^1$ and $4f^o$ levels. With increasing positive ΔH_s, the energy between $4f^1$ and $4f^o$ decreases, at large positive ΔH_s $4f^1$ might be located above the Fermi level. The calculated ΔH_s values vary in the large range from -1.2 eV to +1.8 eV for the systems investigated (Fig.11). As documented in Fig.11, nearly all Ce^{3+} systems are located at negative ΔH_s values, all Ce^{4+} systems are located at positive ΔH_s values and the intermediate valence systems are located between the Ce^{3+} and Ce^{4+} systems. Thus the magnetic behavior of the Ce systems is strongly correlated with the ΔH_s values. The following major results can be inferred from Fig.11:

(i) The formation of local moments of isolated Ce ions in elements is essentially determined by the difference of the heats of solution be-

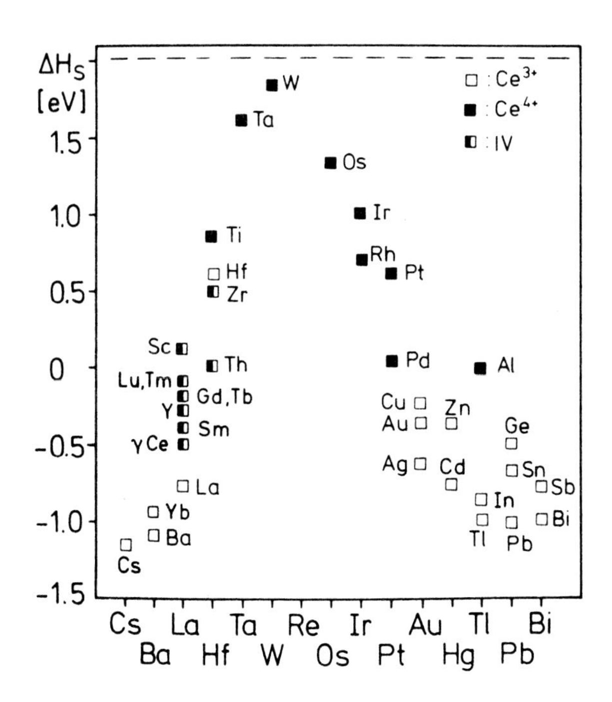

Fig. 11: Difference of the heats of solution between Ce^{4+} and Ce^{3+} ions in elements versus the groups of the periodic table. The magnetic behavior of isolated Ce ions in the various elements is indicated by the symbols ■ : Ce^{4+}, □ : Ce^{3+}, and ◧ : intermediate valence. The observed β values for some of the systems are located at the borderline regions between Ce^{3+} and IV or between Ce^{4+} and IV. Such details are not important for the conclusions given in this work.

tween Ce^{3+} and Ce^{4+}.

(ii) The transition $4f^1 \rightarrow 4f^0$ occurs at $\Delta H_s \sim 0$ eV, suggesting that the $4f^1$ level in both dilute and concentrated intermediate valence systems is located close to the Fermi level. Previous statements that the $4f^1$ level in concentrated Ce systems e.g. in Ce metal is located at about 2 eV[25] below E_F have to be questioned. Within our treatment such an assignment would lead to the expectation that the transition $4f^1 \rightarrow 4f^0$ should occur around $\Delta H_s \sim +2$ eV (dashed line in Fig.11). However we found that in very good approximation Ce^{4+} does exist in dilute systems. The important alternative, that the $4f^1$ electron in Ce^{4+} systems is completely delocalized will be discussed in the last chapter.

IV. 4f SPIN DYNAMICS OF ISOLATED IONS IN SIMPLE METALS

Suppose that the 4f spin fluctuates due to dynamic interactions with the metallic environment e.g. the exchange interaction (see Fig.2). A fluctuating 4f spin produces a fluctuating high magnetic field, B(0), at the nucleus, which in turn causes a relaxation of the nuclear alignment. Following Abragam and Pound[5,26] the magnetic nuclear spin rate is given by

$$\tau_N^{-1}(\text{magn}) = 2(\mu_N/\hbar)^2\ J^{-1}(J+1)\ g_N^2\ B^2(0)\ \tau_J \quad (5)$$

where τ_J is the average correlation time of the atomic spin fluctuation. The validity of Eq.(5) requires $\omega\,\tau_J \ll 1$ which is well satisfied under usual conditions. Eq.(5) represents the basic relation for the sensitivity of the TDPAD method to the 4f spin relaxation rate τ_J^{-1}. In principle τ_J^{-1} rates in the large range 10^9–10^{15} s^{-1} are observable by use of appropriate nuclear states in appropriate systems. Thus TDPAD falls just in the gap between ESR and Mössbauer effect ($\tau_J^{-1} < 10^9\ s^{-1}$) and X-ray photoemission spectroscopy measurements ($\tau_J^{-1} > 10^{15}\ s^{-1}$). In some essential aspects the TDPAD method is complementary to magnetic neutron scattering from which 4f spin rates in concentrated systems might be deduced from the linewidth $\Gamma/2$[27]. TDPAD can be applied to measure the 4f spin dynamics for both extremely dilute and concentrated 4f systems. Of fundamental importance is the possibility to study the 4f spin dynamics of isolated rare earth ions in both Kondo-like systems and intermediate valence systems.

IV.1 4f SPIN DYNAMICS IN DILUTE KONDO-LIKE SYSTEMS

We have successfully applied all nuclear probes listed in Table I to investigate the 4f spin dynamics in various nearly integral valence systems. Many of these results remain to be published.

For an illustrative example we want to concentrate on the 4f spin dynamics of isolated Ce and Nd ions in liquid tin. As derived from the local susceptibility (Fig.7), Ce<u>Sn</u> and Nd<u>Sn</u> are both nearly stable 3^+ systems. The R(t) spectra both exhibit a clear exponential damping from which τ_N^{-1} can be extracted.

Before applying Eq.(5) to extract the 4f spin rate one has to consider possible additional contributions to the τ_N^{-1} observed. First of all one can estimate that the nuclear quadrupole relaxation rates arising from the electric field gradients of the 4f ions are negligibly small compared to the paramagnetic relaxation given by Eq.(5). This is valid for all rare earth systems, if the $|g_N|$ values are not extremely small. Also negligible are the nuclear magnetic and quadrupolar rates arising from interactions with the liquid environment. In liquid metals such rates are known[7)] to be much smaller than the τ_N^{-1} values observed here. For solid 4f systems quadrupolar contributions to the damping of R(t) cannot be neglected in general (see IV.2 and Refs.10,16,18).

Thus the choice of liquid hosts allows a direct extraction of τ_J^{-1} from the τ_N^{-1} observed by Eq. (5). Furthermore the influence of crystal field effects on the magnetic response of the liquid systems seems to be negligible, which can be formulated as k_BT large compared to the crystal field splitting. Additionally the liquid state permits measurements in alloys over a wide range of concentration which has been applied in Ref. 13. The τ_J^{-1} and τ_N values for some Ce and Nd systems are given in Fig.12.

As discussed in more detail in Refs.11,13 the following major results can be obtained from these experiments:

(i) The 4f spin rates are dominated by strong mixing exchange interaction in all systems investigated. TDPAD allows the study of the strength of the mixing exchange coupling for isolated 4f ions in simple s,p, and d metals.

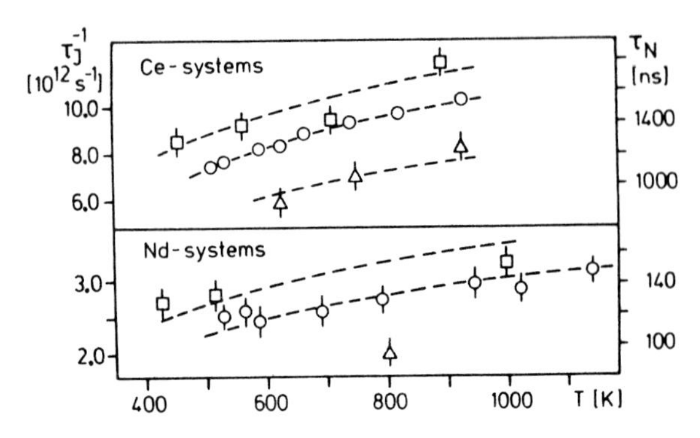

Fig. 12: T-dependence of τ_J^{-1} and τ_N for ^{136}Ce and ^{138}Nd in the liquids $Sn_{57}Bi_{43}$ (squares), Sn (circles) and $Sn_{30}Au_{70}$ (triangles). The dashed lines represent fits which are explained in Ref.13.

(ii) These investigations seem to include the first observation of the Kondo effect in liquid metals and the first observation of Kondo-like behavior in any Nd system. Studies of the Kondo effect are possible at high T, partly because of the large degeneracy factor of the 4f level.

(iii) One is led to argue that the 4f levels, not only for Ce systems but also for Pr and Nd (and perphaps for all light rare earth) systems are close to the Fermi level.

Lastly we want to comment on the suggestion[23,28)] that phonon relaxation contributes to the ESR linewidth in the system Nd<u>LaRh$_2$</u>. If one accepts that the 4f spin rates for Nd<u>LaRh$_2$</u> and Nd<u>Sn</u> (see point (i) in Ref.13) are both dominated by mixing exchange interaction, this suggestion has to be questioned seriously. As shown in Ref.29 similar experimental findings in the Kondo system Yb<u>Au</u> can be analyzed in terms of mixing exchange without any contribution of phonon relaxation. As far as we are aware, phonon relaxation in metallic systems has never been identified by experiment in a reliable manner, which is consistent with the results of this work.

IV.2 4f SPIN DYNAMICS IN DILUTE INTERMEDIATE VALENCE SYSTEMS

The 4f spin fluctuation rate τ_f^{-1} is one of the central parameters in models for valence fluctuations. Hitherto this quantity has been measured in several concentrated rare earth compounds mainly by neutron scattering experiments [2,27]. In this contribution we present the first experimental information about the spin dynamics of an extremely dilute mixed valence impurity.

As discussed above the magnetic part of the nuclear spin relaxation rate τ_N^{-1}(magn) is sensitive to the 4f spin rate τ_f^{-1}. We have performed a series of TDPAD experiments to extract τ_N^{-1}(magn) from the measured $\tau_N^{-1}=\tau_N^{-1}(\text{magn})+\tau_N^{-1}(\text{quadr.})$. The quadrupolar damping in cubic Ce<u>Th</u> might arise from interactions of the nuclear Q moment with small electric field gradients produced by radiation damage around the subsitutional Ce recoil. In a first step we have simultaneously implanted the 10^+, 82 ns isomer of ^{138}Ce and the $19/2^-$, 74 ns isomer of ^{139}Ce into Th metal following the reactions ^{130}Te (^{13}C,xn) (see Fig.13). This experiment yields τ_N^{-1}(138), τ_N^{-1}(139) and the ratio τ_N^{-1}(magn,138)/τ_N^{-1}(magn,139) which is equal to the ratios of the nuclear g_N factors squared. Secondly we have implanted the same two isomers into Pt and Ir metal. Since the static and dynamic response in both systems Ce<u>Pt</u> and Ce<u>Ir</u> correspond in very good approximation to tetravalent Ce ions, we are able to deduce the ratio τ_N^{-1}(quadr.,138)/τ_N^{-1}(quadr.,139). The thus extracted τ_N(magn) values for ^{139}Ce are $\sim$ 500 ns and depend only moderately on temperature in the range 20 to 400 K.

If one applies Eq.(5) to the intermediate valence case also, the spin fluctuation rate for isolated Ce ions in Th is found to be $\tau_f^{-1} \sim 20\cdot10^{12}\ s^{-1}$. The τ_f^{-1}-data reflect a broad maximum around 200 K consistent with theoretical predictions of Schlottmann [30]. We plan to improve the statistics of this experiment in order to measure the important T dependence of τ_f^{-1} more clearly.

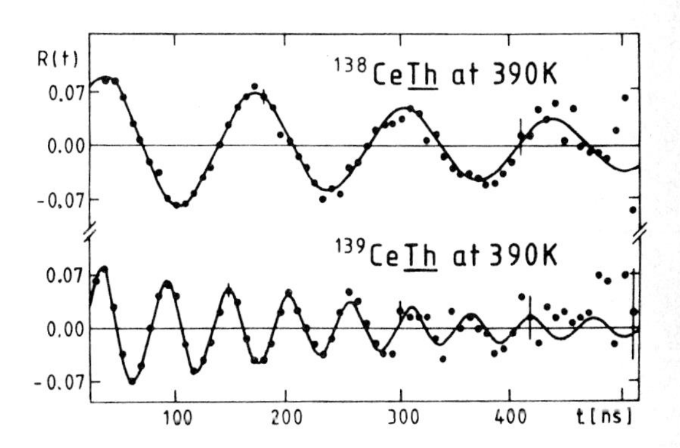

Fig. 13: Spin rotation pattern of isolated ^{138}Ce and ^{139}Ce ions in Th. The difference in frequencies reflects the difference in the g_N factors (see Table I). The shorter relaxation time for ^{139}Ce<u>Th</u> compared to ^{138}Ce<u>Th</u> reflects the larger $\tau_N^{-1}(\text{magn}) \propto g_N^2$ contribution to τ_N^{-1}(139).

The magnitude of the spin fluctuation rate is roughly comparable to the results in concentrated Ce systems [2,27,31] excluding systems in the α phase. With use of $\tau_f\cdot\Delta E = \hbar$ the observed rate corresponds to a linewidth of $\Delta E/k_B$=150 K. These findings support the conclusion that the phenomena of valence fluctuations are mainly governed by the single-ion-behavior.

Besides Ce<u>Th</u>, we have started to investigate the 4f spin dynamics for e.g. Ce in Sc, Y, and Ce in rare earth metals. Most of these systems reflect temperature driven intermediate valence phenomena. Microscopic investigations of the dynamics of the single mixed valence ion in elements are of particular importance for further progress in the field of intermediate valence.

We also have started to investigate the local susceptibility and the 4f spin dynamics in concentrated Ce systems e.g. in γ-Ce and $Ce_{90}Th_{10}$ alloys. The system γ-Ce allows a direct comparison of the 4f spin dynamics as observed by

TDPAD and magnetic neutron scattering. Our preliminary result $\tau_f^{-1}=15\cdot10^{12}$ s^{-1} for γ Ce at 360 K is consistent with the n-scattering result of Rainford et al.[32].

V. SUMMARY AND OUTLOOK

(i) Heavy ion reactions and especially the implantation technique allow the production of very simple 4f systems e.g. isolated rare earth ions in elements. The TDPAD method permits microscopic studies of the local susceptibility and 4f spin dynamics for nearly all 4f ions in liquid and solid metals[33] over wide temperature ranges. Of particular importance is the possibility of measuring high 4f spin rates in the range 10^9-10^{15} s^{-1} for extremely dilute rare earth systems. All these features together represent considerable progress in the field of microscopic observations of the single-ion-behavior of 4f instabilities in metals.

(ii) Concentrated 4f systems can be investigated by TDPAD, also. This offers a wide field of applications, especially to get further insight into the basic problem: isolated versus collective contributions to the phenomena of 4f instabilities. It is of some advantage if one can compare results for dilute and concentrated systems obtained by the same method.

(iii) Ce^{4+} exists in dilute systems.

(iv) The formation of local moments for isolated Ce ions in elements is essentially governed by the difference of the heats of solution between Ce^{3+} and Ce^{4+}. The extension of this type of experiment to other 4f systems e.g. recoil implantation of Pr, Nd, Sm, Eu, Tm, and Yb ions into elements seems to be straightforward.

(v) The findings

a) for Ce in elements the $4f^1 \rightarrow 4f^0$ transitions occurs at $\Delta H_s \sim 0$ eV (Fig.11)

b) the valence transitions for Pr<u>La</u>[34] and Pr<u>Pd</u> seem to be of the $4f^2 \rightarrow 4f^1$ type (see Ref.13)

c) the 4f spin rate for Nd<u>Sn</u> reflects strong hybridization of the $4f^3$ level (Fig.12)

all consistently indicate that the 4f levels are close to the Fermi levels. In view of the findings b), c) it seems to be unlikely that the $4f^1 \rightarrow 4f^0$ transitions for both isolated and concentrated Ce systems can be explained by an extreme broadening of the $4f^1$ linewidth so that the $4f^1$ electron delocalizes. At least for light rare earth metals the interpretation of XPS spectra and various theoretical predictions for the location of the 4f levels (e.g. Ref.25) have to be questioned seriously. These considerations imply the possibility of valence transitions for Nd systems. To test this possibility we plan to study the fate of Nd (and Pm) ions which are implanted into hosts with small metallic radii and high densities of states at E_F. Such hosts are roughly equivalent to hosts with large ΔH_s values[18].

(vi) It has been demonstrated in some detail that the 4f spin dynamics can be affectively investigated in simple Kondo-like systems. These results include the first observation of Kondo-like behavior in liquid metals and for Nd systems. Experiments of this type address to the fundamental problem of the 4f-ce hybridization for dilute 4f ions in simple s,p, and d metals.

(vii) One of the most promising features of TDPAD is the possibility of studying the local susceptibility and 4f spin dynamics in extremely dilute intermediate valence systems. Up to now we have found ten metallic elements in which single Ce ions reflect intermediate valence. As described for the system Ce<u>Th</u> in some detail, one has to conclude that the phenomena of intermediate valence are mainly governed by the single-ion-behavior.

We expect that TDPAD can be applied quite generally to intermediate valence systems containing e.g. Ce, Pr, Nd (see point (v)), Sm, Eu, Tb, Tm, and Yb ions. Appropriate nuclear isomers in all these ions (see Table I for light rare earth ions) can be reached by heavy ion reactions and additionally the technique of implantation can be used to produce the systems of physical interest.

It is our pleasure to thank P. Schlottmann and D. Wohlleben for valuable discussions. We would like to acknowledge the hospitality and support of the Bereich Kern- und Strahlenphysik, Hahn-Meitner-Institut, Berlin. This work was strongly supported by the Deutsche Forschungsgemeinschaft (Sonderforschungsbereich 161).

REFERENCES

1 D.K. Wohlleben, in Valence Fluctuations in Solids, edited by L.M. Falicov, W. Hanke, and M.B. Maple (North Holland 1981) p. 1
2 J.M. Lawrence, P.S. Riseborough, and R.D. Parks, Rep.Prog.Phys. 44, 1 (1981)
3 M.B. Maple, L.E. DeLong, and B.C. Sales, in Handbook on the Physics and Chemistry of Rare Earths, edited by K.A. Gschneidner and L. Eyring (North-Holland, Amsterdam, 1979), Vol. 1, p. 797
4 W. Franz, F. Steglich, W. Zell, D. Wohlleben, and F. Pobell, Phys. Rev. Lett. 45, 64 (1980)
5 H. Frauenfelder and R.M. Steffen, in Alpha-Beta and Gamma-Ray Spectroscopy, edited by K. Siegbahn (North-Holland, Amsterdam, 1966), Vol. 2
6 E. Recknagel, Rev.Roum.Phys. 17, 473 (1972)
7 D. Riegel, Phys. Scripta 11, 228 (1975)
8 J. Christiansen, H.E. Mahnke, E. Recknagel, D. Riegel, G. Weyer, and W. Witthuhn; Phys. Rev. C1, 613 (1970)
9 D. Riegel, N. Bräuer, B. Focke, K. Nishiyama and E. Matthias, Hyperfine Interactions 3, 1 (1977)
10 H.J. Barth, G. Netz, K. Nishiyama, and D. Riegel, Phys. Rev. Lett. 45, 1015 (1980)
11 H.J. Barth, G. Netz, K. Nishiyama, and D. Riegel, Hyperfine Interactions 10,811(1981)
12 M. Luszik-Bhadra, H.J. Barth, H.J. Brocksch, G. Netz, D. Riegel, and H.H. Bertschat, Phys. Rev. Lett. 47, 871 (1981)
13 D. Riegel, Phys. Rev. Lett. 48, 516 (1982)
14 H.J. Barth, G. Netz, K. Nishiyama, D. Riegel, Proc. V. Int. Conf. on Hyperfine Interactions, Berlin 1980, Contr. A6 and to be published
15 W. Klinger, R. Böhm, W. Engel, W. Sandner, R. Seeböck, and W. Witthuhn, Z. Phys. A290, 227 (1979)
16 H.J. Barth, K. Nishiyama, and D. Riegel, Phys. Lett. 77A, 365 (1980)
17 See e.g. many contributions to the IV and V Int. Conf. on Hyperfine Interactions, published in Hyperfine Interactions Vol. 4 (1978) and Vol. 10 (1981)
18 H.J. Barth, M. Luszik-Bhadra, G. Netz, and D. Riegel, to be published
19 G. Günther and I. Lindgren, in Perturbed Angular Correlations, edited by E. Karlsson, E. Matthias, and K. Siegbahn (North-Holland, Amsterdam 1964), p. 357
20 B. Bleaney in Magnetic Properties of Rare Earth Metals, edited by R.J. Elliot (Plenum Press 1972) p. 383
21 A.J. Freeman and P.J. Desclaux, J. Magn. Magn. Mater. 12, 11 (1979)
22 I. Nowik, B.D. Dunlap and J.H. Wernick, Phys. Rev. B8, 238 (1973), and references therein
23 G.E. Barberis, D. Davidov, J.P. Donoso, C. Rettori, J.F. Suassuma, and H.D. Dokter, Phys. Rev. B19, 5495 (1979)
24 A.R. Miedema, P.F. de Chatel, and F.R. de Boer, Physica 100B, 1 (1980)
25 B. Johansson, Phys. Rev. B20, 1315 (1979), P.F. de Chatel, and F.R. de Boer in Valence Fluctuations in Solids, edited by L.M. Falicov, W. Hanke, and M.B. Maple (North Holland 1981) p. 377
26 A. Abragam and R.V. Pound, Phys. Rev. 92 (1953) 943
27 M. Loewenhaupt and E. Holland-Moritz, J. Appl. Phys. 50, 7456 (1979)
28 A. Dodds, J. Sanny, and R. Orbach, Phys. Rev. B18, 1016 (1978)
29 F. Gonzales-Jiminez, B. Cornut, and B. Coqblin, Phys. Rev. B11, 4674 (1975)
30 P. Schlottmann, this conference
31 B.H. Grier, S.M. Shapiro, C.F. Majkrzak, and R.D. Parks, Phys. Rev. Lett. 45, 666 (1980)
32 B.D. Rainford and V.T. Nguyen, J. Phys. (Paris), Colloq. 40, C5-262 (1979)
33 We also have applied TDPAD to investigate the fate of Ce recoils in solid and liquid insulators:
H.J. Barth, G. Netz, K. Nishiyama, and D. Riegel, Proc. V. Int. Conf. on Hyperfine Interactions, Berlin 1980, Contr. G12 and to be published
34 J. Wittig, Phys. Rev. Lett. 46, 1431 (1981)

Valence Instabilities
P. Wachter and H. Boppart (eds.)

ABSENCE OF MAGNETIC ORDER IN Yb MONOPNICTIDES

H.R. Ott, F. Hulliger and H. Rudigier
Laboratorium für Festkörperphysik
ETH Hönggerberg, 8093 Zürich, Switzerland

Measurements of various low temperature thermal properties of Yb monopnictides reveal the absence of cooperative magnetic order in YbN, YbP, YbAs and YbSb down to 12 mK, although at room temperature the Yb ions seem to adopt their trivalent $4f^{13}$ configuration in these materials. Another striking feature of these substances is a strongly enhanced low temperature specific heat anomaly which cannot be explained with a simple single-ion model behaviour.

1. INTRODUCTION

One very obvious consequence of valence instabilities in materials containing rare-earth ions is usually (with still the only exception of some Tm(Se/Te) compounds) the absence of magnetic order at low temperatures. An increasing amount of experimental work on Ce compounds, however, indicates that even in materials undergoing cooperative magnetic order, the ordered moments are either drastically reduced, as e.g. in CeTe[1], or even partially quenched, leading to complex magnetic phase diagrams, as was observed e.g. in CeSb[2,3]. In other Ce compounds, the absence of magnetic order is not necessarily due to a valence fluctuation of the Ce ions but rather a consequence of strong correlation effects, as in the case of $CeAl_3$[4] and probably also $CeCu_2Si_2$[5].

Physical properties similar to those of Ce compounds with nominally one 4f electron or less per Ce ion may also be observed in substances containing Yb, with one possible 4f hole per ion. Most such investigations on Yb compounds have been made on Yb intermetallics[6] where the 4f hole interacts with conduction electrons from other metallic constituents of the compound and may be partially filled by them. Here we give a preliminary report of a study of the Yb monopnictide series, whose Ce analogues are either mixed valent (CeN[7]) or undergo magnetic ordering at temperatures between 5 and 25 K[2,3,8].

2. SAMPLES AND EXPERIMENT

The usual two-stage procedure was applied to prepare poly- and monocrystalline Yb monopnictide samples. In the case of YbN (the polycrystalline YbN sample was made available by Ch. Zürcher) the reaction was carried out in an open molybdenum crucible at a nitrogen pressure of 1.5 atm., while with P, As and Sb, closed silica tubes were used. Prereacted powders were pressed and heated in sealed tungsten crucibles and kept for one or two weeks at temperatures slightly below the melting points, or below the peritectic temperature in the case of YbSb. Using these techniques, YbBi does not form and probably high pressure has to be applied for its synthesis.

The magnetic susceptibility of YbN, YbP, YbAs and YbSb was measured with a sample-moving magnetometer between 1.5 and 300 K. Below 1 K, the low-frequency a.c. suceptibility of the same substances was measured down to 12 mK in a dilution refrigerator with an external magnetic field amplitude of less than 0.1 Oe. Magnetization M(H) was measured on all four substances at 1.6 K in fields up to 70 kOe. Specific heat, thermal expansion and magnetostriction was measured on YbN, YbP and YbAs between 1.5 and 20 K. and finally, also the electrical resistivity was measured for YbP and YbAs between 1.5 K and room temperature.

3. RESULTS AND DISCUSSION

In Fig. 1 we show the temperature dependence of the inverse susceptibilities $\chi^{-1}(T)$ between 1.5 and 300 K as obtained for all four substances. It may be seen, that in all these Yb monopnictides, the general temperature dependence of χ is the same, χ approaching a value of about 0.03 emu/mole at low temperatures in all four cases. Measurements of the a.c. susceptibilities down to 12 mK reveal that χ remains small and no indication of cooperative magnetic ordering is observed. Our data for YbSb are in fair agreement with previously published results[9].

By comparison with the corresponding values of the whole rare-earth monopnictide series[10], the room temperature lattice constants of our substances indicate that, at this temperature, the Yb ions are in their trivalent configuration. Previous inelastic neutron scattering experiments[11] revealed distinct excitations between crystal-electric-field (CEF) split energy levels

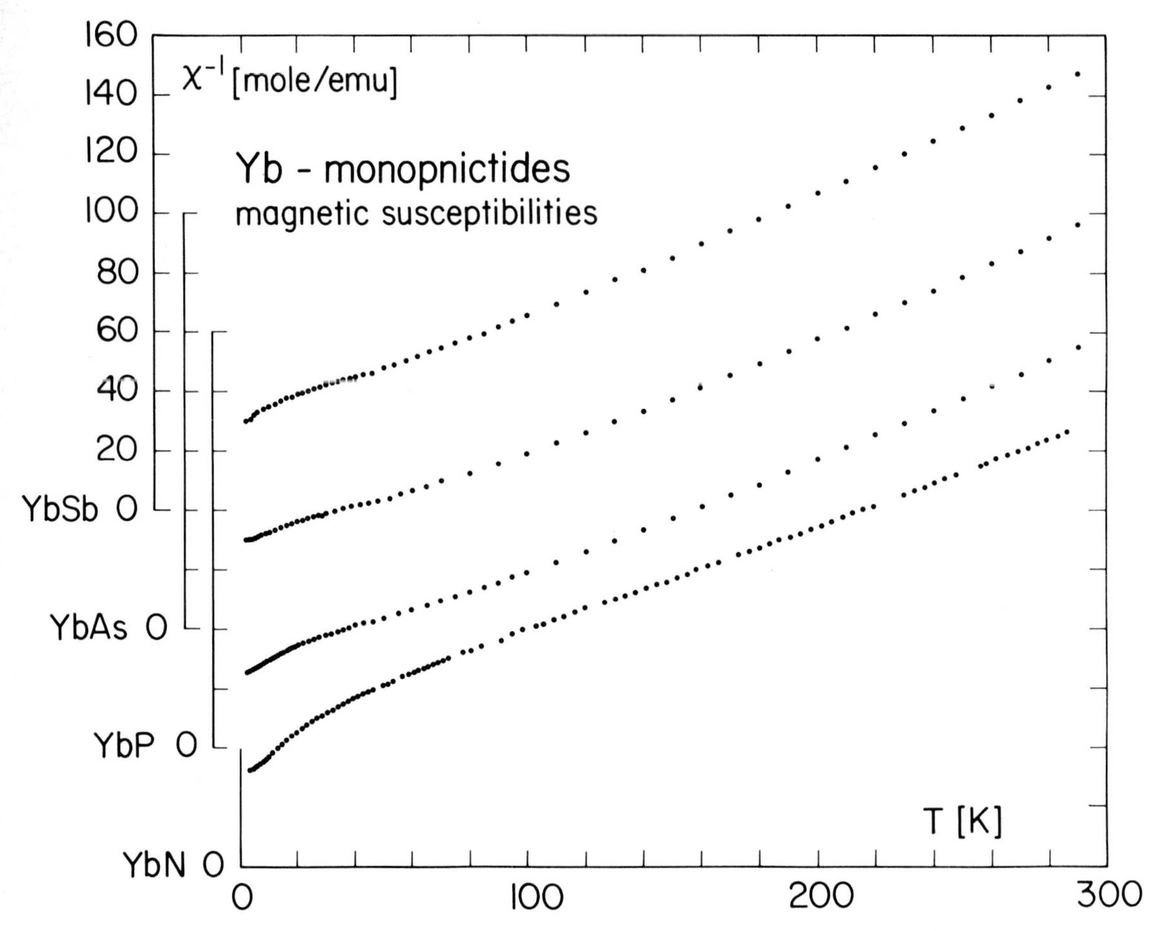

Figure 1 : Temperature dependence of the inverse magnetic susceptibilities χ^{-1} for YbN, YbP, YbAs and YbSb between 1.5 and 300 K.

of the $4f^{13}$, J = 7/2 ground-state multiplet in the case of YbP, and the resulting CEF energy level scheme gave a Γ_6 doublet ground state with a quartet Γ_8 and a doublet Γ_7 as the excited states at 228 K and 365 K, respectively. In what follows, we discuss some of the properties of YbP, using this level scheme.

Fig. 2 shows $\chi^{-1}(T)$ for YbP. The solid line is calculated from the above level scheme, assuming an effective moment $p_{eff} = 4.30\ \mu_B$/Yb and a paramagnetic Curie temperature $\Theta_p = -11.5$ K. It may be seen that these assumptions give an excellent fit to the experimental data above 80 K and hence confirm the trivalent configuration of the Yb ions in this temperature range. The value of the effective moment is within 4% of that of the free-ion moment. Below about 70 K, $\chi^{-1}(T)$ deviates considerably from the calculated curve, indicating an appreciable decrease of the susceptibility below the expected values, presuming a well defined Γ_6 doublet ground state. This behaviour shows quite clearly, that a simple single-ion description is no longer valid.

Further support for this statement is obtained from our low-temperature specific-heat measurements. In Fig. 3 we show, as an example, the difference of the experimental data obtained for YbP and LuP, respectively, namely $\Delta c = c_p(\text{YbP}) - c_p(\text{LuP})$ as a function of temperature between 1.5 K and 20 K. Also shown is the calculated expected magnetic contribution to the specific heat from the above mentioned CEF split 4f electron energy levels, which is virtually zero over almost the entire temperature range. The experimental data reveal a much larger specific heat with a broad maximum between 4 and 5 K. An approximate linear extrapolation of Δc to T = 0 K serves to estimate the entropy associated with the observed anomaly. Integrating from T = 0 to the temperature where Δc has

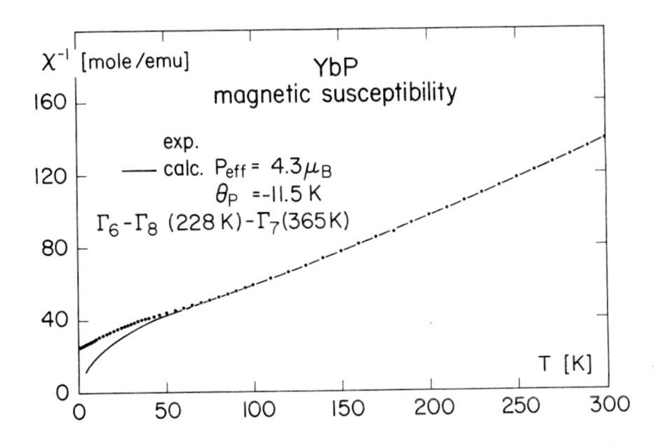

Figure 2 : Inverse magnetic susceptibility $\chi^{-1}(T)$ of YbP from experiment and calculated using the indicated CEF energy level scheme.

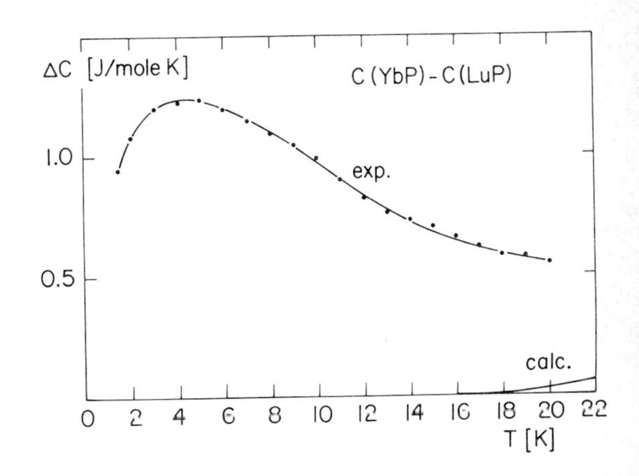

Figure 3 : Experimental results for c_p(YbP)-c_p(LuP) between 1.5 and 20 K. The solid line through the dots is to guide the eye. The calculated line is explained in the text.

its maximum, we obtain S = 2.25 J/mole K, which is about 40% of R ln2, the entropy associated with a doublet ground state. Similar features are found in YbN and YbAs.

Such anomalies have previously been found in non-ordering Ce compounds, especially $CeAl_3$[4,12] and it is, for reasons mentioned in the introduction, not surprising to find a similar behaviour in Yb compounds. In both cases, a relatively simple single-ion model describes the high-temperature properties, such as the magnetic susceptibility χ quite well, but is obviously inadequate at lower temperatures.

In YbP, the magnetization M at 1.6 K increases linearly with increasing external magnetic field and reaches, with a slight bending, about 0.45 μ_B/Yb at 70 kOe, as compared to the expected saturation moment of a Γ_6 ground state of 1.33 μ_B/Yb. Contrary to $CeAl_3$, however, the thermal expansion of the Yb monopnictides is rather small below 15 K. We also note that the magnetostriction is virtually temperature independent and of the order of less than 10^{-7} kOe^{-2} between 1.5 and 15 K. The general features of the electrical resistivity between 1.5 and 300 K are metallic in character with residual resistance ratios of 3 to 5.

4. CONCLUSIONS

It is obvious from this selected collection of data on various low-temperature properties of Yb monopnictides that a similar behaviour as previously found in some Ce compounds is also present in this series of substances. in both cases, the evidence for trivalent Ce or Yb ions in the materials considered disappears when the substances are cooled to helium temperature. The most obvious anomalies associated with this behaviour are a) the absence of cooperative magnetic order and b) an excess specific heat the origin of which is not clear at present. Additional investigations, particularly further experiments below 1 K, which are presently under way, and measurements involving spectroscopy techniques will hopefully help to clarify the situation. A more extensive account of this work will be published later.

ACKNOWLEDGEMENT

We thank Dr. Ch. Zürcher for the generous supply of some YbN material and Mrs. I. Zarbach for helping with the preparation of the substances. Financial support of the Schweizerische Nationalfonds is also gratefully acknowledged.

REFERENCES

(1) Ott, H.R., Kjems, J.K. and Hulliger, F., Phys.Rev.Letters 42 (1979) 1378.
(2) Fischer, P., Lebech, B., Meier, G., Rainford, B.D. and Vogt, O., J. Phys. C 11 (1978) 345.
(3) Rossat-Mignod, J., Burlet, P., Villain, J., Bartholin, H., Wang Tcheng-Si, Florence, D. and Vogt, O., Phys.Rev. B16 (1977) 440.
(4) Andres, K., Graebner, J.E. and Ott, H.R., Phys.Rev.Letters 35 (1975) 1779.
(5) Steglich, F., Aarts, J., Bredl, C.D., Lieke, W., Meschede, D., Franz, W. and Schäfer, H., Phys.Rev.Letters 43 (1979) 1892.

(6) Klaasse, J.C.P., de Boer, F.R. and deChâtel, P.F., Physica 106B (1981) 178.
(7) Baer, Y. and Zürcher, Ch., Phys.Rev.Letters 39 (1977) 956.
(8) Hulliger, F. and Ott, H.R., Z. Phys. B 29 (1978) 47.
(9) Bodnar, R.E., Steinfink, H. and Narasinhan,K. S.V.L., J. Appl. Phys. 39 (1968) 1485.
(10) Hulliger, F., J. Mag. Magn. Mat. 8 (1978) 183.
(11) Birgeneau, R.J., Bucher, E., Maita, J.P., Passell, L. and Turberfield, K.C., Phys.Rev. B 8 (1978) 5345.
(12) van Maaren, M.H., Buschow, K.H.J. and van Daal, H.J., Solid State Communications 9 (1971) 1981.

Valence Instabilities
P. Wachter and H. Boppart (eds.)

$CeSi_2$: A NEW INTERMEDIATE-VALENT COMPOUND

W.H.Dijkman, A.C.Moleman, E.Kesseler, F.R. de Boer and P.F. de Châtel,

Natuurkundig Laboratorium, Universiteit van Amsterdam,
Valckenierstraat 65, 1018 XE Amsterdam, The Netherlands

We report thermal expansion, magnetic susceptibility, high-field magnetization, heat capacity and resistivity measurements on the tetragonal compound $CeSi_2$. Although no easily identifiable anomaly is observed in the lattice-parameter data, the magnetic measurements inevitably lead to the conclusion that Ce is in the intermediate-valent state in this compound.

1. INTRODUCTION

In the series of rare-earth disilicides, $CeSi_2$ is found between superconducting $LaSi_2$ and ferromagnetic $PrSi_2$. Recent magnetization and specific-heat measurements have led to the conclusion that $CeSi_2$ is another compound in which Ce is found in the mixed-valent state |1,2|. In this communication we shall give some of the results presented in ref. 1, which, on the one hand, substantiate the above conclusion, and, on the other hand, suggest that in some respects this compound is unique among the cerium compounds showing mixed valence.

2. CRYSTAL STRUCTURE

Both $LaSi_2$ and $CeSi_2$ have been observed to crystallize in two closely related structures, the tetragonal $ThSi_2$ structure and the orthorhombic $GdSi_2$ structure. Both structures have several interesting features: warped silicon chains can be identified along the axes in the basal plane, and the Si sublattice can also be seen to have a '3-dimensional graphite structure', that is, a 3-dimensional network of distorted hexagonal rings |3|. Both of these features might have some bearing on the interesting superconducting properties of $LaSi_2$ and related compounds |4|. From the point of view of the electronic state of the Ce ions, however, it might be more relevant to point out that the $ThSi_2$ structure can be constructed by stacking appropriately four identical layers, each containing a square lattice of Th atoms. In $CeSi_2$, at room temperature the distance between Ce atoms in the basal plane (a = 4.192 Å) slightly exceeds the one between near neighbour Ce atoms in adjacent layers (4.058 Å).

Two important features of the $ThSi_2$ structure emerge from the above description: that all RE atoms occupy equivalent sites and that the point symmetry at these sites is very low.

The $GdSi_2$ structure can be obtained from the $ThSi_2$ one by an orthorhombic distortion. According to |5| $LaSi_2$ and $CeSi_2$ have the orthorhombic structure with $(a-b)/\frac{1}{2}(a+b)$ = .026 and .014, respectively.

3. SAMPLE PREPARATION

Ingots of $LaSi_2$ and $CeSi_2$ were prepared by melting together stoichiometric amounts of the constituent elements in an arc furnace. While heating up the reactants using a moderate current, the reaction suddenly started at about 800°C, thereby heating up the sample to approximately 1600°C for a few seconds. Homogenization was achieved by repeated remelting. Subsequently, samples were cast in a cylindrical copper crucible. Weight losses were negligible in this process. Analysis of the Debye-Scherrer patterns taken on the $LaSi_2$ and $CeSi_2$ ingots showed unequivocally that only the undistorted $ThSi_2$ structure was present. The ingots had a light grey metallic colour and their visual appearance did not change after exposure to air for several months.

4. MAGNETIC PROPERTIES

Measurements of the magnetic susceptibility of $CeSi_2$ in the temperature range 73-473 K were published by Ferro Ruggiero and Olcese |6|, who found the disilicide to be exceptional among the Ce-Si compounds in that the effective moment deduced from a Curie-Weiss fit is anomalously large, 2.94 μ_B. The same fit gives a paramagnetic Curie temperature of θ = 299 K, a remarkably high value in view of the absence of any phase transition. As this puzzling feature could be an

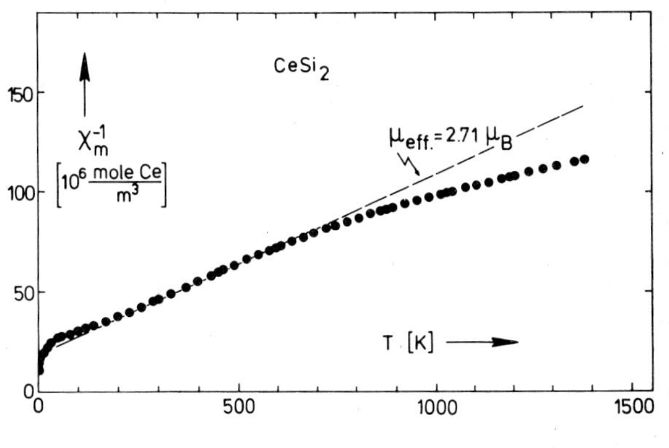

Fig. 1

indication for the existence of an intermediate-valent state in this compound, we have undertaken susceptibility measurements between 1.5 and 1400 K, to determine the range of validity of the Curie-Weiss law. The reciprocal susceptibility, given in fig. 1, shows that a Curie-Weiss law is approximately obeyed in the temperature range 200-600 K. Both the effective moment and the paramagnetic Curie temperature are found to be lower than in ref. 6. Below 200 K there is a deviation from the straight line just like in the case of $CeSn_3$ |7|, suggesting intermediate-valence behaviour or possibly crystal field effects.

The susceptibility becomes approximately constant at 54.0×10^{-9} m^3/mole Ce, from 1.5 - 10 K after subtraction of a Curie contribution equivalent to that of 0.7% Ce ions, each having its full J=5/2 magnetic moment of $2.54\mu_B$. The identification of the low-temperature upturn in the susceptibility with 'misplaced' Ce ions gains further support from the susceptibility of $CeSi_{2.06}$ where the impurity analysis gives only 0.2% trivalent Ce ions. Addition of more excess Si (we have tried $CeSi_{2.12}$) does not change the remaining upturn, suggesting that not misplaced Ce ions but true impurities (presumably other RE metals present in Ce) are responsible for the residual anomaly. Also, annealing at 800 or 1250°C of any of the above-mentioned compounds does not produce the desired result of reducing the 'impurity' contribution. Yashima et al. |2| found no impurity contribution to the low-temperature susceptibility and were able to fit the measured values by $\chi(T) = \chi(0)\left[1+(T/T')^2\right]$ up to almost 30 K, with $\chi(0) = 54.4\times10^9$ m^3/mole and T' = 49 K. No details of their sample preparation procedure are known to us.

The magnetic isotherms at 4.2 and 1.4 K, as determined in the Amsterdam High-Field Magnet, are presented in fig. 2. From the saturating part, visible as a curvature in the isotherm and evaluated by extrapolating the straight, high-field part to B=0, we can estimate the impurity contribution at 0.8% trivalent Ce ions in the stoichiometric compound, in reasonable agreement with the figure derived from the low-temperature susceptibility upturn. After subtraction of this impurity contribution we find that the magnetization is an approximately linear function of the magnetic induction up to 20 T, with a slope of $\chi = 50.6\times10^9$ m^3/mole, deviating by 8% from the constant low-temperature susceptibility as obtained after correction for the 'impurity' contribution. This discrepancy is comparable with differences between results on different samples in the same magnetometer. The assumption that such differences are due to anisotropy was confirmed by experiments in which a sample was remeasured in the pendulum magnetometer after various rotations of the sample holder. The susceptibility was observed to change by amounts up to 18% upon reorientation. Changes in the demagnetizing factor or in the positioning of the sample can be safely excluded as sources of this effect. The need for magnetic measurements on single crystals is obvious, in fact, in order to produce a well-defined magnetic isotherm, a single crystal seems absolutely necessary.

5. HEAT CAPACITY

In fig. 3 the results of heat capacity measurements performed on $LaSi_2$ and $CeSi_2$ are shown. If we assume that the low-temperature upturn in the C/T vs T^2 curve of $CeSi_2$ is caused by trivalent Ce 'impurities', whose magnetic moments are blocked by some weak effective field, we arrive at an amount of 0.9%, reasonably close to the estimates based on magnetic measurements.

In the temperature range investigated, 1.4-28 K, there is no maximum in the heat capacity, which excludes magnetic ordering or crystal-field transitions in this temperature range. The electronic heat capacity coefficient of $LaSi_2$ can be simply determined to be (6 ± .5) mJ/K^2 mole La with θ_D = (252 ± 10) K. The low-temperature upturn hampers the direct determination of γ and θ_D for $CeSi_2$. In fig. 3 the high-temperature

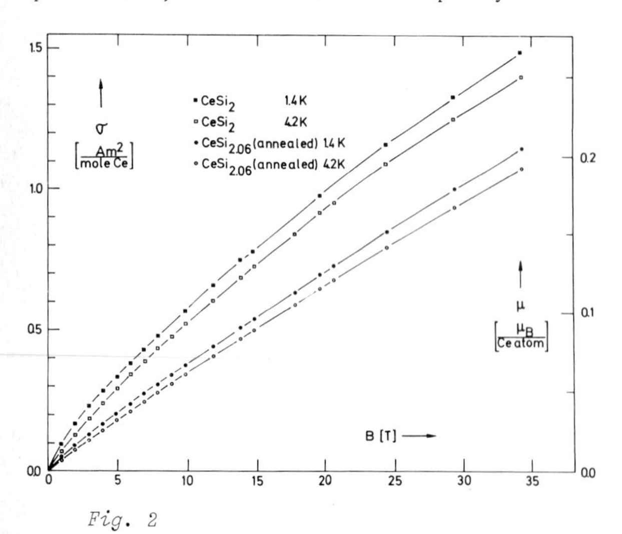

Fig. 2

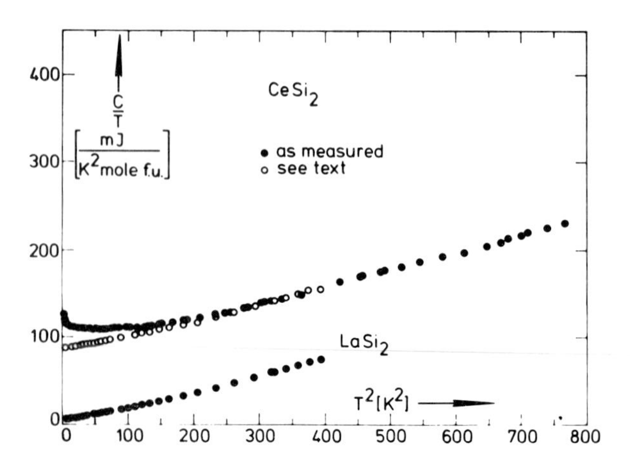

Fig. 3

(T > 20 K) results are extrapolated by shifting the $LaSi_2$ curve (open circles) in such a way that between 16 and 20 K it overlaps with the $CeSi_2$ curve (full circles). This extrapolation gives $\gamma = (87 \pm 1)$ mJ/K^2 mole Ce for $CeSi_2$ and assumes the same Debye temperature for both compounds.

6. ELECTRICAL RESISTIVITY

The electrical resistivities of $LaSi_2$ and $CeSi_2$ have been measured between 1.2 and 300 K and are presented in figs 4 and 5, respectively. A phonon contribution seems to give an adequate description of the temperature-dependent part of the resistivity of $LaSi_2$, which can be described by the Grüneisen-Bloch relation, but the Debye temperature $\theta_R = 425$ K is considerably higher than the θ_D value derived from the heat capacity.

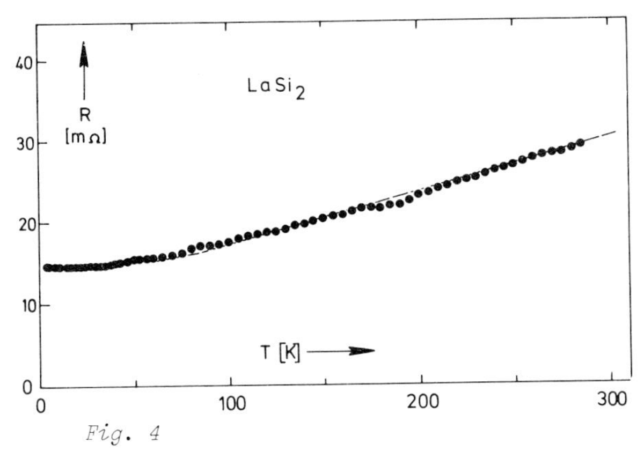

Fig. 4

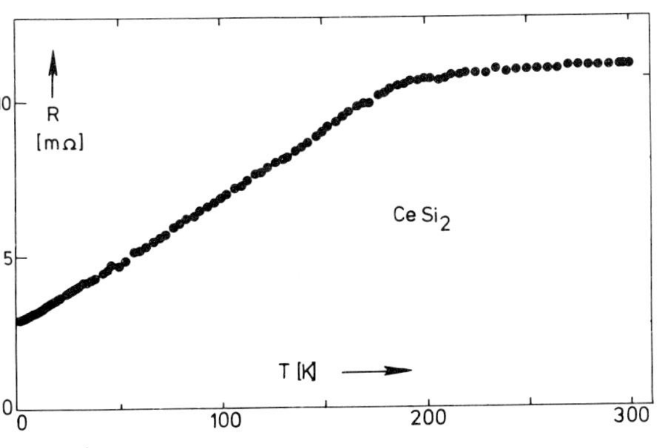

Fig. 5

The most striking feature of the resistivity of $CeSi_2$ (fig. 5) is that it decreases linearly down to the lowest measured temperature, reaching the value of 1.05×10^{-6} Ωm at 1.2 K, without any sign of a tendency to become constant at lower temperatures. The high-temperature slope is presumably due to electron-phonon scattering. The steep decrease below 200 K could indicate that the spin-disorder scattering is eliminated below this temperature. In view of the uncertainty in the cross section of the samples, due to microcracks, only resistances are given in figs 4 and 5.

7. THERMAL EXPANSION

Although the atomic volume of $CeSi_2$ at room temperature is not anomalous with respect to the other, isostructural rare-earth disilicides, in view of the indications from magnetic measurements of a valence instability, lattice-parameter

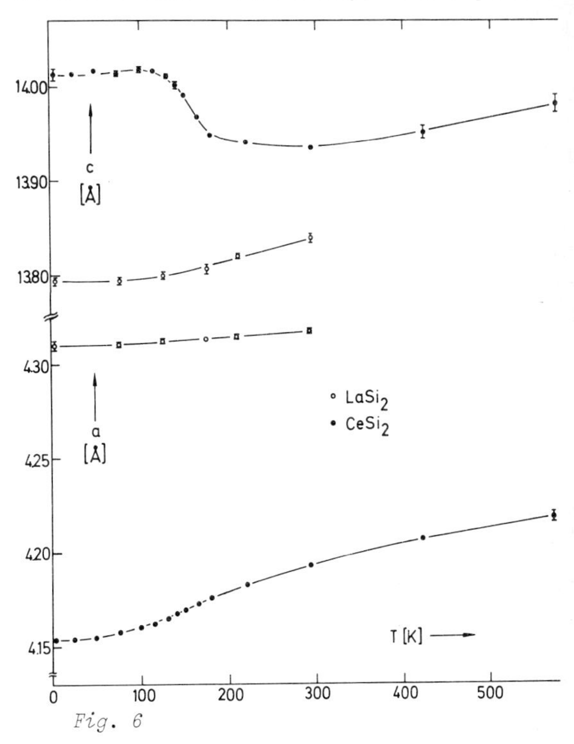

Fig. 6

measurements are of interest. We have measured the lattice parameters of $LaSi_2$ and $CeSi_2$ as functions of the temperature; the results are presented in fig. 6. The thermal expansion anomalies in the parameters a and c between 100 and 200 K do not cancel each other, so that a slight kink results in the temperature dependence of the molar volume. It is clear from fig. 6 that a more pronounced shape anomaly occurs in this temperature range.

8. Gd SUBSTITUTION

As the magnetic properties of $CeSi_2$ do not exclude itinerant magnetism (see section 9 below), the 'giant moment test' was carried out on this compound. A sample with 1% of the Ce substituted by Gd was prepared, which also proved to have the $ThSi_2$ structure, with a trace of a second phase. In the intermediate valent compound YbCuAl, Gd substitution produced an additional Curie-Weiss contribution to the susceptibility, without any indication of giant-moment formation |8|. Our results lead to the same conclusion for $CeSi_2$, although a strict additivity of the host and impurity susceptibilities could not be established. More precisely, the low-temperature susceptibility of $Ce_{0.99}Gd_{0.01}Si_2$ could be decomposed to give a constant plus a Curie-Weiss contribution, the latter corresponding to 0.9 at.% trivalent Ce 'impurities' plus 1.16 at.% Gd, assuming an effective moment of 7.94 μ_B/Gd atom. However, after subtraction of this Curie-Weiss contribution, the remaining susceptibility is 33% lower than that of $CeSi_2$ at low temperature, and gives a higher effective moment than $CeSi_2$ has in the 150-700 K temperature range, where the Curie-Weiss law is obeyed. Despite these difficulties, we consider these measurements conclusive in the sense that they give clear evidence for the absence of giant moments in $Ce_{0.99}Gd_{0.01}Si_2$.

9. DISCUSSION

The results described in the present paper show that $CeSi_2$ shares some properties with mixed-valence compounds, most notably

- (i) a breakdown of the Curie-Weiss law at low temperatures without any indication of ordering;
- (ii) a very high electronic specific-heat coefficient.

On the other hand, unlike in most mixed-valence compounds, we find here

- (iii) no marked anomaly in the atomic volume;
- (iv) no maximum in the susceptibility.

That the latter feature is not an artifact of our correction procedure is clear from the fact that Yashima et al. |2| have been able to determine the negative coefficient of the quadratic temperature dependence of the as-measured susceptibility. A further deviation from the behaviour of known mixed-valence systems is

- (v) an anomalously large R ratio.

The R ratio has been defined by Newns and Hewson |9| as $R = \pi^2 k_B^2 \chi(0)/\gamma g^2 \mu_0 \mu_B^2 J(J+1)$; its value should be 1 for mixed-valence systems and 1.2 for Kondo systems involving Ce. Yashima et al. |2| found R = 1.41, our γ value being lower, we find R = 1.65. The local Fermi liquid approach |9| fails here to account for the temperature dependence of the susceptibility, which is not surprising, in view of (iv).

While the features (iii) to (v) distinguish $CeSi_2$ from most mixed-valence compounds, all observations (i) to (v) are consistent with itinerant-electron magnetism. An interpretation of the results in terms of itinerant electrons gains further support from an Arrott-plot representation of the high-field magnetization, which, after correction for trivalent Ce "impurities' results in a straight line |1|.

The measurements on Gd-substituted $CeSi_2$ described in section 8 are, of course, not conclusive regarding the itinerant character of the magnetic carriers in $CeSi_2$. The absence of giant moments can be due to a lack of coupling between the Gd moments and the matrix, and does not necessarily exclude the long-range polarisability characteristic of exchange-enhanced itinerant-electron paramagnets.

The constancy of the low-temperature susceptibility may be associated with mixed valency or itinerant-electron paramagnetism (which are not necessarily mutually exclusive categories). Whichever is the case, it is clear that the characteristic temperature in this context is much lower than the paramagnetic Curie temperature. The latter is unlikely to be directly related to the low-energy excitations responsible for the high γ value and the quadratic temperature dependence of χ that prevails up to 30 K |2|. It is more natural to associate the high value of θ and the deviations from the Curie-Weiss law that set in at a temperature as high as 200 K (fig. 1) with two other anomalies at elevated temperatures: the strong temperature dependence of the resistivity (fig. 5) and the anomalous thermal expansion (fig. 6). As the latter amounts to only a minor volume anomaly, while it is accompanied by a pronounced shape effect, crystal fields are more likely to be responsible for it than a valence instability. The strong indication we found for a substantial anisotropy in the low-temperature magnetization and the peculiarities of the crystal structure spelled out in section 2 support this assumption.

In conclusion, at low temperatures (below 30 K) $CeSi_2$ shows some of the characteristics of mixed valency and can also be seen as an itinerant-electron paramagnet. On the other hand, crystal field effects appear to play an essential role between 100 and 200 K. The presence of two well separated energy scales is a unique feature of this material among the known mixed-valent Ce compounds. If our conclusion that crystal fields play an important role above 100 K is correct, $CeSi_2$ should be an ideal candidate for a study of the interplay between crystal field effects and valence instability.

REFERENCES

|1| Dijkman, W.H., thesis, University of Amsterdam, 1982.

|2| Yashima, H., Satoh, T., Mori, H., Watanabe, D. and Ohtsuka, T., to be published in Solid State Communications.

|3| Satoh, T. and Asada, Y., J.Phys.Soc. Japan 27 (1969) 1463.

|4| Sawada, A., and Satoh, T., J. of Low Temp. Phys. 30 (1978) 455.
Satoh, T. and Asada, Y., J.Phys.Soc. Japan 28 (1970) 263.

|5| Iandelli A. and Palenzona A. in Handbook on the Physics and Chemistry of Rare Earths (Gschneidner Jr., K.A. and Eyring, L. eds) North-Holland, 2 (1979) 1-55.

|6| Ferro Ruggiero, A. and Olcese, G.L., Atti. Accad. Naz. Lincei 37 (1964) 169.

|7| Dijkman, W.H., De Boer, F.R., De Châtel, P.F. and Aarts, J., J. Magn. Magn. Mat. 15-18 (1980) 970.

|8| Mattens, W.C.M., De Châtel, P.F., Moleman, A.C. and De Boer, F.R., Physica 96B (1970)138.

|9| Newns, D.M. and Newson, A.C., J. Phys. F 10 (1980) 2429

Valence Instabilities
P. Wachter and H. Boppart (eds.)
© North-Holland Publishing Company, 1982

MAGNETIC PROPERTIES AND GROUND STATE IN $CeMg_3$ AND $CeInAg_2$

R.M. Galera*, J. Pierre* and A.P. Murani†

*Laboratoire Louis Néel, C.N.R.S., 166X, 38042 Grenoble-cédex, France
†Institut Laue-Langevin, 156X, 38042 Grenoble-cédex, France

Susceptibility and electrical resistivity measurements, neutron diffraction and inelastic neutron scattering experiments have been performed on RMg_3 and $RInAg_2$ compounds. They show evidence that $CeMg_3$ and $CeInAg_2$ are Kondo systems. A comparison is made with other compounds, such as CeMg.

Rare earth intermetallics RMg_3 and $RInAg_2$ have crystal structures of DO_3 type and $L2_1$ type (Heusler phases) respectively. They may be derived from RMg or RAg compounds (B_2 or CsCl phases) by replacing half of the rare earth (R) atoms by Mg or In respectively. Lattice parameters for Ce compounds are in agreement with a trivalent state (Table I) (1,2).

The magnetic susceptibility of $CeMg_3$ and $CeInAg_2$ follows a Curie-Weiss law above 70 and 5 K respectively, with a Curie constant close to that of trivalent cerium (1). Deviations are observed at lower temperatures for $CeMg_3$, due to crystal field splitting Δ = 190 K as measured by inelastic neutrons cattering. Both compounds have a Γ_7 doublet ground state, whereas CeMg and CeAg have the Γ_8 ground state (3). Fourth order crystal field (CEF) parameters are of larger magnitude than in corresponding Pr and Nd compounds (2).

Table I : Comparison between properties of $CeMg_3$, $CeInAg_2$ and CeMg.

	CeMg	$CeMg_3$	$CeInAg_2$
a (Å)	3.901	7.444	7.108
T_N (K)	19.5	4.1	2.7
CEF splitting Δ (K)	190	195	18
Ground state	Γ_8	Γ_7	Γ_7
$j\,n(E_F)$ from ρ	-0.045	-0.04	-0.08
$\lvert j_{ex}\,n(E_F)\rvert$ from γ_{qe}	0.3	0.3	-
μ_{AF} (μ_B) at T (K)	1.85 (4.2 K)	0.59 (2 K)	0.97 (2.2 K)

$CeMg_3$ and $CeInAg_2$ order antiferromagnetically at 4.1 and 2.7 K respectively. The molecular field constants J(0) for spin-spin interactions could be deduced from the analysis of the susceptibility, they are found 3 to 5 times larger than for corresponding Pr or Nd compounds (2,4). Similarly, in the ordered antiferromagnetic state, the interactions J(Q) deduced from the Néel temperature are much larger than for Nd compounds ($PrMg_3$ and $PrInAg_2$ do not order due to a non-magnetic CEF ground state).

The magnetic contributions to the electrical resistivity have been obtained by subtracting the resistivity of La compounds. They cannot be interpreted in terms of the spin-disorder resistivity for a normal rare earth. Instead, they decrease as ln(T) at high temperature which is a typical feature of the Kondo effect. The same behaviour was also observed for CeMg (2). No resistivity minimum occurs for these concentrated compounds ; however, a minimum is observed at 25 K for $Y_{.8}Ce_{.2}Mg$, together with a very high residual resistivity (65 μΩxcm at 4.2 K), which indicates a strong scattering of conduction electrons by Ce atoms. Thus we anticipate that $CeMg_3$, $CeInAg_2$ and probably CeMg also could be dense Kondo systems. The analysis of the resistivity has been performed following the theory of Cornut and Coqblin (5) for a Kondo impurity in presence of crystal field effects. It enables to evaluate the factor $j\,n(E_F)$, where j is the 4f-band interaction in the Coqblin-Schriffer Hamiltonian. $j\,n(E_F)$ ranges from -0.04 for $CeMg_3$ to -0.08 for $CeInAg_2$ (Table I).

The specific heat has been measured on $LaMg_3$, $CeMg_3$, $NdMg_3$ at the "Service des Basses Températures" (Nuclear Center of Grenoble) (figure 1). The Debye temperature does not change appreciably between La and Ce, which justifies the separation of the magnetic resistivity by simple subtraction of $LaMg_3$ contribution. The electronic contribution to the specific heat $(C_v/T)_o$ is 3.7 $mJ/mole.K^2$ for $LaMg_3$, and could not be determined for Ce and Nd, as the experiments were limited to T > 1.8 K. Previous measurements (3) gave $(C_v/T)_o$ = 35 $mJ/mole.K^2$ for CeMg, a much larger value than for LaMg (6.8 $mJ/mole.K^2$).

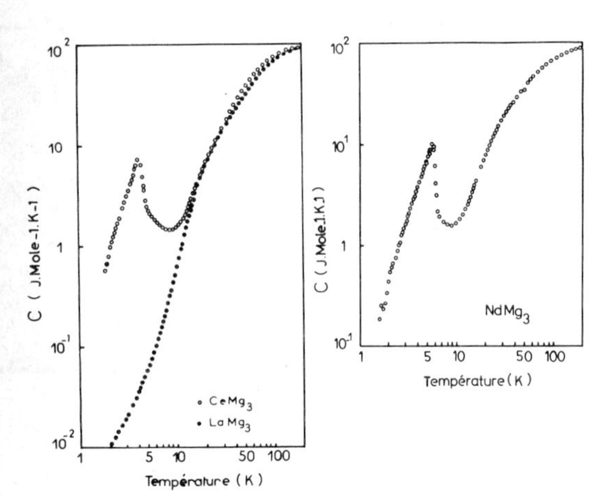

Figure 1 : Specific heat for $LaMg_3$, $CeMg_3$ and $NdMg_3$.

The specific heat for $CeMg_3$ and $NdMg_3$ have a λ-type anomaly at T_N = 4.1 and 6 K respectively. For $CeMg_3$, the magnetic entropy at T_N is only 0.7 x R ln (2), the full entropy of the doublet being recovered at about 25 K. The tail of the magnetic specific heat above T_N can be fitted by a Schottky-type anomaly corresponding to a splitting $\delta \simeq 6$ K for the Γ_7 state. It may be attributed either to short range order or rather to the Kondo coupling, as thought to be the case in $CeAl_2$ (6).

We have determined the magnetic structures of $CeMg_3$, $NdMg_3$ and $CeInAg_2$ by powder neutron diffraction at the Institut Laue-Langevin (I.L.L.) (4). For the Ce f.c.c. sublattice, AF order occurs for both compounds within the chemical cell, with a propagation vector Q = (0, 0, 1) 2Π/a, whereas that for NdMg3 is Q = (1, 1, 1) Π/a, and $NdInAg_2$ has some kind of modulated structure. This change in magnetic order indicates a change in nature of indirect interactions for cerium and neodymium compounds. The observed moment of Ce in CeMg3 (0.59 μ_B at 2 K) is significantly lower (figure 2) than the theoretical value expected from the Γ_7 ground state, which for Ce may be due to the Kondo effect.

We have studied the relaxation of quasielastic magnetic excitations for CeMg (3), $La_{.8}Ce_{.2}Mg$, $Y_{.8}Ce_{.2}Mg$ at T < 300 K and for $CeMg_3$ between 8 and 800 K (figure 3).

For all compounds, the quasielastic linewidth γ_{qe} tends towards a finite value γ_o as $T \rightarrow 0$ (Table II). This residual linewidth appears to be independent of indirect magnetic interactions, considering the different dilutions and ordering temperatures of Ce ions. It has also the same order of magnitude as for other Kondo compounds like $CeAl_2$ (7) or $CeAl_3$ (8). Its value seems representative of the Kondo temperature. Let us note that this width is in agreement with the "splitting" observed by specific heat measurements.

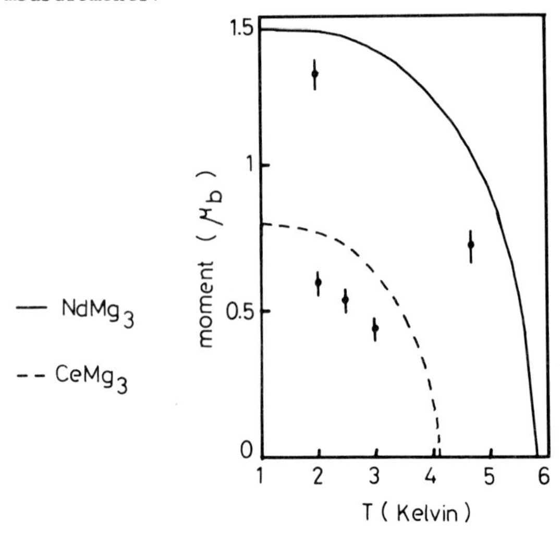

Figure 2 : Temperature dependence of the ordered moment for $CeMg_3$ and $NdMg_3$. The solid and dashed curves give the values calculated within a molecular field model.

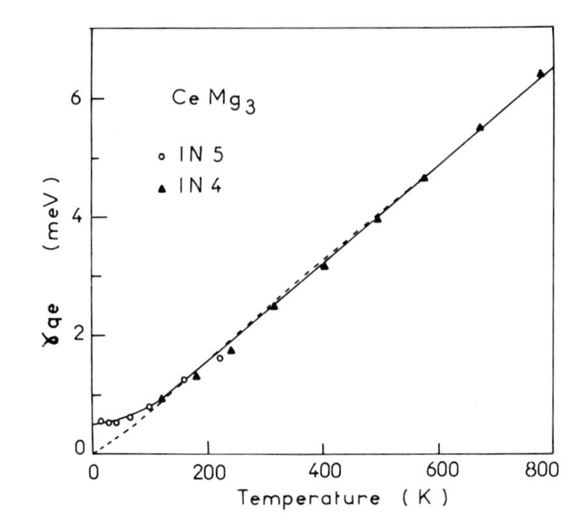

Figure 3 : Temperature dependence of the quasielastic linewidth γ_{qe} for $CeMg_3$. The dashed line is a fit using Becker et al theory (9) with $|j_{ex}\, n(E_F)| = 0.3$.

For $CeMg_3$, γ_{qe} increases linearly with temperature above 200 K. The fit of this linewidth (dashed line on figure 3) has been obtained using expressions derived by Becker et al (9) for the damping of excitations due to the coupling with conduction electrons, taking care of transitions within the Γ_7 and Γ_8 crystal field states. The theory has been developed for the case of a normal 4$\underline{f}$-$\underline{c}$onduction band coupling, where $\mathcal{H} = -(g-1)\, j_{ex}\, \vec{S}.\vec{s}$. The best fit gives

Table II : Residual quasielastic linewidths γ_{qe} and ordering temperatures T_N for some cerium compounds.

Compounds	γ_{qe} (0) (meV)	T_N (K)	Reference
CeMg	0.55	19.5	(3)
$La_{.8}Ce_{.2}Mg$	0.6	< 5	This work
$Y_{.8}Ce_{.2}Mg$	0.8	< 5	This work
$CeMg_3$	0.5	4.1	This work
$CeAl_2$	0.5	3.8	(6)
$CeAl_3$	0.5	-	(7)

$|j_{ex}\, n(E_F)| \simeq 0.3$, which is larger than obtained from resistivity analysis within the Cornut-Coqblin theory (5). We note however that the starting Hamiltonian is not the same for the two cases, and the theory of relaxation (8) may not take into account all the scattering channels which are present with the Kondo model (5). Finally, spin fluctuations are completely suppressed in the ordered range, where we observe for $CeMg_3$ a magnon branch with a strong dispersion.

In conclusion, $CeMg_3$ and $CeInAg_2$ have the features of dense Kondo systems. In particular, $CeMg_3$ behaves very similarly to $CeAl_2$.

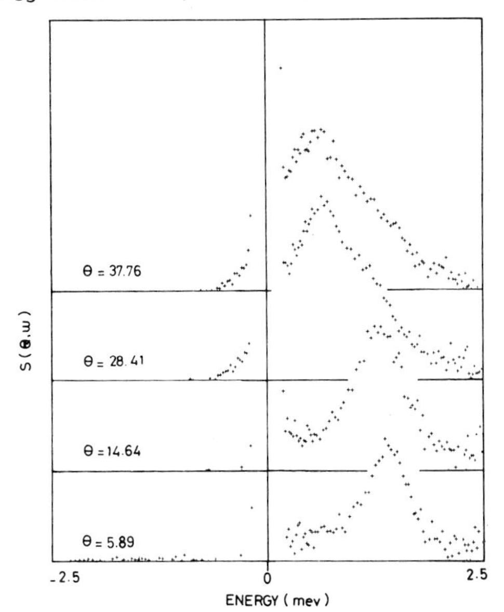

Figure 4. Inelastic spectra for $CeMg_3$ at 2K showing dispersion of magnons.

REFERENCES

1. Pierre, J, Murani AP and Galera RM, J. Phys. F. 11 (1981) 679.
2. Galera RM, Murani AP and Pierre J, J. Magn. Magn. Mat. 23 (1981) 317.
3. Pierre and Murani AP, in Crow JE et al (eds.), Crystalline Electric Field and Structural Effects in f Electrons Systems (Plenum Press, 1980).
4. Galera RM, Pierre J and Pannetier J, J. Phys. F (1982).
5. Cornut B and Coqblin B., Phys. Rev. B5 (1972) 4541.
6. Aarts J, de Boer FR, Horn S, Steglich F, and Meschede D, in Falicov LM et al (eds.), Valence fluctuations in solids (North Holland Publishing Co, 1981).
7. Loewenhaupt M and Holland-Moritz E, J. Appl. Phys. 50 (1979) 7456 and J. Mang. Magn. Mat. 14 (1979) 227.
8. Murani AP, Knorr K, Buschow KHJ, Benoît A and Flouquet J, Sol. Stat. Commun. 36 (1980) 523.
9. Becker KW, Fulde P and Keller J, Z. Physik B 28 (1977) 9.

Valence Instabilities
P. Wachter and H. Boppart (eds.)

EVOLUTION OF MAGNETISM IN $TmSe_{0.6}Te_{0.4}$ UNDER PRESSURE

B. Batlogg,[1] H. Boppart,[2] E. Kaldis,[2] D. B. McWhan[1]
and P. Wachter[2]

[1]Bell Laboratories, Murray Hill, New Jersey
[2]Laboratorium für Festkörperphysik ETH, Zürich, Switzerland

The ac susceptibility and the volume change of $TmSe_{0.6}Te_{0.4}$ have been measured across the pressure-induced first-order valence transformation ($p \sim 7$ kbar at 4K). In the mixed valence phase above 7 kbar, the magnetic ordering temperature T_0 decreases with $dT_0/dp = -0.5$K/kbar. A general magnetic phase diagram of T_0 vs valence is deduced for Tm chalcogenides. With increasing valence mixing, T_0 first decreases rapidly as a result of moment destabilization and then increases below a valence of 2.8 and peaks around 2.5. This is ascribed mainly to double-exchange coupling and moment growth in the strongly mixed valent regime.

INTRODUCTION

The continued effort to understand the Tm-based mixed valence compounds is based on the unique situation that both ionic configurations are magnetic. This results in a sequence of magnetic structures with increasing Tm valence from ferromagnetic ($TmSe_{.8}Te_{.2}$)(1) to type I antiferromagnetic (TmSe)(2) to type II antiferromagnetic (TmS)(3). Recent theoretical studies suggest that the separation in energy of the singlet ground state from the first excited magnetic state is much smaller if both ionic configurations are magnetic than if one is nonmagnetic.(4) To gain insight into the mechanism which determines the magnetic state, the Tm valence has been varied chemically by changing the stoichiometry of TmSe(5,6,7) or by alloying TmSe with TmTe or TmS.(1,8,9) As there are always unknown chemical effects, e.g. vacancy concentration or fine-scale segregation, the more direct way to study the effect of valence on the magnetic properties is by the application of pressure as has been done in TmSe and TmS. (10-14) In this paper we present the results of ac susceptibility measurements on $TmSe_{0.6}Te_{0.4}$. This alloy undergoes a first order transition at P=7 kbar from a nonmagnetic to an apparently ferromagnetic phase. We combine the results of chemical and pressure studies to show that the magnetic ordering temperatures of all Tm based mixed valence compounds fall on a single curve when plotted against Tm valence with a deep minimum at a valence of $V \sim 2.8$. This minimum suggests a competition between ferromagnetic intersite coupling and moment destabilization with the former dominating at V=2.5 and the latter at $V \sim 3$.

EXPERIMENTS

The ac magnetic susceptibility was measured with a traditional phase sensitive technique at $\sim$ 95 Hz with the ac field being $\sim$ 8 Oe. The coils containing the sample were pressurized in a BeCu clamp with a Teflon cell and isoamyl alcohol was used as pressure transmitting medium. At room temperature the pressure was determined by a Manganin gauge and the superconducting transition of Sn served as a pressure calibrant below 4K. The pressure dependence of the volume was determined at RT using the strain-gauge technique.

The single crystals of $TmSe_{0.6}Te_{0.4}$ transform discontinuously from a semiconductor to a metal at $\sim$ 4 kbar and RT. The associated volume collapse amounts to 5.5% and the p-V curve exhibits a very pronounced rounding (high compressibility) in both phases.

The ac susceptibility, χ, at various pressures is shown in Fig. 1. The base line is shifted proportional to the pressure so that the evolution of $\chi(T)$ can easily be followed. Below p=6.8 kbar the susceptibility is very small (few percents of the scale in Fig. 1) and slightly

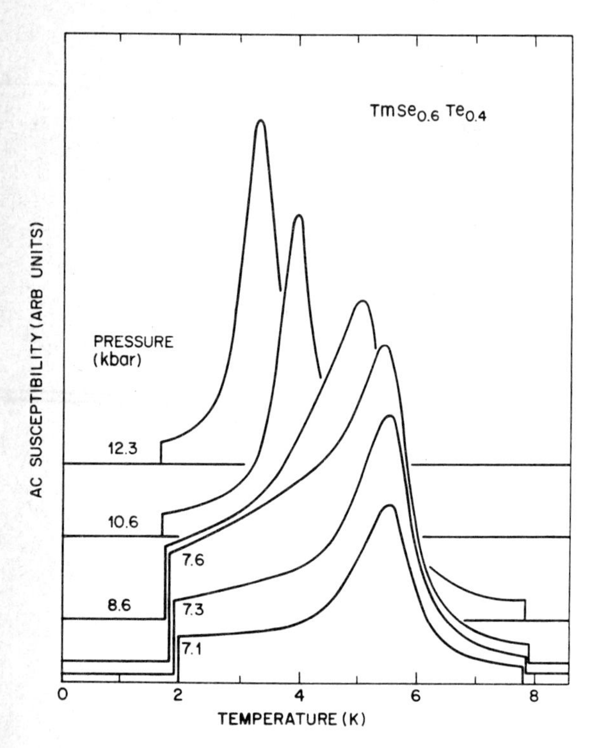

Fig. 1 Evolution of the ac susceptibility of $TmSe_{0.6}Te_{0.4}$ as the valence mixing is varied by pressure.

increases on decreasing temperature, indicating an ordering point well below 1.6K, the lowest temperature reached in this experiment. Raising the pressure by only 0.3 kbar to 7.1 kbar results in a drastic change of $\chi(T)$ which we conclude to be caused by the same phase transformation as observed at RT and P $\sim$ 4 kbar. The peak in $\chi(T)$ remains at 5.6K up to 7.6 kbar and then shifts to lower temperature at higher pressures, reaching 3.25K at 12.3 kbar. The shape of the $\chi(T)$ curve in the range from 7 to 8.6 kbar is similar to those observed at 1 atm. in the ferromagnetic TmSe-TmTe samples. There is a change in the shape of the $\chi(T)$ curves at higher pressures and the comparison with other Tm based alloys given below leads us to speculate that with increasing pressure there may be a sequence from weakly-magnetic (ordering temp. < 1K) to ferromagnetic to antiferromagnetic ground states in $TmSe_{0.6}Te_{0.4}$.

The variation of the ordering temperature as a function of pressure is shown in detail in Fig. 2 (for simplicity we take the peak temperature as the ordering point). Just after the transition we observe the highest ordering temperature for any mixed valent Tm compound. The pressure dependence of T_N in TmSe is also shown in Fig. 2. The reason for the different behavior is the presence of a small energy gap at E_F in TmSe,(15-17) which apparently requires higher pressures (> $\sim$ 25 kbar) to induce a valence change at low T. (12,13,15)

A GENERALIZED MAGNETIC PHASE DIAGRAM

In the following we present for the first time a magnetic phase diagram for a concentrated Tm system which demonstrates the interplay between strong valence mixing and magnetism. Included is also the regime of "almost pure" trivalency.

Again the primary parameter is the degree of valence mixing, which in principle can be deduced from various experimental quantities. We are well aware of the controversy about the reliability of the various methods. In the present study, however, which involves comparison of data from several sources, the lattice parameter turned out to be the only generally available valence-related quantity. Accordingly we have derived the valence of Tm from the

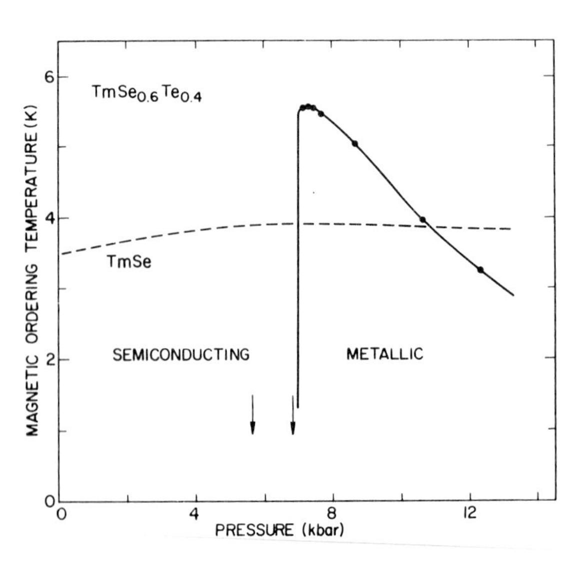

Fig. 2 Pressure dependence of the magnetic ordering temperature in $TmSe_{0.6}Te_{0.4}$ and TmSe.

lattice parameter and found exceptionally good agreement between the different sets of data. The pressure-volume measurements on $TmSe_{0.6}Te_{0.4}$ indicate that the first order transition ends at a valence of 2.53.(18) The valence gradually increases under pressure and is 2.74 at 12.3 kbar. In Fig. 3 the magnetic ordering temperature is shown vs the valence. The full line connects the points determined in this experiment. The broken line represents a smooth curve through all the other which have been obtained over the last years by several research groups working on Tm_xSe, TmSe-TmTe and TmS-TmSe.(2-6) The scatter of the data is remarkably small up to a valence of ∿ 2.9. In particular the ordering temperatures of Tm_xSe with x being close to 1 lie just on the extrapolation of the full line for $TmSe_{0.6}Te_{0.4}$ under pressure. As the trivalent limit is approached, some uncertainty in the determination of the valence is encountered. Nevertheless, it becomes clear that the ordering temperature rises steeply close to trivalency. This is further emphasized by the arrows which indicate the direct pressure measurement of T_N for TmS.(10,14) The overall behavior is characterized by two features: (a) a rapid drop of T_0 as the mixed valence regime is entered from the trivalent border and (b) a gradual increase of T_0 as the valence is further reduced below ∿ 2.85, leading to a broad maximum around v ∿ 2.5. We ascribe the initial drop of T_0 to a combined effect of a reduction of both the moments and the coupling between them via the conduction electrons. This is due to the 4f level approaching the Fermi energy. In the strongly mixed valent regime, the real charge transfer between the 4f shells provides an additional, ferromagnetic intersite coupling. This "double-exchange"(19) is strongest for maximum valence mixing. The observed increase in T_0 and even the shape of the T_0-v curve are in agreement with this mechanism.

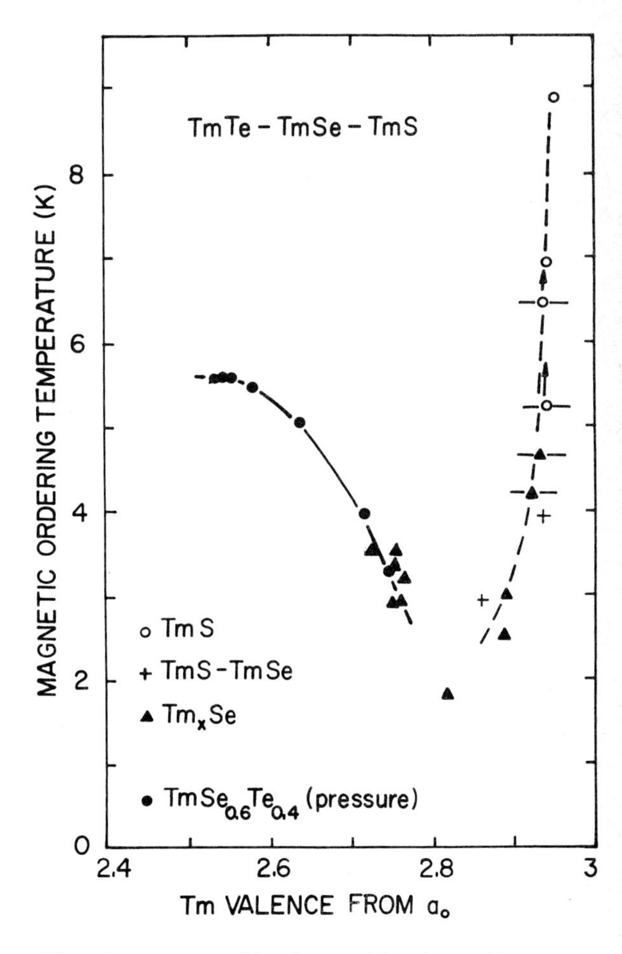

Fig. 3 A generalized magnetic phase diagram for mixed valent Tm chalcogenides.

REFERENCES

[1] Batlogg, B., Ott, H.R., and Wachter, P., Phys. Rev. Lett. 42, 278 (1979).

[2] Bjerrum Møller, H., Shapiro, S.M., and Birgeneau, R.J., Phys. Rev. Lett. 39, 1021 (1977).

[3] Koehler, W.C., Moon, R.M., and Holtzberg, F., J. Appl. Phys. 50, 1975 (1979).

[4] Yafet, Y., and Varma, C.M., Bull. Am. Phys. Soc. 27, 277 (1982).
Varma, C.M., Schlüter, M., and Yafet, Y., these proceedings.

[5] Batlogg, B., Ott, H.R., Kaldis, E., Thöni, W., and Wachter, P., Phys. Rev. B19, 247 (1979).

[6] Holtzberg, F., Penney, T., and Tournier, R., J. de Physique 40, C5-314 (1979).

[7] Haen, P., and Lapierre, F., in Valence Fluctuations in Solids, L. M. Falicov, W. Hanke and M. B. Maple, eds. (North Holland, Amsterdam, 1981).

[8] Batlogg, B., Phys. Rev. B23, 650 (1981).

[9] Köbler, U., Fischer, K., Brickmann, K., and Lustfeld, H., J. Magn. Magn. Mat. 24, 34 (1981).

[10] Guertin, R.P., Foner, S., and Missell, F.P., Phys. Rev. Lett. 37, 529 (1976).

[11] Missell, F.P., Foner, S., and Guertin, R.P., in "Val. Instab. and Rel. Narrow Band Phen." Parks, R.D. (ed.) Plenum Press, New York, 1977, p. 491

[12] Chouteau, G., Holtzberg, F., Peña, O., Penney, T., and Tournier, R., J. de Physique 40, C5-361 (1979).

[13] Vettier, C., Flouquet, J., Mignot, J.M., and Holtzberg, F., J. Magn. Magn. Mat. 15-18, 987 (1980).

[14] Lapierre, F., Haen, P., Cogblin, B., Ribault, M., and Holtzberg, F., Proc. LT16.

[15] Ribault, M., Flouquet, J., Haen, P., Lapierre, F., Mignot, J.M., and Holtzberg, F., Phys. Rev. Lett. 45, 1295 (1980).

[16] Frankowski, I. and Wachter, P., Solid State Comm. 41, 577 (1982)

[17] Güntherodt, G., Thompson, W.A., Holtzberg, F., and Fisk, Z., these proceedings

[18] Boppart, H., and Wachter, P., to be published

[19] Varma, C.M., Solid State Comm. 30, 537 (1979)

Valence Instabilities
P. Wachter and H. Boppart (eds.)

PRESSURE DEPENDENCE OF THE MAGNETIC SUSCEPTIBILITY OF SOME INTERMETALLIC Ce-COMPOUNDS

W. Zell, K. Keulerz, P. Weidner, B. Roden and D. Wohlleben

II. Physikalisches Institut
der Universität zu Köln
5000 Köln 41
W. Germany

The pressure dependence of the magnetic susceptibility is reported for $CePt_2$, CeRhPt, $Ce(Rh_{0.7}Pt_{0.3})_2$, $CeRh_2$, $CePd_3$, $CeSn_3$, $CeBe_{13}$ and $CeNi_2Ge_2$ at 300 K, and between 300 and 4 K for CeRhPt, $Ce(Rh_{0.7}Pt_{0.3})_2$ and $CeNi_2Ge_2$. According to this pressure dependence all systems show intermediate valence except $CePt_2$. $d\ln\chi/dp$ strongly increases with decreasing temperature but saturates at the fluctuation temperature.

In the $Ce(Rh_xPt_{1-x})_2$ system volume reduction by applied pressure has nearly ten times less effect on the susceptibility than volume reduction by alloying.

INTRODUCTION

The pressure dependence of the susceptibility of intermediate valent rare earth-systems is very large compared to that of other paramagnetic metals. Pressure favours the configuration with the smaller volume ($4f^n$ + e) over the one with the larger volume ($4f^{n+1}$). This has two consequences: First, electrons are transferred from the 4f-shell into the conduction-band and therefore the valence is increased, and second, the configuration $4f^{n+1}$ is destabilized.

In the case of Ce $4f^{n+1}$ is the magnetic $4f^1$ configuration. Therefore pressure decreases the number of magnetic electrons and increases the magnetic fluctuation temperature in Ce systems; Both effects decrease the magnetic susceptibility. The same reasoning leads to an increase of the susceptibility of Yb-systems under pressure. The sign of the pressure dependence is indeed as predicted for both types of ions: A negative $\partial\ln\chi/\partial p$ has been reported for γ-Ce, α-Ce (1) and $CeSn_3$ (2), and a positive $\partial\ln\chi/\partial p$ for a number of Yb-intermetallics (3,4,5). The magnitude of $\partial\ln\chi/\partial p$ gives information on the rate of change of the valence and of the fluctuation temperature. For a number of Yb-compounds this quantity shows a maximum at valence 2.5 at room temperature with a strong tendency to go to zero at integral valences (5). This behaviour suggests that near half integral valence, the valence change dominates $d\ln\chi/dp$ because electron transfer by pressure is most efficient there. Recently it was suggested on the basis of L_{III}-absorption measurements (6) that tetravalent Ce compounds do not exist and that the valence of most Ce intermetallic compounds is less than 3.3. If this were true large values of $\partial\ln\chi/\partial p$, comparable to those of Yb-compounds near valence 2.5 should not occur. On the other hand Ce compounds which are traditionally thought to be tetravalent should show appreciable pressure dependence since their valence would be intermediate while they should show no pressure dependence if they were tetravalent. Thus the pressure dependence of the susceptibility provides particularly important information on the question of the nonexistence of tetravalent Ce-compounds.

EXPERIMENTAL PROCEDURE AND RESULTS

The samples were molten in an arc-furnace or an induction-furnace. Weight losses were less than 0.5 at %. Only single phases could be detected in the x-ray diffraction patterns. The measurements were performed in a Faraday magnetometer at temperatures between 1,5 and 300 K and in magnetic fields up to 22 T. For the measurements under pressure the samples were powdered, mixed with teflon powder and enclosed in a teflon cylinder, which was placed inside a small pressure clamp made from CuBe. For details see (7,8,9).

Fig. 1 shows the pressure dependence of the susceptibility of $CeBe_{13}$ at room temperature. The values refer to the total susceptibility of $CeBe_{13}$. Note the linearity within experimental scatter and the reversibility of the data. This curve is exemplary for all other investigated compounds. In Table 1 we give the starting value of the total susceptibility, the structure, the value of the susceptibility of the reference and the relative pressure dependence of the susceptibility, $\partial\ln\chi/\partial p$. This quantity was obtained by dividing the slope of curves like the one in Fig. 1 by the difference between the starting susceptibility and the reference susceptibility. This procedure assumes that the observed dependence is entirely due to that of the 4f-shell. The relative errors given in Table 1 for $\partial\ln\chi/\partial p$ are due to measuring statistics in all cases except for $CeNi_2Ge_2$ where the signal scattered as function of pressure outside of experimental error, in a manner very similar to our findings on YbCuAl (5).

Isobaric susceptibilities as function of temperature are shown in Figs. 2-4. Fig. 2 shows the temperature dependence of the susceptibility of $CeNi_2Ge_2$ at ambient pressure and at 1.2 GPa.

Table 1

substance	structure	χ(300 K) (10^{-3} emu/mole)	reference	$\chi_{ref.}$(300 K) (10^{-3} emu/mole)	$-d(\ln\chi)/dp$ (GPa^{-1})	B(GPa)	$\frac{d(\ln \chi)}{d(\ln V)}$
$CePt_2$	$MgCu_2$	2.32	$LaPt_2/HfPt_2$	-0.032	0.000 ± 0.002	-	-
CeRhPt	$MgCu_2$	1.99	LaRhPt	0.157	0.038 ± 0.002	-	-
$Ce(Rh_{.7}Pt_{.3})_2$	$MgCu_2$	1.59	-	-	0.060 ± 0.005	-	-
$CeRh_2$	$MgCu_2$	0.82	$LaRh_2/HfRh_2$	0.061	0.050 ± 0.005	70(18)	3.5
$CePd_3$	Cu_3Au	1.2	YPd_3(15)	-0.061	0.043 ± 0.004	102.5 (19)	4.4
$CeSn_3$	Cu_3Au	1.6	$LaSn_3$(16)	0.2	0.047 ± 0.005	54.3 (19)	2.6
					0.09 ± 0.04(2)		4.3
$CeBe_{13}$	$NaZn_{13}$	1.76	$LaBe_{13}$(17)	0.024	0.097 ± 0.002	88.5(21)	8.6
$CeNi_2Ge_2$	$ThCr_2Si_2$	2.49	$LaNi_2Ge_2$	0.28	0.070 ± 0.006	42 (20)	3
α-Ce	fcc	0.56(1GPa)	-	-	0.2 ± 0.05 (1)	>40(24)	>8
γ-Ce	fcc	2.25	-	-	0.1 ± 0.03 (1)	24(22)	2.4

The pressure was applied at room temperature and relaxed somewhat during the cool down procedure because of the differential contractions of the different clamp materials. In every case the pressure was measured at He-temperature by superconductive manometry, i.e. by inductively measuring the transition temperature of a small piece of tin enclosed in the pressurized cylinder. A pressure reduction of 0.3 to 0.35 GPa was found.

Also shown in Figs. 2-4, is the temperature dependence of $\partial \ln\chi/\partial p$. These curves were calculated from the "isobars" at p=0 and at $0.9 < p < 1.2$ GPa by assuming linearity of the pressure dependence of the susceptibility (as seen directly in Fig. 1), and that the pressure reduction inside the clamp during the cool down period in the high pressure run is a linear function of the temperature. The relative pressure dependence was calculated by dividing $\Delta\chi/p(T)$ by the difference between the zero pressure susceptibility and the susceptibility of the reference also shown in Figs. 2 and 3, except for $Ce(Rh_{0,7}Pt_{0,3})_2$ where no reference was measured.

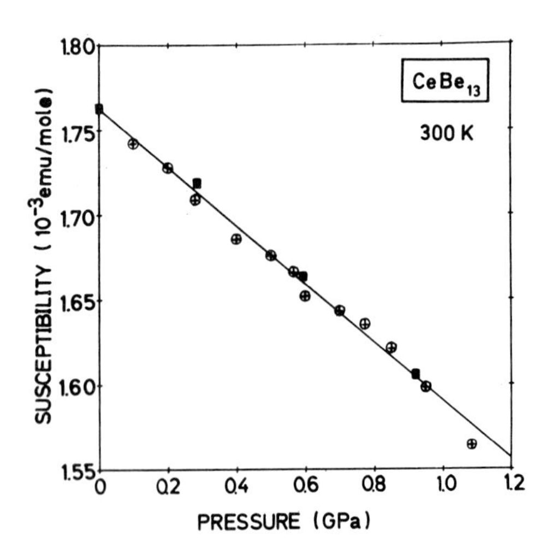

Fig. 1. Room temperature susceptibility of $CeBe_{13}$ vs pressure
(⊕ increasing, ■ decreasing pressure)

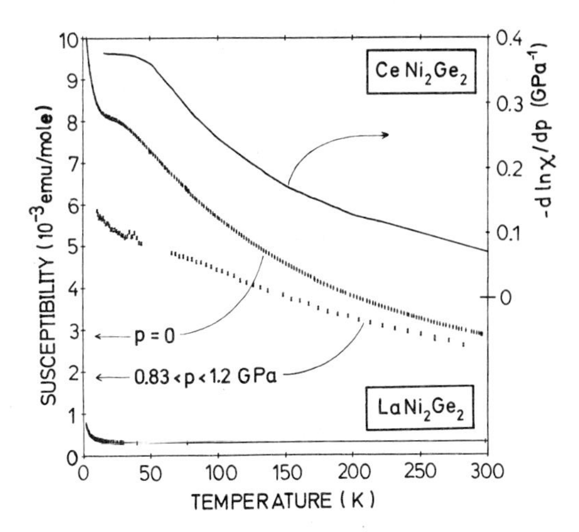

Fig. 2 Susceptiblity of $CeNi_2Ge_2$ vs temperature at atmospheric (+) and high (x) pressure. Bottom curve (+): $\chi(T)$ of $LaNi_2Ge_2$. Top curve: relative pressure dependence $d\ln\chi/dp$ vs temperature.

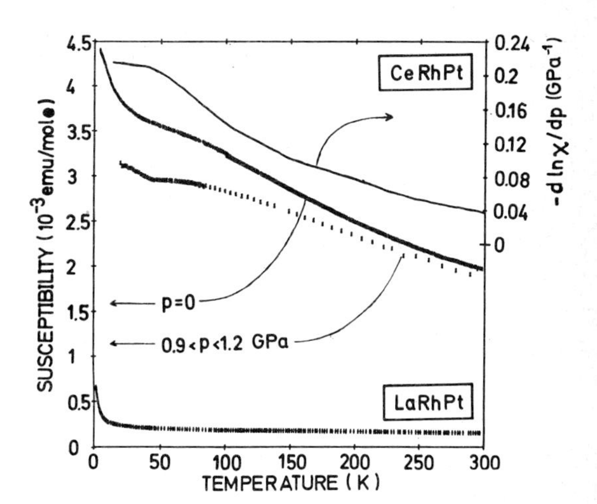

Fig. 3 Susceptibility of CeRhPt vs temperature at atmospheric (+) and high (x) pressure. Bottom curve (+): $\chi(T)$ of LaRhPt. Top curve: relative pressure dependence $d\ln\chi/dp$ vs temperature.

DISCUSSION

In Table 1 we have collected the data taken during this work and also $\partial\ln\chi/\partial p$ as measured previously on γ-Ce and α-Ce (1) and on $CeSn_3$ (2). The first observation is that the room temperature susceptibility does not reach the value $\chi(4f^1) = 2.69.10^{-3}$ emu/mole in any of the systems. In fact most of these susceptibilities are smaller than 6/7 of χ $(4f^1)$. The second observation is that the differences in the room temperature susceptibilities of the listed systems are minor. Only α-Ce and $CeRh_2$ clearly fall out of line from the others, in that their values are four times smaller than $\chi(4f^1)$. If one adds to such small values the observation of no temperature dependence below 300 K, one can understand why some researchers have suspected a total loss of 4f magnetization, i.e. tetravalence (10) in such systems. Of course even for α-Ce and $CeRh_2$ the room temperature susceptibility is still 4 and 10 times larger than that of the reference compound. This speaks against such a total loss as noted long ago. Table 1 now adds further independent arguments for fractional valence in the column for $\partial\ln\chi/\partial p$. We note that the only compound without noticeable pressure dependence outside of experimental error is $CePt_2$. For all other compounds $\partial\ln\chi/\partial p$ lies between 0.2 and 0.04 GPa^{-1}, at least ten times larger than the experimental error. Since $CePt_2$ orders magnetically, it ought to possess the most stable 4f shell (integral valence) and therefore, $\partial\ln\chi/\partial p \to 0$. If $CeRh_2$ were tetravalent it should behave similarly. The large $\partial\ln\chi/\partial p$ observed in this compound therefore speaks against tetravalence, and for intermediate valence. It seems that the relative pressure dependence of the susceptibility will be a good test for or against tetra-

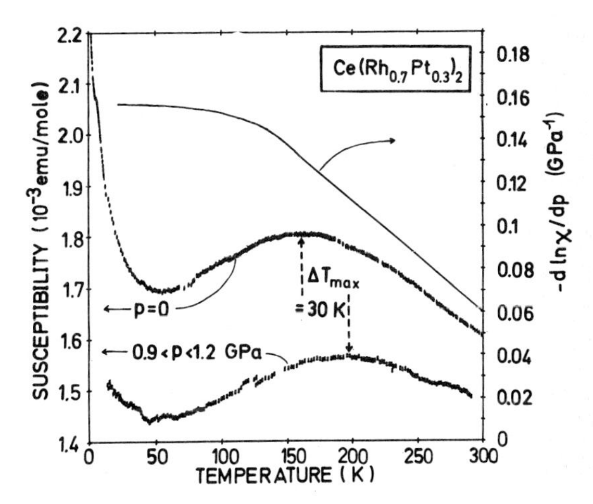

Fig. 4 Susceptibility of $Ce(Rh_{0.7}Pt_{0.3})_2$ vs temperature at atmospheric (+) and high (x) pressure. Top curve: relative pressure dependence $d\ln\chi/dp$ vs temperature.

valent Ce-compounds.

After having disposed of any tetravalent candidates among the systems in Table 1 the next question is whether one can learn anything about the valence from the absolute value of $\partial\ln\chi/\partial p$ or better of $\partial\ln\chi/\partial\ln V$.

The large value of $\partial\ln\chi/\partial p$ for α-Ce and γ-Ce relative to the others is probably due to the high compressibility of Ce-ions in the metallic element compared to other systems. It is common practice to correct such differences by comparing volume rather than pressure dependence. This can be done by calculating $\partial ln\chi/\partial lnV \equiv -B\partial ln\chi/\partial p$ with B, the bulk modulus. In the last two columns of Table 1 we give the bulk modulus where known, and the resulting $\partial ln\chi/\partial lnV$. Viewed via the volume dependence, the sequence of the systems on a scale of increasing numbers has changed, compared with the pressure dependence. $CeRh_2$ is indistinguishable from the others qualitatively.

The pressure dependence of the susceptibility of $YbAl_2$ at 300 K is very strong ($\partial ln\chi/\partial p = 0.37\ GPa^{-1}$ (3,11) in line with the observation that the valence of this compound, 2.4, is close to 2.5, where the maximum pressure dependence is expected (5). With the compressibility of $YbAl_2$ reported recently (12) ($B \simeq 47.5$ GPa), we find $\partial ln\chi/\partial lnV \simeq 18$ which is 4 times larger than that of $CePd_3$. This finding is consistent with a recent analysis of L_{III}-absorption edges which puts the valence of $CePd_3$ to 3.23 at 300 K, away from $\simeq 3.5$ where it is usually put in the older analyses (6). In other words, the relative pressure dependence of the susceptibility of $CePd_3$ seems too small for valence 3.5 if one takes Yb-compounds as a guide.

In Figs 2-4 $\partial ln\chi/\partial p$ is observed to increase strongly with decreasing temperature by factors of 5 for $CeNi_2Ge_2$ and CeRhPt and by 2.4 for $Ce(Rh_{0.7}Pt_{0.3})_2$. Moreover the values of $\partial ln\chi/\partial p$ seem to saturate near the susceptibility maximum or near the point of inflection of $\chi(T)$. This behaviour is quite analagous to that observed in YbCuAl (4) and $YbInAu_2$ (5) and seems to be a general property of mixed valent compounds. Because of the saturation one may regard $\partial ln\chi/\partial p$ at low temperatures as a true ground state property. Moreover since this behaviour is observed in YbCuAl, a compound whose valence is close to integral even at low temperatures ($V \simeq 2.8$), the large value of $\partial ln\chi/\partial p$ observed at the lowest temperature seems to be due more to a change of fluctuation temperature than of valence. If this is true, and if one may regard the fluctuation temperature as the only characteristic energy parameter in this limit, then we may write $\partial ln\chi/\partial lnV = -\partial lnT_f/\partial lnV = \Omega_g$ where Ω_g is the Grüneisen parameter (13). We thus find for $CeNi_2Ge_2$, $\Omega_g = 15$ which is somewhat larger than found for $CePd_3$ and $CeSn_3$ ($\Omega_g \simeq 10$) but smaller than for $CeAl_3$ ($\Omega_g = 50$, $T > 1K$).

The position of the maximum of the susceptibility is often taken as a rough measure of the fluctuation temperature. In Fig. 4 we see that approximately 1 GPa shifts the maximum from $T_{max} = 165$ K to $T_{max} = 195$ K, i.e. by 18% upwards. This value is in accordance with the observed $\partial ln\chi/\partial p = 15\%/GPa$ near T=0 which gives support to the idea that $\partial ln\chi/\partial p = -\partial lnT_f/\partial p$ here.

Recently the susceptibility of the alloy system $Ce(Pt_{1-x}Rh_x)_2$ was investigated (14) as a function of concentration from x=0 to x=1. Simultaneously the volume of the unit cell was measured as function of concentration by x-ray diffraction. The unit cell of $CeRh_2$ ($V = 430\ \AA^3$) is smaller than that of $CePt_2$ ($V = 462\ \AA^3$) and therefore the effect of alloying (increasing x) at atmospheric pressure is comparable to the effect of applying pressure from the outside. It is interesting to ask whether the volume reduction by pressure has the same effect on the magnetic properties as the volume reduction by alloying. Qualitatively this is obviously true because increasing p and increasing x both decrease the susceptibility and shift the maximum towards higher temperatures as observed here and in (14). Speaking more quantitatively a concentration shift of 10% corresponds to an applied pressure of 5GPa, if one compares shifts of susceptibility maxima. A still more exact number can be obtained by comparing the directly measured $\partial ln\chi/\partial p$ of $CeRh_2$ given here with $\chi^{-1} \cdot \Delta\chi/\Delta x$ in the system $Ce(Rh_xPt_{1-x})_2$ for the step from x=1 to x=0.9 at room temperature in (14). One again finds that $\Delta x = 0.1$ corresponds to p=5 GPa. From this and the lattice parameter data as function of x, as given in (14), one would infer a bulk modulus of 700 GPa for $CeRh_2$, if there were a 1:1 correspondence between the susceptibility and the volume. This is an unrealistically large value. In fact the measured compressibility is near 70 GPa (18). We therefore find that volume reduction by alloying changes the susceptibility about 10 times faster than volume reduction by pressure.

A similar comparison of the effects of volume reduction by pressure and by alloying can be made for YbCuAl where data are available for the pressure dependence of YbCuAl (4) and for the concentration dependence of the susceptibility for the systems $Yb_{1-x}Y_xCuAl$ and $Yb_{1-x}Sc_xCuAl$ (23) both at low temperature. Here we find that alloying with Sc increases the susceptibility twice as fast as pressure for the same volume change. Apparently $CeRh_2$ behaves quite abnormally.

ACKNOWLEDGEMENTS:

We thank W. Boksch, R. Pott and S. Stähr for providing us with some of the samples and S. Wood for skilful practical help. This work was supported by the Deutsche Forschungs Gemeinschaft through SFB 125.

REFERENCES

(1) M.R. MacPherson, G.E. Everett, D. Wohlleben and M.B. Maple
Phys. Rev. Letters 26, 20 (1971)

(2) J. Beille, D. Bloch, J. Voiron and G. Parisot,
Physica 86-88B, 231 (1977)

(3) B.C. Sales and D. Wohlleben
Phys. Rev. Letters 35, 1243 (1975)

(4) W.C.M. Mattens, H. Hölscher, G.J.M. Tuin,
A.C. Moleman and F.R. de Boer
J. Magn. Mat 15-18 982 (1980)

(5) W. Zell, R. Pott, B. Roden and D. Wohlleben
Solid State Commun. 40 751 (1981)

(6) Valence Fluctuations in Solids, L.M. Falicov,
W. Hanke and M.B. Maple (eds.)
North Holland Publishing Co. (1981)
K.R. Bauchspieß, W.K. Boksch, E. Holland-Moritz, H. Launois, R. Pott and D.
Wohlleben pg. 417
and D. Wohlleben, ibid., pg. 1

(7) D. Wohlleben and M.B. Maple
Rev. Sci. Inst. 42 1573 (1971)

(8) H.-W.Ludwigs, B. Roden, H. Schöpgens and
W. Zell
to be published

(9) W. Zell, Ph.D. Thesis,
University of Cologne (1981)

(10) G.L. Olcese,
Boll. Sci. Fac. Chim. Ind. Bol. 24, 165
(1966)

(11) B.C. Sales, Ph.D. Thesis, University of
California, San Diego (1974)

(12) T. Penney, B. Barbara, R.L. Melcher,
T.S. Plaskett, H.E. King, Jr. and
S.J. LaPlaca, pg. 341 in Ref. 6

(13) R. Takke, U. Niksch, W. Assmus, B. Lüthi,
R. Pott, R. Schefzyk and D.K. Wohlleben,
Z. Phys. B 44, 33 (1981)

(14) P. Weidner, G.E. Barberis, D. Davidov,
I. Felner, L.C. Gupta and B. Roden,
to be published in Sol. St. Comm.

(15) H.-W. Ludwigs, Ph.D. Thesis, University
of Cologne (1981)

(16) A.M. Toxen, R.J. Gambino and L.B. Welsh
Phys. Rev. B8, 90 (1973)

(17) J.-P. Kappler, Thèse d'Etat, Strasbourg
(1980)

(18) G. Güntherodt, priv. comm.

(19) R. Takke, W. Assmus, B. Lühti, T. Goto
and K. Andres in Crystalline Electric
Field and Structural Effects in f-Electron
Systems, J.E. Crow, R.P. Guertin and
T.W. Mihalisin (eds.) Plenum Press (1980)
p. 321

(20) M. Barth, Diplomarbeit, Köln (1981)

(21) D. Lenz, priv. comm.

(22) F.F. Voronov, L.F. Vereshchagin and
V.A. Goncharova,
Sov. Phys. Doklady 135, 1280 (1960)

(23) W.C.M. Mattens et al. this conference

(24) E. Franceschi, G.L. Olcese
Phys. Rev. Lett. 22, 1299 (1969)

Valence Instabilities
P. Wachter and H. Boppart (eds.)

MAGNETIC AND NONMAGNETIC STATE OF Ce-Si SYSTEM
---FERROMAGNETIC DENSE KONDO BEHAVIOR AND SPIN-FLUCTUATION EFFECTS---

Takeo Satoh, Hideo Yashima and Hiroshi Mori

Department of Physics, Faculty of Science, Tohoku University,
Sendai 980, Japan

Measurements of the specific heat and the magnetic susceptibility on $CeSi_x(1.55 \leq x \leq 2.00)$ revealed a nonmagnetic magnetic transition at around x=1.83. For the magnetically ordered system ($x \leq 1.80$), magnetization measurements were also made. We observed a large reduction of the magnetic moment of Ce atom and also the reduction of the magnetic entropy. On the basis of these results, we propose a dense Kondo model to interpret our Ce-Si system. This is the first example of a ferromagnetic dense Kondo system. Even for the nonmagnetic system ($1.85 \leq x \leq 2.00$), there are several aspects which reflect the fact that the system is a dense system. Discussions are presented within the framework of the paramagnon model.

A valence fluctuation phenomenon is believed to occur due to the proximity of the 4f level to the Fermi energy level. There have been many efforts to find and study systems in which one can control the relative position of these two levels.

We found [1] that $CeSi_2$ shows a nonmagnetic behavior at low temperatures while $CeGe_2$ undergoes a sharp magnetic transition at T_c=7K. These are members of α-$ThSi_2$ type intermetallic compounds, the crystal structure of which is shown in Fig.1. We performed [2] a systematic investigation on $CeSi_x$ in the composition range $1.55 \leq x \leq 2.00$, where the system remains the same crystal structure. A critical examination of the results of density and lattice-parameter measurements [3] in the Si-deficit samples unambiguously shows that the non-stoichiometry results from Si vacancies.

Magnetic susceptibilities were measured in the temperature range from 4.2K to 300K. The results are shown in Fig.2. Specific heat measurements were made in the temperature region from 0.1K to 70K. The results are given in Fig.3 and Fig.4. We can summarize the main features of our results as follows.

(i) In the composition range $1.85 \leq x \leq 2.00$, the system is nonmagnetic at low temperatures. The low temperature specific heat can be expressed very well with the equation $C=\gamma T$. The value of γ is very large compared with that of $LaGe_2$ or $LaSi_2$ [4]. The low temperature magnetic susceptibility can be expressed well with the equation $\chi(T) = \chi(0)\ (1 + aT^2)$.

(ii) For the composition $x \leq 1.80$, the system undergoes a sharp magnetic transition around 10K. A ferromagnetic ordered state is indicated by the divergence of the susceptibility at T_c. The magnetic entropy ΔS below T_c can be calculated from the specific heat data. It is noted that ΔS is much smaller than $R\ell n2$ whereas for $CeGe_2$ it nearly attains a full value expected for a ground doublet.

The values of γ, $\chi(0)$, a, T_c and ΔS are listed in Tab.I.

Magnetization measurements were made on $CeSi_{1.80}$, $CeSi_{1.70}$ and $CeGe_2$ [5]. The results are shown in Fig.5. A remarkable feature to be noted is the fact that, for the two silicides, the ordered magnetic moment ($\simeq 0.3\mu_B$/Ce atom) is much smaller than that of $CeGe_2$. The tetragonal crystal field splits the ground multiplet J=5/2 into three doublets, none of which is expected to exhibit an ordered moment smaller than 0.71 μ_B. The reduction in the magnetic moment and the magnetic entropy observed for $CeSi_{1.80}$ and $CeSi_{1.70}$ suggests a Kondo effect. In the present case, the Ce ion forms a periodic array and one has to consider a dense Kondo system, in which the Kondo effect and the RKKY interaction coexist. If this is the true picture of $CeSi_{1.80}$ and $CeSi_{1.70}$, our system is the first example of a ferromagnetic Kondo system.

To substantiate our picture we make an estimation of the Kondo temperature, T_k, of our system. Expressions for the single-impurity problem are used throughout because there is no established theoretical treatment for a dense Kondo system. From the saturation moment M_s, T_k is obtained with the relation

$$M_s = M_0/\sqrt{1+(T_k/T_c)^2} \qquad (1)$$

in analogy with Ishii's formula for the induced magnetic moment by the applied field [6]. In eq.(1), M_0 means a saturation moment without the Kondo effect and we tentatively adopt the value $M_0=0.71\mu_B$, because the level pattern in the Ce-Si system is still uncertain. The specific heat data below T_c can be used to estimate T_k as follows. ΔS is reduced from $R\ell n2$ by an amount already exhausted above T_c by the Kondo effect. A Kondo effect, when only a ground doublet is involved at low temperatures, also exhausts an entropy of $R\ell n2$. As an analytical expression for the Kondo-effect specific heat anomaly is not available, we replace it with a two-level Schottky specific heat with an energy splitting k_BT_k. Then ΔS is expressed as

$$\Delta S=R\{\ell n(1+\exp(-T_k/T_c))+\frac{T_k}{T_c}\frac{\exp(-T_k/T_c)}{1+\exp(-T_k/T_c)}\} \quad . \qquad (2)$$

Next we extend our picture of the dense Kondo state to the nonmagnetic $CeSi_x$. Here we consider the Kondo effect to dominate the RKKY interaction, thereby favoring a nonmagnetic ground state. The theoretical ground state susceptibility $\chi(0)$ is expressed as

$$\chi(0) = N_A\cdot(g_{eff}\mu_B)^2J(J+1)/3k_BT_k \qquad (3)$$

where N_A is Avogadro's number and J=1/2. We tentatively adopt the effective g-value of the Γ_7 doublet in the cubic crystal field, $g_{eff}=10/7$. Another way of obtaining T_k is to use an expression [7]

$$\chi(T) = \chi(0)\{1-(\pi^2/3)(T/T_k)^2\} \quad . \qquad (4)$$

The estimation of T_k is also made from γ using Yoshimori's expression [8] for S = 1/2

$$\gamma = \pi^2N_Ak_B/6T_k \quad . \qquad (5)$$

The γ value of the magnetically ordered system was obtained from the low temperature specific heat far below T_c (see the inserted figure in Fig.4).

The values of T_k obtained with these various formula are listed in Tab.II and plotted in Fig.5. Considering that we have applied the expression for the single-impurity Kondo effect and also that spin-orbit coupling is not fully taken into account, the agreement between the T_k's deduced from different physical quantities should be regarded rather good, except the T_k's obtained with eq.(5). The deviation of the T_k's obtained from eq.(5) suggests that the variation of γ value with Si concentration x reflects the fact that the system is a dense system. The general trend that T_k's obtained from $\chi(0)$ are smaller than those from other quantities is reasonable because $\chi(0)$ is enhanced by a ferromagnetic interaction.

In Fig.6 γ is plotted as a function as a function of x. As can be seen, γ increases very rapidly when x approaches the critical value, either from the nonmagnetic region or from the magnetic region. In Fig.7 γ is plotted as a function of $\chi(0)$. We have an approximate relation

$$\gamma \propto \ell n\chi(0) \qquad (6)$$

Relation(6) is a well-known behavior predicted for a strongly exchange-enhanced system [9,10]. The trend shown in Fig.6 is similar to the one obtained theoretically for an itinerant electron system near a ferromagnetic instability [11]. These observations motivate us to analyze our data in the framework of the spin-fluctuation model [12]. A spin-fluctuation model for the dense Kondo system may be considered in analogy with Nozières' Fermi liquid picture [13] for the single-impurity Kondo problem. Electrons in the conduction band are interacting with each other by virtually polarizing the singlet-state Ce ions in a periodic array.

We try to interpret the large value of γ as a paramagnon effect. Denoting the quantities without the paramagnon effect with a subscript o, we have

$$\gamma=(m^*/m_o^*)\gamma_o \quad (7) \qquad \text{and} \qquad \chi(0)=S\chi_o(0) \qquad (8)$$

where m* means the effective mass of the conduction electron and S is the Stoner factor. We define the characteristic temperatures

$$T_{SF}=3\lambda/2\chi(0) \quad (9) \qquad \text{and} \qquad T_F(\gamma)=\pi^2Nk_B/2\gamma \qquad (10)$$

where N is the number of conduction electrons and λ is a Curie constant for which we assume the value of the free electron system, $\lambda=N\mu^2_B/k_B$. Then we have

$$(T_{SF}/T_F(\gamma))S = m^*/m_o^* \quad . \qquad (11)$$

We adopt the uniform enhanced paramagnon model [9,10] because there is no theoretical treatment for the spin-fluctuation effect in the dense Kondo system. Assuming a parabolic band [10] we have

$$(T_{SF}/T_F(\gamma))S = 1+(9/2)(1-1/S)^2\ell n(1+S/3). \qquad (12)$$

This can be solved graphically, then γ_o is calculated. Results are given in Tab.II. It is noted that the value of γ_o thus obtained is comparable with the electronic specific heat coefficient of $LaGe_2$ or $LaSi_2$.

Next we will see whether there is a $T^3\ell nT$ term in the specific heat of $CeSi_x$. We take a formula

$$C/T = \gamma+BT^2+DT^2\ell n(T/\Theta_c) \quad . \qquad (13)$$

We assume that the lattice specific heat of $CeSi_x$ is the same as that of $LaSi_2$, so that $B=2x10^{-4}J\cdot K^{-4}\cdot mole^{-1}$. The parameters D and Θ_c are determined to fit the data at the bottom part of the C/T vs. T curve. An example is shown in Fig.9 for $CeSi_{1.85}$. The parameters obtained are listed in Tab.II. As is evident in Fig.9, eq.(13) cannot fit the data at low temperatures although it reproduces the general feature of the experiment, that is, the decrease of C/T with increasing temperature.

In order to get a further insight into the present system, the high-field magnetization measurements [14] and also the dilution experiments [15] are in progress.

References

[1]H.Yashima,T.Satoh,H.Mori,D.Watanabe and T.Ohtsuka; Solid State Commun.41 (1982) 1.
[2]H.Yashima and T.Satoh;ibid.(to be published).
[3]The lattice parameters of $CeSi_2$ given in [1] are wrong and those in [2] are correct.
[4]A.Sawada and T.Satoh;J.Low Temp.Phys.30(1978) 455.
[5]H.Yashima,H.Mori,T.Satoh and K.Kohn; Solid State Commun. (to be published).
[6]H.Ishii; Progr.Theor.Phys. 40 (1968) 201.
[7]K.D.Schotte and V.Schotte; Phys. Lett. 55A (1975) 38.
[8]A.Yoshimori;Progr.Theor.Phys.55 (1976) 67.
[9]S.Doniach and S.Engelsberg; Phys.Rev.Lett. 17 (1966) 750.

[10]J.R.Schrieffer; J.Appl.Phys.,39(1968) 642.
[11]W.F.Brinkman and S.Engelsberg; Phys.Rev.169 (1968) 417, K.Makoshi and T.Moriya;J.Phys. Soc.Japan, 35 (1975) 10.
[12]H.Yashima,N.Sato,H.Mori and T.Satoh; Solid State Commun. (to be published).
[13]P.Nozières;J.Low Temp.Phys.17 (1974) 31.
[14]H.Yashima,T.Satoh,T.Harada,T.Sakakibara and M.Date; to be published.
[15]H.Mori,H.Yashima and T.Satoh; to be published.

Table I

	γ $(\frac{mJ}{mole\cdot k^2})$	$\chi(0)\times10^2$ $(\frac{emu}{mole})$	$a \times 10^3$ (K^{-2})	T_c (K)	ΔS $(\frac{J}{mole\cdot K})$	T_{SF} (K)	$T_F(\gamma)$ $(10^3 K)$
X = 2.00	104	0.42	-0.42			670	1.95
X = 1.90	151	0.70	-0.55			408	1.38
X = 1.85	234	4.0	-2.7			72	0.90
X = 1.80	230			9.0	3.41		
X = 1.70	75			10.9	4.53		
$CeGe_2$				7.0	5.74		

Table II

X	T_{k1} (K)	T_{k2} (K)	T_{k3} (K)	T_{k4} (K)	T_{k5} (K)	S	γ_0 $(\frac{mJ}{mole\cdot k^2})$	$D \times 10^{-4}$ $(\frac{J}{mole\cdot k^4})$	Θ_c (K)
2.00			69	89	132	34	9.0	2.05	34
1.90			41	77	91	43	11.8	3.37	33
1.85			7	35	58	262	11.1	6.96	33
1.80	17	16			59				
1.70	17	13			182				

T_{k1};eq.(1), T_{k2};eq.(2), T_{k3};eq.(3), T_{k4};eq.(4), T_{k5};eq.(5).

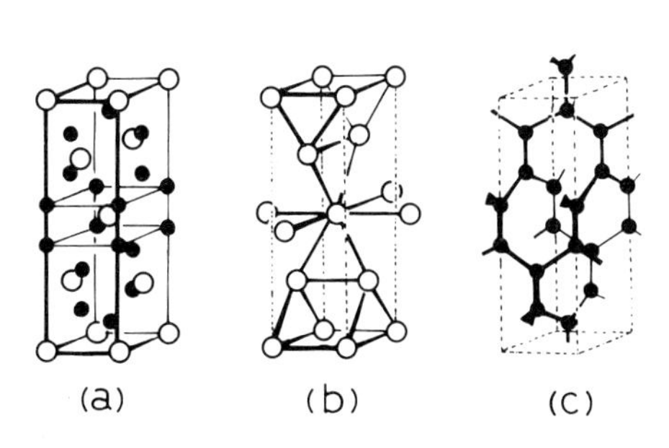

Fig.1 (a)Crystal structure of $CeSi_2$
(b)Sublattice of Ce atoms
(c)Sublattice of Si atoms

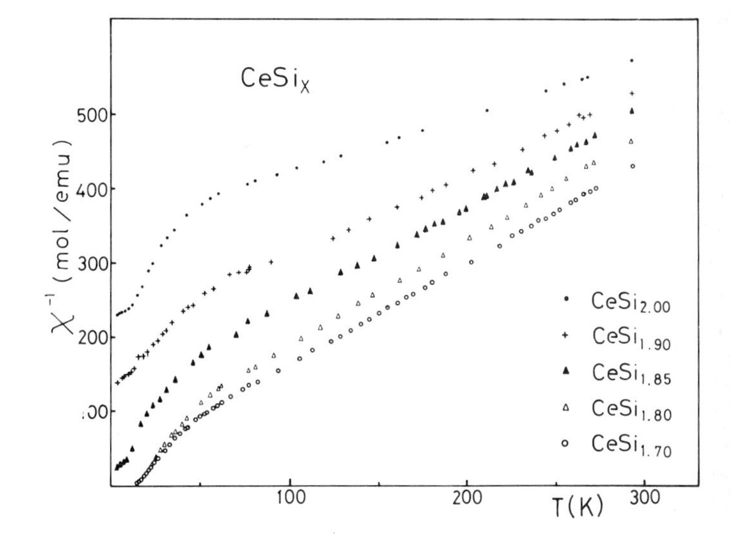

Fig.2 Inverse susceptibility χ^{-1} vs. T plot for $CeSi_x$.

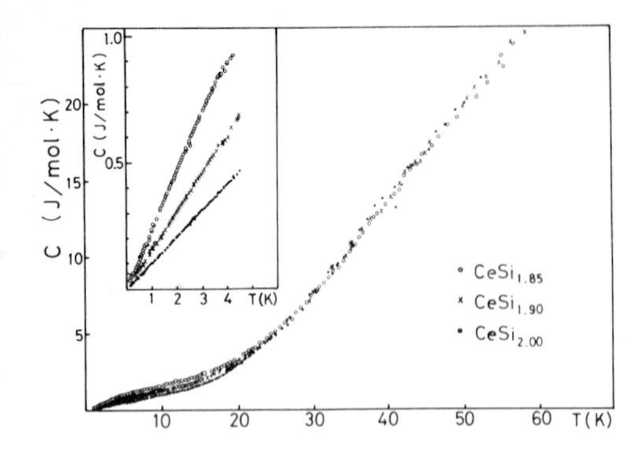

Fig.3 Specific heat C vs.T plot for x=2.00, 1.90 and 1.85. The inserted figure is for the low temperature region.

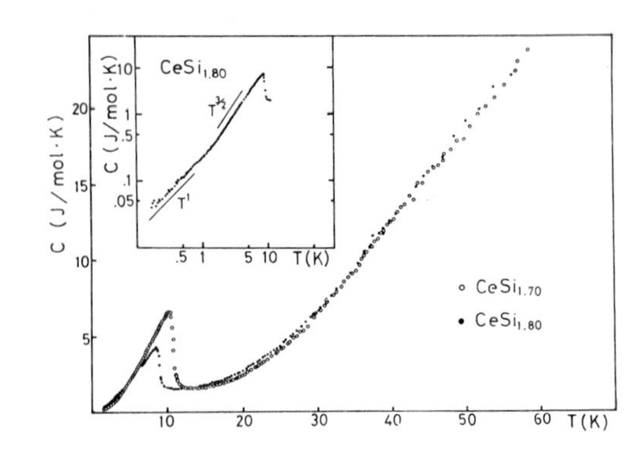

Fig.4 C vs.T plot for x=1.80 and 1.70. The inserted figure is a double logarithmic plot for x=1.80.

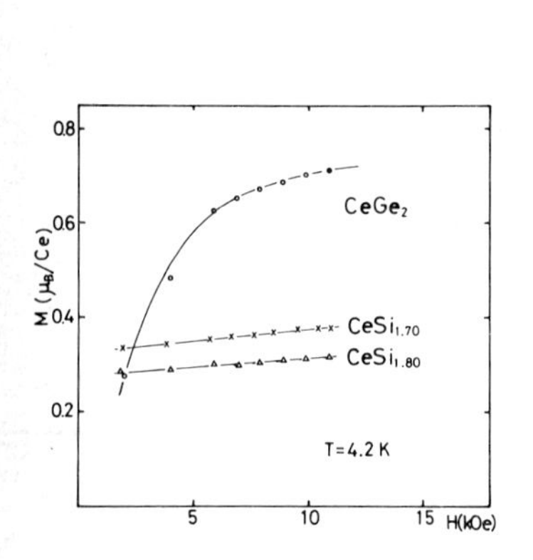

Fig.5 Magnetization at 4.2K as a function of the applied field.

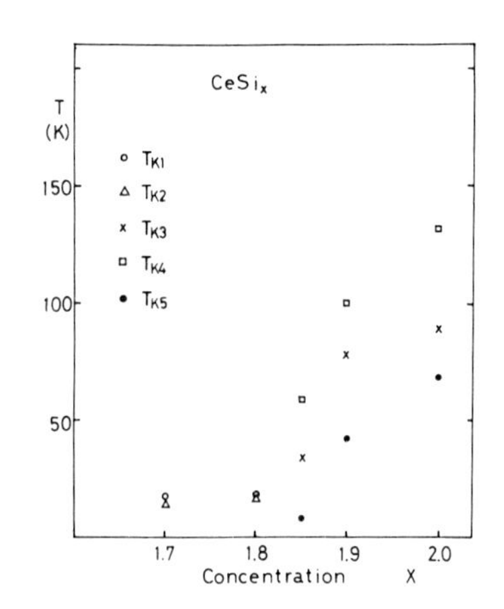

Fig.6 Kondo temperature deduced from various quantities.

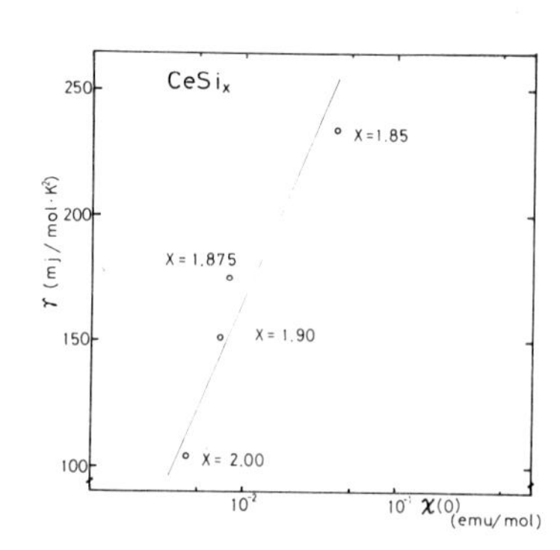

Fig.7 γ vs. $\ln\chi(0)$ for non-magnetic $CeSi_x$.

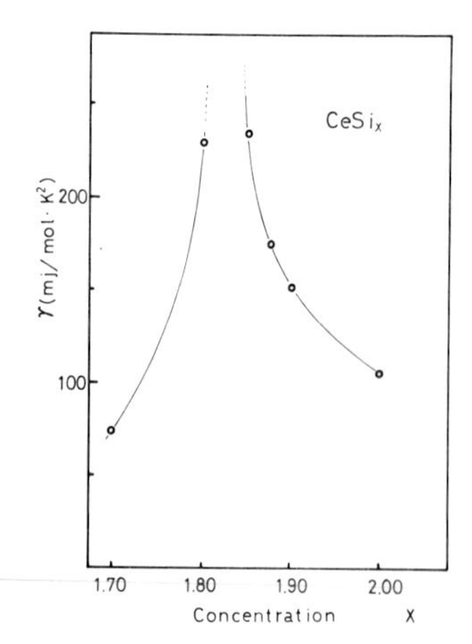

Fig.8 γ is plotted as a function of x.

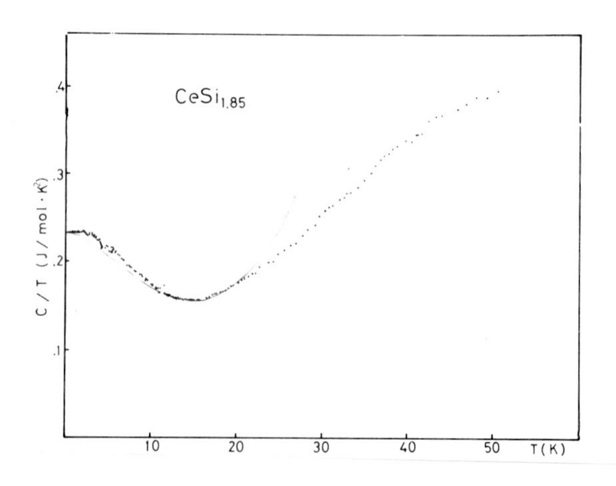

Fig.9 C/T vs.T plot for $CeSi_{1.85}$. The curve represents eq.(13) with the parameters given in Tab.II.

Valence Instabilities
P. Wachter and H. Boppart (eds.)

MOMENT INSTABILITIES AND MAGNETIC ORDERING IN $CePd_2Si_2$, $CeAu_2Si_2$ and $CeAg_2Si_2$

V. MURGAI, S. RAAEN, L.C. GUPTA* and R.D. PARKS

Department of Physics
Polytechnic Institute of New York
333 Jay Street, Brooklyn, N.Y. 11201

Magnetic susceptibility measurements in the temperature range $1.5 \lesssim T \lesssim 300K$ and high field magnetization studies at 4.2 K for fields up to 21 Tesla have been performed on the nearly integral valent systems $CePd_2Si_2$, $CeAu_2Si_2$ and $CeAg_2Si_2$. In addition, electrical resistivity measurements have been made in the temperature range $1.5 \lesssim T \lesssim 300$ K on $CePd_2Si_2$ and $CeAu_2Si_2$. All three systems exhibit magnetic transitions between 8 and 10 K which appear to be antiferromagnetic in $CePd_2Si_2$ and $CeAu_2Si_2$ and ferri-magnetic in $CeAg_2Si_2$, the ordering in all cases being probably of the metamagnetic type. Kondo sideband effects apparent in the resistivity implies the importance of moment instability effects in $CePd_2Si_2$. The smallness of the high field magnetization (H = 21 Tesla) at 4.2 K suggests either the presence of large magnetic anisotropy or the presence of strong spin fluctuations at high fields in $CePd_2Si_2$ and $CeAg_2Si_2$.

1. INTRODUCTION

A number of cerium-based compounds which are nearly integral, but yet exhibit clear manifestations of 4f-instabilities (either valence or Kondo-like fluctuations) exhibit exotic ground states. In Ce_3Al_{11} and $CeAl_2$ the ground state is a moment-modulated antiferromagnetic state in which the 4f instability leads to a position-dependent suppression of the magnetic moment (see, e.g., Ref.1). In $CeAl_3$, which has been labelled an archetypal Kondo lattice system, the moment is completely suppressed which leads to a non-magnetic ground state with large Fermi liquid effects.) In $CeCu_2Si_2$ the ground state is superconducting in spite of, or, alternately, because of, heavy fermions generated by interactions between the 4f-electrons and band electrons.(2) The purpose of the present study was to search for similarly exotic behavior in the systems CeR_2Si_2, with R=Pd, Au and Ag, which are sister compounds of $CeCu_2Si_2$. These systems crystallize in the tetragonal $ThCr_2Si_2$-type structure, wherein the Ce-atoms reside on planes well separated by layers composed of the R and Si atoms.

2. MAGNETIC SUSCEPTIBILITY

The samples were prepared by arc melting the pure materials on a water-cooled copper hearth in an argon atmosphere, followed by a 4-5 day anneal at 800 C. X-ray diffraction studies confirmed that the samples were single phase, to within the resolution of the technique, viz., $\sim$5%.

The temperature dependence of the low field susceptibility and its inverse are shown in Figs. 1, 2 and 3 for the three systems. In the temperature region 100-300 K approximately Curie-Weiss behavior is apparent for all three systems, the Curie-Weiss constants being given (within 5-10%) by $\theta = +75$ K, $+12$ K and $+15$ K for R=Pd, Au and Ag respectively. The departure from Curie-Weiss behavior near T=100 K for R=Pd and Ag is due probably to crystal field (CF) effects. However, the absence of departure from Curie-Weiss behavior in the case of $CeAu_2Si_2$ should not be taken as evidence for the absence of crystal field effects in that system. For example, in $CeCu_2Si_2$, the susceptibility is Curie-Weiss-like (3) in the temperature range $75 \lesssim T \lesssim 300$ K in the presence of 140 K and 364 K splittings between the ground state and the first and second excited CF states respectively.(4) The largeness of the Curie-Weiss constant for $CePd_2Si_2$, compared with its ordering temperature, suggests that it is dominated by spin fluctuation effects rather than Ce-Ce exchange coupling. Evidence for the importance of spin fluctuations in $CePd_2Si_2$ is further suggested by the resistivity measurements as discussed below.

The inserts of Figs. 1, 2 and 3 show an enlarged view of the susceptibility near the magnetic transitions. The cusp-like shapes in $CePd_2Si_2$ and $CeAu_2Si_2$ suggest anti-ferromagnetic transitions, whereas the mesa-shaped anomaly, together with observed magnetic hysteresis effects below the ordering temperature, suggests either a ferri- or ferromagnetic ground state for $CeAg_2Si_2$, the former being more obviously consistent with the positive Curie-Weiss constant. The shape of the susceptibility anomaly at T_N in $CePd_2Si_2$ (inset of Fig. 1) is that expected for a simple antiferromagnet since it mimics the shape of the Fourier transform of the spin-spin correlation function for a non-ordering wave vector, i.e. with an inflection point at T_N and a rounded maximum above.(5) Fisher and Burford(6) have suggested that in less simple situations, as in metamagnets, this same qualitative shape might persist. On the other hand, the anomaly seen in

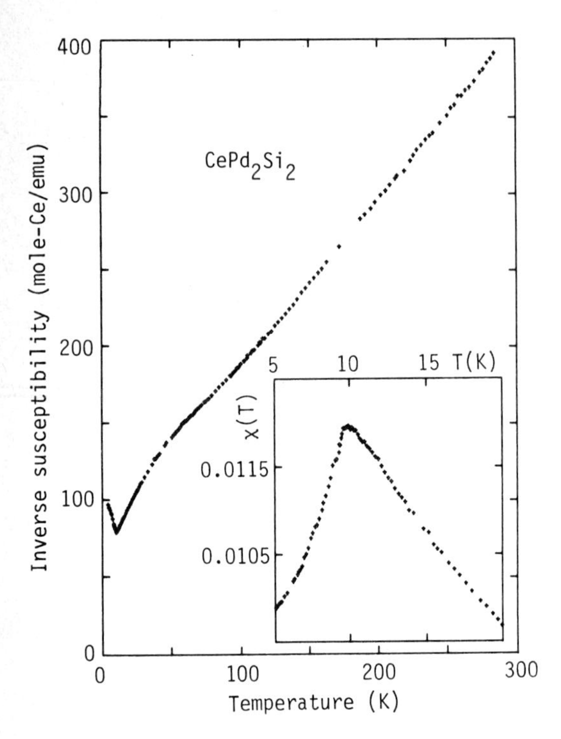

Fig. 1. Inverse magnetic susceptibility of $CePd_2Si_2$. Inset shows expanded view of susceptibility.

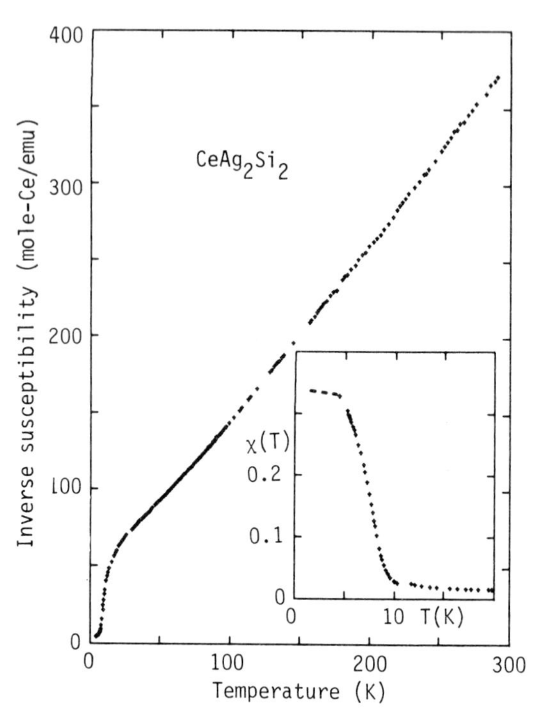

Fig. 3. Inverse magnetic susceptibility of $CeAg_2Si_2$. Inset shows expanded view of susceptibility; dashed line is visual extrapolation.

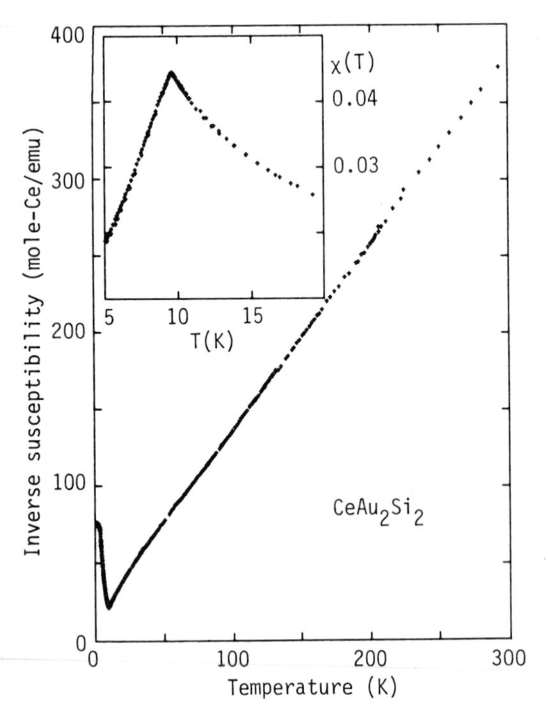

Fig. 2. Inverse magnetic susceptibility of $CeAu_2Si_2$. Inset shows expanded view of susceptibility.

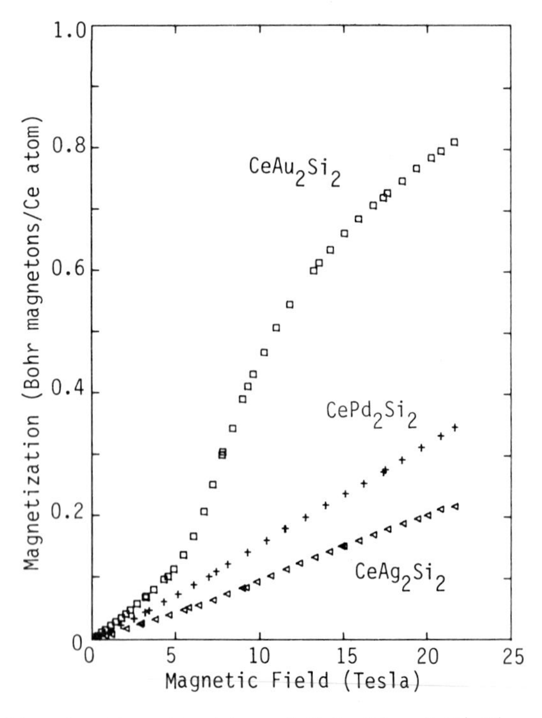

Fig. 4. Magnetization of $CeAu_2Si_2$, $CePd_2Si_2$ and $CeAg_2Si_2$ at 4.2 K in fields up to 21 Tesla.

$CeAu_2Si_2$ (Fig. 2) does not have this shape; rather, its sharp cusplike shape qualitatively mimics the anomalies seen in (cubic) $CeIn_3$ (7) and (orthorhombic) CeAl.(8)

Neutron scattering studies have established the mean field nature of the magnetic order near T_N in both $CeIn_3$(7) and CeAl(8); and, mean field behavior prescribes a cusp-like shape for the susceptibility anomaly. Lawrence and Shapiro(7) suggested that the mean field behavior in $CeIn_3$ might arise from the presence of the moment instabilities. On the other hand, for the case of CeAl where moment instability effects appear to be unimportant, it was suggested (8) that the observed mean field behavior might reflect a somewhat longer than usual coherence length. These ideas do not appear to map onto the present results. It would be surprising if the force range in $CeAu_2Si_2$ were longer than that in its isomorph, $CePd_2Si_2$, yet the susceptibility maximum in $CeAu_2Si_2$ has a mean field shape, whereas in $CePd_2Si_2$ it has the ordinary shape associated with the presence of critical fluctuations. Moreover, from the resistivity and high temperature susceptibility results, moment-instability effects appear to be important in $CePd_2Si_2$, but unimportant in $CeAu_2Si_2$. In CeAl, where the ground state is a metamagnet with a ++-- alternation of signs, the susceptibility near T_N is dominated by short range ferromagnetic fluctuations as clearly evident in the coarse-grained χ^{-1} versus T behavior.(8) To a lesser extent this is also the case for $CePd_2Si_2$ as is evident in the χ^{-1} plot in Fig. 1. The presence of such fluctuations could be expected to severely alter the conventional shape of the susceptibility anomaly at T_N.

3. HIGH FIELD STUDY

The field dependence of the magnetization at 4.2 K and for fields to 21 Tesla is shown in Fig. 4 for the three systems. Qualitatively, the behavior exhibited by $CeAu_2Si_2$ is that expected for a metamagnet for $T \lesssim T_t$, where T_t is the tricritical point (see, e.g., Ref. 9). The nearly constant and small magnetic susceptibility ($\partial M/\partial H$) for fields up to 21 Tesla, observed for $CePd_2Si_2$ and $CeAg_2Si_2$, can be reconciled as conventional metamagnetic behavior only if $T \ll T_t$ and the magnetic anisotropy energy is large compared to kT_N. Alternatively, if spin fluctuations resulting from the 4f instability dominated the magnetic behavior for large H, this could explain the Pauli-like susceptibility which extends up to 21 Tesla. Further experiments, currently underway, to measure the temperature dependence of M(H) should resolve the issue.

4. ELECTRICAL RESISTIVITY

The temperature dependence of the electrical resistivity is dramatically different in $CePd_2Si_2$ than in $CeAu_2Si_2$ as evident in Figs. 5 and 6. The double bump structure observed for

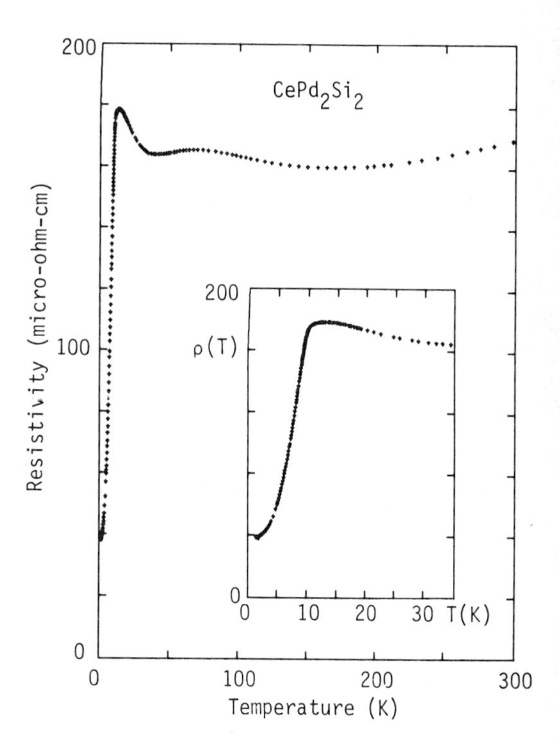

Fig. 5. Electrical resistivity of $CePd_2Si_2$. Inset shows expanded view.

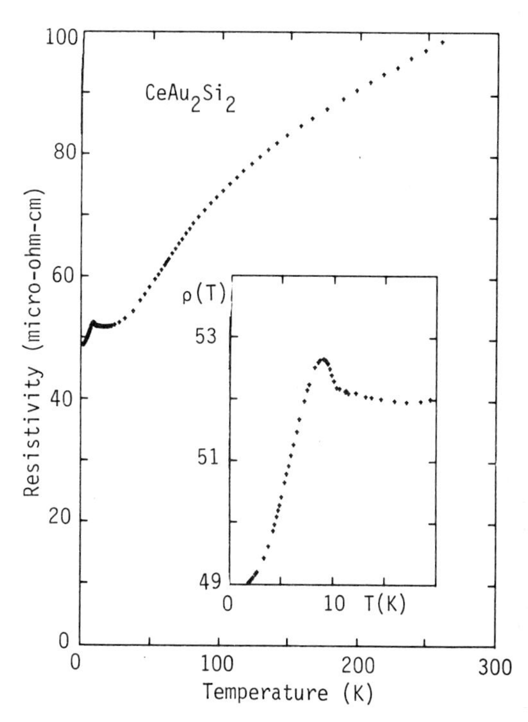

Fig. 6. Electrical resistivity of $CeAu_2Si_2$. Inset shows expanded view.

$CePd_2Si_2$ is qualitatively similar to that observed for $CeCu_2Si_2$,(10) which is the behavior expected when both Kondo fluctuations and crystal field effects are of comparable importance.(10, 11) These are known as Kondo sideband effects because they originate from "sideband" resonances in the conduction electron scattering at discrete energies displaced from the Fermi energy (see, e.g., Ref.11). The only qualitative difference between the shape for $CePd_2Si_2$ and $CeCu_2Si_2$ is that in the former case the $CeCu_2Si_2$-like behavior is cut-short by the magnetic phase transition. In the case of $CeCu_2Si_2$ the maximum value of the low temperature resistivity as in $CePd_2Si_2$ is close to Mott's unitarity limit.(10) Focusing on the magnetic transition itself (inset of Fig. 5), the temperature derivative of $\rho(T)$ in the vicinity of T_N is positive, which is the signature for an antiferromagnetic transition (whether or not in the metamagnetic sense) only when the new Brillouin wall created by the magnetic periodicity does not intersect pieces of the Fermi surface which are important in conduction (see,e.g., Ref. 12). The behavior of $CeAu_2Si_2$ is strikingly different in that there are no signs of strong Kondo scattering as in the cases of $CePd_2Si_2$ and $CeCu_2Si_2$, which is consistent with the Curie-Weiss constant in the susceptibility being roughly equal to, rather than larger than, the ordering temperature, in contradistinction to the situation in $CePd_2Si_2$ and $CeCu_2Si_2$. It is tempting to speculate that the R-derived 4d bands play an important role here through 4d-4f hybridization effects. The resistance near T_N in $CeAu_2Si_2$ exhibits a negative temperature derivative which results from superzone gaps created by the intersection of the "magnetic" domain walls with the Fermi surface, such effects being common in antiferromagnets (see, e.g., Ref.13) and their analogs, order-disorder systems (see, e.g., Ref. 14).

SUMMARY

In summary, the three systems $CePd_2Si_2$,$CeAu_2Si_2$ and $CeAg_2Si_2$ all undergo magnetic phase transitions in the temperature range 8-10 K, which appear to be either antiferromagnetic or ferrimagnetic, probably in the metamagnetic sense, the latter being consistent with the layered structure of the magnetic atoms. Only in the case of $CePd_2Si_2$ is it apparent that the energy of the spin fluctuations is large compared to the ordering energy (kT_N). This, as well as the striking difference in the resistivity anomaly at T_N between $CePd_2Si_2$ and $CeAu_2Si_2$, implies that the band structure, and probably 4f-4d hybridization effects, depends strongly upon the atom chosen for the R sites in the structure, CeR_2Si_2. This preliminary study sets the stage for detailed neutron scattering studies, currently underway, designed to decipher the topology of the magnetic ground states of the three systems under study.

ACKNOWLEDGMENT

We wish to thank Jon Lawrence for a number of very useful discussions. This work was supported by the National Science Foundation. The high field studies were performed at the Francis Bitter National Magnet Laboratory, Massachusetts Institute of Technology, which is supported by the National Science Foundation. We are particularly grateful for the helpful assistance given us by the staff of the Magnet Lab.

REFERENCES

* Permanent address: Tata Institute of Fundamental Research, Bombay, 400005, India.

1. J.M. Lawrence, P.S. Riseborough and R.D. Parks, Rep. Prog. Phys. 44, 1 (1981) and references cited therein.

2. F. Steglich, J. Aarts, C.D. Bredl, W.Licke, D. Meschede, W. Franz and H. Schäfer, Phys. Rev. Lett. 43, 1892 (1979).

3. B.C. Sales and R. Viswanathan, J. Low Temp. Phys. 23, 449 (1976).

4. S. Horn, E. Holland-Moritz, M. Loewenhaupt, F. Steglich, H. Scheuer, A. Benoit and J. Flouquet, Phys. Rev. B 23, 3171 (1981).

5. M.E. Fisher, Phil. Mag. 7, 1731 (1962).

6. M.E. Fisher and R.J. Burford, Phys. Rev. 156, 583 (1967).

7. J.M. Lawrence and S.M. Shapiro, Phys. Rev. B 22, 4379 (1980).

8. J.M. Lawrence, S.M. Shapiro and K. Parvin, to be published.

9. E. Stryjewski and Giordano, Adv. Phys. 26, 487 (1977).

10. W. Franz, A. Griessel, F. Steglich and D. Wohlleben, Z. Physik B 31, 7 (1978).

11. B. Cornut and B. Coqblin, Phys. Rev. B 5, 4541 (1972).

12. R.D. Parks, A.I.P. Conf. Proc. 5, 630 (1971).

13. R.A. Craven and R.D. Parks, Phys. Rev. Lett. 31, 383 (1973).

14. D.P. Chakraborty and R.D. Parks, Phys. Rev. B 18, 6195 (1978).

Valence Instabilities
P. Wachter and H. Boppart (eds.)

DIAGRAMMATIC CALCULATION OF THE MAGNETIC SUSCEPTIBILITY USING A RECENT APPROACH TO THE THEORY OF INTERMEDIATE VALENCE SYSTEM

G.Morandi, Istituto di Fisica and Gruppo Nazionale di Struttura della Materia, Modena, Italy

E.Galleani d'Agliano, F.Napoli and E.Ottaviani, Istituto di Scienze Fisiche and Gruppo Nazionale di Struttura della Materia, Genova, Italy

A diagrammatic expansion is set up for the model proposed by Foglio et al. (1) for intermediate valence systems. From the lowest order diagrams which contribute to the one-particle self-energy we derive an integral equation for the irreducible vertex of the Bethe-Salpeter equation for the dynamic magnetic susceptibility χ_q . Solving this equation we derive an expression for χ_q which agrees with the results of the above mentioned authors in the static uniform limit.
Finally, the valence of the system, which bears in this model a simple relation with the dynamic susceptibility χ_q , can also be evaluated using our technique.

A new approach to the problem of intermediate valence has recently been proposed by Foglio et al. (1),(2),(3). Although the theory proposed by these authors was referred in principle to a general situation of two $(4f)^n$ and $(4f)^{n+1}$ configurations mixed with the conduction band states, we shall restrict our discussion to the simpler and more specific case of a periodic Anderson-type Hamiltonian:

$$\begin{aligned} H &= H_0 + H_1 \\ H_0 &= \sum_{i\sigma} \varepsilon_f n_{i\sigma}^{(f)} + \sum_{i\sigma} \varepsilon_c n_{i\sigma}^{(c)} + \frac{1}{2}\sum_{i\sigma} U n_{i\sigma}^{(f)} n_{i\bar{\sigma}}^{(f)} + \\ &\quad + \sum_{i\sigma} (V c_{i\sigma}^{+} f_{i\sigma} + h.c.) \\ H_1 &= \sum_{\sigma, i\neq j} t_{ij} c_{i\sigma}^{+} c_{j\sigma} \end{aligned} \tag{1}$$

H_0 describes an admixture of localized f- and conduction electrons, with a (screened) Coulomb repulsion U among f-electrons and a hybridization term V. t_{ij} is the hopping integral for conduction electrons, which is assumed to be non zero (and equal to $-|t|$) only between nearest neighbours sites.
The idea of the above mentioned authors is first to diagonalize H_0 in the limit $U \to \infty$ and in the full Hilbert space it acts on, and then to project the entire Hamiltonian onto the manifold of low-lying states which for $\varepsilon_f > \varepsilon_c$ turns out to be a four-dimensional one, whose basis vectors (and eigenvalues of H_0) are given in table I.
The soundness of the whole approach rests of course on the assumption that the manifold of low-lying states is only weakly coupled to the rest of the Hilbert space. In terms of the parameters of the model this amounts to assume $|t| \ll [(\varepsilon_f-\varepsilon_c)^2+4V^2]^{1/2}$.
Introducing new fermion operators $\gamma_{i\sigma}, \gamma_{i\sigma}^{+}$ defined by: $\gamma_{i\sigma}^{+}|0\rangle = |d_\sigma\rangle$, $\gamma_{i\uparrow}^{+}\gamma_{i\downarrow}^{+}|0\rangle = |s\rangle$

TABLE I

Eigenvectors and eigenvalues of H_0 of lower energy

Number of electrons	Eigenvectors of H_0 [a]	Energy
0	$\|0\rangle = \|0;0\rangle$	0
1	$\|d_\uparrow\rangle = \cos\theta\|0;\uparrow\rangle + \mathrm{sen}\,\theta\|+;0\rangle$ $\|d_\downarrow\rangle = \cos\theta\|0;\downarrow\rangle + \mathrm{sen}\,\theta\|-;0\rangle$ $\theta = tg^{-1}(-2V/(\varepsilon_f+\sqrt{\varepsilon_f^2+4V^2}))$	$E_d = \frac{\varepsilon_f - \sqrt{\varepsilon_f^2+4V^2}}{2}$
2	$\|s\rangle = \cos\phi\|0;\uparrow\downarrow\rangle + \frac{\mathrm{sen}\,\phi}{\sqrt{2}} \cdot (\|-;\uparrow\rangle - \|+;\downarrow\rangle)$ $\phi = tg^{-1}(-2\sqrt{2}V/(\varepsilon_f+\sqrt{\varepsilon_f^2+8V^2}))$	$E_s = \frac{\varepsilon_f - \sqrt{\varepsilon_f^2+8V^2}}{2}$

[a] Occupation of spin states of f-electrons (+,-) and of c-electrons (↑,↓) are indicated on left and right sides of kets. The zero in energy scale is set at ε_c.

and the corresponding number operator $\nu_{i\sigma} = \gamma_{i\sigma}^{+}\gamma_{i\sigma}$ the projected Hamiltonian turns out to be given by (4):

$$\tilde{H} = \tilde{H}_0 + \tilde{H}_1 + \tilde{H}_2 \tag{2}$$

where

$$\begin{aligned} \tilde{H}_0 &= \sum_{i\sigma} E_d \nu_{i\sigma} + \cos^2\theta \sum_{\sigma, i\neq j} t_{ij} \gamma_{i\sigma}^{+}\gamma_{j\sigma} \\ \tilde{H}_1 &= \frac{1}{2} I \sum_{i\sigma} \nu_{i\sigma}\nu_{i\bar{\sigma}} + a\cos^2\theta \sum_{\sigma, i\neq j} t_{ij} \gamma_{i\sigma}^{+}\gamma_{j\sigma}(\nu_{i\bar{\sigma}} + \nu_{j\bar{\sigma}}) \\ \tilde{H}_2 &= a^2\cos^2\theta \sum_{\sigma i\neq j} t_{ij} \gamma_{i\sigma}^{+}\gamma_{j\sigma}\nu_{i\bar{\sigma}}\nu_{j\bar{\sigma}} \end{aligned}$$

and

$$a = \cos\phi\,(1 + \tfrac{1}{\sqrt{2}}\,\mathrm{tg}\,\theta\,\mathrm{tg}\,\phi) - 1$$

$$I = E_s - 2E_d \qquad (3)$$

The one-body term $\tilde{H}_0$ gives a band-like dispersion given by:

$$\varepsilon(\underline{k}) = E_d + \cos^2\theta\,|t|\,S(\underline{k}) \qquad (4)$$

with

$$S(\underline{k}) = -\sum_{\underline{\delta}\,(n.n.)} e^{i\underline{k}\cdot\underline{\delta}}$$

We will need in the following the expressions in terms of the new operators, of the conduction-electron occupation number and of the total magnetic moment at any site i. They are:

$$n_i^{(c)} = \sum_\sigma n_{i\sigma}^{(c)} = \sum_\sigma \{\cos^2\theta\,\nu_{i\sigma} + (\mathrm{sen}^2\theta - \frac{\mathrm{sen}^2\phi}{2})\,\nu_{i\sigma}\nu_{i\bar{\sigma}}\} \qquad (5)$$

$$m_i^{(+)} = m_{xi} + i m_{yi} = \frac{\mu_B}{2}(g_c\cos^2\theta + g_f\,\mathrm{sen}^2\theta)\,\gamma_{i\uparrow}^{\dagger}\gamma_{i\downarrow} \qquad (6)$$

where the two gyromagnetic ratios g_c, g_f will be assumed to be equal in the following.

We derive now a simple relation between the dynamic transverse susceptibility

$$\chi_q = \mathrm{F.T.}\{i\langle T(m_i^{(+)}(t)\,m_j(0))\rangle\}\;;\quad q \equiv (\underline{q},\omega) \qquad (7)$$

and "valence" of the system $\langle n_i^{(c)}\rangle$: if we note that $\langle \nu_{i\uparrow}\nu_{i\downarrow}\rangle = \langle \nu_{i\uparrow}\rangle - \langle m_i^{(+)} m_i^{(-)}\rangle$ and we insert this in eq.(5) we easily get:

$$\langle n_i^{(c)}\rangle = \sum_\sigma \Big[(1 - \frac{\mathrm{sen}^2\phi}{2})\langle \nu_{i\sigma}\rangle - (\mathrm{sen}^2\theta - \frac{\mathrm{sen}^2\phi}{2})\cdot \frac{1}{N}\sum_{\underline{q}}\int \frac{d\omega}{2\pi i}\Big(\chi_q/(\tfrac{1}{2}g\mu_B)^2\Big)\Big] \qquad (8)$$

In order to derive an equation for χ_q we need to set up a diagrammatic expansion for the

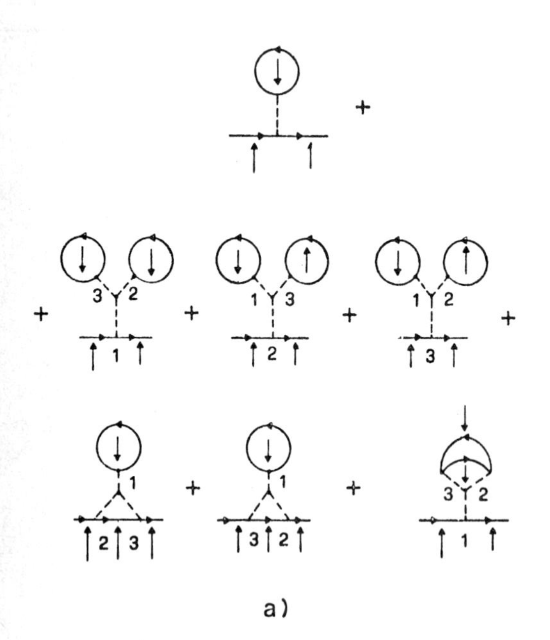

b)

c)

$$= I + a\cos^2\theta\,|t|\,[S(\underline{k}) + S(\underline{k}') + S(\underline{k}'') + S(\underline{k}''')]$$

$$= a^2\cos^2\theta\,|t|\,S(\underline{k}_1 + \underline{k}_3 - \underline{k}_2)$$

Fig. 1 - a) Normal self-energy diagrams.
b) Self-energy diagrams with spin-flip due to the external magnetic field
c) The bare vertices. Total momentum conservation is required.

Hamiltonian (2), treating $\tilde{H}_1$ and $\tilde{H}_2$ as a perturbation. The Feynman diagrams for the self-energy in the 4-momentum space and in the Hartree-Fock approximation are given in Figs. (1a,b). The diagrams are written up to first-order in an external magnetic field. Those in Fig. 1b) vanish in zero external field, and will be of use later for the evaluation of χ_q.

In zero field (and in the paramagnetic region) the self-energy is given by:

$$\Sigma_{HF}(\underline{k}) = \frac{I\nu}{2} - 2a\cos^2\theta\,|t|\,z\tau(1+a\frac{\nu}{2}) + \tag{9}$$
$$+ a\cos^2\theta\,|t|\,(\nu + a\frac{\nu^2}{4} - 3a\tau^2)\,S(\underline{k})$$

where:

$\nu = \sum_\sigma \langle \nu_{i\sigma} \rangle$; z=number of nearest neighbors

and

$$\tau \equiv \tau_\sigma = \frac{1}{N}\sum_{\underline{k}} e^{i\underline{k}\cdot\underline{\delta}} f_{\underline{k}\sigma} \; ; \quad f_{\underline{k}\sigma} = \langle \nu_{\underline{k}\sigma} \rangle \tag{10}$$

$\underline{\delta}$ is anyone of the lattice vectors connecting nearest-neighbors. At equilibrium, the occupation numbers $f_{\underline{k}\sigma}$ will have the same symmetry as the lattice, so τ is in fact independent on which vector $\underline{\delta}$ is chosen. The quasi-particle energies are now given by:

$$\begin{aligned} \varepsilon_{HF}(\underline{k}) &= \varepsilon(\underline{k}) + \Sigma_{HF}(\underline{k}) \\ &= \tilde{E}_d + W\cos^2\theta\,|t|\,S(\underline{k}) \end{aligned} \tag{11}$$

with

$$W = 1 + a\nu + a^2\frac{\nu^2}{4} - 3a^2\tau^2 \,. \tag{12}$$

Apart from an unimportant shift of the band bottom the main effect of the many-body interaction resides in the renormalization of the band width by a factor W, eq.(12). This is in agreement with the results of Foglio et al.(3).

The irreducible vertex for the Bethe-Salpeter equation for the transverse dynamic susceptibility is obtained in a standard way from the "anomalous" diagrams of Fig. 1b) by cutting the internal lines in all possible ways and then setting the external field equal to zero. We find the following integral equation:

$$\chi_q(\underline{k}) = \chi_q^{(0)}(\underline{k})\Big[1 + \tilde{I}_q \frac{1}{N}\sum_{\underline{k}'} \chi_q(\underline{k}') + \tag{13}$$
$$+ \frac{1}{N}\sum_{\underline{k}'}(A_q(\underline{k}) + A_q(\underline{k}'))\chi_q(\underline{k}') + \frac{1}{N}\sum_{\underline{k}'} B_q(\underline{k},\underline{k}')\chi_q(\underline{k}')\Big]$$

where

$$\chi_q = \frac{1}{N}\sum_{\underline{k}} \chi_q(\underline{k})$$

and

$$\chi_q^{(0)}(\underline{k}) = \frac{f_{\underline{k}\sigma} - f_{\underline{k}+\underline{q}\bar{\sigma}}}{\varepsilon_{\underline{k}+\underline{q}} - \varepsilon_{\underline{k}} - \omega - i\eta\,\mathrm{sgn}\,\omega} \; ; \; (\eta = 0^+)$$

with

$$\tilde{I}_q = I - 2a^2\cos^2\theta\,|t|\,\tau\,S(\underline{q})$$

$$A_q(\underline{k}) = a\cos^2\theta\,|t|\,(1+a\frac{\nu}{2})(S(\underline{k}) + S(\underline{k}-\underline{q}))$$

$$B_q(\underline{k},\underline{k}') = -2a^2\cos^2\theta\,|t|\,\tau\,S(\underline{k}+\underline{k}'-\underline{q})$$

For $\underline{q}\to 0$, $\omega/|\underline{q}|\to 0$, and using a simple symmetry argument (cfr. the discussion after eqs.(10)), eq.(13) can be solved exactly for the static uniform susceptibility, yielding:

$$\chi = \frac{\chi^{(0)}}{1 - \left[I + 2a\cos^2\theta\,|t|\left((2+a\nu)S_0 + az\tau(1+\frac{S_0^2}{z^2})\right)\right]\chi^{(0)}}$$
$$\chi^{(0)} = g(S_0)/\cos^2\theta\,|t|\,W \tag{14}$$

where S_0 is the value of $S(\underline{k})$ at the Fermi surface and $g(S_0)$ is the bare density of states per one spin direction in dimensionless variables. The equation for the para-ferromagnetic instability, $\chi^{-1} = 0$ is then given by

$$I = \cos^2\theta\,|t|\left[\frac{W}{g(S_0)} - 2a(2+a\nu) - 2a^2z\tau(1+\frac{S_0^2}{z^2})\right] \tag{15}$$

The numerical results obtained from eq.(15) for a simple cubic structure and for a rectangular density of states are in agreement with those of Foglio et al. (2).

Let discuss eq.(15) in two limiting cases:
i) $V \gg \varepsilon_f$; in this limit eq.(15) gives:

$$V g(S_0) \sim |t| \tag{16}$$

which means that the system is magnetic within most of the range of applicability of the theory ($V \gg |t|$), except for S_o very near to the band edges (i.e. $\nu \sim 0$ or $\nu \sim 2$). ii) $\varepsilon_f \gg V$; here one finds:

$$V g(S_o) \simeq \frac{|t|}{2} \left(\frac{\varepsilon_f}{V}\right)^3 \tag{17}$$

i.e. the system is non magnetic except for very low $|t|$'s.

We now turn to the evaluation of the full dynamic susceptibility. The last term of eq.(13) contains the nonseparable kernel $B_q(\underline{k},\underline{k}')$ which is however proportional to $a^2\tau$; the latter quantity is very small, of order 10^{-3}, within the range of applicability of the theory. As it can only give rise to small corrections (cfr. also eq.(15)) it will be altogether neglected. The remaining equation can be solved at once, yielding:

$$\chi_q = \chi_q^{(o)}(1-\tilde{I}_q\chi_q^{(o)})\left\{\left[(1-\tilde{I}_q\chi_q^{(o)}) - \frac{a}{2}\cos^2\theta\,|t|\,(2+a\nu)\eta_q\right]^2 - \frac{a^2}{4}\cos^4\theta\,|t|^2(2+a\nu)^2\rho_q\chi_q^{(o)}\right\}^{-1} \tag{18}$$

where

$$\chi_q^{(o)} = \frac{1}{N}\sum_{\underline{k}} \chi_q^{(o)}(\underline{k}) \tag{19a}$$

$$\eta_q = \frac{1}{N}\sum_{\underline{k}} [S(\underline{k}) + S(\underline{k}-q)]\,\chi_q^{(o)}(\underline{k}) \tag{19b}$$

$$\rho_q = \frac{1}{N}\sum_{\underline{k}} [S(\underline{k}) + S(\underline{k}-q)]^2\chi_q^{(o)}(\underline{k}) \tag{19c}$$

Now in the sums of eqs.(19 b,c) the most relevant contributions come from the region where $S(\underline{k}) \sim S(\underline{k}-q) \sim S_o$ so that by approximating (5) η_q and ρ_q as:

$$\eta_q \simeq 2S_o\chi_q^{(o)} \;;\; \rho_q \simeq 4S_o^2\chi_q^{(o)} \tag{20}$$

we obtain $\eta_q^2 \simeq \rho_q\chi_q^{(o)}$ and hence:

$$\chi_q \simeq \frac{\chi_q^{(o)}}{1-[\tilde{I}_q + 4a\cos^2\theta\,|t|\,S_o(1+a\nu/2)]\chi_q^{(o)}} \tag{21}$$

Eq. (21) is an expression similar to what one would obtain in the RPA from a Hubbard-type hamiltonian with a renormalized (q-dependent) two-particle interaction:

$$\bar{I} = \tilde{I}_q + 4a\cos^2\theta\,|t|\,S_o(1+a\nu/2) \tag{22}$$

Coming back now to eq.(8) and setting

$$\Delta\chi_q = (\chi_q - \chi_q^{(o)})/(\tfrac{1}{2}g\mu_B)^2 \tag{23}$$

we see that the expression for the valence can be written:

$$\langle n_i^{(c)}\rangle = \sum_\sigma\left[\left(1-\frac{\mathrm{sen}^2\phi}{2}\right)\nu_{i\sigma} - \left(\mathrm{sen}^2\theta - \frac{\mathrm{sen}^2\phi}{2}\right)\cdot\left(\langle\nu_{i\sigma}\rangle(1-\langle\nu_{i\bar\sigma}\rangle) + \frac{1}{N}\sum_q\int\frac{d\omega}{2\pi i}\Delta\chi_q\right)\right] \tag{25}$$

where we have used the (obvious) result:

$$\frac{1}{N}\sum_q\int\frac{d\omega}{2\pi i}\chi_q^{(o)} = \left(\frac{g\mu_B}{2}\right)^2\langle\nu_{i\sigma}\rangle(1-\langle\nu_{i\bar\sigma}\rangle) \tag{25}$$

Noting that $\langle\nu_{i\sigma}\rangle = \nu/2$ in the paramagnetic region eq.(24) becomes:

$$\langle n_i^{(c)}\rangle = \nu\cos^2\theta + \left(\mathrm{sen}^2\theta - \frac{\mathrm{sen}^2\phi}{2}\right)\cdot\left(\nu^2/2 - 2\frac{1}{N}\sum_q\int\frac{d\omega}{2\pi i}\Delta\chi_q\right) \tag{26}$$

This result suggests that valence fluctuations may be driven, and enhanced, by spin fluctuations near the ferromagnetic instability. The detailed evaluation of the integral in the last term of eq.(26) will be reported elsewhere, in a forthcoming paper (6).

REFERENCES AND FOOTNOTES

(1) Foglio, M.E. and Falicov, L.M., Phys. Rev. B20, 4554 (1979)

(2) Foglio, M.E., Balseiro, C.A. and Falicov,L.M. Phys. Rev. B20, 4560 (1979)

(3) Foglio, M.E. and Falicov, L.M., Phys. Rev. B21, 4154 (1980)

(4) See Ref. 1 for more details

(5) Let's remark that the approximation of eq.(20) becomes exact in cases of high symmetry, such as a simple cubic lattice and a half filled band.

(6) Ottaviani, E., Napoli, F., Galleani d'Agliano, E. and Morandi, G., to be published

Valence Instabilities
P. Wachter and H. Boppart (eds.)

THE EFFECT OF La AND Y SUBSTITUTION ON THE MAGNETIC PROPERTIES OF $CeIn_3$

W.H. Dijkman, W.H. de Groot, F.R. de Boer and P.F. de Châtel,

Natuurkundig Laboratorium, Universiteit van Amsterdam,
Valckenierstraat 65, 1018 XE Amsterdam, The Netherlands

$CeIn_3$ is known to be a 'concentrated Kondo system', that is, its properties give indications of the instability of the trivalent state of cerium. To test the stability of the valence state of Ce in this compound, we have prepared La- and Y-substituted quasibinary compounds and performed lattice-constant, susceptibility and magnetization measurements.

1. INTRODUCTION

If it were not for the observation of a resistivity minimum at 175 K |1,2|, $CeIn_3$ could be thought of as a straightforward antiferromagnet. Heat capacity |2,3| and susceptibility |4,5| measurements lead to a picture of stable moments on the Ce atoms, constrained by a cubic crystal field. The crystal-field splitting is Δ/k_B = (155±30) K, the Γ_7 doublet being the ground state. At 10.1 K antiferromagnetic ordering sets in, with an ordered moment |6| of $(0.65\pm0.1)\mu_B$, somewhat smaller than, but not inconsistent with the saturation moment of the Γ_7 doublet ($0.71\mu_B$).

Like $CeIn_3$, $CeSn_3$ also crystallizes in the Cu_3Au structure. The latter being one of the well known mixed-valent Ce compounds, quasibinary $Ce(In,Sn)_3$ compounds have been extensively used in the study of valence instabilities |7-10|. As expected, mixed valence is observed on the Sn-rich side, while local-environment effects blur the picture at higher In concentrations |2,11|. Such effects are less likely to occur in quasibinaries where the substitution takes place outside the nearest-neighbour shell of Ce atoms. Indeed, in Y- and La-substituted $CeSn_3$ |11,12| no indication of inhomogeneous behaviour was seen.

In the present communication we give the results of experiments in which the stability of the valence of cerium in $CeIn_3$ was tested by substituting some of the Ce by La or Y. Both La and Y are expected to be chemically very similar to trivalent Ce, but the atomic volumes of the three metals are rather different. The atomic volume of La (22.5 cm^3/mole) exceeds that of both trivalent (21.6 cm^3/mole) and tetravalent (16.0 cm^3/mole) Ce; whereas the atomic volume of Y (19.9 cm^3/mole) falls between those of the two Ce valence states |13|. On the basis of 'lattice pressure' arguments one would expect thus that the effects of La and Y substitution be opposite. Of these, Y substitution is expected to destabilize the trivalent state of Ce.

2. SAMPLE PREPARATION

Quasibinary $(Ce,La)In_3$ and $(Ce,Y)In_3$ compounds were prepared by argon arc melting. A series of $(La,Y)In_3$ compounds was also made, to serve as dummies for the $(Ce,Y)In_3$ system. Details of the sample preparation procedure are given in ref.11.

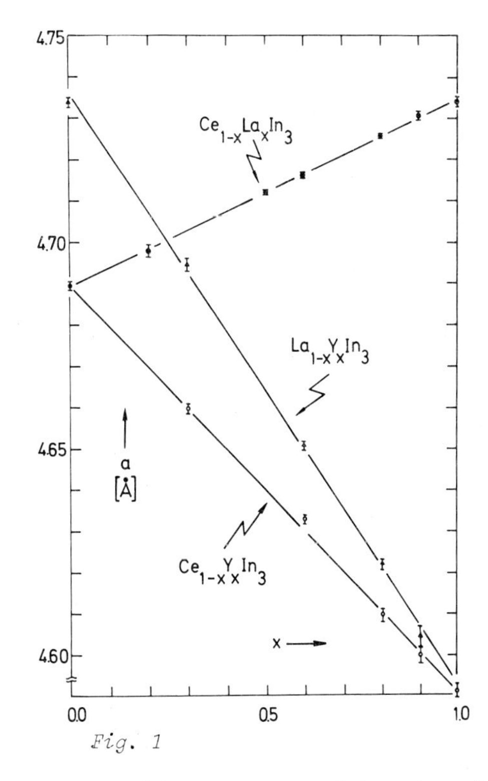

Fig. 1

All samples were found to crystallize in the Cu_3Au structure; no second phase could be detected by X-ray diffraction. The room-temperature lattice parameters as functions of composition are shown in fig. 1. In all three systems, Vegard's law is seen to be obeyed.

3. MAGNETIC PROPERTIES

The contrasting effect of La and Y substitution on the valence state of Ce in $CeIn_3$ is clearly seen from the magnetic properties. While no rigorous proof can be given, there is very strong

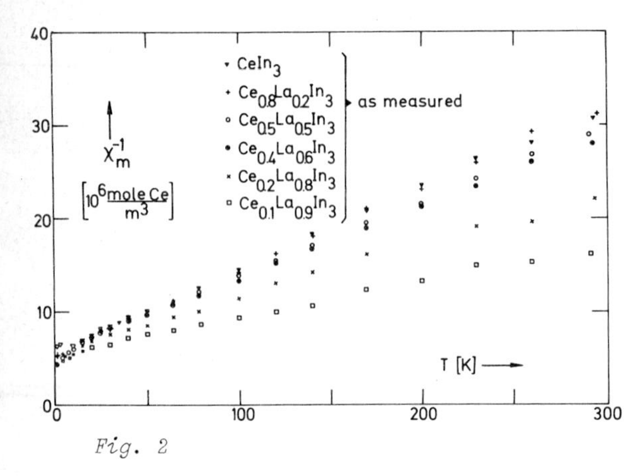

Fig. 2

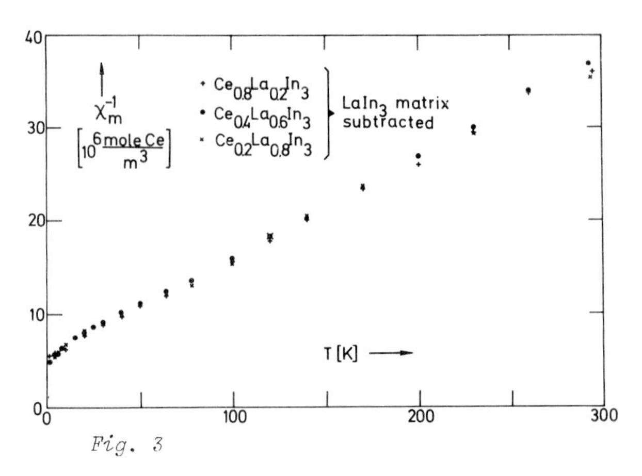

Fig. 3

indication that La substitution only affects the interactions between Ce moments, while Y substitution leads to a mixed-valent state of Ce. This emerges from the temperature dependence of the susceptibility and especially from the high-field magnetization, which was measured in fields up to 20 T.

The temperature dependence of the reciprocal susceptibility of a number of $(Ce,La)In_3$ compounds is shown in fig. 2. To extract the 4f contribution, the susceptibility of $LaIn_3$, which was found to be slightly temperature dependent, has been subtacted from the measured susceptibility values. The resulting curves are shown in fig. 3 for three compositions: the susceptibility per Ce atom is seen to be practically independent of composition. The same is true for the effective moments ($\sim 2.45\ \mu_B$) and paramagnetic Curie temperatures (~ 50 K) derived from the straight portions, between 50 and 300 K, of these curves.

The susceptibility vs temperature curves of a number of $(Ce,Y)In_3$ samples are shown in fig. 4. The low-temperature part of the curves for the Y-rich compounds can be analysed as a sum of a constant value (79 and 97×10^{-9} m^3/mole Ce for 80 and 90% Y, respectively) and a Curie term, C/T, with C corresponding to less than 1% of the Ce atoms acting as magnetic 'impurities', with an effective moment of $2.54\ \mu_B$. To obtain the 4f contribution, we subtracted from the impurity-corrected susceptibility the susceptibility of the corresponding $(La,Y)In_3$ compounds, which was also corrected for a low-temperature impurity contribution. Unlike in the case of the $(Ce,La)In_3$ system, the resulting susceptibility curves do not coincide (fig. 5). The 4f contribution determined in this way clearly becomes constant at low temperatures in the two samples richest in Y. After the two correction procedures and normalization to mole Ce atoms, the resolution does not suffice to establish whether the susceptibility goes through a maximum.

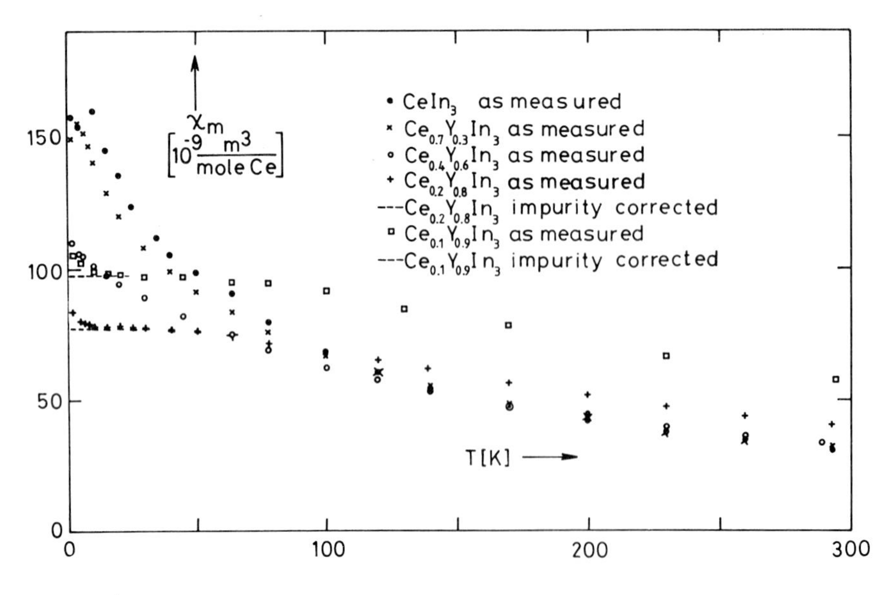

Fig. 4

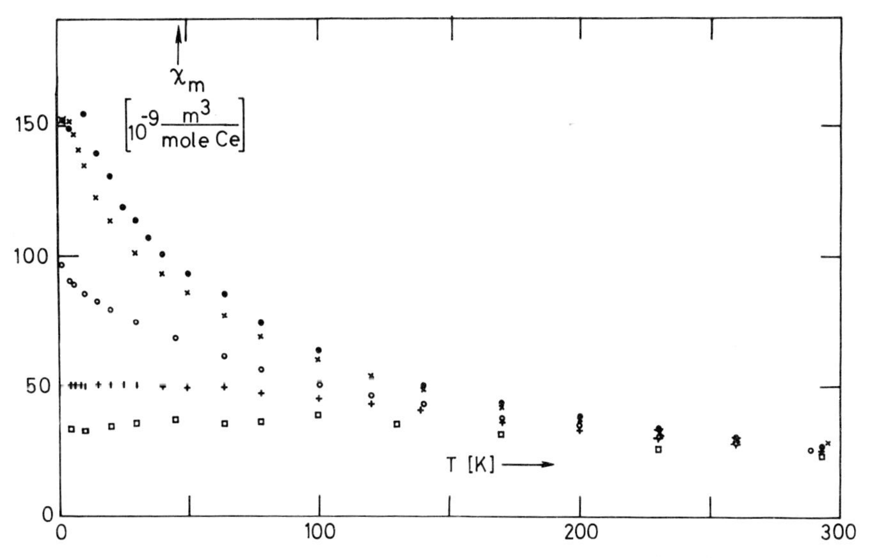

Fig. 5 (See fig. 4 for the meaning of the symbols)

The magnetization curves of (Ce,La)In_3, measured at 4.2 K in fields up to 20 T, are shown in fig. 6. The slight concentration dependence suggested by this figure is considerably changed and reduced in amplitude, if a 'matrix' magnetization, taken to be equal to that of $LaIn_3$, is subtracted from all measured values (fig. 7). A similar convergence cannot be achieved with (Ce,Y)In_3 (figs 8 and 9), where the approximate concentration independence of the magnetization breaks down beyond 30% Y. At 90% Y, the magnetization per Ce atom is reduced by a factor of four compared to the one in the Ce-rich compounds, and the slight curvature observed in the latter (as well as in all (Ce,La)In_3 samples) is not present. The slope of this straight magnetization curve agrees within 1% with the low temperature susceptibility represented in fig. 5 |10|.

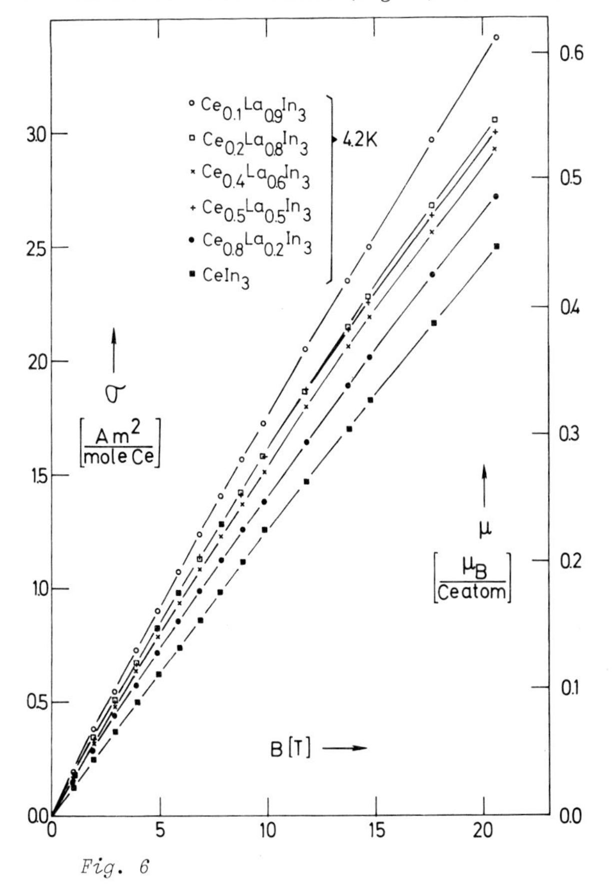

Fig. 6

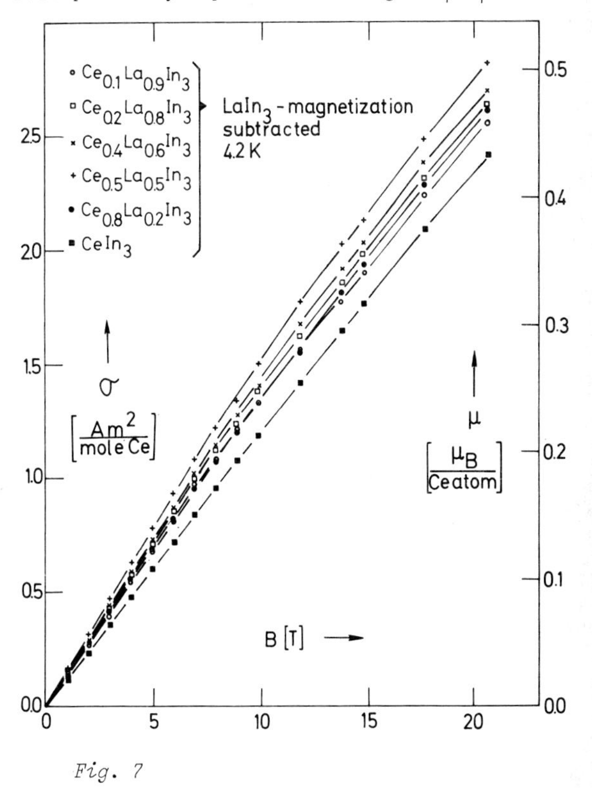

Fig. 7

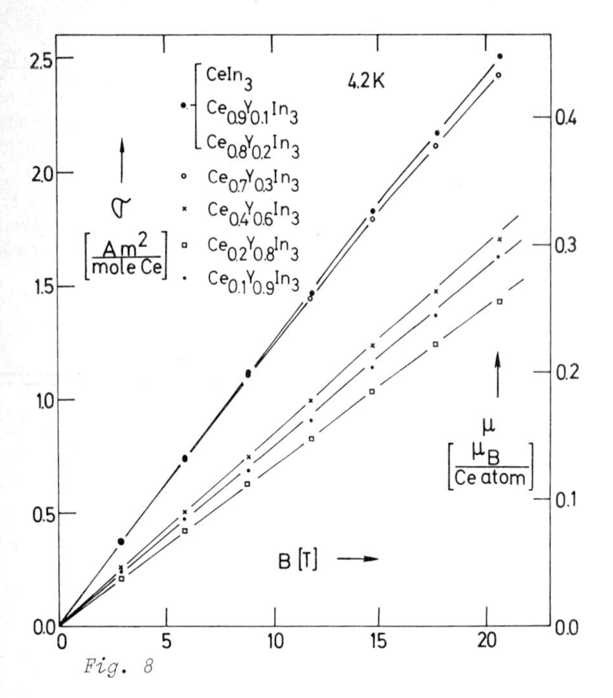

Fig. 8

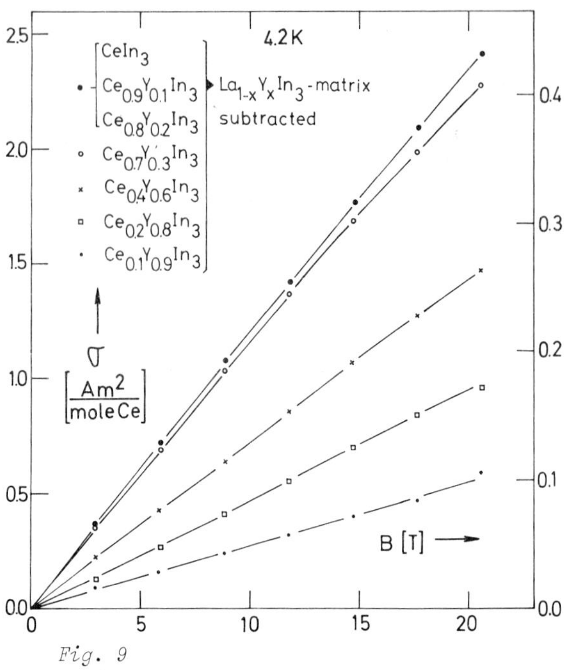

Fig. 9

4. DISCUSSION

It is clear from the above brief description of the magnetic properties of $CeIn_3$-based quasibinaries that in the $(Ce,La)In_3$ system Ce is trivalent throughout the concentration range. The importance of the factors that influence the magnetic behaviour below 50 K (crystal fields, Kondo effect) does not diminish or increase with La substitution. On the other hand, Y substitution appears to destabilize the trivalent state of Ce, leading to characteristic mixed-valent behaviour at the Y-rich end of the $(Ce,Y)In_3$ system.

Destabilization of the trivalent state of Ce upon Y substitution would be expected on the basis of 'lattice pressure' arguments. While such arguments are seldom reliable at a quantitative level, here again, they seem to lead to correct qualitative predictions. However, the fact that Vegard's law is obeyed at room temperature can be seen as evidence that the volume taken up by each Ce atom is independent of concentration. Clearly, other factors (e.g., electronegativity) may also be at play.

Similarly contrasting effects of La and Y substitution have been observed before in $CeAl_2$ |14|. The similarity is not limited to the appearance of mixed-valent behaviour in Y-rich compounds, it is also present in the Ce-rich compounds, as is evident from the rate at which antiferromagnetism is suppressed. In both systems, Y substitution is found to have a moredetrimental effect on antiferromagnetism, which is expected if one assumes that La substitution merely influences the interaction between Ce moments, whereas Y substitution also acts to destabilize the moments.

Summarizing, the magnetic properties of $(Ce,La)In_3$ and $(Ce,Y)In_3$ quasibinary compounds provide sufficient evidence to conclude that La substitution leads to rather uninteresting 'dilution' effects, while Y substitution results in a low-concentration mixed-valent system which deserves further investigation.

REFERENCES

|1| Daal, H.J. van and Buschow, K.H.J., Phys. Stat. Sol. (a) 3 (1970) 853.

|2| Elenbaas, R., thesis, University of Amsterdam (1980).

|3| Diepen, A.M. van, Craig, R.S. and Wallace, W.E., J. Phys. Chem. Solids 32 (1971) 1867.

|4| Tsuchida and Wallace, W.F., J. Chem. Phys. 43 (1965) 3811.

|5| Lethuillier, P., thesis, University of Grenoble (1976).

|6| Lawrence, J.M. and Shapiro, S.M., Phys. Rev. B22 (1980) 4379.

|7| Lawrence, J. and Murphy, D., Phys. Rev. Letters 40 (1978) 961.

|8| Lawrence, J., Phys. Rev. B20 (1979 3770.

|9| Béal-Monod, M.T. and Lawrence, J.M., Phys. Rev. B21 (1980) 5400.

|10| Dijkman, W.H., Boer, F.R. de, Châtel, P.F. de and Aarts, J., J. Magn. Magn. Mat. 15-18 (1980) 970.

|11| Dijkman, W.H., thesis, University of Amsterdam (1982).

|12| Dijkman, W.H., Boer, F.R. de and Châtel, P.F. de, Physica B98 (1980) 271.

|13| Châtel, P.F. de and Boer, F.R. de, in Valence Fluctuations in Solids (Falicov, L.M., Hanke, W. and Maple M.B., eds) North-Holland Publ. Co. (1981) Amsterdam, p. 377.

|14| Aarts, J., Boer, F.R. de, Horn, S, Steglich, F. and Meschede, D., ibid.p. 301.

Valence Instabilities
P. Wachter and H. Boppart (eds.)

NEUTRON DIFFRACTION EVIDENCE FOR LONG-RANGE FERROMAGNETIC MOMENT OF Tm IN INTERMEDIATE-VALENT $TmSe_{1-x}Te_x$ ($0.1 \lesssim x \lesssim 0.2$)

P. Fischer and W. Hälg, Institut für Reaktortechnik, ETH,
CH-5303 Würenlingen

P. Schobinger-Papamantellos, Institut für Kristallographie
und Petrographie, ETH,
CH-8092 Zürich

H. Boppart, E. Kaldis and P. Wachter, Laboratorium für Festkörperphysik, ETH,
CH-8093 Zürich

Elastic neutron-scattering investigations were performed on polycrystalline and single-crystal samples of intermediate-valent $TmSe_{1-x}Te_x$ ($0.1 \lesssim x \lesssim 0.2$) in the temperature range from 1.3 to 293 K. In contrast to antiferromagnetic TmSe and TmTe all samples exhibit long-range ferromagnetism in zero external magnetic field, with Curie points $T_C \geqq 3.4$ K and ordered magnetic moments $\mu_{Tm} \leqq 1.6(1)\ \mu_B$. No antiferromagnetic Bragg peaks were detected. The present results accomplish bulk magnetic measurements and support Varma's theory of mixed-valence behaviour of TmSe based on double-exchange coupling.

1. INTRODUCTION

The stoichiometric intermediate-valent metal $Tm^{2.8+}Se$ with NaCl structure has an antiferromagnetic ground state, corresponding to a fcc type I configuration, a Néel temperature T_N of 3.2 K and an ordered magnetic moment μ_{Tm} = 1.7 to 1.9 μ_B (1,2). Nonstoichiometric Tm_xSe (x<1) tends to Tm^{3+} (3) and stabilizes fcc type II antiferromagnetism (1,4). Also the semiconductor TmTe with valency 2+ of Tm appears to be antiferromagnetic, but the Néel point is considerably lower: T_N = 0.21 K (5). On the other hand macroscopic measurements of magnetic properties of intermediate-valent $TmSe_{1-x}Te_x$ ($0.09 \leqq x \leqq 0.17$) (6÷8) show the remarkable existence of a net spontaneous magnetization (0.1 to 0.8 μ_B per Tm ion for x = 0.09 and 0.17 respectively). Also at the latter composition the paramagnetic Curie temperature is still negative, indicating a certain degree of antiferromagnetic interactions. The substitution of Se by Te increases the lattice constant and reduced the valency of Tm towards 2+ (9), yielding ~2.7 at x = 0.17 (6). This change enhances ferromagnetic interactions via double exchange (10). In order to determine the resulting magnetic structures of $TmSe_{1-x}Te_x$ ($x \lesssim 0.2$) neutron diffraction investigations were performed.

2. EXPERIMENTAL

Table 1 contains details concerning the investigated samples (enclosed in cylindrical vanadium containers under helium atmosphere). The neutron measurements were performed on a two-axis diffractometer with tilting detector, situated at reactor Saphir, Würenlingen. At temperatures T $\leqq$ 4.2 K the samples were immersed in liquid helium. Neutron wavelengths λ = 2.339 and 1.070 Å were used for the powder- and single-crystal studies respectively. The investigated single crystal had dimensions 1.5x1.2x5 mm^3, and the vertical (001) axis was parallel to the longest edge. Absorption corrections were applied only to powder intensities. Single crystal data were corrected for $\lambda/2$ contaminations. Scattering lengths b_{Tm}=0.72 (10^{-12}cm), b_{Se}=0.80 and b_{Te}=0.58 as well as relativistic neutron-magnetic form factors were used in dipole approximation. Results of the data evaluation are summarized in Tab. 1. In the case of the single crystal study an overall temperature factor B = 0.2(1)Å^2 and an agreement ratio $R_{|F|^2}$ = 4.2 % were obtained for 19 independent reflections measured at 4.2 K (concerning powder measurements cf. Fig. 2).

3. LONG-RANGE FERROMAGNETIC ORDERING

The appearance of a narrow 111 Bragg peak of considerable intensity in the 1.5 K neutron diffraction pattern of polycrystalline $TmSe_{0.84}$-$Te_{0.16}$ is shown in Fig. 1. It proves the existence of a long-range ferromagnetic moment of Tm in the intermediate valence system $TmSe_{1-x}Te_x$, in agreement with (6÷8). From the absence of antiferromagnetic peaks and because of the large ordered moment observed at 1.5 K: μ_{Tm} = 1.6(1)μ_B, similar to antiferromagnetic TmSe (1,2), we conclude that the sample is ferromagnetic at low temperatures in zero external magnetic field. On the other hand one cannot exclude weak antiferromagnetic intensity, e.g. due to short-range correlations, which might be hidden in the background. Fig. 2 illustrates the good agreement between observed and calculated profile intensities (Cf. also Tab. 1. The negative value of the overall temperature factor B is presumably caused by residual errors in absorption correction.) Fig. 3 shows the temperature dependence of ferromagnetic intensity measured on the powder specimens. Apparently the Curie point T_C

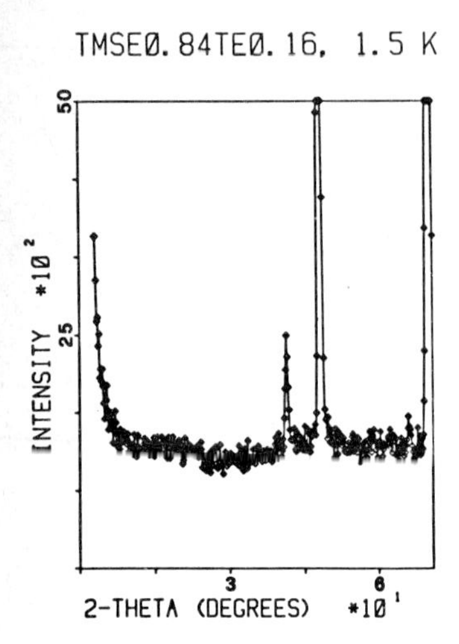

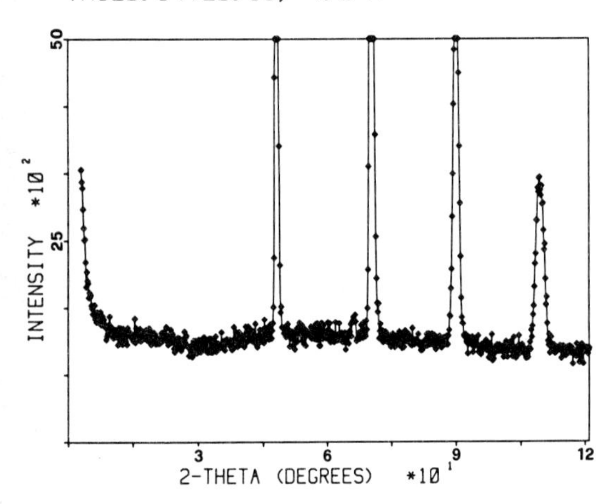

Figure 1: Neutron diffraction patterns of polycrystalline $TmSe_{0.84}Te_{0.16}$: a) at 1.5 K in the ferromagnetic state, b) at 4.2 K corresponding to paramagnetism. The weak impurity peak at scattering angle $2\theta = 66.4^{o}$ was also observed at high temperatures (64 K).

increases from 3.7(1) K to above 4.2 K with increasing concentration of Te, in reasonable agreement with (6÷8).

In none of the measured $TmSe_{1-x}Te_x$ samples antiferromagnetic peaks were detected, although in the case of the single crystal various scans were performed at the lowest temperature. This single crystal (x = 0.17) yields the lowest T_C = 3.4(1) K (cf. Fig. 4 and Tab. 1). Because of possible extinction effects in the case of large nuclear single-crystal intensities (yielding μ_{Tm} = $0.9\mu_B$) the value of the ordered moment μ_{Tm}=0.6

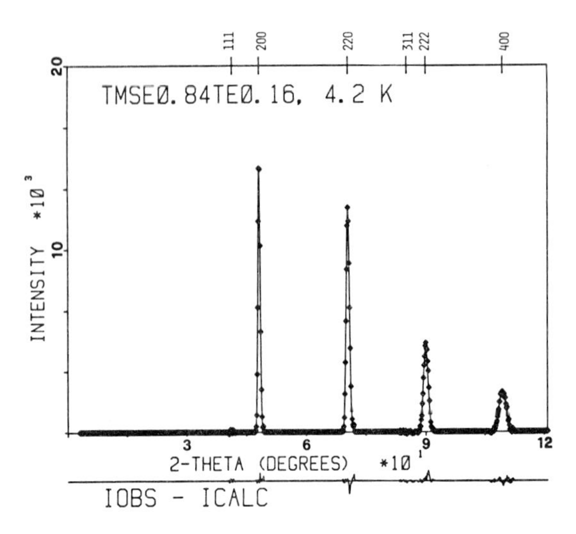

Figure 2: Comparison of observed and calculated profile neutron intensities of $TmSe_{0.84}Te_{0.16}$ powder at 4.2 K. R_n = 2.2 % (at 1.5 K: R_n = 2.7 %, R_m = 2.6 %), a_o = 5.731 Å, B = -1.1(1) Å².

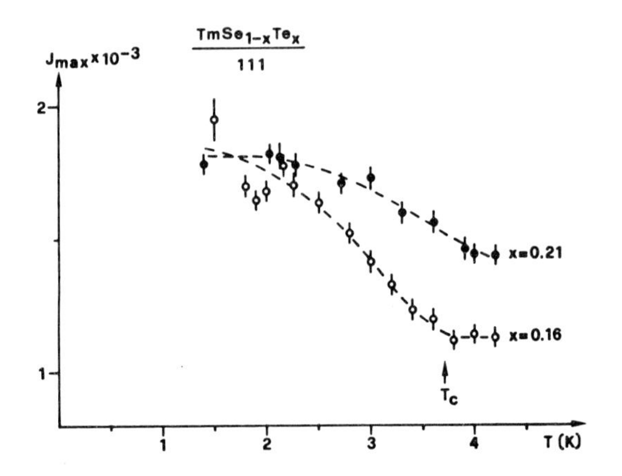

Figure 3: Temperature dependence of maximum neutron intensity of Bragg peak 111 of polycrystalline $TmSe_{1-x}Te_x$.

(3) μ_B is less accurate than for the powder samples. The magnetic origin of the additional neutron intensity observed at low temperatures is evident from a comparison of observed and calculated form factors shown in Fig. 5, which agree within experimental errors. As the halfwidths of Bragg peaks (e.g. 111) of both single crystal and powder samples did not increase significantly on cooling, the ferromagnetic ordering is of long range.

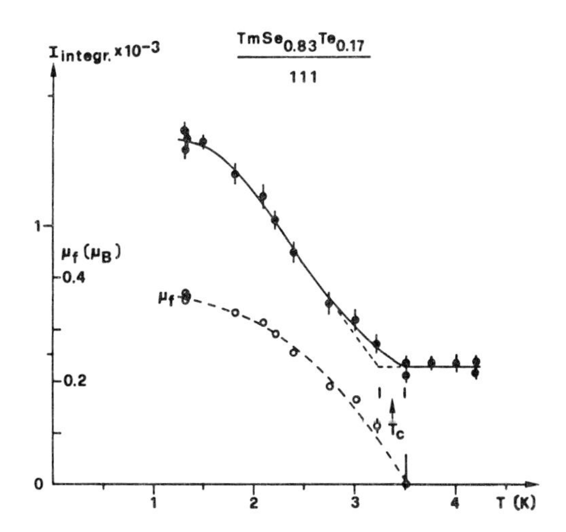

Figure 4: Temperature dependence of integrated neutron intensity of reflection 111 and ordered ferromagnetic moment μ_f of Tm (calculated from weak nuclear intensities) of single crystal $TmSe_{0.83}Te_{0.17}$.

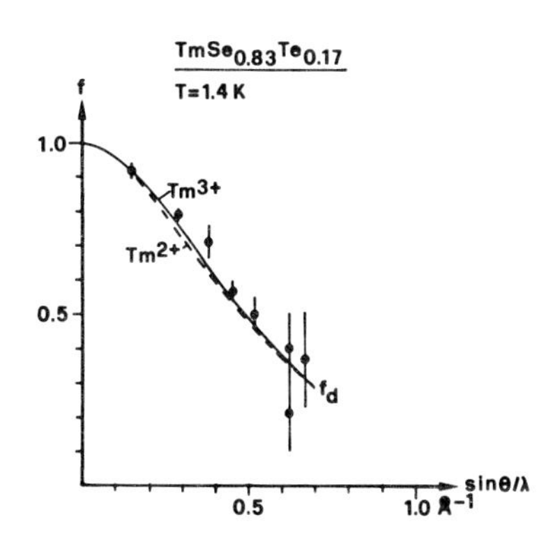

Figure 5: Comparison of observed and calculated neutron-magnetic form factor (dipole approximation) of $TmSe_{0.83}Te_{0.17}$ single crystal.

4. BULK MAGNETIC MEASUREMENTS

On the same single crystal, which was used for the neutron investigations, measurements of magnetic susceptibility and magnetization (cf. Fig. 6,7) were performed. In agreement with the neutron results a net spontaneous magnetization of 0.5(1) μ_B/Tm ion is measured at low temperatures (cf. Fig. 7). Also the temperature dependence of the magnetization (Fig. 6) resembles the ordered moment determined by neutron diffraction (Fig. 4). The peak in the initial magnetic susceptibility suggests domain effects or nonnegligible antiferromagnetic interactions. This could

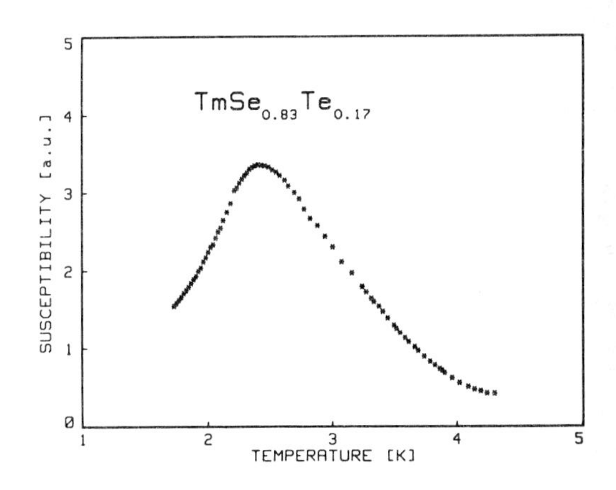

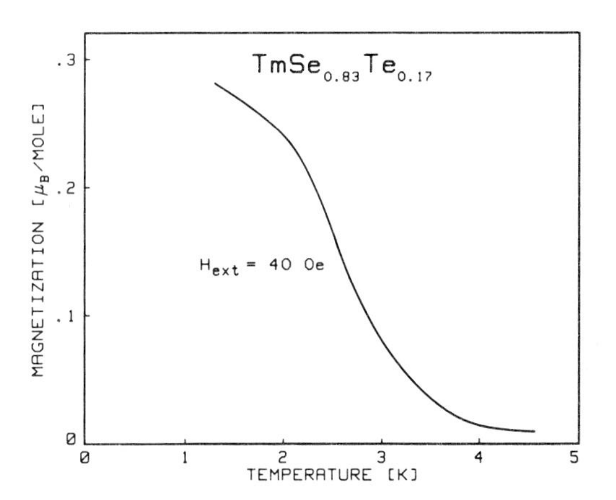

Figure 6: Temperature dependence of a) initial magnetic susceptibility measured in an ac field of 10 Oe and b) magnetization of $TmSe_{0.83}Te_{0.17}$ single crystal.

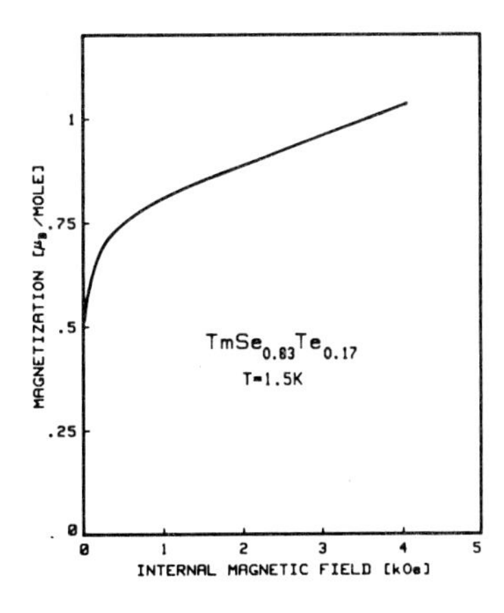

Figure 7: Magnetic field dependence of magnetization (parallel to direction (001)) of $TmSe_{0.83}Te_{0.17}$ at 1.5 K.

Table I: Magnetic properties of Tm_ySe and $TmSe_{1-x}Te_x$.

SC = single crystal, P = powder, A = lattice constant, θ_p = paramagnetic Curie temperature, T_N = Néel temperature, T_C = Curie temperature, μ = ordered magnetic moment, N = neutron scattering, BM = bulk magnetic measurement,*:estimated from lattice constant,**:not single phase (miscibility gap)

x	y	sample	a(Å)	valence of Tm	θ_p (K)	T_N, T_C (K)	magnetic ordering	$\mu_{Tm}(\mu_B)$	method of investigation	ref.
	1	SC	5.71	2.8		3.2	antifmg. fcc I	1.7(2)	N	(1)
	1		5.71	2.8		3.1	"	1.9	N	(2)
	<1	P	5.683	2.8		3.5, 40	magnetic short-range above 3.5 K	2.3	N	(4)
	0.87	SC	5.64	3.0		~4.7	antifmg. fcc II	~0.5	N	(1,2)
0.09						3		0.3	BM	(6÷8)
0.16*		P	5.760			3.7(1)	ferromagnet	1.6(1)	N	
0.17*		SC	5.783			3.4(1)	ferromagnet	0.6(3)	N	
0.17		SC	5.783					0.5(1)	BM	
0.17				~2.7	-17	5.5		0.8	BM	(6÷8)
{0.21*		P**	5.828			≧4.2	ferromagnet	1.2(1)	N	
0.40			6.002							
1			6.364÷ 6.049	2		0.21	antifmg.		BM	(5)

be due to canting of the predominantly ferromagnetic structure or might be associated with antiferromagnetic short-range correlations because of structural defects (such as statistical Se/Te distribution, vacancies, nonstoichiometry, strains).

5. CONCLUSIONS

Similar to bulk magnetic measurements (6÷8) the present neutron diffraction investigations prove the existence of long-range ferromagnetic ordering at zero magnetic field in intermediate-valent $TmSe_{1-x}Te_x$ ($0.1 \lesssim x \lesssim 0.2$). These experimental results support Varma's theory (10) of mixed-valence in TmSe, i.e. ferromagnetism in $TmSe_{1-x}Te_x$ appears to be stabilized by double exchange as consequence of reduced valency of Tm at $x \gtrsim 0.1$.

REFERENCES

(1) Bjerrum Møller H., Shapiro S.M. and Birgeneau R.J., Phys.Rev.Lett. 39 (1977) 1021-1025

Shapiro S.M., Bjerrum Møller H., Axe J.D., Birgeneau R.J. and Bucher E., J.Appl.Phys. 49 (1978) 2101-2106

(2) Vettier C., Flouquet J., Mignot J.M. and Holtzberg F., J.Magn.Magn.Mat. 15-18 (1980) 987-988

(3) Batlogg B., Ott H.R., Kaldis E., Thöni W. and Wachter P., Phys.Rev.B 19 (1979) 247-259

(4) Debray D., Decker D.L., Sougi M., Kahn R. and Buévoz J.L., J.de Phys. 40-C5 (1979) 358

(5) Bucher E., Andres K., di Salvo F.J., Maita J.P., Gossard A.C., Cooper A.S. and Hull G.W., Jr., Phys.Rev.B 11 (1975) 500-513

(6) Batlogg B., Ott H.R. and Wachter P., Phys.Rev.Lett. 42 (1979) 278-281

(7) Batlogg B., Kaldis E. and Wachter P., J.de Phys. 40-C5 (1979) 370-371

(8) Batlogg B., J.Magn.Magn.Mat. 15-18 (1980) 939-941

(9) Kaldis E., Fritzler B., Jilek E. and Wisard A., J.de Phys. 40-C5 (1979) 366-369

Kaldis E. and Fritzler B., J.de Phys. 41-C5 (1980) 135-142

(10) Varma C.M., Solid State Commun. 30 (1979) 537-539

Valence Instabilities
P. Wachter and H. Boppart (eds.)

EXCHANGE INTERACTION IN CERIUM COMPOUNDS

Cesar Proetto[§] and Arturo López

Centro Atómico Bariloche[*] - Instituto Balseiro[#]
8400 Bariloche - Argentina

The exchange interactions between Ce ions caused by hybridization of f-levels to conduction electron states in some Ce compounds are considered in detail [1]. It is shown that the treatment in which these interactions are derived from a multi-site Anderson model through a second order canonical transformation does not include all possible contributions. The correct procedure requires a fourth order canonical transformation [2] and the result coincides with that derived in perturbation theory. Using a free-electron-like band we obtained closed expressions for the range function, which differs from the RKKY function and approaches it only asymptotically. As the localized level moves towards the Fermi level and before entering the full mixed valent regime, the short range behaviour of both functions is different. This means that the interactions for nearest and next nearest neighbors might be determined by a new range function [3]. It is shown that the processes leading to exchange interactions in the hybridization models are not only mediated through the conduction electron spin susceptibility, as results from the second order canonical transformation but also through more complex charge exchange mechanisms. Our results agree qualitatively with those obtained in reference [4].

The effective exchange coupling which is throught to explain complex magnetic orderings in Ce compounds originates in the hybridization between localized and conduction electron states. These fourth order interactions give rise to anisotropic alignements [1] and were first considered by Coqblin and Schrieffer [5], applying a canonical transformation to the Anderson model Hamiltonian. In that way a range function of the RKKY type is obtained.

Such approach does not include all fourth order contributions to the exchange energy and the correct procedure requires a fourth order canonical transformation [2] which takes into account both conduction electron spin polarization and charge fluctuation contributions.

The closed expression for the range function J(R) can be explicitly evaluate for a free electron conduction band.

Results show that the range function approaches the RKKY function only asymptotically. For short distances (first nearest neighbors) both functions are out of phase and differ in amplitude, the RKKY function being usually larger. Any fit to experimental data should take this refined range function into account, especially for the nearest neighbors [3].

We derived our results by means of a canonical transformation, carried out to fourth order in the mixing parameter V, applied to a periodic Anderson model. The aim at each step is to eliminate odd powers of V, as for the Schrieffer-Wolff transformation. The procedure gives rise to an effective Hamiltonin of the form $H_{eff} = H_o+H^{(2)}+H^{(4)}+\ldots$ where H_o describes the f-orbitals and the conduction band without mixing. The correct result for the fourth order intersite exchange has to be calculated using $H^{(2)} + H^{(4)}$ as a perturbation on H_o and coincides with the fourth order result from Ref. 6 .

Assuming a constant V, the range function is given by

$$J(R) = \frac{V^4}{N}\sum_q e^{i\vec{q}\cdot\vec{k}}\sum_k \frac{1-f_k}{(E-\varepsilon_k)^2}\left[\frac{f_{k+q}}{\varepsilon_{k+q}-\varepsilon_k} + \frac{1-f_{k+q}}{\varepsilon_{k+q}-E}\right] \quad (1)$$

where f_k is the zero temperature Fermi function and E is the energy of the localized f electrons. The first term in brackets contains contributions from both $H^{(2)}$ and $H^{(4)}$ which involve spin flip scattering of a conduction electron at some intermediate step and therefore gives rise to intersite interactions mediated by the conduction electron spin polarizability. The second term in the square bracket contains contributions only from $H^{(4)}$ which at some intermediate step have simultaneous promotion of two localized electrons to the conduction band, and is thus related to charge fluctuations. For a free electron band ε_k this gives

$$J(R) = \frac{\Gamma^2}{\rho_o\varepsilon_F}\left\{\frac{\sin(2k_FR)}{R^2(k_o^2-k_F^2)} + \frac{\cos(2k_oR)}{k_oR}\Big[C_i[2(k_F-k_o)R] - C_i[2(k_F+k_o)R]\Big] \quad (2)\right.$$
$$\left. - \frac{\sin(2k_oR)}{k_oR}\Big[si[2(k_F-k_o)R] + si[2(k_F+k_o)R]\Big]\right\}$$

where si(x) and Ci(x) are the integral sine and cosine functions. $\Gamma = \pi\rho_o V^2$, ρ_o is the density of states at the fermi energy and $k_o=(E/2m)^{1/2}$. J(R) behaves asymptotically as

$$\frac{1}{(E-\varepsilon_F)^2}\left[\frac{\cos 2k_FR}{(2k_FR)^3} - \frac{E+3\varepsilon_F}{E-\varepsilon_F}\cdot\frac{\sin(2k_FR)}{(2k_FR)^4}\right] \qquad (3)$$

indicating that the closer the localized level is from the fermi energy the more the range function differs from RKKY.

For small R the dominant term is the first one, which decays as $1/R^2$. Similar behaviour has been obtained by H.Keiter and N.Grewe [4] and by A.M.Tsvelik [7]. Our result, Eq.(2) is strictly fourth order in V and can be simply computed for different values of the parameters.

The general behaviour is given in Fig.1.

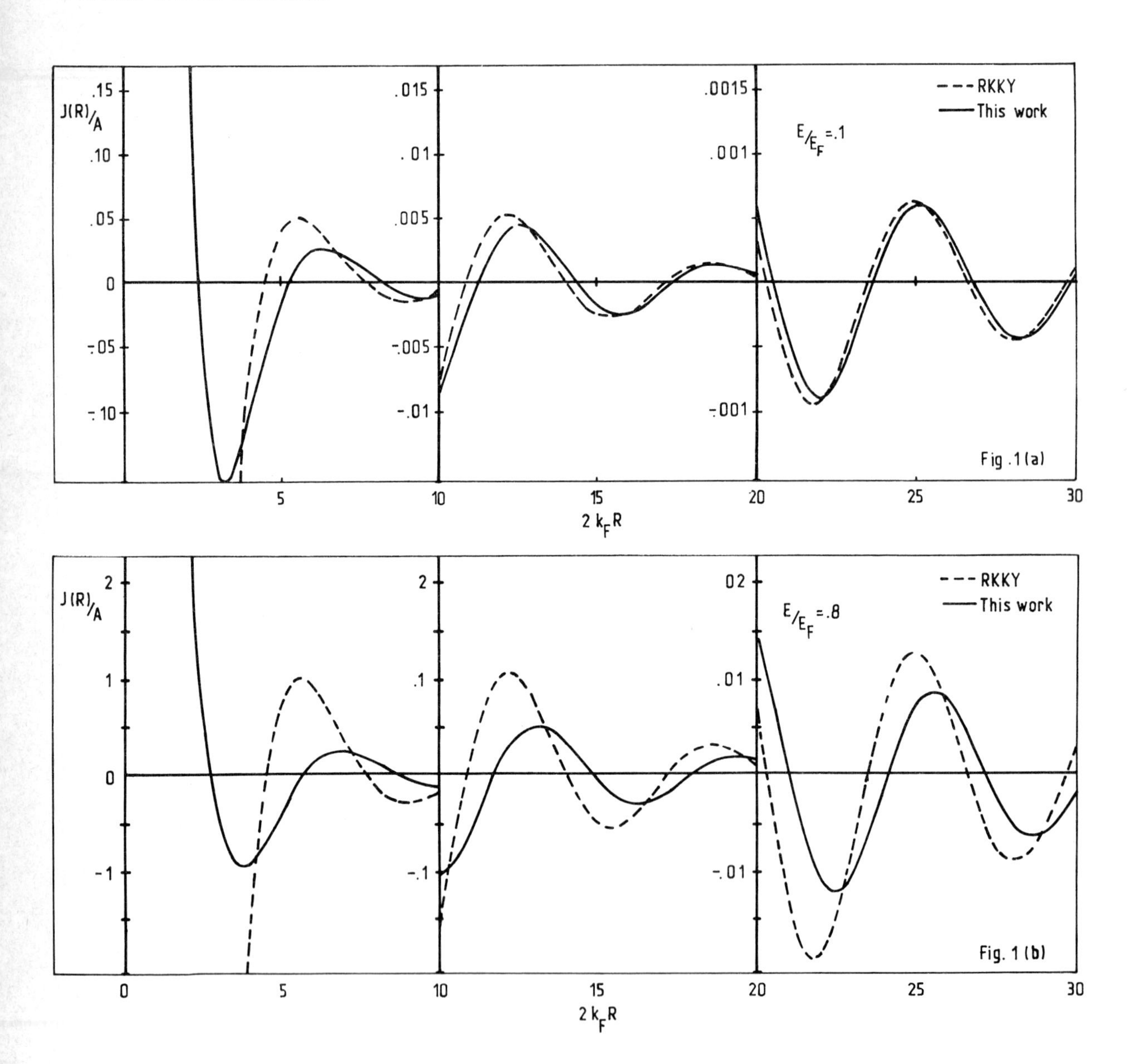

Figure 1: Range function J(R) and comparison with RKKY function for a) E/ε_F = .1 and b) E/ε_F = .8. Notice the change in scale for each figure. It is seen that the difference between the improved range function and the RKKY behaviour increases as E approaches the Fermi energy. The constant $A = \Gamma^2/\rho_o\varepsilon_F^2$.

We want to point out in closing that the canonical transformation (or fourth order perturbation) implies that the magnetic exchange in Ce compounds contains two contributions. One is a spin polarization of the conduction electron gas, as suggested by the equivalence to Kondo Hamiltonian obtained by Schrieffer and Wolff, and the other a charge fluctuation effect which is not contained in the latter.

REFERENCES

§ CONICET (Consejo Nacional de Investigaciones Científicas y Técnicas).
* Comisión Nacional de Energía Atómica.
# Comisión Nacional de Energía Atómica and Universidad Nacional de Cuyo.

1. D.Yang and B.R.Cooper; J.Appl.Phys.52, 2234 (1981); R.Siemann and B.R.Cooper, Phys.Rev. Lett. 44, 1015 (1980).
2. C.Proetto and A.López, Phys.Rev.B24, 3031 (1981).
3. J.Villain, M.B.Gordon and J.Rossat-Mignod, Physica 102B, 262 (1980).
4. H.Keiter and N.Grewe, in Valence Fluctuations in Solids, L.M.Falicov, W.Hanke and M.B.Maple (eds.), (North Holland Publishing Company, Amsterdam, 1981).
5. B.Coqblin and J.R.Schrieffer, Phys.Rev.185, 847 (1969).
6. C.E.T.Goncalves da Silva and L.M.Falicov, J.Phys.C.5, 63 (1972).
7. A.M .Tsvelik, Sov.Phys.JETP 49, 1142 (1979).

Valence Instabilities
P. Wachter and H. Boppart (eds.)

A NEUTRON SCATTERING INVESTIGATION OF THE MAGNETIC PHASE DIAGRAM OF CeB_6

J.M. Effantin, P. Burlet, J. Rossat-Mignod, S. Kunii and T. Kasuya*

Laboratoire de Diffraction Neutronique
Département de Recherche Fondamentale
Centre d'Etudes Nucléaires
85 X, 38041 Grenoble Cédex, France

*Departement of Physics, Tohoku University, Sendai, Japan

Neutron diffraction experiments have been performed on a CeB_6 single crystal with a magnetic field applied along various symmetry directions to investigate the complex magnetic phase diagram of this interesting compound. In the ordered state a strong field dependent anisotropy has been observed. At low temperature for a magnetic field applied along <111> a new phase III' has been discovered between phase III (low temperature and low field phase) and phase II (intermediate phase). In phase III the magnetic structure is indeed composed by two double-$\vec{k}$ structures. The Fourier component $\vec{m}_k$ is perpendicular to $\vec{k}$, i.e. $\vec{m}_k$//[1$\bar{1}$0] for $\vec{k}$=[1/4 1/4 0] and $\vec{k}$'=[1/4 1/4 1/2].

1. INTRODUCTION

Cerium hexaboride, with a simple cubic structure of the CaB_6 type, is a very interesting magnetic compound in the sense that it shows the most typical "dense Kondo behaviour" (1,2). Magnetization (3,4), specific heat (5), resistivity (6) and NMR (7) measurements have revealed very complex magnetic properties and in particular an unusual magnetic phase diagram. Three distinct phases have been reported: i) a paramagnetic high temperature phase I, ii) an intermediate phase II, iii) a low temperature antiferromagnetic phase III. However many uncertainties remain about this phase diagram itself as well as the knowledge of the magnetic structures of the different phases. Indeed recent NMR (8), magnetization (9) and magnetostriction (10) measurements indicate a more complex behaviour, especially at the transition between phase III and phase II. Previous neutron diffraction measurements (11,12,4) have confirmed the main feature of the phase diagram and in the phase III the magnetic ordering has been shown to be described by two distinct wave vectors $\vec{k}$=<1/4 1/4 0> and $\vec{k}$'=<1/4 1/4 1/2>. Several models of structures can be proposed from these experiments, therefore more precise neutron studies have been undertaken to clarify the nature of the magnetic order and the orientation of the magnetic moments. The experimental conditions will be first briefly described, then additional results in zero field will be given. The magnetic phase diagram has been explored for various magnetic field directions and a special attention has been devoted to the study of the motion of magnetic domains. All these experimental results will be finally discussed.

2. EXPERIMENTAL

Neutron diffraction experiments were carried out at the reactor Silce at the C.E.N. Grenoble on the DN_3 spectrometer equipped with a moving up counter arm. This allows us to perform scans along any reciprocal directions which is needed to study the domain distribution. The neutron wavelength was 2.4 Å obtained from a graphite monochromator or 0.99 Å obtained from a copper monochromator for experiments with an applied field.
Single crystals of CeB_6 with 99% enriched ^{11}B were prepared by the floating zone method (3). A crystal with a shape of a sphere of 6 mm in diameter was placed inside a cryomagnetic system which supplies a vertical magnetic field up to 100 kOe and a temperature varying from 1.3 K to 300 K (13).

3. ZERO FIELD RESULTS

Additional experiments in zero field have been performed on single crystals oriented with a vertical [001] and [1$\bar{1}$1] directions in order to get more informations than in the previous study (11). Indeed we observe that the superlattice peaks associated to the wave vectors $\vec{k}$=<1/4 1/4 0> and $\vec{k}$'=<1/4 1/4 1/2> disappear simultaneously at T_0=2.35 ± 0.05 K, i.e. at the transition between phase III and phase II. In a first experiment (11) an ordering temperature of 2.8 K was reported, the origin of this discrepancy is not actually understood and it may come from a technical problem. Then the interpretation given for the phase II - phase III transition must be ruled out.
In phase II (2.35 K < T < 2.9 K) up to now no evidence of any superlattice peak has been obtained. Moreover the phase II - phase III transition is a second order transition and then T_0=2.35 K looks like a Néel temperature.
In phase III, previous results (11,12) do not allow us to distinguish between a helical or a sine wave modulation because the scattering vectors were confined within a (1$\bar{1}$0) plane. Therefore measurements with a [1$\bar{1}$1] vertical axis were performed. Table I contains the intensities of the superlattice magnetic peaks associated to the wave vector $\vec{k}_1$=[1/4 1/4 0]

Table 1 : Comparison between observed and calculated magnetic intensities in zero magnetic field at T=1.48 K.

$\vec{h}$	$I_{obs}(\vec{h})$	$I_{obs}(\vec{h})/P(\vec{h})$	$S(\vec{h})$	$I_{cal}(\vec{h})$
1/4 1/4 0	720 ± 85	327	1	755
3/4 -1/4 0	85 ± 65	62	0.2	95
5/4 1/4 0	120 ± 60	206	0.69	140
1/4 1/4 1	300 ± 60	389	1	265

measured at T=1.48 K in a multi-domain state obtained by cooling down from 4.2 K.
The intensities of a superlattice peak defined by a scattering vector $\vec{h}=\vec{H}+\vec{k}$, where $\vec{H}$ is a reciprocal lattice vector and $\vec{k}$ the wave vector of the modulation, can be written as

$$I(\vec{h})=N\cdot(0.27)^2(A_k^2/4)\cdot P(\vec{h})\cdot S(\vec{h})=K.P(\vec{h}).S(\vec{h})$$

N is the normalization constant, A_k is the amplitude of the Fourier component $\vec{m}_k$ and $P(\vec{h})=A(\vec{h})L(\vec{h})f^2(\vec{h})$ involves the absorption, the Lorentz and the form factors. $S(\vec{h})$ depends on the relative orientation of $\vec{h}$ with respect to the moment direction. In table 1 we can see that this factor, proportional to the ratio $I(\vec{h})/P(\vec{h})$, depends strongly on the orientation of the scattering vector within the (001) plane. Therefore the helical model is ruled out because it cannot explain the ratio observed for the [3/4 -1/4 0] and [1/4 1/4 0] intensities. Thus the ordering must correspond to a sine wave modulation and the Fourier component can be written as,

$$\vec{m}_k = \frac{A_k}{2} e^{i\phi_k} \vec{u}_k ,$$

where $\vec{u}_k$ is a unit vector defining the polarization of the modulation. Then the factor depending on the moment orientation is given by $S(\vec{h})=1-(\vec{u}_k\cdot\vec{h})^2/\vec{h}^2$. The small value of this factor for $\vec{h}$=[3/4 -1/4 0] can be accounted for only if $\vec{u}_k$ is nearly perpendicular to the wave vector $\vec{k}_1$=[1/4 1/4 0], i.e. close to a [1$\bar{1}$0] direction. For such a direction the calculated intensities reported in table 1 are in satisfactory agreement with experimental values. So the sine wave modulations associated to both $\vec{k}_1$=[1/4 1/4 0] and $\vec{k}_1'$=[1/4 1/4 1/2] have a transverse polarization along the [1$\bar{1}$0] direction but they have slightly different amplitudes ($A_{k'}/A_k$=1.16 ± 0.12).

4. STUDY OF THE (H,T) PHASE DIAGRAM

4.1. Magnetic field applied along [1$\bar{1}$1]

In this experiment the crystal was oriented inside the cryomagnet with a [1$\bar{1}$1] direction vertical, in fact a small misorientation of about two degrees existed between [1$\bar{1}$1] and the magnetic field direction giving $H_z > H_x$ and H_y. This misorientation was large enough to break the threefold symmetry. Then, among the twelve wave vectors $\vec{k}$=<1/4 1/4 0> existing in zero field, only four remains when the field is increased : $\vec{k}_1$=±[1/4 1/4 0] and $\vec{k}_2$=±[1/4 -1/4 0].

A similar behaviour is observed for the second wave vector $\vec{k}'$=<1/4 1/4 1/2>. As shown in figure 1, at T=1.40 K, above H=6 kOe a single domain state is reached and the intensity of the remaining peaks is about 2.4 times larger than in zero field. A factor slightly smaller than 3 can be explained by the increase of the ferromagnetic component. Intensity measurements confirm that the Fourier components $\vec{m}_{k_1}$ and $\vec{m}_{k'_1}$ and the ratio of their amplitude are the same as described in section 3. Then up to H=6 kOe the magnetic field moves only the domains existing in phase III. By increasing more the field a very sharp transition occurs at H_{c_1}=10.5 kOe (at T=1.40 K) where the magnetic reflections [1/4 1/4 0] and [1/4 1/4 1/2] abruptly vanish. The ratio of the intensities of the remaining reflections [1/4 -1/4 0] and [1/4 -1/4 1/2] is strongly modified. Then above H_{c_1} the intensity of these two reflections decreases smoothly up to a critical field H_{c_2}=15 kOe (at T=1.40 K). Thus between H_{c_1} and H_{c_2} a new phase, called phase III', exists in the magnetic phase diagram. In decreasing field a similar behaviour has been observed, a small hysteresis of about 1.5 kOe affects H_{c_1}.

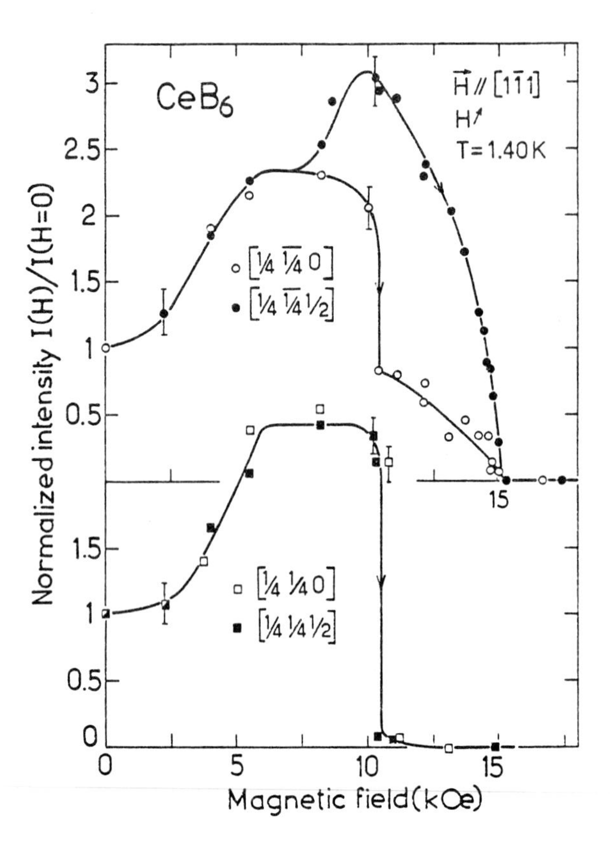

Figure 1 : Magnetic field dependence of the normalized intensities of the magnetic reflections [1/4 1/4 0], [1/4 1/4 1/2], [1/4 -1/4 0] and [1/4 -1/4 1/2].

Table 2 : Comparison between observed and calculated intensities measured in phase III' at T=1.4 K with an applied field H=10 kOe.

$\vec{h}$	$I_{obs}(\vec{h})$	$I_{obs}(\vec{h})/P(\vec{h})$	$S(\vec{h})$	$I_{cal}(\vec{h})$
1/4 -1/4 0	1850±75	477	1	1883
3/4 1/4 0	155±75	153	.20	97
-1/4 1/4 1	300±65	443	1	328
	K=485±50			
1/4 -1/4 -1/2	2900±100	2354	1	2772
1/4 3/4 1/2	590±80	770	.43	744
1/4 -5/4 -1/2	870±60	1695	.73	843
	K'=2251±100			

We deduce from intensity measurements (table 2) that in phase III' the Fourier components $\vec{m}_{k_2}$ and $\vec{m}_{k_2'}$, associated to $\vec{k}_2$=[1/4 -1/4 0] and $\vec{k}_2'$=[1/4 -1/4 1/2], have the same orientation as in zero field, i.e. are parallel to [110], but the ratio of their amplitudes is modified : $A_{k_2'}/A_{k_2}=2.1 \pm 0.1$.

4.2. Magnetic field applied along [001]

A magnetic field applied along [001] induces a domain redistribution which begins in very low field (H∿1.5 kOe). At H=4 kOe and T=1.35 K only remain the four superlattice peaks associated to $\vec{k}_1$=[1/4 1/4 0], $\vec{k}_2$=[1/4 -1/4 0] $\vec{k}_1'$= [1/4 1/4 1/2] and $\vec{k}_2'$= [1/4 -1/4 1/2]. An additional experiment has been performed by tilting the field towards $\vec{k}_2$ ($\vec{H}$//[0.13 0.13 1]), but it was not possible to differenciate $\vec{k}_1$ from $\vec{k}_2$. Then for H > 4 kOe a single domain state is reached with the same magnetic structure as that determined in zero field. By increasing more the field the intensities of the four magnetic peaks decrease smoothly and identically, so in figure 3 we report the behaviour of only one of them. Up to H=22 kOe where the intensities vanish it was not possible to differentiate the four wave vectors. In decreasing field a small hysteresis of about 1 kOe exists.

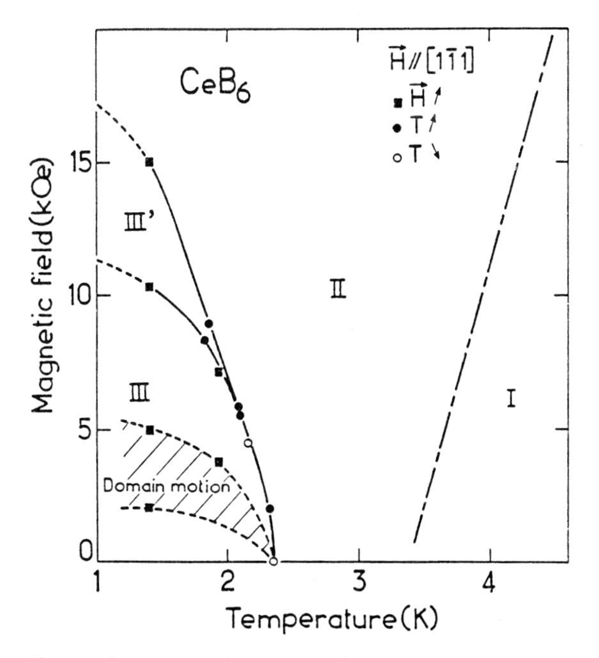

Figure 2 : Magnetic phase diagram obtained for a magnetic field applied along [1$\bar{1}$1].

4.3. Magnetic field applied along [011]

When the magnetic field is applied along [011] at T=1.50 K and H=7 kOe we observed the magnetic peaks associated to the eight wave vectors $\vec{k}_1$=[1/4 1/4 0], $\vec{k}_1'$= [1/4 1/4 1/2], $\vec{k}_2$=[1/4 -1/4 0], $\vec{k}_2'$= [1/4 -1/4 1/2] and $\vec{k}_3$=[1/4 0 1/4], $\vec{k}_3'$= [1/4 1/2 1/4], $\vec{k}_4$=[1/4 0 -1/4] and $\vec{k}_4'$= [1/4 1/2 -1/4]. However if we tilt the field away from [011] along [0 1 1.2] its remains only the four wave vectors $\vec{k}_1$, $\vec{k}_1'$, $\vec{k}_2$ and $\vec{k}_2'$. By increasing the field an abrupt transition occurs at H=11 kOe for T=1.5 K; a precise investigation of the phase diagram is in progress.

5. DISCUSSION

The neutron diffraction results with magnetic field applied along various directions have brought some important informations about the complex magnetic behaviour of CeB_6. First of all the magnetic peaks associated with the two non equivalent wave vectors $\vec{k}$=<1/4 1/4 0> and $\vec{k}'$=<1/4 1/4 1/2> appear and vanish always exactly at the same temperature and at the same field. So the description of the magnetic structure requires to combine at least the Fourier components $\vec{m}_{k_1}$ and $\vec{m}_{k_1'}$, the resulting structure will be called a $\vec{k}.\vec{k}'$ structure. It must be emphasized that it is not an usual 2-$\vec{k}$ structure in the sense that the wave vectors

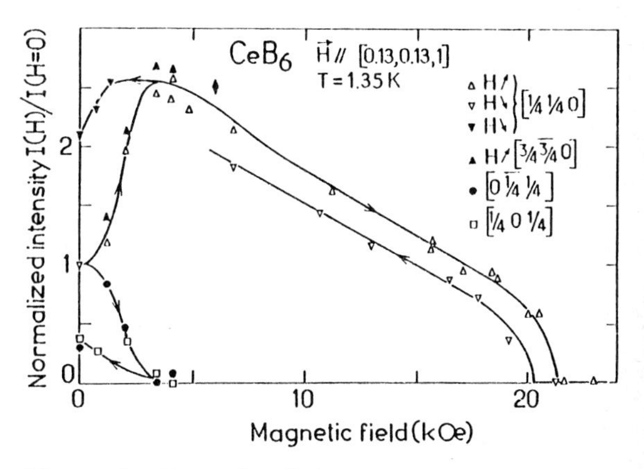

Figure 3 : Magnetic field dependence of the normalized intensities. The field was increased up to 25 kOe and then decreased for the [1/4 1/4 0] satellite. Whereas for the satellites [1/4 1/4 0], [3/4 -3/4 0], [0 -1/4 1/4] and [1/4 0 1/4] the field was increased only to 6 kOe and then decreased.

involved do not belong to the same star. Then to explain the coupling between $\vec{m}_k$ and $\vec{m}_{k'}$ a large fourth order term must be present in the Hamiltonian which implies that the ground state is not a well isolated Γ_7 doublet as inelastic neutron scattering experiments would suggest (4).

Intensity measurements show that in this antiferromagnetic $\vec{k}.\vec{k}'$ structure the magnetic moments are all parallel, along <110> direction perpendicular to $\vec{k}$ and $\vec{k}'$: as example for $\vec{k}_1$=[1/4 1/4 0] and $\vec{k}_1'$= [1/4 1/4 1/2] they lie along [1$\bar{1}$0]. However the exact magnetic structure cannot be deduced because neutron experiments do not give the value of the phase of the two Fourier components (for a discussion of this problem see ref. (11)). Among the different models only one gives a structure with nearly equal magnetic moments on each cerium ion, it corresponds to a ++-- sequence within a (001) plane propagating along a [110] direction, in the neighbouring plane the sequence is shifted by $\pi/4$, i.e.: +--+. As the star of $\{\vec{k}\}$ contains 12 members, a fundamental question is to know if the ordering is described by a single component or by several. The behaviour for $\vec{H}$//[1$\bar{1}$1], [100] and [110] indicates clearly that the wave vectors $\vec{k}_1$=±[1/4 1/4 0], $\vec{k}_3$= ± [1/4 0 1/4] and $\vec{k}_5$=±[0 1/4 1/4] are associated to three domains. For $\vec{k}_1$=[1/4 1/4 0] and $\vec{k}_2$=[1/4 -1/4 0] the situation is not so simple because a field applied along [1$\bar{1}$1] is unable to distinguish these two wave vectors as expected for a single-$\vec{k}$ order. Moreover at H=10.5 kOe an abrupt transition occurs which cannot be explained by a domain redistribution. Therefore the magnetic ordering in phase III must be described by four wave vector ($\vec{k}_1$, $\vec{k}_1'$ and $\vec{k}_2$, $\vec{k}_2'$ for example) and the structure is then a double $\vec{k}.\vec{k}'$structure. Examples of double-$\vec{k}$ structures have been shown to exist in UAs and UP (14) but such a situation of a double-$\vec{k}$ structure with two distinct wave vectors is rather unique. As $\vec{m}_{k_1}$//[1$\bar{1}$0] and $\vec{m}_{k_2}$//[110] are orthogonal we obtain a non collinear ordering with moments within the (001) plane, the moments being close to <100> or <110> directions depending on the choice of the phases. Such a planar structure explains quite well the domain behaviour observed for the various field directions investigated. The exchange interaction in this compound must be very unusual to stabilize such a double-$\vec{k}$ structure which needs again a four or higher order terms in the Hamiltonian.

When the field is increased along the [1$\bar{1}$1] direction, involving a large component in the plane of the moments, the double-$\vec{k}.\vec{k}'$ structure is destroyed to give rise to a collinear single-$\vec{k}.\vec{k}'$ structure which is that of phase III'.

It must be emphasized that phase III' contains two kinds of cerium ions with different moment values. Only the Fourier components which are perpendicular to the field remain (i.e. $\vec{m}_{k_2}$ and $\vec{m}_{k_2'}$), as it is usual for antiferromagnets. Thus by increasing the field we observe a canting of the moments along the field direction. A similar canting behaviour occurs for phase III when the field is nearly perpendicular to the plane of the moments i.e. close to a <001> axis.

Whatever the field direction for H > 25 kOe, the antiferromagnetic components vanish, then the ordering in phase II seems to correspond to a ferromagnetic alignment of the moment along the applied field. In low field in phase II (at T=2.2 K and H=5.5 kOe along [1$\bar{1}$1] we check all the symmetry points of the Brillouin zone, but we fail to detect any magnetic contribution. If there is an ordering the antiferromagnetic component must be smaller than 0.07 μ_B. Indeed in phase II an ordering must exist but it might be not of dipolar type to understand why the antiferromagnetic ordering takes place at T_Q=2.35 K via a second order phase transition while the structure involves two distinct wave vectors.

REFERENCES

(1) Winzer K. and Felsch W., J. de Phys. 39, C6 (1978) 832-834.
(2) Komatsubara T., Suzuki T., Kawakami M., Kunii S., Fujita T., Isikawa Y., Takase A., Kojima K., Suzuki M., Aoki Y., Takigahara K. and Kasuya T., J. Magn. Magn. Mat. 15-18 (1980) 963-964.
(3) Kawakami M., Kunii S., Komatsubara T. and Kasuya T., Solid State Commun. 36 (1980) 435-439.
(4) Horn S., Steglich F., Loewenhaupt M., Scheuer H., Felsch W. and Winzer K., Z. Phys. B 42 (1981) 125-134.
(5) Fujita T., Suzuki M., Komatsubara T., Kunii S., Kasuya T. and Ohtsuka T., Solid State Commun. 35 (1980) 569-572.
(6) Takase A., Kojima K., Komatsubara T. and Kasuya T., Solid State Commun. 36 (1980) 461-464.
(7) Kawakami M., Kunii S., Mizuno K., Sugita M., Kasuya T. and Kume K., J. Phys. Soc. of Japan 50 (1981) 432-437.
(8) Kawakami M., Mizuno K., Kunii S., Kasuya T., Enokuya H. and Kume K., "Magnetic Field Dependence of ^{11}B NMR in CeB_6" to be published.
(9) Sato N., Kunii S. and Kasuya T., private communication.
(10) Nakajima T., Kunii S., Komatsubara T. and Kasuja T., J. Magn. Magn. Mat. 15-18 (1980) 967-968; and recent data: to be published.
(11) Rossat-Mignod J., Burlet P., Kasuya T., Kunii S. and Komatsubara T., Solid Sate Commun. 39 (1981) 471-478.
(12) Burlet P., Rossat-Mignod J., Effantin J.M., Kasuya T., Kunii S. and Komatsubara T., Intern. 3M Conference, Atlanta, Georgia (1981).
(13) Claudet G., Gravil B., Rossat-Mignod J. and Burgess S., Cryogenics, December (1981) 711-714.
(14) Rossat-Mignod J., Burlet P., Quezel S., Vogt O., IV Int. Conf. on Crystal-Field and Structural Effects in f-Electron Systems, Wroclaw (1981), to be published by Plenum Press.

MIXED-VALENCE PROPERTIES

Valence Instabilities
P. Wachter and H. Boppart (eds.)

THERMAL PROPERTIES OF Eu_3S_4: ORDER-DISORDER TRANSITION

R.Pott[+] and G.Güntherodt[+]
II. Physikalisches Institut, Universität zu Köln, 5000 Köln 41, Fed. Rep. of Germany

and

W.Wichelhaus[&] and M.Ohl
Max-Planck-Institut für Festkörperforschung, 7000 Stuttgart 80, Fed. Rep. of Germany

and

H.Bach
Institut für Experimentalphysik IV, Ruhr-Universität Bochum, 4600 Bochum, Fed. Rep. of Germany

Capacitive thermal expansion, specific heat and electrical resistivity measurements on single crystals of Eu_3S_4 and La_3S_4 have established for Eu_3S_4 a first order, nonstructural phase transition near T_t=186 K and the entropy change $\Delta S=(9\pm0.9)$ J/Mole K of an order-disorder transition with a configurational entropy Rln3. The transition exhibits $\Delta V/V=9\cdot10^{-4}$ and $d\ln T_t/dp=d\ln E_a/dp=-8.7\cdot10^{-11}$ Pa^{-1}, i.e a proportionality between T_t and the activation energy E_a of the resistivity.

INTRODUCTION

Homogeneous mixed valence compounds have attracted much attention in recent years because of valence fluctuations, which in the inhomogeneous mixed valence counterpart are due to thermally activated hopping between equivalent cation sites /1/. In the latter class of materials the basic questions concerning charge ordering at low temperatures, associated lattice distortions and order-disorder phase transitions have not been answered unambiguously to date /2/, mostly due to problems with stoichiometry /3/. These questions are also of relevance to intersite couplings /4/ in homogeneous mixed valence systems: Lowering of the ground state energy due to charge fluctuations /5/ or spatial correlations of rare earth ions /6/, Wigner lattice formation /7/, and qualitative changes in Fermi surface /8/.
The prominent inhomogeneous mixed valent compound Eu_3S_4, with equivalent Eu sites at room temperature, but inequivalent Eu^{2+} and Eu^{3+} sites at low temperature /9/, represents an ideal test case because of the well defined change in configurational entropy for a complete order-disorder transition /10/. However, the small entropy change observed /10/ as well as the lack of conclusive evidence from x-ray data about structural changes at the phase transition near 175 K /11/ did not support the previously suggested (Eu^{2+}-Eu^{3+}) charge order-disorder transition /12/.
We report on measurements of (capacitive) thermal expansion, specific heat and electrical resistivity for the first time on single crystals of Eu_3S_4, compared with those of La_3S_4.

EXPERIMENTAL PROCEDURE AND RESULTS

The relative lentgh change $\Delta L/L$ has been measured by a capacitive method /13/ between 1.5 and 380 K (sensitivity 10^{-8}, accuracy ±2%). This very sensitive method could be applied for the first time in the temperature range around T_t, which is usually not easily accessible. The linear thermal expansion coefficient α has been calculated by a Spline fit to $\Delta L/L$ /14/ (sensitivity of 10^{-8} K^{-1}, accuracy ±2%). The specific heat C_p (20<T<350 K) has been measured by a calorimeter described elsewhere /15/ (accuracy ±2%).
The preparation of the single crystals of La_3S_4 and Eu_3S_4 with lattice parameters a=8.724±0.001 Å and a=8.532±0.002 Å at 300 K, respectively, is described elsewhere /16/,/17/.
In Fig.1 we show $\Delta L/L$ of the unoriented Eu_3S_4 single crystal (length 5.63 mm)

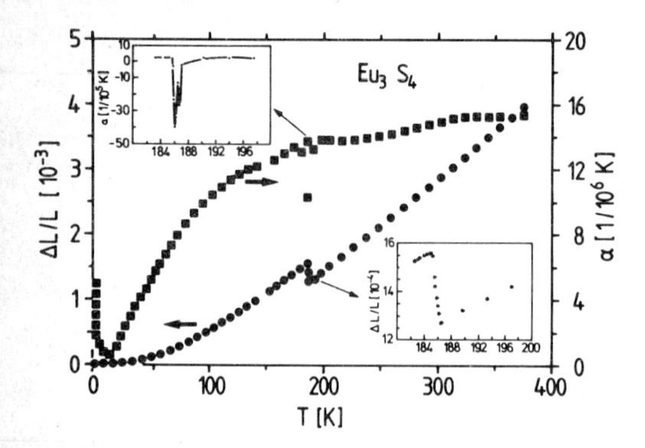

Fig.1: Relative length change $\Delta L/L$ and linear thermal expansion coefficient α of Eu_3S_4 single crystal (unoriented) as function of temperature.

between 1.5 and 380 K. There is a sharp discontinuity at T_t=186 K. α shows a monotonic continuation above and below T_t. The anomaly in α near T=3.8 K is attributed to the ferromagnetic ordering of Eu_3S_4 /10/. The jump in $\Delta L/L$ (insert of Fig.1 (bottom)) with increasing temperature is $-3\cdot10^{-4}$ with a width of only about 1 K.
Fig.2 shows $\Delta L/L$ and α of the unoriented single crystal (length 6.04 mm) of La_3S_4. Contrary to Eu_3S_4, $\Delta L/L$ is a continuous function of temperature near the structural phase transition /18/ (T_M=102 K) and α is not monotonically continued from below to above T_M.

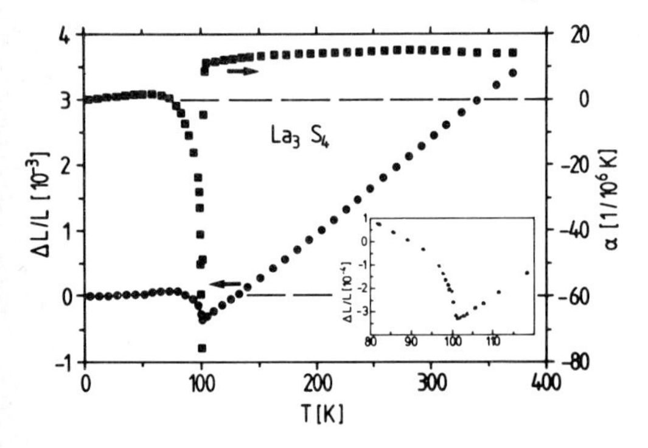

Fig.2: Relative length change $\Delta L/L$ and linear thermal expansion coefficient α of La_3S_4 single crystal (unoriented) as function of temperature.

The specific heat of the Eu_3S_4 single crystal (weight 0.4932 g) exhibits a giant anomaly (Fig.3(a)) near 186 K. The insert of Fig.3(a) shows a well pronounced double peaked structure near 185.5 K and 186.3 K with an overall width of only about 1 K. Note the similarity in C_p and α (see insert of Fig.1 (top)). No significant hysteresis could be detected contrary to the 5 K hysteresis reported by Davis et al. /11/. The entropy of the anomaly is about $\Delta S_1=(5.2\pm0.1)$ J/Mole K. As can be seen in Fig.3(b) there is a step in C_p of Eu_3S_4 for $T>T_t$. Unfortunately, for the determination of the background we cannot consider C_p of La_3S_4, since the phonon spectra of this superconductor and the mixed valent Eu_3S_4 seem to be very different. Therefore in order to use the Debye formula as a reasonable background approximation for $T>T_t$ we have calculated C_v of Eu_3S_4 using $C_v=C_p-9TV\alpha^2B$ (B=92 GPa /19/). The asymptotic limit near T_t of the strongly temperature dependent $\Theta_D(T)$ is Θ_D=180 K, which has been used in the above Debye fit (dashed line). The corresponding excess entropy for $T_t<T<400$ K is about $\Delta S_2=(3.8\pm0.8)$ J/Mole K. The error is mainly due to the uncertainties in Θ_D and accounts also for the uncertainties in the extrapolation $\Delta C_v/T \to 0$ near 400 K. Hence the total entropy is $\Delta S=\Delta S_1+\Delta S_2=(9\pm0.9)$ J/Mole K.

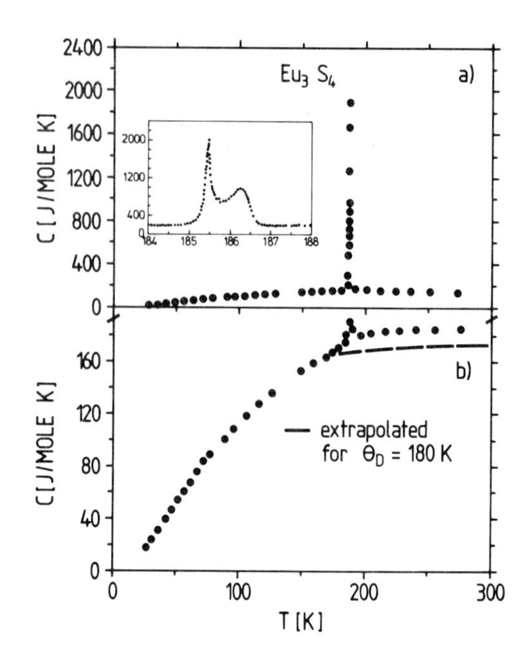

Fig.3: Specific heat C_p of Eu_3S_4 single crystal as function of temperature on full (a) and expanded (b) scale. Dashed line in (b) is the Debye fit extrapolated for T>186 K.

We have also measured the specific heat of several agglomerated single crystals (0.5 - 1mm in diameter) of Eu_3S_4, prepared by the same method. They show one or two peaks with $183.0<T_t<188.3$ K. Note that only C_p shows a variation in

T_t, whereas the lattice parameter of all samples is a=8.532±0.001 Å. Hence C_p is a much more sensitive method to determine the quality of a Eu_3S_4 sample than the lattice parameter.
The specific electrical resistivity ρ of our semiconducting Eu_3S_4 single crystal shows a discontinuity near T_t= 186 K without any significant hysteresis. With decreasing temperature ρ increases at T_t by about a factor of 50. The activation energy E_a is (0.32 ±0.01) eV below and (0.16±0.005) eV above T_t.

DISCUSSION

The discontinuity of $\Delta L/L$ of Eu_3S_4 near 186 K (Fig.1) shows for the first time that the phase transition is definitely of first order. This could not be seen by previous x-ray studies showing a rather poorly resolved transition with peak broadening and loss of intensity upon cooling through the phase transition /11/. On the other hand, La_3S_4 does not show such a discontinuity in $\Delta L/L$ (Fig.2), i.e. the structural phase transition of La_3S_4 is of second order, contrary to the conclusions in the literature /18/,/20/.
The linear thermal expansion coefficient α of our unoriented single crystal of La_3S_4 exhibits a step-like behavior for temperatures below and above T_M /21/. Since α of Eu_3S_4 shows a uniform continuity between 4 and 400 K, except for the peak at T_t (see Fig.1), the phase transition of Eu_3S_4 is not dominantly a structural one, i.e. there is no significant displacive effect on the Eu and S positions. This is consistent with x-ray /11/ and recent neutron diffraction measurements /22/, which show that the change of the atom positions is extremely small, indicated by a small, probably orthorhombic distortion /22/ for $T<T_t$.
From our above results we can calculate the pressure dependence of T_t of Eu_3S_4 using the Clausius-Clapeyron equation $dT_t/dp=\Delta V/\Delta S$. ΔS and ΔV are the entropy and the volume change, respectively, at the phase transition. With ΔS=5.2 J/Mole K we obtain $dT_t/dp=-(1.6\pm0.1)\,10^{-8}$ K/Pa, i.e. we predict that T_t decreases with increasing pressure. Contrary to Eu_3S_4, La_3S_4 is known to have a positive pressure coefficient $dT_M/dp=+(1.6\pm.15)\,10^{-8}$ K/Pa /23/.
The measured total entropy ΔS=9 J/Mole K equals about Rln3. The configurational entropy of an order-disorder transition where three configurations (one Eu^{2+} and two Eu^{3+}) are involved is Rln3 /10/. Hence the first order transition of Eu_3S_4 is of order-disorder type. Note that our entropy is by a factor of three larger than the value reported previously /10/. We attribute this difference to our better sample quality. Furthermore, the entropy (0.9 J/Mole K) of the C_p anomaly of La_3S_4 near T_M=102 K is ten times smaller than that of Eu_3S_4, indicating again the different nature of the phase transition (second order, structural; $dT_M/dp>0$) of La_3S_4 compared to Eu_3S_4 (first order; $dT_t/dp<0$).
The step-like behavior of C_p of Eu_3S_4 for $T>T_t$ (see Fig.3(b)) is assumed to be a short range order effect, extending roughly up to about 400 K. Such a contribution was predicted in mean field theory /24/. However, the reported sign is different. A similar double peaked structure like that of the C_p anomaly of Eu_3S_4 (insert Fig.3(a)) has also been observed in Fe_3O_4 and has been attributed to residual strain effects /25/. The number of peaks, their positions and relative widths depend on the thermal treatment of Fe_3O_4. While this possibility cannot be ruled out completely for Eu_3S_4, temperature cycling between 4 and 400 K did not show any significant change in position, width and height of the C_p anomaly. Another explanation of the double peaked structure of C_p and of α of Eu_3S_4 could be a certain inhomogeneity of our single crystal, showing up in two slightly different transition temperatures within 1 K.
The activation energy of ρ of our Eu_3S_4 single crystal (E_a=0.16 eV) for $T>T_t$ is the same as all values published so far /9/, /26/ and appears to be an intrinsic property. On the other hand, the magnitude of E_a for $T<T_t$ seems to be correlated with that of T_t, amounting to E_a=0.32 eV, T_t=186 K for our single crystal as compared to E_a=0.21 eV, T_t=175 K reported in the literature /26/. Assuming a proportionality between E_a and T_t we obtain
$d\ln T_t/dp = d\ln E_a/dp = -8.7\cdot10^{-11}\ Pa^{-1}$,
which is in excellent agreement with the pressure coefficient of E_a found in Mößbauer spectroscopy /27/. Such a proportionality $E_a \sim T_t$ has been found in Fe_3O_4 and is suggestive of a Mott-Wigner transition in Eu_3S_4 as well, consistent with the low temperature charge ordering.
The electrical resistivity of Eu_3S_4 may be also due to a polaron hopping process /26/. The activation energy E_a^* associated with this process can be assumed as proportional to $1/\omega_0^2$, where ω_0 is an optical phonon frequency /28/. By using the Grüneisen formula we obtain

$d\ln E_a^*/dp = -2d\ln\omega_0/dp = -6\alpha V/C = -4.8\cdot 10^{-11}$ Pa^{-1}, which again is in reasonable agreement with $d\ln E_a/dp$.

We would like to emphasize that our results reveal a very strong similarity between Eu_3S_4 and Fe_3O_4, concerning first order phase transitions near T_t and T_V (Verwey temperature), respectively, and associated entropy ($\Delta S = \Delta S_1 + \Delta S_2$), volume and resistivity changes, pressure coefficients of T_t and T_V and absence of thermal hysteresis.

ACKNOWLEDGEMENTS

We thank B.Schoch for sample preparation, W.Boksch, J.Haag, R.Kuhlmann and G.Neumann for help with the experiments, R.Schefzyk and D.Wohlleben for critical comments, P.Fazekas for fruitful discussions and A.Rabenau for support.

+ Part of this work was supported by the Deutsche Forschungsgemeinschaft, SFB 125.

& Present adress: Henkel KGaA, 4000 Düsseldorf 1, Fed. Rep. of Germany

REFERENCES

/1/ For a recent review see: "Valence fluctuations in Solids", ed. by L.M.Falicov, W.Hanke, and M.B.Maple, North-Holland, Amsterdam, 1981

/2/ N.F.Mott, Phil. Mag. B 42, 325 (1980), and following papers

/3/ F.Holtzberg, Phil. Mag. B 42, 491 (1980)

/4/ T.V.Ramakrishnan, in Ref. /1/, pg.13

/5/ C.M.Varma amd M.Schlüter, in Ref. /1/, pg.37

/6/ A.Sakurai and P.Schlottmann, Sol. St. Commun. 27, 991 (1978)

/7/ T.Kasuya et al., in Ref /1/, pg. 215

/8/ R.M.Martin and J.W.Allen, J. Appl. Phys. 50, 7561 (1979)

/9/ O.Berkooz, M.Malamud and S.Shtrikmann, Sol. St. Commun. 6, 185 (1968)

/10/ O.Massenet, J.M.D.Coey and F.Holtzberg, J. de Physique 37, Coll. C4-297 (1976)

/11/ H.H.Davis, I.Bransky and N.M.Tallan, J. Less Comm. Met. 22, 193 (1970)

/12/ F.L.Carter, J. Sol. St. Chem. 5, 300 (1972)

/13/ R.Pott and R.Schefzyk, to be published

/14/ R.Schefzyk, Diploma thesis, Universität Köln (1980)

/15/ A.Junod, J. Phys. E: Sci. Instr. 12, 945 (1979)

/16/ P.J.Ford, W.A.Lambson, A.J.Miller, G.A.Saunders, H.Bach and S.Methfessel, J.Phys. C: Sol. St. Phys. 13, L697 (1980)

/17/ G.Güntherodt and W.Wichelhaus, Sol. St. Commun. 39, 1147 (1981)

/18/ P.D.Dernier, E.Bucher and L.P.Longinotti, J. of Sol. St. Chem. 15, 203 (1972)

/19/ A.Werner, private communication

/20/ K.Westerholt, H.Bach and S.Methfessel, Sol. St. Commun. 36, 431 (1980)

/21/ The probability of accidentally obtaining a uniform continuity of for temperatures from below to above a structural phase transition for an unoriented single crystal is considered to be quite low.

/22/ K.R.A.Ziebeck, P.J.Brown and W.Wichelhaus, to be published

/23/ R.N.Shelton, A.R.Moodenbaugh, P.D.Dernier and B.T.Matthias, Mat. Res. Bull. 10, 1111 (1975)

/24/ J.L.Moran-Lopez and P.Schlottmann, Phys.Rev. B22, 1912 (1979)

/25/ M.Matsui, S.Todo and S.Chikazumi, J. of the Phys. Soc. of Jap. 42, 1517 (1976)

/26/ I.Bransky, N.M.Tallan and A.Z.Hed, J. Appl. Phys. 41, 1787 (1970)

/27/ J.Röhler and G.Kaindl, Sol. St. Commun. 36, 1055 (1980)

/28/ D.Adler, J.Feinlieb, H.Brooks and W.Paul, Phys. Rev. 155, 851 (1967)

Valence Instabilities
P. Wachter and H. Boppart (eds.)

MÖSSBAUER STUDY OF QUASI-PARTICLE SPECTRUM IN MAGNETITE

A.A. Hirsch

The Lidow Chair in Solid State Physics, Department of Physics,
Technion - Israel Institute of Technology
Haifa, Israel

The recoil-free absorption fraction in magnetite shows a series of dips at characteristic transition temperatures below 150 K, and are indicative of a complex charge-lattice coupling process. These results are ascribed to a spectrum of thermally excited charge complexes which behave as quasi-particles. A multitude of valence instabilities is considered in which four kinds of oxidation states are involved with charges of 3^+, 2^+, 4^+ and 1^+. A model is suggested for the electric conductivity which deals with the growth of the quasi-particle densities with rising temperatures, taking into account the probability for these densities to oscillate at phonon frequency.

1. INTRODUCTION

The interatomic charge transfer in magnetite (Fe_3O_4) at 119 K suggested by Verwey is the first prototype of valence instabilities in a mixed compound. During four decades this idea has stimulated studies of hopping conductivity and metal-insulator transitions. Valence instabilities could be viewed as phase transitions closely related to the lattice dynamics. Soft modes can be inferred from the predominantly displacive character of these transitions and can be associated with electric polarization effects.

2. CHARGE-LATTICE COUPLING

Figure 1 exhibits a sequence of dips in the recoil-free absorption fraction at transition temperatures T_i which may suggest the existence of a multitude of soft modes in the low-temperature phase of magnetite. This fraction is a measure of the Debye-Waller factor (f) and is given by the area under the Mössbauer patterns reduced to the "off-resonance" counts [1]. It is thought that these dips are associated with a spectrum of thermally excited charge complexes of delocalized electrons. Such excitations were viewed in a previous note as quasi-particles which nucleate by quantum mechanical tunneling [2]. The thermal energies associated with the collective excitations are formally represented here by

$$kT_i = \hbar\omega_o(n + \tfrac{1}{2}) - \hbar\omega_o z(n + \tfrac{1}{2})^2 \qquad (1)$$

The values of T_i indicated in Fig. 1 by the bar-diagram correspond to the adjusted parameters ω_o=2.66x10^{12}sec^{-1} and z = 0.035. This formal representation resembles the energy-quantization in the case of a Morse potential. The anharmonicity expressed by the last term in (1) may stress the need to explain the critical slowing-down in terms of charge-lattice coupling.

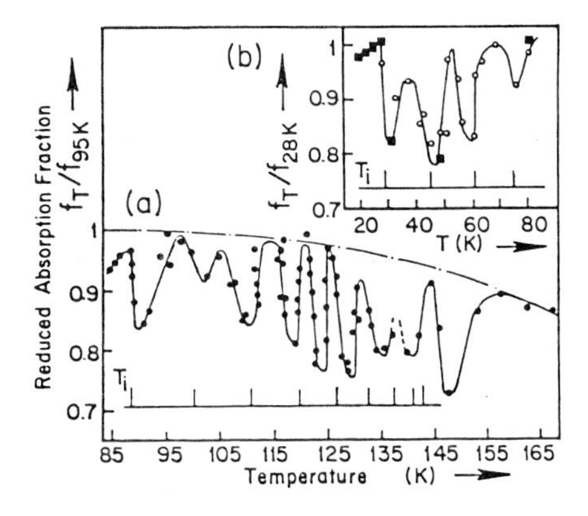

Fig. 1. Mössbauer absorption fraction as a function of temperature.

We consider a lattice which is linearly coupled to a quasi-particle, and express the configurational energy by

$$E = \sum_{k=1}^{k=N_i} \tfrac{1}{2}m_k\, q_k^2\, \omega_s^2 - B\,Q \qquad (2)$$

where m_k and q_k are the effective mass and displacement of a hole associated with the jump of an electron, while ω_s is the frequency of the phonon mode. The configurational parameter Q is a function of temperature and represents an effective single displacement attributed to the total charge of the quasi-particle. We introduce $Q^2 = \langle q_k^2\rangle N_i$ assuming that each m_k equals m. The mean value of the number of jumping electrons of a quasi-particle N_i, we call the quasi-particle density. The configurational crossover will occur when E has a minimum at which $Q = Q_o$. The thermally introduced distortion of the system requires an energy

$$kT_i = (B/2)Q_o = (B^2/2m)/\omega_s^2 \quad (3)$$

Taking $<q_k^2>$ as a constant, one replaces (3) by

$$T_i = (T_i)_{max}\,[N_i/(N_i)_{max}]^{\frac{1}{2}} \quad (4)$$

The experimental data supports (4) provided that N_i changes in a step-wise manner, so that for T>43 K the ratio $(N_i)_{max}/N_i$ obtains integral values from 12 to 1, while $(T_i)_{max}$ = 150 K. The Mössbauer line-intensity rearrangements, which are indicative of valence fluctuations, become detectable at about 10-15 K and persist till 150 K. The minimal jumping distance of the electrons estimated from the uncertainty principle is $R_{min} = \hbar/(2mk(T_i)_{max})^{\frac{1}{2}}$ = 17.6 Å. This result shows that an electron cannot jump between two sites belonging to the same double unit cell (a_o = 2a = 16.8 Å for magnetite). Therefore, we define N_i by the number of jumping electrons in the volume of the double unit cell. The development of the quasi-particles starts at a finite temperature T_o at which N_i is at least equal to 2. By making use of (4) one gets T_o = 10.8 K.

The isotope effect on the metal-insulator transition in magnetite at 119 K and in V_2O_3 at 160 K, may suggest that the phonon-mode mass (M) at low temperatures results mainly from the light oxygen atoms. Because of $\omega_s^2 \propto 1/M$, one derives from Eq. (3) $\Delta T_i = T_i X\,(18-16)/16$, where X is the concentration of the ^{18}O atoms. For example, with X = 0.43 for magnetite and X = 0.11 for V_2O_3 one gets for ΔT_i values of 6.3 K and 2.2 K, respectively. These estimates are in good agreement with the values of 6.1±0.05 K and 2.3±0.1 K observed by Terukov et al. [3,4].

2. VALENCE STATES

The analysis of our Mössbauer patterns suggests that magnetite at absolute zero contains four oxidation states Fe(3), Fe(2), Fe(1) and Fe(4) which are in concentration ratios of 3:1:1:1, respectively. We consider, at least formally, that these states have charge densities of 3^+, 2^+, 1^+ and 4^+. A spectrum of quasi-particles with densities changing in 12 steps could be a consequence of six two-stage valence-fluctuation reactions which saturate at transition temperatures T_i. In order for such reactions to take place the energy of the transferred electron in the initial and final states must be identical. In our case, this could happen when one or more electrons jump simultaneously either from one site to the other, or from both sites. The intermediate states could have integral or non-integral valences. Valence fluctuations are strongly coupled to the lattice vibrations owing to the distortions introduced by the altered ionic volumes. The valence-fluctuation reactions are linked together, so that at a given T_i some reactions may jointly persist, while others not.

(A)

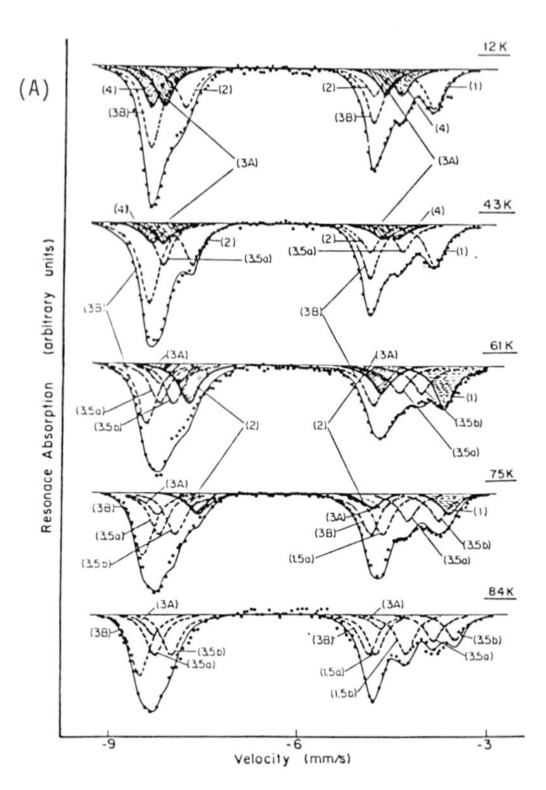

(B)

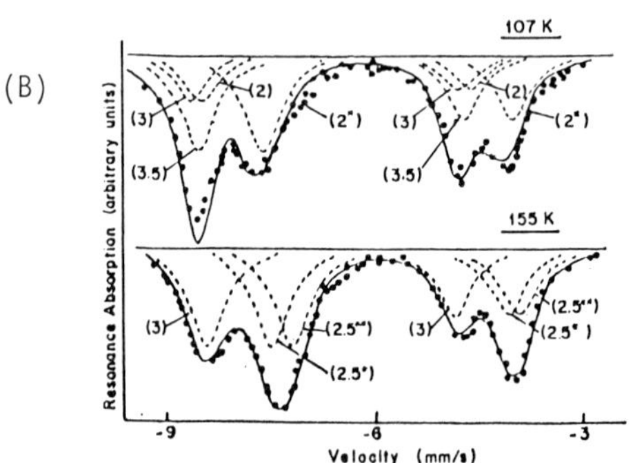

Fig. 2. The first two absorption areas of Mössbauer patterns:

(A) synthetic stoichiometric magnetite

(B) natural magnetite crystal.

Table 1 shows the pairs of charge-exchanging ions involved in the valence instabilities at the various transition temperatures.

The consecutive stages in the formation of intermediate valence states can be followed in Fig. 2 where the spectral components of the first two absorption areas are shown. The Mössbauer patterns at 155 K reveal the existence of two kinds of intermediate states having the same

TABLE I.

T_i (K)	Pairs of charge-exchanging ions -e	-2e	-5e	N_i
43.4	$(4^+,3^+)$			32
61.3	$2(4^+,3^+)$			64
75	$2(4^+,3^+)$ $(2^+,1^+)$			96
86.7	$2(4^+,3^+)$ $2(2^+,1^+)$			128
97.1	$2(4^+,3^+)$ $(2^+,1^+)$	$(3^+,1^+)$		160
106.3	$2(4^+,3^+)$	$2(3^+,1^+)$		192
114.7	$(4^+,3^+)$	$2(3^+,1^+)$ $(4^+,2^+)$		224
122.4		$2(3^+,1^+)$ $2(4^+,2^+)$		256
129.8		$(3^+,1^+)$ $(4^+,2^+)$	$(4^+,1^+)$	288
137			$2(4^+,1^+)$	320
143.5	$(3^+,2^+)$		$2(4^+,1^+)$	356
150	$2(3^+,2^+)$		$2(4^+,1^+)$	384

charge densities of 2.5^+. Evans [5] was the first to point out that the broadening in the absorption areas of the intermediate states results from two sub-spectra having different quadrupole splittings. The densities N_i in Table I are calculated by making use of the number of transferred electrons in each reaction taking into consideration that the double unit cell contains 192 Fe atoms. The intermediate ionic states have quenched orbital moments as concluded from hyperfine field measurements. The Fe(1) and Fe(4) states are located at B-sites. Thus the chemical formula of magnetite is

$$Fe(3)_{0.5}Fe(2)_{0.5}[Fe(1)_{0.5}Fe(4)_{0.5}Fe(3)]O_4$$

The time-averaged moment at A-sites, as follows from this formula, is lower by 10% than that of the Fe^{3+} ion. Moreover, the ratio of the moment at A-sites to that at B-sites equals 1.059. These estimates are in excellent agreement with recent data obtained by neutron scattering techniques [6].

3. ELECTRIC CONDUCTIVITY

The quasi-particles are considered as being trapped by potential barriers of height kT_i and width equal to the jumping distance R of the electrons. The electrons will escape from the trap by quantum mechanical tunneling at a rate $\Omega_q \propto \exp(-2R/\hbar)(2mN_ikT_i)^{\frac{1}{2}}$. On the other hand owing to thermal agitation, the electrons can enter into the trap at the rate $\Omega_c \propto \exp(-E_a/kT)$. Because of these opposing tendencies the density of the quasi-particle may oscillate at a natural ground frequency Ω_o. We represent Ω_o by the minimum value of the geometrical mean-value of the rates Ω_q and Ω_c. Thus $\Omega_o = (\Omega_q\Omega_c)^{\frac{1}{2}}_{min}$. We point out that E_a expresses the energy of a dipole-dipole interaction, so that

$$E_a = 2\, e^2 d^2/R^3 \tag{5}$$

where d is a characteristic polarization displacement of atomic size. The magnitude of d was estimated as equal to 1.5 Å, with $E_a = k(T_i)_{max}$ and $R = R_{min}$ inserted in (5). It is proposed in the present work that only those electrons which are forced to oscillate at the phonon frequency Ω_{ph} can participate in steady electric transport. The fractional change in the amplitude of the charge-density oscillations of the quasi-particles and the phonon occupancy $\langle\mathcal{N}\rangle$ define a transport probability $P_{tr} = \langle\mathcal{N}\rangle(\Omega_{ph}/\Omega_o)^2$. The electric conductivity is given by $\sigma = \mu e\, n_e \langle\mathcal{N}\rangle(\Omega_{ph}/\Omega_o)^2$ where μ is the electric charge mobility, e is the elementary charge, n_e is the concentration of the delocalized electrons ($n_e = N_i/a_o^3$). With the above assumptions one gets for the jumping distance of the electrons

$$R \propto 1/T^{\frac{1}{4}}, \tag{6}$$

and for the conductivity

$$\log\sigma = \text{const.} + \log\mu + \log\langle\mathcal{N}\rangle + \log N_i + K_o/T^{\frac{1}{4}} \tag{7}$$

where $K_o = 0.765(2\, e^2 d^2 \alpha^3/k)^{\frac{1}{4}}$ with $\alpha = (2/\hbar)[2mk(T_i)_{max}(N_i)_{max}]^{\frac{1}{2}}[T_i/(T_i)_{max}]^s$, while $s = 3/2$ for $T \geq 10.8$ K, and $s = 1/2$ for $T < 10.8$ K. The development of the quasi-particles begins at finite temperature at which the quantum tunneling rate Ω_q starts losing dominance over the thermally agitated rate Ω_c. This temperature T_o is determined by equating $\Omega_q = \Omega_c$. One finds $T_o = (T_i)_{max}/2^{4/5}(N_i)^{2/5}_{max} = 10.2$ K. The idea of quantum tunneling nucleation was applied by Stauffer [7] in the case of spin clusters.

The plot (A), in Fig. 3 shows the dependence of $\log\sigma$ on $1/T^{\frac{1}{4}}$, as expected from (7), when one neglects the changes of $\langle\mathcal{N}\rangle$ and μ with temperature. The branch above 10.8 K was drawn through the indicated points which were calculated for the various transition temperatures T_i. The branch below 10.8 K was calculated under the assumption that N_i remains constant and equal to 2. The first branch of this plot is in a good quantitative agreement with recent experimental evidence between 40 K and room temperature [8]. The insert in Fig. 3 shows that the computed ratio $K_o/T^{\frac{1}{4}}$ has a minimum at 10.8 K and predicts a possible increase in the conductivity at liquid helium temperatures. Such an effect was discovered a decade ago by Drabble et al. [9]. Moreover, the illustrated plot (B) shows that the experimental results of these investigators could be described rather well by our model when one also takes into consideration the contribution of $\langle\mathcal{N}\rangle$ and μ to the electric conductivity. In computing the plot (B) we assumed that $\langle\mathcal{N}\rangle \propto T$ below 43 K, whereas $\langle\mathcal{N}\rangle \propto T^4$ when the lowest temperatures are approached. The charge mobility μ is here considered to originate in scattering of the electrons by the localized holes. This mechanism is analogous to that of the charge mobility of a semiconductor with ionic impurities for which $\mu \propto T^{3/2}/Z^2 N_I$, where N_I is the ionic impurity concentration, while Z is the impurity charge. Replacing N_I by $N_i \propto T^2$ [on account of (4)] one obtains

$$\mu \propto 1/T^{\frac{1}{2}} \qquad (8)$$

with Z=1 for holes. The variations of μ with T given by (8) play a minor role in the plot (B) in which the conductivity changes over 15 orders of magnitude.

The Verwey transition is manifested in the model outlined here by a local increase in the slope of log σ versus $1/T^{\frac{1}{4}}$ curve, rather than by a discontinuity as usually assumed. However, a sudden increase in σ at the Verwey temperature may be expected if the space charge of the delocalized electrons is large enough to provide a substantial screening of the holes. A simplified estimate shows that this could happen when $4\pi(r_c/a_o)^3N_i/3$ approaches unity. With the quasi-particle density N_i = 256, as well as with r_c = 1.73 A, both corresponding to T_i = 122.4 K, one obtains for the above term a value of 0.95.

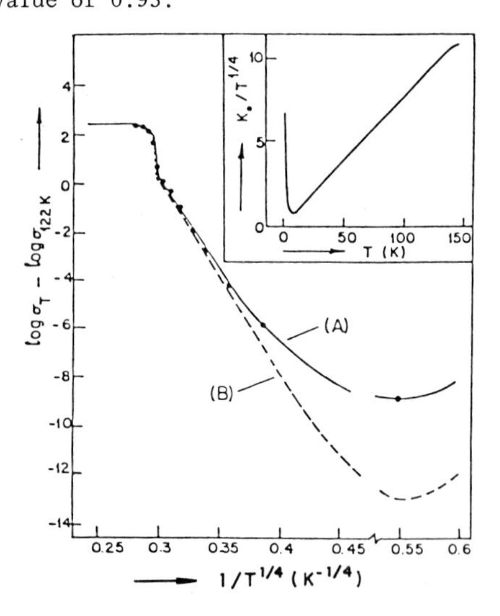

Fig. 3. Electrical Conductivity

In conclusion, it may be useful to note that there is growing evidence recently of magnetic effects which seem to be strongly correlated to the sequence of valence-fluctuation reactions. As an example one may give the spontaneous reversals in the magnetization [10] which are accompanied by line-intensity rearrangements in the Mossbauer patterns, and which coincide with the discrete dips in the recoil-free fraction [11].

ACKNOWLEDGEMENT

The author would like to acknowledge the support of the Technion Fund for Promotion of Research.

REFERENCES

[1] Hirsch, A.A., Phil. Mag. B42 (1980) 427.

[2] Hirsch, A.A. and Galeczki, G., J. Phys. Paris 40 (1979) C2-230.

[3] Terukov, E.I., Reichelt, W., Ihle, D. and Oppermann, H., Phys. Stat. Sol.(b)95 (1979) 491.

[4] Terukov, E.I., Reichelt, W., Wolf, M., Hemschik, H. and Oppermann, H., Phys. Stat. Sol. (a)48 (1978) 377.

[5] Evans, B.J.,AIP Conf. Proc. 5 (1972) 296.

[6] Rakhecha, V.C., Chakravarthy, R. and Satia Murty, N.S., J. Phys. Paris 38 (1977) C1-107.

[7] Stauffer, D., Solid State Commun. 18 (1976) 533.

[8] Graener, H., Rosenberg, M., Wall T.E. and Jones M.R.B., Phil. Mag. B40 (1979) 380.

[9] Drabble, J.R., White, T.D. and Hooper, R.M., Solid State Commun. 9 (1971) 275.

[10] Buckwald, R.A., Hirsch, A.A., Cabib, D. and Callen, E., Phys. Rev. Letters 35 (1975) 878.

[11] Galeczki, G., Buckwald, R.A. and Hirsch, A.A., Solid State Commun. 23 (1977) 201.

Valence Instabilities
P. Wachter and H. Boppart (eds.)

RAMAN SCATTERING IN Sm_3S_4 AND OTHER COMPOUNDS WITH Th_3P_4 STRUCTURE

I. Mörke, G. Travaglini, P. Wachter

Laboratorium f. Festkörperphysik, ETH Zürich
8093 Zürich, Switzerland

Sm_3S_4 is generally considered to be a mixed valent material which crystallizes in the Th_3P_4 structure and has Sm^{2+} and Sm^{3+} ions on equivalent lattice sites. Thermally induced hopping of electrons between these ions is expected. The lattice constant of the system Sm_2S_3-Sm_3S_4-SmS exhibits a jump near the composition Sm_3S_4. We have measured the Raman spectrum of a Sm_3S_4 single crystal and compared it to those of Sm_2S_3, Th_3P_4, Ce_3S_4 and La_3Se_4, all having the same crystal structure, as well as to other Sm_3S_4 with different lattice constants. Furthermore the infrared reflectivity has been investigated, since in this crystal structure infrared active phonon modes are Raman active as well. A new kind of valence mixing is proposed.

Sm_3S_4 crystallizes in the Th_3P_4 structure and is considered to be a mixed valent compound with divalent and trivalent samarium ions on equivalent lattice sites. The results of Mössbauer experiments at room temperature (1) show an isomer shift in this compound which can only be interpreted by thermally induced hopping of electrons between Sm^{2+} and Sm^{3+} or by means of intermediate valent samarium ions. Another problem in Sm_3S_4 is the dependence of the lattice constant on the chemical composition. It was found that the lattice constant of the system Sm_2S_3-Sm_3S_4-SmS exhibits a sudden increase or "jump" from 8.516 Å to 8.545 Å near stoichiometric Sm_3S_4 (2,3,4). In addition it has been shown (4) from the enthalpy of solution that the most stable Sm_3S_4 compounds are stoichiometric samples with a lattice constant around a = 8.516 Å. Sm_3S_4 samples with other lattice parameter have a somewhat different composition and show different physical behaviour, for example in the magnetic susceptibility (6,7,8). For this reason we investigated the Raman and infrared spectra of Sm_3S_4 with varying lattice constant and compared the results to those of other compounds with Th_3P_4 crystal structure.

In this crystal structure (T_d^6 or $I\bar{4}3d$) there are four formula units per unit cell and the irreducible representation of the lattice vibrations at the Γ-point (q = 0) are given by $A_1+2A_2+3E+5F_1+5F_2$. Therefore, there are nine Raman active modes ($A_1+3E+5F_2$) and five infrared active modes ($5F_2$). In this structure the atoms are not situated at sites with inversion

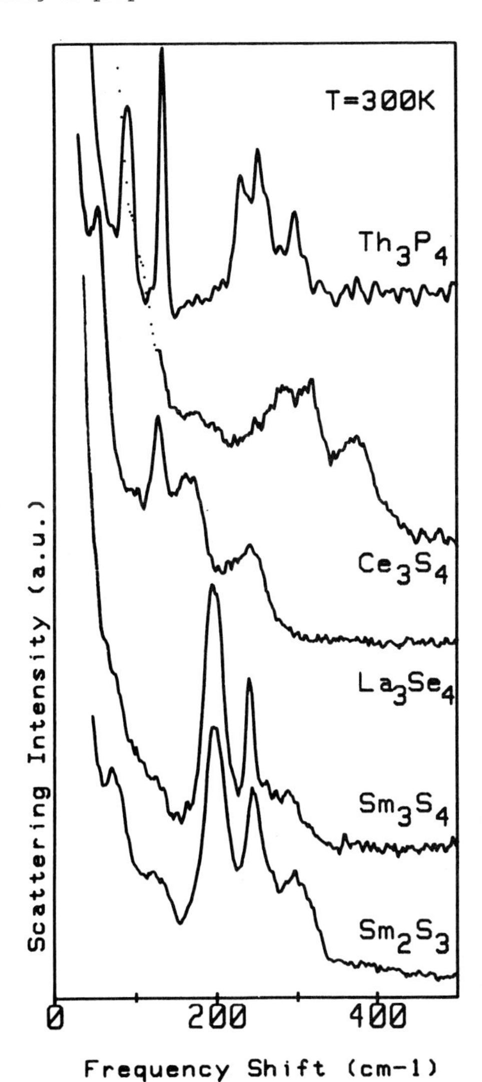

Figure 1: Raman spectra of different compounds with Th_3P_4 crystal structure.

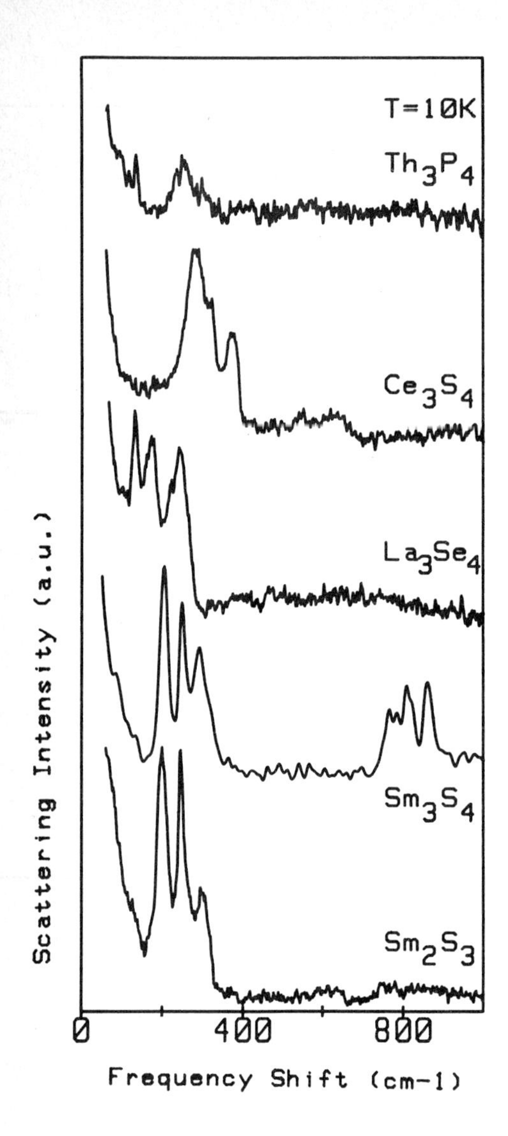

Figure 2: Low temperature Raman spectra of different compounds with Th_3P_4 structure.

symmetry, so infrared active phonon modes are Raman active as well.

The experimental conditions for all Raman spectra shown in this paper are: oblique backscattering geometry; laser wavelength: 5145 Å; resolution: 8 cm^{-1}. In figure 1 we show the room temperature Raman spectra of five different compounds with Th_3P_4 structure. First of all Th_3P_4 which is a semiconductor with a gap of 0.43 eV (9); Ce_3S_4 is metallic as well as La_3Se_4 which in addition is a superconductor with a T_c of 8.6 K (10). Sm_3S_4 and Sm_2S_3 are semiconductors. Sm_2S_3 has a Th_3P_4 defect structure with all sulfur sites filled and randomly distributed vacancies on the samarium sublattice. One should write Sm_2S_3 as $Sm_{2.66}\Box_{0.33}S_4$. All samples were polycrystalline except a stoichiometric Sm_3S_4 which was a 5x2x2 mm^3 single crystal with a = 8.5205 Å.

The Raman spectra show typically three broad bands between 150 cm^{-1} and 400 cm^{-1} and two week excitations below 150 cm^{-1}, except for Th_3P_4 where the latter are very strong. It is difficult to resolve all nine modes, part of them being present only as shoulders in the broad bands or being hidden in the Rayleigh tail of the spectrum. Since we see in all five compounds the same pattern, these excitations are classified as first order phonon scattering.

In figure 2 the same spectra are shown but for a temperature of 10 K and a wave number range of 1000 cm^{-1}. It is to be noted that only in Th_3P_4 the integrated scattering intensity decreases with temperature as one would expect for first order phonon modes at these frequencies. There are no drastic changes in either of these spectra when decreasing the temperature. Only in Sm_3S_4 peaks near 800 cm^{-1} which are very weak at room temperature become strongly pronounced, at low temperatures. These features are due to electronic Raman scattering from intra 4f excitations of the divalent samarium ions. The ground state of Sm^{2+} $(4f^6)$ is ^7F_o and over a larger energy scale, one can observe excitations to the different ^7F_J (J = 1....6) levels as shown in figure 3. The transition energies in the free Sm^{2+} ion are indicated above the spectrum (11). The fine structure of the different excitations is due to crystal field splitting since the Sm site is of low symmetry (S_4). We have to point out that the $^7F_o \rightarrow {}^7F_1$ transition could not be identified unambiguously (but see below) since all the peaks in the vicinity of 290 cm^{-1}, have been found in the spectrum of

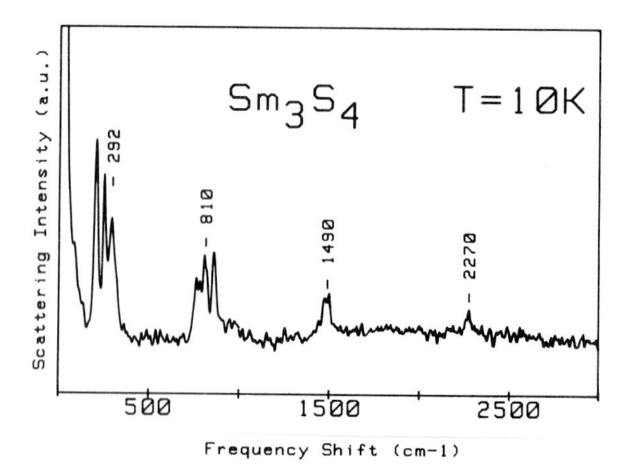

Figure 3: Raman spectrum of Sm_3S_4. The free ion values of the electronic transitions are indicated.

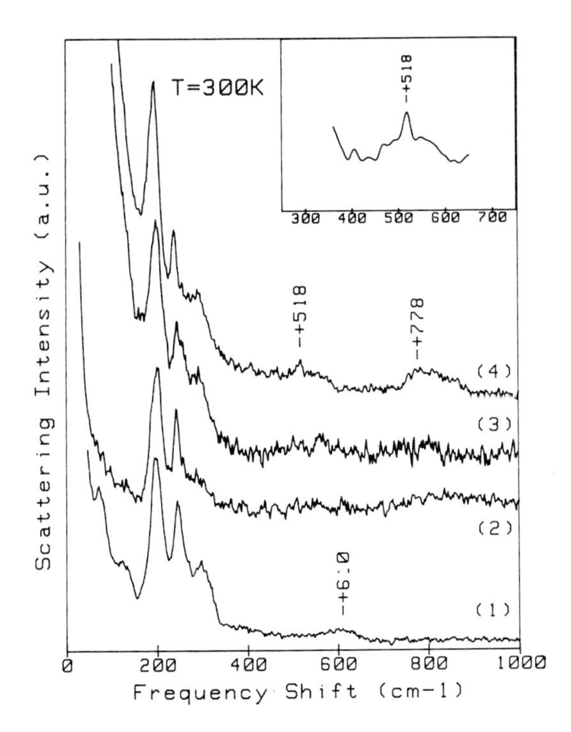

Figure 4: Raman spectrum of Sm_2S_3 (1) and different Sm_3S_4 (2) - (4) samples.

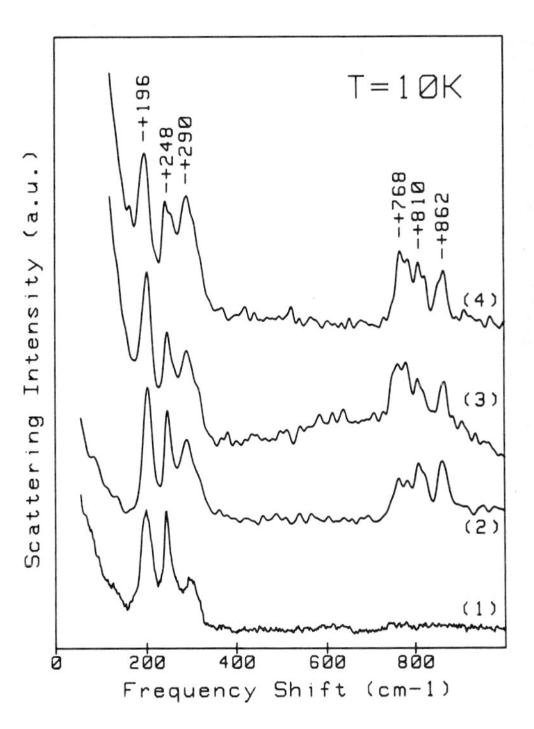

Figure 5: Low temperature Raman spectra of Sm_2S_3 (1) and different Sm_3S_4 (2) - (4) samples.

Sm_2S_3 as well. None of the peaks near 290 cm^{-1} decreases in frequency with decreasing temperature as has been shown for the $^7F_o \rightarrow {}^7F_1$ transition in the samarium monochalcogenides (12). We now discuss the spectra of different Sm_3S_4 within the range of the "jump" of the lattice parameter (4) (fig. 4). As already mentioned, only sample (2) is a single crystal. The lattice constants of the samples are:

(1)	Sm_2S_3	a = 8.447 Å
(2)	Sm_3S_4	a = 8.5205 Å
(3)	Sm_3S_4	a = 8.528 Å
(4)	Sm_3S_4	a = 8.546 Å

As one can easily see there are no changes in the phonon part of the spectra by going from Sm_2S_3 to Sm_3S_4 or between the different Sm_3S_4 compounds at room temperature. However, in Sm_2S_3 (sample 1) there is a peak at 610 cm^{-1} which might be associated with a two phonon spectrum. Since similar two phonon spectra in Sm_3S_4 are absent we rather tend to relate this peak with sulphur vibrations in the defect structure of $Sm_{2.66}\square_{0.33}S_4$. In the Sm_3S_4 sample with the highest lattice parameter (sample 4) there is an additional peak at 518 cm^{-1} which has been measured separately as shown in the insert of figure 4. This peak might be due to the slight stoichiometry deviation which occurs by varying the lattice constant from 8.516 Å (stoichiometry) to 8.546 Å (2-4) or due to the appearance of other crystallographic structures (13). In figure 5 the same spectra are shown at 10 K. It is significant that in Sm_3S_4 compared to 300 K (fig. 4), the 290 cm^{-1} peak is enhanced relatively to the other phonon peaks, mostly for sample (4) with the highest lattice constant, while the relative intensities remain rather the same in Sm_2S_3.

The infrared measurements show the expected five phonon modes as seen in the reflectivity of Sm_2S_3 and Sm_3S_4 at room temperature (fig. 6). Three modes are clearly resolved, the two others are only shoulders. A Kramers-Kronig analysis of these spectra leads to the following transverse optical modes at 125*, 167, 180*, 226, ∿256* cm^{-1} and at 122*, 156, 187*, 226, ∿248* cm^{-1} in Sm_2S_3 and Sm_3S_4, respectively. In Th_3P_4 only three modes have been found in this energy range at 225*, 238 and 256* cm^{-1} (14). The modes with an * could be identified in the Raman spectra as well.

Concerning the "jump" in the lattice constant near stoichiometric Sm_3S_4 we consider that all Sm_3S_4 samples are semiconducting and exactly or nearly stoichiometric compounds. Thus there should not be a significant increase in the amount of Sm^{3+} when reducing the lattice parameter elsewhise one would create free carriers in the 5d band with metallic conductivity. We rather propose that the reduction of the lattice

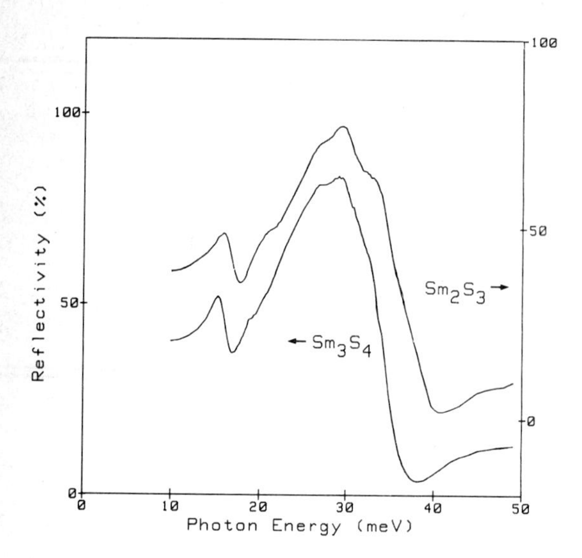

Figure 6: Infrared reflectivity of Sm_2S_3 and Sm_3S_4 at room temperature.

parameter with its concomitant increase in ionic binding energy is partially compensated with an electronic excitation in which a $4f^6$ electron is excited into a localized 5d orbit ($4f^6 \rightarrow 4f^5 5d$) (5) in the sense of an exciton. The radius of such a configuration is mainly that of a $4f^5$ shell in spite of the fact that Sm in the localized $4f^5 5d$ configuration is formally divalent Sm^{2+}. In this sense we write the formula Sm_3S_4 in the following manner:

$$Sm_3S_4: [\underset{\substack{4f^6 \\ ^7F_o}}{Sm^{2+}_{1-x}} + \underset{\substack{4f^5 5d \\ ^7K_4}}{Sm^{2+}_x} + \underset{\substack{4f^5 \\ ^6H_{5/2}}}{Sm^{3+}_2}]\, S^{2-}_4$$

The samples with x = 0 have the highest lattice constant, are slightly off stoichiometric, have additional Raman defect peaks and traces of other crystallographic phases (13). The samples with the lowest lattice constants are stoichiometric, and are most stable (4).

This model is further supported by susceptibility measurements of Sm_3S_4 of compositions within the lattice jump and Sm_2S_3 (15). The susceptibility is a sum of the individual Sm^{2+} and Sm^{3+} contributions (6,7,8,16). For the larger lattice constants ($x \rightarrow 0$) around 8.545 Å only the $4f^6$ and $4f^5$ configurations have to be taken into account (7,8,16), for the smaller lattice constants around 8.520 Å x is of the order of 0.25(5) and the $4f^5 5d$ configuration with 5d spin parallel to the ^7F_o state yields a $^7K_4$ state (11) which is manifest in the susceptibility (6,15).

A susceptibility $\chi(Sm_3S_4) - \chi(Sm_2S_3)$ for samples with $x \rightarrow 0$ and a = 8.546 Å yields a van Vleck term of Sm^{2+} at $T \rightarrow 0$ of $\chi = 8L\mu_B^2 / E_1 = 7 \cdot 10^{-3}$ emu $Mole^{-1}$ with $E_1 = {}^7F_1 - {}^7F_o = 425$ K = 294 cm^{-1}.

It now becomes clear where in the Raman spectrum of Sm_3S_4 we have to look for the ^7F_o-$^7F_1$ transition. It will be most pronounced for samples with $x \rightarrow 0$, a = 8.546 Å because these compounds (although slightly defect) have the highest $Sm^{2+}(^7F_o)$ concentration. Second, we should look in the low temperature spectra, because the electronic Raman effect gets enhanced, compared to the phonon spectra. Looking at figure 5 one observes that the peak at 290 cm^{-1} gets more and more pronounced in going from sample 2 → 4 ($x \sim 0.25 \rightarrow x \sim 0$). A difference spectrum Sm_3S_4 (4) - Sm_2S_3 indeed leaves only one pronounced structure at 290 cm^{-1}. The same is true for the difference spectrum of Sm_3S_4 (4) (10 K) - Sm_3S_4 (4) (300 K) since the electronic transition grows at low temperatures. Thus also the $^7F_o \rightarrow {}^7F_1$ transition in Sm_3S_4 could be identified.

It is quite remarkable that the $4f^5 5d$ state does not result in conducting 5d electrons. However with $x \sim 0.25$ and a total Sm^{2+} concentration of 1/3, the $4f^5 5d$ excitonic state is only present with about 1/12, which is below the percolation limit. However, it is possible that the outer 5d electron exchanges orbits with neighbouring $4f^6$ ions thus resulting in a new intermediate valent $4f^5 5d$-$4f^6$ configuration. Nevertheless a conventional hopping effect between this state and the $4f^5$ (Sm^{3+}) state may occur (17). In any case we do not expect a charge ordering in such a compound with $x \sim 0.25$ and indeed a high resolution X-ray investigation (Weissenberg camera) at 4.2 K has not given a deviation from the cubic Th_3P_4 structure. In conclusion it is thus the first time to be shown that a minute variation in composition near stoichiometry spontaneously performs a "valence transition" in "squeezing out" a $4f^6$ electron into an outer 5d orbit, thus gaining a net binding energy in reducing the lattice constant.

ACKNOWLEDGEMENT

We would like to thank Drs. E. Kaldis, F. Hulliger (La_3Se_4) and Z. Henkie (Th_3P_4) for providing the samples and Dr. M. Ziegler for the low temperature X-ray work.

REFERENCES

1) Eibschütz, M., Cohen, R.L., Buehler, E. and Wernick, J.H., Phys. Rev. B6 (1972) 18
2) Kaldis, E., J. Less-Common Metals 76, (1980) 163

3) Kaldis, E., Spychiger, H., Fritzler, B. and Jilek, E., The Rare Earths in Modern Science and Technology, Vol.3, Mc Carthy et al., Plenum Press 1982
4) Spychiger, H. and Kaldis, E., this conference proceedings
5) Heim, H. and Bärnighausen, H., Acta Cryst. B34, (1978) 2084
6) Wachter, P., Phys. Lett. 58A (1976) 484
7) Coey, J.M.D., Cornut, B., Holtzberg, F. and von Molnar, S., J. Appl. Phys. 50, (1979) 1923
8) Ochiai, A., Suzuki, T. and Kasuya, T., J. de Physique 41, (1980) C5-71
9) Schoenes, J. and Küng, M., Proc. of the 11e Journées des actinides, May 25-27, 1981, Jesolo Lido, Italy, p.227
10) Bozorth, R.M., Holtzberg, F. and Methfessel, S., Phys. Rev. Lett. 14 (1965) 952
11) Dupont, A., J. Opt. Soc. Am. 57, (1967) 867
12) Nathan, M.I., Holtzberg, F., Smith,Jr., J.E., Torrance, J.B. and Tsang, J.C., Phys. Rev. Lett. 37 (1975) 467
13) Bärnighausen, H., private communication
14) Schoenes, J., Küng, M. and Henkie, Z., private communication, to be published
15) to be published
16) von Molnar, S., Holtzberg, F., Benoit, A., Brigger, A., Floquet, J. and Tholence, J.L., this conference proceedings
17) Batlogg, B., Kaldis, E., Schlegel, A., von Schulthess, G. and Wachter, P., Solid State Commun. 19, (1976) 673

Valence Instabilities
P. Wachter and H. Boppart (eds.)

COMMENTS ON THE ELECTRONIC STRUCTURE AND MAGNETIC ORDER IN Sm_2S_3-Sm_3S_4

S. von Molnar and F. Holtzberg
IBM Thomas J. Watson Research Center, Yorktown Heights, New York 10598

A. Benoit, A. Briggs, J. Flouquet and J. L. Tholence
CRTBT, CNRS BP 166X, 38042 Grenoble Cedex, France

The specific heat and magnetization of small single crystals of Sm_2S_3 and Sm_3S_4 have been extended to lower temperature to resolve the controversy concerning the interpretation of earlier data above 2K. The new data indicate magnetic transitions for both compounds with susceptibility cusps, T_{cusp}, of 1K and .5K for Sm_2S_3 and Sm_3S_4 respectively. Specific heats exhibit rounded peaks at temperatures approximately .3K higher than the corresponding magnetic transitions. The entropy associated with the specific heat anomaly in both cases is close to Rln2. The combination of magnetic and thermal data appear to confirm a crystal field model for Sm^{3+} in which the Γ_7 doublet is lowest. In addition new high resolution magnetic measurements are in agreement with earlier results supporting the view that the susceptibility of Sm_3S_4 in the paramagnetic region is well approximated by a simple addition of the contributing Sm^{2+} and Sm^{3+} ions.

INTRODUCTION

The cubic Th_3P_4 structure can accommodate samarium and sulfur with varying stoichiometry having the extremal compositions Sm_2S_3 or Sm_3S_4. If Sm is assumed to be stable in both the di and trivalent states, more precise formulae for the extremal are $Sm_2^{(3+)}S_3^{(2-)}$ and $Sm_2^{(3+)}Sm_1^{(2+)}S_4^{(2-)}$. The latter is defined as a mixed-valent material, in contrast to the intermediate valence materials in which the stable rare earth configuration is not integer valent.

One consequence of mixed valence is, neglecting magnetic interactions between ions, that the susceptibility of Sm_3S_4 should be well described by the addition of two Sm^{3+} ions and one Sm^{2+} ion per mole of compound. Another is that the low temperature thermal and magnetic properties of both compounds should be dominated by contributions from Sm^{3+}, since Sm^{2+} is a Van Vleck ion.

An early study by Wachter[1] concluded that the Sm^{3+} contribution to the susceptibility shows an anomaly between 50 and 100 K and that the Sm^{3+} ion is split by a crystal field into three doublets, with the Γ_7 lowest in energy. The study implied also that antiferromagnetic order might occur near 1 K. Thermal and magnetic measurements by Holtzberg *et al.*[2] and Coey *et al.*[3] on various Th_3P_4 type compounds containing Sm^{3+} ions suggested that the susceptibility of Sm_3S_4 is not anomalous, is a simple addition of ionic contributions, and that the Γ_8 quartet (split into two doublets by a weak axial crystal field component) is lowest. A subsequent analysis of susceptibility of Sm_3Se_4 by Ochiai *et al.*[4] concluded that either crystal field model could be valid. A more recent study of Sm_3Se_4 and Sm_3Te_4 by Sugita *et al.*[5], however, supports the crystal field assignment in which Γ_7 is lowest and again suggests that some kind of ordering occurs at temperatures below ~ 1.5K.

A major shortcoming of earlier experiments is that they were performed at temperatures above 1.5K (with the exception of one high field magnetization measurement at .6K and reference to preliminary specific heat data in Ref. 5). The present communication presents detailed low temperature thermal (~ .4K – 4K) and magnetic (~ .05K – 4K) data on Sm_2S_3 and Sm_3S_4 in magnetic fields up to 50kOe. These results confirm the crystal field model placing the Γ_7 doublet lowest in energy. In addition, high resolution susceptibility measurements between 4 and 300K are consistent with a simple algebraic addition of the various contributions. No evidence of an anomaly near 60K[1] is found.

EXPERIMENTAL RESULTS

The various samples and measurement techniques applied to them is given in Table I. Specific heat measurements were carried out in a He^3 cryostat using a relaxation technique described by Forgan and Nedjat.[6] Results for 5.34 mg of Sm_2S_3 are

Table I

Sample No.	Lattice Constant (Å)	Magnetization θ(K)	T_{cusp}(K)	$C_m \left\{\frac{\text{emuK}}{\text{mole formula wt}}\right\}$
74 I 54 Sm_2S_3	8.451 ± .002	−2	~1	39×10^{-3}
74 II 66 Sm_3S_4	8.543 ± .002	−.9	.5	23×10^{-3}
74 I 34 Sm_3S_4	8.547 ± 9.002	High T susceptibility (4 – 300K)		

Specific Heat	$T_{cusp}(H=0)$ (K)	$T_{cusp}(H=58kOe)$ (K)	$S = \int_0^{5K} (C/T)dT$ (joules/mole Sm^{3+}K)
74 I 54 Sm_2S_3	1.4 ± 0.5	≳1.5	5.6 ± .2
74 II 66 Sm_3S_4	.8 ± .05	~1	5.8 ± .2

shown in Fig. 1, the continuous curve being the specific heat, C, in joules per mole of Sm^{3+} per degree as a function of temperature in zero applied field. The dashed line is the same plot in an applied field H_A = 58kOe. The dotted curve represents the calculated specific heat of a Schottky anomaly derived from a two level system, both levels having the same degeneracy and being separated in energy by 3.33 K.

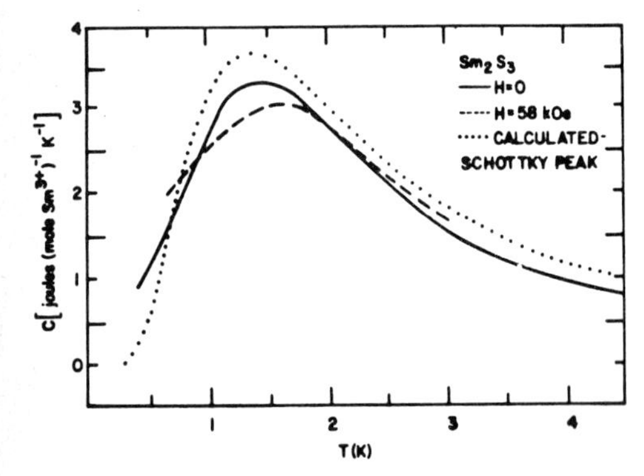

Fig. 1 Heat capacity per mole of Sm^{3+} ion of Sm_2S_3 as a function of temperature. The solid and dashed lines present measurements in 0 and 58 kOe respectively. The dotted curve is a Schottky anomaly calculated for a two level system separated by 3.3K.

For Sm_3S_4 the data are not consistent with Schottky behavior. The apparent agreement with Sm_2S_3 data is, as we shall see, coincidental, the peak having its origin in magnetic order. The entropy, $S = \int \frac{C}{T} dT$, integrated to 5K, in which an exponential dependence of C upon T has been assumed below the lowest measuring temperature, .45K, yields S = 5.6 ± .2joules/mole K^2, a value very close to Rln2 = 5.76 joules/mole K^2.

Similar results for 21.2 mg of Sm_3S_4 are shown in Fig. 2. Again the H_A = 0 curve looks vaguely like, but is smaller in magnitude than, the calculated Schottky anomaly. The experimental entropy S = 5.8 ± .2 joules/mole K^2. In both materials the application of a large 58 kOe magnetic field has only a slight broadening effect. The experimental results are summarized in the last three entries on Table I.

Clearly either a magnetically ordered Γ_7 doublet or a Γ_8 quartet slightly split into two doublets by a noncubic contribution to the field gradient can lead to an Rln 2 entropy. In order to determine which crystal field state is populated at lowest temperature, magnetization measurements are necessary. Representative magnetic data are shown in Figs. 3, 4 and 5. More detailed descriptions of both measurements and results will be given elsewhere.[7] For the purposes of the present discussion the primary results of the low temperature magnetic measurements (see Figs. 3 and 4) are the following: (a) the susceptibility exhibits a peak in both compounds at somewhat lower temperatures than the corresponding specific heat maximum; (b) the solid symbols indicate the presence of an isothermal remanent magnetization, i.e., the remanence obtained after having applied a magnetic field at constant temperature. Other salient parameters, such as θ, the paramagnetic Curie temperature, and C_m, the Curie constant, which was derived from χT vs. T plots, are given in Table I.

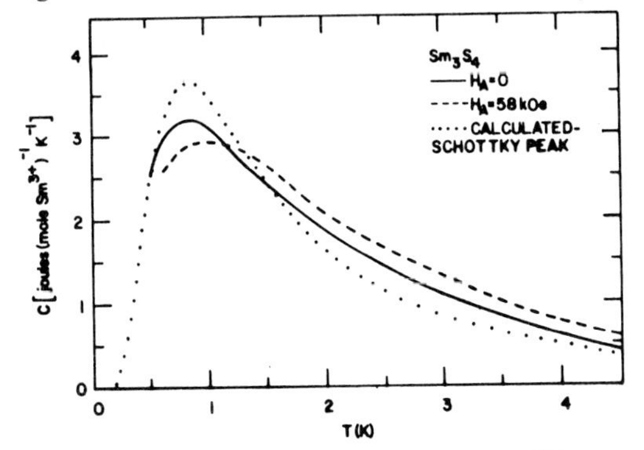

Fig. 2 Heat capacity per mole of Sm^{3+} ion of Sm_3S_4 as a function of temperature. The solid and dashed lines represent measurements in 0 and 58 kOe respectively. The dotted curve is a Schottky anomaly calculated for a two level system separated by 2K.

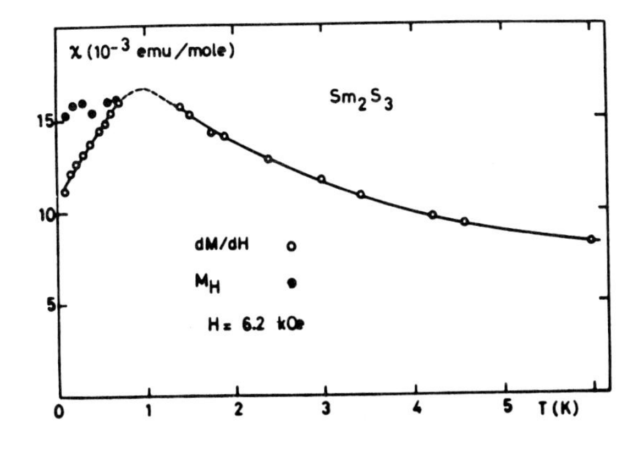

Fig. 3 Susceptibility per mole of compound of Sm_2S_3. The open circles represent the slopes dM/dH, the closed circles M÷H evaluated at 6.2 kOe. The differences between the curves indicate the presence of an irreversible magnetization. Although experimental points are missing in the critical region, T_{cusp} is clearly in the vicinity of 1K.

The Curie constant, C_m, may be calculated for the various Kramers doublets of the Sm^{3+} ion through the relation $C_m = .125 \times p_{eff}^2$ where $p_{eff}^2 = g_{eff}^2 \times \frac{3}{4}$ and g_{eff} is related to the magnetic moment $M = gJ_{eff}\mu_B = g_{eff} \times \frac{1}{2}\mu_B$. One may obtain J_{eff} following the work of Lea *et al.*[8] For two Γ_7 doublets per formula unit this yields $2C_m = 0.042$ emu (mole formula unit)$^{-1}$K in reasonable agreement with the experimental values 0.035 and 0.023 emu (mole formula unit)$^{-1}$K found for Sm_2S_3 and Sm_3S_4 respectively. Similar calculations for the other doublets give 0.015 and 0.206 emu (mole formula unit)$^{-1}$K. The experimental C_m results are consequently consistent with either level scheme, but require a large axial crystal field splitting of the Γ_8 quartet by $\geq$ 20K, since the integrated entropy up to 5 K is only Rln 2. A splitting of 20 K appears to us unlikely, however, and we conclude that Γ_7 is lowest.

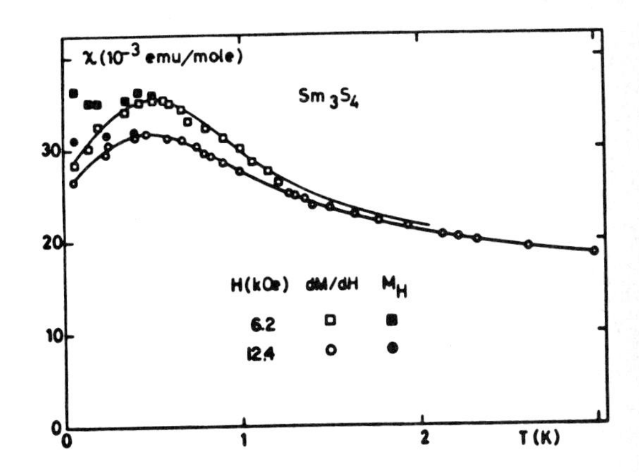

Fig. 4 Susceptibility per moe of compound of Sm_3S_4. The open circles and squares represent the slopes of dM/dH, the solid symbols M÷H evaluated at 6.2 kOe and 12.4 kOe.

Finally, we turn our attention to the high temperature (4.5-300K) susceptibility. Two separate experiments on both large (~ 100 mg; solid points) and small (~6 mg; open circles) single crystals are shown in Fig. 5, where we have plotted the algebraic difference between Sm_3S_4 and Sm_2S_3. Two different samples of Sm_3S_4, 74 II 66 and 75 I 34 were used. Although there are small variations in the absolute value of $\Delta\chi$, which are most probably due to a small amount of second phase found in the X-ray pattern for Sm_3S_4 (75 I 34) or variable filling of the vacancies, the results clearly represent a typical van Vleck susceptibility with no apparent anomalies.[1] The solid line represents data by Bucher *et al.* on di-valent SmTe,[9] from which Birgeneau *et al.*[10] estimated a very small antiferromagnetic exchange interaction of 1.6 ± 1K,[11] almost all of the contribution to the van Vleck term arising from the free ion spin-orbit splitting of 415 ± 6K. The mixed valent model, whereby the total susceptibility of Sm_3S_4 is a simple addition of the contributions from the di and trivalent ions, is therefore justified.

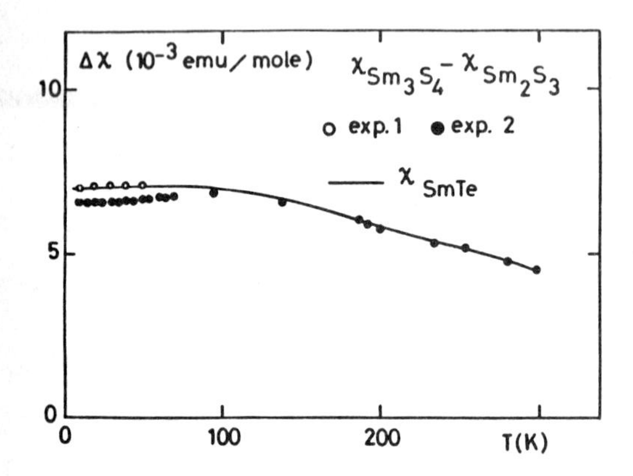

Fig. 5 The algebraic difference, $\Delta\chi$, between the measured susceptibilities of Sm_2S_3 and Sm_3S_4. Open circles are SQUID measurements on milligram size samples 74I54 and 75I34. The solid points are SQUID measurements (4.5K<T<70K) and Faraday Balance measurements (70<T<300K) on larger samples 74I54 and 74II66. The solid line is data on divalent SmTe, see Ref. 9.

DISCUSSION AND CONCLUSION

One clearcut conclusion emerges from the thermal and magnetic data collected below 4K. The coexistence of some form of magnetic order in both compounds concomitant with an integrated entropy approximately equal to Rln 2 rules out the possibility of having Γ_8 lowest. If that were the case an entropy closer to Rln 4 would have been expected. The existence of further crystal field contributions to the specific heat below .45K or near but larger than 5 K, without some noticeable precursor, seems highly unlikely. The low temperature Curie constant also is consistent with the conclusion that the Γ_7 doublet lies lowest.

The nature of the magnetic transition is, however, less clear. The magnetic ordering temperatures are seen to be lower than the thermal data would indicate. In addition the application of the 58kOe magnetic field has relatively little effect on the specific heat, despite the fact that the energy of a Sm^{3+} ion with Γ_7 lowest in this field is ~1K, comparable to the paramagnetic Curie temperature. Furthermore, both compounds display an isothermal remanence which is uncharacteristic of a good antiferromagnet. Both of these observations suggest that Sm_2S_3 and Sm_3S_4 are highly disordered magnets or concentrated spin glasses.[12] This description is consistent with the apparent lack of charge ordering and leads to a model in which the Sm^{2+} and Sm^{3+} ions are randomly distributed throughout the lattice.

Acknowledgments

We thank A. Torressen for his technical assistance and particularly for the fabrication of the bolometers which made the thermal measurements possible. We also thank H. Lilienthal for his help with the magnetic measurements. One of the authors (S. v. M.) gratefully acknowledges the kind hospitality of the low temperature laboratory of the CRNS, Grenoble, where all of the low temperature measurements were performed.

References

1. P. Wachter, Phys. Letters **58A**, 484 (1976)
2. F. Holtzberg, J. M. D. Coey, S. von Molnar and B. Cornut, J.A.P. **49**, 2098 (1978).
3. J. M. D. Coey, B. Cornut, F. Holtzberg and S. von Molnar, **50**, 1923 (1979).
4. A. Ochiai, T. Suzuki and T. Kasuya, J. de Physique **41**, C5-71 (1980).
5. M. Sugita, S. Kunii, K. Takegehara, N. Sato, T. Sakakibara, P. J. Markowski, M. Fujioka, M. Date and T. Kasuya, to be published in The Proceedings of the IVth International Conference on Crystal Fields and Structural Effects in f-Electrons, Sept. 22-25, 1981, Wroclaw, Poland.
6. E. M. Forgan and S. Nedjat, Rev. Sci. Instrum. **51**, 411 (1980).
7. To be published.
8. K. R. Lea, M. J. M. Leask and W. P. Wolf, J. Phys. Chem. Solids **23**, 1381 (1962).
9. E. Bucher, V. Narayanamurti and A.Yayaraman, J. Appl. Phys. **42**, 1741 (1971).
10. R. J. Birgeneau, E. Bucher, L. W. Rupp, Jr., and W. M. Walsh, Jr., Phys. Rev. **B5**, 3412 (1972).
11. An incorrect value of -10K was quoted in Ref. 3.
12. W. E. Fogle, J. D. Boyer, N. E. Phillips and J. Van Curen, Phys. Rev. Lett. **49**, 352(1981)

Valence Instabilities
P. Wachter and H. Boppart (eds.)

ANOMALOUS VALENCE CHANGE IN THE Sm_3S_4-Sm_2S_3 AND Ce_3S_4-Ce_2S_3 SYSTEMS

H. Spychiger, E. Kaldis, E. Jilek

Laboratorium für Festkörperphysik, ETH Zürich
8093 Zürich, Switzerland

The lattice constant of Sm_3S_4 shows a strong, composition dependend, anomaly. This is due to an increase of the Sm^{3+}-contribution indicating a 4f-5d interaction in addition to the heterogeneous mixed valence. Enthalpies of formation as a function of nonstoichiometry in the Sm_3S_4-Sm_2S_3 system show that the solid solution splitts up into three phases, each having, possibly another valence state. The T-x phase diagram of the Sm-S system has been measured, showing several high-temperature phase transitions. A comparison with the Ce_3S_4-Ce_2S_3 shows that this system behaves differently.

1. THE Sm_3S_4 PROBLEM

Sm_3S_4-Sm_2S_3 seems an ideal mixed crystal series for the controlled change of the valence of samarium from +2.66 to +3.00, as a function of the nonstoichiometry. A detailed study of the lattice constants (1) and the heats of formation (2,3) as a function of nonstoichiometry, as well as of the T-x phase diagram (3,4) made clear that this system has some unexpected and fascinating properties indicating a combination of heterogeneous and homogeneous mixed valence. The valence induced thermodynamic instability is discussed also in this volume (5). Raman and magnetic measurements are presented in (6). In this paper we discuss some material properties which reflect the unexpected valence anomalies of this system and make a first comparison with Ce_3S_4-Ce_2S_3.

The first chemical and structural studies of these compounds are due to the Flahaut group (7). This work gave also some information on thermodynamic properties: the m.p. of Sm_2S_3 ∿ 1780 C and the incongruent evaporation of Sm_2S_3 were correct, the m.p. of Sm_3S_4 with 1800 C too low. The lattice constant of one intermediate composition gave the impression that this system follows the Vegard's law. Up to now this opinion was supported by the fact that both Sm_3S_4 and Sm_2S_3 have the bcc structure of Th_3P_4 ($I\bar{4}3d, T_d^6$). Decreasing Sm/S ratio results to the formation of cationic vacancies till the composition $Sm^{3.0+}_{2.67}\square_{0.33}S_4$ (∿ 5% vacancies) corresponding to the formula Sm_2S_3 is reached. This has been shown for the first time, with polycrystalline samples in the Ce_3S_4-Ce_2S_3 system (8). However, for the corresponding Sm-system the validity of this structural model up to the Sm_2S_3 composition has been doubted (9). A structure refinement performed on single crystals (17) showed that Sm_3S_4 does have an undistorted Th_3P_4 structure with statistically distributed Sm^{2+} and Sm^{3+}-sites, and that probably also Sm_2S_3 follows the above model.

An investigation of the transport and optical properties of Sm_3S_4 (10) performed on single crystals (a=8.5198Å) confirmed the hopping mechanism proposed earlier (11), but showed for the first time the paradoxon of Sm_3S_4:

The energy difference between $4f^5$ and $4f^6$ levels is large (U=4.5 eV) so that the trivalent samarium level overlaps with the 3p-band of sulfur. Thus the upper part of the band structure of Sm_3S_4 is almost the same with SmS. As a result of it, Sm_3S_4 is a semiconductor (ΔE ∿ 0.14eV) with hopping conductivity (characteristic of heterogeneously mixed valence) but the 4f-states and the 5d-band are very near to the Fermi energy (characteristic of homogeneously mixed valence). From these results a superposition of the two types of mixed valence could be expected. The investigations of the magnetic susceptibility which followed gave contradicting results: single crystals with low lattice constant (a = 8.5198 Å) gave evidence for valence instability (12) crystals with large lattice constant a = 8.549 Å gave 25% higher susceptibility which was interpreted as the sum of the divalent and trivalent contributions (13).

Table I shows some values of the lattice constants of Sm_3S_4 published upto 1979.

Table I

Compound	a [A°]	Reference
Sm_3S_4	8.556	7a,7b
	8.549	13
	8.544	14
	8.543	15
	8.528	16
	8.523	17
	8.5198	10,12

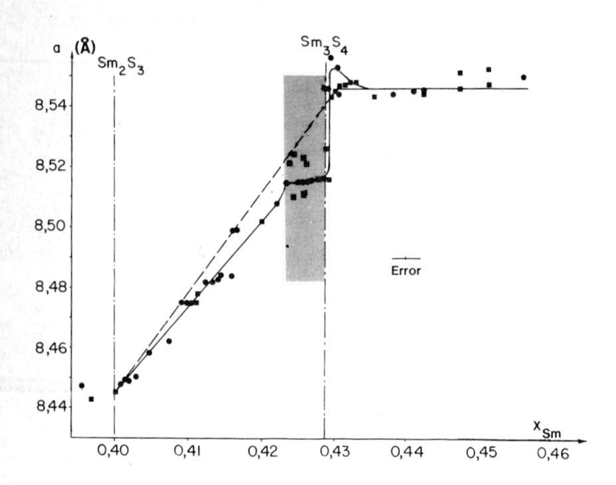

Fig. 1: Lattice constant of the Th_3P_4-structure vs chemical composition in the solid solution Sm_3S_4-Sm_2S_3. The dashed line follows the Vegard's law. ■ samples heated only up to 1600 C.

To make sure that our crystals did not have low lattice constant due to a large concentration of vacancies, we have performed density measurements. They showed a concentration of only 1.7% Schottky vacancies (1), which is typical for crystals grown at high-temperatures (18) and can be reduced by subsequent annealing.

In view of the large variation of the values of Table I ($\Delta a/a \simeq 0.43‰$) we decided that the lattice constant of Sm_3S_4 is probably influenced from the nonstoichiometry, a fact which could open the possibility of chemical manipulation of the samarium valence, like in the case of TmSe (18).

2. THE LATTICE CONSTANT ANOMALY

Figure 1 shows the anomalous dependence of the lattice constant on composition of Sm_3S_4-Sm_2S_3 solid solution. An abrupt change of the lattice constant appears around the stoichiometric Sm_3S_4. It shows clearly that relative changes up to $\Delta a/a \simeq 0.40‰$ appear as a result of deviations from the stoichiometry. This unexpected contraction of the lattice of stoichiometric Sm_3S_4 was explained (1) with a decrease of the divalent samarium concentration due to a change of the valence state. In view of that, it was anticipated (1) that both magnetic susceptibilities were correct, the smaller one (12) applying to stoichiometric Sm_3S_4 crystals and the larger one (13) to overstoichiometric material.

Another possible reason for the anomaly of Fig. 1, could be of course a structural phase transition. However, the high degree of perfection attested to the stoichiometric Sm_3S_4 crystals with low lattice constant (a = 8.523 Å), by the structure refinement (17) removed this possibility.

With increasing lattice constant of the Th_3P_4-structure in the overstoichiometric range, traces of SmS appear in the X-ray powder diffraction patterns. This is understood from the phase diagram (Fig.2) which shows the existence of a eutectic between Sm_3S_4 and SmS. In addition the backreflections become diffuse. A possible reason may be the disorder introduced from the high temperature phase transitions (Fig. 2). To clarify all the structural problems a detailed structural investigation as a function of nonstoichiometry is on the way now (19).

The results of Fig. 1 triggered the question about the thermodynamic stability of the two different valence states of Sm_3S_4 characterized by the small and the large lattice constants (2). To answer this question the enthalpy of formation was measured by solution calorimetry as a function of nonstoichiometry (Fig. 4 in ref.5). It was found that stoichiometric Sm_3S_4 with the small lattice constant is 11% more stable than the overstoichiometric compound. We may conclude, therefore, that the lattice contraction increases appreciably the thermodynamic stability of Sm_3S_4.

3. THE T-x PHASE DIAGRAM OF THE Sm-S SYSTEM

In view of the interesting mixed valent properties of Sm_3S_4-Sm_2S_3 and SmS, and particularly of the possiblity to change the physical properties by controlling the nonstoichiometry, it was found necessary to study the phase relationships in the Sm-S system. Figure 2 shows the up to now existing results from DTA measurements in sealed tungsten capsules (20) and from P-T measurements in Knudsen cells (3). The latter are discussed in the next section. A eutectic between Sm_3S_4 and SmS is followed by the solid solution Sm_3S_4-Sm_2S_3. Several unknown phase transitions appear at higher temperatures. The melting points of the three important samarium-sulfur compounds are given in Table II.

Table II

Mole fraction x_{Sm}	Compound	Melting point (°C)
0.5000	SmS	≃ 2300
0.4286	Sm_3S_4	1950
0.4000	Sm_2S_3	1770

In the solid solution range, the number of measurements is not yet enough to ensure the existence at high temperatures of anomalies corresponding to the instabilites found by calorimetry (Fig. 4, ref. 5) at the compositions x_{Sm}=0.422

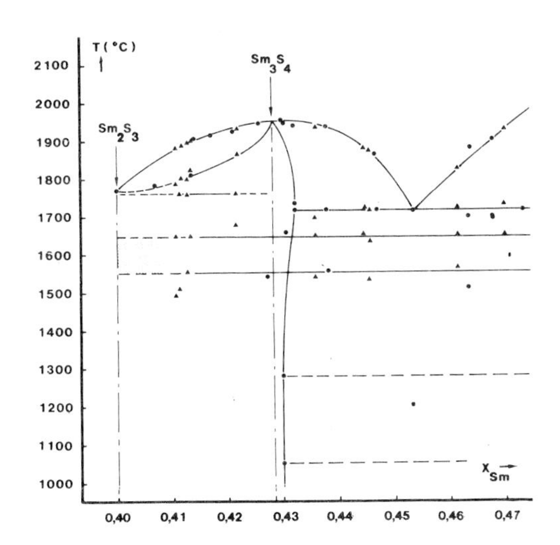

Fig. 2: T-x phase diagram of the Sm-S system • DTA measurements, ▲ P-T measurements (see text). x_{Sm}: molefraction of samarium

and 0.411. Deviations of the solidus and liquidus lines from the lense shape of an ideal solution can be expected due to repulsive interactions of different valence states. However, in view of the phase transitions and of a possible entropy stabilization, at high temperatures such deviations are not a priori necessary.

As it can be seen from Fig. 2 the high-temperature phase transitions are traced mainly by pressure measurements as they are probably too slow for the DTA. In view of the many structures which have been discussed in the literature (9,21) for Sm_2S_3 one may expect some structural transitions.

4. CONGRUENT EVAPORATION OF Sm_3S_4

The abrupt change of the lattice constant of overstoichiometric Sm_3S_4 in a very small composition range, makes the preparation of such samples very difficult. A possible preparation method would be to evaporate samples in vacuum. To this end knowledge of the congruently evaporating composition is necessary. The literature gives two contradicting informations: the incongruent evaporation of Sm_2S_3 leading to Sm_3S_4 (7a) and the incongruent evaporation SmS → Sm_3S_4 → ... leading to more sulfur rich compositions, probably to Sm_2S_3 (14). In view of this discrepancy, we have performed evaporation measurements in order to determine the congruent evaporating composition. Starting with SmS or a sulfur rich component of the Sm_3S_4-Sm_2S_3 solid solutions one reaches after evaporation the vicinity of the Sm_3S_4-SmS phase boundary. We can conclude, therefore, that this is the congruently evaporating composition. The reason for the erroneous conclusion in (14) is that no chemical analysis of the final product has been made. The sulfur rich composition of the solid solution, Sm_3S_4-Sm_2S_3 was estimated from the lattice constant under the assumption of a Vegard's law dependence. In view of the dependence shown in Fig.1, such assumption leads inevitably to an overestimation of the sulfur content. We can, therefore, conclude that the mass spectrometric measurements of the vapor pressure in (14) were not responsible for this error.

Using the incongruent evaporation it was possible to prepare by the same starting batch all intermediate lattice constants (1). A disadvantage of this method is that handling with the open crucible inevitably leads to slight contamination. To avoid this difficulty and composition shifts at the liquidus temperature, a series of polycrystalline samples has been prepared by sintering at 1600 C (■ Fig. 1).

The evaporation has been also used in order to locate the various phase transitions and give a first information about the P-T-x phase diagram of the Sm-S system.

Due to the exponential dependence of the vapor pressure on the enthalpy of evaporation, which changes at each phase transition, breaks in the P-T curve give the transition temperature as shown in Fig. 2 (▲▲▲). It is clear from Fig. 2 that there is close agreement between the DTA and the P-T measurements, the latter being more sensitive to the phase transition. Unfortunately, this method is sensitive only above 1500 C.

5. COMPARISON WITH THE Ce_3S_4-Ce_2S_3 SYSTEM

The results of the above investigations, make also the rest of the solid-solutions (with Th_3P_4-structure), in which a valence change may occur, very interesting. We have selected the system Ce_3S_4- Ce_2S_3 not only for this reason, but also because of the general interest in cerium compounds.

Optical and transport measurements have been published in the literature (22) for Ce_2S_3 and Ce_3S_4, indicating that the former is a semiconductor and the latter a metal. In view of the crude method of the sample preparation and the lack even of lattice constants measurements, the assignement of the measured properties to a certain chemical composition may be erroneous in this work. Single crystals recently grown from several compositions in our laboratory do not show any difference in colour, which is similar to that of the samarium sulfides Sm_3S_4-Sm_2S_3 i.e. black-gray. Optical measurements will soon clear this point.

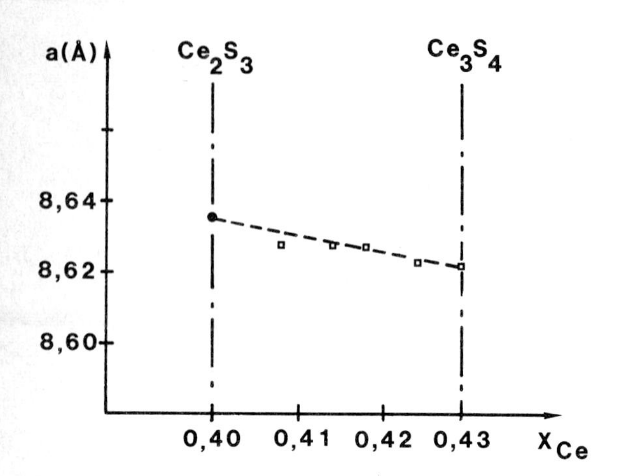

Fig. 3: Lattice constants vs composition for the Ce_3S_4-Ce_2S_3 solid solutions. • after (8).

Figure 3 shows some first results of the dependence of lattice constant on composition. Two striking characteristics appear when comparing with Fig. 1. The change of the lattice constant is 10-times less for the cerium system and the sign of the change is opposite. Thus by incorporating vacancies in the Th_3P_4-structure of Ce_3S_4 (8) a slight expansion takes place instead of contraction. This could be the effect of a valence change overcompensating the vacancy effect. A semiconducting Ce_2S_3 would be in agreement with trivalent cerium in this compound and a metallic Ce_3S_4 with mixed valent or mainly fourvalent cerium. An open question in this case would be the existence of a miscibility gap which normally appears in such solid solutions (5). Up to now no evidence for such gap is existing. It must be pointed out that also in CeN the lattice constant is insensitiv in composition changes (5).

6. CONCLUSIONS

The scattering of the lattice constants of Sm_3S_4 in the literature, is the result of a very pronounced dependence on nonstoichiometry. The strong lattice contraction of the stoichiometric Sm_3S_4 is due to a decrease of the divalent Sm-contribution indicating a contribution of homogeneously mixed valence in the heterogeneously mixed-valent Sm_3S_4.

Enthalpies of formation at room temperature as a function of nonstoichiometry, show that the solid solution Sm_3S_4-Sm_2S_3 splits up into three phases. Probably each of these phases has different valence states. Stoichiometric Sm_3S_4 has a small lattice constant ($a \simeq 8.516$Å) and is 12% more stable than samples with large lattice constants.

First measurements of the T-x phase diagram show a eutectic in the Sm_3S_4-SmS system with $T_{eut} \simeq 1715$ C and a solid solution in the Sm_3S_4-Sm_2S_3 system. The melting points of the compounds are SmS $\simeq$ 2300 C; Sm_3S_4: 1950 C and Sm_2S_3: 1770 C. Several phase transitions appear at high temperatures.

The lattice constants of the system Ce_3S_4-Ce_2S_3 behave quite differently. Like CeN, the lattice constant is almost insensitive to changes in composition.

REFERENCES

1) Kaldis, E., J. Less Common Metals 76 (1980) 163
2) Kaldis, E., Spychiger, H., Fritzler, B., Jilek, E., in "The Rare Earths in Modern Science and Technology" Vol. 3, McCarthy et al. eds., Plenum Press (1982)
3) Spychiger, H., Diploma Thesis, Laboratorium für Festkörperphysik, ETHZ, (1981)
4) Spychiger, H., Kaldis, E., Helvetica Physica Acta (in press)
5) Kaldis, E., Fritzler, B., Spychiger, H., Jilek, E., this proceedings
6) Mörke, I., Wachter, P., this proceedings
7a) Picon, M., Domange, L., Flahaut, J., Guitard, M., Patrie, M., Bull. Soc. Chim. (1960) 221
7b) Picon, M., Flahaut, J., C.R. Acad. Sci. 243 (1956) 2074
8) Zachariasen, W.H., Acta Cryst. 2 (1940) 57
9) Sleight, A.W., Prewitt, C.T., Inorganic Chemistry 7 (1968) 2282
10) Batlogg, B., Kaldis, E., Schlegel, A., Schulthess, G.V., Wachter, P., Solid State Commun. 19 (1976) 673
11) Smirnov, I.A., Parfen'evea, L.S. Khusmutdinova, V.Ya., Segeeva, V.M. Sov.Phys.Solid.State 14 (1972) 2412
12) Wachter, P., Solid State Commun. 19 (1976) 673
13) Coey, J.M., Cornut, B., Holtzberg, F., von Molnar, S., J.Appl.Phys. 50 (1979) 1923
14) Holtzberg, F., Frisch, M.A., Rev. Chim. Min. 10 (1973) 355
15) Leger, L.M., Weil, G., Smirnov, I.A., Panina, L.K., Demina, M.A., sov.Phys.Solid.State 19 (1977) 175
17) Heim, H., Bärnighausen, H., Acta Cryst. Sect. B, 34 (1978) 2084
18) ref. 3 in ref.5
18a) ref. 7 in ref.5
19) Bärnighausen, H., et al. (to be published)
20) ref.7 in Fritzler, Kaldis, this proceedings
21) Eliseev, A., Russ.Inorg.Chem. 23 (1978) 328
22) Kurnick, S.W., Meyer, C., J. Phys. Solids 25, (1964) 115

Valence Instabilities
P. Wachter and H. Boppart (eds.)

MÖSSBAUER STUDY OF THE INTERMEDIATE VALENCE STATE OF IRON IN $FeNb_{1-x}W_xO_4$

F. Grandjean and A. Gérard
Institut de Physique B5, Université de Liège
B-4000 Sart Tilman, Belgique.

Mössbauer spectra of $FeNb_{1-x}W_xO_4$ with $0<x<1$ between 77 and 600 K are discussed here. The temperature quadrupole splitting variation of $FeWO_4$ is explained by an activated electron delocalization in agreement with the semiconducting behavior of this compound. Intermediate valence state for iron is observed by Mössbauer spectroscopy in $FeNb_{1-x}W_xO_4$ with $0<x<1$.

1. INTRODUCTION

$FeWO_4$ and $FeNbO_4$, below 1358 K, crystallize with the monoclinic wolframite structure (1). This structure has zigzag chains along the c-axis which consist of FeO_6 and $W(Nb)O_6$ octahedra sharing two edges. The usual cation valence states in $FeNbO_4$ and $FeWO_4$ are respectively (Fe^{3+}, Nb^{5+}) and (Fe^{2+}, W^{6+}). Therefore, it is expected that in the solid solutions, $FeNb_{1-x}W_xO_4$, Fe^{2+} and Fe^{3+} ions, being at the center of oxygen octahedra sharing edges will be in an intermediate valence state. Noda et al.(1) have studied the magnetic and electric properties and the Mössbauer spectra of this series and have concluded to the existence of a mixed valence state of iron ions.

Because their Mössbauer data were not least-squares fitted and were only recorded at 77 and 300 K, we carried out Mössbauer measurements for x=0, 0.2, 0.4, 0.5, 0.6, 0.8, 1.0 from 77 up to 600 K and we discuss them here.

2. EXPERIMENTAL

The samples were prepared from Fe_2O_3, Fe, Nb_2O_5, WO_3 powders(purity 99,9%). Stoichiometric amounts were mixed for each composition and sealed under vacuum in a quartz tube which was then heated up to 1300 K for 1 day and quenched from 900 K into water.

Unit cell parameters were determined from X-ray powder diffraction patterns and are given in table 1. They are in good agreement with Noda's values.

The Mössbauer spectrometer is of constant acceleration type with a ^{57}Co in Rh source. The isomer shifts are refered to Rh matrix. The low temperature spectra were recorded in a variable temperature Thor Cryogenics cryostat. The sample temperature was measured with an Au-Fe thermocouple within an accuracy of 0.25 K. The high temperature spectra were carried in a furnace built in the laboratory, the temperature being controlled and measured by a chromel-alumel thermocouple with an accuracy of 0.1 K.

TABLE 1. Unit cell parameters for $Fe(Nb_{1-x}W_x)O_4$

x	a(Å)	b(Å)	c(Å)
0	4.998(6)	5.619(4)	4.653(5)
0.2	4.994(3)	5.637(2)	4.656(3)
0.4	4.996(4)	5.656(4)	4.663(3)
0.5	4.99 (2)	5.684(9)	4.68 (2)
0.6	5.009(8)	5.689(7)	4.70 (1)
0.8	5.01 (1)	5.700(9)	4.72 (1)
1.0	4.957(4)	5.707(5)	4.728(5)

3. MÖSSBAUER RESULTS

$FeNbO_4$ is characterized by a quadrupole doublet with the hyperfine parameters given in table 2. Except for the second order Doppler shift of the isomer shift, there is no temperature change down to 77 K.

$FeWO_4$ Mössbauer spectrum is a quadrupole doublet from 77 up to 300 K. In this temperature range, quadrupole splitting and isomer shift decrease with temperature. At 77 K, we found a quadrupole splitting of 1.85 mm/s and Noda et al.(1) of 2.17 mm/s. Our value seems in agreement with the value observed in $Fe_{1-x}Mn_xWO_4$ series (2,3): 1.82 mm/s at 150 K and 1.80 mm/s at 4.2 K. The isomer shift at 77 K is equal to 1.12 mm/s and at 300 K to 0.97 mm/s, i.e. a temperature variation amounting to $5.6\ 10^{-4}$ mm/s.deg which is reasonable for the second order Doppler shift (4). The quadrupole splitting variation is shown in Fig.1.

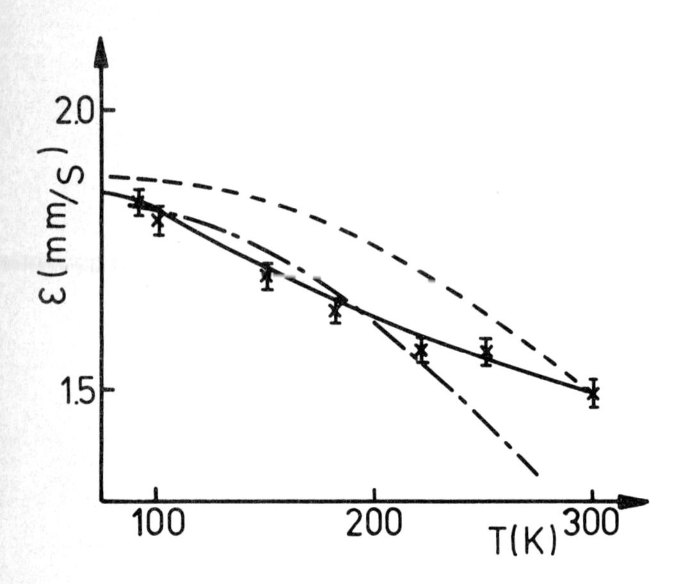

Fig.1 : $FeWO_4$ quadrupole splitting versus temperature. x: experimental points ; ---Ingall's model(5) ($\Delta_1=\Delta_2=766K$),-·--·Ingall's model(5) ($\Delta_1=600K$, $\Delta_2=900K$),— ref.6 model, ($\Delta=343K$).

Such a variation could be due to an increasing population of an excited orbital level. We have applied Ingall's model(5) to an Fe^{2+} ion in an oxygen octahedron with a tetragonal distortion. The dashed (---) curve in Fig.1 is the function

$$\varepsilon(T)=1.88\ \frac{1-e^{-\Delta/kT}}{1+2e^{-\Delta/kT}} \quad \text{with } \Delta_1=\Delta_2=\Delta=766K$$

The curve (-··-·) is formula (15) of ref. 5 with $\Delta_1=600K$, $\Delta_2=900K$. With such a model, it was impossible to account for the experimental points.

Another explanation of the quadrupole splitting variation is the delocalization of one electron with temperature. The model which has been successfully applied to $FeSb_2$ (6) gives the solid curve in Fig.1, corresponding to the following equation :

$$\varepsilon(T)=\frac{\varepsilon_{2+}+\varepsilon_{3+}\ e^{-\Delta/kT}}{1+e^{-\Delta/kT}}$$

with $\varepsilon_{2+}=1.85$ mm/s, $\varepsilon_{3+}=0.40$ mm/s, $\Delta=343$ K. The agreement with the experimental points is very satisfactory.

Furthermore, this electron delocalization would explain the semiconducting behavior of $FeWO_4$ observed by Noda et al.(1), but their activation energy of 0.17 eV∿2000 K is not in agreement with $\Delta=343$ K found here.

The Mössbauer spectra of the solid solutions $FeNb_{1-x}W_xO_4$, $0<x<1$, result in first approximation from the superposition of $FeWO_4$ and $FeNbO_4$ doublets. Fig.2 shows for example, spectra at 300 K and 77 K for x=0.4.

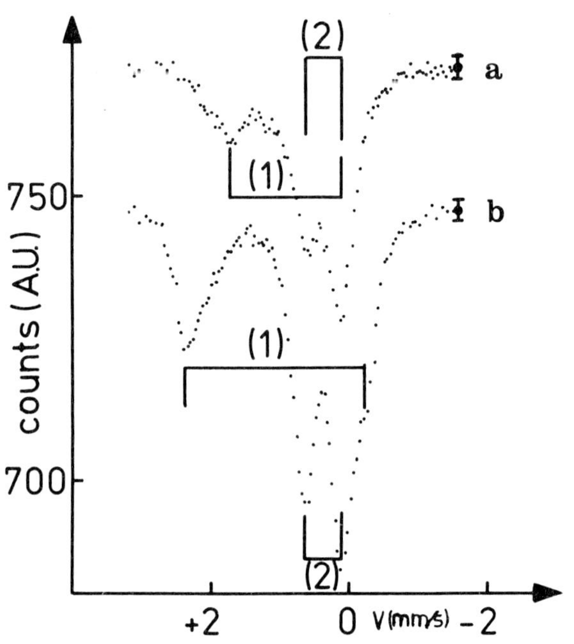

Fig.2 : Mössbauer spectra of $FeNb_{0.6}W_{0.4}O_4$. (a) at 300K, (b) at 77K.

The spectra of the whole series were first least-squares fitted by the superposition of two doublets due to $FeWO_4$ and $FeNbO_4$ with relative intensities in agreement with the chemical composition, between 77 and 300 K. At 77 K, the 2 doublets (D) fit is rather satisfactory, but as temperature increases, as shown in Fig.2, the high velocity line shifts towards zero velocity and broadens. The same behavior is observed for x=0.2, 0.4 and 0.5, for x=0.6 and 0.8, it is not so obvious. The 2D fit does not account very well for that broad absorption around 1.0 mm/s.

It may be noted that the spectra, we observe for the $FeNb_{1-x}W_xO_4$ series are very similar to those observed by Lotgering and van Diepen (7) in the series

$$Zn^{2+}\left[Zn^{2+}_{(1-x)/2}\ Ti^{4+}_{(1+x)/2}\ Fe^{3+}_{1-x}\ Fe^{2+}_{x}\right]O_4.$$

They have interpreted the broad absorption between Fe^{2+} and Fe^{3+} peaks as due to an electron exchange between Fe^{2+} and Fe^{3+} ions. As we shall see, our interpretation is slightly different.
In order to account for that broad absorption, the spectra were fitted by superposing three doublets, the third doublet representing iron ions with an intermediate valence state. The isomer shift variation with temperature of this doublet amounts to 2.4 10^{-3}mm/s.deg for x=0.4 and 1.5 10^{-3}mm/s.deg, for x=0.6. Such large variations are characteristic of average valence variations for iron with temperature.

Table 2 gives the hyperfine parameters given by the 3D fit for all samples at 77 and 300 K and for x=0.4 for intermediate temperatures. For x=0.8, the 3D fit was impossible at 77 K because of the small amount of Fe^{3+}, at 300 K, the results are not satisfactory for the 3+ parameters.

For x=0.6, 0.4 and 0.2, the 3D fit between 77 and 300 K shows that the 3+ intensity does not vary with temperature and is equal to the 3+ fraction in the sample. Thus, 3+ ions do not participate to an electron exchange with 2+ ions as proposed by Lotgering and van Diepen (7) for their series and by Noda et al.(1) for the series $FeNb_{1-x}W_xO_4$.

TABLE 2. Hyperfine parameters for the series $FeNb_{1-x}W_xO_4$

x	T (K)	ε_1	δ_1	ε_2	δ_2	ε_3	δ_3
0	77			0.40	0.37		
	300			0.39	0.27		
0.2	77	2.65	1.06	0.48	0.37	2.12	1.05
	300	1.47	0.90	0.46	0.30	0.72	0.46
0.4	77	2.61	1.06	0.57	0.37	1.88	1.10
	100	2.62	1.06	0.57	0.36	1.94	1.11
	180	2.19	1.03	0.51	0.36	1.68	0.92
	220	1.95	1.10	0.53	0.33	1.60	0.89
	300	1.58	0.93	0.51	0.34	1.13	0.55
0.5	77	2.68	1.12	0.53	0.41	2.09	1.07
	300	1.60	0.90	0.46	0.32	1.09	0.49
0.6	77	2.42	1.08	0.49	0.41	1.78	1.00
	300	1.59	0.96	0.40	0.30	1.52	0.67
0.8	77	2.11	1.13	0.50	0.10		
	300	1.57	1.01	0.26	0.13	1.17	0.48
1.0	77	1.85	1.12				
	300	1.50	0.97				

All values are in mm/s, isomer shifts, δ_i are refered to Rh matrix, errors are ± 0.02 mm/s.

As already mentionned in ref.(1), the 2+ quadrupole splitting is larger for x≠1 than for x=1 at 77 K. It seems that the 2+ character is reinforced by the presence of the 3+ ions which hinder the electron delocalization observed in $FeWO_4$. Nevertheless, increasing temperature activates the delocalization which arises the third doublet.

Above 300 K, 2+ and 3+ lines are so close from each other that fitting the spectra with 3D is impossible. A 2D fit for x=0.5 at 573 K gives the following hyperfine parameters: ε_1=1.09 mm/s, δ_1=0.63 mm/s, ε_2=0.40 mm/s, δ_2=0.28 mm/s.

4. CONCLUSIONS

Among different analyses of the temperature dependence of the quadrupole splitting of $FeWO_4$ between 77 and 300 K, one is in agreement with the semiconducting behavior previously described by Noda et al.(1), but the activation energy deduced from the Mössbauer results is not in agreement with that computed from resistivity measurements.

The solid solutions $FeNb_{1-x}W_xO_4$ for x=0.2 0.4 and 0.6 behave like $FeWO_4$ and an intermediate valence state of iron is observed by Mössbauer spectroscopy.

REFERENCES

(1) Y. Noda, M.Shimada, M.Koisumi, F. Kanamaru, J. Solid St. Chem.28 (1979) 379.
(2) Ch.Klein, R.Geller, J. de Phys.Coll. 35 (1974) C6-589.
(3) J.J.Kunrath, C.S.Muller, A.Vasquez, Hyp. Int.V, Proc., North Holland Ed. (1981) 1013.
(4) J.Steger, E.Kostiner, J. Solid St. Chem. 5 (1972) 131.
(5) R. Ingalls, Phys. Rev. 133 (1964) A787.
(6) A.Gérard, F.Grandjean, J. Phys. Chem. Sol. 36 (1975) 1365.
(7) F.K.Lotgering, A.M. van Diepen, J. Phys. Chem. Sol. 38 (1977) 565.

Valence Instabilities
P. Wachter and H. Boppart (eds.)

CONFERENCE SUMMARY

S. K. Sinha

Solid State Science Division, Argonne National Labortory

Argonne, Illinois 60439

It has been obvious to everyone at this Conference, I think, that the field of Intermediate Valence (IV) continues to be, as was described in the opening talk, "lively but confusing"--beset as it is with unanswered questions and general disagreement among "experts". It is also obvious that this is because the problem of IV is a very difficult one--indeed, one might describe it as the last frontier of magnetism. One need only think of how long it took to solve the Kondo problem, and then reflect that the latter involved correlation effects at only one impurity site, to understand the magnitude of the difficulty. Nevertheless, steady progress seems to have been made, and some of it is apparent from the results which have been presented at this Conference.

There have been obviously newer and more definitive studies of the more familiar IV compounds and in addition several fascinating new materials have been added to the list. In fact, one of the remarkable features of this meeting seems to be the rate at which new IV (and the closely related "Dense Kondo"-type) systems are being discovered and studied. Thus we have heard about the family of compounds based on $EuPd_2Si_2$ (various aspects of which were reported on by Croft, Gupta, Parks, Murgai, Batlogg, Jayaraman, Schneider, Reihl and Martensson). These are interesting for several reasons, one of which is that $Eu[(Au)Pd]_2Si_2$ appears to order antiferromagnetically for a certain range of Au concentration (one of the few IV compounds to do so) but in this case with one of the valence configurations being non-magnetic, unlike TmSe. Then there is the most extraordinary superconductor $CeCu_2Si_2$, reported by Steglich et al. (confirmed also by measurements reported by Ishikawa et al.) which appears to have superconducting electron masses of around 220 m_e, as deduced from the magnitude of the specific heat anomaly at T_c and the extremely large initial slope of the critical field vs. temperature. Other experiments suggest that Ce in this compound is almost fully trivalent. The question thus arises--where are these very "heavy fermions" at E_F coming from? The answer to this question may shed a great deal of light on so-called "dense Kondo systems". Related compounds of the form CeX_2Si_2 (x = Pd, Ag, Au) on the other hand, show magnetic ordering, as reported by Murgai et al. Experimental evidence was presented for IV behavior in $CeSi_2$ (Dijkman et al.), $CeMn_6Al_6$ (Nowik & Felner), in the Yb pnictides (Ott et al.), YbSi and other Yb compounds (Jaccard & Sierro), and for dense Kondo behavior and magnetic ordering in $CeMg_3$ and $CeInAg_2$ (Galera et al.). We also heard an account of detailed experimental evidence for "Kondo lattice" behavior, accompanied by magnetic ordering, in CeB_6 from Kasuya. Detailed magnetic structure studies on CeB_6 were also reported by Rossat-Mignod et al. From all of this, it is clear that we are groping our way to distinguishing conceptually between three different kinds of systems--dilute magnetic alloys (or what one might call "true" Kondo systems where the f-levels are clearly buried well below the Fermi level--such systems would almost certainly order if the spin system became dense, as in ordinary rare earth metals), "true" IV systems, where the Fermi level is pinned at or close to the f-level, and "dense Kondo systems" (or what I would prefer to call "Kondo-like" systems), where the f-level is slightly buried below E_F, but where the hybridization "tail" of the f-state resonance is still finite at the Fermi level. Such systems mimic many of the properties of "true" IV compounds & also of Kondo systems. (Examples are Ce compounds where the Ce ion is practically trivalent.) The experiments also show that such systems may not order at all, or may order with a reduced moment. The theory for such systems is in a primitive state as yet, although it is possible that in such cases a high density of "f-like" states at E_F may arise from a collective many-body resonance, similar to the Abrikosov-Suhl resonance for the single impurity problem. (Theoretical predictions of such a resonance at E_F arising from spin fluctuation effects were presented independently by Lacroix, Baumgartel and Muller-Hartmann, and Fedro and this author). The question then arises as to why many of these systems do not order magnetically. Conventional Kondo theory would predict this can only happen if the effective exchange J becomes of the order of the conduction electron band-width--a very large quantity indeed. However, the lack of magnetic ordering or reduction in moment may be due to the tendency of f-conduction hybridization processes to favor a singlet ground state, as shown by Ramakrishnan in his talk. (Real-space renormalization group and finite-cell calculations presented here by Jullien for a one-dimensional Kondo lattice shed some light on this problem by finding a non-magnetic ground state for the case of two electrons per site.) Several investigations to find out whether IV is a single ion or collective property were presented at this meeting, in the form of dilute alloy studies. Results presented on Y(Tm)Se by Mignot et al. and on neutron scattering from La(Ce)-Th alloys by Holden et al. seemed to indicate that the dilute ions were in fact in an IV state, indicating a single-ion-like behavior. On the other hand, collective effects are also clearly important, as witnessed by the much more dramatic IV transition in $EuPd_2Si_2$ compared to $La(Eu)Pd_2Si_2$ (reported by Croft et al.), and the disappearance of IV behavior in $(Ce_{1-x}Y_x)Pd_3$ for $x > 0.4$ (reported by Kappler et al.). It seems likely that collective f-ion coupling to the lattice may be of importance in many cases, a point emphasized by Jayaraman. A beautiful new technique for studying very dilute systems was presented by Riegel et al. This was the time-differential perturbed x-ray distribution (TDPAD) method which consisted of using a heavy-ion accelerator to both produce and implant isolated rare-earth ions such as Ce in host matrices such as Sn, and simultaneously excite and align the nuclear isomer state so that the 4f spin dynamics can be studied. From the observed Larmor frequencies, the valence of the Ce ion can be inferred and from the decay, the spin relaxation times obtained. The results were mostly as expected, namely, Ce in hosts which do not form IV compounds with it was mainly trivalent or tetravalent, while in hosts which form IV compounds it appeared to have intermediate valence, but with one or two surprises (e.g., Ce in Al was found to be tetravalent).

It is by, by now, well-known that the ionic size change during configuration fluctuations has a large effect on the lattice and on the phonons in IV compounds. It is also well-known that charge fluctuations coupled to the phonons can lead to significant mode softening effects. Phenomenological models incorporating these microscopic concepts were reviewed by Bilz, who showed that one could get a unified picture of both the phonon dispersion curves and the Raman scattering data by incorporating "breathing" degrees of freedom on the rare earth ion. A novel aspect of the model presented by Bilz was the non-linear coupling to the "breathing" deformation on the f-shell, resulting in a "bound phonon" state inside the phonon gap between the acoustic and optic branches. He ascribed the flat mode seen in the neutron scattering data of Mook et al. on $Sm_{1-x}Y_xS$ (which had previously been thought to be a mass-defect-type local mode due to the Y) to this effect. Guntherodt presented convincing evidence, based on Raman scattering from several IV compounds, that this flat mode in the gap is fairly universal and does not scale with mass as a localized mass-defect mode. His interpretation of this mode (or group of excitations) appeared however to be more in terms of a strong polaron-type electron-phonon coupling than in terms of non-linear effects. The interesting and unusual property of a negative elastic constant C_{12}, which has been observed for several IV compounds so far, was also demonstrated for $TmSe_{0.32}Te_{0.68}$ under pressure by Boppart et al. One might almost say it is a hallmark of IV materials, although CeB_6 (a "dense Kondo" system) also exhibits this effect. A convincing microscopic explanation seems to be lacking.

Surprisingly, there was very little in the way of inelastic neutron scattering studies of IV compounds. Perhaps this is because in the past, such studies have generally revealed only a single broad quasi-elastic line which has not led to the wealth of information which INS studies usually provide. Nevertheless, Holland-Moritz et al. presented evidence from careful INS studies on $YbCu_2Si_2$ that this broad quasi-elastic line may in fact be due to inelastic transitions between broadened crystal field levels. Similar conclusions were arrived at by Furrer et al. for TmSe. It is not clear to me whether one can categorically invoke crystal-field states rather than think in terms of peaks due to splitting and broadening of a (2J+1)-fold multiplet on hybridization with a conduction band (or bands). Obviously the relative magnitude of the CEF splitting relative to the hybridization splitting is important. I would suggest, as a test, a careful search for such inelastic transitions in a pure S-state ion, for which the purely CEF splitting should be vanishingly small, but not the hybridization.

The question of "gaps" in IV materials received plenty of attention and some experimental clarification here. It seems clear from presentations by von Molnar et al., Frankowski and Wachter, and Guntherodt et al., that a gap does indeed exist in SmB_6, as evidenced by optical and tunnelling spectroscopy methods. The values for the gap from the different methods agree reasonably well, but there appears to be no consenous as to what "gap" is really being measured--is it an "f-d hybridization gap" (as Mott proposed in order to explain the low-temperature transport properties), or some other kind of gap (e.g., due to a charge density wave or Wigner ordering, as suggested by Kasuya)? Similar questions arose in connection with gaps observed by tunnelling spectroscopy in TmSe and $TmSe_{1-x}Te_x$. The gap disappears when the material is made ferromagnetic (either by cooling below T_c for the latter compounds or by applying external magnetic fields in the case of TmSe). The temptation in the case of TmSe is to ascribe the gap to the antiferromagnetic ordering. However, one cannot then account for its observation in paramagnetic $TmSe_{1-x}Te_x$ by point-contact tunnelling spectroscopy. On the other hand, f-d hybridization gaps should generally look asymmetric, whereas the results showed very symmetrical gaps. A question which needs to be answered here is which kind of tunnelling spectroscopy is sensitive to the ordinary density of electronic states (for example, the gap in an ordinary semiconductor)? Such tunnelling should certainly be able to see f-d hydridization gaps, if they exist.

We turn now to the somewhat confusing question of what information is obtained regarding IV from X-ray spectroscopies such as photoemission and X-ray edge spectroscopy. It is tempting to interpret peaks "below" the Fermi energy seen in photoemission experiments in terms of "f-like" densities of states at those energies. In fact, the photoemission spectra from many IV compounds involving Eu and Yb ions are fairly consistent with this, showing one f-like feature very near E_F, as expected. However, this simple interpretation breaks down in the case of compounds containing Cerium. We saw in the presentations of Parks et al., Hufner and Steiner, and of Gudat, with great regularity, the emergence of the "double-humped Cerium camel" with one peak about 2 eV below E_F and another slightly below E_F regardless of the apparent valence of the cerium! (CeO_2 is a glaring example.) For instance, there is only a subtle change in the relative magnitude of these two peaks as one goes through the $\gamma \rightarrow \alpha$ transition in Cerium (contrary to intuitive expectations based on a promotional model of the transition). There were basically two types of explanation offered for the double peak in Ce photoemission. One was that the lower energy peak corresponds to an f° final state (unscreened f-hole) and the one near E_F to an f^1 state (i.e., screened by a quasi-localized 4f electron). This point of view was taken by Hufner & Gudat and a mathematical model attempting to incorporate this idea is given in the paper by Liu and Ho. Fuggle et al. pointed out that one could obtain information about the degree of f-localization by examining the core-level XPS lineshapes. Schoenes presented optical reflectivity measurements for $CePd_3$ which appeared to confirm that the $4f^1$ state lay very close to E_F. The other proposal was that of Schlueter & Varma who, on the basis of local density functional calculations, find Ce to have a bi-modal character to its 4f state, i.e., they find two types of stable (or metastable) 4f states in Ce and thereby suggest that the double peaks in Ce arise due to excitations out of these two "resonances". It is clear that further theoretical work and a detailed comparison with experiment is required to understand what are the many-body states between which transitions are being observed in X-ray spectroscopic experiments, (in view of the very large photon incident energies required to excite f-states) before one can safely correlate this information with information about the IV ground state and low lying excitations obtained from thermodynamic and magnetic measurements. It is as well to bear in mind that although the "density of states" (defined as the spectrum of poles of the one-electron Green's function) is a helpful concept, it is really only physically meaningful within the one-electron picture as far as these higher-energy dynamical measurements are concerned. Mathematically, of course, the spectral function can in principle be used to obtain rigorously the thermodynamic expectation value of many observables, but there is no reason that it has to be the same quantity observed in photoemission experiments.

I think it is clear that there is a need for a "gentler" spectroscopy to probe the joint density of states around the Fermi energy in these materials. With the development of synchrotron sources and spallation neutron sources, direct scattering measurements of these transitions may soon be available.

What about the state of the general theory of IV systems? Special mention should be made of a model theory by Wohlleben et al. which is based on the fact that there are two characteristic energies in the IV problem--a "spin fluctuation" energy, (or rather temperature T_f), and a characteristic excitation energy E_x. Both are functions of temperature T and pressure P (or perhaps other variables such as concentration). Using a variety of experimental data, Wohlleben showed that T_f was reasonably well represented by a universal function of E_x regardless of T and P. Based on this, a single thermodynamic theory can then be derived to account for a great deal of the systematics observed as a function of T and P.

Turning to "first-principles"-type theories, the talks of Grewe, Keiter and Entel on diagrammatic many-body perturbation approaches impressed me both with the formidable difficulties involved in this approach and the patience and ingenuity with which these difficulties are being attacked! The single IV impurity problem (similar to the approach outlined by Ramakrishnan) appears to be in reasonable shape and we have some suggestive results for the concentrated case, but it may be some time before there exist meaningful results for the full Anderson lattice, which can be compared with experiment. A warning note was sounded by Stevens, however, who felt some of the assumptions in the method of calculating expectation values of the partition function in the Keiter-Kimball perturbation theory are questionable.

Varma and Schlueter presented a real-space renormalization group theory of the 1-D Anderson Lattice, which in principle allows a detailed examination of the ground state and low lying excited states of the system. They find that in the "Kondo-like" regime, antiferromagnetic correlations develop, while a non-magnetic state exists in the true IV regime. The method is somewhat similar conceptionally to the calculation of Jullien et al., but different in the way the actual calculations are carried out. These exact numerical solutions are important in both testing our conceptual understanding of the nature of the many-body states and for comparing with necessarily approximate analytic solutions. (The restriction to 1-D, however, may be important since there is evidence from the 1-D Hubbard model, for instance, that the nature of the solutions may be somewhat special.)

We also were presented with an exact form (although not the exact solution) for the one-electron Green's function for both the conduction and localized f-electrons (Brandow) which should be useful to compare against in a derivation of the latter quantities using equation-of-motion and decoupling techniques, and further investigations (Fazekas) of the Stevens-Brandow variational ansatz to the ground state of an IV system. This variational ansatz explicitly builds in the non-magnetic singlet discussed earlier as forming the lowest energy state once hybridization is introduced. An interesting analysis of many-body spin-fluctuation effects on the electron self-energy in de Haas-van Alphen experiments (Riseborough) indicated that the field dependence of the dHvA mass should differ drastically from the specific heat electronic mass. Finally, (de Menezes and Troper, and Fedro and this author) ideas were put forward, suggesting that there was an instability at low temperatures against formation of a Cooper-paired ground state, involving f-d pairing at opposite ends of the Fermi surface. This pairing instability can arise either due to the usual phonon-exchage mechanism, or through an effective attractive interaction via correlation effects (similar to Schrieffer-Wolff negative exchange, but in this case with quasi-mobile f-electrons). The physical implications of this paired state (which is undoubtedly non-magnetic) still need to be explored.

The more puzzles there are in a field the more "alive" and exciting the field is, and it simply remains to summarize this conference as one which, I think, was indeed exciting and useful for most of us, and to thank Peter Wachter and his colleagues for having performed this valuable service to the community.

AUTHOR INDEX